COLD SPRING HARBOR SYMPOSIA
ON QUANTITATIVE BIOLOGY

VOLUME LXVI

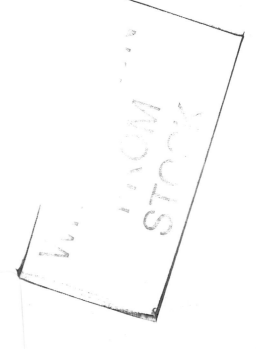

COLD SPRING HARBOR SYMPOSIA ON QUANTITATIVE BIOLOGY

VOLUME LXVI

The Ribosome

COLD SPRING HARBOR LABORATORY PRESS
2001

COLD SPRING HARBOR SYMPOSIA ON QUANTITATIVE BIOLOGY VOLUME LXVI

©2001 by Cold Spring Harbor Laboratory Press
Cold Spring Harbor, New York
International Standard Book Number 0-87969-619-2 (cloth)
International Standard Book Number 0-87969-620-6 (paper)
International Standard Serial Number 0091-7451
Library of Congress Catalog Card Number 34-8174

COLD SPRING HARBOR SYMPOSIA ON QUANTITATIVE BIOLOGY
Founded in 1933 by
REGINALD G. HARRIS
Director of the Biological Laboratory 1924 to 1936
Previous Symposia Volumes

Front Cover (*Paperback*): Large subunit (RNA ribbon, orange; protein, blue) (see N. Ban et al. 2000, *Science*); small subunit (RNA ribbon, yellow; proteins, green) (see B.T. Wimberly et al. 2000, *Nature*); tRNA (space-filled atoms: A-site, purple; P-site, gray; E-site, red). Image courtesy of J.L. Hansen, Yale University, New Haven, Connecticut (see Hansen et al., this volume).

All Cold Spring Harbor Laboratory Press publications may be ordered directly from Cold Spring Harbor Laboratory Press, 500 Sunnyside Boulevard, Woodbury, NY 11797-2924. Phone: 1-800-843-4388 in Continental U.S. and Canada. All other locations: (516) 422-4100. FAX: (516) 422-4097. E-mail: cshpress@cshl.org. For a complete catalog of all Cold Spring Harbor Laboratory Press publications, visit our World Wide Web Site http://www.cshlpress.com/

Symposium Participants

ABO, TATSUHIKO, Div. of Biological Science, Nagoya University Graduate School of Science, Nagoya, Japan

AGRAWAL, RAJENDRA, Dept. of Biomedical Sciences, Wadsworth Center, State University of New York, Albany

ALI, IRAJ, Dept. of Biochemistry, University of Cambridge, Cambridge, United Kingdom

AMALDI, FRANCESCO, Dept. of Biology, University of Rome, Rome, Italy

AMSON, ROBERT, Molecular Engines Laboratories, Paris, France

ANDINO, RAUL, Dept. of Microbiology and Immunology, University of California, San Francisco

ANTHONY, RICHARD, Dept. of Applied Biology and Biomedical Engineering, Rose-Hulman Institute of Technology, Terre Haute, Indiana

AOKI, HIROYUKI, Banting and Best Department of Medical Research, University of Toronto, Toronto, Canada

ASANO, KATSURA, Div. of Biology, Kansas State University, Manhattan, Kansas

ATKINS, JOHN, Dept. of Human Genetics, University of Utah, Salt Lake City

BARTA, ANDREA, Institute of Medical Biochemistry, University of Vienna, Vienna, Austria

BASILE-BORGIA, ANNETTE, Dept. of Biological Sciences, Lehigh University, Bethlehem, Pennsylvania

BAYFIELD, MARK, Dept. of Molecular Biology, Cell Biology, and Biochemistry, Brown University, Providence, Rhode Island

BEALES, LUCY, Dept. of Microbiology, University of Leeds, Leeds, United Kingdom

BECKMANN, ROLAND, Institute for Biochemistry, Humboldt University, Berlin, Germany

BÉLANGER, FRANÇOIS, Dept. of Biochemistry, University of Montréal, Montréal, Québec, Canada

BERARDI, ALESSANDRA, Dept. of Cellular Biotechnology and Hematology, University of Rome, Rome, Italy

BHASKER, RAMAN, Dept. of Biochemistry, Case Western Reserve University, Cleveland, Ohio

BIFFO, STEFANO, Dept. of Molecular Medicine and Pathology, Fondazione Centro San Raffaele, Milan, Italy

BJÖRK, GLENN, Dept. of Microbiology, Umeå University, Umeå, Sweden

BLINN, JAMES, Computer-Aided Drug Discovery Unit, Pharmacia Corp., Kalamazoo, Michigan

BOBKOVA, EKATERINA, Antimicrobial Unit, DuPont Pharmaceuticals Co., Wilmington, Delaware

BODDEKER, NINA, Gilead Sciences, Foster City, California

BOGDANOV, ALEXEY, Dept. of Chemistry, A.N. Belozersky Institute of Physico Chemical Biology, Moscow State University, Moscow, Russia

BRAKIER-GINGRAS, LEA, Dept. of Biochemistry, University of Montréal, Montréal, Québec, Canada

BRANDI, LETIZIA, Dept. of Biology MCA, University of Camerino, Camerino, Italy

BRANLANT, CHRISTIANE, Unité Mixte de Recherche, Centre National de la Recherche Scientifique, Nancy University, Vandoeuvre-les-Nancy, France

BRÉGEON, DAMIEN, Dept. of Molecular, Evolutive and Medical Genetics, School of Medicine, University of Paris, Paris, France

BRIERLEY, IAN, Dept. of Pathology, University of Cambridge, Cambridge, United Kingdom

BRIMACOMBE, RICHARD, Max-Planck-Institut für Moleculare Genetik, Berlin, Germany

BRODERSEN, DITLEV, Dept. of Structural Studies, MRC Laboratory of Molecular Biology, Cambridge, United Kingdom

BURLEY, STEPHEN, Dept. of Molecular Biophysics, Rockefeller University, New York, New York

BUSHELL, MARTIN, Dept. of Microbiology and Immunology, Stanford University, Stanford, California

CAMERON, DALE, School of Microbiology and Immunology, University of New South Wales, Kensington, New South Wales, Australia

CARTER, ANDREW, Dept. of Structural Studies, MRC Laboratory of Molecular Biology, Cambridge, United Kingdom

CASPAR, DONALD, Institute of Molecular Biophysics, Florida State University, Tallahassee, Florida

CHAMPNEY, W. SCOTT, Dept. of Biochemistry and Molecular Biology, East Tennessee State University, Johnson City, Tennessee

CHARPENTIER, BRUNO, Unité Mixte de Recherche, Centre National de la Recherche Scientifique, Nancy University, Vandoeuvre-les-Nancy, France

CHEN, PENG, Dept. of Microbiology, Umeå University, Umeå, Sweden

CHERNYAEVA, NATALYA, Dept. of Molecular Genetics, M.D. Anderson Cancer Center, University of Texas, Houston

CHONG, JEAN-LEON, Dept. of Molecular Genetics, Ohio State University, Columbus, Ohio

CHRISTENSEN, ALBERT, Dept. of Cell and Developmental Biology, University of Michigan, Ann Arbor

CLARK, PATRICIA, Dept. of Biology, Massachusetts Institute of Technology, Cambridge, Massachusetts

COCHELLA, LUISA, Dept. of Molecular Biology and Genetics, School of Medicine, Johns Hopkins University, Baltimore, Maryland

COOPERMAN, BARRY, Dept. of Chemistry, University of Pennsylvania, Philadelphia

CUKRAS, ANTHONY, Dept. of Molecular Biology and Genetics, Johns Hopkins University, Baltimore, Maryland

CULVER, GLORIA, Dept. of Biochemistry, Biophysics, and Molecular Biology, Iowa State University, Ames, Iowa

DAHLBERG, ALBERT, Dept. of Biology and Medicine, Brown University, Providence, Rhode Island

DAHLBERG, JAMES, Dept. of Biomolecular Chemistry, University of Wisconsin, Madison

DALLAS, ANNE, Dept. of Biology, University of California, Santa Cruz

DAS GUPTA, CHANCHAL, Dept. of Biophysics and Molecular Biology, University College of Science, Calcutta, India

DEL CAMPO, MARK, Dept. of Biochemistry and Molecular Biology, School of Medicine, University of Miami, Miami, Florida

DEO, RAHUL, Lab. of Molecular Biophysics, Rockefeller University, New York, New York

DEVER, THOMAS, Lab. of Eukaryotic Gene Regulation, National Institute of Child Health and Human Development, National Institutes of Health, Bethesda, Maryland

DHOLAKIA, JAYDEV, Dept. of Biochemistry and Molecular Biology, School of Medicine, University of Louisville, Louisville, Kentucky

DINMAN, JONATHAN, Dept. of Molecular Genetics and Microbiology, Robert Wood Johnson Medical School-UMDNJ, Piscataway, New Jersey

DORNER, SILKE, Institute of Medical Biochemistry, University of Vienna, Vienna, Austria

DOROVKOV, MAXIM, Dept. of Pharmacology, Robert Wood Johnson Medical School-UMDNJ, Piscataway, New Jersey

DORYWALSKA, MAGDALENA, Dept. of Structural Biology, Stanford University, Stanford, California

DOUDNA, JENNIFER, Dept. of Molecular Biophysics and Biochemistry, Howard Hughes Medical Institute, Yale University, New Haven, Connecticut

DRUZINA, ZHANNA, Dept. of Chemistry, University of Pennsylvania, Philadelphia

DUARTE, CARLOS, Dept. of Biochemistry and Molecular Biophysics, Columbia University, New York, New York

DURBIN, JIM, Structural, Analytical, and Medicinal Chemistry Unit, Pharmacia Corp., Kalamazoo, Michigan

EASTMAN, STEPHEN, Dept. of Biological Sciences, Lehigh University, Bethlehem, Pennsylvania

EBERT, JOAN, Cold Spring Harbor Laboratory Press, Cold Spring Harbor Laboratory, Woodbury, New York

ELLIS, STEVEN, Dept. of Biochemistry, University of Louisville, Louisville, Kentucky

EVERSOLE, ASHLEY, Dept. of Molecular Biophysics and Biochemistry, Yale University, New Haven, Connecticut

FARABAUGH, PHILIP, Dept. of Biological Sciences, University of Maryland-Baltimore County, Baltimore

FAYET, OLIVIER, Dept. of Microbiology and Genetics, Centre National de la Recherche Scientifique, Toulouse, France

FEINBERG, JASON, Dept. of Chemistry and Biochemistry, University of California at San Diego, La Jolla

FELDEN, BRICE, Dept. of Pharmaceutical Biochemistry, University of Rennes, Rennes, France

FLANAGAN, JOHN, Dept. of Biochemistry and Biophysics, Health Science Center, Texas A&M University, College Station, Texas

FRANK, JOACHIM, Dept. of Biomedical Sciences, Howard Hughes Medical Institute, Wadsworth Center, State University of New York, Albany

FREDRICK, KURT, Dept. of Biology, University of California, Santa Cruz

FRIEW, YESHITILA, Dept. of Microbiology, Meharry Medical College, Nashville, Tennessee

FROSHAUER, SUSAN, Rib-X Pharmaceuticals, Inc., Guilford, Connecticut

FUCHS, ECKART, Dept. of Molecular Genetics, University of Baden-Württemberg, Heidelberg, Germany

GANOZA, CLELIA, Banting and Best Department of Medical Research, University of Toronto, Toronto, Canada

GAVIRAGHI, CRISTINA, Dept. of Molecular Medicine and Pathology, Fondazione San Raffaele del Monte Tabor, Milan, Italy

GEBALLE, ADAM, Dept. of Human Biology, Fred Hutchinson Cancer Research Center, Seattle, Washington

GERBI, SUSAN, Dept. of Molecular Cell Biology, BioMed Division, Brown University, Providence, Rhode Island

GESTELAND, HARRIET, Salt Lake City, Utah

GESTELAND, RAYMOND, Dept. of Human Genetics, Howard Hughes Medical Institute, University of Utah, Salt Lake City

GHOSH, SHUBHENDU, Dept. of Molecular Genetics and Microbiology, School of Medicine, University of Massachusetts, Worcester

GHOSH, SRIKANTA, Dept. of Chemistry and Biochemistry, University of California at San Diego, La Jolla

GILMOUR, RAYMOND, Infectious Diseases Unit, Eli Lilly and Co., Indianapolis, Indiana

GLASFELD, ELIZABETH, Infectious Disease Unit, Wyeth-Ayerst Research, Pearl River, New York

GOLDMAN, EMANUEL, Dept. of Microbiology and Molecular Genetics, New Jersey Medical School-UMDNJ, Newark, New Jersey

GONZALO, PHILIPPE, Lab. de Biologie Médicale, Unité Mixte de Recherche, Centre National de la Recherche Scientifique, Lyon, France

GOTTESMAN, SUSAN, Lab. of Molecular Biology, National Cancer Institute, National Institutes of Health, Bethesda, Maryland

GREEN, RACHEL, Dept. of Molecular Biology and Genetics, Johns Hopkins University, Baltimore, Maryland

GREEN, RUSSELL, Astbury Centre for Structural Molecular Biology, University of Leeds, United Kingdom

GREGORY, STEVEN, Dept. of Molecular and Cell Biology and Biochemistry, Brown University, Providence, Rhode Island

GRIFFITHS, ANTHONY, Dept. of Biological Chemistry and Molecular Pharmacology, Harvard Medical School, Boston, Massachusetts

GRODZICKER, TERRI, Cold Spring Harbor Laboratory, Cold Spring Harbor, New York

GROFT, CAROLINE, Dept. of Molecular Biophysics, Rockefeller University, New York, New York

GROSS, JOHN, Dept. of Biological Chemistry and Molecular Pharmacology, Harvard Medical School, Boston, Massachusetts

GUALERZI, CLAUDIO, Dept. of Biology MCA, University of Camerino, Camerino, Italy

HAMIRALLY, SOFIA, Dept. of Pathology, University of Cambridge, Cambridge, United Kingdom

HANNA, RAVEN, Dept. of Physics, University of California, Berkeley

HANSEN, JEFFREY, Dept. of Molecular Biophysics and Biochemistry, Yale University, New Haven, Connecticut

HARGER, JASON, Dept. of Molecular Genetics and Microbiology, Robert Wood Johnson Medical School-UMDNJ, Piscataway, New Jersey

HARTMAN, HYMAN, Dept. of Biology, Massachusetts Institute of Technology, Cambridge, Massachusetts

HEDENSTIERNA, KLAS, Dept. of Molecular Genetics, M.D. Anderson Cancer Center, University of Texas, Houston

HELLEN, CHRISTOPHER, Dept. of Microbiology and Immunology, Health Science Center, State University of New York, Brooklyn

HENDERSON, ALLEN, Dept. of Biological Chemistry, University of California, Davis

HENTZE, MATTHIAS, Gene Expression Programme, European Molecular Biology Laboratory, Heidelberg, Germany

HERMANN, THOMAS, Computational Chemistry and Structure Unit, Anadys Pharmaceuticals, Inc., San Diego, California

HERR, WINSHIP, Cold Spring Harbor Laboratory, Cold Spring Harbor, New York

HICKERSON, ROBYN, Dept. of Biology, University of California, Santa Cruz

HINNEBUSCH, ALAN, Lab. of Eukaryotic Gene Regulation, National Institute of Child Health and Human Development, National Institutes of Health, Bethesda, Maryland

HIROKAWA, GO, Dept. of Pharmaceutical Sciences, Chiba University, Chiba, Japan

HIRSHON, JORDON, Dept. of Biology, Long Island University, Brooklyn, New York

HOHN, THOMAS, Friedrich-Miescher-Institute, Basel, Switzerland

HORI-TAKEMOTO, CHIE, Genomic Sciences Center, RIKEN Yokohama Institute, Yokohama, Japan

HORNEMANN, ULFERT, Dept. of Pharmaceutical Sciences, School of Pharmacy, University of Wisconsin, Madison

INADA, TOSHIFUMI, Div. of Biological Science, Nagoya University, Nagoya, Japan

INAGAKI, YUJI, Dept. of Biochemistry and Molecular Biology, Dalhousie University, Halifax, Nova Scotia, Canada

IPPOLITO, JOSEPH, Dept. of Molecular Biophysics and Biochemistry, Yale University, New Haven, Connecticut

JACKSON, RICHARD, Dept. of Biochemistry, University of Cambridge, Cambridge, United Kingdom

JACOBSON, ALLAN, Dept. of Molecular Genetics and Microbiology, School of Medicine, University of Massachusetts, Worcester

JAYASEKERA, MAITHRI, Bacterial Genomics Unit, Johnson & Johnson, San Diego, California

JIAO, XINFU, Dept. of Cell Biology and Neuroscience, Rutgers University, Piscataway, New Jersey

JOHNSON, ARLEN, Dept. of Molecular Genetics and Microbiology, University of Texas, Austin

JOHNSON, ARTHUR, Health Science Center, Texas A&M University, College Station, Texas

JOHNSON, TONNY, Research and Development Unit, Promega Corp., Madison, Wisconsin

JOSEPH, SIMPSON, Dept. of Chemistry and Biochemistry, University of California at San Diego, La Jolla

JUBIN, RONALD, Antiviral Therapy Unit, Schering-Plough Research Institute, Kenilworth, New Jersey

KAJI, AKIRA, Dept. of Microbiology, School of Medicine, University of Pennsylvania, Philadelphia

KAJI, HIDEKO, Dept. of Biochemistry and Molecular Pharmacology, Thomas Jefferson University, Philadelphia, Pennsylvania

KAMINISHI, TATSUYA, Dept. of Biophysics and Biochemistry, University of Tokyo, Tokyo, Japan

KARLSEN, JESPER, Dept. of Molecular Biophysics and Biochemistry, Yale University, New Haven, Connecticut

KAVRAN, JENNIFER, Dept. of Molecular Biophysics and Biochemistry, Yale University, New Haven, Connecticut

KESSL, JACQUES, Dept. of Biochemistry, Dartmouth Medical School, Hanover, New Hampshire

KIEL, MICHAEL, Dept. of Microbiology, University of Pennsylvania, Philadelphia

KIM, BYUNG-DONG, Dept. of Plant Sciences, Seoul National University, Suwon, South Korea

KIMBALL, SCOT, Dept. of Physiology, College of Medicine, Pennsylvania State University, Hershey, Pennsylvania

KINZY, TERRI, Dept. of Molecular Genetics and Microbiology, Robert Wood Johnson Medical School-UMDNJ, Piscataway, New Jersey

KIRILLOV, STANISLAV, Dept. of Structural and Molecular Biochemistry, North Carolina State University, Raleigh, North Carolina

KLEIN, DANIEL, Dept. of Molecular Biophysics and Biochemistry, Yale University, New Haven, Connecticut

KOC, EMINE, Dept. of Chemistry, University of North Carolina, Chapel Hill

KOEHRER, CAROLINE, Dept. of Biology, Massachusetts Institute of Technology, Cambridge, Massachusetts

KOLUPAEVA, VICTORIA, Dept. of Microbiology, Health Science Center, State University of New York, Brooklyn

KORNEEVA, NADEJDA, Dept. of Microbiology and Immunology, Health Science Center, State University of New York, Brooklyn

KRUPPA, JOACHIM, Dept. of Molecular Cell Biology, Hamburg University, Hamburg, Germany

LA TEANA, ANNA, Institute of Biochemistry, University of Ancona, Ancona, Italy

LAKE, JAMES, Molecular Biology Institute, University of California, Los Angeles

LANCASTER, LAURA, Dept. of Biology, University of California, Santa Cruz

LE ROY, FLORENCE, Dept. of Molecular Genetics and Microbiology, Robert Wood Johnson Medical School-UMDNJ, Piscataway, New Jersey

LIEBERMAN, KATE, Dept. of Biology, University of California, Santa Cruz

LIIV, AIVAR, Estonian Biocentre, Tartu University, Tartu, Estonia

LINDER, PATRICK, Dept. of Medical Biochemistry, University of Geneva, Geneva, Switzerland

LIVINGSTONE, MARK, Cell Signaling Technology, Inc., Beverly, Massachusetts

LOMAKIN, IVAN, Dept. of Microbiology and Immunology, Health Science Center, State University of New York, Brooklyn

LONDEI, PAOLA, Dept. Biochimica Medica e Biologia Medica, University of Bari, Bari, Italy

LONG, JOHN, Dept. of Structural Biology, Message Pharmaceuticals, Malvern, Pennsylvania

LONG, KATHERINE, Institute of Molecular Biology, University of Copenhagen, Copenhagen, Denmark

LORENI, FABRIZIO, Dept. of Biology, University of Rome, Rome, Italy

MADDOCK, JANINE, Dept. of Biology, University of Michigan, Ann Arbor

MADEN, BARRY, School of Biological Sciences, University of Liverpool, Liverpool, United Kingdom

MAGUIRE, BRUCE, Dept. of Biochemistry and Molecular Biology, University of Massachusetts, Amherst

MANKIN, ALEXANDER, Center for Pharmaceutical Biotechnology, University of Illinois, Chicago

MANSELL, JOHN, Thyroid Division, Brigham and Women's Hospital, Harvard Medical School, Harvard Institutes of Medicine, Boston, Massachusetts

MAQUAT, LYNNE, Dept. of Biochemistry and Biophysics, School of Medicine, University of Rochester, Rochester, New York

MARCH, PAUL, School of Microbiology and Immunology, University of New South Wales, Kensington, New South Wales, Australia

MARZI, STEFANO, Div. of Biological Sciences, University of Montana, Missoula

MATHEWS, MICHAEL, Dept. of Biochemistry and Molecular Biology, New Jersey Medical School-UMDNJ, Newark, New Jersey

MATSUO, HIROSHI, Dept. of Biological Chemistry and Molecular Pharmacology, Harvard Medical School, Boston, Massachusetts

MAYER, CHRISTINE, Dept. of Biology, Massachusetts Institute of Technology, Cambridge, Massachusetts

MCCARTY, GREGORY, Dept. of Biological Sciences, University of Maryland-Baltimore County, Baltimore

MCCROSKEY, MARK, Protein Science Unit, Pharmacia Corp., Kalamazoo, Michigan

MEIER, TOM, Dept. of Anatomy and Structural Biology, Albert Einstein College of Medicine, Bronx, New York

MERRYMAN, CHUCK, Whitehead Institute, Massachusetts Institute of Technology, Cambridge, Massachusetts

MESKAUSKAS, ARTURAS, Dept. of Molecular Genetics and Microbiology, Robert Wood Johnson Medical School-UMDNJ, Piscataway, New Jersey

MEYUHAS, ODED, Dept. of Biochemistry, Hebrew University-Hadassah Medical School, Jerusalem, Israel

MOORE, PETER, Dept. of Chemistry, Yale University, New Haven, Connecticut

MUGNIER, PIERRE, CRVA-Infectious Disease Group, Aventis Pharma, Vitry-sur-Seine, France

MURGOLA, EMANUEL, Dept. of Molecular Genetics, M.D. Anderson Cancer Center, University of Texas, Houston

MUSUNURU, KIRAN, Lab. of Molecular Neuro-Oncology, Rockefeller University, New York, New York

MUTH, GREG, Dept. of Molecular Biophysics and Biochemistry, Yale University, New Haven, Connecticut

NAKAMURA, YOSHIKAZU, Dept. of Basic Medical Sciences, Institute of Medical Science, University of Tokyo, Tokyo, Japan

NAPTHINE, SAWSAN, Dept. of Pathology, University of Cambridge, Cambridge, United Kingdom

NAUTRUP-PEDERSEN, GITTE, Institute of Molecular and Structural Biology, Åarhus University, Åarhus, Denmark

NIELSEN, KLAUS, Lab. of Eukaryotic Gene Regulation, National Institute of Child Health and Human Development, National Institutes of Health, Bethesda, Maryland

NIERHAUS, KNUD, AG-Ribosomen, Max-Planck-Institut für Molekulare Genetik, Berlin, Germany

NIX, JAY, Dept. of Biology, University of California, Santa Cruz

NOLLER, HARRY, Center for Molecular Biology of RNA, University of California, Santa Cruz

NOMURA, MASAYASU, Dept. of Biological Chemistry, University of California, Irvine

NYBORG, JENS, Institute of Molecular and Structural Biology, Åarhus University, Åarhus, Denmark

OAKES, MELANIE, Dept. of Biological Chemistry, University of California, Irvine

O'BRIEN, THOMAS, Dept. of Biochemistry and Molecular Biology, University of Florida, Gainesville

O'CONNOR, MICHAEL, Dept. of Molecular and Cell Biology, Brown University, Providence, Rhode Island

OFENGAND, JAMES, Dept. of Biochemistry and Molecular Biology, School of Medicine, University of Miami, Miami, Florida

OGATA, CRAIG, Howard Hughes Medical Institute, Brookhaven National Laboratory, Upton, New York

OGLE, JAMES, Structural Studies Division, MRC Laboratory of Molecular Biology, Cambridge, United Kingdom

PAPE, TILLMANN, Dept. of Biochemistry, Imperial College, London, United Kingdom

PE'ERY, TSAFI, Dept. of Biochemistry and Molecular Biology, New Jersey Medical School-UMDNJ, Newark, New Jersey

PEDERSON, THORU, Dept. of Biochemistry and Molecular Biology, School of Medicine, University of Massachusetts, Worcester

PELTZ, STUART, Dept. of Molecular Genetics and Microbiology, Robert Wood Johnson Medical School-UMDNJ, Piscataway, New Jersey

PESTOVA, TATYANA, Dept. of Microbiology and Immunology, Health Science Center, State University of New York, Brooklyn

PHELPS, STEVEN, Dept. of Chemistry and Biochemistry, University of California at San Diego, La Jolla

POLACEK, NORBERT, Center for Pharmaceutical Biotechnology, University of Illinois, Chicago

POLLOCK, MILA, Cold Spring Harbor Laboratory, Cold Spring Harbor, New York

POOGGIN, MIKHAIL, Friedrich-Miescher-Institute, Basel, Switzerland

POOL, MARTIN, ZMBH, University of Heidelberg, Heidelberg, Germany

RAJBHANDARY, UTTAM, Dept. of Biology, Massachusetts Institute of Technology, Cambridge, Massachusetts

RAMAKRISHNAN, VENKI, MRC Laboratory of Molecular Biology, Cambridge, United Kingdom

RAMARAO, RACHANA, Dept. of Pathology, University of Cambridge, Cambridge, United Kingdom

RICH, ALEXANDER, Dept. of Biology, Massachusetts Institute of Technology, Cambridge, Massachusetts

RICHTER, JOEL, Dept. of Molecular Genetics and Microbiology, School of Medicine, University of Massachusetts, Worcester

ROBERT, FRANCIS, Dept. of Biochemistry, University of Montréal, Montréal, Québec, Canada

ROBINSON, PHILIP, Molecular Medicine Unit, St. James's University Hospital, University of Leeds, Leeds, United Kingdom

ROLL-MECAK, ANTONINA, Dept. of Molecular Biophysics, Rockefeller University, New York, New York

RON, DAVID, Skirball Institute, School of Medicine, New York University, New York, New York

ROSENFELD, AMY, Dept. of Microbiology, Columbia University, New York, New York

RUGGERO, DAVIDE, Dept. of Human Genetics, Memorial Sloan-Kettering Cancer Center, New York, New York

RYAZANOV, ALEXEY, Dept. of Pharmacology, Robert Wood Johnson Medical School-UMDNJ, Piscataway, New Jersey

SACHS, MATTHEW, Dept. of Biochemistry and Molecular Biology, Oregon Graduate Institute, Portland, Oregon

SANBONMATSU, KEVIN, Los Alamos National Laboratory, Los Alamos, New Mexico

SANYAL, SUPARNA, Dept. of Molecular Biophysics, Lund University, Lund, Sweden

SARNOW, PETER, Dept. of Microbiology and Immunology, Stanford University, Stanford, California

SATTLEGGER, EVELYN, Lab. of Gene Regulation and Development, National Institute of Child Health and Human Development, National Institutes of Health, Bethesda, Maryland

SCHEUNER, DONALYN, Dept. of Biological Chemistry, Howard Hughes Medical Institute, University of Michigan, Ann Arbor

SCHIMMEL, PAUL, Skaggs Institute for Chemical Biology, The Scripps Research Institute, La Jolla, California

SCHMEING, MARTIN, Dept. of Molecular Biophysics and Biochemistry, Yale University, New Haven, Connecticut

SCHNEIDER, ROBERT, Dept. of Microbiology, School of Medicine, New York University, New York, New York

SCHRODER, PATRICIA, Dept. of Molecular and Cell Biology, Brandeis University, Waltham, Massachusetts

SEO, HYUK-SOO, Dept. of Chemistry, University of Pennsylvania, Philadelphia

SHIN, BYUNG-SIK, Lab. of Eukaryotic Gene Regulation, National Institute of Child Health and Human Development, National Institutes of Health, Bethesda, Maryland

SHINABARGER, DEAN, Infectious Diseases Unit, Pharmacia Corp., Kalamazoo, Michigan

SHUMAN, STEWART, Molecular Biology Program, Memorial Sloan-Kettering Cancer Center, New York, New York

SIEGEL, VIVIAN, *Cell* and *Molecular Cell*, Cell Press, Cambridge, Massachusetts

SIMONSON, ANNE, Dept. of Molecular, Cell, and Developmental Biology, Molecular Biology Institute, University of California, Los Angeles

SKINNER, GARY, Dept. of Physics, University of Arizona, Tucson

SÖLL, DIETER, Dept. of Molecular Biophysics and Biochemistry, Yale University, New Haven, Connecticut

SOMANCHI, ARAVIND, Dept. of Cell Biology, The Scripps Research Institute, La Jolla, California

SONENBERG, NAHUM, Dept. of Biochemistry, McGill University, Montréal, Québec, Canada

SOUTHWORTH, DANIEL, Dept. of Molecular Biology and Genetics, Johns Hopkins University, Baltimore, Maryland

SPAHN, CHRISTIAN, Howard Hughes Medical Institute, Wadsworth Center, State University of New York, Albany

SPIRIN, ALEXANDER, Dept. of Biochemistry, Institute of Protein Research, Russian Academy of Sciences, Moscow, Russia

SPREMULLI, LINDA, Dept. of Chemistry, University of North Carolina, Chapel Hill

STAHL, GUILLAUME, Dept. of Biological Sciences, University of Maryland-Baltimore County, Baltimore

STARCK, SHELLEY, Dept. of Chemistry and Chemical Engineering, California Institute of Technology, Pasadena, California

STATHOPOULOS, CONSTANTINOS, Dept. of Molecular Biophysics and Biochemistry, Yale University, New Haven, Connecticut

STEITZ, THOMAS, Dept. of Molecular Biophysics and Biochemistry, Howard Hughes Medical Institute, Yale University, New Haven, Connecticut

STELZL, ULRICH, Dept. of Biochemistry and Biophysics, Memorial Sloan-Kettering Cancer Center, New York, New York

STEWART, DAVID, Cold Spring Harbor Laboratory, Cold Spring Harbor, New York

STILLMAN, BRUCE, Cold Spring Harbor Laboratory, Cold Spring Harbor, New York

STROBEL, SCOTT, Dept. of Molecular Biophysics and Biochemistry, Yale University, New Haven, Connecticut

SUTCLIFFE, JOYCE, Antibacterials Unit, Pfizer Global Research and Development, Groton, Connecticut

SWANEY, STEVEN, Infectious Diseases Research Unit, Pharmacia Corp., Kalamazoo, Michigan

SZEP, SZILVIA, Dept. of Chemistry, Yale University, New Haven, Connecticut

TABB, AMY, Dept. of Biochemistry and Molecular Biology, University of Louisville, Louisville, Kentucky

TAN, NANCY, Molecular Medicine Unit, St. James's University Hospital, University of Leeds, Leeds, United Kingdom

TAN, ZHONGPING, Department of Chemistry, Columbia University, New York, New York

TANAKA, TATSUO, Department of Biochemisty, School of Medicine, University of the Ryukyus, Okinawa, Japan

TELERMAN, ADAM, Molecular Engines Laboratories, Paris, France

TENSON, TANEL, Institute of Molecular and Cell Biology, Tartu University, Tartu, Estonia

TERAOKA, YOSHIKA, AG-Ribosomen, Max-Planck-Institut für Molekulare Genetik, Berlin, Germany

THOMAS, ADRI, Dept. of Developmental Biology, Utrecht University, Utrecht, The Netherlands

THOMPSON, JILL, Dept. of Molecular and Cellular Biology and Biochemistry, Brown University, Providence, Rhode Island

TISSIÈRES, ALFRED, Geneva, Switzerland

TISSIÈRES, VIRGINIA, Geneva, Switzerland

TOKIMATSU, HIROAKI, Dept. of Physics, Osaka Medical College, Takatsuki, Japan

TREVATHAN, MEGAN, Dept. of Molecular Biology, The Scripps Research Institute, La Jolla, California

TRIMAN, KATHLEEN, Dept. of Biology, Franklin and Marshall College, Lancaster, Pennsylvania

TRUMPOWER, BERNARD, Dept. of Biochemistry, Dartmouth Medical School, Hanover, New Hampshire

TUMER, NILGUN, Biotechnology Center, Rutgers University, New Brunswick, New Jersey

UHLENBECK, OLKE, Dept. of Chemistry and Biochemistry, University of Colorado, Boulder

URBONAVICIUS, JAUNIUS, Dept. of Microbiology, Umeå University, Umeå, Sweden

VAN HEEL, MARIN, Dept. of Biochemistry, Imperial College of Science, Technology and Medicine, London, United Kingdom

VAN 'T RIET, JAN, Dept. of Biochemistry and Molecular Biology, BioCentrum, Free University, Amsterdam, The Netherlands

VARSHNEY, UMESH, Dept. of Microbiology and Cell Biology, Indian Institute of Science, Bangalore, India

VOERDERWULBECKE, SONJA, Dept. of Biochemistry and Molecular Biology, University of Freiburg, Freiburg, Germany

WADA, AKIRA, Dept. of Physics, Osaka Medical College, Takatsuki, Japan

WAGNER, GERHARD, Dept. of Biological Chemistry and Molecular Pharmacology, Harvard Medical School, Boston, Massachusetts

WARD, CHRISTINE, Drug Discovery Unit, Bacterial Genomics Group, R.W. Johnson Pharmaceutical Research Institute, San Diego, California

WARE, VASSIE, Dept. of Biological Sciences, Lehigh University, Bethlehem, Pennsylvania

WARNER, JONATHAN, Dept. of Cell Biology, Albert Einstein College of Medicine, Bronx, New York

WELCH, ELLEN, Dept. of Biology, PTC Therapeutics, Inc., South Plainfield, New Jersey

WICKENS, MARVIN, Dept. of Biochemistry, University of Wisconsin, Madison

WILLIAMSON, JAMIE, Dept. of Molecular Biology, The Scripps Research Institute, La Jolla, California

WILSON, EMMA, *Trends in Biochemical Sciences*, Elsevier Science, London, United Kingdom

WIMBERLY, BRIAN, Structural Analytical and Medicinal Chemistry Unit, Pharmacia Corp., Kalamazoo, Michigan

WITKOWSKI, JAN, Banbury Center, Cold Spring Harbor Laboratory, Cold Spring Harbor, New York

WOLLENZIEN, PAUL, Dept. of Molecular and Structural Biochemistry, North Carolina State University, Raleigh, North Carolina

YARUS, MICHAEL, Dept. of Biology, University of Colorado, Boulder

YOKOYAMA, SHIGEYUKI, Dept. of Biophysics and Biochemistry, University of Tokyo, Tokyo, Japan

YONATH, ADA, Dept. of Structural Biology, Weizmann Institute of Science, Rehovot, Israel

YOSHIDA, HIDEJI, Dept. of Physics, Osaka Medical College, Takatsuki, Japan

YUSUPOV, MARAT, Institut de Génétique et de Biologie Moléculaire et Cellulaire, Centre National de la Recherche Scientifique, Strasbourg, France

ZARIVACH, RAZ, Dept. of Structural Biology, Weizmann Institute of Science, Rehovot, Israel

ZENGEL, JANICE, Dept. of Biological Sciences, University of Maryland-Baltimore County, Baltimore

ZHANG, AIXIA, National Institute of Child Health and Human Development, National Institutes of Health, Bethesda, Maryland

ZIMMERMANN, ROBERT, Dept. of Biochemistry and Molecular Biology, University of Massachusetts, Amherst

First row: C. Hellen, M. Mathews; K. Nierhaus, G. Skinner, Y. Teraoka
Second row: A. Dahlberg, J. Warner
Third row: M. Oakes, J. Maddock; R. Brimacombe, M. van Heel
Fourth row: A. Kaji, A. Spirin; R. Zimmermann, B. Maguire

First row: E. Murgola, Z. Druzina; R. Agrawal, S. Sanyl, A. Kaji
Second row: A. Johnson, L. Spremulli; A. Thomas, M. Yusupov, J. Ofengand
Third row: M. Yarus, V. Ramakrishnan
Fourth row: F. Amaldi, T. Hohn, O. Meyuhas; A. Spirin, M. Nomura, P. Moore

First row: T. Meier, D. Ron; R. Beckmann, J. Frank, D. Stewart; N. Sonenberg, A. Spirin
Second row: R. Gesteland, B. Stillman, J. Atkins; O. Uhlenbeck, V. Ramakrishnan
Third row: A. La Teana, F. Amaldi, C. Gualerzi; T. O'Brien and rapt audience

First row: M. Nomura, O. Uhlenbeck; J. Nyborg, H. Hartman
Second row: S. Gottesman, A. Hinnebusch; R. Bhasker, M. Wickens
Third row: L. Maquat, R. Jackson; T. Steitz, A. Dahlberg; M. Yarus, U. RajBhandary

First row: A. Jacobson, E. Welch, F. La Roy, S. Ghosh; J. Richter, S. Goldman
Second row: W. Herr, H. Noller; R. Jackson, J. Doudna
Third row: Banquet; E. Fuchs, H. Hartman

First row: E. Wilson, M. Hentze; S. Napthine, S. Hamirally; R. Brimacombe
Second row: P. Moore; A. Tissières, T. Pedersen
Third row: S. Swaney, A. Carter; S. Shuman, P. Linder

Foreword

How proteins are made became a major focus following the discovery of the double-helical structure of DNA in 1953. Once it was realized that the genetic code existed within the order of base pairs of DNA, it became of immediate importance to understand how that information was converted into the sequences of amino acids that defined individual proteins. The solution of this problem, which includes deciphering the genetic code, is one of the great achievements in the science that is now known as molecular biology. Many of these achievements have been recorded and celebrated at previous Symposia in this series.

Protein synthesis occurs within the ribosome, a molecular machine comprising both protein and RNA that in eukaryotes is first assembled in the nucleus of a cell. RNA also functions as the adaptor in the form of tRNA that brings the individual amino acids to the ribosome. Ribosomes themselves associate with other molecular machineries, such as the machinery that directs proteins to the cell surface. It was once thought that figuring out the molecular details of how protein synthesis occurs would long be an insurmountable problem. But dramatic events in structural biology in recent years have changed this view and are the reason for selecting the ribosome as the topic for this year's Symposium.

High-resolution images of the structure of a ribosome not only tell us how beautiful this machine is, but reveal the surprising finding that peptide-bond formation is directed by RNA. Although this had been hinted at for some time, it was difficult to envision without seeing the ribosome structure. The structure of the ribosome, either as its individual parts or as a whole, has also shown how the newly minted peptide chain emerges and, in the case of some proteins, how it interacts with other molecular machinery that exports proteins to the cell surface.

First and foremost, I am most appreciative of the help from David Stewart, Director of our very efficient meetings and courses division, in organizing this meeting. I am also particularly grateful to Peter Moore, Tom Steitz, Michael Mathews, Alan Hinnebusch, and Susan Gerbi for advice on areas of research to be covered. The meeting started with a talk by Jim Watson about his early and important role in the ribosome field while on the faculty at Harvard. Later followed outstanding introductory presentations by Tom Steitz, Venki Ramakrishnan, Nahum Sonenberg, and Roland Beckmann. The Friday evening Reginald Harris Lecture was masterfully presented by Harry Noller, and the Dorcas Cummings Memorial Lecture to our friends and neighbors was beautifully presented by Venki Ramakrishnan. There were 60 oral presentations and 136 poster presentations over the course of the meeting, which was held from May 31 to June 4. On the final day, Peter Moore presented us with a very thoughtful summary that is formalized at the end of this volume.

Particularly pleasing at this meeting was the return to Cold Spring Harbor Laboratory of many of the principal contributors to the early days of ribosome research who reminisced about experiments that contributed to understanding, at a most fundamental level, the central dogma of biology. They included Alfred Tissières, Donald Caspar, and Alex Rich.

As always, the staff at the Laboratory meetings and courses office made this Symposium pleasant for all who attended, and even for the organizer. Those talented people at the Cold Spring Harbor Laboratory Press, particularly Joan Ebert and Patricia Barker, ensured that the volume reflected the high quality of the science presented. Finally, I wish to acknowledge and thank those who financially supported this important meeting, including the National Cancer Institute, which provided a multi-year grant, and the Corporate Sponsors who are listed on the following page.

Bruce Stillman
February 2002

This meeting was funded in part by the **National Cancer Institute**, a branch of the **National Institutes of Health**.

Contributions from the following companies provide core support for the Cold Spring Harbor meetings program.

Corporate Benefactors

Pfizer Inc.

Corporate Sponsors

Abbott Bioresearch Center, Inc.
Amgen Inc.
Applied Biosystems
Aventis Pharma AG
BioVentures, Inc.
Bristol-Myers Squibb Company
Chiron Corporation
Chugai Research Institute for Molecular Medicine, Inc.
Cogene BioTech Ventures, Ltd.
Diagnostic Products Corporation
DuPont Pharmaceuticals Company
Eli Lilly and Company
Forest Laboratories

Genentech, Inc.
Genetics Institute
GlaxoSmithKline
Hoffmann-LaRoche Inc.
Johnson & Johnson
Kyowa Hakko Kogyo Co., Ltd.
Merck Research Laboratories
New England BioLabs, Inc.
Novartis Pharma Research
OSI Pharmaceuticals, Inc.
Pall Corporation
Pharmacia Corporation
Research Genetics, Inc.
Schering-Plough Corporation

Plant Corporate Associates

Monsanto Company
Pioneer Hi-Bred International, Inc.

Torrey Mesa Research Institute
Westvaco Corporation

Corporate Affiliates

Affymetrix, Inc.

Ceptyr, Inc.

Corporate Contributors

Alexis Corporation
Biogen, Inc.
Digital Gene Technologies, Inc.
DoubleTwist, Inc.
Epicentre Technology
GeneMachine
Genomica Corporation
Immuno-Rx, Inc.
Incyte Genomics
InforMax

Invitrogen Corporation
LabBook, Inc.
Lexicon Genetics, Inc.
Proteome, Inc.
Pyrosequencing
Qiagen
Silicon Genetics
Transgenomics, Inc.
ZymoGenetics, Inc.

Foundations

Albert B. Sabin Vaccine Institute, Inc.

Contents

Frameshifting, Recoding, Shunting, IRES

Messenger RNA and Translational Control

Initiation

Elongation and Termination

Regulation of Protein Synthesis

Ribosome Biogenesis

RNA Structure and the Roots of Protein Synthesis

A. RICH

Department of Biology, Massachusetts Institute of Technology,
Cambridge, Massachusetts 02139

When we think about scientific progress in the past, compared to progress in the future, there is a singular asymmetry. Many of the discoveries of the past look obvious. We ask, Why did it take people so long to uncover what today seems obvious? On the other hand, when we look to the future in research, it is unclear, clouded with many uncertainties. We cannot foresee which path will turn out to be correct. Here I look back and describe early research on RNA structure and protein biosynthesis, starting in an era before the term "molecular biology" was used.

This account is by no means a review of early work in protein biosynthesis. That would be an exhaustive enterprise of far greater length. Here, I focus largely on the work of my colleagues and myself dealing with selective problems of RNA structure and protein biosynthesis. Many of the results described here are part of current conventional understanding regarding RNA structure and protein biosynthesis, and as such they appear obvious. At the time of their discovery, however, they were not entirely obvious. This account is written in the context of the current important milestone, determination of the three-dimensional structure of the ribosome.

CAN RNA FORM A DOUBLE HELIX?

My research work on RNA structure started in the early 1950s when nothing was known of its three-dimensional structure, and the question asked was whether RNA could form a double helix comparable to the one described by Watson and Crick (1953). In their paper, they suggested that the double helix could not form with RNA molecules because of the van der Waals crowding associated with the additional oxygen atom on the ribose ring. The issue of double helix formation was further complicated by the fact that it was not known at that time whether RNA was linear or branched, since the additional hydroxyl group on the ribose ring represented a potential branch point. While we were both postdoctoral fellows at Cal Tech in 1953 and 1954, Jim Watson and I tried to make oriented fibers of RNA to see whether their X-ray diffraction pattern was helical. The results were inconclusive (Rich and Watson 1954a,b). On moving to the National Institutes of Health (NIH) later in 1954, I continued to work on the problem. No significant progress was made until the polynucleotide phosphorylase enzyme was discovered by Grunberg-Manago in Ochoa's laboratory (Grunberg-Manago et al. 1955). That enzyme made it possible to make very long strands of polyribonucleotides, and with it, structural investigations became more productive.

David Davies joined me at NIH to work on RNA structure. In 1956, we mixed solutions of polyriboadenylic acid (poly rA) and polyribouridylic acid (poly rU) and discovered a remarkable transformation revealed by X-ray studies. These two molecules actually reacted with each other to produce a double helix! A brief note was sent to the Journal of the American Chemical Society (JACS) (Fig. 1) in June of that year (Rich and Davies 1956), describing the work and a preliminary interpretation of the pattern. The diffraction pattern clearly indi-

[Reprinted from the Journal of the American Chemical Society, **78**, 3548 (1956).]

A NEW TWO STRANDED HELICAL STRUCTURE: POLYADENYLIC ACID AND POLYURIDYLIC ACID

Sir:

While studying the X-ray diffraction patterns of synthetic nucleotide polymers, we mixed together the sodium salts of polyadenylic acid and polyuridylic acid.[1] There resulted a very rapid increase in viscosity as well as the drop in the optical density at 260 mμ which was reported recently by Warner.[2] From this viscous solution, tough, glassy fibers can be drawn which are negatively birefringent, $\Delta n = -0.10$.

These fibers produce a well-oriented X-ray diffraction pattern with a distribution of intensity which is characteristically helical.

. . . .

These results show for the first time that it is possible for the ribonucleic acid (RNA) backbone to assume a configuration not unlike that found in DNA, using the same complementarity in the base pairs.

. . . .

Finally, we would like to point out that this method for forming a two-stranded helical molecule by simply mixing two substances can be used for a variety of studies directed toward an understanding of the formation of helical molecules utilizing specific interactions.

SECTION ON PHYSICAL CHEMISTRY
NATIONAL INSTITUTE OF MENTAL HEALTH   ALEXANDER RICH
BETHESDA 14, MARYLAND   DAVID R. DAVIES

RECEIVED JUNE 8, 1956

Figure 1. Brief excerpt from the publication showing that RNA can form a double helix. This is also the first hybridization reaction. The following figures also have excerpts from publications.

cated the formation of a helical complex that was not present in either of the two individual polymers. Furthermore, the pattern had significant differences compared with those produced from DNA fiber diffraction patterns. In particular, the first layer line was very strong, whereas in the DNA diffraction patterns, the first layer line was weak. It was recognized that this difference implied that, viewed from the axis, the angular separation between the two chains was smaller. This could be due to having the two chains parallel or having the two chains antiparallel but with a larger radius. Subsequent work revealed the latter was correct.

This result, which seems so obvious today, generated a great deal of skepticism at the time. While walking down a long corridor at NIH, I met Herman Kalckar, an eminent Danish biochemist. During our conversation, I mentioned that we discovered that poly rA and poly rU formed a double helix. Kalckar was incredulous. "You mean without an enzyme?" he asked. His attitude was justified, since the only double helix known at that time was one made with the DNA polymerase enzyme that Arthur Kornberg had purified (Kornberg et al. 1956). Other critics thought it was highly unlikely that polymers containing over a thousand nucleotides would be able to disentangle themselves and form a regular structure. They believed it would be hopelessly entangled.

Two weeks after sending off the 1956 JACS note, I wrote a letter to my former postdoctoral mentor, Linus Pauling, describing these results. The letter reveals a sense of incredulity on my part that this reaction could happen and that it was "completely reproducible."

This was the first demonstration that RNA molecules could form a double helix. It was also the first hybridization reaction and, as noted in the JACS letter (Fig. 1), we pointed out that "this method of forming a two-stranded helical molecule" utilizing specific interactions could be used for a variety of studies.

An important method for studying the nucleic acids was measuring their absorbance in the ultraviolet. For some time, it had been known that polymerization of nucleotides resulted in a decrease in absorbance at 260 mμ. Likewise, it had been reported that adding poly rA to poly rU led to hypochromism (Warner 1956). Although the mechanism of hypochromicity was not understood at the time, it was a useful tool for analysis.

Further insight into the reaction of poly rA and poly rU was obtained by carefully measuring hypochromicity in mixtures of varying composition. In work carried out with Gary Felsenfeld (Fig. 2), the hypochromicity fell to a very sharp minimum at a 1:1 mole ratio (Felsenfeld and Rich 1957). This implied that the system was dynamic— the molecules came together to form a helical duplex, but they then disassembled and reassembled so that ultimately all of the gaps between adjacent molecules were closed. This was a dynamic picture of nucleic acid molecules rapidly associating and dissociating, a view that was different from the prevailing view. Before this experiment, macromolecular nucleic acids were regarded as somewhat immobile. The only available mental picture of the structure was the double helix formulated by Wat-

VOL. 26 (1957)   BIOCHIMICA ET BIOPHYSICA ACTA   457

STUDIES ON THE FORMATION ·OF
TWO- AND THREE-STRANDED POLYRIBONUCLEOTIDES

GARY FELSENFELD AND ALEXANDER RICH
National Institute of Mental Health, National Institutes of Health, Bethesda, Md. (U.S.A.)

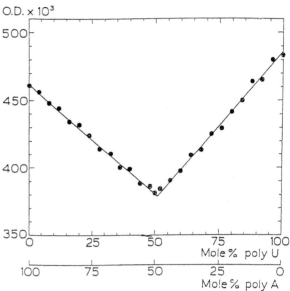

The O.D. at 259 mμ of mixtures of polyadenylic acid and polyuridylic acid. The optical densities were measured within 15 min of mixing. All solutions are in 0.1 M NaCl, 0.01 M glycyl-glycine, pH 7.4, T = 25° C. The least-square lines are shown.

Figure 2. Sharpness of the minimum in the hypochromism of the reaction between poly rA and poly rU meant that the molecules were combining and separating until they formed a duplex without gaps between the base pairs.

son and Crick. Most people thought about the nucleic acids in static terms, whereas the reaction between poly rA and poly rU suggested a more dynamic interpretation of nucleic acid molecules. This led to a significant change in thinking, and it expressed itself in a variety of other systems, such as recombination in which the movements of nucleic acid molecules play a crucial role.

TRIPLE HELICES AS WELL

A year later, it was discovered that the addition of divalent cations such as magnesium would change the picture dramatically (Felsenfeld et al. 1957), leading to the formation of a three-stranded molecule (Fig. 3). We concluded that a second strand of polyuridylic acid bound in the major groove of the poly rA–poly rU duplex. Addition of this strand did not increase the diameter of the molecule and neatly accounted for the 50% increase in sedimentation constant. It was further proposed that the second poly rU strand bound to the N6 and N7 of adenine using two hydrogen bonds to uracil. This proposal was considerably strengthened 2 years later by the X-ray anal-

[Reprinted from the Journal of the American Chemical Society, **79**, 2023 (1957).]

FORMATION OF A THREE-STRANDED POLYNUCLEOTIDE MOLECULE

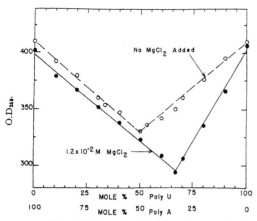

The optical density of various mixtures of polyadenylic acid and polyuridylic acid. Optical densities were measured two hours after mixing. All solutions are in 0.1 M NaCl, 0.01 M glycylglycine, pH 7.4, T = 25°.

We have interpreted these results to mean that the original two-stranded molecule has taken on a third strand of poly-U which fills the helical groove in the (A + U) complex. We have built models to show that such a three-stranded complex is possible with divalent cations neutralizing the charges on the phosphate groups and the additional uracil hydrogen bonded to either one or both bases of the adenine–uracil pair. It should be noted that there is only one position whereby the new uracil residue can make two strong hydrogen bonds, namely, by bonding uracil O_6 and N_1 to adenine N_{10} and N_7. The addition of the third strand need not involve an increase in radius or helical pitch of the molecule, and could therefore account for the increase of approximately 50% in sedimentation coefficient.

SECTION ON PHYSICAL CHEMISTRY   G. FELSENFELD
NATIONAL INSTITUTE OF MENTAL HEALTH
BETHESDA 14, MARYLAND   DAVID R. DAVIES
    ALEXANDER RICH
RECEIVED MARCH 21, 1957

Figure 3. The first nucleic acid triplex.

ysis of the co-crystal containing 9-methyl adenine and 1-methyl thymine by Karst Hoogsteen (1959). The 1957 discovery of the RNA triplex was the precursor of many subsequent studies dealing with nucleic acid triplexes. Furthermore, it pointed out the structural complexity inherent in RNA molecules.

Over the next several years, a variety of polynucleotide interactions were studied (Rich 1957, 1958a,b; Davies and Rich 1958), leading to the formation of two- and three-stranded molecules. Fiber X-ray diffraction and other studies led to the conclusion that the molecules all had bases on the inside and the sugar phosphate chains on the outside. The bases were generally held in place by Watson-Crick or alternative types of hydrogen bonding involving at least two hydrogen bonds.

HOW DOES DNA "MAKE" RNA?

By 1960, several investigators had shown that crude preparations of an RNA polymerase activity could incorporate ribonucleotides into RNA using a DNA template, but the mechanism was not at all clear. It was widely believed that information transfer went from DNA to RNA. But how did that occur (Rich 1959)? The availability of chemically synthesized oligomers of poly deoxythymidylic acid (poly dT) made it possible to study this experimentally (Tener et al. 1958). It was known that the RNA backbone was significantly different from the DNA backbone due to the 2′OH in RNA, so it was not obvious that they could combine. Nonetheless, we showed that these two molecules could accommodate each other to form a hybrid helix containing one strand of poly dT and one strand of poly rA (Rich 1960), as seen from hypochromism (Fig. 4) and other studies. This was the first experimental demonstration of a hybrid helix, and the discovery of messenger RNA was still one year in the future. It immediately provided experimental support for a model of how DNA could "make" RNA, using complementary base-pairing, as in DNA replication. A year later in 1961, experiments by J. Hurwitz with a purified RNA polymerase preparation demonstrated that this was the mechanism underlying information transfer from DNA to RNA (Furth et al. 1961). The reaction between poly dT and poly rA was the first experimental demonstration that the two different backbones could adapt to each other in this method of information transfer. The reaction was also the first hybridization of a DNA molecule with an RNA molecule. The same hybridization is widely used today in the purification of eukaryotic mRNA by hybridizing poly dT to their poly rA tails.

In 1960, Marmur, Doty, and their colleagues demonstrated that it was possible to renature naturally occurring denatured DNA duplexes by incubating them at an intermediate temperature that would allow the single strands to anneal together with the correct sequence (Doty et al. 1960; Marmur and Lane 1960). A year later, this annealing method was also adopted to form DNA–RNA hybrids in viral systems (Hall and Spiegelman 1961).

SINGLE-CRYSTAL X-RAY DIFFRACTION

X-ray diffraction studies of nucleic acid fibers were carried out extensively by M. Wilkins, R. Franklin, and colleagues in the 1950s and 1960s, even though it was realized that the limitations of such studies were enormous. In fiber X-ray diffraction, a rather small number of reflections are registered. However, the number of variables needed to define the structure (at least 3N, where N is the number of atoms) is so great that it was clear that fiber diffraction could not "prove" a structure. It could only say that a particular conformation was compatible with the limited diffraction data from fibers. During this period, the number of single-crystal X-ray diffraction studies was increasing, many of them involving co-crystals of purines and pyrimidine derivatives such as those initiated by Karst Hoogsteen (1959). These studies were

Reprinted from the Proceedings of the NATIONAL ACADEMY OF SCIENCES
Vol. 46, No. 8, pp. 1044–1053.   August, 1960.

A HYBRID HELIX CONTAINING BOTH DEOXYRIBOSE AND RIBOSE
POLYNUCLEOTIDES AND ITS RELATION TO THE TRANSFER OF
INFORMATION BETWEEN THE NUCLEIC ACIDS

By ALEXANDER RICH

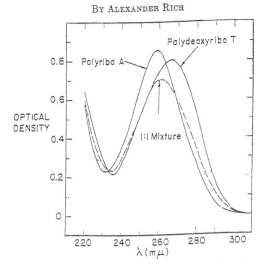

The spectra of polyriboadenylic acid, polydeoxyribothymidylic acid and a 1:1 mixture. All solutions are in 0.6 M NaCl, 0.01 M sodium cacodylate at pH 6.9.   T = 24°C.

Conclusions.—In this paper we have presented evidence which shows that it is possible to have a hybrid helix in which one strand with a DNA backbone can be made to wrap around another strand with an RNA backbone in such a manner that the strands are held together by complementary hydrogen bonds formed between the purine and pyrimidine residues.   This finding may have relevance in the process whereby the genetic information or nucleotide sequence in DNA is transferred to an RNA molecule and suggests experiments which should be carried out to search for the existence of such transfer mechanisms in intact cellular systems.

Figure 4. Discovery of a DNA–RNA hybrid helix.

useful in obtaining information about components of nucleic acid structure. For example, a co-crystal of cytosine and guanine derivatives first demonstrated conclusively that these were held together by three hydrogen bonds (Sobell et al. 1963), not two, as initially suggested by Watson and Crick. Linus Pauling had already strongly emphasized this point based on general structural considerations (Pauling and Covey 1956). Several co-crystals were solved of derivatives of adenine and uracil or adenine and thymine during this period in my laboratory (Mathews and Rich 1964; Katz et al. 1965; Tomita et al. 1967), as well as in others (Voet and Rich 1970). However, a disturbing trend emerged from these studies; namely, all of them were held together by Hoogsteen base-pairing involving N7 and N6 or adenine, and none of them had the Watson-Crick base pairs involving the adenine N6 and N1 atoms. This led some investigators to suggest that the double helix might be held together by Hoogsteen pairing, which would involve protonation of the cytosine residues if two hydrogen bonds were used in

connecting cytosine to the O6 and N7 of guanine. The calculated diffraction pattern of such a helix had many similarities to that predicted by a double helix held together by Watson-Crick base pairs, even though the fit was not good (Arnott et al. 1965). However, the question remained: What is the real structure of the double helix?

THE DOUBLE HELIX AT ATOMIC RESOLUTION

The first single-crystal structures of a double helix were solved in 1973 in my laboratory. This was before it was possible to synthesize and obtain oligonucleotides in significant quantities suitable for crystallographic experiments. However, we succeeded in crystallizing two dinucleoside phosphates, the RNA oligomers GpC (Day et al. 1973) and ApU (Rosenberg et al. 1973). The structures are illustrated in Figure 5. The significant point in this analysis was that the resolution of the diffraction pattern was 0.8 Å. This atomic resolution allowed us to

(*Reprinted from Nature, Vol.* 243, *No.* 5403, *pp.* 150–154.
May 18, 1973)

Double Helix at Atomic Resolution

JOHN M. ROSENBERG, NADRIAN C. SEEMAN,
JUNG JA PARK KIM, F. L. SUDDATH,
HUGH B. NICHOLAS* & ALEXANDER RICH

Department of Biology, Massachusetts Institute of Technology, Cambridge, Massachusetts 02139

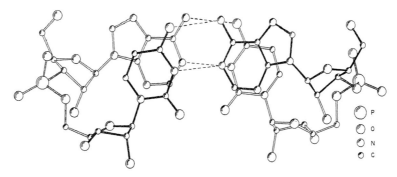

Proc. Nat. Acad. Sci. USA
Vol. 70, No. 3, pp. 849–853, March 1973

**A Crystalline Fragment of the Double Helix: The Structure of the Dinucleoside
Phosphate Guanylyl-3',5'-Cytidine**

ROBERTA OGILVIE DAY*, NADRIAN C. SEEMAN, JOHN M. ROSENBERG, AND ALEXANDER RICH

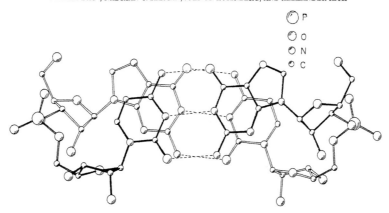

Figure 5. First crystalline nucleic acid duplexes, solved at 0.8 Å resolution. (*Top*) *Nature* cover showing the double helix at atomic resolution: ApU. (*Bottom*) The structure of GpC.

visualize not only the sugar phosphate backbone in the form of a double helix, but also the positions of ions and water molecules. It could be shown that extending the structure using the symmetry of the two base pairs made it possible to generate RNA double helices that were quite similar to the structures that had been deduced from studies of double-helical fibers of RNA. The bond angles and distances from these structures provided the library of acceptable angles and distances and, in addition, gave rise to the nomenclature for identifying torsion angles in the sugar phosphate backbone.

The GpC structure had the anticipated base pairs connected by three hydrogen bonds. However, the ApU structure showed for the first time that Watson-Crick base pairs formed when the molecule was constrained in a double he-

lix, as opposed to the Hoogsteen base pairs that were favored in the single-crystal complexes of adenine with uracil derivatives. I mailed preprints of these to several people, including Jim Watson. He phoned me, saying that after having read the ApU manuscript, he had his first good night's sleep in 20 years! This indicated to him that the uncertainty about the organization of the double helix was resolved. The significance of the double helix at atomic resolution was recognized by the editors of *Nature* who, in their "News and Views" commentary, called it the "missing link" and recognized that "the many pearls offered" helped resolve one of the big uncertainties in nucleic acid structure ("News and Views" [1973] *Nature, 243: 114*).

These structures capped the effort that I had started some 20 years earlier which had been initiated in earnest in 1956 with the recognition that poly rA and poly rU would form a double helix. Here, at last, was the demonstration at atomic resolution of the details of that structure. High-resolution crystallographic analysis of larger fragments of the double helix (DNA or RNA) did not emerge until almost a decade later with the availability of chemically synthesized and purified oligonucleotides, available in large enough quantities to permit single-crystal diffraction analysis.

A single-crystal X-ray structure of a hybrid helix did not appear until 1982 when my colleagues and I solved the structure of a DNA–RNA hybrid linked to double-helical DNA (Wang et al. 1982). This was 22 years after the hybrid helix was first observed (Rich 1960). It showed that the dilemma of two different backbone conformations was resolved by having the DNA strand adopt the RNA duplex conformation. This had been inferred from fiber diffraction studies and has remained a constant feature, reflecting the relative conformational flexibility of the DNA backbone compared with the less flexible RNA strand.

DISCOVERY OF POLYSOMES IN PROTEIN BIOSYNTHESIS

Several experiments in 1961 revealed the presence of a rapidly metabolizing fraction of RNA called messenger RNA that was made on DNA and had the ability to attach itself to ribosomal particles, where it determined the sequence of amino acids (Brenner et al. 1961; Gros et al. 1961). Through the work of Paul Zamecnik, Mahlon Hoagland, and associates, it was known that amino acids were activated by transfer RNA (called soluble RNA at the time) in protein synthesis (Hoagland et al. 1957). Evidence supporting the triplet code (Crick et al. 1961) suggested that very large messenger RNA molecules would be formed. For example, the 150 amino acids in hemoglobin chains required 450 nucleotides or an mRNA over 1500 Å long if the bases were stacked. This was much larger than the ribosomes in which protein synthesis occurred. However, this left many puzzling questions. Some of these questions were resolved by the discovery that a cluster of ribosomes acted on the same mRNA molecule in protein synthesis. Experiments were carried out in my laboratory using rabbit reticulocytes, which can be lysed by gentle methods and synthesize only hemoglobin. (Warner et al. 1962, 1963). Sedimenting the lysate on a sucrose density gradient revealed that the polymerized radioactive amino acids were found in a rapidly sedimenting segment, not with the individual ribosomes. Although rapidly sedimenting "heavy" ribosomes had been seen previously, their nature was not clear (Risebrough et al. 1962). The nature of the rapidly sedimenting reticulocytic fraction was revealed in an electron microscopic study (Warner et al. 1962) in which it was shown that the predominant species synthesizing hemoglobin consisted of a cluster of five ribosomes acting on the same messenger RNA strand (Fig. 6). These were readily dissociated by a brief exposure to ribonuclease. The thin messenger strand could be observed running

between the ribosomes, both in shadowed electron microscopic preparations and in preparations negatively stained with uranyl acetate (Warner et al. 1962; Slayter et al. 1963).

The interpretation of these results was that the ribosomal particles begin protein synthesis by attaching at one end of the messenger RNA strand and then move along it as the polypeptide chain elongates (Fig. 7) (Goodman and Rich 1963; Warner et al. 1963). Experiments by Howard Dintzis had already demonstrated that protein synthesis starts at the amino terminus of the chain (Dintzis 1961). As the ribosome continued along the strand, the peptide chain became steadily larger and detached at the end when the ribosome released the polypeptide chain and then separated from the message (Fig. 7B). The model suggested that longer messenger RNA strands coding for larger polypeptide chains would have larger polysomes. It also provided a natural mechanism for explaining the coordinate synthesis of several peptide chains in unison, since the message could be polycistronic, and the ribosomes moving along the chain would synthesize one polypeptide chain after the other, ensuring a stoichiometric synthesis (Warner et al. 1963).

The discovery of polysomes changed thinking in the field of protein biosynthesis in several ways. First, the electron microscope pictures offered a graphic visualization of the assembly mechanism, implying that ribosomes move along the message and the polypeptide chain elongates. Although there had been limited discussion of the movement of ribosomes along the message, images of the polysomes clearly indicated a dynamic assembly structure, strongly reinforcing the movement of ribosomes.

Further experiments were done with polysomes to investigate this mechanism. Experiments carried out with labeled ribosomes in vitro (Goodman and Rich 1963) showed that these ribosomes could attach to the end of polysomes, while at the same time, further incubation led to the release of completed polypeptide chains and individual ribosomes (Fig. 7B). Although the movement of ribosomes over messenger RNA seems obvious now, this notion required a reordering of the thinking of people in the field of protein biosynthesis. There was some resistance to these ideas for the first year or so, but subsequent experiments in my laboratory and several other laboratories showed the universality of the mechanism and stimulated biochemists to focus their experiments on the enzymatic machinery required to provide the energy needed for ribosomal movement.

Furthermore, the isolation of polysomes on a sucrose gradient meant that the ribosomes actually involved in protein synthesis could be separated from the large number of monomer ribosomes at the top of the gradient that are not active in protein synthesis. The ability to separate active from inactive ribosomes made it possible to uncover further aspects of the protein's synthetic mechanism. For example, it was generally assumed that there was one growing polypeptide chain per ribosome. The ability to isolate active ribosomes made it possible to measure that explicitly and demonstrate experimentally that there is one growing polypeptide chain per ribosome

ELECTRON MICROSCOPE STUDIES OF RIBOSOMAL
CLUSTERS SYNTHESIZING HEMOGLOBIN

Jonathan R. Warner, Alexander Rich, Cecil E. Hall

Reprinted from Science, December 28, 1962, Vol. 138, No. 3548, pages 1399-1403
Copyright © 1962 by the American Association for the Advancement of Science

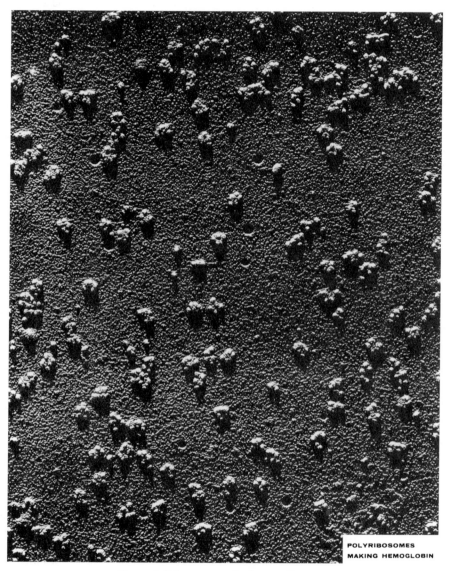

Figure 6. *Science* cover of polysome discovery.

(Warner and Rich 1964b). Of far greater interest, however, was our ability to analyze the number of transfer RNA (tRNA) molecules per active ribosome, as this had a direct bearing on the mechanism of protein synthesis.

TWO tRNA MOLECULES PER RIBOSOME: IDENTIFICATION OF THE A SITE AND P SITE

The polysome experiments were carried out with rabbit reticulocytes, cells that had already lost their nucleus and were no longer synthesizing RNA. This made it pos-

sible for my student Jonathan Warner to do a rather clean analysis of the system by incubating the reticulocytes with radioactive adenine or cytosine (Warner and Rich 1964a). These molecules could penetrate through the reticulocyte cell wall into the interior, where they were incorporated into ribonucleoside triphosphates. Although RNA was not synthesized, the 5′ CCA ends of tRNA molecules were continually cleaved off and then enzymatically re-added to tRNA. The radioactive nucleotides were thus utilized by the CCA-adding enzyme to label the ends of tRNA molecules. Because of the absence of the nucleus, messenger RNA and ribosomal RNA were not

A

Reprinted from the Proceedings of the National Academy of Sciences
Vol. 49, No. 1, pp. 122-129. January, 1963.

A MULTIPLE RIBOSOMAL STRUCTURE IN PROTEIN SYNTHESIS

By Jonathan R. Warner,* Paul M. Knopf,† and Alexander Rich

Department of Biology, Massachusetts Institute of Technology

A tentative view which we can adopt is that ribosomal particles begin protein synthesis by attaching to one end of a messenger RNA strand and then move along it as the polypeptide chain lengthens. If the messenger strip is very long, a correspondingly greater number of ribosomes will be attached at any one time. When the ribosomes reach the end of the strand, they release the polypeptide chain as well as the RNA. Thus, the released ribosome would contribute to the pool of inactive 76S monomers which we observe. . . . .

In addition, the mechanism described above raises the possibility that related groups of proteins are all manufactured at the same time. Thus, one genetic locus in the β-galactosidase system controls the production of a series of enzymes. If one long piece of messenger RNA was produced containing the information for all of these enzymes, ribosomes flowing over this strip might then produce the entire group of enzymes, one after another.

B

(Reprinted from Nature, Vol. 199, No. 4891, pp. 318-322,
July 27, 1963)

MECHANISM OF POLYRIBOSOME
ACTION DURING PROTEIN
SYNTHESIS

By Howard M. Goodman and
Prof. Alexander Rich

Department of Biology, Massachusetts Institute of Technology,
Cambridge, Mass.

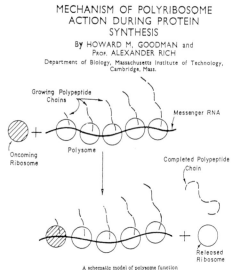

Figure 7. Polysomes synthesizing hemoglobin. Interpretation of polysome mechanism. Experimental demonstration of polysome action.

labeled. Analysis of the number of tRNAs (called soluble RNA or sRNA in that era) led to the demonstration that there were two tRNA molecules bound to the ribosome which is part of a polysome active in protein synthesis, but only one tRNA was found in the inactive ribosomes at the top of the gradient. It had been shown previously by several investigators that ribosomes would bind one tRNA in an exchangeable fashion (Takanami 1962; Cannon et al. 1963). On the basis of these analyses, attempts were made to formulate a mechanism whereby a single tRNA would be involved in protein synthesis on inactive ribosomes (Gilbert 1963). However, it was clear that the mechanism seemed implausible. With the presence of two tRNA molecules, it was possible to formulate a different interpretation. As shown in the excerpt in Figure 8, we postulated that the two tRNA-binding sites are adjacent. One of them we called Site A, which bound aminoacyl tRNA, and the other Site P, which bound peptidyl tRNA. We envisioned these two sites acting in a coordinated manner to transfer the growing polypeptide chain, and the movement of the messenger RNA codon from Site A to Site P constituted the basis of ribosomal movement relative to the messenger strand. This statement, formulated in 1964, rapidly became the standard description for interpreting ribosomal movement and tRNA activity in protein synthesis. Determination of the three-dimensional structure of the ribosome has provided a structural basis for understanding how this mechanism works.

Reprinted from the Proceedings of the National Academy of Sciences
Vol. 51, No. 6, pp. 1134-1141. June, 1964.

*THE NUMBER OF SOLUBLE RNA MOLECULES ON
RETICULOCYTE POLYRIBOSOMES**

By Jonathan R. Warner and Alexander Rich

We postulate that the two sRNA sites on the ribosomes are adjacent. Site A has the property of loosely binding activated sRNA molecules in the absence of messenger RNA. It is the exchange site. When messenger RNA is present, it binds sRNA selectively using the codon nucleotides. Site P next to it has the property of holding the sRNA with the growing polypeptide chain; this sRNA occupies the adjacent codon in the messenger RNA strand. Site P is not filled in the absence of messenger RNA. The mechanism of protein synthesis may thus consist of three coordinated actions: (1) the formation of the peptide bond with the transfer of the growing polypeptide chain to the sRNA initially occupying site A; (2) the displacement of the sRNA from site P and the transfer of the sRNA from site A to site P together with its newly attached polypeptide chain; (3) the simultaneous translation of the messenger RNA codon from site A to site P so that the sRNA remains attached to the same codon as it is being transferred. This movement from site A to site P would be the basis for the movement of the ribosomes relative to the messenger strand.

Figure 8. Discovery that ribosomes active in protein synthesis have two tRNA molecules compared to one on inactive ribosomes. Interpretation of the mechanism of protein synthesis with description of ribosomal A site and P site.

This description of protein synthesis would be modified only slightly in recent years by the observation that, when the tRNA in the P site has had its peptide chain transferred to the adjacent tRNA, it does not leave the ribosome promptly. Rather, it tarries slightly at a site now called the exit or E site, where it may be part of the machinery for triggering the opening of the new A site for the next tRNA (Geigenmuller and Nierhaus 1990). In protein biosynthesis, the ribosome only has two tRNA molecules at any one time, since the tRNA molecules occupying the P site and the Exit site do not permit an additional tRNA to come into the A site. It is interesting, however, that in crystallographic experiments, it is possible to force the ribosome to actually occupy all three sites by providing an abnormally high concentration of tRNA molecules (Cate et al. 1999).

The statement describing the movement of tRNA molecules in the ribosome postulated a simultaneous engagement of the transfer RNA molecule to the messenger RNA, and at the same time, it is the substrate for the peptidyl transferase enzyme that transfers the peptide chain from the tRNA in the P site onto the α amino acid of the aminoacyl tRNA in the A site. The structural and geometric background for understanding the distance separating where these two events occur had to await determination of the three-dimensional folding of transfer RNA molecules.

A PORTION OF THE NASCENT POLYPEPTIDE CHAIN IS BURIED IN THE RIBOSOME

In 1967, together with Leonard Malkin, we asked to what extent the growing polypeptide chain in rabbit reticulocyte polysomes was accessible to externally added proteolytic enzymes (Malkin and Rich 1967). These experiments demonstrated that there was a resistant fragment shielded by the ribosome. It was the most recently synthesized segment, and it was attached to the tRNA molecule. By using sizing experiments, it was shown that the shielded segment contained 30–35 amino acids. In an

extended conformation at 3.6 Å/amino acid, this implies a protected segment of about 100 Å in length (Fig. 9). The structure protecting the chain came to be referred to as the polypeptide tunnel.

It is of great interest that the recent three-dimensional structure determination of the 50S subunit at high resolution has revealed a tunnel of that approximate size, but one that has very interesting differences in its diameter that are likely to have functional significance (Nissen et al. 2000). In our early experiments, the existence of a tunnel or shielded segment could be described along with its approximate length, but no further information was available. We could, however, draw the conclusion that enzymes which modify growing polypeptide chains would not have access to the chains until they were greater than 30–35 Å in length.

The large shielded section of polypeptide probably stabilized its attachment to the ribosome. For very large polypeptide chains, it suggested the possibility that the chains could start to fold in a native conformation while they were still attached to the ribosome. This had been observed at an earlier date as active β-galactosidase could be detected on bacterial polysomes synthesizing that protein (Kiho and Rich 1964).

THE RIBOSOMAL PEPTIDYL TRANSFERASE CAN MAKE ESTER LINKAGES AS WELL AS PEPTIDE LINKAGES

The first indication that the ribosomal peptidyl transferase could catalyze the formation of esters as well as peptides was obtained in experiments with analogs of puromycin in which the α amino group was replaced by a hydroxyl group (Fahnestock et al. 1970). This study revealed that many of the characteristics of ester formation were similar to those found in the peptide formation of the puromycin reaction itself. Ester formation could be monitored directly because of the sensitivity of esters to cleavage by mild alkaline hydrolysis. These experiments encouraged us to incorporate an ester linkage in the viral coat

J. Mol. Biol. (1967) **26**, 329–346

Partial Resistance of Nascent Polypeptide Chains to
Proteolytic Digestion due to Ribosomal Shielding

Leonard I. Malkin and Alexander Rich

The pulse-labeling experiments show that the resistant portions of the nascent polypeptide chain are those which are the most recently synthesized, at the end of the chain adjacent to the sRNA molecule. Thus, this end is presumably buried within the ribosomal structure.

. . . .

The section of nascent polypeptide chain resistant to hydrolysis contains an estimated 30 to 35 amino acids. In an unfolded, extended configuration, amino acids occupy 3·6 Å per residue. In this configuration, the fragment would thus produce an extended polypeptide chain of approximately 100 Å in length.

Figure 9. First evidence of a ribosomal exit tunnel for the growing polypeptide chain.

(Reprinted from Nature New Biology, Vol. 229, No. 1, pp. 8-10, January 6, 1971)

Synthesis by Ribosomes of Viral Coat Protein containing Ester Linkages

by

STEPHEN FAHNESTOCK & ALEXANDER RICH

Department of Biology, Massachusetts Institute of Technology, Cambridge, Massachusetts

Peptidyl transferase, the ribosomal enzyme which synthesizes peptide bonds, can catalyse the formation of ester linkages between α-hydroxyacyl-tRNA and aminoacyl-tRNA when ribosomes carry out a specific translation of phage R17 RNA *in vitro*.

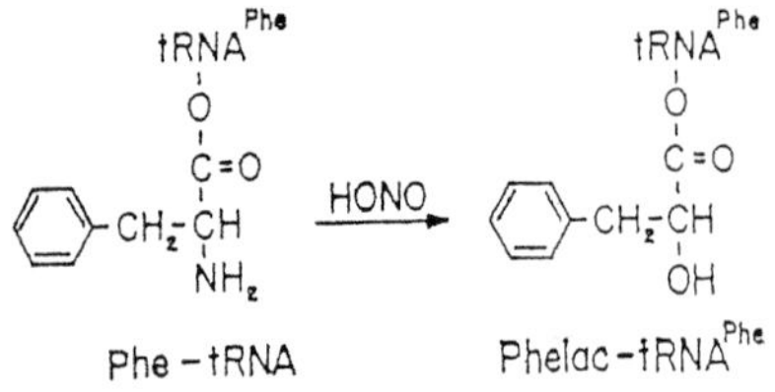

Conversion of phenylalanyl-tRNA to phenyllactyl-tRNA.

Reprinted from
23 July 1971, Volume 173, pp. 340-343

SCIENCE

Ribosome-Catalyzed Polyester Formation

Abstract. Deamination of phenylalanyl–transfer RNA with nitrous acid yields the α-hydroxyacyl analog, phenyllactyl–transfer RNA. When this is incubated in a protein-synthesizing system directed by polyuridylic acid, it yields an acid-precipitable, alkali-labile polyester of phenyllactic acid.

Figure 10. Ribosome peptidyl transferase can form ester bonds as well as peptide bonds.

protein of the R17 bacteriophage by providing all of the amino acids needed in a segment of the viral coat protein, except for one which was provided by chemically modified phenylalanyl tRNA[phe]. Treatment with HONO yielded an α-hydroxy phenyllactyl tRNA (Fahnestock and Rich 1971a). The coat protein was synthesized with an ester linkage specifically introduced at one position in the chain (Fig. 10). The position could be identified explicitly by hydrolyzing the synthesized chain in the presence of alkali and identifying the fragments produced by the reaction.

This experiment carried out in 1971 was the first example of a nonnatural residue incorporated into a protein. In recent years, a large number of nonnatural residues have been incorporated into proteins, and this field is now burgeoning to produce many proteins with specialized features.

Finally, experiments were carried out incorporating an α-hydroxyl phenyllactyl tRNA[phe] in an in vitro synthesis directed by polyuridylic acid (Fahnestock and Rich 1971b). This yielded ribosome-catalyzed polyester formation in which every residue contained ester linkages rather than peptide linkages. It was the first completely nonnatural polymer made by ribosomes directed by mRNA.

These observations were of interest because they pointed out some fundamental features of the peptidyl transferase reaction. The mechanism is usually interpreted as involving a nucleophilic attack by an α amino group on the carboxyl group of an adjacent peptidyl tRNA. However, the nucleophilic attack of an α hydroxyl group was also capable of generating transfer with the formation of an ester linkage. It is clear that the incorporation of esters

into proteins is not desirable because of their inherent lability when they are exposed to pH changes. Aminoacylating enzymes are thus very stringent in excluding them. However, it is likely that mutant auxotrophs could be made in which ester linkages can be incorporated into biological systems, given our present detailed knowledge of the mode of action of the tRNA aminoacyl synthetases and their associated proofreading activities.

THREE-DIMENSIONAL STRUCTURE OF tRNA

As methods for purifying tRNA improved during the 1960s, there was an increase in the number of attempts to form single crystals. This effort was very frustrating because it was easy to fail, and most people involved in the effort failed repeatedly. In 1968, working with Sung Hou Kim in my laboratory, we were able to obtain single crystals of *E.coli* tRNA[phe] (Kim and Rich 1968). Three other groups also obtained single crystals of various tRNAs in that same year, and all of these crystals were poor, in that they were somewhat disordered and the resolution was limited. Our earliest crystals diffracted to ~20 Å. By the next year we were able to get crystals that diffracted to 6 or 7 Å resolution, and a study of the three-dimensional Patterson function using the 12 Å data from the crystals of *E.coli* tRNA[Fmet] yielded approximate molecular dimensions of 80 × 25 × 35 Å (Kim and Rich 1969). These crystals represented progress of a sort, but at the same time, the frustration was great because they were not suitable for solving the structure of the molecule.

We spent the next two years looking at many different purified tRNA preparations and explored many different crystallization procedures. By 1971, we reached an exciting turning point: Yeast tRNA^phe could be crystallized in a simple orthorhombic unit cell with a resolution of 2.3 Å (Kim et al. 1971)! These were the first crystals of tRNA suitable for analysis. The key event in making crystals of this resolution was the incorporation in the crystallization mix of spermine, a naturally occurring polyamine. The spermine bound specifically to yeast tRNA^phe and stabilized it so that it made a high-resolution crystal (Fig. 11). This was an important discovery at the time. The stabilization effect of spermine on yeast tRNA^phe made it possible to form good crystals in other lattices as well. How-

A

Reprinted from
Proc. Nat. Acad. Sci. USA
Vol. 68, No. 4, pp. 841–845, April 1971

High-Resolution X-Ray Diffraction Patterns of Crystalline Transfer RNA That Show Helical Regions

SUNG HOU KIM, GARY QUIGLEY, F. L. SUDDATH, AND ALEXANDER RICH

Department of Biology, Massachusetts Institute of Technology, Cambridge, Mass. 02139

Communicated February 1, 1971

ABSTRACT   Yeast phenylalanyl transfer RNA crystallizes in a simple orthorhombic unit cell (a = 33.2, b = 56.1, c = 161 Å), and the crystal yields an x-ray diffraction pattern with a resolution of 2.3 Å. From an analysis of the packing in the unit cell it is concluded that the molecular dimensions are approximately 80 by 33 by 28 Å. The diffraction pattern viewed along the a-axis has a distribution characteristic of double-helical nucleic acids. However, this distribution is not found when the pattern is viewed along the b-axis. This has been interpreted as indicating that the double-helical portions of the transfer RNA molecule are approximately half a helical turn in length, and therefore can contain 4–7 base pairs. These results are consistent with the cloverleaf formulation of transfer RNA secondary structure.

B

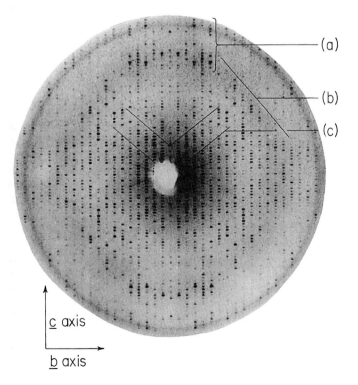

Figure 11. Discovery that spermine-stabilized yeast tRNA^phe produces a high-resolution, highly ordered crystal.

ever, spermine would not necessarily stabilize all tRNA molecules in a similar way.

Analysis of the crystal diffraction pattern showed that it had a characteristic helical distribution of diffracting intensities when viewed in one direction (Fig. 11) but did not show a helical distribution when viewed at right angles. This was taken as evidence that short helical segments containing 4–7 base pairs were found in the molecule, a result entirely consistent with the cloverleaf folding of tRNA molecules as identified by Holley and colleagues after sequencing the first tRNA molecule (Holley et al. 1965). This discovery opened the door to the ultimate solution of the structure of yeast tRNA^{phe}.

TRACING THE FOLDING OF THE POLYNUCLEOTIDE BACKBONE IN YEAST tRNA^{phe}

Myoglobin was the first protein whose three-dimensional structure was solved. The structure was revealed at various levels of resolution. An important event was the tracing of the polypeptide chain which showed how the myoglobin molecule is organized as a series of α-helical and single-stranded regions folded together. This was our first glimpse of how a protein molecule is folded, even though the high-resolution structure had not been completed.

The structure of yeast tRNA^{phe} was revealed in a similar, gradual way. Crystallographic research moved more slowly in the early 1970s than today. Computers were primitive; advanced area detectors, cryo-crystallography, and synchrotron beams were things in the future. However, before work could continue, heavy-atom derivatives had to be discovered that would be useful for phasing the diffraction pattern of a crystalline nucleic acid molecule. This had never been done before, and it took considerable time to discover appropriate derivatives. Three different types of heavy atoms were developed containing platinum, osmium, or samarium ions. The osmium residue was very important since it was known to form complexes with ribonucleotides involving both the 2′ and 3′ hydroxyl groups. Only one pair of *cis* hydroxyl groups was found at the 3′ CCA end of the tRNA molecule. In order to gain some appreciation of the geometry, the structure of an osmium–adenosine complex was solved which enabled us to visualize the interaction (Conn et al. 1974). The single osmium derivative in the tRNA crystal made it possible to identify the 3′ hydroxyl end of the tRNA chain (Kim et al. 1973). The samarium ions were very useful, and they occupied more than one site. The platinum residue was somewhat less valuable since it was only useful for 5.5 Å data. An interim electron density map at 5.5 Å (Kim et al. 1972) made it possible to uncover the external shape of portions of the molecule. However, the true shape of the molecule was not revealed until a map at 4 Å was visualized in 1973 (Kim et al. 1973).

At 4 Å resolution, peaks were seen throughout the electron density map that were due to the electron-dense phosphate groups. We knew a great deal about the distance constraints between adjacent phosphate groups in a polynucleotide chain, and this made it possible to look for peaks between 5 and 7 Å apart. Tracing the chain led to the discovery that the tRNA molecule had an unusual L-shape. The CCA acceptor helix was colinear with the T pseudo-U helix, and it is almost at right angles to the anticodon stem which is colinear with the dehydro U stem. The molecule had the shape of an "L," with the amino acid acceptor 3′ hydroxyl group at one end of the L and the anticodon loop at the other. At the corner of the L, there was a complex folding of the T pseudo-U and dihydro U loops. Figure 12 shows a perspective diagram of the L-shaped molecule, as well as a sample of the electron density map showing double-helical regions in which the intense peaks associated with phosphate groups are separated from each other by a region of lower electron density due to the base pairs.

The L-shaped folding of the tRNA polynucleotide chain was a dramatic and surprising discovery. Because of the constraints in the cloverleaf folding of the molecule, several models had been proposed for the folding of tRNA molecules. All of them were wrong. No one had anticipated that the molecule would organize in this fashion. Even at 4 Å resolution, this folding was compatible with much experimental data concerning tRNA molecules. For example, it was known that photoactivation of *E. coli* tRNA^{val} resulted in the formation of a photo dimer involving the 4-thio-U residue in position 8 and the cytosine in position 13. In the 4 Å folding of the polynucleotide chain, these two bases were in close proximity, and the distance between the phosphate groups of these two residues was short enough to allow formation of the photo dimer (Kim et al. 1973).

The L-shaped tRNA molecule with the folding shown in Figure 12 is now a standard feature of molecular biology, having been found in virtually all tRNA molecules, even when they are complexed to aminoacyl synthetase enzymes. The significance of the folding is twofold. First, it reveals that the 3′ acceptor end is over 70 Å away from the anticodon loop, which has implications for understanding the interaction between tRNA molecules and tRNA aminoacyl synthetases. Second and most important, it suggests that the interaction of the tRNA molecules with the message occurs at one end of the L, whereas the segment responsible for forming the peptide bond is considerably removed from the site. This makes it possible to have great specificity with many interactions at either end of the molecule due to this separation. These features have been incorporated into our present view of protein synthesis in interpreting the three-dimensional structure of ribosomes and the movement of tRNA molecules inside the ribosome. Indeed, a proposal was made regarding the movement of tRNA molecules in the ribosome using the 4 Å folding (Rich 1974). This proposal incorporated some features of the movement that is now seen in the three-dimensional structure of the ribosome.

Today, we are accustomed to seeing a variety of complex ribonucleotide molecules in which double-helical segments and single-chain segments are juxtaposed to make complex structures with a variety of functions, especially in ribozymes. However, the beginning of our

Three-Dimensional Structure of Yeast Phenylalanine Transfer RNA: Folding of the Polynucleotide Chain

SCIENCE
19 January 1973, Volume 179, pp. 285-288

S. H. Kim, G. J. Quigley, F. L. Suddath, A. McPherson, D. Sneden, J. J Kim, J. Weinzierl and Alexander Rich

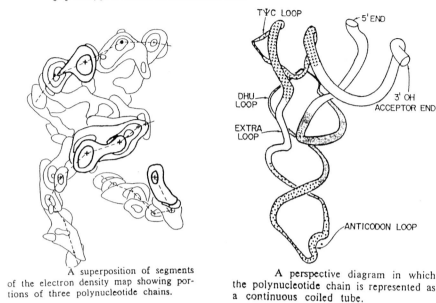

A superposition of segments of the electron density map showing portions of three polynucleotide chains.

A perspective diagram in which the polynucleotide chain is represented as a continuous coiled tube.

Figure 12. Tracing the backbone of yeast tRNA^phe at 4 Å resolution reveals the L-shaped polypeptide chain folding.

understanding of the manner in which complex polynucleotide chains can fold started with this first tracing of yeast tRNA^phe visualized at 4 Å resolution.

This tracing was seen in more detail a year later in our 3 Å analysis of the folding of yeast tRNA^phe in the orthorhombic lattice (Fig. 13) (Kim et al. 1974). Simultaneously, Aaron Klug and colleagues published the 3 Å structure of the same spermine-stabilized yeast tRNA^phe in a monoclinic lattice (Robertus et al. 1974). This also confirmed the same folding of the polynucleotide chain, even though the lattice was different. These 3 Å structures were very similar and revealed in great detail the manner in which base-pairing of nucleotides, both in the double-helical regions and in the single-stranded regions, stabilizes the three-dimensional fold of the molecule. The folding was held together by a variety of hydrogen-bonding interactions, including many in the nonhelical regions of the molecule. These hydrogen-bonding interactions included the formation of triplexes and a variety of interactions beyond that seen in Watson-Crick base-pairing. Recognition of the importance of these alternative types of hydrogen bonds explained why the model builders of that period, trying to anticipate the structure of tRNA, were all incorrect in their conclusions. The main reason was that they relied excessively on Watson-Crick base-pair interactions and did not recognize the stabilizing effect of many other types of hydrogen bonds.

The L-shaped folding was predicted to be a general conformation found in all tRNA molecules (Kim et al. 1974b). Subsequent work has amply verified the relative constancy of the hydrogen-bonding networks (Quigley et al. 1975; Quigley and Rich 1976).

We have known of the L-shaped folding of tRNA molecules since 1973. The full understanding of why this particular folding was robust, and the manner in which modified nucleotides are important to this folding, is still a work in progress. However, for understanding the interaction of tRNA molecules with the ribosome during protein synthesis, it is the L-shaped folding that provides the central information regarding its interactions and movements in protein synthesis.

CONCLUDING COMMENTS

In this short account, I have focused on the work of my colleagues and myself in uncovering some basic aspects of RNA structure and protein biosynthesis over a period of 20 years (1954–1974). This report may be useful to younger investigators who entered the research stream more recently. The work carried out over that period by many investigators brought us to the point of understanding the broad outlines of protein biosynthesis. However, the detailed mechanism could not be understood until the

Reprinted from
2 August 1974, Volume 185, pp. 435-440

SCIENCE

Three-Dimensional Tertiary Structure of Yeast

Phenylalanine Transfer RNA

S. H. Kim, F. L. Suddath, G. J. Quigley, A. McPherson,
J. L. Sussman, A. H. J. Wang, N. C. Seeman and Alexander Rich

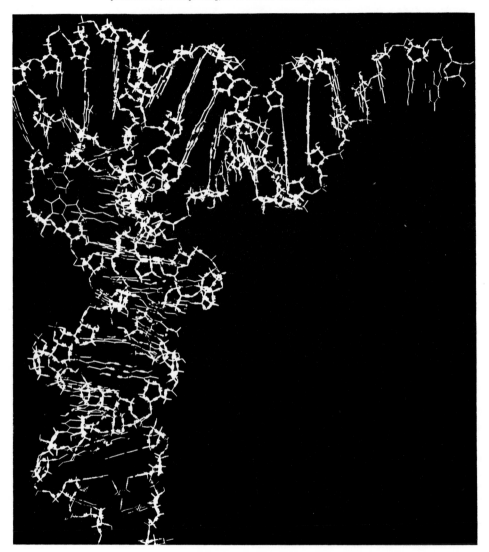

Figure 13. *Science* cover with the 3 Å structure of yeast tRNA[phe].

more recent developments elucidating the three-dimensional structure of the ribosome. The earlier work set the stage for understanding the mechanisms that we now are beginning to envision in more detail in the machinery of protein synthesis.

of students, postdoctoral fellows, and collaborators. We shared together the speculations and the random groping that is part of the research enterprise, as well as the joy of discovery.

ACKNOWLEDGMENTS

It is important to note that the contributions described here are in large part due to the work of a talented group

REFERENCES

Arnott S., Wilkins M.H.F., Hamilton L.D., and Langridge R. 1965. Fourier synthesis studies of lithium DNA. III. Hoogsteen models. *J. Mol. Biol.* **11:** 391.
Brenner S., Jacob F., and Meselson M. 1961. An unstable inter-

mediate carrying information from genes to ribosomes for protein synthesis. *Nature* **190:** 576.

Cannon M., Krug R., and Gilbert W. 1963. The binding of sRNA by *Escherichia coli* ribosomes. *J. Mol. Biol.* **7:** 360.

Cate J.H., Yusupov M.M., Yusupova G.Z., Earnest T.N., and Noller H.F. 1999. X-ray crystal structures of 70S ribosome functional complexes. *Science* **285:** 2095.

Conn J.F., Kim J.J., Suddath F.L., Blattman P., and Rich A. 1974. Crystal and molecular structure of an osmium bispyridine ester of adenosine. *J. Am. Chem. Soc.* **96:** 7152.

Crick F.H.C., Barnett L., Brenner S., and Watts-Tobin R.J. 1961. General nature of the genetic code for proteins. *Nature* **192:** 1227.

Davies D.R. and Rich A. 1958. The formation of a helical complex between polyinosinic acid and polycytidylic acid. *J. Am. Chem. Soc.* **80:** 1003.

Day R.O., Seeman N.C., Rosenberg J.M., and Rich A. 1973. A crystalline fragment of the double helix: The structure of the dinucleoside phosphate guanylyl-3′,5′-cytidine. *Proc. Natl. Acad. Sci.* **70:** 849.

Dintzis H. 1961. Assembly of the peptide chains of hemoglobin. *Proc. Natl. Acad. Sci.* **47:** 247.

Doty P., Marmur J., Eigner J., and Schildkraut C. 1960. Strand separation and specific recombination in deoxyribonucleic acids: Physical chemical studies. *Proc. Natl. Acad. Sci.* **46:** 461.

Fahnestock S. and Rich A. 1971a. Synthesis by ribosomes of viral coat protein containing ester linkages. *Nat. New Biol.* **229:** 8.

———. 1971b. Ribosome catalyzed polyester formation. *Science* **173:** 340.

Fahnestock S., Neumann H., Shashoua V., and Rich A. 1970. Ribosome-catalyzed ester formation. *Biochemistry* **9:** 2477.

Felsenfeld G. and Rich A. 1957. Studies on the formation of two- and three-stranded poly ribonucleotides. *Biochim. Biophys. Acta* **26:** 457.

Felsenfeld G., Davies D.R., and Rich A. 1957. Formation of a three-stranded polynucleotide molecule. *J. Am. Chem. Soc.* **79:** 2023.

Furth J.J., Hurwitz J., and Goldmann M. 1961. The directing role of DNA in RNA synthesis. *Biochem. Biophys. Res. Commun.* **4:** 362.

Geigenmuller U. and Nierhaus K.H. 1990. Significance of the third tRNA binding site, the E site, on *E. coli* ribosomes for the accuracy of translation: An occupied E site prevents the binding of non-cognate aminoacyl-tRNA to the A site. *EMBO J.* **13:** 4527.

Gilbert W. 1963. Polypeptide synthesis in *Escherichia coli*. II. The polypeptide chain and S-RNA. *J. Mol. Biol.* **6:** 389.

Goodman H.M. and Rich A. 1963. Mechanism of polyribosomal action during protein synthesis. *Nature* **199:** 318.

Gros F., Hiatt H., Gilbert W., Kurland C.G., Rissebrough R.W., and Watson J.D. 1961. Unstable ribonucleic acid revealed by pulse labelling of *Escherichia coli*. *Nature* **190:** 581.

Grunberg-Manago M., Ortiz P.J., and Ochoa S. 1955. Enzymatic synthesis of nucleic acidlike polynucleotides. *Science* **122:** 907.

Hall B.D. and Spiegelman S. 1961. Sequence complementarity of T2-DNA and T2-specific RNA. *Proc. Natl. Acad. Sci.* **47:** 137.

Hoagland M.B., Zamecnik P.C., and Stephenson M.L. 1957. Intermediate reactions in protein biosynthesis. *Biochim. Biophys. Acta* **24:** 215.

Holley R.W., Apgar J., Everett G.A., Madison J.T., Marguisse M., Merrill S.H., Penwick J.R., and Zamir A. 1965. Structure of a ribonucleic acid. *Science* **147:** 1462.

Hoogsteen K. 1959. The crystal and molecular structure of a hydrogen-bonded complex between 1-methylthymine and 9-methyladenine. *Acta Crystallogr.* **12:** 822.

Katz L., Tomita K., and Rich A. 1965. The molecular structure of the crystalline complex ethyladenine: Methyl-bromouracil. *J. Mol. Biol.* **13:** 340.

Kiho Y. and Rich A. 1964. Induced enzyme formed on bacterial polyribosomes. *Proc. Natl. Acad. Sci.* **51:** 111.

Kim S.-H. and Rich A. 1968. Single crystals of transfer RNA: An X-ray diffraction study. *Science* **162:** 1381.

———. 1969. Crystalline transfer RNA: The three-dimensional Patterson function at 12-angstrom resolution. *Science* **166:** 1621.

Kim S.-H., Quigley G., Suddath F.L., and Rich A. 1971. High resolution X-ray diffraction patterns of tRNA crystals showing helical regions of the molecule. *Proc. Natl. Acad. Sci.* **68:** 841.

Kim S.-H., Quigley G.J., Suddath F.L., McPherson A., Sneden D., Kim J.J., Weinzierl J., and Rich A. 1973. Three-dimensional structure of yeast phenylalanine transfer RNA: Folding of the polynucleotide chain. *Science* **179:** 285.

Kim S.-H., Suddath F.L., Quigley G.J., McPherson A., Sussman J.L., Wang A.H.-J., Seeman N.C., and Rich A. 1974a. Three-dimensional tertiary structure of yeast phenylalanine transfer RNA. *Science* **185:** 435.

Kim S.-H., Sussman J.L., Suddath F.L., Quigley G.J., McPherson A., Wang A.H.-J., Seeman N.C., and Rich A. 1974b. The general structure of transfer RNA molecules. *Proc. Natl. Acad. Sci.* **71:** 4970.

Kim S.-H., Quigley G., Suddath F.L., McPherson A., Sneden D., Kim J.J., Weinzierl J., Blattmann P., and Rich A. 1972. Three-dimensional structure of yeast phenylalanine transfer RNA: Shape of the molecule at 5.5 Å resolution. *Proc. Natl. Acad. Sci.* **69:** 3746.

Kornberg A., Lehman I.R., Bessman M.J., and Simms E.S. 1956. Enzymic synthesis of deoxyribonucleic acid. *Biochim. Biophys. Acta* **21:** 197.

Malkin L.I. and Rich A. 1967. Partial resistance of nascent polypeptide chains to proteolytic digestion due to ribosomal shielding. *J. Mol. Biol.* **26:** 329.

Marmur J. and Lane D. 1960. Strand separation and specific recombination in deoxyribonucleic acids: Biological studies. *Proc. Natl. Acad. Sci.* **46:** 453.

Mathews F.S. and Rich A. 1964. The molecular structure of a hydrogen bonded complex of N-ethyl adenine and N-methyl uracil. *J. Mol. Biol.* **8:** 89.

Nissen P., Hansen J., Ban N., Moore P.B., and Steitz T.A. 2000. The structural basis of ribosome activity in peptide bond synthesis. *Science* **289:** 821.

Pauling L. and Corey R.B. 1956. Specific hydrogen-bond formation between pyrimidines and purines in deoxyribonucleic acids. *Arch. Biochem. Biophys.* **65:** 164.

Quigley G.J. and Rich A. 1976. Structural domains of a transfer RNA molecule. *Science* **194:** 796.

Quigley G.J., Wang A.H.-J., Seeman N.C., Suddath F.L., Rich A., Sussman J.L., and Kim S.H. 1975. Hydrogen bonding in yeast phenylalanine transfer RNA. *Proc. Natl. Acad. Sci.* **72:** 4866.

Rich A. 1957. The structure of synthetic polyribonucleotides and the spontaneous formation of a new two-stranded helical molecule. In *The chemical basis of heredity* (ed. W.D. McElroy and B. Glass), p. 557. Johns Hopkins University Press, Baltimore, Maryland.

———. 1958a. Formation of two- and three-stranded helical molecules by polyinosinic acid and polyadenylic acid. *Nature* **181:** 521.

———. 1958b. The molecular structure of polyinosinic acid. *Biochim. Biophys. Acta* **29:** 502.

———. 1959. An analysis of the relation between DNA and RNA. *Ann. N.Y. Acad. Sci.* **81:** 709.

———. 1960. A hybrid helix containing both deoxyribose and ribose polynucleotides and its relation to the transfer of information between the nucleic acids. *Proc. Natl. Acad. Sci.* **46:** 1044.

———. 1974. How transfer RNA may move inside the ribosome. In *Ribosomes* (ed. M. Nomura et al.), p. 871. Cold Spring Harbor Laboratory, Cold Spring Harbor, New York.

Rich A. and Davies D.R. 1956a. A new two-stranded helical structure: Polyadenylic acid and polyuridylic acid. *J. Am. Chem. Soc.* **78:** 3548.

Rich A. and Watson J.D. 1954a. Some relations between DNA and RNA. *Proc. Natl. Acad. Sci.* **40:** 759.

————. 1954b. Physical studies on ribonucleic acid. *Nature* **173:** 995.

Risebrough R.W., Tissières A., and Watson J.D. 1962. Messenger-RNA attachment to active ribosomes. *Proc. Natl. Acad. Sci.* **48:** 430.

Robertus J.D., Ladner J.E., Finch J.T., Rhodes D., Brown R.S., Clark B.F.C., and Klug A. 1974. Structure of yeast phenylalanine tRNA at 3 Å resolution. *Nature* **250:** 546.

Rosenberg J.M., Seeman N.C., Kim J.J.P., Suddath F.L., Nicholas H.B., and Rich A. 1973. Double helix at atomic resolution. *Nature* **243:** 150.

Slayter H.S., Warner J.R., Rich A., and Hall C.E. 1963. The visualization of polyribosomal structure. *J. Mol. Biol.* **7:** 652.

Sobell H.M., Tomita K., and Rich A. 1963. The crystal structure of an intermolecular complex containing a guanine and cytosine derivative. *Proc. Natl. Acad. Sci.* **49:** 885.

Takanami M. 1962. Transfer of amino acids from soluble ribonucleic acid to ribosome. III. Further studies on the interaction between ribosome and soluble ribonucleic acid. *Biochim. Biophys. Acta* **61:** 432.

Tener G.M., Khorana H.G., Markham R., and Pol E.H. 1958. Studies on polynucleotides. II. The synthesis and characterization of linear and cyclic thymidine oligonucleotides. *J. Am. Chem. Soc.* **80:** 6223.

Tomita K., Katz L., and Rich A. 1967. Crystal structure of the intermolecular complex 9-ethyladenine: 1-methyl-5-fluorouracil. *J. Mol. Biol.* **30:** 545.

Voet D. and Rich A. 1970. The crystal structures of purine, pyrimidines and their intermolecular complexes. *Prog. Nucleic Acid Res. Mol. Biol.* **10:** 183.

Wang A.H.-J., Fujii S., van Boom J.H., van der Marel G.A., van Boeckel S.A.A., and Rich A. 1982. Molecular structure of r(GCG)d(TATACGC): A DNA-RNA hybrid helix joined to double helical DNA. *Nature* **299:** 601.

Warner J.R. and Rich A. 1964a. The number of soluble RNA molecules on reticulocyte polyribosomes. *Proc. Natl. Acad. Sci.* **51:** 1134.

————. 1964b. The number of growing polypeptide chains on reticulocyte polyribosomes. *J. Mol. Biol.* **10:** 202.

Warner J.R., Knopf P.M., and Rich A. 1963. A multiple ribosomal structure in protein synthesis. *Proc. Natl. Acad. Sci.* **149:** 122.

Warner J.R., Rich A., and Hall C.E. 1962. Electron microscope studies of ribosomal clusters synthesizing hemoglobin. *Science* **138:** 1399.

Warner R.C. 1956. Ultraviolet spectra of enzymatically synthesized polynucleotides. *Fed. Proc.* **15:** 379.

Watson J.D. and Crick F.H.C. 1953. A structure for deoxyribose nucleic acid. *Nature* **171:** 738.

Atomic Structures of the 30S Subunit and Its Complexes with Ligands and Antibiotics

D.E. Brodersen, A.P. Carter, W.M. Clemons, Jr., R.J. Morgan-Warren,
F.V. Murphy IV, J.M. Ogle, M.J. Tarry, B.T. Wimberly, and V. Ramakrishnan
MRC Laboratory of Molecular Biology, Cambridge CB2 2QH, United Kingdom

The two subunits that make up the ribosome have both distinct and cooperative functions. The 30S ribosomal subunit binds messenger RNA (mRNA) and is involved in the selection of cognate transfer RNA (tRNA) by monitoring codon–anticodon base-pairing during the decoding process. The 50S subunit catalyzes peptide-bond formation. Both subunits work in concert to move tRNAs and mRNAs relative to the ribosome in translocation, and both are the target of a large number of naturally occurring antibiotics. Thus, useful information about the mechanism of translation can be gleaned from structures of both individual subunits and the intact ribosome. In this paper, we describe our work on the determination of the atomic structure of the 30S ribosomal subunit and its complexes with RNA ligands, antibiotics, and initiation factor IF1. The results provide structural insights into how the ribosome recognizes cognate tRNA and discriminates against near-cognate tRNA. They also provide a structural basis for understanding the action of various antibiotics that target the 30S subunit.

Knowledge of the atomic structure of the 30S is the culmination of decades of work on its structure and function. The state of the ribosome field two years ago has been summarized in a recent book (Garrett et al. 2000). The initial breakthrough in the goal toward a crystal structure of the ribosome was the crystallization of 50S subunits suitable for diffraction studies by Yonath and coworkers (Yonath et al. 1980). Although these original crystals did not diffract very well, over the course of several years, improved crystals of the 50S subunit were obtained that eventually resulted in diffraction to 3 Å resolution (von Böhlen et al. 1991). Another important advance was the introduction of cryocrystallography to facilitate data collection (Garman and Schneider 1997). Following the demonstration by Hope that cryocrystallography could significantly reduce radiation damage on various proteins (Hope 1988), Hope, Yonath, and coworkers showed that it also reduced damage to ribosome crystals (Hope et al. 1989). Even if better crystals had been obtained in the early 1980s, the technology available then was insufficient to determine the structure of a ribosomal subunit to high resolution. Important developments in synchrotron X-ray sources, instrumentation, and computation (Helliwell 1998; Hendrickson 2000) were all essential for the current progress.

CRYSTALLIZATION OF THE 30S SUBUNIT

Crystals of both the 30S subunit and 70S ribosomes from *Thermus thermophilus* that were suitable for structural studies were first reported by the Puschino group (Trakhanov et al. 1987; Yusupov et al. 1987). These crystals used 2-methyl-2,4-pentanediol (MPD) as the precipitant, and shortly afterward, crystals of the 30S in a mixture of ethanol and ethylbutanol were reported by the Yonath group (Glotz et al. 1987). The 30S subunit crystals in MPD diffracted to about 10–12 Å resolution initially (Yonath et al. 1988; Yusupov et al. 1988), and by 1995, their diffraction limit had only improved to about 7–8 Å resolution (Schluenzen et al. 1995). However, by that time, advances in single-particle reconstruction methods in cryo-electron microscopy had shown that the 30S subunit was conformationally different in the active, inactive, and 50S-bound states (Frank et al. 1995; Lata et al. 1996). This suggested to us that conformational variability of the 30S subunit could be the reason for the poor diffraction reported for its crystals, and we reasoned that, by analogy with smaller enzymes, binding of substrates or cofactors might result in complexes that were conformationally homogeneous and therefore yield better crystals.

As a starting point, we used the original crystallization conditions reported by the Puschino group. To our surprise, a straightforward optimization of the original conditions resulted in crystals that diffracted to beyond 3 Å at third-generation synchrotron sources (Clemons et al. 1999; Wimberly et al. 2000). Unlike improved crystals of the 30S subunit produced by the Yonath group (Yonath et al. 1998; Tocilj et al. 1999), our crystals do not require soaking in tungsten clusters and heat treatment to result in diffraction to high resolution, and the reason they diffract so well will become apparent below.

STRUCTURE DETERMINATION

Just three years ago, it was not clear that the crystallographic phase problem could be solved for asymmetric units as large as a ribosomal subunit. It has long been known that heavy atom clusters of tungsten or tantalum could be used for obtaining phase information for large molecules (Blundell and Johnson 1976). They were used in the structure determination of large complexes such as ribulose biphosphate carboxylase and the nucleosome core particle (O'Halloran et al. 1987; Andersson et al. 1989; Knablein et al. 1997). The use of clusters was also suggested for ribosomes (Thygesen et al. 1996). The first concrete demonstration that such clusters could yield phase information for ribosome-sized problems came when they could be visualized directly in difference-Patterson maps of the 50S subunit and resulted in electron density maps that showed unambiguous right-handed helices corresponding to A-form RNA duplexes (Ban et al. 1998).

At the same time, anomalous scattering has seen increasing use in structure determination, especially through the use of the *m*ulti-wavelength *a*nomalous *d*iffraction (MAD) technique on crystals of proteins labeled with selenomethionine (Hendrickson et al. 1990; Hendrickson 1991). It was shown by us that treatment of MAD as a special case of standard multiple isomorphous replacement (MIR) resulted in excellent phases despite the nominally very small signal from anomalous scattering (Ramakrishnan et al. 1993; Ramakrishnan and Biou 1997). A test calculation shows that a fully selenomethionylated ribosome would have too little signal to be useful (Clemons et al. 2001). However, the LIII edges of many elements typically have 3–4 times the anomalous scattering of the K-edge of selenium, and were previously used for both protein (Kahn et al. 1985; Weis et al. 1991) and RNA (Cate et al. 1996) structure determination. Calculations based on the original equation by Crick and Magdoff (1956) showed that even 10–20 sites per 30S subunit would result in a measurable signal (Clemons et al. 2001), offering an alternative route to the phasing problem. This conclusion appears to have been reached independently by the various groups working on ribosome crystallography. Our atomic structure of the 30S, as well as that of the 50S (Ban et al. 2000) and the 70S (Cate et al. 1999), were all solved primarily using LIII edges of compounds such as osmium or iridium hexammine, although in each case, heavy atom clusters were very useful in the determination of low-resolution phases.

The structure of the 30S in our laboratory was determined in two steps: Initially, a 5.5 Å resolution structure was determined, which implicitly described the correct molecular packing in the crystal (Clemons et al. 1999). Even at this resolution, we were able to interpret about a third of the structure because of the large amount of prior structural and biochemical data on the 30S subunit. We were able to place proteins whose structures had been determined in isolation, deduce the fold and location of the previously unknown protein S20, and determine the fold of the entire central domain of 16S RNA and the relative location of the functionally important helix 44 along the interface of the subunit.

Last year, we published the complete atomic structure of the 30S subunit (Wimberly et al. 2000). Crystallographic details of the determination of the 30S structure have been published recently (Clemons et al. 2001).

GLOBAL ARCHITECTURE

The final refined atomic model of the 30S subunit at 3 Å resolution contains 1513 out of 1521 nucleotides of 16S

RNA and all 20 polypeptide chains present in the crystal (S2-S20, THX), with 98% of the sequence built into the molecule (Wimberly et al. 2000). The missing regions occur exclusively at the termini of the RNA and proteins.

The secondary structure of *Thermus* 16S RNA is very close to that predicted on the basis of phylogenetic comparisons (Gutell 1996), which is shown in Figure 1 along with the standard Brimacombe numbering for the various double helical elements (Glotz and Brimacombe 1980; Mueller et al. 1997). Figure 1 also indicates the insertions and deletions relative to the *Escherichia coli* 16S RNA sequence, and both sets of numberings are shown to aid biochemical analysis. The secondary structure domains of 16S RNA form distinct domains in three dimensions (Fig. 2a), so that the 5′ domain forms the body, the central domain forms the platform, the 3′ major domain forms the head, and the 3′ minor domain consists largely of two helices that lie on the intersubunit interface. This is different from the structure of 23S RNA, where the division of secondary structure domains appears to be more arbitrary and where the secondary structure domains are closely intertwined in three dimensions. The three-dimensional structure of 16S RNA has been described briefly (Wimberly et al. 2000), and a detailed analysis is in progress.

The shape of the 30S is very similar to that seen by electron microscopy. In addition to the classic features of the head, shoulder, and platform, one can also see the "beak" and "spur" that were observed using cryoelectron microscopy (Frank et al. 1995; Stark et al. 1995). From a comparison of the various electron microscopic images of the 30S subunit (Frank et al. 1995; Lata et al. 1996), it appears that the crystal structure is closer to the 50S-bound form than to either the active or inactive conformations of the isolated subunit.

The overall shape and gross morphological features of the 30S subunit are largely determined by the RNA component (cf. Figs. 2a and 2c). The RNA contains more than 50 regular helices. Many of the regions that were considered to be single-stranded loops in the secondary structure are in fact irregular extensions of regular double helices, so the molecule can be thought of in three dimensions as largely helical. Neighboring helical elements are often coaxially stacked; e.g., helices 16 and 17 in the 5′ domain or helices 21, 22, and 23 in the central domain. These coaxially stacked helices are packed using primarily three types of interactions, all of which use the minor groove as the packing interface (Wimberly et al. 2000). A notable feature is the use of highly conserved adenines to make tertiary interactions by binding in the minor groove of an adjacent double-helical section of RNA. As described below, this feature is also used in decoding.

The proteins mainly decorate the back and periphery of the 30S subunit (Fig. 2b), whereas the interface surface consists largely of ribosomal RNA (Fig. 2c). A glaring exception to this rule is S12, which is located near the decoding site at the interface. Proteins S13 and S19 also lie mainly on the 50S side of the head.

Many of the proteins have highly elongated extensions or internal loops (Fig. 3). A particularly dramatic example is S12, which is the only protein that lies directly at the intersubunit interface, but has a long amino-terminal

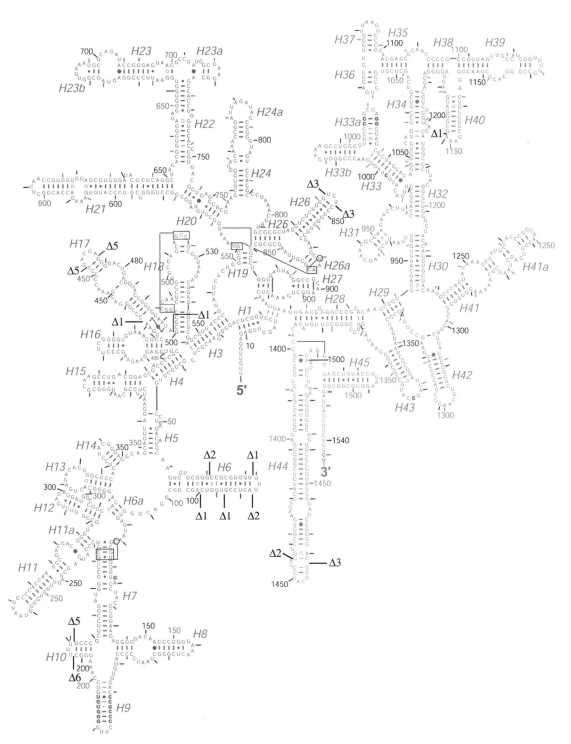

Figure 1. Secondary structure diagram of 16S RNA from *Thermus thermophilus*, colored by domain (5′ domain *red*, central domain *green*, 3′ major domain *orange*, and 3′ minor domain *blue*). Insertions in the *Thermus* sequence relative to *E. coli* occur in the 5′ and 3′ major domains and are shown in green. Deletions are indicated with black tick marks and a "Δ*n*" to show the deletion of *n* nucleotides. The *T. thermophilus* numbering is shown by tick marks for every 10th nucleotide and labels for every 50 nucleotides, colored as for each domain. Corresponding *E. coli* numbering and tick marks are shown in black throughout. (Modified, with permission, from http://www.rna.icmb.utexas.edu; see Gutell 1996.)

extension that threads through the 30S subunit and makes contact with S8 and S17 on the back side. In general, these extensions are highly basic, make intimate contacts with ribosomal RNA, and help with the assembly and folding of the 30S subunit. Similar extensions have also been observed in the 50S structure (Ban et al. 2000).

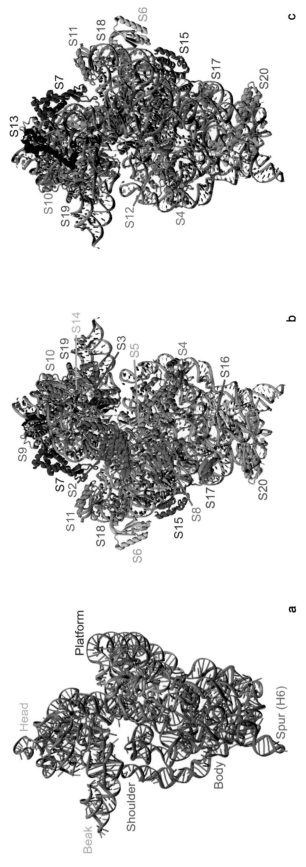

Figure 2. (*a*) Tertiary structure of 16S RNA in the 30S subunit, colored according to domains as in Fig. 1. (*b, c*) Overviews of the back and interface sides of the 30S subunit, respectively. (Panel *a* reprinted, with permission, from Wimberly et al. 2000 [copyright *Nature*].)

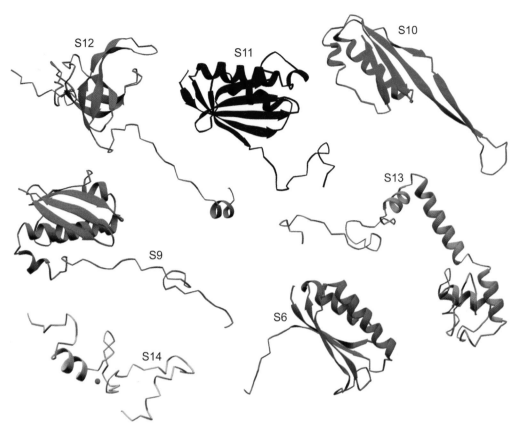

Figure 3. Protein structure gallery. Examples of structures of 30S proteins with long extensions. Of these, only the structure of S6 was determined in isolation. S14 contains a bound zinc ion which is shown as a green sphere.

Interestingly, proteins characterized as primary binders in the 30S assembly map are all globular and lack the long extensions. These proteins generally bind at multihelix junctions, such as S4 at the 5-way junction of helices 3, 4, 16, 17, and 18; S7 at the junction of helices 29, 30, 41, and 42; and S15 at the junction of helices 20, 21, and 22. The primary binders are very likely to be involved in condensing 16S RNA around multihelix junctions in agreement with a great deal of prior data (Nowotny and Nierhaus 1988; Agalarov and Williamson 2000). Finally, proteins with similar or even identical folds, such as S6, S10, and S11, bind RNA in different ways, suggesting that it is difficult to make inferences about RNA binding from topology.

COMPARISON WITH AN INDEPENDENT 30S STRUCTURE

The same month that our structure was published, an independent 3.3 Å structure of the 30S subunit from the Max Planck/Weizmann group was reported (Schluenzen et al. 2000). A comparison reveals a very similar overall conformation of the 30S subunit in the two structures. Thus, there appears to be little evidence to support the claim that heating in the presence of tungsten clusters leads to a functionally more active form of the particle (Tocilj et al. 1999; Schluenzen et al. 2000). Although the

global path of the RNA backbone is in agreement with ours, there are many significant differences in the details. In the Max Planck/Weizmann structure, regions of RNA such as the 560 and 967 loops are missing, irregular conformations of RNA are often modeled differently, and the RNA registry differs from ours in several regions, with these regional differences accounting for about 300 out of 1500 nucleotides. The root mean square deviation (rmsd) between equivalent phosphorus atoms in these regions of discrepancy is 6.1 Å compared to 3.6 Å for the whole RNA. In the Max Planck/Weizmann structure, the previously unsolved proteins are modeled as Cα traces with unidentified residues and are often noncontiguous and incomplete. Major differences in proteins fall into the categories of missing domains (e.g., S2, S4), missing extensions or loops (e.g., S7, S9, S13), or incompatible topologies (e.g., S11, S12, S17). Part of their S3 corresponds to our S14, and S20 has completely reversed polypeptide chain direction. We attribute most of these discrepancies to differences in interpretation of the electron density rather than genuine differences in conformation. Support for this view comes from the observation that recent structures of the 30S subunit published by the Max Planck/Weizmann group are in very close agreement with the structure we originally published about 8 months earlier, with an rmsd between phosphorus atoms of about 1 Å (Pioletti et al. 2001). Any differences that remain could be the result of slightly different crystallization conditions as well as the fact that tungsten clusters are part of the Max Planck/Weizmann structure, but not of ours.

BINDING SITES FOR tRNA AND mRNA

A precise analysis of the tRNA- and mRNA-binding sites was made easier by the occurrence of two fortuitous interactions. The spur (helix 6) from a neighboring molecule in the crystal is inserted into the P site of the 30S subunit and mimics P-site tRNA (Carter et al. 2000). Interestingly, the 3′ end of 16S RNA is folded back into the mRNA-binding cleft, and nucleotides are visible in both the P- and the E-site codons. The terminal three bases of this "pseudo-message" form noncanonical base pairs with the "anticodon" from the spur. Thus, the form of the 30S we crystallized appears to have mimics for tRNA in the P site and mRNA in the P and E sites. In hindsight, it is clear why this crystal form diffracts well even in the absence of bound ligands: It effectively mimics a ligand-bound form through crystal contacts. It is also interesting that in the absence of added mRNA, the 3′ end of 16S RNA is folded back into the 30S cleft in crystals of 70S ribosomes also (see Noller et al., this volume). This suggests that the binding of the 3′ end of 16S RNA into its own mRNA-binding cleft in the 30S subunit is not just a crystal-packing artifact. Rather, the intramolecular burial of the 3′ end of 16S RNA into the message-binding cleft may be energetically more favorable than binding mRNA for entropic reasons, except when extensive additional compensating interactions are made with mRNA; e.g., in base-pairing of the 3′ end with the Shine-Dalgarno sequence. This feature may therefore have the benefit of making initiation more specific.

The 7.8 Å structure of the 70S with mRNA and tRNA identified two elements of the 30S subunit, helix 27 and helix 44 (Cate et al. 1999). Using these two elements, we were able to superimpose the 70S structure onto our atomic model for the 30S subunit to arrive at a model for the interaction of tRNA with the 30S subunit (Fig. 4a). The superposition provided additional evidence that the spur and 3′ end of 16S RNA were good mimics of P-site tRNA and mRNA, respectively, as discussed above. In addition, it allowed us to deduce details of the interactions of tRNA and mRNA with the 30S subunit (Carter et al. 2000). As an example, details of the interaction of the P-site tRNA mimic and codon with the 30S subunit are shown in Figure 4b.

As had been revealed by previous biochemical data, the P-site tRNA makes extensive interactions with the 30S subunit. On the other hand, the A site is shallow and wide, which is consistent with both its lower affinity for tRNA and the need to allow tRNA to rotate significantly from its initial conformation as part of the ternary complex with EF-Tu (Stark et al. 1997) to its eventual conformation after GTP hydrolysis and accommodation, in which the acceptor arm of tRNA swings into the peptidyl transferase site. Additionally, the structure shows that helix 44, helix 34, protein S12, and the 530 loop are all part

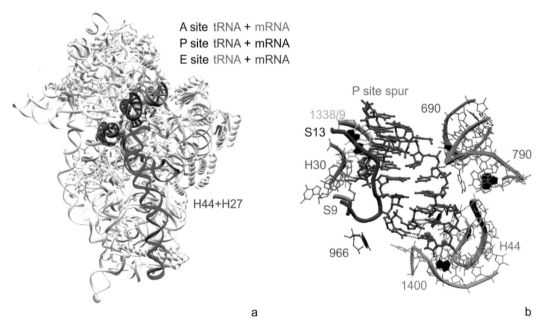

a   b

Figure 4. tRNA-binding sites of the small subunit. (*a*) Overview of the three tRNA molecules modeled onto the 30S structure using the 7.8 Å structure of 70S in complex with tRNA and mRNA. The two 16S elements used for the superposition (helices 44 and 27) are shown in green. (*b*) Details of the interactions in the P site made by the spur (helix 6) from a neighboring subunit in the crystal that mimics tRNA. The 3′ end of 16S RNA is folded into the cleft of the 30S and mimics mRNA (*dark blue*). The spur is shown in red, and the elements of 16S RNA and ribosomal proteins are labeled and colored independently for clarity. Dotted yellow lines indicate base-pairing between the spur and the codon mimic (*dark blue*), and dotted black lines indicate hydrogen bonds. Bases with black filling have previously been implicated in P-site tRNA or mRNA binding. (Reprinted, with permission, from Carter et al. 2000 [copyright *Nature*].)

of the A site where decoding occurs. This confirms the suggestion by Brimacombe and coworkers based on cross-linking experiments with mRNA that these elements are close to the decoding site (Dontsova et al. 1992).

The tRNA-binding sites have several striking features that were previously either unsuspected or only hinted at. Whereas the A and P sites consist mainly of RNA elements, the E-site tRNA includes extensive interactions with proteins S7 and S11. A second unexpected finding was that long tendrils of extended polypeptide chains from S9 and S13 make their way into the tRNA-binding sites. In particular, the lengths of the S9 tail and its terminal lysine residue are highly conserved, suggesting that it makes a crucial interaction with tRNA.

The tRNA interactions deduced from the 30S structure are consistent with a large body of biochemical data and have subsequently been confirmed by the 5.5 Å resolution structure of the 70S ribosome with mRNA and tRNA (Yusupov et al. 2001), as well as our complex of mRNA and an anticodon stem-loop of tRNA with the 30S subunit (Ogle et al. 2001).

ROLE OF THE 30S SUBUNIT IN DECODING

A central role of the 30S is decoding, or selection of the correct aminoacylated tRNA for participation in peptidyl transferase. The discrimination inherent in codon–anticodon base-pairing leads to a free-energy difference of only 2–3 kcal/mole between cognate tRNA and near-cognate tRNA (which has a single mismatch). This difference would predict an error rate of about 10^{-2}, which is 1–2 orders of magnitude higher than the observed error rate of translation (Kurland 1992).

Ever since the discovery that antibiotics can increase the error rate of translation, it has been proposed that the ribosome can modulate the fidelity of translation by directly monitoring codon–anticodon base-pairing (Davies et al. 1964). The site where this monitoring occurs is referred to as the decoding site, and by definition, it must be part of the environment of A-site tRNA. Two possible mechanisms have been proposed to account for the improved accuracy. The first involves recognition by the ribosome of the geometry of correct codon–anticodon base-pairing in the same way that enzymes recognize the geometry of the correct substrate and discriminate against incorrect ones (Eigen and De Maeyer 1966; Potapov 1982; Thompson and Dix 1982). In an alternative mechanism, selection of the correct tRNA would be based on a kinetic proofreading scheme, in which the energy discrimination inherent in codon–anticodon pairing is used twice, by having an initial selection and a "proofreading" step that are separated by an irreversible step (Hopfield 1974; Ninio 1975). In the case of the ribosome, the irreversible step would be the hydrolysis of GTP by EF-Tu. Thus, an error rate of 10^{-2} at each step could result in a total error rate as low as 10^{-4} if each step was allowed to reach equilibrium. Because of the need for speed, the overall accuracy is reduced from the maximum possible, but it is still sufficient to account for the error rate of pro-

tein synthesis. Therefore, in the context of proofreading schemes, it is not clear that recognition of codon–anticodon pairing geometry by the ribosome is even necessary. However, when a slowly hydrolyzable GTP analog was used, it was shown that even a single selection step could theoretically have a discrimination sufficient to account for the accuracy of protein synthesis, provided equilibrium could be reached, which gave evidence for the ability of the ribosome to recognize the geometry of codon–anticodon base-pairing (Thompson and Karim 1982).

LOCATION OF THE DECODING SITE

Where is the decoding site in the 30S subunit? Data from footprinting (Moazed and Noller 1990), cross-linking (Dontsova et al. 1992), and genetics (O'Connor et al. 1995, 1997) have implicated the region that includes helix 44, the 530 loop, and helix 34 in the ribosomal A site. These regions are all close together in the structure. More recent work on decoding has focused on two universally conserved residues, A1492 and A1493, which are part of an internal loop in helix 44. These residues are essential for both viability and A-site tRNA binding (Yoshizawa et al. 1999). This internal loop is also the binding site for the antibiotic paromomycin and related aminoglycosides, which increase the error rate of protein synthesis. In a nuclear magnetic resonance (NMR) structure of paromomycin bound to a fragment of helix 44, it was shown that the drug displaces A1492 and A1493 toward the minor groove of helix 44 (Fourmy et al. 1996). On the basis of biochemical experiments, it was concluded that the N1 of the two bases directly interacted with the 2′ OH of mRNA (Yoshizawa et al. 1999), and that the bases were involved in recognition of the shape of the codon–anticodon helix via minor-groove interactions. However, the same data could be used to arrive at a completely different model that involved interaction of the codon–anticodon helix with the major groove of helix 44 (VanLoock et al. 1999). In any case, when the 7.8 Å structure of the 70S ribosome with tRNA was determined, modeling the NMR structure of the internal loop with or without paromomycin into the 70S electron density suggested that the adenines were too far away from the codon–anticodon helix to have a direct role in decoding (Cate et al. 1999).

These discrepancies began to be resolved when the structure of the 30S subunit complexed with spectinomycin, streptomycin, and paromomycin showed that rather than displace them into the minor groove in a way that would preserve the A1493–A1408 base pair reported in the NMR structure (Fourmy et al. 1996), paromomycin in fact flipped A1492 and A1493 completely out of the internal loop of the helix so that they pointed into the A site (Carter et al. 2000). Modeling of A-site tRNA and mRNA showed that these bases would be in a position to bind directly in the minor groove of the codon–anticodon helix. However, they would do so in a way that would place the N1 of the adenines pointing toward tRNA rather than interacting with the 2′ OH of mRNA, as had previously been proposed (Yoshizawa et al. 1999). There are several examples of highly conserved adenines in 16S RNA that form tertiary interactions with the minor groove of A-form helices. These interactions involve hydrogen bonding of the adenines with the 2′ OH of both strands of the minor groove in a manner that would be sensitive to the shape and width of the groove. On the basis of this analysis, it was proposed that the binding of cognate tRNA would induce a similar conformational change in A1492 and A1493 to the one induced by paromomycin. The energetic cost of the conformational change would be paid for by compensating interactions between the adenines and the codon–anticodon helix. In the case of near-cognate tRNA, the fit would not be perfect, so that the induced changes would not be energetically favorable. However, paromomycin, by partially paying the energetic cost of the induced changes, would promote the binding of near-cognate tRNA and thus lead to an increased error rate.

RECOGNITION OF CODON–ANTICODON BASE-PAIRING BY THE RIBOSOME

The interactions by which the ribosome senses codon–anticodon pairing at the A site were determined experimentally by directly soaking mRNA in the form of a uridine hexanucleotide, as well as the anticodon stem-loop of its cognate phenylalanyl-tRNA, into crystals of the 30S subunit (Ogle et al. 2001). This work directly revealed the codon and anticodon stem-loop in the A site. In doing so, it showed that our earlier modeling of the A-site tRNA was reasonably accurate, and also that a sharp kink exists in the mRNA between the A- and P-site codons, as could be expected if each were to be part of an A-form helix with its corresponding anticodon.

The binding of cognate tRNA induces global domain movements in the 30S subunit. It also causes local changes around the decoding site. In the absence of tRNA or paromomycin, A1492 and A1493 are stacked inside helix 44, and G530 is in the *syn* conformation (Fig. 5a). The binding of paromomycin causes A1492 and A1493 to flip out of the internal loop of helix 44, while leaving G530 in the *syn* conformation (Fig. 5b). On tRNA binding, A1492 and A1493 flip out of helix 44 as we had predicted, but also cause G530 on the other side of the codon–anticodon helix to switch from a *syn* to an *anti* conformation (Fig. 5c). A comparison of Figures 5b and 5c shows that, apart from having no effect on G530, paromomycin alone does not induce precisely the same conformational changes in A1492 and A1493. This suggests that the antibiotic facilitates incorporation of near-cognate tRNA by contributing in part toward the energetic cost of changes required for recognition, rather than by creating a fully preformed binding site. Since the adenines are largely disordered in the native structure as judged by a very high B factor, but well ordered in the complexes with paromomycin or tRNA, there must be some additional entropic cost in the changes induced by binding.

A1493 makes intimate contacts with the first base pair of the codon–anticodon helix (Fig. 6a), whereas A1492 and G530 together span the minor groove of the second base pair and interact with each other via their N1 positions (Fig. 6c). The third base pair is not as closely monitored, with only the codon nucleotide making interactions with the ribosome (Fig. 6d). This is consistent with the requirement of the genetic code that the third position be able to accommodate noncanonical base pairs such as a GU wobble (Crick 1966; Yokoyama and Nishimura 1995). Moreover, the observation that the ribosome makes extensive interactions with the codon base but not the anticodon base is consistent with previous analysis suggesting that the codon base conformation is more restricted than that of the anticodon (Yokoyama and Nishimura 1995). These experiments show how the ribosome recognizes the geometry of codon–anticodon base-pairing, consistent with the looser requirement at the

Figure 5. Overview of A-site tRNA binding. Discrete states of the 30S A site deduced from three different crystal structures of (*a*) native, unbound 30S, (*b*) the 30S in complex with paromomycin, and (*c*) the 30S in complex with mRNA and cognate tRNA. In all panels, tRNA ASL is gold, A-site mRNA codon purple, P-site mRNA codon green, protein S12 brown, and important bases involved in conformational changes red. Magnesium ions are shown as magenta spheres. (*a*) The A site of the native 30S subunit (Wimberly et al. 2000). A1492/3 (*red, right*) are stacked in the center of helix 44, and G530 is in its *syn* conformation. C1054 is shown in the upper left corner. (*b*) In the paromomycin-bound 30S (Carter et al. 2000), the bases of A1492 and A1493 have been pushed out into the A site by the antibiotic binding to the center of helix 44 (*yellow sticks, lower right*). (*c*) In the 30S:ASL:U6 complex, A1492/3 flip out to monitor the codon–anticodon interaction and G530 has switched to its *anti* conformation to stabilize the interaction of the second position. Two Mg ions are visible in the place where A1492/3 are located in the native structure. (Reprinted, with permission, from Ogle et al. 2001 [copyright AAAS].)

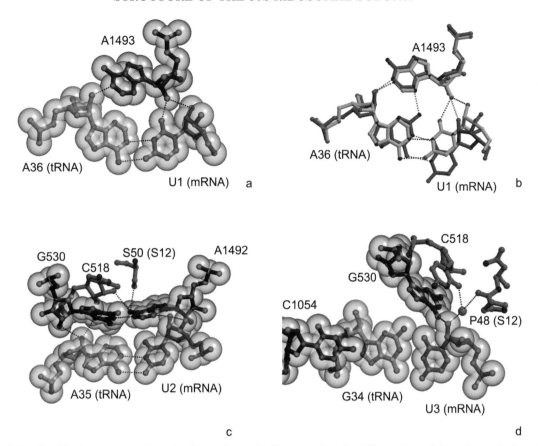

Figure 6. Details of the interactions at the A site. Interactions at the first, second, and wobble positions of the codon–anticodon mini-helix. In panels *a*, *c*, and *d*, the codon (U6) is shown with purple sticks, the anticodon (ASL) with gold sticks, and the monitoring bases with brown sticks. (*a*) At the first position, the minor groove of the A36:U1 base pair is monitored by A1493 (*brown*). (*b*) A super-position of the base pairs in the first position with cognate (*gray*) and near-cognate (*brown*) tRNA. (*c*) At the second position, G530 and A1492 (both *brown*) act in concert to monitor the base pair A35:U2. (*d*) The Watson-Crick-like base-pairing in the wobble position of the 30S:ASL:U6 complex with C1054 (*brown*) stacking against the ribose of A35. The codon–anticodon interaction at the wobble position is stabilized by a Mg-mediated interaction with C518 and residues P48 and N49 from protein S12 (*gray sticks*). (Panels *a*, *c*, and *d* reprinted, with permission, from Ogle et al. 2001 [copyright AAAS].)

wobble position. The structure explains previous chemical modification data and also rationalizes the requirements for adenines at 1492 and 1493 and a guanine at 530. Only purines would be able to span a minor groove, but a guanine at 1492 or 1493 could not have the same conformation as either adenine because of a steric clash with the minor groove from its NH2 at the N2 position. Guanines also would not be able to make the analogous hydrogen bonds through their N1 positions. Similarly, only guanine at G530 would be able to form the observed bond with A1492 through the N1 position. Finally, because DNA has no 2′ OH group and, also, DNA–RNA hybrids have a different groove width and shape, the structure explains the classic observation that DNA is a poor template for translation but is less so in the presence of the aminoglycoside neomycin (McCarthy and Holland 1965). It also explains the inability of the ribosome to bind tRNA in the A site when the codon consists of DNA rather than RNA (Potapov et al. 1995).

Recently, we have also determined the structure of the anticodon stem-loop of near-cognate tRNA in complex with the 30S subunit. This near-cognate tRNA, which codes for leucine, has a GAG anticodon, and hence has a GU wobble in both the first and third positions using the poly-U mRNA. In the absence of paromomycin, no density is visible for tRNA in the A site, whereas weak density is visible for mRNA. In the presence of paromomycin, the structure is very similar to that of cognate tRNA. However, because the first base pair of the codon–anticodon helix is a GU wobble pair rather than a Watson-Crick base pair, the fit of A1493 with the minor groove of this base pair is distorted and hence energetically less favorable. Interestingly, in comparison with the cognate case, it is the codon base rather than the anticodon that has moved to accommodate the wobble pair at the first position (Fig. 6b).

These experiments show that the ribosome recognizes cognate tRNA through an induced-fit mechanism that monitors the geometry of the minor groove of codon–anticodon base-pairing. In doing so, the ribosome uses principles found in other systems. The interactions of A1493 and A1492 with the minor groove are identical to those classified as type I and type II A minor interactions (Nissen et al. 2001), which are found in tertiary interactions in the group I intron, and both 23S RNA and 16S RNA. Much of the free energy of the interaction appears to arise from the shape complementarity of the adenines to the minor groove (Doherty et al. 2001). The principles of minor groove recognition by surface complementarity and sequence-independent hydrogen bonding to ensure fidelity of base-pairing are also employed by DNA and RNA polymerases (Doublie et al. 1998; Kiefer et al. 1998; Cheetham and Steitz 1999).

Since our experiments involve just the anticodon stem-loop of tRNA in the context of the 30S subunit, a natural question concerns what state these structures represent along the reaction pathway. The conformation of the anticodon stem-loop is virtually identical to that of the A-site tRNA in the 70S structure (Yusupov et al. 2001), suggesting that it represents a post-accommodation state of tRNA. Our structure therefore has no direct bearing on the initial recognition step in which a ternary complex of EF-Tu with tRNA and GTP binds to the ribosome. However, we note that the protections observed on the binding of the ternary complex of EF-Tu with tRNA to the ribosome are essentially the same as those observed on nonenzymatic tRNA binding (Powers and Noller 1994), and both are consistent with our structure. This suggests that despite a large change in the orientation of the tRNA, many of the interactions of the ribosome with the codon–anticodon helix observed here probably occur in the initial recognition step and persist through GTP hydrolysis to accommodation. It is clear that A1492 and A1493 are in a conformationally variable region of the 30S subunit and possibly can rotate with the tRNA after the initial binding. This hypothesis needs to be tested experimentally by determining a high-resolution structure of the ternary complex with the ribosome.

PROOFREADING

The basis of proofreading is that the initial binding of tRNA occurs as part of its ternary complex with EF-Tu, and the complex can dissociate at this step. Following GTP hydrolysis by EF-Tu, the aminoacylated end of the tRNA is free to move into the peptidyl transferase site in the accommodation step. During this process, which is part of the proofreading step, tRNA has a second chance to dissociate. The broad consensus is that proofreading is an important step in ensuring the accuracy of protein synthesis (Thompson and Stone 1977; Ruusala et al. 1982; Rodnina and Wintermeyer 2001). However, this view is not unanimous, and Nierhaus and coworkers have proposed an alternative model that excludes proofreading (Nierhaus 1990). Although recognition of the geometry of codon–anticodon base-pairing is an essential ingredient of this alternative model, we note that our demonstration that such recognition occurs does not preclude a proofreading step.

There are several lines of evidence for a proofreading step, the most important of which is that the number of molecules of GTP hydrolyzed per amino acid incorporated is much higher for near-cognate tRNA than for cognate tRNA, showing that near-cognate tRNA can trigger GTP hydrolysis but is usually rejected subsequently and only occasionally incorporated (Thompson and Stone 1977; Ruusala et al. 1982). In addition, it is known that

mutations in the ribosome not directly at the decoding site, e.g., in the tRNA but distant from the anticodon loop (Hirsh 1971), 23S RNA (O'Connor and Dahlberg 1993; Bilgin and Ehrenberg 1994), or EF-Tu (Tapio and Isaksson 1988), can affect accuracy. This argues that considerations other than geometric recognition of codon–anticodon base-pairing, such as the kinetics of various steps in the proofreading hypothesis, also play an essential role in the accuracy of translation.

An important insight into how the ribosome carries out proofreading comes from studies of the antibiotic streptomycin, which is known to reduce the accuracy of protein synthesis (Davies et al. 1964), by affecting primarily the proofreading step (Ruusala and Kurland 1984). Mutants resistant to streptomycin are generally hyperaccurate, whereas ribosome ambiguity (*ram*) mutants, often found as suppressors of streptomycin dependence, are more error-prone than wild-type ribosomes. Moreover, the effects of streptomycin binding and *ram* mutations are not additive, suggesting that their effects arise from a similar mechanism (Ruusala and Kurland 1984).

Restrictive mutations are usually found in protein S12, whereas *ram* mutants usually occur on proteins S4 and S5. However, recently both *ram* and restrictive phenotypes have been generated by mutations in helix 27 of 16S RNA (Lodmell and Dahlberg 1997). Chemical probing studies show that *ram* and restrictive ribosomes have altered chemical reactivity in portions of helices 27 and 44 of 16S RNA (Allen and Noller 1989). More recent chemical probing (Lodmell and Dahlberg 1997) and cryo-electron microscopy (Gabashvili et al. 1999) provide evidence that the *ram* and restrictive ribosomes may have altered conformations throughout the 30S subunit.

A model has been proposed in which the ribosome switches between *ram* and restrictive states even during normal translation (Lodmell and Dahlberg 1997). After hydrolysis of GTP by EF-Tu, the ribosome enters a restrictive conformation that favors dissociation of tRNAs. This is followed by a transition to another state in which the tRNA is stably bound and can become available for peptide-bond formation. Near-cognate tRNAs are more likely to dissociate from the restrictive conformation, whereas cognate tRNAs are more likely to progress to peptide-bond formation. Restrictive mutations alter the energetics of the ribosome so that the lifetime of the restrictive state is prolonged, and hence incorrect tRNAs are more likely to be rejected. In contrast, *ram* mutations "destabilize" this restrictive state, so that more tRNAs (even near-cognate ones) are likely to stay bound to the ribosome.

The structure of the 30S subunit shows the locations of the streptomycin-binding site as well as mutations that affect translational accuracy. The environment of helix 27 in the 30S crystal structure is shown in Figure 7a. The bases with altered chemical reactivity in the *ram* and restrictive states (Allen and Noller 1989) lie at the interface between helix 27 and helix 44. In our structure, bases 888–890 of helix 27 are present in the S-turn conformation thought to be characteristic of the *ram* state (Lodmell and Dahlberg 1997). As shown in Figure 7b, the *ram* mutations that occur in S4 and S5 lie mainly at the interface between the two proteins (Clemons et al. 1999; Wimberly

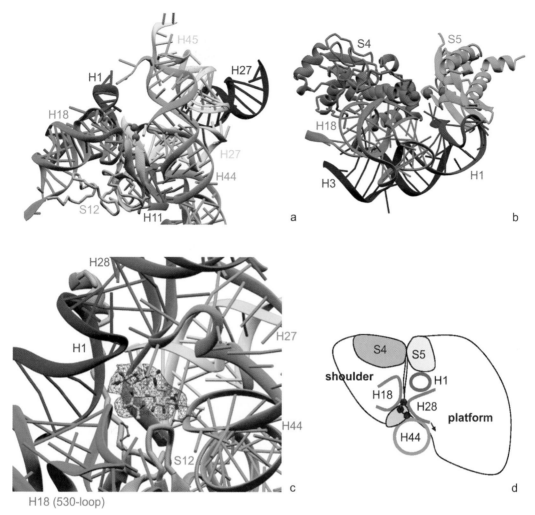

Figure 7. The effect of streptomycin. (*a*) The environment of helix 27 (*yellow*) in the 30S subunit. Red balls highlight sites of restrictive mutations in S12, red sticks show A1492 and A1493. (*b*) The helix 1–helix 18 interaction surface is contiguous with the S4-S5 surface. The locations of *ram* mutations in S4 and S5 are highlighted in red, as are the locations of residues 8 and 26 in S4 whose chemical reactivity changes in *ram* mutants. (*c*) The streptomycin-binding site, showing the interactions with helix 27, the 530-loop (helix 18), helix 44, and ribosomal protein S12. Streptomycin is shown in a stick representation along with the difference Fourier density (*red*). (*d*) Schematic horizontal cross-section of the 30S in the streptomycin-binding region showing the approximate location of the elements involved in controlling the *ram*/restrictive equilibrium. S12 is shown in yellow. (Panels *a–c* reprinted, with permission, from Carter et al. 2000 [copyright *Nature*].)

et al. 2000). This interface is contiguous with an RNA–RNA interface between helix 1 and the 530 loop (helix 18). Together, they represent a hybrid protein–RNA interface that could be disrupted during conformational changes in the 30S subunit. Streptomycin binds tightly in a pocket where it interacts with the phosphate backbone from four regions of 16S RNA (helices 1, 18, 27/28, and 44) as well as the highly conserved Lys-45 (*E. coli* Lys-42) of S12 (Fig. 7c) (Carter et al. 2000). All of these regions lie at the interface between the shoulder and platform domains of the 30S subunit (Fig. 7d).

Together, these observations implicate movement at this interface in the conformational changes associated with proofreading. At one end of the interface, mutations in S4 and S5 that disrupt their interaction lead to the *ram* phenotype, probably by destabilizing the restrictive state and thus accelerating the transition to a *ram* state. In contrast, mutations in S12 that disrupt interactions at the

other end of the interface probably prolong the lifetime of the restrictive state, whereas streptomycin, which stabilizes interactions in the S12 region, suppresses the effect of these mutations (Kurland et al. 1996). In the extreme case of streptomycin-dependence mutations in S12, the transition to the *ram* state is so unfavorable that streptomycin is actually required for function. Interestingly, the single known mutation in S12 that leads to streptomycin resistance without a hyperaccurate phenotype is K45R (*E. coli* K42). This lysine makes direct hydrogen bonds with two OH groups on streptomycin as well as a salt bridge to A913. Its mutation to an arginine would have the effect of disrupting streptomycin binding but would not disrupt its interactions with the ribosome, since the arginine could make the same salt bridge to A913.

Without knowing the structure of the restrictive and *ram* states, it is difficult to predict how streptomycin and accuracy mutations would affect the transition between them. Moreover, it remains unclear how the changes in conformation observed in the mutants relate to conformational changes in the ribosome during the elongation cycle. In addition to structural work, the *ram* and restrictive states need to be placed rigorously in a kinetic pathway from initial selection to peptidyl transferase. Although the details of proofreading remain to be elucidated, the movement between the platform and shoulder domains of the 30S subunit that has been observed recently on cognate tRNA binding or IF1 binding (Carter et al. 2001; Ogle et al. 2001) supports the proposal that movement between these domains is involved.

COMPLEXES WITH ANTIBIOTICS

Many antibiotics act by binding to the large or small ribosomal subunit, and an enormous amount of work has gone into elucidating the biochemical basis for their action (Spahn and Prescott 1996). Most antibiotics appear to bind to the RNA component of the ribosome, although many resistance mutations map to ribosomal proteins.

Despite decades of biochemical work, it has not been possible until recently to obtain direct structural information on the binding of antibiotics to ribosomes because of the lack of a high-resolution structure of either subunit. A notable exception was the site in helix 44 that contains an internal loop implicated in decoding. This region was shown to bind antibiotics as an isolated fragment (Purohit and Stern 1994) and was used to determine the structures of paromomycin and gentamicin in complex with RNA (Fourmy et al. 1996; Yoshizawa et al. 1998).

With the determination of atomic structures for ribosomal subunits, the situation has changed dramatically. It is

now possible to determine the structures of complexes of antibiotics with either ribosomal subunit by cocrystallization or by soaking the native crystals in antibiotic. Refinement of the coordinates of the native structure against diffraction data from the antibiotic complex directly reveals density corresponding to the antibiotic molecules in difference Fourier maps. Using this approach, we have determined the structure of the 30S subunit in complex simultaneously with streptomycin, spectinomycin, and paromomycin (Carter et al. 2000) and also separately with tetracycline, hygromycin B, and pactamycin (Brodersen et al. 2000). The locations of these antibiotics in the 30S structure are shown in Figure 8. These structures shed light on the mechanism of action of the antibiotics, as discussed above for paromomycin and streptomycin in the context of decoding and proofreading. We briefly mention below the results for the other antibiotics.

Tetracycline

Tetracycline is known to have a single strong binding site in the entire ribosome, located in the 30S subunit, in addition to several weaker binding sites on both subunits (Epe et al. 1987; Kolesnikov et al. 1996), but only the strong binding site is responsible for its inhibitory action during translation (Buck and Cooperman 1990). We observe two tetracycline sites in the 30S subunit. The strong or primary binding site makes contacts with an irregular minor groove of helix 34 and peripherally with bases 964–967 in helix 31. This position places it in the A site of the 30S subunit, in a position that would sterically clash with A-site tRNA after accommodation. The structure thus explains the observation that tetracycline bind-

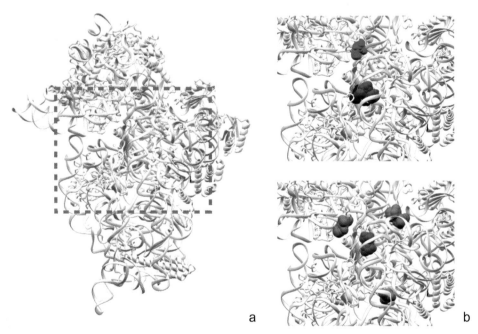

a   b

Figure 8. Antibiotics bound to the 30S subunit. (*a*) Overview of the 30S subunit outlining the central region in which the antibiotics we have studied bind. (*b*) The central region expanded to show the locations of streptomycin (*blue*), spectinomycin (*green*), paromomycin (*red*), tetracycline (*orange*), pactamycin (*cyan*), and hygromycin B (*magenta*).

ing prevents the binding of A-site tRNA but allows GTP hydrolysis by EF-Tu (Gordon 1969). It is also consistent with extensive data on positions on the tetracycline molecule that cannot be modified without losing antibiotic activity (Chopra 1985). These positions are all involved in making interactions with the 30S subunit, whereas the positions where modifications are allowed are not. We also see a second binding site for tetracycline between helix 27 and helix 11 that is consistent with biochemical footprinting data (Moazed and Noller 1987). It is not clear that this second site has any relevance for the action of tetracycline. However, helix 27 is known to be the site of an accuracy switch in the ribosome (as discussed above in the section on proofreading), and this site may have some synergistic effect with the primary site at helix 34.

Recently, a second structure of the complex of the 30S subunit with tetracycline was published by the Max Planck/Weizmann group (Pioletti et al. 2001). This complex, at 4.5 Å resolution, shows six binding sites for tetracycline. Only one of these, the primary binding site at helix 34, is common to both structures. A site observed near helix 27 is distinct from our secondary site near that location and appears not to be consistent with protection data (Moazed and Noller 1987). Other sites observed in the Max Planck/Weizmann structure are consistent with various biochemical data on affinity labeling by tetracycline analogs. However, despite a nominally higher resolution and concentration of tetracycline, we see no signs of these additional sites in our difference Fourier maps. It is not clear whether the differences arise from the data or from the actual soaking times and conditions. In any case, our current opinion is that the minor sites are not relevant physiologically, and the site observed in both structures, namely the A site, is responsible for the effect of tetracycline on protein synthesis.

Pactamycin, Hygromycin B, and Spectinomycin

Pactamycin binds to the ribosomal E site, where it displaces mRNA and could affect initiation by interfering with interaction with either the Shine-Dalgarno sequence or initiation factors (Brodersen et al. 2000). Hygromycin B binds to the major groove of helix 44 near the P site (Brodersen et al. 2000). This region of helix 44 appears to move during translocation (Frank and Agrawal 2000), and the structure provides a rationale for how hygromycin B may inhibit its flexibility and thus translocation.

Spectinomycin binds to a minor groove of helix 34 that appears to be very close to a hinge point between the head and body (Carter et al. 2000). Its binding site is consistent with biochemical and mutational data (Sigmund et al. 1984; Brink et al. 1994). By potentially inhibiting conformational changes in the head, spectinomycin could have the effect of inhibiting translocation as observed.

STUDIES ON INITIATION

During initiation, the 30S subunit binds three initiation factors, IF1, IF2, and IF3, along with initiator tRNA. As a first step in understanding the structure of the initiation complex, we have determined the structure of IF1 bound to the 30S subunit. The structure shows that IF1 interacts directly with A1492 and A1493 in the A site, and its binding would sterically prevent the binding of A-site tRNA. Both of these are consistent with earlier biochemical data (Moazed et al. 1995; Dahlquist and Puglisi 2000). The binding of IF1 causes a change in the conformation of helix 44 that is propagated to result in global domain movements, and since IF1 is known to affect the rate but not extent of dissociation of 30S and 50S (Godefroy-Colburn et al. 1975), it is likely that the IF1 complex with the 30S represents a transition state that has a lowered activation energy barrier for subunit association or dissociation.

CONCLUSIONS AND FUTURE PROSPECTS

These are exciting times to be working on ribosome structure. The recent atomic structures of the 30S and 50S, and the subsequent lower resolution but complete molecular model of the 70S with tRNA, have changed the way that ribosome biology will be studied. The structure of the 30S has given us insights into tRNA and mRNA recognition, decoding, and antibiotic function, as well as interactions with initiation factor IF1. The atomic structures of both subunits have been invaluable in the molecular interpretation of the 5.5 Å resolution structure of the 70S ribosome. Nevertheless, a large number of important questions remain, even in the context of the bacterial ribosome, let alone its more complicated eukaryotic counterpart. These range from the sequence of events during initiation and termination to an understanding of how GTP hydrolysis by the various factors is triggered by the ribosome and how this energy is used in various steps such as proofreading or translocation. We also need to learn much more about the molecular details of the conformational changes in the ribosome during translation.

ACKNOWLEDGMENTS

This work was supported by the Medical Research Council (UK) and National Institutes of Health grant GM-44973 (to S.W. White and V.R.). D.E.B. was the recipient of a Human Frontier Science Program postdoctoral fellowship, and W.M.C. was the recipient of a National Institutes of Health predoctoral fellowship. We dedicate this to the memory of Professor Paul Sigler, who facilitated this work at a crucial juncture.

REFERENCES

Agalarov S.C. and Williamson J.R. 2000. A hierarchy of RNA subdomains in assembly of the central domain of the 30 S ribosomal subunit. *RNA* **6**: 402.

Allen P.N. and Noller H.F. 1989. Mutations in ribosomal proteins S4 and S12 influence the higher order structure of 16 S ribosomal RNA. *J. Mol. Biol.* **208**: 457.

Andersson I., Knight S., Schneider G., Lindqvist Y., Lundqvist T., Brändén C.-I., and Lorimer G.H. 1989. Crystal structure of the active-site of ribulose-bisphosphate carboxylase. *Nature* **337**: 229.

Ban N., Nissen P., Hansen J., Moore P.B., and Steitz T.A. 2000.

The complete atomic structure of the large ribosomal subunit at 2.4 Å resolution. *Science* **289:** 905.

Ban N., Freeborn B., Nissen P., Penczek P., Grassucci R.A., Sweet R., Frank J., Moore P.B., and Steitz T.A. 1998. A 9 Å resolution X-ray crystallographic map of the large ribosomal subunit. *Cell* **93:** 1105.

Bilgin N. and Ehrenberg M. 1994. Mutations in 23 S ribosomal RNA perturb transfer RNA selection and can lead to streptomycin dependence. *J. Mol. Biol.* **235:** 813.

Blundell T.L. and Johnson L.N. 1976. *Protein crystallography.* Academic Press, New York.

Brink M.F., Brink G., Verbeet M.P., and de Boer H.A. 1994. Spectinomycin interacts specifically with the residues G1064 and C1192 in 16S rRNA, thereby potentially freezing this molecule into an inactive conformation. *Nucleic Acids Res.* **22:** 325.

Brodersen D.E., Clemons W.M., Carter A.P., Morgan-Warren R.J., Wimberly B.T., and Ramakrishnan V. 2000. The structural basis for the action of the antibiotics tetracycline, pactamycin, and hygromycin B on the 30S ribosomal subunit. *Cell* **103:** 1143.

Buck M.A. and Cooperman B.S. 1990. Single protein omission reconstitution studies of tetracycline binding to the 30S subunit of *Escherichia coli* ribosomes. *Biochemistry* **29:** 5374.

Carter A.P., Clemons W.M., Jr., Brodersen D.E., Morgan-Warren R., Wimberly B.T., and Ramakrishnan V. 2000. Functional insights from the structure of the 30S ribosomal subunit and its interaction with antibiotics. *Nature* **407:** 340.

Carter A.P., Clemons W.M., Jr., Brodersen D.E., Morgan-Warren R.J., Hartsch T., Wimberly B.T., and Ramakrishnan V. 2001. Crystal structure of an initiation factor bound to the 30S ribosomal subunit. *Science* **291:** 498.

Cate J.H., Yusupov M.M., Yusupova G.Z., Earnest T.N., and Noller H.F. 1999. X-ray crystal structures of 70S ribosome functional complexes (comments). *Science* **285:** 2095.

Cate J.H., Gooding A.R., Podell E., Zhou K., Golden B.L., Kundrot C.E., Cech T.R., and Doudna J.A. 1996. Crystal structure of a group I ribozyme domain: Principles of RNA packing. *Science* **273:** 1678.

Cheetham G.M. and Steitz T.A. 1999. Structure of a transcribing T7 RNA polymerase initiation complex. *Science* **286:** 2305.

Chopra I. 1985. Mode of action of the tetracyclines and the nature of bacterial resistance to them. *Handb. Exp. Pharmacol.* **78:** 317.

Clemons W.M., Jr., May J.L.C., Wimberly B.T., McCutcheon J.P., Capel M., and Ramakrishnan V. 1999. Structure of a bacterial 30S ribosomal subunit at 5.5 Å resolution. *Nature* **400:** 833.

Clemons W.M., Jr., Brodersen D.E., McCutcheon J.P., May J.L.C., Carter A.P., Morgan-Warren R.J., Wimberly B.T., and Ramakrishnan V. 2001. Crystal structure of the 30S ribosomal subunit from *Thermus thermophilus*: Purification, crystallization and structure determination. *J. Mol. Biol.* **310:** 827.

Crick F.H.C. 1966. Codon-anticodon pairing:The wobble hypothesis. *J. Mol. Biol.* **19:** 548.

Crick F.H.C. and Magdoff B.S. 1956. The theory of the method of isomorphous replacement for protein crystals. I. *Acta Crystallogr.* **9:** 901.

Dahlquist K.D. and Puglisi J.D. 2000. Interaction of translation initiation factor IF1 with the *E. coli* ribosomal A site. *J. Mol. Biol.* **299:** 1.

Davies J., Gilbert W., and Gorini L. 1964. Streptomycin, suppression, and the code. *Proc. Natl. Acad. Sci.* **51:** 883.

Doherty E.A., Batey R.T., Masquida B., and Doudna J.A. 2001. A universal mode of helix packing in RNA. *Nat. Struct. Biol.* **8:** 339.

Dontsova O., Dokudovskaya S., Kopylov A., Bogdanov A., Rinke-Appel J., Junke N., and Brimacombe R. 1992. Three widely separated positions in the 16S RNA lie in or close to the ribosomal decoding region; a site-directed cross-linking study with mRNA analogues. *EMBO J.* **11:** 3105.

Doublie S., Tabor S., Long A.M., Richardson C.C., and Ellenberger T. 1998. Crystal structure of a bacteriophage T7 DNA replication complex at 2.2 Å resolution. *Nature* **391:** 251.

Eigen M. and De Maeyer L. 1966. Chemical means of information storage and readout in biological systems. *Naturwissenschaften* **53:** 50.

Epe B., Woolley P., and Hornig H. 1987. Competition between tetracycline and tRNA at both P and A sites of the ribosome of *Escherichia coli. FEBS Lett.* **213:** 443.

Fourmy D., Recht M.I., Blanchard S.C., and Puglisi J.D. 1996. Structure of the A site of *Escherichia coli* 16S ribosomal RNA complexed with an aminoglycoside antibiotic. *Science* **274:** 1367.

Frank J. and Agrawal R.K. 2000. A ratchet-like inter-subunit reorganization of the ribosome during translocation. *Nature* **406:** 319.

Frank J., Zhu J., Penczek P., Li Y., Srivastava S., Verschoor A., Radermacher M., Grassucci R., Lata R.K., and Agrawal R.K. 1995. A model of protein synthesis based on cryo-electron microscopy of the *E. coli* ribosome. *Nature* **376:** 441.

Gabashvili I.S., Agrawal R.K., Grassucci R., Squires C.L., Dahlberg A.E., and Frank J. 1999. Major rearrangements in the 70S ribosomal 3D structure caused by a conformational switch in 16S ribosomal RNA. *EMBO J.* **18:** 6501.

Garman E. and Schneider. T.R. 1997. Macromolecular cryocrystallography. *J. Appl. Crystallogr.* **30:** 211.

Garrett R.A., Douthwaite S.R., Liljas A., Matheson A.T., Moore P.B., and Noller H.F. 2000. *The ribosome: Structure, function, antibiotics and cellular interactions.* ASM Press, Washington, D.C.

Glotz C. and Brimacombe R. 1980. An experimentally-derived model for the secondary structure of the 16S ribosomal RNA from *Escherichia coli. Nucleic Acids Res.* **8:** 2377.

Glotz C., Müssig J., Gewitz H.S., Makowski I., Arad T., Yonath A., and Wittmann H.G. 1987. Three-dimensional crystals of ribosomes and their subunits from eu- and archaebacteria. *Biochem. Int.* **15:** 953.

Godefroy-Colburn T., Wolfe A.D., Dondon J., Grunberg-Manago M., Dessen P., and Pantaloni D. 1975. Light-scattering studies showing the effect of initiation factors on the reversible dissociation of *Escherichia coli* ribosomes. *J. Mol. Biol.* **94:** 461.

Gordon J. 1969. Hydrolysis of guanosine 5′-triphosphate associated wh binding of aminoacyl transfer ribonucleic acid to ribosomes. *J. Biol. Chem.* **244:** 5680.

Gutell R.R. 1996. Comparative sequence analysis and the structure of 16S and 23S rRNA. In *Ribosomal RNA: Structure, evolution, processing, and function in protein biosynthesis* (ed. A.E. Dahlberg and R.A. Zimmermann), p. 111. CRC Press, Boca Raton, Florida.

Helliwell J.R. 1998. Synchrotron radiation facilities. *Nat. Struct. Biol.* (suppl.) **5:** 614.

Hendrickson W.A. 1991. Determination of macromolecular structures from anomalous diffraction of synchrotron radiation. *Science* **254:** 51.

———. 2000. Synchrotron crystallography. *Trends Biochem. Sci.* **25:** 637.

Hendrickson W.A., Horton J.R., and LeMaster D.M. 1990. Selenomethionyl proteins produced for analysis by multiwavelength anomalous diffraction (MAD): A vehicle for direct determination of three-dimensional structure. *EMBO J.* **9:** 1665.

Hirsh D. 1971. Tryptophan transfer RNA as the UGA suppressor. *J. Mol. Biol.* **58:** 439.

Hope H. 1988. Cryocrystallography of biological macromolecules: A generally applicable method. *Acta Crystallogr. B* **44:** 22.

Hope H., Frolow F., von Böhlen K., Makowski I., Kratky C., Halfon Y., Danz H., Webster P., Bartels K.S., Wittmann H.G., and Yonath A. 1989. Cryocrystallography of ribosomal particles. *Acta Crystallogr. B* **45:** 190.

Hopfield J.J. 1974. Kinetic proofreading: A new mechanism for reducing errors in biosynthetic processes requiring high specificity. *Proc. Natl. Acad. Sci.* **71:** 4135.

Kahn R., Fourme R., Bosshard R., Chiadmi M., Risler J.L., Dideberg O., and Wery J.P. 1985. Crystal structure study of *Opsanus* tau parvalbumin by multiwavelength anomalous diffraction. *FEBS Lett.* **179:** 133.

Kiefer J.R., Mao C., Braman J.C., and Beese L.S. 1998. Visualizing DNA replication in a catalytically active *Bacillus* DNA polymerase crystal. *Nature* **391:** 304.

Knablein J., Neuefeind T., Schneider F., Bergner A., Messerschmidt A., Lowe J., Steipe B., and Huber R. 1997. Ta6Br(2+)12, a tool for phase determination of large biological assemblies by X-ray crystallography. *J. Mol. Biol.* **270:** 1.

Kolesnikov I.V., Protasova N.Y., and A.T. Gudkov. 1996. Tetracyclines induce changes in accessibility of ribosomal proteins to proteases. *Biochimie* **78:** 868.

Kurland C.G. 1992. Translational accuracy and the fitness of bacteria. *Annu. Rev. Genet.* **26:** 29.

Kurland C.G., Hughes D., and Ehrenberg M. 1996. Limitations of translational accuracy. In Escherichia coli *and* Salmonella typhimurium: *Cellular and molecular biology,* 2nd edition (ed. F.C. Neidhardt et al.), p. 979. ASM Press, Washington, D.C.

Lata K.R., Agrawal R.K., Penczek P., Grassucci R., Zhu J., and Frank J. 1996. Three-dimensional reconstruction of the *Escherichia coli* 30S ribosomal subunit in ice. *J. Mol. Biol.* **262:** 43.

Lodmell J.S. and Dahlberg A.E. 1997. A conformational switch in *Escherichia coli* 16S ribosomal RNA during decoding of messenger RNA. *Science* **277:** 1262.

McCarthy B.J. and Holland J.J. 1965. Denatured DNA as a direct template for in vitro protein synthesis. *Proc. Natl. Acad. Sci.* **54:** 880.

Moazed D. and Noller H.F. 1987. Interaction of antibiotics with functional sites in 16S ribosomal RNA. *Nature* **327:** 389.

———. 1990. Binding of tRNA to the ribosomal A and P sites protects two distinct sets of nucleotides in 16 S rRNA. *J. Mol. Biol.* **211:** 135.

Moazed D., Samaha R.R., Gualerzi C., and Noller H.F. 1995. Specific protection of 16 S rRNA by translational initiation factors. *J. Mol. Biol.* **248:** 207.

Mueller F., Stark H., van Heel M., Rinke-Appel J., and Brimacombe R. 1997. A new model for the three-dimensional folding of *Escherichia coli* 16 S ribosomal RNA. III. The topography of the functional centre. *J. Mol. Biol.* **271:** 566.

Nierhaus K.H. 1990. The allosteric three-site model for the ribosomal elongation cycle: Features and future. *Biochemistry* **29:** 4997.

Ninio J. 1975. Kinetic amplification of enzyme discrimination. *Biochimie* **57:** 587.

Nissen P., Ippolito J.A., Ban N., Moore P.B., and Steitz T.A. 2001. RNA tertiary interactions in the large ribosomal subunit: The A-minor motif. *Proc. Natl. Acad. Sci.* **98:** 4899.

Nowotny V. and Nierhaus K.H. 1988. Assembly of the 30S subunit from *Escherichia coli* ribosomes occurs via two assembly domains which are initiated by S4 and S7. *Biochemistry* **27:** 7051.

O'Connor M. and Dahlberg A.E. 1993. Mutations at U2555, a tRNA-protected base in 23S rRNA, affect translational fidelity. *Proc. Natl. Acad. Sci.* **90:** 9214.

O'Connor M., Thomas C.L., Zimmermann R.A., and Dahlberg A.E. 1997. Decoding fidelity at the ribosomal A and P sites: Influence of mutations in three different regions of the decoding domain in 16S rRNA. *Nucleic Acids Res.* **25:** 1185.

O'Connor M., Brunelli C.A., Firpo M.A., Gregory S.T., Lieberman K.R., Lodmell J.S., Moine H., Van Ryk D.I., and Dahlberg A.E. 1995. Genetic probes of ribosomal RNA function. *Biochem. Cell Biol.* **73:** 859.

Ogle J.M., Brodersen D.E., Clemons W.M., Jr., Tarry M.J., Carter A.P., and Ramakrishnan V. 2001. Recognition of cognate transfer RNA by the 30S ribosomal subunit. *Science* **292:** 897.

O'Halloran T.V., Lippard S.J., Richmond T.J., and Klug A. 1987. Multiple heavy-atom reagents for macromolecular X-ray structure determination. Application to the nucleosome core particle. *J. Mol. Biol.* **194:** 705.

Pioletti M., Schlunzen F., Harms J., Zarivach R., Gluhmann M., Avila H., Bashan A., Bartels H., Auerbach T., Jacobi C., Hartsch T., Yonath A., and Franceschi F. 2001. Crystal structures of complexes of the small ribosomal subunit with tetra-

cycline, edeine and IF3. *EMBO J.* **20:** 1829.

Potapov A.P. 1982. A stereospecific mechanism for the aminoacyl-tRNA selection at the ribosome. *FEBS Lett.* **146:** 5.

Potapov A.P., Triana-Alonso F.J., and Nierhaus K.H. 1995. Ribosomal decoding processes at codons in the A or P sites depend differently on 2′-OH groups. *J. Biol. Chem.* **270:** 17680.

Powers T. and Noller H.F. 1994. Selective perturbation of G530 of 16 S rRNA by translational miscoding agents and a streptomycin-dependence mutation in protein S12. *J. Mol. Biol.* **235:** 156.

Purohit P. and Stern S. 1994. Interactions of a small RNA with antibiotic and RNA ligands of the 30S subunit. *Nature* **370:** 659.

Ramakrishnan V. and Biou V. 1997. Treatment of multiwavelength anomalous diffraction data as a special case of multiple isomorphous replacement. *Methods Enzymol.* **276:** 538.

Ramakrishnan V., Finch J.T., Graziano V., Lee P.L., and Sweet R.M. 1993. Crystal structure of globular domain of histone H5 and its implications for nucleosome binding. *Nature* **362:** 219.

Rodnina M.V. and Wintermeyer W. 2001. Ribosome fidelity: tRNA discrimination, proofreading and induced fit. *Trends Biochem. Sci.* **26:** 124.

Ruusala T. and Kurland C.G. 1984. Streptomycin preferentially perturbs ribosomal proofreading. *Mol. Gen. Genet.* **198:** 100.

Ruusala T., Ehrenberg M., and Kurland C.G. 1982. Is there proofreading during polypeptide synthesis? *EMBO J.* **1:** 741.

Schluenzen F., Tocilj A., Zarivach R., Harms J., Gluehmann M., Janell D., Bashan A., Bartels H., Agmon I., Franceschi F., and Yonath A. 2000. Structure of functionally activated small ribosomal subunit at 3.3 Å resolution. *Cell* **102:** 615.

Schluenzen F., Hansen H.A.S., Thygesen J., Bennett W.S., Volkmann N., Levin I., Harms J., Bartles H., Zaytzev-Bashan A., Berkovitch-Yellin Z., Sagi I., Fransceschi F., Krumbholtz S., Geva M., Weinstein S., Agmon I., Boddeker N., Morlang S., Sharon R., Dribin A., Maltz E., Peretz M., Weinrich V., and Yonath A. 1995. A milestone in ribosomal crystallography: The construction of preliminary electron density maps at intermediate resolution. *Biochem. Cell Biol.* **73:** 739.

Sigmund C.D., Ettayebi M., and Morgan E.A. 1984. Antibiotic resistance mutations in 16S and 23S ribosomal RNA genes of *Escherichia coli. Nucleic Acids Res.* **12:** 4653.

Spahn C.M. and Prescott C.D. 1996. Throwing a spanner in the works: Antibiotics and the translation apparatus. *J. Mol. Med.* **74:** 423.

Stark H., Rodnina M.V., Rinke-Appel J., Brimacombe R., Wintermeyer W., and van Heel M. 1997. Visualization of elongation factor Tu on the *Escherichia coli* ribosome. *Nature* **389:** 403.

Stark H., Mueller F., Orlova E.V., Schatz M., Dube P., Erdemir T., Zemlin F., Brimacombe R., and van Heel M. 1995. The 70S *Escherichia coli* ribosome at 23 Å resolution: Fitting the ribosomal RNA. *Structure* **3:** 815.

Tapio S. and Isaksson L.A. 1988. Antagonistic effects of mutant elongation factor Tu and ribosomal protein S12 on control of translational accuracy, suppression and cellular growth. *Biochimie* **70:** 273.

Thompson R.C. and Dix D.B. 1982. Accuracy of protein biosynthesis. A kinetic study of the reaction of poly(U)-programmed ribosomes with a leucyl-tRNA2-elongation factor Tu-GTP complex. *J. Biol. Chem.* **257:** 6677.

Thompson R.C. and Karim A.M. 1982. The accuracy of protein biosynthesis is limited by its speed: High fidelity selection by ribosomes of aminoacyl-tRNA ternary complexes containing GTPγS. *Proc. Natl. Acad. Sci.* **79:** 4922.

Thompson R.C. and Stone P.J. 1977. Proofreading of the codon-anticodon interaction on ribosomes. *Proc. Natl. Acad. Sci.* **74:** 198.

Thygesen J., Weinstein S., Franceschi F., and Yonath A. 1996. The suitability of multi-metal clusters for phasing in crystallography of large macromolecular assemblies. *Structure* **4:** 513.

Tocilj A., Schlunzen F., Janell D., Gluhmann M., Hansen H.A., Harms J., Bashan, Bartels H., Agmon I., Franceschi F., and

Yonath A. 1999. The small ribosomal subunit from *Thermus thermophilus* at 4.5 Å resolution: Pattern fittings and the identification of a functional site. *Proc. Natl. Acad. Sci.* **96:** 14252.

Trakhanov S.D., Yusupov M.M., Agalarov S.C., Garber M.B., Ryazantsev S.N., Tischenko S.V., and Shirokov V.A. 1987. Crystallization of 70 S ribosomes and 30 S ribosomal subunits from *Thermus thermophilus*. *FEBS Lett.* **220:** 319.

VanLoock M.S., Easterwood T.R., and Harvey S.C. 1999. Major groove binding of the tRNA/mRNA complex to the 16 S ribosomal RNA decoding site. *J. Mol. Biol.* **285:** 2069.

von Böhlen K., Makowski I., Hansen H.A.S., Bartels H., Berkovitch-Yellin Z., Zaytzev-Bashan A., Meyer S., Paulke C., Franceschi F., and Yonath A. 1991. Characterization and preliminary attempts for derivatization of crystals of large ribosomal subunits from *Haloarcula marismortui* diffracting to 3 Å resolution. *J. Mol. Biol.* **222:** 11.

Weis W.I., Kahn R., Fourme R., Drickamer K., and Hendrickson W.A. 1991. Structure of the calcium-dependent lectin domain from a rat mannose-binding protein determined by MAD phasing. *Science* **254:** 1608.

Wimberly B.T., Brodersen D.E., Clemons W.M., Jr., Morgan-Warren R., von Rhein C., Hartsch T., and Ramakrishnan V. 2000. Structure of the 30S ribosomal subunit. *Nature* **407:** 327.

Yokoyama S. and Nishimura S. 1995. Modified nucleosides and codon recognition. In *tRNA: Structure, biosynthesis and function* (ed. D. Söll and U. RajBhandary), p. 207. ASM Press, Washington, D.C.

Yonath A., Mussig J., Tesche B., Lorenz S., Erdmann V.A., and Wittmann H.G. 1980. Crystallization of the large ribosomal subunits from *Bacillus stearothermophilus*. *Biochem. Int.* **1:** 428.

Yonath A., Glotz C., Gewitz H.S., Bartels K.S., von Böhlen K., Makowski L., and Wittmann H.G. 1988. Characterization of crystals of small ribosomal subunits. *J. Mol. Biol.* **203:** 831.

Yonath A., Harms J., Hansen H.A., Bashan A., Schlunzen F., Levin I., Koelln I., Tocilj A., Agmon I., Peretz M., Bartels H., Bennett W.S., Krumbholz S., Janell D., Weinstein S., Auerbach T., Avila H., Piolleti M., Morlang S., and Franceschi F. 1998. Crystallographic studies on the ribosome, a large macromolecular assembly exhibiting severe nonisomorphism, extreme beam sensitivity and no internal symmetry. *Acta Crystallogr. A* **54:** 945.

Yoshizawa S., Fourmy D., and Puglisi J.D. 1998. Structural origins of gentamicin antibiotic action. *EMBO J.* **17:** 6437.

———. 1999. Recognition of the codon-anticodon helix by ribosomal RNA. *Science* **285:** 1722.

Yusupov M.M., Trakhanov S.D., Ryazantsev S.N., and Garber M.B. 1988. A new crystalline form of 30 S ribosomal subunits from *Thermus thermophilus*. *FEBS Lett.* **238:** 113.

Yusupov M.M., Yusupova G.Z., Baucom A., Lieberman K., Earnest T.N., Cate J.H., and Noller H.F. 2001. Crystal structure of the ribosome at 5.5 Å resolution. *Science* **292:** 883.

Yusupov M.M., Trakhanov S.D., Barynin V.V., Borovyagin V.L., Garber M.B., Sedelnikova O.M., Tishchenko S.V., and Shirokov V.A. 1987. Crystallization of 30S ribosomal subunits from *T. thermophilus*. *Dokl. Akad. Nauk USSR* **292:** 1271.

Progress Toward an Understanding of the Structure and Enzymatic Mechanism of the Large Ribosomal Subunit

J.L. Hansen,* T.M. Schmeing,* D.J. Klein,* J.A. Ippolito,* N. Ban,*¶ P. Nissen,*§
B. Freeborn,† P.B. Moore,*† and T.A. Steitz*†‡
*Departments of *Molecular Biophysics and Biochemistry and †Chemistry, Yale University, and
‡Howard Hughes Medical Institute, New Haven, Connecticut 06520-8114*

The ribosome was discovered in the 1950s (Tissières 1974), and by 1969, the last time it was the subject of a Cold Spring Harbor Symposium, its role in messenger RNA-directed protein synthesis was well understood. Its return to the Cold Spring Harbor stage in 2001 was precipitated, in part, by a series of field-transforming publications, the first of which appeared in the summer of 2000. In August 2000, an atomic structure of the large ribosomal subunit from *Haloarcula marismortui* derived from a 2.4 Å resolution electron density map was published (Ban et al. 2000; Nissen et al. 2000). A few weeks later, a partially interpreted 3.3 Å resolution electron density map of the small ribosomal subunit from *Thermus thermophilus* appeared (Schluenzen et al. 2000), and a few weeks after that, a fully interpreted atomic structure for the same subunit derived from a 3.05 Å resolution map (Carter et al. 2000; Wimberly et al. 2000). Most recently, in April 2001, an atomic model was published for the 70S ribosome from *T. thermophilus* that is based on a 5.5 Å resolution electron density map interpreted using the high-resolution structures of the two subunits (Yusupov et al. 2001). These papers, collectively, converted the ribosome field from one perpetually vexed by the absence of structural information to one in which the structural information is more than sufficient to formulate functional hypotheses.

It has been about two decades since the first crystals of ribosomes were grown by Yonath and Wittmann (Yonath et al. 1980, 1982) and by the group at Pushchino (Trakhanov et al. 1987) and a few more years since the first atomic structure of a spherical virus the size of a ribosome was determined (Harrison et al. 1978). However, only in the last 3 or 4 years has the rapid progress been apparent that led to the publications on ribosome structure mentioned above. What accounts for this rapid progress after so many years?

ADVANCES CRITICAL TO THE RIBOSOME STRUCTURE DETERMINATION

Crystals of the *H. marismortui* large ribosomal subunit were first grown by Yonath, Wittmann, and colleagues

(Shevack et al. 1985), and their quality improved in stages, until by 1991 (von Böhlen et al. 1991) crystals could be obtained that diffract to 3 Å resolution. However, these crystals had a number of flaws that made them unsuitable for high-resolution structure determination. Some of their defects were described by Yonath et al., (1998). They include "hardly any isomorphism" (crystal to crystal), "very high X-ray beam sensitivity, deformed, elongated, nonuniform spot shape, very high mosaicity, and problematic crystal shape and rigidity." Harms et al. (1999) relate an "... unfavorable crystal habit (plates, made of sliding layers, reaching typically up to 0.5 mm² with an average thickness of a few microns in the direction of the C axis), and ... variations in the C axis length (567–570 Å) as a function of irradiation." We encountered all of these problems, and over the course of time, discovered a new one; these crystals are frequently twinned in the insidious manner described below (Ban et al. 1999).

Over the course of about 4 years, we cured most of these crystal pathologies. The maximum resolution of the observed diffraction pattern was extended to 2.2 Å resolution, allowing data measurement to 2.4 Å resolution, and crystal thickness was increased to between 0.1 and 0.2 mm using a reverse extraction procedure (Ban et al. 2000), which also increased their strength and singularity. Crystal-to-crystal isomorphism was improved and twinning was suppressed by the use of an appropriate stabilizing solution (Ban et al. 2000). The end result was crystals of the large ribosomal subunit suitable for structure determination at high resolution.

Once crystallographically suitable crystals have been obtained, the major barrier to solving a crystal structure is determining the phases that are associated with its thousands of X-ray diffraction amplitudes. Although the ribosome is no larger than the viruses whose structures were solved in the late 1970s and early 1980s, the lack of internal symmetry in the ribosome precludes the use of averaging methods for phase (map) improvement that have been so essential for virus structure determination. The only method that could be used for the ribosome was heavy-atom multiple isomorphous replacement combined with anomalous scattering. To obtain measurable diffraction intensity changes from the binding of heavy atoms to crystals of such a large asymmetric assembly, either a large number (50–100) of single heavy atoms or a smaller number (1–10) of heavy-atom cluster compounds

Present addresses: ¶Institute for Molecular Biology and Biophysics, ETH Hönggerberg, Zurich CH-8093, Switzerland; §Department of Molecular and Structural Biology, Aarhus University, Aarhus DK-8000, Denmark.

need to be bound and their positions located in the crystal. The crucial step in obtaining an interpretable map is determining the positions of the bound heavy atoms. This was apparently not successfully accomplished by Yonath and coworkers in the mid- to late 1990s, since the electron density maps they reported then did not contain recognizable features (Schlünzen et al. 1995; Yonath and Franceschi 1998; Yonath et al. 1998; Harms et al. 1999). Not only did these maps lack the continuous electron density features characteristic of RNA and protein, but solvent-flattened maps of *H. marismortui* large subunit were interpreted as indicating a packing of subunits within the unit cell that differs from the actual packing (Ban et al. 1998, 2000).

Because we thought that large numbers of single heavy atoms would be very difficult, if not impossible, to locate in *H. marismortui* large ribosomal subunit crystals by difference Patterson methods, we first used an 18-tungsten-atom cluster compound, which at very low resolution (20 Å) scatters almost as a super atom having 18 × 74 tungsten electrons. To make certain that the tungsten cluster positions determined by difference Patterson methods were correct, we phased the low-resolution X-ray diffraction data by molecular replacement using a 20 Å resolution cryo-electron microscopic reconstruction of the *H. marismortui* 50S subunit provided by J. Frank and colleagues, and calculated a difference electron density map for the tungsten derivative. It had a 7 σ peak at the position deduced from difference Patterson maps (Ban et al. 1998). (Electron microscopically derived phases were *not* combined with X-ray-derived phases for the calculation of the 9 Å map published in 1998 [Ban et al. 1998], contrary to opinions expressed by others [Yonath et al. 1998].) A similar procedure was followed subsequently in structural work on the 70S ribosome (Cate et al. 1999), and similar heavy-atom cluster compounds and refinement approaches were used in the 30S structural work at low resolution (Wimberly et al. 1999).

Molecular replacement using the 20 Å resolution electron microscope reconstruction was also important because it enabled us to identify an intermittent crystal twinning problem that plagued our efforts to determine the structure of the *H. marismortui* 50S subunit, and perhaps the earlier efforts of others. Twinned crystals have a small shift in subunit positions in the unit cell that causes a space group change from orthorhombic C222₁ to monoclinic P2₁, but leaves cell dimensions nearly unchanged and gives diffraction patterns that appear to be orthorhombic because of twinning (Ban et al. 1999).

Phasing and electron density map calculation for the large ribosomal subunit progressed from very low resolution (16 Å in fall 1997) to progressively higher resolutions (9 Å in spring 1998) (Ban et al. 1998), 7 Å in fall 1998 and 5.5 Å in spring of 1999 (Fig. 1) (Ban et al. 1999). At each stage, recognizable structural features could be seen in all experimentally phased electron density maps. The first electron density map published of any ribosome or ribosomal subunit that showed features clearly interpretable in molecular terms was our 9 Å resolution map (Fig. 1a) that showed the right-handed but ir-

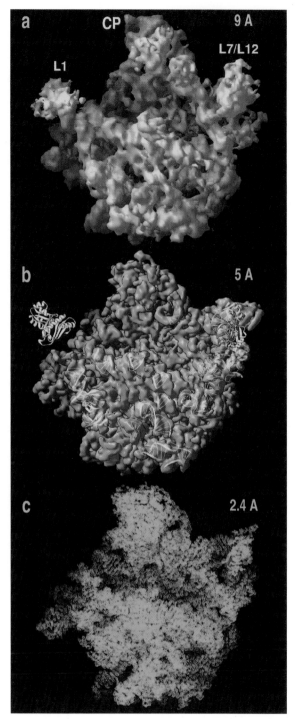

Figure 1. The crown view of the experimental electron density maps of *Haloarcula marismortui* 50S ribosomal subunit at (*a*) 9 Å resolution (Ban et al. 1998), (*b*) 5 Å resolution (Ban et al. 1999), and (*c*) 2.4 Å resolution (Ban et al. 2000). Although density corresponding to the L1 stalk is visible at 9 Å resolution, it disappears at higher resolution.

regular twist of RNA duplex rods (Ban et al. 1998). Because we began at low resolution, we could obtain the data needed using the low intensity, but highly accessible X12B beam line at Brookhaven National Laboratory.

This made it possible to sort out crystal problems and to identify good heavy-atom derivatives very rapidly.

Since the heavy-atom cluster compounds did not have much phasing power beyond about 5 Å resolution, we turned to multiple-site, single heavy-atom derivatives such as osmium pentamine and iridium hexamine. Using the low-resolution-phase information produced by the cluster compounds, the 100 or so binding sites of these compounds could easily be located by difference Fourier maps at 5 Å resolution. In October 1999, during a single 4-day trip to beam line X25 at Brookhaven, we obtained data sets on the now well-characterized native and heavy-atom derivative crystals that extended to a resolution of 3.1 Å. This was our first trip to a high-intensity, insertion device, synchrotron beam line, and it occurred 4 years (!) after the start of the project. These data enabled us to calculate an electron density map phased to 3.1 Å resolution of such quality that all the electron density corresponding to RNA could be interpreted in terms of an atomic model. This work was completed by February 2000. A spring trip to the APS at Argonne National Laboratory allowed data measurement and extension of phasing to 2.4 Å resolution (Ban et al. 2000). Particularly important for the superb quality of this experimentally phased electron density map was the well-collimated, high-intensity, 80 μ diameter X-ray beam available at beam line ID19, and the fine pixel size of the relatively large CCD detector installed there. The solvent flipping density modification program as implemented in the CNS proved remarkably effective at improving both the quality and resolution of the heavy-atom phased maps (Fig. 1c).

STRUCTURE REFINEMENT

The coordinates of the structure have now been refined against the 2.4 Å resolution data. The crystallographic R-factor calculated from the structure published in August 2000 was 25.2%, and the free R-factor was 26.1% (Ban et al. 2000). Currently, the refined structure contains 3622 amino acids, 2847 nucleotides, about 7800 water molecules, 118 Mg^{++} ions, and a handful of Cl^-, K^+, and Cd^{++} ions. The crystallographic R-factor is 18–19%, and the free R-factor is 22.2% to 2.4 Å resolution (Klein et al., in prep.).

The reason that proteins L1, L12, L11, and most of L10, as well as about 5% of the nucleotides in the subunit, are not included in the high-resolution structure is crystallographic, not biochemical. The subunits used for crystallization are highly active in poly(U)-directed phenylalanine incorporation, and subunits recovered from crystals after many weeks of incubation at 19°C remain active (B. Freeborn, unpubl.), which could only be true if these proteins are present. Most of the proteins and nucleotides of RNA not included in the structure are associated with the two lateral protuberances of the subunit, which are so mobile in these crystals that the corresponding electron density is too smeared out to interpret at high resolution. L1, the protein component of the left lateral protuberance (Fig. 1), accounts for 211 of the amino acids

missing at high resolution, but it was indeed visible in the experimental electron density at 9 Å resolution (Fig. 1a). Most of the remaining missing density is accounted for by proteins that bind to the right lateral protuberance: The four copies of L12 (460 amino acids) have never been observed in our maps at any resolution, nor has most of L10 (319 amino acids). The carboxy-terminal half of L11 (161 amino acids) was easily positioned at 5 Å resolution, but its electron density is not distinct enough to interpret in detail at high resolution. Interestingly, the electron density for both protuberances is well enough defined in 70S ribosome crystals to be interpreted, albeit at 5.5 Å resolution (Yusupov et al. 2001). Presumably, their positions become more fixed when the large subunit couples with the small subunit. Furthermore, the position of the L1 protuberance changes significantly in the 70S structure compared to its position in the isolated 50S, as viewed at 9 Å resolution.

RNA ORGANIZATION AND TERTIARY STRUCTURE STABILIZATION

The 23S rRNA can be divided into six domains on the basis of its secondary structure (Noller et al. 1981), and 5S rRNA can be thought of as the subunit's seventh rRNA domain. Being assemblies of rod-like helices, the shapes of all these domains, like those of most other RNA domains, are highly irregular. Nevertheless, they fit together like the pieces of a jigsaw puzzle to form a compact object, the overall shape of which is essentially that of the entire subunit. The interactions between rRNA domains in the large subunit are so extensive and so intimate that it is impossible to tell where one ends and the next begins by visual inspection; the RNA structure of the subunit is monolithic. It appears morphologically to contain a single RNA domain. In this respect, the large subunit differs qualitatively from the small subunit, whose RNA secondary structure domains constitute distinct morphological domains of the intact subunit.

It should not be surprising that an RNA the size of 23S rRNA can form a globular, monolithic structure like this. To first order, RNA structures are assemblies of helical stem-loops, and there is nothing about the physical properties of RNA helices that limits the size of those assemblies. Nor is it obvious that the domains of a large RNA defined by its secondary structure must correspond to morphological subdivisions of the folded molecule. These considerations suggest that division of the small subunit into distinct morphological domains is related to its function, and there is strong evidence to suggest that relative motion of the domains of the small subunit is essential for protein synthesis.

Three kinds of interactions stabilize the tertiary structure of 23S and 5S rRNA: (1) Mg^{++} bridges, (2) RNA–RNA interactions that are largely of two types: (a) long-range base pairs and (b) a newly identified interaction called the A-minor motif, and (3) RNA–protein cross-links. Mg^{++} bridges are found in places where the

23S rRNA fold brings two or more phosphate groups into close proximity with their nonbridging oxygens positioned so that they can function as inner- or outer-shell ligands for a common Mg^{++}. It has been known for years that Mg^{++} bridges can stabilize RNA secondary structure (e.g., tRNA), and thus obvious that under appropriate circumstances, they can stabilize tertiary structure also. About 65 of the 108 Mg^{++} ions identified thus far in the *H. marismortui* large ribosomal subunit stabilize the tertiary structure of 23S rRNA (D. Klein, unpubl.). In addition to single Mg^{++} ions bridging sugar-phosphate backbones that are remote in the sequence, there are clusters of two or three Mg^{++} ions doing likewise (Fig. 2).

There has long been persuasive evidence for the existence of base-pairing between nucleotides associated with different secondary structure elements in rRNAs (Gutell 1996). In addition to stabilizing rRNA tertiary structure, these interactions ensure its specificity. There are about 100 such long-range base pairs in *H. marismortui* 23S rRNA; the exact number depends on how close paired partners may be in the molecule's secondary structure and still be counted as "long range" (Ban et al. 2000).

Adenines are disproportionately abundant in the nonhelical sequences of 23S rRNA, and many of the A's in such sequences are conserved among the three kingdoms (Ware et al. 1983; Gutell et al. 1993). An examination of the role played by these residues in the *H. marismortui* large ribosomal subunit has led to the discovery that many of them are involved in a previously unidentified tertiary interaction that we call the "A-minor motif" (Nissen et al. 2001). The A in an A-minor motif inserts its smooth, minor groove face into the minor groove of a base pair in a helix, often a GC pair, where it forms hydrogen bonds with one or both of its 2' OH groups. Often two or three consecutive A's in a single-stranded region interact with successive base pairs in a helix this way (Fig. 3). A-minor motifs have functional as well as structural significance. For example, the 3'-terminal adenines of tRNAs bound in either the A site or the P site engage in A-minor interactions with 23S rRNA base pairs in the peptidyl transferase region of the large ribosomal subunit. There are 186 A-minor interactions in the *H. marismortui* large ribosomal subunit, and 68 of them involve A's that are conserved across all three kingdoms.

RNA–PROTEIN INTERACTIONS

The K-turn

The 27 proteins whose structures are known in situ as a result of the determination of the crystal structure of the large ribosomal subunit from *H. marismortui* interact with the RNA sequences to which they bind in idiosyncratic ways. Paradoxically, the search for regularities in protein–RNA interactions uncovered instead a new RNA

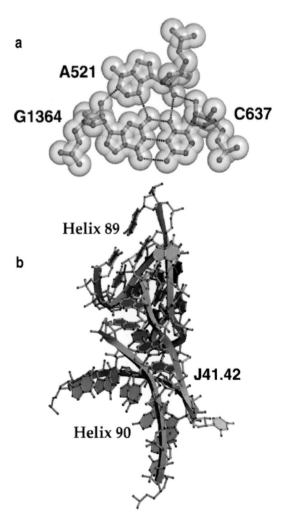

Figure 3. rRNA tertiary structure stabilization by A-minor motifs (Nissen et al. 2001) in which the minor groove face of an A, often in a single-stranded region, is inserted into the minor groove of duplex RNA frequently in patches of 2 to 6 successive and stacked A's. (*a*) The fit of the minor groove edge of an A (*yellow*) into the minor groove of a G-C pair to form a type I A-minor interaction; (*b*) the cross-linking of two rRNA duplexes by a largely single-stranded rRNA containing A's making A-minor interactions. A-minor A's are red.

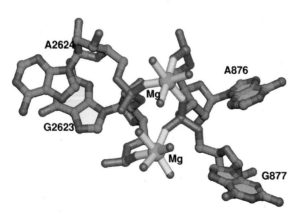

Figure 2. A magnesium ion cluster (*green* atoms and *yellow* bonds) that allows the close approach of rRNA backbone phosphates remote in the rRNA sequence.

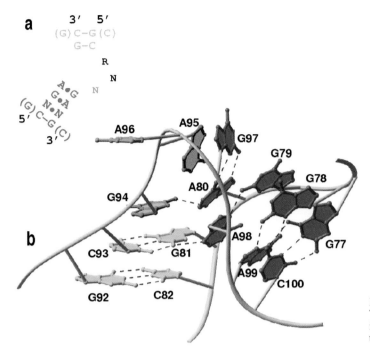

Figure 4. The kink-turn motif (Klein et al. 2001). (*a*) The consensus sequence derived from the 8 K-turns in the ribosome. (*b*) The structure of a typical K-turn.

secondary structure motif that we call the "kink-turn" or "K-turn" (Klein et al. 2001). The consensus sequence for the K-turns in *H. marismortui* 23S rRNA is shown in Figure 4a, and the conformation they assume is displayed in Figure 4b. There are six K-turns in the 23S rRNA of *H. marismortui*, making them about as abundant as bulged-G motifs (Leontis and Westhof 1998; Ban et al. 2000). They came to our attention in part because the globular domains of about a third of the proteins in the large subunit associate with them. Given the distinctive character of the K-turn, the interactions proteins make with them are surprisingly varied, which results in some having more than one protein bound to them. Nonhomologous proteins interact with K-turns in completely different ways, whereas homologous proteins interact in nearly identical ways. The K-turn is also found in 16S rRNA, which has two that associate with proteins, and examples of K-turn/protein complexes have been found in snRNPs and in mRNA/ribosomal protein complexes.

Protein Tails

Twenty-seven of the 31 proteins in the large ribosomal subunit of *H. marismortui* are evident in the 2.4 Å resolution electron density map in more or less complete form, and a small fragment of L10 has also been identified. With the exception of L39e, they all contain at least one globular domain, and many of their folds are similar to folds known in non-ribosomal proteins. The globular regions of all large-subunit proteins are partially exposed to solvent on one side and interact extensively with RNA on the other, thereby serving to both stabilize RNA domain structure and cross-link different RNA domains.

Twelve proteins include at least one sequence of significant length that has an extended nonglobular structure. Viewed in isolation from the rest of the ribosome, the conformations of these "protein tails," which are about 26 mole percent arginine plus lysine, look like random coil, and indeed, it is inconceivable that the conformations they display in the ribosome would be stable in isolation. Their appearances notwithstanding, there is nothing accidental about them; the sequences of ribosomal protein tails are even more conserved than the sequences of their globular domains. The protein tails extend into the interior of the ribosome, filling gaps between RNA helices, and they interact intimately and specifically with RNA groups over their entire lengths. Like the interactions of the globular protein domains, these interactions also play an important role in stabilizing the conformation of the intact particle.

ABSENCE OF RIBOSOMAL PROTEINS FROM THE SITE OF PEPTIDE-BOND FORMATION

The site of peptide-bond synthesis in the large subunit form has been located from the crystal structures of complexes between 50S subunits and substrates, products, or intermediate analogs and was found to consist entirely of RNA (Nissen et al. 2000). The two substrate analogs studied initially were puromycin derivatives: CCdA-p-puromycin, the transition state or intermediate analog devised by Yarus and coworkers (Welch et al. 1995), and an RNA stem-loop having the acceptor stem sequence of tyrosyl tRNA which is extended on its 3′ end with the sequence CC-puromycin (Fig. 5). The CC sequence of the

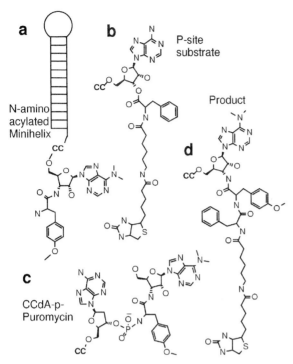

Figure 5. Chemical structures of the substrate, intermediate, and product analogs used in these studies. (*a*) An A-site substrate analog consisting of an RNA hairpin loop with C-C-puromycin at its 3′ terminus (Nissen et al. 2000); (*b*) a P-site substrate analog, N-acetylated C-C- puromycin, which binds to the P site in the presence of sparsomycin (T.M. Schmeing et al., in prep.); (*c*) the transition-state analog CCdA-P-puro (Welch et al. 1995); (*d*) the product of the reactions of the ribosome-catalyzed reaction of compound *b* with puromycin (T.M. Schmeing et al., in prep.).

Yarus inhibitor forms base pairs with G's in the P-loop, as expected (Samaha et al. 1995), and one C of the CC sequence of the tRNA-like analog base-pairs with the A-loop, again as expected (Green et al. 1998). When bound to the ribosome, both analogs are surrounded by 23S rRNA nucleotides that are components of the central loop of domain 5, again consistent with earlier data (Moazed and Noller 1989).

One of the most important results that has emerged from the ribosome structures published so far is the observation that there is no protein in the ribosome's peptidyl transferase site (Nissen et al. 2000). This result is significant because it means that peptide-bond formation is catalyzed entirely by RNA, and that RNA catalysts must have evolved before protein enzymes.

Amino acid sequences are lacking three of the proteins in the large ribosomal subunit of *H. marismortui*, and one of them happens to be one of the four proteins that comes close to the peptidyl transferase site. This protein was identified as L10e by matching the sequence inferred for it from electron density maps with sequences for ribosomal proteins from related organisms (Ban et al. 2000). The experimental map was originally interpreted as indicating that in *H. marismortui* the two-stranded β loop of L10e reaching toward the peptidyl transferase site is 11 amino acids shorter than it is in *all* the archaeal L10e's whose sequences are known. There was no electron density for the additional residues, and the electron density at the end of the loop in question seemed to indicate the existence of a tight turn at that point. It is obvious in the refined map that there is a discontinuity at the point where the electron density ends, not a tight turn. In *H. marismortui*, the sequence of L10e probably includes the 11 residues in question, but they are too mobile to be visualized. The possibility that any of these residues might reach into the peptidyl transferase site at some point during the protein synthesis cycle has been excluded by model building, which shows that even fully extended, the sequence could not reach that far.

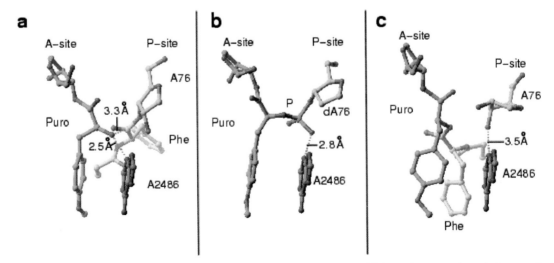

Figure 6. Positions of the (*a*) substrates, (*b*) intermediate, and (*c*) products of the peptidyl transferase reaction relative to each other and to the N3 of A2486 (2451). The A-site substrate is green and the ribose of A76 of the P-site substrate is blue, and its peptide portion is yellow. A2486 is orange. The distances are provided from the N3 of A2486 (2451) to the α-amino of the A-site substrate (2.5 Å), to the phosphate oxygen of the transition-state analog (2.8 Å), and to the 3′ OH of the P-site product (3.3 Å).

A PROPOSAL FOR THE MECHANISM OF THE PEPTIDYL TRANSFERASE REACTION

One way that the 23S rRNA promotes catalysis of peptide-bond formation is by positioning the α-amino group of the amino acid esterified to the CCA-end of tRNA bound in the ribosomal A site so that it can carry out a nucleophilic attack on the carbonyl carbon of the ester bond that links nascent polypeptides to a P-site-bound tRNA that has also been appropriately positioned by interactions with rRNA. It has been argued previously that a juxtaposition of this kind might be all that is required to cause the formation of peptide bonds at the physiologically appropriate rate (Nierhaus et al. 1980). The favorable entropic component arising from the optimal positioning of substrates surely makes as important a contribution to catalysis in the ribosome as it does in other enzymes (Koshland 1971; Page and Jencks 1971). The question to be assessed with this as with any other enzyme is how much additional catalytic efficiency, if any, is provided by acid/base catalysis and/or transition-state stabilization.

There are structural features of the peptidyl transferase active site which suggest that it may also enhance the rate of peptide-bond formation chemically. In the *H. marismortui* large subunit co-crystal structure with CCdA-p-puro, the N3 of A2486 (2451 in *E. coli*) is 2.8 Å from one of the non-bridging oxygens of the phosphate group, which is an analog of the tetrahedral intermediate formed during the attack of the α-amino group of an incoming amino acid on the carbonyl group of a peptidyl tRNA ester bond (Fig. 6a). The distance between the N3 and the phosphate oxygen suggests that they are hydrogen-bonded. Since both would be expected to be unprotonated at pH 5.8, which is the pH of the crystals, this observation implies the N3 of A2486 (2451) is protonated, and it could only be so if its pKa were much higher than normal. If its pKa is unusually high, then A2486(2451) could function as a general acid/base during peptide-bond formation. We have proposed a mechanism for peptide-bond formation in which A2486 (2451) plays a role similar to that played by the active-site histidine of serine proteases in the reverse of the acylation step (Nissen et al. 2000).

More recently, the structures have been determined and refined for the *H. marismortui* large subunit complexed with analogs of both substrates and products bound at the A and P sites (Fig. 5) (Schmeing and Hansen, unpubl.). These structures show that the α-amino group of the A-site-bound substrate is within hydrogen-bonding distance of the N3 of A2486 and only 3.6 Å from the carbonyl carbon of the ester linkage of the P-site substrate that is attacked. Additionally, the 3′ OH of A76 of the P-site product is close to the N3 of A2486. Thus, all of the substrate atoms involved directly in the formation of the peptide bond appear capable of interacting with the N3 of A2486 (2451) at some point in the reaction (Fig. 6). Furthermore, there are no other ribosomal atoms in contact with atoms of the nascent peptide bond in either the substrate, intermediate, or product analog complexes. Consequently, taken together, these structures imply that only A2486

could be involved in chemical catalysis of peptide-bond formation, if indeed there is any chemical catalysis at all.

Finally, since concerns have been expressed that the structure of the 50S ribosomal subunit seen in our crystals corresponds to that of an inactive conformer (Diedrich et al. 2000), it is important to point out that we have established that the 50S ribosomal subunits examined are catalyzing peptide-bond formation in the crystalline state (M. Schmeing, unpubl.). Using substrates developed by S. Strobel and his colleagues that make it possible to carry out the fragment reaction in the absence of alcohol (Muth et al. 2000), the activity of our crystals has been demonstrated both by measuring product formation in crystal suspensions biochemically and by demonstrating crystallographically that when substrates are added to these crystals, product appears bound to the active site. Thus,

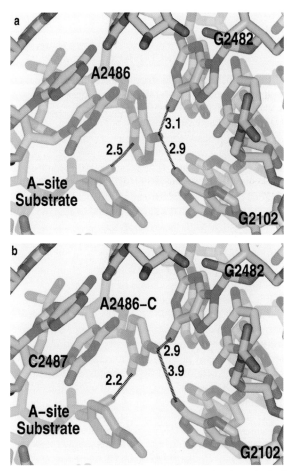

Figure 7. Interactions of A2486 (2451) and a hypothetical A2486C mutant with surrounding A-site substrate and surrounding 23S rRNA nucleotides. Both the wild-type and mutant residue at position 2486 (2451) are shown in yellow. (*a*) The interaction observed between A2486 (2451) and the α-amino group of an A-site substrate (*green*) is highlighted. Also shown are the hydrogen-bonding interactions made by A2486 (2451) with G2482 (2447) and G2101 (2061); (*b*) hypothetical model of A2486C (2451) derived from the observed structure by replacing the wild-type A with a C in exactly the same orientation. Most of the wild-type hydrogen-bonding interactions shown in part *a* of this figure are unaffected by this base change.

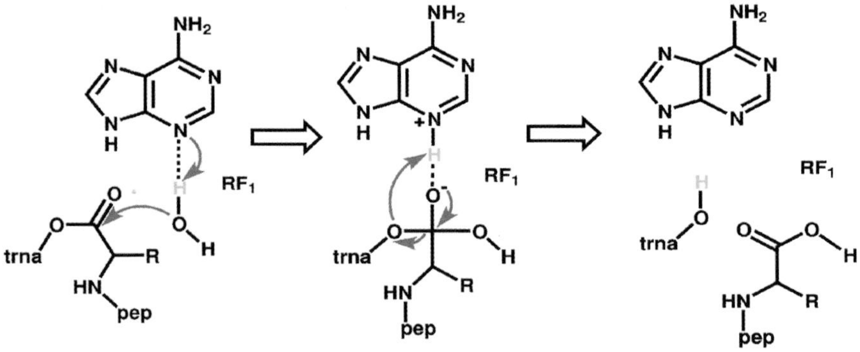

Figure 8. A possible role for A2486 (2451) in catalysis of the deacylation of peptidyl tRNA.

the structures of the 50S ribosomal subunit and its substrate, intermediate, and product complexes discussed above are relevant to an understanding of the mechanism of ribosome-catalyzed peptide-bond synthesis.

DISCUSSION

These biochemical and structural results are consistent with the hypothesis that A2486 (2451) plays a role in peptide-bond formation (as well as in the hydrolytic release of the completed polypeptide chain), but the magnitude of its contribution remains to be determined. Among the issues to be clarified is whether the proton removed from the α-amino group by A2486 (2451) is the third proton of its NH_3^+ form or the second proton of its NH_2 form (Barta et al. 2001). The maximum extent of chemical catalysis expected if the former is the case is tenfold or less, which is not inconsistent with the existing kinetic analyses of reconstituted 50S subunits mutated in A2451 (Polacek et al. 2001; Thompson et al. 2001; Gregory et al.; Kim et al.; both this volume). If the latter is true, the contribution of A2486 (2451) to catalysis could be much larger.

It is interesting and unexpected that replacement of either A2486 (2451) or G2482 (2447) by any other base has only a modest effect on activity, given the complex base-pairing interactions in which they engage and their proximity to the site of peptide-bond formation (Fig. 7a). Assayed using reconstituted 50S subunits, substitution of A2486 (2451) by C, U, or G results in only a threefold to tenfold reduction of activity (Polacek et al. 2001; Thompson et al. 2001). Even if A2486 (2451) contributes nothing to catalysis, one might have anticipated that such base substitutions would result in structural distortions to the peptide synthesis site and thus would have a larger effect. However, when base substitutions are modeled into the structure, insights result that may explain this anomaly. First, G, C, and U all have a hydrogen-bond acceptor (N3 for G and O2 for C or U) in the same place as the N3 of A. Thus, if the bases in A2486 (2451) mutants occupy the same position as the wild-type A, there will be no change in the potential for hydrogen-bonding with the attacking α-NH_2 group. Second, some of the hydrogen bonds between A2486 (2451) and G2482 (2447) and G2102 (2061), which

hold A2486 (2451) in place, can still be made in A2486 (2451) mutants. Substitution of A by C appears the easiest to accommodate (Fig. 7) and substitution by G the hardest. It would be very interesting to obtain crystal structures for these mutated 50S subunits, because they would make it possible to understand why these mutations do not disrupt the peptidyl transferase center. Indeed, until these structures are examined, it may be difficult to interpret the functional effects of these 23S rRNA mutations.

There are four observations pertaining to the mechanism of peptidyl transfer that at present do not appear to be convincingly explained by a single hypothesis. First, all atoms involved in the formation of the nascent peptide bond are in contact with the N3 of A2486 (2451). Second, A2486 (2451) is universally conserved in the cytoplasmic ribosomes of all three kingdoms. Third, mutation of A2486 leads to a dominant lethal phenotype (Muth et al. 2000; Thompson et al. 2001). Fourth, mutation of A2486 (2451) results in only a modest decrease in peptidyl transferase activity using reconstituted 50S subunits (Polacek et al. 2001; Thompson et al. 2001). The first three observations imply an important role for A2486 in peptidyl transfer, whereas the last observation suggests its role is modest. Further biochemical and structure studies will have to be done to resolve the paradox that these observations appear to present.

Several possibilities need further exploration. First, the rate of peptide-bond synthesis in the reconstituted wild-type 50S subunit is greatly reduced from that of in vivo-made ribosomes (Polacek et al. 2001; Thompson et al. 2001). Does that reflect a reduced catalytic contribution from A2486 in the reconstituted 50S ribosome? Second, are the rates being measured those of the chemical step in the sequence of events that leads to peptide-bond formation, which is the only step in the process that should be fully sensitive to the identity of the base at position 2486 (2451)? Finally, the hydrolytic release of the completed protein in the termination step of protein synthesis must be catalyzed, since peptidyl-tRNA is quite stable in isolation. The only obvious candidate for catalyzing the ester hydrolysis required is A2486 (2451) (Fig. 8). Indeed, the dominant lethal phenotype of A2486 (2451) mutants could arise from a failure of those mutant ribosomes to catalyze this hydrolysis.

ACKNOWLEDGMENTS

We are grateful for the help provided by R. Sweet, M. Becker, M. Capel, and L. Berman at the National Synchrotron Light Source (Brookhaven), and by A. Joachimiak, S. Ginell, and R. Sanishvili at the Advanced Photon Source (Argonne). We thank the Strobel laboratory for providing the substrates that enabled us to measure the peptide-bond-forming activity of our crystals. This work was supported by National Institutes of Health grants GM-22778 to T.A.S. and GM-54216 to P.B.M., as well as by a grant from the Agouron Institute.

REFERENCES

Ban N., Nissen P., Hansen J., Moore P.B., and Steitz T.A. 2000. The complete atomic structure of the large ribosomal subunit at 2.4 Å resolution. *Science* **289:** 905.

Ban N., Nissen P., Hansen J., Capel M., Moore P.B., and Steitz T.A. 1999. Placement of protein and RNA structures into a 5 Å-resolution map of the 50S ribosomal subunit. *Nature* **400:** 841.

Ban N., Freeborn B., Nissen P., Penczek P., Grassucci R.A., Sweet R., Frank J., Moore P.B., and Steitz T.A. 1998. A 9 Å resolution X-ray crystallographic map of the large ribosomal subunit. *Cell* **93:** 1105.

Barta A., Dorner S., and Polacek N. 2001. Mechanism of ribosomal peptide bond formation. *Science* **291:** 203.

Carter A.P., Clemons W.M., Brodersen D.E., Morgan-Warren R.J., Wimberly B.T., and Ramakrishnan V. 2000. Functional insights from the structure of the 30S ribosomal subunit and its interactions with antibiotics. *Nature* **407:** 340.

Cate J.H., Yusupov M.M., Yusupova G.Z., Earnest T.N., and Noller H.F. 1999. X-ray crystal structures of 70S ribosome functional complexes. *Science* **285:** 2095.

Diedrich G., Spahn C.M.T., Stelzl U., Schafer M.A., Wooten T., Bochkariov D.E., Cooperman B.S., Traut R.R., and Nierhaus K.H. 2000. Ribosomal protein L2 is involved in the association of the ribosomal subunits, tRNA binding to the A and P sites and peptidyl transfer. *EMBO J.* **19:** 5241.

Green R., Switzer C., and Noller H.F. 1998. Ribosome-catalyzed peptide-bond formation with an A-site substrate covalently linked to 23S ribosomal RNA. *Science* **280:** 286.

Gutell R.R. 1996. Comparative sequence analysis and the structure of 16S and 23S rRNA. In *Ribosomal RNA. Structure, evolution, processing and function in protein biosynthesis* (ed. A. Dahlberg and R. Zimmerman), p. 111. CRC Press, Boca Raton, Florida.

Gutell R.R., Gray M.W., and Schnare M.N. 1993. A compilation of large subunit (23S-like and 28S-like) ribosomal RNA structures. *Nucleic Acids Res.* **21:** 3055.

Harms J., Tocilj A., Levin I., Agmom I., Stark H., Kolln I., van Heel M., Cuff M., Schlünzen F., Bashan A., Franceschi F., and Yonath A. 1999. Elucidating the medium-resolution structure of ribosomal particles: An interplay between electron cryo-microscopy and X-ray crystallography. *Struct. Fold. Des.* **7:** 931.

Harrison S.C., Olson A.J., Schutt C.E., and Winkler F.K. 1978. Tomato bushy stunt virus at 2.9 Å resolution. *Nature* **276:** 368.

Klein D., Schmeing T.M., Moore P.B., and Steitz T.A. 2001. The kink-turn: A new RNA secondary structure motif. *EMBO J.* **20:** 4214.

Koshland D.E. 1971. Molecular basis of enzyme catalysis and control. *Pure Appl. Chem.* **25:** 119.

Leontis N.B. and Westhof E. 1998. A common motif organizes the structure of multihelix loops in 16S and 23S rRNA. *J. Mol. Biol.* **283:** 571.

Muth G.W., Ortoleva-Donnelly L., and Strobel S.A. 2000. A single adenosine with a neutral pKa in the ribosomal peptidyl transferase center. *Science* **289:** 947.

Moazed D. and Noller H.F. 1989. Interaction of tRNA with 23S rRNA in the ribosomal A,P, and E sites. *Cell* **57:** 585.

Nierhaus K.H., Schulze H., and Cooperman B.S. 1980. Molecular mechanisms of the ribosomal peptyl transferase center. *Biochem Int.* **1:** 185.

Nissen P., Ban N., Hansen J., Moore P.B., and Steitz T.A. 2000. The structural basis of ribosome activity in peptide bond synthesis. *Science* **289:** 920.

Nissen P., Ippolito J.A., Ban N., Moore P.B., and Steitz T.A. 2001. RNA tertiary interactions in the large ribosomal subunit: The A-minor motif. *Proc. Natl. Acad. Sci.* **98:** 4899.

Noller H.F., Kop J., Wheaton V., Brosius J., Gutell R.R., Kopylov A.M., Dohme F., Herr W., Stahl D.A., Gupta R., and Waese C.R. 1981. Secondary structure model for 23S ribosomal RNA. *Nucleic Acids Res.* **9:** 6167.

Page M.I. and Jencks W.P. 1971. Aminolysis of acetylimidizaole and rate accelerations caused by intramolecular catalysis. *Fed. Proc.* **30:** 1240.

Polacek N., Gaynor M., Yassin A., and Mankin A.S. 2001. Ribosomal peptidyl transferase can withstand mutations at the putative catalytic nucleotide. *Nature* **411:** 498.

Samaha R.R., Green R., and Noller H.F. 1995. A base pair between tRNA and 23S rRNA in the peptidyl transferase centre of the ribosome. *Nature* **377:** 309.

Schluenzen F., Tocilj A., Zarivach R., Harms J., Gluehmann M., Janell D., Bashan A., Bartles H., Agmon I., Franceschi F., and Yonath A. 2000. Structure of functionally activated small ribosomal subunit at 3.3 Å resolution. *Cell* **102:** 615.

Schlünzen F., Hansen H.A.S., Thygesen J., Bennett W.S., Volkmann N., Levin I., Harms J., Bartels H., Zaytzev-Bashan A., Berkovitch-Yellin Z., Sagi I., Fransceschi F., Krumbholtz S., Geva M., Weinstein S., Agmon I., Boddeker N., Morlang S., Sharon R., Dribin A., Maltz E., Peretz M., Weinrich V., and Yonath A. 1995. A milestone in ribosomal crystallography: The construction of preliminary electron density maps at intermediate resolution. *Biochem. Cell Biol.* **73:** 739.

Shevack A., Gewitz H.S., Hennemann B., Yonath A., and Wittmann H.G. 1985. Characterization and crystallization of ribosomal particles from Halobacterium marismortui. *FEBS Lett.* **184:** 2468.

Thompson J., Kim D.F., O'Connor M., Lieberman K.R., Bayfield M.A., Gregory S.T., Green R., Noller H.F., and Dahlberg A.E. 2001. Analysis of mutations at residues A2451 and G2447 of 23S rRNA in the peptidyltransferase active site of the 50S ribosomal subunit. *Proc. Natl. Acad. Sci.* **98:** 9002.

Tissières A. 1974. Ribosome research: Historical background. In *Ribosomes* (ed. M. Nomura et al.), p. 3. Cold Spring Harbor Laboratory, Cold Spring Harbor, New York.

Trakhanov S.D., Yusupov M.M., Agalarov S.C., Garber M.B., Rayazantsev S.N., Tishenko S.V., and Shirokov V.A. 1987. Crystallization of 70S ribosomes and 30S ribosomal subunits from *Thermus thermophilus*. *FEBS Lett.* **220:** 319.

von Böhlen K., Makowski I., Hansen H.A.S., Bartels H., Berkovitch-Yellin Z., Zaytzev-Bashan A., Meyer S., Paulke C., Franceschi F., and Yonath A. 1991. Characterization and preliminary attempts for derivitization of crystals of large ribosomal subunits from *Haloacrcula marismortui* diffracting to 3 Å resolution. *J. Mol. Biol.* **222:** 11.

Ware V.C., Tague B.W., Clark C.G., Gourse R.L., Brand R.C., and Gerbi S.A. 1983. Sequence analysis of 28S ribosomal DNA from the amphibian *Xenopus laevis*. *Nucleic Acid Res.* **22:** 7795.

Welch M., Chastang J., and Yarus M. 1995. An inhibitor of ribosomal peptidyl transferase using transition-state analogy. *Biochemistry* **34:** 385.

Wimberly B.T., Guymon R., McCutcheon J.P., Jr., White S.W., and Ramakrishnan V.R. 1999. A detailed view of a ribosomal active site: The structure of the L11-RNA complex. *Cell* **97:** 491.

Wimberly B.T., Brodersen D.E., Clemons W.M., Morgan-Warren R.J., Carter A.P., Vonrhein C., Hartsch T., and Ramakrishnan V. 2000. Structure of the 30S ribosomal subunit. *Nature* **407:** 327.

Yonath A. and Franceschi F. 1998. Functional universality and evolutionary diversity: Insights from the structure of the ribosome. *Structure* **6:** 679.

Yonath A., Mussig J., and Wittmann H.G. 1982. Parameters for crystal growth of ribosomal subunits. *J. Cell. Biochemistry* **19:** 145.

Yonath A., Mussig J., Tesche B., Lorenz S., Erdmann V.A., and Wittmann H.G. 1980. Crystallization of the large ribosomal subunits from *Bacillus stearothermophilus. Biochem. Int.* **1:** 428.

Yonath A., Harms J., Hansen H.A., Bashan A., Schlünzen F., Levin I., Koelln I., Tocilj A., Agmon I., Peretz M., Bartels H., Bennett W.S., Krumbholz S., Janell D., Weinstein S., Auerbach T., Avila H., Pioletti M., Morlang S., and Franceschi F. 1998. Crystallographic studies on the ribosome, a large macromolecular assembly exhibiting severe nonisomorphism, extreme sensitivity and no internal symmetry. *Acta Crystallogr. A* **54:** 945.

Yusupov M.M., Yusupova G.Z., Baucom A., Lieberman K., Earnest T.N., Cate J.H.D., and Noller H.F. 2001. Crystal structure of the ribosome at 5.5 Å resolution. *Science* **292:** 883.

High-resolution Structures of Ribosomal Subunits: Initiation, Inhibition, and Conformational Variability

A. Bashan,* I. Agmon,* R. Zarivach,* F. Schluenzen,† J. Harms,† M. Pioletti,†¶
H. Bartels,† M. Gluehmann,† H. Hansen,† T. Auerbach,†¶
F. Franceschi,‡ and A. Yonath*†

*Department of Structural Biology, Weizmann Institute of Science, Rehovot, Israel; † Max-Planck-Research Unit
for Ribosomal Structure, Hamburg, Germany; ‡ Max-Planck-Institute for Molecular Genetics, Berlin, Germany;
¶ Department of Biology, Chemistry, Pharmacology, Free University of Berlin, Germany

The recent impressive progress in ribosomal crystallography has yielded insights into the mechanism of protein biosynthesis. Analysis of the high-resolution structures (Ban et al. 2000; Schluenzen et al. 2000; Wimberly et al. 2000) has led to the identification of dynamic aspects of this process and highlighted strategies adopted by the ribosomes for maintaining their structural integrity and for their survival under extreme conditions (Gluehmann et al. 2001; Harms et al. 2001). Naturally, the crystallographic studies have expanded far beyond the presentation of still pictures and are rapidly progressing toward the elucidation of snapshots describing specific functional stages during the biosynthetic process. Structures of complexes with analogs of transfer RNA and messenger RNA (Weinstein et al. 1999; Auerbach et al. 2000; Brodersen et al. 2000; Yusupov et al. 2001), compounds believed to be substrate analogs (Nissen et al. 2000), translation initiation factors (Carter et al. 2001; Pioletti et al. 2001), and antibiotics (Brodersen et al. 2000; Carter et al. 2000; Pioletti et al. 2001; Schlunzen et al. 2001) are rapidly emerging.

The ribosome is a precisely engineered molecular machine performing an intricate multistep process that requires smooth and rapid switches between different conformations. As such, it contains structural elements that allow global motions together with local rearrangements that create a defined sequence of events at the functional centers. Large-scale movements were detected by cryoelectron microscopy (Frank et al. 1995; Stark et al. 1995; Gabashvili et al. 1999), by surface RNA probing (Alexander et al. 1994), by monitoring the ribosomal activity, by numerous attempts at crystallization (see, e.g., Berkovitch-Yellin et al. 1992), and by the analysis of the high-resolution structures (Schluenzen et al. 2000; Harms et al. 2001; Ogle et al. 2001; Pioletti et al. 2001).

The small ribosomal subunit (30S in prokaryotes) is heavily involved in decoding and translocation—the dynamic aspects of protein biosynthesis—and its significant conformational variability has been correlated with its function. Analysis of the crystal structures of this subunit indicated its mobile structural elements (Gluehmann et al. 2001). Consequently, special efforts were made to identify (Wimberly et al. 2000) or to promote (Tocilj et al. 1999; Carter et al. 2000, 2001; Schluenzen et al. 2000; Pi-

oletti et al. 2001) selected conformations within its crystals. The structures of complexes of this subunit with initiation factors, antibiotics, and mRNA or tRNA analogs showed that the decoding process is accomplished mainly by the 16S ribosomal RNA, and that both the proteins and the RNA features involved in the dynamic functions can assume various conformations.

The large subunit (50S in prokaryotes) is responsible for peptide-bond formation. It is known to show less conformational variability than that found for the small one, but significant mobility can be assigned to some of its features, especially those directly involved in its functions. Both subunits may undergo reversible alterations between active and inactive conformations that may be induced by the environmental conditions. We have previously shown that only functionally active ribosomal particles yield crystals and that the dissolved crystallized material is usually highly active when tested under near-physiological conditions (Berkovitch-Yellin et al. 1992). Nevertheless, within the crystals, the ribosomes may assume a non-active conformation, if maintained under far-from-physiological conditions. Thus, there are reasons to believe that the 2.4 Å structure of the large ribosomal subunit from *Haloarcula marismortui* (H50S), which was determined under far-from-physiological conditions (Ban et al. 2000), reflects less active conformations.

In this paper, we describe our analyses on the structures of the two ribosomal subunits in several conformational states. These studies indicate the strategies that the ribosome adopts for enhancing and directing the binding of factors and substrates. They may also show how the ribosome takes advantage of the built-in flexibility of its components for preventing nonproductive interactions.

THE SMALL RIBOSOMAL SUBUNIT

The small ribosomal subunit (30S) is responsible for the decoding of the genetic information and plays a key role in the initiation phase of protein synthesis. The refined 3.2 Å structure of the functionally activated form of this subunit from *Thermus thermophilus* contains >99% of its 16S RNA chain and most of the amino acids of the subunit's 20 proteins (Schluenzen et al. 2000; Pioletti et al. 2001). The overall fold of the RNA chain, as traced in our map, is in almost perfect agreement with the suggested two-dimensional diagram, based on phylogenetic, protection, and cross-link studies (Gutell 1996). The global architecture of the small subunit that emerged from our crystallographic studies hints at its inherent dynamics, as it is built of loosely attached domains, radiating from one region at the active center of this particle, where the decoding is performed. We identified the elements that form the entrance to the mRNA channel and are able to close it by a latch-like mechanism (Fig. 1) (Schluenzen et al. 2000). This analysis led us to suggest an interconnected network of features that could allow a concerted movement of the subunit during translocation (for review, see Ramakrishnan and Moore 2001).

Analysis of all the available structures of the small subunit (Schluenzen et al. 2000; Wimberly et al. 2000; Pioletti et al. 2001; Yusupov et al. 2001) shows that the 16S RNA is extensively involved in the decoding process. Nevertheless, some proteins are essential for several steps

during the biosynthetic process. Almost all ribosomal proteins are built of globular domains located on the solvent side of the particle. These are connected to extended tails or loops buried within the interior of the particle and seem to stabilize the complex RNA fold. However, the tails of a few proteins are pointing into the solution and are less engaged in RNA contacts. As shown below and in Pioletti et al. (2001), some of these may make crucial contributions to the efficient binding of nonribosomal factors participating in the process of protein biosynthesis.

The Initiation Complex

The small ribosomal subunit is the main player in the initiation of protein biosynthesis. This step has a very important role in governing the accurate setting of the reading frame, as it facilitates the identification of the start codon of mRNA. The mechanisms whereby prokaryotic ribosomes engage mRNA and select the start site are provided by special sequence signals. The initiator mRNA in prokaryotes includes, along with the start codon, an upstream purine-rich sequence (called Shine-Dalgarno [SD]). This pairs with a complementary region in the 16S RNA (called anti-SD), at its 3′-end, thus anchoring the mRNA chains. In the high-resolution structures of the 30S subunit, the anti-SD region is located on the solvent side of the platform, the region that also contains a large part of the E (exit) site.

This intricate process requires the formation of an initiation complex, which in prokaryotes contains the small subunit, mRNA, three initiation factors (IF1, IF2-GTP, and IF3), and initiator tRNA. IF3 plays multiple roles in the formation of this complex. It influences the binding of the other ligands and acts as a fidelity factor by destabilizing noncanonical codon–anticodon interactions. It also selects the start mRNA codon (Sussman et al. 1996) and the correct initiator tRNA to be positioned at the P site (in prokaryotes, the f-met-tRNA). It stabilizes the binding of the fMet-tRNA/IF2 complex to the 30S subunit and discriminates against leaderless mRNA chains (Tedin et al. 1999). IF3 acts as an antiassociation factor because it binds with a high affinity to the 30S subunit and shifts the dissociation equilibrium of the 70S ribosome toward free subunits, thus maintaining a pool of 30S (Grunberg-Manago et al. 1975). It also seems to suppress secondary structure elements in mRNA and to be involved in mRNA shift (La Teana et al. 1995).

IF3 is a small basic protein of about 20 kD. It consists of carboxy- and amino-terminal domains (IF3C and IF3N) connected by a rather long lysine-rich linker region. The structure of the entire protein has not been determined, but nuclear magnetic resonance (NMR) (Garcia et al. 1995a, b) and X-ray structures of the amino- and carboxy-terminal domains have been reported (Biou et al. 1995; Kycia et al. 1995). The interdomain linker appears as a rigid α helix in the crystal structure (of it and IF3N). The NMR studies, however, showed that even under physiological conditions, the linker is partially unfolded and displays flexibility (Kycia et al. 1995; Moreau et al.

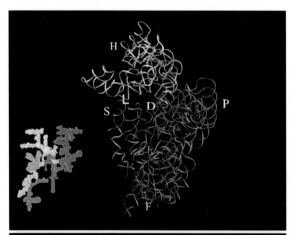

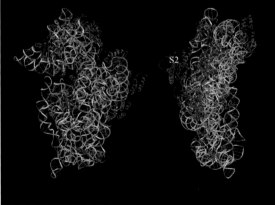

Figure 1. (*Top left*) Secondary structure of the 16S RNA (Gutell 1996). (*Top right*) The small subunit seen from the interface side (the side facing the large subunit in the 70S ribosome). The colors of the domains follow the colors of the secondary structure. The major subdivisions are labeled: (H) head, (S) shoulder, (P) platform, (F) foot. L represents the latch. (*Bottom*) Front (*left*) and side (*right*) views of the 30S structure, showing the proteins in red and the RNA in silver (ribbon backbone and simple lines for base pairs). Protein S2 is highlighted.

1997; Hua and Raleigh 1998a,b; de Cock et al. 1999). Subsequently, the interdomain distances vary between 25 Å and 65 Å, and, as seen below and in Pioletti et al. (2001), the flexibility and ability of the linker to alter its fold are related to the functional roles of IF3.

A double mutation, G1530/A1531 to A1530/G1531, reduces IF3 binding to the 30S subunit tenfold and enhances its affinity to the 70S ribosome, so that it does not promote the dissociation of the assembled ribosome. IF3 was cross-linked to helices H25, H26, and H45 (Ehresmann et al. 1986), produced footprints in the vicinity of the P site; and enhanced the reactivity and/or altered the cross-linking pattern of specific bases on H45 (Shapkina et al. 2000). Mutations in helices H20, H23a, and H24 affect the binding of IF3, and ribosomes lacking S2 have reduced affinity to IF3 (Tapprich et al. 1989; Prescott and Dahlberg 1990; Tedin et al. 1999), but no S2-IF3 cross-links have been reported. The proteins that interact directly with IF3 include S7, S11, S12 S13, S18, and S19 (MacKeen et al. 1980).

It was found that IF3C binds firmly to the ribosome, whereas IF3N and the interdomain linker are loosely attached (Weiel and Hershey 1981; Sette et al. 1999). It has been suggested that the carboxy-terminal domain of IF3 (IF3C) performs many of the tasks assigned to the entire IF3 molecule, preventing the association of the 30S with the 50S subunit and contributing to the dissociation of the entire ribosome (Hershey 1987). IF3C was also shown to influence the formation of the initiation complex. The ability of IF3 to discriminate noncanonical initiation codons, or to verify codon–anticodon complementarity, has been attributed mainly to IF3N (Bruhns and Gualerzi 1980).

IF3 Stretches from the P-site tRNA to the Vicinity of the Anti-Shine-Dalgarno Region

Using crystals of T30S in complex with IF3C, we found that IF3C binds to the 30S particle at the upper end of the platform on the solvent side (Fig. 2), close to the anti-SD region of the 16S rRNA (Pioletti et al. 2001). In the absence of IF3, the region hosting IF3C includes a void of size and shape almost identical to the volume occupied by IF3C (Fig. 2), as determined for IF3C from *Bacillus stearothermophilus* in isolation (Biou et al. 1995). This location reconfirms the results of NMR and mutagenesis of the IF3 molecule (Sette et al. 1999) and is compatible with the effect of the double mutations 1503 1531 (Firpo et al. 1996). It is also consistent with almost all the cross-links, footprints, and protection patterns that were reported for the *Escherichia coli* system (MacKeen et al. 1980; Moazed et al. 1995; Sacerdot et al. 1999), except for protein S12. The bound IF3C is wrapped by residues 7–21 of the central loop of protein S18, interacts with residues 21–27 of the flexible amino terminus of protein S2, with residues 153–156 of the carboxy-terminal end of protein S7, with residues 87–96 of protein S11, and with H23, H26, and H45. Support for this placement, and for the mechanism inferred from it, is also provided

by the analysis of the mode of action and the location of edeine (see Fig. 4), an antibiotic agent that interferes with the initiation process (Pioletti et al. 2001).

We docked the IF3N and the interdomain linker using the program MOLFIT (Eisenstein et al. 1997) by scanning the surface of the whole 30S subunit for possible binding sites. The only location found by this procedure is in close proximity to the P site. In parallel, we fitted IF3N manually, satisfying its proposed function, the constraints posed by the position of IF3C, and the existing biochemical data. The position found manually is almost identical to that identified by the computed search. In this position IF3N interacts with H28, H29, H31, H34, and H44 (via nucleotides 924–927 and 1381–1387 of H28, 1341 of H29, 966–968 of H31, 1062–1064 of H34, and 1398–1400 of H44). Thus, IF3N contacts all helices known to be involved in the peptidyl-tRNA binding and could affect the RNA cross-links C967 x C1400 and C1402 x C1501, as reported by Shapkina et al. (2000).

IF3 Discriminates Initiator tRNA by Space Exclusion and Prevents Subunit Association by Affecting the Conformational Mobility of the Small Subunit

The location of IF3C that we observed suggests that the binding of IF3C to the 30S subunit influences the mobility of H45, close to the anchoring site for the SD sequence. IF3 at this site could affect the conformational mobility of the platform that leads to the association of the two ribosomal subunits, consistent with biochemical observations. The spatial proximity of the IF3C-binding site to the anti-SD region suggests a connection between IF3 and the interactions of the mRNAs with the anti-SD region. These interactions could suppress the change in the conformational dynamics induced by IF3, thus allowing subunit association. The connection between the double mutation of G1530/A1531 to A1530/G1531 and the reduced IF3 binding to the 30S subunit, together with the enhanced affinity of IF3 to the 70S ribosomes, supports this hypothesis.

The binding of IF3C on the solvent side of the upper platform sheds light on the initial step of protein biosynthesis, which involves the detachment of the Shine-Dalgarno sequence. It has been suggested that this region is involved in the displacement of the platform that accompanies the translocation (Gabashvili et al. 1999), as part of the combined head-platform-shoulder conformational changes. The binding of IF3C and the hybridization of the anti-SD sequence are likely to limit the mobility of this region. The detachment of the SD anchor, required at the beginning of the translocation process, allows the platform to regain its conformational mobility.

The bound IF3N leaves a limited, albeit sufficient, space for P-site tRNA. Only small conformational changes are required for simultaneous binding of IF3N, mRNA, and the P-site tRNA. Docking of the P-site tRNA to the 30S-IF3N-bound structure led to close contacts between residues 31–35 of the tRNA and the amino-terminal

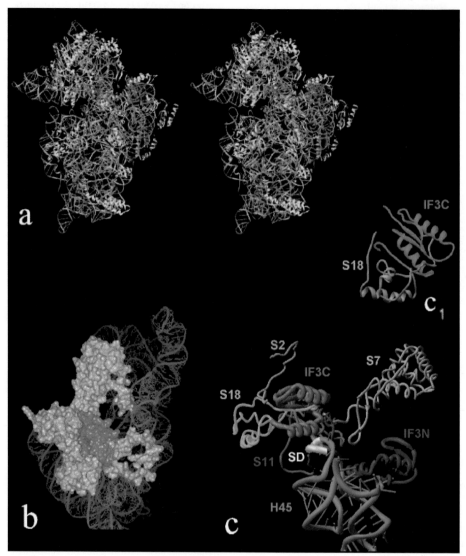

Figure 2. The binding of IF3 to T30S. IF3C is shown is red and IF3N and the interdomain linker in blue. (*a*) Stereo view of the small subunit seen from the interface side (as in Fig. 1, top) with IF3 binding site. (*b*) Space-filling representation of IF3C within its vicinity when bound to T30S. Ribosomal proteins are shown in gray, the RNA chain in blue. (*c*) Close-up of the vicinity of IF3. The RNA features and the proteins are marked by their numbers. (SD) Anti Shine-Dalgarno region. (*c1*) Close-up showing the contacts of S18 and IF3C.

end of IF3N, shedding light on the mechanism of IF3N-mediated discrimination of noncanonical initiation codons. Thus, it seems that the influence of IF3N on initiator tRNA binding is based on space exclusion principles, rather than on specific codon–anticodon complementarity rules, as suggested earlier (Meinnel et al. 1999).

Only indirect contacts exist between IF3N and IF3C, via the curved connection formed by the interdomain linker that wraps around the platform toward the neck. Various mutations, insertions, and deletions that cause significant modifications in the length of the linker do not have major effects on the efficiency of IF3, indicating that the linker maintains its flexibility while IF3 is bound to the 30S subunit. Consequently, it can act as a transmitting

strap between the two domains and can indirectly affect the conformation of the P site and induce its specificity for tRNA-fMet (de Cock et al. 1999). Similarly, the structural changes in IF3 could trigger conformational changes within the 30S subunit that are required for initiating the biosynthetic process and may also lead to a suppression of secondary structure elements in the mRNA. Thus, our structure is consistent with the proposal that the linker maintains its flexibility when IF3 is bound to the 30S subunit and that the flexibility and the ability of the linker region to alter its fold are related to the function of IF3.

The position of IF3 suggests two additional pathways for the transmission of information between the P site and the solvent side of the platform. The flexible protein S2

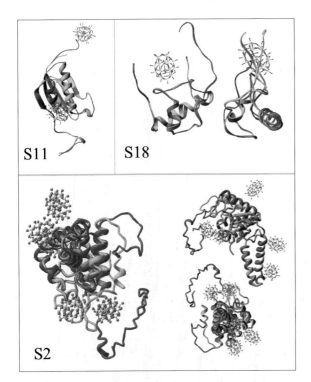

Figure 3. Conformations of proteins S2, S11, and S18 in the presence (*blue*) and the absence (*cyan*) of W18 clusters (the W atoms are shown as red balls). Note the similarity between protein S18-IF3C and S18-W18 contacts (see also Fig. 2).

(Figs. 1, 2, and 3) that interacts with the head and the body of T30S may be an appropriate candidate, consistent with the reciprocal relationship between its existence and the efficiency of IF3 (Tedin et al. 1999). The mRNA itself may also provide a long-range information channel, as successful binding of the cognate tRNA at the P site should be followed by the disruption of the hybridization at the anti-SD region.

In summary, the conformation structure of the small ribosomal subunit that was determined by us is similar to the conformation of this subunit within the initiation complex. The localization of IF3C on the 30S subunit and the docking of IF3N and of the linker region provide a connection between the functions of IF3 and the existing cross-linking, protection, and mutagenesis data. It also explains the correlation between the binding of IF3 to the small ribosomal subunit and the mRNA requirement to interact with the anti-SD region of 16S rRNA for initiating correct translation. The discriminatory function exerted by IF3 against noncanonical codons of the initiator tRNA appears to result from space-exclusion principles, rather than by specific codon–anticodon complementarity rules, since the binding of IF3N leaves limited, albeit sufficient, space for the P-site tRNA. The location of IF3C in our map indicates that the anti-association activity of IF3 is not due to physical blockage of the intersubunit interface, but rather to a change in the conformational dynamics of the subunit.

Agents Interfering with the Initiation Process

Analysis of the structure of the complex of T30S with edeine, a universal initiation inhibitor (Odon et al. 1978; Altamura et al. 1988), supports the mechanism suggested above for the initiation process. We found that edeine binds in the platform between the loop of helix 24 and helix 45 (Fig. 4) (Pioletti et al. 2001). In this position, it would not alter the binding of IF3C, but might well affect the binding of the IF3 intersubunit linker and of IF3N. At the same time, it could influence the mobility of the particle, the interaction of the 3′ end with IF3C and the interactions between the 30S and 50S subunits.

Edeine is a peptide-like antibiotic agent, produced by a strain of *Bacillus brevis*. It contains a spermidine-type moiety at its carboxy-terminal end and a β-tyrosine residue at its amino-terminal end (Kurylo-Borowska 1975). It protects a subset of 16S rRNA nucleotides that are also protected by P-site tRNA (Moazed and Noller 1987; Woodcock et al. 1991). It also shares protections with the antibiotics kasugamycin (bases A794, G926) and pactamycin (bases G693, C795) (Woodcock et al. 1991; Mankin 1997). Mutations in G791 and A792 (H24) reduce association of the 30S and 50S subunits, and an A792 mutant is associated with loss of IF3 binding (Tapprich et al. 1989; Santer et al. 1990). G926 (H28) interacts with the tRNA bound at the P site and is protected by edeine (Woodcock et al. 1991). In addition, mutations in U1498 impair A-site function and enhance tRNA-fMet selectivity (Ringquist et al. 1993), whereas mutations in G1505 increase the levels of stop codon readthrough and frameshifting (O'Connor et al. 1995, 1997).

We found that edeine binds to nucleotides of H24, H28, and H44 at the core of the decoding region, and it is a close neighbor of H45. By physically linking these four helices, critical for tRNA, IF3, and mRNA binding, edeine could lock the small subunit into a fixed configuration (Fig. 4) and hinder the conformational changes that accompany the translation process (Gabashvili et al. 1999; VanLoock et al. 2000).

We also found that the binding of edeine to the 30S subunit induces the formation of a base pair between C795 at the loop of H24 and G693 at the loop of H23. G693 has been shown to be protected when edeine is bound (Woodcock et al. 1991). H23 plays an important role in the binding of the carboxy-terminal domain of IF3, and nucleotides 787–795 of H24 are directly involved in subunit association (Tapprich and Hill 1986). This newly induced G693–C795 base pair alters the mRNA path and would impose constraints on the mobility of the platform, hence interfering with the initiation. Thus, our data suggest that the initiation process is the main target of this universal antibiotic and that edeine induces an allosteric change by the formation of a new base pair—an important new principle of antibiotic action that fits nicely with our suggested mechanism of the initiation step.

Independent studies show that pactamycin, an antibiotic agent that shares a protection pattern with edeine, bridges H23b and H24a, the same helices that are linked by the new base pair that is induced by edeine (Brodersen et al.

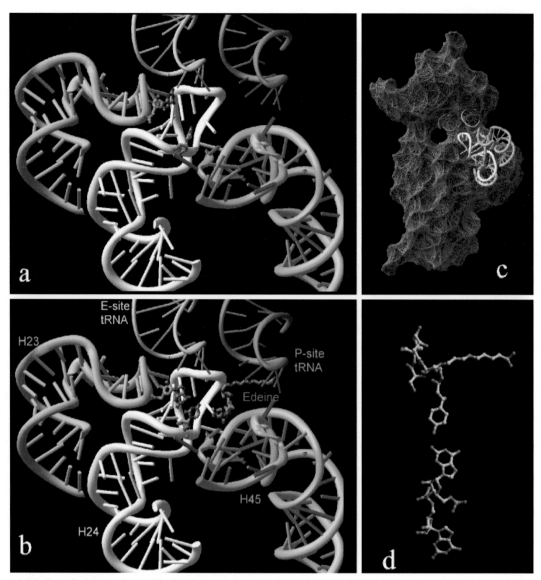

Figure 4. Binding of edeine to the small subunit. Detailed views of the edeine-binding site are shown in *a* and *b*. For both, the assignments of the different ribosomal components are shown in *b*. (*a*) The native conformation. (*b*) The same site once edeine is bound. Note the newly formed base pair. (*c*) The small subunit shown at about 75° rotation (around the vertical axis of the particle) compared to the view of Fig. 1, top. The mRNA channel is clearly seen. (*d*) The pseudo base pair formed by edeine (*top*) upon interaction with the 30S subunit.

2000). This agent is known to interfere with the initiation process, and it is likely that besides reducing the mobility of the platform by locking these two helices, it also alters the mRNA path at the E site. Like edeine, pactamycin interacts with the extended loop of S7—the upper border of the path of the exiting mRNA/tRNA complex—and its mode of interaction suggests that it may interfere with the pairing of the SD sequence or prevent it.

Tetracycline, a multisite antibiotic agent, also binds to protein S7. This is one of the less-occupied sites among the six binding sites of tetracycline that were characterized by us (Pioletti et al. 2001), but it was detected in several biochemical studies. Its influence, if any, at this site is minor, compared to its primary activity—preventing

A-site tRNA binding. Nevertheless, it may induce conformational changes in S7, a protein that plays an important role in initiation and in translocation. S7 is one of the primary rRNA-binding proteins (Held et al. 1974) and is known as one of the proteins that initiate the assembly of the 30S subunit (Nowotny and Nierhaus 1988) in vitro. Therefore, the binding of tetracycline to it could disturb the early assembly steps of new 30S particles, contributing to the overall inhibitory effect of tetracycline.

On the Universality of Initiation

The universal effect of edeine on initiation implies that the main structural elements important for the initiation process are conserved in all kingdoms (Odon et al. 1978). Analysis of our results shows that the rRNA bases defining the edeine-binding site are conserved in chloroplasts, mitochondria, and the three phylogenetic domains. In particular, the effect of edeine on the mRNA path—preventing hybridization of the incoming mRNA—is achieved by edeine's interactions with G926 and G693, two conserved nucleotides. Thus, G926 (H28) has been photo-crosslinked in *E. coli* to position +2 of the mRNA (Sergiev et al. 1997) and to position +1 in human ribosomes (Demeshkina et al. 2000), and G693 has been photo-crosslinked in *E. coli* to positions –1/–3 of the mRNA and in human ribosomes to position –3 of the mRNA (Demeshkina et al. 2000).

Cryo-EM reconstruction of a complex of the small ribosomal subunit from *T. thermophilus* with IF3 from *Thermotoga* localized IF3C at the subunit interface, suggesting that the anti-association activity of IF3 is the product of physical blockage at the interface between the two subunits (McCutcheon et al. 1999). On the other hand, EM studies on rat liver 40S in complex with the eukaryotic initiation factor 3 (eIF3) located eIF3 on the solvent side of the upper edge of the platform (Srivastava et al. 1992), in a region comparable to our findings. In this location, IF3 seems to perform its anti-association activity by effecting the conformational mobility of the small ribosomal subunit—in particular, suppressing the conformational mobility of the platform, essential for association of the two ribosomal subunits. Some aspects of the initiation process of protein biosyntheses were found to be different in eukaryotic and prokaryotic systems (Hershey et al. 1996). Nevertheless, neither of them indicates different locations of IF3. The consistency between our results and the location of the eukaryotic initiation factor may indicate that the main concepts underlying the initiation process and governing the anti-association properties of the initiation complex have been evolutionarily conserved.

Structural Basis for the Tight Binding of IF3C

Most of the extended regions of the ribosomal proteins are buried within RNA features and are believed to stabilize or even assist in shaping the intricate ribosomal structure. In addition to these, there are many flexible tails that are pointing toward the solution, or lie loosely on the ribosomal surface. The amino-terminal ends of S18 and of S2, and the carboxy-terminal ends of S7 and S11, show strikingly different conformations when comparing our structures of T30S (Schluenzen et al. 2000; Pioletti et al. 2001) with that independently determined (Wimberly et al. 2000), presumably related to our crystal stabilization with a heteropolytungstate cluster containing 18 atoms of W (Dawson 1953). We used this cluster to minimize the conformational heterogeneity and limit the mobility of the crystallized T30S subunits (Tocilj et al. 1999; Schluenzen et al. 2000). This procedure was employed for native crystals as well as for complexes of T30S with compounds that facilitate or inhibit protein biosynthesis,

mRNA analogs, initiation factors, and antibiotics. The preparation of the complexes was found to be more efficient if the crystals were soaked in solutions containing the nonribosomal compounds at elevated temperatures, following the routine heat-activation procedure (Zamir et al. 1971), and once the functional complexes were formed, the crystals were cooled down to room temperature and then were treated with the clusters.

Analysis of the structure of the tungstenated crystals (Schluenzen et al. 2000, Pioletti et al. 2001) showed that the extensions that point out into the solution are more ordered than their counterparts in the non-W18-treated crystals (Wimberly et al. 2000). Most of these protein tails bind W18 clusters, creating an extensive network of contacts between their lysines and arginines and the acidic clusters (Fig. 3), suggesting that these tails are able to act as tentacles that enhance the binding of non-ribosomal compounds participating in the process. Once the binding is no longer required, owing to their inherent flexibility, the protein tails can stretch out and release the compounds. Comparing the structure of the tungstenated 30S subunits with that of the complex with carboxy-terminal domain of IF3 shows that the W18 cluster imitates IF3C (Figs. 2 and 3). Indeed, in competition experiments it was found that crystals that were treated with W18 prior to soaking in solutions containing IF3C failed to bind IF3C. This explains why no major conformational changes were observed between the tungstenated and IF3C-bound 30S subunits, contrary to the conformational changes observed while binding IF3 to 30S at its free conformation (Gabashvili et al. 1999; McCutcheon et al. 1999). We therefore conclude that the reference structure of the 30S ribosomal subunit, as determined by using W18-treated crystals, mimics that of the small subunit at the initiation stage.

In summary, the analysis of the structure of the complex of T30S with IF3C not only sheds light on the nature of the binding and action of this factor, but also indicates that the exterior protein tails have important functional tasks that benefit from their significant flexibility. These tasks are quite different from those assigned to the protein extensions in the interior of the particle, which are assumed to be involved mainly in the stabilization of the ribosome structure.

THE LARGE RIBOSOMAL SUBUNIT

Flexibility, Functional Activity, and Apparent Disorder

In ribosomal crystallography, the key to high-resolution data was to crystallize the relatively robust ribosomal particles, assuming that they deteriorate less while being prepared and therefore are expected to yield more homogeneous starting materials for crystallization. The ribosomes from the extreme halophile, *Haloarcula marismortui*, the bacteria that live in the Dead Sea, were found suitable. These bacteria have developed a sophisticated system to accumulate enormous amounts of KCl (3–5 M), although the medium contains only millimolar amounts of it (Table 1) (Ginzburg et al. 1970). Indeed, the func-

Table 1. Concentration of Ions within the Cells
of *H. marismortui*

	Early log	Late log	Stationary
K in cells:	3.7–5.0 M	3.7–4.0 M	3.7–4.0 M
Na in cells:	1.2–3.0 M	1.6–2.1 M	0.5–0.7 M

(Modified from Ginzburg et al. 1970)

tional activity of these ribosomes was found to be directly linked to the concentrations of the potassium ion in the reaction mixture (Fig. 5).

Initially, we grew the crystals of the 50S subunits from this bacterium (H50S) under conditions mimicking the in situ conditions at the bacterial log period. In these experiments, crystals were grown in solutions containing 0.5 M ammonium chloride and 3 M potassium chloride. Under these conditions, nucleation occurred rapidly and yielded small disordered crystals. Consequently, we developed a procedure for crystallization at the lowest potassium concentration required for maintaining the integrity of the subunits (1.2 M KCl), although under these conditions the halophilic ribosomes have only marginal activity (Shevack et al. 1985). Once the crystals grew, we transferred them to solutions containing around 3 M KCl, so that the crystallized particles could rearrange into their active conformation. Post-crystallization rearrangement of ribosomes has also been induced in crystals of the small ribosomal subunits from *T. thermophilus*, either by W18 treatment or by binding initiation factors (Carter et al. 2001; Pioletti et al. 2001).

This procedure yielded crystals showing functional activity and diffracting beyond 2.5 Å resolution (von Bohlen et al. 1991; Yonath et al. 1998). However, the high potassium concentration within these crystals (2.8–3.0 M) caused severe problems in the course of structure determination. The combination of severe non-isomorphism, apparent twinning, high radiation sensitivity,

unstable cell constants, nonuniform mosaic spread, and uneven reflection shape hampered the collection of data usable for structure determination. As these problems became less tolerable at higher resolution, the structure determination under near-physiological conditions stalled at resolutions lower than 5 Å (Yonath et al. 1998; Ban et al. 1999; Harms et al. 1999; Weinstein et al. 1999).

Improved crystals were obtained by drastic reduction of the salt concentration in their stabilization solution, and the exchange of a high concentration of KCl for a relatively low concentration of NaCl (called here L-Na). These far-from-physiological conditions yielded a structure at 2.4 Å resolution (Ban et al. 2000) and even allowed the binding of compounds believed to be substrate analogs, such as CCdA-phosphate-puromycin (Welch et al. 1995; Nissen et al. 2000). Thus, in contrast to the wealth of crystallographic information already obtained for factors and antibiotic-bound complexes of the small subunit, no reports have appeared of high-resolution crystallographic studies of H50S in such complexes. This is consistent with previous observations, showing that extreme halophiles are resistant to most of the antibiotic agents that inhibit bacterial and eukaryotic ribosomes, even when tested under their physiological conditions (Mankin and Garrett 1991). Clearly, the less active conformations are likely to show even lower binding affinities for natural substrates like acylated-tRNA molecules and antibiotics.

The nucleotides that could not be traced in the electron density map obtained from the L-Na crystals, since there was no density that could account for them, were considered to be disordered. These account for less than 10% of the total structure but comprise a significant portion of regions involved in functional activity. They include helix H1, the distal end of helix H38, the tip of H69, the entire regions of H43-H44 and H76-H78. Among these, the L1 stalk (H76-H78 with their bound protein L1) and the L12 stalk (H43-H44 and their bound proteins L10 and L12) create the prominent features of the typical shape of the large ribosomal subunit. These are two lateral protuberances that are readily observed in all EM models and reconstructions, using negative staining, dark-field, or cryo-EM reconstruction. They were also detected in electron density maps obtained from crystals of H50S that were grown and maintained under near-physiological conditions (Ban et al. 1998; Yonath et al. 1998), albeit at lower resolution.

All the untraceable regions of the RNA, together with the proteins that bind to them, contribute to the process of protein biosynthesis. H38 and H69 form intersubunit bridges within the assembled ribosome and interact with the tRNA molecules. L12 and L10 are involved in the contacts with the translocational factors and in factor-dependent GTPase activity (Chandra Sanyal and Liljas 2000). L11 is known to be involved in elongation factor activities (Cundliffe et al. 1979). L1 is a translational repressor binding mRNA (Nikonov et al. 1996), and its absence has a negative effect on the rate of protein synthesis (Subraminian and Dabbs 1980). Since these regions were detected in the maps of the assembled 70S ribo-

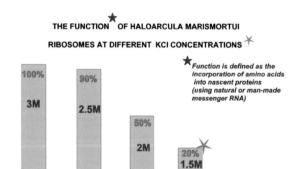

Figure 5. Functional activity of the ribosomes from *H. marismortui* at different potassium concentrations. Activity was checked by the synthesis of polypeptides and by the incorporation of 50S into 70S. In both cases, the ribosomal particles underwent heat activation at 55° for 40 minutes, and homo- or heteronucleotides served as mRNA chains.

some, their absence in the L-Na electron density of H50S stimulated the notion that the functionally important regions are disordered in the free particles, and become fixed in place upon subunit association (Yusupov et al. 2001).

All four proteins (L1, L10, L11, L12) that were not observed in the L-Na map match the list of proteins that we detached selectively from halophilic ribosomes (Franceschi et al. 1994). Evidently, these proteins are less well bound to the core of the large subunit, but it is not clear whether they are indeed disordered, as suggested by Ban et al. (2000), or partially or fully removed from the large subunit. Interestingly, the conditions used for the stabilization of the L-Na crystals are similar to those developed by us for the detachment of these proteins from the ribosome, in solution (Franceschi et al. 1994). It is conceivable that these loosely held proteins were partially or fully detached once the crystals were exposed to the far-from-physiological conditions. The free proteins could then float within the unusually large and continuous solvent regions of these crystal forms (Yonath et al. 1998).

Conformational Variability as a Functional Tool

Recently, we optimized the experimental conditions and minimized the harm caused by the high potassium concentration. Using data collected from crystals of H50S that were grown and kept under conditions mimicking the physiological environment throughout their life span (called here H-K crystals), we constructed a 3.6 Å electron density map. Phases were obtained from anomalous dispersion of several heavy atoms (Harms et al. 1999; Bashan et al. 2000) in combination with crystal averaging and molecular replacement studies based on the 2.4 Å resolution structure (I. Agmon et al., in prep.). Because of the difficulties caused by the high potassium content, the resolution of these studies is somewhat lower than that obtained for the L-Na crystals. Nevertheless, the electron density map is interpretable and enabled a rather detailed comparison between the two structures, aimed at the identification of conformational elements that are required for functional activity in order to reveal how the inherent flexibility of specific features is being exploited by the ribosome to enhance productive, and prevent nonproductive, interactions.

In general, the skeleton of the H-K structure is similar to that determined at low NaCl. However, slight but distinct changes in inter-helix packing, and consequently in the compactness of the structure, were observed. In addition, significant parts of the RNA regions that were not detected in the L-Na map could be traced in the H-K map. These include the tip of H38, which forms the A-site finger and the intersubunit B1a bridge, and H42–H44, the helices forming the GTPase center. We also observed significant discrepancies in the locations of the globular parts of the proteins L10e (L16 in *E. coli*), L37e, L24e, L23, L44e, and L6 in the two maps. Among these, proteins L16 and L6 are of particular interest. L16 is known to be involved in protein biosynthesis in a yet-unknown

way. Protein L6 belongs to the group of five proteins that can be stoichiometrically detached from the particle. It is the only one among this group (L1, L6, L7/12, L10, and L11) that is seen in the L-Na map.

The conformations of almost all the proteins in the H-K structure are different from those observed in L-Na. Some of the differences are subtle, localized within interdomain loops. In order to fit these proteins, minor adjustments were required, and it is likely that such rearrangements may occur after changes in the environment (e.g., salt concentration, type of ion). More interesting are the differences in the locations and the conformations of the termini extensions. Notable are L24, L16 (*E. coli* numbering), and L2. The latter seems to play a key role in facilitating the formation of the peptide bond and in binding both A- and P-tRNA. It was found to increase the hydrolysis rate of peptidyl-tRNA. Mutations in its highly conserved His-229 are fully or conditionally lethal (for review, see Uhlein et al. 1998; Khaitovich and Mankin 1999; Diedrich et al. 2000), and its affinity labeling with an analog of chloramphenicol—an antibiotic known to inhibit protein biosynthesis—caused an irreversible loss of peptidyl-transferase activity (Sonenberg et al. 1973). It was also found to resist extensive digestion with potent proteases in combination with phenol treatment, a procedure that disrupted all other ribosomal proteins (Noller et al. 1992).

Figure 6 shows that the carboxy-terminal extension of L2 loops toward the interior of the particle in the L-Na structure. In the H-K structure, however, there is no density for it, indicating its flexibility. It could therefore extend toward the tRNA molecules located in its vicinity (Yusupov et al. 2001). In this way, it may enhance the binding of the tRNA molecules and effect their accurate positioning as long as they are involved in the peptide-

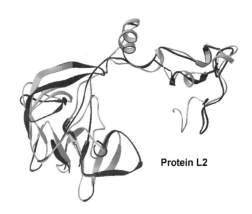

Protein L2

Figure 6. Protein L2 as in the L-Na (in *cyan*) and H-K (in *blue*) structures. Note the differences in the general fold. The carboxyl terminus of the L-Na structure loops toward the main part of the protein, whereas in the H-K structure, there is no density for it for the last 15 residues, indicating its potential flexibility.

bond formation, then, by a small conformational change, relieve the tight binding and allow translocation, acting in a fashion similar to that observed for S18 upon binding of IF3C (Figs. 2 and 4) (Pioletti et al. 2001). This may be the reason for the tight binding of L2 to the large subunit, and for the difficulties encountered in attempts to separate it from the RNA core (Noller et al. 1992).

One conclusion that can be drawn from both structural studies of the large ribosomal subunit *H. marismortui* is that the 2.4 Å L-Na structure represents a conformation that differs from that of the native particle. The absence of almost all the structural features of the 50S subunit that are involved in noncatalytic functional aspects of protein biosynthesis in the 2.4 Å structure of H50S pose an additional shortcoming. These features were not seen in the electron density map and, therefore, were assumed to be disordered. Biochemical, functional, and EM studies indicated that these features are inherently flexible. However, flexibility is not necessarily synonymous with disorder. Thus, flexible parts of molecules or assemblies may assume several well-defined conformations, and the switch from one conformation to another is related to their functional state. In such cases, well-diffracting crystals may be obtained from one of the specific conformations. Therefore, large features that are disordered in the 2.4 Å L-Na H50S map may indicate a special ribosomal strategy to avoid inefficient binding of nonribosomal components. As mentioned above, a similar discriminative binding fashion was detected for the binding of IF3 by the small subunit: The amino-terminal tail of protein S18 is disordered unless it binds the carboxy-terminal domain of IF3 or its mimics.

H. marismortui is an archaea bearing low compatibility with *E. coli*, the species yielding most of our knowledge on ribosomes. Despite the suitability of the ribosomes from *H. marismortui* for high-resolution crystallography, they have not become a subject of many biochemical studies. Consequently, only a small part of the vast amount of data accumulated over almost half a century of ribosomal research can be related directly to its structure. In addition, the antibiotics from the macrolide family hardly bind to the halophilic ribosomes, since a key nucleotide, A2508, is a guanine in their 23S RNA. They are also rather resistant to other antibiotic agents (Mankin and Garret 1991), even under suitable conditions. Thus, it is not surprising that, contrary to the wealth of crystallographic information already obtained about binding of factors and antibiotics to the small subunit (Brodersen et al. 2000; Carter et al. 2000, 2001; Ogle et al. 2001; Pioletti et al. 2001), so far only complexes of H50S with materials believed to represent substrate analogs were found to be suitable for high-resolution crystallographic studies (Nissen et al. 2000). Furthermore, despite extensive studies exploiting these complexes (Nissen et al. 2000), the structural basis for peptidyl-transferase activity is still not well understood (Barta et al. 2001). Thus, in contrast to the strict requirements for antibiotic binding, all nucleotides believed to be crucial for the catalytic activity according to the suggested mechanism (Nissen et al. 2000) could be mutated with little or no effect on peptide-bond formation in vitro (Polacek et al. 2001) and in vivo (Thompson et al. 2001).

For these reasons, we initiated crystallographic studies on the large ribosomal subunit from a eubacterial source that shows a high homology with *T. thermophilus* and *E. coli*. Crystals of the 50S particles from *Deinococcus radiodurans* (D50S) were grown and maintained under conditions that are almost identical to the in situ environment. These crystals, as well as those grown from complexes of these subunits with antibiotics, diffract to higher than 2.9 Å resolution, are relatively stable in the X-ray beam, and yield crystallographic data of very high quality. Thus, they provide an excellent system to investigate antibiotic binding and functional flexibility (Harms et al. 2001; Schlunzen et al. 2001).

The structure of the large ribosomal subunit from this source is significantly more ordered than that of H50S, and many of the functional relevant features that are disordered in H50S are well resolved in D50S. These include the intersubunit bridges, such as those formed by L69 (the bridge to the decoding center), the upper half of H38 (called the A-site finger or the intersubunit B1a bridge) (Fig. 7), the L1 arm (helices 76–78), the GTPase center (helices H42–H44 and protein L11). All are flexible, and most assume orientations that differ, to varying extents, from those seen in the 70S ribosomes, as determined at 5.5 Å (Yusupov et al. 2001).

Of interest is protein L27, which was shown to be important for the biosynthetic process (Sonenberg et al. 1973: Wower et al. 1993). As seen in Figure 8, its location in D50S is consistent with its footprinting in *E. coli* (A.S. Mankin, pers. comm.). In this location it can reach the area between the P and A sites, whereas in *H. marismortui,* its homolog (called H21e) folds toward the interior, consistent with the hypothesis that the tails of the ribosomal proteins that bind factors and substrates fold backward when conditions are not suitable for productive protein biosynthesis.

Many unexpected developments occurred during our long-lasting involvement in ribosomal crystallography. One of them is reported here. Thus, owing to crystallographic difficulties, not only the structures of two conformations of the large subunit from *H. marismortui* are now available for comparative studies, but also a detailed structure of the large ribosomal subunit from a mesophilic source has been determined. Consequently, the gate is opened to detailed comparisons between free and 70S-bound large subunit, to thorough implementation of the biochemical knowledge obtained for *E. coli* ribosomes, and perhaps also to the design of potent antibiotics.

ACKNOWLEDGMENTS

Thanks are given J.M. Lehn for indispensable advice; M. Pope for the tungsten clusters; R. Wimmer for recommending the *D. radiodurans* ribosome; M. Wilchek, W. Traub, and A. Mankin for critical discussions; and to all members of the ribosome-crystallography groups at the

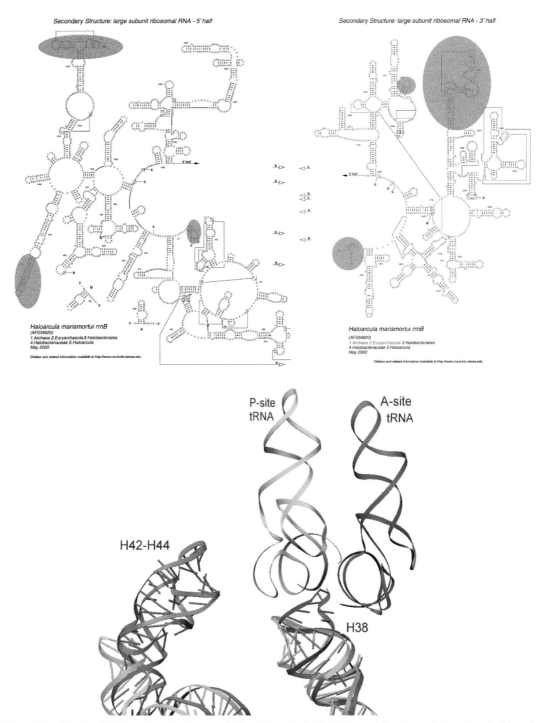

Figure 7. (*Top*) The disordered RNA regions in structure obtained from the L-Na H50S crystals (Ban et al. 2000), marked in red on the 5′ half and in cyan on the 3′ half of the secondary structure diagram of the 23S RNA of *H. marismortui*. (*Bottom*) The conformations and locations of two of these RNA regions in the D50S structure are shown in red. For comparison, the portions of the RNA helices that could be traced in the L-Na structure are shown in green. Note the proximity of H38 to the docked A- and P-tRNA molecules.

Max-Planck Research Unit in Hamburg, the Max-Planck-Institute for Molecular Genetics in Berlin, and the Weizmann Institute in Rehovot, who contributed to these studies. These studies could not have been performed without the cooperation and assistance of the staff of the synchrotron radiation facilities at EMBL and MPG at DESY; ID14/2&4 at EMBL/ESRF, and ID19/APS/ANL. The Max-Planck Society, the U.S. National Institutes of Health (GM-34360), the German Ministry for Science and Technology (Bundesministerium für Bildung, Wissenschaft, Forschung und Technologie grant 05-641EA), and the Kimmelman Center for Macromolecular Assembly at the Weizmann Institute provided support. A.Y. holds the Martin S. Kimmel Professorial Chair.

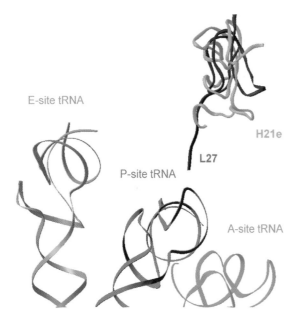

E-site tRNA

H21e

L27

P-site tRNA

A-site tRNA

Figure 8. Comparison between D50S protein L27 (in *blue*) and HmL21e (in *light green*), the protein that was found to occupy the same location in the L-Na structure. The amino-terminal end of HmL21e folds away from the tRNA-binding sites, whereas that of EcCL27 almost reaches them.

REFERENCES

Alexander R.W., Muralikrishna P., and Cooperman B.S. 1994. Ribosomal components neighboring the conserved 518-533-loop of 16S ribosomal-RNA in 30S subunits. *Biochemistry* **33:** 12109.

Altamura S., Sanz J.L., Amils R., Cammarano P., and Londei P. 1988. The antibiotic sensitivity spectra of ribosomes from the thermoproteales phylogenetic depth and distribution of antibiotic binding sites. *Syst. Appl. Microbiol.* **10:** 218.

Auerbach T., Pioletti M., Avila H., Anagnostopoulos K., Weinstein S., Franceschi F., and Yonath A. 2000. Genetic and biochemical manipulations of the small ribosomal subunit from *Thermus thermophilus* HB8. *J. Biomol. Struct. Dyn.* **17:** 617.

Ban N., Nissen P., Hansen J., Moore, P.B., and Steitz T.A. 2000. The complete atomic structure of the large ribosomal subunit at 2.4 Å resolution. *Science* **289:** 905.

Ban N., Nissen P., Hansen J., Capel M., Moore P., and Steitz T.A. 1999. Placement of protein and RNA structures into a 5 Å resolution map of the 50S ribosomal subunit. *Nature* **400:** 841.

Ban N., Freeborn B., Nissen P., Penczek P., Grassucci R.A., Sweet R., Frank J., Moore P.B., and Steitz T.A. 1998. A 9 Å resolution X-ray crystallographic map of the large ribosomal subunit. *Cell* **93:** 1105.

Barta A., Dorner S., and Polacek N. 2001. Mechanism of ribosomal peptide bond formation. *Science* **291:** 203.

Bashan A., Pioletti M., Bartels H., Janell D., Schlunzen F., Gluehmann M., Levin I., Harms J., Hansen H.A.S., Tocilj A., Auerbach T., Avila H., Simitsopoulou M., Peretz M., Bennett W.S., Agmon I., Kessler M., Weinstein S., Franceschi F., and Yonath A. 2000. Identification of selected ribosomal components in crystallographic maps of prokaryotic ribosomal subunits at medium resolution. In *The ribosome: Structure, function, antibiotics and cellular interactions* (ed. R. Garrett et al.), p. 21. ASM Press, Washington, D.C.

Berkovitch-Yellin Z., Bennett W.S., and Yonath A. 1992. Aspects in structural studies on ribosomes. *Crit. Rev. Biochem. Mol. Biol.* **27:** 403.

Biou V., Shu F., and Ramakrishnan V. 1995. X-ray crystallography shows that translational initiation factor IF3 consists of

two compact alpha/beta domains linked by an alpha-helix. *EMBO J.* **14:** 4056.

Brodersen D.E., Clemons W.M., Jr., Carter A.P. Morgan-Warren R.J., Wimberly B.T., and Ramakrishnan V. 2000. The structural basis for the action of the antibiotics tetracycline, pactamycin, and hygromycin B on the 30S ribosomal subunit. *Cell* **103:** 1143.

Bruhns J. and Gualerzi C. 1980. Structure–function relationship in *Escherichia coli* initiation factors: Role of tyrosine residues in ribosomal binding and functional activity of IF-3. *Biochemistry* **19:** 1670.

Carter A.P., Clemons W.M., Brodersen D.E., Morgan-Warren R.J., Wimberly B.T., and Ramakrishnan V. 2000. Functional insights from the structure of the 30S ribosomal subunit and its interactions with antibiotics. *Nature* **407:** 340.

Carter A.P., Clemons W.M., Jr., Brodersen D.E., Morgan-Warren R.J., Hartsch T., Wimberly B.T., and Ramakrishnan V. 2001. Crystal structure of an initiation factor bound to the 30S ribosomal subunit. *Science* **291:** 498.

Chandra Sanyal S. and Liljas A. 2000. The end of the beginning: Structural studies of ribosomal proteins. *Curr. Opin. Struct. Biol.* **10:** 633.

Cundliffe E., Dixon P., Stark M., Stoffler G., Ehrlich R., Stoffler-Meilicke M., and Cannon M. 1979. Ribosomes in thiostrepton-resistant mutants of *Bacillus megaterium* lacking a single 50 S subunit protein. *J. Mol. Biol.* **132:** 235.

Dawson B. 1953. The structure of the 9(18)-heteropoly anion in potassium 9(18)-tungstophosphate, $K_6[(P_2W_{18})O_{62}]\ 14H_2O$. *Acta Crystallogr.* **6:** 113.

de Cock E., Springer M., and Dardel F. 1999. The interdomain linker of *Escherichia coli* initiation factor IF3: A possible trigger of translation initiation specificity. *Mol. Microbiol.* **32:** 193.

Demeshkina N., Repkova M., Ven'yaminova A., Graifer D., and Karpova G. 2000. Nucleotides of 18S rRNA surrounding mRNA codons at the human ribosomal A, P, and E sites: A crosslinking study with mRNA analogs carrying an aryl azide group at either the uracil or the guanine residue. *RNA* **6:** 1727.

Diedrich G., Spahn C.M., Stelzl U., Schafer M.A., Wooten T., Bochkariov D.E., Cooperman B.S., Traut R.R., and Nierhaus K.H. 2000. Ribosomal protein L2 is involved in the association of the ribosomal subunits, tRNA binding to A and P sites and peptidyl transfer. *EMBO J.* **19:** 5241.

Ehresmann C., Moine H., Mougel M., Dondon J., Grunberg-Manago M., Ebel J.P., and Ehresmann B. 1986. Cross-linking of initiation-factor IF3 to *Escherichia coli* 30S ribosomal-subunit by transdiamminedichloroplatinum(II): Characterization of 2 cross-linking sites in 16S ribosomal-RNA; a possible way of functioning for IF3. *Nucleic Acids Res.* **14:** 4803.

Eisenstein M., Shariv I., Koren G., Friesem A.A., and Katchalski-Katzir E. 1997. Modeling supra-molecular helices: Extension of the molecular surface recognition algorithm and application to the protein coat of the tobacco mosaic virus. *J. Mol. Biol.* **266:** 135.

Firpo M.A., Connelly M.B., Goss D.J., and Dahlberg A.E. 1996. Mutations at two invariant nucleotides in the 3´-minor domain of *Escherichia coli* 16S rRNA affecting translational initiation and initiation factor 3 function. *J. Biol. Chem.* **271:** 4693.

Franceschi F., Sagi I., Boeddeker N., Evers U., Arndt E., Paulke C., Hasenbank R., Laschever M., Glotz C., Piefke J., Muessig J., Weinstein S., and Yonath A. 1994. Crystallography, biochemical and genetics studies on halophilic ribosomes. *Syst. Appl. Microbiol.* **16:** 697.

Frank J., Zhu J., Penczek P., Li Y.H., Srivastava S., Verschoor A., Radermacher M., Grassucci R., Lata R.K., and Agrawal R.K. 1995. A model of protein synthesis based on cryo-electron microscopy of the *E. coli* ribosome. *Nature* **376:** 441.

Gabashvili I.S., Agrawal R.K., Grassucci R., and Frank J. 1999. Structure and structural variations of the *Escherichia coli* 30 S ribosomal subunit as revealed by three-dimensional cryo-electron microscopy. *J. Mol. Biol.* **286:** 1285.

Garcia C., Fortier P., Blanquet S., Lallemand J.Y., and Dardel F. 1995a. ^{1}H and ^{15}N resonance assignment and structure of the N-terminal domain of *Escherichia coli* initiation factor 3. *Eur.*

J. Biochem. **228**: 395.

———. 1995b. Solution structure of the ribosome-binding domain of *E. coli* translation initiation factor IF3. Homology with U1A protein of the eukaryotic spliceosome. *J. Mol. Biol.* **254**: 247.

Ginzburg M., Sacks L., and Ginzburg B.Z. 1970. Ion metabolism in *Halobacterium. J. Gen. Physiol.* **55**: 178.

Gluehmann M., Harms J., Zarivach R., Bashan A., Schluenzen F., Bartels H., Agmon I., Rosenblum G., Pioletti M., Auerbach T., Avila H., Hansen H.A.S., Franceschi F., and Yonath A. 2001. Ribosomal crystallography: From poorly diffracting micro-crystals to high resolution structures. (in press).

Grunberg-Manago M., Dessen P., Pantaloni D., Godefroy-Colburn T., Wolfe A.D., and Dondon J. 1975. Light-scattering studies showing the effect of initiation factors on the reversible dissociation of *Escherichia coli* ribosomes. *J. Mol. Biol.* **94**: 461.

Gutell R. 1996. Comparative sequence analysis and the structure of 16S and 23S rRNA. In *Ribosomal RNA: Structure, evolution, processing and function in protein biosynthesis* (ed. R.A. Dahlberg), p. 111. CRC Press, Boca Raton, Florida.

Harms J., Schlunzen F., Zarivach R., Bashan A., Gat S., Agmon I., Bartels H., Franceschi F., and Yonath A. 2001. High resolution structure of the large ribosomal subunit from a mesophilic eubacterium. *Cell* (in press).

Harms J., Tocilj A., Levin I., Agmon I., Stark H., Koelln I., van Heel M., Cuff M., Schluenzen F., Bashan A., Franceschi F., and Yonath A. 1999. Elucidating the medium-resolution structure of ribosomal particles: An interplay between electron cryo-microscopy and X-ray crystallograhy. *Struct. Fold. Des.* **7**: 931.

Held W.A., Ballou B., Mizushima S., and Nomura M. 1974. Assembly mapping of 30S ribosomal-proteins from *Escherichia coli*. Further studies. *J. Biol. Chem.* **249**: 3103.

Hershey J.W. 1987. Protein synthesis. In Escherichia coli *and* Salmonella typhimurium: *Cellular and molecular biology* (ed. F. Neidhardt et al.), p. 613. ASM Press, Washington, D.C.

Hershey J.W., Asano K., Naranda T., Vornlocher H.P., Hanachi P., and Merrick W.C. 1996. Conservation and diversity in the structure of translation initiation factor EIF3 from humans and yeast. *Biochimie* **78**: 903.

Hua Y.X. and Raleigh D.P. 1998a. Conformational analysis of the interdomain linker of the central homology region of chloroplast initiation factor IF3 supports a structural model of two compact domains connected by a flexible tether. *FEBS Lett.* **433**: 153.

———. 1998b. On the global architecture of initiation factor IF3: A comparative study of the linker regions from the *Escherichia coli* protein and the *Bacillus stearothermophilus* protein. *J. Mol. Biol.* **278**: 871.

Khaitovich P. and Mankin A.S. 1999. Effect of antibiotics on large ribosomal subunit assembly reveals possible function of 5 S rRNA. *J. Mol. Biol.* **291**: 1025.

Kurylo-Borowska Z. 1975. Biosynthesis of edeine. II. Localization of edeine synthetase within *Bacillus brevis* Vm4. *Biochim. Biophys. Acta* **399**: 31.

Kycia J.H., Biou V., Shu F., Gerchman S.E., Graziano V., and Ramakrishnan V. 1995. Prokaryotic translation initiation factor IF3 is an elongated protein consisting of two crystallizable domains. *Biochemistry* **34**: 6183.

La Teana A., Gualerzi C.O., and Brimacombe R. 1995. From stand-by to decoding site. Adjustment of the mRNA on the 30S ribosomal subunit under the influence of the initiation factors. *RNA* **1**: 772.

MacKeen L.A., Kahan L., Wahba A.J., and Schwartz I. 1980. Photochemical cross-linking of initiation factor-III to *Escherichia coli* 30S ribosomal-subunits. *J. Biol. Chem.* **255**: 526.

Mankin A.S. 1997. Pactamycin resistance mutations in functional sites of 16S rRNA. *J. Mol. Biol.* **274**: 8.

Mankin A.S. and Garrett R.A. 1991. Chloramphenicol resistance mutations in the single 23S rRNA gene of archaeon Halobacterium halobium. *J. Bacteriol.* **173**: 3559.

McCutcheon J.P., Agrawal R.K., Philips S.M., Grassucci R.A.,

Gerchman S.E., Clemons W.M., Ramakrishnan V., and Frank J. 1999. Location of translational initiation factor IF3 on the small ribosomal subunit. *Proc. Natl. Acad. Sci.* **96**: 4301.

Meinnel T., Sacerdot C., Graffe M., Blanquet S., and Springer M. 1999. Discrimination by *Escherichia coli* initiation factor IF3 against initiation on non-canonical codons relies on complementarity rules. *J. Mol. Biol.* **290**: 825.

Moazed D. and Noller H.F. 1987. Interaction of antibiotics with functional sites in 16S ribosomal RNA. *Nature* **327**: 389.

Moazed D., Samaha R.R., Gualerzi C., and Noller H.F. 1995. Specific protection of 16 S rRNA by translational initiation factors. *J. Mol. Biol.* **248**: 207.

Moreau M., de Cock E., Fortier P.-L., Garcia C., Albaret C., Blanquet S., Lallemand J.-Y., and Dardel F. 1997. Heteronuclear NMR studies of *E. coli* translation initiation factor IF3. Evidence that the inter-domain region is disordered in solution. *J. Mol. Biol.* **266**: 15.

Nikonov S., Nevskaya N., Eliseikina I., Fomenkova N., Nikulin A., Ossina N., Garber M., Jonsson B.H., Briand C., Al-Karadaghi S., Svensson A., Aevarsson A., and Liljas A. 1996. Crystal structure of the RNA binding ribosomal protein L1 from *Thermus thermophilus*. *EMBO J.* **15**: 1350.

Nissen P., Hansen J., Ban N., Moore P.B., and Steitz T.A. 2000. The structural basis of ribosome activity in peptide bond synthesis. *Science* **289**: 920.

Noller H.F., Hoffarth V., and Zimniak L. 1992. Unusual resistance of peptidyl transferase to protein extraction procedures. *Science* **256**: 1416.

Nowotny V. and Nierhaus K.H. 1988. Assembly of the 30s subunit from *Escherichia coli* ribosomes occurs via 2 assembly domains which are initiated by S4 and S7. *Biochemistry* **27**: 7051.

O'Connor M., Thomas C.L., Zimmermann R.A., and Dahlberg A.E. 1997. Decoding fidelity at the ribosomal A and P sites: Influence of mutations in three different regions of the decoding domain in 16S rRNA. *Nucleic Acids Res.* **25**: 1185.

O'Connor M., Brunelli C.A., Firpo M.A., Gregory S.T., Lieberman K.R., Lodmell J.S., Moine H., VanRyk D.I., and Dahlberg A.E. 1995. Genetic probes of ribosomal RNA function. *Biochem. Cell. Biol.* **73**: 859.

Odon O.W., Kramer G., Henderson A.B., Pinphanichakarn P., and Hardesty B. 1978. GTP hydrolysis during methionyl-tRNAf binding to 40 S ribosomal subunits and the site of edeine inhibition. *J. Biol. Chem.* **253**: 1807.

Ogle J.M., Brodersen D.E., Clemons W.M., Jr., Tarry M.J., Carter A.P., and Ramakrishnan V. 2001. Recognition of cognate transfer RNA by the 30S ribosomal subunit. *Science* **292**: 897.

Pioletti M., Schlunzen F., Harms J., Zarivach R., Gluhmann M., Avila H., Bashan A., Bartels H., Auerbach T., Jacobi C., Hartsch T., Yonath A., and Franceschi F. 2001. Crystal structures of complexes of the small ribosomal subunit with tetracycline, edeine and IF3. *EMBO J.* **20**: 1829.

Polacek N., Gaynor M., Yassin A., and Mankin A.S. 2001. Ribosomal peptidyl transferase can withstand mutations at the putative catalytic nucleotide. *Nature* **411**: 498.

Prescott C.D. and Dahlberg A.E. 1990. A single base change at 726 in 16S rRNA radically alters the pattern of proteins synthesized in vivo. *EMBO J.* **9**: 289.

Ramakrishnan V. and Moore P.B. 2001. Atomic structures at last: The ribosome in 2000. *Curr. Opin. Struct. Biol.* **11**: 144.

Ringquist S., Cunningham P., Weitzmann C., Formenoy L., Pleij C., Ofengand J., and Gold L. 1993. Translation initiation complex-formation with 30S ribosomal particles mutated at conserved positions in the 3'-minor domain of 16S RNA. *J. Mol. Biol.* **234**: 14.

Sacerdot C., de Cock E., Engst K., Graffe M., Dardel F., and Springer M. 1999. Mutations that alter initiation codon discrimination by *Escherichia coli* initiation factor IF3. *J. Mol. Biol.* **288**: 803.

Santer M., Bennett-Guerrero E., Byahatti S., Czarnecki S., O'-Connell D., Meyer M., Khoury J., Cheng X., Schwartz I., and McLaughlin J. 1990. Base changes at position-792 of *Escherichia coli* 16s ribosomal-RNA affect assembly of 70S ribosomes. *Proc. Natl. Acad. Sci.* **87:** 3700.

Schluenzen F., Tocilj A., Zarivach R., Harms J., Gluehmann M., Janell D., Bashan A., Bartels H., Agmon I., Franceschi F., and Yonath A. 2000. Structure of functionally activated small ribosomal subunit at 3.3 Å resolution. *Cell* **102:** 615.

Schlunzen F., Zarivach R., Harms J., Bashan J., Tocilj A., Albrecht R., Yonath A., and Franceschi F. 2001. Structural basis for the interaction of chloramphenicol, clindamycin and macrolides with the peptidyl transferase center in eubacteria. *Nature.* **413:** 814.

Sergiev P.V., Lavrik I.N., Wlasoff V.A., Dokudovskaya S.S., Dontsova O.A., Bogdanov A.A., and Brimacombe R. 1997. The path of mRNA through the bacterial ribosome: A site-directed crosslinking study using new photoreactive derivatives of guanosine and uridine. *RNA* **3:** 464.

Sette M., Spurio R., Van Tilborg P., Gualerzi C.O., and Boelens R. 1999. Identification of the ribosome binding sites of translation initiation factor IF3 by multidimensional heteronuclear NMR spectroscopy. *RNA* **5:** 82.

Shapkina T.G., Dolan M.A., Babin P., and Wollenzien P. 2000. Initiation factor 3-induced structural changes in the 30 S ribosomal subunit and in complexes containing tRNA(f)(Met) and mRNA. *J. Mol. Biol.* **299:** 615.

Shevack A., Gewitz H.S., Hennemann B., Yonath A., and Wittmann H.G. 1985. Characterization and crystallization of ribosomal practical from *Halobacterium marismortui. FEBS Lett.* **184:** 68.

Sonenberg N., Wilchek M., and Zamir A. 1973. Mapping of *Escherichia coli* ribosomal components involved in peptidyl transferase activity. *Proc. Natl. Acad. Sci.* **70:** 1423.

Srivastava S., Verschoor A., and Frank J. 1992. Eukaryotic initiation factor-III does not prevent association through physical blockage of the ribosomal-subunit interface. *J. Mol. Biol.* **226:** 301.

Stark H., Mueller F., Orlova E.V., Schatz M., Dube P., Erdemir T., Zemlin F., Brimacombe R., and van Heel M. 1995. The 70S *Escherichia coli* ribosome at 23 Å resolution: Fitting the ribosomal RNA. *Structure* **3:** 815.

Subramanian A.R. and Dabbs E.R. 1980. Functional studies on ribosomes lacking protein L1 from mutant *Escherichia coli. Eur. J. Biochem.* **112:** 425.

Sussman J.K., Simons E.L., and Simons R.W. 1996. *E. coli* translation initiation factor 3 discriminates the initiation codon in vivo. *Mol. Microbiol.* **21:** 347.

Tapprich W.E. and Hill W.E. 1986. Involvement of bases 787-795 of *Escherichia coli* 16S ribosomal RNA in ribosomal subunit association. *Proc. Nat. Acad. Sci.* **83:** 556.

Tapprich W.E., Goss D.J., and Dahlberg A.E. 1989. Mutation at position 791 in *Escherichia coli* 16S ribosomal RNA affects processes involved in the initiation of protein synthesis. *Proc. Natl. Acad. Sci.* **86:** 4927.

Tedin K., Moll I., Grill S., Resch A., Graschopf A., Gualerzi C.O., and Blasi U. 1999. Translation initiation factor 3 antagonizes authentic start codon selection on leaderless mRNAs. *Mol. Microbiol.* **31:** 67.

Thompson J., Kim D.F., O'Connor M., Lieberman K.R., Bayfield M.A., Gregory S.T., Green R., Noller H.F., and Dahlberg A.E. 2001. Analysis of mutations at residues A2451 and G2447 of 23S rRNA in the peptidyltransferase active site of the 50S ribosomal subunit. *Proc. Natl. Acad. Sci.* **98:** 9002.

Tocilj A., Schlunzen F., Janell D., Gluhmann M., Hansen H.A., Harms J., Bashan A., Bartels H., Agmon I., Franceschi F., and Yonath A. 1999. The small ribosomal subunit from *Thermus thermophilus* at 4.5 Å resolution: Pattern fittings and the identification of a functional site. *Proc. Natl. Acad. Sci.* **96:** 14252.

Uhlein M., Weglohner W., Urlaub H., and Wittmann-Liebold B. 1998. Functional implications of ribosomal protein L2 in protein biosynthesis as shown by in vivo replacement studies. *Biochem. J.* **331:** 423.

VanLoock M.S., Agrawal R.K., Gabashvili I.S., Qi L., Frank J., and Harvey S.C. 2000. Movement of the decoding region of the 16 S ribosomal RNA accompanies tRNA translocation. *J. Mol. Biol.* **304:** 507.

von Bohlen K., Makowski I., Hansen H.A., Bartels H., Berkovitch-Yellin Z., Zaytzev-Bashan A., Meyer S., Paulke C., Franceschi F., and Yonath A. 1991. Characterization and preliminary attempts for derivatization of crystals of large ribosomal subunits from *Haloarcula marismortui* diffracting to 3 Å resolution. *J. Mol. Biol.* **222:** 11.

Weiel J. and Hershey J.W. 1981. Fluorescence polarization studies of the interaction of *Escherichia coli* protein synthesis initiation factor 3 with 30S ribosomal subunits. *Biochemistry* **20:** 5859.

Weinstein S., Jahn W., Glotz C., Schluenzen F., Levin I., Janell D., Harms J., Koelln I., Hansen H., Gluehmann M., Bennett W., Bartels H., Bashan A., Agmon I., Kessler M., Pioletti M., Avila H., Anagnostopoulos K., Peretz M., Auerbach T., Franceschi F., and Yonath A. 1999. Metal compounds as tools for the construction and the interpretation of medium-resolution maps of ribosomal particles. *J. Struct. Biol.* **127:** 141.

Welch M., Chastang J., and Yarus M. 1995. An inhibitor of ribosomal peptidyl transferase using transition-state analogy. *Biochemistry* **34:** 385.

Wimberly B.T., Brodersen D.E., Clemons W.M., Jr., Morgan-Warren R.J., Carter A.P., Vonrhein C., Hartsch, T., and Ramakrishnan V. 2000. Structure of the 30S ribosomal subunit. *Nature* **407:** 327.

Woodcock J., Moazed D., Cannon M., Davies J., and Noller H.F. 1991. Interaction of antibiotics with A- and P-site-specific bases in 16S ribosomal RNA. *EMBO J.* **10:** 3099.

Wower J., Sylvers L.A., Rosen K.V., Hixson S.S., and Zimmermann R.A. 1993. A model for the tRNA binding sites on the *E. coli* ribosome In *The translational apparatus: Structure, function, regulation, evolution* (ed. K.H. Nierhaus et al.), p. 445. Plenum Press, New York.

Yonath A., Harms J., Hansen H.A.S., Bashan A., Schlunzen F., Levin I., Koelln I., Agmon I., Peretz M., Bartels H., Bennett W.S., Krumbholz S., Janell D., Weinstein S., Auerbach T., Avila H., Piolleti M., Morlang S., and Franceschi F. 1998. Crystallographic studies on the ribosome, a large macromolecular assembly exhibiting severe nonisomorphism, extreme beam sensitivity and no internal symmetry. *Acta Crystallogr. A* **54:** 945.

Yusupov M.M., Yusupova G.Z., Baucom A., Lieberman K., Earnest T.N., Cate J.H., and Noller H.F. 2001. Crystal structure of the ribosome at 5.5 Å resolution. *Science* **292:** 883.

Zamir A., Miskin R., and Elson D. 1971. Inactivation and reactivation of ribosomal subunits: Amino acyl transfer RNA binding activity of the 30S subunit from *E. coli. J. Mol. Biol.* **60:** 347.

Structure of the Ribosome at 5.5 Å Resolution and Its Interactions with Functional Ligands

H.F. Noller,* M.M. Yusupov,† G.Z. Yusupova,† A. Baucom,* K. Lieberman,*
L. Lancaster,* A. Dallas,* K. Fredrick,* T.N. Earnest,‡ and J.H.D. Cate¶

*Center for Molecular Biology of RNA, Sinsheimer Laboratories, University of California at Santa Cruz,
Santa Cruz, California 95064; ‡Berkeley Center for Structural Biology, Physical Biosciences Division,
Lawrence Berkeley National Laboratory, Berkeley, California 94720; ¶Whitehead Institute,
Cambridge, Massachusetts 01242

Ribosomology as a field was launched in the late 1950s, largely by the pioneering efforts of Tissières, working in the Watson laboratory at Harvard, who did the first physical characterizations of the *E. coli* ribosome and its subunits (Tissières and Watson 1958). During the subsequent discoveries of the genetic code and the outlines of the mechanism of translation, the ribosome was variously viewed as a passive surface upon which the tRNAs, mRNAs, and factors carried out the functional steps of protein synthesis, or, at the opposite extreme, as an enormously complex macromolecular machine whose structural and functional elucidation would be all but hopeless. As it became clear that the ribosome was not at all inert, tremendous effort was put into the isolation and characterization of the ribosomal proteins, the presumed determinants of ribosome function. By the late 1960s, discussions over the molecular origins of life had already uncovered the "ribosome paradox": How could the first ribosomes have evolved if the ribosome itself depended on proteins for its function? Crick's solution to this chicken-or-the-egg problem was to ask whether the first ribosomes might have been made of RNA (Crick 1968). Evidence supporting this possibility began to emerge in the early 1970s. Kethoxal modification of 30S ribosomal subunits was shown to inactivate their ability to bind tRNA (Noller and Chaires 1972). Prior binding of tRNA specifically protected against inactivation, suggesting that the target of modification was the binding site itself. Reconstitution and labeling experiments showed that inactivation was due to modification of a handful of guanine bases in 16S rRNA; the proteins from the modified subunits were fully active. These experiments pointed to the possibility (but did not prove) that parts of 16S rRNA were directly involved in binding tRNA to the ribosome.

However, proposals that ribosomal RNA was responsible for, or even a participant in, the function of protein synthesis, did not receive widespread support until the discovery of ribozymes about a decade later (Kruger et al. 1982; Guerrier-Takada et al. 1983). An exception was the acceptance of the special role of the tail of 16S rRNA in translational start-site selection (Shine and Dalgarno 1974; Steitz and Jakes 1975). By the late 1990s, arguments against a functional role for rRNA had largely been buried beneath an avalanche of biochemical, biophysical, genetic, and phylogenetic evidence (for review, see Green and Noller 1997). Any lingering doubts have now been removed by the spectacular high-resolution crystal structures of the ribosomal subunits that have emerged in the past two years (Ban et al. 2000; Schluenzen et al. 2000; Wimberly et al. 2000); as simply put by Nissen et al. (2000), "the ribosome is a ribozyme." Although the ribosomal proteins do indeed appear to participate in functional interactions, their relation to the rRNA clearly reveals them as "molecular arrivistes," structurally and functionally playing supporting roles to a process clearly dominated by, and essentially based on, RNA. Crick's conjecture is all but proven.

In this paper, we summarize the results of X-ray crystallographic studies on complete ribosomes, cocrystallized with mRNA and tRNAs, at resolutions of up to 5.5 Å (Cate et al. 1999; Yusupov et al. 2001). These structures reveal the nature of the interactions between the ribosome and the A-, P-, and E-site tRNAs, and show the details of how the two ribosomal subunits interact with each other to form the 70S ribosome. We also describe the path of mRNA through the ribosome (Yusupova et al. 2001) and a proposal for the position of initiation factor IF3 (Dallas and Noller 2001). Finally, we address the problem of how the mRNA and tRNAs are translocated, in the context of the molecular structure of the ribosome. The details of the structures of the rRNAs and ribosomal proteins and their interactions are presented elsewhere in this volume in the contexts of the high-resolution structures of the 30S and 50S subunits (Hansen et al., Bashan et al., Brodersen et al.).

STRUCTURE OF THE 70S RIBOSOME

Thermus thermophilus 70S ribosomes were co-crystallized with a synthetic mRNA analog and tRNAs bound to the P and E sites as described earlier (Cate et al. 1999), and their diffraction improved to 5 Å resolution. Experi-

†Present address: UPR 9004 de Biologie et de Genomiques Structurales du CNRS, IGBMC B.P. 163, 1 rue L.Fries, 67404 Illkirch Cedex-CU de Strasbourg, France.

mental phases to 7.5 Å obtained from multi-wavelength anomalous dispersion (MAD) experiments were extended to an effective resolution of 5.5 Å by use of density modification (Brünger et al. 1998). The quality of the phases was confirmed by the electron density of the bound P and E tRNAs, which provided internal standards of known structure. At 5.5 Å, the RNA backbones could be traced with high confidence, and proteins of known structure could be fitted readily to the electron density. Using 70S complexes crystallized with and without A-site tRNA, we obtained a 7 Å Fourier difference map that

provided the position of the A-site tRNA. Although final interpretation of our electron density maps was greatly facilitated by the availability of the high-resolution subunit structures (Ban et al. 2000; Schluenzen et al. 2000; Wimberly et al. 2000), the quality of our maps was sufficient to allow a reasonable initial fit of the 16S rRNA chain (overall rmsd = 5.7 Å) guided by biochemical and phylogenetic constraints but independent of any high-resolution structural information.

Figure 1 shows the structure of the 70S ribosome from three different orthogonal views. The 2.5 MD particle has

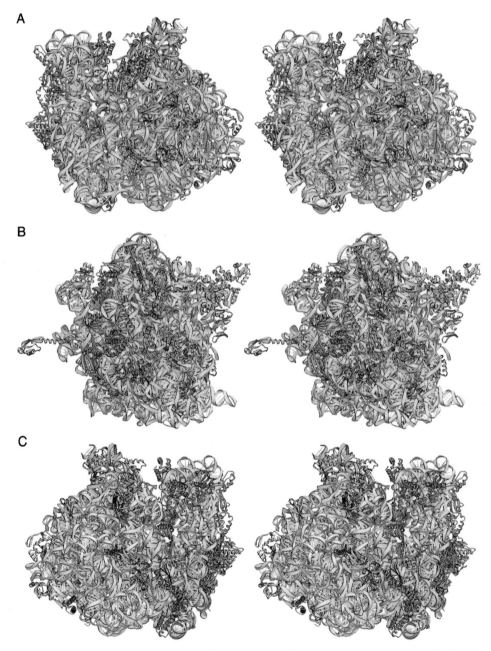

Figure 1. Structure of the 70S ribosome from *Thermus thermophilus* in a functional complex with an mRNA fragment and tRNAs bound to the A, P, and E sites (Yusupov et al. 2001). Stereo ribbon drawings are shown from (*A*) the A-site view, with the 30S subunit on the left and the 50S subunit on the right; (*B*) the 30S solvent view; and (*C*) the E-site view, with the 30S subunit on the right and the 50S subunit on the left. The molecular components are 16S rRNA (*cyan*), 23S rRNA (*gray*), 5S rRNA (*blue*), the small subunit proteins (*dark blue*), the large subunit proteins (*magenta*), and tRNAs bound to the A (*yellow*), P (*orange*), and E (*red*) sites.

an overall diameter of about 250 Å. From the right-hand side (Fig. 1A), the anticodon end of the A-site tRNA (gold) is visible in the near end of the subunit interface cavity, viewed through the large funnel-shaped opening where elongation factors EF-Tu and EF-G interact with the ribosome. The "standard view" (Fig. 1B) from the solvent face of the 30S subunit shows its head, body, platform, and neck features and their corresponding 16S rRNA (cyan) and protein (blue) components. In the background, parts of the 50S subunit are visible in the "crown" view, with its 23S rRNA (gray), 5S rRNA (top; blue), and 50S subunit proteins (magenta). Protein L9 can be seen at the left, extending more than 50 Å beyond the surface of the 50S subunit proper. In the left-hand view (Fig. 1C), close approach of the platform of the 30S subunit with the 50S subunit at the interface is evident. The E-site tRNA (red) is visible at the near side of the interface cavity, partly shielded from view by L1 and its RNA-binding site, which appear to block the path for its exit from the ribosome.

Viewed from the interface (Fig. 2), fewer proteins are visible on the 30S and 50S subunits, and they are located mainly around the periphery, leaving large exposed surfaces of ribosomal RNA. The three tRNAs are aligned on

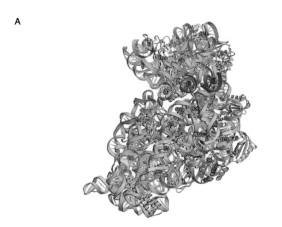

A

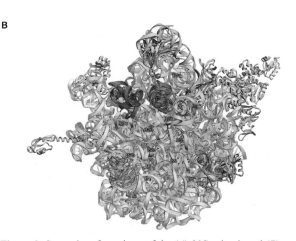

B

Figure 2. Stereo interface views of the (*A*) 30S subunit and (*B*) 50S subunit (Yusupov et al. 2001). RNA–RNA intersubunit contacts are shown in magenta, protein contacts are shown in yellow. The three tRNAs (*B*) or their anticodon stem-loops (*A*) are shown in yellow (A tRNA), orange (P tRNA), or red (E tRNA).

the 30S subunit with their anticodon ends bound in the RNA-rich groove between the head, body, and platform (Fig. 2A). The rest of the tRNAs, including their D stems, elbows, and acceptor arms, interact with the 50S subunit (Fig. 2B). The acceptor arms of the A and P tRNAs point downward into the peptidyl transferase cavity, whereas the E-tRNA acceptor arm is directed into a separate cleft next to the L1 ridge. The tRNA-binding-site neighborhoods are dominated by rRNA, as are the interface contact surfaces.

INTERACTIONS ACROSS THE SUBUNIT INTERFACE

Intersubunit contacts were first visualized as discrete bridges in low-resolution cryoelectron microscopy (cryo-EM) studies and more recently at 11.5 Å resolution by Frank and coworkers (Frank et al. 1995; Gabashvili et al. 2000). In our earlier 7.8 Å structure (Cate et al. 1999), at least ten individual bridges could be resolved. At 5.5 Å, all of the molecular components involved in the intersubunit contacts can be identified, including two additional protein-containing bridges. As inferred from earlier chemical probing (Merryman et al. 1999a,b) and modification-interference (Herr et al. 1979) studies, most of the bridge contacts involve rRNA.

Figure 2A shows the 30S bridge contacts, viewed from the interface, with the anticodon stem-loops of the A, P, and E tRNAs in their respective 30S subunit-binding sites. The distribution of RNA–RNA versus RNA–protein or protein–protein contacts is striking; the RNA–RNA contacts (magenta) are centrally located on the platform and penultimate stem, directly abutting the tRNA-binding sites. In contrast, contacts involving proteins (yellow) are peripherally located, more distal from the functional sites. On the 50S subunit side (Fig. 2B), the RNA–RNA contacts are again central, forming a triangular patch across the front surface of the interface wall that separates the peptidyl transferase and E sites from the interface cavity.

The molecular contacts forming the twelve intersubunit bridges have been described in detail by Yusupov et al. (2001). Multiple contacts can be seen in the electron density map for many of the bridges, giving a total of more than 30 individual interactions. The RNA components of the bridges come almost exclusively from the central and 3′ minor domains of 16S rRNA and from domains II and IV of 23S rRNA. The most common types of RNA–RNA contacts are minor groove–minor groove interactions, although major groove, loop, and backbone contacts are also found. The bridge proteins (S13, S15, L2, L5, L14, and L19) use a wide variety of RNA features for recognition, including major groove, minor groove, backbone, and loop elements.

tRNA–RIBOSOME INTERACTIONS

All three tRNAs are shared between the two ribosomal subunits in a similar way; their anticodon stem-loops are bound by the 30S subunit, and contacts with the rest of the tRNA—D stem, elbow, and acceptor arm—are made by the 50S subunit. The anticodon stem-loops point into the groove separating the head from the rest of the 30S subunit.

In the 30S A site (Fig. 3A), the geometry of the codon–anticodon interaction is monitored by G530, A1492, and A1493, as shown by Ogle et al. (2001) in their high-resolution structure. These are the same three bases that were most strongly implicated in 30S A-site function by chemical footprinting and mutational analysis (Moazed and Noller 1986, 1990; Powers and Noller 1990; Yoshizawa et al. 1999). In addition, contacts are made by protein S12, which has long been known to have a strong influence on the decoding function of the ribosomal A site (Gorini 1971). The bulged base C1054, mutations in which have been shown to suppress UGA nonsense mutations (Murgola et al. 1988), projects toward the apex of the A-tRNA anticodon loop. The 30S A site is rather open, with few nonspecific contacts between the ribosome and the tRNA, presumably to maximize codon-specific binding.

In contrast, the 30S P site (Fig. 3B) is more closed, making several nonspecific contacts with the backbone and minor groove of the anticodon stem. In addition, a set of interactions appears to stabilize P-site codon–anticodon pairing: The base G966, interacting with the anticodon backbone at position 34 of the tRNA, and the backbone of nucleotide U1498, interacting with the backbone of position 1 of the P codon, appear to clamp the codon and anticodon together. The P-site wobble base pair appears to be stabilized by a stacking interaction with C1400, an interaction that was predicted nearly 20 years ago by Ofengand, Zimmermann, and their coworkers (Prince et al. 1982). Again, many of the bases identified by tRNA footprinting experiments (Moazed and Noller 1986, 1990) are found to make interactions with the tRNA; some, however, including the "class III" bases (Moazed and Noller 1987), are protected indirectly, presumably by tRNA-induced conformational changes. Two proteins, S9 and S13, interact with the P-tRNA, both via their extended, basic carboxy-terminal tails, which interact with the anticodon stem-loop in a way that suggests that they function as sophisticated polyamines, bolstering the RNA–RNA interactions.

A model for the interaction of initiation factor IF-3 with 30S subunits has been derived using constraints obtained from hydroxyl radical footprinting of IF-3 and its individual domains on 16S rRNA, and from directed hydroxyl radical probing studies using Fe(II) tethered to various positions on the surface of IF-3 (A. Dallas and

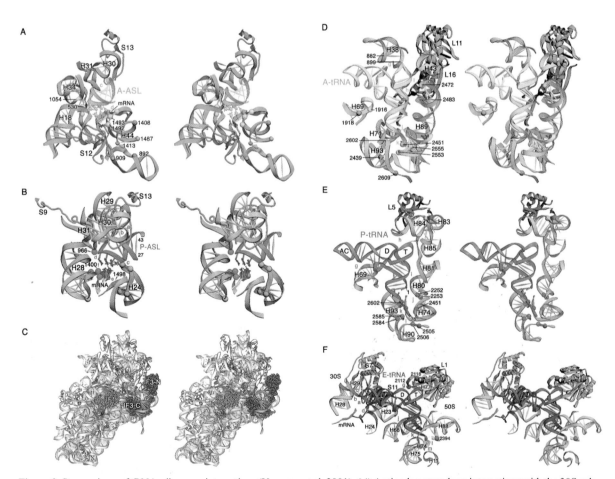

Figure 3. Stereo views of tRNA–ribosome interactions (Yusupov et al. 2001). (*A*) Anticodon stem-loop interactions with the 30S subunit A site; (*B*) anticodon stem-loop interactions with the 30S P site; (*C*) model for the interaction of IF3 with the 30S subunit (A. Dallas and H.G. Noller, in prep.); (*D*) interaction of tRNA with the 50S A site and (*E*) the 50S P site; (*F*) interaction of tRNA with the ribosomal E site.

H.F. Noller, in prep.). These constraints place the C-domain of IF-3 at the subunit interface surface of the platform of the 30S subunit (Fig. 3C), almost precisely in the position of helix 69 of 23S rRNA. This result accounts for the dissociation activity of IF-3, which would result from disruption of several intersubunit bridges. The N-domain is positioned in the E site, where it may prevent binding of tRNA to the E site during the initiation process. Also shown in Figure 3C is the crystallographically determined position of initiation factor IF-1 (Carter et al. 2001), which blocks the 30S A site, leaving only the P site available for tRNA binding. Discrimination of the initiator tRNA, another IF-3 function, has been shown to depend on the run of three G-C base pairs that is found uniquely at the bottom of the anticodon stem in initiator tRNAs (Mandal et al. 1996). In our structure of the complex of the 70S ribosome containing initiator tRNA, G1338 and A1339 are closely juxtaposed with these same base pairs in the minor groove of the anticodon stem. A modest rotation of the head of the 30S subunit could bring these two bases in close contact with the minor groove to allow discrimination between the initiator and elongator Met tRNAs, by a mechanism reminiscent of the one proposed by Ramakrishnan and coworkers for discrimination of A-site tRNAs. In this case, the role of IF-3 could be to perturb the position of the head of the 30S subunit during initiation. Indeed, cryo-EM difference maps of IF-3 complexes (McCutcheon et al. 1999) provide direct evidence for movement of the head of the subunit that is compatible with such a proposal.

The anticodon stem-loop of the E tRNA is also bound by the 30S subunit, something that was not predicted from the base-specific footprinting studies, probably because the interactions with 16S rRNA are mainly via its sugar-phosphate backbone. As for the A and P sites, there are also protein–tRNA interactions in the E site. In this case, the β hairpin and carboxy-terminal α helix of protein S7 interact with the minor groove surface of the anticodon loop. A possible role of S7 could be to disrupt codon–anticodon interaction in the 30S E site. Recent studies show that deletion of either the β hairpin or the carboxy-terminal helix of S7 influences the efficiency and accuracy of EF-G-dependent translocation (K. Fredrick, unpubl.).

Among the most functionally intriguing RNA features of the ribosome is the universally conserved helix 69 from domain IV of 23S rRNA. The minor groove side of its hairpin loop interacts with the top of the penultimate stem of 16S rRNA to form bridge B2A; the major groove side of its loop interacts with the minor groove of the D stem of the A tRNA (Fig. 3D); and the minor groove of its stem interacts with the minor groove of the D stem of the P tRNA (Fig. 3E). As discussed below, helix 69 is a prime candidate for involvement in ribosomal dynamics associated with translocation.

The elbows and acceptor arms of the three tRNAs interact mainly with 23S rRNA, but in each case, the elbow region makes contact with a ribosomal protein: L16 in the A site, L5 in the P site, and L1 in the E site (Fig. 3D,E,F). Again, most of the base-specific protections correspond to direct interactions between 23S rRNA and tRNA, whereas the remaining ones are indicative of tRNA-dependent conformational changes. The acceptor ends of the A and P tRNAs are inserted into a deep cavity in the 50S subunit that contains the peptidyl transferase catalytic site, which is made up exclusively of RNA, as shown by the high-resolution 50S subunit structure (Ban et al. 2000). Details of the interactions between the ends of the tRNAs and the peptidyl transferase site can be inferred from the high-resolution structures of complexes between the 50S subunit and model substrates or transition-state analogs (Nissen et al. 2000), described by Moore, Steitz, and coworkers elsewhere in this volume (Hansen et al.). There appear to be some differences in the conformation of the 23S rRNA in the peptidyl transferase region between our 70S ribosomal complexes and the 50S structures, localized mainly around positions 2450 and 2505, and in the P loop (helix 80). These differences could be due to factors such as the presence of the 30S subunit or complete tRNAs in the 70S structures, or to the different ionic conditions in the two different kinds of crystals. Because of the lower resolution of the 70S structure, these differences should be regarded as tentative for the time being.

In contrast, the acceptor arm and CCA end of the E tRNA are oriented away from the A and P tRNAs and inserted into a separate cavity near protein L1 (Fig. 3F). The blockage of the exit path for the E tRNA by protein S7 and by L1 and its rRNA-binding site requires that one or both of these structures move to allow release of the deacylated tRNA. Rotation of the head of the 30S subunit (see below) and the observed mobility of the L1 region of yeast 80S ribosomes (Gomez-Lorenzo et al. 2000) support both possibilities.

PATH OF mRNA THROUGH THE RIBOSOME

Using ribosome–tRNA complexes formed in the presence or absence of mRNA, we obtained Fourier difference maps of three different mRNAs bound to the ribosome (Yusupova et al. 2001). This allowed us to directly observe the path of the mRNA through the ribosome (Fig. 4A). It winds around the neck of the 30S subunit, in a groove between the head and the body of the particle. The Shine-Dalgarno helix is bound to a cleft between the head and the rear of the platform, and contacts both proteins S11 and S18, as well as several elements of 16S rRNA. On the interface side of the subunit, only about eight nucleotides, centered on the junction between the A and P codons, are exposed to the interface cavity. Interactions between the mRNA and ribosome in the interface region are almost exclusively with 16S rRNA, except for some contacts with protein S12 in the A site. The phosphate group of nucleotide 1401 is positioned directly in the path of the mRNA downstream from the P codon, forcing the mRNA backbone to kink between the two codons.

Upstream from the P codon, the mRNA passes through a short upstream tunnel, bounded by the neck and platform of the subunit and the β hairpin and carboxy-terminal helix of S7. Downstream from the A codon, the mRNA passes through a larger tunnel, formed between the head and shoulder of the subunit. On the interface side, the tunnel is made almost entirely of RNA. In contrast, the solvent side of the tunnel is composed of proteins S3, S4, and S5, which form a ring around the mRNA where it enters the ribosome (Fig. 4B). The problem of how the ribosome unwinds helical structures such as hairpin stem-loops in mRNA has largely remained unaddressed, and may be related to the observation that downstream pseudoknots and helices stimulate –1 translational frameshifting (Brierley et al. 1989; Alam et al. 1999). We have proposed a possible helicase mechanism, based on the arrangement of structural features in the downstream tunnel. Protein S3 is bound to the head of the particle, whereas S4 and S5 are bound to the body. If the head of the subunit rotates during translocation, as appears to be indicated by cryo-EM reconstruction experiments (Agrawal et al. 1999), the two strands of a hairpin stem could be pulled apart mechanically by movement of the head relative to the body, coupled to the translocation of tRNA and mRNA.

STRUCTURAL INTERPRETATION OF THE TRANSLOCATIONAL CYCLE

Figure 5A shows the overall relative geometry of the A, P, and E tRNAs and the mRNA as they are positioned

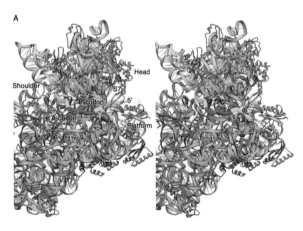

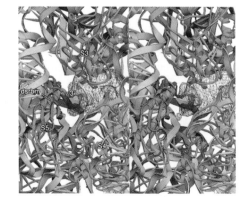

Figure 4. (*A*) Stereo view of the path of messenger RNA through the ribosome, viewed from the interface side of the 30S subunit. (*B*) Stereo view of the downstream mRNA tunnel. (dsrbd) Double-stranded RNA-binding domain in protein S5.

in the 70S ribosome crystals. Their specific contacts with the ribosome indicate that they are in their "classic" (A/A, P/P and E/E), rather than hybrid, binding states (Moazed and Noller 1989). The planes of the A and P tRNAs form an included angle of 26°, and the P and E tRNAs an angle of 46°. Simultaneous reading of the adjacent A and P codons is accommodated by a kink in the mRNA backbone of about 45° between the A and P codons (Fig. 5A). The distances between corresponding positions of the three tRNAs is a measure of the magnitude of the movement of tRNA during translocation. Thus, the anticodon end of tRNA moves about 28 Å between the 30S A and P sites, and 20 Å between the P and E sites. Because of the rotation of the plane of the tRNA, the elbow moves through much larger distances of 40 and 55 Å, as it transits from A to P to E.

Our current understanding of the hybrid-states model (Moazed and Noller 1989) is shown schematically in Figure 5B. Experimental evidence from several laboratories over the past decade has led to the introduction of some modifications to the minimal model. First, the aforementioned crystallographic evidence necessitates participation of a 30S E site in the mechanism. Second, evidence for an "accommodation" step following release of EF-Tu (Pape et al. 1999) raises the possibility that proofreading of the incoming aminoacyl tRNA could take place during this step; possibly, the accommodation process could involve regulation of peptidyl transferase activity, permitting only the cognate aminoacyl tRNA to participate in peptide-bond formation. Third, several lines of evidence (Green et al. 1998; M. Rodnina and S. Joseph, unpubl.) have convincingly demonstrated that movement from the A/A to A/P and P/P to P/E states occurs sequentially, rather than concertedly, with peptide-bond formation. Therefore, a separate state in which the peptidyl tRNA occupies the A/A state has been introduced (Fig. 5B).

Extensive evidence has accumulated in support of the essential feature of the hybrid-states model, that the tRNAs move independently with respect to the two ribosomal subunits, first on the 50S subunit and then on the 30S subunit (coupled to mRNA movement). Direct structural observation of the A/P and P/E states has occurred in cryo-EM reconstructions (Agrawal et al. 2000). The A/T state, in which the incoming aminoacyl tRNA is still bound to EF-Tu, has also been observed by cryo-EM studies (Stark et al. 1997).

Figure 5C shows a three-dimensional interpretation of the hybrid-states translocational cycle. Here, the orientations of the classic-state tRNAs (A/A, P/P, and E/E) are represented by those that we have directly observed crystallographically. The positions of the A/P and P/E hybrid-state tRNAs were modeled starting with the classic-state tRNAs, fixing the positions of their anticodon ends, and rotating them as rigid bodies to dock their respective acceptor ends in the 50S subunit. The resulting models bear close resemblance to the low-resolution structures observed experimentally by cryo-EM (Agrawal et al. 2000). The A/T tRNA was modeled in two steps: First, the structure of EF-G (Czworkowski et al. 1994) was docked on the 70S ribosome structure using constraints from foot-

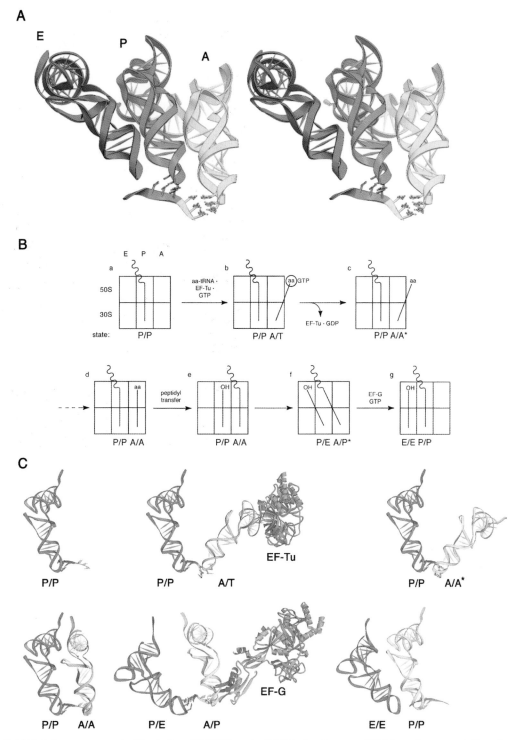

Figure 5. (*A*) Relative arrangement of the A, P, and E tRNAs and mRNA in the ribosome. (*B*) Schematic representation of an updated version of the hybrid-states model (Moazed and Noller 1989) for the translational cycle. (*C*) A three-dimensional representation of the movements of tRNA through the hybrid-states cycle. (*A*, Reprinted, with permission, from Yusupov et al. 2001 [copyright AAAS].)

printing and directed hydroxyl radical probing. Second, the structure of the EF-Tu–tRNA–GTP ternary complex (Nissen et al. 1995) was docked on EF-G by virtue of their homologous G domains. The result is again in good agreement with the position of the ternary complex deter-mined by cryo-EM (Stark et al. 1997). A striking obser-vation is that the distance traversed by the acceptor end of the aminoacyl tRNA in going from the A/T to the A/A states is on the order of 70 Å, roughly the overall dimen-sions of the tRNA itself.

CLUES TO THE MECHANISM OF
tRNA MOVEMENT

The dynamic nature of translation implies that the ribosome contains moving parts that enable its function (Spirin 1970). The translocation step of protein synthesis inescapably requires movements of 20 Å or more by the tRNAs, as they move from the A to P to E sites. It seems unlikely that such movements would not be matched by corresponding structural rearrangements of the ribosome (Wilson and Noller 1998). The hybrid-states model carries the implication that the mechanism of translocation involves relative movement of the 30S and 50S subunits, or of particular structural domains or substructures of the two subunits (Moazed and Noller 1989). This notion is reinforced by the observation that many of the nucleotides implicated in tRNA–ribosome interactions by biochemical and genetic experiments are adjacent to nucleotides involved in subunit association (Merryman et al. 1999a,b). The crystal structure in fact shows that the tRNAs directly contact intersubunit bridges, at least some of which are believed to be dynamic elements of the ribosome. For example, among the structural elements that are disordered in the high-resolution 50S subunit structure are the bridges B1a, B1b, and B2a. Disorder is informative in that it identifies specific molecular features of the ribosome that are capable of independent motion, at least under conditions prevailing in the crystal, and thus are candidates for participation in ribosomal dynamics.

Figure 6 shows the features directly surrounding the A and P tRNAs at the subunit interface, viewed from the two opposite interface sides. The two tRNAs are sandwiched between bridges B1a and B1b at the top, and B2a at the bottom. The intersubunit contacts for all three of these bridges are disordered in the 50S crystal structure (Ban et al. 2000), suggesting that all three are dynamic elements. On the 30S side (Fig. 6B), the tRNAs are sandwiched between the head and the tops of the penultimate stem and platform, all of which show conformational differences between the free 30S subunits and 70S ribosomes (Yusupov et al. 2001), again suggesting that they are capable of movement during translation. Moreover, the fact that these potentially dynamic elements all interact with each other across the subunit interface points to the likelihood that their respective movements are coordinated. Thus, movement of bridges B1a and B1b would be coupled to rotation of the head, and movement of bridge B2a to movement of the penultimate stem and platform. In fact, low-resolution cryo-EM images of the pre- and post-translocation states of *E. coli* ribosomes (Agrawal et al. 1999) are consistent with such a coordinated movement.

A potentially important clue to the mechanism of translocation comes from tRNA modification-interference studies by Feinberg and Joseph (2001). Their studies show that introduction of a single 2′-O-methyl group at position 71 of P tRNA abolishes EF-G-dependent translocation. Interestingly, the sole interaction between the ribosome and position 71 of tRNA occurs in the 50S E site, indicating that the effect of the methyl group must be on the P/E state. This finding is consistent with a ki-

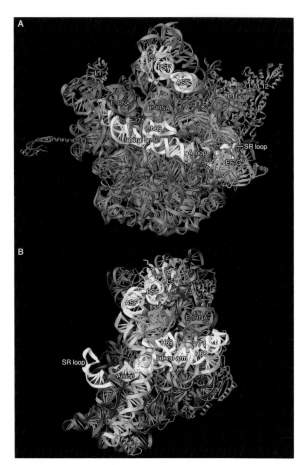

Figure 6. rRNA elements surrounding the A and P tRNAs at the subunit interface. The position of ribose 71 of the E tRNA is shown by a red sphere. See the text for details.

netic analysis that has shown the importance of hybrid-state formation for EF-G-dependent translocation (Semenkov et al. 2000). The mechanism by which ribose 71 methylation inhibits translocation must be indirect, since the nearest approach of EF-G is about 70 Å away, and its catalytic center is over 100 Å away. One possibility is that the effect is mediated through 23S rRNA. Ribose 71 contacts helix 68 of 23S rRNA, which is at the far left end of the lateral arm of domain IV that lies horizontally across the top of the subunit interface surface of the 50S subunit (Fig. 6A). The lateral arm is a continuously coaxially stacked system of canonical and noncanonical helices that traverse the interface. Its far right-hand end terminates in a hairpin loop that makes an A-minor interaction with the stem of the A loop. The single-stranded loop at the base of the A loop in turn interacts with the minor groove of the sarcin/ricin loop, which has been directly implicated in the GTPase function of EF-G (Hausner et al. 1987). In addition, helix 69, which directly contacts both the A and P tRNAs in their respective D stems, is connected to the conserved, noncanonical helix in the middle of the lateral arm of domain IV (Fig. 6A). These structural clues present a circumstantial case for the involvement of specific elements of ribosomal RNA in the mechanism of translocation. The availability of detailed structural information means that we now have a basis for formulating concrete molecular models for this most intriguing of biological mechanisms, and for testing them experimentally.

ACKNOWLEDGMENTS

We thank Abraham Szöke, Hanna Szöke, Sanjoy Ray, Bill Scott, James Murray, and Chuck Wilson for stimulating discussions; Robin Gutell and Francois Michel for generously providing unpublished sequence analysis results and alignments, and for discussions; Gerry McDermott, Keith Henderson, Jay Nix, and Li-Wei Hung for advice and help with data collection; and Simpson Joseph for communicating unpublished results. Three-dimensional model renderings were generated using RIBBONS (Carson 1997). This work was supported by grants from the National Institutes of Health and the Agouron Institute (to H.F.N.), from the W.M. Keck Foundation (to the Center for Molecular Biology of RNA), from the Whitehead Institute for Biomedical Research (to J.H.D.C.), and from a Damon Runyon-Walter Winchell postdoctoral fellowship and a Searle Scholar Award to J.H.D.C. X-ray data were collected at the Berkeley Center for Structural Biology beamline 5.0.2, which is supported by the Office of Biological and Environmental Research of the Department of Energy and by grants from the National Institutes of Health Institute of General Medical Sciences, and the Agouron Institute (to T.N.E.).

REFERENCES

Agrawal R.K., Heagle A.B., Penczek P., Grassucci R.A., and Frank J. 1999. EF-G-dependent GTP hydrolysis induces translocation accompanied by large conformational changes in the 70S ribosome. *Nat. Struct. Biol.* **6:** 643.

Agrawal R.K., Spahn C.M., Penczek P., Grassucci R.A., Nierhaus K.H., and Frank J. 2000. Visualization of tRNA movements on the *Escherichia coli* 70S ribosome during the elongation cycle. *J. Cell Biol.* **150:** 447.

Alam S.L., Atkins J.F., and Gesteland R.F. 1999. Programmed ribosomal frameshifting: Much ado about knotting! *Proc. Natl. Acad. Sci.* **96:** 14177.

Ban N., Nissen P., Hansen J., Moore P.B., and Steitz T.A. 2000. The complete atomic structure of the large ribosomal subunit at 2.4 Å resolution (see comments). *Science* **289:** 905.

Brierley I., Digard P., and Inglis S.C. 1989. Characterization of an efficient coronavirus ribosomal frameshifting signal: Requirement for an RNA pseudoknot. *Cell* **57:** 537.

Brünger A.T., Adams P.D., Clore G.M., DeLano W.L., Gros P., Grosse-Kunstleve R.W., Jiang J.S., Kuszewski J., Nilges M., Pannu N.S., Read R.J., Rice L.M., Simonson T., and Warren G.L. 1998. Crystallography & NMR system: A new software suite for macromolecular structure determination. *Acta Crystallogr. D Biol. Crystallogr.* **54:** 905.

Carson M. 1997. Ribbons. *Methods Enzymol.* **277B:** 493.

Carter A.P., Clemons W.M., Jr., Brodersen D.E., Morgan-Warren R.J., Hartsch T., Wimberly B.T., and Ramakrishnan V. 2001. Crystal structure of an initiation factor bound to the 30S ribosomal subunit. *Science* **291:** 498.

Cate J.H., Yusupov M.M., Yusupova G.Z., Earnest T.N., and Noller H.F. 1999. X-ray crystal structures of 70S ribosome functional complexes. *Science* **285:** 2095.

Crick F.H.C. 1968. *The origin of the genetic code. J. Mol. Biol.*

38: 367.

Czworkowski J., Wang J., Steitz T.A., and Moore P.B. 1994. The crystal structure of elongation factor G complexed with GDP, at 2.7 Å resolution. *EMBO J.* **13:** 3661.

Dallas A. and Noller H.F. 2001. Interaction of translation initiation factor 3 with the 30S ribosomal subunit. *Mol. Cell.* **98:** 855.

Feinberg J. and Joseph S. 2001. Identification of molecular interactions between P site tRNA and the ribosome essential for translocation. *Proc. Natl. Acad. Sci.* **98:** 11120.

Frank J., Verschoor A., Li Y., Zhu J., Lata R.K., Radermacher M., Penczek P., Grassucci R., Agrawal R.K., and Srivastava S. 1995. A model of the translational apparatus based on a three-dimensional reconstruction of the *Escherichia coli* ribosome. *Biochem. Cell Biol.* **73:** 757.

Gabashvili I.S., Agrawal R.K., Spahn C.M., Grassucci R.A., Svergun D.I., Frank J., and Penczek P. 2000. Solution structure of the *E. coli* 70S ribosome at 11.5 Å resolution. *Cell* **100:** 537.

Gomez-Lorenzo M.G., Spahn C.M., Agrawal R.K., Agrawal R.A. Grassucci, Penczek P., Chakraburtty K., Ballesta J.P., Lavandera J.L., Garcia-Bustos J.F., and Frank J. 2000. Three-dimensional cryo-electron microscopy localization of EF2 in the *Saccharomyces cerevisiae* 80S ribosome at 17.5 Å resolution. *EMBO J.* **19:** 2710.

Gorini L. 1971. Ribosomal discrimination of tRNAs. *Nature* **234:** 261.

Green R. and Noller H.F. 1997. Ribosomes and translation. *Annu. Rev. Biochem.* **66:** 679.

Green R., Switzer C., and Noller H.F. 1998. Ribosome-catalyzed peptide-bond formation with an A-site substrate covalently linked to 23S ribosomal RNA. *Science* **280:** 286.

Guerrier-Takada C., Gardiner K., Marsh T., Pace N., and Altman S. 1983. The RNA moiety of ribonuclease P is the catalytic subunit of the enzyme. *Cell* **35:** 849.

Hausner T.P., Atmadja J., and Nierhaus K.H. 1987. Evidence that the G2661 region of 23S rRNA is located at the ribosomal binding sites of both elongation factors. *Biochimie* **69:** 911.

Herr W., Chapman N.M., and Noller H.F. 1979. Mechanism of ribosomal subunit association: Discrimination of specific sites in 16 S RNA essential for association activity. *J. Mol. Biol.* **130:** 433.

Kruger K., Grabowski P.J., Zaug A.J., Sands J., Gottschling D.E., and Cech T.R. 1982. Self-splicing RNA: Autoexcision and autocyclization of the ribosomal RNA intervening sequence of Tetrahymena. *Cell* **31:** 147.

Mandal N., Mangroo D., Dalluge J.J., McCloskey J.A., and Rajbhandary U.L. 1996. Role of the three consecutive G:C base pairs conserved in the anticodon stem of initiator tRNAs in initiation of protein synthesis in *Escherichia coli*. *RNA* **2:** 473.

McCutcheon J.P., Agrawal R.K., Philips S.M., Grassucci R.A., Gerchman S.E., Clemons W.M., Jr., Ramakrishnan V., and Frank J. 1999. Location of translational initiation factor IF3 on the small ribosomal subunit. *Proc. Natl. Acad. Sci.* **96:** 4301.

Merryman C., Moazed D., Daubresse G., and Noller H.F. 1999a. Nucleotides in 23S rRNA protected by the association of 30S and 50S ribosomal subunits. *J. Mol. Biol.* **285:** 107.

Merryman C., Moazed D., McWhirter J., and Noller H.F. 1999b. Nucleotides in 16S rRNA protected by the association of 30S and 50S ribosomal subunits. *J. Mol. Biol.* **285:** 97.

Moazed D. and Noller H.F. 1986. Transfer RNA shields specific nucleotides in 16S ribosomal RNA from attack by chemical probes. *Cell* **47:** 985.

———. 1987. Interaction of antibiotics with functional sites in 16S ribosomal RNA. *Nature* **327:** 389.

———. 1989. Intermediate states in the movement of transfer RNA in the ribosome. *Nature* **342:** 142.

———. 1990. Binding of tRNA to the ribosomal A and P sites protects two distinct sets of nucleotides in 16S rRNA. *J. Mol. Biol.* **211:** 135.

Murgola E.J., Hijazi K.A., Goringer H.U., and Dahlberg A.E. 1988. Mutant 16S ribosomal RNA: A codon-specific translational suppressor. *Proc. Natl. Acad. Sci.* **85:** 4162.

Nissen P., Hansen J., Ban N., Moore P.B., and Steitz T.A. 2000. The structural basis of ribosome activity in peptide bond synthesis (see comments). *Science* **289:** 920.

Nissen P., Kjeldgaard M., Thirup S., Polekhina G., Reshetnikova L., Clark B.F.C., and Nyborg J. 1995. Crystal structure of the ternary complex of Phe-tRNA[Phe], EF-Tu, and a GTP analog. *Science* **270:** 1464.

Noller H.F. and Chaires J.B. 1972. Functional modification of 16S ribosomal RNA by kethoxal. *Proc. Natl. Acad. Sci.* **69:** 3113.

Ogle J.M., Brodersen D.E., Clemons W.M., Tarry M.J., Carter A.P., and Ramakrishnan V. 2001. Recognition of cognate transfer RNA by the 30S ribosomal subunit. *Science* **292:** 897.

Pape T., Wintermeyer W., and Rodnina M. 1999. Induced fit in initial selection and proofreading of aminoacyl-tRNA on the ribosome. *EMBO J.* **18:** 3800.

Powers T. and Noller H.F. 1990. Dominant lethal mutations in a conserved loop in 16S rRNA. *Proc. Natl. Acad. Sci.* **87:** 1042.

Prince J.B., Taylor B.H., Thurlow D.L., Ofengand J., and Zimmermann R.A. 1982. Covalent crosslinking of tRNA[Val] to 16S RNA at the ribosomal P site: Identification of crosslinked residues. *Proc. Natl. Acad. Sci.* **79:** 5450.

Schluenzen F., Tocilj A., Zarivach R., Harms J., Gluehmann M., Janell D., Bashan A., Bartels H., Agmon I., Franceschi F., and Yonath A. 2000. Structure of functionally activated small ribosomal subunit at 3.3 angstroms resolution. *Cell* **102:** 615.

Semenkov Y.P., Rodnina M.V., and Wintermeyer W. 2000. Energetic contribution of tRNA hybrid state formation to translocation catalysis on the ribosome. *Nat. Struct. Biol.* **7:** 1027.

Shine J. and Dalgarno L. 1974. The 3´-terminal sequence of *E. coli* 16S ribosomal RNA complementarity to nonsense triplets and ribosome binding sites. *Proc. Natl. Acad. Sci.* **71:** 1342.

Spirin A.S. 1970. A model of the functioning ribosome: Locking and unlocking of the ribosome subparticles. *Cold Spring Harbor Symp. Quant. Biol.* **34:** 197.

Stark H., Rodnina M.V., Rinke-Appel J., Brimacombe R., Wintermeyer W., and van Heel M. 1997. Visualization of elongation factor Tu on the *Escherichia coli* ribosome. *Nature* **389:** 403.

Steitz J.A. and Jakes K. 1975. How ribosomes select initiator regions in mRNA: Base pair formation between the 3´ terminus of 16S rRNA and the mRNA during initiation of protein synthesis in *Escherichia coli*. *Proc. Natl. Acad. Sci.* **72:** 4734.

Tissières A. and Watson J.D. 1958. Ribonucleoprotein particles from *E. coli*. *Nature* **182:** 778.

Wilson K. and Noller H.F. 1998. Molecular movement inside the translational engine. *Cell* **92:** 337.

Wimberly B.T., Brodersen D.E., Clemons W.M., Jr., Morgan-Warren R.J., Carter A.P., Vonrhein C., Hartsch T., and Ramakrishnan V. 2000. Structure of the 30S ribosomal subunit. *Nature* **407:** 327.

Yoshizawa S., Fourmy D., and Puglisi J.D. 1999. Recognition of the codon-anticodon helix by ribosomal RNA. *Science* **285:** 1722.

Yusupov M., Yusupova G., Baucom A., Lieberman K., Earnest T.N., Cate J.H., and Noller H.F. 2001. Crystal structure of the ribosome at 5.5 Å resolution. *Science* **292:** 883.

Yusupova G.Z., Yusupov M., Cate J.H.D., and Noller H.F. 2001. The path of messenger RNA through the ribosome. *Cell* **106:** 233.

Ratchet-like Movements between the Two Ribosomal Subunits: Their Implications in Elongation Factor Recognition and tRNA Translocation

J. Frank*† AND R.K. Agrawal†‡

*Howard Hughes Medical Institute, Health Research, Inc. at the Wadsworth Center; † Department of Biomedical Sciences, State University of New York at Albany; ‡Wadsworth Center, New York State Department of Health, Albany, New York 12201-0509

Three-dimensional cryo-electron microscopy (3D cryo-EM) is uniquely suited to provide information about dynamic changes of the ribosome structure in the course of translation. Usually, a functional ribosome–ligand complex is stalled by the addition of an antibiotic that is known to inhibit a biochemical reaction at a particular intermediate state (see Spahn and Prescott 1996) or substitution of high-energy nucleotide triphosphate compounds, such as GTP, by its nonhydrolyzable analogs. In most cases, such compounds bind to the protein factors to provide a conformation recognizable by the ribosome. Many of the changes are on such a scale that they are easily detected and measured by fitting of X-ray structures or by comparison of three-dimensional density maps showing the ribosome in different states. For example, on the basis of cryo-EM maps with resolutions ranging from 17 Å to 20 Å, EF-G bound to the ribosome in the presence of fusidic acid (Agrawal et al. 1998, 1999b; Stark et al. 2000) was shown to be in a different conformation than in the X-ray structures, both in its nucleotide-free state (Aevarsson et al. 1994) and in complex with GDP (Czworkowski et al. 1994; al-Karadaghi et al. 1996), and the cryo-EM maps allowed a quantitative description of the movements of individual domains (Agrawal et al. 1998, 1999b; Wriggers et al. 2000). An 8 Å movement of the top portion of helix 44 of 16S rRNA within the smaller 30S subunit was detected by comparing cryo-EM maps of ribosome complexes in the pre- and post-translocational state (VanLoock et al. 2000). Another most recent example is the observation of a variation in the geometry of the entrance of the tunnel that conducts the polypeptide chain within the 50S subunit with different functional states (Gabashvili et al. 2001).

The idea about conformational changes of the ribosome is not new. Many biochemical and biophysical studies in the past decades have suggested that the translating ribosome undergoes reversible conformational transitions between the pre- and post-translocational states (see, e.g., Burma et al. 1986; Spirin et al. 1987; Möller 1990; Noller 1991; Wool et al. 1992; Nagel and Voigt 1993; Nierhaus et al. 1995; Agrawal and Burma 1996; Lodmell and Dahlberg 1997; see Agrawal et al. 1999a and references therein). These changes have been studied at both the global and local levels. Only now, with

the advent of 3D cryo-EM technique, have we begun to directly observe conformational changes in defined functional states of the ribosome (see Agrawal and Frank 1999) and are in a position to pinpoint some of these changes in greater detail.

Here we specifically focus on a large-scale conformational change of the ribosome that accompanies translocation and involves a relative movement of the two ribosomal subunits as a whole. After examining the rationale for this movement in the translocation process, we look at the wider context and pose a more general question: Could the change in the geometry of subunit association be related to the discrimination between one factor and the other during the course of the elongation cycle?

DOMAIN ARCHITECTURE OF THE RIBOSOME

The ribosome, as depicted by recent X-ray studies of individual ribosomal subunits (Ban et al. 2000; Schlünzen et al. 2000; Wimberly et al. 2000), is a highly complex structure with complex dynamic properties. Systems of this complexity have a large number of degrees of freedom, which prohibits the description of all modes by methods of molecular mechanics without some simplification. The most obvious simplification is to break the structure down into more or less solid "blocks." Such blocks were already identified by Spirin (1985) in a discussion of the possible dynamic behavior of the ribosome. In the case of the small subunit, it is known that the four major secondary-structure domains are reflected in autonomous architectural blocks: head, shoulder, platform, and body. In the isolated small subunit, these blocks appear in varying orientations (Lata et al. 1996; Agrawal et al. 1999b; Gabashvili et al. 1999). Upon association of the subunits, head and platform were seen to change their positions (Lata et al. 1996; Agrawal et al. 1999a; Gabashvili et al. 1999). In the 70S ribosome, the shoulder and head move to change the cleft they enclose into a channel (Lata et al. 1996; Agrawal et al. 1999a; Schlünzen et al. 2000) that conducts the mRNA into the ribosome (Frank et al. 1995, 2000). In contrast, the large subunit is formed as a tightly integrated meshwork of RNA and protein that behaves like one solid block (Ban et al. 2000). Still, three of the 50S subunit's peripheral features—the protrusions that determine the characteristic crown appearance—seem to have a life of their own: the L1 stalk (Agrawal et al. 1999a; Gomez-Lorenzo et al. 2000), the central protuberance, and the L7/L12-stalk region (Traut et al. 1995; Malhotra et al. 1998; Agrawal et al. 1999a).

The most important division of the ribosome into blocks is defined by the subunits themselves, however, which are connected by a number of bridges (Frank et al. 1995; Lata et al. 1996; Cate et al. 1999; Gabashvili et al. 2000; Yusupov et al. 2001). Early on, the two-subunits composition of the ribosome, a universal feature, gave rise to the speculation that the subunits must move relative to each other in the course of elongation (Bretscher 1968; Spirin 1968). A detailed explanation for the rationale underlying this hypothesis is from Spirin (1985):

> For what purpose has nature placed one part of each substrate (and product) molecule on one subunit and the other part on the other subunit? Why is the boundary between the large ribosomal blocks the place where the events of the elongation cycle, including translocation, are played out? Simple logic implies that the relative movement of the two subunits (and, perhaps, of the large blocks of each subunit) is somehow required for function.

Whereas Bretscher (1968) left open the question whether translational or rotational movement was employed, Spirin (1968) initially considered a model where the subunits are alternately locked and unlocked. Recent observations by cryo-EM have shown direct evidence of such global subunit movements accompanying translocation events. The two ribosomal subunits are seen to rotate relative to each other, as blocks, in response to EF-G binding in the GTP state and (according to a separate experiment) subsequent hydrolysis (Frank and Agrawal 2000). The movement is reminiscent of a ratchet, and therefore the term *ratchet movement* is used in this paper for brevity. In the following, we first summarize these observations, then attempt to bring the results into the larger context of ribosome–ligand binding.

THE RATCHET MOVEMENT

A major movement of the small subunit, relative to the large subunit, was observed when comparing the cryo-EM maps of three complexes (Agrawal et al. 1999b): (A) fMet-tRNA•70S (= control), (B) EF-G•GMPP(CH$_2$)P•70S, and (C) EF-G•GDP•fusidic acid•70S. The main component of the movement is a rotation around an axis that is perpendicular to the solvent view (Fig. 1). Manual superimposition (Frank and Agrawal 2000) gave the angles of ~6° for A versus B (i.e., the small subunit rotates counterclockwise) and ~3° clockwise for B versus C. This corresponds to the following sequence of events: (1) Upon EF-G binding in the GTP state, the small subunit rotates counterclockwise by ~6°; (2) upon GTP hydrolysis, and in the presence of fusidic acid, the small subunit rotates clockwise by ~3°. Without fusidic acid, the small subunit rotates all the way back into the control state, as can be inferred from a comparison of cryo-EM maps of ribosome in pre- and post-translocational states (Agrawal et al. 2000b). The presence of fusidic acid, which is known to block the release of EF-G in the GDP state from the ribosome (Willie et al. 1975), apparently locks the ribosome into an intermediate state that follows immediately after translocation.

A more detailed analysis, using computational fitting, was done by Spahn and coworkers as part of a binding study of Tet(O), a protein that has ~51% sequence similarity with EF-G and is found specifically in tetracycline-resistant bacteria. In this measurement, the movement between states A and B was characterized by two angles, corresponding to two successive rotations around two orthogonal axes: first –2.7° around the anticodon stem helix, then 4.3° around the acceptor stem helix of the P-site tRNA. Tet(O), having high sequence homology with EF-G, proved to bind to the ribosome in a very similar way

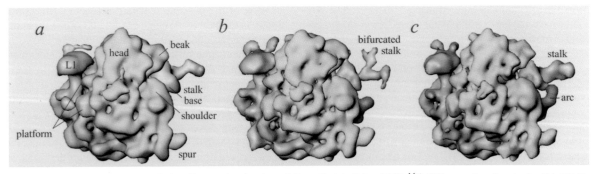

Figure 1. Ratchet movement of the ribosomal subunits of *E. coli.* (*a*) fMet-tRNA$_f^{Met}$•70S complex (control); (*b*) EF-G•GMPP(CH$_2$)P•70S complex (GTP state); (*c*) EF-G•GDP•fusidic acid•70S (GDP state). In *b*, the small subunit is seen to be rotated counterclockwise by ~6° relative to its position in *a*. In *c*, it is rotated by ~3° relative to *a*. 30S and 50S subunits are shown in *yellow* and *blue*, respectively. EF-G density is shown in *red*. A large conformational change in the L7/L12-stalk and stalk-base regions of the 50S subunit in different functional states is quite obvious. Arc in panel *c* points to a connection between the G'domain of EF-G and the amino-terminal domain of protein L11 (R.K. Agrawal et al., in prep.). Other landmarks of the two subunits are self-explanatory. (Reprinted, with permission, from Frank and Agrawal 2000 [copyright MacMillan].)

(Fig. 2), apparently utilizing some of the same binding sites. The analysis of the cryo-EM map of a complex of the 70S ribosome and Tet(O) revealed, however, that the subunits had the same relative position as in the control. Thus, Tet(O) binding *does not* induce the kinds of movements that are observed with EF-G. What might be significant in explaining this difference in ribosome response, in view of the following discussion of factor recognition, is that Tet(O) binds to the ribosome in the post-translocational state, where it competes with the binding of the ternary complex, not with the binding of EF-G. Indeed, an earlier check (I. Gabashvili, unpubl.) of the subunit constellation for the ribosome bound with the ternary complex (Agrawal et al. 2000a) had also failed to find evidence for a ratchet movement.

In a separate cryo-EM study (Stark et al. 2000), the ribosome was frozen in a conformational state by adding an antibiotic, thiostrepton. The mechanisms of action of thiostrepton and fusidic acid in inhibiting the translocation are completely different. Thiostrepton inhibits the reaction by binding to the GTPase-associated center (see Wimberly et al. 1999; Agrawal et al. 2001) of the ribosome, apparently preventing the proper interaction of this center with EF-G, whereas fusidic acid interacts with EF-G after it has bound to the ribosome and prevents its dissociation from the ribosome. In the study by Stark et al. (2000), EF-G was reported to be bound to the ribosome in a position completely different from the binding positions of EF-G both in the GDP and GTP forms described above. It should be noted, however, that in this study a large portion of the stalk-base region of the ribosome density was wrongly attributed to the density corresponding to domains I and II of EF-G. What is important here is that no evidence for a ratchet-like movement was found. Proper interaction of EF-G with the ribosome is probably required to induce this movement.

In our work, the large-scale movement of the subunits is accompanied by a reorganization of the inner frameworks of the two subunits. The most radical changes of architecture are seen in the 30S subunit: The head changes position, and the geometry of entrance and exit channels for mRNA varies—the openings become wide upon EF-G binding and narrow in the control (Frank and Agrawal 2000). This change in geometry is just as expected, since during translocation, the mRNA must be able to freely move, along with the base-paired anticodon ends of A- and P-site tRNAs, but on the other hand must be in a fixed position during decoding, for error-free codon–anticodon pairing. Post-translocational narrowing of the mRNA channel may help not only in fixation of the mRNA codon, but also in preventing the unnecessary exposure of additional nucleotides on the 3′ side of the mRNA codon at the ribosomal A site.

In contrast to the 30S subunit, the architecture of the 50S subunit appears to vary to a lesser extent, except for the obvious conformational changes in the L7/L12-stalk and stalk-base regions, central protuberance (CP), and the group of extended bridges B1a, B1b, and B1c that connect the CP to the head of the 30S subunit. The structural components associated with CP move in complicated ways to accommodate the movement of the small subunit.

Whereas most of the intersubunit bridges formed between the lower body regions of 30S and 50S subunits involve specific regions of 16S and 23S rRNAs, bridges between the head of the 30S subunit and CP of the 50S subunit have greater participation of ribosomal proteins. (It should be noted that in the cryo-EM reconstruction of the *Escherichia coli* 70S ribosome [Gabashvili et al. 2000], three distinct bridges between the 30S subunit head and the 50S subunit CP can be seen, whereas the X-ray structure of the *Thermus thermophilus* 70S ribosome [Yusupov et al. 2001] shows only two bridges, B1a and B1b. Comparison of the maps shows that the bridge identified as B1b in the X-ray study actually corresponds to bridge B1c identified in the cryo-EM work. In the following, we use the nomenclature of the cryo-EM work to refer to these bridges.) For example, bridge B1a is formed by protein S13 from the 30S and the tip of helix 38 of 23S rRNA from the 50S, whereas bridge B1c involves the amino-terminal domain of protein S13 and protein L5

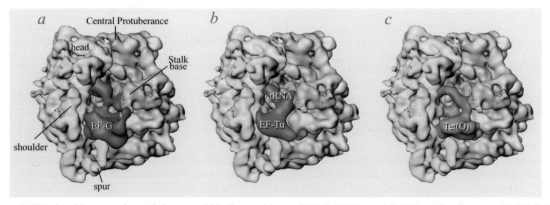

Figure 2. Side-by-side comparison of shapes and binding positions of EF-G, EF-Tu, and Tet(O) on the ribosome. (*a*) EF-G (*red*, adapted from Agrawal et al. 1999b); (*b*) aa-tRNA-EF-Tu-GDP-kirromycin complex (*orange*, adapted from Agrawal et al. 2000a); and (*c*) Tet(O) (*magenta*, adapted from Spahn et al. 2001). Experimental densities of all three factors are superimposed onto the 11.5 Å resolution cryo-EM map of the *E. coli* 70S ribosome (Gabashvili et al. 2000) with 30S (*yellow*) and 50S (*blue*) subunits identified.

from the 50S subunit (see, e.g., Yusupov et al. 2001). In the cryo-EM map, the additional bridge B1b, located between bridges B1a and B1c, apparently comprises rRNA on the 50S subunit side and protein S13 on the 30S side. Thus, protein S13 is the major player on the 30S subunit side. During the ratchet-like movement, the bridges B1a and B1c appear to be completely disrupted, while two components of the bridges B1b and B1c from the 50S side undergo a large structural reorganization and merge with the components of bridge B1b from the 30S side.

ROLE OF THE RATCHET MOVEMENT IN TRANSLOCATION

The cryo-EM study showed that the direction of the movement of the 30S subunit with respect to the 50S subunit, upon EF-G•GTP binding (Agrawal et al. 1999b; Frank and Agrawal 2000), is in the expected direction of the movement of tRNAs from A and P sites to P and E sites, respectively. This raised the following question: Does tRNA movement occur during the movement of the subunit or as a separate event? We have found that the largest movement of the 30S subunit takes place upon binding of EF-G in the GTP state, when a nonhydrolyzable GTP analog is used, and translocation of tRNA is apparently a separate event, associated with EF-G-dependent GTP hydrolysis. It should be noted that the 30S subunit head is most strongly affected by the movement. Since the anticodon ends of the tRNAs are tightly associated with the decoding region of the 30S subunit, and both the anticodon loops and arms make several intimate contacts with the various components of the 30S subunit head (Malhotra et al. 1998; Cate et al. 1999; Agrawal et al. 2000b; Yusupov et al. 2001), the ratchet-like movement may have the purpose of destabilizing the tRNA–ribosome interactions, in preparation for the actual translocation. Thus, destabilization of various tRNA–ribosome interactions could begin with the interaction of EF-G with the ribosome, but the actual translocation would be facilitated only by GTP hydrolysis.

The initial destabilization of the tRNA–ribosome interaction due to intersubunit reorganization might affect not only the contacts with the 30S subunit, but also those with the 50S subunit. Specifically, it might contribute to the partial puromycin reactivity, the biochemical measure of translocation, to the A-site tRNA even in the GTP state (see, e.g., Modolell et al. 1975). The puromycin reaction would require the CCA end of the A-site tRNA to move away from the A-site to the P-site region, or alternatively, would require a corresponding opposite movement of the peptidyl transferase center of the ribosome. In this context, it should be noted that a cryo-EM study (Agrawal et al. 2000b) has indicated a small movement (~6 Å) of the CCA end of the dipeptidyl-tRNA toward the P site upon peptide-bond formation. The movement of the CCA end of the tRNA was also indicated earlier by chemical protection experiments (Moazed and Noller 1989). The atomic structure of the isolated 50S subunit from an archaeon (Ban et al. 2000) shows that the peptidyl transferase center, where the CCA-end analogs of A- and P-site tRNAs bind (Nissen et al. 2000), is a compact structure, leaving little room for a structural reorganization in that region. However, the ratchet-type intersubunit movement would be able to account for a change in the lateral position of the CCA relative to the peptidyl transferase center of a magnitude similar to the change associated with the peptide-bond formation. This might explain the occurrence of puromycin reactions prior to complete tRNA translocation. Evidence of periodic conformational change in the 23S rRNA, near the peptidyl transferase center (Polacek et al. 2000), also indicates that the core of the 50S subunit moves during the translocation process.

It is important to note that ribosomes can translocate tRNAs even in the absence of EF-G (Pestka 1969; Gavrilova et al. 1976), indicating that the entire machinery required for tRNA translocation is present within the ribosome. However, the rate of translocation in the absence of EF-G is very low (see Rodnina et al. 1997). It is likely that EF-G promotes the intrinsic translocation ability of the ribosome by triggering an existing switch on the ribosome through its interaction with ribosomal binding sites. We have identified the binding sites of EF-G on the ribosome from an analysis of various EF-G•70S complexes (Table 1). It appears that helix 69 of the 23S rRNA, on which the anticodon arms of both A- and P-site tRNAs reside in the pre-translocational state (Agrawal et al. 2000b), as well as the anticodon arm of the P-site tRNA and the tip of domain IV of EF-G in the post-translocational state (Figs. 3 and 4), plays a pivotal role in translocation. Helix 69 is a flexible structure that is disordered in the X-ray map of the isolated 50S subunit (Ban et al. 2000) but gets stabilized in the 70S ribosome, as clearly seen in both cryo-EM (Gabashvili et al. 2000) and X-ray (Yusupov et al. 2001) maps. Helix 69 is also involved in the formation of one of the most crucial intersubunit bridges, B2a (see Gabashvili et al. 2000) with the 30S subunit, involving the tip of helix 44, the decoding site, of the 16S rRNA. We have previously shown that the tip of helix 44 moves in the expected direction of the tRNA movement, in response to the EF-G-dependent GTP hydrolysis (VanLoock et al. 2000). A similar movement of helix 44 was also observed in an X-ray study of initiation factor 1 (IF1) binding to the 30S subunit (Carter et al. 2001), indicating that helix 44 is involved in a variety of ribosome functions. It is likely that the movement of helix 44 and associated movement or conformational changes in helix 69 are directly involved in coordination of translocation between the two ribosomal subunits.

FACTOR BINDING AND RECOGNITION

A number of factors bind to the same region of the ribosome in a closely similar way. EF-G•GDP and ternary complex aa-tRNA•EF-Tu•GTP, known to be molecular mimics (Nissen et al. 1995), were found to interact with a number of binding sites scattered over some regions of the 30S and 50S subunits (see, e.g., Moazed et al. 1988; Stark et al. 1997; Agrawal et al. 1998, 1999b, 2000a; Wilson and Noller 1998.) Analysis of the cryo-EM maps suggested that the binding sites of the two factors are in close mutual proximities (Agrawal et al. 2000a,b). The most detailed data are available on the interaction between

Table 1. List of Molecular Contacts of EF-G on 30S and 50S Subunits,[a] and Classification of Function into "Anchor" and "Work"

EF-G domains	Ribosomal subunit	Subunit component[b]	Function
I (G)	50S	H95 (α-sarcin/ricin loop),[c]	anchor
		tip of H91	anchor
I (G′)	50S	arc formation with amino-terminal domain of protein L11	work[d]
II	30S	h4, h5, h15;[e]	anchor
		base of h17 (GTP state)	work
		base of h3, h4 &h18 (GDP state)[e]	work
III	30S	protein S12	anchor/work
IV	30S	A-site tRNA,[f,g,h] tip of h44[i,j]	work
		h30	work
	50S	tip of H69	work
V	50S	tip of H43[d] and H89	work[d]

[a]Contact sites were inferred by docking the X-ray structures of 30S (Wimberly et al. 2000) and 50S (Ban et al. 2000) subunits into the densities corresponding to 30S and 50S subunits, respectively, of the cryo-EM maps of 70S-EF-G complexes (Agrawal et al. 1999b).

[b]h and H indicate helices of the 16S and 23S rRNAs, respectively.

References: [c]Wriggers et al. 2000; [d]Agrawal et al. (2001); [e]Wilson and Noller (1998); [f]Agrawal et al. (1998); [g]Agrawal et al. (1999b); [h]Agrawal et al. (2000a); [i]Frank and Agrawal (2000); [j]VanLoock et al. (2000).

EF-G and the 70S ribosome of *E. coli* (Table 1, Fig. 4). Other large molecules that interact with the ribosome, such as release factor RF3 (see, e.g., Nakamura et al. 2000) and the protection protein Tet(O) (in certain tetracycline-resistant bacteria; see Spahn et al. 2001), appear to use at least a subset of the same binding sites. Tet(M), a close homolog of Tet(O), is likely part of the growing family of ribosome-binding G proteins. In eukaryotes, the proteins EF2, EF1α, and eRF3 play a similar role in translocation, decoding, and release, and cryo-EM has already confirmed that EF2 is in a very similar binding position as EF-G (Gomez-Lorenzo et al. 2000). To date, no X-ray structure of any elongation factor–ribosome complex has been solved, but cryo-EM maps of functional complexes, when combined with docking of X-ray structures of ribosomal subunits and elongation factors, have

provided a means to study locations and dynamics of ribosomal binding sites (see, e.g., Agrawal et al. 2001).

One unresolved question in understanding the molecular events of translation is how the factors are selected in the course of elongation. Decoding and translocation are events that must strictly alternate (Fig. 5). If molecular mimicry would imply that both elongation factors could compete for the same binding sites, then the temporal sequence of events, requiring alternate binding of the two factors, could not be easily explained. Some of these questions were poignantly raised by Peter Moore (Moore 1995) at the time the molecular similarity was discovered.

The key to the selection of factors may lie in the fact that they interact with both subunits, and thus, that the condition for simultaneous multiple contacts is dependent on the precise constellation of 30S and 50S binding sites

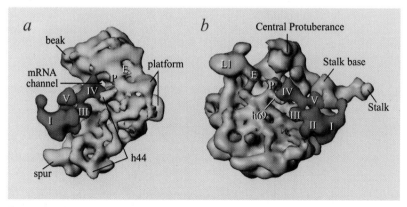

Figure 3. Positions of EF-G domains in the EF-G•GDP•fusidic acid•70S•(tRNA)₂ complex (Agrawal et al. 1999b) relative to the two ribosomal subunits. (*a*) 30S and (*b*) 50S subunits are shown from the subunit interface side. The subunit maps were created by computationally separating the cryo-EM map. Densities corresponding to P- and E-site tRNAs (marked as P and E, respectively) are divided between 30S and 50S subunits. EF-G domains are identified in Roman numbers based on the X-ray structure (see Fig. 4) (Czworkowski et al. 1994). Tip of domain IV and a T-loop side of the anticodon arm of the P-site tRNA directly interact with helix 69.

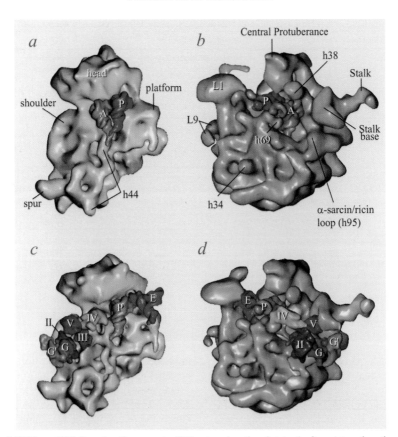

Figure 4. Locations of tRNAs and EF-G on the ribosome. (*a, b*) Pre-translocational state; (*c,d*) post-translocational state. Atomic models of tRNA and EF-G were fitted into the corresponding cryo-EM densities and filtered to the resolution of cryo-EM maps. The tRNA positions are adapted from Agrawal et al. (2000b), and the EF-G position in the GDP state from Agrawal et al. (1999b). A- and P-site tRNAs (*pink* and *green*, respectively) in panels *a* and *b* have moved to the P- and E-sites, respectively, in panels *c* and *d*, after EF-G-dependent translocation. Color codes of EF-G domains: (*magenta*) G domain; (*brown*) G´ domain; (*blue*) domain II; (*forest green*) domain III; (*yellow*) domain IV; and (*red*) domain V. Other landmarks: (L1) L1-protein region; (L9) protein L9; (h44) 16S rRNA helix number 44; (h34, h38, h69 and h95), various 23S rRNA helices.

(Fig. 5) (see Agrawal et al. 2000a,b). The fact that this constellation changes in a reorientation of the subunits makes it possible that distinctly different matches can be made at different time points along the elongation cycle. Because of the large size of the ribosome, even a small reorientation—much smaller than the one observed in response to EF-G binding in the GTP state—will change the geometry of binding interactions significantly.

More specifically, the observed ratchet movement of the two subunits implies that a given set of m binding sites on the small subunit $\{s_1, s_2, ... , s_m\}$ and n binding sites on the large subunit $\{l_1, l_2, ..., l_n\}$ will in the course of elongation change to form different constellations $\{R_i \{s_1\}, R_i \{s_2\}, ... R_i \{s_m\}; l_1, l_2, ...l_n\}$. Here R_i denotes a series of different coordinate transformations characterizing the positions of the small subunit relative to its position in the control. (The use of a general formulation with m and n binding sites is prudent at this point, since cryo-EM maps do not resolve the broader contacts.) Considering the highly coupled nature of the mechanical framework making up the ribosome, it is conceivable that this series of transformations can be parameterized with a single parameter. (This could be an angle indicating the relative rotation around an axis joining the subunits, but it could also

be a parameter of a more complex movement entailing both rotation and translation.) In that case, the possible constellations have the important property that they are ordered along a single one-dimensional continuum.

A quick look at the elongation factors indicates that the hypothesis requires a more careful formulation: Clearly, some of the molecular contacts between the factors and the ribosome are not made until after the factor is bound—they are not to be counted among the sites making structural connections. Therefore, it will be useful to distinguish between "anchor sites" and "work binding sites" (see Table 1).

For factor binding to occur, the constellation of anchor sites on the factor must correspond to the constellation of anchor sites on the ribosome. This binding condition will be fulfilled for one unique ribosome conformation (Fig. 6). In contrast to the anchor sites, the constellation of the work binding sites will not be decisive for the binding. Subsequently, as a result of the anchor binding interactions, the ribosome or the factor or both may change conformation. It is during this phase where the work binding sites will engage. Eventually, these changes will result in the destabilization of the binding interactions and a release of the factor. In the case of elongation factor G, a

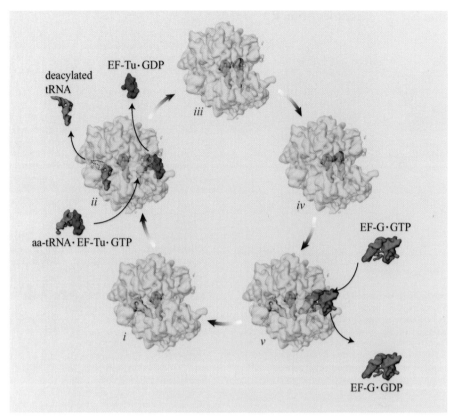

Figure 5. Positions of tRNAs and elongation factors during the elongation cycle, as inferred by cryo-EM of functionally defined states (for details, see Agrawal et al. 2000b). 30S and 50S subunits are shown in transparent yellow and blue colors, respectively, from the top view; i.e., with the head of the 30S subunit and the central protuberance of the 50S subunit facing the viewer. (*i*) Post-translocational state with P-(*green*) and E-(*solid yellow*) site tRNAs; (*ii*) an intermediate state, in which the E-site tRNA has moved to a new position (shown in *brown*, orientation not defined; for details, see Agrawal et al. 2000b) and a new aa-tRNA-EF-Tu-GTP ternary complex binds to the ribosome; (*iii*) pre-translocational state with A-(*pink*) and P-(*green*) site tRNAs; (*iv*) an intermediate, peptide-bond formation state, in which the CCA end of the A-site tRNA (*purple*, designated as the A_{pep} state of the tRNA; for details, see Agrawal et al. 2000b) has moved toward the CCA end of the P-site tRNA (*green*); and (*v*) interaction of EF-G (*solid blue*)–GTP with the ribosome to promote the translocation of A_{pep} state and P-site tRNAs (*purple* and *green* from state *iv*) to P- and E-sites (*green* and *solid yellow*, respectively). (Adapted, with permission, from Agrawal et al. 2000b [copyright Rockefeller University Press].)

tabulation (Table 1) shows that there are at least two anchor sites on each of the two subunits. In addition, there is one work binding site on the 30S subunit, and two such sites on the 50S subunit. These sites and their approximate locations have been derived by analysis of cryo-EM maps (Agrawal et al. 1999b).

Provided the general scheme underlying ribosome–factor interaction proves correct, what would be its rationale? One aspect is that the ribosome would be able to change its chemical appearance in a very economic way. Instead of providing multiple sets of binding sites, a subset of which would have to be activated at a given time to be receptive for one of the factors, the ribosome will present a single set of sites in different constellations. Another aspect is that, once recognized and bound, the factor will act as a brace, stabilizing the ribosome in a particular conformation that is required for the factor's function.

What remains to be seen is whether all movements of subunit domains containing anchor sites are accompanied by an intersubunit rotation. In a cryo-EM study of pre- and post-translocational complexes (Agrawal et al. 2000b), we found that the shoulder region of the 30S subunit shifts toward the stalk-base region of the 50S subunit, apparently narrowing the gap between the two subunits on the factor entry side, although we found no clear indication for an associated ratchet-like intersubunit movement. Nevertheless, this result proves that EF-G and EF-Tu recognize completely different structural constellations. Mesters et al. (1994) observed a significant positive cooperativity between EF-Tu- and EF-G-dependent GTPase activities of the ribosome. Their study also indicated that EF-G and EF-Tu recognize different conformers of the ribosome and that interaction with one elongation factor stimulates a conformation that is favorable for the binding of the other elongation factor and vice versa.

CONCLUSIONS

Despite the progress in the elucidation of ribosomal structure, and numerous observations facilitated by cryo-EM, of binding interactions and conformational changes of the factors, and associated conformational changes of the ribosome, the sequence and causation of events in the translation process are still a mystery. To get a step closer

Conformation R₁ Conformation R₂

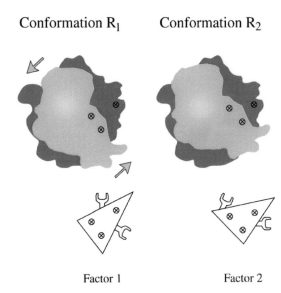

Factor 1   Factor 2

Figure 6. Diagrammatic presentation of the proposed link between ratchet movement and factor recognition. A factor will bind only if its constellation of anchor sites matches the ratchet-dependent constellation of ribosomal anchor sites. Shown are two constellations of ribosomal anchor sites, obtained as a result of different subunit rotations, and two different factors that have matching anchor sites. In addition to the anchor sites (marked by a cross in a circle), some work binding sites are also shown (marked by a spanner).

to the solution, it is necessary to obtain more detailed information, at higher resolution, on the dynamics of the process. Trapping of transitional states by physical means (spray-freezing or use of caged compounds) in conjunction with 3D cryo-EM may yield such information once these methods have been worked out (see Frank 2001).

ACKNOWLEDGMENTS

This work was supported by grants from the National Institutes of Health (R37 GM-29169, R01 GM-55440, and P41 RR0121919 to J.F., and R01 GM-61576 to R.K.A.). We thank Jens Nyborg for discussions and Yu Chen and Michael Watters for assistance with the preparation of the illustrations.

REFERENCES

Aevarsson A., Brazhnikov E., Garber M., Zheltonosova J., Chirgadze Y., al-Karadaghi S., Svensson L.A., and Liljas A. 1994. Three-dimensional structure of the ribosomal translocase: Elongation factor G from *Thermus thermophilus*. *EMBO J.* **13:** 3669.

Agrawal R.K. and Burma D.P. 1996. Sites of ribosomal RNAs involved in the subunit association of tight and loose couple ribosomes. *J. Biol. Chem.* **271:** 21285.

Agrawal R.K. and Frank J. 1999. Structural studies of the translational apparatus. *Curr. Opin. Struct. Biol.* **9:** 215.

Agrawal R.K., Heagle A.B., and Frank J. 2000a. Studies of elongation factor G-dependent tRNA translocation by three-dimensional cryo-electron microscopy. In *The ribosome: Structure, function, antibiotics, and cellular interactions* (ed. R.A. Garrett et al.), p. 53. ASM Press, Washington, D.C.

Agrawal R.K., Lata K.R., and Frank J. 1999a. Conformational variability in *E. coli* 70S ribosome as revealed by 3D cryo-electron microscopy. *Int. J. Biochem. Cell Biol.* **31:** 243.

Agrawal R.K., Penczek P., Grassucci R.A., and Frank J. 1998. Visualization of elongation factor G on the *Escherichia coli* 70S ribosome: The mechanism of translocation. *Proc. Natl. Acad. Sci.* **95:** 6134.

Agrawal R.K., Heagle A.B., Penczek P., Grassucci R.A., and Frank J. 1999b. EF-G-dependent GTP hydrolysis induces translocation accompanied by large conformational changes in the 70S ribosome. *Nat. Struct. Biol.* **6:** 643.

Agrawal R.K., Linde J., Sengupta J., Nierhaus K.H., and Frank J. 2001. Localization of L11 protein on the ribosome and elucidation of its involvement in EF-G-dependent translocation. *J. Mol. Biol.* **311:** 777.

Agrawal R.K., Spahn C.M.T., Penczek P., Grassucci R.A., Nierhaus K.H., and Frank J. 2000b. Visualization of tRNA movements on the *Escherichia coli* 70S ribosome during the elongation cycle. *J. Cell Biol.* **150:** 447.

al-Karadaghi S., Aevarsson A., Garber M., Zheltonosova J., and Liljas A. 1996. The structure of elongation factor G in complex with GDP: Conformational flexibility and nucleotide exchange. *Structure* **4:** 555.

Ban N., Nissen P., Hansen J., Moore P.B., and Steitz T.A. 2000. The complete atomic structure of the large ribosomal subunit at 2.4 Å resolution. *Science* **289:** 905.

Bretscher M.S. 1968. Translocation in protein synthesis: A hybrid structure model. *Nature* **218:** 675.

Burma D.P., Srivastava S., Srivastava A.K., Mahanti S., and Dash D. 1986. Conformational change of 50S ribosomes during protein synthesis. In *Structure, function and genetics of ribosomes* (ed. B. Hardesty and G. Kramer), p. 438. Springer-Verlag, New York.

Carter A.P., Clemons W.M., Jr., Brodersen D.E., Morgan-Warren R.J., Hartsch T., Wimberly B.T., and Ramakrishnan V. 2001. Crystal structure of an initiation factor bound to the 30S ribosomal subunit. *Science* **291:** 498.

Cate J.H., Yusupov M.M., Yusupova G.Z., Earnest T.N., and Noller H.F. 1999. X-ray crystal structures of 70S ribosome functional complexes. *Science* **285:** 2095.

Czworkowski J., Wang J., Steitz T.A., and Moore P.B. 1994. The crystal structure of elongation factor G complexed with GDP, at 2.7 Å resolution. *EMBO J.* **13:** 3661.

Frank J. 2001. Ribosomal dynamics explored by cryo-electron microscopy. *Methods* (in press).

Frank J. and Agrawal R.K. 2000. A ratchet-like inter-subunit reorganization of the ribosome during translocation. *Nature* **406:** 318.

Frank J., Penczek P., Grassucci R.A., Heagle A., Spahn C.M.T., and Agrawal R.K. 2000. Cryo-electron microscopy of the translational apparatus: Experimental evidence for the paths of mRNA, tRNA, and the polypeptide chain. In *The ribosome: Structure, function, antibiotics, and cellular interactions* (ed. R.A. Garrett et al.), p. 45. ASM Press, Washington, D.C.

Frank J., Zhu J., Penczek P., Li Y., Srivastava S., Verschoor A., Radermacher M., Grassucci R., Lata R.K., and Agrawal R.K. 1995. A model of protein synthesis based on cryo-electron microscopy of the *E. coli* ribosome. *Nature* **376:** 441.

Gabashvili I.S., Agrawal R.K., Grassucci R.A., and Frank J. 1999. Structure and structural variations of the *Escherichia coli* 30S ribosomal subunit as revealed by three-dimensional cryo-electron microscopy. *J. Mol. Biol.* **286:** 1285.

Gabashvili I.S., Agrawal R.K., Spahn C.M.T., Grassucci R.A., Frank J., and Penczek P. 2000. Solution structure of the *E.coli* 70S ribosome at 11.5 Å resolution. *Cell* **100:** 537.

Gabashvili I.S., Gregory S.T., Valle M., Grassucci R., Worbs M., Wahl M.C., Dahlberg, A.E., and Frank J. 2001. The polypeptide tunnel system in the ribosome and its gating in erythromycin resistant mutants of L4 and L22. *Mol. Cell* **8:** 181.

Gavrilova L.P., Kostiashkina O.E., Koteliansky V.E., Rutkevitch N.M., and Spirin A.S. 1976. Factor-free ("non-enzymic") and factor-dependent systems of translation of polypuridylic acid by *Escherichia coli* ribosomes. *J. Mol. Biol.* **101:** 537.

Gomez-Lorenzo M.G., Spahn C.M.T., Agrawal R.K., Grassucci R.A., Penczek P., Chakraburtty K., Ballesta J.P.G., Lavandera J.L., Garcia-Bustos J.F., and Frank J. 2000. Three-dimensional cryo-electron microscopy localization of EF2 in the *Saccharomyces cerevisiae* 80S ribosome at 17.5 Å resolution. *EMBO J.* **19:** 2710.

Lata K.R., Agrawal R.K., Penczek P., Grassucci R., Zhu J., and Frank J. 1996. Three-dimensional reconstruction of the *Escherichia coli* 30 S ribosomal subunit in ice. *J. Mol. Biol.* **262:** 43.

Lodmell J.S. and Dahlberg A.E. 1997. A conformational switch in *Escherichia coli* 16S ribosomal RNA during decoding of messenger RNA. *Science* **277:** 1262.

Malhotra A., Penczek P., Agrawal R.K., Gabashvili I.S., Grassucci R.A., Junemann R., Burkhardt N., Nierhaus K.H., and Frank J. 1998. *Escherichia coli* 70 S ribosome at 15 Å resolution by cryo-electron microscopy: Localization of fMet-tRNA$_f^{Met}$ and fitting of L1 protein. *J. Mol. Biol.* **280:** 103.

Mesters J.R., Potapov A.P., de Graaf J.M., and Kraal B. 1994. Synergism between the GTPase activities of EF-Tu.GTP and EF-G.GTP on empty ribosomes: Elongation factors as stimulators of the ribosomal oscillation between two conformations. *J. Mol. Biol.* **242:** 644.

Moazed D. and Noller H.F. 1989. Intermediate states in the movement of transfer RNA in the ribosome. *Nature* **342:** 142.

Moazed D., Robertson J.M., and Noller H.F. 1988. Interaction of elongation factors EF-G and EF-Tu with a conserved loop in 23S RNA. *Nature* **334:** 362.

Modolell J., Girbes T., and Vazquez D. 1975. Ribosomal translocation promoted by guanylylimido diphosphate and guanylylmethylene diphosphonate. *FEBS Lett.* **60:** 109.

Moller W. 1990. Hypothesis: Ribosomal protein L12 drives rotational movement of tRNA. In *The ribosome: Structure, function, and evolution* (ed. W.E. Hill et al.), p. 380. American Society for Microbiology, Washington, D.C.

Moore P.B. 1995. Molecular mimicry in protein synthesis? *Science* **270:** 1453.

Nagel K. and Voigt J. 1993. Regulation of the uncoupled GTPase activity of elongation factor G (EF-G) by the conformations of the ribosomal subunits. *Biochim. Biophys. Acta* **1174:** 153.

Nakamura Y., Ito K., and Ehrenberg M. 2000. Mimicry grasps reality in translation termination. *Cell* **101:** 349.

Nierhaus K.H., Beyer D., Dabrowski M., Schafer M.A., Spahn C.M., Wadzack J., Bittner J.U., Burkhardt N., Diedrich G., and Junemann R. 1995. The elongating ribosome: Structural and functional aspects. *Biochem. Cell Biol.* **73:** 1011.

Nissen P., Hansen J., Ban N., Moore P.B., and Steitz T.A. 2000. The structural basis of ribosome activity in peptide bond synthesis. *Science* **289:** 920.

Nissen P., Kjeldgaard M., Thirup S., Polekhina G., Reshetnikova L., Clark B.F.C., and Nyborg J. 1995. Crystal structure of the ternary complex of Phe-tRNA ^{Phe}, EF-TU, and a GTP analog. *Science* **270:** 1464.

Noller H.F. 1991. Ribosomal RNA and translation. *Annu. Rev. Biochem.* **60:** 191.

Pestka S. 1969. Studies on the formation of transfer ribonucleic acid-ribosome complexes. VI. Oligopeptide synthesis and translocation on ribosome in the presence and absence of transfer factors. *J. Biol. Chem.* **244:** 1533.

Polacek N., Patzke S., Nierhaus K.H., and Barta A. 2000. Periodic conformational changes in rRNA: Monitoring the dynamics of translating ribosomes. *Mol. Cell* **6:** 159.

Rodnina M.A., Savelsbergh A., Katunin V.I., and Wintermeyer W. 1997. Hydrolysis of GTP by elongation factor G drives tRNA movement on the ribosome. *Nature* **385:** 37.

Schlünzen F., Tocilj A., Zarivach R., Harms J., Glühmann M., Janell D., Bashan A., Bartels H., Agmon I., Franceschi F., and Yonath A. 2000. Structure of functionally activated small ribosomal subunit at 3.3 Å resolution. *Cell* **102:** 615.

Spahn C.M. and Prescott C.D. 1996. Throwing a spanner in the works: Antibiotics and the translation apparatus. *J. Mol. Med.* **74:** 423.

Spahn C.M.T., Blaha G., Agrawal R.K., Penczek P., Grassucci R.A., Trieber C.A., Connell S.R., Taylor D.E., Nierhaus K.H., and Frank J. 2001. Localization of the tetrecycline resistance protein Tet(O) on the ribosome and the inhibition mechanism of tetracycline. *Mol. Cell* **7:** 1037.

Spirin A.S. 1968. On the mechanism of ribosome function. The hypothesis of locking-unlocking of subparticles. *Dokl. Akad. Nauk. USSR* **179:** 1467.

———. 1985. Ribosomal translocation: Facts and models. *Prog. Nucleic Acid Res. Mol. Biol.* **32:** 75.

Spirin A.S., Baranov V.I., Polubesov G.S., Serdyuk I.N., and May R.P. 1987. Translocation makes the ribosome less compact. *J. Mol. Biol.* **194:** 119.

Stark H., Rodnina M.V., Wieden H.J., van Heel M., and Wintermeyer W. 2000. Large-scale movement of elongation factor G and extensive conformational change of the ribosome during translocation. *Cell* **100:** 301.

Stark H., Rodnina M.V., Rinke-Appel J., Brimacombe R., Wintermeyer W., and van Heel M. 1997. Visualization of elongation factor Tu on the *Escherichia coli* ribosome. *Nature* **389:** 403.

Traut R.R., Dey D., Bochkarlov D.E., Oleinikov A.V., Jokhadze G.G., Hamman B., and Jameson D. 1995. Location and domain structure of *Escherichia coli* ribosomal protein L7/L12: Site-specific cysteine crosslinking and attachment of fluorescent probes. *Biochim. Biophys. Acta* **73:** 949.

VanLoock M.S., Agrawal R.K., Gabashvili I.S., Qi L., Frank J., and Harvey S.C. 2000. Movement of the decoding region of the 16S ribosomal RNA accompanies tRNA translocation. *J. Mol. Biol.* **304:** 507.

Willie G.R., Richman N., Godtfredson W.O., and Bodley J.W. 1975. Some characteristics of and structural requirements for the interaction of 24, 25-dihydrofusidic acid with ribosome elongation factor G complexes. *Biochemistry* **14:** 1713.

Wilson K.W. and Noller H.F. 1998. Mapping the position of EF-G in the ribosome by directed hydroxyl radical probing. *Cell* **92:** 131.

Wimberly B.T., Guymon R., McCutcheon J.P., White S.W., and Ramakrishnan V. 1999. A detailed view of a ribosomal active site: The structure of the L11-RNA complex. *Cell* **97:** 423.

Wimberly B.T., Brodersen D.E., Clemons W.M., Jr., Morgan-Warren R.J., Carter A.P., von Rhein C., Hartsch T., and Ramakrishnan V. 2000. Structure of the 30S ribosomal subunit. *Nature* **407:** 327.

Wool I.G., Gluck A., and Endo Y. 1992. Ribotoxin recognition of ribosomal RNA and proposal for the mechanism of translocation. *Trends Biochem. Sci.* **17:** 266.

Wriggers W., Agrawal R.K., Drew D.L., McCammon A., and Frank J. 2000. Domain motions of EF-G bound to the 70S ribosome: Insights from a hand-shaking between multi-resolution structures. *Biophys. J.* **79:** 1670.

Yusupov M.M., Yusupova G.Z., Baucom A., Lieberman K., Earnest T.N., Cate J.N., and Noller H.F. 2001. Crystal structure of the ribosome at 5.5 Å resolution. *Science* **10:** 1126.

Do Single (Ribosome) Molecules Phase Themselves?

M. VAN HEEL

*Imperial College of Science, Technology and Medicine, Department of Biological Sciences,
Biochemistry Building, London, SW7 2AY, United Kingdom*

Whole new avenues of research have been opened by the flexibility and speed of electron microscopy of single molecules embedded in vitreous ice (single-particle cryo-EM). Different functional states of the ribosome can now readily be imaged in three dimensions at resolutions of 10–20 Å, as was first demonstrated by the visualization of elongation factor Tu on the *Escherichia coli* ribosome (Stark et al. 1997a). Single-particle cryo-EM already regularly yields three-dimensional structures at sub-nanometer resolution such as the 7.5 Å map of the 50S ribosomal subunit of *E. coli* (Matadeen et al. 1999). However, the recent atomic-resolution X-ray structure of whole ribosomal subunits shows that more can always be learned from a structure at atomic resolution than at the intermediate resolution level of 5–20 Å. There is also more to be learned from the X-ray crystallographic methodologies for improving single-particle cryo-EM techniques. The standard X-ray crystallographic solvent-flattening approach, when applied to noncrystalline objects in cryo-EM, may prove so powerful that it could help extend the realm of single-particle analyses to atomic resolution.

Single-particle cryo-EM is coming of age. At the 5–10 Å resolution levels one can now achieve α helices and, in favorable cases, even β sheets can be visualized in large macromolecular complexes. The current situation in cryo-EM of large biomolecular assemblies is reminiscent of the early days of protein crystallography when the α helices of myoglobin and hemoglobin were first seen. Despite the high inherent instrumental resolution of the electron microscope, the harsh high-vacuum environment and the ionizing electron radiation have hampered the elucidation of biological structures since the invention of the instrument. In the early 1980s, Jacques Dubochet and his group at the EMBL in Heidelberg brought the cryo-EM specimen preparation technique to maturity (Dubochet et al. 1988). A thin layer of solution (say, spanning a small hole in a carbon foil) is rapidly cooled to liquid nitrogen temperatures by freeze-plunging into cooled liquid ethane/propane, causing the solution to solidify into a vitreous phase of water which, for all practical considerations, behaves just like water. This technique brought us the first well-preserved images of macromolecular assemblies (Adrian et al. 1984).

For calculating three-dimensional (3D) structures from such images, good algorithms are required, in combination with powerful computers. The development of computational tools for processing electron microscopic images started mainly in the Cambridge research group around Aaron Klug some 30 years ago (DeRosier and Klug 1968; Crowther 1971). The techniques allowed us to calculate the structures of icosahedral viruses to resolution levels of 7–9 Å (Böttcher et al. 1997; Conway et al. 1997), and more recently to 5.9 Å (van Heel et al. 2000) or even 4.5 Å (A. Patwardhan et al., in prep.). The highest resolution hitherto achieved for an entirely symmetrical structure is the above-mentioned 7.5 Å for the 50S large ribosomal subunit of *E. coli* (Matadeen et al. 1999). When compared to X-ray crystallography, an unsurpassed strength of the single-particle cryo-EM approach is its ability to image different functional states of biological complexes such as the ribosome (Stark et al. 1997a,b; Agrawal et al. 1998). X-ray crystallography, in turn, excels in yielding atomic resolution structures (of complexes that form stable crystals) such as the eye-catching recent structures of ribosomal subunits (Ban et al. 2000; Schluenzen et al. 2000; Wimberly et al. 2000). Electron microscopy has yielded atomic-resolution structures based on images of 2D crystals (Henderson et al. 1990). However, only a handful of structures have so far been solved by electron crystallography. It would clearly be an ideal situation if one were able to produce atomic-resolution structures based on single-particle images alone without the limitation associated with using two- or three-dimensional crystals.

WHAT LIMITS SINGLE-PARTICLE ANALYSIS TODAY?

Modern electron microscopes operating with a field-emission gun at an acceleration voltage of 200–300 kV and equipped with a stage that can hold the sample at liquid nitrogen, or even at liquid helium, temperatures produce images of stable inorganic samples with information beyond 2 Å (Zemlin et al. 1996). Such resolution levels are more than sufficient to elucidate atomic-resolution biological structures. It is thus not the microscope that limits us in the achievable resolution. Biological molecules start disintegrating under the electron beam long before enough electrons are collected to create a high-resolution, low-noise image. Cooling the specimens helps to keep the biological molecules intact during the data collection. However, the tolerable exposure levels for biological macromolecules remain—by some three orders of magnitude—lower than what one would need to directly see ~3 Å information in the images. Achieving high-resolution structures of biological materials is thus a battle against noise and requires the averaging of large numbers of extremely noisy molecular images (cf. van Heel et al. 2000). We cannot afford to even look at the molecules before collecting the images, and thus, we cannot focus the images precisely. Although it is possible to correct the effects of different defoci (contrast transfer function [CTF] correction) to a certain extent, the residual errors in the defocus determination cause the high-resolution phases in the molecular images to be unreliable. Upon averaging images taken at slightly different defoci (unavoidably), the high-resolution information will tend to be averaged out. One of the main problems in single-particle cryo-EM is thus to average the high-frequency information coherently. It is already not simple to precisely determine the defocus parameters for each area of the micrograph. Moreover, focus differences exist between the bottom and the top of the particle and of the ice layer, making some intrinsic defocus variations within the data. The net effect of these uncertainties is a smearing out of the high-frequency image details ("defocus envelope"; Van Heel et al. 2000). The focus differences within a particle are also known as the "curvature of the Ewald sphere" in Fourier space (DeRosier 2000).

One key technique that allows us to continuously improve the resolution of the results is multivariate statistical analysis (MSA) data compression and classification (van Heel and Frank 1981; van Heel 1984, 1989). Whatever causes statistically significant differences within the data set can be, in principle, pinpointed by the MSA techniques. Despite the extreme noise levels, these techniques allow us to discriminate between molecular images in slightly different orientations and/or at slightly different defocus conditions. One can normally only assign Euler orientations (by "angular reconstruction"; van Heel 1987; van Heel et al. 2000) after noise reduction is achieved by averaging the members of the various MSA classes. While chiseling away the final Angstroms in our quest toward structures interpretable in terms of their atomic coordinates (better than ~3.5 Å), it is time to take a step back and look at the data refinement techniques in routine use in X-ray crystallography to see what good they could bring in single-particle cryo-EM.

WHAT CAN CRYO-EM LEARN FROM X-RAY CRYSTALLOGRAPHY?

In X-ray crystallography, the direct observables are the diffraction amplitudes of the crystals. The relative phases of the diffraction peaks are lost upon data registration ("phase problem"), and the phases must subsequently be retrieved by various techniques like the classic use of heavy-metal derivatives. Once the phases in X-ray crystallography are found to a sufficiently high level of resolution, "density modification" tricks such as noncrystallographic averaging and solvent flattening can be applied to improve the maps (Wang 1985; Abrahams and Leslie 1996). Averaging of subunits related by noncrystallographic symmetry (due to pointgroup symmetry axes not all coinciding with crystallographic axes) is a very powerful technique that helps boost the resolution and the consistency of the electron-density maps. The approach has, for example, been instrumental for solving the X-ray structures of icosahedral viruses.

In electron microscopy, the observables are projection *images* of the macromolecules, which include both amplitude and phase information. Thus, the X-ray crystallographic phase problem is nonexistent in electron microscopy. The price one pays in electron microscopy is that the very noisy molecular images need to first be aligned with respect to each other. That is to say, the molecular images in the data set need to be given a common phase origin in order to "build a crystal in the computer." In X-ray crystallography, the actual averaging over all molecules takes place in the (analog) diffraction experiment. If there are "alignment" problems between the molecules in the crystal, the effect will be that the crystal will not diffract (or not diffract to high resolution), and one then just speaks of a "bad" crystal. For the phases to be consistent within the single-particle data set, not only do the images need to be aligned with respect to a common 3D origin, but we also need to eliminate the frequency-dependent phase differences within the data set that are due to the defocus (CTF) variations among the molecular images. Such phase consistency problems do not have a direct equivalent in X-ray crystallography, where the phases are not measured directly in the first place.

Where can we still improve on single-particle cryo-EM using the techniques first developed in X-ray crystallography? The angular reconstitution approach (van Heel 1987) already fully exploits the pointgroup symmetry of the assembly: The higher the degree of symmetry in an oligomeric assembly, the easier that data processing becomes (cf. Orlova et al. 1997). For example, each individual image of an icosahedral virus contains 60 copies of the asymmetric unit of the highly regular capsid. With such high degrees of symmetry, it is much easier to find the orientation of the particle with respect to the common 3D phase origin of the data set. In X-ray crystallography it is often not possible to directly exploit the full pointgroup symmetry of the oligomer during the crystallographic analysis per se because not all the pointgroup symmetry axes necessarily coincide with the crystal axes. What one can do, however, after having found a preliminary solution for the phases of crystal, is to indeed average the real-space electron densities over the independent copies of the monomers in the crystal's unit cell. In single-particle cryo-EM, this type of averaging over the asymmetric units of the homo-oligomer is intrinsically built into the current processing and leaves no space for improvements. That situation could be quite different for solvent flattening.

SOLVENT FLATTENING

In X-ray crystallographic solvent flattening (Wang 1985; Abrahams and Leslie 1996), one exploits the a priori knowledge that, outside the space occupied by the molecule, there should only be the unstructured density of the solvent within the unit cell. Once a preliminary so-

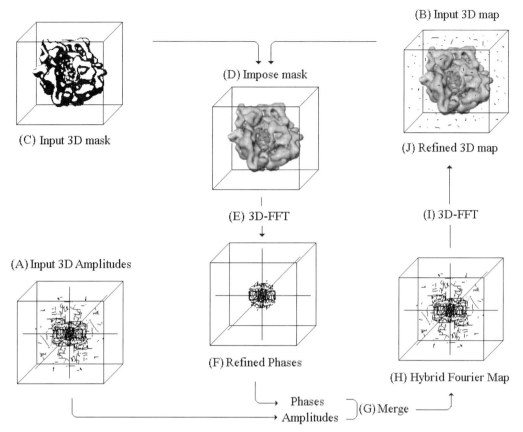

Figure 1. Single-particle solvent flattening (SPSF). Overall scheme of the proposed SPSF procedure modeled after similar procedures in use in X-ray crystallography (see text for a more detailed explanation): (*A*) One starting point for the SPSF procedure is an input volume of experimental 3D amplitudes that are assumed to extend to high resolution. (*B*) The other starting point for the SPSF analysis is a preliminary 3D map (containing both the low- and medium-resolution amplitudes and phases but no reliable high-resolution phases). (*C*) A 3D mask is calculated from the 3D map (*B*), which defines the outer contour of the particle and separates the molecule from the solvent environment (background noise). (*D*) The 3D mask (*C*) is imposed onto the input 3D map (*B*) or (*J*), respectively, in order to suppress the residual modulations present in the solvent (background) area. (*E*) The 3D Fourier transform (FFT) of the masked map (*D*) is calculated. Both phases and amplitudes of the resulting transform are consistent with the limited extend (boundary) of the particle. (*F*) The amplitudes of the 3D FFT (*E*) are discarded; only the refined phases of *E* are used in the subsequent processing steps. (*G*) The refined phases (*F*) and the input 3D amplitudes (*B*) are merged. (*H*) The result is a hybrid 3F Fourier map. (*I*) The hybrid 3D Fourier map is 3D Fourier-transformed back to real space. (*J*) The result is a refined 3D map with reliable phases up to a higher resolution than in the starting 3D map (*A*). If specified convergence criteria are fulfilled, the process is stopped. Otherwise, the refined 3D map (*J*) is re-entered into the process (*D*).

lution for the crystal's structure exists, one can refine the electron density by explicitly forcing the densities outside the known molecular envelope to approach zero. The modified real-space densities are then Fourier-transformed to 3D Fourier space yielding complex (amplitudes plus phases) structure factors. One then proceeds to refine the complex structure factors by combining these phases (modified by the imposed real-space constraints) with the experimental amplitudes. When these modified 3D structure factors are Fourier-transformed back to real space, the resulting densities are again no longer entirely contained within the molecular boundary. This is because the refined phases are based on the masked electron density map, but that is not true for the experimental amplitudes. By repeating the procedure iteratively, as illustrated in Figure 1 (in its single-particle equivalent), one can iteratively refine the electron density map. X-ray crystallographic solvent flattening works best when the

solvent content of the crystals is high. The higher the solvent content, the more stringent the boundary conditions represented by the imposed flatness of the solvent areas become, and thus the more helpful the technique is for improving the X-ray maps. There is, of course, an upper limit to the solvent content that a crystal can have because—and this is the very nature of crystals—the molecules in the crystal must be close to each other (touch) to form the crystal bonds.

The individual molecules in single-particle cryo-EM preparations are surrounded by frozen solvent. Separating the resulting 3D maps from the residual noise in the background is a routine operation that can be performed manually or automatically. This editing out the surroundings is applied often to remove noise prior to visualization of final results. At that stage of the processing, the solvent area surrounding the structure is of course no longer the vitrified solvent, but rather a computational box, the size of which can be chosen freely by the researcher. In contrast to the X-ray crystallographic experiment where the level of solvent content is limited by the necessity of physical contact between the molecules in the crystal, the solvent content in a single-particle experiment can be chosen freely. The idea of improving 3D map structure determined by cryo-EM by editing out the surroundings has already been put into practical use in various forms beyond the level of simply polishing the final results. For example, with a 3D reconstruction available, one can calculate the possible maximal extent of any 2D molecular image (for which one has determined the orientation parameters) and create a precise 2D mask around that molecular image to remove all background in the original micrographs. The thus-better-masked molecular images will yield better 3D reconstruction results. Such improvement procedures have, however, never been incorporated into a systematic method for improving the cryo-EM maps.

SINGLE-PARTICLE SOLVENT FLATTENING

The purpose of this paper is to make a case for exploiting solvent-flattening procedures for solving structures by single-particle cryo-EM. As the starting point of this discussion, let us assume that we have already performed a standard 3D reconstruction based on a large number of noisy molecular electron microscopic images. In particular, we assume that we have achieved at least an approximate correction of the CTF such that the structural information is no longer spread over an environment of the individual molecules but is actually concentrated within the particle boundary (cf. van Heel et al. 2000). With the availability of a (preliminary) 3D reconstruction, we also know approximately the 3D boundaries of the molecular complex and we can thus define the residual noise area around the 3D molecular density to be solvent and then proceed to iteratively remove those modulations from the data. Apart from the preliminary 3D reconstruction and the 3D mask separating the molecule from its surroundings, we need a source of amplitude information. The preliminary 3D map contains amplitude information, but the basic idea of the refinement approaches in X-ray crystallography is that one has reliable experimental amplitudes available beyond the resolution level for which reliable phases are available. For now, let us assume that that is also the case in the single-particle cryo-EM case (see discussion) and now proceed with model calculations to illustrate the feasibility of phase extension in single-particle processing.

Asymmetrical Particles:
50S *E. coli* Subunit

As a test data set for illustrating the SPSF procedures, the earlier 7.5 Å 3D reconstruction of the large ribosomal subunit of *E. coli* (Matadeen et al. 1999) is used (Fig. 2A). Note that this resolution quote is actually irrelevant in this context, since the object is simply used as a test object; i.e., as a 3D array of numbers. The highest spatial frequency possible in the test data set (the Nyquist frequency) is arbitrarily set to unity. Although the absolute resolution of the original map is irrelevant for the current experiment, the relative power spectrum distribution (Fig. 2A) is important because it can be used to help interpret the success of the operation (see below). From this test object we first generate the three different types of input volumes (Fig. 1) required for the SPSF procedures: (1) a map with only the low-resolution phases (and amplitudes) in place, (2) a 3D mask delineating the molecule, and (3) a 3D volume containing all the Fourier-space amplitudes of the test object.

The low-pass-filtered version of this map (Fig. 2B) that is used here as a starting model for the refinements is generated from the test data (Fig. 2A) by removing all the amplitudes and phases ("structure factors") beyond one-sixth of the Nyquist frequency using a sharp spherical mask. The effects of this sharp mask can be directly appreciated from the "fuzziness" of the sections (Fig. 2B) and from the Fourier shell correlation curve (FSC; Harauz and van Heel 1986) depicted in Figure 2B. (The FSC measures the similarity between two 3D maps as a normalized correlation coefficient between the structure factors within corresponding frequency shells in Fourier space.) The FSC curve drops to zero rapidly at one-sixth of the Nyquist frequency and then continues to oscillate around that value up to the Nyquist frequency. The 3-σ threshold curve for an asymmetric particle (Orlova et al. 1997) is intersected at exactly one-sixth of the Nyquist frequency, as expected. Upon refinement with the SPSF scheme (Fig. 1), the FSC curve improves dramatically (Fig. 2C) and now intersects the 3-σ threshold curve beyond half (0.5) of the Nyquist frequency. Since most of the power spectrum distribution is concentrated in the region below the 0.5 value (Fig. 2A), it is clear that the phases have indeed been largely recovered; that is, wherever there was enough power to make them sufficiently important. This experiment shows that it is possible to significantly extend the phases in single-particle cryo-EM, even for asymmetrical particles, provided that reliable experimental amplitudes are available beyond the extent of the experimental phases. Of course, we need to define exactly what "reliable" means. The model data were completely free of noise in this experiment (other than for the limitations in precision of the computational representations: all performed in 32-bit floating-point format), and to be able to assess the stability of the procedures, we must at least introduce experimental errors in the form of noise into the model calculations (see below).

Symmetrical Particles:
Lumbricus terrestris Hemoglobin

Let us now repeat this model experiment with a 3D reconstruction of the D6-symmetric giant worm hemoglobin of *L. terrestris* (Fig. 3A) (cf. van Heel et al. 2000). In this case, rather than starting with a low-pass-filtered version of the test object, let us start the iterative refinements with a 3D volume containing densities stemming purely from a random-number generator. The random noise data are then imposed on D6 symmetry using

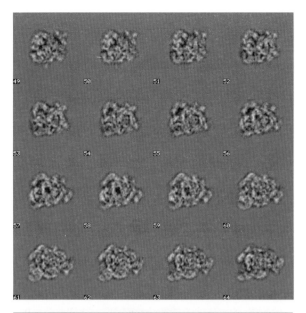

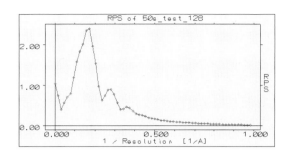

(*A*) Sections from a 3D reconstruction of the 50S ribosomal subunit of *E. coli*. This map is used as a test object for a solvent-flattening experiment for asymmetrical particles. The curve illustrates the rotationally averaged power spectrum distribution of this test object.

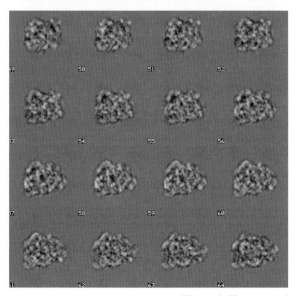

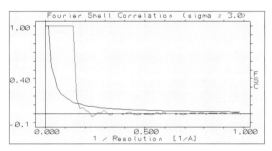

(*B*) Low-pass-filtered (hard spherical mask to 0.15 times the Nyquist frequency) version of input image used as starting point for the solvent-flattening test. The Fourier shell correlation between the test object (*A*) and the filtered starting model reflects the sharp Fourier-space mask.

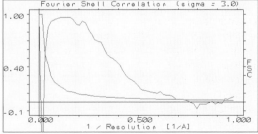

(*C*) Using a few hundred iterations of the algorithm depicted in Fig. 1, a significant part of the low-resolution phases could be restored. The total variance represented in the high-frequency data components that have not reached statistical significance represents less than a few percent of the total variance in the original map.

Figure 2. Phase extension of 50S ribosomal subunits.

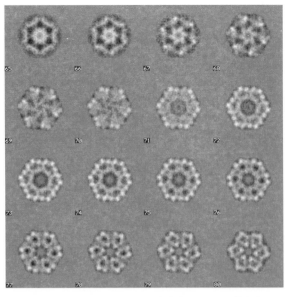

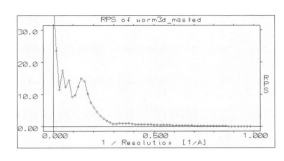

(*A*) A 3D reconstruction of the giant hemoglobin of *Lumbricus terrestris* (van Heel et al. 2000) is used as a test object for solvent flattening. Worm hemoglobin has D6 pointgroup symmetry; only some sections of the 128^3 map are depicted. The molecule occupies less than 15% of the available 3D volume; its rotational power spectrum distribution is depicted in the curve in arbitrary units.

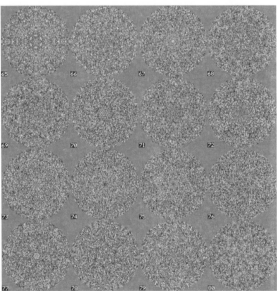

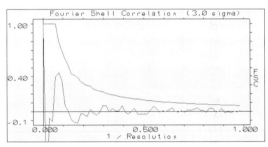

(*B*) Starting model for phasing: random noise density with imposed D6 symmetry. The Fourier shell correlation (Harauz and van Heel 1986) between the test object (*A*) and this symmetrized-random-noise starting model obviously shows no correlations above the 3-σ significance curve. The low-resolution statistical fluctuations are largely due to the small number of voxels in the low-resolution shells close to the origin.

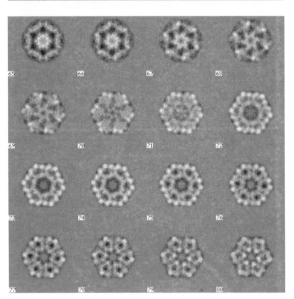

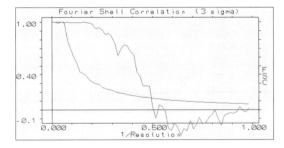

(*C*) Regeneration of the original phases is illustrated in the sections shown here and in the FSC between the original map and the phase-extended map. The map was inverted in contrast since the iterative algorithm had flipped the contrast (the inverse of a map has the same amplitude spectrum).

Figure 3. Creating phases out of thin air.

the appropriate IMAGIC command (van Heel et al. 1996). As expected, the FSC between the D6-symmetrized random noise and the original *L. terrestris* hemoglobin model never crosses the threshold curve (Fig. 3B), which, in this case, is corrected for the 12-fold redundancy of the D6-pointgroup symmetry by multiplication with a √12 factor (Orlova et al. 1997). The FSC curve fluctuates around zero for all special frequencies. Note that close to the origin, the fluctuations around zero have a much larger amplitude than do those close to the Nyquist frequency. This is indeed expected from the "√N" type of fluctuations in the FSC measurement caused by the smaller number of Fourier space voxels in a shell close to the Fourier space origin. This change in statistical fluctuations as a function of N is the reason one needs threshold curves rather than a fixed threshold value criterion. The lack of appreciation of this fact has unfortunately led to endless discussions/misunderstandings in the use of the FSC criterion (cf. Orlova et al. 1997).

The results of the phase extension in this example are quite counterintuitive (Fig. 3C). Despite having thrown out all the phases at the start of the model experiments, the phases are restored to excellent precision in all spatial frequency ranges everywhere (in Fourier space) where the test data contain a significant amount of power. The imposing of D6 symmetry, repeated regularly during the iterative solvent-flattening procedures, clearly improves the convergence of the overall process. Again, the test data were noise-free and there are thus no internal inconsistencies or errors in this computer experiment, other than the errors introduced by the limited precision of the data representation in the computer.

Asymmetrical Particles with Added Noise: 50S *E. coli* Subunit

For the experiments to be somewhat more realistic, it is important to investigate the stability of the numerical procedures in the presence of (simulated) experimental errors. Experimental errors are primarily expected to affect the experimental amplitudes, the main source of information in this procedure for retrieving the desired high-resolution phases. Using the *E. coli* 50S ribosomal-subunit data as a starting point, the power spectrum of a volume containing only Gaussian random white noise was added to the input amplitude spectrum (Fig. 1A). The half-width value of the modulations in real space was chosen to be about one-tenth of the largest modulations in the 50S map, yet the random white noise fills the entire calculation volume, whereas the 50S test data fill only the densities within the 3D mask (Fig. 1C). The starting point for the phase extension was the same low-pass-filtered 50S map (Fig. 1B) used for the noise-free experiment discussed above.

The results of this experiment (Fig. 4) indicate that significant levels of phase extension remain possible in the presence of noise. The resolution in the final map is about twice that present in the starting low-resolution map. It is interesting to compare the FSC curve (Fig. 4A) with the distribution of the power in both the test model and the

random white noise (Fig. 4B). The results indicate that, for the chosen parameters, the SPSF is capable of retrieving the phases reasonably well in those areas of the 3D Fourier space where the signal-to-noise ratio (SNR) exceeds unity.

DISCUSSION: WHY DOES SOLVENT FLATTENING WORK SO WELL?

The two programs needed to apply solvent-flattening procedures (cf. Wang 1985; Abrahams and Leslie 1996) are an automatic masking program, which defines where to find the object of interest inside the reconstruction volume, and a program that will go through the actual SPSF iterations as depicted in Figure 1. The new SPSF programs were implemented in the context of the IMAGIC software system (van Heel et al. 1996). The automatic masking program is based on determining the local variance in a 3D environment of each point (voxel) in the reconstruction volume, whereby only the variances in relevant frequency ranges are considered, similar to an earlier algorithm aimed at detecting objects in noisy images (van Heel 1982). An interesting observation is that typically only about 10–15% of a typical reconstruction volume actually contains the molecule. One tends to place the 2D molecular images in a frame that, in terms of available area, covers about four times the area occupied by the molecular projection; as a consequence, only about one-eighth of the 3D reconstruction volume will be filled. This implies that even under conventional circumstances, the largest part of the reconstruction volume is available for solvent flattening.

As a consequence of the volume being only partially (about one-eighth) filled, in Fourier space the amplitude and phase information will be correlated over a volume of about eight voxels around each Fourier-space voxel (the correlation volume is the reciprocal of the relative volume filled by the object). As mentioned above, in X-ray crystallography, the higher the solvent content of the crystals, the better the solvent-flattening phase refinements work. In single-particle cryo-EM, we can make the box around our particles as large as we want and can thus choose the solvent content as high as we want, thus extending the Fourier-space correlation volume at will. Many implementation details that cannot all be discussed here influence the convergence of the procedure. One important aspect is that, following the X-ray crystallographic tradition, we start at the low-resolution end of Fourier space and slowly include the high-resolution information. The increase in the Fourier space radius, per iteration, is chosen small such that they approximately remain within the correlation volume. These are important because the phase consistency within the correlation volume is the main driving force of the procedure.

While we were developing the new programs for SPSF on cryo-EM test data (Figs. 2 and 3), the quality of the results came as quite a surprise: The structures of biological macromolecules can be reconstructed using (almost) only the amplitudes of the molecular images and (almost) ignoring their phases. In the 50S ribosomal subunit test (Fig. 2), an increase in resolution by a factor of about two

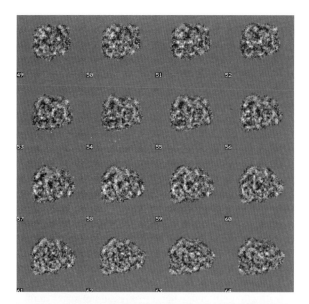

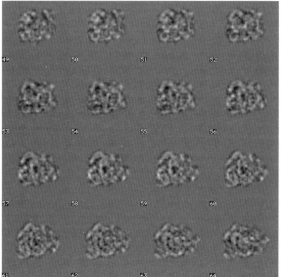

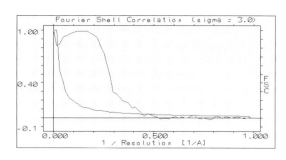

(A) A SPSF 3D reconstruction of the 50S ribosomal subunit of *E. coli* after adding (the power spectrum of) random white noise to the exact amplitude spectrum used as input for the solvent-flattening procedure. The FSC plot shows that the resolution has increased by at least a factor of 2 compared to that of the low-pass-filtered starting model (Fig. 2B).

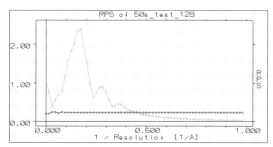

(B) The rotational power spectrum of the 50S *E. coli* test object (Fig. 2A) shown together with that of the added random white noise. By low-pass filtering the noisy SPSF result (A) to a resolution of ~40% of the Nyquist frequency, a relatively good quality map (*shown*) can still be obtained. For more details see the text.

Figure 4. The influence of experimental noise.

was achieved by the solvent-flattening procedures. This corresponds to an approximately eightfold (2^3) increase in information content that is visibly reflected in the results. Since in the worm hemoglobin example (Fig. 3) none of the original phases were used, the increase in available information content (as defined by the FSC) is phenomenal. (Strictly speaking, the relevant information must already have been coded in the known mask, and in the known amplitudes.) By exploiting its D6 pointgroup symmetry, the structure converged toward a solution much more quickly than did the entirely asymmetrical 50S ribosomal subunit. In the course of the iterations, asymmetrical particles tended to drift somewhat against the edges of the mask, leading to increased errors. Measures to prevent such shifts in real space (phase ramps in Fourier space) have yet to be taken. Various other ideas commonly used in such iterative refinement schemes, like exploiting positivity constraints, have yet to be tested. The results are counterintuitive in that the consen-

sus in the cryo-EM field is that the phases contain virtually all the relevant information. The tests show that even when the 3D amplitude spectrum is corrupted with noise, retrieving the phase information from the amplitude information can remain possible.

There have been earlier iterative Fourier-space/real-space refinement schemes in electron microscopy. Known as "maximum entropy" or "projection onto convex sets" approaches, these techniques were proposed to improve the quality of 3D tomographic reconstructions (cf. Carazo and Carrascosa 1987). The purpose of these approaches is to exploit constraints such as positivity of the density or the limited spatial extent of the object to improve the quality of the maps and, in particular, to compensate for the effects of the missing cone. However, these proposals were never aimed at improving single-particle 3D maps. In tomography one cannot exploit known amplitude information to retrieve phases because the amplitude of the data in the missing cone is also missing. The earlier proposals were never motivated by the fundamental aspect of single-particle cryo-EM, namely that of the free choice of the solvent content. None of these methods has to my knowledge been shown in model experiments to actually restore the complex structure factors computationally removed from an artificially created missing cone.

The SPSF approach contains important elements of an old school of thinking. A whole body of publications exists on reconstructing the complex wave front behind an object from the image plane and the diffraction plane amplitudes (cf. Gerchberg and Saxton 1972; Huiser and Ferwerda 1976; Saxton 1978). Solvent-flattening algorithms (cf. Wang 1985; Abrahams and Leslie 1996) can be seen as a special case of such procedures. It has also been known for some time (Sayre 1952; Szöke 2001) that the phase problem in X-ray crystallography is a consequence of the critical sampling of the diffraction pattern that could in principle be overcome if a complete coverage of the amplitude information (including the information from between diffraction spots) were available. Recently, the use of free-electron lasers was considered for collecting diffraction patterns of single molecules, which would then be phased by oversampling (Neutze et al. 2000). Indeed, in a most recent computer experiment, Miao et al. (2001) have shown that it is possible to reconstruct the structure of a rubisco monomer from simulated single-particle X-ray diffraction patterns alone. Clearly, there is an intimate relationship between these single-particle diffraction (SPD) calculations and the SPSF procedures proposed above. The diffraction method relies on collecting X-ray laser diffraction data from single molecules, isolated from any background using a mass spectrometer (Neutze et al. 2000). In the SPSF approach, in contrast, one relies rather on images for collecting high-resolution amplitude information. The method is based on the fact that amplitude information in the images is more readily accessible than the wildly fluctuating high-resolution phases. The image information needed for the SPSF approach is readily available today, whereas the development of the first X-ray laser may take another decade.

THE FUTURE IS FLAT!

SPSF is a new technique, the details of which are still largely not understood. It will clearly take some time to digest the details of the new SPSF procedures, and various open questions remain. For example, what kind of information is represented by the real-space mask/envelope? The better the resolution of a preliminary 3D reconstruction, the better we can define the mask, and that again directly influences the quality of the phase extension by solvent flattening. Note that the handedness of the mask information directly makes the reconstructions converge toward the correct handedness, despite the fact that the correct and the mirrored 3D maps are equally compatible with the experimental amplitudes. The mask, for the time being, will be the result of a "conventional" single-particle 3D reconstruction and thus of conventionally determined phases. The mask obviously represents shape information

about the object which *phase* information will probably be first obtained from conventional single-particle analysis (and may later stem from earlier rounds of the SPSF iterations). It is thus (for the time being) more appropriate to think of the SPSF procedures as phase extension procedures rather than as ab initio phasing approaches.

The model calculations indicate that the approach could become very important indeed. Because of the oversampling aspects of single-particle processing, the SPSF cryo-EM approach may become much more powerful than its X-ray crystallographic predecessor. How can we put these new ideas to work for us? The real gain is to be expected from the fact that it is simpler to determine the high-resolution amplitudes from the molecular images in the micrographs than it is to achieve a perfect synchronization of their high-resolution phases. To determine the high-resolution phases of single-particle data we must find the perfect (translational + rotational) alignment between the molecular images, and we must precisely determine their individual defocus parameters. In contrast, the high-resolution image amplitudes are not sensitive to phase fluctuations at all. Note, however, that this idea is in opposition to the widespread opinion in this field that the high-resolution phases are easier to obtain than the corresponding amplitudes. The SPSF procedures may be implemented in real space as a 3D reconstruction of the 3D auto-correlation function (ACF) or of the 3D self-correlation function (SCF) (van Heel et al. 1992) of the molecules, or, alternatively, in Fourier space by assembling the 3D amplitude spectrum of the particles. Since there already are cryo-EM structures at ~5–7 Å resolution, with the anticipated additional resolution boost we could soon witness the long-anticipated breakthrough of single cryo-EM to atomic resolution.

CONCLUSIONS

Single-particle cryo-EM is a rapidly maturing technique elbowing its way into the world of high-resolution structural biology. Solvent flattening may significantly contribute toward boosting the resolution levels achievable by single-particle cryo-EM into a regime, compatible with an atomic interpretation of macromolecules and their complexes. Let us see how quickly this new technique matures and whether it will live up to my high levels of expectation.

ACKNOWLEDGMENTS

I thank Ardan Patwardhan, Janos Hajdu, Abraham Szöke, Edgar Weckert, Michael Schatz, and Fritz Zemlin for stimulating discussions. I am also indebted to my Ph.D. supervisor Eddy Ferwerda for introducing me—more than 25 years ago—to the theoretical concepts at hand. The work was supported by grants from the European Community (especially: BIO4-CT98-0377) and the Biotechnology and Biological Sciences Research Council (BBSRC). Our electron microscopes have been funded by Higher Education Funding Council for England/BBSRC, with matched funding from FEI Electron Optics (Philips) and GlaxoSmithKlein.

REFERENCES

Abrahams J.P. and Leslie A.G. 1996. Methods used in the structure determination of bovine mitochondrial F1 ATPase. *Acta Crystallogr. D* **52:** 30.

Adrian M., Dubochet J., Lepault J., and McDowall A.W. 1984. Cryo-electron microscopy of viruses. *Nature* **308:** 32.

Agrawal R.K., Penczek P., Grassucci R.A., and Frank J. 1998. Visualization of the elongation factor G on *E. coli* 70S ribosome: The mechanism of translation. *Proc. Natl. Acad. Sci.* **95:** 6134.

Ban N., Nissen P., Hansen J., Moore P.B., and Steitz T.A. 2000. The complete atomic structure of the large ribosomal subunit at 2.4 Å resolution. *Science* **289:** 905.

Böttcher B., Wynne S.A., and Crowther R.A. 1997. Determination of the fold of the core protein of hepatitis B virus by electron cryomicroscopy. *Nature* **386:** 88.

Carazo M. and Carrascosa J.L. 1987. Information recovery in missing angular data cases: An approach by the convex projections method in three dimensions. *J. Microsc.* **145:** 23.

Conway J., Cheng N., Wingfield P.T., Stahl S.J., and Steven A.C. 1997. Visualisation of a 4-helix bundle in the hepatitis B virus capsid by cryo-electron microscopy. *Nature* **385:** 91.

Crowther R.A. 1971. Procedures for three-dimensional reconstruction of spherical viruses by Fourier synthesis from electron micrographs. *Philos. Trans. R. Soc. Lond. B Biol. Sci.* **261:** 221.

DeRosier D.J. 2000. Correction of high-resolution data for curvature of the Ewald sphere. *Ultramicroscopy* **81:** 83.

DeRosier D.J. and Klug A. 1968. Reconstruction of three-dimensional structures from electron micrographs. *Nature* **217:** 130.

Dubochet J., Adrian M., Chang J.-J., Homo J.-C., Lepault J., McDowall A., and Schultz P. 1988. Cryo-electron microscopy of vitrified specimens. *Q. Rev. Biophys.* **21:** 129.

Gerchberg R.W. and Saxton W.O. 1972. A practical algorithm for the determination of phase from image and diffraction plane pictures. *Optik* **35:** 237.

Harauz G. and van Heel M. 1986. Exact filters for general geometry three-dimensional reconstruction. *Optik* **73:** 146.

Henderson R., Baldwin J.M., Ceska T.A., Zemlin F., Beckmann E., and Downing K.H. 1990. Model for the structure of bacteriorhodopsin based on high-resolution electron cryomicroscopy. *J. Mol. Biol.* **213:** 899.

Huiser A.M.J. and Ferwerda H.A. 1976. The problem of phase retrieval in light and electron microscopy of strong objects. 2. On the uniqueness and stability of object reconstruction procedures using two defocused images. *Opt. Acta* **23:** 445.

Matadeen R., Patwardhan A., Gowen B., Orlova E.V., Pape T., Mueller F., Brimacombe R., and van Heel M. 1999. The *E. coli* large ribosomal subunit at 7.5 Å resolution. *Structure* **7:** 1575.

Miao J., Hodgson K.O., and Sayre D. 2001. An approach to three-dimensional structures of biomolecules by using single-molecule diffraction images. *Proc. Natl. Acad. Sci.* **98:** 6641.

Neutze R., Wouts R., van der Spoel D., Weckert E., and Hajdu J. 2000. Potential for biomolecular imaging with femtosecond X-ray pulses. *Nature* **406:** 752.

Orlova E.V., Dube P., Harris J.R., Beckman E., Zemlin F., Markl J., and van Heel M. 1997. Structure of keyhole limpet hemocyanin type 1 (KLH1) at 15 Å resolution by electron cryomicroscopy and angular reconstitution. *J. Mol. Biol.* **271:** 417.

Saxton W.O. 1978. *Computer techniques for image processing in electron microscopy.* Academic Press, New York.

Sayre D. 1952. The squaring method: A new method for phase determination. *Acta Crystallogr.* **5:** 843.

Schluenzen F., Tocilj A., Zarivach R., Harms J., Gluehmann M., Janell D., Bashan A., Bartels H., Agmon I., Franceschi F., and Yonath A. 2000. Structure of functionally activated small ribosomal subunit at 3.3 Å resolution. *Cell* **102:** 615.

Stark H., Rodnina M.V., Rinke-Appel J., Brimacombe R., Wintermeyer W., and van Heel M. 1997a. Visualisation of elongation factor Tu on the *Escherichia coli* ribosome. *Nature* **389:** 403.

Stark H., Orlova E.V., Rinke-Appel J., Jünke N., Mueller F., Rodnina M., Wintermeyer W., Brimacombe R., and van Heel M. 1997b. Arrangement of tRNAs in pre- and post-translocational ribosomes revealed by electron cryomicroscopy. *Cell* **88:** 19.

Szöke A. 2001. Diffraction of partially coherent X-rays and the crystallographic phase problem. *Acta Crystallogr. A* **57:** 586.

van Heel M. 1982. Detection of objects in quantum noise limited images. *Ultramicroscopy* **8:** 331.

———. 1984. Multivariate statistical classification of noisy images (randomly oriented biological macromolecules). *Ultramicroscopy* **13:** 165.

———. 1987. Angular reconstitution: A posteriori assignment of projection directions for 3D reconstruction. *Ultramicroscopy* **21:** 111.

———. 1989. Classification of very large electron microscopical image data sets. *Optik* **82:** 114.

van Heel M. and Frank J. 1981. Use of multivariate statistics in analysing the images of biological macromolecules. *Ultramicroscopy* **6:** 187.

van Heel M., Schatz M., and Orlova E.V. 1992. Correlation functions revisited. *Ultramicroscopy* **46:** 304.

van Heel M., Harauz G., Orlova E.V., Schmidt R., and Schatz M. 1996. A new generation of the IMAGIC image processing system. *J. Struct. Biol.* **116:** 17.

van Heel M., Gowen B., Matadeen R., Orlova E.V., Finn R., Pape T., Cohen D., Stark H., Schmidt R., Schatz M., and Patwardhan A. 2000. Single-particle cryo electron microscopy: Towards atomic resolution. *Q. Rev. Biophys.* **33:** 307.

Wang B.-C. 1985. Resolution of phase ambiguity in macromolecular crystallography. *Methods Enzymol.* **115:** 148.

Wimberly B.T., Brodersen D.E., Clemons W.M., Morgan-Warren R.J., Carter A.P., Vonrhein C., Hartsch T., and Ramakrishnan V. 2000. The structure of the 30 S ribosomal subunit. *Nature* **407:** 327.

Zemlin F., Beckmann E., and van der Mast K.D. 1996. A 200 kV electron microscope with Schottky field emitter and a helium-cooled super-conducting objective lens. *Ultramicroscopy* **3:** 227.

Correlating the X-Ray Structures for Halo- and Thermophilic Ribosomal Subunits with Biochemical Data for the *Escherichia coli* Ribosome

P. Sergiev,* A. Leonov,* S. Dokudovskaya,* O. Shpanchenko,*
O. Dontsova,* A. Bogdanov,* J. Rinke-Appel,† F. Mueller,† M. Osswald,†
K. von Knoblauch,† and R. Brimacombe†

Department of Chemistry of Natural Compounds and Belozersky Institute, Moscow State University, Moscow 119899, Russia; †Max-Planck-Institut für Molekulare Genetik, D-14195 Berlin, Germany

The ribosome is a ribonucleoprotein particle of outstanding complexity, and during the last three decades an unprecedented variety of chemical and biochemical methods has been applied to the study of its structure in solution (for reviews, see Brimacombe 1995; Green and Noller 1997). Moreover, a number of these methods were developed specifically for the investigation of the ribosome (Bogdanov et al. 2000). The recent publication of X-ray structures for both the 30S and 50S ribosomal subunits at resolutions of 2.4–3.5 Å (Ban et al. 2000; Schluenzen et al. 2000; Wimberly et al. 2000) now offers a unique opportunity for comparing the crystallographic data with the biochemical data; none of the biochemical methods is ideal, and each embodies its own inherent assumptions and potential pitfalls. The purpose of this review is to make a retrospective assessment of the various methods that have been applied to the ribosome, so as to determine which techniques have provided the most reliable data in terms of the X-ray structures. Such an assessment should prove useful for the further development of these methods, both for the study of the different functional states of the highly dynamic ribosomal structures and for investigations of ribosomal complexes or other types of ribosomes for which crystallographic structures are not likely to become available in the near future. Furthermore, this information should also be of interest to scientists working with other complex ribonucleoprotein systems, to enable them to select the most reliable methods for their research. Sergiev et al. (2001) have very recently made an extensive and detailed analysis of the methods that have been used for investigating inter- or intra-RNA interactions within the ribosome. Here, we summarize this latter analysis, and extend it to cover the biochemical techniques that have been applied to the study of RNA–protein interactions.

There are two potential problems in the comparison of the biochemical and X-ray data. First, there may be conformational differences between the ribosomal structure in solution (especially as a reflection of different functional states) and the crystal structures of the ribosomal subunits. Such differences could arise either from the flexibility of the ribosomal structure in solution or from inter-

molecular contacts within the crystals. However, as Sergiev et al. (2001) have already demonstrated, the quality of the correlation between the biochemical and X-ray data is in general clearly dependent on the particular biochemical method under consideration. In consequence, a simple comparison of the distances between ribosomal components that are expected from the biochemical data to be close together with the corresponding distances in the X-ray structures allows us to make an objective judgment as to the reliability of the method concerned, although some discrepancies may indeed be the result of conformational changes or flexibilities.

The second potential problem stems from the fact that the X-ray analyses were made with 30S ribosomal subunits from the thermophile *Thermus thermophilus* (Schluenzen et al. 2000; Wimberly et al. 2000) and with 50S subunits from the halophile *Haloarcula marismortui* (Ban et al. 2000), whereas the great majority of the biochemical studies have been made with ribosomes from the eubacterium *E. coli* (Brimacombe 1995; Bogdanov et al. 2000). In the latter case, the best available structural information has come from cryo-electron microscopy (cryo-EM; Matadeen et al. 1999; Gabashvili et al. 2000), and the transposition of the high-resolution X-ray structures into the *E. coli* cryo-EM structures at somewhat lower resolution is currently being made in our laboratories. As a first step in this procedure, the X-ray structures for both ribosomal proteins and rRNA have been fitted as rigid bodies into a cryo-EM map of the *E. coli* 70S ribosome, and the 16S and 23S rRNA sequences from *T. thermophilus* and *H. marismortui*, respectively, have been "converted" as far as possible into those of *E. coli*. For obvious reasons, in our assessment of the biochemical methods we do not consider data relating to those parts of the rRNA molecules where a direct secondary structure equivalence between *T. thermophilus* or *H. marismortui* and *E. coli* does not exist, or data relating to regions where the X-ray structures were disordered. To avoid confusion, *E. coli* numbering is used throughout the following discussion, the rRNA helices being numbered as in Mueller and Brimacombe (1997a) for the 16S rRNA and in Mueller et al. (2000) for the 23S and 5S rRNAs.

CROSS-LINKING BY DIRECT UV-IRRADIATION OR THE USE OF SIMPLE BIFUNCTIONAL REAGENTS

One of the earliest methods to be applied to the ribosome with a view to determining neighborhoods or contacts within the structures was the simple exposure of isolated ribosomal subunits to UV light at 254 nm, or treatment of the subunits with symmetrical bifunctional reagents such as *bis*-(2-chloroethyl)-methylamine ("nitrogen mustard"; summarized in Brimacombe 1995; Mueller and Brimacombe 1997a,b; Mueller et al. 2000). These treatments can in principle lead to the formation of protein–protein, protein–RNA, or RNA–RNA cross-links, and Figure 1 illustrates the locations of intra-rRNA cross-links within the X-ray structures of 23S (Fig. 1a) and 16S (Fig. 1b) rRNA that were localized by these simple methods (cf. Sergiev et al. 2001). In each case, to help orient the reader, the rRNA molecule is shown within the cryo-EM contour of the corresponding subunit, excised from the cryo-EM density of the *E. coli* 70S ribosome (Brimacombe et al. 2000). It can be seen that the oligonucleotides identified as encompassing the cross-link sites are in every case (with one exception, see below) closely neighbored within the ribosomal subunits. Other cross-links were also observed in these experiments, corresponding to sites that would be expected to be neighbors from the primary or secondary structures of the 16S or 23S rRNA; these are not included in Figure 1.

There are three cases in Figure 1 in which the same nucleotide (or oligonucleotide) was involved in two different cross-links. In the 23S rRNA, nucleotides 2030–2032 were linked in different experiments both to nucleotides 571 and 2054–2055. Similarly, in the 16S rRNA, nucleotide 31 was linked to nucleotides 48 and 306, and nucleotide 894 to 244 and 1468. The latter cross-link (894 to 1468) is the only example where the cross-linked partners lie somewhat apart in the X-ray structure, and this raises the question of the methods used for identification of the cross-link sites. In most of the cross-links shown in Figure 1 (those identified in our laboratories), the cross-linked rRNA was first subjected to partial digestion with nuclease, and the cross-link site identification was then made with the help of classic "fingerprint" oligonucleotide analysis. This technique is very accurate, but has the disadvantage that the cross-link can often only be localized to within a few nucleotides, rather than to one precise nucleotide. In contrast, Wilms et al. (1997) used primer extension analysis to localize UV-induced cross-links within the 16S rRNA. However, as has been discussed in detail previously (Brimacombe 1995), the primer extension method, although it has the potential of pinpointing a cross-link site to a single nucleotide, can also lead to serious artifacts. This is because the reverse transcription reaction is carried out in the presence of the bulky cross-linked partner, with the result that primer extension stops or pauses may be observed at positions that have nothing to do with the cross-link site per se. This will be a recurring theme in the following sections of this article, and the apparent cross-link between nucleotides 894 and 1468 could represent such an artifact.

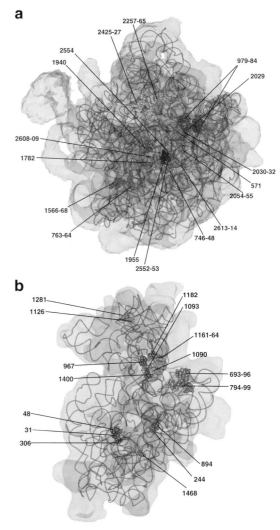

Figure 1. Locations of intra-RNA cross-link sites in the 50S (*a*) or 30S (*b*) ribosomal subunit. The cross-links are those that were induced by direct UV-irradiation or treatment of the ribosomal subunits with simple bifunctional reagents. The figures show the atomic structures of the rRNA molecules (23S and 5S rRNA in *a*, 16S in *b*) as gray backbone tubes fitted to cryo-EM reconstructions of the respective subunits, rendered as transparent contours. The view in each case is from the interface side of the subunit. The pairs of nucleotides (or oligonucleotides) identified at each cross-link site are shown as nucleotides in CPK format, displayed in the same color (e.g., nucleotide 1782 in 23S rRNA was cross-linked to nucleotides 2608–2609, shown in *orange*). See text for references and further details.

These simple cross-linking procedures, in particular the use of UV irradiation at 254 nm, were often criticized on the grounds that such treatment could seriously disturb the ribosomal structure. The data of Figure 1, perhaps surprisingly, indicate that this criticism is groundless, and suggest that the observed cross-links were formed before any serious damage to the structure had occurred. It is also possible that any cross-linking from damaged ribosomes takes place in a random manner that only results in low background levels of spurious cross-links.

The locations of RNA–protein cross-link sites on 23S or 16S rRNA in relation to the atomic structures of the corresponding ribosomal proteins within the 50S or 30S subunits are illustrated in Figures 2 and 3. Here the cross-links were for the most part induced by treatment with hetero-bifunctional reagents (summarized in Mueller and Brimacombe 1997b; Mueller et al. 2000), but the analysis of the cross-link sites on the rRNA was in all cases still made by classic oligonucleotide fingerprinting (cf. Fig. 1). All of the cross-link identifications, with the exception of that from protein L2 to nucleotide 1963 of the 23S rRNA (Fig. 2a) (Thiede et al. 1998), were made in our laboratories.

Figure 2a shows cross-link sites to proteins located on the interface side of the 50S subunit (further examples, from the solvent side of the subunit, are given in Fig. 5, below), and it can be seen that—as with the intra-rRNA

cross-link data of Figure 1—there is a high degree of correspondence between the positions of the proteins and the positions of the oligonucleotides identified as their cross-linked partners. As an extrapolation of this correspondence, it is noteworthy that in the X-ray structure of the *H. marismortui* 50S subunit (Ban et al. 2000), a number of protein structures were mapped which have no identifiable equivalent in *E. coli*. Two of these proteins, L21e and L44, are located in the central protuberance of the 50S subunit, and a stereo close-up view of this area of the subunit is shown in Figure 2b, together with the cross-link sites on the 23S rRNA found for *E. coli* proteins L33 (one cross-link site) and L27 (four sites, two in helix h81 and two in h85, respectively). L33 and L27 have no apparent equivalent in *H. marismortui*, but the positions of their cross-link sites (Fig. 2b) suggest that L33 could correspond to L44, and L27 to L21e.

Similar data for the 30S subunit are shown in Figure 3a, where again the degree of correspondence between protein positions and cross-link sites is very high. Here it is noteworthy that two of the proteins, S8 and S9, have multiple cross-link sites on the 16S rRNA. In the case of S8 there are three sites, all of which lie within a single helix (h21) of the rRNA molecule. With S9, however, the two cross-link sites are two different regions of the 16S rRNA, position 954 lying in h30 and positions 1130–1131 in h39 on the opposite side of the head of the 30S subunit. Protein S9 is one of the ribosomal proteins that carry a long nonglobular extension, and whereas the cross-link site at nucleotides 1130–1131 is located adjacent to the globular center of the protein, the site at nucleotide 954 lies close to the end of the nonglobular extension. It is easy to see how results of this kind caused problems in model-building studies on the 16S rRNA (see, e.g., Stern et al. 1988; Mueller and Brimacombe 1997b), which attempted to bring such pairs of sites into proximity in the 16S structure, working on the assumption that the ribosomal proteins are essentially globular.

Two cross-link sites to different regions of the 16S rRNA were also found for protein S7 (Tanaka et al. 1998), and in this case the cross-linked amino acid positions on the protein could be identified as well (Möller et al. 1978; Urlaub et al. 1997). These data are illustrated as a stereo close-up in Figure 3b, from which it can be seen that in both cases the nucleotide and amino acid components of the cross-link are indeed in direct contact with one another in the X-ray structure of the 30S subunit; the Met-115/1240 cross-link lies at the upper left-hand corner of the orange backbone tube for protein S7 (Fig. 3b) and the Lys-75/1377-78 cross-link on the lower left-hand side. Figure 3b also shows the corresponding footprinting data on the 16S rRNA for protein S7, the footprinting approach being the subject of the following section.

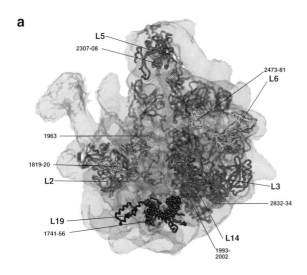

a

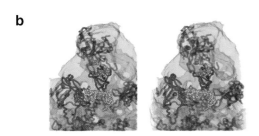

b

Figure 2. Locations of RNA–protein cross-link sites in the 50S subunit. In *a* a transparent contour of the 50S subunit is shown as in Fig. 1a, this time with the atomic structures of the ribosomal proteins rendered as gray backbone tubes. Nucleotides (or oligonucleotides) identified in cross-links to particular proteins are displayed in CPK format, with the corresponding protein backbone highlighted in the same color. In *b* a close-up stereo view of the central protuberance region of the 50S subunit is shown (in the same orientation as *a*). The oligonucleotide cross-linked to protein L33 is displayed in green, and those cross-linked to 27 in helices h81 and h85 in pink and orange, respectively. The *H. marismortui* proteins L44 and L21e (which have no equivalent in *E. coli*) are highlighted in green and red, respectively. See text for explanation.

RNA–PROTEIN FOOTPRINTING

Many different footprinting methods have been used in a number of laboratories for the study of RNA–protein interactions in the ribosome. In the 30S subunit, most of the

a

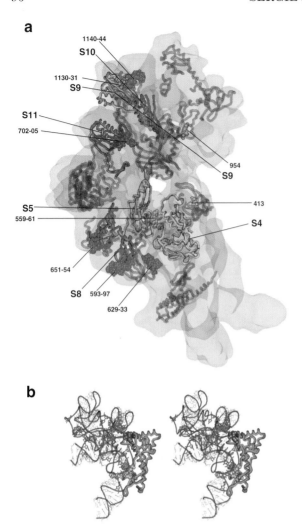

b

Figure 3. Locations of RNA–protein cross-link sites (and foot-print sites) in the 30S subunit. In *a,* the cross-link sites are displayed in the same manner as that of Fig. 2a. In the view shown, the 50S subunit would lie on the right. Note that protein S9 (*orange*), for which there are two widely separated cross-link sites (at nucleotides 1130–1131 and 954, respectively), has a long nonglobular extension partly concealed behind protein S10 (*green*). In *b*, the cross-link and footprint data for protein S7 are shown as a close-up stereo view; the subunit has been rotated by ~180° about the *y* axis with respect to the view shown in *a*. The protein is displayed as an orange tube with the amino acids in *T. thermophilus* corresponding to Lys-75 and Met-115 in *E. coli* highlighted in CPK format in dark pink and dark red, respectively. The corresponding cross-linked nucleotides in 16S rRNA (1377–1378 and 1240) are also in CPK format, in lighter pink and red, respectively. The area of the 16S rRNA carrying footprint sites (helices h28–h30 and h41–h43) is displayed in wire-frame format with gray backbone tubes; base-specific footprint sites are shown as nucleotides in blue "stick" format, Fe(II)-EDTA footprint sites in green.

data has been produced by Noller's group with the help of two different types of footprinting analyses, namely the use of base-specific probes or of hydroxyl radicals generated from Fe(II)-EDTA (summarized by Powers and Noller 1995). The footprinting data for ribosomal protein S7 illustrated in Figure 3b show that, whereas the RNA and protein components of the cross-link sites are in direct

contact as just mentioned above, the footprint sites from both techniques are spread more deeply into the 16S rRNA, away from the protein. The primer extension technique is universally used for the determination of footprint sites on the rRNA, and—in contrast to its application in the analysis of cross-link sites—there is no reason to doubt its accuracy for this purpose. It therefore seems likely that binding of protein S7 to the16S rRNA causes some local rearrangement of the latter at positions that are not in direct contact with the protein; this would explain the distribution of footprint sites in Figure 3b.

Figures 4 and 5 show some further examples of different types of footprinting data. Figures 4a and 4b give the base-specific and Fe(II)-EDTA footprints for proteins S12 and S20, respectively, within the 30S subunit (Powers and Noller 1995). In both cases the Fe(II)-EDTA footprint sites are in close contact with the protein, whereas the base-specific sites are more broadly distributed, as predicted by Powers and Noller (1995). Protein S20 is particularly interesting for two reasons. First, the location of this protein in the 30S subunit was determined incorrectly by the neutron scattering method (Capel et al. 1988) but correctly by immunoelectron microscopy (Schwedler et al. 1993). Second, since protein S20 is well known to be identical to protein L26 (i.e., it is found on both 30S and 50S subunits after dissociation of the 70S ribosome), one would expect it to have significant contacts to the 23S as well as to the 16S rRNA; however, this does not appear to be the case, and the mysterious behavior of S20 remains to be clarified. (There are no cross-linking data for either S12 or S20.)

The corresponding data for protein S4 are included in Figure 4c. Here it should be noted that base-specific footprint sites were observed in helices h23a and h27 (Powers and Noller 1995; not shown in Fig. 4c); both of these helices are remote from protein S4 in the 30S subunit, thus providing further support for the contention that the Fe(II)-EDTA method is more reliable. Figure 4c also shows the sites of scission in the 16S rRNA observed from Fe(II)-EDTA tethered to a specific position (amino acid 31) of S4 (Heilek et al.1995); these sites lie in reasonable proximity to the protein. In contrast, the sites of scission in 23S rRNA from Fe(II)-EDTA tethered to a number of specific positions in protein L9 cover a wide area of the 23S molecule (Lieberman et al. 2000). These data are illustrated in Figure 4d. The *H. marismortui* 50S subunit does not contain an equivalent protein to L9 (Ban et al. 2000), but the position of the latter protein in *E. coli* is very clear from cryo-EM studies (Matadeen et al. 1999). The Fe(II)-EDTA was tethered to sites in L9 lying predominantly within the long central α helix of the protein, but it can be seen from Figure 4d that the spread of footprint or scission sites observed, which cover four of the six secondary structural domains of the 23S rRNA, extends well beyond the limit of 44 Å predicted for this method (Joseph et al. 1997). On the other hand, in the same series of experiments (Lieberman et al. 2000) the "simple" Fe(II)-EDTA or base-specific footprinting approaches gave strong footprints at the junction of helices h76 and h79, in close proximity to the protein (cf. Fig. 4d). In similar studies with protein S5 (Culver et al.

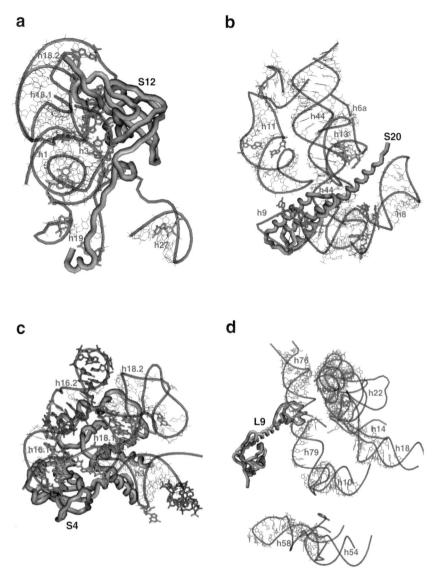

Figure 4. Examples of RNA–protein footprint data, and of probing experiments with tethered Fe(II)-EDTA. In each part of the figure, the protein concerned is represented as an orange backbone tube, and the areas of the rRNA (16S or 23S) carrying the footprint or tethered Fe(II)-EDTA sites are displayed in wire-frame format with gray backbone tubes as in Fig. 3b. Helix names (in the 16S or 23S rRNA, as appropriate) are marked. (*a*) Data for protein S12. Base-specific footprint sites in the 16S rRNA are highlighted as blue "stick" nucleotides, Fe(II)-EDTA footprint sites as green stick nucleotides. (*b*) Data for protein S20, with footprint sites highlighted as in *a*. (*c*) Data for protein S4. Here the footprint sites are marked as in *a* or *b*, and in addition, the sites of cleavage from tethered Fe(II)-EDTA at position 31 of S4 are highlighted in magenta. (*d*) Sites of cleavage in 23S rRNA from tethered Fe(II)-EDTA at various positions in protein L9 are highlighted in green. See text for references.

1999), tethered Fe(II)-EDTA scissions were even found in the 23S rRNA, again at a considerable distance from the protein. Our impression is that data from the tethered Fe(II)-EDTA method are dependent on whether the tethered residue has an "unobstructed view" of its target sites on the rRNA, rather than being a direct measure of the distance of the target from the tether.

Figures 5a, b, and c show some examples of RNA–protein footprints from the 50S subunit reported by Garrett's group (summarized by Egebjerg et al. 1991), and these footprints are compared with the corresponding RNA–protein cross-link sites (Osswald et al. 1990). Here

the footprinted regions of the 23S rRNA were compiled from a variety of different probes, including the use of base-specific reagents and various types of nuclease, and the results are presented as "areas" of the 23S molecule rather than as sets of individual nucleotides as in Figure 4a–c. The data for protein L3 (Fig. 5a) demonstrate that the footprinted area lies very close to the protein, as does also the RNA–protein cross-link site (cf. Fig. 2a), although different helices of the 23S rRNA are involved (h94 and h100, respectively). The same holds true for protein L24 (Fig. 5b), where two distinct footprinted regions were observed, in helices h3/h24 and h18/20; taken

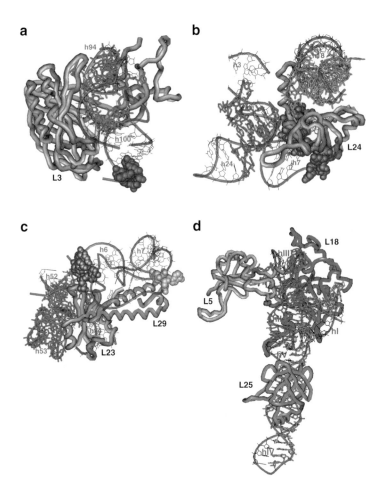

Figure 5. Comparison of RNA–protein footprint and cross-link data in the 50S subunit. Proteins and appropriate areas of rRNA are displayed as in Fig. 4. (*a*) Data for protein L3. Footprint sites are highlighted as green "stick" nucleotides, and the oligonucleotide identified as containing the cross-link site is indicated by the red CPK nucleotides. (*b*) Data for protein L24. Cross-link and footprint sites are denoted as in *a*, but with the two distinct footprinted regions of the 23S rRNA (helices h3, h24 and helices h18, h20) being distinguished as green and blue stick nucleotides, respectively. (*c*) Data for proteins L23 and L29. The green stick nucleotides represent the footprint site for L23, and the red CPK nucleotides the cross-link site to the latter. The cross-link site for L29 is shown in pink (there being no footprint data for this protein). (*d*) Footprint data for proteins L5, L18, and L25 complexed with 5S rRNA. Each of the three proteins is shown as a backbone tube, with the corresponding footprint sites on the 5S rRNA highlighted in the same color. See text for references.

together with the cross-link site in h7, the protein is effectively "surrounded," and this is a good example of how the footprint and cross-link data complement each other. Figure 5c shows data for the neighboring proteins L23 and L29. L23 has footprints in domain III of the 23S rRNA, whereas the cross-link site in h6 is in domain I, and lies on the opposite side of the protein. L29 (for which there are no footprint data) has a cross-link site in the closely adjacent helix h7. It should be noted that there is a further cross-link to L23 in helix h9 of the 23S rRNA, but since the latter site lies in a region where the secondary structure of the *H. marismortui* is ambiguous in relation to that of *E. coli* (Matadeen et al. 2001), this cross-link is not included in Figure 5c.

The final example in this section (Fig. 5d) gives the footprinting data of Huber and Wool (1984), which were obtained from isolated 5S rRNA–protein complexes using α-sarcin as a probe for nuclease sensitivity. It can be seen that the footprint data for proteins L18 and L25 are in good agreement with the in situ arrangement of 5S rRNA within the 50S subunit, but that the corresponding results for the third 5S rRNA-binding protein, L5, are less satisfactory. This is perhaps not surprising, as L5 was found to be only capable of binding to the 5S molecule in the presence of L18, and the L5 footprint was accordingly deduced by "subtraction" of the footprint for L18 (Huber and Wool 1984).

CROSS-LINKING TO THE RIBOSOME OF IN-VITRO-TRANSCRIBED 5S rRNA CARRYING PHOTOREAGENTS

The incorporation of photoactive nucleotides into short RNA molecules such as the 5S rRNA can easily be achieved by in vitro transcription. The most commonly used analog, 4-thiouridine, can be directly incorporated in this way (Dontsova et al. 1994), or alternatively, analogs carrying amino groups can be used, which are derivatized after the transcription with a suitable photoactive reagent such as a diazirine (Sergiev et al. 1998). In a similar approach, photoreactive groups have also been incorporated into selected regions of the 5S molecule by in vitro transcription of 5S rRNA fragments, followed by ligation (Osswald and Brimacombe 1999). The modified 5S rRNA transcripts are subsequently reconstituted into the 50S subunit, and cross-linking is induced by irradiation at wavelengths above 300 nm, which leaves the natural RNA bases undamaged. Cross-link sites on the 23S rRNA were analyzed by first localizing the cross-linked region as closely as possible by site-directed cleavage with ribonuclease H, followed by reverse transcription. The corresponding modified nucleotides in the 5S rRNA involved in the cross-links were determined by oligonucleotide fingerprinting (Dontsova et al. 1994). The results obtained from all these experiments are summarized in Figure 6a.

In those experiments in which the modified residues were distributed throughout the 5S rRNA molecule, all of the cross-links identified were to 5S residue U89, which has been shown by nuclear magnetic resonance (NMR) studies to be particularly exposed in the *E. coli* 5S molecule (Dallas and Moore 1997). This nucleotide is shown in blue in Figure 6a, and the cross-linked targets on the 23S rRNA are displayed in green. The figure shows that the nucleotide residues from domain II (in h39 and h41–42) which were cross-linked to the 5S rRNA form a cluster around U89. At the same time, although ribonuclease H was used as just noted to locate the cross-linked regions as closely as possible, the identification of the precise cross-linked nucleotides within these regions by primer extension can still be problematic and subject to artifact. This is exemplified by the supposed cross-link to nucleotide 2475 in helix h89. The ribonuclease H analysis localized this cross-link site to the region of h89 (shown as a blue backbone tube in Fig. 6a), and the base of this helix is indeed close to residue U89 of the 5S rRNA. On the other hand, nucleotide 2475, which was "identified" in the primer extension analysis, is at the loop-end of the helix, rather far away from U89. Similar difficulties were experienced in the primer extension analyses of cross-links from 6S rRNA carrying modifications at specific positions within helices hII–hIII (the red nucleotides in Fig. 6a), and in this case only the regions found from the ribonuclease H digestions are displayed, as pink and red backbone tubes. The cross-linked region

encompassing helices h83–h84 (pink) can be seen to lie close to the modified 5S nucleotides, whereas the corresponding region in h38 (red) is farther away; h38 (the "A-site finger") is, however, known to be a flexible part of the 50S structure (Matadeen et al. 1999).

It is worth noting that the cross-linking data, combined with results obtained by immune electron microscopy, enabled us to correctly predict the overall arrangement of the 5S rRNA in the 50S ribosomal subunit (Dontsova et al. 1994). These data indicated directly for the first time that the peptidyl transferase- and GTPase-associated regions of the 23S rRNA are in close proximity in the 50S subunit, and pointed to a possible participation of 5S rRNA in the interaction between these two functional domains. Furthermore, the technique of protection of phosphorothioate internucleotide bonds from iodine-induced cleavage has also given reliable results when applied to the 5S rRNA; those internucleotide bonds in the latter that are protected by the small subunit (Shpanchenko et al. 1998) are clustered in the X-ray structure at the interface surface of the large subunit (data not shown).

CROSS-LINKING OF RIBOSOMES CARRYING PHOTOREACTIVE GROUPS AT INTERNAL SITES IN THE rRNA

The construction of intact 16S or 23S rRNA molecules containing photoreactive reagents at desired specific in-

Figure 6. Examples of site-directed cross-linking in rRNA. (*a*) Close-up of the central protuberance area of the 50S subunit (viewed from the L7/L12 side) showing the cross-linking data for 5S rRNA. The 5S rRNA is displayed as an orange backbone tube, with nucleotides in its helical regions added in wire-frame format. Cross-links from a thiouridine residue or from diazirine derivatives at position U89 of the 5S molecule (highlighted as a *blue* CPK nucleotide) to positions in the 23S rRNA in the region of helices h39, h89, or h41–42 are marked by the green CPK nucleotides. In the case of the cross-link at position 2475 within h89, the whole helix is in addition displayed as a blue backbone tube. Cross-links in h38 or in the h83–h84 region of the 23S rRNA from thiouridines at positions 40, 48, 55, or 65 of the 5S rRNA (*red* CPK nucleotides) are indicated by the rRNA backbone tubes displayed in red and pink, respectively. See text for further explanation. (*b*) Nucleotides in 23S rRNA cross-linked from photoreactive complementary oligodeoxynucleotide probes. The 50S subunit and the 23S rRNA are displayed as in Fig. 1a. Cross-linked nucleotides from a probe at position 2252 (*red* CPK nucleotide) are shown as orange CPK nucleotides, and those from a probe at position 2604 (*blue* CPK nucleotide) as green CPK nucleotides. (Nucleotides that were cross-linked at positions proximal to the probe site itself are not shown.) (*c*) Nucleotides in 16S rRNA cross-linked from a reagent attached to the rRNA by oligonucleotide-directed psoralen photoaddition. The 30S subunit and the 16S rRNA are displayed as in Fig. 1b. The position of the probe (at nucleotides 788–789) is indicated by the red CPK nucleotides, and the positions of the resulting cross-link sites by the orange CPK nucleotides. See text for references.

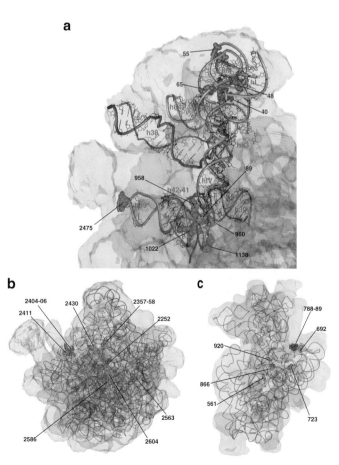

ternal positions is a very difficult task. Nevertheless, a number of attempts have been made to develop methods for the investigation of the environment of individual nucleotide residues within the large rRNA molecules, involving the use of either 5′-modified rRNA, fragments of rRNA, or oligonucleotides complementary to the rRNA. The results of these attempts have been discussed in detail by Sergiev et al. (2001; see also Leonov et al. 1999) and accordingly are only very briefly summarized here.

Photoreactive labels have been attached to the 5′ terminus of recombinant 16S rRNAs truncated at their 5′ end (Juzumiene and Wollenzien 2000), and also to the 5′ terminus of a 16S rRNA fragment encompassing its 3′ major domain (Montpetit et al. 1998), in order to investigate the 3D environment of the 16S molecule in the neighborhood of these 5′-end regions. In some of the recombinant 16S rRNA variants used, the 5′-end deletion was extended as far as the functionally important central pseudoknot region. Although some 16S rRNA residues could be identified as being in contact with the 5′ terminus concerned, the level of correlation between these data and the X-ray structure of the 30S subunit is in general disappointingly low.

Several laboratories have made use of rRNA molecules with an artificially introduced break at a single internucleotide bond for their research. The free 5′ or 3′ ends generated in this way were exploited as attachment sites either for photoactive reagents or for reagents catalyzing the cleavage of spatially proximal internucleotide bonds. In Noller's laboratory (Newcomb and Noller 1999), 16S rRNA fragments corresponding to nucleotides 1–423 and 424–1542 were used for the reconstitution of the 30S subunit, and the tethered Fe(II)-EDTA reagent (cf. Fig. 4c, d) was attached to the 5′ end of the second fragment. Comparison of the results of this work with the X-ray structure shows that the nucleotide residues found to be cleaved by this process were indeed restricted to the distance of 40–50 Å from nucleotide 424, in agreement with the estimated diffusion distance of the free hydroxyl radicals generated (Joseph et al. 1997). However, this large distance is not very informative, even for a particle the size of the ribosome. In our laboratories, we attached a photoreagent with a cross-linking range of ~10 Å to cleavage sites in the 16S or 23S rRNA (Baranov et al. 1997, 1998) in the hope that this would lead to more precise information. Unfortunately, this proved not to be the case; the nucleotides identified as being proximal to the cleavage sites are significantly separated in the crystal structures, most probably as a result of disturbance of the second and tertiary structure of the rRNA in the vicinity of these cleavages.

Another often-used approach has been the application of oligodeoxyribonucleotides complementary to selected single-stranded regions of the rRNA, carrying either photoreagents (see, e.g., Bukhtiyarov et al. 1999; Vladimirov et al. 2000) or cleavage-inducing reagents (Muth et al. 1999) at their termini. However, the binding of such oligonucleotides forces the complementary sequence on the rRNA into a helical conformation and must a priori disturb the local structure. Furthermore, it is clear from the crystal structures (Ban et al. 2000; Schluenzen et al. 2000) that practically all the "single-stranded" regions of the rRNA are themselves highly structured, and it is therefore not surprising that the results of this type of experiment do not correlate well with the X-ray structures. An example, in which photoreactive probes were attached to the complementary oligodeoxynucleotides under study (Bukhtiyarov et al. 1999; Vladimirov et al. 2000), is illustrated in Figure 6b. An interesting attempt to circumvent this problem was developed in Wollenzien's laboratory (Mundus and Wollenzien 1998). Here, an oligonucleotide carrying a psoralen residue was used to direct this residue by photochemical addition to its complementary region on the rRNA. After removal of the oligonucleotide by DNase treatment, a "secondary" photoreagent, azidophenacyl bromide, was attached to the psoralen moiety. As a result, an intact 16S rRNA molecule was obtained carrying the photoreactive azido group at nucleotide 788. The results are illustrated in Figure 6c, from which it can be seen that the correlation with the X-ray structure is again disappointing. How far this is a weakness of the method itself, or whether this is another example of the poor performance of the primer extension method in the analysis of the cross-linked sites, is impossible to assess.

THE STUDY OF rRNA–LIGAND INTERACTIONS

The interaction of the ribosome with its ligands, primarily mRNA and tRNA, has been the subject of many studies, and very recently an X-ray structure of the *T. thermophilus* 70S ribosome carrying tRNAs and mRNA has been published at 5.5 Å resolution (Yusupov et al. 2001). Moreover, in the corresponding structure of the 50S ribosomal subunit from *H. marismortui* (Ban et al. 2000), analogs of the 3′ ends of tRNA molecules in a state imitating the peptide transfer were observed at atomic resolution (Nissen et al. 2000). The positions of tRNA and mRNA have also been deduced from relatively low resolution structures obtained by cryo-EM (Stark et al. 1997a,b; Agrawal et al. 1999). Here again, a detailed comparison of the biochemical data with the atomic structures has been made by Sergiev et al. (2001). Recently, the three-dimensional structures of complexes of ribosomal subunits with more than half a dozen different antibiotics have been analyzed by X-ray crystallography (Brodersen et al. 2000; Carter et al. 2000; Pioletti et al. 2001); on the whole, the X-ray data were found to be in excellent agreement with the chemical probing and cross-linking results. In the following sections we restrict ourselves to a brief summary of the data relating to ribosomal contacts with tRNA, mRNA, and the growing peptide chain.

Contacts between tRNA and rRNA

The interaction of the CCA end of tRNA with the ribosomal A and P sites has been studied by a compensatory mutation approach, which predicted a base-pairing between nucleotide C-74 of the P-site-bound tRNA with residue G2252 of the 23S rRNA (Samaha et al. 1995) and, similarly, a pairing between C75 of the A-site-bound tRNA and G2553 of the 23S rRNA (Kim and Green 1999). Both of these predictions have proved to be correct. Here, as in the rest of this paper, we are more concerned with the footprinting and cross-linking type of approach, and examples of these data are illustrated in Figure 7.

A number of nucleotides in the 23S and 16S rRNA are protected by tRNA from chemical modification (Moazed and Noller 1989, 1990; Bocchetta et al. 2001), and those

that relate to the A- or P-site tRNA are given in Figure 7a and b, respectively. Nucleotides whose modification interferes with tRNA binding (von Ahsen and Noller 1995) form a subset of those that are protected by the tRNAs. It is possible that the protection of some nucleotides by tRNA could be mediated by other components of the translation apparatus, which change their location upon tRNA binding. For example, nucleotide 532 of the 16S rRNA (Fig. 7a) is separated from the major nucleotide cluster protected by P-site-bound tRNA; it is known that this nucleotide is in direct contact with mRNA (Dontsova et al. 1992; see below), so it may well be that its protection reflects a tRNA-dependent docking of mRNA to the

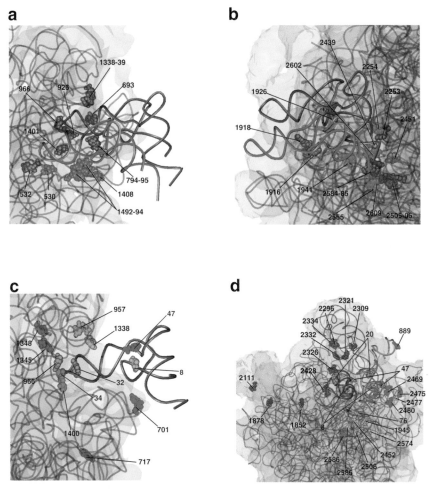

Figure 7. Footprints and site-directed cross-links from tRNA. In each part of the figure, the 30S or 50S subunits, and the 16S or 23S rRNA, are displayed as in Fig. 1. A-site and P-site tRNA molecules are shown as green and red backbone tubes, respectively. (*a*) Footprint sites from tRNA on the 16S rRNA. The A-site and P-site footprints are marked by the green and red CPK nucleotides, respectively. (*b*) Footprint sites from tRNA on the 23S rRNA. The sites are marked as in *a*, with sites that are footprinted by both A- and P-site tRNA being additionally marked in orange. (*c*) Site-directed cross-links from P-site tRNA on the 16S rRNA. The cross-link from the naturally photoreactive residue at position 34 of the tRNA to nucleotide 1400 is indicated by the blue CPK nucleotides. The cross-links from photoreactive residues introduced at positions 8, 32, and 47 of the tRNA are indicated by CPK residues (both on the tRNA and on the 16S rRNA) in the same color. (*d*) Site-directed cross-links from P- or A-site tRNA to the 23S rRNA. The P-site specific cross-link from position 76 of the tRNA to nucleotide 1945 is marked by the pink CPK nucleotides. Cross-links from photoreactive residues at positions 8, 20, or 47 in the elbow region of the tRNA (marked on the A- and P-site tRNA molecules by orange CPK nucleotides) are denoted by red CPK nucleotides (P site) or green CPK nucleotides (A site). Yellow CPK nucleotides denote cross-linked positions that were not specific to either the A or the P site. The loop-proximal region of helix 38 (which was cross-linked from all three elbow positions of the A-site tRNA) is highlighted by the blue backbone tube. See text for references.

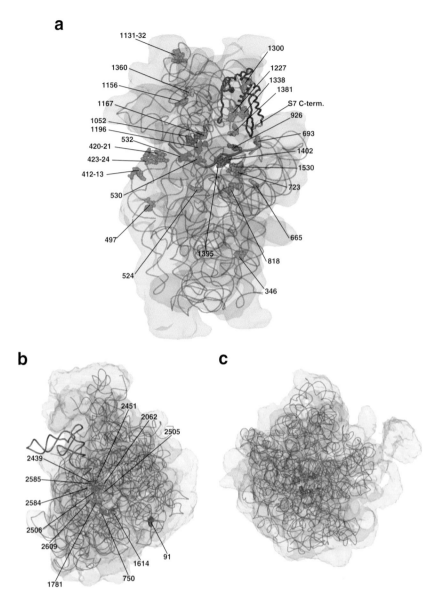

Figure 8. Cross-links from mRNA or from the amino terminus of nascent peptides. The 30S or 50S subunits, and the 16S or 23S rRNA, are displayed as in Fig. 1. (*a*) Cross-links from mRNA. The putative path of the mRNA is indicated by the colored tube, whereby the orange segment represents the upstream region of the mRNA (residues –6 to –1), the red segment the P- and A-site codons (residues +1 to +6), and the green segment the downstream region (residues +7 to +12). Cross-links identified in our laboratories to these three mRNA regions are marked by CPK nucleotides on the 16S rRNA in the same colors. A cross-link to the carboxyl terminus of protein S7 (*blue* backbone tube) is similarly indicated by the orange marking. The blue CPK nucleotides are sites of cross-linking from mRNA identified by Wollenzien and colleagues. (*b*) Cross-links from the amino acid residue of aminoacyl tRNA, or from the amino terminus of nascent peptides of different lengths. The red tRNA molecule is at the ribosomal P site, and the sites close to its CCA end represented by the CPK nucleotides in different colors are cross-links from the aminoacyl residue. The CPK nucleotides at positions 2609, 1781, 750, 1614, and 91 indicate cross-links from the amino terminus of peptides with increasing length. (*c*) The same data as in *b* (without the tRNA molecule), with the 50S subunit rotated so as to show a view along the peptide tunnel, with the exit site closest to the reader. (The L1 protuberance can be seen on the upper right side of the figure.) See text for references and further explanation.

decoding center. With this exception, all of the 16S rRNA nucleotides protected by tRNA bound to the A or P sites form compact clusters surrounding the respective tRNA molecule (Fig. 7a). A similar situation is observed in the 50S subunit (Fig. 7b), where only nucleotide 2254 is distant from the major A-site-associated group. Here, the proximity of this nucleotide to the P-site-protected clus-

ter could be indicative of an allosteric influence of A-site occupation at the P site.

In contrast, the situation with regard to cross-links between tRNA and the rRNA is in general less satisfactory. Figure 7c shows site-specific cross-links that have been reported from the P-site tRNA to 16S rRNA, and it can be seen that, whereas the cross-links from the tRNA anticodon loop at positions 34 (Prince et al. 1982) and 32 (Döring et al. 1994) are consistent with the atomic structure of the 16S rRNA, those from the elbow region of the molecule at positions 8 or 47 (Rinke-Appel et al. 1995) are not. This is most likely due to the tRNA slipping from one binding site to another under the conditions of cross-linking, but it is also possible that the attachment of the cross-linking agent to the tRNA position concerned was not as specific as we had believed in these experiments. Again, a similar situation is found in the 50S subunit (Fig. 7d). In this case, the cross-link from P-site tRNA position 76 to nucleotide 1945 (Wower et al. 1989, 2000) is consistent with the atomic structure of the 23S rRNA, but the cross-links from the elbow region (positions 8, 20, or 47) of either P- or A-site tRNA (Rinke-Appel et al. 1995) show a wide spread of targets, with those from the P site extending toward the expected E-site location (on the left in Fig. 7d; cf. Stark et al. 1997b; Agrawal et al. 1999) and those from the A site extending toward the "pre-A site" location (on the right in Fig. 7d; cf. Stark et al. 1997a). Furthermore, a number of cross-link sites were observed (yellow in Fig. 7d) for which there was no apparent specificity for A- or P-site tRNA binding. It is noteworthy that the supposed A-site-specific cross-links include sites in helices h38 and h89 (nucleotides 2469–2480 in the latter case), both of these helices also being targets for cross-linking from the 5S rRNA (see Fig. 6a).

The tRNA environment in the ribosome has also been studied using reagents inducing internucleotide bond cleavage. Copper–phenanthroline complexes attached to tRNA were used to study the ribosomal P site (see, e.g., Bullard et al. 1995), and similarly, Noller's group used Fe(II)-EDTA derivatives of tRNA anticodon analogs for structural studies on the ribosomal A and P sites (Joseph et al. 1997). The analysis of these data by Sergiev et al. (2001) showed that the nucleotides cleaved by hydroxyl radicals emitted by the A-site-bound anticodon analogs are located close to each other and are in agreement with other data relating to the ribosomal A site, within the framework of the 40–50 Å resolution estimated for the tethered Fe(II)-EDTA method (Joseph et al. 1997). On the other hand, the corresponding data for the P site showed cleavages occurring at larger distances than expected. It seems likely that in the latter case hydroxyl radicals released into the intersubunit cavity could diffuse to all unprotected residues protruding into the space between the subunits. Such an extensive level of internucleotide bond cleavage would not take place in the case of the A-site-bound anticodon analogs, since here the intersubunit cavity was partially blocked by the P-site tRNA. In contrast, the method of site-directed cleavage induced by copper–phenanthroline derivatives (Bullard et al. 1995) shows a disappointingly low level of specificity, with the target sites being spread apparently randomly throughout the ribosome (see Sergiev et al. 2001).

Contacts between mRNA and rRNA

There are no footprinting data relating directly to the mRNA, but its environment in the ribosome has been studied extensively using a set of photoreactive nucleotides incorporated into mRNA analogs by in vitro transcription (see, e.g., Wollenzien et al. 1991; Banghu and Wollenzien 1992; Dontsova et al. 1992; Sergiev et al. 1997; cf. the 5S rRNA, above). The results are illustrated in Figure 8a, in which the putative path of the mRNA (based on the location of the two codons binding the A- and P-site rRNAs) is indicated by the colored tube. The nucleotides identified in our laboratories (Dontsova et al. 1992; Rinke-Appel et al. 1993; Sergiev et al. 1997) as being cross-linked from the downstream region of the mRNA (residues +1 to +12) correlate well with the X-ray structure of the 30S subunit, but this is not the case with the corresponding cross-links from the spacer region between the Shine-Dalgarno sequence and the AUG codon (Rinke-Appel et al. 1994); the cross-links to 16S rRNA nucleotides 665 and (especially) 1360 are not close to the mRNA path and cannot be explained even by a high flexibility of the mRNA spacer region. On the other hand, the cross-link from the latter region to the carboxyl terminus of protein S7 (Greuer et al. 1999) is plausible (Fig. 8a). We cannot offer any explanation for this discrepancy.

Figure 8a also includes the set of cross-links from mRNA to the 16S rRNA identified in Wollenzien's laboratory (Wollenzien et al. 1991; Banghu and Wollenzien 1992). In these experiments the precise residues of the mRNA that were involved in the individual cross-links were not determined, and furthermore, the cross-link sites were identified by reverse transcription only, without a preliminary localization by oligonucleotide-directed ribonuclease H digestion. Figure 8a shows that the cross-link sites found are distributed throughout the 30S subunit.

Contacts between the Nascent Peptide and rRNA

The path of the growing peptide within the 50S ribosomal subunit has been studied by cross-linking using photoreactive derivatives attached either to the aminoacyl residue of aminoacyl-tRNA (Steiner et al. 1988; Mitchell et al. 1993; Stade et al. 1995) or to the amino terminus of peptides of different lengths synthesized in situ on the ribosome (Stade et al. 1995; Choi and Brimacombe 1998). The results of these experiments are shown in Figure 8,b and c. The cross-links from the aminoacyl residue lie—as would be expected—close to the 3′ terminus of the tRNA, whereas those from the amino terminus of peptides of increasing length penetrate progressively further into the peptide tunnel (Fig. 8b), all of the sites being within or very close to the tunnel itself (Fig. 8c). As discussed in a recent review (Brimacombe 2000), the correlation of these cross-link data with the X-ray structure of the 50S subunit (Ban et al. 2000) provides in retrospect the most direct evidence that the nascent peptide does indeed pass through the tunnel. It is interesting to compare this high degree of agreement between the nascent peptide cross-links and the X-ray structure with the much less satisfactory level of agreement found with the cross-links from tRNA described above (Fig. 7c, d), since the cross-linking technology employed was very similar in both cases. It seems likely that flexibility of the nascent peptide chain in the 50S subunit (cf. Choi and Brimacombe 1998) is confined to the limits of the tunnel, whereas, in contrast, the tRNA molecules—as suggested above—can potentially move about on the ribosomal surface. In other words, it is probably the mobility of the ligand concerned under the conditions of cross-linking that leads to an overly wide distribution of cross-link sites, rather than a defect in the cross-linking methodology per se.

CONCLUSIONS

It should be clear from the foregoing discussion that the quality of the biochemical data in relation to the X-ray structures of the ribosomal subunits covers as wide a range as do the biochemical techniques themselves. If any generalization at all is possible, it is the perhaps surprising one that the older and simpler techniques have proved on the whole to be more reliable than the newer, more sophisticated methodologies. The set of methods based on the chemical modification of nucleotides—footprinting— gives reliable information, although it must be borne in mind that footprinting data will always include effects arising from induced fits or allosteric changes caused by binding of the protein or ligand concerned, and do not necessarily reflect direct contacts. In the case of the footprinting methods based on Fe(II)-EDTA cleavage, the "simple" footprinting approach (Powers and Noller 1995) has given more useful results than the tethered Fe(II)-EDTA method (see, e.g., Lieberman et al. 2000). On the other hand, the copper–phenanthroline cleavage technique (Bullard et al. 1995) does not appear to be reliable. For the investigation of the 3D structures of the large rRNA molecules, there are unfortunately still no good methods for the introduction of a photoreactive residue at internal positions with a view to inducing site-specific cross-links within the molecules (Sergiev et al. 2001). Here the best cross-linking methods—as is also the case with RNA–protein cross-linking—are unquestionably the direct "old fashioned" methods (Figs. 1–3), and it is regrettable that a prerequisite for the application of these methods is the availability of highly radioactive in-vivo-labeled substrates, which nobody nowadays would want to prepare. One of the main reasons for the success of these old methods was the use of oligonucleotide fingerprinting techniques for the identification of the cross-link sites, and we end by once again stressing the importance of using oligonucleotide-directed cleavage with ribonuclease H, or some similarly appropriate method, for the preliminary localization of cross-link sites; primer extension analyses alone can be very misleading.

ACKNOWLEDGMENTS

The authors gratefully acknowledge financial support from the Volkswagen Foundation (I/74 598), the Russian Foundation for Basic Research (99-04-49070 and 99-04-49054), the Deutsche Forschungsgemeinschaft (Br 632/4-1) and the Alexander von Humboldt Foundation (Research Award to A.B.).

REFERENCES

Agrawal R.K., Penczek P., Grassucci R.A., Burkhardt N., Nierhaus K.H., and Frank J. 1999. Effect of buffer conditions on the position of tRNA on the 70S ribosome as visualized by cryoelectron microscopy. *J. Biol. Chem.* **274:** 8723.

Ban N., Nissen P., Hansen J., Moore P.B., and Steitz T.A. 2000. The complete atomic structure of the large ribosomal subunit at 2.4 Å resolution. *Science* **289:** 905.

Banghu R. and Wollenzien P. 1992. The mRNA binding track in the *E. coli* ribosome for mRNAs of different sequences. *Biochemistry* **31:** 5937.

Baranov P.V., Gurvich O.L., Bogdanov A.A., Brimacombe R., and Dontsova O.A. 1998. New features of 23S ribosomal RNA folding: The long helix 41-42 makes a "U-turn" inside the ribosome. *RNA* **4:** 658.

Baranov P.V., Dokudovskaya S.S., Oretskaya T.S., Dontsova O.A., Bogdanov A.A., and Brimacombe R. 1997. A new technique for the characterization of long-range tertiary contacts in large RNA molecules: Insertion of a photolabel at a selected position in 16S rRNA within the *E. coli* ribosome. *Nucleic Acids Res.* **25:** 2266.

Bocchetta M., Xiong L., Shah S. and Mankin A.S. 2001. Interactions between 23S rRNA and tRNA in the ribosomal E site. *RNA* **7:** 54.

Bogdanov A.A., Sergiev P.V., Lavrik I.N., Spanchenko O.V., Leonov A.A., and Dontsova O.A. 2000. Modern site-directed cross-linking approaches: Implication for ribosome structure and functions. In *The ribosome: Structure, function, antibiotics and cellular interactions* (ed. R.A. Garrett et al.), p. 245. ASM Press, Washington, D.C.

Brimacombe R. 1995. The structure of ribosomal RNA: A three-dimensional jigsaw puzzle. *Eur. J. Biochem.* **230:** 365.

———. 2000. The bacterial ribosome at atomic resolution. *Structure* **8:** R195.

Brimacombe R., Greuer B., Mueller F., Osswald M., Rinke-Appel J., and Sommer I. 2000. Three-dimensional organization of the bacterial ribosome and its subunits: Transition from low-resolution models to high-resolution structures. In *The ribosome: Structure, function, antibiotics and cellular interactions* (R.A. Garrett et al.), p. 151. ASM Press, Washington D.C.

Brodersen D.E., Clemons W.M., Carter A.P., Morgan-Warren R.J., Wimberly B.T., and Ramakrishnan V. 2000. The structural basis for the action of the antibiotics tetracycline, pactamycin, and hygromycin B on the 30S ribosomal subunit. *Cell* **103:** 1143.

Bukhtiyarov Y., Druzina Z., and Cooperman B.S. 1999. Identification of 23S rRNA nucleotides neighboring the P-loop in the *E. coli* 50S subunit. *Nucleic Acids Res.* **27:** 4376.

Bullard J.M., van Waes M.A., Bucklin D.J., and Hill W.E. 1995. Regions of 23S ribosomal RNA proximal to transfer RNA bound at the P and E sites. *J. Mol. Biol.* **252:** 572.

Capel M.S., Kjeldgaard M., Engelman D.M., and Moore P.B. 1988. Positions of S2, S13, S16, S17, S19 and S21 in the 30S ribosomal subunit of *E. coli. J. Mol. Biol.* **200:** 65.

Carter A.P., Clemons W.M., Brodersen D.E., Morgan-Warren R.J., Wimberly B.T., and Ramakrishnan V. 2000. Functional insights from the structure of the 30S ribosomal subunit and its interactions with antibiotics. *Nature* **407:** 340.

Choi K.M. and Brimacombe R. 1998. The path of the growing peptide chain through the 23S rRNA in the 50S ribosomal subunit; a comparative cross-linking study with three different peptide families. *Nucleic Acids Res.* **26:** 887.

Culver G.M., Heilek G.M., and Noller H.F. 1999. Probing the rRNA environment of ribosomal protein S5 across the subunit interface and inside the 30S subunit using tethered Fe(II). *J. Mol. Biol.* **286:** 355.

Dallas A. and Moore P.B. 1997. The loop E-loop D region of *E. coli* 5S rRNA: The solution structure reveals an unusual loop that may be important for binding ribosomal proteins. *Structure* **5:** 1639.

Dontsova O., Dokudovskaya S., Kopylov A., Bogdanov A.A., Rinke-Appel J., Jünke N., and Brimacombe R. 1992. Three widely separated positions in the 16S RNA lie in or close to the ribosomal decoding region; a site-directed cross-linking study with mRNA analogues. *EMBO J.* **11**: 3105.

Dontsova O., Tishkov V., Dokudovskaya S., Bogdanov A., Döring T., Rinke-Appel J., Thamm S., Greuer B., and Brimacombe R. 1994. Stem-loop IV of 5S rRNA lies close to the peptidyl transferase center. *Proc. Natl. Acad. Sci.* **91**: 4125.

Döring T., Mitchell P., Osswald M., Bochkariov D., and Brimacombe R. 1994. The decoding region of 16S RNA; a cross-linking study of the ribosomal A, P and E sites using tRNA derivatized at position 32 in the anticodon loop. *EMBO J.* **13**: 2677.

Egebjerg J., Christiansen J., and Garrett R.A. 1991. Attachment sites of primary binding proteins L1, L2 and L23 on 23S ribosomal RNA of *E. coli*. *J. Mol. Biol.* **222**: 251.

Gabashvili I.S., Agrawal R.K., Spahn C.M.T., Grassucci R.A., Svergun D.I., Frank J., and Penczek P. 2000. Solution structure of the *E. coli* ribosome at 11.5 Å resolution. *Cell* **100**: 537.

Green R. and Noller H.F. 1997. Ribosomes and translation. *Annu. Rev. Biochem.* **66**: 679.

Greuer B., Thiede B., and Brimacombe R. 1999. The cross-link from the upstream region of mRNA to ribosomal protein S7 is located in the C-terminal peptide; experimental verification of a prediction from modelling studies. *RNA* **5**: 1521.

Heilek G., Marusek R., Meares C.F., and Noller H.F. 1995. Direct hydroxyl radical probing of 16S rRNA using Fe(II) tethered to ribosomal protein S4. *Proc. Natl. Acad. Sci.* **92**: 1113.

Huber P.W. and Wool I.G. 1984. Nuclease protection analysis of ribonucleoprotein complexes: Use of the cytotoxic ribonuclease α-sarcin to determine the binding sites for *E. coli* ribosomal proteins L5, L18 and L25 on 5S rRNA. *Proc. Natl. Acad. Sci.* **81**: 322.

Joseph S., Weiser B., and Noller H.F. 1997. Mapping the inside of the ribosome with an RNA helical ruler. *Science* **278**: 1093.

Juzumiene D.I. and Wollenzien P. 2000. Organization of the 16S rRNA around its 5′ terminus determined by photochemical cross-linking in the 30S ribosomal subunit. *RNA* **6**: 26.

Kim D.F. and Green R. 1999. Base-pairing between 23S rRNA and tRNA in the ribosomal A site. *Mol. Cell* **4**: 859.

Leonov A.A., Sergiev P.V., Dontsova O.A., and Bogdanov A.A. 1999. Directed introduction of photoaffinity reagents in internal segments of RNA. *Mol. Biol.* **33**: 1063.

Lieberman K.R., Firpo M.A., Herr A.J., Nguyenle T., Atkins J.F., Gesteland R.F., and Noller H.F. 2000. The 23S rRNA environment of ribosomal protein L9 in the 50S ribosomal subunit. *J. Mol. Biol.* **297**: 1129.

Matadeen R., Patwardhan A., Gowen B., Orlova E., Stark H., Pape T., Cuff M., Mueller F., Brimacombe R., and van Heel M. 1999. The *E. coli* large ribosomal subunit at 7.5 Å resolution. *Structure* **7**: 1575.

Matadeen R., Sergiev P., Leonov A., Pape T., van der Sluis E., Mueller F., Osswald M., von Knoblauch K., Brimacombe R., Bogdanov A., van Heel M., and Dontsova O. 2001. Direct localization by cryo-electron microscopy of secondary structural elements in *E. coli* which differ from the corresponding regions in *H. marismortui*. *J. Mol. Biol.* **307**: 1341.

Mitchell P., Stade K., Osswald M., and Brimacombe R. 1993. Site-directed cross-linking studies on the *E. coli* tRNA-ribosome complex; determination of sites labelled with an aromatic azide attached to the variable loop or aminoacyl group of tRNA. *Nucleic Acids Res.* **21**: 887.

Moazed D. and Noller H.F. 1989. Interaction of tRNA with 23S rRNA in the ribosomal A, P and E sites. *Cell* **57**: 585.

———. 1990. Binding of tRNA to the ribosomal A and P sites protects two distinct sets of nucleotides in 16S rRNA. *J. Mol. Biol.* **211**: 135.

Möller K., Zwieb C., and Brimacombe R. 1978. Identification of the oligonucleotide and oligopeptide involved in an RNA-protein cross-link induced by UV-irradiation of *E. coli* 30S ribosomal subunits. *J. Mol. Biol.* **126**: 489.

Montpetit A., Payant C., Nolan J.M., and Brakier-Gingras L. 1998. Analysis of the conformation of the 3′ major domain of *E. coli* 16S ribosomal RNA using site-directed photoaffinity

cross-linking. *RNA* **4**: 1455.

Mueller F. and Brimacombe R. 1997a. A new model for the three-dimensional folding of *E. coli* 16S ribosomal RNA. I. Fitting the RNA to a 3D electron microscopic map at 20 Å. *J. Mol. Biol.* **271**: 524.

———. 1997b. A new model for the three-dimensional folding of *E. coli* 16S ribosomal RNA. II. The RNA-protein interaction data. *J. Mol. Biol.* **271**: 545.

Mueller F., Sommer I., Baranov P., Matadeen R., Stoldt M., Wöhnert J., Görlach M., van Heel M., and Brimacombe R. 2000. The 3D arrangement of the 23S and 5S rRNA in the *E. coli* 50S ribosomal subunit based on a cryo-electron microscopic reconstruction at 7.5 Å resolution. *J. Mol. Biol.* **298**: 35.

Mundus D. and Wollenzien P. 1998. Neighborhood of 16S rRNA nucleotides U788/U789 in the 30S ribosomal subunit determined by site-directed cross-linking. *RNA* **4**: 1373.

Muth G.W., Hennelly S.P., and Hill W.E. 1999. Positions in the 30S ribosomal subunit proximal to the 790 loop as determined by phenathroline cleavage. *RNA* **5**: 856.

Newcomb L.F. and Noller H.F. 1999. Directed hydroxyl radical probing of 16S rRNA in the ribosome: Spatial proximity of RNA elements of the 3′ and 5′ domains. *RNA* **5**: 849.

Nissen P., Hansen J., Ban N., Moore P.B., and Steitz T.A. 2000. The structural basis of ribosome activity in peptide bond synthesis. *Science* **289**: 920.

Osswald M. and Brimacombe R. 1999. The environment of 5S rRNA in the ribosome: Cross-links to 23S rRNA from sites within helices II and III of the 5S molecule. *Nucleic Acids Res.* **27**: 2283.

Osswald M., Greuer B. and Brimacombe R. 1990. Localization of a series of RNA-protein cross-link sites in the 23S and 5S ribosomal RNA from *E. coli*, induced by treatment of 50S subunits with three different bifunctional reagents. *Nucleic Acids Res.* **18**: 6755.

Pioletti M., Schlünzen F., Harms J., Zarivach R., Glühmann M., Avila H., Bashan A., Bartels H., Auerbach T., Jacobi C., Hartsch T., Yonath A., and Franceschi F. 2001. Crystal structures of complexes of the small ribosomal subunit with tetracycline, edeine and If3. *EMBO J.* **20**: 1829.

Powers T. and Noller H.F. 1995. Hydroxyl radical footprinting of ribosomal proteins on 16S rRNA. *RNA* **1**: 194.

Prince J.B., Taylor B.H., Thurlow D.L., Ofengand J., and Zimmermann R.A. 1982. Covalent cross-linking of tRNA$_1^{Val}$ to 16S RNA at the ribosomal P site; identification of cross-linked residues. *Proc. Natl. Acad. Sci.* **79**: 5450.

Rinke-Appel J., Jünke N., Osswald M., and Brimacombe R. 1995. The ribosomal environment of tRNA; cross-links to rRNA from positions 8 and 20:1 in the central fold of tRNA located at the A, P or E site. *RNA* **1**: 1018.

Rinke-Appel J., Jünke N., Brimacombe R., Dokudovskaya S., Dontsova O., and Bogdanov A. 1993. Site-directed cross-linking of mRNA analogues to 16S ribosomal RNA; a complete scan of cross-links from all positions between +1 and +16 on the mRNA, downstream from the decoding site. *Nucleic Acids Res.* **21**: 2853.

Rinke-Appel J., Jünke N., Brimacombe R., Lavrik I., Dokudovskaya S., Dontsova A., and Bogdanov A. 1994. Contacts between 16S ribosomal RNA and mRNA, within the spacer region separating the AUG initiator codon and the Shine-Dalgarno sequence; a site-directed cross-linking study. *Nucleic Acids Res.* **22**: 3018.

Samaha R.R., Green R., and Noller H.F. 1995. A base-pair between tRNA and 23S rRNA in the peptidyl transferase center of the ribosome. *Nature* **377**: 309.

Schluenzen F., Tocilj A., Zarivach R., Harms J., Gluehmann M., Janell D., Bashan A., Bartels H., Agmon I., Franceschi F., and Yonath A. 2000. Structure of functionally activated small ribosomal subunit at 3.3 Å resolution. *Cell* **102**: 615.

Schwedler G., Albrecht-Erlich R., and Rak K.H. 1993. Immunoelectron microscopic localization of proteins BS8, BS9, BS20, BL3 and BL21 on the surface of 30S and 50S subunits from *B. stearothermophilus*. *Eur. J. Biochem.* **217**: 361.

Sergiev P.V., Dontsova O.A., and Bogdanov A.A. 2001. Chemical methods in the structural study of the ribosome: Judgement day. *Mol. Biol.* **35:** 472.

Sergiev P., Dokudovskaya S., Romanova E., Topin A., Bogdanov A., Brimacombe R., and Dontsova O. 1998. The environment of 5S rRNA in the ribosome: Cross-links to the GTPase-associated area of 23S rRNA. *Nucleic Acids Res.* **26:** 2519.

Sergiev P.V., Lavrik I.N., Wlasoff V.A., Dokudovskaya S.S., Dontsova O.A., Bogdanov A.A., and Brimacombe R. 1997. The path of mRNA through the bacterial ribosome; a site-directed cross-linking study using new photoreactive derivatives of guanosine and uridine. *RNA* **3:** 464.

Shpanchenko O.V., Dontsova O.A., Bogdanov A.A., and Nierhaus K.H. 1998. Structure of 5S rRNA within the *E. coli* ribosome. Iodine-induced cleavage patterns of phosphothioate derivatives. *RNA* **4:** 1154.

Stade K., Jünke N., and Brimacombe R. 1995. Mapping the path of the nascent peptide chain through the 23S rRNA in the 50S ribosomal subunit. *Nucleic Acids Res.* **23:** 2371.

Stark H., Rodnina M., Rinke-Appel J., Brimacombe R., Wintermeyer W., and van Heel M. 1997a. Visualization of elongation factor Tu on the *Escherichia coli* ribosome. *Nature* **389:** 403.

Stark H., Orlova E.V., Rinke-Appel J., Jünke N., Mueller F., Rodnina M., Wintermeyer W., Brimacombe R., and van Heel M. 1997b. Arrangement of tRNAs in pre- and post-translocational ribosomes revealed by electron cryomicroscopy. *Cell* **88:** 19.

Steiner G., Kuechler E., and Barta A. 1988. Photo-affinity labelling at the peptidyl transferase centre reveals two different positions for the A and P sites in domain V of 23S rRNA. *EMBO J.* **7:** 3949.

Stern S., Weiser B., and Noller H.F. 1988. Model for the three-dimensional folding of 16S ribosomal RNA. *J. Mol. Biol.* **204:** 447.

Tanaka I., Nakagawa A., Hosaka H., Wakatsuki S., Mueller F., and Brimacombe R. 1998. Matching the crystallographic structure of ribosomal RNA. *RNA* **4:** 542.

Thiede B., Urlaub H., Neubauer H., Grelle G., and Wittmann-Liebold B. 1998. Precise determination of RNA-protein contact sites in the 50S ribosomal subunit of *E. coli*. *Biochem. J.* **334:** 39.

Urlaub H., Thiede B., Müller E.C., Brimacombe R., and Wittmann-Liebold B. 1997. Identification and sequence analysis of contact sites between ribosomal proteins and rRNA in *E. coli* 30S subunits by a new approach using matrix-assisted laser desorption/ionization-mass spectrometry combined with N-terminal microsequencing. *J. Biol. Chem.* **272:** 14547.

Vladimirov S.N., Druzina Z., Wang R., and Cooperman B.S. 2000. Identification of 50S components neighboring 23S rRNA nucleotides A2448 and U2604 within the peptidyl transferase center of *E. coli* ribosomes. *Biochemistry* **39:** 183.

Von Ahsen U. and Noller H.F. 1995. Identification of bases in 16S rRNA essential for tRNA binding at the 30S ribosomal P site. *Science* **267:** 234.

Wilms C., Noah J.W., Zhong D., and Wollenzien P. 1997. Exact determination of UV-induced crosslinks in 16S ribosomal RNA in 30S ribosomal subunits. *RNA* **3:** 602.

Wimberly B.T., Brodersen D.E., Clemons W.M., Morgan-Warren R.J., Carter A.P., Vohrhein C., Hartsch T., and Ramakrishnan V. 2000. Structure of the 30S ribosomal subunit. *Nature* **407:** 327.

Wollenzien P., Expert-Bezançon A., and Favre A. 1991. Sites of contact of mRNA with 16S rRNA and 23S rRNA in the *E. coli* ribosome. *Biochemistry* **30:** 1788.

Wower J., Hixson S.S., and Zimmermann R.A. 1989. Labeling the peptidyl transferase center of the *E. coli* ribosome with photoreactive tRNA[Phe] derivatives containing azidoadenosine at the 3′ end of the acceptor arm; a new model of the tRNA-ribosome complex. *Proc. Natl. Acad. Sci.* **86:** 5232.

Wower J., Kirillov S., Wower I., Guven S., Hixson S., and Zimmermann R.A. 2000. Transit of tRNA through the *E. coli* ribosome: Cross-linking of the 3′ end of tRNA to specific nucleotides of the 23S ribosomal RNA at the A, P and E sites. *J. Biol. Chem.* **275:** 37887.

Yusupov M.M., Yusupova G.Z., Baucom A., Liebermann K., Earnest T.N., Cate J.H.D., and Noller H.F. 2001. Crystal structure of the ribosome at 5.5 Å resolution. *Science* **292:** 883.

Probing Ribosome Structure and Function by Mutagenesis

S.T. Gregory, M.A. Bayfield, M. O'Connor, J. Thompson, and A.E. Dahlberg
Department of Molecular Biology, Cell Biology and Biochemistry, Brown University,
Providence, Rhode Island 02912

Recent ribosome structure studies provide a new framework within which to interpret results from decades of genetic and biochemical experiments, and indicate numerous new experimental strategies to investigate how the ribosome works in atomic detail. The first high-resolution crystal structures of ribosomal subunits will eventually be supplanted by structures of more ambitious ribosome complexes, representing several, perhaps even most, of the intermediate stages of protein synthesis. In this context, high-resolution structures of ribosomes bearing biochemically well-defined mutations will provide additional insight into questions regarding the translation mechanism. Cryo-electron microscopic (cryo-EM) reconstructions of ribosome complexes will complement higher-resolution studies and will also continue to make important contributions to our understanding of how mutations perturb the global conformation of the ribosome. Ultimately, a blend of classic genetics, biochemistry, X-ray crystallography, and cryo-EM will guide our investigations into the underlying mechanisms of protein synthesis. Here we describe several aspects of ribosome function currently being addressed in our laboratory by a combination of genetics, biochemistry, and structural biology. These include dynamic structural changes during protein synthesis, the role of intersubunit bridges, and the mechanism of peptide-bond formation.

STRUCTURAL DYNAMICS OF THE RIBOSOME

The notion that the ribosome's structure is dynamic has become an article of faith in the ribosome community. Structure-probing experiments have hinted at conformational rearrangements of rRNA during the Zamir-Elson active–inactive transition (Moazed et al. 1986) and upon tRNA or antibiotic binding or subunit association (Moazed and Noller 1987), whereas structure probing with metal ions has revealed conformational changes in the large subunit accompanying translocation (Polacek et al. 2000). Recent structural studies provide the most compelling support for this notion. Conformational changes occurring during EF-G binding (Agrawal et al. 1999), followed by an intersubunit ratcheting motion during translocation (Frank and Agrawal 2000), have been observed by cryo-EM. Such intersubunit movement is predicted by the hybrid states model (Moazed and Noller 1989). Large-scale conformational rearrangements

within 40S subunits upon binding an IRES element have also been observed by cryo-EM (Spahn et al. 2001). Chemical probing (Gregory and Dahlberg 1999a) and cryo-EM reconstructions (Gabashvili et al. 2001) of ribosomes bearing mutations in ribosomal proteins L4 or L22 suggest the existence of a dynamic gating mechanism in the polypeptide exit channel. At atomic resolution, movement of specific nucleotide residues upon tRNA or antibiotic binding (Ogle et al. 2001), or initiation factor binding (Carter et al. 2001), has been observed by X-ray crystallography.

Our laboratory has acquired genetic evidence of a conformational rearrangement occurring within the 30S subunit (Fig. 1) (Lodmell and Dahlberg 1997). The model, derived in part from these experiments, includes the cooperation of ribosomal proteins S4, S5, and S12, and helix 27 (H27) of 16S rRNA in establishing two distinct conformations of the 30S subunit that are thought to be analogous to conformations adopted by ribosomes from *ram* S4 and restrictive S12 mutants (Allen and Noller 1989). Current models predict that the conformations adopted by *ram* and restrictive ribosomes are related to two distinct functional states of the ribosome.

The fact that mutations in ribosomal proteins S4 and S12 and in 16S rRNA H18 and H27 confer streptomycin resistance in various organisms, coupled with footprinting by streptomycin of one of the bases involved in the H27 triplet switch, suggests that the switch is most likely involved in some aspect of protein synthesis affected by streptomycin, and further, that streptomycin acts by inhibiting that step. This notion is not entirely helpful in elucidating the functional role of the switch, since streptomycin, and mutations in S4, S5, and S12 for that matter, affect multiple aspects of protein synthesis including tRNA selection, tRNA-binding affinity, and translocation (for references, see Gregory et al. 2001). The possibility that the switch also affects multiple functions makes biochemically defining its role difficult.

Cryo-EM reconstructions of 70S ribosomes containing H27 mutations reveal large conformational rearrangements in both 30S and 50S subunits (Fig. 1) (Gabashvili et al. 1999). These results affect the switch model in several ways. First, they confirm the prediction that the H27 mutant ribosomes adopt different conformations, although the magnitude of the difference was unanticipated. Second, the global nature of the effects indicates

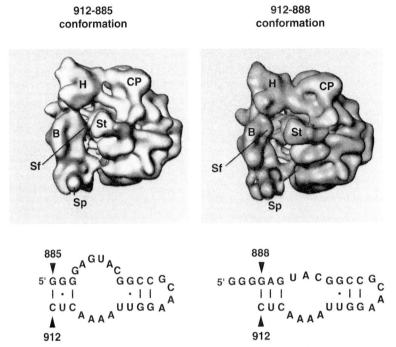

Figure 1. The H27 conformational switch model. Cryo-EM reconstructions of 70S ribosomes containing mutations in H27 (Gabashvili et al. 1999) show large conformational rearrangements (*top*). Ribosomes bear the mutations C912G and G885U (*left*) or C912G (*right*). The two proposed conformations of H27 (912–885 and 912–888) (*bottom*).

that conformational changes involving the putative *ram*-restrictive transition are propagated throughout the 70S ribosome. The cross-subunit influence observed in vacant 70S ribosomes demonstrates the existence of intersubunit communication independent of tRNA and drives speculation that such communication may play a role in ribosome function during the translation cycle.

One current goal of our laboratory is to examine the conformations adopted by restrictive and *ram* ribosomes at high resolution. For this purpose, we have identified mutants of *Thermus thermophilus* with alterations of ribosomal protein S12 (Fig. 2) (Gregory et al. 2001) and are attempting to solve the structures of ribosomes from these mutants by X-ray crystallography. *T. thermophilus* has proven to be an excellent source of ribosomes for crystallization studies (Moore 2001; Ramakrishnan and Moore 2001) and has the additional advantage of being amenable to genetic manipulations. Mutants of *T. thermophilus* have been identified that express streptomycin-resistance, streptomycin-dependence, or streptomycin-pseudodependence phenotypes. All the mutations we have identified are located within two loop structures of S12 that interact with H18, H27, and H44, three structural elements associated with decoding (see Fig. 2). These mutations are similar, and in many instances identical, to mutations identified in *Escherichia coli* and other mesophiles, giving added validity to the use of X-ray structures derived from *T. thermophilus* ribosomes to interpret functional data obtained with *E. coli* ribosomes (see Gregory et al. 2001). Some of the *T. thermophilus* mutants are almost certainly hyperaccurate or "restrictive," and the triplet switch model (Lodmell and

Dahlberg 1997) predicts that H27 will appear in the 912-888 base-pairing configuration in these ribosomes (Fig. 1). We have also isolated a number of suppressors of

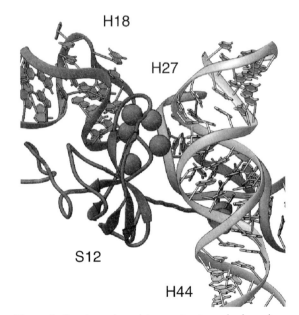

Figure 2. Streptomycin-resistance, streptomycin-dependence, and streptomycin-pseudodependence mutations in *Thermus thermophilus* ribosomal protein S12 (Gregory et al. 2001). Shown are ribosomal protein S12 and 16S rRNA H18, H27, and H44 as observed from the subunit interface side of the crystal structure of the *T. thermophilus* 30S subunit (PDB 1FJF; Wimberly et al. 2000). Sites of mutations are indicated by spheres.

streptomycin-dependence that are likely to promote the opposing *ram* configuration. These might include *ram* mutations at the S4–S5 interface, which have been suggested to exert their influence via H18 (Carter et al. 2000). Structure determination of *ram* and restrictive ribosomes is expected to reveal other structural changes that occur in the functional states mimicked by these two opposing conformations, and to give definitive insight into their role in ribosome function.

INTERSUBUNIT BRIDGES, tRNA CONTACTS, AND TRANSLATIONAL FIDELITY

High-resolution ribosome structures will certainly have a major impact on the interpretation of results from mutational studies. Mutations affecting translational fidelity have been identified in most, if not all, of the functional centers of the ribosome. One unanticipated revelation emerging from structural studies thus far is that many *ram* mutations arise in intersubunit bridges. Still other *ram* mutations occur within structures of the large subunit now known to be in intimate contact with tRNA at some time during the translation cycle.

Intersubunit bridges were first observed in cryo-EM reconstructions of 70S particles (Frank et al. 1995; Gabashvili et al. 2000), and the RNA and protein structures that comprise them were recently identified in the 5.5 Å 70S crystal structure (Yusupov et al. 2001). These bridges are likely to perform crucial functions above and beyond maintaining subunit–subunit association. They are now being viewed as likely candidates for central players in subunit and substrate movement during translocation (Frank and Agrawal 2000) and perhaps at other steps of the elongation cycle as well. One intriguing possibility is that intersubunit bridges act to transfer, between subunits, information regarding the binding states of tRNAs and accessory factors. The structure of the *T. thermophilus* 70S ribosome complexed with tRNA and mRNA also suggests that the ribosome and the interactions between its subunits are dynamic in nature (Yusopov et al. 2001). Notably, RNA helical elements comprising intersubunit bridges seen in the 70S structure are disordered in the isolated *Haloarcula marismortui* 50S subunit structure (Ban et al. 2000), suggesting that they are inherently flexible and would allow relative movement of the two subunits as envisioned by the hybrid states model (Moazed and Noller 1989). Support for this notion is derived from the identification of 50S subunit mutations that perturb the fidelity of decoding (for review, see O'Connor et al. 1995) and the realization that many of these mutations are in structures comprising intersubunit bridges.

Selection for frameshift suppressor mutations gives rise to *ram* mutations in H69 of 23S rRNA (O'Connor and Dahlberg 1995). H69 participates in two bridging interactions and makes direct contact with the anticodon stem-loops of A-site and P-site tRNA (Yusupov et al. 2001). In bridge B2a, H69 interacts with the 1408–1410 and 1495 segment of H44. Mutations at the 1409–1491

base pair also affect decoding accuracy (Gregory and Dahlberg 1995). Selection of mutants resistant to the initiation inhibitor kasugamycin (Vila-Sanjurjo et al. 1999) produces 16S rRNA mutations at A794 in H24 and at A1519 in H45. Together with H69, these elements are seen in the 70S crystal structure to combine to form bridge B2b (Yusupov et al. 2001).

Another structure, H62, participates in two bridges, B5 and B6 (Yusupov et al. 2001). Mutations in H62 act as second-site suppressors of the deleterious mutation U2555A (M. O'Connor and A.E. Dahlberg, unpubl.). We now know that H62 mutations act by reducing subunit association, thereby removing ribosomes carrying the U2555A mutation from the translating pool. This result clearly indicates a role for H62 in subunit association, although it is unclear whether this is its sole function.

An interesting finding from genetic selections and site-directed mutagenesis experiments is the large number of fidelity mutations arising in large-subunit RNA structures interacting with tRNA. The P loop (H80) and A loop (H92) are known from structure-probing experiments to interact with the acceptor end of tRNA (Samaha et al. 1995; Kim and Green 1999), interactions that are confirmed by the *Haloarcula* 50S subunit structure (Nissen et al. 2000). Mutations at U2555 in the A loop are obtained in genetic selections for suppressors of a frameshift mutation in *trpE* of *E. coli* and exhibit a generalized *ram* phenotype (O'Connor and Dahlberg 1993). Mutations at G2251, G2252, and G2253 in the P loop also exert effects on fidelity in vivo (Gregory et al. 1994) and perturb the conformation of both P- and A-site nucleotides as assessed by base-specific chemical probing (Gregory and Dahlberg 1999b). Although it remains uncertain exactly how mutations in these two loops generate accuracy defects, these observations are consistent with the existence of frameshift suppressor mutations at C74 of tRNA^{Val} (O'Connor et al. 1993). Mutated CCA fragments show decreased binding to the ribosome and A- and P-loop mutations also affect substrate binding (Samaha et al. 1995; Kim and Green 1999). Nucleotides in H89 or in the central loop of domain V are in close proximity to, or make contact with, the acceptor end of tRNA, and mutations in H89 (O'Connor and Dahlberg 1995) or the central loop (Saarma and Remme 1992; Thompson et al. 2001) affect accuracy. Together, these data suggest that obstructing tRNA–ribosome contacts enhances various kinds of translational errors. It is also clear from all these studies that the accuracy of tRNA selection is a highly cooperative process not confined to a simple active site.

What is now needed is a series of precise biochemical characterizations of the mutants presently in hand in order to determine the exact nature of their influence on the accuracy of tRNA selection. The existing high-resolution structures provide a context in which to interpret such results. Furthermore, mutant ribosomes affected in bridging interactions are ideal candidates for examination by cryo-EM and X-ray crystallography and are expected to provide important clues to the mechanism of translocation, tRNA selection, and intersubunit communication.

INVESTIGATIONS INTO PEPTIDE-BOND FORMATION

The peptidyl-transferase center and the peptidyl-transferase reaction have been subjected to intense scrutiny for many years (for review, see Lieberman and Dahlberg 1995). Early biochemical studies clearly pinpointed the sufficiency of the large ribosomal subunit for the catalysis of this reaction using minimal substrates under the appropriate conditions (Monro 1967). Despite being the simplest activity of the ribosome, the peptidyl-transferase reaction's precise mechanism and the identity of active-site catalytic residues have been difficult to establish. Early reconstitution studies narrowed the components potentially responsible for catalysis to include 23S rRNA and a handful of ribosomal proteins (Schulze and Nierhaus 1982). Examination of active protein-depleted particles (Noller et al. 1992) excluded all but 23S rRNA, L2, and L3 (Khaitovich et al. 1999). Although consistent with models proposing an RNA-catalyzed reaction, this result is also interpretable in light of models proposing the involvement of a histidine residue of L2 (Diedrich et al. 2000). This issue now appears to be settled. The recent X-ray crystallographically derived structure of the *H. marismortui* 50S subunit with substrate analogs or a transition-state analog seemingly eliminates any amino acid residue from the vicinity of the peptidyl transferase active site (Nissen et al. 2000). Rather, the structure reveals the proximity of the N3 position of residue A2451 of 23S rRNA as the closest functional group to the site of reaction. Dimethyl sulfate (DMS) modification experiments with *E. coli* ribosomes have been interpreted as indicating a near-neutral pKa of the N3 of A2451; this observation, combined with the crystallographic evidence of a close proximity of the N3 of this base to the putative tetrahedral reaction intermediate, has led to the proposal of an acid–base mechanism of catalysis (Muth et al. 2000; Nissen et al. 2000). To explain an aberrant pKa of A2451 N3, a charge relay mechanism involving residues G2061, G2447, A2450, and A2451 has been hypothesized (Nissen et al. 2000).

Several observations make this model extremely attractive in the context of the crystallographic evidence. Perhaps first and foremost, the identity of A2451 is absolutely conserved without exception (R.R. Gutell et al., in prep.), and, not surprisingly, mutations at this residue produce a dominant lethal phenotype (Muth et al. 2000; Thompson et al. 2001), possibly suggesting a critical role for this residue in ribosome function. Finally, the pH dependence of DMS reactivity of A2451 is consistent with an increased pKa of the residue, conceivably at the N3 position (Muth et al. 2000).

The proposed acid–base mechanism suggests several experiments that might potentially facilitate its verification. Mutations in any one of the residues implicated in the charge-relay system should have dramatic, if not catastrophic, consequences for peptide-bond formation, depending, of course, on how critical the acid–base mechanism is for catalysis. Such a test has been performed. Unexpectedly, the results obtained from these experiments do not support the existence of the charge-relay system and raise doubts regarding a central role for an acid–base mechanism. Ribosomes bearing mutations at A2451 are active in protein synthesis in an in-vitro-coupled transcription–translation system (Fig. 3) (Thompson et al. 2001) and, moreover, synthesis is resistant to inhibition by chloramphenicol, essentially validating the identification of a mitochondrial chloramphenicol-resistance

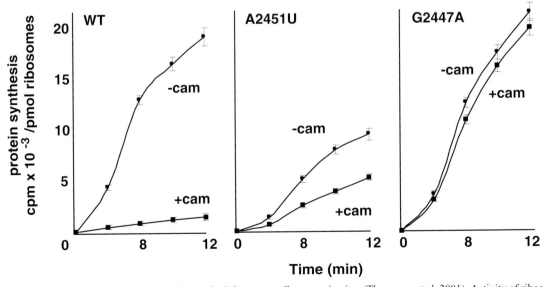

Figure 3. Chloramphenicol-resistant protein synthesis by mutant ribosomes in vitro (Thompson et al. 2001). Activity of ribosomes in a coupled in vitro transcription–translation system with β-lactamase mRNA as template for protein synthesis measured as TCA-precipitable [^{35}S]methionine incorporation. Polypeptide synthesis, TCA-precipitable ^{35}S cpm x 10^{-3}. (WT) Wild-type; (A2451U) a mixed population of wild-type and A2451U mutant ribosomes; (G2447A) a pure population of G2447A mutant ribosomes; (cam) chloramphenicol.

mutation at A2451 (Kearsey and Craig 1981). The processiveness of protein synthesis is demonstrated by the production of full-length β-lactamase in this system. Production of full-length β-galactosidase in vivo is indicated by increased suppression of a frameshift or a nonsense mutation by mutant ribosomes. Reconstituted *Bacillus stearothermophilus* ribosomes with substitutions at A2451 are active in the fragment reaction in vitro. Similarly, ribosomes mutated at G2447, a residue proposed to participate in the charge-relay network (Nissen et al. 2000), are active in protein synthesis and peptide-bond formation, and are resistant to chloramphenicol in vitro (Fig. 3) (Thompson et al. 2001). Interestingly, the G2447A mutation is not lethal. Antibiotic-resistance mutations have also been found at G2447 (Dujon 1980; Hummel and Böck 1987). Similar results have been observed by other investigators (Polacek et al. 2001). It would seem that peptidyl transferase can tolerate mutations at residues which are almost certainly direct participants in the peptide-bond formation process.

The dominant lethal phenotypes of mutations at A2451 (Muth et al. 2000; Nissen et al. 2001) have been presented as a powerful argument for a critical role of this residue in ribosome function. Such arguments should be judged with some caution. There exists a profound distinction between our ability to biochemically assess effects of mutations on single steps of protein synthesis and our capacity to understand how these effects might ultimately impact cellular physiology. In fact, it could reasonably be argued that we have yet to fully understand the mechanism of lethality of any ribosomal mutation (although a case can be made for streptomycin-dependent S12 mutations). One complicating factor is the necessity to express lethally mutant ribosomes only transiently with a background of wild-type ribosomes. Under such conditions, lethality cannot simply be accounted for by functional inactivity. A very large number of lethal mutations have been identified in both subunits in site-directed mutagenesis studies, and invariably, mutant ribosomes are found in translating polysomes. Second-site suppressors of lethal mutations often act by excluding mutant ribosomes from polysomes via defects in subunit association or subunit assembly. It would seem that lethality requires active protein synthesis by mutant ribosomes, at least when they are produced in mixed populations with wild-type ribosomes. By analogy, it has long been known that the bactericidal action of aminoglycoside antibiotics requires active protein synthesis and can be abrogated by chloramphenicol, a peptidyl transferase inhibitor (Jawetz et al. 1952). Furthermore, mutations conferring resistance to peptidyl transferase inhibitors are dominant, demonstrating that inhibition of peptidyl transferase activity of sensitive ribosomes in a mixed population is not lethal. The correct prediction, therefore, is that mutations which abolish peptide-bond formation should not be lethal provided there is a background of wild-type ribosomes. The lethal phenotype is itself an indicator of active protein synthesis by A2451 mutant ribosomes, and by logical extension, active peptide-bond formation. These results themselves do not exclude the proposed mechanism, but simply demand that an additional mechanism(s) not involving direct participation of A2451 in the chemistry of peptide-bond formation also be considered.

Perhaps the greatest difficulty for a general acid–base mechanism is our finding regarding the postulated shift in pKa of A2451 (Muth et al. 2000). It would appear that in our hands the aberrant DMS reactivity of A2451 occurs only in ribosomes that are in an inactive conformation (Fig. 4) (Bayfield et al. 2001) as defined by criteria established by Miskin, Zamir, and Elson (Miskin et al. 1970). The Zamir-Elson active–inactive conformational transition in the 30S subunit results from depletion of monovalent cations and can be reversed by a heat activation in the presence of monovalent cations. A similar transition also occurs in the 50S subunit and, importantly, precise correlations exist between monovalent cations and competence for peptide-bond formation (Miskin et al. 1970). We have discovered that activation of either 50S subunits or 70S ribosomes by heating to 37°C in the presence of monovalent cations eliminates the purported pH-dependent DMS modification shift; under such conditions, A2451 is unreactive to DMS (Fig. 4a) (Bayfield et al. 2001). Ribosomes in the Zamir-Elson inactive conformation are inactive in peptide-bond formation; therefore, the conditions under which the proposed pKa shift is observed are those in which peptidyl transferase is inactive (Fig. 4). In addition to a pH-dependent DMS reactivity of A2451 in the inactive conformation, several "antishifts" are observed (Bayfield et al. 2001). These residues exhibit enhanced DMS reactivity at lower pH, which of course cannot result from an increased pKa, but instead is indicative of a pH-dependent conformational change. Finally, mutation of G2447, a base proposed to be essential to the charge-relay system, has no effect on the pH dependence of DMS reactivity of A2451. Our interpretation of these data is that A2451 does not have a perturbed pKa and that the N3 of that residue does not directly participate in the chemistry of peptide-bond formation. At the very least, the pH-dependent DMS modification of A2451 cannot be used as an indicator of a perturbed pKa of that residue. It seems reasonable to conclude that the N1 is available to react with DMS in an inactive conformation that is distinct from that observed in the *H. marismortui* 50S subunit structure. The chemical probing data are therefore consistent with the assertion that the *H. marismortui* 50S structure represents an active conformation in which the N1 of A2451 is hydrogen-bonded and unavailable for DMS modification.

In summary, the above-described experimental results raise serious questions about the proposed acid–base catalytic mechanism of peptide-bond formation (Nissen et al. 2000). If such a mechanism exists, it cannot be the sole mechanism, and it remains to be determined what contribution such a mechanism might make to the rate of catalysis.

IS THERE A CATALYTIC RESIDUE IN THE PEPTIDYL TRANSFERASE?

Our experimental results do not support an essential role for A2451 as the lone peptidyl transferase catalytic residue; the precise nature of the transpeptidation mechaa)

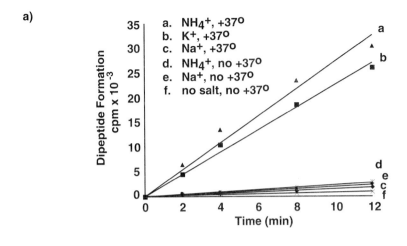

b)

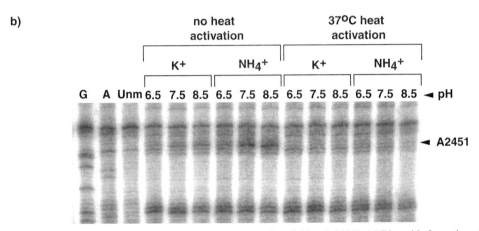

Figure 4. The active–inactive transition and DMS modification of A2451 (Bayfield et al. 2001). (*a*) Dipeptide formation of ribosomes treated as indicated. (NH_4^+) Ribosomes treated with 100 mM NH_4Cl; (K^+), ribosomes treated with 150 mM KCl; (37°C) 10 minutes heat activation. (*b*) DMS modification at the indicated pH of ribosomes treated as in *a*. (A2451) Position of RT stop corresponding to A2451.

nism remains unresolved. The existence of a single catalytic residue has not been rigorously demonstrated, and perhaps future efforts are best guided by the question of whether a residue with such a property even exists. Two of the residues implicated by the *H. marismortui* structure to be near the site of catalysis can be mutated without complete loss of activity. The acid–base model (Muth et al. 2000; Nissen et al. 2000) or other models derived from it do not exclude substrate alignment by the ribosome as a contributing and perhaps primary mechanism (Moore 2001). It is possible that a firmly established catalytic mechanism will emerge only after more detailed kinetic analyses and high-resolution structure determinations of ribosomes mutated in the active site.

CONCLUSIONS

Recent crystallographic and cryo-EM studies of the ribosome are having a tremendous impact on the ribosome field, not only in the reinterpretation of results obtained by other methodologies, but also in the way in which investigators will consider how to tackle the many questions still remaining. They will also, to a very large ex-

tent, dictate which questions will be asked. It is clear that the structural dynamics of the ribosome facilitate protein synthesis in a fundamental manner that is only superficially understood. The ability to solve high-resolution structures of ribosomes in various stages of translation will allow this issue to be addressed in a meaningful way, and with reasonable hope for the development of some unifying concepts. It is also almost certain that combining genetics, biochemistry, and structural biology will lead to rapid progress not likely to emerge from any one approach alone.

The importance of intersubunit bridging interactions for ribosome function is apparent. Although it is obvious that such structures are ideal target sites for mutagenesis studies, it is not surprising that many such mutations have already been identified or constructed. It is perhaps suspected by many, if not most, investigators in the field that such bridging interactions are the key to understanding both subunit communication and the mechanism of tRNA movement through the ribosome.

Even with high-resolution structures and 40 years of biochemistry and genetics, the most fundamental mechanisms of the ribosome are yet to be understood. Perhaps the most elusive remains the catalysis of peptide-bond formation. Despite the seeming simplicity of the chemistry of the reaction, efforts to dissect the ribosome's role have caused many laboratories a great deal of angst. Exploring one of the most fundamental reactions in biology merits nothing less than the most comprehensive and rigorous experimentation.

ACKNOWLEDGMENTS

The authors are grateful to Sun-Thorn Pond-Tor for technical assistance and good humor, to Judy Nathanson for assistance in manuscript preparation, and to George Q. Pennable for prudence and perspicacity. This work was supported by a grant GM-19756 from the National Institutes of Health to A.E.D.

REFERENCES

Agrawal R.K., Heagle A.B., Penczek P., Grassucci R.A., and Frank J. 1999. EF-G-dependent GTP hydrolysis induces translocation accompanied by large conformational changes in the 70S ribosome. *Nat. Struct. Biol.* **6:** 643.

Allen P.N. and Noller H.F. 1989. Mutations in ribosomal proteins S4 and S12 influence the higher order structure of 16S ribosomal RNA. *J. Mol. Biol.* **208:** 457.

Ban N., Nissen P., Hansen J., Moore P.B., and Steitz T.A. 2000. The complete atomic structure of the large ribosomal subunit at 2.4 Å resolution. *Science* **289:** 905.

Bayfield M.A., Dahlberg A.E., Schulmeister U., Dorner S., and Barta A. 2001. A conformational change in the ribosomal peptidyl transferase center upon active/inactive transition. *Proc. Natl. Acad. Sci.* **98:** 10096.

Carter A.P., Clemons W.M., Brodersen D.E., Morgan-Warren R.J., Wimberly B.T., and Ramakrishnan V. 2000. Functional insights from the structure of the 30S ribosomal subunit and its interaction with antibiotics. *Nature* **407:** 340.

Carter A.P., Clemons W.M., Brodersen D.E., Morgan-Warren R.J., Hartsch T., Wimberly B.T., and Ramakrishnan V. 2001. Crystal structure of an initiation factor bound to the 30S ribosomal subunit. *Science* **291:** 498.

Diedrich G., Spahn C.M., Stelzl U., Schafer M.A., Wooten T., Bochkariov D.E., Cooperman B.S.,Traut R.R., and Nierhaus K.H. 2000. Ribosomal protein L2 is involved in the association of the ribosomal subunits, tRNA binding to A and P sites and peptidyl transfer. *EMBO J.* **19:** 5241.

Dujon B. 1980. Sequence of the intron and flanking exons of the mitochondrial 21S rRNA gene of yeast strains having different alleles at the omega and *rib-1* loci. *Cell* **20:** 185.

Frank J. and Agrawal R.K. 2000. A ratchet-like inter-subunit reorganization of the ribosome during translocation. *Nature* **406:** 318.

Frank J., Verschoor A., Li Y., Zhu J., Lata R.K., Radermacher M., Penczek P., Grassucci R., Agrawal R.K., and Srivastava S. 1995. A model of the translational apparatus based on a three-dimensional reconstruction of the *Escherichia coli* ribosome. *Biochem. Cell Biol.* **73:** 757.

Gabashvili I.S., Agrawal R.K., Grassucci R., Squires C.L., Dahlberg A.E., and Frank J. 1999. Major rearrangements in the 70S ribosomal 3D structure caused by a conformational switch in 16S ribosomal RNA. *EMBO J.* **18:** 6501.

Gabashvili I.S., Agrawal R.K., Spahn C.M., Grassucci R.A., Svergun D.I., Frank J., and Penczek P. 2000. Solution structure of the *E. coli* 70S ribosome at 11.5 Å resolution. *Cell* **100:** 537.

Gabashvili I.S., Gregory S.T., Valle M., Grassucci R., Worbs M., Wahl M.C., Dahlberg A.E., and Frank J. 2001. The polypeptide tunnel system in the ribosome and its gating in erythromycin resistance mutants of L4 and L22. *Mol. Cell* **8:** 181.

Gregory S.T. and Dahlberg A.E. 1995. Nonsense suppressor and

antisuppressor mutations at the 1409-1491 base pair in the decoding region of *Escherichia coli* 16S rRNA. *Nucleic Acids Res.* **23:** 4234.

———. 1999a. Erythromycin resistance mutations in ribosomal proteins L22 and L4 perturb the higher order structure of 23S ribosomal RNA. *J. Mol. Biol.* **289:** 827.

———. 1999b. Mutations in the conserved P loop perturb the conformation of two structural elements in the peptidyl transferase center of 23 S ribosomal RNA. *J. Mol. Biol.* **285:** 1475.

Gregory S.T., Cate J.H.D., and Dahlberg A.E. 2001. Streptomycin-resistant and streptomycin-dependent mutants of the extreme thermophile *Thermus thermophilus*. *J. Mol. Biol.* **309:** 333.

Gregory S.T., Lieberman K.R., and Dahlberg A.E. 1994. Mutations in the peptidyl transferase region of *E. coli* 23S rRNA affecting translational accuracy. *Nucleic Acids Res.* **22:** 279.

Hummel H. and Böck A. 1987. 23S ribosomal RNA mutations in halobacteria conferring resistance to the anti-80S ribosome targeted antibiotic anisomycin. *Nucleic Acids Res.* **15:** 2431.

Jawetz E., Gunnison J.B., and Bruff J.B. 1952. Studies on antibiotic synergism and antagonism. *J. Bacteriol.* **64:** 29.

Kearsey S.E. and Craig I.W. 1981. Altered ribosomal RNA genes in mitochondria from mammalian cells with chloramphenicol resistance. *Nature* **290:** 607.

Khaitovitch P., Mankin A.S., Green R., Lancaster L., and Noller H.F. 1999. Characterization of functionally active subribosomal particles from *Thermus aquaticus*. *Proc. Natl. Acad. Sci.* **96:** 85.

Kim D.F. and Green R. 1999. Base-pairing between 23S rRNA and tRNA in the ribosomal A site. *Mol. Cell* **4:** 859.

Lieberman K.R. and Dahlberg A.E. 1995. Ribosome-catalyzed peptide-bond formation. *Progr. Nucleic Acids Res.* **50:** 1.

Lodmell J.S. and Dahlberg A.E. 1997. A conformational switch in *Escherichia coli* 16S ribosomal RNA during decoding of messenger RNA. *Science* **277:** 1262.

Miskin R., Zamir A., and Elson D. 1970. Inactivation and reactivation of ribosomal subunits: The peptidyl transferase activity of the 50S subunit of *Escherichia coli*. *J. Mol. Biol.* **54:** 355.

Moazed D. and Noller H.F. 1987. Interaction of antibiotics with functional sites in 16S ribosomal RNA. *Nature* **327:** 389.

———. 1989. Intermediate states in the movement of transfer RNA in the ribosome. *Nature* **342:**142.

Moazed D., Van Stolk B.J., Douthwaite S., and Noller H.F. 1986. Interconversion of active and inactive 30 S ribosomal subunits is accompanied by a conformational change in the decoding region of 16 S rRNA. *J. Mol. Biol.* **191:** 483.

Monro R.E. 1967. Catalysis of peptide bond formation by 50 S ribosomal subunits from *Escherichia coli*. *J. Mol. Biol.* **26:** 147.

Moore P.B. 2001. The ribosome at atomic resolution. *Biochemistry* **40:** 3243.

Muth G.W., Ortoleva-Donnelly L., and Strobel S.A. 2000. A single adenosine with a neutral pKa in the ribosomal peptidyl transferase center. *Science* **289:** 947.

Nissen P., Hansen J., Ban N., Moore P.B., and Steitz T.A. 2000. The structural basis of ribosome activity in peptide bond synthesis. *Science* **289:** 920.

Nissen P., Hansen J., Muth G.W., Ban N., Moore P.B., Strobel S.A., and Steitz T.A. 2001. Mechanism of ribosomal peptide bond formation. *Science* **291:** 203a.

Noller H.F., Hoffarth V., and Zimniak L. 1992. Unusual resistance of peptidyl transferase to protein extraction procedures. *Science* **256:** 1416.

O'Connor M. and Dahlberg A.E. 1993. Mutations at U2555, a tRNA-protected base in 23S rRNA, affect translational fidelity. *Proc. Natl. Acad. Sci.* **90:** 9214.

———. 1995. The involvement of two distinct regions of 23S ribosomal RNA in tRNA selection. *J. Mol. Biol.* **254:** 838.

O'Connor M., Willis, N.M., Bossi L., Gesteland R.F., and Atkins J. 1993. Functional tRNAs with altered 3′ ends. *EMBO J.* **12:** 2559.

O'Connor M., Brunelli C.A., Firpo M.A., Gregory S.T., Lieberman K.R., Lodmell J.S., Moine H., Van Ryk D.I., and Dahlberg A.E. 1995. Genetic probes of ribosomal RNA function. *Biochem. Cell Biol.* **73:** 859.

Ogle J.M., Brodersen D.E., Clemons W.M., Tarry M.J., Carter
    A.P., and Ramakrishnan V. 2001. Recognition of cognate
    transfer RNA by the 30S ribosomal subunit. *Science* **292:**
    897.
Polacek N., Gaynor M., Yassin A., and Mankin A.S. 2001. Ri-
    bosomal peptidyl transferase can withstand mutations at the
    putative catalytic nucleotide. *Nature* **411:** 498.
Polacek N., Patzke S., Nierhaus K.H., and Barta A. 2000. Peri-
    odic conformational changes in rRNA: Monitoring the dy-
    namics of translating ribosomes. *Mol. Cell* **6:** 159.
Ramakrishnan V. and Moore P.B. 2001. Atomic structures at
    last: The ribosome in 2000. *Curr. Opin. Struct. Biol.* **11:** 144.
Saarma U. and Remme J. 1992. Novel mutants of 23S RNA:
    Characterization of functional properties. *Nucleic Acids Res.*
    20: 3147.
Samaha R.R., Green R., and Noller H.F. 1995. A base pair be-
    tween tRNA and 23S rRNA in the peptidyl transferase centre
    of the ribosome. *Nature* **377:** 309.
Schulze H. and Nierhaus K.H. 1982. Minimal set of ribosomal
    components for reconstitution of the peptidyltransferase ac-

tivity. *EMBO J.* **1:** 609.
Spahn C.M., Kieft J.S., Grassucci R.A., Penczek P.A., Zhou K.,
    Doudna J.A., and Frank J. 2001. Hepatitis C virus IRES RNA-
    induced changes in the conformation of the 40S ribosomal
    subunit. *Science* **291:** 1959.
Thompson J., Kim D.F., O'Connor M., Lieberman K.R., Bay-
    field M.A., Gregory S.T., Green R., Noller H.F., and
    Dahlberg A.E. 2001. Analysis of mutations at residues A2451
    and G2447 of 23S rRNA in the peptidyl transferase active site
    of the 50S ribosomal subunit. *Proc. Natl. Acad. Sci.* **98:** 9002.
Vila-Sanjurjo A., Squires C.L., and Dahlberg A.E. 1999. Isola-
    tion of kasugamycin resistant mutants in the 16S ribosomal
    RNA of *Escherichia coli*. *J. Mol. Biol.* **293:** 1.
Wimberly B.T., Brodersen D.E., Clemons W.M., Morgan-War-
    ren R.J., Carter A.P., Vonrhein C., Hartsch T., and Rama-
    krishnan V. 2000. The structure of the 30S ribosomal subunit.
    Nature **407:** 327.
Yusupov M.M., Yusupova G.Z., Baucom A., Lieberman K.,
    Earnest T.N., Cate J.H., and Noller H.F. 2001. Crystal struc-
    ture of the ribosome at 5.5 Å resolution. *Science* **292:** 883.

Exploring the Mechanism of the Peptidyl Transfer Reaction by Chemical Footprinting

S.A. STROBEL, G.W. MUTH, AND L. CHEN

Department of Molecular Biophysics and Biochemistry and Department of Chemistry, Yale University, New Haven, Connecticut 06520-8114

The most important step of protein biosynthesis from a chemical point of view is formation of the peptide bond (Chladek and Sprinzl 1985). This reaction is accomplished within the large 50S subunit of the ribosome using aminoacyl tRNA and peptidyl tRNAs as substrates (Noller 1991). The reaction takes place via the transfer of a peptide from a peptidyl tRNA bound in the P site to the α-amino group of an aminoacyl tRNA bound in the A site. The A and P sites are each specified by base-pairing interactions between the 3′-terminal CCA end of the tRNAs and complementary sequences in the 23S rRNA (Chladek and Sprinzl 1985; Samaha et al. 1995; Kim and Green 1999; Nissen et al. 2000). The peptidyl transfer reaction involves nucleophilic attack by the α-amino group in the A site on the ester linking the peptide to the P-site tRNA. The transacylation reaction is expected to proceed through two transition states separated by a short-lived tetrahedral intermediate (Hegazi et al. 1978). The reaction is catalyzed within the peptidyl transferase center, which is composed of a section of 23S rRNA domain V that is extremely well conserved throughout biology (Fig. 1A).

In this paper, we review mechanistic possibilities for how the peptidyl transfer reaction is catalyzed, including early biochemical work and mechanistic implications from the crystal structure of the 50S ribosomal subunit. We also discuss chemical footprinting data within the peptidyl transferase center on ribosomes from bacterial, archaebacterial, and eukaryotic sources. We describe three independent lines of evidence to indicate that the DMS pH titration of the active-site residue, A2451 (Fig. 1A), reflects conformational flexibility within the peptidyl transferase center rather than protonation of A2451.

MECHANISTIC PROPOSALS FOR THE PEPTIDYL TRANSFER REACTION

Two fundamentally different hypotheses for the mechanism of peptide-bond formation were summarized 20 years ago by Nierhaus et al. (1980). According to the first model, the peptidyl transferase serves solely as a template to align the A-site and P-site substrates. In the second model, the peptidyl transferase acts as a template, but further stabilizes the chemical transition state by a general acid-base mechanism similar to that employed in the acylation reaction of serine proteases (Kraut 1977).

PEPTIDYL TRANSFERASE AS A TEMPLATE

Since peptidyl transfer is simply the aminolysis of the peptidyl-tRNA ester bond by the α-amino group of an aminoacyl tRNA, the reaction is thermodynamically favored. Nierhaus et al. (1980) argued, based on model unimolecular reactions, that the required rate of peptide-bond formation in vivo can be achieved when the

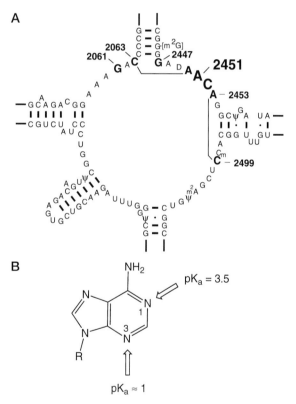

Figure 1. (*A*) RNA sequence and secondary structure within the *E. coli* peptidyl transferase center (Gutell et al. 1993). A2451 is highlighted, as are other nucleotides discussed in the course of this study. All of the highlighted nucleotides are conserved between *E. coli*, *H. marismortui*, and *S. cerevisiae* rRNA except A2453 and C2499, which are both U's in the *S. cerevisiae* rRNA (Gutell 2000). The A•C pairs that flank A2451 are indicated by bent lines. (*B*) Chemical structure of adenosine with the N1 and N3 imino groups numbered and their pKa values indicated (Seela et al. 1998; Muth et al. 2000).

nucleophile and the electrophile are appropriately juxta-posed. A relevant example of a rapid "uncatalyzed" transacylation reaction is provided by the isomerizaion of the aminoacyl residue from the 2′ OH to the 3′ NH_2 of tRNA-CCA3′-NH_2, which can proceed at an efficient rate (Chladek and Sprinzl 1985).

Hecht and coworkers provided indirect evidence for a template-only mechanism using unnatural aminoacyl-tRNA analogs (Hecht 1992). They reported that a P-site substrate (N-[chloroacetyl]-phenylalanyl-tRNA[Phe]) containing two electrophilic centers could react at either position with the nucleophilic α-amino group of the A-site tRNA (Roesser et al. 1986). Similarly, an A-site substrate ([S])-α-hydrazino-phenylalanyl-tRNA[Phe]] with two nucleophilic centers formed peptide bonds using either reactive position (Killian et al. 1998). The ability of these substrates to react indiscriminately implies that there is significant flexibility within the peptidyl transferase center.

PEPTIDYL TRANSFERASE AS AN ENZYME

Other observations argue that the ribosome may also catalyze the reaction via a general acid-base mechanism similar to serine proteases (Nierhaus et al. 1980). At the time of the Nierhaus et al. proposal in 1980, they were focusing on a protein-based catalyst involving an ionizable histidine (His) residue at the active site. They suggested that the aminoacyl-tRNA α-amino group is deprotonated by this His (general base catalysis) and the tetrahedral intermediate is activated by proton donation from the His to the leaving group 3′ oxyanion of the peptidyl tRNA (general acid catalysis).

Early biochemical evidence to support a mechanism involving an ionizable functional group such as histidine with a near-neutral pK_a included the following observations: (1) The peptidyl transferase reaction has a substantial pH dependence with an acidity constant (pK_a) of approximately 7.2 (Maden and Monro 1968; Pestka 1972). (2) Ribosomal inactivation by ethoxyformic anhydride, which was thought to react with an active-site His, had a pH maximum at 7.0 (Baxter and Zahid 1978). (3) Photochemical inactivation of the ribosome with Rose Bengal dye followed first-order kinetics with a pH optimum of 7.5, consistent with an active-site residue with a neutral pK_a (Wan et al. 1975). (4) The peptidyl transferase is inhibited by phenylboric acid, a specific inhibitor of serine proteases (Cerna and Rychlik 1980). (5) A preliminary measure of peptide-bond formation in H_2O versus D_2O showed a solvent isotope effect of 1.5, suggesting that a general acid-base catalysis mechanism is rate-limiting for peptide-bond formation (see Fig. 3 in Nierhaus et al. 1980).

MECHANISTIC CLUES FROM THE 50S RIBOSOMAL CRYSTAL STRUCTURE

The structure of the 50S ribosomal subunit from *Haloarcula marismortui* was determined at 2.4 Å resolu-tion in a collaborative effort between the laboratories of Professors Peter Moore and Thomas Steitz (Ban et al. 2000). To identify the peptidyl transfer center of the ribosome, Nissen et al. utilized a transition-state analog of peptidyl transfer reaction (Welch et al. 1995; Nissen et al. 2000). This antibiotic is a puromycin derivative that contains a CCdAp (P-site analog) coupled via a phosphoramidate linkage to the puromycin α-amino group (A-site analog). Interestingly, A2486 (*E. coli* A2451) is within hydrogen-bonding distance of the phosphoramidate mimic of the tetrahedral intermediate despite the fact that neither the imino group nor the nonbridging oxygen should be protonated at neutral pH (Fig. 2A). Because this oxygen is expected to mimic the negatively charged oxyanion of the tetrahedral intermediate, a protonated A2451 may provide charge stabilization of the intermediate. On the basis of its placement in the active site, Nissen et al. also suggested that A2451 could play a role as a general acid-base catalyst, such as was originally ascribed to the putative histidine in the Nierhaus model (Nierhaus et al. 1980; Nissen et al. 2000). Thus, A2451 in the ribosome might promote peptidyl transfer by any or all of three strategies: deprotonation of the α-amino nucleophile prior to or during the first transition state (general base catalysis), stabilization of the oxyanion in the tetrahedral intermediate, and/or protonation of the 3′ oxyanion of the P-site tRNA during the second transition state (general acid catalysis). A schematic that incorporates all three of these possibilities is shown in Figure 2B.

During early stages of refinement on the inhibitor-bound 50S ribosome crystals, it became clear that the nucleotide bases rather than the RNA phosphate backbone are closest to the active site (Ban et al. 2000). This suggested that the ribosome might utilize an RNA heterocyclic base to catalyze peptide-bond formation. Although the structure clearly indicated that there were no histidines in the active site, none of the functional groups in RNA normally has a pK_a near the neutral pH that would be required for a general acid-base mechanism (Saenger 1984). Therefore, we postulated that the pK_a of the active-site residue might be significantly perturbed in a manner analogous to that seen in protein enzymes and proposed for an essential C within the hepatitis delta virus ribozyme (Fersht 1985; Ferre-D'Amare et al. 1998; Perrotta et al. 1999; Nakano et al. 2000). On the basis of the unperturbed pK_a values of the nucleobases, the two nucleotides most likely to have such an effect are A and C with pK_a values of 3.5 and 4.2, respectively (Saenger 1984). Both of these functional groups can be alkylated with dimethyl sulfate (DMS) to generate an adduct that creates a stop during reverse transcription (Lawley and Brookes 1963; Stern et al. 1988). We postulated that if the pK_a of an active-site nucleotide were perturbed, it would lead to increased protonation at neutral pH, resulting in protection from DMS methylation (Connell and Yarus 1994). However, as the pH is raised above the nucleotide's pK_a, it would become susceptible to DMS modification. In this manner, we attempted to identify the nucleotide at the ribosomal active site and measure its pK_a.

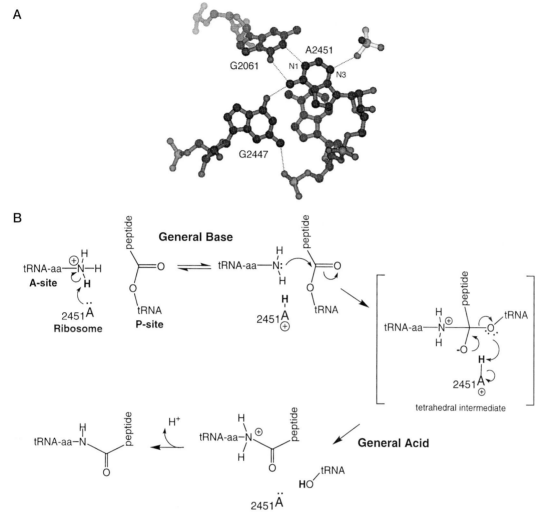

Figure 2. (*A*) Ribosome active site bound to a phosphoramidate inhibitor that mimics the tetrahedral intermediate. (Reprinted, with permission, from Nissen et al. 2000 [copyright AAAS].) (*B*) Possible catalytic mechanism of the peptidyl transfer reaction. (Adapted, with permission, from Barta et al. 2000; Muth et al. 2000; Nissen et al. 2000.)

CALIBRATION OF THE DMS APPROACH USING A NUCLEOSIDE ANALOG

pH-dependent DMS modification was used to measure the pK_a values of four nucleotides within an RNA aptamer selected for its ability to bind arginine and guanosine (Connell and Yarus 1994); however, the method had not been calibrated on an RNA where the acidity constant was determined by an independent approach. To validate the method, we measured the pK_a of 3-deazaadenosine (3dA), whose remaining N1 imino group has a spectrophotometrically determined value of 7.0 (Minakawa et al. 1999). Rates of DMS modification of A or 3dA were measured from pH 5.5 to 8.0 (Fig. 3). As expected, the rate of DMS reactivity with A was constant at all pH values tested, consistent with its acidic pK_a (Fig. 1B). In contrast, the rate of 3dA methylation was dramatically dependent on the pH of the solution (Fig. 3). The log of the methylation rate was plotted versus pH, resulting in an apparent pK_a of 6.4 ± 0.1. This is within half a pH unit of

the spectroscopically determined value, which was measured at substantially different ionic strength (Minakawa et al. 1999). This simple system demonstrates that DMS reactivity can be used to approximate a nucleotide's pK_a.

DMS MAPPING OF *E. COLI* RIBOSOMES

We performed DMS modification analysis on intact *E. coli* 50S ribosomal subunits as a function of pH from 4.5 to 8.6 (Fig. 4). The extent of methylation was determined for each of the solvent-accessible A and C residues within domain V (nucleotides 2043–2625, *E. coli* numbering), which biochemical evidence suggests harbors the peptidyl transferase center (Green and Noller 1997). All the A's and C's throughout this region showed a similar level of DMS reactivity at all pH values tested, with the single exception of A2451. The extent of A2451 DMS reactivity was plotted versus pH to give a calculated apparent pK_a of 7.6 ± 0.2 (Fig. 4). This estimate is close to the re-

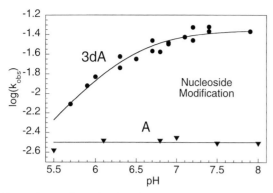

Figure 3. pK_a determination of 3-deazaadenosine as measured by the pH dependence of its reactivity with DMS. The log of the observed methylation rates of nucleosides A (*triangles*) and 3dA (*circles*) are plotted versus pH. Individual rates of DMS reactivity were calculated by plotting the change in UV absorbance (300 nm) as a function of reaction time. Although the overall level of DMS reactivity is greater for 3dA than for A, only the 3dA reactivity changes as a function of pH. The data indicate a pK_a of 6.4 ± 0.2 for 3-deazaadenosine.

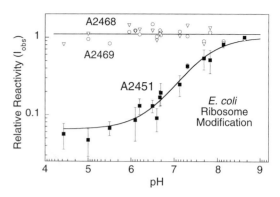

Figure 4. pH dependence of A2451 DMS reactivity in *E. coli* ribosomes. Plot of the extent of DMS modification for three A's (A2468, *open circles*; A2469, *open triangles*; A2451, *closed squares*) within the peptidyl transferase region of 50S ribosomal subunits as a function of pH. The data were normalized for lane loading and the extent of primer extension and scaled such that the maximum observed reactivity is defined as 1.0. Each point is an average of five replicates at each pH, and the standard deviation is indicated with error bars. The data indicate an apparent pK_a for *E. coli* A2451 of 7.6 ± 0.2.

ported value for the peptidyl transferase center based on the pH dependence of the fragment reaction (7.7 ± 0.3 in Maden and Monro 1968; 7.2 ± 0.1 in Pestka 1972).

A2451 is located in the central loop of domain V (Fig. 1). It has been shown by DMS footprinting of peptidyl tRNAs, aminoacyl tRNAs, and antibiotics to be within the peptidyl transferase center (Moazed and Noller 1987, 1989, 1991; Vannuffel et al. 1994). A2451 is universally conserved across all three kingdoms of life, including rRNA of mitochondria and chloroplasts (Gutell 1996, 2000). Upon further refinement of the 50S ribosomal structure, we learned that A2451 (A2486 in *H. marismortui*) is immediately adjacent to the phosphoramidate mimic of the tetrahedral intermediate in the peptidyl transfer center (Nissen et al. 2000).

A2451 IS ESSENTIAL FOR RIBOSOMAL FUNCTION IN VIVO

Although A2451 is universally conserved, we were surprised to find that the functional importance of A2451 in the ribosome had not been systematically investigated by mutagenesis (Triman 1999), although one early study reported that spontaneous mutation of A2451 (*E. coli* numbering) to U in rat mitochondrial rRNA appeared to confer chloramphenicol resistance to 3T3 cells in culture (Kearsey and Craig 1981). To investigate this further, we mutated the active-site A to either C, G, or U and expressed the mutant ribosomes under the control of a temperature-sensitive promoter (Brosius et al. 1981; Douthwaite et al. 1989). At 30°C, where the mutant rRNA is not expressed, *E. coli* containing either the wild type or any one of the three mutant plasmids grew equally well. However, expression of all three mutant 23S rRNAs at 42°C resulted in a dominant lethal phenotype. Lethality and complete conservation of this residue imply that A2451 is important for ribosomal function and is consistent with a potential role in catalyzing the peptidyl transfer reaction.

CONCERNS REGARDING THE INTERPRETATION OF THE DMS MODIFICATION DATA IN *E. COLI*

The biochemistry appeared to fit nicely with the crystallography, but three uncertainties remained. First, the preferred site of adenosine DMS modification is the N1 position (Lawley and Brookes 1963), yet the N3 imino group is presented into the active-site cleft by the ribosome (Figs. 1B and 2A) (Nissen et al. 2000). Second, although pK_a values approaching neutrality have been observed for the N1 imino group (Cai and Tinoco 1996), the N3 pK_a is estimated to be 2–3 units lower (Fig. 1B) (Seela et al. 1998; Muth et al. 2000). There is no other precedent for its perturbation, especially to a value as high as 7.6. Third, pH-dependent DMS modification cannot rigorously distinguish between protection induced by protonation (a proton footprint) and protection induced by increased conformational flexibility (a structural footprint). In addressing these issues, it has become clear that the DMS footprint reflects a pH-dependent conformational change rather than being a direct measure of the A2451 pK_a.

CHEMICAL FOOTPRINTING OF *H. MARISMORTUI* RIBOSOMES

We hypothesized that if a perturbed pK_a at A2451 is a conserved feature of the peptidyl transferase center, then a similar pH-dependent DMS pattern should be observed across phylogeny, including eukaryotic and archaebacterial ribosomal sources. The 50S ribosomal crystal structure was determined for *H. marismortui*, a halophilic archaebacterium (Ban et al. 2000). We probed these ribosomes to allow direct comparison between the mapping data and the crystal structure (Muth et al. 2001). The extent of A2451 methylation was monitored between pH 5 and 9.5, and the sites of modification were revealed by reverse transcription. In contrast to the observations made in *E. coli* (Muth et al. 2000), the nucleotide equivalent to A2451 in *H. marismortui* ribosomes was DMS-unreactive at high pH (≥8.0), but it became about fivefold more reactive as the pH was reduced (Fig. 5). Such an inverted DMS modification pattern is inconsistent with a base protonation event, because the reactivity should be maximized in the deprotonated form at high pH and minimized at low pH where the nucleophilic imino group is blocked by protonation. For example, this pattern is opposite that observed in model reactions with the nucleoside 3-deazaadenosine where the pK_a measured by DMS modification closely matched that determined spectroscopically (Minakawa et al. 1999; Muth et al. 2000). Because the *H. marismortui* ribosomes are the same as those used for crystallization, the results can be compared directly to the structure without need for extrapolation. The inverted pattern implies that A2451 becomes more solvent-exposed at low pH, even if the exposure is only transient. This is not the conformation observed in either the apoenzyme or the inhibitor-complexed ribosomes (Ban et al. 2000; Nissen et al. 2000).

MODIFICATION OF YEAST RIBOSOMES

Given that two systems displayed contradictory modification patterns, we next explored the pH dependence of DMS modification within a eukaryotic ribosome (Muth et al. 2001). 60S ribosomes were isolated from *Saccharomyces cerevisiae* and subjected to DMS modification between pH 5.5 and 9.0. This resulted in a modification pattern different from that seen in either the *E. coli* or the *H. marismortui* ribosomal active sites (Muth et al. 2000). In the region of the sequence from 2300 to 2493, three nucleotides (A2478, A2482, and A2392) showed a pH-dependent increase in their DMS reactivity (data not shown), but A2451 was only modestly reactive and its re-

activity did not change with pH (Fig. 6). Instead, C2452, which neighbors A2451 and is also universally conserved throughout phylogeny (Gutell 2000; Gutell et al. 2000), showed increased DMS reactivity with increased pH (Fig. 6). C2452 is stacked on the active-site A within the 50S ribosomal structure (Ban et al. 2000). Modification followed a standard pH profile in that C2452 was unreactive at low pH and became increasingly reactive as the pH was raised, resulting in an apparent pK_a of 7.2. A telling feature of this observation is that the only nucleophilic position on C is the N3, which is relatively solvent-inaccessible within the *H. marismortui* crystal structure without a structural rearrangement.

DEFINING THE POSITION OF A2451 MODIFICATION WITHIN *E. COLI* RIBOSOMES BY DIMROTH REARRANGEMENT

It is difficult to explain the irregular pH dependence of nucleotide modification within archaebacteria and eukaryotic peptidyl transferase centers as anything other than a pH-dependent conformation change. This caused us to revisit the question as to which imino group (N1 or N3) on A2451 was modified by DMS within *E. coli* ribosomes. N3 is solvent-exposed in the peptidyl transferase center (Fig. 2A) (Ban et al. 2000), so DMS modification at N3 would be consistent with a pK_a perturbation. It would explain the apparent hydrogen bond observed between A2451 and the nonbridging phosphate of the phosphoramidate inhibitor (Nissen et al. 2000). In contrast, modification at N1 could only be explained by a conformational change from that observed in the crystal structure, because the A2451 N1 is involved in a hydrogen bond to G2061 (Ban et al. 2000).

We utilized selective Dimroth rearrangement of the 1-methyladenosine adduct to distinguish between N1 and N3 methylation of adenosine in a sequence-specific manner (Fig. 7) (Macon and Wolfenden 1968; Fujii et al.

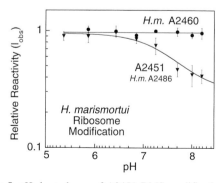

Figure 5. pH dependence of A2451 DMS modification in *H. marismortui* ribosomes. The extent of A2451 (*H.m.* A2486) and *H.m.* 2460 methylation was normalized as described in Fig. 4 and plotted versus pH (*H.m.* 2460, *circles* and A2451, *triangles*). The data are the average of three experiments, and the standard deviations are indicated with error bars.

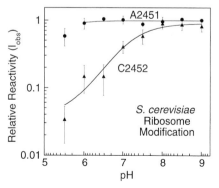

Figure 6. pH-independent modification of A2451 and pH-dependent modification of C2452 in yeast *S. cerevisiae* 60S ribosomes. Data were normalized as described in Fig. 4 and plotted versus pH (A2451, *circles* and C2452, *triangles*). The data indicate an apparent pK_a for *S. cerevisiae* C2452 of 7.3. Each point is an average of seven independent experiments.

1989; Muth et al. 2001). Methylation of N1 or N3 results in opening of the six-membered ring at neutral to slightly alkaline pH (Macon and Wolfenden 1968; Saito and Fujii 1979; Fujii et al. 1989). This process is quite slow for the nucleoside 1-methyladenosine at neutral pH, but at pH 9 at 25°C, it occurs with a half-life of about 1 day (Macon and Wolfenden 1968) . Once the ring is opened, it rapidly undergoes a Dimroth rearrangement to produce the stable 6-methyladenosine product (Fig. 7) (Jones and Robins 1963). From the perspective of the reverse transcriptase that is used to map sites of modification, this rearrangement converts an N-1 RT stop into a readthrough position. The N3-methylated adduct also undergoes ring opening to form the *N*-methylformamido-imidazole derivative at neutral to alkaline pH, but it does not undergo a subsequent rearrangement before ring closure (Fig. 7) (Saito and Fujii 1979; Fujii et al. 1989). Instead, it recircularizes back to 3-methyladenosine and reaches a uniform equilibrium between the two adducts (Saito and Fujii 1979). Both the methylated and the ring-opened adducts could cause transcriptional termination (Saito and Fujii 1979; Muth et al. 2000). We reasoned that retention of the N3 methyl group and migration of the N1 methyl group upon alkaline treatment of the modified RNA could provide a straightforward approach to determine whether the N1 or N3 imino group of a given A is methylated within an RNA sequence.

The position of adenosine heterocyclic ring methylation within *E. coli* 23S rRNA (residues 2478–2451) was investigated using conditions that promote Dimroth rearrangement of the N1-methylated adenosines. DMS-modified (pH 7.9) 23S rRNA from *E. coli* was isolated and subjected to alkaline (pH 9.0) incubation at 37°C for various times up to 60 hours. A2451 modification was then assayed by reverse transcription. The intensities of several DMS-specific N-1 RT bands within the ladder (A2468, A2469, A2476, and A2478) were progressively attenuated as a function of the alkaline incubation time relative to the intensity of the non-DMS-specific stops at G2448-m^2G2445 (Fig. 8, lanes 2, 4, 6, and 8). The attenuation does not appear to be due to RNA hydrolysis, because the overall background intensity at non-DMS-specific sites does not increase significantly as would be expected if the RNA were being progressively degraded (compare Fig. 8, lanes 1, 3, 5, and 7). The signal attenuation suggests that these four positions have undergone a Dimroth rearrangement and therefore are at least partially modified at the N1 position of the base. N1 DMS modification at A2468, A2469, A2476, and A2478 is consistent with the 50S ribosomal crystal structure.

If the level of signal attenuation seen for A2468, A2469, A2476, and A2478 is that expected for a nucleotide modified at N1, how does the pattern compare to that at A2451? The intensity of the A2451 N-1 RT stop is attenuated at least as much as that of the other A's in this region (Fig. 8). About half of the signal was lost after 20 hours (compare lanes 2 and 4, Fig. 8), and the band intensity had fallen to that of the no-DMS control after 60 hours (compare lanes 7 and 8, Fig. 8). This suggests that a substantial fraction of the DMS-modified A2451 can undergo a Dimroth rearrangement, which implies that within *E. coli* ribosomes methylation at A2451 occurred at the more common N1 position rather than at the relatively unusual N3 group. Since the A2451 N1 is hydrogen-bonded to the N1 of G2061, the DMS effect observed in *E. coli* at A2451 is more likely to reflect a pH-dependent conformational flexibility than the pK_a of A2451's N3 imino group.

Thus, large ribosomal subunits obtained from organisms spanning the three phylogenetic kingdoms each display a different DMS modification pattern within the peptidyl transferase core, and the data are most consistent with pH-dependent conformational flexibility. This is not altogether surprising upon examination of the crystal structure. Even though A2451 is involved in an array of hydrogen bonds that do not appear to require unusual pro-

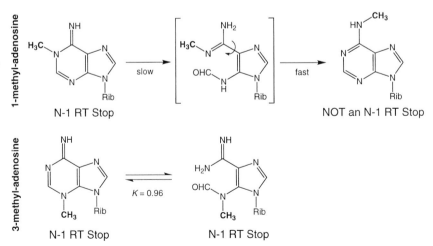

Figure 7. Scheme for hydrolytic ring opening of N1 and N3 methylated adenosine adducts. Ring opening of the N1 adduct is followed by a Dimroth rearrangement to produce the N6-methylated product (Jones and Robins 1963; Macon and Wolfenden 1968). The N3 adduct also undergoes ring opening, but it does not undergo rearrangement (Saito and Fujii 1979; Fujii et al. 1989). The outcome is that only the N1 adduct is converted into a reverse transcriptase readable modification upon mild alkaline incubation, whereas the N3 adduct remains a reverse transcriptase stop.

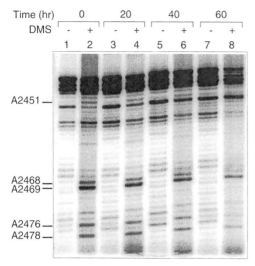

Figure 8. Dimroth analysis of DMS-modified 23S rRNA isolated from *E. coli* 50S subunits. Autoradiogram of reverse-transcribed 23S rRNA (nucleotides 2443–2480) incubated under conditions conducive to a Dimroth rearrangement of N1 methylated adenosine residues. The presence or absence of DMS and the length of the alkaline incubation are indicated above each lane. The bands corresponding to the N-1 RT stops for individual nucleotides are labeled to the left. Band intensities corresponding to adenosine residues 2468, 2469, 2476, 2478, and A2451 are attenuated as a function of time in a manner consistent with a Dimroth rearrangement.

tonation, A2451 and C2452 are flanked on either side by two A•C wobble pairs that should be highly pH-sensitive (Fig. 1) (Ban et al. 2000). The nucleotide 5′ of the active-site residue, A2450, forms a wobble pair with C2063, and these residues are conserved in all three organisms. In *E. coli* and *H. marismortui* ribosomes, the nucleotide on the 3′ side, A2453, forms a wobble pair with C2499. In a typical A•C wobble pair, the C O2 forms a hydrogen bond to the A•N1, which is protonated (Cai and Tinoco 1996). Based on nuclear magnetic resonance (NMR) characterization of such A•C wobble pairs, the pK_a of the A is perturbed to a value approaching neutrality (6.0–6.5). The wobble geometries of the A•C pairs observed crystallographically within the peptidyl transferase center are consistent with an A N1 to C O2 interaction (Ban et al. 2000). Although they should be mostly protonated at pH 5.8, the condition under which the 50S ribosomes were determined, they are unlikely to be protonated at the slightly basic pH (7.5–8.5) where the peptidyl transferase center displays maximal activity (Pestka 1972). Deprotonation of either or both wobble pairs at neutral to alkaline pH might create local rearrangements in the active site.

IMPLICATIONS FOR THE PEPTIDYL TRANSFERASE REACTION MECHANISM

Although these results indicate that DMS modification of A2451 in the peptidyl transferase center cannot be taken as evidence for or against the N3 pK_a perturbation, it remains possible that the ribosome promotes catalysis by a general acid-base mechanism utilizing A2451. How-

ever, arguments for such a mechanism currently rest entirely upon functional group proximity within the crystal structure.

Although the DMS data do not speak directly to the pK_a of A2451, they do indicate that something in the active site is ionized with a near-neutral pK_a. Furthermore, the macroscopic pK_a of the peptidyl transferase reaction (pK_a 7.2–7.6) suggests that ionization of a functional group in the substrate or the enzyme is the rate-limiting step of the reaction and that the deprotonated form of the ribosomal-substrate complex is active (Pestka 1972). An obvious candidate is the nucleophilic amine of the A-site substrate, since its pK_a closely matches that of the reaction. However, Fahnestock and Rich reported that the macroscopic pK_a of the peptidyl transfer reaction was unchanged when the α-amino group of the A-site tRNA was replaced with a hydroxyl group, even though this raised the nucleophilic pK_a by at least three pH units (Fahnestock et al. 1970). If the macroscopic pK_a is not that of the nucleophile, then there must be an ionizable group in the ribosome whose deprotonation constitutes the rate-limiting step in the peptidyl transferase reaction. This might be A2451, but it could also be another group involved in a catalytic or a conformational step in the reaction profile.

On the basis of its crystal structure, A2451 has been proposed to stabilize the oxyanion in the tetrahedral intermediate, and it is possible that the A2451 pK_a could be perturbed, even if only transiently. This may result from juxtaposition of the negatively charged oxyanion, which is in proximity to the A2451 N3 during the short-lived tetrahedral intermediate. Unfortunately, the only evidence for this perturbation comes from the close approach ($\approx$3 Å) of the A2451 N3 with the phosphoramidate mimic of the tetrahedral intermediate (Nissen et al. 2000). Because protons cannot be detected by macromolecular crystallography, this and other hypotheses regarding the catalytic reaction mechanism await further biochemical evidence.

One mechanistic alternative that has not been given much consideration is that A2451 N3 may be a hydrogen-bond acceptor to the α-amino group in the A site. In this scenario, the N3 group would help orient the α-amino group for attack. An N3 to α-amino group hydrogen bond is consistent with the A-site substrate-bound state observed crystallographically (Nissen et al. 2000), and it may explain the in vitro mutagenesis data recently reported for the peptidyl transferase active site (Polacek et al. 2001). Using reconstituted 50S ribosomes from in vitro transcribed *Thermus aquaticus* 23S rRNA, Polacek et al. mutated A2451 to U, C, or G and observed only modest changes in activity (2- to 50-fold, depending on the assay conditions). Although greater care was taken to measure relative reaction rates and to measure initial velocities, similar results were obtained by Thompson et al. (2001) using a *Bacillus stearothermophilus* in vitro reconstitution system. It is worth noting that despite making all three possible mutations at A2451, none of these changes actually alters the hydrogen-bond-accepting character of the functional group of interest. An unprotonated N3 is retained between A and G, whereas the O2 carbonyl of C and U occupies an equivalent spatial location and can also act as a hydrogen-bond acceptor. It is not clear how G, C, and U could be accommodated in the greater overall network of interactions that are seemingly specific to A, but poor accommodation may explain why the mutations are dominant lethal in vivo where the catalytic rates are substantially faster than observed in the in vitro reconstitution assays. Since traditional mutagenesis does not summarily change the N3 group of interest, it appears that chemical mutagenesis may be necessary to determine the catalytic role of this functional group.

ACKNOWLEDGMENTS

We thank Betty Freeborn and Peter B. Moore for the gift of *H. marismortui* and *S. cerevisiae* ribosomes, and Dieter Söll and Thomas Steitz for helpful discussion. This work was supported by an American Cancer Society postdoctoral fellowship to G.W.M. and National Institutes of Health grant GM-54839 to S.A.S.

REFERENCES

Ban N., Nissen P., Hansen J., Moore P., and Steitz T. 2000. The complete atomic structure of the large ribosomal subunit at 2.4 Å resolution. *Science* **289:** 905.

Barta A., Dorner S., Polacek N., Berg J.M., Lorsch J.R., Nissen P., Hansen J., Muth G.W., Ban N., Moore P.B., Strobel S.A., and Steitz T.A. 2000. Mechanism of ribosomal peptide bond formation (see technical comments). *Science* **291:** 203.

Baxter R.M. and Zahid N.D. 1978. The modification of the peptidyl transferase activity of 50S ribosomal subunits, LiCl-split proteins and L16 ribosomal protein by ethoxyformic anhydride. *Eur. J. Biochem.* **91:** 49.

Brosius J., Dull T.J., Sleeter D.D., and Noller H.F. 1981. Gene organization and primary structure of a ribosomal RNA operon from *Escherichia coli. J. Mol. Biol.* **148:** 107.

Cai Z. and Tinoco I. 1996. Solution structure of loop A from the hairpin ribozyme from tobacco ringspot virus satellite. *Biochemistry* **35:** 6026.

Cerna J. and Rychlik I. 1980. Phenylboric acids—A new group of peptidyl transferase inhibitors. *FEBS Lett.* **119:** 343.

Chladek S. and Sprinzl M. 1985. The 3′-end of tRNA and its role in protein biosynthesis. *Angew. Chem. Int. Ed. Engl.* **24:** 371.

Connell G.J. and Yarus M. 1994. RNAs with dual specificity and dual RNAs with similar specificity. *Science* **264:** 1137.

Douthwaite S., Powers T., Lee J.Y., and Noller H.F. 1989. Defining the structural requirements for a helix in 23 S ribosomal RNA that confers erythromycin resistance. *J. Mol. Biol.* **209:** 655.

Fahnestock S., Neumann H., Shashoua V., and Rich A. 1970. Ribosome-catalyzed ester formation. *Biochemistry* **9:** 2477.

Ferre-D'Amare A.R., Zhou K., and Doudna J.A. 1998. Crystal structure of a hepatitis delta virus ribozyme. *Nature* **395:** 567.

Fersht A. 1985. *Enzyme structure and mechanism.* W.H. Freeman and Company, New York.

Fujii T., Saito T., and Nakasaka T. 1989. Purines. XXXIV. 3-Methyladenosine and 3-methyl-2′-deoxadenosine: Their synthesis, glycosidic hydrolysis, and ring fission. *Chem. Pharm. Bull.* **37:** 2601.

Green R. and Noller H.F. 1997. Ribosomes and translation. *Annu. Rev. Biochem.* **66:** 679.

Gutell R.R. 1996. Comparative sequence analysis and the structure of 16S and 23S rRNA. In *Ribosomal RNA: Structure, evolution, processing, and function in protein biosynthesis* (ed. R.A. Zimmermann and A.E. Dahlberg) p. 111, CRC Press, Boca Raton, Florida.

———. http://www.rna.icmb.utexas.edu/CSI/fts.html.

Gutell R.R., Gray M.W., and Schnare M.N. 1993. A compilation of large subunit (23S and 23S-like) ribosomal RNA structures: 1993. *Nucleic Acids Res.* **21:** 3055.

Gutell R.R., Cannone J.J., Shang Z., Du Y., and Serra M.J. 2000. A story: Unpaired adenosine bases in ribosomal RNAs. *J. Mol. Biol.* **304:** 335.

Hecht S.M. 1992. Probing the synthetic capabilities of a center of biochemical catalysis. *Acc. Chem. Res.* **25:** 545.

Hegazi M.F., Quinn D.M., and Schowen R.L. 1978. Transition-state properties in acyl and methyl transfer. In *Transition states of biochemical processes* (ed. R.D. Gandour and R.L. Schowen), p. 355. Plenum Press, New York.

Jones J.W. and Robins R.K. 1963. Purine nucleosides. III. Methylation studies of certain naturally occurring purine nucleosides. *J. Am. Chem. Soc.* **85:** 193.

Kearsey S.E. and Craig I.W. 1981. Altered ribosomal RNA genes in mitochondria from mammalian cells with chloramphenicol resistance. *Nature* **290:** 607.

Killian J.A., Van Cleve M.D., Shayo Y.F., and Hecht S.M. 1998. Ribosome-mediated incorporation of hydrazinophenylalanine into modified peptide and protein analogues. *J. Am. Chem. Soc.* **120:** 3032.

Kim D.F. and Green R. 1999. Base-pairing between 23S rRNA and tRNA in the ribosomal A site. *Mol. Cell* **4:** 859.

Kraut J. 1977. Serine proteases: Structure and mechanism of catalysis. *Annu. Rev. Biochem.* **46:** 331.

Lawley P. and Brookes P. 1963. Further studies on the alkylation of nucleic acids and their constituent nucleotides. *Biochem. J.* **89:** 127.

Macon J.B. and Wolfenden R. 1968. 1-Methyladenosine. Dimroth rearrangement and reversible reduction. *Biochemistry* **7:** 3453.

Maden B. and Monro R. 1968. Ribosome-catalyzed peptidyl transfer. Effects of cations and pH value. *Eur. J. Biochem.* **6:** 309.

Minakawa N., Kojima N., and Matsuda A. 1999. Nucleosides and nucleotides. 184. Synthesis and conformation investigation of anti-fixed 3-deaza-3-halopurine ribonucleosides. *J. Org. Chem.* **64:** 7158.

Moazed D. and Noller H.F. 1987. Chloramphenicol, erythromycin, carbomycin and vernamycin B protect overlapping sites in the peptidyl transferase region of 23S ribosomal RNA. *Biochimie* **69:** 879.

———. 1989. Interaction of tRNA with 23S rRNA in the ribosomal A, P, and E sites. *Cell* **57:** 585.

———. 1991. Sites of interaction of the CCA end of peptidyl-tRNA with 23S rRNA. *Proc. Natl. Acad. Sci.* **88:** 3725.

Muth G.W., Ortoleva-Donnelly L., and Strobel S.A. 2000. A single adenosine with a neutral pKa in the ribosomal peptidyl transferase center. *Science* **289:** 947.

Muth G.W., Chen L., Kosek A.B., and Strobel S.A. 2001. pH-dependent conformational flexibility within the ribosomal peptidyl transferase center. *RNA* **7:** 1403.

Nakano S., Chadalavada D.M., and Bevilacqua P.C. 2000. General acid-base catalysis in the mechanism of a hepatitis delta virus ribozyme. *Science* **287:** 1493.

Nierhaus K.H., Schulze H., and Cooperman B.S. 1980. Molecular mechanisms of the ribosomal peptidyltransferase center. *Biochem. Int.* **1:** 185.

Nissen P., Hansen J., Ban N., Moore P., and Steitz T. 2000. The structural basis of ribosome activity in peptide bond synthesis. *Science* **289:** 920.

Noller H.F. 1991. Ribosomal RNA and translation. *Annu. Rev. Biochem.* **60:** 191.

Perrotta A.T., Shih I., and Been M.D. 1999. Imidazole rescue of a cytosine mutation in a self-cleaving ribozyme. *Science* **286:** 123.

Pestka S. 1972. Peptidyl-puromycin synthesis on polyribosomes from *Escherichia coli. Proc. Natl. Acad. Sci.* **69:** 624.

Polacek N., Gaynor M., Yassin A., and Mankin A.S. 2001. Ribosomal peptidyl transferase can withstand mutations at the putative catalytic nucleotide. *Nature* **411:** 498.

Roesser J.R., Chorghade M.S., and Hecht S.M. 1986. Ribosome-catalyzed formation of an abnormal peptide analogue. *Bio-*

chemistry **25:** 6361.

Saenger W. 1984. Principles of nucleic acid structure. Springer-Verlag, New York.

Saito T. and Fujii T. 1979. Synthesis and hydrolysis of 3-methyladenosine. *J. Chem. Soc. Chem. Commun.* **1979:** 135.

Samaha R.R., Green R., and Noller H.F. 1995. A base pair between tRNA and 23S rRNA in the peptidyl transferase centre of the ribosome. *Nature* **377:** 309.

Seela F., Debelak H., Usman N., Burgin A., and Beigelman L. 1998. 1-Deazaadenosine: Synthesis and activity of base-modified hammerhead ribozymes. *Nucleic Acids Res* **26:** 1010.

Stern S., Moazed D., and Noller H.F. 1988. Structural analysis of RNA using chemical and enzymatic probing monitored by primer extension. *Methods Enzymol.* **164:** 481.

Thompson J., Kim D.F., O'Connor M., Lieberman K.R., Bayfield M.A., Gregory S.T., Green R., Noller H.F., and Dahlberg A.E. 2001. Analysis of mutations at residues A2451 and G2447 of 23S rRNA in the peptidyl transferase active site of the 50S ribosomal subunit. *Proc. Natl. Acad. Sci.* **98:** 9002.

Triman K.L. 1999. Mutational analysis of 23S ribosomal RNA structure and function in *Escherichia coli. Adv. Genet.* **41:** 157.

Vannuffel P., Di Giambattista M., and Cocito C. 1994. Chemical probing of a virginiamycin M-promoted conformational change of the peptidyl-transferase domain. *Nucleic Acids Res.* **22:** 4449.

Wan K.K., Zahid N.D., and Baxter R.M. 1975. The photochemical inactivation of peptidyl transferase activity. *Eur. J. Biochem.* **58:** 397.

Welch M., Chastang J., and Yarus M. 1995. An inhibitor of ribosomal peptidyl transferase using transition-state analogy. *Biochemistry* **34:** 385.

Analysis of the Active Site of the Ribosome by Site-directed Mutagenesis

D.F. Kim, K. Semrad, and R. Green

Howard Hughes Medical Institute, Johns Hopkins University School of Medicine, Baltimore, Maryland 21205

Atomic-resolution X-ray crystallographic data of the large subunit of the ribosome complexed with a transition-state analog (TSA) for peptide-bond formation has revealed the identity of the components present near the active site for peptide-bond formation (Nissen et al. 2000). Consistent with evolutionary considerations, the active site of the ribosome is composed exclusively of highly conserved 23S rRNA nucleotides. Strikingly, biochemical and genetic approaches had accurately predicted most of its components (for review, see Lieberman and Dahlberg 1995; Green and Noller 1997).

Highly conserved bases scattered over hundreds of nucleotides in the primary sequence come together in three dimensions to promote peptide-bond formation. First, the CCA ends of the peptidyl and aminoacyl tRNAs are oriented in the active site by Watson-Crick base-pairing with two separate loops (the P loop and the A loop, respectively) in domain V of the 23S rRNA. Two of the three Watson-Crick interactions seen in the structure had previously been predicted on the basis of in vitro genetic approaches (Samaha et al. 1995; Kim and Green 1999), these data serving now to validate the arrangement of active-site components identified crystallographically by "soaks" of a variety of substrate analogs into the crystal lattice. Next, the terminal adenosines of both tRNA substrates are oriented by specific interaction between their N1 positions and 2′ hydroxyl groups located on 23S rRNA nucleotides. Finally, the 23S rRNA nucleotide in the active site most proximal to the tetrahedral phosphoramide linkage of a bound transition-state analog is A2451. In the structure with the bound TSA, the N3 position of A2451 is approximately 3.4 Å from the nonbridging oxygen atom of the transition-state analog and approximately 4.2 Å from the nucleophilic amine mimic. On the basis of the high-resolution structure of the apoenzyme, there is proposed to be a network of hydrogen-bonding interactions connecting G2447, via a K^+ ion (and G2061), to A2451 and from there to the TSA. However, the exact arrangement of the active site differs slightly between the unbound (apoenzyme) and the lower-resolution TSA-bound structures of the large subunit (see Fig. 1). The potassium ion critical in establishing this network is not seen in the TSA-bound structure (Nissen et al. 2000).

As in any active site, the substrates are intimately surrounded by enzymatic moieties that provide a protected environment where chemistry can occur. The high-resolution structure provides us with one snapshot of how these components are arranged with respect to one an-

other. Even with this detailed structural information, an understanding of how these RNA nucleotides conspire in the catalysis of peptide-bond formation will require substantial biochemical characterization.

Of immediate interest in the large subunit structure is the fact that the N3 position of A2451 is within hydrogen-bonding distance of the phosphoramide oxygen of the bound transition-state analog. This observation led to speculation that an unanticipated proton might be found there even though the unperturbed pK_a of either of these moieties is far from neutral or from pH 6.2 (the pH at which the structure was solved). A series of biochemical experiments revealed that A2451 has an unusual dimethyl sulfate (DMS) reactivity profile consistent with a substantially perturbed pK_a (Muth et al. 2000) that might be critical for the mechanism of peptide-bond formation. More recent biochemical experiments suggest that pH-dependent conformational changes in the active site can

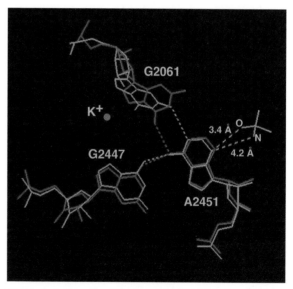

Figure 1. Active site of the large subunit of the ribosome from *H. marismortui.* Shown in red is the structure of the apoenzyme with observed K^+ ion; in green is the structure of the TSA-bound structure in the same region and the tetrahedral phosphoramide group. Hydrogen bonds between G2447, G2061, and A2451 were placed automatically by QUANTA. The dashed lines between the N3 of A2451 and the nonbridging oxygen and the nitrogen of the phosphoramide include the distances between these positions in the TSA-bound structure. Significant differences between these two structures are readily seen.

explain much of the earlier data. Bayfield et al. (2001) demonstrated that A2451 becomes largely inaccessible to DMS modification when ribosomes are converted to their most active state. Regardless of whether A2451 has a perturbed pK_a, its N3 position remains the most proximal moiety to the critical tetrahedral phosphoramide of the transition-state analog, and it thus remains a candidate for direct involvement in active-site chemistry.

Initial proposals suggested that A2451 might act as a general base to deprotonate the amine of the aminoacyl tRNA (Nissen et al. 2000), whereas revised proposals suggested that A2451 might instead deprotonate the incoming ammonium form of the aminoacyl tRNA (pK_a of approximately 7) (Berg and Lorsch 2001). It was also proposed that A2451 might function to stabilize the developing negative charge in the transition state in a manner analogous to the oxyanion hole of the serine proteases (Nissen et al. 2000). These different models predict different relative contributions by nucleotide A2451 to the rate of peptide-bond formation (Berg and Lorsch 2001). We have used a site-directed mutagenesis approach to begin to address these questions by asking what the specific contribution of nucleotides A2451 and G2447 might be to the rates of peptide-bond formation.

MATERIALS AND METHODS

Bacterial strains and plasmids. Site-directed mutagenesis of A2451 of *Bacillus stearothermophilus* and *Escherichia coli* 23S rRNAs for in vitro transcription was performed using the Quikchange procedure by Stratagene (on plasmids pBS23S-T7 and pCW, respectively). G2447 mutant ribosomes were prepared as described from the Δ7 prrn strain MC250 (Δ*rrnE* Δ*rrnB* Δ*rrnH* Δ*rrnG::cat* Δ*rrnA* Δ*rrnD::cat* Δ*rrnC::cat rpsL121 recA56*/ prrnS12 p70*)* (Thompson et al. 2001).

In vitro reconstitution of B. stearothermophilus ***and E. coli*** 50S ***subunits.*** *B. stearothermophilus* 50S subunits were reconstituted from in vitro transcripts (wild-type and mutant versions) of 23S rRNA as described previously (Green and Noller 1999). *E. coli* 50S subunits were reconstituted from in vitro transcripts of 23S rRNA using modifications of the original protocol for natural rRNA (Dohme and Nierhaus 1976). 1.25 A_{260} of 23S rRNA transcript and 0.05 A_{260} of 5S rRNA were incubated with 1 equivalence unit of TP50 (Dohme and Nierhaus 1976) in 20 mM Tris-HCl (pH 7.4), 4 mM MgOAc, 400 mM NH$_4$Cl, 0.2 mM EDTA, and 5 mM β-mercaptoethanol at 44°C for 20 minutes. In addition to these standard components, 750 mM trimethylamine oxide (TMAO) and 0.5 mM HMR 3647 (a ketolide antibiotic; Khaitovich and Mankin 1999) were included (Semrad and Green 2002). The magnesium concentration was then raised to 20 mM and the reaction was incubated at 50°C for 90 minutes.

Peptidyl transferase reactions. N-protected aminoacylated tRNA substrates and fragments and radioactively labeled [^{32}P]cytidyl-puromycin (CPm) was prepared as described previously (Kim and Green 1999). Assays A, B, and C were performed as described previously (Kim

and Green 1999; Thompson et al. 2001). Chloramphenicol resistance was tested by incubating the reaction mixture with 62.5 μM chloramphenicol prior to the addition of the puromycin substrate. Peptidyl hydrolase (PTH) was prepared in the laboratory by cloning the *E. coli* gene into the pQE70 Qiagen vector and isolating the protein on a Ni^{++} affinity column (data not shown). In reactions including PTH, the enzyme was added simultaneously with the puromycin substrate to initiate the peptidyl transfer reaction. Independent experiments (data not shown) confirmed that PTH completely deacylates N-acetyl-Phe-tRNA$_{Phe}$ in less than 30 seconds. Rapid kinetic reactions were performed for assay B with native *E. coli* ribosomes in 20 mM Tris (pH 7.5), 100 mM NH$_4$Cl, and 20 mM MgCl$_2$. Reactions were started by mixing in a KinTek apparatus equal volumes of preformed 70S ribosome-N-Ac-[^{35}S]-Met-tRNA$_{Met}$-gene 32 mRNA complex in 1× buffer with puromycin (2 mM) in 1× buffer to yield final concentrations of 0.5 μM and 1 mM, respectively. Reactions were quenched in the same apparatus with an equal volume of 0.2 N KOH and quantitated as described previously (Kim and Green 1999).

RESULTS

Analysis of Mutant Ribosome Populations

Straightforward analysis of the effect of specific mutations on the rates of catalysis of peptide bond-formation depends on the generation of pure populations of mutant ribosomes. 23S rRNA genes carrying mutations at nucleotide G2447 or A2451 were expressed in the Δ7 prrn strain of *E. coli* in which all of the ribosomal RNA operons have been deleted so that growth depends solely on the plasmid-borne (in this case mutant) version of the rRNA gene (Asai et al. 1999; O'Connor et al. 2001). G2447A and G2447C mutant ribosomes were able to support growth of this strain and allowed us to isolate pure populations of these mutant ribosomes. In contrast, mutations at A2451 (to C, G, and U) do not support growth of the Δ7 prrn strains of *E. coli* and are dominant lethal when coexpressed with the wild-type version of the 23S rRNA gene in the λ pL conditional system (Gourse et al. 1985). Extensive characterization of the in vivo properties of the various mutant ribosomes is presented in another paper by Dahlberg and colleagues (Gregory et al., this volume). Here, for the biochemical analysis of peptidyl transferase activity, G2447 mutant ribosomes were isolated from the appropriate strain of *E. coli*, whereas A2451 mutant ribosomes were generated using an in vitro reconstitution approach. We have previously optimized conditions for the in vitro reconstitution of *B. stearothermophilus* 50S subunits (Green and Noller 1999) and have used this system for a number of studies (Green and Noller 1999; Kim and Green 1999). Recently, we have also made substantial improvements in the *E. coli* reconstitution protocol allowing characterization of mutant *E. coli* ribosomes as well. Both of these reconstitution systems were used to analyze A2451 mutant ribosomes.

In both reconstitution protocols, in-vitro-transcribed 23S rRNA is combined with 5S rRNA and total proteins from the 50S subunit (Fahnestock et al. 1974; Dohme and Nierhaus 1976) and allowed to incubate under specific conditions that result in maximal incorporation of the 23S rRNA into functional particles. A number of years ago, we learned that in vitro reconstitution of *E. coli* 50S subunits from in-vitro-transcribed 23S rRNA was extremely inefficient relative to the high efficiency observed with natural rRNA. We then identified a specific region of *E. coli* 23S rRNA where posttranscriptional modifications were apparently critical for efficient reconstitution (Green and Noller 1996). We subsequently found that the reconstitution efficiency was substantially better for the *B. stearothermophilus* 50S subunit (than for *E. coli*) when using in-vitro-transcribed 23S rRNA (Green and Noller 1999). Recently, we have made significant improvements in the reconstitution efficiency of *E. coli* in vitro transcripts by adding two different low-molecular-weight compounds, a ketolide antibiotic HMR3647 (Khaitovich and Mankin 1999) and the "chemical chaperone" trimethylamine oxide (TMAO) (Semrad and Green 2002). In either of these systems, reconstitution efficiency is on the order of 10% (i.e., the K_{obs} from the reconstitution reaction is 10% as fast as the rate with a natural 50S subunit).

With these active reconstituted particles, it is possible to look at the effect of mutations in the active site of the ribosome and to ask specific questions about the contribution of these nucleotides to the observed rates of peptide-bond formation. It should be kept in mind that the population being analyzed in these reconstituted systems is certainly heterogeneous, which constrains meaningful physical analysis of these populations; in addition, complications arise in data interpretation. For example, the activity in a population of ribosomes with a reconstitution efficiency of 10% could reflect a large fraction of the species in the population having less than normal levels of activity, or a smaller fraction of the population having considerable activity. Despite these limitations, the re-

constituted particles can be exploited to analyze certain ribosomal functions. rRNA mutations may have effects on a number of steps, including reconstitution efficiency, substrate binding, conformational changes, or catalysis. These possibilities can often be distinguished by biochemical and kinetic approaches.

Single-turnover Peptidyl Transferase Assays

A number of sensitive peptidyl transferase assays can be used to compare the rates of catalysis of wild-type and mutant ribosome populations. The "fragment reaction" is often used to measure peptidyl transferase activity (assay A, Fig. 2) (Monro and Marcker 1967). This reaction looks at the peptidyl transferase activity of 50S subunits in a buffer containing high salt, magnesium, and an organic solvent (33%). The substrates for this reaction are generally minimal tRNA fragments such as CACCA-*N*-acetyl-methionine in the peptidyl site and puromycin in the aminoacyl site. The organic solvent apparently allows the binding of such minimal substrates to the P site of the ribosome, as the solvent is not required for promoting the puromycin reactivity of isolated 50S subunits if they are isolated with peptidyl tRNA already bound (Traut and Monro 1964). The product of the fragment reaction is *N*-acetyl-methionine-puromycin and can be resolved and quantitated following organic extraction and/or paper electrophoresis (assay A, Fig 2). This assay is appealing in its simplicity, requiring no 30S subunits, messenger RNA, or intact tRNA substrates. Because mimimal substrates are used, binding to the P site generally limits the rate of catalysis. This feature was particularly useful in the identification of binding interactions in the ribosome critical in orienting the CCA ends of the P-site substrate (Samaha et al. 1995).

More recently, we have focused on assays using intact P-site tRNA substrates, 70S ribosomes, messenger RNA, and various puromycin (Pm) derivatives as the A-site sub-

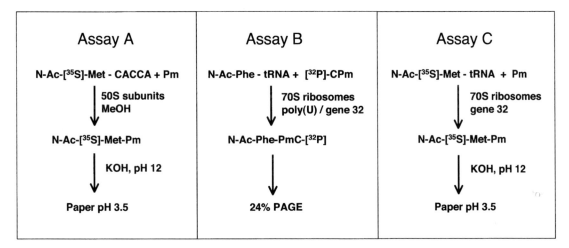

Figure 2. Peptidyl transferase assays. The various substrates, products, ribosomes, and mRNAs for three different peptidyl transferase reactions are shown. These assays have previously been described in detail (Monro and Marcker 1967; Green and Noller 1996; Kim and Green 1999).

strates (assays B and C, Fig. 2) (Kim and Green 1999; Thompson et al. 2001). The use of 70S ribosomes eliminates the need for high salt, magnesium, and organic solvents for catalysis and creates a situation where single-turnover kinetics are easily observed. Here, two basic assays are used with 70S ribosomes. In the first assay, 70S ribosomes are prebound with either poly(U) or gene 32 messenger RNA. N-acetyl-Phe-tRNA$_{Phe}$ is then bound in the P site of the ribosome, and the peptidyl transferase reaction is initiated with small amounts of a radioactively labeled [^{32}P]CPm derivative. The substrates and product ([^{32}P]CPm-N-acetyl-Phe) of this reaction are readily resolved on 24% acrylamide/5 M urea gels and quantitated (assay B) (Kim and Green 1999). In a second assay, 70S ribosomes are prebound with gene 32 mRNA and N-acetyl-[^{35}S]Met-tRNA$_{Met}$ in the P site, and the reaction is initiated with saturating amounts of puromycin. The substrates and products from this reaction are resolved using paper electrophoresis (assay C) (Kim and Green 1999). These assays are similar, except that in one case the A-site substrate is present at subsaturating concentrations and in the latter it is saturating. These two assays can in principle distinguish between A-site binding defects and defects in a step after tRNA binding (e.g., chemistry).

In the 70S ribosome assays (Fig. 2), it is possible to look at single-turnover peptidyl transferase kinetics to compare rate constants. We have confirmed in assays B and C that the P-site substrates, N-acetyl-Met-tRNA$_{Met}$ or N-acetyl-Phe-tRNA$_{Phe}$, are present in saturating concentrations relative to both mutant and wild-type ribosomes. In addition, in assay C, we confirmed that 1 mM puromycin was saturating for both the mutant and wild-type ribosomes. Finally, we were interested in confirming that the reaction rates being measured represented single-turnover peptidyl transfer events. If multiple turnover events are being measured, observed rate differences can reflect binding, product release, *or* catalytic events. In the case of the A2451 and G2447 mutant ribosomes, we were most interested in looking at the effects of these mutations on the chemical steps of catalysis.

In both assays, intact N-acetylated tRNAs are used as P-site substrates. Following peptidyl transfer, the deacylated P-site substrate likely moves to a P/E state (Moazed and Noller 1989), and it is the stability of this binding state that determines whether the rates being observed represent single-turnover events. In the case where the off-rate of the P/E-bound deacylated tRNA is fast relative to the time scale of the assay, multiple turnovers of the ribosome could be observed. Assays B and C were both run under conditions of high magnesium where the off-rates of deacylated tRNA were expected to be slow relative to the time scale of the assay. However, the binding affinities of N-acetylated tRNAs for reconstituted ribosomes have not been previously characterized.

To experimentally address this issue, a chase experiment was performed using the enzyme peptidyl hydrolase. Peptidyl hydrolase deacylates unbound N-acetylated aminoacyl tRNAs, thus eliminating the possibility of secondary binding and catalytic events. Free in solution, the peptidyl hydrolase completely deacylated N-acetyl-Phe-

tRNA$_{Phe}$ in less than 30 seconds (data not shown). The chase reaction was performed by adding PTH simultaneously with the [^{32}P]CPm substrate. A comparison of this observed rate to the rate of the reaction in the absence of PTH indicated that there was no effect of PTH on catalysis (Fig. 3) measured over a period of approximately 30 minutes with wild-type *E. coli* ribosomes. Similar results were obtained with G2447A mutant *E. coli* ribosomes and with *B. stearothermophilus* reconstituted wild-type and A2451C mutant ribosomes (data not shown). These data establish that the 70S reactions that we have developed measure single-turnover peptidyl transfer events. In contrast, the fragment reaction (assay A) likely measures multiple-turnover reaction kinetics, thus making it more difficult to determine whether the actual rates of catalysis are affected by a given nucleotide change in the active site.

Measuring the Chemical Step of Peptide-bond Formation

Ultimately, to understand the mechanism of catalysis in the active site of the ribosome, it is critical that the assay measures the chemical step of the reaction rather than, for example, binding or conformational changes preceding catalysis. The known in vivo rate of translation is approximately 20 amino acids/sec, indicating that peptidyl transfer must be at least equally fast. Interestingly, although the literature reports rate constants of peptide-bond formation of ~100 per second with intact aminoacyl tRNA substrates (loaded enzymatically by EF-Tu) (Pape et al. 1999), reactions analogous to those presented here

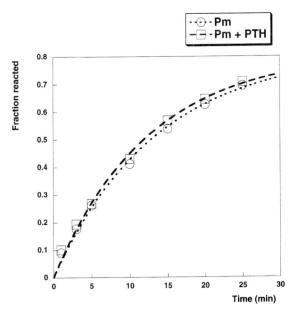

Figure 3. Peptidyl hydrolase chase experiment. In assay B using wild-type ribosomes, peptidyl transferase reactions were initiated either with the addition of [^{32}P]CPm substrate or with a mixture of [^{32}P]CPm and peptidyl hydrolase enzyme. Time points were taken, and the fractional product ([^{32}P]CPm-N-Ac-Phe) generated as a function of time (in minutes) is plotted. The rate of the reaction was unaffected by the peptidyl hydrolase chase.

using puromycin as an A-site analog are reported to have rate constants in the range of several per minute (Synetos and Coutsogeorgopoulos 1987). We have been interested in understanding the apparent discrepancies between these results and in establishing an experimental system where the chemistry of peptidyl transfer can be dissected.

Initial measurements of first-order rate constants (assay C) with native ribosomes indicate that peptidyl transfer rate constants substantially exceeding 20 per second can be measured with puromycin as an A-site substrate under standard conditions. In the initial fastest phase of the reaction, up to 30% of the ribosomal population catalyzes peptide-bond formation with a rate constant of ~75 per second at 20°C (>100/sec at 37°C) (Fig. 4). A kinetically slower population of ribosomes converts most of the remainder of the N-Ac-Met-tRNA$_{Met}$ substrate to product. We are currently trying to understand the distinctions between these ribosomal populations and whether they can be interconverted. Future experiments will aim at more extensive analysis of the pH dependence of the peptidyl transferase reaction and at modification of the A- and P-site substrates to explore mechanistic details.

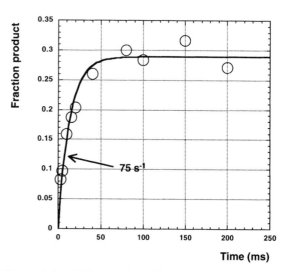

Figure 4. Rapid kinetics of peptide-bond formation by wild-type ribosomes. The rate constant (75/sec) of wild-type *E. coli* ribosomes in assay B at 20°C was determined using a rapid quench system to measure early time points in the reaction profile. Fractional product (N-Ac-[^{35}S]Met-Pm) generated as a function of time (in milliseconds) is plotted.

A2451 Mutant Analysis

The effects of mutations at A2451 were measured using the three different peptidyl transferase assays (described above). A2451 mutant ribosomes were generated using in vitro reconstitution reactions in both the *B. stearothermophilus* and the *E. coli* systems. Reconstitutions were performed as described previously for *B. stearothermophilus* using in-vitro-transcribed 23S rRNAs, 5S rRNA, and TP50 (total ribosomal proteins isolated from the 50S subunit). For the peptidyl transferase reaction with A2451 mutant reconstituted ribosomes, 70S particles were assembled at 45°C by adding an equimolar amount of highly purified 30S subunits from *E. coli* to the reconstitution reaction with saturating amounts of mRNA template (poly(U) or gene 32 mRNA) and an equimolar amount of *N*-acetylated P-site tRNA substrate (*N*-acetyl-Phe-tRNA$_{Phe}$ for assay B or *N*-acetyl-Met-tRNA$_{Met}$ for assay C). The reaction was then started with the addition of either sub-stoichiometric amounts of the radiolabeled puromycin derivative [^{32}P]CPm (0.04 μM) or with saturating amounts of free puromycin (1 mM). Reaction products were analyzed as described previously (Kim and Green 1999), and rates were determined from the linear portion of the reaction profile.

Both assays yielded similar results with the A2451 mutant ribosomes. In assay B where the A-site substrate is limiting and in assay C where the puromycin derivative is saturating, the A2451C, G, and U mutant ribosomes had effects on the rates of the reaction ranging from 3- to 10-fold (Table 1). In assay C, increases in puromycin concentration had no effect on the relative rates of catalysis by these ribosomes. The consistency of these results is striking and suggests that these mutant ribosomes are not defective in A-site substrate binding since no compensation of the deleterious effect was seen when saturating levels of substrate were used.

To extend the analysis of A2451, mutations were analogously generated in *E. coli* ribosomes using the optimized in vitro reconstitution protocol (Semrad and Green 2002). Peptidyl transferase activity for these ribosomes was first assessed using the traditional "fragment assay" that follows the catalytic activity of isolated 50S subunits incubated with minimal P- and A-site substrates

Table 1. Relative Peptidyl Transferase Rates of A2451 Mutant Ribosomes

| | | Assay B | | Assay C | Assay A | Assay B |
| | Recon. | *B. stearo* | *B. stearo* | *B. stearo* | *E. coli* | *E. coli* |
	Cam	(−)	(+)	(−)	(−)	(−)
A2451 (wt)		**1.0**	0.01	**1.0**	**1.0**	**1.0**
A2451C		0.32	0.19	0.25	0.02	1.4
A2451G		0.09	0.03	0.07	0.04	0.5
A2451U		0.31	0.21	0.2	0.03	0.6

Initial rates were determined from time points in the linear range of the reactions and normalized to the rate of the wild type for each set and assay. The normalized rates (in bold type) are shown for assays performed (+) with, or (−) without, 62.5 μM chloramphenicol (Cam). The 2451 mutants were assayed by assay A, B, or C (as indicated) using either 50S subunits or 70S ribosomes reconstituted (Recon.) in vitro from transcripts of *B. stearothermophilus* or *E. coli* 23S rRNA (as indicated). Data represent the average of multiple experiments. Standard deviations were less than 50% of the average values derived from multiple experiments.

(N-acetyl-[^{35}S]Met-CACCA and puromycin, respectively) (assay A). A2451C, G, and U mutant ribosomes had undetectable levels of activity in the fragment reaction (down > 50-fold). In contrast, when the same reconstitution reactions were tested in the assay utilizing an intact P-site tRNA substrate, sub-stoichiometric [^{32}P]CPm and assembled 70S ribosomal particles (assay B), these ribosomes had levels of activity that were indistinguishable from that seen for wild-type reconstituted ribosomes. The simplest interpretation of these results is that the A2451 mutant ribosome species have decreased P-site binding affinity relative to the wild-type version. These deficiencies are most evident in the fragment reaction (assay A) with minimal P- and A-site substrates and are masked in assays utilizing intact P-site tRNA substrates (assays B and C) where the 30S ribosomal subunit binds tightly to the anticodon of the tRNA guiding the acceptor end of the tRNA into the active site for catalysis. Previous analysis of mutations at G2252 in the 23S rRNA (which forms part of the P loop) yielded similar patterns (Samaha et al. 1995).

G2447 Mutant Analysis

We next looked at the effects of mutations at G2447 on peptidyl transferase activity. Two different mutations at G2447 (to A or C) permitted growth of the Δ7 prrn strains of *E. coli* (Gregory et al., this volume), allowing the isolation of pure mutant ribosomes from *E. coli* for use in in vitro assays. As seen in Table 2, G2447A mutant ribosomes have reduced levels of peptidyl transferase activity (~14-fold) in assay B where the A-site substrate is present at sub-stoichiometric levels. In contrast, in assay C where A-site substrate is supplied at a saturating concentration, G2447A ribosomes have activity levels comparable to that of wild-type ribosomes. These data suggest that the G2447A ribosomes have an A-site binding deficiency. The high levels of activity observed in assay C for G2447A mutant ribosomes are consistent with the observation that ribosomes with mutations at this position are sufficient to support life in the Δ7 prrn strains of *E. coli*. We are currently using rapid kinetic approaches to more thoroughly characterize these mutant ribosomes.

Chloramphenicol Resistance of Mutant Ribosomes

The peptidyl transferase activity of wild-type ribosomes, either natural or reconstituted, is severely inhibited by micromolar concentrations of chloramphenicol. In contrast, ribosomes with mutations at either A2451 or G2447 were largely insensitive to the same level of chloramphenicol (Tables 1 and 2). These data are consistent with the fact that chloramphenicol-resistant mutants were observed in vivo at both of these positions (Dujon 1980).

DISCUSSION

Atomic resolution crystallographic data of the large ribosomal subunit of *H. marismortui* led to specific models

Table 2. Relative Peptidyl Transferase Rates of G2447 Mutant Ribosomes

	Cam	Assay B		Assay C
		(−)	(+)	(−)
G2447 (wt)		**1.0**	0.01	**1.0**
G2447A		0.07	0.06	1.1

Initial rates were determined from time points in the linear range of the reactions and normalized to the rate of the wild type for each set and assay. The normalized rates (in bold type) are shown for assays performed (+) with, or (−) without, 62.5 μM chloramphenicol (Cam). The 2447 mutants were assayed by assay B or C (as indicated) using 70S ribosomes prepared from *E. coli* strain MC250 carrying solely mutant ribosomes. Data represent the average of multiple experiments. Standard deviations were less than 50% of the average values derived from multiple experiments.

as to how the nucleotide components of the active site might be directly involved in catalysis of peptide-bond formation (Nissen et al. 2000). The universally conserved nucleotide A2451 was proposed to be intimately involved in promoting peptide-bond formation by (1) acting as a general base or (2) stabilizing developing negative charge in the transition state. G2447 was seen to be involved in a network of hydrogen-bonding interactions that was proposed to delocalize charge in the active site and potentially to perturb the pK_a of A2451. Here we present a mutational analysis of these two nucleotides in an attempt to test these predictions.

Mutations incorporated at A2451 (C, G, and U) resulted in dominant lethality when coexpressed with wild-type versions of the 23S rRNA gene, and these mutations do not support growth in the Δ7 prrn strains of *E. coli*. For the analysis of A2451 mutants, ribosomes were reconstituted in vitro from in-vitro-transcribed rRNA and TP50 (total proteins isolated from the 50S subunit). In contrast, mutations at G2447 were viable in the Δ7 prrn strains of *E. coli*, allowing the isolation of pure mutant in-vivo-derived ribosomes. These data are covered in detail in Gregory et al. (this volume).

The simplest interpretation of these in vivo properties is that nucleotide A2451 is more critical to the function of the ribosome than is G2447. Dominant lethality is generally interpreted in the context of the polyribosomes characteristic of bacterial translation. A partially functional ribosome (e.g., a "slow" ribosome) in a polyribosome can prevent wild-type ribosomes from completing translation of a gene and re-entering the active pool, thus resulting in a dominant phenotype. In this interpretation, markedly deficient ribosomes unable to engage a message (those that, e.g., have an assembly or initiation defect) may not have a *dominant* phenotype, although they are in one sense more defective than the simply slow class. Dominant phenotypes could result from a variety of translation or related defects and so should be interpreted with caution. Arguably, modest effects on the overall rate of translation of all messages in a cell (e.g., 3-fold effects) may well be sufficient to cause a dominant lethal phenotype.

Dominant lethality has been associated with mutation of a number of highly conserved nucleotides in the rRNA, including mutations in the A and P loops and in multiple sites in the central ring of domain V of 23S rRNA (Samaha et al. 1995; Porse et al. 1996; Green et al. 1997; Kim and Green 1999), as well as in functionally interesting regions of the small ribosomal subunit (Powers and Noller 1990). Transducing a dominant lethal phenotype to knowledge of the underlying biochemical defect requires careful analysis in defined in vitro systems.

In the single-turnover peptidyl transferase assays that we describe and characterize here, A2451 and G2447 mutations had surprisingly modest effects (Thompson et al. 2001; Gregory et al., this volume). Although the G2447A E. coli mutant ribosomes were partially defective in the assay utilizing sub-stoichiometric A-site substrate (assay B), the deficiency was suppressed by saturating A-site substrate concentrations. These data suggest that the deficiency of these mutant ribosomes was related to A-site binding defects. The fact that G2447A ribosomes support life in the Δ7 prrn strains of E. coli suggests that the deficiencies of these ribosomes are not substantial in vivo in the context of intact tRNA substrates being loaded enzymatically into the A site by EF-Tu.

A2451 mutant ribosomes reconstituted in two different systems (B. stearothermophilus and E. coli) yielded similar overall results. In a fragment reaction, where both P- and A-site substrates are minimal, A2451 mutant ribosomes from E. coli had substantially diminished activity (at least 50-fold). However, such severely compromised activity was not observed in either the E. coli or the B. stearothermophilus reconstitutions where intact P-site substrates and 30S subunits were used. In either assay B or C (with differing amounts of A-site substrate), the mutant ribosomes were at most 3- to 10-fold less active in the catalysis of peptide-bond formation. These data suggest that A2451 mutant ribosomes have reduced capacity for binding minimal P-site substrates in the absence of their interaction with the 30S subunit and mRNA, but that catalysis proceeds largely unimpaired when substrates are bound. These data are also consistent with results reported by Mankin and colleagues (Polacek et al. 2001).

On a visceral level, these results are surprising. How can universally conserved nucleotides in the heart of the active site of the ribosome have so little effect on the catalysis of peptide-bond formation? If we think about the details, however, perhaps these observations are less surprising. The chemistry of peptide-bond formation is straightforward—the attack of a nucleophilic primary amine on a highly activated ester linkage. Solution studies performed by Weber and Orgel using 2′(3′)-O-(glycyl)-adenosine-5′-(O-methylphosphate) (as a donor substrate) and serine ethyl ester (as the acceptor substrate) yielded a substantial second-order rate constant of ~0.1/M/min (Weber and Orgel 1979). Thus, in 30 minutes, about half of the glycyl donor was converted to product when the serine ester was present at a concentration of 0.4 M.

What might the ribosome do to catalyze peptide-bond formation? It has long been argued that positioning by the ribosome of the tRNA substrates will account for much of the observed rates of catalysis (Nierhaus et al. 1980). Inspection of the highly ordered active site poising the CCA ends of the tRNAs for catalysis makes it clear that this

prediction is most likely true. The reported data are consistent, however, with several of the models presented for A2451. If it functions to stabilize developing negative charge in the transition state or to deprotonate the ammonium form of the aminoacyl tRNA substrate, A2451 might be predicted to have mere 10- to 100-fold effects on the observed rates of catalysis (Berg and Lorsch 2001). Similar effects have been observed for the serine protease family when, for example, elements of the oxyanion hole are disrupted (Nicolas et al. 1996). At a minimum, to distinguish between the various models will require more thorough analysis of the pH dependence of the reaction in both mutant and wild-type ribosomes. Such experiments will be most useful if it is established that peptidyl transfer chemistry is rate-limiting in the observed reaction profiles.

A final note of caution is urged for interpretation of the data presented here. These are not the first site-directed mutations to be incorporated into the active site of the ribosome. Mutations at U2506, U2580, U2584, U2585, and A2602, highly conserved nucleotides in the active-site inner shell, and even at nucleotides G2252 and G2553 involved in direct interaction with the conserved cytidines of the A- and P-site substrates, all have relatively minor effects (from 2- to 20-fold) on the observed rates of peptide-bond formation (Samaha et al. 1995; Porse et al. 1996; Green et al. 1997; Green and Noller 1999; Kim and Green 1999). It is not yet clear whether the ribosome is more refractive to mutations in the active site than other enzymes or whether the most critical nucleotides have not yet been identified. Perhaps what we observe in the existing atomic resolution structure is not the most active conformer, and the atomic level structures of other states will lead us to these critical nucleotides. In the meantime, improved methodologies for generating and analyzing mutant ribosomes will make biochemical analysis that much more revealing.

ACKNOWLEDGMENTS

We thank A. Dahlberg, J. Thompson, M. O'Connor, M. Bayfield, S. Gregory, K. Lieberman, and H. Noller for collaborations on the analysis of the active site; J. Lorsch for critical discussions and comments on the manuscript; and A. Cukras for help with figures. Funding for this research was provided by National Institutes of Health grant GM-59425-01 to R.G.

REFERENCES

Asai T., Condon C., Voulgaris J., Zaporojets D., Shen B., Al-Omar M., Squires C., and Squires C.L. 1999. Construction and initial characterization of *Escherichia coli* strains with few or no intact chromosomal rRNA operons. *J. Bacteriol.* **181:** 3803.

Bayfield M.A., Dahlberg A.E., Schulmeister U., Dorner S., and Barta A. 2001. A conformational change in the ribosomal peptidyl transferase center upon active/inactive transition. *Proc. Natl. Acad. Sci.* **98:** 10096.

Berg J.M. and Lorsch J.R. 2001. Mechanism of ribosomal peptide bond formation. *Science* **291:** 203.

Dohme F. and Nierhaus K.H. 1976. Total reconstitution and assembly of 50 S subunits from *Escherichia coli* ribosomes in vitro. *J. Mol. Biol.* **107:** 585.

Dujon B. 1980. Sequence of the intron and flanking exons of the mitochondrial 21S rRNA gene of yeast strains having different alleles at the omega and rib-1 loci. *Cell* **20:** 185.

Fahnestock S., Erdmann V., and Nomura M. 1974. Reconstitution of 50 S ribosomal subunits from *Bacillus stearothermophilus*. *Methods Enzymol.* **30:** 554.

Gourse R.L., Takebe Y., Sharrock R.A., and Nomura M. 1985. Feedback regulation of rRNA and tRNA synthesis and accumulation of free ribosomes after conditional expression of rRNA genes. *Proc. Natl. Acad. Sci.* **82:** 1069.

Green R. and Noller H.F. 1996. In vitro complementation analysis localizes 23S rRNA posttranscriptional modifications that are required for *Escherichia coli* 50S ribosomal subunit assembly and function. *RNA* **2:** 1011.

———. 1997. Ribosomes and translation. *Annu. Rev. Biochem.* **66:** 679.

———. 1999. Reconstitution of functional 50S ribosomes from in vitro transcripts of *Bacillus stearothermophilus* 23S rRNA. *Biochemistry* **38:** 1772.

Green R., Samaha R.R., and Noller H.F. 1997. Mutations at nucleotides G2251 and U2585 of 23 S rRNA perturb the peptidyl transferase center of the ribosome. *J. Mol. Biol.* **266:** 40.

Khaitovich P. and Mankin A.S. 1999. Effect of antibiotics on large ribosomal subunit assembly reveals possible function of 5 S rRNA. *J. Mol. Biol.* **291:** 1025.

Kim D.F. and Green R. 1999. Base-pairing between 23S rRNA and tRNA in a site. *Mol. Cell* **4:** 859.

Lieberman K.R. and Dahlberg A.E. 1995. Ribosome-catalyzed peptide-bond formation. *Prog. Nucleic Acid Res. Mol. Biol.* **50:** 1.

Moazed D. and Noller H.F. 1989. Intermediate states in the movement of transfer RNA in the ribosome. *Nature* **342:** 142.

Monro R.E. and Marcker K.A. 1967. Ribosome-catalysed reaction of puromycin with a formylmethionine-containing oligonucleotide. *J. Mol. Biol.* **25:** 347.

Muth G.W., Ortoleva-Donnelly L., and Strobel S.A. 2000. A single adenosine with a neutral pKa in the ribosomal peptidyl transferase center (see comments). *Science* **289:** 947.

Nicolas A., Egmond M., Verrips C.T., de Vlieg J., Longhi S., Cambillau C., and Martinez C. 1996. Contribution of cutinase

serine 42 side chain to the stabilization of the oxyanion transition state. *Biochemistry* **35:** 398.

Nierhaus K.H., Schulze H., and Cooperman B.S. 1980. Molecular mechanisms of the ribosomal peptidyltransferase center. *Biochem. Int.* **1:** 185.

Nissen P., Hansen J., Ban N., Moore P.B., and Steitz T.A. 2000. The structural basis of ribosome activity in peptide bond synthesis (comments). *Science* **289:** 920.

O'Connor M., Lee W.M., Mankad A., Squires C.L., and Dahlberg A.E. 2001. Mutagenesis of the peptidyltransferase center of 23S rRNA: The invariant U2449 is dispensable. *Nucleic Acids Res* **29:** 710.

Pape T., Wintermeyer W., and Rodnina M.W. 1999. Induced fit in initial selection and proofreading of aminoacyl-tRNA on the ribosome. *EMBO J.* **18:** 3800.

Polacek N., Gaynor M., Yassin A., and Mankin A.S. 2001. Ribosomal peptidyl transferase can withstand mutations at the putative catalytic nucleotide. *Nature* **411:** 498.

Porse B.T., Thi-Ngoc H.P., and Garrett R.A. 1996. The donor substrate site within the peptidyl transferase loop of 23 S rRNA and its putative interactions with the CCA-end of N-blocked aminoacyl-tRNA(Phe). *J. Mol. Biol.* **264:** 472.

Powers T. and Noller H.F. 1990. Dominant lethal mutations in a conserved loop in 16S rRNA. *Proc. Natl. Acad. Sci.* **87:** 1042.

Samaha R.R., Green R., and Noller H.F. 1995. A base pair between tRNA and 23S rRNA in the peptidyl transferase centre of the ribosome (erratum appears in *Nature* [1995] **378:** 419). *Nature* **377:** 309.

Semrad K. and Green R. 2002. Osmolytes stimulate the reconstitution of functional 50S ribosomes from in vitro transcripts of *Escherichia coli* 23S rRNA. *RNA* (in press).

Synetos D. and Coutsogeorgopoulos C. 1987. Studies on the catalytic rate constant of ribosomal peptidyltransferase. *Biochim Biophys. Acta* **923:** 275.

Thompson J. Kim D.F., O'Connor M., Lieberman K.R., Bayfield M.A., Gregory S.T., Green R., Noller H.F., and Dahlberg A.E. 2001. Analysis of mutations at residues A2451 and G2447 of 23S rRNA in the peptidyl transferase active site of the 50S ribosomal subunit. *Proc. Natl. Acad. Sci.* **98:** 9002.

Traut R.R. and Monro R.E. 1964. The puromycin reaction and its relation to protein synthesis. *J. Mol. Biol.* **10:** 63.

Weber A.L. and Orgel L.E. 1979. The formation of dipeptides from amino acids and the 2′(3′)-glycyl ester of an adenylate. *J. Mol. Evol.* **13:** 185.

Codon Recognition and Decoding: The Transorientation Hypothesis

A.B. SIMONSON AND J.A. LAKE

Molecular Biology Institute and MCD Biology, University of California, Los Angeles, California 90095

Protein synthesis is a complex, multistep process comprising initiation, elongation, and termination steps (Garret et al. 2000). Elongation itself consists of two principal phases. In the first, or decoding phase, a codon of messenger RNA is matched with its cognate tRNA, and the amino acid carried by the tRNA is added to the growing protein chain. In the second, the mRNA and the peptidyl tRNA are translocated exactly one codon to make room for the next tRNA. In the decoding step, elongation factor Tu (EF-Tu) binds to the ribosome as a ternary complex (EF-Tu•GTP•aminoacyl tRNA). Upon recognition of the cognate codon, EF-Tu switches conformations and helps place the aminoacyl tRNA in the A site. It is thought that the high accuracy of translation results from the combination of an initial codon-reading step and a second proofreading step following GTP hydrolysis (Hopfield 1971; Pape et al. 1998). Models also exist for codon recognition that do not require a separate proofreading step (Nierhaus 1990).

Much is known about the locations of the A-, P-, and E-tRNA-binding sites. All three of these sites have been localized on the ribosome using cryo-electron microscopy (cryo-EM) (Agrawal et al. 1996) and atomic resolution X-ray diffraction methods (Carter et al. 2000; Yusupov et al. 2001). In addition, the binding sites of EF-Tu in the presence of the antibiotic kirromycin (Stark et al. 1997), and of EF-G in the presence of the antibiotic fusidic acid (Agrawal et al. 1998), have been mapped by cryo-EM. In contrast, the molecular details of the decoding step are difficult to obtain because the ternary complex binds only transiently to programmed ribosomes in the absence of antibiotics.

Here a molecular mechanism for the decoding step, the transorientation hypothesis, is described. Our lab had hints of this model years earlier based on immunoelectron microscopy (Lake 1977). It wasn't quite right, but with the aid of the new ribosome structures determined over the last several years (Ban et al. 2000; Schluenzen et al. 2000; Wimberly et al. 2000; Yusupov et al. 2001), the early model was modified until it was completely consistent with the X-ray structures. The proposed mechanism is compatible with the known structures, conformations, and functions of the ribosome and its component parts, including tRNAs and EF-Tu.

OVERVIEW OF TRANSORIENTATION

Codon recognition and decoding starts with ternary complex binding to a newly identified decoding, or D, site and ends with transfer of the nascent protein chain to the incoming aminoacyl tRNA (aa-tRNA). Transorientation refers to the rotation of an aa-tRNA from the D site to the A site, just as translocation refers to the movement of a peptidyl tRNA from the A site to the P site. The model utilizes a conformational switch between two tRNA anticodon loop structures, one in which the anticodon is stacked on the 3′ side of the anticodon loop (3′ stack) and one in which the anticodon is stacked on the 5′ side of the loop (5′ stack). Through this switching mechanism, the anticodon loop acts like a three-dimensional hinge that allows aa-tRNAs to rotate from the D site to the A site. The tRNA rotation is coupled to the large EF-Tu conformational changes that result from GTP hydrolysis.

The rotation of an aa-tRNA during transorientation is illustrated in two, approximately orthogonal, projections of the small subunit in Figures 1a and 1b. In frames 1, an incoming tRNA approaches the D site with its anticodon in the 3′ stacked conformation. In the initial binding of the ternary complex to the D site, the aa-tRNA anticodon stem-loop fits into a channel between the head domain of the 30S and the 530-loop. The acceptor stem points up and toward the 50S subunit, at an angle of roughly 45° with respect to the approximate plane separating the two subunits, so that EF-Tu•GTP is adjacent to the GTPase-associated region of the 23S rRNA.

During entry into the D site, the anticodon switches into the 5′ stack (frames 2) where it can pair with the codon. Initial recognition of the anticodon by its cognate codon induces GTP hydrolysis, EF-Tu conformational changes, and transorientation (frames 3–6), during which the tRNA is rotated about the relatively fixed anticodon until the aa-tRNA occupies a pre-A site, where it can be accommodated into the A site (frames 7).

KEY ISSUES FOR TRANSORIENTATION

The 5′ stacked conformation of the tRNA anticodon loop is a cornerstone of the transorientation hypothesis, since it is essential for rotating an aa-tRNA from the D

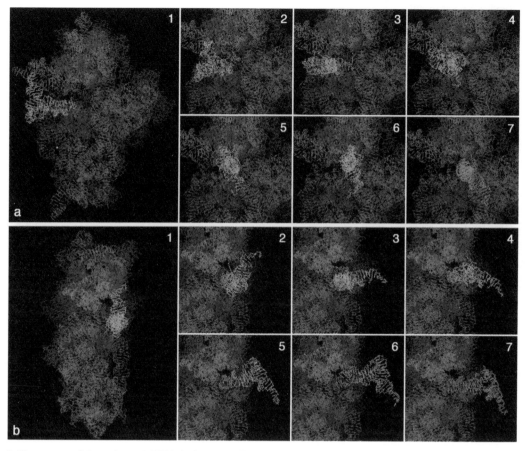

Figure 1. Movements of the aminoacyl tRNA during transorientation. Transfer RNAs are shown in orange and the 30S rRNA and proteins are shown in light and dark purple, respectively. The asymmetric projection of the 30S subunit, as viewed from the 50S subunit, is shown in *a*. The quasisymmetric projection of the 30S subunit is shown in *b*. In the orientation in *b*, the 50S subunit (not shown) would be on the right. The numbers refer to various stages of transorientation. In frames 2, the tRNA is docked in the D site in the 5′ stack. Frames 3–6 illustrate tRNA movement during the EF-Tu conformational switch. Frame 7 shows the tRNA (*dark orange*) bound in the A site determined by X-ray and cryo-EM studies (Agrawal et al. 1996; Carter et al. 2000; Yusupov et al. 2001). (Adapted from Simonson and Lake 2002.)

site to the A site, around a fixed codon–anticodon pair. The 5′ and the 3′ stacked anticodon conformations are represented schematically in Figure 2. In the 3′ stack, a U-turn breaks the helical stack at U_{33} so that nucleotides 34–38 continue the 3′ strand of the acceptor stem. In the 5′ stack, nucleotides 32–36 are stacked to continue the 5′ strand of the acceptor stem. No direct X-ray structural evidence for the 5′ stacked conformation exists. X-ray diffraction studies of individual tRNAs (Quigley et al. 1974; Ladner et al. 1975), of EF-Tu in the ternary complex (Nissen et al. 1995), and of tRNA synthetase complexes (Eiler et al. 1999) have consistently revealed tRNAs in the 3′ stacked conformation, although the anticodon loop is considerably distorted in the synthetase structure. A number of key issues, discussed below, must be addressed if transorientation is to be a viable hypothesis.

Can the 5′ Stacked Conformation Be Built?

Although the 5′ stacked tRNA has not been observed in X-ray-derived structures, the 5′ stacked structure can be constructed using the bond angles, bond distances, and

sugar pucker characteristic of A-form RNAs. For reference, the 3′ stacked tRNA, an intermediate between the 3′ and 5′ stacked tRNAs, and a 5′ stacked tRNA, are shown

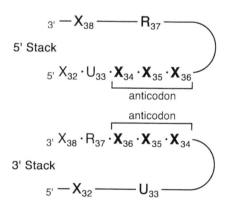

Figure 2. Schematic diagram of anticodon loop conformations in the 5′ stacked structure (*top*) and in the 3′ stacked structure (*bottom*). (Adapted from Simonson and Lake 2002.)

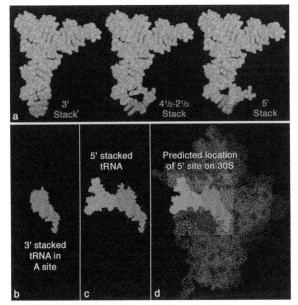

Figure 3. Structures of tRNAs and their orientation on the 30S subunit. The tRNAs are orange, except that the anticodon is magenta and the modification of nucleotide A_{37} is cyan. Anticodon conformations shown in *a* are the 3′ stack, at the left; an intermediate conformation, the 4 1/2–2 1/2 stack, and the 5′ stack, at the right. The relative orientations of the D and A tRNA-binding sites are shown in *b–d*. In all three frames, the direction of the codon is fixed, so that the orientations of the tRNAs are determined only by the conformations of the anticodon loops. The small subunit is purple in *b–d*. (Adapted from Simonson and Lake 2002.)

in Figure 3a, from left to right, respectively. A characteristic of the 5′ stacked structure is that the anticodon, shown in magenta, stacks against base U_{33} and continues the helical strand. As a result, the anticodon in the 5′ stacked tRNA is approximately perpendicular to the anticodon in the 3′ stacked tRNA. Structures intermediate between the 3′ and 5′ stacked forms, such as the 4 1/2–2 1/2 stacked tRNA shown in the middle of Figure 3a, were useful prototypes for the 5′ stacked molecule, and may have a role in –1 frameshifting (Atkins et al. 2000).

Where Must a 5′ Stacked tRNA Bind in Order to Read a Codon in the A Site?

Messenger RNA moves one codon during translocation and, as judged by RNase protection experiments, moves very little during transorientation (Gupta et al. 1971; Thach and Thach 1971). Thus, the initial aa-tRNA-binding site can be determined, in principle, by working backward from the A site, using the A-site codon orientation as a guide, as shown in Figure 3, b–d. A tRNA bound to the A site in the 3′ stack is shown in Figure 3b, as it would appear viewed from the 50S subunit. For comparison, both the 3′ and the 5′ stacked tRNAs paired with the A-site codon are shown in Figure 3c. In this orientation, the 5′ stacked tRNA is accommodated nearly perfectly by its counterpart, the 30S-binding site shown in Figure 3d.

This putative binding site, the D site, accommodates the variable hypermodified purine, R_{37} (shown in cyan in Fig. 3d), in the 5′ stacked structure within a groove formed between parallel RNA helices 34 and 32.

Why Can't a 3′ Stacked tRNA Function in the Decoding Site?

A tRNA in the 3′ stack is shown as it approaches the D site in Figure 2a, frame 1. In this conformation, the complementary codon (5′ to 3′) would need to point from left to right. This is nearly opposite the direction that the codon is known to point (Carter et al. 2000; Yusupov et al. 2001), and therefore would require a drastic movement of the message to effect base-pairing with the anticodon. Given that nuclease protection experiments designed to monitor mRNA movement during elongation detected no codon movement during transorientation (Gupta et al. 1971; Thach and Thach 1971), it would seem impossible for a 3′ stacked tRNA to base-pair with the codon while in the D site. In contrast, if the tRNA in the D site assumes the 5′ stacked conformation, Figure 1a, frame 2, it is in a position to pair with the A-site codon, as identified by X-ray diffraction studies (Carter et al. 2000; Yusupov et al. 2001).

Is There Evidence from the Literature for a 5′ Anticodon Stack?

Nuclease protection experiments suggest that the 5′ stack exists in solution. Nuclease S_1 hydrolyzes single-stranded, but not double-stranded, nucleic acids and is a sensitive probe of base stacking. Harada and Dahlberg (1975) investigated the cutting of elongator tRNAs at pH 4.5 and found that the primary cut was at the 3′ side of the anticodon, implying a 5′ stack for essentially all of the intact molecules. Some heterogeneity of cleaving was found, with cleavages toward the 5′ side of the anticodon probably occurring secondarily, after the initial cut. Tal (1975) independently found similar results at pH 7.5 using a *Neurospora* endonuclease with similar specificity for polynucleotides. Wrede, Woo, and Rich (1979) compared the S_1 cleavage patterns of initiator and elongator tRNAs. For elongator tRNAs they found similar results as those of Harada and Dahlberg, consistent with the 5′ anticodon stack; but for initiator tRNAs, which enter directly into the P site and would not need the 5′ stacked conformation, they found a unique pattern lacking the characteristic cut expected for the 5′ stacked conformation. A recent NMR comparison of initiator and elongator anticodon loop structures (Schweisguth and Moore 1997) showed "a more flexible structure" in the elongator tRNA than in the initiator tRNA, again consistent with switches between the 3′ and 5′ stacked conformations in the elongator tRNA; however, these studies were performed with anticodon stem-loops lacking modified bases. Thus, there is a considerable body of experimental data in the literature consistent with the existence of the 5′ stacked anticodon conformation in elongator tRNAs.

What Is the Mechanism for Generating the 5′ Stack within the Ribosome?

In double-stranded B-form DNA, sequences of pyrimidines (Y) followed by a purine (R) are known as helix breakers, since they are characterized by high roll angles and tend to interrupt helices (Dickerson and Chiu 1997). Similarly, single-stranded A-form RNA helices have decreasing rigidity according to the surface area available for stacking between adjacent bases in the order: $_5RY_{3'}$ $>_5RR_{3'} > _5YY_{3'} > _5YR_{3'}$, where $_5YR_{3'}$, the least stable, is a helix breaker. The anticodon is flanked on the 5′ side by a universally conserved pyrimidine, U_{33}, and on the 3′ side by a purine, R_{37}, which is often hypermodified. As a consequence, flexible "hot spots," indicated by caret signs, enclose the anticodon: $U_{33} \wedge X_{34} X_{35} X_{36} \wedge R_{37}$. These hot spots are rationalized as follows. If the nucleotide at position 34 is a purine, then the dinucleotide $Y_{33} R_{34}$ is a helix breaker, and if nucleotide 34 is a Y, then $Y_{33} Y_{34}$ is the second most flexible dinucleotide. Similarly, on the other side of the anticodon, if nucleotide 36 of the anticodon is a pyrimidine, then $Y_{36} R_{37}$ is the helix breaker. However, if position 36 is a purine, then $R_{36} R_{37}$ is a strong, although not the strongest, dinucleotide.

A clear-cut correlation between tRNA nucleotide 36 (R or Y) and modification of R_{37} suggests that R_{37} modification could be part of a mechanism that functions to convert the 3′ stacked tRNA to the 5′ stacked conformer in the event that a purine is at position 36. In analyzing phylogenetically diverse eukaryotic and prokaryotic elongator tRNAs (Gauss et al. 1979), we found when R_{37} is modified, the sequence of nucleotide 36 is just as likely to be a pyrimidine (30 sequences) as a purine (32 sequences). However, if R_{37} is not modified, then nucleotide 36 is very likely to be a pyrimidine (22 sequences) and unlikely to be a purine (3 sequences). By the Chi-Squared test, the null hypothesis (that R_{37} modification is independent of whether nucleotide 36 is a purine or a pyrimidine) is highly unlikely, $P<.001$ (df = 2). This finding suggests R_{37} could have a role in facilitating the switch from the 3′ to 5′ stacked conformation.

When Phe-tRNA^{Phe} is docked in the D site in the 5′ stacked configuration, the hypermodified base at position 37, yW_{37}, is proximate to several 30S components, potentially implicating them in the 3′ to 5′ switch. These include nucleotides G_{953} and G_{954} (universally conserved in 30S helix 30) separated by 3–4 Å; G_{963} and A_{964} (universally conserved, except for fungi and animals, in 30S helix 31) separated by 5–7 Å; and amino acids K_{120}, K_{121}, K_{122} (near the carboxyl terminus in protein S13).

Can the 70S Accommodate the Ternary Complex Bound in the D Site?

The ternary complex docked in the D site fits onto the 70S ribosome extremely well. There is no instance of steric hindrance to its binding, with one intriguing exception. The L11–rRNA complex, thought to regulate entry into the ribosome during its different functional states, displays significant steric overlap with EF-Tu and is discussed below.

The proposed binding of the ternary complex to the 70S structure (Yusupov et al. 2001) is shown in Figure 4. The ternary complex is shown in the asymmetric 30S projection, as viewed from the 50S subunit, in Figure 4, a–c, and in the nonoverlap projection of the 70S ribosome in Figure 4, d–f. To orient the reader, a tRNA in the D site is shown at reduced magnification in Figure 4, a and d.

Inclusion of the ternary complex likewise causes no steric occlusion, Figure 4, b and e. When the L7/L12 protein complex from the 70S structure is included, Figure 4, c and f, L7/L12 encloses domain I of EF-Tu, without overlap or other interference. Proteins L7/L12 form a conspicuous structural feature of the ribosome, the L7/L12 stalk (Strycharz et al. 1978), thought to be related to factor binding, but not essential for GTPase activity (Briones et al. 1998).

What Is the Role of L11 in the Transorientation Model?

Ribosomal protein L11 and 23S rRNA nucleotides 1051–1108 form a well-ordered ribonucleoprotein complex. It is thought that the conformational state of this complex may regulate elongation, through switches involving RNA, protein, or both (Cundliffe 1986). The importance of this highly conserved ribonucleoprotein complex is underscored by the observation that it is a target for thiostrepton-antibiotics that disrupt elongation factor function. Recent evidence suggests that L11 itself undergoes conformational changes which are inhibited by thiostrepton and micrococcin antibiotics (Markus et al. 1997; Wimberly et al. 1999), possibly by preventing EF-G release (Wimberly et al. 1999).

The structure of the complex in isolation has been solved by X-ray crystallography (Wimberly et al. 1999). L11 consists of a carboxy-terminal domain, which interacts directly with the 23S rRNA, and an amino-terminal domain, which is tethered to the carboxyl domain by a flexible linker. Mutations in the L11 gene that resulted in thiostrepton resistance corresponded to protein residues that lie along a narrow cleft between the RNA and the amino-terminal domain of the protein, suggesting that thiostrepton binds in the cleft and blocks a putative conformational switch by the L11–RNA complex (Wimberly et al. 1999).

With the ternary complex docked in the D site, there is extensive clash between EF-Tu domain II and the carboxy-terminal domain of L11, and considerable overlap between the amino-terminal domain of L11 and EF-Tu domain III (Fig. 5a). Even with L11 removed, significant overlap still remains between EF-Tu domain III and 23S rRNA helix 1057–1081 (Fig. 5b). Thus, the potential clash between EF-Tu and the L11–RNA complex is predicted to occur independently of whether L11 is present. Given that L11 is not required for viability in E. coli (Stoeffler et al. 1980), the overlap between EF-Tu and helix 1057–1081 suggests that the switch may involve RNA as well as the protein.

This may represent a mechanism for controlling the access of ternary complex to the ribosome. A large cavity, sufficiently large to hold L11 and helix 1057–1081, is

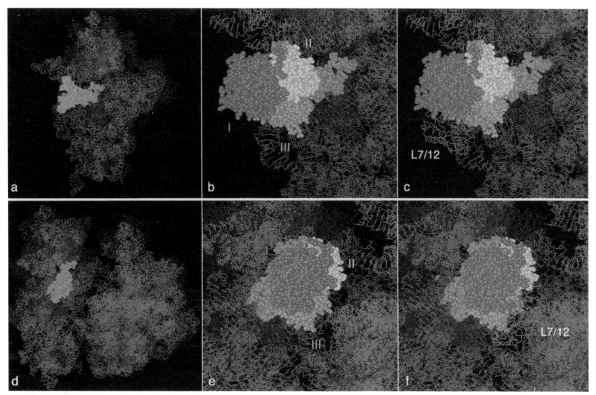

Figure 4. The 70S-binding site for the ternary complex, with an aa-tRNA docked in the D site. EF-Tu domain I is magenta, domain II is yellow, and domain III is cyan. The 30S subunit is purple, the tRNA is orange, and 50S subunit proteins L7/L12 are green and red. Frames *a–c* show the 30S subunit in the asymmetric projection, as viewed from the 50S subunit, and frames *d–f* show the interface view of the 70S ribosomes. Frames *a* and *d* are to orient the reader and show only the aa-tRNA in the D site, whereas *b–c* and *e–f* show the entire ternary complex. In *b* and *e*, ribosomal proteins L7/L12 are missing, whereas in frames *c* and *f*, L7/L12 are included. The binding was tested and analyzed using the 70S structure of Yusupov et al. (2001). However, since only coordinates of α carbons and phosphate atoms were provided in the 70S map, the images illustrated here are from the 30S (Wimberly et al. 2000) and 50S (Ban et al. 2000) structures aligned using the 70S α carbons and phosphates. L7/L12 in *c* and *f* is taken from Yusupov et al. (2001). (Adapted from Simonson and Lake 2002.)

present just above helix 1086–1102 (i.e., toward the viewer) in Figure 5. This cavity is bounded on the right by L16. If the L11 complex were to act as a gatekeeper, then the cavity would be large enough to accommodate the complex when the gate was open. Alternatively, ternary complex binding may simply push the gate open.

What Is the Role of EF-Tu in Transorientation?

EF-Tu is an integral part of the decoding process and functions to bring the aa-tRNA to the ribosome. It is a three-domain, GTP-binding protein that undergoes dramatic conformational changes upon the hydrolysis of GTP. Domain I binds GTP and is connected to domains II and III through a linker to domain II. Domains II and III form a long, cylindrically shaped structure in the ternary complex that contacts the acceptor end and the T loop of a bound aa-tRNA, respectively. As a result of GTP hydrolysis, domain II moves distally from domain I, and domains II and III rotate 90° about their long axis (Abel et al. 1996; Polekhina et al. 1996; Song et al. 1999). Only after GTP hydrolysis does the incoming aa-tRNA enter the A site, suggesting that movements of EF-Tu might physically move aa-tRNAs from the D site to the A site.

It is not essential that EF-Tu move the aa-tRNA all the way from the D site to the A site. The movement need only switch the anticodon conformation past the transition-state energy barrier separating the 5′ stack from the 3′ stack, thereby allowing anticodon–codon stacking to position the aa-tRNA into the A site. Because the point at which the aa-tRNA is released from the ternary complex is unknown (Nissen et al. 1995), there are multiple ways in which EF-Tu could bring an aa-tRNA to the A site from the D site.

One possibility is that EF-Tu moves the aa-tRNA only part of the way to the A site. Following GTP hydrolysis, domain II is pushed away from domain I so that the aminoacyl end of the attached tRNA is moved toward the A site. This push is in the direction needed to break the codon–anticodon helix on the 3′ side of U_{33}. The push of domain II, by itself, may be sufficient to surmount the 5′ to 3′ transition state, since in factor-free protein synthesis (Pestka 1968; Gavrilova and Spirin 1971) the transition happens spontaneously, although slowly.

Another possibility is that EF-Tu may not release the tRNA until the GTP to GDP conformational transition is nearly completed. This would require the motions of EF-Tu to be constrained so that tRNA would maintain its contact with the mRNA through the anticodon hinge dur-

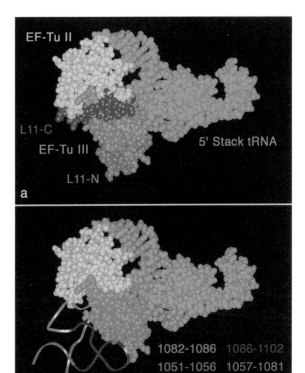

Figure 5. Overlap of the ternary complex with the 50S L11–RNA complex. EF-Tu domains II and III, and aa-tRNA in the 5′ stacked conformation, are bound as a ternary complex in the D site, viewed from the 50S subunit. (*Orange*) 5′-stack aa-tRNA; (*cyan*) EF-Tu domain III; (*yellow*) EF-Tu domain II. The 30S subunit and EF-Tu domain I are not shown. (*a*) There is significant overlap between the carboxy-terminal domain of L11 (L11-C) and EF-Tu domain II, as well as some overlap between the amino-terminal domain of L11 (L11-N) and EF-Tu domain III. (*Purple*) L11-C; (*pink*) L11-N. (*b*) EF-Tu domain III and a small portion of EF-Tu domain II exhibit steric overlap in the 1067 loop of the 23S rRNA. (*Red*) 1056 stem; (*pumpkin*) 1086 hairpin; (*dark blue*) 1097 stem-loop; (*green*) 1067 stem-loop. (Adapted from Simonson and Lake 2002.)

ing transorientation. In practice, this can be accomplished by permitting EF-Tu domains II and III to rotate around the linker connecting them to domain II. This connector is a well-conserved region of the EF-Tu sequence and has the sequence $_{200}$PEPERAIDKPFLLP$_{213}$ in *E. coli*. In a collection of 53 EF-Tu sequences representing the diversity of prokaryotes and eukaryotes, prolines are found at every position between residues 200 and 205 in at least some organisms, and prolines are present almost universally at positions 209 and 213. Thus, only a few amino acids in this region might allow the desired rotation of domains II and III about the linker. We found that small modifications of the ψ angles at positions 203 and 204, and of ϕ and ψ angles at positions 206, 207, and 210, would keep the anticodon in contact with the message during transorientation, without causing clash. Furthermore, the starting and ending ϕ and ψ angles and the paths connecting them are comfortably within the acceptable range for Ramachandran plots of EF-Tu (Polekhina et al.

1996). The starting and ending conformations of EF-Tu and the aa-tRNA are detailed elsewhere (Simonson and Lake 2002.). Remarkably, in this alternative, EF-Tu is carried to the very edge of the A site without clash within EF-Tu, or between EF-Tu and the aa-tRNA or the 70S ribosome. In the present state of our knowledge, this alternative is more likely.

Polekhina et al. (1996) suggested that the function of the EF-Tu switch I might be to "physically dislocate domains II and III" from domain I. This makes sense in the context of the constrained model for EF-Tu movement, since physical separation of the domains is required to allow the rotation of domains II and III with respect to domain I, without clashing. Similarly, the kirromycin family of antibiotics functions by linking domain III to domain I (Abdulkarim et al. 1994; Vogeley et al. 2001). Again, it makes sense that the linking could inhibit transorientation, since it would likewise prevent the rotation of domains II and III with respect to domain I.

How Does the Transorientation Hypothesis Fit with Proofreading?

The decoding of mRNA requires the recognition of each codon by the correct aa-tRNA. Because the difference in binding energy between a correct codon–anticodon interaction and an incorrect one is too small to account for the high accuracy of translation (Kurland 1992), it is thought that the message may be read twice, before and after GTP hydrolysis.

The multiple kinetic steps in codon recognition have been identified and characterized (Pape et al. 1998). These steps parallel the structural steps in the transorientation hypothesis (Fig. 6). The first step, initial binding, corresponds to the reversible binding of the ternary complex to the decoding site. Ternary complex binding is followed by the reversible switching of the anticodon conformation to the 5′ stack and by codon matching. Successful codon recognition triggers a series of irreversible steps involving GTP activation, GTP hydrolysis, and, ultimately, EF-Tu conformational change. Since EF-Tu conformational change may test the codon–anticodon interaction, this step might provide an opportunity to proofread the codon, and tRNA rejection could potentially occur at this point. Similarly, the process of accommodation during which the aa-tRNA leaves EF-Tu•GDP and enters the A site may also provide an opportunity for proofreading. This would be reasonable, given the impressive functional and structural data on antibiotic involvement in misreading and fidelity (Carter et al. 2000). Finally, peptidyl transfer completes the transorientation minicycle and irreversibly prepares the ribosome for the start of translocation.

The wobble hypothesis (Crick 1966) predicts that alternative codon–anticodon base-pairings are more acceptable at the 3′ end of a codon than at other positions. Because the observed wobble is the product of separate decoding and proofreading steps, it is possible that the patterns associated with each reading may differ significantly from the wobble pattern. Reading in the 5′ stacked

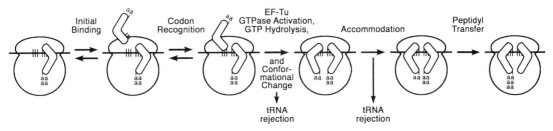

Figure 6. Schematic representation of the ribosome, mRNA, and tRNAs during decoding. The figure illustrates the correspondence between the structural transitions in the transorientation hypothesis and the steps in the kinetic proofreading mechanism observed by Pape et al. (1998). See the text for details. (Adapted from Simonson and Lake 2002.)

conformation could produce a distinctly different pattern than reading in the 3′ stacked conformation. During the initiation of protein synthesis, start codons (thought to be read in the P site) are presumably read by a tRNA in the 3′ stacked conformation. They exhibit a 5′ wobble, in which AUG, GUG, and sometimes CUG are acceptable start codons. Perhaps the proofreading step (occurring in the 3′ stack) also generates 5′ wobble. Specifically designed kinetic experiments on near-cognate tRNAs that mispair at either the 5′ or 3′ anticodon positions may reveal additional details about the process of translation.

CONCLUSIONS

The transorientation hypothesis generates numerous testable predictions. It is startling that it is consistent with so many details of ribosome structure. It is striking that a tRNA in the 5′ stacked conformation, when base-paired with the A-site codon, is matched nearly perfectly by its counterpart D site. The observed S1 nuclease digests of tRNAs very closely match those predicted by the 5′ stacked loop conformation. The parallels observed between the structural details of transorientation and the energetics of codon recognition and proofreading were also unforeseen. Finally, it is remarkable that the experimentally observed conformational changes of EF-Tu are such that they direct an incoming aa-tRNA from the D site to the A site.

Transorientation is a very specific hypothesis, hence there are many ways in which it may be falsified. This makes the hypothesis eminently testable; i.e., it has high Popperian verifiability. One should not take the structures presented here as being comparable to X-ray or NMR structures; they are not. Rather, they should be regarded as the first steps taken toward understanding the functioning of a fascinating machinery, which will surely occupy many of us for several decades.

Finally, it is interesting that transorientation is dominated and shaped by the structure and the movements of the tRNA molecule. The ribosome acts as a facilitator, but the tRNA is the keystone in the mechanism. One can well imagine how protein synthesis could have evolved, perhaps beginning from only mRNAs and tRNAs, and possibly prototypical ribosomal subunits. As such, this mechanism is consistent with the view that protein synthesis may have originated in a preexisting RNA World (Zaug and Cech 1986; Noller et al. 1992).

ACKNOWLEDGMENTS

This work was supported by grants from the National Institutes of Health, from the Astrobiology Institute, and from the Department of Energy to J.A.L. We thank D. Eisenberg and R. Dickerson for advice, and for access to graphics terminals and software. Correspondence and requests for materials should be addressed to J.A.L. (E-mail: Lake@mbi.ucla.edu) or A.B.S (E-mail: Asimonso@ucla.edu). Coordinates for the 5′ stacked tRNA will be deposited in the Protein Data Bank (see Simonson and Lake 2002.).

REFERENCES

Abdulkarim F., Liljas L., and Hughes D. 1994. Mutations to kirromycin resistance occur in the interface of domains I and III of EF-Tu•GTP. *FEBS Lett.* **352:** 118.

Abel K., Yoder M.D., Hilgenfeld R., and Jurnak F. 1996. An alpha to beta conformational switch in EF-Tu. *Structure* **4:** 1153.

Agrawal R.K., Penczek P., Grassucci R.A., and Frank J. 1998. Visualization of elongation factor G on the *Escherichia coli* ribosome: The mechanism of translocation. *Proc. Natl. Acad. Sci.* **95:** 6134.

Agrawal R.K., Penczek P., Grassucci R.A., Li Y., Leith A., Nierhaus K.H., and Frank J. 1996. Direct visualization of A-, P-, and E-site transfer RNAs in the *E. coli* ribosome. *Science* **271:** 1000.

Atkins J.F., Herr A.J., Massire C., O'Connor M., Ivanov I., and Gesteland R.F. 2000. Poking a hole in the sanctity of the triplet code: Inferences for framing. In *the ribosome: Structure, function, antibiotics, and cellular interactions* (ed. R.A. Garrett et al.), p. 369. ASM Press, Washington, D.C.

Ban N., Nissen P., Hansen J., Moore P.B., and Steitz T.A. 2000. The complete atomic structure of the large ribosomal subunit at 2.4 angstroms resolution. *Science* **289:** 905.

Briones E., Briones C., Remacha M., and Ballesta J.P. 1998. The GTPase center protein L12 is required for correct ribosomal stalk assembly but not for *Saccharomyces cerevisiae* viability. *J. Biol. Chem.* **273:** 31956.

Carter A.P. Clemons W.M., Brodersen D.E., Morgan-Warren R.J., Wimberly B.T., and Ramakrishnan V. 2000. Functional insights from the structure of the 30S ribosomal subunit and its interactions with antibiotics. *Nature* **407:** 340.

Crick F.H. 1966. Codon-anticodon pairing: The wobble hypothesis. *J. Mol. Biol.* **19:** 548.

Cundliffe E. 1986. Involvement of specific portions of rRNA in defined ribosomal functions: A study utilizing antibiotics. In *Structure, function and genetics of ribosomes* (ed. B. Hardesty and G. Kramer), p. 586. Springer-Verlag, New York.

Dickerson R.E. and Chiu T.K. 1997. Helix bending as a factor in protein/DNA recognition. *Biopolymers* **44:** 361.

Eiler S., Dock-Bregeon A., Moulinier L., Thierry J.C., and

Moras D. 1999. Synthesis of aspartyl-tRNA^{Asp} in *Escherichia coli*—A snapshot of the second step. *EMBO J.* **18**: 6532.

Garrett R.A., Douthwaite S.R., Liljas A., Matheson A.T., Moore P.B., and Noller H.F., Eds. 2000. *The ribosome: Structure, function, antibiotics, and cellular interactions.* ASM Press, Washington, D.C.

Gauss D.H., Gruter F., and Sprinzl M. 1979. Compilation of tRNA sequences. In *Transfer RNA: Structure, properties, and recognition* (ed. P.R. Schimmel et al.), p. 521. Cold Spring Harbor Laboratory, Cold Spring Harbor, New York.

Gavrilova L.P. and Spirin A.S. 1971. Stimulation of "non-enzymic" translocation in ribosomes by *p*-chloromercuribenzoate. *FEBS Lett.* **17**: 324.

Gupta S. L., Waterson J., Sopori M.L., Weissman S.M., and Lengyel P. 1971. Movement of the ribosome along the messenger ribonucleic acid during protein synthesis. *Biochemistry* **10**: 4410.

Harada F. and Dahlberg J.E. 1975. Specific cleavage of tRNA by nuclease S1. *Nucleic Acids Res.* **2**: 865.

Hopfield J.J. 1971. Kinetic proofreading: A new mechanism for reducing errors in biosynthetic processes requiring high specificity. *Proc. Natl. Acad. Sci.* **71**: 4135.

Kurland C.G. 1992. Translational accuracy and the fitness of bacteria. *Annu. Rev. Genet.* **26**: 29.

Ladner J.E., Jack A., Robertus J.D., Brown R.S., Rhodes D., Clark B.F., and Klug A. 1975. Structure of yeast phenylalanine transfer RNA at 2.5 Å resolution. *Proc. Natl. Acad. Sci.* **72**: 4414.

Lake J.A. 1977. Amino-acyl tRNA binding at the recognition site is the first step of the elongation cycle of protein synthesis. *Proc. Natl. Acad. Sci.* **74**: 1903.

Markus M.A., Hinck A.P., Huang S., Draper D.E., and Torchia D.A. 1997. High resolution solution structure of ribosomal protein L11-C76, a helical protein with a flexible loop that becomes structured upon binding to RNA. *Nat. Struct. Biol.* **4**: 70.

Nierhaus K.H. 1990. The allosteric three-site model for the ribosomal elongation cycle: Features and future. *Biochemistry* **29**: 4997.

Nissen P., Kjeldgaard M., Thirup S., Polekhina G., Reshetnikova L., Clark B.F.C., and Nyborg J. 1995. Crystal structure of the ternary complex of Phe-tRNA^{Phe}, EF-Tu, and a GTP analog. *Science* **270**: 1464.

Noller H.F., Hoffarth V., and Zimniak L. 1992. Unusual resistance of peptidyl transferase to protein extraction procedures. *Science* **256**: 1416.

Pape T., Wintermeyer W., and Rodnina M.V. 1998. Complete kinetic mechanism of elongation factor Tu-dependent binding of aminoacyl-tRNA to the A site of the *E. coli* ribosome. *EMBO J.* **17**: 7490.

Pestka S. 1968. Studies on the formation of transfer ribonucleic acid-ribosome complexes. *J. Biol. Chem.* **243**: 2810.

Polekhina G., Thirup S., Kjeldgaard M., Nissen P., Lippmann C., and Nyborg J. 1996. Helix unwinding in the effector region of elongation factor EF-Tu-GDP. *Structure* **4**: 1141.

Quigley G.J., Suddath F.L., McPherson A., Kim J.J., Sneden D., and Rich A. 1974. The molecular structure of yeast phenylalanine transfer RNA in monoclinic crystals. *Proc. Natl. Acad. Sci.* **71**: 2146.

Schluenzen F. Tocilj A., Zarivach R., Harms J., Gluehmann L., Janell D., Bashan A., Bartels H., Agmon I., Franceschi F., and Yonath A. 2000. Structure of functionally activated small ribosomal subunit at 3.3 Å resolution. *Cell* **102**: 615.

Schweisguth D.C. and Moore P.B. 1997. On the conformation of the anticodon loops of initiator and elongator methionine tRNAs. *J. Mol. Biol.* **267**: 505.

Simonson A.B. and Lake J.A. 2002. The transorientation hypothesis for tRNA proofreading during protein synthesis. *Nature* (in press).

Song H., Parsons M.R., Rowsell S., Leonard G., and Phillips S.E. 1999. Crystal structure of intact elongation factor EF-Tu from *Escherichia coli* in GDP conformation at 2.05 Å resolution. *J. Mol. Biol.* **285**: 1245.

Stark H., Rodnina M.V., Rinke-Appel J., Brimacombe R., Wintermeyer W., and van Heel M. 1997. Visualization of elongation factor Tu on the *Escherichia coli* ribosome. *Nature* **389**: 403.

Stoeffler G., Cundliffe E., Stoeffler-Meilicke M., and Dabbs E.R. 1980. Mutants of *Escherichia coli* lacking ribosomal protein L11. *J. Biol. Chem.* **255**: 10517.

Strycharz W.A., Nomura M., and Lake J.A. 1978. Ribosomal proteins L7/L12 localized at a single region of the large subunit by immune electron microscopy. *J. Mol. Biol.* **126**: 123.

Tal J. 1975. The cleavage of transfer RNA by a single strand specific endonuclease from *Neurospora crassa*. *Nucleic Acids Res.* **2**: 1073.

Thach S.S. and Thach R.E. 1971. Translocation of messenger RNA and accommodation of fMet-tRNA. *Proc. Natl. Acad. Sci.* **68**: 1791.

Vogeley L., Palm G.J., Mesters J.R., and Hilgenfeld R. 2001. Conformational change of elongation factor Tu (EF-Tu) induced by antibiotic binding: Crystal structure of the complex between EF-Tu•GDP and aurodox. *J. Biol. Chem.* **276**: 17149.

Wimberly B.T., Guymon R., McCutcheon J.P., White S.W., and Ramakrishnan V. 1999. A detailed view of a ribosomal active site: The structure of the L11-RNA complex. *Cell* **97**: 491.

Wimberly B.T., Brodersen D.E., Clemons W.M., Jr., Morgan-Warren R.J., Carter A.P., Vonrhein C., Hartsch T., and Ramakrishnan V. 2000. Structure of the 30S ribosomal subunit. *Nature* **407**: 327.

Wrede P., Woo N.H., and Rich A. 1979. Initiator tRNAs have a unique anticodon loop conformation. *Proc. Natl. Acad. Sci.* **76**: 3289.

Yusupov M.M., Yusupova G.Z., Baucom A., Lieberman K., Earnest T.N., Cate J.H.D., and Noller H.F. 2001. Crystal structure of the ribosome at 5.5 Å resolution. *Science* **292**: 883.

Zaug A.J. and Cech T.R. 1986. The intervening sequence RNA of *Tetrahymena* is an enzyme. *Science* **231**: 470.

Features and Functions of the Ribosomal E Site

G. Blaha and K.H. Nierhaus

Max-Planck-Institut für Molekulare Genetik, AG Ribosomen, D-14195 Berlin, Germany

According to a general observation, the scientific community receives a new discovery in three stages. At first, the general opinion is "It's wrong," later "It's right, but not important," and finally "It's important, but not new." Concerning the acceptance of the E site, we are now at the transition toward the third stage.

The first hints toward a third tRNA-binding site on ribosomes were provided by Noll and his coworker in 1965 studying eukaryotic ribosomes. They saturated polysomes with ^{32}P-labeled tRNA and concluded from their results that the ribosome contains at least two, possibly three, tRNA-binding sites (Wettstein and Noll 1965). The first solid evidence was the saturation of programmed 70S ribosomes with three deacylated tRNAs that was presented in 1981 with *Escherichia coli* ribosomes (Rheinberger et al. 1981), a finding that was confirmed by others (Grajevskaja et al. 1982; Lill et al. 1984). Later, a third tRNA-binding site was found in ribosomes from all three evolutionary domains, viz., in addition to the eubacterial ribosomes from *E. coli* in archaeal ribosomes (Saruyama and Nierhaus 1986) and in eukaryotes (yeast as low eukaryotes [Triana-Alonso et al. 1995] and rabbit liver as high eukaryotes [El'skaya et al. 1997]). A report that rabbit liver ribosomes contain four tRNA-binding sites (Rodnina and Wintermeyer 1992) could not be confirmed (El'skaya et al. 1997). The hybrid-site model of the elongation brought an inflation of the number of binding states into the discussion; six and more tRNA-binding states were postulated by this model (Moazed and Noller 1989), which was based on footprinting the rRNAs of ribosomes after binding tRNA to the various sites. Indeed, cryo-electron microscopy (cryo-EM) has revealed a hybrid site for a deacylated tRNA on programmed ribosomes, but only under unfavorable buffer conditions; under near in vivo conditions, a deacylated tRNA was seen in the classic P site (Agrawal et al. 1999), suggesting that a hybrid site might be a buffer artifact. Indeed, a systematic analysis of tRNA-binding sites with authentic functional states of the elongating ribosomes has not revealed a hybrid site (Agrawal et al. 2000). Furthermore, in ribosome crystals bearing tRNAs and anticodon–stem-loop structures, a deacylated tRNA was found in the classic P site (Yusupov et al. 2001). This finding violates a diagnostic feature of the hybrid-site model according to which a deacylated tRNA is never found in the P site during protein synthesis, but rather in the P/E hybrid site, viz., the tRNA contacts the 30S at the P site and the 50S at the E site. Therefore, we do not consider the hybrid-site model here (for a detailed discussion of the current models of the elongation cycles, see Spahn and Nierhaus 1998).

An additional position for a deacylated tRNA was reported and called E2 position (Agrawal et al. 2000). This position collides with the universal E site, and thus, occupation of the E site and E2 position are mutually exclusive. This finding explains why programmed ribosomes can be saturated with three and not more tRNAs (for review, see Nierhaus 1990). At the E2 position, the tRNA clings with its inner elbow toward the L1 protein and has no further significant interactions with the ribosome. Considering the fact that L1 is not an essential ribosomal protein, since viable mutants exist that lack L1 (Subramanian and Dabbs 1980), it is clear that the E2 position is not a tRNA-binding site of the same rank as the three binding sites, A, P, and E. It is thought that the tRNA might remain for a short while in the E2 position on the way out of the ribosome after being released from the E site (Agrawal et al. 2000).

It follows that three tRNA-binding sites are a universal feature of ribosomes. The A site is the first site a tRNA contacts in the form of an aminoacyl tRNA (A for aminoacyl tRNA-binding site; see Fig. 1A for the reactions of one elongation cycle). At this site, decoding takes place, and after that the aminoacyl tRNA fully enters the A site, now determining a ribosomal state that is called the pre-translocational state (PRE state). Peptide-bond formation transfers the nascent peptide chain from the peptidyl tRNA at the P site to the aminoacyl tRNA, leaving a deacylated tRNA at the P site. The result is that now the peptidyl tRNA lengthened by one amino acid resides at the A site. The translocation reaction shifts the peptidyl tRNA to the P site (P for peptidyl tRNA) and the deacylated tRNA to the E site (E for exit) and establishes the ribosomal post-translocational state (POST state). The E site is specific for deacylated tRNA. It follows that the tRNAs move completely through the ribosome in two elongation cycles going successively through the A, P, and E sites.

The relative tRNA locations of the three ribosomal sites were assessed in the eubacterium *E. coli* with authentic states of the elongating ribosome by cryo-EM, as well as deduced from crystals of programmed 70S ribosomes of *Thermus thermophilus*. Both the mutual arrangement of the tRNAs and their ribosomal sites coincide remarkably well (Fig. 1B). A systematic analysis by cryo-EM allowed for the first time the reconstruction of the tRNA movement through the ribosome during protein synthesis (Fig. 1C).

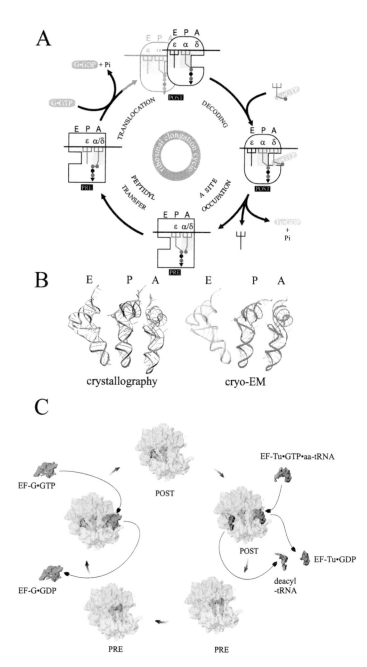

Figure 1. The elongation cycle and the mutual tRNA arrangements on the ribosome. (*A*) The elongation cycle of the ribosome according to the α-ε model (Dabrowski et al. 1998). According to this model, a mobile domain of the ribosome exists that binds tightly both tRNAs in the PRE state at A and P sites and transports them during the translocation reaction to the P and E sites, thus establishing the POST state. The decoding center is thought to be fixed at the A site, since decoding takes place at the POST state when the mobile domain with the tRNAs should be at the P and E sites. (*B*) The mutual arrangements of tRNAs at the A, P, and E sites (from right to left) according to an X-ray analysis of crystals of ribosome complexes that contain deacylated tRNAs (*left side*, resolution of 5.5–7.5 Å; Yusupov et al. 2001) and a cryo-EM study of authentic PRE and POST states of elongating ribosomes (*right side*, resolution at 17 Å; Agrawal et al. 2000). (*B*, Adapted, with permission, from Yusupov et al. 2001.) (*C*) The latter study allowed a full reconstruction of the tRNA movement through the ribosome during an elongation cycle. The two PRE states at the bottom represent the situation before and after peptide-bond formation (*right* and *left*, respectively). (*C*, Adapted, with permission, from Agrawal et al. 2000.)

Here we review the knowledge accumulated about the features and functions of the E site. All observations discussed here were made under buffer conditions that mimic those found in vivo and allow a protein-synthesis performance concerning rate and accuracy of almost in vivo perfection, whereas under conventional buffer conditions both rate and accuracy are inferior to the corresponding in vivo parameters by more than an order of magnitude. A striking correspondence is found between in vitro results gathered under near in vivo conditions and those observed with native polysomes, in contrast to many observations collected under conventional buffer conditions (for review, see Nierhaus et al. 2000).

E-SITE FEATURE 1: DEACYLATED tRNA IS BOUND AT THE E SITE IN A STABLE FASHION

Under unfavorable buffer conditions, the dissociation rate of a tRNA at the E site was reported to be 0.3/second (Robertson et al. 1986a), whereas the elongation rate was reported to be ten times faster (3/second) determined under comparable conditions (Robertson et al. 1986b). This difference between rates already indicates a relatively stable E-site occupation. Under near in vivo conditions, the stability of a tRNA at the E site is even more pronounced: The dissociation rate is strikingly slower and evidently has to be measured in hours rather than in seconds, since POST states can be isolated via overnight centrifugations through sucrose cushions without loss of any deacylated tRNA from the E site (see, e.g., Wadzack et al. 1997). Likewise, native polysomes isolated with a procedure that lasts longer than 24 hours contain an occupied E site almost quantitatively (Remme et al. 1989). These observations correspond well with the crystal structure of tRNAs bound to programmed ribosomes, where the E-site tRNA seems to be enclosed in the ribosomal matrix (Yusupov et al. 2001). It is clear that a stable binding has to be taken into account when ribosomal functions are discussed. Some reports claim weak interactions of tRNA at the E site; these contradictory results are explained by the use of unfavorable buffer conditions (for references and discussion, see Nierhaus et al. 2000).

E-SITE FEATURE 2: RECIPROCAL LINKAGE BETWEEN THE A AND E SITES

After translocation, the ribosome harbors two tRNAs, a peptidyl tRNA at the P site and a deacylated tRNA at the E site. The construction of this ribosomal state allows an easy demonstration of the reciprocal linkage between the A and the E sites that flank the P site: The rate of the A-site occupation is virtually identical with that of the E-site release. This was demonstrated with eubacterial ribosomes (*E. coli*, Rheinberger and Nierhaus 1986b) as well as archaeal (Saruyama and Nierhaus 1986) and eukaryotic ribosomes

(yeast ribosomes, Triana-Alonso et al. 1995). The experiment with the latter ribosomes is shown in Figure 2A. Three different codons were present in the A, P, and E sites; correspondingly, a ^{32}P-labeled tRNA was present at the E site and a [^{14}C]Val-tRNA at the P site of the posttranslocational ribosome. Upon addition of a ternary complex containing EF-1, GTP, and [^{3}H]Phe-tRNA, the rate of the A-site occupation was the same as the rate of the E-site release, whereas the intervening Val-tRNA at the P site did not react at all. The reaction depended on the presence of elongation factor 3, EF-3, a factor specific to higher fungi and essential for viability. This factor is an ATPase, in contrast to all other nucleotide-binding elongation factors, which are exclusively G proteins. Therefore, it appears that yeast ribosomes bind deacylated tRNA so tightly to the E site that the reciprocal mechanism prevents an occupation of the A site in the absence of EF-3 and ATP. EF-3 is an E-site factor in that it opens the E site, thus allowing quantitative A-site occupation and E-site release (Fig. 2B) (Triana-Alonso et al. 1995).

Recently, this reciprocal linkage between A and E sites was also demonstrated with rat liver ribosomes (T. Budkevich et al., unpubl.). Figure 3A shows the simplest presentation of the reciprocal linkage between the A and E sites during the elongation cycle. The consequence is that statistically two tRNAs should be found on polysomes irrespective of whether they are in the PRE or POST state. This expectation could be confirmed with *E. coli* polysomes (Remme et al. 1989).

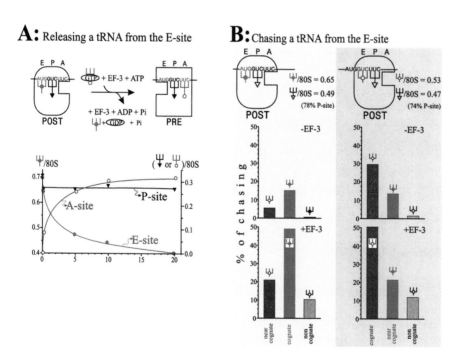

Figure 2. Reciprocal effects at A and E sites and the function of the third elongation factor EF-3 in higher fungi (here yeast; Triana-Alonso et al. 1995). (*A*) In the presence of EF-3 and ATP, the A site can be occupied, and with the same rate the deacylated tRNA is released from the E site. The binding values of the acylated tRNA at the intervening P site do not move. (*B*) The chasing of deacylated ^{32}P labeled-tRNA from the E site with various unlabeled tRNAs is only efficient if the chasing substrate is cognate to the codon at the E site, viz., if this tRNA can undergo codon–anticodon interaction. This observation is strong evidence for codon–anticodon interaction at this site. (Modified, with permission, from Triana-Alonso et al. 1995.)

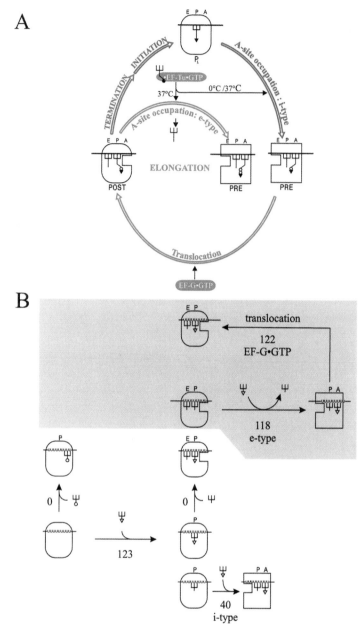

Figure 3. The two main states of the elongating ribosome and the various ribosomal states that can be thermodynamically distinguished. (*A*) In the PRE and POST states, the ribosome can bind only two tRNAs in a stable fashion. (*A*, Modified, with permission, from Hausner et al. 1988.) (*B*) Activation energies for the various tRNA-binding steps determined at the polyamine buffer in the presence of 3 mM Mg^{++} (Schilling-Bartetzko et al. 1992a). The states relevant for the elongation cycle have a blue background. The numbers indicate activation energies (kJ/mole). For further explanation, see text. (*B*, Modified, with permission, from Schilling-Bartetzko et al. 1992a.)

E-SITE FEATURE 3: CODON–ANTICODON INTERACTION AT THE RIBOSOMAL E SITE

This feature was demonstrated with two experimental strategies. (1) The E site can be quantitatively filled with a deacylated tRNA only if the ligand is a cognate tRNA; i.e., the tRNA must have an anticodon complementary to the codon at the E site (Gnirke et al. 1989). (2) A deacylated labeled tRNA can be chased effectively out of the E site exclusively with cognate nonlabeled tRNA. This has been shown with both *E. coli* (Rheinberger and Nierhaus 1986a; Rheinberger et al. 1986) and yeast ribosomes (Triana-Alonso et al. 1995); the experiment with the latter is shown as an example in Figure 2B. Two posttranslational complexes were constructed; in one case an AUG codon and the cognate ^{32}P-labeled tRNA^{Met} were at the E

site, whereas in the other case a GUC codon and the cognate ^{32}P-labeled tRNA^{Val} were located in the E site. In both cases only the cognate nonlabeled tRNA could effectively chase the ^{32}P-labeled tRNA. In other words, only a tRNA with an anticodon complementary to a codon at the E site could effectively chase a tRNA out of this site, clearly indicating the existence and importance of codon–anticodon interaction at this site. Similar results were also obtained with *E. coli* ribosomes (Rheinberger and Nierhaus 1986a; Rheinberger et al. 1986), and these data collectively indicate that codon–anticodon interaction significantly contributes to the binding energy of the E-site-bound tRNA.

The anticodons of the tRNAs at the P and E sites are at least as near to each other as those of the tRNAs at A and P sites, suggesting that the anticodons of the tRNAs at the

P and E sites form a duplex with the corresponding codons as they do at A and P sites (Fig. 1B). Furthermore, the E-site codon seems to be near the anticodon region of the E-site tRNA in the crystal structure of the programmed 70S ribosomes carrying tRNAs at all three sites (probably the three sites are not filled simultaneously, but rather the presented pictures are statistical averages; Yusupov et al. 2001), although the E-site codon of the mRNA is difficult to position unequivocally at a resolution of 5.5 Å. Nevertheless, the authors state that "normal codon–anticodon interaction is absent" at the E site, although they admit a possible contact between the bases of the middle positions. We note that the tRNA at the E site of the crystal is a non-cognate (a tRNA^{Met}), although the mRNA fragment contained only one AUG that is required for the tRNA^{Met} binding to the P site (see Cate et al. 1999). It follows that (1) the tRNA–70S complex of the crystal does not represent a state that occurs during protein synthesis and (2) this tRNA at the E site per se cannot undergo codon–anticodon interaction, making it prohibitively difficult to draw any conclusion as to whether or not a tRNA at the E site can undergo codon–anticodon interaction during the ribosomal elongation cycle. Only the crystal structure of an authentic posttranslocational complex will bring a conclusive answer to this issue.

HIGH ACTIVATION ENERGIES SEPARATE PRE- AND POSTTRANSLOCATIONAL COMPLEXES

The pre- and posttranslocational states of the ribosome are stable states, since they are separated by high activation energy barriers. A systematic analysis revealed the activation energies shown in Figure 3B. Both A-site binding and translocation during an elongation cycle require about 90 kJ/mole under conventional buffer conditions. These relatively high values are even higher at ionic conditions mimicking the relevant values for concentrations of Mg^{++} ions, monovalent ions, and polyamines in vivo and amount to about 120 kJ/mole (Schilling-Bartetzko et al. 1992a), indicating that both PRE and POST states are better defined at near in vivo conditions. The high-energy barrier is reduced by elongation factors similar to the action of enzymes, but in addition, the factors determine the direction of reaction, whereas enzymes accelerate the reaction rate only until equilibrium is reached. Therefore, all organisms require two elongation factors (Schilling-Bartetzko et al. 1992a), one for the transition pre- to posttranslocational state (EF-G in eubacteria and EF-2 in archaea and eukarya), and one for the transition post- to pretranslocational state (EF-Tu in eubacteria and EF-1 in archaea and eukarya). Only higher fungi such as yeast or *Candida albicans* have a third elongation factor EF-3, an ATPase, that is required for the release of the E-site-bound tRNA upon A-site occupation (see above; Triana-Alonso et al. 1995).

An important consequence of the high activation energy barrier is that PRE and POST states are the two main states between which the ribosome oscillates during protein synthesis (Fig. 3A). Even the empty ribosome can still adopt two main conformations that are reminiscent of the PRE and the POST states, as demonstrated when the mutual stimulation of the uncoupled GTPase activities (i.e., uncoupled from protein synthesis) of the elongation factors EF-Tu and EF-G was analyzed (Mesters et al. 1994).

STABLE E-SITE OCCUPATION IS AN ESSENTIAL FEATURE OF THE POSTTRANSLOCATIONAL STATE AND INFLUENCES THE INHIBITION PATTERNS OF SOME ANTIBIOTICS

When the activation energy of A-site binding was determined, a striking dependence on the E-site occupation was revealed: With occupied P and E sites, a value was found that was three times larger than that determined with a free E site, viz., 120 kJ/mole versus 40 kJ/mole (see Fig. 3B) (AcPhe-tRNA binding at 3 mM; Schilling-Bartetzko et al. 1992a). In other words, the high-energy barrier in the transition between POST and PRE states is only observed if the E site is occupied, and an occupied E site is essential to define the POST state. An A-site occupation with a free E site occurs only once during protein synthesis, viz., just after the initiation of protein synthesis, where an fMet-tRNA binds to the P site and initiates protein synthesis (A-site occupation of the *i*-type, *i* for initiation). All ensuing A-site occupations occur with an occupied E site (Fig. 3A) (A-site occupation of the *e*-type, *e* for elongation; Hausner et al. 1988).

A-site occupations of the i- and e-types behave strikingly differently with respect to the effects of some antibiotics. For example, all aminoglycosides, viz., streptomycin and the gentamycin, kanamycin, and neomycin families, hardly interfere with the A-site occupation of the i-type (E-site-free), whereas they completely block the A-site occupation of the e-type (Hausner et al. 1988). This effect is common to all aminoglycosides in addition to the misreading effect, and is probably responsible for the bactericidal effect of this large group of antibiotics. Note that the A-site effects of the aminoglycosides are stronger than those of the tetracyclines (Fig. 4), the classic A-site blocker. However, tetracyclines inhibit the A-site occupation of both types, whereas the aminoglycosides interfere specifically with that of the e-type.

In this respect, the inhibition patterns of the drugs viomycin and thiostrepton are also noteworthy. Both drugs are known to block the factor-dependent GTPase activity and are potent inhibitors of the translocation reaction (for references, see Hausner et al. 1988). However, as shown in Figure 4, both drugs also interfere strongly with the A-site occupation of the e-type; i.e., they block the transitions between the POST and the PRE states in both directions equally well. In contrast, only weak effects are seen with the A-site occupation of the i-type.

A consequence of the strong differences seen between the two types of A-site occupation is that the two types have to be tested separately for an assessment of the inhibition spectrum of a drug.

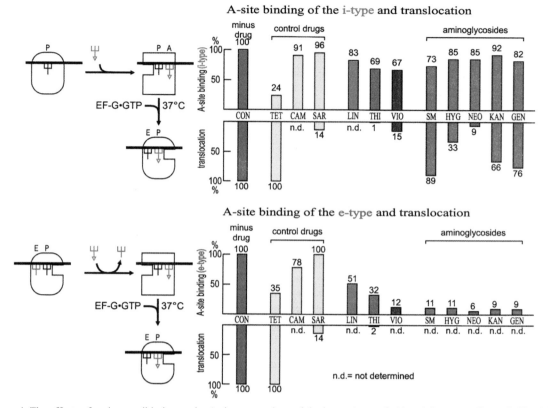

Figure 4. The effects of various antibiotics on the A-site occupations of the i-type (*upper half*) and the e-type (*lower half*). (CON) Control; (TET) tetracycline; (CAM) chloramphenicol; (SAR) α-sarcin; (LIN) lincomycin; (THI) thiostrepton; (VIO) viomycin; (SM) streptomycin; (HYG) hygromycin; (NEO) neomycin; (KAN) kanamycin; (GEN) gentamycin. (Modified, with permission, from Hausner et al. 1988.)

IMPORTANCE OF THE RECIPROCAL LINKAGE BETWEEN THE A AND E SITES

At least three functions of the E site can be envisaged: (1) The release of the deacylated tRNA from the E site is actively promoted and does not occur passively, (2) the low affinity of the A site induced by an occupied E site in the POST state plays an important role during the selection of the correct aminoacyl tRNA at the A site, and (3) codon–anticodon interaction at the E site seems to be an important element for the maintenance of the reading frame during protein synthesis.

Active Release of the Deacylated tRNA from the E Site

The stable binding of a deacylated tRNA at the E site mentioned above means that the half-life of a deacylated tRNA at the E site (several hours) surpasses the period of time required for an elongation cycle (~50 msec) by more than five orders of magnitude. Therefore, the release of a deacylated tRNA from the E site must be an active process and cannot occur via a simple diffusion process as considered by some authors (see, e.g., Rodnina et al. 1997).

A possible mechanism for such an active release is suggested by the α-ε model, according to which a movable

ribosomal domain binds two tRNAs and transports them from the A and P sites to the P and E sites, respectively (Fig. 1A, translocation reaction). In the POST state the decoding occurs. After a successful decoding the movable domain swings back and leaves the E-site region without tRNA-binding capacity (see Fig. 1A, A-site occupation). This triggers the release of the deacylated tRNA that dissociates from the ribosome, probably with a transient stop at the E2 position (Fig. 1C).

The Low Affinity of the A Site Induced by an Occupied E Site of the POST State Plays an Important Role during the Selection of the Correct Aminoacyl tRNA at the A Site

Let us consider briefly the three types of molecular recognition (Fig. 5) (see also Nierhaus 1993). The first type is typical for enzymes that recognize small substrates by binding a site of the substrate that is distinct from related substrates. The consequence is that the binding energy is roughly identical to the discrimination energy and allows an accuracy of binding that is between 10^{-3} and 10^{-6} (i.e., one wrong binding of 10^3 to 10^6 correct binding reactions, respectively), which ensures that metabolism does not degenerate into a chaos. Type II is qualitatively the same as type I, but here the binding area

Type I **Type II** **Type III**

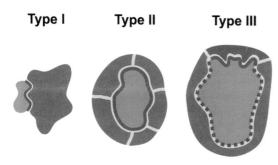

Figure 5. The three types of molecular recognition (Nierhaus 1993). (*Pink*) Bound substrate; (*red line*) discriminatory binding region; (*green dashed line*) nondiscriminatory binding region; (*blue*) binding component. For explanation, see text. (Modified, with permission, from Nierhaus 1993.)

is almost the whole surface of the molecule, resulting in a huge binding that equals discrimination energy. This case applies for ribosomal proteins that are bound mainly inside the ribosome, coat proteins of virus heads, or subunits of multisubunit enzymes and provides the energy for complicated assembly processes of the corresponding quaternary structures.

Type III presents an unfavorable case, where the binding energy is large and the discrimination energy only a small fraction of the total energy. Selection of the correct ternary complex at the decoding center of the small subunit is such a case: The discrimination region is only the anticodon, and the contacts between tRNA and ribosomes outside the anticodon and between EF-Tu and the ribosome are the same for all 41 different ternary complexes (Komine et al. 1990; two ternary complexes are defined as different when their anticodons differ). Note that EF-Tu within the ternary complex is identical in all ternary complexes and thus contributes to the nondiscriminatory fraction of the binding energy. EF-Tu increases the affinity to the A site by two orders of magnitude over that of the naked aminoacyl tRNA (Schilling-Bartetzko et al. 1992b).

Since the discrimination energy of codon–anticodon interaction can be exploited only under equilibrium conditions, the selection procedure must take a long time in order to harvest the discrimination potential, as the discrimination energy is only a small fraction of the total energy (Fig. 6A, left half). It follows that protein synthesis should be either fast and inaccurate (the equilibrium cannot be reached) or slow and accurate. That is not what we see in protein synthesis; instead, it is relatively fast (10–20 amino acids are incorporated into the growing peptide chain per one ribosome in one second; Bremer and Dennis 1987) *and* precise (about 1 wrong amino acid per 3000 incorporations; Bouadloun et al. 1983).

The E site provides a solution for this selection riddle. Occupation of this site by a deacylated tRNA induces a low-affinity state at the A site, and the hypothesis is that this low-affinity site allows just codon–anticodon interaction restricting the ternary complex–ribosome interaction to the discriminatory region only, thus changing the unfavorable type III into the favorable type I. The result is that only cognate tRNA (with a complementary anti-

codon to the A-site codon) and near-cognate tRNAs (anticodon similar to the cognate one) compete for the decoding center, whereas the majority of the ternary complexes, namely the non-cognate complexes, are out of the game (Fig. 6A, right half). For them the A site does not exist, since their anticodon cannot interact with the codon presented at the A site, and interactions outside the anticodon are not possible. The selection problem is, therefore, reduced by an order of magnitude: Instead of 1 out of 41 ternary complexes, only 1 out of 4–6 ternary complexes has to be selected. This situation is similar to that of a polymerase that selects 1 out of 4 nucleotides and does so with a precision of 1 error in more than 50,000 incorporations without even taking a proofreading mechanism into account (Blank et al. 1986; Libby et al. 1989). This hypothesis explains why the non-cognate ternary complexes do not show any ribosome-triggered GTPase activity and why the respective amino acids are never misincorporated into proteins.

Figure 6B shows a test of this hypothesis: When the E site is not occupied and thus representing a high-affinity state of the A site of poly(U) programmed ribosomes, even the non-cognate aspartic acid coded by GAC/U is incorporated into dipeptides as demonstrated by an HPLC analysis, whereas with a deacylated tRNA^{Phe} at the E site exclusively cognate AcPhe-Phe was observed (Geigenmüller and Nierhaus 1990). Obviously, an occupied E site triggers a low A-site affinity that abolishes the interference with non-cognate ternary complexes.

Figure 7 shows another detail. In experiment 1 the results demonstrated in Figure 6B are reproduced: A free E site allows an incorporation of non-cognate aspartic acid residues indicated by a large error of 0.72%, an occupied E site shows only background values of aspartic acid incorporation (0.05%). However, binding a near-cognate tRNA^{Leu} (codon UUG/A) to the E site does not prevent aspartic acid incorporation. In other words, codon–anticodon interaction seems to be the trigger for the ribosome to flip into the POST state and to establish a low-affinity A site for proper selection of the ternary complexes.

These in vitro data are supported by in vivo data. A tRNA with an altered CCA-3′ end does not bind well to the E site (Lill et al. 1988). Therefore, a tRNA^{Val} at the E site with a mutation at the CCA-3′ end cannot bind well to this site, and triggers an increased error at the A site (O'Connor et al. 1993). These results suggest that the deacylated tRNA^{Val} prematurely falls off the E site, and that the now-vacant E site induces a high-affinity A site that is prone to misincorporations.

According to the reciprocal relationship between E and A sites, the first occupation of the A site after initiation (FMet-tRNA at the P site, E site free; A-site occupation of the i-type) should have an increased error as compared to the following A-site occupations (e-type occupation). A detailed analysis revealed that in eubacterial ribosomes, the Shine-Dalgarno interaction between mRNA and the 3′ end of the 16S rRNA can functionally replace the interaction of the E-site codon of the mRNA with the anticodon of the corresponding tRNA (V. DiGiacco et al., unpubl.).

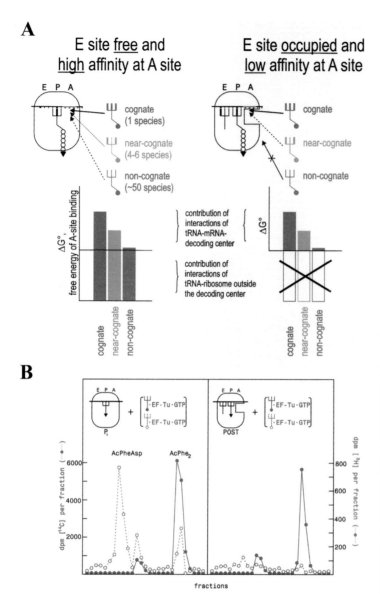

A

E site <u>free</u> and high affinity at A site

cognate (1 species)

near-cognate (4-6 species)

non-cognate (~50 species)

contribution of interactions of tRNA-mRNA-decoding center

contribution of interactions of tRNA-ribosome outside the decoding center

$\Delta G°$, free energy of A-site binding

cognate near-cognate non-cognate

E site <u>occupied</u> and <u>low</u> affinity at A site

cognate

near-cognate

non-cognate

$\Delta G°$

cognate near-cognate non-cognate

B

P_i

POST

$+ \begin{bmatrix} EF\text{-}Tu\cdot GTP \\ EF\text{-}Tu\cdot GTP \end{bmatrix}$

$+ \begin{bmatrix} EF\text{-}Tu\cdot GTP \\ EF\text{-}Tu\cdot GTP \end{bmatrix}$

AcPheAsp   AcPhe$_2$

dpm [^{14}C] per fraction (○—○)

dpm [^{3}H] per fraction (●—●)

fractions

Figure 6. The role of the E site for the accuracy of decoding. (*A, left half*) If the E site is free, the A site is in a high-affinity state at which all 41 different ternary complexes interfere with the selection of the cognate complex. (*A, right half*) The E site is occupied, inducing a low-affinity state of the A site. This low-affinity state restricts the interaction of the ternary complex to the decoding center, thus mainly allowing discriminatory codon–anticodon interactions but preventing nondiscriminatory interaction outside the anticodon. (*B*) A test of the above outlined hypothesis with poly(U) programmed ribosomes (Geigenmüller and Nierhaus 1990). When the E site is free (*left half*), the A site adopts a high-affinity state. This allows the incorporation of even non-cognate amino acids, in this case aspartic acid that is coded for by the codon GAC/U (HPLC analysis of the formed dipeptides). With an occupied E site (*right half*), the incorporation of aspartic acid is abolished. (Modified, with permission, from Geigenmüller et al. 1990.)

The fact that amino acids of non-cognate ternary complexes are never incorporated into proteins is explained by the reciprocal linkage between A and E sites. What remains is the decoding and selection between cognate versus near-cognate ternary complexes. Concerning the molecular mechanism of decoding, two alternative models exist. The first assumes that the stability of the codon–anticodon duplex is used for the selection process. Since this mechanism can explain an accuracy of not much better than 1:10 (Grosjean et al. 1978), excessive proofreading was advocated, where the discrimination potential of codon–anticodon interaction was exploited several times (Ehrenberg et al. 1986; Kurland 1987a,b). Alternatively, the correctness of the partial Watson-Crick structure could be sensed by the decoding center, where, e.g., the correct position of the sugar pucker (2′OH in the duplex RNA) would contribute to the discrimination energy. The two models were tested using a heteropolymeric mRNA of 41 nucleotides containing in the middle the three codons AUG-UUC-GUC coding for methionine, phenylalanine and valine, respectively, where the mRNA contained either a canonical ribo-codon UUC or a 2′-deoxyribo codon dUdUdC coding for phenylalanine (F or dF codon, respectively; Fig. 8; Potapov et al. 1995). The idea was that a comparison between a ribo-codon and 2′-deoxyribo codon could discriminate between the two models, since base-pairing in RNA–RNA duplexes is almost identical to that in DNA–RNA hybrids, which also adopt an overall A form typical for RNA duplexes. However, the sugar puckering in the DNA strand of hybrids is quite different from that of RNA duplexes (Hall 1993). The first model predicts that the RNA–RNA duplex and the DNA–RNA hybrid would be equally accepted in the decoding process, whereas according to the second model, a striking difference is expected.

Figure 8 demonstrates some results relevant in this discussion. F or dF codon at the P site has no influence on extent and affinity for AcPhe-tRNA binding to this site

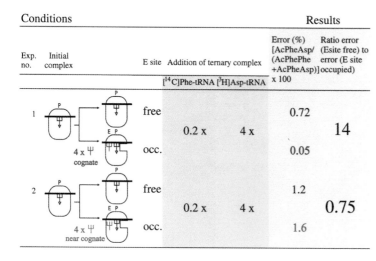

Conditions				Results	
Exp. no.	Initial complex	E site	Addition of ternary complex	Error (%) [AcPheAsp/ (AcPhePhe +AcPheAsp)] x 100	Ratio error (Esite free) to error (E site occupied)
			[^{14}C]Phe-tRNA [^{3}H]Asp-tRNA		
1		free		0.72	
			0.2 x 4 x		14
		occ.		0.05	
2		free		1.2	
			0.2 x 4 x		0.75
		occ.		1.6	

Figure 7. Accuracy of aminoacyl tRNA selection at the A site and E site occupation. A cognate deacylated tRNA has to be at the E site in order to avoid the incorporation of non-cognate Asp in poly(U) programmed 70S ribosomes (experiment 1). A near-cognate deacylated tRNA^{Leu} (codon UUG/A) is not sufficient (experiment 2), implying that a tRNA at the E site has to undergo codon–anticodon interaction in order to induce the low–affinity A site that is prerequisite for noninterference of non-cognate aminoacyl tRNAs at the A site. (Modified, with permission, from Geigenmüller and Nierhaus 1990.)

(experiment 2). However, a striking difference was found at the A site (experiment 3). Considering the A-site fractions found in the presence of UUC or dUdUdC, the lack of the 2′OH groups in the A-site codon reduces the binding to this site by a factor of 5; the affinity was reduced by a factor of 10. These results identify an important role for the 2′OH groups of the codon at the A site, in contrast to that at the P site, and strongly favor the second model where proofreading plays a minor role, if any.

Recently, this issue was settled by an X-ray structure of programmed 30S subunits carrying an anticodon stem-loop structure at the A site (Ogle et al. 2001). Indeed, the correct positions of the 2′OH groups of the first and second base pair of the codon–anticodon duplex are sensed by the decoding center. The elements of the latter are the universally conserved bases A1492 and A1493 of helix 44, the universally conserved bases G530 and C518, and protein S12, mutations of which are known to improve the accuracy of the ribosome (Ozaki et al. 1969; Breckenridge and Gorini 1970; Funatsu et al. 1972 and references therein). A1493 alone clings into the minor groove of the first base pair of the codon–anticodon duplex and forms H bonds to both nucleotides of the base pair plus an H bond to the O2 of the uracil present in this position (Fig. 9A). Even the latter H bond is not sequence-specific, since O2 of pyrimidines and N3 of purines adopt equivalent positions in the minor groove of Watson-Crick base pairs, because both are H acceptors (Seeman et al. 1976). The ribosomal groups sensing the 2′OH groups of the second base pair in the middle of the codon–anticodon duplex seem to be even more rigidly fixed. Here, two universally conserved bases are involved, namely A1492 and G530. Either one forms equivalent H bonds to one of the bases of the duplex; for example, the 2′OH of one base of the duplex forms two H bonds with N3 and the 2′OH group of A1492. The same is true for the other duplex base and G530 (Fig. 9B). Both A1492 and G530 are backed up by C518 and S12 (Ser-50), resulting in a rigid structure that senses correct positions of the 2′OH groups of the middle pair of the codon–anticodon duplex. The recognition of the

third base pair is less restricted and allows the formation of wobble pairs (Ogle et al. 2001).

The binding of A residues into the minor groove of a Watson-Crick base pair seems to be a general recognition mode. It is seen frequently in the rRNA within the ribosome and seems to play an important role in stabilizing the tertiary structure of the rRNA. Four different types can be distinguished (Nissen et al. 2001): The most important ones are type I and type II, which are represented in the recognition of the first and second base pair of codon–anticodon interaction (Fig. 9, A and B, respectively).

Codon–Anticodon Interaction at the E Site Seems to Be Important to Maintain the Reading Frame

The simple feedback regulation of the RF2 synthesis might shed some light on this aspect of the E site. The

Exp. no.	binding state	mRNA	binding of AcPhe-tRNA[pmol]	binding site	binding ratio
1	E P A	no	2.2	p	
2	E P A	EFV	2.9	P	1
		E(dF)V	3.0	P	
3	E P A	EFV	3.0	A 85%	5
		E(dF)V	1.0	(A 52%)	

Figure 8. Importance of the 2′OH groups of the codons at P and A sites for binding of acylated tRNAs. A deoxy-codon at the P site (dF) does not seriously affect tRNA binding to this site, in contrast to A-site binding, where the presence of a deoxy-codon severely impairs tRNA binding. (Modified, with permission, from Potapov et al. 1995.)

A   **B**

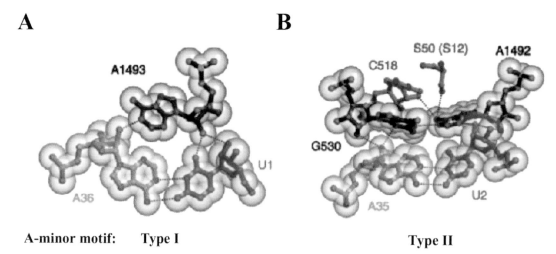

A-minor motif: Type I   **Type II**

Figure 9. Decoding mechanism at the first and second base pair of codon–anticodon interaction at the A site (*A* and *B*, respectively) as deduced from crystal of a programmed 30S subunit with a cognate anticodon stem-loop structure at the A site (Ogle et al. 2001). (*A*) U1-A36 is the first base pair of codon–anticodon interaction at the A site, where U1 is the first base of the A-site codon and A36 the 3′ base of the tRNA anticodon. A1493 is from 16S rRNA. (*<None>*) U2-A35 is the second base pair. The other elements are from the 16S rRNA or the ribosomal protein S12. 2′OH groups of the duplex are the main elements of the recognition process. These recognition modes represent two types of four so-called A-minor motifs, where A residues are packed into the minor grooves of base pairs (Nissen et al. 2001). Note that here G530 instead of an A represents an "A-minor motif." (Reprinted, with permission, from Ogle et al. 2001 [copyright AAAS].)

26th codon position of the RF2 mRNA is taken by a stop-codon UGA that is recognized by RF2. If the concentration of RF2 in the cell is sufficient, the stop codon will be recognized and the translation of the RF2 mRNA will abort after the synthesis of a 25-mer peptide that is quickly degraded. However, if there is a shortage of RF2 in the cell, a +1 frameshift will occur with an astonishing efficiency of up to 100% (Donly et al. 1990), whereas during normal translation, the error frequency of frameshifts is not more than 1 case in 30,000 amino acid incorporations (Jorgensen and Kurland 1990); i.e. the frameshift on the RF2 mRNA that is necessary for the complete synthesis of RF2 occurs with a frequency that is more than four orders of magnitude larger than normal. What kind of mechanism is switched off to abandon the strict maintenance of the reading frame?

What is special with this UGA? It has a Shine-Dalgarno upstream, and what has been overlooked in a number of publications is that the Shine-Dalgarno sequence extends into the E site and occupies the first base of the E-site codon. Our hypothesis is that the collision between Shine-Dalgarno sequence and the E-site codon triggers the release of the deacylated tRNA from this site and leaves the ribosome with just the peptidyl tRNA at the P site. This situation does not occur during elongation and might facilitate a slip of the ribosome on the mRNA, thus inducing with a high probability a +1 frameshift. This hypothesis can be tested, and if true, it not only gives a satisfying explanation for the extreme frameshift frequency during the translation of the RF2 mRNA, but also suggests that the codon–anticodon interaction at the E site has the important role of maintaining the reading frame during translation.

There is also in vivo evidence for codon–anticodon interaction at the E site that influences frameshifts. Heptameric sequences such as X XXY YYZ can perform a –1

frameshift with good efficiency. An ingeniously designed experiment has revealed that –1 frameshifts at such slippery sequences occur *after* translocation in vivo (Horsfield et al. 1995), indicating that both tRNAs at the P and E sites during the POST state undergo codon–anticodon interactions. If duplex formation only at the P site would be adequate for a frameshift event rather than at P *and* E sites, a 4-mer rather than a 7-mer sequence would be sufficient for such a frameshift, but a tetrameric slippery sequence has not been observed up to now.

CONCLUSIONS

We have seen that the role of the E site is not limited to simple functions by, e.g., passively facilitating the release of the tRNA from the high-affinity P site, but rather the deacylated tRNA is ejected actively from this site. The E site actively participates in the decoding process by preventing the non-cognate aminoacyl tRNAs from interfering with the selection of the cognate aminoacyl tRNA and probably plays an important role for holding the reading frame during translation.

ACKNOWLEDGMENTS

We thank Sean Connell for help and discussion. Support was provided by the German Ministry for Sciences and Technology (Bundesministerium für Bildung, Wissenschaft, Forschung und Technologie, grant BMBF 0312552).

REFERENCES

Agrawal R.K., Penczek P., Grassucci R.A., Burkhardt N., Nierhaus K.H., and Frank J. 1999. Effect of buffer conditions on the position of tRNA on the 70S ribosome as visualized by cryo-electron microscopy. *J. Biol. Chem.* **274:** 8723.

Agrawal R.K., Spahn C.M.T., Penczek P., Grassucci R.A., Nierhaus K.H., and Frank J. 2000. Visualization of tRNA movements on the *Escherichia coli* 70S ribosome during the elongation cycle. *J. Cell Biol.* **150:** 447.

Blank A., Gallant J.A., Burgess R.R., and Loeb L.A. 1986. An RNA polymerase mutant with reduced accuracy of chain elongation. *Biochemistry* **25:** 5920.

Bouadloun F., Donner D., and Kurland C.G. 1983. Codon-specific missense errors in vivo. *EMBO J.* **2:** 1351.

Breckenridge L. and Gorini L. 1970. Genetic analysis of streptomycin resistance in *Escherichia coli*. *Genetics* **65:** 9.

Bremer H. and Dennis P.P. 1987. Modulation of chemical composition and other parameters of the cell by growth rate. In *Escherichia coli and Salmonella typhimurium: Cellular and molecular biology* (ed. F.C. Neidhardt et al.), p. 1527. ASM Press, Washington, D.C.

Cate J.H., Yusupov M.M., Yusupova G.Z., Earnest T.N., and Noller H.F. 1999. X-ray crystal structures of 70S ribosome functional complexes. *Science* **285:** 2095.

Dabrowski M., Spahn C.M.T., Schäfer M.A., Patzke S., and Nierhaus K.H. 1998. Contact patterns of tRNAs do not change during ribosomal translocation. *J. Biol. Chem.* **273:** 32793.

Donly B.C., Edgar C.D., Adamski F.M., and Tate W.P. 1990. Frameshift autoregulation in the gene for *Escherichia coli* release factor-2: Partly functional mutants result in frameshift enhancement. *Nucleic Acids Res.* **18:** 6517.

Ehrenberg M., Andersson D., Mohman K., Jelenc P., Ruusala T., and Kurland C.G. 1986. Ribosomal proteins tune rate and accuracy in translation. In *Structure, function and genetics of ribosomes* (ed. B. Hardesty et al.), p. 573. Springer, New York.

El'skaya A.V., Ovcharenko G.V., Palchevskii S.S., Petrushenko Z.M., Triana-Alonso F.J., and Nierhaus K.H. 1997. Three tRNA binding sites in rabbit liver ribosomes and role of the intrinsic ATPase in 80S ribosomes from higher eukaryotes. *Biochemistry* **36:** 10492.

Funatsu G., Nierhaus K., and Wittmann H. G. 1972. Ribosomal proteins. XXXVII. Determination of allele types of amino acid exchanges in protein S12 of three streptomycin-resistant mutants of *Escherichia coli*. *Biochim. Biophys. Acta* **287:** 282.

Geigenmüller U. and Nierhaus K.H. 1990. Significance of the third tRNA binding site, the E site, on *E. coli* ribosomes for the accuracy of translation: An occupied E site prevents the binding of non-cognate aminoacyl-transfer RNA to the A site. *EMBO J.* **9:** 4527.

Gnirke A., Geigenmüller U., Rheinberger H.-J., and Nierhaus K.H. 1989. The allosteric three-site model for the ribosomal elongation cycle. *J. Biol. Chem.* **264:** 7291.

Grajevskaja R.A., Ivanov Y.V., and Saminsky E.S. 1982. 70S ribsosomes of *Escherichia coli* have an additional site for deacylated tRNA. *Eur. J. Biochem.* **128:** 47.

Grosjean H.J., Dehenau S., and Crothers D.M. 1978. On the physical basis for ambiguity in genetic coding interactions. *Proc. Natl. Acad. Sci.* **75:** 610.

Hall K.B. 1993. NMR spectroscopy of DNA/RNA hybrids. *Curr. Opin. Struct. Biol.* **3:** 336.

Hausner T.P., Geigenmüller U., and Nierhaus K.H. 1988. The allosteric three site model for the ribosomal elongation cycle. New insights into the inhibition mechanisms of aminoglycosides, thiostrepton, and viomycin. *J. Biol. Chem.* **263:** 13103.

Horsfield J.A., Wilson D.N., Mannering S.A., Adamski F.M., and Tate W.P. 1995. Prokaryotic ribosomes recode the HIV-1 gag-pol-1 frameshift sequence by an E/P site post-translocation simultaneous slippage mechanism. *Nucleic Acids Res.* **23:** 1487.

Jorgensen F. and Kurland C.G. 1990. Processivity errors of gene expression in *Escherichia coli*. *J. Mol. Biol.* **215:** 511.

Komine Y., Adachi T., Inokuchi H., and Ozeki H. 1990. Genomic organization and physical mapping of the transfer RNA genes in *Escherichia coli* K12. *J. Mol. Biol.* **212:** 579.

Kurland C.G. 1987a. Strategies for efficiency and accuracy in gene expression. 1. *Trends Biochem. Sci.* **12:** 126.

——. 1987b. Strategies for efficiency and accuracy in gene expression. 2. Growth optimized ribosomes. *Trends Biochem. Sci.* **12:** 169.

Libby R.T., Nelson J.L., Calvo J.M., and Gallant J.A. 1989. Transcriptional proofreading in *Escherichia coli*. *EMBO J.* **8:** 3153.

Lill R., Robertson J.M., and Wintermeyer W. 1984. tRNA binding sites of ribosomes from *Escherichia coli*. *Biochemistry* **23:** 6710.

Lill R., Lepier A., Schwägele F., Sprinzl M., Vogt H., and Wintermeyer W. 1988. Specific recognition of the 3′-terminal adenosine of tRNAPhe in the exit site of *Escherichia coli* ribosomes. *J. Mol. Biol.* **203:** 699.

Mesters J.R., Potapov A.P., de Graaf J.M., and Kraal B. 1994. Synergism between the GTPase activities of EF-Tu.GTP and EF-G.GTP on empty ribosomes. Elongation factors as stimulators of the ribosomal oscillation between two conformations. *J. Mol. Biol.* **242:** 644.

Moazed D. and Noller H.F. 1989. Intermediate states in the movement of transfer RNA in the ribosome. *Nature* **342:** 142.

Nierhaus K.H. 1990. The allosteric three-site model for the ribosomal elongation cycle: Features and future. *Biochemistry* **29:** 4997.

——. 1993. Solution of the ribosomal riddle: How the ribosome selects the correct aminoacyl-tRNA out of 41 similar contestants. *Mol. Microbiol.* **9:** 661.

Nierhaus K.H., Spahn C.M.T., Burkhardt N., Dabrowski M., Diedrich G., Einfeldt E., Kamp D., Marquez V., Patzke S., Schäfer M.A., Stelzl U., Blaha G., Willumeit R., and Stuhrmann H.B. 2000. Ribosomal elongation cycle. In *The ribosome: Structure, function, antibiotics, and cellular interactions* (ed. R.A. Garrett et al.), p. 319. ASM Press, Washington D.C.

Nissen P., Ippolito J.A., Ban N., Moore P.B., and Steitz T.A. 2001. RNA tertiary interactions in the large ribosomal subunit: The A-minor motif. *Proc. Natl. Acad. Sci.* **98:** 4899.

O'Connor M., Willis N.M., Bossi L., Gesteland R.F., and Atkins J.F. 1993. Functional tRNAs with altered 3′ ends. *EMBO J.* **12:** 2559.

Ogle J.M., Brodersen D.E., Clemons W.M., Jr., Tarry M.J., Carter A.P., and Ramakrishnan V. 2001. Recognition of cognate transfer RNA by the 30S ribosomal subunit. *Science* **292:** 897.

Ozaki M., Mizushima S., and Nomura M. 1969. Identification and functional characterization of the protein controlled by the streptomycin-resistant locus in *E. coli*. *Nature* **222:** 333.

Potapov A.P., Triana-Alonso F.J., and Nierhaus K.H. 1995. Ribosomal decoding processes at codons in the A or P sites depend differently on 2′-OH groups. *J. Biol. Chem.* **270:** 17680.

Remme J., Margus T., Villems R., and Nierhaus K.H. 1989. The third ribosomal tRNA-binding site, the E site, is occupied in native polysomes. *Eur. J. Biochem.* **183:** 281.

Rheinberger H.-J. and Nierhaus K.H. 1986a. Adjacent codon-anticodon interactions of both tRNAs present at the ribosomal A and P or P and E sites. *FEBS Lett.* **204:** 97.

——. 1986b. Allosteric interactions between the ribosomal transfer RNA-binding sites A and E. *J. Biol. Chem.* **261:** 9133.

Rheinberger H.-J., Sternbach H., and Nierhaus K.H. 1981. Three tRNA binding sites on *E. coli* ribosomes. *Proc. Natl. Acad. Sci.* **78:** 5310.

——. 1986. Codon-anticodon interaction at the ribosomal E site. *J. Biol. Chem.* **261:** 9140.

Robertson J.M., Paulsen H., and Wintermeyer W. 1986a. Pre-steady-state kinetics of ribosomal translocation. *J. Mol. Biol.* **192:** 351.

Robertson J.M., Urbanke C., Chinali G., Wintermeyer W., and Parmeggiani A. 1986b. Mechanism of ribosomal translocation. Translocation limits the rate of *Escherichia coli* elongation factor G-promoted GTP hydrolysis. *J. Mol. Biol.* **189:** 653.

Rodnina M.V. and Wintermeyer W. 1992. 2 transfer RNA-binding sites in addition to A-site and P-site on eukaryotic ribosomes. *J. Mol. Biol.* **228:** 450.

Rodnina M.V., Savelsbergh A., Katunin V.I., and Wintermeyer

W. 1997. Hydrolysis of GTP by elongation factor G drives tRNA movement on the ribosome. *Nature* **385:** 37.

Saruyama H. and Nierhaus K.H. 1986. Evidence that the three-site model for ribosomal elongation cycle is also valid in the archaebacterium *Halobacterium halobium. Mol. Gen. Genet.* **204:** 221.

Schilling-Bartetzko S., Bartetzko A., and Nierhaus K.H. 1992a. Kinetic and thermodynamic parameters for transfer RNA binding to the ribosome and for the translocation reaction. *J. Biol. Chem.* **267:** 4703.

Schilling-Bartetzko S., Franceschi F., Sternbach H., and Nierhaus K.H. 1992b. Apparent association constants of transfer RNAs for the ribosomal A-site, P-site, and E-site. *J. Biol. Chem.* **267:** 4693.

Seeman N.C., Rosenberg J.M., and Rich A. 1976. Sequence-specific recognition of double helical nucleic acids by proteins. *Proc. Natl. Acad. Sci.* **73:** 804.

Spahn C.M.T. and Nierhaus K.H. 1998. Models of the elongation cycle: An evaluation. *Biol. Chem.* **379:** 753.

Subramanian A.R. and Dabbs E.R. 1980. Functional studies on ribosomes lacking protein L1 from mutant *Escherichia coli. Eur. J. Biochem.* **112:** 425.

Triana-Alonso F.J., Chakraburtty K., and Nierhaus K.H. 1995. The elongation factor 3 unique in higher fungi and essential for protein biosynthesis is an E site factor. *J. Biol. Chem.* **270:** 20473.

Wadzack J., Burkhardt N., Jünemann R., Diedrich G., Nierhaus K.H., Frank J., Penczek P., Meerwinck W., Schmitt M., Willumeit R., and Stuhrmann H.B. 1997. Direct localization of the tRNAs within the elongating ribosome by means of neutron scattering (proton-spin contrast-variation). *J. Mol. Biol.* **266:** 343.

Wettstein F.O. and Noll H. 1965. Binding of transfer ribonucleic acid to ribosomes engaged in protein synthesis: Number and properties of ribosomal binding sites. *J. Mol. Biol.* **11:** 35.

Yusupov M.M., Yusupova G.Z., Baucom A., Lieberman K., Earnest T.N., Cate J.H., and Noller H.F. 2001. Crystal structure of the ribosome at 5.5 Å resolution. *Science* **292:** 883.

Pseudouridines and Pseudouridine Synthases of the Ribosome

J. OFENGAND,* A. MALHOTRA,* J. REMME,† N.S. GUTGSELL,* M. DEL CAMPO,*
S. JEAN-CHARLES,* L. PEIL,† AND Y. KAYA*

*Department of Biochemistry and Molecular Biology, University of Miami School of Medicine, Miami,
Florida 33101; †Department of Molecular Biology, Tartu University, Tartu, Estonia

Pseudouridine (5-ribosyl uracil, Ψ) is the single most abundant modified nucleoside found in ribosomal RNA (Table 1). This is true even when each ribosomal subunit is considered separately, with the exception of some prokaryotes which appear to have little or no Ψ in their small subunits (see Table 2). Although the totals of all 2′-O-methyl modifications are approximately equal in amount, clearly this would not be true for each of the Am, Cm, Um, or Gm modified nucleosides. Similarly, the methylated bases in the table consist of a variety of different modified residues. Ψ is made from uridine residues in an oligo- or polynucleotide by enzymatic cleavage of the N-glycosyl bond, rotation of the uracil ring (apparently while still enzyme-bound) so that C-5 occupies the position previously held by N-1, followed by re-formation of the glycosyl link as a C–C bond. This process is energetically favored as the reaction goes to apparent completion in the absence of any external energy source or cofactors. Since Ψ formation only takes place with selected uridines that are in a preformed polynucleotide chain, a major question is the mechanism of selection of particular uridines for conversion to Ψ. A second major issue is what purpose is served by this reaction.

Ψ is found only in those RNA molecules whose tertiary structure is integral to their function; for example, rRNA, tRNA, tmRNA, snRNA, and snoRNA. The most obvious consequence of Ψ formation is the creation of a new H-bond donor. Therefore, it may be that the main function of Ψ is to act as a structural stabilizer. One might envision Nature as testing out the various RNA structures and, where a bit too much flexibility exists, adding in a Ψ or two to tighten things up by increasing RNA–RNA contacts or, in some cases, perhaps by improving rRNA–protein interaction. Another possibility worth consideration is that Ψ adds an additional element of potential interaction with functionally important ligands such as, in the case of rRNA in the ribosome, tRNA and mRNA.

Ψ in ribosomal RNA has been the subject of several recent reviews (Ofengand and Fournier 1998; Bachellerie et al. 2000; Ofengand and Rudd 2000). A more general review of Ψ in RNA is also available (Charette and Gray 2000).

NUMBER AND LOCATION OF PSEUDOURIDINE RESIDUES IN RIBOSOMAL RNA

Number of Pseudouridines in rRNA

The number of Ψ in known rRNAs is listed in Table 2. 5S RNAs are not included, although Ψ is present in a few cases (Szymanski et al. 2000). The number of Ψ varies widely from organism to organism, increasing as the organism becomes more complex. Some small-subunit RNAs have none, as in the case of two of the archaeal organisms (*Halobacter halobium* and *Haloferax volcanii*), one as in the two eubacteria (*Escherichia coli* and *Bacillus subtilis*), or up to ~40 in the higher eukaryotes. All the large-subunit RNAs have four or more Ψ with the higher eukaryotes having ~55. Even organelle large-subunit RNAs have at least one Ψ. It appears that although the small subunit can make do without Ψ, this is not the case for the large subunit. The vast increase in number in eukaryotes compared to the prokaryotes is also noteworthy and is not due to the increase in size of the RNAs in eukaryotes. When the number of Ψ is expressed as a percentage of either the U residues or total nucleotides, there is a clear approximately fivefold increase in the percentage of Ψ in the higher eukaryotes. As expected from its lesser complexity, the value for yeast is intermediate between the two extremes. The reason(s) for this disparity in Ψ usage is (are) unknown.

Table 1. Number of Modified Nucleosides in Representative Ribosomal RNAs

Organism	SSU			LSU		
	base methyl	2′-OMe ribose	Ψ plus Ψ*	base methyl	2′-OMe ribose	Ψ plus Ψ*
Escherichia coli	11	1	1	10	3	10
Yeast	3	18	14	6	37	30
Xenopus laevis	4	33	~44	5	62–64	~52
Mouse/human	4	40	~36	5	63–65	57/55

Ψ*, modified Ψ.

Table 2. Number of Pseudouridine Residues in Small and Large Subunit Ribosomal RNAs

Organism	No. of Ψ residues	Ψ (% of U)	Ψ (% of all residues)	How determined (% RNA sequenced)	Reference
Small subunit cytoplasm					
Escherichia coli	1	0.3	0.06	seq. (100)	a
Bacillus subtilis	1	0.3	0.06	seq. (15)	b
Halobacter halobium	0	—	—	seq. (70)	c
Haloferax volcanii	0	—	—	total	d
Sulfolobus solfataricus	4–5	1.8–2.2	0.28–0.35	total	e
Saccharomyces cerevisiae	14‡	2.7	0.78	seq. (98)	c
Xenopus laevis	~44	10.7	2.4	total	f
Mus musculus	~36	8.9	1.9	seq. (80)	f
Homo sapiens	~36	8.6	1.8	seq. (80)	f
Large subunit cytoplasm					
Escherichia coli	10*	1.7	0.34	seq. (99)	g
Bacillus subtilis	5*	0.9	0.17	seq. (56)	h
Halobacter halobium	4	0.7	0.14	seq. (54)	h
Sulfolobus solfataricus	4	0.8	0.13	total	e
Sulfolobus acidocaldarius	6	1.2	0.20	seq. (20)	i
Saccharomyces cerevisiae	30	3.5	0.88	seq. (99)	j
Drosophila melanogaster	57	4.8	1.44	seq. (59)	h
Xenopus laevis	~52	7.9	1.27	total	f
Mus musculus	57	7.3	1.21	seq. (46)	h
Homo sapiens	55	7.3	1.09	seq. (41)	h
Large subunit mitochondria					
Saccharomyces cerevisiae	1	0.1	0.03	seq. (99)	j
Mus musculus	1	0.2	0.06	seq. (17)	h
Homo sapiens	1	0.3	0.06	seq. (41)	h
Trypanosoma brucei	6	0.8	0.35	seq. (98)	h
Large subunit chloroplasts					
Zea mays	4*	0.7	0.14	seq. (29)	h
5.8 S RNA					
Saccharomyces cerevisiae	1	2.2	0.63	seq. (100)	k
Mus musculus	2	5.7	1.3	seq. (100)	l
Homo sapiens	2	5.5	1.3	seq. (100)	l

‡This value includes the hypermodified pseudouridine, $m^1acp^3\Psi1189$.
*Includes the $m^3\Psi1916$ in *E. coli* and the equivalent residues in *B. subtilis* and *Z. mays*.
[a]Bakin et al. 1994b; [b]Niu and Ofengand 1999; [c]Bakin and Ofengand 1995; [d]Kowalak et al. 2000; [e]Noon et al. 1998; [f]Maden 1990; [g]Bakin and Ofengand 1993; [h]Ofengand and Bakin 1997; [i]Massenet et al. 1999; [j]Bakin et al. 1994a; [k]Rubin 1973; [l]Nazar et al. 1976; Khan and Maden 1977.

Location of the Pseudouridines in the Ribosome

Small subunit (SSU) RNA. Many of the Ψ listed in Table 2 have been located in their RNA sequence, mostly by the use of a sequencing method for Ψ (Bakin and Ofengand 1993). They are mapped on the secondary structure of *Thermus thermophilus* SSU RNA (Fig. 1) and on the three-dimensional structure of Wimberly et al. (2000) (see also Fig. 3A, below). As noted in the figure legend, 11 of the mammalian Ψ and 3 of the yeast Ψ could not be mapped. Many of the sites shown are not U in the figure because the sequence varies from organism to organism even though the core secondary structure remains the same. Although all of the Ψ sites have been put at the equivalent positions in the *T. thermophilus* 30S subunit, exact positional correspondence of the RNA chains in the ribosomes of the different species is not assured. However, given the overall conservation of structure and function of ribosomes, any deviation is likely to be small.

Whereas *E. coli* has but one Ψ and *B. subtilis* also has one at the same site, the Ψ in yeast and mammals are numerous and distributed throughout the RNA and the ribosome. A number of the sites are shared by yeast and mammals, but each also has sites unique to itself. Despite the fact that ~40 Ψ sites are known in mammals, none are at

or even near the site used by the bacteria. The wide distribution of Ψ residues in the 30S ribosome (see Fig. 3a), the lack of any clustering around the decoding site, and the fact that some organisms have no Ψ at all do not permit any universal or even simple interpretation of the geographic localization of Ψ in the small subunit.

Large subunit (LSU) RNA. The large subunit Ψ in Table 2 have been similarly mapped on the secondary structure of the *Haloarcula marismortui* LSU RNA (Fig. 2) and on the three-dimensional structure of Ban et al. (2000) (Fig. 3b, c). Included are the incomplete data for the Ψ of *H. marismortui*, showing both those Ψ definitely located as well as those that are likely but not yet certain. Because only 17% of the RNA has been sequenced so far, these sites should not be considered the complete set.

The 5′ half of the RNA in prokaryotes is remarkably free of Ψ except for the two in *E. coli*. The eukaryotic Ψ cluster in the vicinity of helices 33, 35, 35.1, 37, 38, and 39 of domain II with a few in helices 6 and 7 of domain I, leaving large areas of the 5′half free of Ψ in all of the studied organisms. Many more Ψ are to be found in the 3′ half of the RNA, comprising domains IV to VI. Interestingly, only the higher eukaryotes have Ψ in domain VI. All four of the prokaryotic Ψ sites of domain IV can

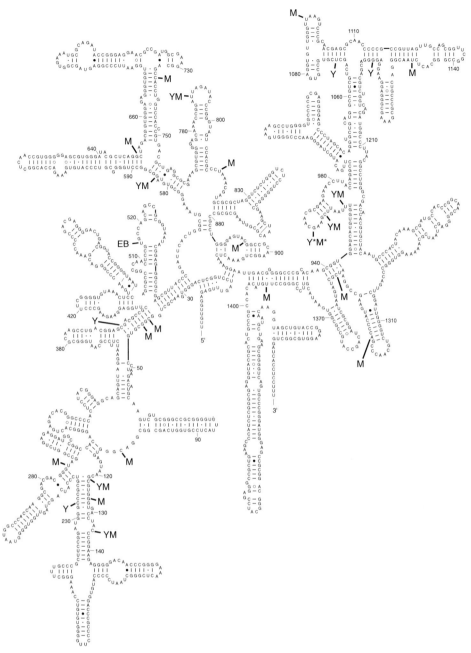

Figure 1. Ψ locations in small subunit RNAs. The secondary structure of *Thermus thermophilus* 16S RNA with the *E. coli* numbering system (Yusupov et al. 2001) is shown with the sites for Ψ in *E. coli* (E), *B. subtilis* (B), *S. cerevisiae* (Y), and mammals (M) obtained from the sources listed in Table 2. Three Y sites and 11 M sites are not shown either because the structure is absent from *T. thermophilus* or because the sites were not located to nucleotide resolution.

be found within a 13-nucleotide segment of helix 69 and its loop. One of these loop positions is universally found, and another is in all organisms studied except for *S. acidocaldarius*. This is the stem-loop that projects out from the 50S subunit at the interface side (Fig. 3b, c) and which has been shown to be directly under the elbow of both A- and P-site-bound tRNAs (Yusupov et al. 2001). It can hardly be a coincidence that these two Ψ sites are the only two in the entire RNA which are so highly conserved. Moreover, prokaryotes with few total Ψ nevertheless

have two or three Ψ in the helix 69 stem-loop. *B. subtilis* has three of its five there, *H. halobium* has two of its four, *Z. mays* three of its four, *E. coli* three of its ten, and *S. acidocaldarius* two of its six. Domain V, which includes the peptidyl transferase center (PTC), is by far the richest overall in Ψ. Of the ten Ψ in *E. coli*, five are at the PTC, the remaining two of *B. subtilis* and *H. halobium* are there, and the single remaining Ψ of *Z. mays* is also there. The detailed distribution (Fig. 2B) shows several sites that are common to a number of organisms. In additional

A

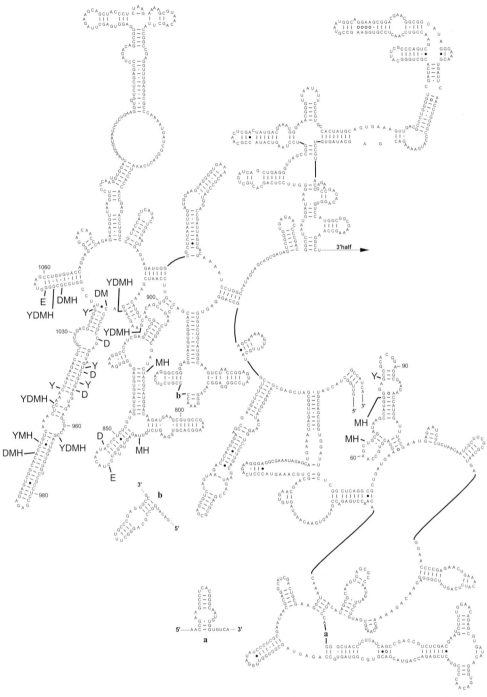

Figure 2. Ψ locations in large subunit RNAs. (*A*) 5′ half; panel (*B*) 3′half. The secondary structure of *Haloarcula marismortui* rrnB 23S RNA (AF034620) from the Gutell Web site (www.rna.icmb.utexas.edu) is shown using *H. marismortui* numbering. The sites for Ψ in *E. coli* (E), *B. subtilis* (B), *Z. mays* (Z), *H. halobium* (A), *S. acidocaldarius* (S), *S. cerevisiae* (Y), *D. melanogaster* (D), *M. musculus* (M), and *H. sapiens* (H) were obtained from the sources listed in Table 2. E* is m³Ψ (Kowalak et al. 1996) and B* and Z* are probably the same (Ofengand and Bakin 1997). Seven Ψ in *D. melanogaster* are not shown because they occur in a large expansion segment absent in *H. marismortui*. Ⓨ , Ψ, definite or probable sites, respectively, in *H. marismortui* (S. Jean-Charles and J. Ofengand, unpubl.). Arrow brackets indicate the segments of *H. marismortui* RNA sequenced for Ψ. (*Figure continues on next page.*)

cases, the exact site is not conserved, but a region 1–2 nucleotides distant or on the opposite side of a helix is in common.

The location of the Ψ in three dimensions is shown in Figure 3,b and c. The 6 Å diameter spheres representing the Ψ ring are somewhat larger than the 4.2 Å uracil ring

for ease of viewing. All of the prokaryotes are grouped in Figure 3b and all of the eukaryotes are in Figure 3c. Since many of the sites are shared by a number of organisms (Fig. 2), the color assigned to each site was determined according to the hierarchy described in the figure legend. There is a striking correspondence in grouping of all the

B

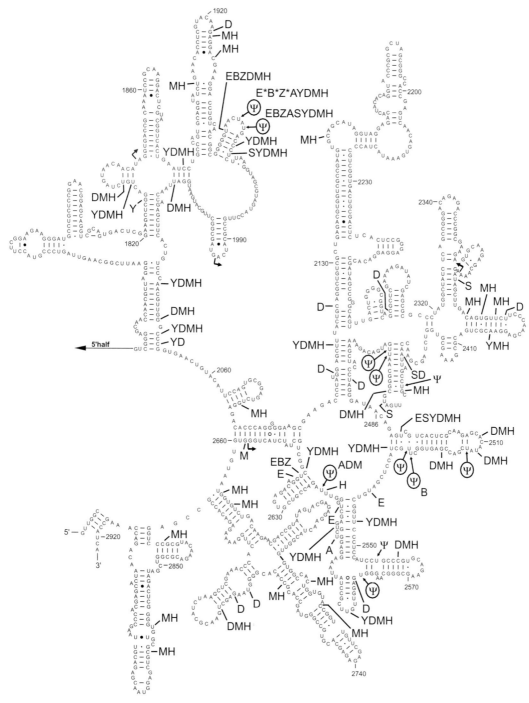

Figure 2. (*Continued. See facing page for legend.*)

prokaryotes. One major cluster is in the vicinity of the A-site tRNA, possibly including the EF-Tu-binding region and part of the peptide exit channel, and a minor cluster, so far only found in two of the three archaeal organisms, is near the top of the E site. In the eukaryotes, the pattern is much more complex. Similar clusters to those seen in the prokaryotes are present, but tend to be overshadowed by the many Ψ located elsewhere in the subunit. Nevertheless, some parts of the subunit have no Ψ. The top, the left side in the interface view, the bottom, and the back of

the subunit, which together comprise considerable material, are not sites for Ψ.

MECHANISM OF BIOSYNTHESIS

Eubacteria

In *E. coli*, the only eubacterium studied, the various Ψ in its rRNA are made by a set of site-specific enzymes, termed synthases. Figure 4 shows the locations of the Ψ in 23S RNA, and the synthases that make them. Not

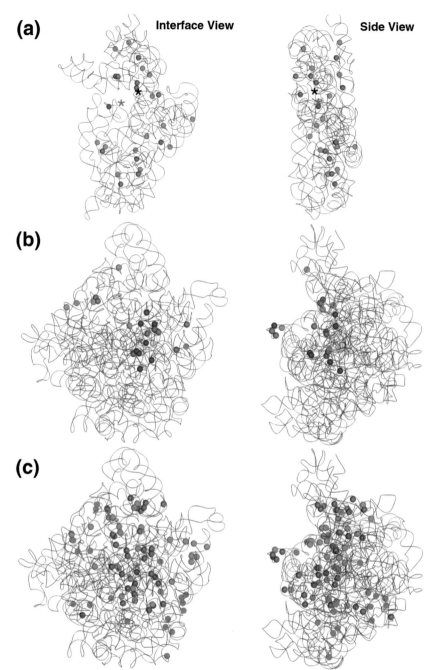

Figure 3. Three-dimensional structure of the RNA in each subunit showing the location of the Ψ residues in all known organisms. Ribosomal subunits are shown with all proteins removed and the phosphodiester backbone represented by a continuous line. The Ψ locations were marked by a 6 Å diameter sphere whose center was located at the glycosyl N of the base in question except for bases 1952, 1956, 1958, and 1962 derived from helix 69 of the large subunit (Yusupov et al. 2001), since only the phosphate coordinates were available. Each panel shows the interface view and an ~90° side view. (*a*) *T. thermophilus* 30S subunit (Wimberly et al. 2000). *E. coli* (*blue*), *S. cerevisiae* (*red*). Only *additional* Ψ sites in mammals (*green*) are shown. The decoding site is marked by G513 (*red star*) for the A site and C1382 (*black star*) for the P site (Ogle et al. 2001; Yusupov et al. 2001). (*b*) *H. marismortui* 50S subunit (Ban et al. 2000) with helix 69 (Yusupov et al. 2001) added. *H. marismortui* Ψ are shown as definite (*red*) or probable (*violet*). Only Ψ at *additional* sites in *E. coli* (*blue*) are shown. One Ψ in *H. halobium* (*yellow*) is new, and four Ψ in *S. acidocaldarius* (*green*) are not found elsewhere. The peptidyl transferase center is marked by a red star at A2486. (*c*) As in *b*, but with Ψ from the eukaryotes as indicated. All Ψ sites in *S. cerevisiae* (*red*) are shown. Only Ψ at *additional* sites in *D. melanogaster* (*green*) are shown. Only Ψ in *M. musculus* (*violet*) *additional* to the first two sets of sites are shown. Only one Ψ in *H. sapiens* (*blue*) is at a site not found in the above organisms. Ψ998 (*green*) and Ψ2154 (*violet*) are absent because these positions are missing from the published structure.

shown is the single Ψ516 in 16S RNA made by the synthase RsuA (Wrzesinski et al. 1995a; Conrad et al. 1999). Each synthase makes its own Ψ(s). There is no case

where more than one synthase makes the same Ψ. This is evident since deletion of each of the synthase genes in turn causes a corresponding absence of specific Ψ in

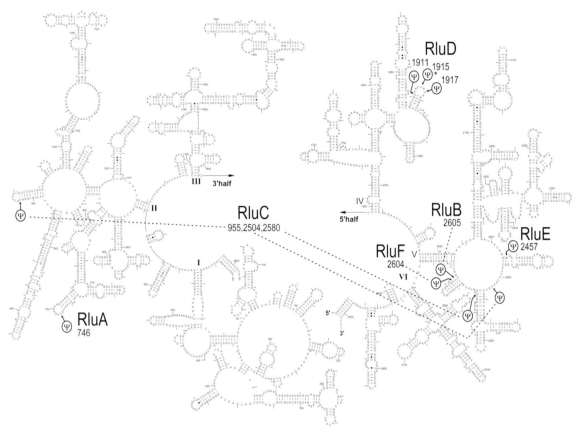

Figure 4. Secondary structure of *E. coli* 23S RNA showing the locations of the ten Ψ in this RNA along with the assignments of the six synthases that form them. RluA (Wrzesinski et al. 1995b; Raychaudhuri et al. 1999); RluB (Del Campo et al. 2001); RluC (Conrad et al. 1998; Huang et al. 1998b); RluD (Huang et al. 1998b; Raychaudhuri et al. 1998); RluE and RluF (Del Campo et al. 2001).

vivo, and all of the Ψ sites can be accounted for by the six synthases shown in the figure. The reverse is not the case, however. Two synthases, RluD and RluC, make three Ψ each. The mechanism of recognition by RluD could be explained by binding to the stem-loop and conversion of all U residues within range, since all of the U residues in this stem-loop are modified until the G•U pair at the base of the stem is reached. RluC, on the other hand, specifically recognizes 3 U residues which share nothing obvious in common—neither primary sequence, secondary structure, nor nearness in the 50S subunit. The recognition mechanism in this case is not understood. RluB and RluF are also interesting from the standpoint of specificity. They recognize adjacent U residues and carry out the same enzymatic reaction, yet are specific for their respective sites. How this specificity is maintained is another open question. RluA is special because it was the first example of a modifying enzyme with dual specificity. It is specific for U746 in 23S RNA but also is specific for U32 in those tRNAs that have Ψ32, and it is the only synthase in the cell able to carry out both reactions. This finding is important because it shows that a substrate under study may not be the only substrate for a particular enzyme in vivo. Therefore, if a functional loss occurs

when a modifying activity is eliminated by mutation, it need not be directly related to the studied substrate. As a case in point, RluA was isolated as an rRNA-modifying enzyme and only subsequently was shown to possess tRNA-modifying activity as well.

Eukaryotes

For rRNA of eukaryotes, site-specificity is determined by guide RNAs acting in conjunction with a (probably) single Ψ synthase protein plus additional auxiliary proteins (Ofengand and Fournier 1998; Bachellerie et al. 2000). These two different ways to achieve specificity in Ψ formation in the rRNA of bacteria and eukaryotes is either a cause or a consequence of the disparate number of Ψ in their respective rRNAs, eukaryotes having up to 10 times the number found in bacteria (Table 2). However, since it seems likely that archaeal organisms with small numbers of Ψ use a guide RNA system (see below), a more probable scenario is that guide RNAs were developed first in Archaea to provide the site-specificity of Ψ formation even when only a few Ψ were made, and only later eukaryotes exploited the simplicity of the system to make more and more Ψ at useful sites in the ribosome.

Archaea

There is no definitive information on whether archaeal organisms use a set of site-specific synthases like eubacteria or guide RNAs like eukaryotes. From the small number of Ψ found in Archaea so far, it might be deduced that only a few site-specific synthases would be needed. On the other hand, analysis of ten archaeal genomes for Ψ synthase homologs failed to reveal such putative synthases. Instead, only a TruA-like one and a guide RNA-associated type like Cbf5 were found (see Table 4). This result, although indirect, strongly argues for guide RNA usage, at least for these ten organisms. Unfortunately, it is not known whether they have a few Ψ like the Archaea in Table 2 or many Ψ like eukaryotes. If guide RNAs are used but there are only a few Ψ sites, it might suggest that guide RNAs were first used in Archaea and only later elaborated in eukaryotes.

THE PSEUDOURIDINE SYNTHASES

Number and Assignment in *E. coli*

There are a total of ten Ψ synthases in *E. coli* as judged by amino acid sequence homology analysis of its genome (Ofengand and Rudd 2000). Seven are accounted for above as rRNA Ψ synthases. The remaining three make Ψ in tRNA. TruA makes Ψ38–40, TruB makes Ψ55, and TruC is responsible for Ψ65 in tRNA^{Asp} and tRNA^{Ile1} (see references in Ofengand and Rudd 2000; Del Campo et al. 2001). Thus, all ten synthases and all known Ψ in rRNA and tRNA have been connected to each other. The two Ψ in tmRNA are probably made by TruB and TruA (Felden et al. 1998). The only exception is Ψ13 in tRNA^{Glu}, which has no corresponding synthase. It may be that this synthase differs so much in sequence homology from the other ten that it is not detectable by current search methods. Another possibility is that two (or more) of the identified synthases are able to form Ψ13 in addition to their other specific reaction. At least two would be required, since deletion individually of all ten Ψ synthases did not cause the loss of Ψ13 (Y. Kaya and J. Ofengand, unpubl.).

Ψ Synthase Classification into Families and an Essential Aspartate in a Conserved Sequence Motif

Conserved sequence motifs in putative Ψ synthases were first described by Koonin (1996) and have been summarized and expanded by Ofengand and Rudd (2000). They serve to subdivide all Ψ synthases into four families, three of which, RsuA, RluA, and TruB, are related, and one, TruA, which is distinct. As shown in Table 3, in *E. coli* the RsuA family synthases all make Ψ in ribosomal RNA but without distinction between 16S and 23S RNAs. Whereas RsuA is specific for 16S RNA, RluB, RluE, and RluF make Ψ in 23S RNA. Similarly, whereas RluA, RluC, and RluD make Ψ in 23S RNA, TruC is a tRNA-specific synthase. TruB and TruA, being the only members of their respective families, define the

Table 3. An Essential Aspartate Residue Present in Motif II of all *E.coli* Ψ Synthases

Protein	Motif II	Aspartate mutant activity (%)	
		in vivo	in vitro
RsuA family			
RsuA	A GRLD I D T T G L V L M	<1[a]	—
RluB	V GRLD V N T C G L L L F	<1[b]	—
RluE	A GRLD R D S E G L L V L	<1[b]	—
RluF	I GRLD K D S Q G L I F L	<1[b]	—
RluA family			
RluA	V HRLD M A T S G V I V V	<1[c]	<2[d]
RluD	V HRLD K D T T G L M V V	<1[e]	<5[e]
RluC	V HRLD R D T S G V L L V	<1[f]	—
TruC	A HRLD R P T S G V L L M	<5[b]	—
TruB family			
TruB	T GALD P L A T G M L P I	<1[g]	<0.1[d]
TruA family			
TruA	A GRTD A G V H G T G Q V	—	<0.1[h]

Conserved nucleotides are underlined. The essential aspartate is in bold.

[a]Conrad et al. 1999; [b]Del Campo et al. 2001; [c]Raychaudhuri et al. 1999; [d]Ramamurthy et al. 1999; [e]Gutgsell et al. 2001; [f]S. Jean-Charles and J. Ofengand, unpubl.; [g]Gutgsell et al. 2000; [h]Huang et al. 1998a.

classes as synthases for tRNA Ψ55 and Ψ38–40, respectively. In yeast (Table 4), the RsuA family is absent, as in all eukaryotes. Of the four members of the RluA family, only one makes an rRNA Ψ, namely Ψ2819 in mitochondrial LSU RNA (Ansmant et al. 2000). The other three make Ψ in tRNA (Ansmant et al. 2001; C. Branlant, pers. comm.). The yeast TruA family has three members that make Ψ in tRNA, one of which is the homolog of *E. coli* TruA, and the TruB family consists of the classic tRNA Ψ55 enzyme plus Cbf5, the ribosomal Ψ synthase that works in conjunction with guide RNAs (see references in Gutgsell et al. 2000; Ofengand and Rudd 2000). The Cbf5 synthase is sufficiently different structurally from the TruB class that it constitutes its own subclass. Two Ψ synthases in mouse have been analyzed. They are homologous to two yeast synthases of the TruA family (Chen and Patton 1999; 2000). On the basis of this limited study, RsuA or RluA family synthases should make Ψ in 16S RNA, 23S RNA, or in tRNA. TruA family members should be limited to tRNA substrates, TruB synthases should make only Ψ55 in tRNA, and Cbf5 members should make only ribosomal Ψ.

All motifs II from all families of all organisms are related to each other (Ofengand and Rudd 2000), as shown in Table 3 for all ten of the *E. coli* synthases. In particular, a conserved aspartate is embedded in motif II. This residue was shown to be essential for activity of TruA and was even proposed to be at the catalytic center (Huang et al. 1998a). Whether this is so or not, the residue is clearly essential for activity of all ten of the *E. coli* synthases. Most were analyzed in vivo by the transformation of deletion strains with rescue plasmids containing the mutated gene, and several were tested in vitro by overexpression and affinity purification of the wild-type and mutant synthases. Similar results were obtained for three yeast synthases (Zebarjadian et al. 1999; Ansmant et al. 2000, 2001). The presence of an essential residue in these syn-

Table 4. Family Distribution of Pseudouridine Synthases in Completely Sequenced Genomes

Organism	Families			
	RsuA	RluA	TruA	TruB
Bacteria				
Aquifex aeolicus	2	1	1	1
Bacillus halodurans	2	2	1	1
Bacillus subtilis	2	3	1	1
Borrelia burgdorferi	1	2	1	1
Buchnera sp.	1	2	1	1
Campylobacter jejuni	2	3	1	1
Caulobacter crescentus	3	3	1	1
Chlamydia muridarum	1	2	1	1
Chlamydia pneumoniae	1	1	1	1
Chlamydia trachomatis	1	2	1	1
Chlamydophila pneumoniae AR39	1	2	1	1
Deinococcus radiodurans	1	3	1	1
Escherichia coli	4	4	1	1
Haemophilus influenzae	3	5	1	1
Helicobacter pylori	1	3	1	0
Lactococcus lactis	3	3	1	1
Mesorhizobium loti	2	2	1	1
Mycobacterium leprae	1	1	1	1
Mycobacterium tuberculosis	1	2	1	1
Mycoplasma genitalium[a]	0	2	1	0
Mycoplasma pneumoniae[a]	0	2	1	0
Mycoplasma pulmonis	1	2	0	1
Neisseria meningitidis MC58	3	4	1	1
Pasteurella multocida PM70	3	5	1	1
Pseudomonas aeruginosa	4	4	1	1
Rickettsia prowazekii	1	2	1	1
Streptococcus pyogenes	3	3	1	1
Synechocystis sp.	2	2	1	1
Thermotoga maritima	1	2	1	1
Treponema pallidum	1	3	1	1
Ureaplasma urealyticum	0	2	1	1
Vibrio cholerae	4	6	1	1
Xylella fastidiosa	2	2	1	1
Archaea				
Archaeoglobus fulgidus	0	0	1	0, 1*
Aeropyrum pernix	0	0	0	0, 1*
Halobacterium sp NRC-1	0	0	1	0, 1*
Methanococcus jannaschii	0	0	1	0, 1*
Methanobacterium thermoautotrophicum[b]	0	0	1	0, 1*
Pyrococcus abyssii	0	0	1	0, 1*
Pyrococcus furiosus	0	0	1	0, 1*
Pyrococcus horikoshii	0	0	1	0, 1*
Thermoplasma acidophilum[b]	0	0	1	0, 1*
Thermoplasma volcanium	0	0	1	0, 1*
Eukaryotes				
Guillardia theta	0	0	0	0, 1*
Saccharomyces cerevisiae	0	4	3	1, 1*
Caenorhabditis elegans	0	1	2	0, 1*
Drosophila melanogaster	0	1–2	3	1, 1*

Genome sequences were obtained from publicly available databases or as described in Ofengand and Rudd (2000). Superscript * in the TruB column indicates the number of ORFs with homology to yeast Cbf5 and which are presumed to function with guide RNAs.

[a]Other *Mycoplasma* strains, *capricolum* and *mycoides*, have Ψ55, the TruB product.

[b]*M.thermoautotrophicum* has Ψ54Ψ55 in tRNA[Gly] and tRNA[Asn]; *T. acidophilum* has Ψ54Ψ55 in tRNA[Metm] and tRNA[Metf].

thases has made it possible to inactivate function by changing a single amino acid, a process that does not appear to alter the amount of the protein produced (Gutgsell et al. 2001). This ability has allowed us to independently study the effect of loss of Ψ from the effect of the loss of the synthase itself (see below).

Ψ Synthases in Sequenced Genomes

Keeping in mind the characteristic family distinctions just described, all publicly available sequenced genomes were analyzed for the kinds of Ψ synthases predicted to be present as an indicator of which Ψ residues might be there (Gutgsell et al. 2000; Ofengand and Rudd 2000). As shown in Table 4, bacteria have a typical complement of synthases, namely a variable number in the RsuA and RluA class, and one each in the TruB and TruA classes. Notably, three organisms lack a TruB enzyme, suggesting either that Ψ55 is missing in the tRNA of these organisms (sequencing of these tRNAs has not been done), or more likely, that a synthase from one of the other three families has taken over this function. The most striking result is that all ten of the sequenced archaeal genomes have at most two synthases, one TruA-like, members of which have not shown any activity on rRNA, and one guide RNA-dependent Cbf5-like. Moreover, *A. pernix* has even dispensed with the TruA synthase. These results suggest that Archaea, at least these ten, use the guide RNA system found in eukaryotes. It is also worth noting that the nucleomorph of *G. theta*, an ultra-streamlined eukaryote (Douglas et al. 2001), also has retained a Cbf5-like synthase, presumably because even it needs some Ψ in its ribosomes.

FUNCTIONAL ROLES FOR PSEUDOURIDINES AND THEIR SYNTHASES

Deletion of Ψ Synthase Genes and Their Effect on Growth

In order to probe the functional role of Ψ in the ribosome, it was necessary to choose both a suitable organism and an appropriate experimental approach. *E. coli* was chosen because it had a sufficient number of Ψ to be interesting but not so many as to be overwhelming. That this organism was easy to manipulate genetically and had the best-characterized ribosomes was also a positive feature. In addition, *E. coli* was predicted to have a set of Ψ synthases with specificity for particular Ψ that would allow deletion of Ψ residues by deletion or disruption of the respective synthase. This, in turn, would reveal the metabolic consequences of the lack of specific Ψ. The prediction was borne out, and as noted above, seven synthases were identified for the eleven ribosomal Ψ in *E. coli*. Inactivation by deletion of six of these genes individually had no effect on exponential growth in rich media at normal temperatures. Only the inactivation of one of them, *rluD*, had a strong effect on growth (Table 5).

Two RluD-minus strains were studied. *rluD*::cam is considered to be the original RluD-minus phenotype generated by a miniTn10 insertion two-thirds of the way along the gene, and Tiny is a pseudorevertant (Gutgsell et al. 2001). The disruption of RluD caused a fivefold drop in growth rate that was only partly rescued by the second site mutation. In both strains, however, supply of the wild-type *rluD* on a plasmid completely restored normal growth. Was this a result of restoration of Ψ1911, Ψ1915, and Ψ1917, the products of RluD activity, or was it due

Table 5. Growth Rate of Tiny and *rluD*::cam Mutants in *E. coli* MG1655 with and without Transformation by Rescue Plasmids

Rescue plasmid	Doubling time (min)			
	experiment I		experiment II	
	wild-type	Tiny strain	wild-type	*rluD*::cam strain
pTrc99A	33 ± 1	51 ± 2	30 ± 1	152 ± 15
pTrc99A/D139D	—	34 ± 2	—	31 ± 2
pTrc99A/D139N	—	33 ± 2	—	29 ± 1
pTrc99A/D139T	—	32 ± 1	—	29 ± 1

Cells were grown in LB medium plus 100 µg/ml carbenicillin at 37°C. Doubling times are the average of two determinations. Data from Gutgsell et al. (2001).

to the lack of the synthase itself? As noted above and in Table 3, mutation of an essential aspartate in RluD, D139, was sufficient to block the enzymatic activity. However, as Table 5 shows, it did not block the ability of D139 mutants of RluD from fully restoring growth. To be sure that no Ψ was being made by the mutant RluD, even though in vitro studies showed that such mutants were inactive (Gutgsell et al. 2001), Ψ sequencing of the RNA from both *rluD*-minus strains rescued with mutant *rluD* genes was done (Fig. 5). Panel A confirms that in the Tiny strain

only the wild-type gene on a plasmid could restore Ψ formation, despite the fact that both mutants restored growth as well as the wild type. Therefore, the growth-restoring ability is not due to the Ψ, but to the synthase protein. The results shown in panel B require further study. Despite being inactive in vitro, the mutant proteins were able to make their Ψ in the *rluD*::cam (referred to as Dust) strain. Possible reasons for this anomalous result have been discussed previously (Gutgsell et al. 2001). Nevertheless, the results with the Tiny strain (panel A) show clearly that growth rescue is possible without Ψ synthesis.

Effect of the Absence of RluD on Ribosome Assembly

Ribosome profiles from cell lysates of the RluD-minus strains and various plasmid-rescued derivatives are shown in Figure 6. Panel A compares the profiles of wild-type and the *rluD*::cam (Dust) mutant. Lack of RluD resulted in a major change in the profile. The 70S were strongly decreased, the 50S increased, there was an appearance of a major new peak at 39S, and the 30S peak shifted to 27S. The 39S peak was derived from a 50S precursor as it contained 50S proteins and precursor 23S

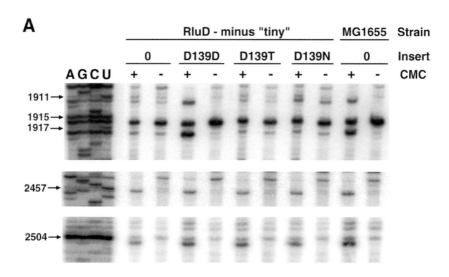

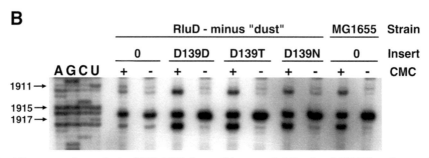

Figure 5. Pseudouridine sequencing analysis of 23S rRNA from wild-type and rluD-minus MG1655 strains transformed with plasmids carrying wild-type or mutant *rluD* genes. (0) No insert; (D139N, D139T, and D139D) mutant and wild-type inserts, respectively. (*A*) IPTG-induced *rluD*-minus Tiny cells; (*B*) uninduced *rluD*::cam (Dust) cells. Positions 1911, 1915, 1917, 2457, and 2504 are indicated. RNA for the A, G, C, U sequencing lanes was from wild-type MG1655 *E. coli*. (Reprinted, with permission, from Gutgsell et al. 2001.)

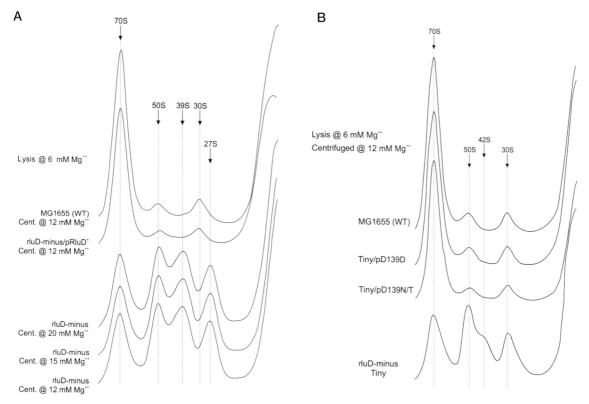

Figure 6. Analysis of ribosomal particles by sucrose gradient centrifugation. Exponentially growing bacterial cells were lysed at 6 mM Mg^{++} and centrifuged through a 10–25% sucrose gradient containing 12–20 mM Mg^{++} as indicated. (*A*) Ribosomes from wild-type (MG1655), *rluD*::cam (Dust), and Dust transformed with a plasmid carrying the wild-type *rluD* gene (rluD-minus/pRluD$^+$). (*B*) Ribosomal particles from the *rluD*-minus Tiny strain with and without wild-type or mutant rescue plasmids. Tiny plus plasmid-borne wild-type *rluD* gene (Tiny/pD139D) or either D139N or D139T mutant *rluD* (pD139N/T). Ribosome profiles of both mutants were identical; only one is shown.

RNA, and it did not shift to 50S even at 20 mM Mg^{++}. It appears to be a trapped 50S precursor. Clearly, this assembly and association defect could explain the severe growth inhibition shown in Table 5. As with the growth defect, complete restoration to wild type was achieved after supply of a wild-type copy of *rluD* on a plasmid. In Figure 6B, a similar experiment with the Tiny strain is shown. Consistent with its faster growth rate, the ribosome profile is less severely disturbed. Still, 70S are reduced, 50S and 30S are increased, and there is a well-defined shoulder at 42S. Assembly is affected here, as is subunit joining. Significantly, a normal ribosome profile was obtained not only when the wild-type gene was added back, but also when either of two mutant genes was used. This result is again in good agreement with the growth results. Taken together, these findings reveal the unexpected result that the RluD synthase itself, independent of its role in making Ψ, plays an important role in 50S subunit assembly.

SUMMARY AND CONCLUSIONS

Ψ are ubiquitous in ribosomal RNA. Eubacteria, Archaea, and eukaryotes all contain Ψ, although their number varies widely, with eukaryotes having the most. The small ribosomal subunit can apparently do without Ψ in some organisms, even though others have as many as 40 or more. Large subunits appear to need at least one Ψ but can have up to 50–60. Ψ is made by a set of site-specific enzymes in eubacteria, and in eukaryotes by a single enzyme complexed with auxiliary proteins and specificity-conferring guide RNAs. The mechanism is not known in Archaea, but based on an analysis of the kinds of Ψ synthases found in sequenced archaeal genomes, it is likely to involve use of guide RNAs. All Ψ synthases can be classified into one of four related groups, virtually all of which have a conserved aspartate residue in a conserved sequence motif. The aspartate is essential for Ψ formation in all twelve synthases examined so far.

When the need for Ψ in *E. coli* was examined, the only synthase whose absence caused a major decrease in growth rate under normal conditions was RluD, the synthase that makes Ψ1911, Ψ1915, and Ψ1917 in the helix 69 end-loop. This growth defect was the result of a major failure in assembly of the large ribosomal subunit. The defect could be prevented by supplying the *rluD* structural gene in *trans*, and also by providing a point mutant gene that made a synthase unable to make Ψ. Therefore, the RluD synthase protein appears to be directly involved in 50S subunit assembly, possibly as an RNA chaperone, and this activity is independent of its ability to form Ψ. This result is not without precedent. Depletion of PET56, a 2′-O-methyltransferase specific for G2251 (*E. coli* numbering) in yeast mitochondria virtually blocks 50S subunit assembly and mitochondrial function (Sirum-Connolly et al. 1995), but the methylation activity of the enzyme is not required (T. Mason, pers. comm.). The absence of FtsJ, a heat shock protein that makes Um2552 in *E. coli*, makes the 50S subunit less stable at 1 mM Mg^{++} (Bügl et al. 2000) and inhibits subunit joining (Caldas et al. 2000), but, in this case, it is not yet known whether the effects are due to the lack of 2′-O-methylation or to the absence of the enzyme itself.

Is there any role for the Ψ residues themselves? First, as noted above, the 3 Ψ made by RluD which cluster in the end-loop of helix 69 are highly conserved, with one being universal (Fig. 2B). In the 70S–tRNA structure (Yusupov et al. 2001), the loop of this helix containing the Ψ supports the anticodon arm of A-site tRNA near its juncture with the amino acid arm. The middle of helix 69 does the same thing for P-site tRNA. Unfortunately, the resolution is not yet sufficient to provide a more precise alignment of the Ψ residues with the other structural elements of the tRNA–ribosome complex so that one cannot yet determine what role, if any, is played by the N-1 H that distinguishes Ψ from U.

Second, and more generally, some Ψ residues in the LSU appear to be near the site of peptide-bond formation or tRNA binding but not actually at it (Fig. 2B) (Nissen et al. 2000; Yusupov et al. 2001). For example, position 2492 is commonly Ψ and is only six residues away from A2486, the A postulated to catalyze peptide-bond formation. Position 2589 is Ψ in all the eukaryotes and is next to 2588, which base-pairs with the C75 of A-site tRNA. Residue 2620, which interacts with the A76 of A-site-bound tRNA, is a Ψ or is next to a Ψ in eukaryotes and Archaea, and is five residues away from Ψ2580 in *E. coli*. A2637, which is between the two CCA ends of P- and A-site tRNA, is near Ψ2639, Ψ2640, and Ψ2641, found in a number of organisms. Residue 2529, which contacts the backbone of A-site tRNA residues 74–76, is near Ψ2527Ψ2528 in *H. marismortui*. Residues 2505–2507, which contact A-site tRNA residues 50–53, are near Ψ2509 in higher eukaryotes, and residues 2517–2519 in contact with A-site tRNA residues 64–65 are within 1–3 nucleotides of Ψ2520 in higher eukaryotes and Ψ2514 in *H. marismortui*. A way to rationalize this might be to invoke the concept suggested in the Introduction that Ψ acts as a molecular glue to hold loose elements in a more rigid configuration. It may well be that this is more important near the site of peptide-bond formation and tRNA binding, accounting for the preponderance of Ψ in this vicinity.

What might be the role of all the other Ψ in eukaryotes? One can only surmise that cells, having once acquired the ability to make Ψ with guide RNAs, took advantage of the system to inexpensively place Ψ wherever an undesirable loose region was found. It might be that in some of these cases, Ψ performs the role played by proteins in other regions, namely that of holding the rRNA in its proper configuration. Confirmation of this hypothesis will have to await structural determination of eukaryotic ribosomes.

Notes Added in Proof

An updated version of the number and location of Ψ in *H. marismortui* 50S can be found in Ofengand (2002). The crystal structure of pseudouridine synthase (TruB) complexed with substrate (Hoang and Ferré-D'Amaré 2001) confirms that the essential aspartate (Table 3) is at the catalytic center.

ACKNOWLEDGMENTS

We thank the Peter Moore laboratory for the gift of *H. marismortui* ribosomal RNA used for sequencing. This work was supported in part by National Institutes of Health grant 58879 (J.O.), in part by the Estonian Science Foundation (J.R.), and in part by the USA National Research Council (NRC) under the Collaboration in Basic Science and Engineering Program (J.O. and J.R.). The contents of this publication do not necessarily reflect the views or policies of the NRC, nor does mention of trade names, commercial products, or organizations imply endorsement by the NRC.

REFERENCES

Ansmant I., Massenet S., Grosjean H., Motorin Y., and Branlant C. 2000. Identification of the *Saccharomyces cerevisiae* RNA:pseudouridine synthase responsible for formation of $Ψ_{2819}$ in 21S mitochondrial ribosomal RNA. *Nucleic Acids Res.* **28**: 1941.

Ansmant I., Motorin Y., Massenet S., Grosjean H., and Branlant C. 2001. Identification and characterization of the tRNA:Ψ31-synthase (Pus6p) of *Saccharomyces cerevisiae*. *J. Biol. Chem.* **276**: 34934.

Bachellerie J.-P., Cavaillé J., and Qu L.-H. 2000. Nucleotide modification of eukaryotic rRNAs: The world of small nucleolar RNA guides revisited. In *The ribosome: Structure, function, antibiotics, and cellular interactions* (ed. R. Garrett et al.), p.191. ASM Press, Washington, D.C.

Bakin A. and Ofengand J. 1993. Four newly located pseudouridylate residues in *Escherichia coli* 23S ribosomal RNA are all at the peptidyl transferase center: Analysis by the application of a new sequencing technique. *Biochemistry* **32**: 9754.

————. 1995. Mapping of the thirteen pseudouridine residues in *Saccharomyces cerevisiae* small subunit ribosomal RNA to nucleotide resolution. *Nucleic Acids Res.* **23**: 3290.

Bakin A., Lane B.G., and Ofengand J. 1994a. Clustering of pseudouridine residues around the peptidyl transferase center of yeast cytoplasmic and mitochondrial ribosomes. *Biochemistry* **33**: 13475.

Bakin A., Kowalak J.A., McCloskey J.A., and Ofengand J. 1994b. The single pseudouridine residue in *Escherichia coli* 16S RNA is located at position 516. *Nucleic Acids Res.* **22**: 3681.

Ban N., Nissen P., Hansen J., Moore P.B., and Steitz T.A. 2000. The complete atomic structure of the large ribosomal subunit at 2.4 Å resolution. *Science* **289**: 905.

Bügl H., Fauman E.B., Staker B.L., Zheng F., Kushner S.R., Saper M.A., Bardwell J.C.A., and Jakob U. 2000. RNA methylation under heat shock control. *Mol. Cell* **6**: 349.

Caldas T., Binet E., Bouloc P., and Richarme G. 2000. Translational defect of *Escherichia coli* mutants deficient in the Um2552 23S ribosomal RNA methyltransferase RrmJ/FTSJ. *Biochem. Biophys. Res. Commun.* **271**: 714.

Charette M. and Gray M.W. 2000. Pseudouridine in RNA: What, where, how, and why. *IUBMB Life* **49**: 341.

Chen J. and Patton J.R. 1999. Cloning and characterization of a mammalian pseudouridine synthase. *RNA* **5:** 409.

————. 2000. Pseudouridine synthase 3 from mouse modifies the anticodon loop of tRNA. *Biochemistry* **39:** 12723.

Conrad J., Sun D., Englund N., and Ofengand J. 1998. The *rluC* gene of *Escherichia coli* codes for a pseudouridine synthase which is solely responsible for synthesis of pseudouridine at positions 955, 2504, and 2580 in 23S ribosomal RNA. *J. Biol. Chem.* **273:** 18562.

Conrad J., Niu L., Rudd K., Lane B.G., and Ofengand J. 1999. 16S ribosomal RNA pseudouridine synthase RsuA of *Escherichia coli:* Deletion, mutation of the conserved Asp102 residue, and sequence comparison among all other pseudouridine synthases. *RNA* **5:** 751.

Del Campo M., Kaya Y., and Ofengand J. 2001. Identification and site of action of the remaining four putative psuedouridine synthases in *Escherichia coli*. *RNA* **7:** 1603.

Douglas S., Zauner S., Fraunholz M., Beaton B., Penny S., Deng L.T., Wu X., Reith M., Cavalier-Smith T., and Maier U.G. 2001. The highly reduced genome of an enslaved algal nucleus. *Nature* **410:** 1091.

Felden B., Hanawa K., Atkins J.F., Himeno H., Muto A., Gesteland R.F., McCloskey J.A., and Crain P.F. 1998. Presence and location of modified nucleotides in *Escherichia coli* tmRNA: Structural mimicry with tRNA acceptor branches. *EMBO J.* **17:** 3188.

Gutgsell N.S., Del Campo M., Raychaudhuri S., and Ofengand J. 2001. A second function for pseudouridine synthases: A point mutant of RluD unable to form pseudouridines 1911, 1915, and 1917 in *Escherichia coli* 23S ribosomal RNA restores normal growth to an RluD-minus strain. *RNA* **7:** 990.

Gutgsell N., Englund N., Niu L., Kaya Y., Lane B.G., and Ofengand J. 2000. Deletion of the *Escherichia coli* pseudouridine synthase gene *truB* blocks formation of pseudouridine 55 in tRNA in vivo, does not affect exponential growth, but confers a strong selective disadvantage in competition with wild-type cells. *RNA* **6:** 1870.

Hoang C. and Ferre-D'Amare A.R. 2001. Cocrystal structure of a tRNA Psi55 pseudouridine synthase. Nucleotide flipping by an RNA-modifying enzyme. *Cell* **107:** 929.

Huang L., Pookanjanatavip M., Gu X., and Santi D.V. 1998a. A conserved aspartate of tRNA pseudouridine synthase is essential for activity and a probable nucleophilic catalyst. *Biochemistry* **37:** 344.

Huang L., Ku J., Pookanjanatavip M., Gu X., Wang D., Greene P.J., and Santi D.V. 1998b. Identification of two *Escherichia coli* pseudouridine synthases that show multisite specificity for 23S RNA. *Biochemistry* **37:** 15951.

Khan M.S.N. and Maden B.E.H. 1977. Nucleotide sequence relationships between vertebrate 5.8 S ribosomal RNAs. *Nucleic Acids Res.* **4:** 2495.

Koonin E.V. 1996. Pseudouridine synthases: Four families of enzymes containing a putative uridine-binding motif also conserved in dUTPases and dCTP deaminases. *Nucleic Acids Res.* **24:** 2411.

Kowalak J.A., Bruenger E., Crane P., and McCloskey J.A. 2000. Identities and phylogenetic comparisons of posttranscriptional modifications in 16 S ribosomal RNA from *Haloferax volcanii*. *J. Biol. Chem.* **275:** 24484.

Kowalak J.A., Bruenger E., Hashizume T., Peltier J.M., Ofengand J., and McCloskey J.A. 1996. Structural characterization of 3-methylpseudouridine in domain IV from *E. coli* 23S ribosomal RNA. *Nucleic Acids Res.* **24:** 688.

Maden B.E.H. 1990. The numerous modified nucleotides in eukaryotic ribosomal RNA. *Prog. Nucleic. Acid Res. Mol. Biol.* **39:** 241.

Massenet S., Ansmant I., Motorin Y., and Branlant C. 1999. The first determination of pseudouridine residues in 23S ribosomal RNA from hyperthermophilic Archaea *Sulfolobus acidocaldarius*. *FEBS Lett.* **462:** 94.

Nazar R.N, Sitz T.O., and Busch H. 1976. Sequence homologies in mammalian 5.8S ribosomal RNA. *Biochemistry* **15:** 505.

Nissen P., Hansen J., Ban N., Moore P.B., and Steitz T.A. 2000. The structural basis of ribosome activity in peptide bond synthesis. *Science* **289:** 920.

Niu L. and Ofengand J. 1999. Cloning and characterization of the 23S RNA pseudouridine 2633 synthase from *Bacillus subtilis*. *Biochemistry* **38:** 629.

Noon K.R., Bruenger E., and McCloskey J.A. 1998. Posttranscriptional modifications in 16S and 23S rRNAs of the archaeal hyperthermophile *Sulfolobus solfataricus*. *J. Bacteriol.* **180:** 2883.

Ofengand J. 2002. Ribosomal RNA pseudouridines and pseudouridine synthases. *FEBS Lett.* (in press).

Ofengand J. and Bakin A. 1997. Mapping to nucleotide resolution of pseudouridine residues in large subunit ribosomal RNAs from representative eukaryotes, prokaryotes, archaebacteria, mitochondria, and chloroplasts. *J. Mol. Biol.* **266:** 246.

Ofengand J. and Fournier M. 1998. The pseudouridine residues of rRNA: Number, location, biosynthesis, and function. In *Modification and editing of RNA.* (ed. H. Grosjean and R. Benne), p. 229. ASM Press, Washington, D.C.

Ofengand J. and Rudd K. 2000. The bacterial, archaeal, and organellar ribosomal RNA pseudouridines and methylated nucleosides and their enzymes. In *The ribosome: Structure, function, antibiotics, and cellular interactions* (ed. R. Garrett et al.), p.175. ASM Press, Washington, D.C.

Ogle J.M., Brodersen D.E., Clemons W.M., Jr., Tarry M.J., Carter A.P., and Ramakrishnan V. 2001. Recognition of cognate transfer RNA by the 30S ribosomal subunit. *Science* **292:** 897.

Ramamurthy V., Swann S.L., Paulson J.L., Spedaliere C.J., and Mueller E.G. 1999. Critical aspartic acid residues in pseudouridine synthases. *J. Biol. Chem.* **274:** 22225.

Raychaudhuri S., Conrad J., Hall B.G., and Ofengand J. 1998. A pseudouridine synthase required for the formation of two universally conserved pseudouridines in ribosomal RNA is essential for normal growth of *Escherichia coli*. *RNA* **4:** 1407.

Raychaudhuri S., Niu L., Conrad J., Lane B.G., and Ofengand J. 1999. Functional effect of deletion and mutation of the *Escherichia coli* ribosomal RNA pseudouridine synthase RluA. *J. Biol. Chem.* **274:** 18880.

Rubin G.M. 1973. The nucleotide sequence of *Saccharomyces cerevisiae* 5.8 S ribosomal ribonucleic acid. *J. Biol. Chem.* **248:** 3860.

Sirum-Connolly K.S., Peltier J.M., Crain P.F., McCloskey J.A., and Mason T.L. 1995. Implications of a functional large ribosomal RNA with only three modified nucleotides. *Biochimie* **77:** 30.

Szymanski M., Barciszewska M.Z., Barciszewski J., and Erdmann V.A. 2000. 5S ribosomal RNA database Y2K. *Nucleic Acids Res.* **28:** 166.

Wimberly B.T., Brodersen D.E., Clemons W.M., Jr., Morgan-Warren R.J., Carter A.P., Vonrhein C., Hartsch T., and Ramakrishnan V. 2000. Structure of the 30S ribosomal subunit. *Nature* **407:** 327.

Wrzesinski J., Bakin A., Nurse K., Lane B.G., and Ofengand J. 1995a. Purification, cloning, and properties of the 16S RNA Ψ516 synthase from *Escherichia coli*. *Biochemistry* **34:** 8904.

Wrzesinski J., Nurse K., Bakin A., Lane B.G., and Ofengand J. 1995b. A dual-specificity pseudouridine synthase: Purification and cloning of a synthase from *Escherichia coli* which is specific for both Ψ746 in 23S RNA and for Ψ32 in tRNA^Phe. *RNA* **1:** 437.

Yusupov M.M., Yusupova G.Z., Baucom A., Lieberman K., Earnest T.N., Cate J.H., and Noller H.F. 2001. Crystal structure of the ribosome at 5.5 Å resolution. *Science* **292:** 883.

Zebarjadian Y., King T., Fournier M.J., Clarke L., and Carbon J. 1999. Point mutations in yeast CBF5 can abolish in vivo pseudouridylation of rRNA. *Mol. Cell. Biol.* **19:** 7461.

Formation of Two Classes of tRNA Synthetases in Relation to Editing Functions and Genetic Code

P. Schimmel and L. Ribas de Pouplana

The Skaggs Institute for Chemical Biology, The Scripps Research Institute, Beckman Center,
La Jolla, California 92037

The genetic code is an algorithm that relates each of the 20 amino acids to specific nucleotide triplets. Its origin is not known, but any efforts to understand the development of the code centers on transfer RNAs and aminoacyl tRNA synthetases. The synthetases catalyze the aminoacylations of tRNAs that establish the algorithm of the code. In the aminoacylation reactions, each amino acid is linked to the tRNA that bears the anticodon triplet for that amino acid.

Typically, there is one tRNA synthetase for each of the 20 amino acids. The 20 enzymes are divided into two classes of 10 enzymes each (Webster et al. 1984; Ludmerer and Schimmel 1987; Cusack et al. 1990; Eriani et al. 1990). These classes are defined by two types of distinct architectures used to construct the catalytic sites. No relationship between these architectures has been identified, nor do sequences of the more than 700 synthetases from various organisms suggest that a common ancestor gave birth to the two distinct core structures. The catalytic domain is thought to represent the ancient, historical enzyme, with determinants for activation of the amino acid (formation of an aminoacyl adenylate) and with insertions for binding to the acceptor helix of the tRNA (Cusack et al. 1993; Delarue and Moras 1993; Hamann and Hou 1995; Martinis and Schimmel 1995; Schimmel and Ribas de Pouplana 1995). Indeed, helical substrates based on tRNA acceptor stems are active for specific aminoacylations by many of the tRNA synthetases (Francklyn and Schimmel 1989; Francklyn et al. 1992; Frugier et al. 1992, 1994; Nureki et al. 1993; Saks and Sampson 1996; Musier-Forsyth and Schimmel 1999). Later in evolution, additional motifs and domains were added to the core structures of the synthetases. These elements facilitated interactions with parts of the tRNA distal to the acceptor end.

Thus, attempts to understand the historical development of the code have some justification in focusing on interactions of the catalytic domains of the synthetases with tRNA acceptor stems. Here we summarize recent results which suggest that such interactions may form the basis for the two classes of tRNA synthetases. At the same time, the editing functions of tRNA synthetases now appear to be critical for refining the code from an ambiguous to a precise state. The timing of the introduction of the domains that encodes the editing functions, and the way that these domains combine with the ancient, catalytic domain, are examined here as a means to understand a critical step in the refinement of the genetic code.

SUBCLASSES AND CROSS-CLASS PAIRINGS OF SYNTHETASES ON ACCEPTOR STEMS

Using a composite of sequences for each of the 20 enzymes, trees for class I and, separately, for class II enzymes can be constructed (Fig. 1). These two trees show that the enzymes in each class can be divided into subclasses, based on grouping those enzymes that have the closest relationships. The largest to the smallest of these subclasses are designated as a, b, and c, respectively. Thus, class Ia is made up of 6 enzymes, class Ib of 3 (including lysyl-tRNA synthetase which with a minor exception [Ibba et al. 1997] is a class II enzyme), and class Ic of 2 enzymes. These subclasses group together many of the enzymes that have specificities for the same types of amino acid side chains.

Class IIa, like class Ia, is made up of six synthetases, with three in class IIb and one in class IIc. Here again, some common features can be seen among the amino acid specificities associated with members of each subclass. These features are shared by their counterparts in the subclasses of class I. For example, subclass IIc contains phenylalanine, which is aromatic, like the two members of subclass Ic. The relationship between amino acids associated with members of Ib and IIb (glutamate versus aspartate, glutamine versus asparagine) is also obvious. As for the large subclasses Ia and Ib, the connections are somewhat looser, perhaps because of the inherent diversity that occurs when more than half of the amino acids are in only two subclasses. Still, the similar shapes of cysteine versus serine and of threonine versus valine are apparent. The hydrophobic nature of alanine and proline of class IIa is held in common with the hydrophobic nature of several of the subclass Ia amino acids.

Interestingly, many of the subclass Ia and IIa enzymes are specific for amino acids that are listed in the first two columns of the table of codons—class Ia amino acids in column 1 where the second base of the triplet is U and class Ib amino acids in column 2 where the second base is C. This and other observations that link groups of codons with specific synthetase subclasses suggest a potential connection between the subclasses and the organization of the genetic code. To understand the possible origin of this connection, specific functional characteristics of members of the two classes are useful to consider. In par-

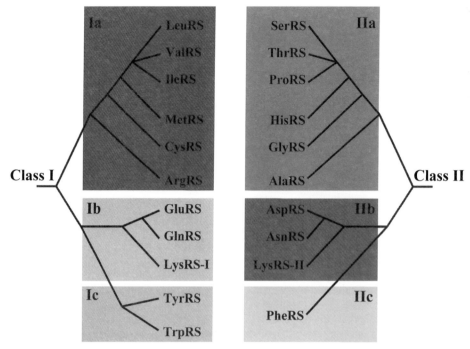

Figure 1. Tree of aminoacyl tRNA synthetases showing the lineage of the six subclasses. Although a class I LysRS has been detected (LysRS-I), this enzyme is typically class II (LysRS-II).

ticular, class I enzymes aminoacylate tRNA on the 2′-OH, whereas those in class II catalyze aminoacylation to the 3′-OH (Eriani et al. 1990). This distinction arises because class I enzymes bind to the minor groove side of the acceptor stem, whereas those in class II bind to the major groove side (Rould et al. 1989; Ruff et al. 1991; Carter 1993; Biou et al. 1994; Cusack 1997).

This difference in binding raised the possibility that synthetases from opposite classes could have sterically compatible dockings on the tRNA acceptor stem. At least one crystal structure of a synthetase–tRNA complex from each of the subclasses has been determined. These structures show significant differences, between subclasses, in the precise way a synthetase fits onto the acceptor stem. At the same time, all members of a given subclass are thought to bind in the same way to the acceptor stem. The differences in the ways that specific enzymes orient on the acceptor stem suggested that pairings of synthetases on acceptor stems would not be arbitrary.

Considering just the class-defining catalytic domain, sterically compatible dockings of two synthetases on one acceptor stem were identified (Ribas de Pouplana and Schimmel 2001). These dockings are not random (Fig. 2), but instead link enzymes in subclass Ia with those in subclass IIa, enzymes in Ib with those in IIb, and link Ic with IIc. Most significant is the coupling of Ic with IIc. In the case of tyrosyl-tRNA synthetase, the enzyme is bound in an unusual way, relative to other class I enzymes. This unique orientation is exactly compensated by the way that phenylalanyl-tRNA synthetase is bound to the acceptor stem. In fact, tyrosyl-tRNA synthetase cannot be paired with any class II enzyme, other than phenylalanyl-tRNA synthetase.

SIGNIFICANCE OF SUBCLASS-SPECIFIC PAIRINGS

The organization of synthetases into subclasses, with similar amino acid types within a given subclass, is consistent with the possibility that the early code was ambiguous or imprecise. For example, the progenitor of subclass Ia may have activated several of the 6 amino acids in that subclass and assigned them more or less equivalently to a group of early codons. (Not all of the 20 contemporary amino acids may have been present, and those that were may have varied in concentration so that some were favored over others.) The result would be "rough" or heterogeneous mixtures of proteins synthesized from the same coding strand of RNA. These rough proteins would have to have had the capability to form crude active sites and RNA-binding domains.

That class-defining domains of tRNA synthetases can in principle make specific, sterically compatible dockings

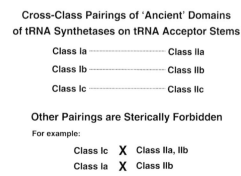

Figure 2. Summary of specific pairings of active-site domains that are possible on tRNA acceptor stems.

on tRNA acceptor stems suggests that early enzymes may have acted as chaperones, to cover and protect the tRNA acceptor stem in an environment (e.g., thermophilic) deleterious to RNA (Ribas de Pouplana and Schimmel 2001). The earliest aminoacylations may have been done by ribozymes (Piccirilli et al. 1992; Illangasekare et al. 1995; Illangasekare and Yarus 1999; Lee et al. 2000), with the chaperone-like synthetases gradually taking over that role. (As the code became less ambiguous, more accurate proteins including synthetases were made that could outperform ribozymes.) Eventually, the synthetases added other domains that interact with other parts of the tRNA, distal to the acceptor stem (Delarue and Moras 1993; Schimmel and Ribas de Pouplana 1995). Addition of these domains would lead to steric clashes and force apart the synthetase pairs.

TRANSITION FROM AN AMBIGUOUS TO A PRECISE CODE

Ambiguity occurs because certain amino acids cannot be bound at active sites with the high specificity demanded by the code, where errors of mistranslation occur at frequencies estimated as less than 0.03% (Loftfield and Vanderjagt 1972). Several examples can be demonstrated within subclasses Ia and Ib. Because the isopropyl side chain of valine fits into a pocket designed for the isobutyl side chain of isoleucine, isoleucyl-tRNA synthetase misactivates valine (to form valyl-adenylate) with a frequency of about one part in 100–200 (Baldwin and Berg 1966; Schmidt and Schimmel 1994). Similarly, valine is isosteric with threonine, so that valyl-tRNA synthetase misactivates threonine (Fersht 1985; Lin and Schimmel 1996). Threonyl-tRNA synthetase activates serine (because its side chain easily fits into the pocket for threonine; Dock-Bregeon et al. 2000), whereas proly-tRNA synthetase activates alanine (Beuning and Musier-Forsyth 2000).

The misactivated adenylates or mis-aminoacyl esters fused to tRNA are cleared by an editing activity. For example, misacylated Val-tRNA^{Ile} and Thr-tRNA^{Val} are deacylated by isoleucyl- and valyl-tRNA synthetases (Eldred and Schimmel 1972; Fersht 1985), respectively. These hydrolytic activities are encoded by a domain (known as CP1) that is inserted into the active site and bulges out from the body of the enzyme as a distinct lobe (Fig. 3) (Schmidt and Schimmel 1995; Lin et al. 1996; Hale and Schimmel 1997; Nureki et al. 1998). The active site is separated from the center for editing by about 30 Å, so that the misactivated adenylate or aminoacyl moiety on tRNA must be translocated from one site to the other. The translocation process requires specific nucleotides within the tRNA that are distinct from those needed for aminoacylation (Hale et al. 1997; Farrow et al. 1999). Translocation is rate-limiting for the editing reaction.

The genetic code can be made ambiguous by disruption of the center for editing. For example, if the center for editing in valyl-tRNA synthetase is disrupted, both valine and threonine will be inserted at a codon for valine (Doring et al. 2001). To test whether the center for editing is essential for cell viability, an editing-defective T222P mutation in the CP1 editing domain of valyl-tRNA syn-

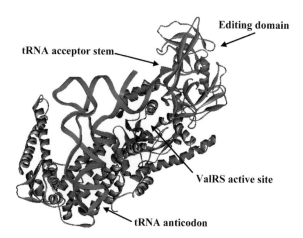

Figure 3. Ribbon diagram of ValRS (*blue*)-tRNA^{Val} (*green*) complex, with editing domain highlighted in red. (Adapted from Fukai et al. 2000.)

thetase was constructed. This mutation enables facile mischarging of tRNA^{Val} with threonine (Fig. 4). As a consequence, exogenously added threonine is toxic to *Escherichia coli* cells bearing a gene for valyl-tRNA synthetase that encodes the T222P substitution.

The T222P mutation in valyl-tRNA synthetase affords an opportunity to determine conditions, if any, where a modern cell can survive when it synthesizes an ensemble of rough proteins. Each protein would be a heterogeneous product resulting from an ambiguous code. For this purpose, cells bearing the T222P mutation in valyl-tRNA synthetase were studied with α-amino butyrate (Abu). This amino acid is a metabolite that is not normally incorporated into proteins. Because its butyl side chain can fit into the pocket for the isopropyl group of valine, Abu is misactivated by valyl-tRNA synthetase and subsequently cleared by the editing reaction. Like threonine, exogenously added Abu is toxic to cells harboring the T222P mutation in valyl-tRNA synthetase. By adjusting the concentration of Abu to just below the threshold

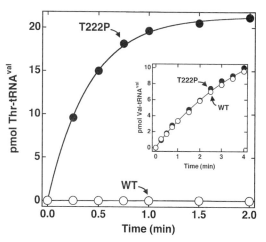

Figure 4. Mischarging of tRNA^{Val} with threonine, using a mutant ValRS that has a T222P mutation in the CP1 editing domain. (Adapted, with permission, from Doring et al. 2001 [copyright AAAS].)

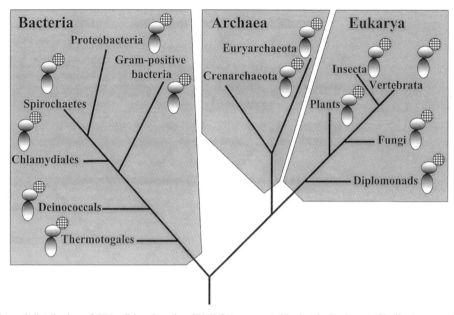

Figure 5. Universal distribution of CP1 editing domain of ValRS (represented by hatched spheres). Similar trees can be constructed for the CP1 domains of the closely related IleRS and LeuRS.

where cell death occurs, proteins that incorporate Abu were synthesized in vivo.

Amino acid analysis of hydrolysates of whole-cell proteins isolated from *E. coli* cells harboring the T222P-encoding chromosomal mutation demonstrated that 24% of the valines were replaced with Abu. Mass spectrometry of tryptic peptides from a digestion of a protein purified from these cells proved that Abu was inserted specifically at valine codons (Doring et al. 2001). Thus, the precision of the modern genetic code is sharply dependent on the editing function of tRNA synthetases. Surprisingly, however, even a modern cell can tolerate a high population of rough, imprecise proteins. This result is consistent with the idea that an early genetic code that was ambiguous could support viable living systems.

APPEARANCE OF EDITING ACTIVITY

Each of the six subclasses of tRNA synthetases arose from a progenitor that eventually differentiated into the several distinct enzymes of the subclass. These six progenitors may have activated the entire set of amino acids associated with their subclass, or activated amino acids similar to those that eventually were settled upon. Under these circumstances, the early code would have a high degree of ambiguity. In the tree of Figure 1, the appearance of isoleucyl-, leucyl-, and valyl-tRNA synthetases is a radiation that split out the three enzymes of subclass Ia. A key step was the introduction of an editing function that enabled each of these enzymes to distinguish between closely similar amino acids.

Distribution analysis of the CP1 editing domain (found in isoleucyl-, leucyl-, and valyl-tRNA synthetases) shows clearly that this domain is conserved throughout evolution and is ancient (Fig. 5). For example, CP1 is found in all three of the domains of life—bacteria, eukarya, and ar-

chaea. *Aquifex aeolicus* is near the base of the bacterial kingdom and harbors the CP1 domain in the three aforementioned tRNA synthetases. Similarly, *Aeropyrum pernix* is near the base of the archaea kingdom and harbors the CP1 domain. This analysis is consistent with the possibility that the CP1 domain was added no later than at or near the time of appearance of the last common ancestor.

If the editing domain was added early and at the time when compatible pairs of synthetase were bound to tRNA acceptor stems, it would have to be added in a way that was sterically possible. The CP1 editing domain fits into a complex of subclass 1a valyl-tRNA synthetase with subclass IIa threonyl-tRNA synthetase (Fig. 6), without causing any steric clashes. (An equivalent complex can

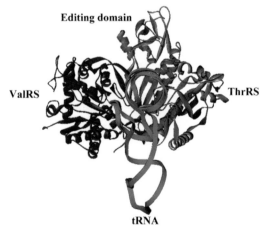

Figure 6. Three-domain complex of catalytic domain of ValRS and ThrRS fit onto acceptor stem, with CP1 domain added in. Structures of ValRS and ThrRS are based on Dock-Bregeon et al. (2000) and Fukai et al. (2000), respectively. The structure of the CP1 domain is based on Fukai et al. (2000).

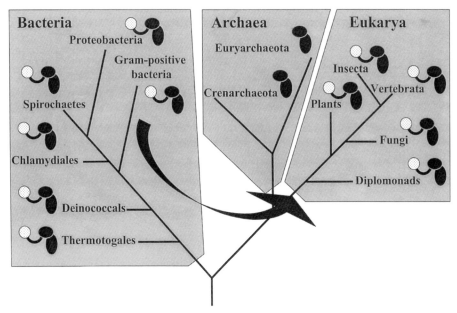

Figure 7. Distribution of editing domain of class II ThrRS (represented by dotted spheres). This domain lacks the universal distribution seen for the CP1 domain of class I enzymes that is shown in Fig. 5.

be made with isoleucyl-tRNA synthetase and its CP1 domain replacing valyl-tRNA synthetase.) Thus, the editing activity could in principle be added to the emerging synthetase at a time when two enzymes were bound to a single acceptor stem.

In contrast to the CP1 editing domain of class I tRNA synthetases, the unique editing domains of the class II threonyl- and proly-tRNA synthetases are not universally distributed. For example, the editing domain of threonyl-tRNA synthetase is absent from available sequences of members of the Archaea kingdom (Fig. 7). Its presence in eubacteria and eukaryotes is consistent with the possibility that lateral gene exchanges facilitated its present-day distribution in these two kingdoms (Stehlin et al. 1998; Woese et al. 2000; Ribas de Pouplana et al. 2001). In the complex of two synthetases bound to a single acceptor stem, the editing domain of ThrRS clashes with that of a class I active site. Thus, the editing domain for at least some class II tRNA synthetases may have been added after the splitting apart of the ancestral pairs.

ACKNOWLEDGMENTS

This work was supported by grants GM-15539 and GM-23562 from the National Institutes of Health and by a fellowship from the National Foundation for Cancer Research. We thank Leslie Nangle for providing a copy of Figure 4.

REFERENCES

Baldwin A.N. and Berg P. 1966. Transfer ribonucleic acid-induced hydrolysis of valyladenylate bound to isoleucyl ribonucleic acid synthetase. *J. Biol. Chem.* **241:** 839.
Beuning P.J. and Musier-Forsyth K. 2000. Hydrolytic editing by a class II aminoacyl-tRNA synthetase. *Proc. Natl. Acad. Sci.* **97:** 8916.
Biou V., Yaremchuk A., Tukalo M., and Cusack S. 1994. The 2.9 Å crystal structure of *T. thermophilus* seryl-tRNA synthetase complexed with tRNA^Ser. *Science* **263:** 1404.
Carter C.W., Jr. 1993. Cognition, mechanism, and evolutionary relationships in aminoacyl-tRNA synthetases. *Annu. Rev. Biochem.* **62:** 715.
Cusack S. 1997. Aminoacyl-tRNA synthetases. *Curr. Opin. Struct. Biol.* **7:** 881.
Cusack S., Berthet-Colominas C., Hartlein M., Nassar N., and Leberman R. 1990. A second class of synthetase structure revealed by X-ray analysis of *Escherichia coli* seryl-tRNA synthetase at 2.5 Å. *Nature* **347:** 249.
Cusack S., Berthet-Colominas C., Biou V., Borel F., Fujinaga M., Hartlein M., Krikliviy I., Nassar N., Price S., Tukalo M. A., Yaremchuk A.D., and Leberman R. 1993. The crystal structure of seryl-tRNA synthetase and its complexes with ATP and tRNA^Ser. In *The translation apparatus* (ed. K.H. Nierhaus et al.), p. 1. Plenum Press, New York.
Delarue M. and Moras D. 1993. The aminoacyl-tRNA synthetase family: Modules at work. *BioEssays* **15:** 675.
Dock-Bregeon A., Sankaranarayanan R., Romby P., Caillet J., Springer M., Rees B., Francklyn C. S., Ehresmann C., and Moras D. 2000. Transfer RNA-mediated editing in threonyl-tRNA synthetase. The class II solution to the double discrimination problem. *Cell* **103:** 877.
Doring V., Mootz H.D., Nangle L.A., Hendrickson T.L., de Crecy-Lagard V., Schimmel P., and Marliere P. 2001. Enlarging the amino acid set of *Escherichia coli* by infiltration of the valine coding pathway. *Science* **292:** 501.
Eldred E.W. and Schimmel P.R. 1972. Rapid deacylation by isoleucyl transfer ribonucleic acid synthetase of isoleucine-specific transfer ribonucleic acid aminoacylated with valine. *J. Biol. Chem.* **247:** 2961.
Eriani G., Delarue M., Poch O., Gangloff J., and Moras D. 1990. Partition of tRNA synthetases into two classes based on mutually exclusive sets of sequence motifs. *Nature* **347:** 203.
Farrow M.A., Nordin B.E., and Schimmel P. 1999. Nucleotide determinants for tRNA-dependent amino acid discrimination by a class I tRNA synthetase. *Biochemistry* **38:** 16898.
Fersht A. 1985. *Enzyme structure and mechanism,* 2nd edition.

W.H. Freeman, New York.

Francklyn C. and Schimmel P. 1989. Aminoacylation of RNA minihelices with alanine. *Nature* **337:** 478.

Francklyn C., Shi J.-P., and Schimmel P. 1992. Overlapping nucleotide determinants for specific aminoacylation of RNA microhelices. *Science* **255:** 1121.

Frugier M., Florentz C., and Giegé R. 1992. Anticodon-independent aminoacylation of an RNA minihelix with valine. *Proc. Natl. Acad. Sci.* **89:** 3990.

———. 1994. Efficient aminoacylation of resected RNA helices by class II aspartyl-tRNA synthetase dependent on a single nucleotide. *EMBO J.* **13:** 2218.

Fukai S., Nureki O., Sekine S., Shimada A., Tao J., Vassylyev D.G., and Yokoyama S. 2000. Structural basis for double-sieve discrimination of L-valine from L-isoleucine and L-threonine by the complex of tRNA(Val) and valyl-tRNA synthetase. *Cell* **103:** 793.

Hale S.P. and Schimmel P. 1997. DNA aptamer targets translational editing motif in a tRNA synthetase. *Tetrahedron* **53:** 11985.

Hale S.P., Auld D.S., Schmidt E., and Schimmel P. 1997. Discrete determinants in transfer RNA for editing and aminoacylation. *Science* **276:** 1250.

Hamann C.S. and Hou Y.-M. 1995. Enzymatic aminoacylation of tRNA acceptor stem helices with cysteine is dependent on a single nucleotide. *Biochemistry* **34:** 6527.

Ibba M., Morgan S., Curnow A.W., Pridmore D.R., Vothknecht U.C., Gardner W., Lin W., Woese C.R., and Söll D. 1997. Euryarchael lysyl-tRNA synthetase: Resemblance to class I synthetases. *Science* **278:** 1119.

Illangasekare M. and Yarus M. 1999. Specific, rapid synthesis of Phe-RNA by RNA. *Proc. Natl. Acad. Sci.* **96:** 5470.

Illangasekare M., Sanchez G., Nickles T., and Yarus M. 1995. Aminoacyl-RNA synthesis catalyzed by an RNA. *Science* **267:** 643.

Lee N., Bessho Y., Wei K., Szostak J.W., and Suga H. 2000. Ribozyme-catalyzed tRNA aminoacylation. *Nat. Struct. Biol.* **7:** 28.

Lin L. and Schimmel P. 1996. Mutational analysis suggests the same design for editing activities of two tRNA synthetases. *Biochemistry* **35:** 5596.

Lin L., Hale S.P., and Schimmel P. 1996. Aminoacylation error correction. *Nature* **384:** 33.

Loftfield R.B. and Vanderjagt D. 1972. The frequency of errors in protein biosynthesis. *Biochem. J.* **128:** 1353.

Ludmerer S.W. and Schimmel P. 1987. Gene for yeast glutamine tRNA synthetase encodes a large amino terminal extension and provides strong confirmation of the signature sequence for a group of aminoacyl tRNA synthetases. *J. Biol. Chem.* **262:** 10801.

Martinis S.A. and Schimmel P. 1995. Small RNA oligonucleotide substrates for specific aminoacylations. In *tRNA: Structure, biosynthesis and function* (ed. D. Söll and U.L. RajBhandary), p. 349. American Society for Microbiology, Washinton, D.C.

Musier-Forsyth K. and Schimmel P. 1999. Atomic determinants for aminoacylation of RNA minihelices and relationship to genetic code. *Accts. Chem. Res.* **32:** 368.

Nureki O., Vassylyev D.G., Tateno M., Shimada A., Nakama T., Fukai S., Konno M., Hendrickson T.L., Schimmel P., and Yokoyama S. 1998. Enzyme structure with two catalytic sites for double-sieve selection of substrate (comments). *Science* **280:** 578.

Nureki O., Niimi T., Muto Y., Kanno H., Kohno T., Muramatsu T., Kawai G., Miyazawa T., Giege R., Florentz C., and Yokoyama S. 1993. Conformational change of tRNA upon interaction of the identity-determinant set with aminoacyl-tRNA synthetase. In *The translation apparatus* (ed. K.H. Nierhaus et al.), p. 59. Plenum Press, New York.

Piccirilli J.A., McConnell T. S., Zaug A.J., Noller H.F., and Cech T.R. 1992. Aminoacyl esterase activity of the *Tetrahymena* ribozyme. *Science* **256:** 1420.

Ribas de Pouplana L. and Schimmel P. 2001. Two classes of tRNA synthetases suggested by sterically compatible dockings on tRNA acceptor stem. *Cell* **104:** 191.

Ribas de Pouplana L., Brown J.R., and Schimmel P. 2001. Structure-based phylogeny of class IIa tRNA synthetases in relation to an unusual biochemistry. *J. Mol. Evol.* **53:** 261.

Rould M.A., Perona J.J., Söll D., and Steitz T.A. 1989. Structure of *E. coli* Glutaminyl-tRNA synthetase complexed with tRNA[Gln] and ATP at 2.8 Å resolution. *Science* **246:** 1135.

Ruff M., Krishnaswamy S., Boeglin M., Poterszman A., Mitschler A., Podjarny A., Rees B., Thierry J. C., and Moras D. 1991. Class II aminoacyl transfer RNA synthetases: Crystal structure of yeast aspartyl-tRNA synthetase complexed with tRNA[Asp]. *Science* **252:** 1682.

Saks M.E. and Sampson J.R. 1996. Variant minihelix RNAs reveal sequence-specific recognition of the helical tRNA[Ser] acceptor stem by *E. coli* seryl-tRNA synthetase. *EMBO J.* **15:** 2843.

Schimmel P. and Ribas de Pouplana L. 1995. Transfer RNA: From minihelix to genetic code. *Cell* **81:** 983.

Schmidt E. and Schimmel P. 1994. Mutational isolation of a sieve for editing in a transfer RNA synthetase. *Science* **264:** 265.

———. 1995. Residues in a class I tRNA synthetase which determine selectivity of amino acid recognition in the context of tRNA. *Biochemistry* **34:** 11204.

Stehlin C., Burke B., Yang F., Liu H., Shiba K., and Musier-Forsyth K. 1998. Species-specific differences in the operational RNA code for aminoacylation of tRNA[Pro]. *Biochemistry* **37:** 8605.

Webster T.A., Tsai H., Kula M., Mackie G.A., and Schimmel P. 1984. Specific sequence homology and three-dimensional structure of an aminoacyl transfer RNA synthetase. *Science* **226:** 1315.

Woese C.R., Olsen G.J., Ibba M., and Söll D. 2000. Aminoacyl-tRNA synthetases, the genetic code, and the evolutionary process. *Microbiol. Mol. Biol. Rev.* **64:** 202.

Structural Basis for Amino Acid and tRNA Recognition by Class I Aminoacyl-tRNA Synthetases

O. Nureki,*†‡ S. Fukai,*† S. Sekine,† A. Shimada,*†‡ T. Terada,*‡ T. Nakama,*†
M. Shirouzu,†‡ D.G. Vassylyev,† and S. Yokoyama*†‡

*Department of Biophysics and Biochemistry, Graduate School of Science, the University of Tokyo, 7-3-1 Hongo,
Bunkyo-ku, Tokyo 113-0033, Japan; †Cellular Signaling Laboratory, RIKEN Harima Institute at SPring-8,
1-1-1 Kohto, Mikazuki-cho, Sayo-gun, Hyogo 679-5148, Japan; and ‡RIKEN Genomic Sciences Center,
1-7-22 Suehiro-cho, Tsurumi, Yokohama 230-0045, Japan

Aminoacyl-tRNA synthetase (aaRS) catalyzes the ester bond formation between the cognate amino acid and tRNA(s). The aminoacylation of tRNA is carried out in a two-step reaction, the formation of an active intermediate (aminoacyl-adenylate or aa-AMP) from the amino acid and ATP, and the transfer of the aminoacyl moiety of the aa-AMP to the 3′ terminal adenosine (A76) of the tRNA. Strict recognition of both the amino acid and the tRNA by the aaRS ensures the correct translation of the genetic code.

Some amino acids are similar to one another, which requires the strict fitting of the cognate amino acid to the binding pocket in the aaRS active site. In the most difficult cases, valine has a branched aliphatic chain highly similar to that of isoleucine and also resembles threonine, even more than isoleucine, with respect to shape and size, regardless of hydrophobicity. Because the differences in the properties of these amino acids are not large enough to achieve sufficiently strict discrimination by ordinary single-step mechanisms, several aaRSs have a sophisticated two-step mechanism for amino acid discrimination: Incorrectly formed aminoacyl tRNA (and aa-AMP) is eliminated in the second step (proofreading or editing) (Fersht et al. 1985; Lin et al. 1996; Dock-Bregeon et al. 2000).

To distinguish the cognate tRNA from noncognate tRNAs, aaRS recognizes nucleotide residues that constitute the "identity set" of the tRNA (Giegé et al. 1998). In most tRNAs, the identity set consists of a small number of elements (or bases) concentrated in the anticodon and/or the acceptor stem (Giegé et al. 1998). aaRS should strictly recognize all of the identity elements, and its identity set may be overlapped partly with some identity sets for different amino acid specificities. Therefore, strict discrimination requires not only the correct recognition of each element, but also the recognition of the proper spatial arrangement of the elements on the common tRNA framework (the two-dimensional cloverleaf model and the three-dimensional L shape). In particular, the anticodon is situated in a flexible single-stranded loop region farthest from the aminoacyl terminus of the tRNA, suggesting that the enzyme has a sophisticated mechanism to locate the anticodon properly on the protein. Therefore, the tertiary structures of tRNAs and aaRSs are essential for strict molecular recognition.

The 20 aaRSs are divided into two structurally unrelated classes, I and II (Eriani et al. 1990), with only one exception (Fabrega et al. 2001). Both classes have 10 members each. Class I aaRSs have a catalytic domain characterized by the Rossmann fold, as well as by the consensus His-Ile-Gly-His (HIGH) and Lys-Met-Ser-Lys-Ser (KMSKS) motifs (Cusack 1995). On the other hand, class II aaRSs have a catalytic domain constructed with an antiparallel β sheet with three signature motifs (Cusack 1995). Class I is further divided into three subclasses, Ia, Ib, and Ic, on the basis of sequence homology and domain architecture (Cusack 1995). Class Ia consists of the isoleucyl-, methionyl-, valyl-, leucyl-, cysteinyl-, and arginyl-tRNA synthetases (IleRS, MetRS, ValRS, LeuRS, CysRS, and ArgRS, respectively); class Ib includes the glutamyl- and glutaminyl-tRNA synthetases (GluRS and GlnRS, respectively); and class Ic is composed of the tyrosyl- and tryptophanyl-tRNA synthetases (TyrRS and TrpRS, respectively). In most archaea, many bacteria, and some organelles, GlnRS is missing and is replaced by GluRS for misacylation of tRNA^{Gln} with Glu, and then by a transamidase for conversion of Glu-tRNA^{Gln} to Gln-tRNA^{Gln} (Tumbula et al. 2000). Furthermore, the lysyl-tRNA synthetases from many archaea and bacteria are of class I (LysRS-I), whereas all other organisms have a class II lysyl-tRNA synthetase (LysRS-II).

Three-dimensional structures of class I aaRSs have been determined for TyrRS (Brick et al. 1989), GlnRS (Rould et al. 1989), GluRS (Nureki et al. 1995), TrpRS (Doublie et al. 1995), IleRS (Nureki et al. 1998), MetRS (Mechulam et al. 1999; Sugiura et al. 2000), ArgRS (Cavarelli et al. 1998; Shimada et al. 2001), ValRS (Fukai et al. 2000), LeuRS (Cusack et al. 2000), and LysRS-I (T. Terada et al., unpubl.) (The CysRS structure has not yet been determined.) For GlnRS, IleRS, GluRS, ArgRS, ValRS, and TyrRS, the tRNA-bound structures have been reported. Class Ic enzymes, TyrRS and TrpRS, form a homodimer and have different tRNA-binding modes from those of class Ia and class Ib enzymes. We have been working mainly on class Ia and class Ib aaRSs from an extreme thermophile, *Thermus thermophilus* HB8. In this paper, the idiosyncratic mechanisms of amino acid and tRNA recognition are described on the basis of our crystal structures of *T. thermophilus* MetRS, IleRS, ValRS, ArgRS, and GluRS, as well as LysRS-I from a hyperthermophilic archaeon, *Pyrococcus horikoshii*, with respect to the combination of the common Rossmann-fold domain and the diversified additional domains.

DOMAIN ARCHITECTURES OF CLASS Ia AND CLASS Ib aaRSs

Figure 1 shows the crystal structures of four class Ia enzymes, MetRS, ArgRS, IleRS, and ValRS. Figure 2 displays those of three class Ib enzymes, GluRS, GlnRS, and LysRS-I. The Rossmann fold (colored gray) has a $\beta_5\alpha_4$ topology, which is well conserved among the class Ia and Ib aaRSs. The amino acid- and ATP-binding sites and one of the two class-I-characterizing motifs, the HIGH sequence, are located in this Rossmann-fold domain. The Rossmann-fold domain is followed by a small domain named the SC-fold domain (colored cyan), which contacts rather widely the inner side of the L-shaped tRNA molecule and displays the other class-I-characterizing motif, the KMSKS sequence. This SC-fold domain is also to locate the carboxy-terminal anticodon-binding domain(s). The class Ia aaRSs have an α-helix bundle domain (colored red) for anticodon recognition. In contrast, GlnRS has β-rich domains, and GluRS and LysRS-I have α-helical domains, including the carboxy-terminal "helix cage" domain, for anticodon recognition (colored blue in

Fig. 2). The structural similarity between GluRS and LysRS-I suggests that these two enzymes have a common ancestor. ArgRS has the amino-terminal domain (colored green) for the recognition of A20 in the D loop. IleRS and ValRS (class Ia) have, on top of the Rossmann-fold domain, the CP1 domain for amino acid editing.

RECOGNITION OF AMINO ACIDS

We have reported the crystal structures of the *T. thermophilus* IleRS in the isoleucine-bound form (Nureki et al. 1998) and the Ile-AMS-bound form (Ile-AMS, 5′-*N*-[*N*-(L-isoleucyl)sulfamoyl] adenosine is a nonhydrolyzable analog of Ile-AMP, and the –O–P–O– linkage between the isoleucyl and adenylate moieties is replaced by a –NH–S–NH– linkage) (Nakama et al. 2001), and that of the *T. thermophilus* ValRS bound with Val-AMS and tRNA^{Val} (Fukai et al. 2000). As shown in Figure 3, Ile-AMS and Val-AMS are bound in deep catalytic clefts of the central Rossmann-fold domains of IleRS and ValRS, respectively. In the IleRS•Ile-AMS complex structure (Fig. 3) (Nakama et al. 2001), the α-amino and α-carbonyl groups of the isoleucyl moiety hydrogen-bond to the main-chain carbonyl group of Pro-46 and the side-chain ε-amino group of Gln-554, respectively. In the IleRS•isoleucine complex structure (Nureki et al. 1998), the isoleucine interacts with IleRS in almost the same manner as that of Ile-AMS, except that the amino group

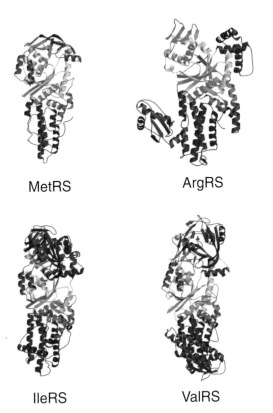

MetRS   ArgRS

IleRS   ValRS

Figure 1. Structures of class Ia aminoacyl-tRNA synthetases, methionyl-, arginyl-, isoleucyl-, and valyl-tRNA synthetases (MetRS, ArgRS, IleRS, and ValRS, respectively) from *Thermus thermophilus* HB8. (Reprinted from O. Nureki et al., in prep.)

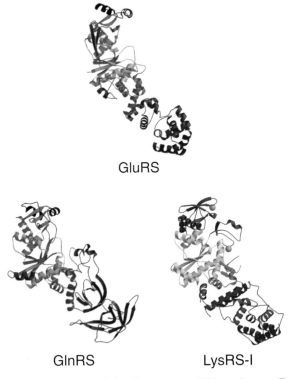

GluRS

GlnRS   LysRS-I

Figure 2. Structures of class Ib aminoacyl-tRNA synthetases, *T. thermophilus* glutamyl-tRNA synthetase (GluRS), *Escherichia coli* glutaminyl-tRNA synthetase (GlnRS), and *Pyrococcus horikoshii* lysyl-tRNA synthetase (LysRS-I). The GlnRS structure is that in a complex with the cognate tRNA. (Reprinted from O. Nureki et al., in prep.)

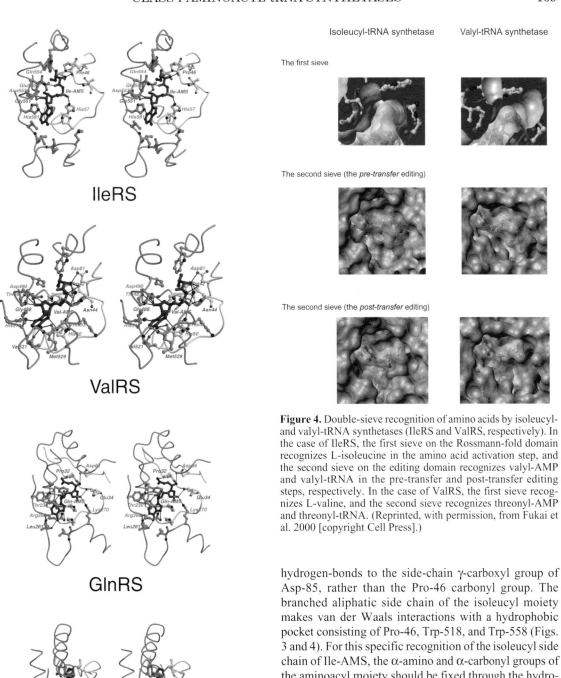

IleRS

ValRS

GlnRS

TyrRS

Figure 3. Comparison of the molecular recognition of the cognate aa-AMP analogs by the Rossmann-fold domains of the *T. thermophilus* IleRS (Nakama et al. 2001), the *T. thermophilus* ValRS (Fukai et al. 2000), the *E. coli* GlnRS (Rath et al. 1998), and the *B. stearothermophilus* TyrRS (Brick et al. 1989). The bound aa-AMP analogs are shown in green. The amino acid residues that recognize the aa-AMP analogs are indicated by ball-and-stick models. (Reprinted, with permission, from Nakama et al. 2001 [copyright ABMB].)

Isoleucyl-tRNA synthetase   Valyl-tRNA synthetase

The first sieve

The second sieve (the *pre-transfer* editing)

The second sieve (the *post-transfer* editing)

Figure 4. Double-sieve recognition of amino acids by isoleucyl- and valyl-tRNA synthetases (IleRS and ValRS, respectively). In the case of IleRS, the first sieve on the Rossmann-fold domain recognizes L-isoleucine in the amino acid activation step, and the second sieve on the editing domain recognizes valyl-AMP and valyl-tRNA in the pre-transfer and post-transfer editing steps, respectively. In the case of ValRS, the first sieve recognizes L-valine, and the second sieve recognizes threonyl-AMP and threonyl-tRNA. (Reprinted, with permission, from Fukai et al. 2000 [copyright Cell Press].)

hydrogen-bonds to the side-chain γ-carboxyl group of Asp-85, rather than the Pro-46 carbonyl group. The branched aliphatic side chain of the isoleucyl moiety makes van der Waals interactions with a hydrophobic pocket consisting of Pro-46, Trp-518, and Trp-558 (Figs. 3 and 4). For this specific recognition of the isoleucyl side chain of Ile-AMS, the α-amino and α-carbonyl groups of the aminoacyl moiety should be fixed through the hydrogen bonds described above.

In the ValRS•Val-AMS complex structure (Fukai et al. 2000), the α-amino group of the valyl moiety of Val-AMS hydrogen-bonds not only with the main-chain carbonyl group of Pro-42, which corresponds to Pro-46 in IleRS, but also with the γ-carboxyl group of Asp-81 and the γ-carbonyl group of Asn-44 (Fig. 3). On the other hand, the α-carbonyl group of the valyl moiety of Val-AMS does not hydrogen-bond with any protein residue. The aliphatic side chain of the valyl moiety fits into a hydrophobic pocket consisting of Pro-41, Pro-42, Trp-456, Ile-491, and Trp-495 of ValRS (Figs. 3 and 4) (Pro-42, Trp-456, and Trp-495 of ValRS correspond to Pro-46, Trp-518, and Trp-558 of IleRS). The hydrophobic pocket of ValRS is significantly smaller than that of IleRS, mainly because of the bulky Pro-41 in ValRS in place of

Gly-45 in IleRS. Therefore, the ValRS pocket is just as large as the valyl moiety and is too small for the isoleucyl moiety, whereas the IleRS pocket is just as large as the isoleucyl moiety. Interestingly, the interactions of the α-amino and α-carbonyl groups of the aminoacyl moiety with the active site of the enzyme are fine-tuned, corresponding to the difference in the side-chain binding mode between the valyl and isoleucyl moieties of Val-AMS and Ile-AMS, respectively.

The aminoacyl-adenylate-bound structures have been reported for the *T. thermophilus* LeuRS (Cusack et al. 2000), the *Escherichia coli* GlnRS (Fig. 3) (Rath et al. 1998), the *Bacillus stearothermophilus* TyrRS (Fig. 3) (Brick et al. 1989), and the *B. stearothermophilus* TrpRS (Doublie et al. 1995; Ilyin et al. 2000). The arginine-bound structures of yeast ArgRS have also been determined (Cavarelli et al. 1998). In the class Ia and Ib aaRSs, the α-amino group of the aminoacyl moiety hydrogen-bonds to the main-chain carbonyl group of the residue corresponding to Pro-46 in the *T. thermophilus* IleRS. Furthermore, the α-amino group of the aminoacyl moiety hydrogen-bonds to the side chain of the aspartic acid residue highly conserved in most of the class Ia and Ib aaRSs. Within this common framework, the interactions of the aminoacyl moiety with the active site are fine-tuned, according to the idiosyncratic recognition of the amino acid side chain. On the other hand, in the class Ic TyrRS (Fig. 3) and TrpRS, the hydrogen-bonding interaction of the α-amino group of the aminoacyl moiety is different from those in the class Ia and Ib enzymes.

The substrate-binding pockets for hydrolysis of incorrectly formed aminoacyl-tRNA (substrate for the "post-transfer" editing) and aminoacyl-AMP (substrate for the "pre-transfer" editing) have been shown by the IleRS•valine complex and ValRS•tRNA^{Val}•Val-AMS complex structures. The valine molecule is bound to the CP1 (or editing) domain in the IleRS•valine complex, whereas 3′-A (A76) of the tRNA^{Val} is bound to the editing domain in the ValRS•tRNA^{Val}•Val-AMS complex. The structures of the editing domains of *T. thermophilus* IleRS and ValRS are similar to each other, and can be superposed (rmsd of C$_\alpha$ atoms 1.75 Å over 157 residues). The superposition of the IleRS and ValRS editing domains allowed us to simply model A76 of tRNA^{Ile} onto the IleRS•valine complex, and the threonyl moiety onto the ValRS•tRNA^{Val}•Val-AMS complex (Fig. 4, middle). The models are likely to represent post-transfer editing by IleRS and ValRS. With regard to post-transfer editing, the valyl-binding pocket of *T. thermophilus* IleRS is formed by the side chains of Thr-233, His-319, and Ala-321, and the main chains of Phe-324 and Gly-325. The valyl-binding pocket of IleRS for post-transfer editing is just as large as valine, but is too small for isoleucine. On the other hand, the threonyl-binding pocket of *T. thermophilus* ValRS is formed by the hydrophilic side chains of Arg-216, Thr-219, Lys-270, Thr-272, Asp-276, and Asp-279. Our model suggests that the γ-COO$^-$ group of Asp-328 hydrogen-bonds with the characteristic γ-OH group of the threonyl moiety. The valyl moiety should be prevented from binding to this hydrophilic pocket, due to

the lack of proper hydrogen-bonding with the δ-O atom of Asp-328. The ValRS and IleRS aminoacyl-binding pockets for post-transfer editing clearly show the difference in the substrate selection manners at the post-transfer editing step. In addition to aminoacyl-binding pockets for post-transfer editing, we pointed out those for "pre-transfer editing" (Fig. 4, bottom), on the assumption that the adenosyl-binding manners of incorrectly formed aminoacyl-tRNA and aminoacyl-AMP are the same. The putative aminoacyl-binding pockets of IleRS and ValRS for pre-transfer editing also reflect the difference in the substrate selection manners.

RECOGNITION OF tRNA

We have determined the crystal structure of the *T. thermophilus* GluRS•tRNA^{Glu} complex (Sekine et al. 2001), as well as the tRNA-free GluRS structure (Nureki et al. 1995). The tRNA^{Glu} anticodon loop maintains the canonical U-turn structure of the anticodon loop and the stacking of the anticodon bases, C34–U35–C36 of tRNA^{Glu}. The anticodon is recognized by the carboxy-terminal α-helical anticodon-binding domains (domains 4 and 5) of GluRS (Fig. 5). In contrast, the two carboxy-terminal domains of GlnRS, which recognize the tRNA^{Gln} anticodon, have completely distinct β-barrel architectures, whereas the anticodon bases are unstacked for independent recognition by GlnRS (Fig. 5) (Rould et al. 1989, 1991). In the GluRS•tRNA^{Glu} complex, C34 and U35 (the first and second bases of the anticodon) are recognized by many amino acid residues of domain 5, or the "α-helix cage" domain (Fig. 3) (Nureki et al. 1995).

In contrast, C36 (the third anticodon base) is recognized by an arginine side chain (Arg-358) protruding from domain 4, through the cytosine-specific hydrogen bonds with O2 and N3 (Fig. 3). This Arg-358 residue is therefore the major determinant for the discrimination of the Glu-type anticodon (^{34}YUC^{36}) from the Glu type (^{34}YUG^{36}), as the bulky guanine base would cause steric hindrance with the arginine side chain. The *T. thermophilus* GluRS is a "discriminating" GluRS, which aminoacylates only tRNA^{Glu}, but not tRNA^{Gln}. On the other hand, "non-discriminating" GluRSs from grampositive bacteria that synthesize gln-tRNA^{Gln} through the transamidation pathway have glutamine at the position corresponding to Arg-358 of the *T. thermophilus* GluRS. In contrast to arginine, the smaller glutamine side chain would not cause steric hindrance with the bulky G36, and its polar side chain may allow recognition of both cytosine and guanine as the third anticodon base (Fig. 5). Actually, the R358Q mutant GluRS aminoacylates the wild-type (C36) and variant (C36G) tRNA^{Glu} molecules with equal efficiency. This mutant GluRS with a relaxed specificity toward both C36 and G36 is a mimic of the nondiscriminating GluRSs, which aminoacylate both tRNA^{Glu} and tRNA^{Gln} with glutamate.

As described above, the structure of the class I LysRS (LysRS-I) is similar to that of the discriminating GluRS. In particular, the α-helix cage domains of GluRS and LysRS-I are highly homologous to each other. Therefore,

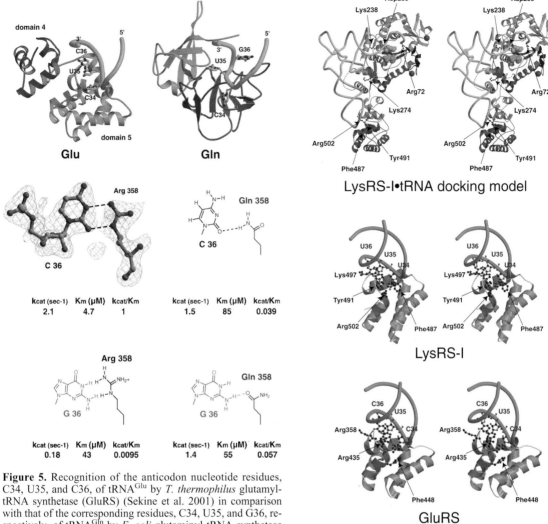

Figure 5. Recognition of the anticodon nucleotide residues, C34, U35, and C36, of tRNA^Glu by *T. thermophilus* glutamyl-tRNA synthetase (GluRS) (Sekine et al. 2001) in comparison with that of the corresponding residues, C34, U35, and G36, respectively, of tRNA^Gln by *E. coli* glutaminyl-tRNA synthetase (GlnRS) (*top*) (Rould et al. 1991). Recognitions of C36 and G36 of the tRNA^Glu wild type and mutant, respectively, by Arg-358 and Gln-358 of the Glu-RS wild type and mutant, respectively (*bottom*) (Sekine et al. 2001). (Reprinted, with permission, from Sekine et al. 2001 [copyright Nature Publishing Group].)

Figure 6. Recognition of the anticodon by the carboxy-terminal, helical domains of class Ib glutamyl- and lysyl-tRNA synthetases (GluRS and LysRS-I, respectively). (Reprinted from T. Terada et al., in prep.)

on the basis of the structure of the GluRS•tRNA^Glu complex, that of the LysRS-I•tRNA^Lys complex was modeled (Fig. 6). Both tRNA^Glu and tRNA^Lys have C34 and U35, which may be recognized by the common α-helix cage domain in the same manner (Fig. 6). On the other hand, the tRNA^Lys-characterizing base, U36, is possibly recognized by the neighboring domain with a topology largely different from that of domain 4 of GluRS.

The anticodon recognition mechanisms were also compared between the ValRS•tRNA^Val (Fukai et al. 2000) and IleRS•tRNA^Ile (Silvian et al. 1999) structures (see Fig. 7). ValRS recognizes all of the valine-accepting tRNA species having the identity elements, A35–C36, in common. These two residues are recognized by the α-helix bundle domain, which characterizes class Ia (Fig. 7). In contrast, G, C, and a modified U in position 34 of the tRNA^Val species are not recognized by ValRS,

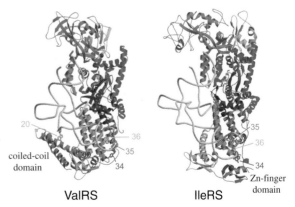

Figure 7. Interactions of *T. thermophilus* ValRS (Fukai et al. 2000) and *Staphylococcus aureus* IleRS (Silvian et al. 1999) with tRNA. (Reprinted from S. Fukai et al., in prep.)

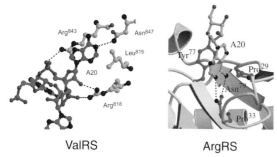

VaIRS   ArgRS

Figure 8. Interactions of *T. thermophilus* VaIRS (crystal structure; Fukai et al. 2000) and ArgRS (model; Shimada et al. 2001) with A20 of tRNA. (Reprinted, with permission, from Shimada et al. 2001 [copyright National Academy of Science].)

which is again in agreement with the crystal structure. On the other hand, the bacterial tRNA[Ile] species have G or lysidine (a modified C; Muramatsu et al. 1988) in position 34 (the first position of the anticodon) and the A35–U36 sequence. Because tRNA[Met] species have the same A35–U36 sequence, IleRS should recognize all of the three nucleotides of the anticodon, in order to discriminate tRNA[Ile] from tRNA[Met]. Correspondingly, IleRS has an additional domain, which is not present in VaIRS. This carboxy-terminal domain has a Zn-finger and actually recognizes the base in position 34 (Fig. 7).

In the case of VaIRS, there is a coiled-coil domain at its carboxyl terminus. The elongated shape of the coiled-coil domain is to support the back of the L-shaped tRNA molecule (Fig. 7) and contact A20 in the D loop (Fig. 8). Aminoacyl-tRNA synthetases bind to the inner side of the L-shaped tRNA molecule, whereas position 20 is near the outer corner of the L shape. Therefore, recognition of position 20 requires an additional domain. The only case where the nucleotide in position 20 is a major element for the tRNA identity is tRNA[Arg]. ArgRS has an additional domain at the amino terminus, which is protruded from the anticodon-recognizing helix bundle domain to the left (Fig. 1). This amino-terminal domain has a β-sheet surface for recognition of the A20 base strictly conserved in most organisms except for yeast. As shown in Figure 8, a model of the A20 recognition was made on the basis of the structures of the yeast ArgRS•tRNA complex (Delagoutte et al. 2000) and the A20-recognizing ArgRS from *T. thermophilus* (Shimada et al. 2001).

ACKNOWLEDGMENT

The authors thank T. Nakayama for her assistance in preparation of the manuscript.

REFERENCES

Brick P., Bhat T.N., and Blow D.M. 1989. Structure of tyrosyl-tRNA synthetase refined at 2.3 Å resolution. Interaction of the enzyme with the tyrosyl adenylate intermediate. *J. Mol. Biol.* **208**: 83.
Cavarelli J., Delagoutte B., Eriani G., Gangloff J., and Moras D. 1998. L-arginine recognition by yeast arginyl-tRNA synthetase. *EMBO J.* **17**: 5438.
Cusack S. 1995. Eleven down and nine to go. *Nat. Struct. Biol.*

2: 824.
Cusack S., Yaremchuk A., and Tukalo M. 2000. The 2 Å crystal structure of leucyl-tRNA synthetase and its complex with a leucyl-adenylate analog. *EMBO J.* **19**: 2351.
Delagoutte B., Moras D., and Cavarelli J. 2000. tRNA aminoacylation by arginyl-tRNA synthetase: Induced conformations during substrates binding. *EMBO J.* **19**: 5599.
Dock-Bregeon A., Sankaranarayanan R., Romby P., Caillet J., Springer M., Rees B., Francklyn C.S., Ehresmann C., and Moras D. 2000. Transfer RNA-mediated editing in threonyl-tRNA synthetase. The class II solution to the double discrimination problem. *Cell* **103**: 877.
Doublie S., Bricogne G., Gilmore C., and Carter C.W., Jr. 1995. Tryptophanyl-tRNA synthetase crystal structure reveals an unexpected homology to tyrosyl-tRNA synthetase. *Structure* **3**: 17.
Eriani G., Delarue M., Poch O., Gangloff J., and Moras D. 1990. Partition of tRNA synthetases into two classes based on mutually exclusive sets of sequence motifs. *Nature* **347**: 203.
Fabrega C., Farrow M.A., Mukhopadhyay B., de Crecy-Lagard V., Ortiz A.R., and Schimmel P. 2001. An aminoacyl tRNA synthetase whose sequence fits into neither of the two known classes. *Nature* **411**: 110.
Fersht A.R. 1985. *Enzyme structure and mechanism.* W.H. Freeman, New York.
Fukai S., Nureki O., Sekine S., Shimada A., Tao J., Vassylyev D.G., and Yokoyama S. 2000. Structural basis for double-sieve discrimination of L-valine from L-isoleucine and L-threonine by the complex of tRNA[Val] and valyl-tRNA synthetase. *Cell* **103**: 793.
Giegé R., Sissler M., and Florentz C. 1998. Universal rules and idiosyncratic features in tRNA identity. *Nucleic Acids Res.* **26**: 5017.
Ilyin V.A., Temple B., Hu M., Li G., Yin Y., Vachette P., and Carter C.W., Jr. 2000. 2.9 Å crystal structure of ligand-free tryptophanyl-tRNA synthetase: Domain movements fragment the adenine nucleotide binding site. *Protein Sci.* **9**: 218.
Lin L., Hale S.P., and Schimmel P. 1996. Aminoacylation error correction. *Nature* **384**: 33.
Mechulam Y., Schmitt E., Maveyraud L., Zelwer C., Nureki O., Yokoyama S., Konno M., and Blanquet S. 1999. Crystal structure of *Escherichia coli* methionyl-tRNA synthetase highlights species-specific features. *J. Mol. Biol.* **294**: 1287.
Muramatsu T., Nishikawa K., Nemoto F., Kuchino Y., Nishimura S., Miyazawa T., and Yokoyama S. 1988. Codon and amino-acid specificities of a transfer RNA are both converted by a single post-transcriptional modification. *Nature* **336**: 179.
Nakama T., Nureki O., and Yokoyama S. 2001. Structural basis for the recognition of isoleucyl-adenylate and an antibiotic, mupirocin, by isoleucyl-tRNA synthetase. *J. Biol. Chem.* **276**: 47387.
Nureki O., Vassylyev D.G., Katayanagi K., Shimizu T., Sekine S., Kigawa T., Miyazawa T., Yokoyama S., and Morikawa K. 1995. Architectures of class-defining and specific domains of glutamyl-tRNA synthetase. *Science* **267**: 1958.
Nureki O., Vassylyev D.G., Tateno M., Shimada A., Nakama T., Fukai S., Konno M., Hendrickson T.L., Schimmel P., and Yokoyama S. 1998. Enzyme structure with two catalytic sites for double-sieve selection of substrate. *Science* **280**: 578.
Rath V.L., Silvian L.F., Beijer B., Sproat B.S., and Steitz T.A. 1998. How glutaminyl-tRNA synthetase selects glutamine. *Structure* **6**: 439.
Rould M.A., Perona J.J., and Steitz T.A. 1991. Structural basis of anticodon loop recognition by glutaminyl-tRNA synthetase. *Nature* **352**: 213.
Rould M.A., Perona J.J., Soll D., and Steitz T.A. 1989. Structure of *E. coli* glutaminyl-tRNA synthetase complexed with tRNA[Gln] and ATP at 2.8 Å resolution. *Science* **246**: 1135.
Sekine S., Nureki O., Shimada A., Vassylyev D.G., and Yokoyama S. 2001. Sturctural basis for anticodon recognition by discriminating glutamyl-tRNA synthetase. *Nat. Struct. Biol.* **8**: 203.
Shimada A., Nureki O., Goto M., Takahashi S., and Yokoyama

S. 2001. Structural and mutational studies of the recognition of the arginine tRNA-specific major identity element, A20, by arginyl-tRNA synthetase. *Proc. Natl. Acad. Sci.* **98:** 13537.

Silvian L.F., Wang J., and Steitz T.A. 1999. Insights into editing from an Ile-tRNA synthetase sructure with tRNA[Ile] and mupirocin. *Science* **285:** 1074.

Sugiura I., Nureki O., Ugaji-Yoshikawa Y., Kuwabara S., Shi-

mada A., Tateno M., Lorber B., Giegé R., Moras D., Yokoyama S., and Konno M. 2000. The 2.0 Å crystal structure of *Thermus thermophilus* methionyl-tRNA synthetase reveals two RNA-binding modules. *Struct. Fold. Des.* **8:** 197.

Tumbula D.L., Becker H.D., Chang W.Z., and Soll D. 2000. Domain-specific recruitment of amide amino acids for protein synthesis. *Nature* **407:** 106.

Aminoacyl-tRNA Synthesis: A Postgenomic Perspective

C. STATHOPOULOS, I. AHEL, K. ALI,† A. AMBROGELLY, H. BECKER, S. BUNJUN,* L. FENG,
S. HERRING, C. JACQUIN-BECKER, H. KOBAYASHI, D. KORENCIC, B. KRETT,
N. MEJLHEDE,‡ B. MIN, H. NAKANO, S. NAMGOONG, C. POLYCARPO, G. RACZNIAK,
J. RINEHART, G. ROSAS-SANDOVAL, B. RUAN, J. SABINA, A. SAUERWALD, H. TOOGOOD,
D. TUMBULA-HANSEN, M. IBBA,‡ AND D. SÖLL*

*Departments of Molecular Biophysics and Biochemistry, *Chemistry, and †Pharmacology, Yale University, New Haven, Connecticut 06520-8114; ‡Center for Biomolecular Recognition, Department of Medical Biochemistry and Genetics, Laboratory B, The Panum Institute, Copenhagen, Denmark*

The ribosome requires a full complement of correctly aminoacylated tRNAs in order to perform mRNA-templated protein synthesis. The aminoacyl-tRNA synthetase (AARS) family of proteins comprises the principal catalysts of aminoacyl-tRNA (AA-tRNA) formation. The ability of an AARS to discriminate between multiple substrates and to accurately attach a particular amino acid to its cognate tRNA species is a universal requirement for the faithful translation of genetic information into the sequence of a polypeptide. In the pregenomic era it was a generally held belief, as originally propounded in the adaptor hypothesis (Crick 1958), that virtually every cell and organelle would contain 20 AARSs, with one particular enzyme responsible for the synthesis of each of the 20 AA-tRNA isoacceptor species required for mRNA translation. The identification in *Escherichia coli* of two distinct AARS classes, each containing 10 members, provided seemingly incontrovertible support for the "20 AARS" facet of the adaptor hypothesis. This view of AA-tRNA synthesis was initially confounded when the genomes of two methanogenic archaea were sequenced and annotated (Bult et al. 1996; Smith et al. 1997). A variety of analyses consistently identified genes encoding only 16 AARSs, raising the question of how these organisms synthesize asparaginyl tRNA (Asn-tRNA), cysteinyl tRNA (Cys-tRNA), glutaminyl tRNA (Gln-tRNA), and lysyl tRNA (Lys-tRNA). The identification of the alternative pathways by which these four AA-tRNAs are made has provided unexpected insight into the structural and functional diversity of AA-tRNA synthesis not envisaged within the structures of the 20 AARS hypothesis. Furthermore, functional genomics and comparative phylogenetic studies have shown that, far from being idiosyncrasies of methanogenic archaea, these alternative pathways of AA-tRNA synthesis are widespread in the living kingdom. To date, three novel activities have been characterized: an unusual class I lysyl-tRNA synthetase for Lys-tRNA formation (Ibba et al. 1997c), tRNA-dependent amidation enzymes for Asn-tRNA and Gln-tRNA synthesis (Tumbula et al. 2000), and a dual-specificity prolyl-tRNA synthetase generating Cys-tRNA and Pro-tRNA (Stathopoulos et al. 2000). Here we describe the current state of knowledge on these unusual routes for AA-tRNA synthesis.

CLASS I LYSYL-tRNA SYNTHETASES

Initial analyses of the genomic sequences of *Methanococcus jannaschii* and *Methanothermobacter thermautotrophicus* failed to identify any open reading frames encoding a lysyl-tRNA synthetase (LysRS). The possibility that LysRS might be absent from these and related archaea was not supported by the observation that cell-free extracts of both *Haloferax volcanii* and *Methanococcus maripaludis* were able to directly attach lysine to tRNA (M. Ibba and A.W. Curnow, unpubl.). This assay was used as the basis for the purification of *M. maripaludis* LysRS, the amino-terminal sequencing of which allowed the cloning of the corresponding gene (originally annotated as *lysS*, now renamed *lysK*; M. Ibba and J. Krzycki, pers. comm.). Heterologous expression of this gene in *E. coli* gave rise to a protein with canonical LysRS activity. The sequence of the *M. maripaludis lysK* gene showed no similarity to known class II LysRS-encoding genes, but was highly homologous to unassigned open reading frames (ORFs) in all of the archaeal genomes that lacked an identifiable LysRS (Ibba et al. 1997c). Interestingly, this group of archaeal *lysK* genes all encoded proteins containing sequence motifs characteristic of class I aminoacyl-tRNA synthetases, in contrast to all other then-known LysRS proteins that were members of class II. Subsequent biochemical studies supported the assignment of the *lysK*-encoded protein as a class I-type LysRS (Ibba et al. 1999).

An aminoacyl-tRNA synthetase of particular substrate specificity will normally belong to either class I or class II regardless of its biological origin, reflecting the ancient evolution of this enzyme family (Woese et al. 2000). The first exception to this rule was found among the lysyl-tRNA synthetases with the discovery of a class I enzyme in certain archaea (Ibba et al. 1997c), all previously characterized members of this family belonging to class II. Subsequent work originating from analysis of whole-

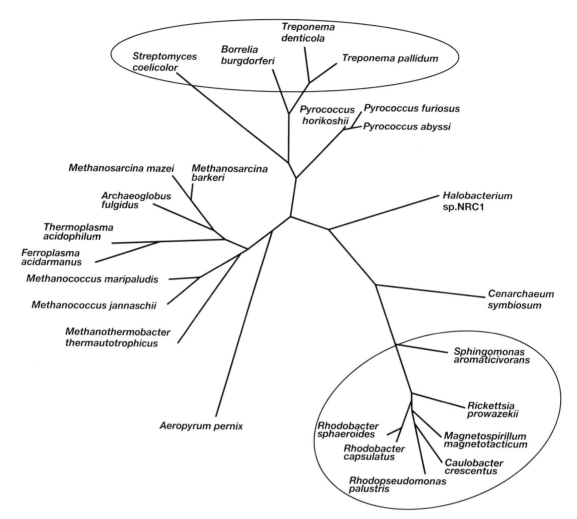

Figure 1. Unrooted phylogeny of LysRS1 proteins. All known examples are from archaea, except those within the ovals, which are bacterial. The phylogeny was constructed by using the maximum likelihood method (5000 puzzling steps) implemented in the program Tree-Puzzle 4.0.2 under the JTT model.

genome sequences showed that the class I-type LysRS (LysRS1) is found in the majority of archaea and a scattering of bacteria, to the exclusion of the more common class II-type protein (LysRS2; Ibba et al. 1997b, Söll et al. 2000). To date, 25 organisms have been identified whose genomes encode LysRS1 (Fig. 1). Only one organism, *Methanosarcina barkeri*, has so far been found to encode both LysRS1 and LysRS2. This organism differentially expresses the corresponding genes in response to changes in growth conditions (J. Krzycki, pers. comm.). Not all archaea possess a class I-type LysRS. For example, the crenarchaeotes *Pyrobaculum aerophilum* and *Sulfolobus solfataricus* both contain only class II-type LysRS proteins. Closer examination of these crenarchaeal sequences shows that they are more similar to each other than to any other LysRS and that they have reasonably high homology with bacterial proteins (e.g., both show >43% identity at the amino acid level to *Thermotoga maritima* LysRS2).

All known eukaryal LysRSs (both cytoplasmic and organellar) are of the class II-type. Despite their lack of sequence similarity, LysRS1 and LysRS2 are able to recognize the same amino acid and tRNA substrates both in vitro and in vivo, providing an example of functional convergence by divergent enzymes (Ibba et al. 1999). Comparison of tRNA^{Lys} recognition by LysRS1 and LysRS2 proteins indicated that although they approach the acceptor stem from different sides, both recognize the discriminator base, the acceptor stem, and the anticodon in tRNA^{Lys}. Moreover, tRNA sequence comparisons do not allow the assignment of specific features for tRNA^{Lys} recognition by either LysRS1 or LysRS2 (with the exception of the bacterial spirochete LysRS1 proteins; Ibba et al. 1999). The cross-reactivity of bacterial tRNAs and archaeal tRNAs with both classes of LysRS is also an indication that the tRNA identity sets are partially overlapping. This, together with phylogenetic analyses of the class I LysRS family, suggests that tRNA^{Lys} may predate at least one of the LysRS families in the evolution of AA-tRNA synthesis (Ribas de Pouplana et al. 1998; Ibba et al. 1999).

tRNA-DEPENDENT AMIDATIONS

AA-tRNA formation proceeds with exquisite specificity regarding both substrates (amino acid and tRNA) catalyzed by the AARSs, some of which have sophisticated editing functions to prevent errors (Jakubowski and Goldman 1992). Among the AARSs, the activities of asparaginyl-tRNA synthetase (AsnRS) or glutaminyl-tRNA synthetase (GlnRS) may be completely absent in an organism. These organisms then harbor one or two nondiscriminating AARSs. Such synthetases have relaxed RNA substrate recognition and can acylate two different tRNAs. For instance, it has been shown that in *Bacillus subtilis* and *M. thermautotrophicus,* a nondiscriminating GluRS acylates efficiently with glutamate the noncognate $tRNA^{Gln}$ in addition to the cognate $tRNA^{Glu}$ to generate $Glu-tRNA^{Gln}$ (Lapointe et al. 1986; Tumbula et al. 2000). Similarly, a nondiscriminating AspRS, often found in archaea, charges both $tRNA^{Asp}$ and $tRNA^{Asn}$ with aspartate. Whereas sequence-based phylogeny has failed to identify structural differences between discriminating and nondiscriminating GluRS, the structural dissimilarity between AspRSs is correlated with functional divergences (Becker et al. 2000a). Bacterial-like AspRSs only aspartylate $tRNA^{Asp}$, whereas archaeal-like AspRSs aspartylate $tRNA^{Asp}$ and $tRNA^{Asn}$ with similar efficiencies (Becker et al. 2000a). The reason for the existence of these enzymes is described below.

tRNA-dependent Amidotransferases Substitute for Two AARSs

In many bacteria and archaea, $Asn-tRNA^{Asn}$ and $Gln-tRNA^{Gln}$ are formed in a two-step indirect pathway involving first a misaminoacylating AspRS or GluRS and then a tRNA amidotransferase (AdT) (Wilcox and Nirenberg 1968; Curnow et al. 1996). In the first step, the mischarged $Asp-tRNA^{Asn}$ (Eq. 1) and $Glu-tRNA^{Gln}$ (Eq. 3) are produced as described above; they constitute the substrates of tRNA-dependent amidotransferase (AdT) enzymes which, in the presence of an amide donor (e.g., glutamine) convert these misacylated tRNAs to $Asn-tRNA^{Asn}$ (Eq. 2) and $Gln-tRNA^{Gln}$ (Eq. 4), respectively:

$$Asp + tRNA^{Asn} + ATP \rightleftharpoons$$
$$Asp-tRNA^{Asn} + AMP + PP_i \qquad (1)$$

$$Asp-tRNA^{Asn} + Asn + ATP \rightleftharpoons$$
$$Asn-tRNA^{Asn} + Asp + ADP + P_i \qquad (2)$$

$$Glu + tRNA^{Gln} + ATP \rightleftharpoons$$
$$Glu-tRNA^{Gln} + AMP + PP_i \qquad (3)$$

$$Glu-tRNA^{Gln} + Gln + ATP \rightleftharpoons$$
$$Gln-tRNA^{Gln} + Glu + ADP + P_i \qquad (4)$$

(In Eq. 2, ATP hydrolysis in this reaction is patterned after what has been shown for Eq. 4.) This is the only route to Asn-tRNA or Gln-tRNA in organisms that lack AsnRS or GlnRS.

Structure and Roles of the tRNA Amidotransferases

Two distinct species-specific AdT enzymes have been described. Thus far, regardless of the source, the heterotrimeric GatCAB amidotransferase forms $Asn-tRNA^{Asn}$ or $Gln-tRNA^{Gln}$ in vitro in the presence of the misaminoacylated substrate (Curnow et al. 1998; Becker et al. 2000b; Tumbula et al. 2000). Therefore, the role of the enzyme in vivo is believed to depend on the presence of the misaminoacylating AspRS or GluRS in the cell, at least in bacteria (see below). However, this has not been verified in vivo by substituting the direct aminoacylation pathway with the transamidation route.

The second class of AdT, the heterodimeric GatDE, has so far been found only in archaea (Tumbula et al. 2000). Unlike GatCAB, GatDE is capable of generating only $Gln-tRNA^{Gln}$. The distribution of the two AdTs in archaea suggests that the role of the archaeal GatCAB is to form $Asn-tRNA^{Asn}$, because GatCAB is present only in archaea that also lack AsnRS (Table 1). Likewise, whereas GlnRS is absent from archaea, GatDE is ubiquitous in this domain, signifying that GatDE is the major $Gln-tRNA^{Gln}$-forming enzyme. However, the presence of both AdTs in several archaea makes GatDE appear redundant in some archaea. Future in vivo studies of the archaeal AdTs may provide the physiological basis for separation of the Asn- and Gln-tRNA-forming pathways in this domain.

Possible Function of Individual Subunits of GatCAB and GatDE

Sequence alignments indicate that the GatA subunit is homologous to amidases and that GatD is similar to type I L-asparaginases (Curnow et al. 1997; Tumbula et al. 2000). Both amidases and asparaginases belong to the glutamine amidotransferase superfamily that catalyzes glutamine hydrolysis and subsequent transfer of ammonia to a wide variety of amide acceptors (for review, see Zalkin and Smith 1998). Therefore, GatCAB and GatDE have each recruited a distinct glutaminase to activate ammonia for the transamidation reaction. The crystal structure of a type II L-asparaginase has been solved, in which the asparaginase catalytic triad Thr-Asp-Lys was found in the aspartate-binding site (Swain et al. 1993; Lubkowski et al. 1996). However, neither GatA nor GatD carries the amino-terminal cysteine characteristic of the Ntn-type glutaminases nor the Cys-His-Glu catalytic triad found in the "Triad" class. Instead, site-directed mutagenesis of the amidase of *Rhodococcus* sp. has demonstrated that the serine and aspartate residues within the amidase signature motif (DTGGS) are important for activity (Kobayashi et al. 1997). Mutational analysis of homologous residues in GatA and GatD should reveal whether they play similar roles in the AdTs.

GatB and GatE are ~30% identical, but a 140-amino-acid domain found in GatE is absent from GatB (Tumbula et al. 2000). The functions of GatB and GatE are unclear, but both are essential for the overall transamidation reac-

Table 1. Presence of Direct and Indirect Routes of Amide Aminoacyl-tRNA Synthesis

| | | | | Presence of the activity or a homologous ORF | | |
| | | | | gatCAB | | |
Organism	Taxonomic group	asnS	glnS	Asp-AdT[a]	Glu-AdT[a]	gatDE
Aeropyrum pernix	Crenarchaeota	–	–	+	+	+
Sulfolobus solfataricus	Crenarchaeota	–	–	+	+	+
Sulfolobus tokodaii	Crenarchaeota	–	–	+	+	+
Archaeoglobus fulgidus	Euryarchaeota	–	–	+	+	+
Halobacterium salinarum	Euryarchaeota	–	–	+	+	+
Methanothermobacter thermautotrophicus[b]	Euryarchaeota	–	–	+	+	+
Methanococcus jannaschii	Euryarchaeota	–	–	+	+	+
Methanosarcina mazei	Euryarchaeota	–	–	+	+	+
Aquifex aeolicus	Aquificales	–	–	+	+	–
Thermotoga maritima	Thermotogales	–	–	+	+	–
Mycobacterium tuberculosis	Firmicutes	–	–	+	+	–
Chlamydia muridarum	Chlamydiales	–	–	+	+	–
Chlamydia trachomatis[b]	Chlamydiales	–	–	+	+	–
Chlamydia pneumoniae	Chlamydiales	–	–	+	+	–
Chlamydophila pneumoniae	Chlamydiales	–	–	+	+	–
Campylobacter jejuni	ε-Proteobacteria	–	–	+	+	–
Helicobacter pylori[b]	ε-Proteobacteria	–	–	+	+	–
Rickettsia prowazekii	α-Proteobacteria	–	–	+	+	–
Pyrobaculum aerophilum	Crenarchaeota	+	–	–	–	+
Pyrococcus abyssi	Euryarchaeota	+	–	–	–	+
Pyrococcus furiosus	Euryarchaeota	+	–	–	–	+
Pyrococcus horikoshii	Euryarchaeota	+	–	–	–	+
Thermoplasma acidophilum	Crenarchaeota	+	–	–	–	+
Bacillus halodurans	Firmicutes	+	–	–	+	–
Bacillus subtilis[b]	Firmicutes	+	–	–	+	–
Mycoplasma genitalium	Firmicutes	+	–	–	+	–
Mycoplasma pneumoniae	Firmicutes	+	–	–	+	–
Ureaplasma urealyticum	Firmicutes	+	–	–	+	–
Clostridium acetobutylicum	Firmicutes	+	–	–	+	–
Lactobacillus delbrueckii	Firmicutes	+	–	–	+	–
Enterococcus faecalis	Firmicutes	+	–	–	+	–
Streptococcus pyogenes[b]	Firmicutes	+	–	–	+	–
Synechocystis sp.	Cyanobacteria	+	–	–	+	–
Borrelia burgdorferi	Spirochaetales	+	–	–	+	–
Treponema pallidum	Spirochaetales	+	–	–	+	–
Neisseria meningitidis	β-Proteobacteria	–	+	+	–	–
Neisseria gonorrhoeae	β-Proteobacteria	–	+	+	–	–
Pseudomonas aeruginosa	γ-Proteobacteria	–	+	+	–	–
Porphyromonas gingivalis	Cytophagales	+	+	–	–	–
Buchnera sp.	γ-Proteobacteria	+	+	–	–	–
Escherichia coli	γ-Proteobacteria	+	+	–	–	–
Haemophilus influenza	γ-Proteobacteria	+	+	–	–	–
Vibrio cholerae	γ-Proteobacteria	+	+	–	–	–
Xylella fastidiosa	γ-Proteobacteria	+	+	–	–	–
Yersinia pestis	γ-Proteobacteria	+	+	–	–	–
Deinococcus radiodurans[b]	Thermus/Deinococcus	+	+	+	–	–
Thermus thermophilus[b]	Thermus/Deinococcus	+	+	+	–	–

This table describes the presence or absence in the genomes of genes for AsnRS (*asnS*), GlnRS (*glnS*), bacterial-type AdT (*gatCAB*), and archaeal-type AdT (*gatDE*) in prokaryotes whose genome sequences have been completed. When the *gatCAB*-encoded enzyme is present, its in vivo Asp-AdT (aspartyl-tRNA amidotransferase) or Glu-AdT (glutamyl-tRNA amidotransferase) activity is indicated by [a]. Except for the few species where the specificity of the GatCAB enzyme has been characterized biochemically (indicated by [b]), the specificity of the AdT is based on presence or lack of AsnRS or GlnRS. When both AsnRS and GlnRS are absent, the AdT is a Glu/Asp-AdT; when GlnRS is absent, the AdT is a Glu-AdT; and when AsnRS is absent, the AdT acts like an Asp-AdT. These different categories are separated by a space in the table. The two last parts of the table represent bacteria where both AsnRS and GlnRS are present. In each part of the table the species are grouped according to their taxonomic nature.

tion, because *B. subtilis* GatCA or *M. thermautotrophicus* GatD alone could not convert Glu-tRNA^{Gln} to Gln-tRNA^{Gln} (Curnow et al. 1997; D. Tumbula and D. Söll, unpubl.). GatB and GatE have been suggested to be important for tRNA recognition (Curnow et al. 1997), but

this proposition presently lacks experimental support. GatC was reported to be important for expression of *B. subtilis* GatA (Curnow et al. 1997), but dispensable for *Thermus thermophilus* GatCAB activity in vitro (Becker and Kern 1998). In addition, deletion of *gatC* in *Heli-*

cobacter pylori was shown to be lethal (Chalker et al. 2001), indicating the necessity of GatC. These apparently conflicting results may be explained by the fact that for GatC the in vitro function is different from that in vivo.

Glutamine and ATP Hydrolysis Are Tightly Coupled to Amidation Reaction

Conversion of Glu-tRNA[Gln] or Asp-tRNA[Asn] to the correctly aminoacylated tRNA is a complex process catalyzed by AdTs. Wilcox (1969) proposed three distinct steps for the AdT reaction (see Fig. 2): hydrolysis of glutamine to generate active ammonia, phosphorylation of Glu-tRNA[Gln], and amidation of the activated tRNA intermediate. This reaction mechanism is similar to that of carbamoyl phosphate synthetase (Raushel et al. 1999). Using *B. subtilis* S100 extract and [γ-^{32}P]ATP, Wilcox demonstrated the accumulation of TCA-precipitable ^{32}P-labeled material before the addition of glutamine. However, the proposed γ-phosphoryl ester of the Glu-tRNA[Gln] intermediate has not been isolated. Attempts to repeat Wilcox's experiment with purified GatCAB failed due to a high level of nonspecific binding of [^{32}P]ATP to nitrocellulose membranes (Horiuchi et al. 2001; H.D. Becker and D. Söll, unpubl.). Therefore, the existence of the proposed intermediate awaits further studies.

Recent kinetic characterization of the *Streptococcus pyogenes* GatCAB enzyme (Horiuchi et al. 2001) has provided some initial glimpses of the reaction mechanism. Using a high-performance liquid chromatography (HPLC)-based assay, the authors demonstrate that both glutaminase and ATPase activities of GatCAB are tightly coupled to the overall transamidation reaction (Fig. 2). In the absence of the misaminoacylated tRNA, the basal glutaminase and ATPase activities of the enzyme were extremely low. However, when Glu-tRNA[Gln] was added, the k_{cat} of the glutaminase reaction was increased ~70-fold while the K_m for glutamine remained unchanged. Similarly, the rate of ATP hydrolysis jumped 400-fold when both Glu-tRNA[Gln] and glutamine were present. These data suggest that binding of the misaminoacylated tRNA may trigger a conformational change, which leads to a structural rearrangement so that active-site residues of glutaminase and ATPase are configured properly for catalysis. In the presence of Glu-tRNA[Gln], binding of adenosine 5′-[γ-thio]triphosphate (ATPγS) stimulated the glutaminase activity of GatCAB to the same level as did ATP. However, hydrolysis of ATPγS was extremely slow, and no transamidation of Glu-tRNA[Gln] was observed. The lack of transamidation using ATPγS is likely due to the slow formation of the thiophosphorylated Glu-tRNA[Gln] intermediate. Therefore, binding of ATP and Glu-tRNA[Gln] is sufficient to induce the glutaminase activity, but ATP hydrolysis is required for the transamidation to proceed. It is hoped that structural studies and biochemical characterization will provide the molecular basis for any allosteric interactions observed in GatCAB.

How Are the Misaminoacylated tRNAs Sequestered?

Although the nondiscriminating AspRS and GluRS enzymes are intrinsic to the tRNA-dependent transamidation pathways, their existence poses an obvious problem for the translation process. How do the misaminoacylated products, Asp-tRNA[Asn] and Glu-tRNA[Gln], avoid misincorporation at asparagine and glutamine codons? It appears that this problem is taken care of by the elongation factor EF-Tu. Stanzel et al. (1994) demonstrated that spinach chloroplast EF-Tu does not interact with the misacylated Glu-tRNA[Gln], but binds Glu-tRNA[Glu]. However, *E. coli* EF-Tu, which in vivo does not encounter this misacylated tRNA species, complexed with both Glu-tRNA species equally well. In comparison, *T. thermophilus* EF-Tu, like its spinach chloroplast counterpart, does discriminate. The substrate of the aspartyl-AdT enzyme, Asp-tRNA[Asn], is not bound to EF-Tu, whereas Asn-tRNA[Asn] is (Becker and Kern 1998). In addition, EF-Tu protected Asn-tRNA[Asn] significantly better from alkaline deacylation than Asp-tRNA[Asn]. The molecular basis for EF-Tu "rejection" of Asp-tRNA[Asn] and Glu-tRNA[Gln] is currently unexplained, as EF-Tu is assumed to recognize only a general AA-tRNA structure (Nissen et al. 1995) in its ability to deliver 20 different AA-tRNAs to the ribosome.

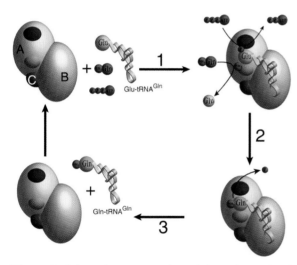

Figure 2. Schematic representation of the AdT-catalyzed transamidation. The AdT subunits (GatA, GatB, and GatC) are identified with bold letters, and only Glu-tRNA[Gln] transamidation is represented. The numbers beside the arrows indicate the order of the intermediate steps of the reaction. ATP is binding to the A subunit in the ATPase site (*purple*), the amide-group donor Gln binds to the glutaminase site (*gray*) also located on the A subunit. *Step 1*: Upon binding of the Glu-tRNA[Gln] in the active site, ATPase activity is stimulated leading to cleavage of ATP into ADP, which is released, and inorganic phosphate (P$_i$), which is transferred on the γ-carboxylic group of the glutamate (Glu) esterifying tRNA[Gln]. Concomitantly, the glutaminase activity is stimulated, forming Glu, which is released, and ammonia that will subsequently be transferred onto the activated intermediate P-γ-Glu-tRNA[Gln]. *Step 2*: Upon formation of Gln-tRNA[Gln], P$_i$ is released prior to dissociation of Gln-tRNA[Gln] from the AdT. *Step 3*: The free enzyme can then begin a new cycle.

A different way of "sequestering" these misacylated tRNAs would involve their direct transfer from the nondiscriminating AARS to the amidotransferase as suggested by Schön et al. (1988). However, evidence for "substrate channeling" is still absent, despite a growing number of large-scale protein–protein interaction studies (Tucker et al. 2001). A recent study (Rain et al. 2001) describing over 1,200 such interactions in *H. pylori*, which contains nondiscriminating AspRS and GluRS enzymes (Tumbula et al. 1999; S. Kareem et al., unpubl.), failed to identify an interaction between any AARS and a subunit of the GatCAB amidotransferase. Whatever the final outcome, the question of how these misaminoacylated tRNAs evade translational incorporation remains an intriguing mystery of the tRNA-dependent transamidation route.

tRNA Amidotransferases Are Common Routes to Amide Aminoacyl tRNA

The tRNA amidotransferase pathway of amide AA-tRNA formation was long considered an exception restricted to Gln-tRNA formation in some gram-positive bacteria (Wilcox and Nirenberg 1968) and organelles (Schön et al. 1988). The recent gene characterizations of the *B. subtilis* GatCAB (Curnow et al. 1997) and archaeal GatDE (Tumbula et al. 2000) enzymes, together with the increasingly large number of genomic sequences, provided a better perspective of the distribution of each pathway among organisms in all three domains of life (Table 1).

For Gln-tRNA formation, all prokaryotes lack GlnRS and use a Glu-tRNA^{Gln} amidotransferase; the exceptions are β- and γ-proteobacteria and the *Thermus/Deinococcus* group, which all have GlnRS (Table 1). Bacteria use the heterotrimeric GatCAB enzyme to catalyze the transamidation step, and archaea utilize the heterodimeric GatDE protein (Tumbula et al. 2000). In eukarya, although Gln-tRNA is exclusively synthesized by GlnRS in the cytoplasm, it is probable that organellar glutaminylation of tRNA is generally achieved by an AdT. In fact, such a pathway has been characterized in several plant chloroplasts (Schön et al. 1988; Jahn et al. 1990; W. Chang and B. Min, unpubl.) and suggested in yeast mitochondria (Martin and Rabinowitz 1984; Kim et al. 1997). To date, only trypanosomatids have been shown to possess a mitochondrial GlnRS activity (Nabholz et al. 1997), although the enzyme and its evolutionary history have yet to be determined.

Sequence-based phylogenies of GlnRS suggest that this enzyme arose from eukarya and was acquired by a very limited number of bacteria via horizontal gene transfer (Lamour et al. 1994; Handy and Doolittle 1999). The rare occurrence of GlnRS among prokaryotes suggests a "poor efficiency" of horizontal gene transfer for which there is no explanation yet. However, in the case of archaea, which all lack GlnRS, we showed that the archaeal tRNA^{Gln} is not a suitable substrate for eukaryal and bacterial GlnRSs; this may have prevented the stable transfer of *glnS*, the gene encoding GlnRS, into these organisms (Tumbula et al. 2000). The same reason might account for the poor representation of *glnS* among bacteria. The acquisition of the *glnS* gene by the few extant bacteria led without exception to the loss of the nondiscriminating property of GluRS, thus blocking the transamidation route to Gln-tRNA^{Gln}, and often the concomitant loss of the *gatCAB* genes.

AsnRS is absent from approximately half of the prokaryotes studied so far (Table 1). Discovery of a tRNA-dependent transamidation pathway for Asn-tRNA formation in the archaeon *Haloferax volcanii* (Curnow et al. 1996) gave the first clue as to how protein synthesis can be achieved in organisms lacking AsnRS. Investigation of the Asn-tRNA formation pathway in *T. thermophilus* and *Deinococcus radiodurans* showed that the indirect route to Asn-tRNA was catalyzed by the same GatCAB enzyme involved in Gln-tRNA synthesis (Curnow et al. 1998; Becker et al. 2000b). Genomic analysis and biochemical studies showed that 8 out of 13 archaea (Tumbula et al. 2000), as well as many bacteria (Table 1), use GatCAB to form Asn-tRNA. Several bacteria, like *H. pylori* (Tumbula et al. 1999) and *Chlamydia trachomatis* (Raczniak et al. 2001), are expected to use GatCAB to generate both Asn-tRNA and Gln-tRNA because these organisms lack both AsnRS and GlnRS (Table 1). In contrast, eukarya appear to synthesize both cytoplasmic and organellar Asn-tRNAs using a cytoplasmic and an organellar-targeted AsnRS, respectively.

AsnRS sequence-based phylogenies only localize the origin of this enzyme to the archaeal genre of AspRS (Woese et al. 2000). It is very likely that AsnRS evolved in the common ancestor before the later split into the three extant domains. AsnRS, unlike GlnRS, has been found to coexist with a GatCAB enzyme that generates the same product. However, this occurs in only a very restricted branch of bacteria, the *Thermus/Deinococcus* group. In this case, the AdT has probably been retained because AsnRS is not autonomous due to a lack of asparagine synthetase, the enzyme responsible for the formation of the free asparagine substrate of AsnRS (Becker and Kern 1998; Curnow et al. 1998). Thus, it is very likely that in this group, the indirect pathway serves as an alternative route to Asn-tRNA under conditions of asparagine depletion.

ALTERNATE ROUTES OF CYS-tRNA SYNTHESIS

Among all the absences of recognizable AARS orthologs from the genomes of many organisms, the most difficult to explain was the lack of genes encoding a canonical cysteinyl-tRNA synthetase (CysRS). In two methanogenic archaea, *M. jannaschii* and *M. thermautotrophicus*, genomic analysis revealed that this enzyme may be completely absent (Bult et al. 1996; Smith at al. 1997). This suggested that a tRNA-dependent pathway, like the one responsible for the formation of selenocysteinyl-tRNA (Commans and Böck 1999), could be involved in the biosynthesis of Cys-tRNA in these organisms. Alternatively, Cys-tRNA could be made by a completely different enzyme (in terms of amino acid sequence), as is the case for Lys-tRNA synthesis by class I LysRS (see above). Biochemical studies with *Methanothermobacter marburgensis* seryl-tRNA synthetase (SerRS) showed that this enzyme was unable to form Cys-tRNA[Ser], the crucial intermediate in a tRNA-dependent amino acid transformation pathway (Kim et al. 1998); thus, the indirect pathway does not exist. Instead, an enzymatic activity that could directly charge cysteine onto tRNA was easily detected by aminoacylation assays (Hamann et al. 1999). Standard biochemical purification of cysteinylation activity from *M. jannaschii* cell extracts led to a homogeneous aminoacyl-tRNA synthetase. Sequence analysis and biochemical activity showed this enzyme to be prolyl-tRNA synthetase (ProRS) (Stathopoulos et al. 2000). Cloning and overexpressing the archaeal *proS* gene in *E. coli* confirmed this unexpected finding, as the purified recombinant protein was able to catalyze the formation of both Pro-tRNA[Pro] and Cys-tRNA[Cys] without cross-reactivity. The archaeal *proS* was also active in vivo as it could rescue the growth of a *cysS* temperature-sensitive *E. coli* strain. The dual specificity of this enzyme was later confirmed independently (Lipman et al. 2000). This enzyme, named ProCysRS (Yarus 2000), is unique among the known AARSs in that it has as substrates two cognate amino acids (proline and cysteine) and two cognate tRNAs (tRNA[Pro] and tRNA[Cys]).

Substrate Specificity of Prolyl-cysteinyl-tRNA Synthetase

Mutational analysis and subsequent biochemical characterization of the reaction steps showed that ProCysRS has overlapping amino acid-binding sites. Replacement of conserved residues that participate in the formation of the binding pocket alters the character of the enzyme for either cysteinylation or prolylation (Stathopoulos et al. 2001). The critical point of the reaction is the requirement for the presence of tRNA[Cys] during cysteinyl-adenylate formation, but not for tRNA[Pro] during proline activation. Thus, the cysteine-binding site is only properly "configured" after tRNA binding (Ibba et al. 2000). This observation, although in dispute (Lipman et al. 2000), clearly indicates a sophisticated mechanism that would halt concomitant proline activation, thus preventing Pro-tRNA[Cys] formation. tRNA-independent proline activation ensures an induced fit of the active site for only Pro-AMP. This prevents the misactivation of smaller amino acids (e.g., alanine) as has been reported for *E. coli* ProRS (Beuning and Musier-Forsyth 2000) and also provides the correct conformation for proline accessibility to the 3′ end of the tRNA (Yaremchuk et al. 2001). This discrimination mechanism is in good agreement with the recent structural analysis of *T. thermophilus* ProRS (Yaremchuk et al. 2000), a bacterial enzyme that is also a dual specificity ProCysRS (Feng et al. 2001). All the above data suggest that this enzyme can form the correct products (Cys-tRNA[Cys] and Pro-tRNA[Pro]) without recourse to ATP-dependent hydrolytic editing, a required reaction in certain proofreading mechanisms (Beuning and Musier-Forsyth 2001).

tRNA Recognition Assures the Fidelity of ProCysRS

One of the major characteristics of all ProCysRS enzymes tested to date is their inability to efficiently charge unmodified tRNA[Cys] synthesized by in vitro transcription (Hamann et al. 1999; Stathopoulos et al. 2000; I. Ahel, unpubl.). It is likely that nucleotide modification(s) in the tRNA is part of the identity elements that are recognized by ProCysRS. The inability to use unmodified tRNAs in these studies will make the determination of the identity elements much more difficult. On the other hand, a modified nucleotide may be crucial for folding the tRNA into the correct conformation required for the proper interaction with the enzyme, resulting in configuring the cysteine-binding site.

Occurrence of Dual Specificity ProCysRS in the Living Kingdom

ProRS enzymes can be divided into two subgroups (Woese et al. 2000; Yaremchuk et al. 2000; Burke et al. 2001): the archaeal genre ProRS, which has given rise to the ProCysRS, and a bacterial-type ProRS, which has a 180-amino-acid insertion domain between motifs II and III. The archaeal genre enzyme also has an idiosyncratic carboxy-terminal extension. A biochemical survey of selected archaea demonstrated CysRS activity of the archaeal ProRS from *M. jannaschii*, *M. thermautotrophicus*, and *M. maripaludis* (Stathopoulos et al. 2000). To date only one dual-specificity eukaryotic ProRS has been characterized, in the deep-rooted eukaryote *Giardia lamblia* (Bunjun et al. 2000), which likely acquired the archeal *proS* by lateral gene transfer (Woese et al. 2000). Biochemical examination of many bacterial ProRSs of the archaeal genre revealed that they also have CysRS activity. Whereas some of these bacterial genes also complemented the temperature-sensitive *E. coli cysS* mutant, it remains to be determined whether they do form Cys-tRNA in their respective organisms, all of which also harbor a canonical CysRS (I. Ahel and C. Stathopoulos, unpubl.).

Prolyl-cysteinyl-tRNA Synthetase as an Indicator of the Origins of Aminoacyl-tRNA Synthesis

The ability of ProCysRS to act in the presence of tRNA[Cys] as a "ribonucleoprotein enzyme" or as a single-specificity ProRS is a direct reflection of the evolutionary history of AARSs. This family of proteins may have evolved from common ancestral enzymes that specified more than one amino acid, first as ribozymes and then evolving as solely proteinaceous catalysts with the recruitment of protein cofactors (Ibba et al. 1997a, 2000). The common origin of these enzymes also explains the chiral symmetry in the architecture of their active sites (Ribas dePouplana and Schimmel 2001). Future studies might explain why enzymes like the unconventional synthetase that forms Cys-tRNA are found only in a very limited number of organisms (i.e., *M. jannaschii*, *T. mar-*

itima, and *D. radiodurans*) (Fàbrega et al. 2001) and never found wider use during the evolution of protein synthesis as a specifier of cysteine codons in translation. Comparative phylogenetic analyses revealed that the canonical CysRS enzymes joined the translational machinery later during evolution (Li et al. 1999), an observation that makes the history of Cys-tRNA formation more puzzling.

CONCLUSIONS

The recent advent of whole-genome sequence analysis has profoundly changed our view of AA-tRNA synthesis. The previous depiction of an evolutionarily conserved system of 20 distinct synthetic pathways can now be seen as an inadequate description of the overall process of AA-tRNA synthesis. Although it is still true that the majority of AA-tRNAs are in fact synthesized the same way in all organisms, a significant minority is made by more than one route, sometimes even in the same organism. The scale of these evolutionary differences is highly variable both from one AA-tRNA synthetic route to another and between different organisms. This raises the question of whether variations in AA-tRNA synthesis reflect an arbitrary flexibility in particular pathways, perhaps arising from undetermined biochemical constraints, or instead, reflect key events in the evolution of contemporary protein synthesis. For example, although it seems reasonable to accept that the transamidation pathways indicate the late recruitment of asparagine and glutamine to coded protein synthesis, their coexistence with the corresponding direct pathways in some organisms suggests a correlation with contemporary amino acid metabolism. Similarly, the need to minimize potential aminoacylation errors resulting from changes in substrate availability during the cell cycle may go some way to explaining the need for alternative routes for Cys-tRNA synthesis. The present uncertainty as to the significance of the alternative routes of AA-tRNA synthesis arises from the overwhelming preponderance of in vitro data, on which most of our present assumptions are based. The possibility of looking inside the cell by revisiting classic genetic approaches with functional genomics tools will now allow the study of AA-tRNA synthesis in a cellular context, which will in turn provide new insights into the physiological constraints that govern AA-tRNA synthesis in vivo.

ACKNOWLEDGMENTS

Work in the authors' laboratories is supported by grants from the National Institute of General Medical Sciences (D.S.), the National Aeronautics and Space Administration (D.S.), the Office of Energy Biosciences of the Department of Energy (D.S.), and the European Commission (QLG-CT-99-00660, M.I.). M.I. gratefully acknowledges the support of the Alfred Benzon Foundation. D.T.H. and L.F. were National Institutes of Health postdoctoral fellows, H.B. was an EMBO postdoctoral fellow.

REFERENCES

Becker H.D. and Kern D. 1998. *Thermus thermophilus:* A link in evolution of the tRNA-dependent amino acid amidation pathways. *Proc. Natl. Acad. Sci.* **95:** 12832.

Becker H.D., Roy H., Moulinier L., Mazauric M.H., Keith G., and Kern D. 2000a. *Thermus thermophilus* contains an eubacterial and an archaebacterial aspartyl-tRNA synthetase. *Biochemistry* **39:** 3216.

Becker H.D., Min B., Jacobi C., Raczniak G., Pelaschier J., Roy H., Klein S., Kern D., and Söll D. 2000b. The heterotrimeric *Thermus thermophilus* Asp-tRNA^{Asn} amidotransferase can also generate Gln-tRNA^{Gln}. *FEBS Lett.* **476:** 140.

Beuning P.J. and Musier-Forsyth K. 2000. Hydrolytic editing by a class II aminoacyl-tRNA synthetase. *Proc. Natl. Acad. Sci.* **97:** 8916.

———. 2001. Species-specific differences in amino acid editing by class II prolyl-tRNA synthetase. *J. Biol. Chem.* **276:** 30779.

Bult C.J., White O., Olsen G.J., Zhou L., Fleischmann R.D., Sutton G.G., Blake J.A., FitzGerald L.M., Clayton R.A., Gocayne J.D., Kerlavage A.R., Dougherty B.A., Tomb J.-F., Adams M.D., Reich C.I., Overbeek R., Kirkness E.F., Weinstock K.G., Merrick J.M., Glodek A., Scott J.L., Geoghagen N.S., Weidman J.F., Furhmann J.L., Nguyen D., Utterback T.R., Kelley J.M., Peterson J.D., Sadow P.W., Hanna M.C., Cotton M.D., Roberts K.M., Hurst M.A., Kaine B.P., Borodovsky M., Klenk H.-P., Fraser C.M., Smith H.O., Woese C.R., and Venter J.C. 1996. Complete genome sequence of the methanogenic archaeon, *Methanococcus jannaschii. Science* **273:** 1058.

Bunjun S., Stathopoulos C., Graham D., Min B., Kitabatake M., Wang A.L., Wang, C.C., Vivarès, C.P., Weiss, L.M., and Söll D. 2000. A dual-specificity aminoacyl-tRNA synthetase in the deep-rooted eukaryote *Giardia lamblia. Proc. Natl. Acad. Sci.* **97:** 12997.

Burke B., Lipman R.S., Shiba K., Musier-Forsyth K., and Hou Y.M. 2001. Divergent adaptation of tRNA recognition by *Methanococcus jannaschii* prolyl-tRNA synthetase. *J. Biol. Chem.* **276:** 20286.

Chalker A.F., Minehart H.W., Hughes N.J., Koretke K.K., Lonetto M.A., Brinkman K.K., Warren P.V., Lupas A., Stanhope M.J., Brown J.R., and Hoffman P.S. 2001. Systematic identification of selective essential genes in *Helicobacter pylori* by genome prioritization and allelic replacement mutagenesis. *J. Bacteriol.* **183:** 1259.

Commans S. and Böck A. 1999. Selenocysteine inserting tRNAs: An overview. *FEMS Microbiol. Rev.* **23:** 335.

Crick F.H.C. 1958. On protein synthesis. *Symp. Soc. Exp. Biol.* **12:** 138.

Curnow A.W., Ibba M., and Söll D. 1996. tRNA-dependent asparagine formation. *Nature* **382:** 589.

Curnow A.W., Tumbula D.L., Pelaschier J.T., Min B., and Söll D. 1998. Glutamyl-tRNA^{Gln} amidotransferase in *Deinococcus radiodurans* may be confined to asparagine biosynthesis. *Proc. Natl. Acad. Sci.* **95:** 12838.

Curnow A.W., Hong K.-W., Yuan R., Kim S.-I., Martins O., Winkler W., Henkin T.M., and Söll D. 1997. Glu-tRNA^{Gln} amidotransferase: A novel heterotrimeric enzyme required for correct decoding of glutamine codons during translation. *Proc. Natl. Acad. Sci.* **94:** 11819.

Fàbrega C., Farrow M.A., Mukhopadhyay B., de Crécy-Lagard V., Ortiz A.R., and Schimmel P. 2001. An aminoacyl tRNA synthetase whose sequence fits into neither of the two known classes. *Nature* **411:** 110.

Feng L., Stathopoulos C., Ahel I., Mitra A., Tumbula-Hansen D., Hartsch T., and Söll D. 2002. Aminoacyl-tRNA formation in the extreme thermophile *Thermus thermophilus. Extremophiles* (in press)

Hamann C.S., Sowers K.R., Lipman R.S., and Hou Y.M. 1999. An archaeal aminoacyl-tRNA synthetase missing from genomic analysis. *J. Bacteriol.* **181:** 5880.

Handy J. and Doolittle R.F. 1999. An attempt to pinpoint the phylogenetic introduction of glutaminyl-tRNA synthetase among bacteria. *J. Mol. Evol.* **49:** 709.

Horiuchi K.Y., Harpel M.R., Shen L., Luo Y., Rogers K.C., and Copeland R.A. 2001. Mechanistic studies of reaction coupling in Glu-tRNA^{Gln} amidotransferase. *Biochemistry* **40:** 6450.

Ibba M., Curnow A.W., and Söll D. 1997a. Aminoacyl-tRNA synthesis: Divergent routes to a common goal. *Trends Biochem. Sci.* **22:** 39.

Ibba M., Bono J.L., Rosa P.A., and Söll D. 1997b. Archaeal-type lysyl-tRNA synthetase in the Lyme disease spirochete *Borrelia burgdorferi. Proc. Natl. Acad. Sci.* **94:** 14383.

Ibba M., Becker H.D., Stathopoulos C., Tumbula D.L., and Söll D. 2000. The adaptor hypothesis revisited. *Trends Biochem. Sci.* **25:** 311.

Ibba M., Losey H.C., Kawarabayasi Y., Kikuch, H., Bunjun S., and Söll D. 1999. Substrate recognition by class I lysyl-tRNA synthetases: A molecular basis for gene displacement. *Proc. Natl. Acad. Sci.* **96:** 418.

Ibba M., Morgan S., Curnow A.W., Pridmore D.R., Vothknecht U.C., Gardner W., Lin W., Woese C.R., and Söll D. 1997c. A euryarchaeal lysyl-tRNA synthetase: Resemblance to class I synthetases. *Science* **278:** 1119.

Jahn D., Kim Y.C., Ishino Y., Chen M.W., and Söll D. 1990. Purification and functional characterization of the Glu-tRNA^{Gln} amidotransferase from *Chlamydomonas reinhardtii. J. Biol. Chem.* **265:** 8059.

Jakubowski H. and Goldman E. 1992. Editing of errors in selection of amino acids for protein synthesis. *Microbiol. Rev.* **56:** 412.

Kim H., Vothknecht U.C., Hedderich R., Celic I., and Söll D. 1998. Sequence divergence of seryl-tRNA synthetases in archaea. *J. Bacteriol.* **180:** 6446.

Kim S.-I., Stange-Thomann N., Martins O., Hong K.-W., Söll D., and Fox T.D. 1997. A nuclear genetic lesion affecting *Saccharomyces cerevisiae* mitochondrial translation is complemented by a homologous *Bacillus* gene. *J. Bacteriol.* **179:** 5625.

Kobayashi M., Fujiwara Y., Goda M., Komeda H., and Shimizu S. 1997. Identification of active sites in amidase: Evolutionary relationship between amide bond- and peptide bond-cleaving enzymes. *Proc. Natl. Acad. Sci.* **94:** 11986.

Lamour V., Quevillon S., Diriong S., N'guyen V.C., Lipinski M., and Mirande M. 1994. Evolution of the Glx-tRNA synthetase family: The glutaminyl enzyme as a case of horizontal gene transfer. *Proc. Natl. Acad. Sci.* **91:** 8670.

Lapointe J., Duplain L., and Proulx M. 1986. A single glutamyl-tRNA synthetase aminoacylates tRNA^{Glu} and tRNA^{Gln} in *Bacillus subtilis* and efficiently misacylates *Escherichia coli* tRNA$^{Gln}_1$ in vitro. *J. Bacteriol.* **165:** 88.

Li T., Graham DE., Stathopoulos C., Haney P.J., Kirn H., Vothknecht U.C., Kitabatake M., Hong K., Eggertsson G., Curnow A.W., Lin W., Celic I., Whitman W., and Söll D. 1999. Cysteinyl-tRNA formation: The last puzzle of aminoacyl-tRNA synthesis. *FEBS Lett.* **462:** 302.

Lipman R.S., Sowers K., and Hou Y.M. 2000. Synthesis of cysteinyl-tRNA^{Cys} by a genome that lacks a normal cysteine-tRNA synthetase. *Biochemistry* **39:** 7792.

Lubkowski J., Palm G.J., Gilliland G.L., Derst C., Rohm K.H., and Wlodawer A. 1996. Crystal structure and amino acid sequence of *Wolinella succinogenes* L-asparaginase. *Eur. J. Biochem.* **241:** 201.

Martin N.C. and Rabinowitz M. 1984. Glu-tRNA^{Gln}: An intermediate in yeast mitochondrial protein synthesis. *Methods Enzymol.* **106:** 152.

Nabholz C.E., Hauser R., and Schneider A. 1997. *Leishmania tarentolae* contains distinct cytosolic and mitochondrial glutaminyl-tRNA synthetase activities. *Proc. Natl. Acad. Sci.* **94:** 7903.

Nissen P., Kjeldgaard M., Thirup S., Polekhina G., Reshetnikova L., Clark B.F., and Nyborg J. 1995. Crystal structure of the ternary complex of Phe-tRNA^{Phe}, EF-Tu, and a GTP analog. *Science* **270:** 1464.

Raczniak G., Ibba M., and Söll D. 2001. Genomics-based identification of targets in pathogenic bacteria for potential therapeutic and diagnostic use. *Toxicology* **160:** 181.

Rain J.C., Selig L., De Reuse H., Battaglia V., Reverdy C., Simon S., Lenzen G., Petel F., Wojcik J., Schachter V., Chemama Y.,

Labigne A., and Legrain P. 2001. The protein-protein interaction map of *Helicobacter pylori. Nature* **409:** 211.

Raushel F.M., Thoden J.B., and Holden H.M. 1999. The amidotransferase family of enzymes: Molecular machines for the production and delivery of ammonia. *Biochemistry* **38:** 7891.

Ribas de Pouplana L. and Schimmel P. 2001. Two classes of tRNA synthetases suggested by sterically compatible dockings on tRNA acceptor stem. *Cell* **104:** 191.

Ribas de Pouplana L., Turner R.J., Steer B.A., and Schimmel P. 1998. Genetic code origins: tRNAs older than their synthetases? *Proc. Natl. Acad. Sci.* **95:** 11295.

Schön A., Kannangara C.G., Gough S., and Söll D. 1988. Protein biosynthesis in organelles requires misaminoacylation of tRNA. *Nature* **331:** 187.

Smith D.R., Doucette-Stamm L.A., Deloughery C., Lee H., Dubois J., Aldredge T., Bashirzadeh R., Blakely D., Cook R., Gilbert K., Harrison D., Hoang L., Keagle P., Lumm W., Pothier B., Qiu D., Spadafora R., Vicaire R., Wang Y., Wierzbowski J., Gibson R., Jiwani N., Caruso A., Bush D., Safer H., Patwell D., Prabhakar S., McDougall S., Shimer G., Goyal A., Pietrokovski S., Church G.M., Daniels C.J., Mao J.-I., Rice P., Nölling J., and Reeve J.N. 1997. Complete genome sequence of *Methanobacterium thermoautotrophicum* ΔH: Functional analysis and comparative genomics. *J. Bacteriol.* **179:** 7135.

Söll D., Becker H.D., Plateau P., Blanquet S., and Ibba M. 2000. Context-dependent anticodon recognition by class I lysyl-tRNA synthetases. *Proc. Natl. Acad. Sci.* **97:** 14224.

Stanzel M., Schön A., and Sprinzl M. 1994. Discrimination against misacylated tRNA by chloroplast elongation factor Tu. *Eur. J. Biochem.* **219:** 435.

Stathopoulos C., Li T., Longman R., Vothknecht U.C., Becker H.D., Ibba M., and Söll D. 2000. One polypeptide with two aminoacyl-tRNA synthetase activities. *Science* **287:** 479.

Stathopoulos C., Jacquin-Becker C., Becker H.D., Li T., Ambrogelly A., Longman R., and Söll D. 2001. *Methanococcus jannaschii* prolyl-cysteinyl-tRNA synthetase possesses overlapping amino acid binding sites. *Biochemistry* **40:** 46.

Swain A.L., Jaskolski M., Housset D., Rao J.K., and Wlodawer A. 1993. Crystal structure of *Escherichia coli* L-asparaginase, an enzyme used in cancer therapy. *Proc. Natl. Acad. Sci.* **90:** 1474.

Tucker C.L., Gera J.F., and Uetz P. 2001. Towards an understanding of complex protein networks. *Trends Cell. Biol.* **11:** 102.

Tumbula D.L., Becker H.D., Chang W.-Z., and Söll D. 2000. Domain-specific recruitment of amide amino acids for protein synthesis. *Nature* **407:** 106.

Tumbula D., Vothknecht U.C., Kim H.-S., Ibba M., Min B., Li T., Pelaschier J., Stathopoulos C., Becker H., and Söll D. 1999. Archaeal aminoacyl-tRNA synthesis: Diversity replaces dogma. *Genetics* **152:** 1269.

Wilcox M. 1969. Gamma-glutamyl phosphate attached to glutamine-specific tRNA: A precursor of glutaminyl-tRNA in *Bacillus subtilis. Eur. J. Biochem.* **11:** 405.

Wilcox M. and Nirenberg M. 1968. Transfer RNA as a cofactor coupling amino acid synthesis with that of protein. *Proc. Natl. Acad. Sci.* **61:** 229.

Woese C.R., Olsen G.J., Ibba M., and Söll D. 2000. Aminoacyl-tRNA synthetases, the genetic code, and the evolutionary process. *Microbiol. Mol. Biol. Rev.* **64:** 202.

Yaremchuk A., Cusack S., and Tukalo M. 2000. Crystal structure of an eukaryote/archaea-like prolyl-tRNA synthetase at 2.4 Å resolution and its complex with tRNA^{Pro}(CGG). *EMBO J.* **19:** 4745.

Yaremchuk A., Tukalo M., Grotli M., and Cusack S. 2001. A succession of substrate induced conformational changes ensures the amino acid specificity of *Thermus thermophilus* prolyl-tRNA synthetase: Comparison with histidyl-tRNA synthetase. *J. Mol. Biol.* **309:** 989.

Yarus M. 2000. Unraveling the riddle of ProCys tRNA synthetase. *Science* **287:** 440.

Zalkin H. and Smith J.L. 1998. Enzymes utilizing glutamine as an amide donor. *Adv. Enzymol. Relat. Areas Mol. Biol.* **72:** 87.

tRNA Conformity

A.D. Wolfson, F.J. LaRiviere,* J.A. Pleiss,† T. Dale, H. Asahara,
and O.C. Uhlenbeck

Department of Chemistry and Biochemistry, University of Colorado, Boulder, Colorado 80309-0215

The 30 or more different transfer RNA molecules present in all cells have a similar size and overall three-dimensional architecture, but differ in nucleotide sequence. Numerous enzymes are involved in the maturation, modification, biochemical function, and degradation of each individual tRNA. These include enzymes that are specific for a small subclass of all tRNAs, such as the aminoacyl-tRNA synthetases and certain tRNA-modifying enzymes, as well as a group of enzymes that interact with nearly all tRNAs, such as RNase P, rT_{54} methylase, elongation factor Tu (EF-Tu), and the ribosome. A challenge for biochemists is to understand how this set of structurally similar RNA molecules can be substrates for both specific and nonspecific enzymes.

The term *tRNA identity* can be defined as the structural properties of a tRNA molecule that make it a substrate for its cognate aminoacyl-tRNA synthetase and not a substrate for the other, noncognate synthetases present in the cell. Extensive in vivo and in vitro experiments with mutant tRNAs developed the concept that a finite number of nucleotides, termed identity nucleotides, present in each tRNA were required for specific interaction with the cognate synthetase (McClain 1994; Saks et al. 1994; Giege et al. 1998). The critical experiment that supported this concept was the "swap" experiment, where proposed identity nucleotides from one tRNA were transplanted into a second tRNA sequence, effecting a change in the identity of the esterified amino acid (Normanly et al. 1986). Although these swap experiments were not always successful, they strongly suggested that a significant part of tRNA identity was defined by a limited number of bases positioned at different sites on a generic tRNA framework. Although co-crystal structures of a number of tRNA•synthetase complexes generally confirmed direct contacts between identity nucleotides and the protein, extensive interactions between the phosphodiester backbone of tRNA and the enzyme were also observed (Rould et al. 1989; Ruff et al. 1991). This has led to the suggestion that backbone contacts could also contribute to tRNA identity (McClain et al. 1998; Nissan et al. 1999).

The term *tRNA conformity* can be defined as the structural properties of elongator tRNAs that make them

equivalent substrates for those enzymes which act on all tRNAs, including those of the translation apparatus. Although the rate of translation of different codons in *E. coli* varies considerably (Curran and Yarus 1989; Sorensen and Pedersen 1991), most of the variation is due to the aminoacyl-tRNA (aa-tRNA) concentration. When their concentration is saturating, all aa-tRNAs pass through the translation machinery at approximately the same rate (Thomas et al. 1988), despite differences in their sequences, differences in the size, charge, and hydrophobicity of the esterified amino acid, and differences in the stability of the codon–anticodon interaction. Perhaps the simplest explanation for tRNA conformity in translation is that EF-Tu and the ribosome interact with one or more of the nine universally conserved nucleotides of tRNA. However, many of the conserved nucleotides can be mutated without dramatically affecting tRNA function either in vivo (Hou and Schimmel 1992) or in vitro (Nazarenko et al. 1994). Thus, it appears that a considerable part of tRNA conformity is achieved in a different way.

tRNA CONFORMITY WITH EF-Tu

When complexed with GTP, EF-Tu binds all aminoacylated elongator tRNAs, but not the initiator or selenocystine tRNAs. A typical dissociation constant (K_D) for an aa-tRNA to bacterial EF-Tu•GTP is on the order of 1 nM (Abrahamson et al. 1985), whereas the K_D for deacylated tRNA is about 1 μM (Janiak et al. 1990), indicating that about one-third of the binding energy is associated with the amino acid. *E. coli* tRNAs exhibit a high degree of conformity with *E. coli* EF-Tu•GTP, as similar K_D values were observed for 27 different aa-tRNAs (Louie et al. 1984; Louie and Jurnak 1985; Ott et al. 1990). However, conformity is not complete, since the different *E. coli* aa-tRNAs ranged in K_D from 0.5 nM for Gln-tRNA^{Gln} to 7 nM for Val-tRNA^{Val}. Because these differences correlate fairly well with tRNA abundance in *E. coli* (Jakubowski 1988), it appears that EF-Tu "homogenizes" differences in tRNA abundance so that uniform binding to the ribosome can be achieved.

The co-crystal structures of EF-Tu•GMPPNP with yeast Phe-tRNA^{Phe} (Nissen et al. 1995) and *E. coli* Cys-tRNA^{Cys} (Nissen et al. 1999) suggest how tRNA conformity is achieved. In both structures, the protein forms hydrogen bonds with both the amino group and the carbonyl of the esterified amino acid, presumably resulting in tighter binding of the aa-tRNA. The protein also binds to

Present addresses: *Brandeis University, Department of Biochemistry, 415 South Street, Waltham, Massachusetts 02454-9110; †University of California-San Francisco, Department of Biochemistry and Biophysics, San Francisco, California 94143-0448.

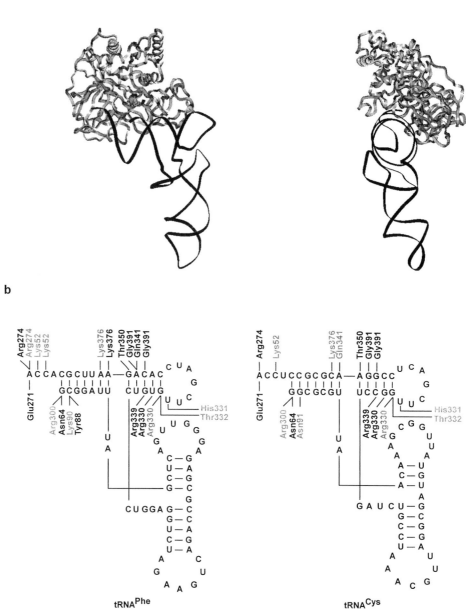

Figure 1. (*a*) Two views of the crystal structure of the complex between *T. thermophilus* EF-Tu•GMPPNP and yeast Phe-tRNA[Phe] (Nissen et al. 1995). (*b*) Summary of contacts between EF-Tu and the phosphodiester backbone of yeast tRNA[Phe] (*left*) and *E. coli* tRNA[Cys] (*right*) deduced by the respective X-ray structures (Nissen et al. 1995, 1999). Amino acids in gray contact one or both of the phosphate oxygens 5′ to the nucleotide, and amino acids in black contact 2′ hydroxyl groups. Most of the contacts use the side chain of the amino acid, but some involve the peptide backbone.

the upper part of the tRNA (Fig. 1a), making contacts with the acceptor stem and T stem, which are both structurally conserved among tRNAs. Interestingly, a close examination of the hydrogen bonds and ion pairs between the tRNA and protein reveal no base-specific contacts at all. Instead, the only direct contacts are made with 2′ hydroxyls and phosphate residues, suggesting a uniform mechanism of binding for all tRNAs. However, somewhat surprisingly, the exact contacts differ slightly between tRNA[Phe] and tRNA[Cys] (Fig. 1b). It is not known whether these differences reflect alternative binding

modes in solution or are the result of different packing constraints in the two crystal lattices (Nissan et al. 1999). Thus, the structural data generally support the concept that EF-Tu binds all aa-tRNAs in a uniform manner despite differences in the amino acids and tRNA sequences.

In contrast with this simple view, there have been several observations suggesting that EF-Tu may have an underlying specificity for both the amino acid and the tRNA body. Changes in the sequence of the acceptor stem of yeast tRNA[Asp] resulted in tenfold differences in EF-Tu affinity (Rudinger et al. 1996). The poor binding of the structurally atypical tRNA^{Sec} to EF-Tu appears to be the result of a particular sequence in the acceptor stem (Rudinger et al. 1996). An important set of observations suggesting that EF-Tu shows amino acid specificity are that EF-Tu•GTP often binds misacylated tRNAs differently than the corresponding correctly acylated form. For example, misacylated Gln-tRNA^{Su+7} binds *E. coli* EF-Tu•GTP about threefold tighter than the correctly acylated Trp-tRNA^{Su+7} counterpart, presumably explaining the more efficient incorporation of Gln into the UAG codons of suppressed proteins (Knowlton and Yarus 1980). A more striking example comes from the substantial number of eubacteria and archaea that lack asparaginyl- and/or glutaminyl-tRNA synthetases. These organisms misacylate their tRNA^{Asn} or tRNA^{Gln} with either aspartate or glutamate and then use a specialized tRNA-amidotransferase to make Asn-tRNA^{Asn} and Gln-tRNA^{Gln} (Ibba and Söll 2000). In these organisms, it is critical that the misacylated Asp-tRNA^{Asn} or Glu-tRNA^{Gln} intermediates are not misincorporated into protein. Two groups have shown that these misacylated tRNAs do not bind EF-Tu (Stanzel et al. 1994; Becker and Kern 1998), suggesting a possible mechanism for avoiding misincorporation but requiring that EF-Tu can somehow differentiate between correctly acylated and misacylated tRNAs. It appears that the mechanism underlying the nearly uniform binding affinity of EF-Tu to the different aa-tRNAs may be more complex than the structural data suggest.

EF-Tu ACHIEVES CONFORMITY BY THERMODYNAMIC COMPENSATION

To examine the binding of EF-Tu•GTP to tRNAs esterified with both correct and incorrect amino acids, we chose four unmodified tRNAs with similar overall architectures (tRNA^{Phe}, tRNA^{Ala}, tRNA^{Val}, and tRNA^{Gln}) and aminoacylated each of them with the cognate and three noncognate amino acids (Phe, Ala, Val, and Gln) (La Riviere et al. 2001). To facilitate misacylation, one or more identity nucleotides for noncognate enzymes were introduced into each tRNA. Mutations were targeted to regions of the tRNA unlikely to affect the interaction with EF-Tu, but also designed so that many of the tRNAs could be aminoacylated by more than one tRNA synthetase. For example, when the G73A, C70U, C34G, U35A, and G36C mutations are introduced into tRNA^{Gln}, the resulting tRNA can be misacylated by either PheRS, ValRS, or AlaRS, thus creating three misacylated versions of tRNA^{Gln}.

The affinities of the four correctly acylated and twelve misacylated tRNAs to *T. thermophilus* EF-Tu were determined using a ribonuclease protection assay (Knowlton and Yarus 1980; Louie et al. 1984). This classic assay makes use of the fact that when ^{3}H-aa-tRNA is treated with a high concentration of pancreatic ribonuclease, the ^{3}H amino acid rapidly becomes acid-soluble unless EF-Tu•GTP is bound, presumably because the protein restricts access to the very ribonuclease-sensitive residues C74 and C75. The assay can be performed in a microtiter

format using a series of EF-Tu•GTP concentrations and a constant low concentration of ^{3}H-aa-tRNA (Pleiss and Uhlenbeck 2001). A plot of fraction-protected versus EF-Tu concentration can be used to calculate a K_D. An alternate assay is to form the complex at a sufficiently high protein concentration to achieve saturation, add ribonuclease, and take time points (Louie and Jurnak 1985). Since any aa-tRNA that dissociates from the complex is rapidly degraded, a dissociation rate constant can be obtained. Although the ribonuclease can cleave other parts of the tRNA in the complex, we find excellent correlation between the two assays for many different aa-tRNAs, suggesting that the value of the association rate constant does not change. Although both assays were used, the dissociation rate assay is easier to perform and more reproducible, since it does not require an independent determination of the fraction of active protein or involve the error-prone serial dilution of protein.

Preliminary binding experiments with the four different tRNAs esterified with their cognate amino acids confirmed the expectation that they all bound EF-Tu•GTP with a similar affinity. Control experiments also confirmed that the mutations introduced to promote misacylation did not affect EF-Tu affinity with the cognate amino acid. However, when the 12 misacylated tRNAs were assayed, many of the resulting affinities were either too tight or too weak to measure under the standard reaction conditions. When $K_D < 1$ nM, it was not possible to obtain a complete binding curve because the lowest protein concentrations needed were far less than a detectable amount of labeled tRNA. In addition, when $K_D < 5$ nM, the dissociation rate became so slow (< 0.03 min^{-1}) that spontaneous hydrolysis of the aminoacyl linkage could occur. Similarly, when $K_D > 100$ nM, the dissociation rate of the complex becomes comparable to the rate of ribonuclease digestion of the free tRNA, compromising the experiment.

In order to relate the K_D for the tight and weak binding misacylated tRNAs to the correctly acylated tRNAs, K_D values were determined for all 16 aa-tRNAs as a function of NH$_4$Cl concentration. Since log K_D increased linearly with increasing log [NH$_4$Cl] for all the aa-tRNAs, it was possible to get accurate data over at least a limited range of ionic strength for all 16. Thus, the tight-binding tRNAs could be measured at high ionic strength and the weak binding tRNAs at low ionic strength. Extrapolation of each data set to a reference ionic strength of 0.5 M NH$_4$Cl permits all 16 aa-tRNAs to be compared directly.

The comparison of the binding of the 16 aa-tRNAs to *T. thermophilus* EF-Tu•GTP can be made in two different ways. In Figure 2a, the ΔG values for each tRNA are shown with the four different esterified amino acids. It can be seen that although the absolute affinities differ, a similar hierarchy of amino acids is present on each tRNA body. Thus, each tRNA esterified with Gln is the tightest, followed by Phe, Val, and Ala. For a given tRNA, the affinities ranged by about 100-fold among these four amino acids. A similar conclusion can be made about the four tRNA bodies esterified to each amino acid (Fig. 2b). In this case, tRNA^{Ala} binds EF-Tu•GTP tighter than the

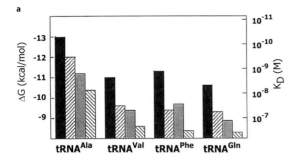

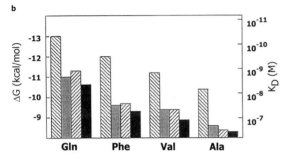

Figure 2. Affinities of aa-tRNA with *T. thermophilus* EF-Tu•GTP (LaRiviere et al. 2001). (*a*) Indicated tRNAs esterified with Gln (*black*), Phe (*left hatch*), Val (*gray*), and Ala (*right hatch*). (*b*) Indicated amino acids esterified to tRNA^Ala (*right hatch*), tRNA^Val (*gray*), tRNA^Phe (*left hatch*), and tRNA^Gln (*black*).

other three tRNAs. This is followed by the similar binding tRNA^Val and tRNA^Phe, with tRNA^Gln binding the weakest. For a given amino acid, the range of affinities among these four relatively similar tRNAs is from 40- to 120-fold. These results suggest that the amino acid and the tRNA body contribute independently to the overall binding energy. This conclusion was confirmed for each pair of tRNAs by showing that the sum of the ΔG values of the two cognate tRNAs was equal to the ΔG values of the same two tRNAs esterified with each other's amino acid.

Examining the data in Figure 2, it is clear that the comparatively uniform binding of the four correctly acylated tRNAs arises from a thermodynamic compensation mechanism where the differing contributions of the

amino acid and the tRNA to the binding affinity are combined to achieve a uniform overall affinity. Thus, a "tight" amino acid such as glutamine is correctly esterified to a "weak" tRNA^Gln, whereas the "weak" amino acid alanine is correctly esterified to the "tight" tRNA^Ala. Therefore, when a tRNA is incorrectly esterified with a noncognate amino acid, the resulting K_D can either be much tighter or much weaker than the correctly acylated counterpart. These effects are quantitatively substantial. For example, when the "weak" Ala is esterified onto the "weak" tRNA^Gln, the resulting affinity is about 60-fold weaker than the corresponding cognate Gln-tRNA^Gln. Similarly, the "tight" Gln esterified onto the "tight" tRNA^Ala has a K_D about 125-fold tighter than the cognate Ala-tRNA^Ala. Thus, the entire range of affinities among the 12 misacylated tRNAs tested here is about 5000-fold (Fig. 3), similar to the difference in affinities observed between acylated and deacylated tRNA^Phe (Janiak et al. 1990).

Our observation that EF-Tu shows quite specific binding for both the tRNA body and amino acid side chain is generally consistent with the previous observations described above. For example, the poor binding of misacylated Glu-tRNA^Gln to EF-Tu (Stanzel et al. 1994) is easily explained if we hypothesize that glutamate is a weak amino acid. Since we have found that tRNA^Gln is a comparatively weak tRNA, Glu-tRNA^Gln would be too weak to bind EF-Tu until its amino acid is converted to the tight-binding glutamine. In other words, thermodynamic compensation may explain how EF-Tu "knows" that the tRNA^Gln is misacylated.

It should be noted that the data set is quite limited so far, with contributions established for only 4 tRNAs and 4 amino acids. An important goal is to investigate the contributions of the remaining 16 amino acids and their esterified tRNAs using a similar approach. An even larger range of affinities for misacylated tRNAs can be anticipated. Another point to note is that the observed thermodynamic compensation is not perfect in the sense that although the 4 correctly acylated tRNAs show similar K_D values, they are not identical but differ by ninefold, similar to the range observed previously for cellular tRNAs (Louie and Jurnak 1985). This suggests that the observed variation cannot be attributed to the absence of modified nucleotides. In addition, we have confirmed previous experiments (Nazarenko et al. 1994) that the

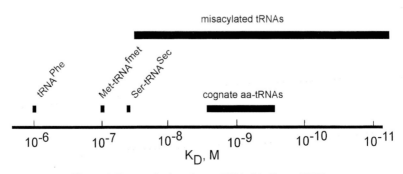

Figure 3. Range of values for aa-tRNAs binding to EF-Tu.

modified versions of tRNA^{Phe}, tRNA^{Val}, tRNA^{Ala}, and tRNA^{Gln} show nearly identical binding affinities to their unmodified counterparts. Thus, the variation may reflect differences in the intracellular concentrations of tRNAs as suggested previously (Jakubowski 1988) or, possibly, may be a consequence of the nonphysiological conditions used in the assays.

Relating Structure to Function

The fact that the amino acid side chain contributes to the affinity of aa-tRNA for EF-Tu•GTP is consistent with the co-crystal structures. The structures around the Phe and Cys side chain in the two co-crystals are shown in Figure 4. The side chain of the esterified amino acid is located in a roomy pocket made up of several peptide backbone residues, the side chains of eight amino acids, and the phosphodiester backbone of C$_{75}$ and A$_{76}$ of the tRNA. Although the overall structures are quite similar, some rearrangement in the pocket can occur depending on the esterified amino acid. In their respective structures, both the Phe and Cys side chains are within van der Waals distance of His-67 and Asn-285. However, in the Phe-tRNA^{Phe} structure, His-67 stacks on top of the Phe side chain, whereas Asn-285 forms a tighter-packing interaction with the Cys side chain in the Cys-tRNA^{Cys} structure. The only large difference between the two structures is the position of Arg-274, which may be a consequence of a nearby crystal lattice contact (Nissen et al. 1999). All in all, differences between the two structures are of the same scale as the differences between EF-Tu•GTP (Kjeldgaard et al. 1993) and EF-Tu•GTP•aa-tRNA, suggesting that the pocket maintains the same functional groups regardless of the presence or identity of the amino acid, yet is flexible enough to accommodate any of the 20 different amino acid side chains.

The observed amino acid hierarchy in Figure 2 can, to some extent, be rationalized by the structure and properties of the binding pocket. Phenylalanine presumably binds tighter than alanine and valine because of the energy it obtains by stacking with His-67. However, the tight binding of glutamine is less clear. Perhaps it makes a hydrogen bond with one of the side chains in the pocket. Glu-226, for example, appears to be close enough to form a hydrogen bond with the Gln side chain. The weaker binding of Glu and Asp compared to Gln and Asn discussed above can be rationalized by the overall negative charge of the amino-acid-binding pocket as calculated by the program GRASP (Nicholls et al. 1991). The contributions of several other side chains seem easy to predict on the basis of their size. Gly, like Ala, should be weak, whereas Trp and Tyr should stack well and therefore be tight. Indeed, the misacylated Tyr-tRNA^{Phe} binds as well as Phe-tRNA^{Phe} (Wagner and Sprinzl 1980), suggesting that both amino acids are tight. Since EF-Tu recognizes the α-amino group in the co-crystal structures, Pro, which lacks the amino group, may be weak. Less easy to predict are Ser, Met, His, Arg, and Ile, which will depend on how electrostatic, hydrophobic, and potential hydrogen bonds contribute to the binding energy.

A corresponding understanding of how the different tRNA bodies contribute differently to the overall binding energy is more difficult to achieve because the protein directly contacts the tRNA entirely through parts of the phosphodiester backbone that are thought to be identical in all tRNAs. One possibility, suggested by the comparison of the tRNA^{Phe} and tRNA^{Cys} structures (Fig. 1), is that the protein makes slightly different contacts with each tRNA, resulting in different affinities. Alternatively, the contacts may be identical for all tRNAs, but may have slightly different individual binding energies. In either case, the differences in affinity among tRNAs are pre-

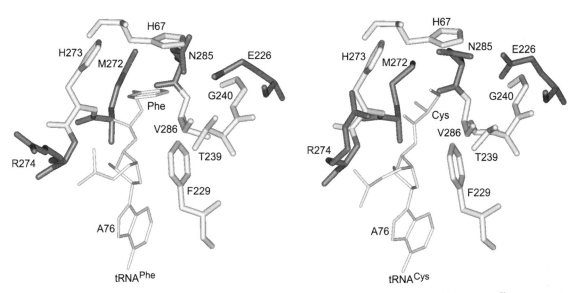

Figure 4. The amino-acid-binding pocket of *T. thermophilus* EF-Tu•GTP from the crystal structures with Phe-tRNA^{Phe} (*left*) and Cys-tRNA^{Cys} (*right*) (Nissen et al. 1995, 1999). The positions of the black amino acid side chains differ in the two structures, whereas the positions of the gray side chains and the aa-tRNA are similar. All the indicated side chains are highly conserved.

sumably the result of small differences in the structure or the dynamic properties of the tRNA backbone that are a consequence of the particular sequences of the acceptor and T helices. This explains why different sequences have been found to modulate EF-Tu affinity (Rudinger et al. 1996). Differential binding of DNA sequences solely through backbone contacts has been observed for several DNA-binding proteins and is often called "indirect readout" (Bareket-Samish et al. 1998; Leonard and Kerppola 1998; Martin et al. 1999).

To better understand how tRNA achieves its binding energy when binding EF-Tu•GTP, 44 different tRNA^{Phe} variants, each containing a single deoxynucleotide residue, were prepared and their affinities to *T. thermophilus* EF-Tu•GTP were determined (Pleiss and Uhlenbeck 2001). The data are summarized in Figure 5 and compared to the positions that contact 2′ hydroxyls in the co-crystal structure. Only five single deoxynucleotide-substituted tRNA^{Phe}s bind EF-Tu•GTP with a weaker binding affinity. Reassuringly, four of these five correspond to positions where a clear hydrogen bond occurs between the protein and the 2′ hydroxyl group. The fifth, at ribose 7, corresponds to an intramolecular hydrogen bond that contributes to the correct folding of the tRNA^{Phe} tertiary structure. Of the remaining 39 deoxynucleotide-substituted tRNAs that do not affect protein binding, most do not contact the protein and, therefore, serve as controls. However, there are five sites where the structure shows a potential hydrogen bond involving a 2′ hydroxyl group, but no change in binding energy was observed when a deoxynucleotide was introduced. In most cases, these "thermodynamically neutral" contacts can be explained by the structure of EF-Tu•GTP without bound aa-tRNA (Kjeldgaard et al. 1993). At these sites, the amino acid side chain involved in the contact is hydro-

gen-bonded to another amino acid on the surface of the free protein, and this bond must be broken when the tRNA is bound. Thus, there is no net free energy derived from forming the hydrogen bond in the complex. These experiments illustrate the need for biochemical experiments to interpret X-ray crystal structures. It will be interesting to evaluate the 2′ hydroxyl contacts in a "strong" tRNA body such as tRNA^{Ala} and see if they differ from those observed with the "weak" tRNA^{Phe}.

The above thermodynamic experiments employed EF-Tu from *T. thermophilus* or the very closely related protein from *T. aquaticus* in the crystal structures. Although no differences between these two proteins were observed, it will be interesting to investigate whether similar thermodynamic compensation is observed for EF-Tu proteins from other organisms. An alignment of 58 different bacterial EF-Tu sequences indicates that the amino acids that make thermodynamically important contacts with tRNA, as well as those that make up the amino-acid-binding pocket are nearly completely conserved. Although it remains possible that more distal amino acids that differ among bacterial EF-Tus could modulate the properties of aminoacyl-tRNA binding, it is reasonable to expect that some degree of thermodynamic compensation will be observed for most species. One interesting exception to the sequence conservation of bacterial EF-Tus is that His-273 in the amino-acid-binding pocket of *T. aquaticus* is a phenylalanine in most bacteria. Does this sequence change result in a different amino acid hierarchy for other bacteria? Another interesting point is that the alignment of eukaryotic sequences with the bacterial sequences reveals a conserved amino-acid-binding pocket, but less conservation in the amino acids that contact the T arm of tRNA. Indeed, a recent analysis of the singly deoxynucleotide-substituted tRNAs with wheat germ elongation

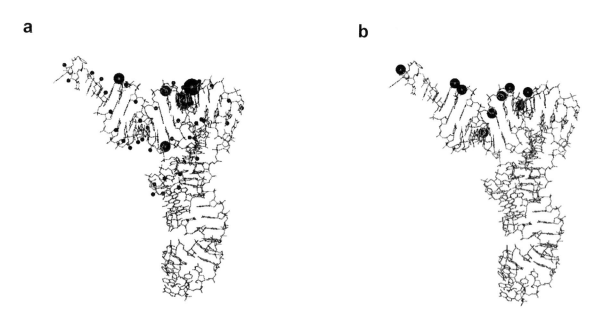

a   **b**

Figure 5. (*a*) Effects of a single deoxynucleotide substitution in yeast tRNA^{Phe} on the affinity of *T. thermophilus* EF-Tu•GTP (Pleiss and Uhlenbeck 2001). Large dots indicate the 2′ positions where a deoxynucleotide weakens K_D, small dots are where there is no effect. (b) Sites on tRNA^{Phe} where EF-Tu contacts 2′ hydroxyl groups in the co-crystal structure.

factor 1α indicates a slightly different pattern of thermodynamically relevant contacts in the T arms (J.A. Pleiss et al., in prep.). Thermodynamic compensation in this eukaryotic protein remains to be investigated.

THERMODYNAMIC COMPENSATION BY EF-Tu AND TRANSLATIONAL ACCURACY

We have shown that the relative lack of specificity of EF-Tu for different cognate aa-tRNAs is the result of a mechanism where the energies of specific binding to the amino acid and the tRNA body are combined in a compensatory manner. From a biophysical perspective, this situation can be viewed as an inevitable consequence of the need for a uniform binding energy and the fact that the amino acid side chain interacts with a pocket in the protein. Indeed, it would be very difficult for all the different amino acid side chains to interact in an asymmetric pocket in a protein without showing at least some difference in binding energy. A similar argument can be made to explain why the different tRNA bodies bind differently. Although the site of EF-Tu binding is essentially helical for all tRNAs, RNA helices can show sequence-dependent structural or dynamic differences (Shen et al. 1995; Holbrook and Kim 1997). Since many contacts are formed over an extensive RNA–protein interface, it is hardly surprising that overall tRNA affinities vary. Indeed, from a broader perspective, when any two macromolecules interact over a large interface, a significant amount of specificity can be expected. If no specificity is observed, some kind of complex compensation mechanism must occur. In other words, it is much easier to arrange a specific macromolecular interaction than a truly nonspecific one.

Given that a thermodynamic compensation mechanism is expected in the EF-Tu•tRNA complex, is there any reason why it evolved in the particular way that it did? That is, why has glutamine evolved to be a tighter amino acid and alanine a weaker one? One possibility is that discrimination by EF-Tu has evolved in such a way as to increase translational accuracy by excluding certain misacylated tRNAs from the ribosome. Misacylated tRNAs are produced by errors made by tRNA synthetases, either by using the wrong amino acid or the wrong tRNA substrate in the reaction. In the well-studied acylation of incorrect amino acids by synthetases, high accuracy is obtained by a double-sieve mechanism (Fersht and Dingwall 1979) where larger incorrect amino acids are excluded because of steric clashes in the amino-acid-binding site (the first sieve) and small incorrect amino acids are hydrolytically deacylated in a second "editing" site (the second sieve). Interestingly, the current limited data suggest that EF-Tu will not bind well to misacylated tRNAs when a small amino acid is esterified to a tRNA body meant for a larger amino acid, such as Ala-tRNA^{Val}. Thus, like the Glu-tRNA^{Gln} discussed above, such misacylated tRNAs may be blocked from translation due to their inability to interact with EF-Tu. Thus, EF-Tu may act as a "third sieve" by discriminating against those misacylated tRNAs that did not get efficiently edited.

One difficulty in proposing that EF-Tu discriminates against misacylated tRNAs is the observation that a substantial fraction of misacylated tRNAs bind EF-Tu more tightly than their correctly acylated counterparts, suggesting that they could get misincorporated into protein with abnormally high efficiency. From the available data, tight-binding aa-tRNAs generally arise when a large amino acid is esterified onto a tRNA meant for a small amino acid, such as Phe-tRNA^{Ala}. Such misacylated tRNAs can be made either by a failure of the first sieve or by aminoacylation of an incorrect tRNA. Considering the efficiency of the first sieve (Fersht and Dingwall 1979) and competition between aminoacyl tRNA synthetases for tRNAs in vivo (Yarus 1972; Swanson et al. 1988; Sherman et al. 1992), both these events are likely to be rare. Nevertheless, if such "tight" misacylated tRNAs do form, they should efficiently bind EF-Tu and be delivered to the ribosome. However, it is possible that such tRNAs will not function well in translation precisely because they bind EF-Tu so tightly. When the EF-Tu•GTP•aa-tRNA complex binds the ribosome, it is likely that many of the thermodynamically relevant protein–RNA contacts found in the free complex will be preserved. Thus, after decoding has occurred and GTP has been hydrolyzed, a tight-binding aa-tRNA may release EF-Tu•GDP and enter the ribosomal A site more slowly than usual, which could affect either the rate of translation or the efficiency of proofreading at near-cognate codons. In other words, the rate of release of an aa-tRNA by EF-Tu•GDP on the ribosome may be an equally important selective pressure for maintaining a uniform affinity between aa-tRNA and EF-Tu. An analysis of translational efficiency and accuracy of such tight-binding tRNAs is, therefore, of high priority.

HOW DOES CONFORMITY WORK ON THE RIBOSOME?

The complicated path of a tRNA molecule through the ribosome involves a substantial number of discrete intermediates where the tRNA makes several contacts, primarily with the rRNA. Known intermediates include the site of ternary complex binding, the classic A, P, and E sites, and at least three "hybrid" sites, but additional intermediates may be revealed as kinetic analysis improves. A molecular description of these sites is beginning to emerge. The 5.5 Å X-ray structures of tRNA in the A, P, and E sites of *T. thermophilus* ribosomes indicate that, aside from the anticodon and the terminal CCA, the majority of the contacts are made to the tRNA backbone, similar to that seen with EF-Tu (Yusupov et al. 2001). Although the description may be incomplete, it is clear that different residues on tRNA contact the ribosome at each site. Experiments evaluating the thermodynamic contribution of the contacts are at a very early stage (Koval'chuke et al. 1991; Dao et al. 1994; von Ahsen et al. 1997; Feinberg and Joseph 2001), but it would not be surprising if different tRNAs bind each site with slightly different thermodynamic details, as we have found with EF-Tu.

How, then, is conformity achieved in translation? Does the situation resemble EF-Tu and subtle differences in the structures of different tRNAs result in "homogenization" of the rate of translation? If this concept is correct, it remains unknown at which site on the ribosome and at which step in the mechanism such conformity is imposed. Among the steps in translation where different tRNAs are likely to act differently are when the codon–anticodon interaction is formed and when it is broken. The energetic and kinetic differences between the different codon–anticodon interactions are likely to be significant. For example, the affinity of tRNA[Phe] with the UUU codon is likely to be much weaker than tRNA[Gly] with the GGG codon, and the corresponding dissociation rate much faster. Are these differences offset by the extensive contacts between the backbone of the tRNA and the sites observed in the ribosome crystal structure? The well-established correlations between the sequence of the anticodon stem and suppressor efficiency (Curran and Yarus 1986; Raftery and Yarus 1987) support this idea. Experiments measuring elemental rates of translation (Thomas et al. 1988; Pape et al. 1998, 1999) using different tRNA bodies with identical anticodons and identical amino acids could help to resolve this issue.

One concept that emerges from this discussion is that part of the selective pressure on the sequences of tRNA helices may be to achieve conformity. Thus, the choice of certain acceptor-arm and T-stem sequences will subtly modulate EF-Tu affinity and perhaps ribosome affinity to offset different amino acid identities. Similarly, the choice of certain anticodon and possibly other sequences will subtly modulate decoding function to offset different codon–anticodon strengths. When these signals are integrated, each tRNA acts equivalently in translation. This view potentially complicates experiments evaluating the suppression activity of mutant tRNAs, where it is generally assumed that any tRNA which aminoacylates well will be used equivalently in translation. Thus, in a typical "swap" experiment where identity nucleotides are introduced into a different tRNA body, one has basically created a misacylated tRNA with unknown consequences in translation. It is possible that this problem is one reason why certain swap experiments were not successful. In addition, since suppression is a rare event in translation, a high suppression efficiency of a tRNA does not necessarily mean that the tRNA functions at optimal efficiency on the ribosome. Thus, analysis of such swapped tRNAs in normal elongation may reveal that they are not as active as the suppression assay suggests.

CONCLUSIONS

It is beginning to appear that the different tRNA molecules are not simply a set of generic adapters connecting amino acid and anticodon, but instead, have evolved their sequences in a way to meet the conflicting demands of having both specific and nonspecific functions. Thus, tRNA sequences must have different anticodons and be sufficiently different so that they can be distinguished by the aa-tRNA synthetases. However, the very structural features that make each tRNA unique make it difficult for it to progress through translation in a uniform manner. As a result, tRNA sequences have evolved to achieve conformity. Our understanding of structural features in tRNA that ensure conformity remains incomplete and will be an interesting topic for future experiments.

REFERENCES

Abrahamson J.K., Laue T.M., Miller D.L., and Johnson A.E. 1985. Direct determination of the association constant between elongation factor Tu X GTP and aminoacyl-tRNA using fluorescence. *Biochemistry* **24:** 692.

Bareket-Samish A., Cohen I., and Haran T.E. 1998. Direct versus indirect readout in the interaction of the trp repressor with noncanonical binding sites. *J. Mol. Biol.* **277:** 1071.

Becker H.D. and Kern D. 1998. *Thermus thermophilus:* A link in evolution of the tRNA-dependent amino acid amidation pathways. *Proc. Natl. Acad. Sci.* **95:** 12832.

Curran J.F. and Yarus M. 1986. Base substitutions in the tRNA anticodon arm do not degrade the accuracy of reading frame maintenance. *Proc. Natl. Acad. Sci.* **83:** 6538.

———. 1989. Rates of aminoacyl-tRNA selection at 29 sense codons in vivo. *J. Mol. Biol.* **209:** 65.

Dao V., Guenther R., Malkiewicz A., Nawrot B., Sochacka E., Kraszewski A., Jankowska J., Everett K., and Agris P.F. 1994. Ribosome binding of DNA analogs of tRNA requires base modifications and supports the "extended anticodon." *Proc. Natl. Acad. Sci.* **91:** 2125.

Feinberg J.S. and Joseph S. 2001. Identification of molecular interactions between P-site tRNA and the ribosome essential for translocation. *Proc. Natl. Acad. Sci.* **98:** 11120.

Fersht A.R. and Dingwall C. 1979. Evidence for the double-sieve editing mechanism in protein synthesis. Steric exclusion of isoleucine by valyl-tRNA synthetases. *Biochemistry* **18:** 2627.

Giege R., Sissler M., and Florentz C. 1998. Universal rules and idiosyncratic features in tRNA identity. *Nucleic Acids Res.* **26:** 5017.

Holbrook S.R. and Kim S.H. 1997. RNA crystallography. *Biopolymers* **44:** 3.

Hou Y.M. and Schimmel P. 1992. Novel transfer RNAs that are active in *Escherichia coli. Biochemistry* **31:** 4157.

Ibba M. and Söll D. 2000. Aminoacyl-tRNA synthesis. *Annu. Rev. Biochem.* **69:** 617.

Jakubowski H. 1988. Negative correlation between the abundance of *Escherichia coli* aminoacyl-tRNA families and their affinities for elongation factor Tu- GTP. *J. Theor. Biol.* **133:** 363.

Janiak F., Dell V.A., Abrahamson J.K., Watson B.S., Miller D.L., and Johnson A.E. 1990. Fluorescence characterization of the interaction of various transfer RNA species with elongation factor Tu.GTP: Evidence for a new functional role for elongation factor Tu in protein biosynthesis. *Biochemistry* **29:** 4268.

Kjeldgaard M., Nissen P., Thirup S., and Nyborg J. 1993. The crystal structure of elongation factor EF-Tu from *Thermus aquaticus* in the GTP conformation. *Structure* **1:** 35.

Knowlton R.G. and Yarus M. 1980. Discrimination between aminoacyl groups on su+7 tRNA by elongation factor Tu. *J. Mol. Biol.* **139:** 721.

Koval'chuke O.V., Potapov A.P., El'skaya A.V., Potapov V.K., Krinetskaya N.F., Dolinnaya N.G., and Shabarova Z.A. 1991. Interaction of ribo- and deoxyriboanalogs of yeast tRNA(Phe) anticodon arm with programmed small ribosomal subunits of *Escherichia coli* and rabbit liver. *Nucleic Acids Res.* **19:** 4199.

LaRiveire F.J., Wolfson A.D., and Uhlenbeck O.C. 2001. Thermodynamic compensation as a mechanism to achieve the uniform binding of aminoacyl-tRNA to elongation factor Tu. *Science* **294:** 165.

Leonard D.A. and Kerppola T.K. 1998. DNA bending determines Fos-Jun heterodimer orientation. *Nat. Struct. Biol.* **5:** 877.

Louie A. and Jurnak F. 1985. Kinetic studies of *Escherichia coli* elongation factor Tu-guanosine 5´- triphosphate-aminoacyltRNA complexes. *Biochemistry* **24:** 6433.

Louie A., Ribeiro N.S., Reid B.R., and Jurnak F. 1984. Relative affinities of all *Escherichia coli* aminoacyl-tRNAs for elongation factor Tu-GTP. *J. Biol. Chem.* **259:** 5010.

Martin A.M., Sam M.D., Reich N.O., and Perona J.J. 1999. Structural and energetic origins of indirect readout in site-specific DNA cleavage by a restriction endonuclease. *Nat. Struct. Biol.* **6:** 269.

McClain W.H. 1994. The tRNA identity problem: Past, present and future. In *tRNA: Structure, biosynthesis and function.* (ed. D. Söll and U.L. RajBhandary), p. 335. ASM Press, Washington, D.C.

McClain W.H., Schneider J., Bhattacharya S., and Gabriel K. 1998. The importance of tRNA backbone-mediated interactions with synthetase for aminoacylation. *Proc. Natl. Acad. Sci.* **95:** 460.

Nazarenko I.A., Harrington K.M., and Uhlenbeck O.C. 1994. Many of the conserved nucleotides of tRNA(Phe) are not essential for ternary complex formation and peptide elongation. *EMBO J.* **13:** 2464.

Nicholls A., Sharp K.A., and Honig B. 1991. Protein folding and association: Insights from the interfacial and thermodynamic properties of hydrocarbons. *Proteins* **11:** 281.

Nissan T.A., Oliphant B., and Perona J.J. 1999. An engineered class I transfer RNA with a class II tertiary fold. *RNA* **5:** 434.

Nissen P., Thirup S., Kjeldgaard M., and Nyborg J. 1999. The crystal structure of Cys-tRNACys-EF-Tu-GDPNP reveals general and specific features in the ternary complex and in tRNA. *Struct. Fold. Des.* **7:** 143.

Nissen P., Kjeldgaard M., Thirup S., Polekhina G., Reshetnikova L., Clark B.F., and Nyborg J. 1995. Crystal structure of the ternary complex of Phe-tRNAPhe, EF-Tu, and a GTP analog. *Science* **270:** 1464.

Normanly J., Ogden R.C., Horvath S.J., and Abelson J.N. 1986. Changing the identity of a transfer RNA. *Nature* **321:** 2163.

Ott G., Schiesswohl M., Kiesewetter S., Forster C., Arnold L., Erdmann V.A., and Sprinzl M. 1990. Ternary complexes of *Escherichia coli* aminoacyl-tRNAs with the elongation factor Tu and GTP: Thermodynamic and structural studies. *Biochim. Biophys. Acta* **1050:** 222.

Pape T., Wintermeyer W., and Rodnina M.V. 1998. Complete kinetic mechanism of elongation factor Tu-dependent binding of aminoacyl-tRNA to the A site of the *E. coli* ribosome. *EMBO J.* **17:** 7490.

———. 1999. Induced fit in initial selection and proofreading of aminoacyl-tRNA on the ribosome. *EMBO J.* **18:** 3800.

Pleiss J.A. and Uhlenbeck O.C. 2001. Identification of thermodynamically relevant interactions between EF-Tu and backbone elements of tRNA. *J. Mol. Biol.* **308:** 895.

Raftery L.A. and Yarus M. 1987. Systematic alterations in the anticodon arm make tRNA(Glu)-Suoc a more efficient suppressor. *EMBO J.* **6:** 1499.

Rould M.A., Perona J.J., Söll D., and Steitz T.A. 1989. Structure of *E. coli* glutaminyl-tRNA synthetase complexed with tRNA(Gln) and ATP at 2.8 Å resolution. *Science* **246:** 1135.

Rudinger J., Hillenbrandt R., Sprinzl M., and Giege R. 1996. Antideterminants present in minihelix(Sec) hinder its recognition by prokaryotic elongation factor Tu. *EMBO J.* **15:** 650.

Ruff M., Krishnaswamy S., Boeglin M., Poterszman A., Mitschler A., Podjarny A., Rees B., Thierry J.C., and Moras D. 1991. Class II aminoacyl transfer RNA synthetases: Crystal structure of yeast aspartyl-tRNA synthetase complexed with tRNA(Asp). *Science* **252:** 1682.

Saks M.E., Sampson J.R., and Abelson J.N. 1994. The transfer RNA identity problem: A search for rules. *Science* **263:** 191.

Shen L.X., Cai Z., and Tinoco I., Jr. 1995. RNA structure at high resolution. *FASEB J.* **9:** 1023.

Sherman J.M., Rogers M.J., and Söll D. 1992. Competition of aminoacyl-tRNA synthetases for tRNA ensures the accuracy of aminoacylation [corrected and republished article originally printed in *Nucleic Acids Res.* [1992] **20:** 1547. *Nucleic Acids Res.* **20:** 2847.

Sorensen M.A. and Pedersen S. 1991. Absolute in vivo translation rates of individual codons in *Escherichia coli.* The two glutamic acid codons GAA and GAG are translated with a threefold difference in rate. *J. Mol. Biol.* **222:** 265.

Stanzel M., Schon A., and Sprinzl M. 1994. Discrimination against misacylated tRNA by chloroplast elongation factor Tu. *Eur. J. Biochem.* **219:** 435.

Swanson R., Hoben P., Sumner-Smith M., Uemura H., Watson L., and Söll D. 1988. Accuracy of in vivo aminoacylation requires proper balance of tRNA and aminoacyl-tRNA synthetase. *Science* **242:** 1548.

Thomas L.K., Dix D.B., and Thompson R.C. 1988. Codon choice and gene expression: Synonymous codons differ in their ability to direct aminoacylated-transfer RNA binding to ribosomes in vitro. *Proc. Natl. Acad. Sci.* **85:** 4242.

von Ahsen U., Green R., Schroeder R., and Noller H.F. 1997. Identification of 2′-hydroxyl groups required for interaction of a tRNA anticodon stem-loop region with the ribosome. *RNA* **3:** 49.

Wagner T. and Sprinzl M. 1980. The complex formation between *Escherichia coli* aminoacyl-tRNA, elongation factor Tu and GTP. The effect of the side-chain of the amino acid linked to tRNA. *Eur. J. Biochem.* **108:** 213.

Yarus M. 1972. Intrinsic precision of aminoacyl-tRNA synthesis enhanced through parallel systems of ligands. *Nat. New Biology* **239:** 106.

Yusupov M.M., Yusupova G.Z., Baucom A., Lieberman K., Earnest T.N., Cate J.H., and Noller H.F. 2001. Crystal structure of the ribosome at 5.5 A resolution. *Science* **292:** 883.

Initiator tRNA and Its Role in Initiation of Protein Synthesis

C. Mayer, A. Stortchevoi, C. Köhrer, U. Varshney, and U.L. RajBhandary

Department of Biology, Massachusetts Institute of Technology, Cambridge, Massachusetts 02139

The initiation step of protein synthesis is quite different from the repetitive steps of elongation that follow it. Initiation involves the assembly of the two ribosomal subunits, mRNA, and the initiator tRNA in a reaction requiring the participation of several initiation factors, and is often the rate-limiting step in protein synthesis (Kozak 1983; Gualerzi and Pon 1990; RajBhandary and Chow 1995; Hershey and Merrick 2000). The initiator tRNA plays an important role in this process and in the selection of the appropriate initiation codon. Because of its special function, an initiator tRNA possesses many highly specific properties that are different from those of elongator tRNAs (RajBhandary 1994). We have been interested in identifying the sequence and/or structural features in the initiator tRNA important for specifying its distinctive properties and in understanding the molecular basis of the specific interactions of the initiator tRNA with the various proteins, the initiation factors, and the ribosome. Here we report on our work with the eubacterial *Escherichia coli* initiator tRNA. We describe briefly the system that we use for in vitro and in vivo functional analyses of mutant initiator tRNAs, a summary of the overall results, and some of our conclusions related to participation of the tRNA in initiation.

STRUCTURAL FEATURES OF *E. COLI* INITIATOR tRNA IMPORTANT FOR ITS OVERALL ACTIVITY

Protein synthesis is initiated with methionine or formylmethionine in all organisms studied to date. Of the two species of methionine tRNAs present in these organisms, the initiator is used for the initiation of protein synthesis, whereas the elongator is used for insertion of methionine into internal peptidic linkages (Clark and Marcker 1965; Kozak 1983). In eubacteria and in eukaryotic organelles such as mitochondria and chloroplasts, the initiator tRNAs are used as formylmethionyl-tRNAs (fMet-tRNA) (Marcker and Sanger 1964; Kozak 1983). In the cytoplasm of eukaryotes and in archaea, they are used as methionyl-tRNAs (Met-tRNA) (Housman et al. 1970; Smith and Marcker 1970; Ramesh and RajBhandary 2001). Figure 1 shows the steps in the utilization of initiator and elongator tRNAs in protein synthesis in eubacteria and highlights the special properties of initiator tRNAs. It shows that everything that happens

to the initiator tRNA subsequent to its aminoacylation is different from what happens to the elongator tRNAs. These steps include: (1) the formylation of the initiator Met-tRNA by the enzyme methionyl-tRNA formyltransferase (MTF) to form fMet-tRNA (Dickerman et al. 1967); (2) the binding of fMet-tRNA to the ribosomal P site in contrast to elongator tRNAs, which bind to the A site; (3) the very poor binding of the initiator Met-tRNA to the elongation factor EF-Tu, which ensures that the initiator tRNA functions exclusively in initiation (Seong and RajBhandary 1987b); and (4) the unique resistance of the initiator fMet-tRNA to peptidyl-tRNA hydrolase (PTH), which ensures the availability of fMet-tRNA for initiation instead of its hydrolysis to fMet and tRNA (Kossel and RajBhandary 1968).

The approach for identifying the features in the initiator tRNA important for specifying its distinctive properties involved mutagenesis of the initiator tRNA gene followed by analysis of function of the mutant initiator tRNAs at each of the steps of protein synthesis in vitro and in vivo. The in vitro work was facilitated greatly by the development of a one-step method for purification of mutant initiator tRNAs, free of wild-type initiator tRNA (Mandal and RajBhandary 1992). The in vivo work was similarly facilitated by the development of two methods: (1) the use of acid-urea polyacrylamide gel electrophoresis to separate the initiator tRNA, Met-tRNA, and fMet-tRNA from each other (Varshney et al. 1991), which allowed a direct analysis of the effect of mutations in the tRNA on its aminoacylation and formylation in vivo (Fig. 2), and (2) the use of a CAU→CUA anticodon sequence mutant (U35A36 mutant) of the initiator tRNA to initiate protein synthesis from UAG instead of AUG as the initiation codon of a reporter chloramphenicol acetyltransferase (CAT) gene (Fig. 3), which provided an assay for in vivo function of mutant tRNAs in initiation (Varshney and RajBhandary 1990). By coupling the mutations in the anticodon with mutations in the main body of the tRNA, it was possible to determine the effect of the latter mutations on the overall activity of the mutant tRNAs in initiation. In addition, because the mutant initiator tRNA reads UAG as a codon, by measuring the activity of the same mutant tRNAs in amber suppression in *E. coli* strains carrying an amber mutation in the chromosomal β-galactosidase gene, it was possible to identify mutations in the initiator tRNA that allow it to bind to EF-Tu and act

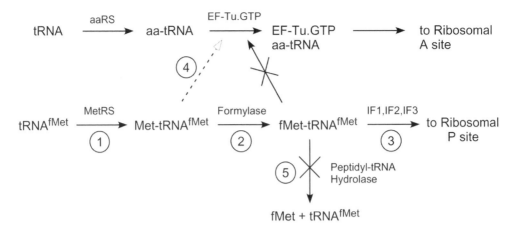

Figure 1. Steps in the utilization of elongator (*top*) and initiator (*bottom*) tRNAs in protein synthesis in eubacteria. (Reprinted, with permission, from RajBhandary and Chow 1995 [copyright ASM Press].)

as an elongator (Seong et al. 1989). Similarly, by comparing the effect of mutations on the properties of the mutant tRNAs in a strain carrying a temperature-sensitive peptidyl-tRNA hydrolase with those in a parent strain, it was possible to identify mutations in the initiator tRNA that make it a substrate for PTH (Atherly and Menninger 1972; Lee et al. 1992).

The combined results of such in vitro and in vivo studies allowed us to identify the main features in the initiator tRNA important for aminoacylation and for specifying each of its distinctive properties (Fig. 4). The results of in vitro and in vivo studies were remarkably consistent (Raj-Bhandary 1994). The three consecutive G:C base pairs conserved in all initiator tRNAs are important for binding of the tRNA to the ribosomal P site (Seong and Raj-Bhandary 1987a). Interestingly, the base mismatch conserved at the end of the acceptor stem in all eubacterial and chloroplast initiator tRNAs (C1xA72 in case of the *E. coli* tRNA) is important for specifying three of its four distinctive properties. The elements critical for identity of the *E. coli* initiator tRNA are clustered mostly in the acceptor stem and in the anticodon stem. This conclusion has been confirmed by introduction of these elements into

E. coli elongator methionine and glutamine tRNAs and showing that the mutant tRNAs are now active in initiation (Guillon et al. 1993; Varshney et al. 1993).

A COMMON MECHANISM FOR INITIATOR–ELONGATOR DISCRIMINATION IN EUBACTERIA AND IN EUKARYOTES

The primary negative determinant blocking activity of the *E. coli* initiator tRNA in elongation is the C1xA72 mismatch at the end of the acceptor stem (Seong and Raj-

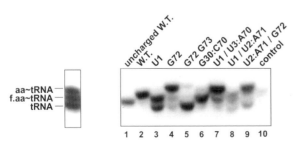

Figure 2. Analysis of the effect of mutations of *E. coli* initiator tRNA on aminoacylation and formylation in vivo. (Reprinted, with permission, from Lee et al. 1992 [copyright National Academy of Science].)

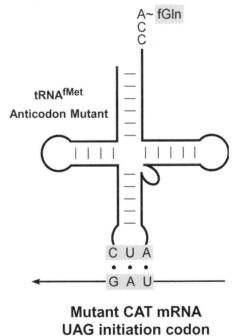

Figure 3. Codon–anticodon pairing between mutant CAT mRNA carrying AUG→UAG initiation codon change and mutant initiator tRNA carrying CAU→CUA anticodon sequence change. (Reprinted, with permission, from RajBhandary and Chow 1995 [copyright ASM Press].)

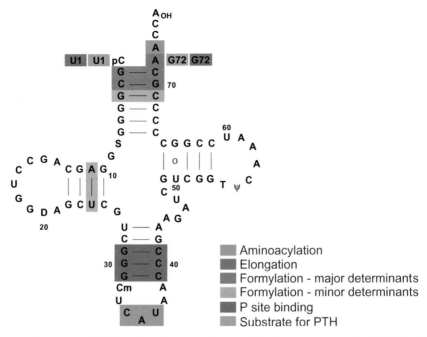

Figure 4. Regions of *E. coli* initiator tRNA important for specifying its various properties highlighted in color.

Bhandary 1987b). This is due to the perturbation of the RNA helical structure caused by the C1xA72 mismatch. Mutations of C1xA72 to either a U1:A72 base pair (U1 mutant) or to a C1:G72 base pair (G72 mutant) produced tRNAs that could bind to EF-Tu and that were active in elongation in vitro and in vivo. The G72 mutant Met-tRNA, with the stronger C1:G72 base pair, bound to EF-Tu tighter than the U1 mutant with the weaker U1:A72 base pair. The G72 mutant was also more active in elongation in vitro than the U1 mutant. However, the activity

of the G72 mutant in elongation was still lower than that of the elongator Met-tRNA. Therefore, we searched for secondary negative determinants in the *E. coli* initiator tRNA, which might affect its activity in elongation, and have identified the U50.G64 wobble base pair, which is conserved near the junction of the extended acceptor stem-TΨC stem helix in many eubacterial initiator tRNAs, as a secondary negative determinant. Mutation of U50.G64 to U50:A64 (A64 mutant) or to C50:G64 (C50 mutant) significantly increased the activity in amber sup-

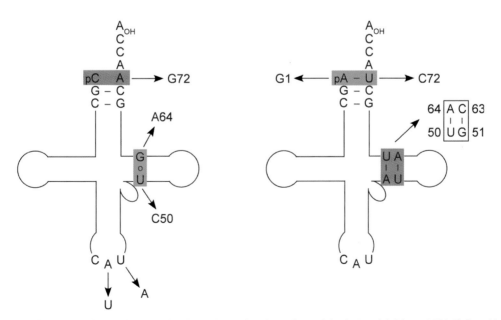

Figure 5. Sites of primary and secondary negative determinants for elongation activity in *E. coli* initiator tRNA (*left*) and human initiator tRNA (*right*) highlighted in color. (*Red*) Primary negative determinant; (*green*) secondary negative determinant. Arrows indicate mutations that confer partial elongation activity on the mutant tRNAs.

Table 1. Effect of Mutation of U50.G64 Wobble Base Pair to C50:G64 or U50:A64 Watson-Crick Base Pairs on Activity of U35A36/G72 Mutant Initiator tRNA in Elongation as Measured by Suppression of an Amber Codon in the β-Galactosidase Gene

Mutant tRNA	Relative β-galactosidase activity
U35A36	.01
U35A36/G72	1
U35A36/G72/C50	2.6
U35A36/G72/A64	5.02

pression in *E. coli* of mutant initiator tRNAs carrying the G72 mutation in the acceptor stem and the U35A36 mutations in the anticodon sequence (Table 1 and Fig. 5, left) (A. Storchevoi et al., unpubl.). The increased activity of the A64 and C50 mutants in elongation is likely due to their increased affinity toward EF-Tu, compared to the tRNA with the U50.G64 wobble base pair, although this remains to be established.

Interestingly, eukaryotic initiator tRNAs also have negative determinants for elongation activity at exactly these sites (Fig. 5, right), except that the primary negative determinant is in the base pair 50:64 and 51:63 region of the TΨC stem, and the secondary negative determinant is the A1:U72 base pair highly conserved at the end of the acceptor stem of eukaryotic initiator tRNAs (Drabkin et al. 1998). Mutation of the A1:U72 base pair in human initiator tRNA to G1:C72 produces a tRNA that has partial activity in elongation in rabbit reticulocyte cell-free extracts. Mutations of A50:U64 and U51:A63 base pairs to U50:A64 and G51:C63 base pairs, respectively, produce a tRNA that is more active in elongation than the G1:C72 mutant. Combination of these TΨC stem mutations with the G1:C72 mutation produces a tRNA that is even more active in elongation than either of the mutants, with the activity of the last mutant approaching that of the human elongator tRNA. Results in mammalian COS1 cells are essentially the same as the results in vitro. It is quite likely that the A50:U64 and U51:A63 base pairs act as negative determinants by perturbing the sugar phosphate backbone in the TΨC stem and thereby blocking the binding of the initiator Met-tRNA to eEF1. The structural perturbation could be due to a weakening of the RNA helix because of the consecutive A:U and U:A base pairs near the junction of the acceptor stem-TΨC stem helix or due to a change in the helical structure, which alters the sugar phosphate backbone conformation and/or position from that of an RNA A helix. In contrast to vertebrate initiator tRNAs, plant and fungal initiator tRNAs have a bulky modification attached to the 2′-hydroxyl of ribose 64 (Simsek and Raj Bhandary 1972; Desgres et al. 1989), which protrudes into the minor groove of the TΨC stem helix and most likely sterically blocks binding to the eukaryotic elongation factor eEF1 (Kiesewetter et al. 1990; Basavappa and Sigler 1991; Forster et al. 1993; Astrom and Bystrom 1994).

The mechanism by which the A1:U72 base pair functions as a secondary negative determinant in eukaryotic initiator tRNAs for elongation could be analogous to that of bases 1 and 72 in the *E. coli* initiator tRNA. There, a mismatch between bases 1 and 72 is the primary negative

determinant that prevents the initiator Met-tRNA from binding to EF-Tu. The *E. coli* initiator tRNA with a C1xA72 mismatch has a structure at the end of the acceptor stem that is distinct from those of normal elongator tRNAs. Whether the presence of a weak A1:U72 base pair in eukaryotic initiator tRNAs allows the initiator Met-tRNAs to adopt a different structure is not known. Interestingly, the only exception to the presence of A1:U72 in eukaryotic initiator tRNA is found in *Schizosaccharomyces pombe*, where it is Ψ1:A72, also a weak base pair.

Figure 6 highlights the location of the negative determinants in the initiator tRNAs for elongation in a ribbon diagram of the three-dimensional structure of tRNA. The importance of tRNA backbone conformation, at the end of the acceptor stem and near the junction of the acceptor stem-TΨC stem extended helix, for binding to the elongation factor can be understood from the co-crystal structures of the *Thermus aquaticus* EF-Tu•GDPNP•aminoacyl-tRNA ternary complexes (Nissen et al. 1995, 1999). The co-crystal structures show intimate contact of the protein with the sugar phosphate backbone of the aminoacyl-tRNA at the end of the acceptor stem and in the minor groove of the TΨC stem region around the sugar phosphate backbones of nucleotides 50–54 and 64–67. In particular, the 2′-OH groups of ribose 63 and 64 are H-bonded to the main-chain carbonyl and amide of Gly-391, respectively, thereby establishing a rigid contact between the protein main chain and the backbone of the TΨC stem of the tRNA (Nissen et al. 1999). Work on tRNA-like structures in plant viral RNAs also suggest contacts between the TΨC stem region and EF-Tu and the eukaryotic elongation factor eEF1 (Joshi et al. 1986). These findings, combined with our results above on *E. coli* and human initiator tRNAs (Seong and RajBhandary 1987b; Drabkin et al. 1998) and those of Bystrom, Sprinzl, and coworkers on the yeast and plant initiator tRNAs (Kiesewetter et al.

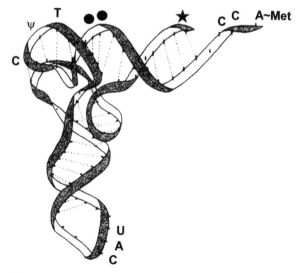

Figure 6. Ribbon diagram of 3D structure of tRNA highlighting the negative determinants for elongation in initiator tRNA.

1990; Forster et al. 1993; Astrom and Bystrom 1994), suggest that the requirements for eEF1 are quite similar to those for EF-Tu and suggest a common mode of binding of the eubacterial and eukaryotic elongation factors to the corresponding aminoacyl-tRNAs.

IMPORTANCE OF AMINO ACID ATTACHED TO THE INITIATOR tRNA IN INITIATION OF PROTEIN SYNTHESIS

The U35A36 mutant of the *E. coli* initiator tRNA, which initiates protein synthesis using UAG (Fig. 3), is a very poor substrate for methionyl-tRNA synthetase (MetRS) and is aminoacylated with glutamine (Schulman and Pelka 1985; Seong et al. 1989). Therefore, protein synthesis in this case is initiated with formylglutamine (Varshney and RajBhandary 1990). The amount of the reporter CAT protein produced in cells expressing the mutant CAT gene and overproducing this mutant initiator tRNA is approximately the same as in cells expressing the wild-type CAT gene and overproducing the wild-type initiator tRNA. However, in cells overproducing the U35A36 mutant initiator tRNA and also overproducing MetRS, the amount of CAT protein produced increased by 2.5-fold (Varshney and RajBhandary 1992). Since there was no overall change in the extent of aminoacylation and formylation of the U35A36 mutant tRNA in cells overproducing MetRS, a likely explanation for increased CAT activity is that some of the mutant tRNA is now aminoacylated with methionine and that tRNA carrying fMet is a better initiator than tRNA carrying fGln. Thus, some component of the initiation machinery, which interacts with the initiator tRNA subsequent to its aminoacylation and formylation, prefers fMet over fGln. We show below that this component is IF2.

To investigate this phenomenon further, we have studied several other anticodon sequence mutants of the initiator tRNA, which are aminoacylated with different amino acids (Pallanck and Schulman 1991), and studied their activities in initiation in vivo. To determine whether there are any steps limiting the activity of these mutants in initiation in vivo, we also analyzed the activity of the mutant tRNAs in cells overproducing the corresponding aminoacyl-tRNA synthetases (aaRS), MTF, IF2, or a combination of these. In parallel, we also followed the extent of aminoacylation and formylation of the mutant tRNAs in vivo. Results of these and other studies show that the amino acid attached to the initiator tRNA is important for recognition of the tRNA by MTF and by IF2.

The results obtained with a number of mutant initiator tRNAs carrying different amino acids are shown in Table 2. With the wild-type tRNA and the wild-type CAT gene, there is not much difference in the amounts of CAT protein produced under different conditions. With the tRNA aminoacylated with Gln also, there is not much difference in the amounts of CAT protein in cells overproducing GlnRS or MTF. However, there is a large increase in cells overproducing GlnRS and IF2 (Mangroo and Raj-Bhandary 1995). This suggests that aminoacylation and formylation of the mutant tRNA is not limiting; what is limiting is the activity of the fGln-tRNA in initiation, most likely because of its poor affinity for IF2. With the tRNA aminoacylated with valine, overproduction of ValRS or MTF leads to increases in CAT activity, overproduction of both leads to an even larger increase, but overproduction of IF2 has no effect. Thus, aminoacylation and formylation of the tRNA are limiting in vivo, but binding to IF2 is not (Wu et al. 1996). With the tRNA aminoacylated with isoleucine, aminoacylation, formylation, and binding of the fIle-tRNA to IF2 are all limiting (C. Mayer et al., unpubl.). The tRNA aminoacylated with phenylalanine is much less active in initiation (Pallanck and Schulman 1991), most likely because it is poorly aminoacylated in vivo and also probably because the fPhe-tRNA binds poorly to IF2 (C. Mayer et al., unpubl.). The above conclusions regarding the limitations of aminoacylation and formylation, based on CAT activities in cell extracts, are also confirmed by analysis of the state of the mutant tRNAs in *E. coli*. For example, with the tRNA aminoacylated with valine, the increases in CAT

Table 2. Effect of Overproduction of aaRS, MTF, and IF2 Either Singly or in Combination on Activity of the Various Mutant tRNAs in Initiation

CAT initiation codon (Initiator tRNA anticodon)	Amino acid attached	Fold increase in CAT activities[a] upon overproduction of							Limiting factors
		—	aaRS[b]	MTF	IF2	aaRS[b] +MTF	MTF +IF2	aaRS[b] +IF2	
AUG (CAU)	Met	1	<1	1	<1	1	<1	n.d.	—
UAG (CUA)	Gln	1	1	1	n.d.	1.2	n.d.	9	IF2
GUC (GAC)	Val	1.6	2.2	3	1.3	5.4	3.4	n.d.	aaRS and MTF
AUC (GAU)	Ile	1.1	2.7	3.2	1.5	5	6.2	n.d.	aaRS, MTF, and IF2
UUC (GAA)	Phe	.2	n.d.	n.d.	.4	n.d.	n.d.	3.3	aaRS and IF2

[a]CAT activity with AUG initiation codon is set as 1. To rule out the possibility of changes in CAT activity due to fluctuations in plasmid copy numbers, the CAT activities in cell extracts were all normalized to β-lactamase activity encoded by the same plasmid. (aaRS) Aminoacyl-tRNA synthetase. (n.d.) Not determined.

[b]The nature of aaRS overproduced (MetRS, GlnRS, ValRS, IleRS, or PheRS) depended on which amino acid was attached to the mutant initiator tRNA.

activities in cells overproducing ValRS, MTF, or both are paralleled by increases in extent of aminoacylation and formylation of the tRNA in vivo (Wu et al. 1996). Similarly, there is a reciprocal correlation between the effect of overproduction of IF2 on initiator activity and binding affinity of the formyl aminoacyl-tRNA toward IF2 as measured by gel-shift analysis (Wu and RajBhandary 1997) and by surface plasmon resonance spectroscopy (Fig. 7) (C. Mayer et al., unpubl.).

These and other results show that the amino acid attached to the initiator tRNA is inspected at least twice, once by MTF (Giege et al. 1973; Li et al. 1996) and once by IF2, the order of preference being Met>Gln~Phe>Val~Ile>Lys for MTF and fMet>fVal>fIle>fGln~fLys for IF2. The important role of the amino acid attached to the tRNA for formylation by MTF is highlighted by the finding that a tRNA, which contains the critical determinants for formylation but which is aminoacylated with lysine, is an extremely poor substrate for MTF in vivo, whereas the same tRNA aminoacylated with methionine is an excellent substrate (Varshney et al. 1993; Li et al. 1996). Thus, methionine may well be the best amino acid for initiation of protein synthesis in *E. coli*. Work with the eukaryotic initiator tRNA suggests that methionine may well be the best amino acid for initiation in eukaryotes also (Wagner et al. 1984; Drabkin and RajBhandary 1998). These findings are consistent with the fact that methionine is used to initiate protein synthesis in all organisms and further suggest that contacts of IF2 or eIF2, respectively, with the formylmethionine or the methionine attached to the initiator tRNA may make a significant energetic contribution toward the discrimination of initiator tRNAs from elongator tRNAs. O.C. Uhlenbeck and coworkers (pers. comm.) have also noted a significant effect of the amino acid attached to the tRNA on the binding affinity of aminoacyl-tRNAs toward EF-Tu.

IS IF2 A CARRIER OF fMet-tRNA TO THE RIBOSOME?

How the fMet-tRNA gets to the ribosome during the assembly of the initiation complex is not established. On the basis of in vitro studies, it was suggested initially that IF2 acts as a carrier of fMet-tRNA to the ribosome in much the same way as EF-Tu does for aminoacyl-tRNAs. However, the binding of fMet-tRNA to IF2 is weak (dissociation constant in the micromolar range) and is destabilized by magnesium ions. On the basis of these findings and the estimated intracellular concentrations of IF2 and its affinity for the 30S ribosome in the presence of the other initiation factors, it has been suggested that the 30S ribosomes in cells are all bound to IF2 and that IF2 selects the fMet-tRNA from the pool of cellular tRNAs after binding to the 30S ribosome rather than binding to fMet-tRNA and carrying it to the ribosome (Gualerzi and Wintermeyer 1986; Zucker and Hershey 1986; Gualerzi and Pon 1990). Our results (Table 2) on the effect of overproduction of IF2 on the increased activity in initiation of fGln-, fIle-, and fPhe-tRNAs are not easily explained if ribosomes are already saturated with IF2 and if IF2 binds first to the ribosome and then to the formyl aminoacyl-tRNAs (fAA-tRNAs). They are more easily explained if IF2 is acting as a carrier of these fAA-tRNAs to the ribosome (Wu and RajBhandary 1997). According to this hypothesis, the fGln-, fIle-, and fPhe-tRNAs are less active than fMet-tRNA because they bind less well to IF2. In cells overproducing IF2, more of these tRNAs would bind to IF2 and be carried to the ribosome. The conclusion that IF2 can form a binary complex with initiator tRNA in vivo is also supported by the work of Blanquet and coworkers (Guillon et al. 1996). The reciprocal correlations between the effect of overproduction of IF2 on activity of initiator fMet-, fVal-, and fGln-tRNAs in cells and their affinities in vitro for IF2 (Fig. 7) (Wu and RajBhandary 1997) are also consistent with this hypothesis. It is clearly desirable to obtain similar data on the relative binding affinities of fIle- and fPhe-mutant initiator tRNAs toward IF2, and this work is in progress.

The above conclusion that IF2 can act as a carrier of initiator tRNA to the ribosome is based on activity of mutant initiator tRNAs in cells overproducing IF2. The possibility that the situation is different in cells not overproducing IF2 cannot, therefore, be ruled out (Guillon et al. 1996). It is important to note, however, that aminoacyl-tRNAs (Kozak 1983), the eukaryotic initiator Met-tRNA (Hershey and Merrick 2000), and the selenocysteine inserting SeCys-tRNA (Forchhammer et al. 1989) are all carried to the ribosome by a protein carrier, EF-Tu or eEF1α, eIF2, and the SelB protein, respectively. This raises the question of why the transport of fMet-tRNA to the ribosome should be an exception. Our results could provide an answer to this otherwise puzzling question.

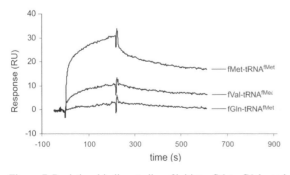

Figure 7. Real-time binding studies of initiator fMet-, fVal-, and fGln-tRNAs to *E. coli* IF2 using surface plasmon resonance in a BIACORE 3000 unit. IF2 was immobilized on the surface of a CM5 chip and the formylaminoacyl-tRNAs were all used at 1 μM concentration.

MOLECULAR BASIS OF RECOGNITION OF INITIATOR tRNA BY MTF

The formylation of initiator Met-tRNA by MTF is important for initiation of protein synthesis in eubacteria. Formylation provides a positive determinant for allowing IF2 to select the initiator fMet-tRNA over other tRNAs (Sundari et al. 1976). Because the amino group of

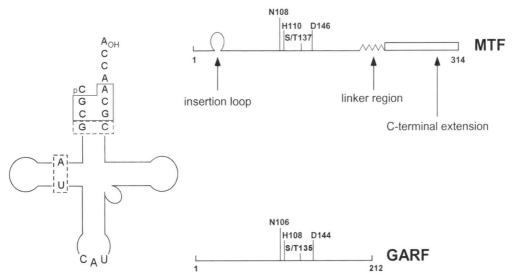

Figure 8. (*Left*) Determinants in the *E. coli* initiator tRNA important for formylation. (*Boxed line*) Major determinants; (*dotted line*) minor determinants. (*Right*) Schematic alignment of MTF (*top*) and GARF (*bottom*) sequences. Arrows indicate sites of insertions in MTF compared to GARF.

aminoacyl-tRNA plays an important role in the binding of aminoacyl-tRNA to EF-Tu (Nissen et al. 1995), formylation also provides an additional negative determinant for blocking the binding of the initiator tRNA to EF-Tu (Guillon et al. 1993). In *E. coli*, mutants of the initiator tRNA that are defective in formylation are essentially inactive in initiation (Varshney et al. 1991). Additionally, an *E. coli* strain carrying disruptions in the gene for MTF grows extremely slowly at 37°C and not at all at 42°C (Guillon et al. 1992a). In view of the importance of formylation of the initiator tRNA for its function in initiation, our laboratory and that of Blanquet have been studying the mechanism of the specific recognition of initiator tRNAs by MTF. The results from our two laboratories using biochemical, genetic, and structural approaches have provided a satisfyingly consistent picture of the molecular mechanism of recognition. In our laboratory, we have used a combination of approaches involving tRNA–protein cross-linking, isolation and analysis of suppressor mutations in MTF, site-specific mutants of MTF, and footprinting experiments on both the tRNA and the protein. Blanquet and coworkers have focused on site-specific mutants of MTF and on structural studies of MTF and the MTF•fMet-tRNA complex.

Table 3. Kinetic Parameters in Formylation of Mutant tRNA[fMet]

tRNA	V_{max}/K_m Wild type
	V_{max}/K_m Mutant
tRNA[fMet]	1.0
G72	495
G72C73	3.7
G72U73	3.7
G72G73	>1000

Requirements in the tRNA for Formylation

As shown above, the sequence and/or structural elements in the tRNA important for formylation by MTF are clustered mostly at the end of the acceptor stem (Fig. 8, left) (Lee et al. 1991, 1992; Guillon et al. 1992b). The key determinants are a mismatch (as found in the wild-type tRNA) or a weak base pair between nucleotides 1 and 72, a G:C base pair between nucleotides 2 and 71, and a C:G base pair between nucleotides 3 and 70. Mutations at the G4:C69 base pair also affect formylation kinetics but have less of an effect than mutations at the above positions. Mutations at position 73 affect formylation minimally except for the G73 mutation. The effect of G73 mutation is quite specific and is, therefore, likely due to negative interactions and/or an altered structure (Lee et al. 1993).

The nature of the mismatch between nucleotides 1 and 72 is less important for formylation than that they must not form a strong Watson-Crick base pair. tRNAs carrying the wild-type CxA mismatch or virtually any mismatch are good substrates, whereas tRNAs carrying strong base pairs, such as C1:G72 or G1:C72, are extremely poor substrates (Guillon et al. 1992b; Lee et al. 1992). These results suggest that nucleotides 1 and 72 must be unpaired for formylation to occur. Interestingly, the severe effect on formylation of a C1:G72 base pair can be compensated for by a secondary mutation of the neighboring base 73 to C or U but not to G (Table 3). The most likely explanation of this result is that the C1:G72 base pair, which is at the end of an RNA helix and may, therefore, have a tendency to "breathe," is stabilized by stacking of the neighboring base A73, on top of the C1:G72 base pair (Lee et al. 1992). Change of A73 to pyrimidine bases could destabilize the C1:G72 base pair due to loss of this stacking interaction and thereby lower the energetic cost of disrupting this base pair. The increased accessibility of C1 in such mutant tRNAs toward sodium bisulfite, which reacts only with cytidine residues that are unpaired and unstacked is consistent with this idea (Lee et al. 1993).

Outside of the acceptor stem, the only mutant with an alteration in kinetic parameters for formylation (K_m up by a factor of about 7) is the A11:U24 to C11:G24 mutant. This finding explains the strict conservation of the purine 11: pyrimidine 24 base pair in eubacterial, mitochondrial, and chloroplast initiator tRNAs (Lee et al. 1991).

Crystal Structure of MTF

The crystal structure of MTF (Schmitt et al. 1996) shows that the enzyme consists of two domains connected by a linker region. The amino-terminal domain structure contains a Rossmann fold, and its α-helix and β-strand topology is strikingly homologous to that of *E. coli* glycinamide ribonucleotide formyltransferase (GARF), which, like MTF, uses N^{10}-formyltetrahydrofolate (fTHF) as a formyl donor in formylation reactions (Fig. 8, right). The amino-terminal domain of MTF also contains the same amino acid residues, Asn-106, His-108, Ser- or Thr-135, and Asp-144, thought to be important for catalysis in GARF (Warren et al. 1996) in exactly the same order and with the same spacing, suggesting that the two classes of enzymes are closely related evolutionarily and that the basic mechanism of formyl group transfer is very similar between them. A notable difference between MTF and GARF sequences is the insertion of a 16-amino-acid sequence in MTF, in the loop region between the second β-strand and the second α helix from the amino terminus (Schmitt et al. 1996). This insertion sequence is found in all MTFs and is always 16 amino acids long in eubacterial MTFs (Ramesh et al. 1997). Another difference between MTF and GARF is the presence of an ~100-amino-acid extension at the carboxyl terminus of MTF. This carboxy-terminal extension, which consists of a β barrel with five antiparallel β strands, is structurally homologous to the anticodon-binding domain of some aminoacyl-tRNA synthetases. It has a positive electrostatic potential on the surface pointing toward the catalytic center of MTF and binds tRNA on its own, although nonspecifically (Schmitt et al. 1996). These findings suggest that the insertion loop and the carboxy-terminal extension together are important for tRNA recognition by MTF, with the carboxy-terminal extension being used mostly for nonspecific binding of tRNA and for orientation of the 3′ end of the initiator Met-tRNA toward the catalytic center of MTF and the insertion loop for the specific recognition of the determinants in the initiator tRNAs.

Topology of Initiator tRNA–MTF Interactions

First evidence that amino acids in the insertion loop play an important role in recognition of the determinants for formylation in the tRNA came from analysis of suppressor mutations in MTF (Ramesh et al. 1997). The expression of CAT genes carrying UAG as the initiation codon is dependent on the presence of an anticodon sequence mutant of the initiator tRNA, which can read the UAG codon (Fig. 3). We showed that G72G73 mutations in the acceptor stem, which removed one of the critical determinants for formylation (Table 3), essentially abolished activity of the mutant tRNA in initiation. As a result, cells carrying the mutant CAT gene and the mutant initiator tRNA do not make CAT and are sensitive to chloramphenicol. Using this system, we isolated suppressor mutations in MTF, which compensate for the formylation defect of the mutant initiator tRNA. The suppressor mutant had a Gly-41 to arginine change in the insertion loop. The mutant with Gly-41 changed to lysine also acts as a suppressor whereas other mutants do not, suggesting that a positive charge at this position is necessary for suppression. In vitro studies with the mutant enzyme(s) using wild-type and mutant tRNAs as substrate show that the suppressor mutation compensates specifically for the strong negative effect of the G72G73 mutation on formylation. These and other results suggest that this region of MTF comes close to the critical determinants for formylation in the acceptor stem of the initiator tRNA and probably interacts directly with them. A positively charged amino acid at position 41 in the insertion loop could facilitate melting of an otherwise stable C1:G72 base pair and thereby promote formylation of the mutant tRNA.

Studies on mutants of MTF within the insertion loop obtained by site-specific mutagenesis provided further support for interaction of amino acids in the insertion loop with the determinants for formylation in the acceptor stem (Ramesh et al. 1998; Schmitt et al. 1998). It was found that the basic amino acids in the insertion loop were important for MTF function; in particular, Arg-42, which is adjacent to Gly-41, was very important. Mutations of Arg-42 to any amino acid except to lysine resulted in an enzyme that was much less active, the Arg-42 to leucine mutant being essentially inactive in vitro and in vivo (V. Ramesh et al., unpubl.). Comparison of activity of the wild-type and Arg-42 mutant enzymes on wild-type and acceptor stem mutants of initiator tRNA suggests that the Arg-42 is interacting directly with the C3:G70 base pair and also to some extent the G2:C71 base pair, which are among the important determinants for formylation (Ramesh et al. 1998).

The crystal structure of MTF and the unique sensitivity of the insertion loop toward proteases indicated that amino acids 40–45 of MTF were unstructured and flexible (Schmitt et al. 1996). We therefore examined whether the very poor activity of the Arg-42 to leucine mutant could be rescued by mutation of the preceding Gly-41 to arginine. The kinetic parameters of the Gly-41 to arginine and Arg-42 to leucine double mutant were essentially the same as that of the Arg-42 to leucine mutant. Thus, the requirement for an arginine at position 42 cannot be fulfilled by an Arg at position 41. This suggests that the insertion loop adopts a defined conformation upon binding to the tRNA and supports the idea of the side chain of

Arg-42 making specific contacts with some of the key determinants for formylation in an induced-fit mechanism (Ramesh et al. 1998).

Further evidence for an induced fit of the insertion loop in MTF came from studies of protection of MTF against proteolytic cleavage by the initiator Met-tRNA substrate. We found that the initiator Met-tRNA protects MTF against trypsin cleavage, whereas a formylation-defective G72 mutant initiator Met-tRNA, which binds to MTF with approximately the same affinity, does not. Additionally, mutants of MTF within the insertion loop, which are defective in formylation, are not protected by the initiator Met-tRNA. These results suggest that a functional MTF•Met-tRNA complex is necessary for protection of the insertion loop of MTF against proteolytic cleavage and suggest that a segment of the insertion loop, which is exposed and unstructured in MTF, undergoes an induced fit in the functional MTF•Met-tRNA complex but not in the nonfunctional complex (Ramesh et al. 1999). The notion of a conformational change in the insertion loop of MTF, subsequent to binding, only in a functional MTF•Met-tRNA complex is analogous to the situation with aminoacyl-tRNA synthetase•tRNA complexes, in which a conformational change is triggered by cognate tRNAs but not by noncognate tRNAs (Schimmel and Söll 1979).

The results of footprint analysis on the effect of MTF on cleavage of Met-tRNA by nucleases show that MTF protects the acceptor stem and the 3′-end region of the tRNA but not the D loop, anticodon stem and loop, or the variable loop. This protection also depends on the formation of a functional MTF•Met-tRNA complex, suggesting that the insertion loop interacts mostly with the acceptor stem of the initiator Met-tRNA, which contains the critical determinants for formylation (Ramesh et al. 1999).

Finally, results of mutagenesis of basic and aromatic amino acids conserved in the carboxy-terminal domain of MTF are consistent with the notion that this region interacts mostly in a nonspecific manner with the tRNA. Mutation of most of the lysine residues to alanine has relatively minor effects, but mutations to glutamic acid have strong effects (Gite et al. 2000).

Co-crystal Structure of the
MTF•fMet-tRNA Complex

The crystal structure of the complex (Schmitt et al. 1998) shows that the enzyme binds to the tRNA on the inside of the L-shape of the tRNA (Fig. 9, top). The enzyme undergoes two substantial conformational changes upon binding to the tRNA. The insertion loop, which is unstructured in the free enzyme, adopts a well-defined structure with many of the amino acids in the loop making contact with the determinants in the acceptor stem (Fig. 9, top; compare green and black ribbons). The carboxy-terminal domain moves as a unit so as to fill the inside of the L-shape of the tRNA (Fig. 9, top; compare red and black ribbons). The contacts of the enzyme to the tRNA are limited to where the main determinants for

formylation are, the acceptor stem and the 3′-end region and the D stem region. The acceptor stem is clamped between the insertion loop and part of the carboxy-terminal domain of MTF. Part of the insertion loop runs parallel to bases 72–76 of the acceptor stem and the 3′-end region. In agreement with the results of suppressor mutants, Gly-41 comes close to A72 and A73 with the backbone NH of Gly-41 forming H-bonds with N1 and N3 of A72 and A73, respectively. The backbone NH of Gly-43 is H-bonded to O6 of G2, another important determinant for formylation. The side chain of Arg-42 is involved in several interactions including H-bonding to O6 and N7 of G70, and the guanidinium group is stacked on top of C69. The presence of a mismatch between bases 1 and 72 at the end of the acceptor stem, a critical determinant for formylation, allows the amino acid residues in the insertion loop to contact the major groove of base pairs 2:71 and 3:70 as well as the phosphate backbone in the 3′-end region. This causes the 3′ end of the tRNA to dip into the active-crevice site (Schmitt et al. 1998).

The carboxy-terminal domain of MTF is seen to interact mostly nonspecifically with the acceptor stem and the D stem. Most of the contacts are between the phosphate backbone and an electropositive channel on the surface of MTF. Base-specific contacts include contact of the side chain of Asn-301 with O2 of U24 and the backbone carbonyl group of Asn-301 with N2 of G12.

Although the co-crystal structure of the MTF•fMet-tRNA complex involves the product of the formylation reaction rather than the substrate, the crystal structure confirms the identification of the determinants for formylation based on biochemical and genetic experiments and is in accord with essentially all of the data generated in our laboratory and that of Blanquet using a combination of biochemical and genetic studies. Figure 9, bottom, shows two examples. Analysis of suppressor mutants of MTF suggested that Gly-41 (red) must come very close to bases A72 and/or A73 (green) of the tRNA (Ramesh et al. 1997). This is seen in the crystal structure of the complex (Fig. 9, bottom left). Similarly, the results of chemical cross-linking of periodate-oxidized tRNA to MTF indicated a cross-link of ribose of A76 specifically to Lys-206 (Gite and RajBhandary 1997). The crystal structure shows that these two residues on the tRNA and MTF are indeed quite close (Fig. 9, bottom right).

Finally, the crystal structure highlighting the roles of the insertion loop and the carboxy-terminal extension confirms our previous conclusion that MTF consists of a modular structure in which polypeptide sequences necessary for tRNA binding and discrimination have been inserted into an ancestral enzyme, capable of binding fTHF and transferring the formyl group to simple acceptors (Ramesh et al. 1997).

ACKNOWLEDGMENTS

This work was supported by grant R37GM17151 from the National Institutes of Health. We thank Annmarie McInnis for patience and care in the preparation of this manuscript.

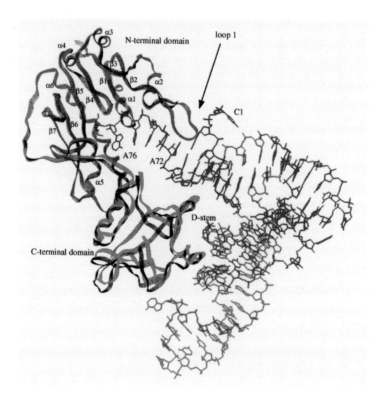

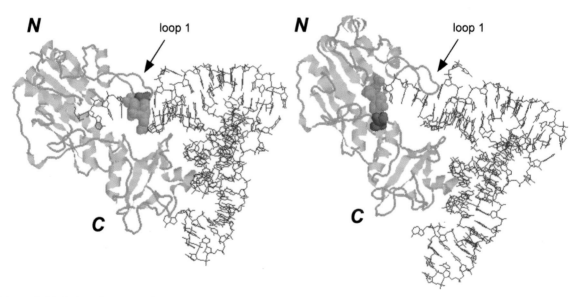

Figure 9. (*Top*) Crystal structure of *E. coli* MTF•fMet-tRNA complex. tRNA in blue wire frame and MTF in ribbon; amino-terminal domain, green; and the linker and the carboxy-terminal domain, red. Also shown is the structure of the free MTF, black ribbon. (Reprinted, with permission, from Schmitt et al. 1998 [copyright Oxford University Press].) (*Bottom*) Diagram highlighting the proximity of Gly-41 (*red*) of MTF and A72 and A73 (*green*) of tRNA on the left and the proximity of Lys-206 (*red*) of MTF and A76 (*green*) of tRNA on the right.

REFERENCES

Astrom S.U. and Bystrom A.S. 1994. Rit1, a tRNA backbone-modifying enzyme that mediates initiator and elongator tRNA discrimination. *Cell* **79:** 535.

Atherly A.G. and Menninger J.R. 1972. Mutant *E. coli* strain with temperature sensitive peptidyl-transfer RNA hydrolase. *Nat. New Biol.* **240:** 245.

Basavappa R. and Sigler P.B. 1991. The 3 Å crystal structure of yeast initiator tRNA: Functional implications in initiator/elongator discrimination. *EMBO J.* **10:** 3105.

Clark B.F.C. and Marcker K.A. 1965. Coding response of N-formyl-methionyl-sRNA to UUG. *Nature* **207:** 1038.

Desgres J., Keith G., Kuo K.C., and Gehrke C.W. 1989. Presence of phosphorylated O-ribosyl-adenosine in T-Ψ-stem of yeast methionine initiator tRNA. *Nucleic Acids Res.* **17:** 865.

Dickerman H.W., Steers E., Jr., Redfield B.G., and Weissbach H. 1967. Methionyl soluble ribonucleic acid transformylase. I. Purification and partial characterization. *J. Biol. Chem.* **242:** 1522.

Drabkin H.J. and RajBhandary U.L. 1998. Initiation of protein synthesis in mammalian cells with codons other than AUG and amino acids other than methionine. *Mol. Cell. Biol.* **18:** 5140.

Drabkin H.J., Estrella M., and RajBhandary U.L. 1998. Initiator-elongator discrimination in vertebrate tRNAs for protein synthesis. *Mol. Cell. Biol.* **18:** 1459.

Forchhammer K., Leinfelder W., and Bock A. 1989. Identification of a novel translation factor necessary for the incorporation of selenocysteine into protein. *Nature* **342:** 453.

Forster C., Chakrabuttty K., and Sprinzl M. 1993. Discrimination between initiation and elongation of protein biosynthesis in yeast: Identity assured by a nucleotide modification in the initiator tRNA. *Nucleic Acids Res.* **21:** 5679.

Giege R., Ebel J.P., and Clark B.F. 1973. Formylation of mischarged *E. coli* tRNA$_f^{Met}$. *FEBS Lett.* **30:** 291.

Gite S. and RajBhandary U.L. 1997. Lysine 207 as the site of cross-linking between the 3'-end of *Escherichia coli* initiator tRNA and methionyl-tRNA formyltransferase. *J. Biol. Chem.* **272:** 5305.

Gite S., Li Y., Ramesh V., and RajBhandary U.L. 2000. *Escherichia coli* methionyl-tRNA formyltransferase: Role of amino acids conserved in the linker region and in the C-terminal domain on the specific recognition of the initiator tRNA. *Biochemistry* **39:** 2218.

Gualerzi C.O. and Pon C.L. 1990. Initiation of mRNA translation in prokaryotes. *Biochemistry* **29:** 5881.

Gualerzi C.O. and Wintermeyer W. 1986. Prokaryotic initiation factor 2 acts at the level of the 30 S ribosomal subunit. *FEBS Lett.* **202:** 1.

Guillon J.M., Mechulam Y., Blanquet S., and Fayat G. 1993. Importance of formylability and anticodon stem sequence to give a tRNA^{Met} an initiator identity in *Escherichia coli*. *J. Bacteriol.* **175:** 4507.

Guillon J.M., Mechulam Y., Schmitter J.M., Blanquet S., and Fayat G. 1992a. Disruption of the gene for Met-tRNA$_f^{Met}$ formyltransferase severely impairs growth of *Escherichia coli*. *J. Bacteriol.* **174:** 4294.

Guillon J.M., Meinnel T., Mechulam Y., Lazennec C., Blanquet S., and Fayat G. 1992b. Nucleotides of tRNA governing the specificity of *Escherichia coli* methionyl-tRNA$_f^{Met}$ formyltransferase. *J. Mol. Biol.* **224:** 359.

Guillon J.M., Heiss S., Soutourina J., Mechulam Y., Laalami S., Grunberg-Manago M., and Blanquet S. 1996. Interplay of methionine tRNAs with translation elongation factor Tu and translation initiation factor 2 in *Escherichia coli*. *J. Biol. Chem.* **271:** 22321.

Hershey J.W.B. and Merrick W.C. 2000. The pathway and mechanism of initiation of protein synthesis. In *Translation control of gene expression* (ed. N. Sonenberg et al.), p. 33. Cold Spring Harbor Laboratory Press, Cold Spring Harbor, New York.

Housman D., Jacobs-Lorena M., RajBhandary U.L., and Lodish H.F. 1970. Initiation of haemoglobin synthesis by methionyl-tRNA. *Nature* **227:** 913.

Joshi R.L., Ravel J.M., and Haenni A.-L. 1986. Interaction of turnip yellow mosaic virus Val-RNA with eukaryotic elongation factor Ef-1a. Search for function. *EMBO J.* **5:** 1143.

Kiesewetter S., Ott G., and Sprinzl M. 1990. The role of modified purine 64 in initiator/elongator discrimination of tRNA$_f^{Met}$ from yeast and wheat germ. *Nucleic Acids Res.* **18:** 4677.

Kossel H. and RajBhandary U.L. 1968. Studies on polynucleotides. LXXXVI. Enzymatic hydrolysis of N- acylaminoacyl-transfer RNA. *J. Mol. Biol.* **35:** 539.

Kozak M. 1983. Comparison of initiation of protein synthesis in procaryotes, eucaryotes, and organelles. *Microbiol. Rev.* **47:** 1.

Lee C.P., Seong B.L., and RajBhandary U.L. 1991. Structural and sequence elements important for recognition of *Escherichia coli* formylmethionine tRNA by methionyl-tRNA transformylase are clustered in the acceptor stem. *J. Biol. Chem.* **266:** 18012.

Lee C.P., Mandal N., Dyson M.R., and RajBhandary U.L. 1993. The discriminator base influences tRNA structure at the end of the acceptor stem and possibly its interaction with proteins. *Proc. Natl. Acad. Sci.* **90:** 7149.

Lee C.P., Dyson M.R., Mandal N., Varshney U., Bahramian B., and RajBhandary U.L. 1992. Striking effects of coupling mutations in the acceptor stem on recognition of tRNAs by *Escherichia coli* Met-tRNA synthetase and Met-tRNA transformylase. *Proc. Natl. Acad. Sci.* **89:** 9262.

Li S., Kumar N.V., Varshney U., and RajBhandary U.L. 1996. Important role of the amino acid attached to tRNA in formylation and in initiation of protein synthesis in *Escherichia coli*. *J. Biol. Chem.* **271:** 1022.

Mandal N. and RajBhandary U.L. 1992. *Escherichia coli* B lacks one of the two initiator tRNA species present in *E. coli* K-12. *J. Bacteriol.* **174:** 7827.

Mangroo D. and RajBhandary U.L. 1995. Mutants of *Escherichia coli* initiator tRNA defective in initiation. Effects of overproduction of methionyl-tRNA transformylase and the initiation factors IF2 and IF3. *J. Biol. Chem.* **270:** 12203.

Marcker K.A. and Sanger F. 1964. N-formyl-methionyl-s-RNA. *J. Mol. Biol.* **8:** 835.

Nissen P., Thirup S., Kjeldgaard M., and Nyborg J. 1999. The crystal structure of Cys-tRNA^{Cys}-EF-Tu-GDPNP reveals general and specific features in the ternary complex and in tRNA. *Struct. Fold. Des.* **7:** 143.

Nissen P., Kjeldgaard M., Thirup S., Polekhina G., Reshetnikova L., Clark B.F., and Nyborg J. 1995. Crystal structure of the ternary complex of Phe-tRNA^{Phe}, EF-Tu, and a GTP analog. *Science* **270:** 1464.

Pallanck L. and Schulman L.H. 1991. Anticodon-dependent aminoacylation of a noncognate tRNA with isoleucine, valine, and phenylalanine in vivo. *Proc. Natl. Acad. Sci.* **88:** 3872.

RajBhandary U.L. 1994. Initiator transfer RNAs. *J. Bacteriol.* **176:** 547.

RajBhandary U.L. and Chow C.M. 1995. Initiator tRNAs and initiation of protein synthesis. In *tRNA: Structure, biosynthesis, and function* (ed. D. Söll and U.L. RajBhandary), p. 511. American Society for Microbiology, Washington, D.C.

Ramesh V. and RajBhandary U.L. 2001. Importance of the anticodon sequence in the aminoacylation of tRNAs by methionyl-tRNA synthetase and by valyl-tRNA synthetase in an Archaebacterium. *J. Biol. Chem.* **276:** 3660.

Ramesh V., Gite S., and RajBhandary U.L. 1998. Functional interaction of an arginine conserved in the sixteen amino acid insertion module of *Escherichia coli* methionyl-tRNA formyltransferase with determinants for formylation in the initiator tRNA. *Biochemistry* **37:** 15925.

Ramesh V., Gite S., Li Y., and RajBhandary U.L. 1997. Suppressor mutations in *Escherichia coli* methionyl-tRNA formyltransferase: Role of a 16-amino acid insertion module in initiator tRNA recognition. *Proc. Natl. Acad. Sci.* **94:** 13524.

Ramesh V., Mayer C., Dyson M.R., Gite S., and RajBhandary U.L. 1999. Induced fit of a peptide loop of methionyl-tRNA formyltransferase triggered by the initiator tRNA substrate. *Proc. Natl. Acad. Sci.* **96:** 875.

Schimmel P.R. and Soll D. 1979. Aminoacyl-tRNA synthetases: General features and recognition of transfer RNAs. *Annu. Rev. Biochem.* **48:** 601.

Schmitt E., Blanquet S., and Mechulam Y. 1996. Structure of crystalline *Escherichia coli* methionyl-tRNA$_f^{Met}$ formyltransferase: Comparison with glycinamide ribonucleotide formyltransferase. *EMBO J.* **15:** 4749.

Schmitt E., Panvert M., Blanquet S., and Mechulam Y. 1998. Crystal structure of methionyl-tRNA$_f^{Met}$ transformylase complexed with the initiator formyl-methionyl-tRNA$_f^{Met}$. *EMBO J.* **17:** 6819.

Schulman L.H. and Pelka H. 1985. In vitro conversion of a methionine to a glutamine-acceptor tRNA. *Biochemistry* **24:** 7309.

Seong B.L. and RajBhandary U.L. 1987a. *Escherichia coli* formylmethionine tRNA: Mutations in GGG:CCC sequence conserved in anticodon stem of initiator tRNAs affect initiation of protein synthesis and conformation of anticodon loop. *Proc. Natl. Acad. Sci.* **84:** 334.

———. 1987b. Mutants of *Escherichia coli* formylmethionine tRNA: A single base change enables initiator tRNA to act as an elongator *in vitro*. *Proc. Natl. Acad. Sci.* **84:** 8859.

Seong B.L., Lee C.P., and RajBhandary U.L. 1989. Suppression of amber codons in vivo as evidence that mutants derived from *Escherichia coli* initiator tRNA can act at the step of elongation in protein synthesis. *J. Biol. Chem.* **264:** 6504.

Simsek M. and RajBhandary U.L. 1972. The primary structure of yeast initiator transfer ribonucleic acid. *Biochem. Biophys. Res. Commun.* **49:** 508.

Smith A.E. and Marcker K.A. 1970. Cytoplasmic methionine transfer RNAs from eukaryotes. *Nature* **226:** 607.

Sundari R.M., Stringer E.A., Schulman L.H., and Maitra U. 1976. Interaction of bacterial initiation factor 2 with initiator tRNA. *J. Biol. Chem.* **251:** 3338.

Varshney U. and RajBhandary U.L. 1990. Initiation of protein synthesis from a termination codon. *Proc. Natl. Acad. Sci.* **87:** 1586.

———. 1992. Role of methionine and formylation of initiator tRNA in initiation of protein synthesis in *Escherichia coli*. *J. Bacteriol.* **174:** 7819.

Varshney U., Lee C.P., and RajBhandary U.L. 1991. Direct analysis of aminoacylation levels of tRNAs in vivo. Application to studying recognition of *Escherichia coli* initiator tRNA mutants by glutaminyl-tRNA synthetase. *J. Biol. Chem.* **266:** 24712.

———. 1993. From elongator tRNA to initiator tRNA. *Proc. Natl. Acad. Sci.* **90:** 2305.

Wagner T., Gross M., and Sigler P.B. 1984. Isoleucyl initiator tRNA does not initiate eucaryotic protein synthesis. *J. Biol. Chem.* **259:** 4706.

Warren M.S., Marolewski A.E., and Benkovic S.J. 1996. A rapid screen of active site mutants in glycinamide ribonucleotide transformylase. *Biochemistry* **35:** 8855.

Wu X.Q. and RajBhandary U.L. 1997. Effect of the amino acid attached to *Escherichia coli* initiator tRNA on its affinity for the initiation factor IF2 and on the IF2 dependence of its binding to the ribosome. *J. Biol. Chem.* **272:** 1891.

Wu X.Q., Iyengar P., and RajBhandary U.L. 1996. Ribosome-initiator tRNA complex as an intermediate in translation initiation in *Escherichia coli* revealed by use of mutant initiator tRNAs and specialized ribosomes. *EMBO J.* **15:** 4734.

Zucker F.H. and Hershey J.W. 1986. Binding of *Escherichia coli* protein synthesis initiation factor IF1 to 30S ribosomal subunits measured by fluorescence polarization. *Biochemistry* **25:** 3682.

On Translation by RNAs Alone

M. Yarus

*Department of Molecular, Cellular and Developmental Biology, University of Colorado,
Boulder, Colorado 80309-0347*

In this paper, I discuss the first protein biosynthesis and ask in particular whether modern experiments bear on the ancient emergence of coded proteins. The crux of this investigation is the premise that there was an RNA World (White 1976; Gilbert 1986). Accordingly, we posit an era immediately preceding the current Nucleoprotein World, when our ancestral biota used a congener of RNA as the principal informational and catalytic molecule.

Given that an RNA World immediately preceded the current biological era, one predicts with some confidence that RNAs invented translation, that is, carried out the first templated synthesis of peptides. This is required in order that specific peptide catalysts displace RNAs as principal biomolecular agents, leading to the nucleoprotein biota dominant now.

We cannot be completely certain of the detailed biochemistry of an ancient RNA translation apparatus, but the reactions of translation are presently universal. Therefore, translation very much as we know it today dates at least to the era before separation of the three modern domains of life: eubacteria, archaea, and eukarya. A conservative hypothesis is that these ancient reactions descend from the immediately preceding translation system, and therefore ancestral translational RNAs would be required to:

1. Activate amino acids—carboxyl activation is required in any case to thermodynamically favor peptide formation

2. Synthesize aminoacyl-RNAs—the current universal connection of amino acids to coding nucleic acids independently argues that aminoacyl-RNAs are relics of an RNA World

3. Form peptide bonds

4. Implement (part of) the genetic code—given reaction 2, enough coding would be required to specify the synthesis of effective peptide catalysts, although some codon assignments to amino acids may be more recent (for review, see Knight et al. 1999).

In the rest of this text, I first show that examples of the four predicted reactions are within the repertoire of modern RNAs. Although I only have space to sketch the isolation and properties of the four sorts of active RNAs, these data show that an experimental proof of principle exists. That is, the chemistry of RNA-like molecules allows a simple RNA-only translation system. This review concludes with a discussion of the implications of this

demonstration, an indication that it may be possible to carry out selection experiments that go beyond proof of principle, and finally, a reflection on the validity of the four assumptions just above.

THE METHOD

In all cases, the findings referred to here are the result of selection-amplification or SELEX experiments (Fig. 1) (Ellington and Szostak 1990; Robertson and Joyce 1990; Tuerk and Gold 1990). In such experiments, 10^{13} to 10^{15} initial RNAs of randomized sequence (derived by transcription of synthetic DNA templates containing randomized tracts with all four nucleotides) are subjected to cycles composed of selection (that is, biochemical fractionation) and then amplification (that is, replication; e.g., cDNA synthesis, PCR, and transcription). The unique ability of nucleic acids to be replicated allows recycling and reapplication of the fractionation as many times as required to purify a rare (≥ 1 part 10^{15}) active sequence. Purified sequences with the initially exceptional property (small circle in Fig. 1) are cloned and thereby made completely homogeneous. Application of similar protocols has for the first time allowed answers to the type of open-ended questions that concern us here. We may now ask:

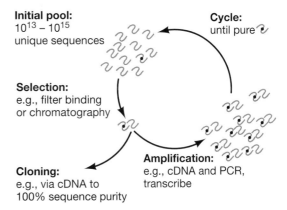

A Selection-Amplification Cycle

Initial pool:
$10^{13} - 10^{15}$
unique sequences

Cycle:
until pure

Selection:
e.g., filter binding
or chromatography

Amplification:
e.g., cDNA and PCR,
transcribe

Cloning:
e.g., via cDNA to
100% sequence purity

Figure 1. Course of isolation of an RNA by selection-amplification experiment. An RNA (*squiggle*) with a rare binding or catalytic activity (*circle*) is being purified.

Is there any RNA (among ≈1 nmole arbitrary sequences) that can carry out any reaction (for which a purifying selection exists)? Such investigation frees us from dependence on modern biological examples and answers a new type of experimental question, virtually unapproachable before.

THE FOUR ESSENTIAL REACTIONS

Amino Acid Activation

The chemistry of amino acid activation, as universally practiced by protein aminoacyl RNA synthetases (aaRS; Ibba and Söll 2000), is based on the formation of a mixed anhydride between AMP phosphate and amino acid carboxyl:

$$\text{aa-C(O)O}^- + \text{ATP} \leftrightarrow \text{AMP-O-C(O)-aa} + \text{PP}_i$$

The selection of this RNA reaction posed substantial difficulties (Yarus and Illangasekare 1999). The desired product is highly reactive (appropriate to a precursor for aa-RNA synthesis) and unstable to hydrolysis ($t_{1/2} < 10$ minutes), even at pH 7 and 0°C. The forward reaction can also be quite slow, because the carboxyl is quite stable on its own and therefore not a good nucleophile. After unsuccessful selections based on varied principles, a selection at lowered pH (=4; slows hydrolysis) with a substrate that yielded a stabilized product (a carboxylic rather than an α-amino acid) yielded a pool containing many independent RNA sequences. Therefore, the reaction itself, although disfavored, is not difficult for RNAs, in the sense that many sequences do it. The selection of the reactive RNAs relied on formation of a disulfide between an RNA-bound carboxylic acid (3-mercaptopropionic acid) and thiopropyl-Sepharose (Fig. 2A).

These RNAs are also able to activate amino acids (Kumar and Yarus 2001). In particular, formation of the mixed anhydride on the 5′ α-phosphate of the pppRNA transcript (RNA KK 13; Fig. 2B) was demonstrated using phenylalanine and leucine:

$$\text{Phe} + \text{pppRNA} \rightarrow \text{Phe-pRNA} + \text{PP}_i$$

Having been selected at lowered pH, the derived RNA was also most active at pH 4–4.5 (leucine $K_M = 48$ mM; k_{cat} (1.1 min^{-1}). Such a pH optimum might pose a problem for assembly of a complete RNA aaRS (performing activation and aa-RNA synthesis) at neutrality using this activation domain. However, this problem is likely to be solvable either by shifting the pH optimum by selection, or by selecting high reaction velocities which, although less than optimal, will still be rapid at pH 7. Activation by formation of these mixed acid anhydrides is presently universal (and therefore ancient). However, selected RNAs can use other chemically activated amino acids (such as cyanomethyl esters; Lee et al. 2000; Saito et al. 2001), and this versatility may have been exploited during evolution.

Aminoacyl-RNA Synthesis

In present-day translation, a cognate aaRS carrying its activated amino acid transfers it to the 2′(3′) terminus of an RNA carrying its anticodon. The synthesis of aa-RNA from activated amino acids has been known for some time to be an RNA reaction (Illangasekare et al. 1995):

$$\text{AMP-aa} + \text{RNA} \leftrightarrow \text{aa-RNA} + \text{AMP}$$

This reaction was first detected by incubating randomized RNA sequences with AMP-Phe. Prospective catalysts that aminoacylated themselves to Phe-RNA were captured by exploiting their chemically unique phenylalanine α-amino groups as sites of substitution with a large hydrophobic tag (the naphthoxy group; Fig. 3A). Hydrophobic-Phe-self-aminoacylating RNAs were then fractionated by HPLC and captured as a trailing peak (Illangasekare et al. 1995).

Study of structural variants of one selected class of active sequences ($K_M = 12$ mM ; $k_{cat} = 1.4$ min^{-1}; Illangasekare et al. 1997) allowed the reduction of a parental 95-mer to only those structures immediately surrounding the active site. RNAs as small as 29 nucleotides continue to form Phe-RNA at a rate similar to larger parental

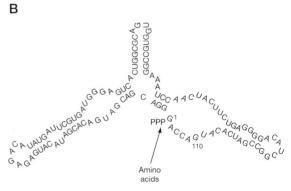

Figure 2. (*A*) Selection for carboxyl activation. An RNA that activates carboxyl groups by forming a mixed anhydride with its own 5′ α-phosphate is purified after incubation with mercaptopropionic acid at lowered pH. RNAs thereby tagged with a sulfhydryl are trapped by disulfide formation on thiopropyl-Sepharose resin. (*B*) An RNA that activates amino acids. RNA KK13, isolated by the selection in *A*, reacts with mercaptopropionic acid, Leu, and Phe to give bona fide activated (-C(O)O-PO$_3$-) products, judged by comparison with synthetic standards confirmed by mass spectroscopy.

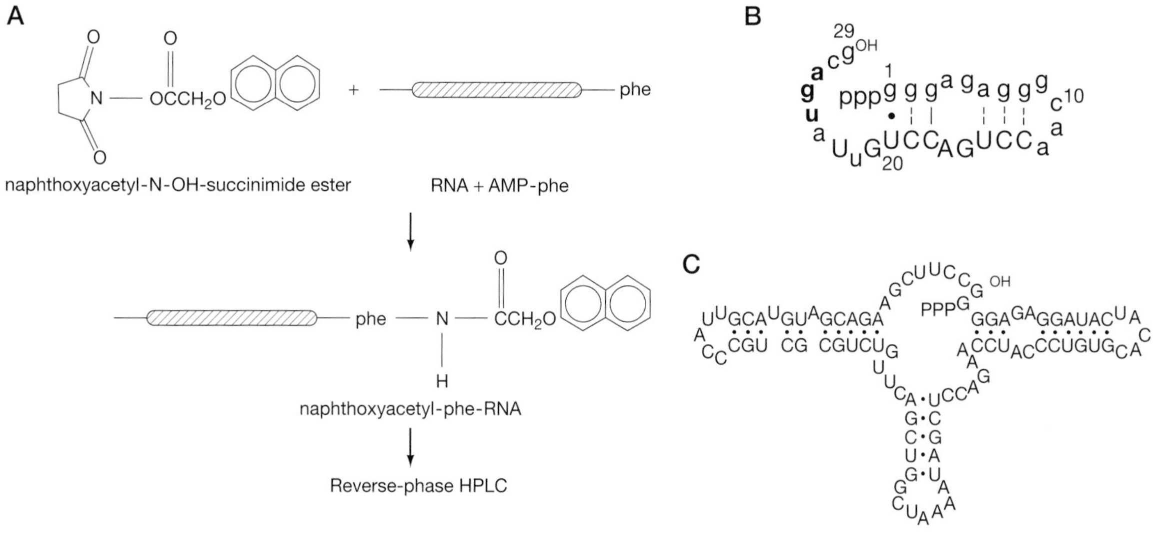

Figure 3. (*A*) Selection for aminoacyl-RNA synthesis. A self-aminoacylating RNA puts the amino acid on its 3′ end when incubated with aa-AMP. The unique α-amino group of the amino acid is tagged with the naphthoxy-group, and these unusually hydrophobic RNAs are isolated by HPLC. (*B*) A miniaturized (29 nucleotide) self-aminoacylating RNA that also catalyzes formation of peptide bonds. Uppercase nucleotides are conserved from the originally selected RNA. Bold lowercase nucleotides were randomized during selection of active small RNAs. The 3′-terminal CG-OH is also an essential sequence (M. Illangasekare et al., unpubl.). (*C*) A self-aminoacylating RNA that is very specific for formation of Phe-RNA and very rapidly reacting (see text).

molecules (Fig. 3B) (Illangasekare and Yarus 1999a). The removal of the 3′ domain of the self-aminoacylating RNAs during this contraction also accelerated a secondary reaction. The 3′ end of Phe-RNA formed initially returns to the active site and accepts a second (and further) phenylalanine to form the peptidyl RNA, Phe-Phe-RNA. Peptide synthesis for different small RNAs is 2- to 13-fold slower than Phe-RNA synthesis. However, this RNA of only 29 nucleotides accelerates two of the three reactions of translation; aminoacyl-RNA synthesis and peptide-bond formation. This finding, even in isolation, strengthens the argument for an ancient RNA translation apparatus. However, it is necessary to keep in mind that the 29-mer is using an unnatural substrate for peptide-bond formation (AMP-Phe) and also that peptide formation is uncoded. Perhaps the simplicity of this RNA catalysis might be taken as evidence for uncoded but RNA-catalyzed peptide synthesis as an evolutionary precursor of more sophisticated systems.

Larger self-aminoacylating RNAs show more rapid and selective catalysis (Fig. 3C) (Illangasekare and Yarus 1999b). A 90-mer RNA with a complex active site makes itself Phe-RNA 300-fold more rapidly ($K_M = 4.7$ mM; k_{cat} = 430 min^{-1}). In fact, the reaction was too rapid for hand measurement at neutrality, and the kinetics quoted here were derived by measurement at lower pH and extrapolation to pH 7.25. The inferred limiting first-order rate, 430 min^{-1}, is faster than modern PheRS proteins and among the fastest ribozymes known from any source. The same RNA, selected using phenylalanine, is highly accurate. For example, it shows a 40,000-fold discrimination against isoleucine in substrate form (Illangasekare and Yarus 1999b). This is more selective than modern PheRS proteins (Freist et al. 1996).

This RNA thus shows that RNAs can outperform modern proteins in the speed and accuracy of aminoacyl transfer. Because modern PheRS is plainly competent to support an evolutionarily successful translation system, this 90-mer suggests strongly that RNAs could once have supported the evolutionarily successful synthesis of aminoacyl-RNA. In fact, if the properties of this RNA can be generalized, RNAs roughly the size of modern tRNA clearly could have both specifically aminoacylated themselves and served as coding adapters. An RNA aminoacylation system can also be broken into an "enzyme" and "aa-RNA" molecules (Saito et al. 2001), which might have succeeded a unified progenitor.

Peptide Synthesis

An argument for aboriginal RNA-catalyzed peptide synthesis can be constructed from selection-amplification of RNAs that perform related reactions (see Yarus 1999). However, below I rely instead on studies of the biological peptidyl transferase and on selected RNAs that are related to the peptidyl transferase.

Transition state analogy. A successful transition state analog (TSA) is a stable compound that resembles the most transient intermediate formed during a chemical reaction. A TSA is complementary to the active site of a catalyst carrying out the reaction mechanism the TSA emulates. Thus, a TSA will bind strongly to that active site. Conversely, a surface selected for complementarity to an accurate TSA will resemble an active site for the same reaction.

Synthetic CCdApPuro (Fig. 4A) was designed as a TSA for the ribosomal peptidyl transferase, assuming direct nucleophilic attack as the peptide-forming mechanism (Welch et al. 1995). The synthetic molecule CCdApPuro has a P-site-directed portion (CCdA), an emulator of the tetrahedral, polar, reacting carbonyl (p) and an A-site-directed moiety (Puro = puromycin). CCdApPuro has served effectively in both senses mentioned above; both as affinity label and as a template for a ribosomal peptidyl transferase active site.

TSA as affinity reagent. CCdApPuro was diffused into crystals of the 50S ribosomal subunit from the archaean *Haloarcula marismortui,* and a high-resolution structure was derived (Ban et al. 2000). One molecule was bound and completely resolved by the crystallographic analysis. In Figure 4B, the small central green lobes are CCdApPuro bound to the active center. Thus, the peptidyl transferase active site was tagged, and the location can be verified—for example, by formation of a previously characterized P-site base pair (Samaha et al. 1995) and by overlap in the A site with other A-site analogs (Nissen et al. 2000). The reacting group is oriented to direct the nascent chain down the 50S "exit tunnel." The site is quite similar in 70S ribosomes, where A- and P-site tRNA 3′ ends appear at 5.5 Å resolution to converge in the same way (Yusupov et al. 2001). These findings confirm that the ribosomal reaction center is constructed to support the direct nucleophilic attack mechanism (Yarus and Welch 2000) embodied in the design of CCdApPuro (Welch et al. 1995). In addition, the 23S rRNA nucleotide A_{2451} (*E. coli* numbering) is hydrogen-bonded to the phosphate oxygen that represents the reacting P-site amino acid. This suggests that there could be catalytic effects beyond simple P-site /A-site proximity during peptide-bond formation (Nissen et al. 2000), perhaps including general acid or base catalysis or direct transition-state stabilization (Muth et al. 2000; Polacek et al. 2001).

Even more relevant to our present purpose, there was no ribosomal protein, only rRNA, within 18.4 Å (about a protein diameter) of the point of the peptide formation reaction (Nissen et al. 2000). In Figure 4B, the yellow ribosomal proteins do not approach the peptidyl transferase site, which is exclusively surrounded by the gray RNA. Thus, the *H. marismortui* peptidyl transferase is necessarily a ribozyme, barring very improbable ribosomal convulsions during peptide formation. Given that ribosomes from all three domains of life bind CCdApPuro (Yarus and Welch 2000), and the almost complete conservation of the binding sequences, peptidyl transferase has likely been a ribozyme since the last common ancestor of life on Earth. Furthermore, a ribozyme this ancient probably dates from the earlier RNA World. However, we provide a more direct argument bearing on this possibility in the next section.

TSA as template. RNAs were isolated from randomized pools, selecting for binding to agarose-linked CCdApPuro and for elution by free CCdApPuro (Welch et al. 1997). The most prevalent CCdApPuro-binding motif, which independently recurred twice in the initial randomized RNAs, was a conserved helix-loop containing 16 invariant nucleotides. These nucleotides did not vary under mutagenesis and reselection for binding, and a very

strong footprint from CCdApPuro confirms that the conserved loop is the binding site for the TSA. Remarkably, the essential loop sequence that was selected in vitro shares an octamer with the peptidyl transferase loop of 23S RNA. The red nucleotides in Figure 4C are the octamer, and the radial numbers show the conservation at each position in ribosomes from the three kingdoms. This ribosomal sequence is almost completely conserved in large subunit RNAs from thousands of sources (Yarus and Welch 2000). Even more remarkably, the conserved ribosomal octamer, one of 65,536 possible octamers, is also the binding site for CCdApPuro in the 50S ribosome crystal (Nissen et al. 2000).

The octamer core of the ribosomal peptidyl transferase ribozyme, including elements of the P site, the A site, and the potentially catalytic nucleotide A_{2451} (Muth et al. 2000; Polacek et al. 2001), emerged directly from randomized, 50-nucleotide, ribonucleotide sequences when affinity for reacting amino acids was selected (Fig. 4D) (Welch et al. 1997). These data show then that the core sequences of the presently universal peptidyl transferase could have emerged from short RNAs in an RNA World. Not only is the peptidyl transferase presently a ribozyme, but it is probably the same ribozyme that has persisted since an appearance long before the three domains of life appeared. This probably is attributable to the fact that peptidyl transferase substrates since very early on have been small ribonucleotides, lightly substituted with amino acids. No protein has been able to displace this ancient ribozyme, likely because an RNA is peculiarly qualified to guide reactants that are, by mass or surface area, predominantly ribonucleotides.

Coding

The genetic code is a set of equations, relating triplet nucleotide sequences to amino acids. If translation originated in an RNA World, some of those coding equations could be chemical ones; simply expressing the outcome(s) of nucleic acid–amino acid interactions (Woese et al. 1967; compare with Crick 1968). Because the code cannot have appeared suddenly in complete and final form, it probably evolved after it became possible to code for peptides. Thus, there may also be a part of the modern code that reflects other kinds of chemistry. Evidence from the code's structure supports, for example, coding assignments based on similarities between the chemistry of amino acid side chains (Freeland and Hurst 1998). In terms of the argument of this manuscript, the necessity is to show that RNA chemistry alone allowed the invention of a substantial ancestral coding table. In fact, a much more explicit and interesting conclusion seems plausible: that some of the modern genetic code was determined entirely by RNA–amino acid interactions, and is therefore a relic of the RNA World.

A POTENTIAL MOLECULAR FOSSIL

The origins of this argument lie in the first discovered side-chain-specific and stereoselective binding site for an

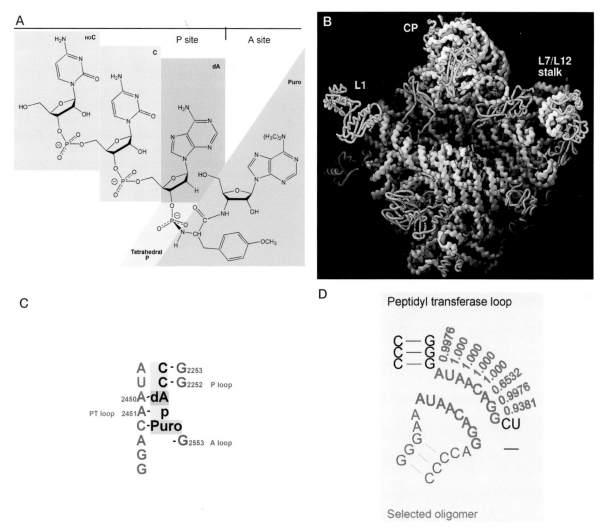

Figure 4. (*A*) A transition state analog, CCdApPuro, directed to the ribosomal peptidyl transferase site. Ribosomal P-site and A-site moieties are indicated. Different sections of the molecule are displayed on backgrounds of distinctive colors, retained in panel *C* to clarify the display. (Reprinted, with permission of Elsevier Science, from Yarus and Welch 2000.) (*B*) The crystallographic structure of the 50S ribosomal subunit, viewed toward the combining face for the 30S particle. Ribosomal proteins are the yellow tubes, the RNA backbone is shown as a light gray zigzag tube, and CCdApPuro bound to the particle is the small, central green lobed structure. (Reprinted, with permission, from Ban et al. 2000 [copyright AAAS].) (*C*) Schematic of the peptidyl transferase center. The red sequence is the selected octamer, and the blue sequences are other nucleotides that interact with reacting aminoacyl-RNAs. CCdAp-Puro, colored as in panel *A*, is shown in its binding site. Dashed lines show interactions of varied kinds, deduced from the crystal structure, with CCdApPuro or reacting aminoacyl-RNAs. Numbering corresponds to *E. coli* 23S rRNA. (Reprinted, with permission of Elsevier Science, from Yarus and Welch 2000.) (*D*) Comparison of the selected CCdApPuro-binding sequence with the conserved sequence of the 23S peptidyl transferase loop. The conserved octamers are in red, and the radial red numbers are the fractional conservation of the associated nucleotide in summed eubacterial, archaeal, and eukaryal rRNA sequences (R.R. Gutell and associates, www.rna.icmb.utexas.edu). (Reprinted, with permission of Elsevier Science, from Yarus and Welch 2000.)

amino acid in an RNA structure. L-Arginine uniquely inhibits self-splicing by the group I intron, competing for the splicing cofactor site usually occupied by a guanine nucleotide (Yarus 1988). In Figure 5 there is a view into the major groove of the site, with intron nucleotides represented as blocks with H-bonding groups labeled. Discovery of this site suggested that specific RNA–amino acid affinities, whose absence inhibited the development of a "stereochemical" (Woese et al. 1967) account of the code's origin, might nevertheless exist. In fact, when the splicing cofactor site was elucidated (Michel et al. 1989),

it had an even more interesting property. The nucleotides closest to bound G and Arg ligands in the group I active center were arginine triplets (Yarus and Christian 1989), and arginine triplets exclusively (nucleotides 263, 264, and 265 in Fig. 5; for review, see Yarus 1993). This further suggested that some coding assignments, including arginine, originated because primordial nucleic acids bound amino acids, and triplet pieces of these binding sites ultimately were adapted to serve as modern codons. An initial translation apparatus composed of such adjacent RNA-binding sites for activated amino acids (DRT the-

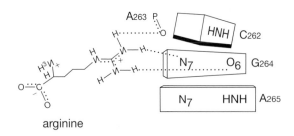

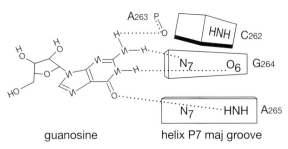

Figure 5. View into the major groove of the group I splicing cofactor site, with numbering as in *Tetrahymena*. Nucleotide bases in this hypothetical model structure are shown as blocks, with H-bonding groups inscribed on them. The dashed lines suggest how G nucleotides and arginine (on the left) can H-bond to occupy the same RNA site. The site is partially composed of the conserved arginine triplet nucleotides at nucleotides 263, 264, and 265. (Reprinted, with permission, from Yarus 1993 [copyright Cold Spring Harbor Laboratory Press].)

ory) has been defended as the simplest possible scheme for initiation of coded peptide synthesis (Yarus 1998), although other possibilities exist (Szathmáry 1999).

EXTENSION OF THE IDEA OF CODONS IN SITES

An unbiased general test of this idea became possible on the invention of selection-amplification. Using this technique, RNA-binding sites for amino acids could be derived ad libitum (Famulok and Szostak 1992; Connell et al. 1993). New sites can be examined to see whether triplets were unexpectedly concentrated within the binding-site nucleotides. In fact, newly selected L-arginine sites were the first to be examined in some detail, and like the natural example in the group I active center, they appeared to disproportionately contain codons (Connell and Yarus 1994). However, an objective statistical test was needed, and was supplied (Knight and Landweber 1998) by comparing the surrounding nucleotides in newly derived amino-acid-binding oligomers to the nucleotides within the amino acid sites. The site nucleotides were resolved by nucleotide protections, modification-interference experiments, and conservation in independent examples of the same sites. The surrounding nucleotides used as controls are in the same RNAs, come from the same randomized pools, via the same selection procedures, but are less (or not) related to the bound amino acid. The validity of chemical criteria for placing nucleotides in or out of the amino-acid-binding site can be supported where an independent three-dimensional structure is known, as for the nuclear magnetic resonance (NMR) structure of an arginine site (Yang et al. 1996; Yarus 2000). This statistical test strongly favors the concentration of arginine codons within arginine sites (Knight and Landweber 1998).

This statistical analysis for arginine binding has been criticized (Ellington et al. 2000), but it appears that the objections raised do not weaken the argument (Knight and Landweber 2000; Illangasekare and Yarus 2002). Moreover, the presently published examples extend far beyond the initial 5 arginine oligomers to 21 amino-acid-binding RNAs for the chemically disparate amino acids arginine, isoleucine (Majerfeld and Yarus 1998), tyrosine (Mannironi et al. 2000), and phenylalanine (Illangasekare and Yarus 2002). These sequences are generally the most prevalent (likely) structures that meet varied selections for amino acid binding. The RNAs comprise 1342 nucleotides in all, with 262 of those classified as site nucleotides by nucleotide protection, interference, and conservation. There is a 2.7-fold excess in the proportion of triplet nucleotides within all binding site nucleotides. The overall probability that these triplet sequences are distributed independently of binding sites is tiny; about 5×10^{-11} in these RNAs taken together (Table 1) (Illangasekare and Yarus 2002). Thus, codons do appear disproportionately within amino-acid-binding sites, and this is exceedingly unlikely to be an experimental accident. However, considering individual cases, phenylalanine does not demonstrably concentrate its triplets, whereas arginine, isoleucine, and tyrosine do so. The idea that some of the code is based directly on interactions between RNA and amino acids is accordingly strongly supported by a large body of data. The extension of such experiments may even provide a minimal list of amino acids whose coding assignments could have been made in an RNA World. As required for RNA-mediated synthesis of catalytic peptides, the amino acids potentially coded via RNA chemistry already seem to be both chemically varied and potentially substantial in number.

PROBABILITY OF AN RNA WORLD

The RNA World hypothesis predicts the existence of translational reactions via RNAs. The predicted reactions in turn have been shown to exist. The fulfillment of these multiple predictions should increase our confidence that the RNA World itself existed.

Table 1. The Combined Probability (Sokal and Rohlf 1995) That Cognate Codons Are Placed Randomly with Respect to Amino-acid-binding Sites in Four Independent Published Populations of Amino-acid-binding RNAs from Selection-amplification Are Considered

Amino acid [# sites]	Probability (P)	$-2 \ln P$ (Sum is chi-squared with 8 df)
Arginine [5]	8.12×10^{-8}	32.7
Isoleucine [5]	6.46×10^{-4}	14.7
Tyrosine [3]	5.43×10^{-3}	10.4
Phenylalanine [8]	2.67×10^{-2}	7.25

Combined $P = 4.8 \times 10^{-11}$ by Fisher's test for multiple experiments

This is especially so because none of the four outcomes seems trivial in any sense. For examples, at the time of the initiation of this work, it was uncertain whether other cases could be found like the group I arginine triplets. Amino acid activation is thermodynamically unfavorable. Aminoacyl-RNA synthesis required carbonyl chemistry before this was demonstrated to be within the ambit of RNA catalysis. It could not have been asserted with confidence, to say the least, that all peptide bonds in modern proteins were formed by a ribozyme.

How seriously should these results be taken as confirmation of the RNA World hypothesis itself? How can we objectively take account of the existence of four varied predicted reactions? There is a rational answer to such questions in the form of Bayes' Theorem (see, e.g., Meyer 1975), which is a prescription for re-evaluation of a hypothesis in the light of new evidence. The theorem, rewritten in a suitable form, is:

$$\frac{P(RNAworld \mid Expts)}{P(RNAworld)} = \frac{P(Expts \mid RNAworld)}{P(Expts)}$$

All P are probabilities for the parenthetical events, and the vertical bars can be read as "given that." The left-hand side is what we want to know; the ratio of probabilities assigned to the hypothesis of the RNA World's existence, before (denominator) and after (numerator) certain experiments (*Expts*) have been performed. The calculation is uninformative (both sides ≈1) for someone who already believed there was a historical RNA World, so we adopt the position of a skeptic. The right-hand side is the probability of the experimental outcomes given that the RNA World existed, divided by the a priori probability that the experiments would have succeeded. The probability of the experimental outcomes given that the RNA World existed is 1. RNA must have performed translational reactions to exit a real ancestral RNA World. The experiments, especially to the skeptic, are reasonably taken as independent. Thus:

$$\frac{P(RNAworld \mid Expts)}{P(RNAworld)} = \frac{1.0}{P^4}$$

where we do not distinguish between P for different experiments, so probability for a positive outcome for all is P^4. Therefore, the crucial quantity is P, the a priori probability of success in a selection for an arbitrary new RNA activity. Although *seleccionistas* will certainly vary in their estimation of P, I suggest 0.2 as a starting point for discussion. In other words, if I set out to select something like a RNA lipid oxidase, one in five such programs would ultimately yield the desired activity (readers are invited to substitute their own estimates). The implication of such a conservative choice is that I should elevate my estimate of the probability of the RNA World 625-fold (0.2^{-4}) on the basis of the results reviewed above. If I thought the RNA World had a probability of 1 in 1000 before, now I am quite sure it existed.

I know that the above discussion may seem abstract or arbitrary (or both). However, Bayes' Theorem captures and quantifies a common experimentalist's intuition.

That is: Confirmation of one prediction of an idea is progress, but it is when you have two or more independent confirmed predictions that one should feel some confidence in a hypothesis. A serviceable summary of this section is, then: The RNA World is due the credence appropriate to a multiply confirmed hypothesis.

SUMMARY OF THE PROOF OF PRINCIPLE

The capabilities of RNA (and by extension RNA congeners) do in fact embrace examples of all four activities required for a simplified translation system. It is therefore chemically plausible that our present translation system descends from one that employed RNA alone. In fact, it would already be defensible to assign this proposition a high probability.

Nonetheless, a critic might take another tack. He could argue instead that we have performed few and partial reactions. There are still many amino acids to be examined. No RNA system has been put together capable of accurately assembling free amino acids into a templated peptide, and so on. However, assembly of components we already possess using selective methods that we already understand can likely fill many of these gaps. Bayes' Theorem further shows that such filling is useful employment. Even without agreement on any particular probability, as our RNA constructions become more elaborate, their probabilities must progressively become smaller. So $1/P^4$ in our form of Bayes' rapidly becomes larger because of the inverse exponent. Accordingly, for a rational skeptic, if successive RNA constructs better approximate real translation, descent from an RNA World quickly advances from provable to proven.

THE EXIGUITY OF MOLECULAR EVOLUTION

The discussion so far has not dealt with an unanticipated resemblance between selection results and biological reality; I turn to this now. Exiguity is a standard dictionary word meaning sparseness or fewness. It was first employed (Yarus and Welch 2000) in this kind of discussion as a simple, nearly assumption-free way of evoking an unexpected quality of this work. That is: In multiple cases where we know the historical path that biological evolution has taken, that choice is reproduced by in vitro selections performed under seemingly arbitrary modern conditions. I have in mind the three amino acids whose most frequent sites concentrate their codons, as well as the emergence of the partial peptidyl transferase sequence by selection for affinity for CCdApPuro. How could this have happened? Almost every selection-amplification yields galaxies of solutions that are never characterized because of limited time and manpower. This is particularly striking for CCdApPuro affinity. The molecule is a 3′-3′ linked tetranucleotide that offers many sites of interaction to another RNA. How is it that an apparent piece of the biological peptidyl transferase was its predominant binding sequence? The exiguous answer is that the possible molecular solutions to a selective problem are so few, or so sparse, or so narrow that there is a substantial chance that modern experiments and ancient biology find related sequences.

Exiguity in this sense would be a stunning piece of good fortune, and should therefore be taken cautiously. It would imply that modern experiments can be plausible as investigations of the ancient course of molecular evolution. Furthermore, we might get most help from exiguity when we most need it—when we cannot recognize the most plausible answer because of lack of modern parallels. It would imply that early evolution must have been deterministic—and this would differ radically from macroevolution, where contingency and irreproducibility are the default hypotheses (see, e.g., Gould 1989).

Most likely, exiguity rests on the unique accessibility of the simplest molecular solutions. It is well known (Ciesiolka et al. 1996; Sabeti et al. 1997) that the simplest RNAs that meet an in vitro selection will probably be most numerous in a selection-amplification pool. Here "simplicity" explicitly means "least fixation of nucleotide identities" to produce an active structure. An RNA site might be physically large, but if it has only a few nucleotides whose identities are crucial, it might also be "simple" in this sense. We may hope that the early course of biology was also a search for simplest solutions, the latter necessarily few in number. However, the above macroevolution example suggests that as evolution proceeds, exiguity will become irrelevant. That is, molecular simplicity cannot remain the primary basis for selection. In fact, recent experiments on selection for active peptides suggest that ATP-binding sites in modern proteins may be too complex to be exiguous (Keefe and Szostak 2001). Therefore, the challenge now is to find the limits of molecular simplicity or exiguity as a useful constraint on biological history.

EXIGUITY AND INITIAL ASSUMPTIONS

The exiguous outcomes of these experiments reflect on the initial assumptions. The four essential reactions were introduced as the most reasonable hypotheses, given evidence for a translation apparatus using these reactions before the separation of the three domains of life. However, (re)selection of the core sequence of the peptidyl transferase potentially revises our reasoning. Ever since the appearance of the modern peptidyl transferase, peptide synthesis has evidently been quite similar. In particular, the modern selected/conserved peptidyl transferase core interacts with CCA-amino acid in the P site, and A-amino acid in the A site (Nissen et al. 2000), and facilitates a tetrahedral reaction intermediate. We can deduce from this that direct nucleophilic substitution between aminoacylated oligoribonucleotides has been the rule since the modern peptidyl transferase appeared, presumably midway in the RNA World. In addition, the multiple demonstration that amino acid sites contain codons suggests that an RNA-amino acid code ancestral to the present one existed at the same era. It is tempting to associate this code with the ribonucleotide moieties of the ancient aminoacyl nucleotides implied by the selection of the core sequence. Thus, selection of RNAs related to peptidyl transferase independently and quite explicitly supports the existence of an ancestral RNA translation apparatus that closely approximates the one originally postulated in the introduction. In fact, aminoacyl-RNA synthesis, peptide formation, and coding have been confirmed. Only the case for a particular ancient amino acid activation reaction now relies entirely on the initial argument from continuity, and in addition, on the demonstration (Kumar and Yarus 2001) that the RNA reaction exists.

ACKNOWLEDGMENTS

Thanks are due Mali Illangasekare, Vasant Jadhav, Rob Knight, Krishna Kumar, Irene Majerfeld, Alexandre Vlassov, and Ico de Zwart for comments on a draft. I am grateful also for support by National Institutes of Health Research grants GM 30881 and 48080 for the work summarized here.

REFERENCES

Ban N., Nissen P., Hansen J., Moore P.B., and Steitz T.A. 2000. The complete atomic structure of the large ribosomal subunit at 2.4 Å resolution. *Science* **289:** 905.

Ciesiolka J., Illangasekare M., Majerfeld I., Nickles T., Welch M., Yarus M., and Zinnen S. 1996. Affinity selection-amplification from randomized ribooligonucleotide pools. *Methods Enzymol.* **267:** 315.

Connell G.J. and Yarus M. 1994. RNAs with dual specificity and dual RNAs with similar specificity. *Science* **264:** 1137.

Connell G.J., Illangasekare M., and Yarus M. 1993. Three small ribo-oligonucleotides with specific arginine sites. *Biochemistry* **32:** 5497.

Crick F.H.C. 1968. The origin of the genetic code. *J. Mol. Biol.* **38:** 367.

Ellington A. and Szostak J. 1990. In vitro selection of RNAs that bind specific ligands. *Nature* **346:** 818.

Ellington A., Khrapov M., and Shaw C.A. 2000. The scene of a frozen accident. *RNA* **6:** 485.

Famulok M. and Szostak J.W. 1992. Stereospecific recognition of tryptophan-agarose by in vitro selected RNA. *J. Am. Chem. Soc.* **114:** 3990.

Freeland S.J. and Hurst L.D. 1998. The genetic code is one in a million. *J. Mol. Evol.* **47:** 238.

Freist W., Sternbach H., and Cramer F. 1996. Phenylalanyl-tRNA synthetase from yeast and its discrimination of 19 amino acids in aminoacylation of tRNA(Phe)-C-C-A and tRNA(Phe)-C-C-A(3′ NH2). *Eur. J. Biochem.* **240:** 526.

Gilbert W. 1986. The RNA world. *Nature* **319:** 618.

Gould S.J. 1989. *Wonderful life: The Burgess Shale and the nature of history.* W.W. Norton, New York.

Ibba M. and Söll D. 2000. Aminoacyl-tRNA synthesis. *Annu. Rev. Biochem.* **69:** 617.

Illangasekare M. and Yarus M. 1999a. A tiny RNA that catalyzes both aminoacyl-RNA and peptidyl-RNA synthesis. *RNA* **5:** 1482.

———. 1999b. Specific, rapid synthesis of phe-RNA by RNA. *Proc. Natl. Acad. Sci.* **96:** 5470.

———. 2002. Phenyl-alanine-binding RNAs and genetic code evolution. *J. Mol. Evol.* **54:** 298.

Illangasekare M., Kovalchuke O., and Yarus M. 1997. Essential structures of a self-aminoacylating RNA. *J. Mol. Biol.* **274:** 519.

Illangasekare M., Sanchez G., Nickles T., and Yarus M. 1995. Aminoacyl-RNA synthesis catalyzed by an RNA. *Science* **267:** 643.

Keefe A.D. and Szostak J.W. 2001. Functional proteins from a random-sequence library. *Nature* **410:** 715.

Knight R.D. and Landweber L.F. 1998. Rhyme or reason: RNA-arginine interactions and the genetic code. *Chem. Biol.* **5:** R215.

―――. 2000. Guilt by association: The arginine case revisited. *RNA* **6:** 499.

Knight R.D., Freeland S.J., and Landweber L.F. 1999. Selection, history and chemistry: The three faces of the genetic code. *Trends Biochem. Sci.* **24:** 241.

Kumar R.K. and Yarus M. 2001. RNA-catalyzed amino acid activation. *Biochemistry* **40:** 6998.

Lee N., Bessho Y., Wei K., Szostak J.W., and Suga H. 2000. Ribozyme catalyzed tRNA aminoacylation. *Nat. Struct. Biol.* **7:** 28.

Majerfeld M. and Yarus M. 1998. Isoleucine:RNA sites with associated coding sequences. *RNA* **4:** 471.

Mannironi C., Scherch C., Fruscoloni P., and Tocchini-Valentini G.P. 2000. Molecular recognition of amino acids by RNA aptamers: The evolution into a tyrosine binder of a dopamine-binding motif. *RNA* **6:** 520.

Meyer S.L. 1975. *Data analysis for scientists and engineers.* p. 131. John Wiley & Sons, New York.

Michel F., Hanna M., Green R., Bartel D.P., and Szostak J.W. 1989. The guanosine binding site of the *Tetrahymena* ribozyme. *Nature* **342:** 391.

Muth G.W., Ortoleva-Donnelly L., and Strobel S.A. 2000. A single adenosine with a neutral pKa in the ribosomal peptidyl transferase center. *Science* **289:** 947.

Nissen P., Hansen J., Ban N., Moore P.B., and Steitz T.A. 2000. The structural basis of ribosome activity in peptide bond synthesis. *Science* **289:** 920.

Polacek N., Gaynor M., Yassin A., and Mankin A.S. 2001. Ribosomal peptidyl transferase can withstand mutations at the putative catalytic nucleotide. *Nature* **411:** 498.

Robertson D.L. and Joyce G.F. 1990. Selection in vitro of an RNA enzyme that specifically cleaves single-stranded DNA. *Nature* **344:** 467.

Sabeti P.C., Unrau P.J., and Bartel D.P. 1997. Accessing rare activities from random RNA sequences: The importance of the length of molecules in the starting pool. *Chem. Biol.* **4:** 767.

Saito H., Kourouklis D., and Suga H. 2001. An in vitro evolved precursor tRNA with aminoacylation activity. *EMBO J.* **20:** 1797.

Samaha R.R., Green R., and Noller H.F. 1995. A base pair between tRNA and 23S rRNA in the peptidyl transferase centre of the ribosome. *Nature* **377:** 309.

Sokal R.R and Rohlf F.J. 1995. *Biometry,* 3rd edition, p. 794. W.H. Freeman, New York.

Szathmáry E. 1999. The origin of the genetic code: Amino acids as cofactors in an RNA world. *Trends Genet.* **15:** 223.

Tuerk C. and Gold L. 1990. Systematic evolution of ligands by exponential enrichment. *Science* **249:** 505.

Welch M., Chastang J., and Yarus M. 1995. An inhibitor of ribosomal peptidyl transferase using transition-state analogy. *Biochemistry* **34:** 385.

Welch M., Majerfeld I., and Yarus M. 1997. 23S rRNA similarity from selection for peptidyl transferase mimicry. *Biochemistry* **36:** 6614.

White H.B., III. 1976. Coenzymes as fossils of an earlier metabolic state. *J. Mol. Evol.* **7:** 101.

Woese C.R., Dugre D.H., Dugre S.A., Kondo M., and Saxinger W.C. 1967. On the fundamental nature and evolution of the genetic code. *Cold Spring Harbor Symp. Quant. Biol.* **31:** 723.

Yang Y., Kochoyan M., Burgstaller P., Westhof E., and Famulok M. 1996. Structural basis of ligand discrimination by two related aptamers resolved by NMR spectroscopy. *Science* **272:** 1343.

Yarus M. 1988. A specific amino acid binding site composed of RNA. *Science* **240:** 1751.

―――. 1993. An RNA-amino acid affinity. In *The RNA world* (ed. R. Gesteland and J. Atkins), p. 205. Cold Spring Harbor Laboratory Press, Cold Spring Harbor, New York.

―――. 1998. Amino acids as RNA ligands: A direct-RNA-template theory for the code's origin. *J. Mol. Evol.* **47:** 109.

―――. 1999. Boundaries for an RNA world. *Curr. Opin. Chem. Biol.* **3:** 260.

―――. 2000. RNA-ligand chemistry: A testable source for the genetic code. *RNA* **6:** 475.

Yarus M. and Christian E.L. 1989. Genetic code origins. *Nature* **342:** 349.

Yarus M. and Illangasekare M. 1999. Aminoacyl-tRNA synthetases and self-acylating ribozymes. In *The RNA world*, 2nd edition (ed. R. Gesteland et al.), p. 183. Cold Spring Harbor Laboratory Press, Cold Spring Harbor, New York.

Yarus M. and Welch M. 2000. Peptidyl transferase: Ancient and exiguous. *Chem. Biol.* **7:** R187.

Yusupov M.M., Yusupova G.Z., Baucom A., Lieberman K., Earnest T.N., Cate J.H.D., and Noller N. 2001. Crystal structure of the ribosome at 5.5 Å resolution. *Science* **292:** 883.

Overriding Standard Decoding: Implications of Recoding for Ribosome Function and Enrichment of Gene Expression

J.F. ATKINS,* P.V. BARANOV,* O. FAYET,† A.J. HERR,* M.T. HOWARD,* I.P. IVANOV,*
S. MATSUFUJI,‡ W.A. MILLER,¶ B. MOORE,* M.F. PRÈRE,†
N.M. WILLS,* J. ZHOU,* AND R.F. GESTELAND*

*Department of Human Genetics, University of Utah, Salt Lake City, Utah 84112-5330;
†Microbiologie et Génétique Moléculaire, CNRS, 31062 Toulouse Cedex, France;
‡Department of Biochemistry 2, Jikei University School of Medicine,
Minato-ku, Tokyo 105-8461, Japan; ¶Plant Pathology Department,
Iowa State University, Ames, Iowa 50011

When the genetic code and the general mechanism of its readout were elucidated, there was no reason to think that decoding would not be uniform—that the code would be read differently at specific sites within coding sequences. However, we now know that ribosomes show remarkable versatility in response to recoding signals. They can shift reading frame in either direction, bypass blocks of nucleotides within mRNA or redefine the meaning of certain codons, usually stop codons. The crucial signals can involve not only specific sequences in the mRNA, but also mRNA structures, specialized translation factors, specialized tRNAs, and even amino acid sequences in the nascent peptide chain.

Some of these recoding events are regulatory so that expression of the novel protein is responsive to specific cues. Other cases lead to the synthesis of two proteins in a set ratio: one the result of standard decoding, the other a novel protein sharing some amino-terminal sequences with the former. In yet other cases, the 21st amino acid selenocysteine is encoded, often at the active site of an enzyme. These events, collectively termed recoding (Gesteland and Atkins 1996), are of interest in their own right as a sophisticated "extra layer" on top of standard decoding; however, they also provide insights into standard decoding. A database (Baranov et al. 2001) at http://recode.genetics.utah.edu/ tabulates the currently known cases.

RIBOSOMAL PROTEIN L9

Initially, decoding was thought to be rigidly triplet and that no single change of the translational apparatus would permit even localized shifts in reading frame. This view changed with the isolation of external suppressors for frameshift mutants in *Salmonella* that would partially restore reading frame (Riyasaty and Atkins 1968; Yourno and Tanemura 1970; for reviews, see Roth 1981 and Magliery et al. 2001) and the finding of frameshift mutant leakiness due to error frameshifting (Atkins et al. 1972; Fox and Weiss-Brummer 1980). One of the two classes of –1 frameshift mutant extragenic suppressors initially isolated were mutants of tRNA$^{Gly}_2$, with the strongest having the mutation U*34 to C. Recent work has shown that the second class, which comprised weaker suppressors, have two mutations (C. Johnston et al., unpubl.). One mutation alters release factor 2 (RF2), the other alters ribosomal protein L9. Together these two mutants permit +2 frameshifting at the sequence *GUG UG*(A) (just as seen previously with a mutant of tRNA$^{Val}_1$, O'Connor et al. 1989). This result suggested a possible relationship between RF2 and L9 that we then investigated in *Escherichia coli*.

The starting point was a strain with a deletion of the gene for L9 and a C40G mutant of tRNA$^{Gly}_2$ that was isolated in a different selection (described below). The strain also contained a –1 frameshift mutation in a gene for a tryptophan biosynthetic enzyme, and selection was for growth in the absence of tryptophan (Fig. 1). One of the three mutants characterized had an altered RF2: Phe-207 to Leu. The WT gene for L9 was then reintroduced into this strain and, surprisingly, this proved to be deleterious. A selection for mutants with restored growth rates of this strain led to the isolation of a new mutation affecting the carboxy-terminal domain of L9. This analysis suggests L9 enhances a deleterious activity inherent to the RF2 mutant (A.J. Herr et al., unpubl.). How this effect is mediated is an intriguing, unresolved question.

A similar release factor 2 mutant, F207T, promotes termination at UAG, in addition to the normal UAA and UGA, and also at UGG, a tryptophan codon (Ito et al. 2000). Mutants at position 207 somehow affect recognition at the ribosomal A site. Even though a direct interaction between L9 and RF2 may not require part of the L9 to be at the A site, it is still difficult to imagine how such an interaction could occur. As described below, the amino terminus of L9 is bound to a distant segment of rRNA. This result from frameshift mutant suppression has to be reconciled with findings about L9 emerging from studies of a recoding event—the programmed bypassing of 50 nucleotides (Fig. 2) (Weiss et al. 1990a; Herr et al. 2000a).

Fifty nucleotides between codons 46 and 47 are not translated in the mRNA for phage T4 gene *60*, which en-

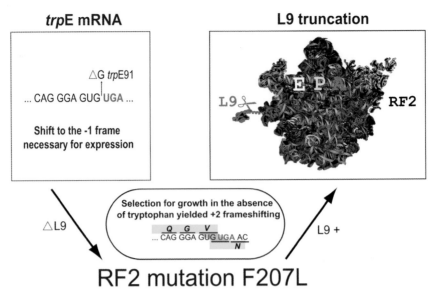

Figure 1. Genetic interaction between L9 and RF2. Selection in the absence of ribosomal protein L9 for frameshifting that would compensate for a frameshift mutant yielded RF2 mutant F207L. A further selection with this RF2 mutant in the presence of WT L9 yielded mutant L9. L9 is shown in red in the right-hand panel.

codes a topoisomerase subunit (Huang et al. 1988). In 50% of the ribosomes translating this mRNA, the anticodon of peptidyl tRNA$^{Gly}_2$ detaches from its cognate GGA codon 46 and the mRNA moves relative to the ribosome-peptidyl tRNA complex until the tRNA$^{Gly}_2$ repairs to the mRNA at GGA located 47 nucleotides downstream (Weiss et al. 1990a; Herr et al. 1999, 2001). Decoding then resumes at the adjacent 3′ codon 47. (A selection for mutants that reduce bypassing yielded the C40G mutant of tRNA$^{Gly}_2$ that was used in the work described above.) Not surprisingly, several signals are required for this remarkable event (Fig. 2); if multiple signals were not involved, standard nonoverlapping triplet decoding might be precariously poised on the brink of disaster. One of the signals is a stem capped by a tetraloop at the 5′ side of the coding gap. Codon 46, the "take-off"

site, is within the 5′ side of the stem. When the length of this stem-loop is extended, bypassing decreases but can be partially restored by mutants of L9 (Herbst et al. 1994; Adamski et al. 1996). This, and further genetic experiments, led to the deduction that the role of L9 is to prevent forward mRNA slippage, but that in the case of gene 60, this restraint can be overridden by the stem-loop structure (Herr et al. 2000b, 2001; see below).

Interestingly, a mutant of what is apparently the L9 gene of *Rhizobium leguminosarum* was isolated a decade ago, although neither the original nor subsequent papers have pointed out the relationship to L9. The mutant was called csn-1, for *cultivar specific nodulation* (Lewis-Henderson and Djordjevic 1991). Mutants of an outer membrane protein are suppressed by csn-1, but the mechanism remains mysterious.

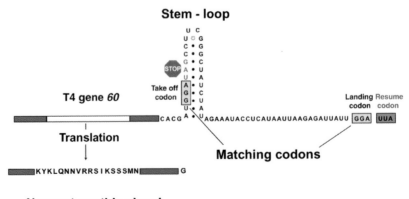

Figure 2. Translational bypassing of 50 nucleotides in decoding phage T4 gene *60*. Following dissociation of "take-off" codon GGA from pairing with the anticodon peptidyl tRNA$^{Gly}_2$, the mRNA slips until pairing is reestablished by the matched landing-site codon adjacent to the site of decoding resumption. Part of the nascent peptide while still in the ribosome, a stop codon immediately 3′ of the take-off codon, and a stem-loop at the start of the coding gap contribute to efficient bypassing.

L9 is a highly elongated ribosomal protein with a long central α-helical region separating globular RNA-binding domains at each end (Hoffman et al. 1996). New crystallographic data have shown that its amino terminus binds to rRNA at the base of the L1 stalk (Yusupov et al. 2001), consistent with numerous earlier studies including cross-linking and immune electron microscopy (EM) (Walleczek et al. 1988), toeprinting (Adamski et al. 1996), and cryo-EM studies (Matadeen et al. 1999). Recent results suggest that L9 exists in two states. In one state, its carboxy-terminal region is bound back to a more central portion of the ribosome (Lieberman et al. 2000; Mueller et al. 2000). In the other form, seen both in crystal-derived structures (Yusupov et al. 2001) and by cryo-EM, the carboxyl terminus extends some 50 Å out and away from the ribosome. Cryo-EM pictures suggest that L9 can go from the "intraribosomal" state to the "projecting" state on binding of EF-G in the GTP state (Spahn et al. 2001). However, even in the intraribosomal state, L9 cannot directly reach the decoding center on the 30S subunit of the same ribosome, where it could interact with RF2 in the A site. How the genetic interaction between L9 and release factor comes about is an intriguing mystery. Could the interaction reflect contact between projecting L9 and the trailing ribosome on the mRNA? In one crystal structure analysis of ribosomes, L9 from one ribosome projects into the A-site side of the 30S subunit of an adjacent ribosome (Yusupova et al. 2001). It is certainly unclear how these behaviors of L9 might relate to its prevention of inappropriate forward mRNA slippage that is somehow overcome by the stem-loop at the start of the gene *60* coding gap. (It is also unclear what performs the L9-equivalent function in eukaryotic cells.)

The role of L9 in translation focuses on ribosome slippage and is distinct from general issues of classic translational fidelity involving the correct identity of the amino acid inserted. Error-restrictive mutants of *E. coli* ribosomal proteins S12, S17, or L6 increase the stringency of A-site selection, whereas error-prone mutants of S4, S5, and L7/L12 decrease accuracy. We have no reason to think that L9 influences the selection of aminoacyl tRNAs.

This opening section on L9 illustrates a connection between work on mutants of translation components that compensate for frameshift mutants and later work on recoding. It also serves to introduce transitory codon–anticodon dissociation within the ribosome. This is explored more systematically in the next section.

RECODING SITES: CODON–ANTICODON INTERACTIONS IN THE A, P, AND E RIBOSOMAL SITES

A-site Occupancy Influences P-site Frameshifting

The great majority of cases of ribosomal frameshifting involve codon–anticodon dissociation, mRNA slippage by a single nucleotide, and mRNA re-pairing to the tRNA anticodon(s) on overlapping codon(s) in a new reading frame. In a number of cases, it is the tRNA in the P site

that shifts relative to the mRNA, but it does so under the influence of the state of the A site.

Dissociation of tRNA followed by re-pairing in +1 frameshifting utilizes the first nucleotide of the next zero-frame codon, and thus, the efficiency of frameshifting is strongly influenced by occupancy of that codon. A stop codon or a rare codon in the ribosomal A site promotes +1 frameshifting (Gallant and Foley 1980; Weiss et al. 1987; Belcourt and Farabaugh 1990), as exemplified by the expression of *E. coli* RF2, eukaryotic antizyme, and *Saccharomyces cerevisiae* retrotransposon Ty 1 (see below). Stop codons are decoded slowly (Freistroffer et al. 1997), as are rare codons by sparse tRNAs. This gives competitive advantage to re-pairing of peptidyl tRNA to the +1 frame codon. The stringency of the re-pairing is less than for initial pairing in the A site, as the proofreading step is absent. The key to efficiency then is in the kinetic details of dissociation and re-pairing of peptidyl tRNA. As discussed below, competition for the first nucleotide of the next zero-frame codon provides an opportunity for regulation of translation. It is easy to imagine how unoccupied codon bases in the A site could be captured by P-site tRNA by just a slip of the mRNA.

In addition to +1 frameshifting occurring in the P site being influenced by occupancy of the A site, –1 frameshifting is also affected. Stop codons can also stimulate shifting of peptidyl tRNA into the –1 frame (Weiss et al. 1987, 1990b; Brault and Miller 1992), as exemplified in the programmed frameshifting for expression of potato virus M (Gramstat et al. 1994). A vacant or "hungry" codon due to limitation of aminoacyl tRNA can also play a similar role and the codon rules have been well described using synthetic constructs (for review, see Gallant et al. 2000). (Even though starvation is a common situation for bacteria in nature, no examples of the use of hungry codons for programmed frameshifting have yet been discovered). The stop codon or vacant codon must be in the A site; again it must be the P-site peptidyl tRNA that shifts, but to shift –1 it must capture the next 5′ base. Most likely this happens by securing the last base in the E site. Some argue that there is pairing between the E-site tRNA and the E-site codon (for review, see Spahn and Nierhaus 1998; Agrawal et al. 2000). If so, the P-site anticodon involved in re-pairing to mRNA must displace at least the last base pair.

Weak A-site Pairing Promotes Tandem Shifts

Other mechanisms for –1 frameshifting involve the tRNA in the ribosomal A site, in which case P-site pairing is affected. This frameshifting is facilitated if the A-site tRNA has relatively weak initial pairing, an important feature often being the strength of pairing with the third codon base.

Shifting of the pairing of this A-site tRNA to the overlapping –1 frame codon often occurs in tandem with that of the P-site tRNA so the two together move to the new frame. More than one model has been advanced for the stage in the ribosome cycle at which this tandem shift occurs (Jacks et al. 1988a; Weiss et al. 1989). The requirement for pairing of both tRNAs in the zero frame and also in the –1 frame constrains the shifty mRNA sequence to X XXY YYZ, where X, Y, and Z may be the same or different nucleotides.

However, only a small subset of the potential heptanucleotides of the form X XXY YYZ are utilized in known cases of programmed frameshifting because of special features of a small subset of tRNAs that can enable tandem shifts. Y is restricted to being A or U. Where it is A, presumably poor anticodon pyrimidine stacking and the potential for just two hydrogen bonds from each of the Us at anticodon positions 36 and 35 contribute to dissociation in the A site. However, poor pairing with the third codon base may be more important. Not only is tRNA anticodon base-34 identity relevant, but its modification status can be pertinent also. This subset of tRNAs is not the same in *E. coli* and mammals, which is reflected by the relative shiftiness of the heptanucleotide sites. In *E. coli* the sequence A AAA AAG is especially shift-prone (Weiss et al. 1989) and is the site of frameshifting for many IS elements, especially members of the IS3 family (Chandler and Fayet 1993), for *dnaX* where it is required for synthesis of a DNA polymerase III subunit, and for several long-tailed double-stranded DNA phages for synthesis of their GT fusion product (Levin et al. 1993; R. Hendrix, pers. comm.). One feature that makes A AAA AAG especially shifty is that fixation in the new frame is facilitated. In the A site, pairing of the sole tRNA^{Lys} (anticodon $^{3'}$UUmnm^{5}s^{2}U$^{5'}$) with AAA in the new –1 frame is stronger than with its initial AAG partner, since pairing with the third codon base, G, is poor (Sundaram et al. 2000). Consequently, introduction of a gene encoding a tRNA^{Lys} that pairs better with AAG reduces frameshifting (Tsuchihashi and Brown 1992). Such a tRNA occurs in mammals where A AAG is not shift-prone. Instead, A AAC is especially shifty and is found in the corresponding position in the shift site of many retroviruses and Coronaviruses (Brierley et al. 1992; Hatfield et al. 1992).

The conclusion that P-site and A-site codons are involved in tandem slippage at the Rous sarcoma virus and infectious bronchitis virus shift sites is inferred from the finding that placing a stop codon immediately 3′ of the shift heptanucleotide had no effect on frameshifting (Jacks et al. 1988a; Brierley et al. 1992). A similar experiment in *E. coli* with the sequence A AAA AAG also shows no effect (Bertrand et al. 2002).

Single Shifts Versus Tandem Shifts

As introduced above, many studies have focused on –1 slippage at sequences other than heptanucleotides (see, e.g., Weiss et al. 1990b), including shifting from UUC back to an overlapping UUU (Atkins et al. 1983), with the presumption that only single slippage in the P site was involved. Having a stop codon in the A site greatly boosts such P-site frameshifting and, in these cases, the identity of 7 mRNA bases is important, e.g., U UUC UAG, but it

is clear that the shift takes place in the P site. In some other cases, discerning whether shifts are tandem or not is difficult given the latitude in re-pairing rules in the P site. Examples of imperfect heptanucleotide sites where the involvement of tandem slippage is unclear include phage T7 gene *10* (G GUU UUC) (Condron et al. 1991) and equine arteritis virus frameshifting (G UUA AAC) (Brierley et al. 1992).

Even with classic heptanucleotide tandem –1 frameshifting, some single slippage is detectable (Jacks et al. 1988b; Weiss et al. 1989). Limiting aminoacyl tRNA for the second codon of a heptanucleotide shift site elevates a frameshift at the first codon of the heptanucleotide which is then presumed to be in the P site (Yelverton et al. 1994). Recently, the issue of whether –1 slippage can occur without P-site slippage has been brought into sharper focus by a study of decoding the *Bacillus subtilis* cytidine deaminase gene (*cdd*) and the eubacterial IS*1222* transposase (Fig. 3) (Mejlhede et al. 1999 and unpubl.) The identity of the base 5′ of the CGA AAG hexanucleotide shift site has little or no influence on *cdd* frameshifting, suggesting that tRNA$_2^{Arg}$ (anticodon $^{3'}$GCI$^{5'}$) which decodes CGA does not slip back, because if it did, only inosine (I):G pairing would be involved when the 5′ mRNA base is G. Instead, weak purine:purine pairing of the inosine with A in the third position of CGA permits tRNA^{Lys} to "rob" this codon base for its re-pairing to mRNA in the –1 frame. A stop codon when present as the adjacent 3′ codon has only a marginal effect on frameshifting, consistent with the anticodon of peptidyl-tRNA^{Lys} being in the A site or, perhaps more likely, undergoing translocation at the time of the frameshift (N. Wills et al., unpubl.). However, this mechanism of frameshifting is not unique to apposition of inosine with adenosine; a subset of other NNA codons can substitute for the CGA and function in an equivalent manner (P.

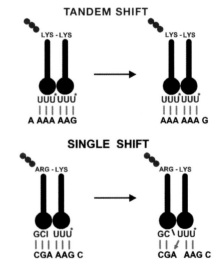

Figure 3. Comparison of frameshifting involving mRNA dissociating from, and re-pairing to, two tRNAs or just to one tRNA. The pre-shift and post-shift pairings are shown (I is inosine. The modified anticodon base designated with * in *E. coli* is mnm^{5}s^{2}U, although some tRNA has mnm^{5}Se^{2}U.)

Licznar et al., unpubl.). The location of the anticodon of the "robbed" tRNA is unknown, but of interest. Either it remains transitorily paired to the first two codon bases or it completely dissociates from the codon at the time the third codon base starts to pair with tRNA^{Lys}.

Relevant to this issue is –2 frameshifting initially seen with synthetic constructs with the sequence GG GGG, where the zero frame is underlined (Weiss et al. 1987, 1990b). However, with this –2 frameshifting and that found when the mammalian antizyme 1 programmed shift cassette is expressed in the heterologous host *S. cerevisiae* (Matsufuji et al. 1996; Ivanov et al. 1998), the anticodon of the shift tRNA is expected to be in the P site when the shift is initiated, since an adjacent 3′ stop codon is important for the frameshifting. (Phage Mu uses –2 frameshifting to express its essential G-T fusion protein, but it does not have a stop codon 3′ adjacent [J. Xu, R. Duda, and R. Hendrix, in prep.].) Further examples doubtless await discovery and should reveal the latitude in 3′ flanking codons. Consideration of frameshifting at runs of G prompts mention of the frameshift mutant leakiness that depends on runs of Gs in the thymidine kinase gene of acyclovir-resistant mutants of Herpesvirus (Horsburgh et al. 1996). Whether the frameshifting occurs at the run of Gs, or is dependent on them for another reason, has not yet been determined, even though the phenomenon is of medical interest.

Shifting without Codon–Anticodon Dissociation: tRNA Balance

Frameshifting in the A site can also occur without mRNA slippage. There is provocative evidence that in the programmed +1 frameshifting used for expression of the yeast transposable element Ty3, neither A-site nor P-site tRNAs dissociate. Instead, it appears that special features of P-site pairing of the third codon base cause occlusion of the first base of the A site such that the incoming tRNA is cognate for the codon of bases 2, 3, and 4 (Farabaugh et al. 1993; Sundararajan et al. 1999). However, an alternative model that involves dissociation and atypical re-pairing in the P site is also viable (Ivanov et al. 2002). (Mutants at the 3′ CCA end of tRNA in the P or E sites can also influence decoding in the A site [O'-Connor et al. 1993].)

The absence of certain tRNAs that form strong Watson-Crick pairing with the third base of particular codons permits the cognate tRNA for that codon to be "shifty" when reading that codon. In contrast to this frameshifting caused by the cognate tRNA, frameshifting by near-cognate tRNA is known in mutant situations (O'Connor 1998; Qian et al. 1998) and –1 frameshifting by tRNAs that are not cognate for the zero or the new frame is known from in vitro studies with the normal tRNA balance (Atkins et al. 1979; Bruce et al. 1986). Whether this is due to doublet decoding in the A site, events before entry to the A site, or nonorthodox tRNA stacking is currently unknown. Irrespective of mechanism, the level of frameshifting in this and several other cases can be strongly affected by perturbations of the balance of

tRNAs (Atkins et al. 1979; Weiss et al. 1988a; O'Connor 1998; Sundararajan et al. 1999; for review, see Gallant et al. 2000). Whether the synthesis of particular tRNAs is induced or repressed to mediate frameshifting at certain developmental stages is an open question, although obvious analogies are known (for instance, the synthesis of certain tRNAs to cope with silk fibroin synthesis).

Two Classes of Heptanucleotide Shift Sites?

Different heptanucleotide sequences respond differently to 3′ adjacent stop codons. With some the effect is benign as described above, with others a decrease in frameshifting occurs. Horsfield et al. (1995) reported that with the heptanucleotide U UUU UUA in *E. coli*, a stop codon at the 3′ adjacent position *reduces* frameshifting. This result languished as an oddity without significant comment from others despite the sequence being the shift site required for expression of HIV-1 GagPol that is the source of viral reverse transcriptase. Although the particular contexts used could conceivably give more efficient termination and competition with frameshifting than most termination contexts, the process is still going to be much slower than reading a sense codon, and so there must be another reason that these heptanucleotides are different from the A-rich ones discussed above. On its own, the U UUU UUA sequence is one of the most shift-prone sequences in both eubacteria (where it is the shift site for several double-stranded DNA phages [J. Xu, R. Duda, and R. Hendrix, in prep.]) and mammalian cells (Brierley et al. 1992). In several cases, the full heptanucleotide has been shown to be required for efficient frameshifting (Jacks et al. 1988a; Weiss et al. 1989; Reil et al. 1993). (UUU pyrimidine is an especially shifty +1 site [Fu and Parker 1994; Schwartz and Curran 1997].) The simplest interpretation is that the stop codon is sensed in the A site, which would imply that the shift occurs in the P and E sites with pairing in both sites (Horsfield et al. 1995). Does the presence of A, which has maximal stacking potential, at anticodon positions 36 and 35 of the tRNAs in *both* the A and P sites diminish dissociation such that translocation to the E and P sites takes place before dissociation occurs? If so, does dissociation from these anticodons then occur unusually readily in these sites with somewhat inefficient subsequent re-pairing? Regardless of the answers, there is apparently a puzzle of why a stop codon in the A site should diminish –1 frameshifting at this sequence in contrast to the stimulation of –1 frameshifting first described by Weiss et al. (1987) with different sequences. (The possibility that release factor binding with some abortive termination step may be relevant in this case stands in contrast to the studies described above and needs investigation.)

As described below, the efficiency of frameshifting at shift-prone sites can be greatly elevated by stimulatory recoding signals present in the mRNA. It was generally assumed that, at least for heptanucleotide –1 simultaneous slippage sites, the recoding signals would be interchangeable within the same organism with an efficiency characteristic of the recoding signal. Extensive cross-switching of recoding signals is only now being performed, but one of us (O.F.) has found that in *E. coli*, U UUU UUN only weakly responds to a particular set of strong –1 frameshift stimulatory signals. This is unlike the response of other heptanucleotides that are inherently less shift-prone. In contrast, in reticulocyte lysates, U UUU UUC is responsive to the IBV 3′ pseudoknot (Brierley et al. 1992). The rules for mixing and matching may be complicated.

Coding Resumption without P-site Pairing

Clearly, codon–anticodon interactions play a key role in moving mRNA through the A, P, and E ribosomal sites, and the nature of the base at the 3′ side of the anticodon helps with framing (Björk et al. 2001). But is tRNA pairing with mRNA only necessary to ensure the correct order of amino acids? Can the mRNA move independently in intact ribosomes or only if linked to tRNA? Ribosomal protein L9 functions to limit mRNA slippage, especially forward slippage. However, in the absence of L9, with the construct shown in Figure 4, multiple sites for re-pairing with mRNA are found (polygamous pairing). In the absence of a single strong pairing partner, a variety of other pairings become evident. Some of the sister pairing partners are near each other, others are at a greater distance. Most codons with no potential for Watson-Crick pairing are excluded from the family of codons to which re-pairing occurs, but one codon, AAU, without any Watson-Crick pairing possibility, permits landing and coding resumption with peptidyl transfer. Of course, it is not surprising that peptidyl transfer can occur without peptidyl-tRNA anticodon pairing, but to have a system where it is observable in vivo on ribosomes is novel. The in vitro

demonstration of peptide elongation on ribosomes without codon–anticodon interactions occurred long ago (Yusupova et al. 1986). Thus, rather than the interest of the present finding being focused on peptidyl transfer, the main issue raised is why coding resumption preferentially occurs at certain sites with little or no Watson-Crick or wobble pairing but not at others.

Although the intricacies of ribosome conformational changes on mRNA movement remain to be discovered, effects mediated through tRNA, and the subtleties of its anticodon–codon interaction, are crucial for framing and its elaboration in recoding.

NASCENT PEPTIDE

Some 30 amino acids of the nascent peptide extend from peptidyl tRNA through an exit channel of the large subunit before reaching the exterior of the ribosome. A select subset of these nascent peptides influences translation activities of the ribosome that has just synthesized them. Examples include an effect on translation of the leader sequence for some bacterial antibiotic-resistance mRNAs, the *E. coli* tryptophanase mRNA (see Gong and Yanofsky 2001), and at least several eukaryotic mRNAs, including one that is responsive to polyamine levels (for review, see Morris and Geballe 2000). The 50-nucleotide translational bypassing in decoding T4 gene *60* is another example (Weiss et al. 1990a; Herr et al. 2000a). The identity of a stretch of 15 amino acids in the growing chain is crucial. When the ribosome is poised to initiate bypassing, this sequence is located 14–30 amino acids from the peptidyl tRNA and thus should still be within the ribosome. When these amino acids are substituted, the efficiency of bypass drops by ninefold (Herr et al. 2001). The same nascent

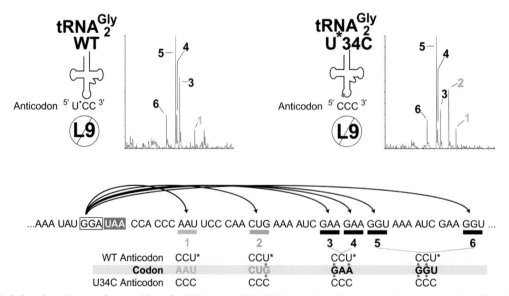

Figure 4. Relaxed requirement for re-pairing of mRNA to peptidyl tRNA permits coding resumption at several sites. The sites of re-pairing, as deduced from mass spectral data of protein product shown at the top of the figure, to the anticodon of WT or a mutant tRNA^Gly are shown together with the possibilities for Watson-Crick pairing. The experiment was performed in the absence of L9, which constrains forward mRNA slippage. (*Blue*) 0 bases involved in Watson-Crick or wobble codon: anticodon pairing; (*green*) 1 base involved in wobble pairing; (*black*) 1 or 2 bases involved in Watson-Crick pairing.

peptide signal can also stimulate –1 frameshifting, and a pause in translation can substantially increase the effect of the nascent peptide signal (Herr et al. 2001). The signal is distributed within the sequence with multiple residues contributing to activity—the maximal effect of altering a single residue is a threefold reduction in activity (for review, see Herr et al. 2000a). The most obvious model for action of the peptide sequence is that the effect is mediated while the peptide is within the exit tunnel that leads from the peptidyl transferase center through the 50S subunit to the far side of the ribosome. However, cross-linking studies indicated that the peptide sequence, like at least some other nascent peptides, can interact with small-subunit components near the intersubunit decoding center (Choi et al. 1998), where it might be able to destabilize peptidyl-tRNA: mRNA pairing (Herr et al. 2001). To date, gene *60* bypassing is the only case where a substantial part of the nascent peptide is known to play a role in recoding. However, the identity of the last two amino acids in nascent chains contributes to termination efficiency (Björnsson et al. 1996), and this is likely to be important for redefinition of stop codons, including that to selenocysteine (Grundner-Culemann et al. 2001).

5′ STIMULATORS ACTING AT THE mRNA LEVEL

5′ recoding elements that function at the RNA level rather than the protein level to stimulate frameshifting are known in one eukaryotic case and a number of eubacterial cases.

The eubacterial cases involve the Shine-Dalgarno (SD) interaction between a sequence near the 3′ end of the 16S rRNA and a complementary mRNA sequence. This Shine-Dalgarno interaction is not exclusive to start codon selection, rather the anti-SD sequence of translating ribosomes is continuously scanning mRNA for potential complementarity. When the rRNA pairs with mRNA, it can stimulate frameshifting at 3′ shift-prone sites, with the spacing between the paired region and the shift site on the mRNA influencing the directionality of frameshifting. In the case of the *E. coli* release factor 2, the required +1 frameshifting used for regulation of expression is stimulated by an SD sequence whose 3′ end is 3 bases 5′ of the +1 shift-prone site (Weiss et al. 1987, 1988b; Curran and Yarus 1988). If one counts the spacing between the mRNA base complementary to a specified base in the Shine-Dalgarno sequence in 16S rRNA, then the spacing between the SD and the shift site is the same for RF2 mRNAs from all 69 organisms sequenced that utilize frameshifting (Fig. 5).

In contrast, an SD sequence that stimulates –1 frameshifting is positioned 9–14 bases 5′ of a tandem, 6AG shift site (Larsen et al. 1994), utilized for frameshifting in decoding *E. coli dnaX* and many IS elements, including IS911 and IS1222. (This anti-SD sequence in 16S rRNA is offset one base 3′ to the anti-SD used for release factor 2 frameshifting.) These observations strongly suggest that the SD interaction 5′ of a shift site can influence the ribosome poised on the shift site and that the spacing determines whether it exerts a push or a pull (Fig. 6).

Placing a second SD sequence with its 3′ end 3 bases before the 5′ end of the SD sequence used for stimulating +1 frameshifting on RF2 mRNA eliminates the stimulatory effect of the natural SD before the shift site (Weiss et al. 1990b), as if formation of rRNA:mRNA pairing stays intact for several steps of chain elongation and prevents recognition of a nearby downstream SD sequence. Perhaps when the minimal spacing between the hybrid formed by the anti-SD sequence and the nearby part of 16S rRNA involved at the decoding site creates conformational tension that pushes forward whereas, just before the hybrid breaks, the effect is a "pull-back." This highlights the general issue that, for the various 5′-element in-

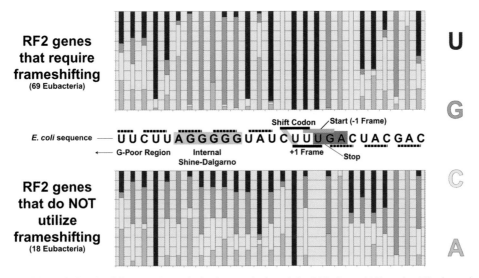

Figure 5. Nucleotide variation in different eubacteria in the proximity of the RF2 frameshifting site. The bacteria that utilize frameshifting for expression of RF2 have a highly conserved Shine-Dalgarno sequence 5′ of the shift site. This SD stimulates +1 frameshifting from CUU to the overlapping UUU by pairing with its complementary sequence in 16S rRNA of *translating* ribosomes.

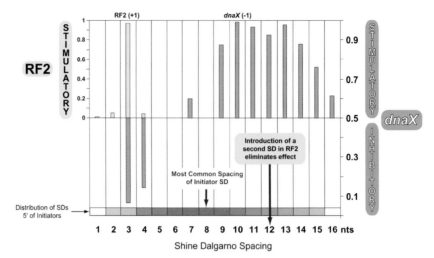

Figure 6. Spacing of the internal Shine-Dalgarno sequence before the shift site influences directionality of frameshifting. For stimulating +1 frameshifting (e.g., RF2 in *yellow*) the spacing has to be 3 nucleotides, whereas for –1 frameshifting (e.g., *dnaX* in *green*) the spacing is optimally 10 or more nucleotides from the shift site. If the ribosome encounters an "extra" SD shortly before reaching the "stimulatory" SD, its recognition of the stimulatory SD is eliminated.

teractions within the ribosome, there may need to be latitude in the mRNA and/or rRNA environment(s) so that there is space to accommodate looping out of sequences.

Despite the absence of a known Shine-Dalgarno-like interaction in eukaryotic cytoplasmic translation, a 5′ element in mRNA is important for effective recoding of most of the antizyme family of mRNAs. The antizyme frameshift is a +1 shift that occurs predominately at UCC UGA, with limited variations of the first codon being known in certain nematodes and fungi (Ivanov et al. 2000a).

Plus-one frameshifting is required for synthesis of all antizymes (Matsufuji et al. 1995; Ivanov et al. 2000a), and an element 5′ of the shift site is involved in stimulating frameshifting with all (except for the likely exception of mammalian antizyme 3 mRNA). Progressive deletion of the 5′ sequence of mammalian antizyme 1 mRNA leads to stepwise reduction of frameshifting efficiency as tested both in vitro and by transfection (see Howard et al. 2001; S. Matsufuji, in prep.), implying a modular nature for the stimulatory element. The 5′ element starts about 45 nucleotides upstream of the frameshift site and ends with the penultimate codon of ORF1. A core sequence, 5′-CCG GGG CCU CGG-3′ located 3–6 codons upstream, is quite conserved and is responsible for a three- to fivefold stimulation. Changing the spacing between the 5′ element and the shift site by one codon in either direction severely diminishes frameshifting.

In *Schizosaccharomyces pombe* antizyme, the 5′ element is short, comprising only 6 nucleotides (Ivanov et al. 2000b). It seems clear that additional modules have been added progressively going up the evolutionary tree. These added elements are also functional in *S. pombe*, providing the opportunity to investigate the mechanism of their action.

The results of mutational analysis of antizyme 5′ elements, together with phylogenetic comparisons, show that they act at the mRNA level but not by intra-mRNA

structure (S. Matsufuji, in prep.). It seems very likely that these sequences directly interact with some part of the ribosome, and rRNA is the most likely target; so far, however, there is no evidence. A direct test of 5′ mRNA–rRNA interactions in frameshifting is likely to come from use of genetic analyses in *S. pombe*, where compensatory changes in sequences may be possible. However, an understanding of this phenomenon of mRNA–ribosome interactions may provide important clues about how translating ribosomes might monitor mRNA sequences.

FLANKING 3′ RECODING SIGNALS

A wide variety of 3′ stimulators flanking recoding sites, ranging from relatively simple sequences to complex structures in the mRNA, affect different types of recoding events. The importance of a 3′-flanking base and 3′ stop codons have been discussed above, and as shown in Figure 5, the base 3′ to a stop codon is universally conserved in the RF2 shift site. Even longer sequences seem to act without forming structures within mRNA. The 15 nucleotides 3′ adjacent to the +1 shift site for the yeast retrotransposon stimulate frameshifting 7.5-fold, probably via pairing with the 530 loop of small-subunit rRNA (Farabaugh et al. 1993; Li et al. 2001). The six bases 3′ of the readthrough site in tobacco mosaic virus in the form CAR YYA are important for redefinition of the stop codon by some unknown mechanism that is unlikely to involve intra-mRNA structure (Skuzeski et al. 1991; Zerfass and Beier 1992; Stahl et al. 1995). This and related sequences are utilized by a very high proportion of the viruses known or suspected to utilize readthrough (L. Harrell et al., in prep.).

Other 3′ elements, from simple stem-loops to complex pseudoknots, act through the folded structures that they form. In one case, *E. coli dnaX* –1 frameshifting, the magnitude of the stimulatory effect of a simple stem-loop ap-

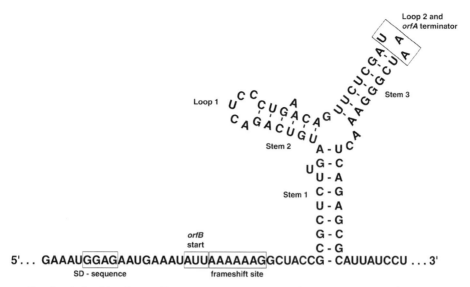

Figure 7. The recoding signals for IS911 frameshifting include an SD sequence 5′ of the shift site and 3′ stem-loops with a three-way junction. Frameshifting, to yield the ORFA:ORFB transframe-encoded transposase, occurs 14 codons before the end of ORFA at an A AAA AAG shift site. A rare AUU start codon 5′ adjacent to the shift site yields a low level of ORFB product alone.

pears wholly dependent on stability of the stem (Larsen et al. 1997). With more complex structures, there is likely to be a combination of stability and specific structural features that generate the stimulatory response, including the branched stem-loop structure important for IS911 frameshifting (Fig. 7) (Rettberg et al. 1999; M.F. Prère et al., in prep.); the stem-loop whose loop sequence is crucial for binding a special elongation factor, SELB, for eubacterial selenocysteine incorporation (Heider et al. 1992); and pseudoknots important for –1 frameshifting (Brierley et al. 1989; ten Dam et al. 1990), +1 frameshifting (Matsufuji et al. 1995), or readthrough (Wills et al. 1991; Feng et al. 1992; Alam et al. 1999).

Only a subset of possible pseudoknots function in recoding. Those that do come in a variety of styles. Even for –1 frameshifting they vary from the small beet western yellow virus pseudoknot with intricate triplex structures (Su et al. 1999) involving what has subsequently been categorized as an A-minor motif (Nissen et al. 2001); to the compact simian retrovirus-1 pseudoknot, which additionally has a ribose zipper motif (Michiels et al. 2001); to the medium-sized pseudoknot, bent due to a wedged base, used for mouse mammary virus *gag-pro* frameshifting (Shen and Tinoco 1995; Kang and Tinoco 1997; Theimer and Giedroc 2000); to the long stem 1 containing pseudoknots lacking the wedged base used for Corona-, Toro- and Arterivirus frameshifting (Liphardt et al. 1999; Napthine et al. 1999).

In some cases, the essential elements of the pseudoknot are just the two stems themselves without identity of the connecting loop sequences being important. This is evident in the case of the antizyme pseudoknot, where evolutionary comparisons of sequences show absolute conservation of the stem sequences in vertebrates but extensive variation in the loop sequences (Ivanov et al. 2000a). Mutational testing of these sequences corroborates the importance of the stems and not of the loops

(although the extreme stem conservation without even a switch of a C-G for a G-C is noteworthy).

The identity of the so-called loop 2 bases of beet western yellow virus pseudoknot that are actually part of triplex structures are experimentally known to be important (Kim et al. 1999). With MuLV *gag* readthrough, the important 3′ pseudoknot has key bases outside the stems—some in the spacer region between the stop codon and the bottom of stem one and some in loop two (Wills et al. 1994). Probing experiments have not revealed any specific structural characteristics involving these bases (Alam et al. 1999), and the possible involvement of some of these bases in specific ribosome interactions has not been addressed.

The pseudoknots that have been tested cause a translational pause of questionable relevance to stimulation of the recoding event, but additional features of pseudoknots are required for frameshifting (Tu et al. 1992; Somogyi et al. 1993; Dinman 1995; Lopinski et al. 2000; Kontos et al. 2001). There is functional specificity in the pseudoknots used for murine leukemia virus stop codon readthrough (Wills et al. 1994) and antizyme +1 frameshifting (S. Matsufuji, unpubl.), and in one heterologous case, spacing from the shift site affects frameshift directionality (Matsufuji et al. 1996).

DISTANT 3′ RECODING SIGNALS

Distant 3′ elements can also be crucial for recoding events. This was first discovered in investigations of the direct encoding of selenocysteine in eukaryotes. The SECIS structure that is essential for redefinition of UGA codons to specify selenocysteine rather than termination is in the 3′UTR (Berry et al. 1993). Embedded in a long sequence with standard base-pairing, SECIS structures have a quartet of non-Watson-Crick base pairs featuring A-N, U-U, G-A, and A-G pairs (Walczak et al. 1996).

This sequence is bound by the protein SBP2 (Copeland et al. 2000, 2001), which in turn binds the protein eEFsec that brings aminoacylated selenocysteine tRNA to the ribosome (Fagegaltier et al. 2000; Tujebajeva et al. 2000b). The finding of the latter was facilitated by studies on Archaeal selenocysteine incorporation that has eukaryotic-like features (Rother et al. 2000). (SBP2 has a motif [Copeland et al. 2001] like that now known to be in proteins that bind kink-turn RNA motifs which contain A-G, G-A noncanonical base pairs [Klein et al. 2001].) How the acylated tRNA is delivered by the SECIS-associated complex to ribosomes is an intriguing unresolved question. Although the great majority of the 30–50 selenoprotein mRNAs in mammals have a single selenocysteine-encoding UGA, rat selenoprotein P mRNA has 10 selenocysteine-specifying UGAs (Hill et al. 1993), and zebrafish selenoprotein P has 17 (Tujebajeva et al. 2000a). Whereas ribosome loading of selenoprotein mRNAs is reduced compared to standard mRNAs, nevertheless the mRNAs are translated by polysomes (Martin and Berry 2001), and studies on the efficiency of the process have begun (Low et al. 2000). Understanding how the far distant signals function is a major challenge but is not unique either to selenocysteine or the recoding field.

Distant recoding signals are also known in redefinition in certain plant viruses where a different standard amino acid rather than selenocysteine is specified. In-frame readthrough of the coat-protein gene stop codon occurs during translation of barley yellow dwarf virus (BYDV) subgenomic RNA 1 (Fig. 8). This requires a nearby element comprising at least six CCXXXX repeats, and a sequence located about 700 nucleotides downstream (Brown et al. 1996). The six bases immediately following the stop codon are conserved, but altering them had no effect on readthrough (Brown et al. 1996). The CCXXXX motif does not begin until 16 nucleotides downstream of the stop codon. No obvious conserved Watson-Crick secondary structure is present in either of the required elements. Potential base-pairing between the distant readthrough element and sequence adjacent to the stop codon was proposed for BYDV (Brown et al. 1996), but it is poorly conserved among other members of the *Luteoviridae* family.

Distant 3′ recoding signals are also known for frameshifting. The first case identified was near the end of the coding sequence for the major coat protein of the phage T7 (Condron et al. 1991; Gabashvili et al. 1997), but it has not been investigated in detail (T7 RNA polymerase transcribes considerably faster than *E. coli* RNA polymerase and faster than ribosomes translate mRNA, and this allows distal sequences to fold back and form structures that will be encountered by translating ribosomes). The main case studied, however, is again in barley yellow dwarf virus. BYDV contains a canonical shifty site, followed by a large bulged stem-loop structure. However, these two elements are not sufficient for

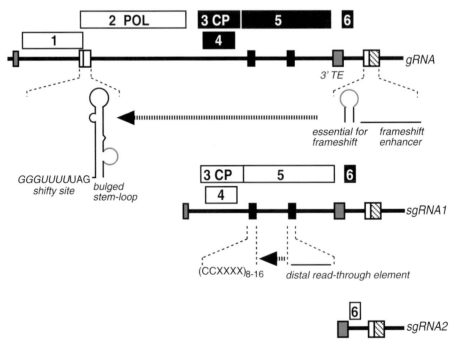

Figure 8. Recoding signals on BYDV RNA. Numbered open boxes indicate ORFs translatable from the indicated RNA (*bold line*). (POL) Polymerase; (CP) coat protein; (gRNA) viral genomic RNA; (sgRNA) subgenomic RNAs generated during virus infection. Open boxes on RNA indicate *cis*-acting signals required for frameshifting, and the hatched box signal enhances frameshifting. The loop sequence in the essential distant frameshift element may interact with the conserved bulged sequences near the 3′ end of the stem-loop adjacent to the shifty site (*gray lines*). (*Solid boxes*) *cis*-Acting sequences essential for readthrough of the CP stop codon. The role for readthrough of the sequence between the CCXXXX repeat and the distal element is unknown. (*Shaded boxes*) 3′-Cap-independent translation element (3′TE) and 5′UTR sequences that interact with it.

the –1 frameshifting required for expression of its polymerase. The missing sequence is located 4 kb downstream of the frameshift site, in the 3′-untranslated region (3′UTR) of the 5.7-kb viral genome (Fig. 8) (Paul et al. 2001). This predicted stem-loop structure has conserved nucleotides that have the potential to base-pair to a bulge in the stem-loop 3′ adjacent to the shifty site. Preliminary mutagenesis data support the requirement for this long-distance interaction to effect frameshifting (J. Barry, pers. comm.). This would form a complex type of pseudoknot or kissing stem-loops.

The general secondary structure of the bulged stem-loop adjacent to the frameshift site is conserved in all members of the *Luteovirus* (BYDV), *Dianthovirus*, and *Umbravirus* genera. The essential role of this structure in frameshifting was shown in red clover necrotic mosaic *dianthovirus* (Kim and Lommel 1998). However, no downstream element was reported. All of the above viruses have a sequence in the 3′UTR capable of forming at least six base pairs to the bulge adjacent to the frameshift site, but, due to the high probability of such downstream sequences occurring by chance, their relevance must be determined experimentally.

How any of the above structures interact with the ribosomes to facilitate frameshifting or readthrough remains a key unsolved mystery in our understanding of recoding. Control of gene expression and replication by very distant elements is an emerging phenomenon among RNA viruses. In addition to recoding and initiation, base-pairing across kilobases can regulate RNA-templated transcription (van Marle et al. 1999) and RNA replication (You et al. 2001) in a diverse range of RNA viruses. The use of such interactions may allow viruses like BYDV to coordinate translation and replication in *cis*, and (via subgenomic RNAs) in *trans* (Wang et al. 1999), to ensure efficient propagation of the genome. In addition to recoding, initiation of BYDV translation is mediated by a distant downstream element. The 3′-cap-independent translation element (3′TE, Fig. 8) is essential for initiation of translation at the 5′-proximal AUGs on genomic and subgenomic RNAs. It acts by direct base-pairing (kissing) of a stem-loop in the 3′TE with one in the 5′UTR (Guo et al. 2001).

ROLES FOR RECODING

Evolution has led to utilization of recoding mechanisms for at least three diverse purposes. First, there are regulatory roles, in several cases autoregulatory roles. An example of the latter is the frameshifting in *E. coli* RF2 decoding where the first open reading frame apparently has no function other than to set up the opportunity for the recoding event that leads to the active product (Craigen and Caskey 1986). RF2 frameshifting efficiency is a competition between termination at a UGA codon and +1 frameshifting whose product is active RF2 protein that catalyzes termination at UGA (Fig. 9).

The frameshifting required for animal antizyme expression is not subject to regulation by antizyme itself, but rather by polyamines whose intracellular level is gov-

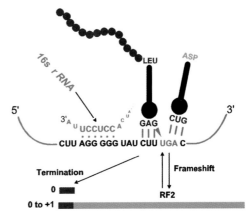

Figure 9. Autoregulatory +1 frameshifting in expression of *E. coli* release factor 2. RF2 mediates termination at the UGA terminator at codon 26 to liberate a peptide that is degraded. Frameshifting occurs by the CUU codon dissociating from peptidyl tRNA, mRNA slipping by one base and the overlapping UUU codon re-pairing to the tRNA. The third base of this codon is the first base of the UGA, and a deficit of RF2 leads to increased shifting to the +1 frame. This frame encodes the bulk of RF2 protein, and so the increased frameshifting leads to greater synthesis of RF2.

erned by antizyme (Matsufuji et al. 1995; Howard et al. 2001; for review, see Coffino 2001). The frameshifting acts as a sensor of free intracellular polyamines, and its efficiency in turn controls polyamine levels by affecting the synthesis of polyamines and their import and export from cells (Fig. 10). The effect on synthesis is mediated by antizyme binding to ornithine decarboxylase, inhibiting it and targeting it for degradation by the 26S proteosome directly without ubiquitination (Murakami et al. 1992). The effect on import and export is mediated by influencing the transporters. There is a single antizyme in the yeast *S. pombe*, in *Caenorhabditis elegans,* and in *Drosophila melanogaster,* but three known antizymes in mammals (for review, see Ivanov et al. 2000a). As more becomes known about the different mammalian antizymes, the distinctions in their regulatory significance will become apparent and will shed light on the differences in frameshifting, especially between that required for the synthesis of antizyme 3 and that required for antizymes 1 and 2, which are much more similar to each other. Despite the differences between the frameshifting involved in the synthesis of antizyme 3 compared to that of antizymes 1 and 2, common features in antizyme frameshifting are apparent. The sequence UGG-UGX (X = G, C or U) immediately upstream of this frameshift site is inferred to have been present in the last common ancestor of all known members of the antizyme gene family, which implies an origin of at least 1 billion years ago.

Examples of the utilization of redefinition for regulatory purposes are only beginning to emerge—as far as is known, the well-known cases of redefinition where a standard amino acid is inserted all yield a set ratio of products. However, regulated stop codon redefinition is known for *Drosophila* kelch, although the recoding signal is unknown (Robinson and Cooley 1997). Whether

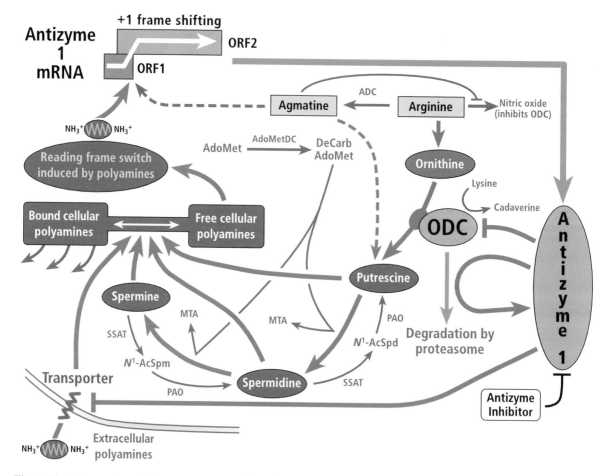

Figure 10. Antizyme frameshifting acts as a sensor for cellular polyamine levels. Increased frameshifting required for synthesis of antizyme occurs with elevated levels of free putrescine, spermidine, spermine, or agmatine. Ornithine decarboxylase (ODC) is required for the synthesis of polyamines and is inhibited and targeted for degradation by antizyme. Antizyme also negatively regulates uptake of polyamines into cells by inhibiting a transporter.

regulation is involved in the redefinition of a UAA stop codon in the *Drosophila* headcase gene is unclear, but in this case, a start has been made in defining the elaborate recoding signal (Steneberg and Samakovlis 2001).

In other cases, recoding responds more generally to metabolic status. Studies with certain sequences not known to be sites used for frameshifting for gene expression have shown an elevation of frameshifting as bacteria enter stationary phase (Barak et al. 1996; Wentzel et al. 1998; Ballesteros et al. 2001). Since nutrient limitation is a common cause for bacteria being in stationary phase in nature, it may well be that some of this frameshifting plays a role in dealing with adversity. This work has been complemented by studies of growth phase influence on the –1 frameshifting used for expression of the transposase of a subset of bacterial insertion sequences (IS elements). In several IS elements, the frequency of transposition is known to be directly related to the frequency of frameshifting (Rettberg et al. 1999; M.F. Prère and O. Fayet, unpubl.). With the complete frameshift regions of IS911 (Rettberg et al. 1999) and IS3 (Sekine et al. 1994), frameshifting increases fivefold from exponential growth to entry to stationary phase (O. Fayet and M.F. Prère, un-

publ.). The effect of this variation on transposition rates is now being tested, but earlier work has shown a general link between growth conditions and IS transposition frequency (Hall 1999).

Second, there are examples where recoding is used to provide a fixed ratio of two functional protein products from one mRNA, although, as expected, the reasons for utilization of a set ratio of the two products are diverse. In several cases, the product of the first coding sequence plays a structural role. Recoding yields a product that also has information from a second ORF that encodes a catalytic function. The recoding efficiency is comparatively low, giving a set ratio of structural product to catalytic product that is relatively high. At least initial fusion of the catalytic product to the structural product may serve a localization function. An example is in decoding retroviral RNA, where either frameshifting or stop codon readthrough is utilized to give a fixed ratio of Gag protein to Gag-Pol or Gag-Pro-Pol (where two shifts in reading frame are required, the efficiency of the first is comparatively high). Processing of the resulting polyproteins leads to a tightly determined ratio of the Gag structural proteins, protease and reverse transcriptase. If the efficiency of these recoding events is altered even modestly in either direction, viability drops (Shehu-Xhilaga et al. 2001), as was shown earlier for the yeast virus L-A (Dinman and Wickner 1992). A different situation exists where frameshifting in a central region of a coding sequence leads quickly to termination in the new frame and production of a product lacking domains present in the product of standard decoding. An example is in the decoding of *dnaX*, where a 50% efficient frameshift two-thirds of the way through the coding sequence yields the γ product. γ and the product of standard decoding are subunits of the major replicative polymerase, DNA polymerase III, where they occur in a 1:1 ratio. Also, as described above, many plant viruses use frameshifting and/or readthrough to express correct levels of the polymerase encoded downstream.

Third, redefinition specifies the 21st encoded amino acid, selenocysteine. Specification of selenocysteine utilizes protein(s) and a tRNA that are not used in standard decoding in addition to the special mRNA structures.

The list of functions at the product level is very likely to expand as more becomes known about bypassing and other forms of recoding. There may also be non-product roles of recoding, although none have yet been clearly characterized. When frameshifting or readthrough brings ribosomes to a region downstream of a gene terminator, on occasion it may not be the product per se that is important, but rather the consequences of ribosome movement on RNA structure.

OVERVIEW

The ability of translating ribosomes to respond to signals in coding sequences, with resultant reprogramming of readout, has been exploited for gene expression and is providing insights to ribosome function. Investigations of the extent to which these mechanisms are utilized is being facilitated by the deluge of genome sequence information, but proteome analysis is badly needed to identify other cases, especially those involving bypassing. As the dramatic advances in ribosome structural knowledge reported in this volume extend to an understanding of the conformational changes during translation, the versatility required for recoding will be revealed and will increase our appreciation of the sophistication of ribosomes and decoding.

ACKNOWLEDGMENTS

We thank Dr. Roger Hendrix for providing unpublished information, and we thank him and our colleagues Cindy Paul, Jennifer Barry, and Darrell Davis for discussions. Our work was supported by grants from National Institutes of Health (GM48152 to J.F.A; GM/DK61200 to R.F.G); Department of Energy (DE-FG03-01ER63132 to R.F.G), U.S. Department of Agriculture (National Research Initiative grant no. 95-37303-1813 to A.M.), Centre Nationale de la Recherche Scientifique (O.F), and Ministère de l'Education Nationale (O.F).

REFERENCES

Adamski F.M., Atkins J.F., and Gesteland R.F. 1996. Ribosomal protein L9 interactions with 23 S rRNA: The use of a translational bypass assay to study the effect of amino acid substitutions. *J. Mol. Biol.* **261:** 357.

Agrawal R.K., Spahn C.M., Penczek P., Grassucci R.A., Nierhaus K.H., and Frank J. 2000. Visualization of tRNA movements on the *Escherichia coli* 70S ribosome during the elongation cycle. *J. Cell. Biol.* **150:** 447.

Alam S.L., Wills N.M., Ingram J.A., Atkins J.F., and Gesteland R.F. 1999. Structural studies of the RNA pseudoknot required for readthrough of the *gag*-termination codon of murine leukemia virus. *J. Mol. Biol.* **288:** 837.

Atkins J.F., Elseviers D., and Gorini L. 1972. Low levels of β-galactosidase in frameshift mutants of *Escherichia coli*. *Proc. Natl. Acad. Sci.* **69:** 1192.

Atkins J.F., Nichols B.P., and Thompson S. 1983. The nucleotide sequence of the first externally suppressible –1 frameshift mutant, and of some nearby leaky frameshift mutants. *EMBO J.* **2:** 1345.

Atkins J.F., Gesteland R.F., Reid B.R., and Anderson C.W. 1979. Normal tRNAs promote ribosomal frameshifting. *Cell* **18:** 1119.

Ballesteros M., Fredriksson A., Henriksson J., and Nyström T. 2001. Bacterial senescence: Protein oxidation in non-proliferating cells is dictated by the accuracy of the ribosomes. *EMBO J.* **20:** 5280.

Barak Z., Gallant J., Lindsley D., Kwieciszewki B., and Heidel D. 1996. Enhanced ribosome frameshifting in stationary phase cells. *J. Mol. Biol.* **263:** 140.

Baranov P.V., Gurvich O.L., Fayet O., Prère M.F., Miller W.A., Gesteland R.F., Atkins J.F., and Giddings M.C. 2001. RECODE: A database of frameshifting, bypassing and codon redefinition utilized for gene expression. *Nucleic. Acids Res.* **29:** 264.

Belcourt M.F. and Farabaugh P.J. 1990. Ribosomal frameshifting in the yeast retrotransposon Ty: tRNAs induce slippage on a 7 nucleotide minimal site. *Cell* **62:** 339.

Berry M.J., Banu L., Harney J.W., and Larsen P.R. 1993. Functional characterization of the eukaryotic SECIS elements which direct selenocysteine insertion at UGA codons. *EMBO J.* **12:** 3315.

Bertrand C., Prère M.F., Gesteland R.F., Atkins J.F., and Fayet O. 2002. Influence of the stacking potential of the base 3′ of tandem shift codons on –1 ribosomal frameshifting used for gene expression. *RNA* (in press).

Björk G.R., Jacobsson K., Nilsson K., Johansson M.J.O., Byström A.S., and Persson O.P. 2001. A primordial tRNA modification required for the evolution of life. *EMBO J.* **20:** 231.

Björnsson A., Mottagui-Tabar S., and Isaksson L.A. 1996. Structure of the C-terminal end of the nascent peptide influences translation termination. *EMBO J.* **15:** 1696.

Brault V. and Miller W.A. 1992. Translational frameshifting mediated by a viral sequence in plant cells. *Proc. Natl. Acad. Sci.* **89:** 2262.

Brierley I., Digard P., and Inglis S.C. 1989. Characterization of an efficient coronavirus ribosomal frameshifting signal: Requirement for an RNA pseudoknot. *Cell* **57:** 537.

Brierley I., Jenner A.J., and Inglis S.C. 1992. Mutational analysis of the "slippery-sequence" component of a coronavirus ribosomal frameshift signal. *J. Mol. Biol.* **227:** 463.

Brown C.M., Dinesh-Kumar S.P., and Miller W.A. 1996. Local and distant sequences are required for efficient readthrough of the barley yellow dwarf virus PAV coat protein gene stop codon. *J. Virol.* **70:** 5884.

Bruce A.G., Atkins J.F., and Gesteland R.F. 1986. tRNA anticodon replacement experiments show that ribosomal frameshifting can be caused by doublet decoding. *Proc. Natl. Acad. Sci.* **83:** 5062.

Chandler M. and Fayet O. 1993. Translational frameshifting in the control of transposition in bacteria. *Mol. Microbiol.* **7:** 497.

Choi K.M., Atkins J.F., Gesteland R.F., and Brimacombe R. 1998. Flexibility of the nascent polypeptide chain within the ribosome. Contacts from the peptide N-terminus to a specific region of the 30S subunit. *Eur. J. Biochem.* **255:** 409.

Coffino P. 2001. Regulation of cellular polyamines by antizyme. *Nat. Rev. Mol. Cell Biol.* **2:** 188.

Condron B.G., Gesteland R.F., and Atkins J.F. 1991. An analysis of sequences stimulating frameshifting in the decoding of gene 10 of bacteriophage T7. *Nucleic Acids Res.* **19:** 5607.

Copeland P.R., Stepanik V.A., and Driscoll D.M. 2001. Insight into mammalian selenocysteine insertion: Domain structure and ribosome binding properties of Sec insertion sequence binding protein 2. *Mol. Cell. Biol.* **21:** 1491.

Copeland P.R., Fletcher J.E., Carlson B.A., Hatfield D.L., and Driscoll D.M. 2000. A novel RNA binding protein, SBP2, is required for the translation of mammalian selenoprotein mRNAs. *EMBO J.* **19:** 306.

Craigen W.J. and Caskey C.T. 1986. Expression of peptide chain release factor 2 requires high-efficiency frameshift. *Nature* **322:** 273.

Curran J.F. and Yarus M. 1988. Use of tRNA suppressors to probe regulation of *Escherichia coli* release factor 2. *J. Mol. Biol.* **203:** 75.

Dinman J.D. 1995. Ribosomal frameshifting in yeast viruses. *Yeast* **11:** 1115.

Dinman J.D. and Wickner R.B. 1992. Ribosomal frameshifting efficiency and the *gag/gag-pol* ratio are critical for yeast M_1 double-stranded RNA virus propagation. *J. Virol.* **66:** 3669.

Fagegaltier D., Hubert N., Yamada K., Mizutani T., Carbon P., and Krol A. 2000. Characterization of mSelB, a novel mammalian elongation factor for selenoprotein translation. *EMBO J.* **19:** 4796.

Farabaugh P.J., Zhao H., and Vimaladithan A. 1993. A novel programmed frameshift expresses the POL3 gene of retrotransposon Ty3 of yeast: Frameshifting without tRNA slippage. *Cell* **74:** 93.

Feng Y.X., Yuan H., Rein A., and Levin J.G. 1992. Bipartite signal for read-through suppression in murine leukemia virus mRNA: An eight-nucleotide purine-rich sequence immediately downstream of the *gag* termination codon followed by an RNA pseudoknot. *J. Virol.* **66:** 5127.

Fox T.D. and Weiss-Brummer B. 1980. Leaky +1 and −1 frameshift mutations at the same site in a yeast mitochondrial gene. *Nature* **288:** 60.

Freistroffer D.V., Pavlov M.Y., MacDougall J., Buckingham R.H., and Ehrenberg M. 1997. Release factor RF3 in *E. coli* accelerates the dissociation of release factors RF1 and RF2 from the ribosome in a GTP-dependent manner. *EMBO J.* **16:** 4126.

Fu C., and Parker J. 1994. A ribosomal frameshifting error during translation of the *argI* mRNA of *Escherichia coli*. *Mol. Gen. Genet.* **243:** 434.

Gabashvili I.S., Khan S.A., Hayes S.J., and Serwer P.1997. Polymorphism of bacteriophage T7. *J. Mol. Biol.* **273:** 658.

Gallant J. and Foley D. 1980. On the causes and prevention of mistranslation. In Ribosomes: Structure, function and genetics (eds. G. Chambliss et al.) p. 615. University Park Press, Baltimore, Maryland.

Gallant J.A., Lindsley D., and Masucci J. 2000. The unbearable lightness of peptidyl-tRNA. In *The ribosome: Structure, function, antibiotics, and cellular interactions* (ed. R.A. Garrett et al.), p. 385. ASM Press, Washington, D.C.

Gesteland R.F. and Atkins J.F. 1996. Recoding: Dynamic reprogramming of translation. *Ann. Rev. Biochem.* **65:** 741.

Gong F. and Yanofsky, C. 2001. Reproducing *tna* operon regulation *in vitro* in an S-30 system. Tryptophan induction inhibits cleavage of TnaC. *J. Biol. Chem.* **276:** 1974.

Gramstat A., Prüfer D., and Rohde W. 1994. The nucleic acid-binding zinc finger protein of potato virus M is translated by internal initiation as well as by ribosomal frameshifting involving a shifty stop codon and a novel mechanism of P-site slippage. *Nucleic Acids Res.* **22:** 3911.

Grundner-Culemann E., Martin G.W., Tujebajeva R., Harney J.W., and Berry M.J. 2001. Interplay between termination and translation machinery in eukaryotic selenoprotein synthesis. *J. Mol. Biol.* **310:** 699.

Guo L., Allen E., and Miller W.A. 2001. Base-pairing between untranslated regions facilitates translation of uncapped, nonpolyadenylated viral RNA. *Mol. Cell* **7:** 1103.

Hall B.G. 1999. Transposable elements as activators of cryptic genes in *E. coli*. *Genetica* **107:** 181.

Hatfield D.L., Levin J.G., Rein A., and Oroszlan S. 1992. Translational suppression in retroviral gene expression. *Adv. Virus Res.* **41:** 193.

Heider J., Baron C., and Böck A. 1992. Coding from a distance: Dissection of the mRNA determinants required for the incorporation of selenocysteine into protein. *EMBO J.* **11:** 3759.

Herbst K.L., Nichols L.M., Gesteland R.F., and Weiss R.B. 1994. A mutation in ribosomal protein L9 affects ribosomal hopping during translation of gene 60 from bacteriophage T4. *Proc. Natl. Acad. Sci.* **91:** 12525.

Herr A.J., Atkins J.F., and Gesteland R.F. 1999. Mutations which alter the elbow region of tRNA^{Gly} $_2$ reduce T4 gene 60 translational bypassing efficiency. *EMBO J.* **18:** 2886.

―――. 2000a. Coupling of open reading frames by translational bypassing. *Annu. Rev. Biochem.* **69:** 343.

Herr A.J., Gesteland R.F., and Atkins J.F. 2000b. One protein from two open reading frames: Mechanism of a 50 nucleotides translational bypass. *EMBO J.* **19:** 2671.

Herr A.J., Nelson C.C., Willis N., Gesteland R.F., and Atkins J.F. 2001. Analysis of the roles of tRNA structure, ribosomal protein L9, and the phage T4 gene 60 bypassing signals during ribosome slippage on mRNA. *J. Mol. Biol.* **309:** 1029.

Hill K.E., Lloyd R.S., and Burk R.F. 1993. Conserved nucleotide sequences in the open reading frame and 3′ untranslated region of selenoprotein P mRNA. *Proc. Natl. Acad. Sci.* **90:** 537.

Hoffman D.W., Cameron C.S., Davies C., White S.W., and Ramakrishnan V. 1996. Ribosomal protein L9: A structure determination by the combined use of X-ray crystallography and NMR spectroscopy. *J. Mol. Biol.* **289:** 223.

Horsburgh B.C., Kollmus H., Hauser H., and Coen D.M. 1996. Translational recoding induced by G-rich mRNA sequences that form unusual structures. *Cell* **86:** 949.

Horsfield J.A., Wilson D.N., Mannering S.A., Adamski F.M., and Tate W.P. 1995. Prokaryotic ribosomes recode the HIV-1 −1 frameshift sequence by an E/P site post-translocation simultaneous slippage mechanism. *Nucleic Acids Res.* **23:** 1487.

Howard M.T., Shirts B.H., Zhou J., Carlson C.L., Matsufuji S., Gesteland R.F., Weeks R.S., and Atkins J.F. 2001. Cell culture analysis of the regulatory frameshift event required for the expression of mammalian antizymes. *Genes Cells* **6:** 931.

Huang W.M., Ao S.Z., Casjens S., Orlandi R., Zeikus R., Weiss R., Winge D., and Fang M. 1988. A persistent untranslated sequence within bacteriophage T4 DNA topoisomerase gene 60. *Science* **239:** 1005.

Ito K., Uno M., and Nakamura Y. 2000. A tripeptide 'anticodon' deciphers stop codons in messenger RNA. *Nature* **403:** 680.

Ivanov I.P., Gesteland R.F., and Atkins J.F. 2000a. Antizyme expression: A subversion of triplet decoding, which is remarkably conserved by evolution, is a sensor for an autoregulatory circuit. *Nucleic Acids Res.* **28:** 3185.

Ivanov I.P., Gesteland R.F., Matsufuji S., and Atkins J.F. 1998. Programmed frameshifting in the synthesis of mammalian antizyme is +1 in mammals, predominantly +1 in fission yeast, but −2 in budding yeast. *RNA* **4:** 1230.

Ivanov I.P., Gurvich O.L., Gesteland R.F., and Atkins J.K. 2002. Recoding: Site- or mRNA-specific alteration of genetic readout utilized for gene expression in *Translation mechanisms* (ed. J. Lapointe and L. Brakier-Gingras). Landes Bioscience, Austin, Texas. (In press).

Ivanov I.P., Matsufuji S., Murakami Y., Gesteland R.F., and Atkins J.F. 2000b. Conservation of polyamine regulation by translational frameshifting from yeast to mammals. *EMBO J.* **19:** 1907.

Jacks T., Madhani H.D., Masiarz F.R., and Varmus H.E. 1988a. Signals for ribosomal frameshifting in the Rous sarcoma virus *gag-pol* region. *Cell* **55:** 447.

Jacks T., Power M.D., Masiarz F.R., Luciw P.A., Barr P.J., and Varmus H.E. 1988b. Characterization of ribosomal frameshifting in HIV-1 *gag-pol* expression. *Nature* **331:** 280.

Kang H.S. and Tinoco I. 1997. A mutant RNA pseudoknot that promotes ribosomal frameshifting in mouse mammary tumor virus. *Nucleic Acids Res.* **25:** 1943.

Kim K.H. and Lommel S.A.1998. Sequence element required for efficient –1 ribosomal frameshifting in red clover necrotic mosaic dianthovirus. *Virology* **250:** 50.

Kim Y.G., Su L., Maas S., O'Neill A., and Rich A. 1999. Specific mutations in a viral RNA pseudoknot drastically change ribosomal frameshifting efficiency. *Proc. Natl. Acad. Sci.* **96:** 14234.

Klein D.J., Schmeing T.M., Moore P.B., and Steitz T.A. 2001. The kink-turn: A new RNA secondary motif. *EMBO J.* **20:** 4214.

Kontos H., Napthine S., and Brierley I. 2001. Ribosomal pausing at a frameshifter RNA psuedoknot is sensitive to reading phase but shows little correlation with frameshift efficiency. *Mol. Cell. Biol.* **21:** 8657.

Larsen B., Gesteland R.F., and Atkins J.F. 1997. Structural probing and mutagenic analysis of the stem-loop required for *Escherichia coli dnaX* ribosomal frameshifting: Programmed efficiency of 50%. *J. Mol. Biol.* **271:** 47.

Larsen B., Wills N.M., Gesteland R.F., and Atkins J.F. 1994. rRNA-mRNA base pairing stimulates a programmed –1 ribosomal frameshift. *J. Bacteriol.* **176:** 6842.

Levin M.E., Hendrix R.W., and Casjens S.R. 1993. A programmed translational frameshift is required for the synthesis of a bacteiophage λ tail assembly protein. *J. Mol. Biol.* **234:** 124.

Lewis-Henderson W.R. and Djordjevic M.A. 1991. A cultivar-specific interaction between *Rhizobium leguminosarum* bv. trifolii and subterranean clover is controlled by *nodM*, other cultivar specificity genes, and a single recessive host gene. *J. Bacteriol.* **173:** 2791.

Li Z., Stahl G., and Farabaugh P.J. 2001. Programmed +1 frameshifting stimulated by complementarity between a downstream mRNA sequence and an error-correcting region of rRNA. *RNA* **7:** 275.

Lieberman K.R., Firpo M.A., Herr A.J., Nguyenle T., Atkins J.F., Gesteland R.F., and Noller H.F. 2000. The 23 S rRNA environment of ribosomal protein L9 in the 50 S ribosomal subunit. *J. Mol. Biol.* **297:** 1129.

Liphardt J., Napthine S., Kontos H., and Brierley I. 1999. Evidence for an RNA pseudoknot loop-helix interaction essential for efficient –1 ribosomal frameshifting. *J. Mol. Biol.* **288:** 321.

Lopinski J.D., Dinman J.D., and Bruenn J.A. 2000. Kinetics of ribosomal pausing during programmed –1 translational frameshifting. *Mol. Cell. Biol.* **20:** 1095.

Low S.C., Grundner-Culemann E., Harney J.W. and Berry M.J. 2000. SECIS-SBP2 interactions dictate selenocysteine incorporation efficiency and selenoprotein hierarchy. *EMBO J.* **19:** 6882.

Magliery T.J., Anderson J.C., and Schultz P.G. 2001. Expanding the genetic code: Selection of efficient suppressors of four-base codons and identification of "shifty" four-base codons with a library approach in *Escherichia coli*. *J. Mol. Biol.* **307:** 755.

Martin G.W. and Berry M.J. 2001. Selenocysteine codons decrease polysome association on endogenous selenoprotein mRNAs. *Genes Cells* **6:** 121.

Matadeen R., Patwardhan A., Gowen B., Orlova E.V., Pape T., Cuff M., Mueller F., Brimacombe R., and van Heel M. 1999. The *Escherichia coli* large ribosomal subunit at 7.5 Å resolution. *Structure* **7:** 1575.

Matsufuji S., Matsufuji T., Wills N.M., Gesteland R.F., and Atkins J.F. 1996. Reading two bases twice: Mammalian antizyme frameshifting in yeast. *EMBO J.* **15:** 1360.

Matsufuji S., Matsufuji T., Miyazaki Y., Murakami Y., Atkins J.F., Gesteland R.F., and Hayashi S. 1995. Autoregulatory frameshifting in decoding mammalian ornithine decarboxy-lase antizyme. *Cell* **80:** 51.

Mejlhede N., Atkins J.F., and Neuhard J. 1999. Ribosomal –1 frameshifting during decoding of *Bacillus subtilis cdd* occurs at the sequence CGA AAG. *J. Bacteriol.* **181:** 2930.

Michiels P.J.A., Versleijen A.A.M., Verlaan P.W., Pleij C.W.A., Hilbers C.W., and Heus H.A. 2001. Solution structure of the pseudoknot of SRV-1 RNA, involved in ribosomal frameshifting. *J. Mol. Biol.* **310:** 1109.

Morris D.R. and Geballe A.P. 2000. Upstream open reading frames as regulators of mRNA translation. *Mol. Cell. Biol.* **20:** 8635.

Mueller F., Sommer I., Baranov P., Matadeen R., Stoldt M., Wöhnert J., Görlach M., van Heel M., and Brimacombe R. 2000. The 3D arrangement of the 23S and 5S rRNA in the *Escherichia coli* 50S ribosomal subunit based on a cryo-electron microscopic reconstruction at 7.5Å resolution. *J. Mol. Biol.* **298:** 35.

Murakami Y., Matsufuji S., Kameji T., Hayashi S., Igarashi K., Tamura T., Tanaka K., and Ichihara A. 1992. Ornithine decarboxylase is degraded by the 26S proteosome without ubiquitination. *Nature* **360:** 597.

Napthine S., Liphardt J., Bloys A., Routledge S., and Brierley I. 1999. The role of pseudoknot stem 1 length in the promotion of efficient –1 ribosomal frameshifting. *J. Mol. Biol.* **288:** 305.

Nissen P., Ippolito J.A., Ban N., Moore P.B., and Steitz T.A. 2001. RNA tertiary interactions in the large ribosomal subunit: The A-minor motif. *Proc. Natl. Acad. Sci.* **98:** 4899.

O'Connor M. 1998. tRNA imbalance promotes –1 frameshifting via near-cognate decoding. *J. Mol. Biol.* **279:** 727.

O'Connor M., Gesteland R.F., and Atkins J.F. 1989. tRNA hopping: Enhancement by an expanded anticodon. *EMBO J.* **8:** 4315.

O'Connor M., Wills N.M., Bossi L., Gesteland R.F., and Atkins J.F. 1993. Functional tRNAs with altered 3´ ends. *EMBO J.* **12:** 2559.

Paul C.P., Barry J.K., Dinesh-Kumar S.P., Brault V., and Miller W.A. 2001. A sequence required for –1 ribosomal frameshifting located four kilobases downstream of the frameshift site. *J. Mol. Biol.* **310:** 987.

Qian Q., Li J.-N., Zhao H., Hagervall T.G., Farabaugh P.J., and Björk G.R. 1998. A new model for phenotypic suppression of frameshift mutations by mutant tRNAs. *Mol. Cell* **1:** 471.

Reil H., Kollmus H., Weidle U.H., and Hauser H. 1993. A heptanucleotide sequence mediates ribosomal frameshifting in mammalian cells. *J. Virol.* **67:** 5579.

Rettberg C.C., Prère M.F., Gesteland R.F., Atkins J.F., and Fayet O. 1999. A three-way junction and constituent stem-loops as the stimulator for programmed –1 frameshifting in bacterial insertion sequence IS911. *J. Mol. Biol.* **286:** 1365.

Riyasaty S. and Atkins J.F. 1968. External suppression of a frameshift mutant in *Salmonella*. *J. Mol. Biol.* **34:** 541.

Robinson D.N. and Cooley L. 1997. Examination of the function of two kelch proteins generated by stop codon suppression. *Development* **124:** 1405.

Roth J.R. 1981. Frameshift suppression. *Cell* **24:** 601.

Rother M., Wilting R., Commans S., and Böck A. 2000. Identification and characterization of the selenocysteine-specific translation factor SelB from the Archaeon *Methanococcus jannaschii*. *J. Mol. Biol.* **299:** 351.

Schwartz R. and Curran J.F. 1997. Analysis of frameshifting at UUU-pyrimidine sites. *Nucleic Acids Res.* **25:** 2005.

Sekine Y., Eisaki N., and Ohtsubo E. 1994. Translational control in production of transposase and in transposition of insertion sequence IS3. *J. Mol. Biol.* **235:** 1406.

Shehu-Xhilaga M., Crowe S.M., and Mak J. 2001. Maintenance of the Gag/Gag-Pol ratio is important for human immunodeficiency virus type 1 RNA dimerization and viral infectivity. *J. Virol.* **75:** 1834.

Shen L.X. and Tinoco I. 1995. The structure of an RNA pseudoknot that causes efficient frameshifting in mouse mammary tumor virus. *J. Mol. Biol.* **247:** 963.

Skuzeski J.M., Nichols L.M., Gesteland R.F., and Atkins J.F. 1991. The signal for a leaky UAG stop codon in several plant viruses includes the two downstream codons. *J. Mol. Biol.* **218,** 365.

Somogyi P., Jenner A.J., Brierley I., and Inglis S.C. 1993. Ribosomal pausing during translation of an RNA pseudoknot. *Mol. Cell. Biol.* **13:** 6931.

Spahn C.M. and Nierhaus K.H. 1998. Models of the elongation cycle: An evaluation. *Biol. Chem.* **379:** 753.

Spahn C.M.T., Blaha G., Agrawal R.K., Penczek P., Grassucci R.A., Trieber C.A., Connell S.R., Taylor D.E., Nierhaus K.H., and Frank J. 2001. Localization of the ribosomal protection protein Tet(O) on the ribosome and the mechanism of tetracycline resistance. *Mol. Cell* **7:**1037.

Stahl G., Bidou L., Rousset J.-P., and Cassan M. 1995. Versatile vectors to study recoding: Conservation of rules between yeast and mammalian cells. *Nucleic. Acids Res.* **23:** 1557.

Steneberg P. and Samakovlis C. 2001. A novel stop codon readthrough mechanism produces functional Headcase protein in *Drosophila* trachea. *EMBO Rep.* **2:** 593.

Su L., Chen L., Egli M., Berger J.M., and Rich A. 1999. Minor groove RNA triplex in the crystal structure of a ribosomal frameshifting viral pseudoknot. *Nat. Struct. Biol.* **6:** 285.

Sundaram M., Durant P.C., and Davis D.R. 2000. Hypermodified nucleosides in the anticodon of tRNA^{Lys} stabilize a canonical U-turn structure. *Biochemistry.* **39:** 12575.

Sundararajan A., Michaud W.A., Qian Q., Stahl G., and Farabaugh P.J. 1999. Near-cognate peptidyl-tRNAs promote +1 programmed translational frameshifting in yeast. *Mol. Cell* **4:** 1005.

ten Dam E.B., Pleij C.W.A., and Bosch L. 1990. RNA pseudoknots; translational frameshifting and readthrough on viral RNAs. *Virus Genes* **4:** 121.

Theimer C.A. and Giedroc D.P. 2000. Contribution of the intercalated adenosine at the helical junction to the stability of the *gag-pro* frameshifting pseudoknot from mouse mammary tumor virus. *RNA* **6:** 409.

Tsuchihashi Z. and Brown P.O. 1992. Sequence requirements for efficient translational frameshifting in the *Escherichia coli dnaX* gene and the role of an unstable interaction between tRNA^{Lys} and an AAG lysine codon. *Genes Dev.* **6:** 511.

Tu C.-L., Tzeng T.-H., and Bruenn J.A. 1992. Ribosomal movement impeded at a pseudoknot required for frameshifting. *Proc. Natl. Acad. Sci.* **89:** 8636.

Tujebajeva R.M., Ransom D.G., Harney J.W., and Berry M.J. 2000a. Expression and characterization of nonmammalian selenoprotein P in the zebrafish, *Danio rerio. Genes Cells* **5:** 897.

Tujebajeva R.M., Copeland P.R., Xu X.M., Carlson B.A., Harney J.W., Driscoll D.M., Hatfield D.L., and Berry M.J. 2000b. Decoding apparatus for eukaryotic selenocysteine insertion. *EMBO Rep.* **1:** 158.

van Marle G., Dobbe J.C., Gultyaev A.P., Luytjes W., Spaan W.J., and Snijder E.J. 1999. Arterivirus discontinuous mRNA transcription is guided by base pairing between sense and antisense transcription-regulating sequences. *Proc. Natl. Acad. Sci.* **96:** 12056.

Walczak R., Westhof E., Carbon P., and Krol A. 1996. A novel RNA structural motif in the selenocysteine insertion element of eukaryotic selenoprotein mRNAs. *RNA* **2:** 367.

Walleczek J., Schuler D., Stoffler-Meilicke M., Brimacombe R., and Stoffler G. 1988. A model for the spatial arrangement of the proteins in the large subunit of the *Escherichia coli* ribosomes. *EMBO J.* **7:** 3571.

Wang S., Guo L., Allen E., and Miller W.A. 1999. A potential mechanism for selective control of cap-independent translation by a viral RNA sequence in *cis* and in *trans. RNA* **5:** 728.

Weiss R.B., Huang W.M., and Dunn D.M. 1990a. A nascent peptide is required for ribosomal bypass of the coding gap in bacteriophage T4 gene *60. Cell* **62:** 117.

Weiss R.B., Dunn D.M., Atkins J.F., and Gesteland R.F. 1987. Slippery runs, shifty stops, backward steps, and forward hops: –2, –1, +1, +2, +5, and +6 ribosomal frameshifting. *Cold Spring Harbor Symp. Quant. Biol.* **52:** 687.

———. 1990b. Ribosomal frameshifting from –2 to +50 nucleotides. *Prog. Nucleic Acid Res. Mol. Biol.* **39:** 159.

Weiss R., Lindsley D., Falahee B., and Gallant J. 1988a. On the mechanism of ribosomal frameshifting at hungry codons. *J. Mol. Biol.* **203:** 403.

Weiss R.B., Dunn D.M., Dahlberg A.E., Atkins J.F., and Gesteland R.F. 1988b. Reading frame switch caused by base-pair formation between the 3′ end of 16S rRNA and the mRNA during elongation of protein synthesis in *Escherichia coli. EMBO J.* **7:** 1503.

Weiss R.B., Dunn D.M., Shuh M., Atkins J.F., and Gesteland R.F. 1989. *E. coli* ribosomes re-phase on retroviral frameshift signals at rates ranging from 2 to 50 percent. *New Biol.* **1:** 159.

Wentzel A.M., Stancek M., and Isaksson L.A. 1998. Growth phase dependent stop codon readthrough and shift of translation reading frame in *Escherichia coli. FEBS Lett.* **421:** 237.

Wills N.M., Gesteland R.F., and Atkins J.F. 1991. Evidence that a downstream pseudoknot is required for translational readthrough of the Moloney murine leukemia virus *gag* stop codon. *Proc. Natl. Acad. Sci.* **88:** 6991.

———. 1994. Pseudoknot-dependent read-through of retroviral *gag* termination codons: Importance of sequences in the spacer and loop 2. *EMBO J.* **13:** 4137.

Wilson K.S., Ito K., Noller H.F., and Nakamura, Y. 2000. Functional sites of interaction between release factor RF1 and the ribosome. *Nat. Struct. Biol.* **7:** 866.

Yelverton E., Lindsley D., Yamauchi P., and Gallant J.A. 1994. The function of a ribosomal frameshifting signal from human immunodeficiency virus-1 in *Escherichia coli. Mol. Microbiol.* **11:** 303.

You S., Falgout B., Markoff L., and Padmanabhan R. 2001. In vitro RNA synthesis from exogenous dengue viral RNA templates requires long range interactions between 5'- and 3'-terminal regions that influence RNA structure. *J. Biol. Chem.* **276:** 15581.

Yourno J. and Tanemura S. 1970. Restoration of in-phase translation by an unlinked suppressor of a frameshift mutation in *Salmonella typhimurium. Nature* **225:** 422.

Yusupova G.Z., Belitsina N.V., and Spirin A.S. 1986. Template-free ribosomal synthesis of polypeptides from aminoacyl-tRNA. Polyphenylalanine synthesis from phenylalanine-tRNA^{Lys}. *FEBS Lett.* **206:** 142.

Yusupova G.Z., Yusupov M.M., Cate J.H., and Noller H.F. 2001. The path of messenger RNA through the ribosome. *Cell* **106:** 233.

Yusupov M.M., Yusupova G.Z., Baucom A., Lieberman K., Earnest T.N., Cate J.H.D., and Noller H.F. 2001. Crystal structure of the ribosome at 5.5 Å resolution. *Science* **292:** 883.

Zerfass K. and Beier H. 1992. Pseudouridine in the anticodon GψA of plant cytoplasmic tRNA^{Tyr} is required for UAG and UAA suppression in the TMV-specific context. *Nucleic Acids Res.* **20:** 5911.

Structure and Function of the Stimulatory RNAs Involved in Programmed Eukaryotic −1 Ribosomal Frameshifting

I. Brierley and S. Pennell*

Division of Virology, Department of Pathology, University of Cambridge, Cambridge CB2 1QP, United Kingdom

The elongation phase of protein synthesis is of necessity a precise process, and mechanisms exist to promote translational fidelity (for review, see Czworkowski and Moore 1996). However, the system must endure a degree of inaccuracy in order that translation may proceed with sufficient speed. Two major kinds of errors have been described: missense, where an incorrect amino acid is incorporated; and processivity, where premature termination of translation or translational frameshifting occurs (Farabaugh and Bjork 1999). Fortunately, these events take place at a relatively low frequency, with frameshift errors the less common (10^{-5} per codon; Kurland 1992). A growing number of examples have been described, however, of highly efficient "programmed" frameshift sites (for review, see Farabaugh 1996, 2000). These ribosomal frameshift signals do not induce "errors" in the classic sense in that the frameshifts generate authentic proteins, are stimulated by specific elements encoded in the mRNA, and occur at frequencies that can approach 100%. For this reason, they are considered more as extensions of the genetic code (recoding sites; Gesteland et al. 1992; Gesteland and Atkins 1996) rather than "natural" errors, although there may be mechanistic similarities between the two (Farabaugh and Bjork 1999). There is considerable interest in how programmed frameshifting occurs, since this may provide insights into normal frame maintenance, tRNA movement, and the unwinding of mRNA secondary structures by ribosomes. This paper focuses on a particular class of programmed frameshift signal, that which directs −1 ribosomal frameshifting, with an emphasis on the stimulatory mRNA elements that enhance the efficiency of the process. The review concentrates mainly on eukaryotic systems, as prokaryotic frameshifting is discussed elsewhere in this volume.

OCCURRENCE AND ROLE OF −1 RIBOSOMAL FRAMESHIFT SIGNALS

Ribosomal frameshift signals have been found in a variety of mRNAs, most commonly in the genomes of eukaryotic RNA viruses and *Escherichia coli* insertion elements (for review, see Chandler and Fayet 1993; Brierley 1995; Dinman 1995; Fütterer and Hohn 1996; Farabaugh 2000). To date, there are no confirmed examples from conventional eukaryotic cellular genes, although computer-assisted database searches have identified a number of candidates (Hammell et al. 1999; Liphardt 1999). The signals for frameshifting are typically contained within a short, contiguous stretch of mRNA (usually less than 100 nucleotides in length) and instruct the ribosome to move from its current frame (arbitrarily zero) to the −1 reading frame (i.e., 5′-ward) at a defined point and to continue translation. Frameshift sites occur almost exclusively at the junction of overlapping coding sequences and allow the product of the downstream coding region, at a certain frequency, to be appended to that of the upstream coding sequence, generating a single, translationally fused, polypeptide. In RNA viruses, frameshifting is usually involved in the expression of replicases; in insertion elements, in the biologically equivalent transposases. For most virus groups the exact role of frameshifting is uncertain, but it is reasonably well understood for the retroviruses, where it allows replicative enzymes (as part of the Pol polyprotein) to be synthesised as a carboxy-terminal extension of the structural proteins (Gag). The product of ribosomal frameshifting, the Gag-Pol polyprotein, is incorporated into virions, and this is an essential step in the virus life cycle, since the reverse transcriptase enzyme (encoded in *pol*) is required for subsequent events. There is growing evidence that modulation of frameshift efficiency can reduce retroviral infectivity, either by eliminating the replicase from virions, or by influencing particle assembly (Dinman and Wickner 1992, Karacostas et al. 1993; Hung et al. 1998; Shehu-Xhilaga et al. 2001). This has important human health implications, because many pathogenic RNA viruses of humans and animals employ frameshifting during their life cycle, including human immunodeficiency virus type 1 (HIV-1), the human T-cell lymphotropic viruses (HTLV-I and -II), and the coronaviruses. A better understanding of both the occurrence and the molecular basis of frameshifting will be required, however, before it can be considered as a genuine antiviral target.

STRUCTURE OF −1 FRAMESHIFT SIGNALS

Eukaryotic ribosomal frameshift signals contain two essential elements: a "slippery" sequence, where the ribosome changes reading frame, and a stimulatory RNA

*Present address: National Institute for Medical Research, Division of Protein Structure, The Ridgeway, Mill Hill, London, NW7 1AA, United Kingdom.

secondary structure, often an RNA pseudoknot, located a few nucleotides downstream (Jacks et al. 1988a; Brierley et al. 1989; ten Dam et al. 1990). A spacer region between the slippery sequence and the stimulatory RNA is also required, and the precise length of this spacer must be maintained for maximal frameshifting efficiency (Brierley et al. 1989, 1992; Kollmus et al. 1994).

The Slippery Sequence

The slippery sequence is a heptanucleotide stretch that contains two homopolymeric triplets and conforms in the vast majority of cases to the motif XXXYYYZ (e.g., UUUAAAC in the coronavirus infectious bronchitis virus (IBV) *1a/1b* signal). In eukaryotic systems, frameshifting at this sequence is thought to occur by "simultaneous slippage" of two ribosome-bound tRNAs, presumably peptidyl and aminoacyl tRNAs, which are translocated from the zero (X XXY YYN) to the −1 phase (XXX YYY) (Jacks et al. 1988a). The homopolymeric nature of the slippery sequence seems to be required to allow the tRNAs to remain base-paired to the mRNA in at least two out of three anticodon positions following the slip. Frameshift assays, largely carried out in vitro, have revealed that the X triplet can be A, C, G, or U, but the Y triplet must be A or U (Jacks et al. 1988b; Dinman et al. 1991; Brierley et al. 1992). In addition to these restrictions, slippery sequences ending in G (XXXAAAG or XXXUUUG) do not function efficiently in in vitro translation systems (Brierley et al. 1992), nor in yeast (Dinman et al. 1991) or mammalian cells (Marczinke et al. 2000). At naturally occurring frameshift sites, of the possible codons which are decoded in the ribosomal A site prior to tRNA slippage (XXXYYYN), only five are represented in eukaryotes: AAC, AAU, UUA, UUC, and UUU (Farabaugh 1996). Together with the in vitro data, it seems that the sequence restrictions observed are a manifestation of the need for the pre-slippage codon–anticodon complex in the A site to be weak enough such that the tRNAs can detach from the codon during the process of frameshifting. G-C pairs are thus avoided.

There is considerable experimental support for the simultaneous-slippage model, particularly from site-directed mutagenesis studies (Jacks et al. 1988a; Dinman et al. 1991; Brierley et al. 1992; Dinman and Wickner 1992), sequencing of trans-frame proteins (Hizi et al. 1987; Jacks et al. 1988a,b; Weiss et al. 1989; Nam et al. 1993), and nucleotide sequence comparisons (Jacks et al. 1988a; ten Dam et al. 1990). The protein sequencing studies indicate that the frameshift occurs at the second codon of the tandem slippery pair; i.e., at that codon decoded in the ribosomal aminoacyl (A) site (XXXYYYN). The importance of the A-site tRNA in frameshifting is also apparent from the mutagenesis studies; point mutations in this region are generally more inhibitory than those that would influence the P-site tRNA. In the simultaneous-slippage model, frameshifting is proposed to take place prior to peptide-bond formation, with the ribosome-bound tRNAs occupying the traditional P and A sites. An updated version advanced by Weiss and colleagues (1989),

taking into account the tRNA hybrid-states model of Moazed and Noller (1989), points out that frameshifting could also occur following peptide-bond formation and prior to, or even during, translocation (see Fig. 1). To date, the precise point in the elongation cycle at which frameshifting occurs remains unestablished, although the effects on frameshifting of various antibiotics suggest that the event occurs prior to translocation and possibly during peptidyl transfer (Dinman et al. 1997, 1998; Tumer et al. 1998; Brunelle et al. 1999; Hudak et al. 2001).

Stimulatory RNA Secondary Structures

Efficient frameshifting requires the presence of a stimulatory RNA structure located a few nucleotides downstream of the slippery sequence. In some cases, this is a stem-loop, but more commonly, an RNA pseudoknot is present. These structures are sometimes referred to as frameshifter stem-loops or frameshifter pseudoknots to distinguish them from related structures in cellular RNAs. The interaction of the ribosome with the stimulatory RNA is thought to pause ribosomes in the act of decoding the slippery sequence, allowing more time for the tRNAs to realign in the −1 reading frame (Jacks et al. 1988a). Our knowledge of the folding of these RNAs has been derived from site-specific mutagenesis, chemical and enzymatic structure probing, and more recently by nuclear magnetic resonance (NMR) and X-ray crystallography. The following sections summarize what is known about the structures of representative examples of the stimulatory RNAs and the key features that influence frameshifting before considering how this fits with models for the process. We begin with frameshifter RNA pseudoknots, since these are the best studied. This topic has recently been the subject of an excellent review (Giedroc et al. 2000).

FRAMESHIFTER RNA PSEUDOKNOTS

Mouse Mammary Tumor Virus *gag/pro*

The pseudoknot at the *gag-pro* junction of MMTV was the first to be analyzed in three dimensions using NMR spectroscopy (Shen and Tinoco 1995). The structure solved was actually a variant termed VPK, which retained the same frameshift efficiency of the wild-type molecule (Chen et al. 1995). Initial experiments with the wild-type pseudoknot gave multiple overlapping signals that could not be disentangled until certain base pairs in each stem of the pseudoknot were flipped from G-C to C-G, to create VPK (Fig. 2). Overall, four converged pseudoknot structures were identified, three of which shared similar free-energy values (within 15%), and a fourth with a free energy some 25–30% higher. The analysis revealed that the VPK pseudoknot has a 5-base-pair stem 1, a 6-base-pair stem 2, a 2-nucleotide loop 1, and a short, 8-nucleotide loop 2. The G-U and A-U base pairs in stem 2 located at the ends of the helix did not give sharp resonances in the NMR spectrum, suggesting that the pairs are only transiently formed or absent. The outstanding feature of the structure is a pronounced interstem kink of approximately 60° from vertical which is brought about by a combination of an unpaired adenosine residue located between the two stems and sterically constrained loops. The stacking of bases in loop 2 stabilizes base-pairing at the bottom of S1 and effectively shortens the loop to the equivalent of 6 nucleotides. This restriction prevents the backbone of loop 2 from adopting an extended conformation, limiting the effective distance by which loop 2 can span across the minor groove, thus causing stem 1 to bend toward stem 2. Together, the short loops hold the kinked stems in place and reinforce the compact nature of the pseudoknot. No long-range nuclear Overhauser enhancements were observed between the loops and stems of VPK, arguing against base triplex interactions. However, it has been suggested from molecular modeling studies (Le et al. 1998) that interactions between loop 2 and stem 1 could occur, as has also been seen at other frameshifter pseudoknots (see below). A combination of experimental approaches, including mutational analysis, structure probing, and NMR spectroscopy of VPK variants (Chen et al. 1995, 1996; Kang et al. 1996; Kang and Tinoco 1997), has revealed a specific requirement for the unpaired adenosine between the

stems, and no other residue will replace it functionally. Indeed, the data are consistent with the view that the specific kinked conformation is the determining feature of frameshift stimulation by this pseudoknot.

Beet Western Yellows Virus *p1/p2*

The frameshift signal in the genomic RNA of BWYV was first described by Veidt and colleagues (1988) and demonstrated experimentally some years later (Garcia et al. 1993). Sequence analysis had suggested that the site contained an RNA pseudoknot (ten Dam et al. 1990), and this was subsequently confirmed by Su and colleagues (1999), who published the crystal structure at 1.6 Å resolution, the first crystal structure of a frameshifter RNA pseudoknot. This was a remarkable achievement and also provided an example of how secondary structure predictions can often be misleading. The BWYV pseudoknot was thought to consist of short, coaxially stacked stems of five (stem 1) and four (stem 2) base pairs connected by short loops of two (loop 1) and six (loop 2) nucleotides, respectively, in classic H-type pseudoknot folding (Pleij et al. 1985; Garcia et al. 1993; Hilbers et al. 1998). In fact, the crystal structure revealed many unusual features and highlighted several changes to the previously assumed

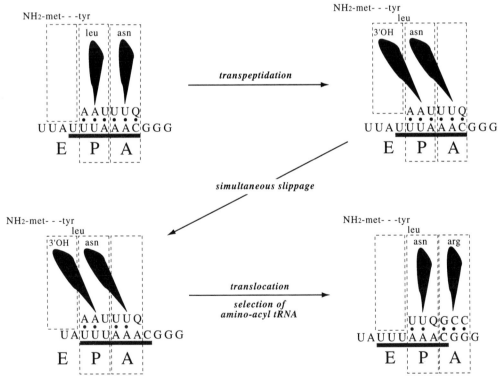

Figure 1. The simultaneous slippage model of programmed –1 ribosomal frameshifting. The model shown is a variant of the original (Jacks et al. 1988a) as proposed by Weiss and colleagues (1989). The first stage shows the amino-acyl (A) and peptidyl (P) tRNAs base-paired to the slippery sequence (UUUAAAC) in the zero frame before transpeptidation. In the second stage, the tRNAs are still in the zero frame but occupy hybrid sites (P-A and E [exit]-P) based on the displacement model for the peptidyl transfer reaction (Moazed and Noller 1989). The third stage shows the tRNAs slipping back by one nucleotide, retaining (in the case of this slippery sequence) two of three anticodon–codon base pairs. In the fourth step, the incoming tRNA[Arg] decodes the –1 frame codon, completing the cycle.

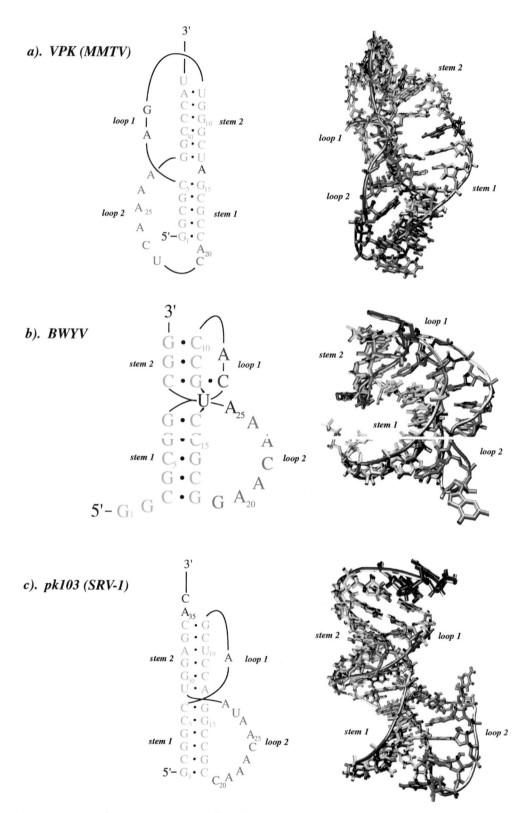

Figure 2. Frameshifter pseudoknots at atomic resolution. Secondary structure representations and atomic structures of the VPK (MMTV) (Shen and Tinoco 1995), BWYV (Su et al. 1999), and pk103 (SRV-1) (Michiels et al. 2001) pseudoknots are shown. Stem bases are in yellow, loop 1 in red, loop 2 in green, and other residues (see text) in blue. Outstanding features include the unpaired, intercalating A14 residue of VPK that contributes to the bend between the two stems, interactions between loop 2 and stem 1 in BWYV and pk103 that result in base triples, and the unpaired U13 A25 residues at the bottom of stem 2 in BWYV. In the latter case, A25 stacks onto loop 2 and U13 is excluded from the helix of stem 2. Accession codes of the atomic coordinates from the Protein Data Bank (www.rcsb.org) are 1RNK (VPK), 437D (BWYV), and 1E95 (pk103). Structural representations were generated using the program VMD (Humphrey et al. 1996; www.ks.uiuc.edu/Research/vmd) and rendered using POV-RAY (www.povray.org).

secondary structure, including a number of unexpected interactions (see Fig. 2). The most striking was the presence of a novel adenosine-rich triplet formed between most of the length of loop 2 and the minor groove of stem 1. A highly conserved 5′-AACAA-3′ motif in this loop has its bases rotated by varying degrees to allow interactions with bases on both strands in the minor groove of stem 1. This is the first time that such an extended RNA triplex has been described. It is an RNA-specific feature, with each interaction involving a hydrogen bond formed from a ribose 2′-hydroxyl group. The first base of loop 2 is the only one that is not involved in the triplex; in fact, it projects away from stem 1, forming a sharp turn facilitated by the possession of a C2′-endo sugar pucker. Surprisingly, only three base pairs are formed in stem 2; the predicted A-U pair is absent and, instead, the U is displaced from the helix and the A is stacked onto loop 2. This stacking, along with the short length of loop 1, induces a rotation of stem 2 relative to stem 1 such that the stems are displaced by approximately 5 Å, with stem 2 tilting some 25° from vertical. Thus, the stems are not coaxially stacked. The geometry of the stems also differs significantly from A-form duplex RNA; the base pairs are largely propeller-twisted, and in stem 1, they are tilted less far from the helix axis than is normal. Another notable feature is a base quadruple interaction formed between C8 of loop 1 (which is inserted into the major groove of stem 2), A25, and the G12-C26 base pair of stem 2. A fascinating statistic concerning the BWYV pseudoknot is that more hydrogen bonds are present in tertiary interactions than are involved in canonical Watson-Crick base pairs (26 versus 24, respectively).

The contribution to frameshifting of the various structural features of the BWYV pseudoknot and a closely related pseudoknot from potato leafroll virus has been investigated by site-directed mutagenesis and in vitro translation (Kim et al. 1999; 2000). These studies are of particular relevance because the influence on pseudoknot conformation of the various base substitutions, insertions, and deletions could be predicted with accuracy. It was found that frameshifting efficiency depended on the maintenance of the conformation of the structure, particularly the specific nucleotide tertiary interactions at the junction between the two stems and the bases involved in the triplex interaction. Unexpectedly, a number of exposed residues not involved in tertiary contacts were also found to affect frameshifting, including U13, which is extruded from the helix; A9 of loop 1, which stacks partially onto stem 2; and finally, a nucleotide(s) at the junction of stem 1 and loop 2. It was proposed that certain specific residues may have dynamic interactions with the ribosome or auxiliary protein factors needed to trigger the –1 slip.

Simian Retrovirus 1 *gag/pro*

An atomic resolution NMR structure of a variant of the SRV-1 pseudoknot has recently been reported (see Fig. 2) (Michiels et al. 2001). As was the case with the MMTV pseudoknot, a variety of alterations to the primary sequence of the SRV-1 pseudoknot were required before the molecule was amenable to high-resolution NMR

analysis. The resulting pseudoknot pk103, which retained most of the functionality of the wild-type structure, was found to have a classic H-type fold and contained a number of interesting features. Most notably, the two stems of the pseudoknot stack upon each other. Earlier secondary structure predictions had indicated that the pseudoknot would resemble that of MMTV (ten Dam et al. 1990) and that the U-A "pair" at the top of stem 1 of the SRV-1 pseudoknot would probably be unpaired, with the A intercalating between the two stems, generating a kinked pseudoknot similar in conformation to that of VPK. However, from the analysis of pk103, and an earlier lower-resolution study of the wild-type pseudoknot (Du et al. 1997), it is clear that the U-A pair at the top of stem 1 is formed. The two base pairs at the junction of stem 1 and stem 2 have a helical twist of approximately 49°, allowing proper alignment and close approach of the three different strands at the junction. In addition to the overwound junction, the structure is somewhat kinked between stem 1 and stem 2, assisting the single adenosine in spanning the major groove of stem 2. The pk103 pseudoknot also has a loop 2–stem 1 triple helical interaction similar to that seen in the BWYV pseudoknot. The interactions involve the minor groove of stem 1 and, in addition to base–base and base–sugar interactions, a ribose zipper motif occurs, the first time such a feature has been seen in a frameshifter pseudoknot. The A and C residues included at the 3′ end of the molecule (blue in Fig. 2) were found to improve the homogeneity of the major conformer in solution, probably a result of stacking effects.

The structure of the SRV-1 pseudoknot was published only recently, and we await the outcome of functional studies informed by this knowledge. Earlier mutational analysis of the wild-type molecule assessed the importance of stem stability and loop lengths (ten Dam et al. 1995) and of the U-A pair at the top of stem 1 (Chen et al. 1996; Sung and Kang 1998). It was found that the predicted stability of stem 1 did not correlate tightly with frameshift efficiency, and a functional requirement for a guanosine-rich 5′ arm at the bottom of this stem was proposed. In contrast to stem 1, the calculated thermodynamic stability of stem 2 mutants correlated qualitatively with the frameshift efficiency of the mutants, although no clear linear relationship was apparent. The primary sequence of the single loop-1 residue was found to be unimportant in frameshifting, and this is consistent with the observation that the loop-1 base does not interact with stem 2 (Michiels et al. 2001). Two observations were at odds with the NMR structure. First, mutational analysis revealed that the sequence and length of loop 2 were unimportant as long as the loop was of sufficient length to span stem 1. This was unexpected, given the evidence for a loop 2–stem 1 triplex interaction revealed by NMR analysis. It may be that the modified loop-2 sequences in the mutants were still able to form such interactions. Second, the identity of the nucleotides at the top of stem 1 and their ability to form a base pair (or not) has apparently no effect on frameshifting efficiency (Chen et al. 1996; Sung and Kang 1998). Thus, although the U-A pair is formed at the top of stem 1, pairing does not appear to be required for function.

Infectious Bronchitis Virus *1a/1b*

The frameshift signal present at the *1a-1b* overlap of the coronavirus IBV was the first shown to require the participation of an RNA pseudoknot (see Fig. 3, panel A) (Brierley et al. 1989). The three-dimensional structure of this molecule has not been solved, and our knowledge is based on mutagenesis studies (Brierley et al. 1989, 1991, 1992) and secondary structure probing (Liphardt et al. 1999; Napthine et al. 1999; Pennell 2001). In comparison to the MMTV, BWYV, and SRV-1 pseudoknots, the IBV pseudoknot is significantly larger, with a longer stem 1 (11 base pairs, cf. 5–7) and loop 2 (32 nucleotides, cf. 7–12). The "extra" length of stem 1 is required for frameshifting; pseudoknots with shorter stems show greatly reduced activity, although longer stems (up to at least 13 base pairs) are tolerated (Napthine et al. 1999). This length requirement seems to be unrelated to the predicted stability of stem 1. Pseudoknots based on the IBV signal have been prepared with stem-1 lengths of less than 11 base pairs, yet highly GC-rich and with greater predicted thermodynamic stabilities than the wild-type stem 1. These variants still show greatly reduced frameshifting efficiency (Napthine et al. 1999). Another feature of stem 1 that contributes to frameshifting is the G-rich 5′ arm at the bottom of the stem. This seems to be an important determinant of functionality, as is also seen with the SRV-1 pseudoknot (ten Dam et al. 1995). In contrast, most of loop 2 is dispensable; it can be shortened to as few as 8 nucleotides without compromising frameshift activity, although further reduction (to 5 nucleotides) severely disrupts frameshifting, probably because the latter loop is too short to span stem 1 (Brierley et al. 1991). The frameshifter pseudoknot present at the *gag/pol* overlap of the *Saccharomyces cerevisiae* L-A virus may represent an example of an IBV-like stimulator with a naturally short loop 2. The predicted secondary structure of the L-A pseudoknot is similar to that of IBV, with a long stem 1 (some 12–13 base pairs in length), but loop 2 is likely to be only 12 nucleotides in length (Dinman and Wickner 1992).

An important question to address is whether IBV pseudoknot-induced frameshifting depends on a kinked conformation and/or tertiary contacts described for other frameshifter pseudoknots. It seems unlikely that the IBV pseudoknot contains a kinked conformation, or at least one brought about by an unpaired, intercalated residue between the stems. It has been noted that the base pair at the top of stem 1 of the wild-type IBV pseudoknot is a relatively weak U-G pair, and if this pair did not form, an unpaired G residue would be present at the stem junction that could lead to a kinked structure (Chen et al. 1996). However, IBV pseudoknot variants exist with more stable C-G or G-C pairs at the top of stem 1 (Brierley et al. 1991; Napthine et al. 1999), and these stimulate frameshifting at slightly higher levels than the wild-type structure, arguing against a role for an unpaired base at the top of stem 1. In addition, both pseudoknot loops can tolerate insertions (Brierley et al. 1991), implying that they are not sterically constrained, as seems to be a requirement for kinking of the VPK pseudoknot. Regarding

triplex interactions between loop 2 and stem 1, again there is little supportive evidence. A specific length or primary sequence of loop 2 appears to be unimportant as long as a minimal length sufficient to span stem 1 is maintained (Brierley et al. 1991). On the surface, therefore, there appear to be functional differences between pseudoknots like that of IBV and the shorter ones described earlier. Confirmation of this will require a high-resolution structure, and efforts to crystallize a minimal version of the IBV pseudoknot with a shortened loop 2 (Brierley et al. 1992) are under way (Pennell 2001).

Liphardt and colleagues (1999) recently investigated the sequence manipulations that are required to bypass the requirement for a long stem 1 in IBV and to convert a short nonfunctional pseudoknot into a highly efficient, kinked frameshifter pseudoknot. The starting point was an inactive IBV-based pseudoknot with 6 base pairs in each stem and a loop 2 of 8 nucleotides. A variety of pseudoknots were constructed that contained, among other changes, an intercalating A residue at the junction between the two pseudoknot stems. However, efficient frameshifting was only observed when the last nucleotide of loop 2 was changed from a G to an A residue (to create the pseudoknot pKA-A; Fig. 3). The preference for adenines at the end of RNA loops has been linked with loop–RNA minor groove interactions in other RNAs (Cate et al. 1996a,b; Kolk et al. 1998), consistent with the possibility that a similar contact was contributing to frameshifting in pKA-A. A mutational analysis of both partners of the proposed interaction, the loop-2 terminal adenine and two GC pairs near the top of stem 1, revealed that the interaction was essential for efficient frameshifting. The specific requirement for an A at the end of loop 2, however, was not seen in pseudoknots with a longer loop 2 (14 nucleotides in length), suggesting that the loop–helix contact may be required only in those pseudoknots with a short loop 2. Whether the loop–helix interaction seen with pKA-A resembles the triplexes observed with the BWYV and pk103 pseudoknots remains to be determined.

Rous Sarcoma Virus *gag/pol*

The frameshift signal in the genomic RNA of RSV was the first documented example of –1 ribosomal frameshifting (Jacks and Varmus 1985). The RSV stimulatory RNA was originally described as a complex stem-loop based on a computer prediction of secondary structure, yet it was clear from deletion analysis that additional information downstream of the stem-loop was required for maximal frameshift efficiency, and a pseudoknot structure was suspected (Jacks et al. 1988a). A combination of structure probing and mutational analysis later revealed that the stimulatory RNA is indeed a complex stem-loop with a long stable stem and two additional stem-loops contained as substructures within the main loop region (Fig. 3, panel B) (Marczinke et al. 1998). An additional interaction occurs between a stretch of 8 nucleotides in the main loop and a region downstream to generate an unusual RNA pseudoknot. If the RSV stimulator is drawn in

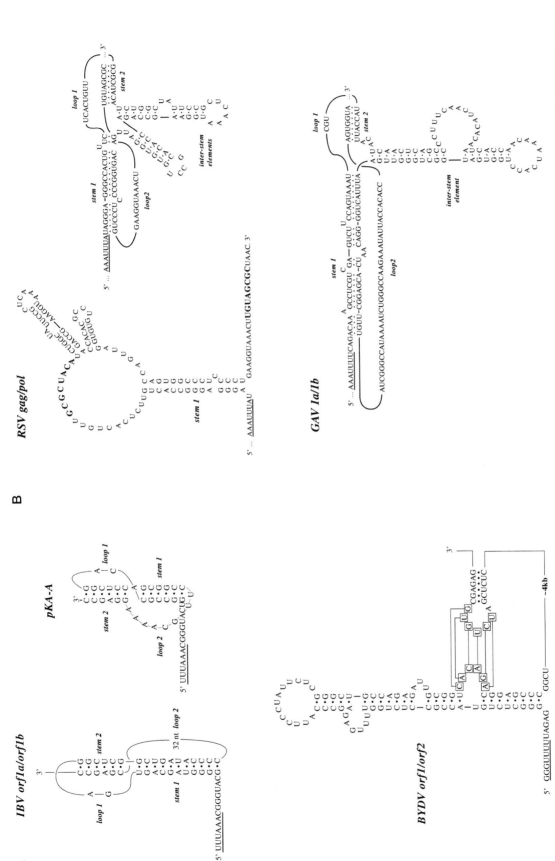

Figure 3. Examples of pseudoknot-containing ribosomal frameshift signals. Shown are signals from (*A*) IBV *1a/1b* (Brierley et al. 1991), pKA-A (Liphardt et al. 2001) and (*B*) RSV *gag/pol* (Jacks et al. 1988a; Marczinke et al. 1998), and GAV *1a/1b* (Cowley et al. 2000). Slippery sequences are underlined. The RSV *gag/pol* signal is shown twice, once as a stem-loop representation (*left*) and again folded in classic H-type pseudoknot conformation (*right*).

the classic conformation, the substructures in the loop are positioned at the junction of the two pseudoknot stems (we refer to these as inter-stem elements). Earlier work with the IBV pseudoknot had indicated that the insertion of bases between the component pseudoknot stems was strongly inhibitory to frameshifting (Brierley et al. 1991), so the predicted organization of the RSV pseudoknot is puzzling. It seems likely, however, that this kind of frameshifter pseudoknot is not uncommon, as a number of other retroviruses appear to use a frameshift signal similar in organization to that employed by RSV (Marczinke et al. 1998). Recently, a frameshift signal has been discovered in the genomic RNA of gill-associated virus (GAV), a prawn pathogen related to the coronaviruses (Cowley et al. 2000). The proposed secondary structure of the GAV stimulatory shows considerable similarity to that of RSV (Fig. 3, panel B), including a stable inter-stem element. It is intriguing that the stimulatory RNA of this invertebrate virus appears to be structurally related to that of a vertebrate retrovirus.

Barley Yellow Dwarf Virus *orf 1/orf 2*

Early studies on the frameshift signal of BYDV had indicated the likely involvement of a bulged stem-loop structure located downstream of a heptanucleotide stretch GGGUUUU (Brault and Miller 1992). When the putative signal was cloned into a heterologous context (i.e., into a reporter gene) to allow frameshifting to be studied, it was found that the response of these elements to mutation was not consistent with those that had been observed at other frameshift sites. For example, the zero-frame termination codon downstream of the slippery sequence was found to be required, and mutations within the slippery sequence that ought to prevent frameshifting showed no obvious effects (Paul et al. 2001). The situation was reversed when the signal was studied within the context of the viral genome. Under these circumstances, the slippery sequence was clearly required for frameshifting, and the termination codon was dispensable in vitro (Paul et al. 2001), although a complete molecular explanation for this reversal is lacking. A remarkable observation made during the course of these studies was that sequences at the 3′ end of the viral genome were required for frameshifting. Indeed, it was proposed that a stretch of bases at the 3′ end of the genome pair with nucleotides in a bulge on the 3′ arm of the BYDV hairpin to generate a long-range pseudoknot (see Fig. 3, panel A). If confirmed, this pseudoknot would possess a loop 2 of about 4 kb and would represent by far the longest loop described in a frameshifter pseudoknot.

FRAMESHIFTER STEM-LOOPS

Stem-loop-containing frameshift signals are found in a variety of animal and plant virus genomes, including the human astrovirus serotype-1 *1a/1b* (HAst-1; Jiang et al. 1993), HTLV-II *gag/pro* (Mador et al. 1989), HIV-1 *gag/pol* (Jacks et al. 1988b), Cocksfoot mottle sobemovirus *2a/2b* (CfMV; Makinen et al. 1995), and red

clover necrotic mosaic virus *p27/p57* (see Fig. 4) (RCNMV; Xiong and Lommel 1989). Of the human viruses, the HAst-1 signal is the simplest, comprising the slippery sequence AAAAAAC and a small G-C-rich stem-loop structure located some six nucleotides downstream. The HTLV-II *gag/pro* signal is similar, with an identical slippery sequence but a longer stem-loop. The HIV-1 frameshift site also has a largely homopolymeric slippery sequence, although in this case UUUUUUA, and a stable hairpin downstream. RNA structure probing has confirmed the presence of the stem-loop at the HAst-1 (Marczinke et al. 1994) and HIV-1 sites (Kang 1998), and there is little evidence to suggest that the loop regions of these hairpins could base-pair elsewhere to form a pseudoknot, although the primary sequences of the loops are well conserved among virus isolates (L. King et al., unpubl.). Du and Hoffman (1997) have proposed that the HIV-1 frameshift region could fold into a pseudoknot, involving pairing between part of the HIV-1 spacer region (5′-GGGAAG-3′) and the top part of the first arm of the stem (3′-UCCUUC-5′; Fig. 4). In this configuration, however, the slippery sequence and pseudoknot would be

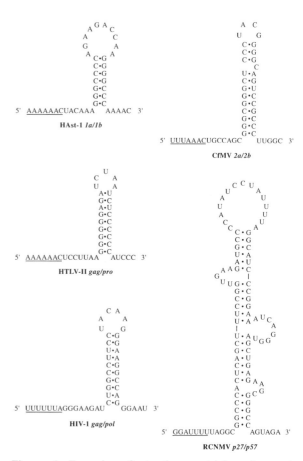

Figure 4. Examples of stem-loop-containing ribosomal frameshift signals. Illustrated are the signals from HAst-1 *1a/1b* (Marczinke et al. 1994), HTLV-II *gag/pro* (ten Dam et al. 1990), HIV-1 *gag/pol* (Jacks et al. 1988b; Kang 1998), CfMV *2a/2b* (Makinen et al. 1995), and RCNMV *p27/p57* (Kim and Lommel 1994). The slippery sequences are underlined.

contiguous on the mRNA, an arrangement that is rarely seen at frameshift sites. The plant virus frameshift signals of CfMV (Lucchesi et al. 2000) and RCNMV (Kim and Lommel 1998) have been characterized by mutational analysis. They differ somewhat from the mammalian stem-loop-containing sites in that the slippery sequences are less homopolymeric and the stems are longer, especially in the case of RCNMV. A potential pseudoknot interaction between the loop of the RCNMV hairpin (5′-UAUCCU) and the large internal bulge in the 3′ arm of the stem (3′-AUGGGA-5′) has been investigated, but no support for such an interaction was obtained (Kim and Lommel 1998). Interestingly, the RCNMV structure closely resembles that of BYDV, so the possibility of a long-range pseudoknot in this virus must be considered. A region capable of pairing to the 3′ bulge of the frameshifter stem-loop is present at the 3′ end of the RNA-1 segment of the viral genome (Xiong and Lommel 1989). However, changing the bases in the aforementioned stem bulge to their Watson-Crick complementary bases does not appear to influence frameshifting or virus replication, arguing against long-range pseudoknot formation (Kim and Lommel 1998).

In terms of RNA structure analysis, frameshifter stem-loops have generally received less experimental attention than pseudoknots, and no high-resolution images are available. As discussed below, there is a need to determine whether stem-loops induce frameshifting by the same pathway as pseudoknots and whether it is possible that frameshifter stem-loops contain unanticipated structural features that warrant analysis.

MECHANISTIC ASPECTS OF RIBOSOMAL FRAMESHIFTING

Despite considerable study, the mechanism of programmed –1 ribosomal frameshifting has remained elusive. In elementary terms it is relatively straightforward; the ribosome encounters the stimulatory RNA structure while in the act of decoding the slippery sequence, something unusual happens, and a proportion of the ribosomes enter the –1 frame. However, the details are entwined with the natural movement processes of the tRNAs and ribosomal subunits, topics about which we have only limited understanding. Presently, there are at least three models of frameshifting, which for brevity can be termed the pausing, unwinding, and protein factor models. These are discussed below under separate headings, but it should be borne in mind that to some extent, they are interlinked, and certain features of stimulatory RNAs have been seized upon as supportive evidence for more than one model.

The Central Thesis Is Ribosomal Pausing

Ribosomal pausing has been at the heart of all models of –1 frameshifting, and indeed, most recoding phenomena. In its simplest form, it increases the time at which ribosomes are held at a recoding site, promoting alternative events that would normally be kinetically unfavorable

(Farabaugh 2000). In the case of –1 frameshifting, pausing at the stimulatory RNA would allow more time for the tRNAs decoding the slippery sequence to realign in the –1 frame (Jacks et al. 1988a). There is considerable experimental evidence that pausing occurs at frameshift signals. Polypeptide intermediates corresponding to ribosomes paused at RNA pseudoknots have been detected at the frameshift sites of IBV (Somogyi et al. 1993) and L-A (Lopinski et al. 2000), and footprinting studies of elongating ribosomes have defined the site of pausing at

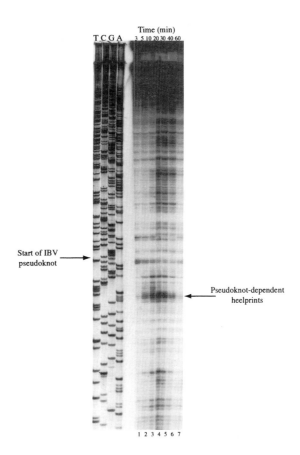

Figure 5. Ribosomal pausing at the minimal IBV pseudoknot. This figure shows a ribosomal pausing assay performed using the heelprinting technique (Wolin and Walter 1988). Full methodological details can be found in Kontos et al. (2001). In brief, an mRNA containing the minimal IBV pseudoknot was translated in vitro (in RRL), and at various times, aliquots were removed, treated sequentially with cycloheximide and micrococcal nuclease, and the ribosome-protected mRNA fragments were purified. The fragments were annealed to a complementary DNA template and mapped by primer extension using T7 DNA polymerase. The gel shows the primer extension reactions and a sequencing ladder generated from the same DNA template. The start of the IBV pseudoknot (5′-GGGG) is indicated, as are the pseudoknot-dependent heelprints that arise as a result of ribosomal pausing at the pseudoknot. The primary sequence of the mRNA upstream of the pseudoknot is shown at the bottom, and the positions of the pseudoknot-dependent heelprints are indicated with arrows (the first four G residues of the pseudoknot are shown in bold type).

the L-A signal (Tu et al. 1992; Lopinski et al. 2000) and more recently at the IBV, SRV-1, and pKA-A signals (Kontos et al. 2001). There is also evidence for pausing at the frameshift site in the *E. coli dnaX* gene, again from the analysis of translational intermediates (Tsuchihashi 1991). The position of paused ribosomes is also highly consistent with a role in the frameshift process. This is shown experimentally in Figure 5. Here, we employed the heelprint assay (Wolin and Walter 1988) to map the position of ribosomes paused at the minimal IBV pseudoknot (Brierley et al. 1992). An in vitro translation reaction (the rabbit reticulocyte lysate system, RRL) was programmed with the pseudoknot-containing mRNA (pPS1a; Somogyi et al. 1993; Kontos et al. 2001) and at set time intervals, aliquots of the translation reaction were withdrawn and ribosome-protected fragments (RPFs) were purified as described previously (Kontos et al. 2001). These were annealed to a complementary single-stranded DNA template along with an end-labeled sequencing primer, and following extension of the sequencing primer using T7 DNA polymerase, which terminates upon encountering an annealed RNA fragment, the site of termination was mapped by running out the primer extension products on a denaturing polyacrylamide gel alongside a sequencing ladder prepared using the same sequencing primer. Since paused ribosomes produce an increased amount of specific RPFs, the T7 DNA polymerase extension products, corresponding to the trailing 5′ edges of the stalled ribosomes from which these RPFs were obtained, appear as more intense species on the gel. As shown in lanes 2–6, among a number of species, a strong heelprint was observed that spanned 4 nucleotides at a position 21–24 nucleotides upstream of the first base (G) of the pseudoknot. This pattern has been seen previously and is indicative of ribosomal pausing at the minimal IBV pseudoknot (Kontos et al. 2001). Given that the length of the RPFs are typically 28–36 nucleotides in this system (Kontos et al. 2001) and that the 5′ edge of the ribosome is positioned 21–24 nucleotides upstream of the first base of the pseudoknot, the 3′ edge can be calculated, on the basis of a mean ribosomal-site size of 32, to be 8–11 nucleotides into the IBV pseudoknot-forming sequence. Heelprinting of ribosomes paused at initiation codons has shown that the 5′ edge of the ribosome is some 12–13 nucleotides from the first base of the AUG (Wolin and Walter 1988). On this basis, in our experiment, the ribosomal P site will be approximately 8–12 nucleotides 5′ of the start of the pseudoknot; i.e., at or around those bases which are in the equivalent position of the slippery sequence in the natural frameshift signal (the pPS1a mRNA does not possess a slippery sequence). Thus, the heelprint data are consistent with a role for pausing in frameshifting, with the paused ribosome being positioned over the slippery sequence while in contact with the pseudoknot. Similar observations have also been made at the L-A frameshift signal (Tu et al. 1992; Lopinski et al. 2000).

One of the great virtues of the pausing model is its ability to accommodate the variety of stimulatory RNAs present at –1 frameshifting signals. Notwithstanding the range of secondary and tertiary features presented to the ribosome, as long as pausing occurs, frameshifting results. Unfortunately, the idea that a pause alone is sufficient to induce frameshifting is highly questionable. Simple provision of a roadblock to ribosomes in the form of stable RNA hairpins (Brierley et al. 1991; Somogyi et al. 1993), a tRNA (Chen et al. 1995), or even different kinds of RNA pseudoknots (Liphardt et al. 1999; Napthine et al. 1999) is not sufficient to bring about frameshifting, and furthermore, non-frameshifting pseudoknots and stem-loops exist that can still pause ribosomes (Tu et al. 1992; Somogyi et al. 1993; Lopinski et al. 2000; Kontos et al. 2001). These experiments, of course, do not rule out a contribution of pausing to the mechanism of frameshifting, since there are no documented examples where frameshifting has occurred in the absence of a detectable pause. Indeed, one cannot ignore the possibility that a precise "kinetic pause" is required for frameshifting (Bidou et al. 1997), which only certain stimulatory RNAs can generate. For example, during a –1 frameshift, two pauses could occur, one productive (in terms of frameshifting) upon initial encounter of the stimulatory RNA structure, and a second, non-productive pause, corresponding to a delay in unwinding the structure *after* the crucial event in frameshifting has already taken place. The magnitude of the initial pause would perhaps influence the extent of the frameshift, whereas the second pause, occurring during the time that a hypothetical ribosomal "unwinding activity" (see below) locates and deals with a secondary structure, would be irrelevant. The pausing assays employed currently would probably not distinguish between two such pausing events, and a detailed analysis of the kinetics of pausing will require further experimentation, including the development of techniques to study translational elongation at the level of individual ribosomes.

The Unwinding Model

Regardless of whether pausing is the sole mediator of frameshifting, a contributor to the process, or simply a by-product of the stimulatory RNA–ribosome interaction, how it is brought about is a matter of considerable interest. One possibility is that the stimulatory RNAs are especially resistant to unwinding by an 80S-associated, hypothetical RNA helicase responsible for unwinding mRNA structures ahead of the decoding center. As outlined above, RNA pseudoknots possess unique features that may be refractory to standard helix unwinding, including the loop-helix triplexes seen in the BWYV and SRV-1 pseudoknots and the requirement for a G-rich stretch at the base of stem 1 in IBV and SRV-1. Modeling studies predict that the triplex (and the base of stem 1) is likely to be one of the first features of the pseudoknot to be encountered by the ribosome (Giedroc et al. 2000), where it could function to stabilize stem 1 and increase the time taken to unwind the structure. As noted several years ago (Draper 1990) and reiterated recently (Michiels et al. 2001), another potential barrier to unwinding is the unusual topology of the pseudoknot at the beginning of stem 1, where in addition to the two base-paired strands, loop 2 adds a third strand in close proximity. Perhaps the ribosome does not deal effectively with this kind of topological arrangement. In one form or another, it has been speculated that the resistance to unwinding sets up a strain in the mRNA that can only be relieved by tRNA movement; i.e., by frameshifting. Although there is no direct experimental evidence to support this hypothesis, it is a compelling one.

Clearly, the relative positions of the ribosome and pseudoknot during decoding of the slippery sequence are of great relevance here. The recent crystal structure of the *T. thermophilus* 70S ribosome complexed with an mRNA fragment has revealed that the distance from the first base of the P-site codon to the point at which the mRNA enters the ribosome is some 13–15 nucleotides (Yusupova et al. 2001). Using the IBV frameshift signal as an example, this would place the first three bases of the pseudoknot in the ribosome during decoding of the slippery sequence. This fits nicely with the observed organization of frameshift signals, with the slippery sequence spaced such that the pseudoknot is in intimate association with the ribosome during decoding. The heelprinting studies outlined above gave a slightly different view of this association in that some 8–11 nucleotides of the pseudoknot are protected from nuclease treatment by a paused ribosome. This apparent discrepancy may arise from inherent limitations of the technique. The ribosomal heelprint is defined by the length of the RPFs, which are generated upon micrococcal nuclease treatment of cycloheximide-treated ribosomes. With this assay, it is difficult to distinguish between a significantly unwound pseudoknot and a largely intact pseudoknot associated with the ribosome, since certain regions, especially the single-stranded loops, would likely remain accessible and be cleaved by the micrococcal nuclease. However, it may simply be that the eukaryotic ribosome is somewhat larger, with more surface-associated proteins (Dube et al. 1998), and the distance from the P site to the mRNA entry site could effectively be larger.

On the basis of the crystal structure of a complex between a short mRNA and a prokaryotic ribosome, Yusupova and colleagues (2001) have suggested a possible mechanism by which secondary structure unwinding can be achieved. At the point where the mRNA enters the ribosome, the RNA is threaded through a narrow tunnel between three proteins, S3, S4, and S5. The tunnel appears to be too shallow to allow a helix to enter, so it is presumably at this point that the RNA is unwound. The authors point out that S4 and S5 are integral to the body of the 30S subunit, whereas S3 is part of the head. If one of the helical strands is bound to S4 and/or S5 and the other to S3, then the rotation of the head that is believed to occur during translocation (Agrawal et al. 1999) could disrupt the helix in a stepwise fashion, three nucleotides at a time, while advancing the mRNA through the ribosome. During frameshifting, the unusual topology of a pseudoknot could interfere with this process, blocking movement of the mRNA and creating a strain that ultimately leads to tRNA movement.

Recently, the issue of pseudoknot unwinding upon encounter of the ribosome has been considered from the perspective of our knowledge of pseudoknot thermodynamics and unfolding kinetics (Giedroc et al. 2000). In brief, these authors argue for distinct unfolding pathways for different pseudoknots. In the first pathway, encounter of the ribosome with the proximal stem 1 and loop 2 of a pseudoknot destabilizes the helical junction and the distal stem 2, leaving a partially folded stem-1 hairpin. This unfolding pathway would occur for short pseudoknots like BWYV and SRV-1 where the stems are "coupled" via the short, often triplex, loop 2. For pseudoknots with a long, stable stem 1 and a long loop 2 (e.g., IBV), where there is little apparent coupling between the two stems, stem 1 would melt first. It should be noted, however, that the unfolding pathway and the rate of unfolding may influence the magnitude of the frameshift event, but would not necessarily trigger the event in the first place.

The Protein Factor Model

When frameshifter pseudoknots were first described, it was proposed that they could act as binding sites for cellular or viral proteins responsible for promoting or regulating the frameshift process (Jacks et al. 1988a; Brierley et al. 1989). To date, however, no such pseudoknot-binding proteins have been unearthed, although examples have been described in other systems (see, e.g., Schlax et al. 2001). Certainly, if such factors exist, they are not easily titratable. The addition of a large molar excess of the SRV-1 pseudoknot to an in vitro translation reaction programmed with an SRV-1 frameshift reporter mRNA has no effect on frameshifting (ten Dam et al. 1994), and the same is true for the IBV pseudoknot (I. Brierley and S.C. Inglis, unpubl.). Furthermore, several laboratories have reported a failure to identify frameshifter pseudoknot-binding factors in searches using traditional band-shift and UV cross-linking technologies. These experiments do not rule out the involvement of a protein or protein complex that is tightly associated with the translation apparatus, perhaps even an integral ribosomal component, since the pseudoknot may be recognized only in the context of the elongating ribosome. For example, a comparison of the VPK pseudoknot with the structure of a tRNA–codon complex has revealed certain similarities, and this has led to the proposal that the pseudoknot could be recognized by a component of the translation apparatus responsible for stabilizing tRNA–codon complexes (Shen and Tinoco 1995). Nevertheless, biochemical approaches have failed to identify such proteins. In the previous section, we speculated that pseudoknots may resist unwinding by virtue of an unusual RNA topology. From the protein factor viewpoint, it has been remarked that pseudoknot action could therefore be mediated because of an *inability* to be recognized by ribosomal components, a model mischievously termed the "no factor" model (ten Dam 1993). However, the apparent requirement for a kinked pseudoknot structure in MMTV and the identification of specific residues of the BWYV pseudoknot that are essential for frameshifting but not involved in tertiary interactions is not inconsistent with specific recognition of a protein component of the translation apparatus.

Although no frameshifter pseudoknot-binding proteins have been described, a growing number of cellular proteins have been identified that can influence frameshifting (Dinman and Wickner 1994). In some cases, the pathways by which the proteins influence frameshifting seem relatively straightforward. Certain mutants of yeast ribosomal protein L5, for example, can stimulate frameshifting at the L-A signal (Meskauskas and Dinman 2001). The L5 protein is likely to be involved in anchoring the peptidyl tRNA to the P site; mutations that compromise this ability will presumably allow the peptidyl tRNA to move around more freely, promoting the frameshift event. Similarly, mutants of yeast ribosomal protein L3, a protein present at the peptidyl transferase center of the ribosome, can also stimulate frameshifting at the L-A site. In this case, the stimulation is thought to be a result of increased ribosomal pausing by ribosomes containing defective, or depleted of, L3 (Peltz et al. 1999). Some elongation factor 1α mutants have also been found to stimulate both −1 and +1 frameshifting. Whether these mutants act by promoting ribosomal pausing or by affecting the natural ribosomal error-correcting mechanisms, however, is a matter of some debate (Dinman and Kinzy 1997; Farabaugh and Vimaladithan 1998). Another contentious area is the involvement of a component of the nonsense-mediated mRNA decay pathway, UPF3, in frameshifting, with evidence for (Cui et al. 1996; Ruiz-Echevarria et al. 1998; Dinman et al. 2000) and against (Bidou et al. 2000; Stahl et al. 2000) a role for this protein.

Models for Stem-loop Stimulators

It is pertinent at this juncture to raise the issue of stimulatory stem-loops and frameshifting mechanisms. Almost all of the models presented above depend to a greater or lesser extent on pseudoknot-specific features, yet it must be borne in mind that some sites have stem-loop stimulators that are not thought to possess unusual structural motifs. How these RNAs stimulate frameshifting has not been addressed in terms of specific models. Historically, the explanation is that stem-loop structures induce only low levels of frameshifting, thus, they need only block elongation long enough to create a modest ribosomal pause, and an elaborate structure is not required to produce this low-level pausing. However, the idea that frameshifter stem-loops generally produce only low levels of frameshifting needs consideration. During the course of this paper, we have avoided comparisons of frameshift efficiencies from different frameshift signals, since it is generally difficult to make meaningful comparisons given that the assays are carried out in different translation systems, both in vitro and in vivo, and with varying flanking mRNA sequences. There is, however, sufficient evidence that some stem-loops can engender frameshift efficiencies of a similar magnitude to those seen at pseudoknot-containing sites. For example, the HIV-1 *gag/pol* frameshift signal induces about 3% frameshifting in mammalian cells (Reil et al. 1993; Kollmus et al. 1994), only some

threefold lower than the IBV frameshift signal (Marczinke et al. 2000). So how do we account for the activity of frameshifter stem-loops? One possibility is that the stem-loop stimulators are in fact pseudoknots that have not been identified yet. Most studies of frameshifter stem-loops have been carried out on regions subcloned from the context of the natural mRNA, and it is possible that some long-range interactions have been overlooked. A second possibility is that the stem-loops themselves possess hitherto uncharacterized, novel structural features that can promote frameshifting. A high-resolution structure of a frameshifter stem-loop would be informative in this regard. Of course, it may be that our models of frameshifting are incomplete. There is evidence to support the idea that stem-loops may act simply to provide a thermodynamic barrier to bring about ribosomal pausing. The frameshift efficiency of the HIV-1 signal in yeast and mammalian cells has been shown to be directly related to the stability of the stem-loop structure, with an almost linear relationship between frameshift efficiency and stability over quite a broad range (Bidou et al. 1997). However, there are also examples of stem-loops that are unable to stimulate efficient frameshifting when placed downstream of slippery sequences. These include an "artificial" stem-loop designed to have a stem length and base composition similar to that of the IBV pseudoknot (Brierley et al. 1991), and most mutant pseudoknots able to form only the first stem have poor activity. Thus, there is more work to be done to understand how stem-loops induce frameshifting. Mutational analysis of such sites has already provided hints that the traditional combination of slippery sequence and hairpin may not be the sole defining feature of the signal, and other elements may contribute. Kim and colleagues (2001) have recently measured the frameshift efficiencies evoked in vitro by a series of HIV-1 *gag/pol*–HTLV-II *gag/pro* chimeras. They defined four elements, namely the slippery sequence, spacer, stem-loop, and a region upstream of the slippery sequence, and mixed these in various combinations to create a range of hybrid sites. Surprisingly, it was found that the regions flanking the slippery sequence and stem-loop could influence frameshifting quite dramatically, possibly by modulating stem-loop unfolding kinetics. The influence of termination codons at or around the frameshift site is another issue. A stop codon placed some 27 nucleotides downstream of the CfMV stem-loop in the −1 frame was found to increase frameshifting some threefold, perhaps by influencing movement of an upstream ribosome poised at the slippery sequence (Lucchesi et al. 2000). It is harder to interpret the reduction in frameshifting observed when a −1 frame stop codon is positioned some 15 nucleotides *upstream* of a chimeric HIV–MMTV signal (Honda and Nishimura 1996).

TRANSFER RNAs AND FRAMESHIFTING

Although this paper has concentrated mainly on stimulatory mRNA structures, it must be emphasized that the tRNAs that decode the slippery sequence are central to the frameshift process, and there is considerable interest in whether these are canonical tRNAs or special "shifty" tRNAs, more prone to frameshift than their "normal" counterparts (Jacks et al. 1988a). In this respect, no novel tRNAs have been described as yet, but it has been noted that all the A-site tRNAs that function in frameshifting are decoded by tRNAs with a highly modified base in the anticodon loop (see Hatfield et al. 1992 and references therein). In tRNA^{Asn} (AAC, AAU), the wobble base is queuosine (Q); in tRNA^{Phe} (UUC, UUU), wyebutoxine (Y) is present just 3′ of the anticodon; in tRNA^{Lys} (AAA, AAG), the wobble base is 5-methylaminomethyl-2-thiouridine (mnm^{5}s^{2}U) (prokaryotes); and in tRNA^{Leu} (UUA) 2-methyl-5-formylcytidine is present at the wobble position. Hatfield and colleagues (1992) have raised the possibility that hypomodified variants of these tRNAs may exist which could act as specific "shifty" tRNAs, since such variants would have a considerably less bulky anticodon and be more free to move around at the decoding site. Support for this hypothesis comes from the observation that purified tRNA^{Phe} populations devoid of the Y modification can stimulate frameshifting at certain slippery sequences in RRL (Carlson et al. 1999, 2001). In contrast, the frameshift capacity of tRNA^{Asn} appears to be uninfluenced by the absence of the Q modification in either prokaryotic (Brierley et al. 1997) or eukaryotic cells (Marczinke et al. 2000), so hypomodification per se is insufficient to stimulate frameshifting. At present, therefore, the role of modified or hypomodified bases in frameshifting is uncertain. Where a role can be ascribed for the hypomodification (e.g., the Y base above), it is not yet known whether the effect is mediated through a change in codon–anticodon stability or by some other route.

CONCLUSIONS

During the last 5 or 6 years, the most notable progress in the field has been the determination of the atomic structure of three frameshifter pseudoknots. Considering that only a handful of naturally occurring RNA structures had been determined prior to 1995, this represents a remarkable achievement. Our knowledge of the structures of the VPK, BWYV, and SRV-1 pseudoknots has highlighted hitherto unexpected features and provided new avenues to explore in the search of the frameshift mechanism(s). However, much remains to be done. Further structural analysis of a broader range of stimulatory RNAs is required, including stem-loop stimulators. Currently, we are attempting to crystallize the minimal IBV pseudoknot to ascertain whether the tertiary features present in the shorter pseudoknots are conserved. It is too early to say whether this information will contribute to our understanding of how frameshifting works. The problem faced by the field is not inconsiderable, a determination of how elongation is subverted without a detailed understanding of how ribosomes achieve normal triplet decoding in the first place. Nevertheless, given the recent progress in crystallographic and cryo-electron microscopy (cryo-EM) techniques for the study of ribosomes and associated factors (detailed extensively in this volume), it is highly likely that additional

ribosome–mRNA complexes will be studied, including mRNAs containing frameshift signals. These experiments should reveal the precise position of the stimulatory RNAs on the ribosome. Ultimately, a complex with a ribosome from a eukaryotic organism will be needed, as there appear to be subtle differences in frameshifting between prokaryotic and eukaryotic systems (Farabaugh 2000). A complete mechanistic description of frameshifting will probably require new experimental approaches. These will include the development of methods to investigate the unwinding (as opposed to melting) behavior of RNA pseudoknots, site-specific cross-linking studies of pseudoknots during the process of translation, and the preparation of cryo-EM images of paused ribosomes. Related studies on translational recoding events like pseudoknot-dependent termination codon suppression (Alam et al. 1999) may also provide comparative insights. These are worthy objectives; it seems likely that the study of frameshifting and recoding phenomena in general will make a significant contribution to our understanding of the natural elongation cycle.

ACKNOWLEDGMENTS

We acknowledge the Biotechnology and Biological Sciences Research Council and the Medical Research Council for financial support and thank Philip Farabaugh for his substantial contributions to our work. I.B. is grateful to the organizers of the 66th Cold Spring Harbor Symposium on the Ribosome for their kind invitation to speak at this meeting, and to Joan Ebert for her patience.

REFERENCES

Agrawal R.K., Heagle A.B., Penczek P., Grassucci R.A., and Frank J. 1999. EF-G-dependent GTP hydrolysis induces translocation accompanied by large conformational changes in the 70S ribosome. *Nat. Struct. Biol.* **6:** 643.

Alam S.L., Wills N.M., Ingram J.A., Atkins J.F., and Gesteland R.F. 1999. Structural studies of the RNA pseudoknot required for readthrough of the *gag*-termination codon of murine leukemia virus. *J. Mol. Biol.* **288:** 837.

Bidou L., Stahl G., Grima B., Liu H., Cassan M., and Rousset J.-P. 1997. In vivo HIV-1 frameshift efficiency is directly related to the stability of the stem-loop stimulatory signal. *RNA* **10:** 1153.

Bidou L., Stahl G., Hatin I., Namy O., Rousset J.-P., and Farabaugh P.J. 2000. Nonsense-mediated decay mutants do not affect programmed –1 frameshifting. *RNA* **6:** 952.

Brault V., and Miller W.A. 1992. Translational frameshifting mediated by a viral sequence in plant cells. *Proc. Natl. Acad. Sci.* **89:** 2262.

Brierley I. 1995. Ribosomal frameshifting on viral RNAs. *J. Gen. Virol.* **76:** 1885.

Brierley I., Digard P., and Inglis S.C. 1989. Characterisation of an efficient coronavirus ribosomal frameshifting signal: Requirement for an RNA pseudoknot. *Cell* **57:** 537.

Brierley I., Jenner A.J., and Inglis S.C. 1992. Mutational analysis of the "slippery sequence" component of a coronavirus ribosomal frameshifting signal. *J. Mol. Biol.* **227:** 463.

Brierley I., Meredith M.R., Bloys A.J., and Hagervall T.G. 1997. Expression of a coronavirus ribosomal frameshift signal in *Escherichia coli*: Influence of tRNA anticodon modification on frameshifting. *J. Mol. Biol.* **270:** 360.

Brierley I., Rolley N.J., Jenner A.J., and Inglis S.C. 1991. Mutational analysis of the RNA pseudoknot component of a coronavirus ribosomal frameshifting signal. *J. Mol. Biol.* **220:** 889.

Brunelle M.N., Payant C., Lemay G., and Brakier-Gingras L. 1999. Expression of the human immunodeficiency virus frameshift signal in a bacterial cell-free system: Influence of an interaction between the ribosome and a stem-loop structure downstream from the slippery site. *Nucleic Acids Res.* **27:** 4783.

Carlson B.A., Kwon S.Y., Chamorro M., Oroszlan S., Hatfield D.L., and Lee B.J. 1999. Transfer RNA modification status influences retroviral ribosomal frameshifting. *Virology* **255:** 2.

Carlson B.A., Mushinski J.F., Henderson D.W., Kwon S.Y., Crain P.F., Lee B.J., and Hatfield D.L. 2001. 1-Methylguanosine in place of Y base at position 37 in phenylalanine tRNA is responsible for its shiftiness in retroviral ribosomal frameshifting. *Virology* **279:** 130.

Cate J.H., Gooding A.R., Podell E., Zhou K., Golden B.L., Kundrot C.E., Cech T.R., and Doudna J.A. 1996a. Crystal structure of a group I ribozyme domain: Principles of RNA packing. *Science* **273:** 1678.

Cate J.H., Gooding A.R., Podell E., Zhou K., Golden B.L., Szewczak A.A., Kundrot C.E., Cech T.R., and Doudna J.A. 1996b. RNA tertiary structure mediation by adenosine platforms. *Science* **273:** 1696.

Chandler M. and Fayet O. 1993. Translational frameshifting in the control of transposition in bacteria. *Mol. Microbiol.* **7:** 497.

Chen X.Y., Kang H.S., Shen L.X., Chamorro M., Varmus H.E., and Tinoco I. 1996. A characteristic bent conformation of RNA pseudoknots promotes –1 frameshifting during translation of retroviral RNA. *J. Mol. Biol.* **260:** 479.

Chen X., Chamorro M., Lee S.I., Shen L.X., Hines J.V., Tinoco I., Jr., and Varmus H.E. 1995. Structural and functional studies of retroviral RNA pseudoknots involved in ribosomal frameshifting: Nucleotides at the junction of the two stems are important for efficient ribosomal frameshifting. *EMBO J.* **14:** 842.

Cowley J.A., Dimmock C.M., Spann K.M., and Walker P.J. 2000. Gill-associated virus of *Penaeus monodon* prawns: An invertebrate virus with ORF1a and ORF1b genes related to arteri- and coronaviruses. *J. Gen. Virol.* **8:** 1473.

Cui Y., Dinman J.D., and Peltz S.W. 1996. Mof4-1 is an allele of the UPF1/IFS2 gene which affects both mRNA turnover and –1 ribosomal frameshifting efficiency. *EMBO J.* **15:** 5726.

Czworkowski J. and Moore P.B. 1996. The elongation phase of protein synthesis. *Prog. Nucleic Acid Res. Mol. Biol.* **54:** 293.

Dinman J.D. 1995. Ribosomal frameshifting in yeast viruses. *Yeast* **11:** 1115.

Dinman J.D. and Kinzy T.G. 1997. Translational misreading: Mutations in translation elongation factor 1-alpha differentially affect programmed ribosomal frameshifting and drug sensitivity. *RNA* **3:** 870.

Dinman J.D. and Wickner R.B. 1992. Ribosomal frameshifting efficiency and *gag/gag-pol* ratio are critical for yeast M1 double-stranded RNA virus propagation. *J. Virol.* **66:** 3669.

———. 1994. Translational maintenance of frame: Mutants of *Saccharomyces cerevisiae* with altered –1 ribosomal frameshifting efficiencies. *Genetics* **136:** 75.

Dinman J.D., Icho T., and Wickner R.B. 1991. A –1 ribosomal frameshift in a double-stranded RNA virus of yeast forms a Gag-Pol fusion protein. *Proc. Natl. Acad. Sci.* **88:** 174.

Dinman J.D., Ruiz-Echevarria M.J., and Peltz S.W. 1998. Translating old drugs into new treatments: Ribosomal frameshifting as a target for antiviral agents. *Trends Biotechnol.* **16:** 190.

Dinman J.D., Ruiz-Echevarria M.J., Czaplinski K., and Peltz S.W. 1997. Peptidyl-transferase inhibitors have antiviral properties by altering programmed –1 ribosomal frameshifting efficiencies: Development of model systems. *Proc. Natl. Acad. Sci.* **94:** 6606.

Dinman J., Ruiz-Echevarria M., Wang W., and Peltz S. 2000. The case for the involvement of the Upf3p in programmed –1 ribosomal frameshifting. *RNA* **6:** 1685.

Draper D.E. 1990. Pseudoknots and the control of protein synthesis. *Curr. Opin. Cell Biol.* **2:** 1099.

Du Z. and Hoffman D.W. 1997. An NMR and mutational study of the pseudoknot within the gene 32 mRNA of bacteriophage T2: Insights into a family of structurally related RNA pseudoknots. *Nucleic Acids Res.* **25:** 1130.

Du Z.H., Holland J.A., Hansen M.R., Giedroc D.P., and Hoffman D.W. 1997. Base-pairings within the RNA pseudoknot associated with the simian retrovirus-1 *gag-pro* frameshift site. *J. Mol. Biol.* **270:** 464.

Dube P., Bacher G., Stark H., Mueller F., Zemlin F., van Heel M., and Brimacombe R. 1998. Correlation of the expansion segments in mammalian rRNA with the fine structure of the 80S ribosome; a cryoelectron microscopic reconstruction of the rabbit reticulocyte ribosome at 21Å resolution. *J. Mol. Biol.* **279:** 403.

Farabaugh P.J. 1996. Programmed translational frameshifting. *Microbiol. Rev.* **60:** 103.

———. 2000. Translational frameshifting: Implications for the mechanism of translational frame maintenance. *Prog. Nucleic Acid Res. Mol. Biol.* **64:** 131.

Farabaugh P.J. and Bjork G.R. 1999. How translational accuracy influences frame maintenance. *EMBO J.* **18:** 1427.

Farabaugh P.J. and Vimaladithan A. 1998. Effect of frameshift-inducing mutants of elongation factor 1-alpha on programmed +1 frameshifting in yeast. *RNA* **4:** 38.

Fütterer J. and Hohn T. 1996. Translation in plants—Rules and exceptions. *Plant Mol. Biol.* **32:** 159.

Garcia A., van Duin J., and Pleij C.W. 1993. Differential response to frameshift signals in eukaryotic and prokaryotic translational systems. *Nucleic Acids Res.* **21:** 401.

Gesteland R.F. and Atkins J.F. 1996. Recoding: Dynamic reprogramming of translation. *Annu. Rev. Biochem.* **65:** 741.

Gesteland R.F., Weiss R.B., and Atkins J.F. 1992. Recoding: Reprogrammed genetic decoding. *Science* **257:** 1640.

Giedroc D.P., Theimer C.A., and Nixon P.L. 2000. Structure, stability and function of RNA pseudoknots involved in stimulating ribosomal frameshifting. *J. Mol. Biol.* **298:** 167.

Hammell A.B., Taylor R.C., Peltz S.W., and Dinman J.D. 1999. Identification of putative programmed –1 ribosomal frameshift signals in large DNA databases. *Genome Res.* **9:** 417.

Hatfield D., Levin J.G., Rein A., and Oroszlan S. 1992. Translational suppression in retroviral gene expression. *Adv. Virus Res.* **41:** 193.

Hilbers C.W., Michiels P.J., and Heus H.A. 1998. New developments in structure determination of pseudoknots. *Biopolymers* **48:** 137.

Hizi A., Henderson L.E., Copeland T.D., Sowder R.C., Hixson C.V., and Oroszlan S. 1987. Characterization of mouse mammary tumor virus *gag-pro* gene products and the ribosomal frameshift site by protein sequencing. *Proc. Natl. Acad. Sci.* **84:** 7041.

Honda A. and Nishimura S. 1996. Suppression of translation frameshift by upstream termination codon. *Biochem. Biophys. Res. Commun.* **221:** 602.

Hudak K.A., Hammell A.B., Yasenchak J., Tumer N.E., and Dinman J.D. 2001. A C-terminal deletion mutant of pokeweed antiviral protein inhibits programmed +1 ribosomal frameshifting and Ty1 retrotransposition without depurinating the sarcin/ricin loop of rRNA. *Virology* **279:** 292.

Humphrey W., Dalke A., and Schulten K. 1996. VMD: Visual molecular dynamics. *J. Mol. Graph.* **14:** 33.

Hung M., Patel P., Davis S., and Green S.R. 1998. Importance of ribosomal frameshifting for human immunodeficiency virus type 1 particle assembly and replication. *J. Virol.* **72:** 4819.

Jacks T. and Varmus H.E. 1985. Expression of the Rous sarcoma virus *pol* gene by ribosomal frameshifting. *Science* **230:** 1237.

Jacks T., Madhani H.D., Masiarz F.R., and Varmus H.E. 1988a. Signals for ribosomal frameshifting in the Rous sarcoma virus *gag-pol* region. *Cell* **55:** 447.

Jacks T., Power M.D., Masiarz F.R., Luciw P.A., Barr P.J., and Varmus H.E. 1988b. Characterization of ribosomal frameshifting in HIV-1 *gag-pol* expression. *Nature* **331:** 280.

Jiang B., Monroe S.S., Koonin E.V., Stine S.E., and Glass R.I. 1993. RNA sequence of astrovirus: Distinctive genomic organization and a putative retrovirus-like ribosomal frameshifting signal that directs the viral replicase synthesis. *Proc. Natl. Acad. Sci.* **90:** 10539.

Kang H. 1998. Direct structural evidence for formation of a stem-loop structure involved in ribosomal frameshifting in human immunodeficiency virus type 1. *Biochim. Biophys. Acta.* **1397:** 73.

Kang H.S. and Tinoco I. 1997. A mutant RNA pseudoknot that promotes ribosomal frameshifting in mouse mammary tumor virus. *Nucleic Acids Res.* **25:** 1943.

Kang H.S., Hines J.V., and Tinoco I. 1996. Conformation of a non-frameshifting RNA pseudoknot from mouse mammary tumor virus. *J. Mol. Biol.* **259:** 135.

Karacostas V., Wolffe E.J., Nagashima K., Gond M.A., and Moss B. 1993. Overexpression of the HIV-1 *gag-pol* polyprotein results in intracellular activation of HIV-1 protease and inhibition of assembly and budding of virus-like particles. *Virology* **193:** 661.

Kim K.H. and Lommel S.A. 1994. Identification and analysis of the site of –1 ribosomal frameshifting in red clover necrotic mosaic virus. *Virology* **200:** 574.

———. 1998. Sequence element required for efficient –1 ribosomal frameshifting in red clover necrotic mosaic dianthovirus. *Virology* **250:** 50.

Kim Y.G., Maas S., and Rich A. 2001. Comparative mutational analysis of *cis*-acting RNA signals for translational frameshifting in HIV-1 and HTLV-2. *Nucleic Acids Res.* **29:** 1125.

Kim Y.G., Maas S., Wang S.C., and Rich A. 2000. Mutational study reveals that tertiary interactions are conserved in ribosomal frameshifting pseudoknots of two luteoviruses. *RNA* **6:** 1157.

Kim Y.G., Su L., Maas S., O'Neill A., and Rich A. 1999. Specific mutations in a viral RNA pseudoknot drastically change ribosomal frameshifting efficiency. *Proc. Natl. Acad. Sci.* **96:** 14234.

Kolk M.H., van der Graaf M., Wijmenga S.S., Pleij C.W.A., Heus H.A., and Hilbers C.W. 1998. NMR structure of a classical pseudoknot: Interplay of single- and double-stranded RNA. *Science* **280:** 434.

Kollmus H., Honigman A., Panet A., and Hauser H. 1994. The sequences of and distance between two *cis*-acting signals determine the efficiency of ribosomal frameshifting in human immunodeficiency virus type 1 and human T-cell leukemia virus type II in vivo. *J. Virol.* **68:** 6087.

Kontos H., Napthine S., and Brierley I. 2001. Ribosomal pausing at a frameshifter RNA pseudoknot is sensitive to reading phase but shows little correlation with frameshift efficiency. *Mol. Cell. Biol.* **21:** 8657.

Kurland C.G. 1992. Translational accuracy and the fitness of bacteria. *Annu. Rev. Genet.* **26:** 29.

Le S.Y., Chen J.H., Pattabiraman N., and Maizel J.V. 1998. Ion-RNA interactions in the RNA pseudoknot of a ribosomal frameshifting site: Molecular modeling studies. *J. Biomol. Struct. Dyn.* **16:** 1.

Liphardt J.T. 1999. "The mechanism of –1 ribosomal frameshifting: experimental and theoretical analysis." Ph.D. thesis, University of Cambridge, United Kingdom.

Liphardt J., Napthine S., Kontos H., and Brierley I. 1999. Evidence for an RNA pseudoknot loop-helix interaction essential for efficient –1 ribosomal frameshifting. *J. Mol. Biol.* **288:** 321.

Lopinski J.D., Dinman J.D., and Bruenn J.A. 2000. Kinetics of ribosomal pausing during programmed –1 ribosomal frameshifting. *Mol. Cell. Biol.* **20:** 1095.

Lucchesi J., Makelainen K., Merits A., Tamm T., and Makinen K. 2000. Regulation of –1 ribosomal frameshifting directed by cocksfoot mottle sobemovirus genome. *Eur. J. Biochem.* **267:** 3523.

Mador N., Panet A., and Honigman A. 1989. Translation of *gag, pro,* and *pol* gene products of human T-cell leukemia virus type 2. *J. Virol.* **63:** 2400.

Makinen K., Tamm T., Naess V., Truve E., Puurand U., Munthe T., and Saarma M. 1995. Characterization of cocksfoot mottle sobemovirus genomic RNA and sequence comparison with related viruses. *J. Gen. Virol.* **76:** 2817.

Marczinke B., Hagervall T., and Brierley I. 2000. The Q-base of asparaginyl-tRNA is dispensible for efficient –1 ribosomal frameshifting in eukaryotes. *J. Mol. Biol.* **295:** 179.

Marczinke B., Fisher R., Vidakovic M., Bloys A.J., and Brierley I. 1998. Secondary structure and mutational analysis of the ribosomal frameshift signal of Rous sarcoma virus. *J. Mol. Biol.* **284:** 205.

Marczinke B., Bloys A.J., Brown T.D., Willcocks M.M., Carter M.J., and Brierley I. 1994. The human astrovirus RNA-dependent RNA polymerase coding region is expressed by ribosomal frameshifting. *J. Virol.* **68:** 5588.

Meskauskas A. and Dinman J.D. 2001. Ribosomal protein L5 helps anchor peptidyl-tRNA to the P-site in *Saccharomyces cerevisiae*. *RNA* **7:** 1084.

Michiels P.J., Versleijen A.A., Verlaan P.W., Pleij C.W., Hilbers C.W., and Heus H.A. 2001. Solution structure of the pseudoknot of SRV–1 RNA, involved in ribosomal frameshifting. *J. Mol. Biol.* **310:** 1109.

Moazed D. and Noller H.F. 1989. Intermediate states in the movement of transfer RNA in the ribosome. *Nature* **342:** 142.

Nam S.H., Copeland T.D., Hatanaka M., and Oroszlan S. 1993. Characterization of ribosomal frameshifting for expression of *pol* gene products of human T-cell leukemia virus type I. *J. Virol.* **67:** 196.

Napthine S., Liphardt J., Bloys A., Routledge S., and Brierley I. 1999. The role of RNA pseudoknot stem 1 length in the promotion of efficient –1 ribosomal frameshifting. *J. Mol. Biol.* **288:** 305.

Paul C.P., Barry J.K., Dinesh-Kumar S.P., Brault V., and Miller W.A. 2001. A sequence required for –1 ribosomal frameshifting located four kilobases downstream of the frameshift site. *J. Mol. Biol.* **310:** 987.

Peltz S.W., Hammell A.B., Cui Y., Yasenchak J., Puljanowski L., and Dinman J.D. 1999. Ribosomal protein L3 mutants alter translational fidelity and promote rapid loss of the yeast killer virus. *Mol. Cell. Biol.* **19:** 384.

Pennell S. 2001. "Structural studies of RNA pseudoknots involved in programmed –1 ribosomal frameshifting." Ph.D. thesis, University of Cambridge, United Kingdom.

Pleij C.W.A., Rietveld K., and Bosch L. 1985. A new principle of RNA folding based on pseudoknotting. *Nucleic Acids. Res.* **13:** 1717.

Reil H., Kollmus H., Weidle U.H., and Hauser H. 1993. A heptanucleotide sequence mediates ribosomal frameshifting in mammalian cells. *J. Virol.* **67:** 5579.

Ruiz-Echevarria M.J., Yasenchak J.M., Han X., Dinman J.D., and Peltz S.W. 1998. The upf3 protein is a component of the surveillance complex that monitors both translation and mRNA turnover and affects viral propagation. *Proc. Natl. Acad. Sci.* **95:** 8721.

Schlax P.J., Xavier K.A., Gluick T.C., and Draper D.E. 2001. Translational repression of the *E. coli* α operon mRNA: Importance of an mRNA conformational switch and a ternary entrapment complex. *J. Biol. Chem.* **276:** 38494.

Shehu-Xhilaga M., Crowe S.M., and Mak J. 2001. Maintenance of the Gag/Gag-Pol ratio is important for human immunodeficiency virus type 1 RNA dimerization and viral infectivity. *J. Virol.* **75:** 1834.

Shen L.X. and Tinoco I. 1995. The structure of an RNA pseudoknot that causes efficient frameshifting in mouse mammary tumor virus. *J. Mol. Biol.* **247:** 963.

Somogyi P., Jenner A.J., Brierley I., and Inglis S.C. 1993. Ribosomal pausing during translation of an RNA pseudoknot. *Mol. Cell. Biol.* **13:** 6931.

Stahl G., Bidou L., Hatin I., Namy O., Rousset J.P., and Farabaugh P. 2000. The case against the involvement of the NMD proteins in programmed frameshifting. *RNA* **6:** 1687.

Su L., Chen L., Egli M., Berger J.M., and Rich A. 1999. Minor groove RNA triplex in the crystal structure of a ribosomal frameshifting viral pseudoknot. *Nat. Struct. Biol.* **6:** 285.

Sung D. and Kang H. 1998. Mutational analysis of the RNA pseudoknot involved in efficient ribosomal frameshifting in simian retrovirus 1. *Nucleic Acids Res.* **26:** 1369.

ten Dam E.B. 1993. "Pseudoknot-dependent ribosomal frameshifting." Ph.D. thesis, University of Leiden, The Netherlands.

ten Dam E.B., Pleij C.W., and Bosch L. 1990. RNA pseudoknots: Translational frameshifting and readthrough on viral RNAs. *Virus Genes* **4:** 121.

Ten Dam E., Verlaan P., and Pleij C. 1995. Analysis of the role of the pseudoknot component in the SRV-1 *gag-pro* ribosomal frameshift signal: Loop lengths and stability of the stem regions. *RNA* **1:** 146.

ten Dam E., Brierley I., Inglis S., and Pleij C. 1994. Identification and analysis of the pseudoknot-containing *gag-pro* ribosomal frameshift signal of simian retrovirus-1. *Nucleic Acids Res.* **22:** 2304.

Tsuchihashi Z. 1991. Translational frameshifting in the *Escherichia coli dnaX* gene *in vitro*. *Nucleic Acids Res.* **19:** 2457.

Tu C., Tzeng T.-H., and Bruenn J.A. 1992. Ribosomal move-

ment impeded at a pseudoknot required for frameshifting. *Proc. Natl. Acad. Sci.* **89:** 8636.

Tumer N.E., Parikh B.A., Li P., and Dinman J.D. 1998. The pokeweed antiviral protein specifically inhibits Ty1-directed +1 ribosomal frameshifting and retrotransposition in *Saccharomyces cerevisiae*. *J. Virol.* **72:** 1036.

Veidt I., Lot H., Leiser M., Scheidecker D., Guilley H., Richards K., and Jonard G. 1988. Nucleotide sequence of beet western yellows virus RNA. *Nucleic Acids Res.* **16:** 9917.

Weiss R.B., Dunn D.M., Shuh M., Atkins J.F., and Gesteland R.F. 1989. *E. coli* ribosomes re-phase on retroviral frameshift signals at rates ranging from 2 to 50 percent. *New Biol.* **1:** 159.

Wolin S.L. and Walter P. 1988. Ribosome pausing and stacking during translation of a eukaryotic mRNA. *EMBO J.* **7:** 3559.

Xiong Z. and Lommel S.A. 1989. The complete nucleotide sequence and genome organization of red clover necrotic mosaic virus RNA-1. *Virology* **171:** 543.

Yusupova G.Z., Yusupov M.M., Cate J.H., and Noller H.F. 2001. The path of messenger RNA through the ribosome. *Cell* **106:** 233.

Programmed +1 Translational Frameshifting in the Yeast *Saccharomyces cerevisiae* Results from Disruption of Translational Error Correction

G. Stahl, S. Ben Salem, Z. Li, G. McCarty, A. Raman, M. Shah, and P.J. Farabaugh
Department of Biological Sciences and Program in Molecular and Cell Biology,
University of Maryland Baltimore County, Baltimore, Maryland 21250

A commonly held belief, especially among beginning students of molecular biology, is that the process of protein synthesis is faithful—the sequence of codons in an mRNA defines exactly the sequence of amino acids in the encoded protein. Of course, like any biological process, protein synthesis is inherently error-prone. Nevertheless, it is important that translation is as nearly faithful as possible to avoid disastrous translational errors. However, increasing accuracy can also be counterproductive; despite the fact that slower translation is more accurate, decreasing the rate of translation can actually limit growth (for review, see Kurland et al. 1996). Therefore, the translational machinery has been subject to opposing selections: to translate rapidly in order to produce more protein, and to translate accurately in order to produce products more faithfully (Kurland et al. 1996). The compromise results in an average of 5×10^{-4} missense errors and fewer than 3×10^{-5} frameshift errors per codon translated (Kurland 1979; Parker 1989).

Correct (cognate), in-frame tRNAs bind to the ribosomal A site and are accepted by the ribosome much more rapidly than are incorrect (non-cognate) or out-of-frame tRNAs. The ribosome blocks errant decoding by noncognate tRNAs by a process called kinetic proofreading (Hopfield 1974; Ninio 1975). Aminoacyl tRNAs (aa-tRNA) bind to the ribosome in a ternary complex (TC) with elongation factor Tu (EF-Tu, or in eukaryotes, EF-1α) and GTP. Cognate TC rarely dissociates from the ribosome, whereas non-cognate TC dissociates rapidly. Binding to the ribosome activates the intrinsic GTPase activity of EF-Tu; after GTP hydrolysis, again cognate aa-tRNA rarely dissociates, whereas non-cognate aa-tRNA does so rapidly. The two kinetic selection steps of this process amplify the preference for cognate aa-tRNA, providing a high level of accuracy.

The ribosome actively participates in the process of selection. First, the ribosome undergoes structural rearrangements in response to cognate aa-tRNA's occupation of the A site that increase the likelihood of its retention (Powers and Noller 1994; Lodmell and Dahlberg 1997). Since non-cognates do not promote these rearrangements, the ribosome in effect selects correct and rejects incorrect tRNAs. Second, cognate aa-tRNAs cause the ribosome to accelerate the rate of GTP hydrolysis, again creating a preference for cognate aa-tRNAs (Pape et al. 1999, 2000).

Translational errors can be stimulated by a variety of means, including directly perturbing the ribosome. Certain antibiotics that bind to the ribosomal decoding sites can reduce the ability of the ribosome to distinguish between correct and incorrect aa-tRNAs. For example, aminoglycoside antibiotics increase translational errors by binding inside helix 44 of the rRNA immediately adjacent to the ribosomal A site (Fourmy et al. 1996; Carter et al. 2000; Vicens and Westhof 2001). Selection of cognate aa-tRNA is thought to occur in part by its inducing a change in the structure of this helix (Ogle et al. 2001). Aminoglycosides cause the same rearrangement, paying the energetic cost of the less favorable binding of non-cognate aa-tRNA, and thus increasing errors. Mutating elements of the rRNA or ribosomal proteins surrounding the decoding site can also affect fidelity. For example, mutations affecting domains of *Escherichia coli* rpS12 increase translational fidelity (for review, see Kurland et al. 1996). Residues in these domains, Pro-44 and Ser-46, form hydrogen bonds stabilizing the rearranged structure of helix 44 (Ogle et al. 2001), suggesting that more stringent A-site selection causes the increased fidelity.

Changes in *trans*-acting factors involved in translation can also decrease translational fidelity. Certain EF-Tu mutants cause increased frameshifting and nonsense readthrough (Hughes et al. 1987; Sandbaken and Culbertson 1988; Vijgenboom and Bosch 1989). Changes to the structure of tRNAs can also decrease the fidelity of their decoding. For example, a mutation in the D-stem of *E. coli* tRNA^Trp allows errant decoding of UGA as Trp (Hirsh 1971) by reducing the rate of tRNA dissociation from the A site (Smith and Yarus 1989a,b).

The mRNA, as a substrate used by the ribosome, can also affect fidelity. The 30S ribosome directly contacts the mRNA in the decoding sites to ensure cognate decoding (Ogle et al. 2001), and large segments of mRNA contact the ribosome on either side of the decoding center (Yusupova et al. 2001). Certain mRNA sequences can reduce fidelity, greatly increasing the frequency of non-canonical events such as readthrough of termination codons or frameshifting. These sequences are termed recoding sites to signify the idea that they alter translational coding rules. The sequences can stimulate these events by altering the kinetics of TC selection (see, e.g., Belcourt and Farabaugh 1990; Farabaugh et al. 1993) or recruitment of other translational factors, e.g., peptide release factor (for review, see Tate and Brown 1992), or possibly by interfering with the function of the accuracy center (Li et al. 2001).

Recoding sites from the Ty family of retrotransposons in the yeast *Saccharomyces cerevisiae* cause the ribosome to efficiently frameshift 1 nucleotide in the downstream direction (+1 frameshifting). Our laboratory has extensively characterized these frameshift sites, which stimulate frameshifting by manipulating the process of translational error correction in several ways. By combining multiple error-promoting effects, the sites are able to achieve highly efficient frameshifting.

PROGRAMMED +1 FRAMESHIFTING IN Ty RETROTRANSPOSONS IN *S. CEREVISIAE*

Ty retrotransposons, like metazoan retroviruses, include a *GAG* and a *POL* gene, which encode the structural and enzymatic proteins required for reverse transcription. (In early work, these genes were respectively termed *TYA* and *TYB* [Clare and Farabaugh 1985]. To clarify the relationship to retroviruses, we use conventional retrovirus nomenclature.) As in retroviruses, the *POL* product is expressed as a Gag-Pol translational fusion protein. In Ty elements, the *POL* open reading frame overlaps the last few dozen nucleotides of the upstream *GAG* gene, shifted into the +1 reading frame. Expression of the Gag-Pol product occurs by +1 programmed translational frameshifting, meaning that the ribosome shifts from the *GAG*, or zero reading frame, to the *POL*, or +1 reading frame, while continuing elongation.

Frameshifting accomplishes a morphogenetic purpose by placing the enzymatic activities of Pol inside a nucleocapsid particle constructed from Gag monomers (Fig. 1). The geometry of the fusion of Gag to Pol means that as the Gag-Pol fusion protein assembles with the much more abundant Gag monomers, the enzymatic activities of the Pol protein are positioned in the interior of the particle in which reverse transcription later takes place. The proportion of Gag to Gag-Pol proteins expressed sensitively modulates the function of these particles (for review, see Dinman 1995; Farabaugh 1995). Increasing (Kawakami et al. 1993) or decreasing (Xu and Boeke 1990) frameshift efficiency can each strongly reduce the frequency of retroviral-like transposition of the elements. The same type of effect is seen in metazoan viruses where fusing *gag* and *pol* genes into one reading frame destroys their ability to replicate (Felsenstein and Goff 1988; Park and Morrow 1992). Dinman and Wickner (1992) proposed that changing the ratio of Gag to Gag-Pol might result in construction of incomplete particles, therefore reducing transposition. Surprisingly, Kawakami et al. (1993) found that changing the ratio of Gag to Gag-Pol interferes with proteolytic processing of these polyprotein precursors, which is essential to their enzymatic

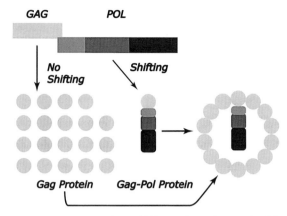

Figure 1. Translational frameshifting and morphogenesis of the Ty virus-like particle. Gag (*circle*) and Gag-Pol (*multiple shapes* representing multiple enzymatic activities) form into the virus-like particle, or nucleocapsid. Frameshifting controls the ratio of Gag to Gag-Pol, shown as ~5%.

function. It is possible that incomplete formation of particles blocks proper processing, but Kawakami et al. (1993) did not see evidence of incomplete particles. In addition, the proteolysis defect was not general, but specifically blocked processing of the Gag-Pol precursor while allowing normal processing of Gag. One model for this phenomenon is that proper processing requires a specific overall structure for the immature particle; changing stoichiometry changes this structure, resulting in inefficient processing. It is significant that all retroviruses maintain a similar ratio of Gag to Gag-Pol between 20:1 and 50:1 (Dickson et al. 1984)—the ratio for Ty retrotransposons is 33:1 (Kawakami et al. 1993). Perhaps all of these viruses share a similar immature particle structure and processing mechanism.

A MISPAIRED PEPTIDYL tRNA IN THE RIBOSOMAL P SITE INDUCES FRAMESHIFTING

The first characterized Ty programmed frameshift site was derived from the Ty1 element (Belcourt and Farabaugh 1990). A 7-nucleotide sequence is both necessary and sufficient to allow up to 40% frameshifting: CUU-AGG-C (shown as codons of the upstream *GAG* gene). The sequence is the site of a stochastic event that results in a +1 shift in frames (Fig. 2a). Ribosomes that encounter the sequence have either of two fates: They may decode the region normally, incorporating Leu-Arg by reading the successive CUU and AGG codons (the upper pathway in the figure). Alternatively, they can encode Leu-Gly by reading the CUU followed by the GGC codon, in effect skipping the intervening A residue (the lower pathway). The ratio of ribosomes undergoing the two events defines the efficiency of frameshifting. The two events compete at the frameshift site. Any decrease in the rate of canonical in-frame decoding causes increased frameshifting, as would an increase in out-of-frame decoding. Programmed frameshift sites employ

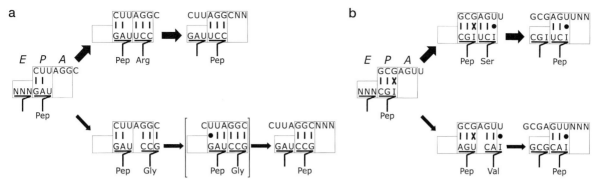

Figure 2. Models for programmed +1 frameshifting in yeast. Thickness of arrows indicates approximate relative probability of the two alternative events that occur at the frameshift site. (*a*) Frameshifting in Ty1 elements. (*b*) Frameshifting in Ty3 elements. See text for explanation.

diverse mechanisms to reduce canonical or increase out-of-frame decoding, often directly disturbing the process of error correction.

Frameshifting at the Ty1 site depends on the unusual tRNA $_{UAG}^{Leu}$, which is the last tRNA to decode in the zero frame. In vitro this tRNA can recognize all six leucine codons (UUR and CUN), although in vivo it only reads the four CUN codons (Weissenbach et al. 1977). This expansive decoding depends on the lack of modification of the wobble U residue of the tRNA. In most cases, such a base would be modified to reduce its pairing to nucleotides other than A; lack of modification actually allows even U•pyrimidine combinations, although perhaps without any hydrogen bonding between the nucleotides. Lagerkvist (1978) termed this type of pairing "2-out-of-3". Notably, frameshifting occurs at a site where tRNA $_{UAG}^{Leu}$ could pair in either the zero (CUU) or +1 (UUA) frames, implying that frameshifting may occur by slippage between these two codons. Introduction of a synthetic tRNA with an AAG anticodon complementary to CUU strongly reduces frameshifting at CUU-AGG-C (Belcourt and Farabaugh 1990), implying the necessity of weak interaction by tRNA $_{UAG}^{Leu}$, as might be expected with a slippage model. Mutations of the mRNA that reduce the ability of the tRNA to pair in the +1 reading frame also reduce frameshifting, further supporting the tRNA slippage model (Belcourt and Farabaugh 1990). Slippage by tRNAs is a mechanism used to explain almost all programmed frameshifts (for review, see Farabaugh 1996). Curran (1993) demonstrated using the *prfB* gene of *E. coli*, which also uses a +1 programmed frameshift mechanism, that frameshift efficiency is directly related to the ability to pair in the shifted frame. The data on the Ty1 site emphasize that slippage may depend equally on weakness of pairing in the zero frame, a phenomenon also observed with –1 frameshifting in the *dnaX* gene of *E. coli* (Tsuchihashi and Brown 1992).

The programmed frameshift site from the Ty3 retrotransposon seems to employ a different mechanism. Frameshifting occurs at the sequence GCG-AGU-U by reading of the GCG codon as alanine, then the GUU as valine. Again, the middle nucleotide of the 7-nucleotide frameshift signal is bypassed but, unlike the Ty1 event,

frameshifting at Ty3 cannot involve tRNA slippage (Farabaugh et al. 1993). In our initial work on the Ty3 site, we proposed that frameshifting would occur after a putative cognate tRNA $_{CGC}^{Ala}$ decodes GCG. This tRNA would be incapable of slippage +1 since it could make no pairs with the +1 frame codon, CGA. In fact, using the metric developed by Curran (1993), tRNA $_{CGC}^{Ala}$ reading GCG has the lowest potential to +1 slippage of any pair of codons in yeast (Vimaladithan and Farabaugh 1994). We suggested, therefore, that frameshifting at the Ty3 site does not require slippage, but occurs by out-of-frame recruitment of the next aa-tRNA (Fig. 2b).

We presumed that there was something unusual about tRNA $_{CGC}^{Ala}$, since only a small minority of codons in yeast shared the ability to cause out-of-frame decoding (Vimaladithan and Farabaugh 1994). Unexpectedly, the sequence of the completed yeast genome revealed that yeast lacked a tRNA $_{CGC}^{Ala}$, meaning that another tRNA^{Ala} isoacceptor must normally decode GCG. It turns out that the absence of the tRNA explains frameshift induction. In the absence of a normal cognate tRNA, the cell is forced to use an isoacceptor, and the abnormal interaction between this alternative tRNA and the mRNA predisposes the ribosome to shift reading frames.

By manipulating tRNA availability, we were able to show that frameshifting at the Ty3 site occurred when GCG was decoded by tRNA $_{IGC}^{Ala}$ (Sundararajan et al. 1999). This tRNA normally decodes for GCU and GCC; the inosine in the wobble position of this tRNA is expected to pair very poorly with guanosine. Crick (1966) originally proposed that inosine could pair with U, C, and A, although more recent work shows that inosine-containing tRNAs decode A-ending codons very inefficiently (Curran 1995). We know that frameshifting occurs when peptidyl tRNA $_{IGC}^{Ala}$ occupies the P site; its unusual I•G wobble interaction suggested a model for frameshifting in which aberrant P-site interactions promoted frame errors at the next codon. Because 11 codons can stimulate +1 frameshifting above background levels (Vimaladithan and Farabaugh 1994), we looked to see whether unusual P-site interactions could explain any of these other cases. Surprisingly, in each case, frameshifting depends on an unusual wobble interaction by such a

Table 1. Normal Cognate-decoding tRNAs and Frameshift-inducing Near-cognates

Codon	Cognate[a]	Near-cognate
CUU	CUU | | • GAG	CUU | | GAU
CUC	CUC | | | GAG	CUC | | GAU
CUG	CUG | | | GAC	CUG | | GAU
CCU	CCU | | • CG I	CCU | | GGU*
CCC	CCC | | • CG I	CCC | | GGU*
CCG	CCG | | | GGC	CCG | | GGU*
CGA	CGA | | | GCU*	CGA | | x GC I
AGG	AGG | | | GAC	AGG | | GAU*
GUG	GUG | | | CAC	GUG | | CAU*
GCG	GCG | | | CGC	GCG | | x CG I
GGG	GGG | | | CCC	GGG GGG | | or | | x CCU* CCG

[a]Symbols: (\|) Watson-Crick; (•) wobble pair; (x) purine•purine clash.

near-cognate tRNA in the P site (Table 1). To clarify our recently criticized terminology, by "near-cognate" we mean any tRNA that makes a less-than-optimal interaction with a codon. For example, tRNA $_{UGC}^{Ala}$ and tRNA $_{IGC}^{Ala}$ are near cognates for GCG; in yeast, a tRNA with a wobble C decodes codons ending in G (Percudani et al. 1997; Sprinzl et al. 1998). In most cases, a weak pyrimidine•pyrimidine interaction causes frameshifting. In a few cases, it is a purine•purine clash, like the I•G clash in the Ty3 frameshift. This suggests that frameshifting does not require a specific and special structure, but rather that it results from a poor interaction between the decoding tRNA and the mRNA.

A TRANSIENT PAUSE REDUCES THE PREFERENCE FOR COGNATE, IN-FRAME DECODING

Translation is a kinetically controlled process that operates far from chemical equilibrium (for review, see Thompson 1988; Kurland et al. 1996). Elongation continues at a rate of from 5 to 20 codons per second (Sorensen and Pedersen 1991), averaging about 12 codons per second (Sorensen et al. 1989). Translation ac-

curacy depends on the idea that the higher rate of acceptance of cognate aa-tRNA by the ribosome means that it nearly always is selected in preference to any other aa-tRNA. One of the ways that mRNAs can stimulate programmed frameshift events is by limiting the rate of continued canonical translation. If the rate of the generally preferred canonical event is slowed sufficiently, the probability of competing noncanonical events (e.g., frameshifts) will be indirectly increased. Programmed +1 frameshifting occurs when a very poorly recognized codon occupies the ribosomal A site, either a sense (Belcourt and Farabaugh 1990; Farabaugh et al. 1993) or nonsense codon (Craigen and Caskey 1986; Donly et al. 1990; Matsufuji et al. 1995). The effect is to pause the ribosome with a peptidyl tRNA in the P site, and the A site unoccupied. Frameshifting efficiency can be increased by further restricting recognition of the A-site codon. Deleting the gene encoding the tRNA that recognizes such a codon increases frameshifting (Kawakami et al. 1993). Presumably, this obliges it to be recognized by a near-cognate isoacceptor, which would do so much less rapidly than would the cognate. By the same reasoning, overexpressing the cognate isoacceptor should accelerate cognate decoding at the frameshift site, which explains why it reduces frameshift efficiency (Belcourt and Farabaugh 1990; Farabaugh et al. 1993). The same effect obtains for frameshifting at poorly recognized termination codons; decreasing the rate of recognition using a partially defective peptide release factor (RF) mutant increases frameshifting (Donly et al. 1990; Bidou et al. 2000), whereas an oversupply of RF reduces frameshifting (Craigen and Caskey 1986; Donly et al. 1990).

The requirement for slow recognition of codon following the last zero-frame codon implies that frameshifting occurs with the tRNA decoding the last zero-frame codon in the ribosomal P site. By manipulating availability of the cognate tRNAs, we showed that +1 frameshifting in yeast also depends on rapid recognition of the first +1 frame codon (Pande et al. 1995). This is true for frameshifting by out-of-frame binding of aa-tRNA, which is expected, because the frameshifting in that case explicitly requires recognition of the first +1 frame codon. Unexpectedly, it is also true for frameshifting dependent on slippage, suggesting that the first +1 frame tRNA may actively participate during slippage of peptidyl tRNA $_{UAG}^{Leu}$ with the +1 frame tRNA actually in the A site as the peptidyl tRNA slips +1. Alternatively, the +1 frame tRNA could act by mass action, favoring the competing reaction in which it participates. This seems an unlikely explanation given the lack of chemical equilibrium, which would be required for this type of effect. Given this effect, we need to rethink the mechanism of +1 frameshifting. The effect of the pause codon was originally interpreted to suggest that frameshifting occurs when the A site is empty. If frameshifting occurs coincident with recognition of the first +1 frame codon, then even if frameshifting occurs by slippage of peptidyl tRNA, the A site must actually be filled during the frameshifting, although perhaps before release of EF-Tu•GDP. As shown above, near-cognate peptidyl tRNA in the P site stimulates frameshifting. Perhaps the near-cognate disrupts the ribosomal mechanism that rejects aa-tRNAs which enter the A site out of the proper reading frame.

The evidence for slippage of peptidyl tRNAs is strong in bacteria; Curran (1993) clearly showed that +1 frameshifting efficiency is roughly proportional to the ability of the peptidyl tRNA to bind in the shifted reading frame. This can only be explained by the necessity for the peptidyl tRNA to dissociate from the zero-frame codon and re-pair with the +1 frame codon, which is what we mean by tRNA slippage. In yeast, we have little direct evidence for tRNA slippage during +1 frameshifting. Frameshift efficiency among the 11 shift-prone codons is actually not correlated with the ability to bind in the +1 frame. These data are consistent with a radical interpretation that slippage is not required for frameshifting in yeast. We note that Curran's experiments used a frameshift site derived from the *prfB* gene of *E. coli*. Frameshifting in this gene requires base-pairing between the Shine-Dalgarno interaction site of the 16S rRNA and a sequence in the mRNA immediately upstream of the frameshift site. Available data suggest that Shine-Dalgarno interactions stress the ribosome to physically force slippage (for review, see Larsen et al. 1995). Perhaps in yeast the absence of this kind of force means that slippage cannot occur, so that the programmed sites in yeast instead may depend on unusual P-site wobble pairs to force frameshifting without slippage.

COMPLEMENTARITY BETWEEN THE rRNA AND mRNA STIMULATES FRAMESHIFTING IN YEAST

The sequence that induces programmed +1 frameshifting in the Ty3 retrotransposon extends beyond the 7 nucleotides at which the shift in frame occurs. The heptameric sequence at the site of frameshifting, GCG-AGU-U, stimulates only very low efficiency frameshifting, 2% (Farabaugh et al. 1993), barely above background efficiency, 0.5% (Vimaladithan and Farabaugh 1994). However, the sequence downstream of the heptamer increases frameshifting about 7.5-fold (Farabaugh et al. 1993). We call this sequence the Ty3 stimulator. Downstream sequences can stimulate increased frameshifting in other systems. Most familiar are the secondary structures immediately downstream of the slippery heptamer responsible for simultaneous slippage –1 frameshifting. The structures are commonly pseudoknots in eukaryotes and hairpin loops in prokaryotes (for review, see ten Dam et al. 1990; Farabaugh et al. 2000). The Ty3 stimulator sequence is not predicted to fold into either structure. A longer sequence could form into a non-standard pseudoknot, but structural mutagenesis of this sequence showed that the ability to form this structure is not necessary for the stimulatory effect (Li et al. 2001). The bacteriophage T4 gene *60* translational bypass site includes a stimulator upstream of the recoding site that functions as a primary protein sequence. We tested whether the primary protein sequence encoded by the

Ty3 stimulator was responsible for its effect, creating mutations that retained the protein encoded in either the zero or +1 reading frame but altered the mRNA sequence. Each of these mutations eliminated stimulation, arguing against the protein model (Li et al. 2001).

Having eliminated the possibility of an mRNA secondary structure or primary protein sequence causing stimulation left only one remaining model for the stimulator: that it functions as a primary mRNA sequence. The simplest model to explain how a primary mRNA sequence could stimulate frameshifting is that it base-pairs to an RNA target. Since the ribosome is largely composed of rRNA, we considered the idea that the stimulator base-paired to a segment of it. A segment of the rRNA complementary to a region of 6 nucleotides in the 5′ half of the stimulator is part of the loop of helix 18, also known as the 530 loop (Fig. 3). Helix 18 has been proposed to be involved in recognition of cognate aa-tRNA in the A site (Powers and Noller 1994). More recently, X-ray crystallography has shown that helix 18 is in intimate contact with the tRNA bound in the A site (Ogle et al. 2001). In fact, two nucleotides, C518 and G530, directly or indirectly interact with tRNA and mRNA nucleotides in the A site; one of these, G530, is within the region predicted to base-pair with the stimulator. Finally, Yusupova et al. (2001) have visualized the path of the mRNA on the face of the 30S ribosome, showing that it passes very near helix 18.

Mutational analysis of the Ty3 stimulator shows that altering the region complementary to helix 18 tends to reduce frameshift stimulation (Li et al. 2001). The correlation is not perfect. Some mutations that might be expected to destabilize pairing have no effect on stimulation, perhaps because non-Watson–Crick pairing is allowed in this complex. The clear trend in the data, however, is that mutations that reduce the ability to pair with helix 18 reduce the effect of the stimulator on frameshifting.

ERROR DISRUPTION DURING PROGRAMMED +1 FRAMESHIFTING IN YEAST

Over the course of the last several years, we have arrived at a rather full understanding of the mechanism underlying programmed +1 frameshifting in yeast. The most critical fact for our understanding frameshifting is that a tRNA that makes a less than optimal interaction with the mRNA occupies the P site during the event. Such

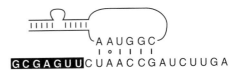

Figure 3. A proposed mRNA•rRNA pairing interaction. Pairing is cartooned between the Ty3 mRNA (*below*) and helix 18 of the 18S rRNA (*above*). White letters on black indicate the heptameric site of frameshifting.

near-cognate decoding is absolutely essential for efficient frameshifting. We have even shown an example in which an mRNA signal that normally cannot induce frameshifting above the background level becomes an extremely efficient frameshift signal when deprived of its cognate tRNA. The cognate tRNA is tRNA$^{Arg}_{CCU}$, which decodes AGG. When the P-site codon of a frameshift signal is changed to AGG, we see very low levels of frameshifting (Vimaladithan and Farabaugh 1994). Deleting the single gene encoding this tRNA causes a massive increase in frameshifting, making it one of the most efficient frameshift signals in that background. The only difference between the two conditions is that tRNA$^{Arg}_{UCU}$ must recognize the AGG codon in the mutant background rather than tRNA$^{Arg}_{CCU}$. This shows that a weak pyrimidine•pyrimidine wobble interaction by peptidyl-tRNA$^{Arg}_{UCU}$ in the P site is sufficient to induce high-efficiency frameshifting.

The other requirements for frameshifting appear to play a less critical role. The slow recognition of the in-frame codon in the A site indirectly increases frameshifting by slowing the competing reaction that leads to continued in-frame decoding. However, in the absence of a special P-site wobble interaction, slowing decoding has little effect on frame maintenance. From our data we see that drastically reducing the rate of recognition of an A-site AGG codon stimulates frameshifting an average of 22-fold on the 10 codons normally read by near-cognate tRNAs. For the other codons apparently read exclusively by cognate tRNAs, the average frameshift efficiency rose only slightly. Thus, pausing alone is insufficient to stimulate significant levels of frameshifting.

How does the state of the wobble pair in the P site disrupt frame maintenance in the A site? With the availability of molecular structures for the ribosomes bound to tRNAs (Ogle et al. 2001; Yusupov et al. 2001) and mRNA (Yusupova et al. 2001), we now have an image of the interaction between the ribosome and cognate tRNAs. From this we can speculate on how a tRNA that differs from a cognate might disturb frame maintenance. Knowing that a normal cognate wobble interaction in the P site is essential to correct framing in the adjacent A site suggests that that interaction constrains the ribosome to reject aa-tRNAs cognate for a codon in other frames. The structure of an 80S ribosome carrying a tRNA in the P site shows that the wobble base pair is bound in a pincer-like grip between two nucleotides of the 16S rRNA, C1400 contacting the wobble base of the tRNA and m^{2}G966 the backbone of the wobble nucleotide in the mRNA (Yusupov et al. 2001). Since the resolution of this structure is only 5.5 Å, we do not know the details of these interactions. A wobble mismatch would be predicted to destabilize these interactions, implying that constraining the wobble pair is essential for proper framing in the A site.

Strangely, the structure of an 80S ribosome shows that the A-site and P-site tRNAs can approach no closer than 10 Å while bound to the mRNA (Yusupov et al. 2001). This is because of an approximately 45° kink in the mRNA between the A-site and P-site codons (Yusupov et al. 2001). The kink forces the phosphoribose backbone

into an extended conformation. No part of the ribosome appears to be in direct contact with the kink, so its conformation must be forced by interactions with the mRNA in the two decoding sites. One of these interactions is the set of pincer contacts with the P-site wobble base pair. Although structures of near-cognate complexes have not been solved, a non-standard interaction at this base pair could disrupt the conformation of the kink, potentially disturbing the structure of the A-site codon•anticodon complex as well. Other genetic evidence suggests that recognition of the A-site codon is affected by the nature of the P-site contact, with even normal non-Watson–Crick wobble interactions in the P site reducing the efficiency of A-site cognate decoding (Kato et al. 1990). As we have seen, the frameshift-inducing near-cognates in yeast do not appear simply to decrease the rate of in-frame cognate decoding and thus increase the duration of a translational pause, but rather to directly promote shifted decoding.

Perhaps, then, the effect of these P-site wobble disruptions is to disrupt the mechanism that the ribosome uses to distinguish in-frame from out-of-frame aa-tRNA complexes in the A site. By implication, the kink between the two decoding sites must be integral to that distinction. How could that work? Given what we now know about cognate recognition in the A site, it seems likely that "shifted decoding" does not mean that the tRNA binds to the mRNA shifted slightly out of the A site (Fig. 4a), but rather that the +1 frame codon must move into the normal location in the A site at which codons are recognized, looping out a nucleotide between the A and P sites (Fig. 4b). Such a conformation might be possible, since the mRNA kink between the sites lies directly above a cavity in the 30S ribosome, which may be large enough to accommodate an extra nucleotide. Alternatively, the extra nucleotide could stack on the end of the A-site codon, extending this pseudohelix one more nucleotide toward the P site. In either of these cases, it may be impossible to maintain the normal interactions between the P-site wobble codon and the ribosome. Perhaps, then, the ribosome senses out-of-frame recognition in the A site by its disruption of the P-site wobble contacts. This would explain how poor non-standard P-site wobble interactions would stimulate frameshifting, since by themselves destabilizing the ribosomal contacts they would reduce the ability of the ribosome to distinguish between correct and incorrect framing in the A site.

The effect of the Ty3 stimulator on frameshifting provides a second example of how a recoding site can directly stimulate frame errors, but this time by directly manipulating the interaction between the A site and the incoming aa-tRNA. Two helix-18 nucleotides interact directly with the codon•anticodon helix in the A site: nucleotide G530 and C518 (Ogle et al. 2001). These two residues are involved in four of the seven H-bond contacts between the ribosome and the A-site codon•anticodon helix (Fig. 5). Pairing between the Ty3 stimulator and helix 18 could disrupt some or all of these bonds. Eliminating these contacts would decrease the stability of tRNA binding to the A site. Paradoxically, one might ex-

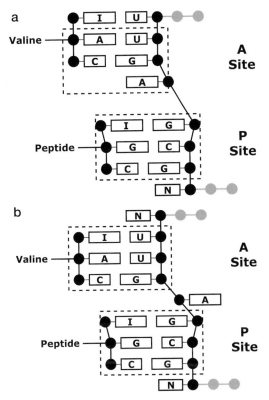

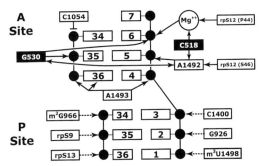

Figure 5. Helix-18 nucleotides make critical contacts in the A site. The A- and P-site codon•anticodon complexes are cartooned as in Fig. 4. Nucleotides and amino acids in direct or indirect contact with the complexes are shown; the two nucleotides of helix 18 are shown as black rectangles. Solid arrows indicate hydrogen bond or metal-mediated interactions with the 2′-OH of ribose (*black balls*) or side groups on bases. A stacking interaction between C1054 and the ribose of the A-site tRNA's wobble nucleotide (base 34) is indicated. Dashed arrows are used where the nature of the interaction with the mRNA or tRNA nucleotide is unknown. The nature of the interactions in the P site remains unknown.

Figure 4. Two models for out-of-frame recognition during frameshifting. The A and P sites are cartooned to show the approximate relative positions of the two decoding sites including the 45° kink between them (Ogle et al. 2001; Yusupov et al. 2001; Yusupova et al. 2001). Black balls represent the phosphoribose backbone, and boxes represent the nucleotides of codons (*right*) and anticodons (*left*). (*a*) Pairing out of the A site. In this model the tRNA aligns with a codon including one base outside the normal A site. An abnormal P-site wobble interaction may predispose the tRNA not to bind to the first position of the A-site codon. No specific mechanism for this effect is implied by the cartoon. (*b*) Out-of-frame pairing to a shifted codon. In this alternative model, +1 codon•anticodon recognition occurs fully in the A site. To allow this, the middle base of the heptameric site must be looped out between the decoding sites. Gray balls represent those nucleotides buried within the small subunit.

pect that to increase discrimination, since it would lead to dissociation of the tRNA from the ribosome, which is how non-cognate aa-tRNAs are rejected. However, if disrupting these contacts leads to dissociation of even cognate aa-tRNAs, the interaction would leave the ribosome

less able to distinguish cognate from non-cognate tRNAs, or perhaps even in-frame from out-of-frame. In such a situation, one might expect a net increase in errors.

The stimulator effect is different from both the effect of pausing induced by a poor in-frame A-site codon and the effect of P-site wobble disruption. Its function is not simply to increase translational pausing at the recoding signal. Rather, the stimulator appears to antagonize the binding of cognate aa-tRNA in the A site. The first evidence for this was that the stimulator has no effect on frameshifting at sites with any of the three nonsense codons in the position of the pause-inducing codon, although it stimulates when any of three sense codons are used instead (Li et al. 2001). This suggested that the stimulator acts by reducing the rate of acceptance of in-frame cognate aa-tRNAs; this explains the nonsense codon effect, since that involves recognition by peptide release factor instead. As a second test of this conclusion, using the frameshift signal GUG-AGG-C, we tested the effect of the Ty3 stimulator in two strains that differed only in the availability of the cognate tRNA for AGG. In the strain KK240, the single structural gene for tRNA$_{CCU}^{Arg}$ has been deleted; KK242 is the corresponding wild type. As described above, in the absence of its cognate tRNA$_{CCU}^{Arg}$,

Table 2. The Ty3 Stimulator Does Not Increase Frameshifting in the Absence of Cognate Decoding

Strain	GUG-AGG-C (–stimulator)		GUG-AGG-C (+ stimulator)		Stimulator effect
	activity	S.E.M. (%)	activity	S.E.M. (%)	
KK242	3.1×10^4	22	1.7×10^5	5	5.4
KK240	3.7×10^7	27	2.6×10^7	5	0.7

Strains KK242 (*HSX1*) and KK240 (*hsx1::HIS3*) differ only in the presence or absence of the gene encoding tRNA$_{CCU}^{Arg}$, *HSX1*. The strains were transformed with a reporter plasmid containing a *lacZ* gene fused by translational frameshifting to a downstream *luc* gene. Frameshifting is directed by the recoding heptamer GUG-AGG-C either with or without the Ty3 stimulator downstream. Comparing the ratio of luciferase, the *luc* product, to β-galactosidase, the *lacZ* product, between constructs gives relative frameshift efficiency.

the near-cognate tRNA$^{Arg}_{UCU}$ must decode AGG; this tRNA carries a wobble modification restricting pairing to codons other than AGA. We found that addition of the stimulator downstream of the recoding site strongly stimulated frameshifting in the wild-type KK242, as expected (Table 2, cf. lines 1 and 2). However, in KK240, the presence of the stimulator had no effect (Table 2, cf. lines 3 and 4). Thus, the stimulator has no effect in a situation in which no cognate aa-tRNA is available, although it strongly stimulates frameshifting on the same sequence when the cognate is available. This suggests that the stimulator does reduce the rate of cognate decoding; the data do not distinguish whether it has a general effect on all decoding in the A site. Surprisingly, then, A-site ribosomal interactions are not necessary for framing, although P-site interactions are. We interpret this to mean that the role of the A site may be to simply test cognate versus non-cognate interactions, independent of frame, and that the P-site tRNA complex constrains the decoding frame.

More evidence that the Ty3 stimulator affects translation accuracy comes from studies of its effect on translation initiation. Recently, we tested whether the presence of the stimulator immediately downstream of the initiation codon could affect either the efficiency or accuracy of initiation. Although it had neither effect in yeast, in *E. coli* we found evidence of increased initiation at non-AUG codons. We tested each of the nine codons mismatched by one base from AUG (NUG, ANG, and AUN) and found a strong increase in initiation at all but the purine substitutions at the middle position, AAG and AGG (G. Stahl et al., in prep.). The stimulator had little or no effect in the absence of an upstream Shine-Dalgarno interaction, showing that its effect was distinct from the docking effect of that interaction. The data are consistent with the idea that rRNA helix 18 also regulates the accuracy of translational initiation. This is surprising, because helix 18 forms part of the A site, and initiation occurs in the P site. These data provide further genetic evidence for functional cross-talk between the two decoding sites to ensure translational accuracy.

IS MANIPULATION OF ERROR MAINTENANCE A GENERAL ASPECT OF PROGRAMMED FRAMESHIFTING?

All of the phenomenology of programmed +1 frameshifting in yeast can be understood as an effect of reduced error correction. This could be a special case, or it could be that programmed frameshifting always results, at least to some extent, from disruption of the ribosomal error correction machinery. We have put forward the hypothesis that all programmed frameshifting, and indeed other types of frame disruption including frameshift suppression by mutant tRNAs, result in part from amplification of the normal rate of translational errors (Farabaugh and Bjork 1999; Farabaugh 2000). This proposal was meant as a provocation, as appropriately noted by Atkins et al. (2000), in the hope that it would stimulate others to consider new ways to understand these perplexing events. For example, is it possible that the stimulatory effect on

simultaneous slippage –1 frameshifting of downstream pseudoknots reflects their interference with a mechanism that ensures that frame is not disrupted during EF-G-catalyzed translocation?

At present, a definitive answer cannot be given to the question posed in the title of this section. However, other than the +1 frameshift signals in yeast and *E. coli*, it should be noted that we do not know precisely how any other element of a programmed recoding site functions to stimulate errors. In the absence of any specific evidence contradicting the hypothesis, it must be considered an open question.

Arguments have been advanced against the hypothesis. Atkins et al. (2000) have argued that there is no direct evidence for a counting mechanism in the ribosome that allows it to maintain a repeating 3-nucleotide reading frame. They conclude that it is likely that the structure of the tRNA•mRNA complex is dominant in determining translational step size. This would easily explain how frameshift suppressor tRNAs cause +1 shifts in-frame (Atkins et al. 2000); most such tRNAs have an expanded anticodon which under the conventional theory (Roth 1981) allows quadruplet reading of the mRNA leading to quadruplet translocation. Qian et al. (1998) recently showed that modification of the anticodon loops of some frameshift proline-inserting suppressor tRNAs makes quadruplet decoding impossible. We suggested that triplet decoding might be obligatory, and that frameshift suppressor tRNAs might cause peptidyl tRNA slippage reminiscent of programmed +1 frameshifting (Farabaugh and Bjork 1999).

The structure of the ribosome seems to support the contention that a triplet reading frame is encoded directly in the ribosome. The ribosome makes multiple stabilizing contacts with the 3 base pairs formed in each of the decoding sites. These data challenge the suggestion that an extra base pair would form in either decoding site without interacting with the ribosome. It is equally difficult to see how 4 base pairs could be accommodated in the space provided on the ribosome for recognition in the A site, as envisioned in the conventional theory of frameshift suppression, since ribosomal components flanking the codon•anticodon helices should clash sterically with any extra base pair. The simplest interpretation of the structural data is that the ribosome accepts only triplet reading, but as we have suggested, it can distinguish cognate triplet reading in frame from the same decoding out of frame.

CONCLUSIONS

Recoding events occur in competition with continued normal decoding. In the yeast *S. cerevisiae*, ribosomes frameshift +1 at mRNA sequences that manipulate the efficiency of translational error control. Manipulation takes several forms, including reducing the rate of recognition of the next in-frame codon (the "pause codon" effect), interfering with retention of cognate, in-frame aa-tRNAs in the A site (the effect of the Ty3 stimulator) and direct interference with a mechanism necessary to distinguish cognate in-frame from cognate out-of-frame decoding (the effect of wobble disruption in the P site). Highly efficient frameshifting, in some cases approaching the efficiency of normal in-frame decoding, can result from the juxtaposition of these three effects at special mRNA signals. In combination, they raise the frequency of translational reading frame errors at least 10,000-fold. We have used this system to provide clues as to how the ribosome identifies each successive codon during the elongation phase of protein synthesis. The fact that the Ty3 stimulator also drastically increases errors during initiation of protein synthesis in *E. coli* shows that the same error correction machinery is involved in selection of tRNAs in both initiation and elongation. Phenotypes associated with this system provide powerful tools for the further genetic and molecular dissection of the process of ribosomal error correction during both phases of protein synthesis.

ACKNOWLEDGMENTS

This work was supported by grants to P.J.F. from the National Institutes of Health (GM-29480 and TW-02211-01) and the Swedish Cancer Foundation (3717-B95-01VAA).

REFERENCES

Atkins J.F., Herr A., Massire C., O'Connor M., Ivanov I., and Gesteland R.F. 2000. Poking a hole in the sanctity of the triplet code: Inferences for framing. In *The ribosome: Structure, function, antibiotics, and cellular interactions* (ed. R. Garrett et al.), p. 369. ASM Press, Washington, D.C.

Belcourt M.F. and Farabaugh P.J. 1990. Ribosomal frameshifting in the yeast retrotransposon Ty: tRNAs induce slippage on a 7 nucleotide minimal site. *Cell* **62:** 339.

Bidou L., Stahl G., Hatin I., Namy O., Rousset J.P., and Farabaugh P.J. 2000. Nonsense-mediated decay mutants do not affect programmed –1 frameshifting. *RNA* **6:** 952.

Carter A.P., Clemons W.M., Brodersen D.E., Morgan-Warren R.J., Wimberly B.T., and Ramakrishnan V. 2000. Functional insights from the structure of the 30S ribosomal subunit and its interactions with antibiotics. *Nature* **407:** 340.

Clare J. and Farabaugh P.J. 1985. Nucleotide sequence of a yeast Ty element: Evidence for an unusual mechanism of gene expression. *Proc. Natl. Acad. Sci.* **82:** 2829.

Craigen W.J. and Caskey C.T. 1986. Expression of peptide chain release factor 2 requires high-efficiency frameshift. *Nature* **322:** 273.

Crick F.H. 1966. Codon-anticodon pairing: The wobble hypothesis. *J. Mol. Biol.* **19:** 548.

Curran J.F. 1993. Analysis of effects of tRNA:message stability on frameshift frequency at the *Escherichia coli* RF2 programmed frameshift site. *Nucleic Acids Res.* **21:** 1837.

———. 1995. Decoding with the A:I wobble pair is inefficient. *Nucleic Acids Res.* **23:** 683.

Dickson C., Eisenman R., Fan H., Hunter E., and Teich N. 1984. Protein biosynthesis and assembly. In *RNA tumor viruses 1/Text* (ed. R. Weiss et al.), p. 527. Cold Spring Harbor Laboratory, Cold Spring Harbor, New York.

Dinman J.D. 1995. Ribosomal frameshifting in yeast viruses. *Yeast* **11:** 1115.

Dinman J.D. and Wickner R.B. 1992. Ribosomal frameshifting efficiency and *gag/gag-pol* ratio are critical for yeast M₁ double-stranded RNA virus propagation. *J. Virol.* **66:** 3669.

Donly B.C., Edgar C.D., Adamski F.M., and Tate W.P. 1990. Frameshift autoregulation in the gene for *Escherichia coli* release factor 2: Partly functional mutants result in frameshift

enhancement. *Nucleic Acids Res.* **18:** 6517.

Farabaugh P. 1995. Post-transcriptional regulation of transposition by Ty retrotransposons of *Saccharomyces cerevisiae*. *J. Biol. Chem.* **270:** 10361.

———. 1996. Programmed translational frameshifting. *Microbiol. Rev.* **60:** 103.

———. 2000. Translational frameshifting: Implications for the mechanism of translational frame maintenance. *Prog. Nucleic Acid Res. Mol. Biol.* **64:** 131.

Farabaugh P.J. and Bjork G.R. 1999. How translational accuracy influences reading frame maintenance. *EMBO J.* **18:** 1427.

Farabaugh P., Qian Q., and Stahl G. 2000. Programmed translational frameshifting, hopping and readthrough of termination codons. In *Translational control of gene expression* (ed. N. Sonenberg et al.), p. 741. Cold Spring Harbor Laboratory Press, Cold Spring Harbor, New York.

Farabaugh P.J., Zhao H., and Vimaladithan A. 1993. A novel programed frameshift expresses the *POL3* gene of retrotransposon Ty3 of yeast: Frameshifting without tRNA slippage. *Cell* **74:** 93.

Felsenstein K. and Goff S. 1988. Expression of the *gag-pol* fusion protein of Moloney murine leukemia virus without *gag* protein does not induce virion formation or proteolytic processing. *J. Virol.* **62:** 2179.

Fourmy D., Recht M.I., Blanchard S.C., and Puglisi J.D. 1996. Structure of the A site of *Escherichia coli* 16S ribosomal RNA complexed with an aminoglycoside antibiotic. *Science* **274:** 1367.

Hirsh D. 1971. Tryptophan transfer RNA as the UGA suppressor. *J. Mol. Biol.* **58:** 439.

Hopfield J. 1974. Kinetic proofreading: A new mechanism for reducing errors in biosynthetic processes requiring high specificity. *Proc. Natl. Acad. Sci.* **71:** 4135.

Hughes D., Atkins J.F., and Thompson S. 1987. Mutants of elongation factor Tu promote ribosomal frameshifting and nonsense readthrough. *EMBO J.* **6:** 4235.

Kato M., Nishikawa K., Uritani M., Miyazaki M., and Takemura S. 1990. The difference in the type of codon-anticodon base pairing at the ribosomal P-site is one of the determinants of the translational rate. *J. Biochem.* **107:** 242.

Kawakami K., Pande S., Faiola B., Moore D.P., Boeke J.D., Farabaugh P.J., Strathern J.N., Nakamura Y., and Garfinkel D.J. 1993. A rare tRNA-Arg(CCU) that regulates Ty1 element ribosomal frameshifting is essential for Ty1 retrotransposition in *Saccharomyces cerevisiae*. *Genetics* **135:** 309.

Kurland C.G. 1979. Reading-frame errors on ribosomes. In *Nonsense mutations and tRNA suppressors* (ed. J.E. Celis and J.D. Smith), p. 97. Academic Press, New York.

Kurland C., Hughes D., and Ehrenberg M. 1996. Limitations of translational accuracy. In *Escherichia coli and Salmonella: Cellular and molecular biology* (ed. F.C. Neidhardt et al.), p. 979. ASM Press, Washington, D.C.

Lagerkvist U. 1978. "Two out of three": An alternative method for codon reading. *Proc. Natl. Acad. Sci.* **75:** 1759.

Larsen B., Peden J., Matsufuji S., Matsufuji T., Brady K., Maldonado R., Wills N.M., Fayet O., Atkins J.F., and Gesteland R.F. 1995. Upstream stimulators for recoding. *Biochem. Cell Biol.* **73:** 1123.

Li Z., Stahl G., and Farabaugh P.J. 2001. Programmed +1 frameshifting stimulated by complementarity between a downstream mRNA sequence and an error-correcting region of rRNA. *RNA* **7:** 275.

Lodmell J.S. and Dahlberg A.E. 1997. A conformational switch in *Escherichia coli* 16S ribosomal RNA during decoding of messenger RNA. *Science* **277:** 1262.

Matsufuji S., Matsufuji T., Miyazaki Y., Murakami Y., Atkins J.F., Gesteland R.F., and Hayashi S. 1995. Autoregulatory frameshifting in decoding mammalian ornithine decarboxylase antizyme. *Cell* **80:** 51.

Ninio J. 1975. Kinetic amplification of enzyme discrimination. *Biochimie* **57:** 587.

Ogle J., Brodersen D., Clemons W.M., Jr., Tarry M., Carter A., and Ramakrishnan V. 2001. Recognition of cognate transfer RNA by the 30S ribosomal subunit. *Science* **292:** 897.

Pande S., Vimaladithan A., Zhao H., and Farabaugh P.J. 1995. Pulling the ribosome out of frame +1 at a programmed frameshift site by cognate binding of aminoacyl-tRNA. *Mol. Cell. Biol.* **15**: 298.

Pape T., Wintermeyer W., and Rodnina M.V. 1999. Induced fit in initial selection and proofreading of aminoacyl-tRNA on the ribosome. *EMBO J.* **18**: 3800.

———. 2000. Conformational switch in the decoding region of 16S rRNA during aminoacyl-tRNA selection on the ribosome. *Nat. Struct. Biol.* **7**: 104.

Park J. and Morrow C.D. 1992. Overexpression of the *gag-pol* precursor from human immunodeficiency virus type 1 proviral genomes results in efficient proteolytic processing in the absence of virion production. *J. Virol.* **65**: 5111.

Parker J. 1989. Errors and alternatives in reading the universal genetic code. *Microbiol. Rev.* **53**: 273.

Percudani R., Pavesi A., and Ottonello S. 1997. Transfer RNA gene redundancy and translational selection in *Saccharomyces cerevisiae. J. Mol. Biol.* **268**: 322.

Powers T. and Noller H.F. 1994. The 530 loop of 16S rRNA: A signal to EF-Tu? *Trends Genet.* **10**: 27.

Qian Q., Li J.N., Zhao H., Hagervall T.G., Farabaugh P.J., and Bjork G.R. 1998. A new model for phenotypic suppression of frameshift mutations by mutant tRNAs. *Mol. Cell* **1**: 471.

Roth J.R. 1981. Frameshift suppression. *Cell* **24**: 601.

Sandbaken M.G. and Culbertson M.R. 1988. Mutations in elongation factor EF-1α affect the frequency of frameshifting and amino acid misincorporation in *Saccharomyces cerevisiae. Genetics* **120**: 923.

Smith D. and Yarus M. 1989a. Transfer RNA structure and coding specificity. I. Evidence that a D-arm mutation reduces tRNA dissociation from the ribosome. *J. Mol. Biol.* **206**: 489.

———. 1989b. Transfer RNA structure and coding specificity. II. A D-arm tertiary interaction that restricts coding range. *J. Mol. Biol.* **206**: 503.

Sorensen M.A. and Pedersen S. 1991. Absolute in vivo translation rates of individual codons in *Escherichia coli*. The two glutamic acid codons GAA and GAG are translated with a threefold difference in rate. *J. Mol. Biol.* **222**: 265.

Sorensen M.A., Kurland C.G., and Pedersen S. 1989. Codon usage determines translation rate in *Escherichia coli. J. Mol.*

Biol. **207**: 365.

Sprinzl M., Horn C., Brown M., Ioudovitch A., and Steinberg S. 1998. Compilation of tRNA sequences and sequences of tRNA genes. *Nucleic Acids Res.* **26**: 148.

Sundararajan A., Michaud W.A., Qian Q., Stahl G., and Farabaugh P.J. 1999. Near-cognate peptidyl-tRNAs promote +1 programmed translational frameshifting in yeast. *Mol. Cell* **4**: 1005.

Tate W.P. and Brown C.M. 1992. Translational termination: "Stop" for protein synthesis or "pause" for regulation of gene expression. *Biochemistry* **31**: 2443.

ten Dam E., Pleij C., and Bosch L. 1990. RNA pseudoknots: Translational frameshifting and readthrough of viral RNAs. *Virus Genes* **4**: 121.

Thompson R. 1988. EFTu provides an internal kinetic standard for translational accuracy. *Trends Biochem. Sci.* **13**: 91.

Tsuchihashi Z. and Brown P.O. 1992. Sequence requirements for efficient translational frameshifting in the *Escherichia coli dnaX* gene and the role of an unstable interaction between tRNA[Lys] and an AAG lysine codon. *Genes Dev.* **6**: 511.

Vicens Q. and Westhof E. 2001. Crystal structure of paromomycin docked into the eubacterial ribosomal decoding A site. *Structure* **9**: 647.

Vijgenboom E. and Bosch L. 1989. Translational frameshifts induced by mutant species of the polypeptide chain elongation factor Tu of *Escherichia coli. J. Biol. Chem.* **264**: 13012.

Vimaladithan A. and Farabaugh P.J. 1994. Special peptidyl-tRNA molecules promote translational frameshifting without slippage. *Mol. Cell. Biol.* **14**: 8107.

Weissenbach J., Dirheimer G., Falcoff R., Sanceau J., and Falcoff E. 1977. Yeast tRNA[Leu] (anticodon U–A–G) translates all six leucine codons in extracts from interferon treated cells. *FEBS Lett.* **82**: 71.

Xu H. and Boeke J.D. 1990. Host genes that influence transposition in yeast: The abundance of a rare tRNA regulates Ty1 transposition frequency. *Proc. Natl. Acad. Sci.* **87**: 8360.

Yusupov M.M., Yusupova G.Z., Baucom A., Lieberman K., Earnest T.N., Cate J.H., and Noller H.F. 2001. Crystal structure of the ribosome at 5.5 Å resolution. *Science* **292**: 883.

Yusupova G., Yusupov M., Cate J., and Noller H. 2001. The path of messenger RNA through the ribosome. *Cell* **106**: 233.

Preferential Translation of Adenovirus mRNAs in Infected Cells

R. Cuesta, Q. Xi, and R.J. Schneider

Department of Microbiology, New York University School of Medicine, New York, New York 10016

Adenovirus (Ad) can infect a wide array of tissue types, but most natural infections are confined to the respiratory tract, gastrointestinal tract, or the conjunctiva (for review, see Schneider 2000). Human Ads contain a double-stranded linear DNA genome 36 kb in size. Ad gene expression is organized into an early phase corresponding to expression of genes involved in DNA replication and cell cycle control, which occurs prior to the onset of viral replication. The viral late phase corresponds to genes activated with or after viral DNA replication. Gene products synthesized during the late phase of infection are generally structural and nonstructural polypeptides that are required in large amounts for assembly of viral capsids. Most Ad late gene expression is initiated from the viral major late promoter (MLP), which synthesizes a large precursor RNA corresponding to 80% of the viral genome. The precursor RNA gives rise to five families of late mRNAs (L1–L5), which are derived by differential splicing and polyadenylation (Fig. 1). Large amounts of mRNAs also encode pIX and pIVa2 proteins during the late phase, but from promoters independent of the MLP. MLP mRNAs all contain a common 5′ noncoding region (5′NCR) 212 nucleotides in length known as the tripartite leader, so called because it is produced by the splicing of three small exons (Berget et al. 1977).

Several hours after Ad enters the late stage of infection, it inhibits the translation of most cellular mRNAs and its own early mRNAs (for review, see Schneider 2000). Cellular mRNAs remain stable, capped, and polyadenylated, and can be translated in in vitro extracts (Thimmappaya et al. 1982), indicating that they remain intact and functional, but that the translation machinery must be altered. Several cellular mRNAs, such as β-tubulin, can escape the Ad block in cellular mRNA export from the nucleus, but they are still unable to translate during late infection (Moore et al. 1987), indicating a very specific block to translation of all but late Ad mRNAs. Most translating mRNAs during the late phase of Ad infection are therefore late Ad mRNAs, indicating a selective inhibition of cellular mRNAs and exclusive translation of Ad late mRNAs. The tripartite leader was shown to be essential for selective translation of mRNAs following Ad inhibition of cellular protein synthesis (Logan and Shenk 1984; Berkner and Sharp 1985). Structural and mutational analyses showed that the tripartite leader contains an unstructured 5′ end of ~25–40 nucleotides, followed by a group

of moderately stable hairpin (stem-loop) structures that possess significant single-stranded loops (Zhang et al. 1989; Dolph et al. 1990). Both the unstructured 5′ end and the hairpin structures were found to be important for the ability of the tripartite leader to direct translation dur-

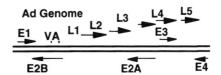

Ad inhibition of eIF4F-dependent translation

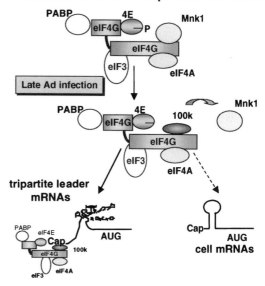

Figure 1. Model for translational inhibition during the late phase of Ad infection. The Ad dsDNA genome is shown, representing early transcription units E1–E4, and the family of late transcripts L1–L5. Inhibition of cell protein synthesis and preferential translation of late Ad mRNAs is also shown. The eIF4F cap-binding complex is depicted with associated eIF3, poly(A)-binding protein (PABP), and the eIF4E kinase, Mnk1. The Ad late L4-100k protein blocks phosphorylation of eIF4E, promoting dissociation of the eIF4E kinase, Mnk1. Cellular mRNA translation is therefore suppressed. The tripartite leader 5′ noncoding region on late Ad mRNAs permits translation when levels of phosphorylated eIF4E are low, either by efficiently recruiting the small amounts of active (eIF4E-phosphorylated) eIF4F, or by utilizing eIF4E/eIF4F containing the viral 100k protein. (Reprinted, with permission, from Schneider 2000.)

ing late Ad infection or during heat shock of cells (Dolph et al. 1988, 1990; Yueh and Schneider 1996). Studies showed that the tripartite leader requires a cap for translation (Dolph et al. 1990; Thomas et al. 1992) and does not function as an internal ribosome entry site (IRES) (Dolph et al. 1990; Jang et al. 1990).

Ad LATE mRNAs TRANSLATE BY RIBOSOME SHUNTING

The tripartite leader directs mRNA translation in poliovirus-infected cells (Castrillo and Carrasco 1987; Dolph et al. 1988), which inhibits cap-dependent mRNA translation by proteolytic degradation of initiation factor eIF4G. Initiation factor eIF4G is a scaffolding protein upon which the cap–initiation complex assembles to direct 40S ribosome subunit addition to capped mRNAs (for review, see Gingras et al. 1999). Thus, the tripartite leader reduces the dependence of Ad late mRNAs, which are capped, on the cap–initiation complex for their translation.

Studies have demonstrated that the tripartite leader directs a novel form of cap-dependent translation initiation known as ribosome shunting. Ribosome shunting has now been described for the cauliflower mosaic virus 35S (CaMV) mRNA (Futterer et al. 1993), Ad late mRNAs (Yueh and Schneider 1996), the Sendai virus Y mRNA (Curran and Kolakofsky 1988; Latorre et al. 1998), papillomavirus E1 mRNA (Remm et al. 1999), and mammalian heat shock protein 70 (Hsp70) mRNA (Yueh and Schneider 2000). The general mechanism for ribosome shunting is not well understood. It has been shown to involve the loading of 40S ribosome subunits to the 5′ end of the capped mRNA, probably via limited scanning, followed by direct translocation of 40S subunits to the downstream initiation codon, directed by poorly defined "shunting elements" (for review, see Hohn et al. 1998). Ribosome shunting is therefore a discontinuous form of ribosome scanning. It is thought that there are two roles for ribosome shunting on mammalian and viral mRNAs. Shunting can extend the coding capacity of an mRNA, as shown for the CaMV 35S mRNA (Futterer et al. 1993), in that it permits ribosomes to skip an upstream open reading frame (ORF) and initiate translation downstream at an internal location on the same mRNA. A second function for ribosome shunting, as shown for late Ad tripartite leader mRNAs (Yueh and Schneider 1996), is to permit selective translation, such as when the cap–initiation complex has been largely inactivated.

A mechanistic understanding of ribosome shunting directed by the Ad tripartite leader 5′NCR is beginning to emerge. The tripartite leader was shown to be inhibited in directing translation at the downstream AUG codon if strong secondary structure, or an AUG codon, is inserted within the first 80 nucleotides of the 5′NCR (Yueh and Schneider 1996). If strong secondary structure is inserted farther 3′, translation is directed by the tripartite leader at normal levels. Therefore, 40S ribosome subunits are thought to contact the first 40 to 80 nucleotides of the tripartite leader in a cap- and scanning-dependent manner. Shunting of 40S ribosome subunits was shown to occur to

a limited distance downstream from the tripartite leader, as the AUG cannot be located farther than ~160 nucleotides from the 3′ end of the leader. The limitation in translation of 40S ribosome subunits is consistent with the location of Ad late AUG codons, which are within 35 nucleotides of the 3′ end of the tripartite leader. The RNA "shunting elements" include hairpin structures downstream of nucleotide 80. In uninfected cells, the tripartite leader directs translation by conventional 5′ scanning of 40S ribosome subunits and also by 40S ribosome shunting, at roughly equal levels (Yueh and Schneider 1996). However, translation initiation by both scanning and shunting ribosomes was found to be incompatible on the same tripartite leader mRNA, probably because scanning ribosomes disrupt the conformation of shunting elements, abolishing the ability to direct ribosome shunting (Yueh and Schneider 1996). Accordingly, in late Ad-infected cells, when the cap–initiation complex is largely inactivated or altered in function (Cuesta et al. 2000), the tripartite leader directs translation exclusively by ribosome shunting (Yueh and Schneider 1996, 2000).

There are certain features in common between ribosome shunting as described for the CaMV 35S mRNA and late Ad tripartite leader mRNAs. CaMV and tripartite leader mRNAs can alternate between scanning and shunting initiation mechanisms. However, on both mRNAs, translation by scanning or shunting does not occur simultaneously, and in fact was shown to be mutually exclusive (Yueh and Schneider 1996; Hohn et al. 1998). This probably indicates that disruption of secondary and tertiary RNA structure also disrupts the function of shunting elements. In CaMV, a short upstream ORF followed by one or more hairpin structures comprises a minimal shunting element (Hemmings-Mieszczak et al. 1997, 1998; Dominguez et al. 1998; Pooggin et al. 1998). The short ORF and downstream stable hairpin structure are thought to provide 40S ribosome subunits to the shunting elements for translocation to a downstream AUG. The unstructured 5′ conformation of the tripartite leader is thought to bind with high affinity to 40S ribosome subunits. Farther downstream in the tripartite leader is a complex set of stable hairpin structures (Dolph et al. 1988; Zhang et al. 1989) which comprise elements that direct ribosome shunting (Yueh and Schneider 2000). However, there is no ORF in the tripartite leader 5′NCR. Instead, the hairpins are thought to both impede 40S ribosome movement, providing 40S subunits to the shunting elements, which might facilitate 40S ribosome recruitment and translocation to a downstream AUG (Yueh and Schneider 1996, 2000). In neither case is it understood how the 40S subunit is directly translocated to the downstream AUG codon for initiation.

Ad INHIBITS CELLULAR PROTEIN SYNTHESIS BY TARGETING eIF4E

It has been shown that Ad inhibits host cell protein synthesis by acting on the cap–initiation complex (see Fig. 1) (for review, see Schneider 2000). The complex contains a core group of translation initiation factors known as eIF4F. eIF4F functions as a cap-dependent RNA helicase that stimulates protein synthesis by unwinding the 5′ end of mRNAs, promoting 40S ribosome binding. Core eIF4F contains the 24-kD cap-binding protein eIF4E, the 45-kD ATP-dependent RNA helicase eIF4A, and the 220-kD molecular adapter protein eIF4G upon which the complex assembles at the cap (for review, see Gingras et al. 1999). eIF4E is phosphorylated by two related MAP kinases known as Mnk1 and Mnk2 (Fukunaga and Hunter 1997; Waskiewicz et al. 1997), either of which is bound to eIF4G (Pyronnet et al. 1999; Waskiewicz et al. 1999). eIF4E binds an amino-terminal region of eIF4G, eIF4A binds a central region, and Mnk1,-2 binds a carboxy-terminal region (for review, see Gingras et al. 1999). Mnk1 phosphorylates eIF4E in vivo only when both are bound to eIF4G (Pyronnet et al. 1999; Waskiewicz et al. 1999). Translation initiation factor eIF3, which recruits the small 40S ribosome subunit, binds the central region of eIF4G. Poly(A)-binding protein (PABP), which promotes cap-dependent and poly(A)$^+$-dependent mRNA translation, also binds to the amino terminus of eIF4G (Tarun et al. 1997; Imataka et al. 1998), potentially circularizing capped and polyadenylated mRNAs (Wells et al. 1998). eIF4G is therefore an adapter protein that couples cap recognition of the mRNA, 5′ mRNA unwinding activity, 40S ribosome subunit loading, and surveillance of the poly(A) tail.

Phosphorylation of eIF4E at Ser209/Thr210 by Mnk1 is thought to stabilize the interaction between the capped mRNA and the initiation complex (Marcotrigiano et al. 1997; Pyronnet et al. 1999; Waskiewicz et al. 1999). Stimulation of cap-dependent mRNA translation by growth factors and other activating stimuli correlates with increased phosphorylation of eIF4E (Gingras et al. 1999), consistent with this mechanistic understanding. Dephosphorylation of eIF4E strongly correlates with inhibition or impairment of cap-dependent mRNA translation, probably by reducing cap-complex affinity or stability with capped mRNA, thereby decreasing ribosome loading (Gingras et al. 1999).

Shutoff of cellular protein synthesis during late Ad infection is associated with a 90–95% block in the phosphorylation of eIF4E (Huang and Schneider 1991; Zhang et al. 1994), with no change in the integrity of eIF4G (Dolph et al. 1988). In addition, several drugs that act on protein kinase activity were shown to block the ability of Ad to shut off cell protein synthesis, which also blocks dephosphorylation of eIF4E (Huang and Schneider 1990; Feigenblum et al. 1998). Dephosphorylation of eIF4E in late Ad-infected cells is therefore strongly associated with viral inhibition of cellular protein synthesis. Shutoff does not involve eIF4E sequestration by 4E-BPs (Feigenblum and Schneider 1996; Gingras and Sonenberg 1997). In fact, Ad induces the increased phosphorylation of 4E-BPs, resulting in release of eIF4E, which is mediated by the E1A early protein. Studies have found that Ad late gene expression is essential for inhibition of eIF4E phosphorylation and host protein synthesis, indicating the requirement for one or more late viral gene products (Zhang et al. 1994).

MECHANISM FOR Ad INHIBITION OF CELLULAR PROTEIN SYNTHESIS: DISPLACEMENT OF Mnk FROM eIF4G

Based on genetic evidence, the Ad late L4-100-kD nonstructural polypeptide has been implicated in preferential translation of viral late mRNAs (Hayes et al. 1990). The 100k protein has both a nuclear and cytoplasmic distribution. 100k protein is involved in morphogenesis of the viral particle in the nucleus (Oosterom-Dragon and Ginsberg 1981). However, 100k protein in the cytoplasm can also be photo-UV-crosslinked specifically to both cellular and viral poly(A)$^+$ RNAs (Adam and Dreyfuss 1987), suggesting a role in translational control. An Ad mutant containing a temperature-sensitive (ts) mutation in 100k protein (Ad2ts1) was found to be reduced in late viral protein synthesis at restrictive temperature, with no defect in viral particle assembly. More recent studies (described below) confirmed that Ad2ts1 does not efficiently translate late Ad mRNAs at restrictive temperature, and also showed that it no longer inhibits host cell protein synthesis.

The importance of eIF4E phosphorylation for translation of cellular mRNAs and the ability of tripartite leader mRNAs to translate in its absence were established with model reporter mRNAs. Two reporter mRNAs were expressed by transfection of cells, one containing β-galactosidase and the Ad late tripartite leader, and the other from a nonviral eIF4F-dependent 5′NCR known as CR3 (Feigenblum and Schneider 1996). To test the requirement for eIF4E phosphorylation, two mutants of eIF4E were developed that cannot be phosphorylated at activating positions Ser-209 and Thr-210. The single mutant contained a Ser-209→Ala substitution, eliminating the

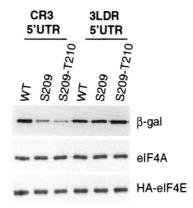

Figure 2. Effect of eIF4E phosphorylation on translation of model Ad late and cellular mRNAs. 293T cells were cotransfected with plasmids expressing β-galactosidase mRNAs containing either the Ad late tripartite leader 5′NCR (3LDR), or an eIF4F-dependent 5′NCR (CR3), and plasmids expressing HA-tagged wild-type (WT) eIF4E, Ser-209→Ala eIF4E, or Ser-209→Ala/Thr-210→Ala eIF4E. At 36 hours posttransfection, equal amounts of cell lysates were resolved by SDS-12%PAGE and immunoblotted using specific antisera for β-galactosidase protein (β-gal), HA-eIF4E, and eIF4A (for protein levels). (Reprinted, with permission, from Cuesta et al. 2000 [copyright Oxford University Press].)

major site of eIF4E phosphorylation. A double mutant in addition contained a Thr-210→Ala mutation to eliminate secondary phosphorylation at this adjacent site (Cuesta et al. 2000). 293T cells were cotransfected with expression vectors for wild-type or mutant eIF4E proteins, and tripartite leader or eIF4F-dependent β-galactosidase reporters. Translation was measured by immunoblot analysis of β-galactosidase reporter protein. Translation of the eIF4F-dependent mRNA was reduced fourfold by expression of the Ala-209 eIF4E mutant, and about eightfold by expression of the eIF4E double mutant (Fig. 2)(Cuesta et al. 2000). Translation of the tripartite leader reporter mRNA was unaffected by expression of either eIF4E mutant compared to wild-type eIF4E. Thus, dephosphorylation of eIF4E can account for inhibition of cellular protein synthesis during late Ad infection.

Since one or more Ad late gene proteins were implicated in inhibition of cellular protein synthesis (Zhang et al. 1994; Feigenblum et al. 1998), and 100k protein was shown to facilitate translation of late viral (tripartite leader) mRNAs, it was determined whether 100k protein interacts with the eIF4F complex (Cuesta et al. 2000). 293 cells were transfected with a vector expressing Flag-100k, eIF4F complexes were isolated by m⁷GTP(cap)-affinity chromatography, and associated proteins were detected by immunoblot analysis (Fig. 3) (Cuesta et al. 2000). Cap-binding complexes purified from cells expressing Flag-100k protein contained eIF4G, eIF4E, and 100k protein, indicating that 100k protein interacts with cap-binding eIF4E/4G complexes. 100k protein interaction was not evident for complexes isolated from control cells, as expected. Studies conducted in vivo and in vitro demonstrated that 100k protein association with the eIF4F complex is coupled to loss of Mnk1 interaction (shown diagramatically in Fig. 1). 293 cells were cotransfected with vectors expressing Flag-100k protein and either GST-, GST-Mnk1, or a catalytic mutant known as GST-T2A2 (Waskiewicz et al. 1999). eIF4G was immunoprecipitated and the association of proteins with eIF4G was determined by immunoblotting (Fig. 4). Binding was evident for endogenous eIF4G to Flag-100k protein which was not present in Flag-transfected cells. eIF4G immunoprecipitates from Flag-100k-expressing cells also contained unchanged amounts of eIF4E, eIF4A, and PABP, but lacked Mnk1 protein (Fig. 4). It was concluded from these and other data (Cuesta et al. 2000) that 100k protein expression is linked to dissociation of Mnk1 from eIF4F/eIF4G.

The interaction partner for 100k protein was shown to be eIF4G, and the site of interaction corresponded to the carboxy-terminal third of the protein, which contains eIF4A- and Mnk1-binding sites (Cuesta et al. 2000). This conclusion was confirmed in vitro using recombinantly expressed and purified proteins corresponding to the amino terminus (N4G), middle (M4G), and carboxyl terminus (C4G) of eIF4G, full-length 100k protein, and Mnk1 (Cuesta et al. 2000). In vitro experiments showed that binding of 100k protein to eIF4G blocks binding of, or evicts, Mnk1 from the eIF4F complexes. Intact eIF4F complexes were isolated from cells by immunoprecipitation (Cuesta et al. 2000), using 293 cells transfected with plasmids expressing GST alone as a control, or GST-Mnk1. Isolated eIF4F complexes were then incubated with purified GST or GST-100k protein, and the effect on eIF4F composition was determined by immunoblot analysis (Fig. 5). Addition of GST alone did not alter the composition of eIF4F, which contained PABP, Mnk1, eIF4A, and eIF4E proteins. In contrast, addition of GST-100k protein resulted in the specific loss of Mnk1-GST from eIF4G (about tenfold reduced). There was also less eIF4A retained and a greater amount of 100k protein bound to eIF4G. Control studies showed equal levels of proteins in whole-cell lysates. 100k protein therefore binds the carboxy-terminal region of eIF4G, resulting in dissociation of Mnk1 protein from cap–initiation complexes and/or blocking its interaction in a competitive manner.

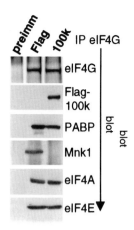

Figure 4. Characterization of 100k protein/eIF4F complexes in the absence of virus infection. 293 cells were transfected with plasmids expressing Flag or Flag-100k protein, endogenous eIF4GI was immunoprecipitated with specific antisera (IP eIF4G) or preimmune sera (preimm), and immunoblotted as shown. (Reprinted, with permission, from Cuesta et al. 2000 [copyright Oxford University Press].)

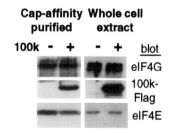

Figure 3. Association of Ad L4-100k protein with eIF4F complexes. 293 cells were either mock transfected or transfected with a cDNA clone expressing Flag-epitope-tagged 100k protein. Equal amounts of lysates were prepared, and eIF4F complexes were purified by cap-affinity chromatography (*left panel*), or directly resolved by SDS-10%PAGE (*right panel*). Proteins were detected by immunoblot analysis using antisera as shown. (Reprinted, with permission, from Cuesta et al. 2000 [copyright Oxford University Press].)

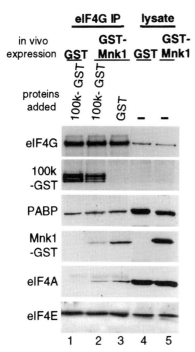

Figure 5. 100k protein binds the carboxyl terminus of eIF4G and evicts Mnk1 protein from eIF4F complexes in vitro. 293 cells were transfected with plasmids expressing GST or GST-Mnk1, extracts were prepared and equal amounts of eIF4F isolated by immunoprecipitation of eIF4G (eIF4G IP). Equal amounts (2 μg) of purified recombinant GST-100k or GST proteins were incubated in vitro with eIF4F complexes, and associated proteins were detected by immunoblot analysis as shown. Equal amounts of whole cell lysates ("lysate") were resolved in lanes *4* and *5* to establish protein levels. (Reprinted, with permission, from Cuesta et al. 2000 [copyright Oxford University Press].)

Dephosphorylation of eIF4E is thought to weaken the interaction of capped mRNAs with eIF4F by eliminating a clamping salt bridge about the capped mRNA (Marcotrigiano et al. 1997). Consequently, capped cellular mRNAs will not translate well in the absence of eIF4E phosphorylation. Nevertheless, Ad late mRNAs are capped, as are cellular mRNAs, and they initiate ribosome binding through a shunting mechanism that still is cap-dependent (for review, see Schneider 2000). Studies were therefore conducted to determine whether there are specific *cis*-acting elements (i.e., shunting elements) in the tripartite leader that permit high-level initiation of translation in the absence of eIF4E phosphorylation (Yueh and Schneider 2000).

RIBOSOME SHUNTING DIRECTED BY THE TRIPARTITE LEADER IS FACILITATED BY *CIS* ELEMENTS COMPLEMENTARY TO 18S rRNA

One significant feature of the tripartite leader that is not generally present in other mRNAs is a group of three elements of split (bipartite) complementarity to both RNA strands of the stem in the 3′ hairpin of 18S rRNA (Fig. 6). There is no established role for the 18S rRNA 3′ hairpin,

which is located on the outer surface of the 40S (small) ribosome subunit. This element is conserved in prokaryotes and eukaryotes, and consistent with its location on the outside surface of the 40S subunit (Green and Noller 1997), it is accessible for interactions with complementary RNA sequences (Moore 1996). The tripartite leader complementarity to the 18S rRNA is well conserved in human and mammalian Ads (Dolph et al. 1990; Yueh and Schneider 1996). Mutational analysis of the tripartite leader was carried out to determine whether the complementarity to 18S rRNA is important for facilitating ribosome shunting in normal and/or eIF4E dephosphorylated cells (Yueh and Schneider 1996) by introducing small, specific deletions. The ability to direct scanning and shunting initiation was compared for wild-type and reporter tripartite leader cDNAs, with or without a stable hairpin at the 3′ end of the tripartite leader, to eliminate ribosome scanning in the latter case. Deletion of complementarities C1–3, when performed individually, had little effect on scanning or shunting translation initiation. Paired deletions of C1 and C2 (mutant D14/B202) reduced shunting slightly (to 60% of the wild-type B202 shunting constructs), without strongly decreasing scanning initiation. Paired deletion of C1 and C3 (mutant D15/B202) reduced shunting initiation 3-fold, whereas paired mutation of C2 and C3 (mutant D16/B202) strongly inhibited shunting by ~20-fold. Similar effects were not observed with mutations outside the boundaries of the 18S rRNA complementarities. There is therefore a functional redundancy and hierarchy in the importance of these elements, in which C2 and C3 are functionally dominant over C1. Triple mutation of C1, C2, and C3 (mutant DL3) did not further reduce ribosome shunting or scanning initiation compared to the C2–C3 mutant. Again, these data are consistent with the redundant and hierarchical functional pattern of these elements.

The importance of tripartite leader elements C1,-2, and -3 in facilitating ribosome shunting during late Ad infection was also established (Yueh and Schneider 2000). 293 cells were transfected with plasmids expressing reporter mRNAs containing the wild-type or mutant tripartite leader 5′NCRs followed by infection with Ad. Reporter translation was determined prior to shutoff of cellular protein synthesis (from 10–16 hours) or after shutoff (from 24–30 hours), by measuring the amount of HBsAg secreted into the medium during this 6-hour period (Table 1). Whereas cytoplasmic levels of mRNAs decreased by only 25% during the late phase of Ad infection (Yueh and Schneider 2000), translation of the wild-type tripartite leader (3LDR), and of a tripartite leader containing the 3′ B202 hairpin (B202) to measure shunting translation, increased by 30–40% despite inhibition of cell protein synthesis. All mutants translated at levels similar to the wild-type tripartite leader mRNAs during early infection. Mutant D14 lacking C1 and C2 translated at about half the level of the B202 control mRNA, and mutant D15 translated at only about 30% the level of B202 during late Ad infection. Mutant D16 lacking C2 and C3 was very poorly translated (~5%) during late infection. These data show that complementarity to 18S rRNA, particularly C2 and C3

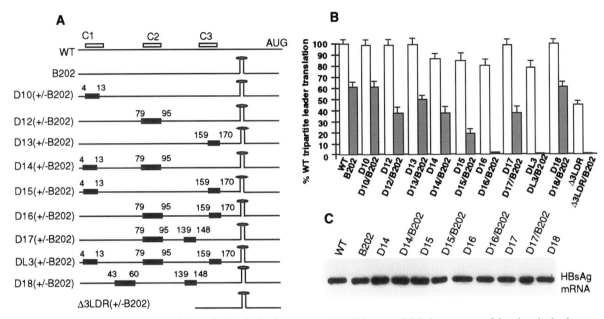

Figure 6. Scanning and shunting translation of tripartite leader mutants. (*A*) Wild type and deletion mutants of the tripartite leader are shown, ± insertion of a stable hairpin structure (B202) in the 3′ end of the NCR to block scanning-dependent initiation. Deletions are indicated by filled boxes, with the position of junction sites indicated. (*B*) ELISA analysis of HBsAg levels obtained from 5 independent transfections of HeLa cells performed at similar efficiencies with the different tripartite leader 5′NCRs. Transfection efficiencies (30–35%) were determined by cotransfection of a green fluorescent protein expression vector (pGFP). Data were normalized relative to the average translational activity of the wild-type tripartite leader (WT). Mutant Δ3LDR contains a 50-nucleotide (*Bam*HI to *Sal*I) fragment of pBluescriptIISK polylinker in place of the tripartite leader. (*C*) Steady-state northern analysis of HBsAg reporter mRNAs obtained from equal numbers of transfected HeLa cells. Blots were hybridized to a ^{32}P-labeled probe prepared from the HBsAg coding region. (Reprinted from Yueh and Schneider 2000.)

in combination, is important for efficient tripartite leader translation by ribosome shunting during late Ad infection.

Although ribosome shunting directed by the tripartite leader can account for translation of most late viral mRNAs after shutoff of cellular protein synthesis, it does not explain how pIX and pIVa2 mRNAs translate. Ad late IVa2 mRNA is synthesized during the intermediate phase of virus infection, it lacks the tripartite leader 5′NCR, but it is still translated during late viral infection, after inhibition of eIF4F-dependent protein synthesis (Hayes et al.

Table 1. Effect of Ad Infection on Ribosome Shunting of Tripartite Leader and IVa2 5′NCRs

	Translation activity	
Construct	10–16 hr p.i. (%)	24–30 hr p.i. (%)
3LDR	100	136
B202	63	105
D12	92	54
D13	100	57
D14	85	29
D15	73	22
D16	71	5
WT IVa2	38	25
IVa2/B202	21	26
mIVa2	30	<2
WT IX	16	0

Translation activity was calculated by normalizing the average ELISA value for HBsAg secretion of different constructs obtained from 5 independent experiments against that of the wild-type tripartite leader (3LDR) at the 10–16-hour (hr) time point postinfection (p.i.). (Taken from Yueh and Schneider 2000.)

1990; Huang and Schneider 1990). The Ad late pIX intermediate-phase mRNA, which also lacks the tripartite leader, is weakly translated during the late phase of infection (Hayes et al. 1990; Huang and Schneider 1990). Inspection of these mRNAs indicated that the pIVa2 5′NCR contains two sequences which are complementary to the 18S rRNA 3′ hairpin and bear an overlapping relationship to those in the tripartite leader (Fig. 7A). No significant complementarity to 18S rRNA exists in the pIX mRNA 5′NCR. pIVa2 and pIX 5′NCRs were tested for ribosome shunting activity when linked to the HBsAg coding region reporter. The B202 hairpin structure was inserted in the 3′ end of the NCR to determine the level of shunting initiation (Fig. 7B). The IVa2 5′NCR directed ribosome shunting activity at approximately one-third that of the tripartite leader (compare B202 and IVa2/B202). Deletion of most of the IVa2 5′NCR, including the two complementarities to the 18S rRNA 3′ hairpin (mutant ΔIVa2), prevented shunting. A mutant deleted in the majority of the two IVa2 complementarities (mIVa2) only very poorly directed ribosome shunting. Overall translation activity of this mutant was reduced by only ~30%, which is the loss of shunting translation. Thus, the IVa2 5′NCR promotes translation initiation by both ribosome shunting and scanning, and likely utilizes complementarity to the 3′ hairpin of 18S rRNA.

The IX mRNA 5′NCR did not translate by ribosome shunting, as it was inhibited by insertion of the B202 hairpin (Fig. 7). This is consistent with the absence of elements complementary to the 18S rRNA 3′ hairpin. It is

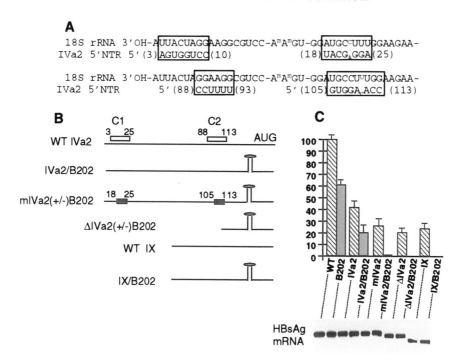

Figure 7. Scanning and shunting translation directed by the Ad5 IVa2 5′NCR. (*A*) Bipartite sequence complementarity of the 18S rRNA 3′ hairpin and the 5′NCR of human Ad5. Complementary elements C1 and C2 (boxed) and nucleotide positions in the IVa2 and 5′NCR are indicated, relative to the mRNA transcription start site. (*B*) Wild type and deletion mutants of the Ad5 IVa2 and IX 5′NCRs are shown, +/− insertion of the B202 stable hairpin structure. The boundaries of deletions in cDNA copies of the 5′NCRs are shown above filled boxes. (*C*) ELISA analysis of HBsAg reporter mRNAs represents an average of 5 independent transfections of HeLa cells at similar efficiencies. The ΔIVa2 mutant is deleted from position 1–112 of the IVa2 5′NCR. Steady-state northern analysis of cytoplasmic reporter HBsAg mRNAs for different constructs was performed using HeLa cells transfected at similar efficiencies. (Reprinted, with permission, from Yueh and Schneider 2000.)

likely that the low translation level of the IX mRNA during the late phase of infection results from the large abundance of the transcript (Huang and Schneider 1990).

CONCLUSIONS

The mechanism by which 40S ribosome shunting is directed by the tripartite leader still remains to be described in molecular detail. On the basis of the present state of understanding, two models can be proposed to describe ribosome shunting (Fig. 8). The "dissociating model" for ribosome shunting proposes that 40S ribosome subunits are loaded onto the 5′ unstructured end of the tripartite leader, provided by 5′ scanning or direct placement from the cap. Due to an intrinsically high off-rate resulting from the stability of the internal hairpin structures and the

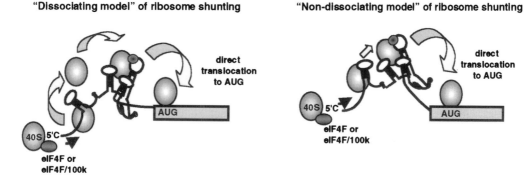

Figure 8. Model for ribosome shunting directed by the tripartite leader during late adenovirus infection. In the "dissociating model" of ribosome shunting, 40S ribosome subunits are thought to be displaced from the tripartite leader due to the inability to unwind the RNA in the absence of normal eIF4F activity. In the "non-dissociating model," 40S ribosome subunits are thought to stall on the mRNA but do not dissociate. The shunting elements would then recover dissociated ribosome subunits, or directly interact with tethered 40S ribosomes, promoting direct translocation to the downstream AUG. 40S ribosomes might interact directly through 18S rRNA:tripartite leader RNA interaction, or via initiation factors recruited to the shunting elements. (Reprinted, with permission, from Schneider 2000.)

loss of eIF4F/RNA unwinding activity, a significant number of 40S subunits might then dissociate from the tripartite leader. The shunting element would then serve the function of recovering these ribosomal subunits by direct binding to the RNA shunting elements through a tripartite leader RNA:18S rRNA hairpin interaction. On the other hand, the shunting element might bind the very limited amount of native eIF4F or eIF4F/100k complex, or specific translation factors such as eIF4G or eIF3, which would recruit 40S subunits to the shunting elements. The translocation step is not at all understood. It might involve direct placement of 40S subunits to the AUG, or upstream of the AUG, followed by limited scanning. The "non-dissociating model" of ribosome shunting proposes that 40S ribosome subunits are loaded onto the 5′ end of the tripartite leader as above, but rather than dissociate, they are blocked from proceeding into the body of the tripartite leader due to dephosphorylation of eIF4E and loss of eIF4F mRNA-unwinding activity. Tethered 40S subunits would then interact with the shunting elements, possibly by RNA:RNA interactions, or by initiation factors recruited to the shunting elements.

ACKNOWLEDGMENTS

This work was supported by a grant from the National Institutes of Health (CA-42357).

REFERENCES

Adam S.A. and Dreyfuss G. 1987. Adenovirus proteins associated with mRNA and hnRNA in infected Hela cells. *J. Virol.* **61:** 3276.

Berget S.M., Moore C., and Sharp P. 1977. Spliced segments at the 5′ terminus of Ad2 late mRNA. *Proc. Natl. Acad. Sci.* **74:** 3171.

Berkner K.E. and Sharp P.A. 1985. Effect of tripartite leader on synthesis of a non-viral protein in an adenovirus 5′ recombinant. *Nucleic Acids Res.* **13:** 841.

Castrillo J.L. and Carrasco L. 1987. Adenovirus late protein synthesis is resistant to the inhibition of translation induced by poliovirus. *J. Biol. Chem.* **262:** 7328.

Cuesta R., Xi K., and Schneider R.J. 2000. Adenovirus specific translation by selective disassembly of cap-initiation complex. *EMBO J.* **19:** 3465.

Curran J. and Kolakofsky D. 1988. Scanning independent ribosomal initiation of the Sendai virus X protein. *EMBO J.* **7:** 2869.

Dolph P.J., Huang J., and Schneider R.J. 1990. Translation by the adenovirus tripartite leader: Elements which determine independence from cap-binding protein complex. *J. Virol.* **64:** 2669.

Dolph P.J., Racaniello V., Villamarin A., Palladino F., and Schneider R.J. 1988. The adenovirus tripartite leader eliminates the requirement for cap binding protein during translation initiation. *J. Virol.* **62:** 2059.

Dominguez D.I., Ryabova L.A., Pooggin M.M., Schmidt-Puchta W., Futterer J., and Hohn T. 1998. Ribosome shunting in cauliflower mosaic virus. Identification of an essential and sufficient structural element. *J. Biol. Chem.* **273:** 3669.

Feigenblum D. and Schneider R.J. 1996. Cap-binding protein (eukaryotic initiation factor 4E) and 4E-inactivating protein BP-1 independently regulate cap-dependent translation. *Mol. Cell. Biol.* **16:** 5450.

Feigenblum D., Walker R., and Schneider R.J. 1998. Adenovirus induction of an interferon-regulatory factor during entry into the late phase of infection. *J. Virol.* **72:** 9257.

Fukunaga R. and Hunter T. 1997. MNK1, a new MAP kinase-activated protein kinase, isolated by a novel expression screening method for identifying protein kinase substrates. *EMBO J.* **16:** 1921.

Futterer J., Kiss-Laszlo Z., and Hohn T. 1993. Nonlinear ribosome migration on cauliflower mosaic virus 35S RNA. *Cell* **73:** 789.

Gingras A. C. and Sonenberg N. 1997. Adenovirus infection inactivates the translational inhibitors 4E-BP1 and 4E-BP2. *Virology* **237:** 182.

Gingras A.-C., Raught B., and Sonenberg N. 1999. eIF4 initiation factors: Effectors of mRNA recruitment to ribosomes and regulators of translation. *Annu. Rev. Biochem.* **68:** 913.

Green R. and Noller H.F. 1997. Ribosomes and translation. *Annu. Rev. Biochem.* **66:** 679.

Hayes B.W., Telling G.C., Myat M.M., Williams J.F., and Flint S.J. 1990. The adenovirus L4 100 kilodalton protein is necessary for efficient translation of viral late mRNA species. *J. Virol.* **64:** 2732.

Hemmings-Mieszczak M., Steger G., and Hohn T. 1997. Alternative structures of the cauliflower mosaic virus 35 S RNA leader: Implications for viral expression and replication. *J. Mol. Biol.* **267:** 1075.

———. 1998. Regulation of CaMV 35 S RNA translation is mediated by a stable hairpin in the leader. *RNA* **4:** 101.

Hohn T., Dominguez D.I., Scharer-Hernandez N., Pooggin M.M., Schmidt-Puchta W., Hemmings-Mieszczak M., and Futterer J. 1998. Ribosome shunting in eukaryotes: What the viruses tell me. In *A look beyond transcription: Mechanisms determining mRNA stability and translation in plants* (ed. J. Bailey-Serres and D.R. Gallie), p. 84. American Society of Plant Physiologists, Rockville, Maryland.

Huang J. and Schneider R.J. 1990. Adenovirus inhibition of cellular protein synthesis is prevented by the drug 2-aminopurine. *Proc. Natl. Acad. Sci.* **87:** 7115.

———. 1991. Adenovirus inhibition of cellular protein synthesis involves inactivation of cap binding protein. *Cell* **65:** 271.

Imataka H., Gradi A., and Sonenberg N. 1998. A newly identified N-terminal amino acid sequence of human eIF4G binds poly(A)-binding protein and functions in poly(A)-dependent translation. *EMBO J.* **17:** 7480.

Jang S.K., Pestova T.V., Hellen C.U.T., Witherall G.W., and Wimmer E. 1990. Cap-independent translation of picornaviral RNAs: Structure and function of internal ribosome entry site. *Enzyme* **44:** 292.

Latorre P., Kolakofsky D., and Curran J. 1998. Sendai virus Y proteins are initiated by a ribosomal shunt. *Mol. Cell. Biol.* **18:** 5021.

Logan J. and Shenk T. 1984. Adenovirus tripartite leader sequence enhances translation of mRNAs late after infection. *Proc. Natl. Acad. Sci.* **81:** 3655.

Marcotrigiano J., Gingras A.C., Sonenberg N., and Burley S.K. 1997. Cocrystal structure of the messenger RNA 5′ cap-binding protein (eIF4E) bound to 7-methyl-GDP. *Cell* **89:** 951.

Moore M., Schaack J., Baim S.B., Morimoto R.I., and Shenk T. 1987. Induced heat shock mRNAs escape the nucleocytoplasmic transport block in adenovirus infected Hela cells. *Mol. Cell. Biol.* **7:** 4505.

Moore P.B. 1996. In *Ribosomal RNA: Structure, evolution, processing, and function in protein biosynthesis* (ed. R.A. Zimmermann and A.E. Dahlberg), p. 199. CRC Press, Boca Raton, Florida.

Oosterom-Dragon E.A. and Ginsberg H.S. 1981. Characterization of two temperature sensitive mutants of type 5 adenovirus with mutations in the 100,000k dalton protein gene. *J. Virol.* **40:** 491.

Pooggin M.M., Hohn T., and Futterer J. 1998. Forced evolution reveals the importance of short open reading frame A and secondary structure in the cauliflower mosaic virus 35S RNA leader. *J. Virol.* **72:** 4157.

Pyronnet S., Imataka H., Gingras A.C., Fukunaga R., Hunter T., and Sonenberg N. 1999. Human eukaryotic translation initiation factor 4G (eIF4G) recruits Mnk1 to phosphorylate eIF4E. *EMBO J.* **18:** 270.

Remm M., Remm A., and Ustav M. 1999. Human papillomavirus type 18 E1 protein is translated from polycistronic mRNA by a discontinuous scanning mechanism. *J. Virol.* **73:** 3062.

Schneider R.J. 2000. Adenovirus inhibition of cellular protein synthesis and preferential translation of viral mRNAs. In *Translational control of gene expression* (ed. N. Sonenberg et al.), p. 901. Cold Spring Harbor Laboratory Press, Cold Spring Harbor, New York.

Tarun S.Z., Wells S.E., and Sachs A.B. 1997. Translation initiation factor eIF-4G mediates in vitro polyA tail dependent translation. *Proc. Natl. Acad. Sci.* **94:** 9046.

Thimmappaya B., Weinberger C., Schneider R.J., and Shenk T. 1982. Adenovirus VA1 RNA is required for efficient translation of viral mRNA at late times after infection. *Cell* **31:** 543.

Thomas A.M., Scheper G.C., Kleijn M., DeBoer M., and Voorma H.O. 1992. Dependence of the adenovirus tripartite leader on the p220 subunit of eukaryotic initation factor 4F during in vitro translation. *Eur. J. Biochem.* **207:** 471.

Waskiewicz A.J., Flynn A., Proud C.G., and Cooper J.A. 1997. Mitogen-activated protein kinases activate the serine/threo-nine kinases Mnk1 and Mnk2. *EMBO J.* **16:** 1909.

Waskiewicz A.J., Johnson J.C., Penn B., Mahalingham M., Kimball S.R., and Cooper J.A. 1999. Phosphorylation of the cap-binding protein eukaryotic translation initiation factor 4E by protein kinase Mnk1 in vivo. *Mol. Cell. Biol.* **19:** 1871.

Wells S.E., Hillner P.E., Vale R.D., and Sachs A.B. 1998. Circularization of mRNA by eukaryotic translation initiation factors. *Mol. Cell* **2:** 135.

Yueh A. and Schneider R.J. 1996. Selective translation by ribosome jumping in adenovirus infected and heat shocked cells. *Genes Dev.* **10:** 1557.

———. 2000. Translation by ribosome shunting on adenovirus and Hsp70 mRNAs facilitated by complementarity to 18S rRNA. *Genes Dev.* **14:** 414.

Zhang Y., Dolph P.J., and Schneider R.J. 1989. Secondary structure analysis of adenovirus tripartite leader. *J. Biol. Chem.* **264:** 10679.

Zhang Y., Feigenblum D., and Schneider R.J. 1994. A late adenovirus factor induces eIF-4E dephosphorylation and inhibition of cell protein synthesis. *J. Virol.* **68:** 7040.

Shunting and Controlled Reinitiation: The Encounter of Cauliflower Mosaic Virus with the Translational Machinery

T. Hohn, H.-S. Park, O. Guerra-Peraza, L. Stavolone, M.M. Pooggin, K. Kobayashi, and L.A. Ryabova

Friedrich-Miescher-Institute, CH-4002 Basel, Switzerland

CaMV belongs to the caulimoviridae, the only group of complex plant pararetroviruses. The replication of caulimoviridae is closely related to that of the animal retroviruses; however, the exception is that their genome accumulates in the host nucleus as an episome rather than as an integrate and that the extracellular virions contain DNA rather than RNA. The genomes of CaMV and other members of caulimoviridae (Fig. 1) contain a module including a Zn-finger-containing capsid (GAG)-, a protease (PR)-, and a reverse transcriptase/RNase H (RT/RH) gene, related to a similar module in retroviruses and retrotransposons. An additional module includes genes typical for plant RNA viruses responsible for cell-to-cell movement (MOV) and insect transmission (for CaMV: aphid transmission, ATF) functions. CaMV also codes for an interesting multifunctional protein, *trans*-activator/viroplasmin (TAV), involved in sublocalization of virus production within viroplasm and translational control.

Viruses with an RNA-based replicative intermediate require signals for replication, which are usually based on primary and secondary structure. Especially when located in a leader or coding region, those structures are often inhibitory to the classic scanning or the translation process, per se. On the other hand, if scanned and translated, those structures will be melted and covered by ribosomes and temporarily lose their specific function. In order to preserve functional replication signals, viruses often use alternative translation strategies, such as internal ribosome entry and shunting.

Most complex retroviruses use alternative splicing and polyprotein processing to produce all viral proteins from one original transcript. CaMV and other plant pararetroviruses such as rice tungro bacilliform virus (RTBV) use alternative splicing and polyprotein processing, too, but CaMV has in addition a second promoter producing a subgenomic mRNA (19S RNA). However, in spite of this, the viral ORFs still outnumber the viral mRNAs, leading to the suggestion that polycistronic translation is involved. In the case of RTBV, the first three ORFs are translated by a leaky scanning mechanism (Fütterer et al. 1997). In CaMV, TAV promotes, for eukaryotes, an unusual reinitiation mechanism.

SHUNTING

The CaMV leader is 600 nucleotides long. It was predicted by computer analysis (Fütterer et al. 1988) and shown by chemical and enzymatic modification analysis (Hemmings-Mieszczak et al. 1997) that its central 480 nucleotides form an energy-rich bulged hairpin (Fig. 2). One can distinguish three stem sections, named 1 to 3 from base to top, with stem section 1 being the strongest section. The leader contains 6 to 7 small open reading frames (sORFs) ranging in size from 2 to 34 codons. The first of these, sORF A, precedes the central hairpin, whereas the others are fully or partially (sORF F) located within the hairpin. It can be predicted that such a leader is very inhibitory to translation, especially because the last of these sORFs (sORF F) overlaps the first large ORF. Baughman and Howell (1988) showed that indeed a reporter gene attached to the leader is poorly translated in transfected plant protoplasts. Fütterer et al. (1989) confirmed this result but pointed out that the remaining translation level is still remarkable. They then showed by deletion analysis that the leader consists of a mosaic of enhancing and inhibiting elements (the flanking regions and the central hairpin, respectively) and suggested that the inhibiting elements are bypassed by "shunting" ribosomes (Fütterer et al. 1990).

A series of transient expression experiments was performed to foster the shunt model (Fig. 3) (Fütterer et al. 1993). The stem structure used by Kozak (1986; "Kozak stem") as a tool to study the scanning process inhibited translation if positioned close to the cap site of the original or extended CaMV leader. This showed that a cap, or at least the 5'-end of the leader, is involved in translation initiation and excludes internal ribosome entry, as shown for picornavirus and other RNA viruses (Jackson 2000). Insertion of the Kozak stem into the flanking region of the leader strongly inhibited translation, whereas insertion into the central part had little effect. This showed that the flanking parts are scanned, while the central stem structure is bypassed (shunted), and ruled out a leaky scanning mechanism. Furthermore, a dicistronic RNA was constructed, whereby a GUS-reporter gene was inserted into

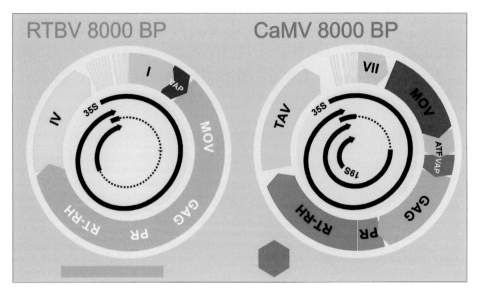

Figure 1. Maps of caulimoviridae: rice tungro bacilliform virus (RTBV) and cauliflower mosaic virus (CaMV).

the central portion of the hairpin and a CAT reporter at the 3′end of the leader. Both ORFs were translated, showing that a subpopulation of ribosomes scans into the central hairpin to reach GUS, while a second subpopulation shunts the hairpin to translate CAT. Finally, shunting was achieved by providing the leader in two parts that could form an interrupted leader by base-pairing. Only both parts together allowed reporter gene expression, finally excluding leaky scanning, since any ribosome scanning into the central hairpin would fall off at the "internal" poly(A) tail, if it got that far.

The CaMV leader also contains a splice donor with a corresponding splice acceptor site within ORF II (Kiss-

Lázló et al. 1995), and splicing out of all inhibitory regions within the leader was a scenario that had to be seriously considered. A construct was therefore made mutating all possible (classic) shunt acceptor consensus sequences within sORF F and further downstream parts of the leader. This had no effect on expression of the reporter gene, showing that classic splicing within the leader did not occur. However, atypical splicing (Tarn and Steitz 1996) could not be fully excluded until an in vitro translation system for shunting was established (Schmidt-Puchta et al. 1997). In this case, the reporter RNA including the CaMV leader was produced in vitro, purified, and translated. Results with this system correlated to what had been observed in the transient expression systems and excluded splicing without any doubt. The in vitro system also facilitated the study of the effect of antisense oligonucleotides directed against various parts of the leader. Antisense oligonucleotides directed against the 5′- and 3′-terminal parts of the leader strongly inhibited translation, whereas antisense oligonucleotides directed against the central part had little effect, again indicating that the central part of the leader is not scanned but shunted (Schmidt-Puchta et al. 1997).

The in vitro translation system also allowed a more finely tuned observation of scanning and shunting details. A new reporter construct was produced by a one-base deletion mutation by which the originally overlapping sORF F and the CAT reporter were fused in-frame. CAT could now be measured not only by its enzymatic activity, but also as a radioactive band in SDS gels, and its mobility indicated whether translation of the fused version started at either the sORF F AUG codon, at AUG- and non-AUG start codons within the sORF region, or at the CAT AUG (Ryabova and Hohn 2000). In fact, most translation events were initiated at the CAT AUG, i.e., within the fusion ORF, very few were initiated at the sORF F AUG and a substantial number were initiated at

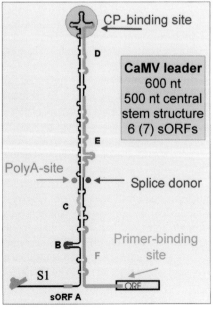

Figure 2. Leader of CaMV RNA.

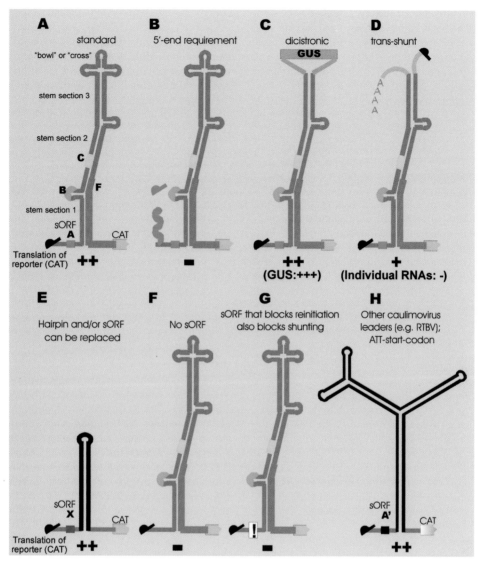

Figure 3. Schematic representation of experiments proving and characterizing shunting. Experiments were performed in transfected plant protoplasts and/or in vitro. (*A*) CaMV leader with the hairpin structure, and some of the sORFs attached to the CAT reporter gene, shown schematically. (*B*) A cap is required; a strong stem near the cap inhibits translation. This argues against internal initiation. (*C*) Strong additional stem structures (not shown) or large ORFs (shown) can be inserted into the central regions of the stem structure without inhibiting shunting of the original reporter gene. This argues against complete scanning of the leader for translation of the second ORF. However, the central ORF is also translated, in this case after scanning. Consequently, the setup allows dicistronic translation. (*D*) Trans-shunting occurs when the leader is provided in two parts. This excludes a purely scanning-based mechanism. (*E*) The CaMV central hairpin can be replaced by other strong stem structures. The sORF can also be replaced with almost any other sORF. The position of the sORF controls both shunting efficiency and the landing site position (not shown). (*F*) An sORF is essential for efficient shunting. However, sORFs starting with non-AUG start codons can also be used (not shown). (*G*) sORFs that block ribosome release also block shunting. This indicates that shunting in CaMV is a modified reinitiation process. (*H*) Other caulimoviridae have adopted a shunting mechanism similar to that of CaMV. In the case of RTBV, a non-AUG start codon is used to initiate the first ORF. In general, an increased efficiency of non-AUG start codon usage is observed after shunting.

three different non-AUG start codons located a few bases downstream from the stem, a region that had been mapped as shunt landing site before (Fütterer et al. 1993). Interestingly, the first ORF of the rice tungro bacilliform virus (RTBV), another member of caulimoviridae using the shunting strategy (see below), starts efficiently with a position-dependent ATT codon located a few bases downstream of the RTBV stem structure (Fütterer et al. 1996). These results indicate that upon landing, the shunt-

ing ribosomes are deficient in factors responsible for correct AUG start codon recognition. A similar behavior was observed in certain cases of internal ribosome entry (Jackson 2000).

The in vitro translation system helped to confirm that the central stem is not only the obstacle that is bypassed by shunting, but also an essential element to promote shunting (Fig. 3) (Dominguez et al. 1998; Hemmings-Mieszczak et al. 1998). The primary sequence of CaMV stem section 1 turned out not to be essential. This stem section could be replaced by stem section 2, by inverted stem section 1, or even by the above-mentioned Kozak stem (Dominguez et al. 1998; Hemmings-Mieszczak and Hohn 1999; Pooggin et al. 2000; Ryabova and Hohn 2000; Ryabova et al. 2000).

Another essential element in the caulimovirus shunting system is sORF A, located ~6 nucleotides in front of the base of the stem (Dominguez et al. 1998; Hemmings-Mieszczak and Hohn 1999). This location allows for some melting of the stem by the translating ribosome. After this melting, the ribosome is in direct contact with a base pair of the stem with the distance to the other site of the stem in fact being a "cat-leap." If the number of nucleotides located between sORF stop codon and stem is reduced, a corresponding number of additional stem base pairs is melted and, as a consequence, the shunt landing site is shifted to a position farther upstream (Ryabova and Hohn 2000). In contrast, with the increase of the number of nucleotides located between sORF stop codon and stem, the landing site is moved farther downstream and the shunting efficiency is strongly diminished.

Shunting is usually not much affected if the sequence of sORF A is altered and when it is amino-terminally elongated to about 30 codons. Longer versions are very inefficient. Generally, the sequence of the sORF has little effect (Pooggin et al. 2001), with the exception of negatively regulatory sORFs such as the AdoMetDC sORF (Hill and Morris 1992; Ryabova and Hohn 2000) or the yeast GCN4 sORF4 (Hinnebusch 1997; Hemmings-Mieszczak et al. 2000; Poogin et al. 2000). The fact that an sORF, which inhibits reinitiation, also inhibits shunting strongly suggests that shunting involves a reinitiation process. It therefore is not surprising that the CaMV TAV protein, which allows reinitiation, also enhances shunting (Pooggin et al. 2000, 2001).

The importance of sORF A in CaMV shunting was not immediately obvious for two reasons. (1) Certain mutations of the sORF A AUG start codon still allowed shunting at the original position because several non-AUG start codons are located in front of and in frame with CaMV sORF A (Dominguez et al. 1998; Ryabova et al. 2000). These can still be used with relatively high efficiency due to the stem structure located downstream of them. Kozak (1990) had shown that downstream secondary structures facilitate recognition of initiator codons by eukaryotic ribosomes. (2) The sORF/stem arrangement is redundant; i.e., a second (Ryabova and Hohn 2000) and maybe even a third sORF/stem arrangement (with sORF B and C, respectively) can be used for shunting.

The sORF A/stem arrangement is, with variations, conserved in most of the caulimoviridae (Pooggin et al. 1999). An exception seems to be the soybean chlorotic mottle virus class, such as the *Cestrum* yellow leaf curling virus (CmYLCV). In this case, a strong stem is positioned directly after the cap, resulting in a strong inhibition of translation (L. Stavolone et al., in prep.). We believe that the caulimoviridae strong stem structure has evolved as an element resisting melting by scanning ribosomes in order to preserve the structure of important *cis*-

functions, e.g., for RNA dimerization, packaging (Guerra-Peraza et al. 2000), and reverse transcription (Pfeiffer and Hohn 1983). In the case of CmYLCV, the leader is either melted and translation occurs, or it is not melted and then used for replication. In the case of the other caulimoviruses, the same molecule can be translated (by shunting) and become sequestered for replication, since the secondary structure of the shunted region is preserved.

The importance of sORF A was further recognized by testing the infectivity of CaMV with mutations in sORF A. In this case, symptoms appeared with much delay, and progeny virus had reverted at first and second sites to reestablish an sORF at the position of the original ORF A. In contrast, mutations of the other sORFs had little effect on infectivity and reverted only slowly, if at all (Pooggin et al. 1998, 2001).

CaMV-related shunting is not confined to protoplast- and wheat germ-translation systems; it also functions in infected plants (Pooggin et al. 2001) and in transgenic plants (Schärer-Hernandez and Hohn 1998), and it is not confined to plant systems, since it also works in the rabbit reticulocyte in vitro translation systems (Ryabova and Hohn 2000) and yeast (Hemmings-Mieszczak et al. 2000).

A type of shunting was also observed in various animal viruses, i.e., Sendai virus (Curran and Kolakofsky 1989; Latorre at al. 1998), budgerigar fledgling disease virus (Li 1996), adenovirus tripartite leader (Yueh and Schneider 1996), and perhaps also in porcine transmissible gastroenteritis coronavirus (TGEV; O'Connor and Brian 2000). It was also suggested for mammalian heat shock protein hsp70 mRNAs (Yueh and Schneider 2000). Shunting in adenovirus and hsp70 mRNAs seems to function by a different mechanism, since an sORF is not involved. Instead, it is facilitated by complementarity of the leader to 18S rRNA (Yueh and Schneider 2000). A direct comparison in our hands showed that, placing either an sORF or the adenovirus-derived 18S complementary region in front of a stem structure has a similar effect on shunting (L.A. Ryabova et al., unpubl.).

Based on the accumulated data on shunting, search criteria can be defined to detect CaMV-like shunting events in eukaryotic mRNAs once databases based on careful mRNA mapping data that can identify the precise start of transcript leaders become available. Such search criteria would be the presence of an sORF closely followed by a strong hairpin, with this arrangement being flanked by relatively unstructured sequences.

REGULATED POLYCISTRONIC TRANSLATION

Historically, three observations indicated that CaMV employs polycistronic translation to express most of its genes from the full-size pregenomic RNA:

1. There is a lack of subgenomic mRNAs. Only one of the viral proteins, the transactivator/viroplasmin (TAV), encoded by ORF VI, is translated from a subgenomic RNA (19S RNA; Odell et al. 1981), the

other six ORFs (ORFs VII, and I–V) are translated from the full-length or a spliced version of the pre-genomic RNA; splicing between a donor within the leader and an acceptor in ORF II yields a subgenomic mRNA for ORFs III–V (Kiss-László et al. 1995).

2. ORFs I–V of CaMV and related viruses are tightly arranged, thus there is no space for promoters or splice signals between the ORFs. Although promoters could overlap coding regions (in fact some of the 35S enhancer regions overlap ORF VI) and splice sites could be located within ORFs, the fact that viral ORFs often overlap by only a single base suggests a role for a translational mechanism linking their expression.

3. Some mutations within ORFs VII and II are polar (Dixon and Hohn 1984; Gronenborn 1987); i.e., they affect expression of the following ORF. Such polar mutants are typical of polycistronic, usually prokaryotic, mRNAs.

To confirm that polycistronic translation of CaMV RNA occurs, dicistronic reporter constructs were expressed in transfected protoplasts (Fig. 4). Such constructs, coding for an RNA consisting of ORF VII and an ORF I::CAT fusion, yielded very little CAT activity if expressed alone or together with any single CaMV ORF from I to V. However, if expressed together with ORF VI (TAV), high levels of CAT activity were obtained; this property of ORF VI was named *trans*-activator. The level of CAT produced depended on the amount of TAV available (Bonneville et al. 1989). Similar results were obtained with RNAs more virus-like in nature from CaMV (Scholthof et al. 1992), figwort mosaic virus (Gowda et al. 1989), and peanut chlorotic streak virus (Maiti et al. 1998). TAV as *trans*-activator has also been observed in yeast (Sha et al. 1995) and in transgenic plants (Zjilstra

and Hohn 1992). In the latter case, plants transgenic for a dicistronic GUS reporter gene were either crossed with plants transgenic for TAV or infected with CaMV. In both cases, activation of GUS activity was observed; in the latter case, the progressive infection of the plant could in fact be followed by observing the extent of GUS staining.

To elucidate the mechanism of TAV action, artificial mRNAs consisting of two reporter ORFs were tested (Fütterer and Hohn 1991, 1992). Again, TAV promoted second cistron translation, showing that specific CaMV primary sequences are not required. Interestingly, however, an sORF present in the leader (as occurs in the natural CaMV 35S RNA) strongly enhanced second cistron translation. Furthermore, insertion of the Kozak stem into the leader, or between the first and second reporter ORFs, strongly inhibited second cistron translation, indicating that TAV promotes reinitiation events.

TAV, as viroplasmin, is the main component of the cytoplasmic inclusion bodies or viroplasm of most caulimoviruses, and these are, especially in early phases of infection, surrounded by masses of polysomes as seen by electron microscopy (Fig. 5) (Kitajima et al. 1969; Lawson and Hearon 1974; Furusawa et al. 1980). Inclusion bodies are also the site of virus protein accumulation, including proteins produced from payload ORFs introduced into CaMV, such as human α-interferon (De Zoeten et al. 1989); of virus assembly (Shepherd et al. 1979); and of reverse transcription (Pfeiffer and Hohn 1983). This suggests that TAV is not only promoting polycistronic CaMV translation, but is also involved in sequestering the translation products and fostering their correct assembly. The observation that TAV is required, in addition to POL and GAG, for virus proliferation in single cells (K. Kobayashi and T. Hohn, in prep.), supports this notion.

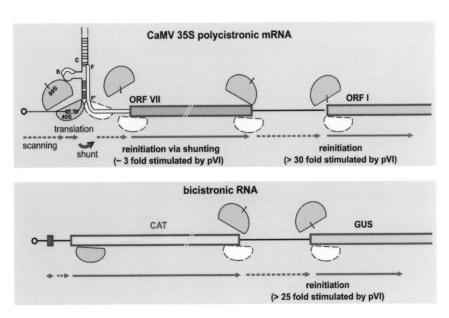

Figure 4. The CaMV *trans*-activator/viroplasmin (TAV) enhances reinitiation after small and large ORFs.

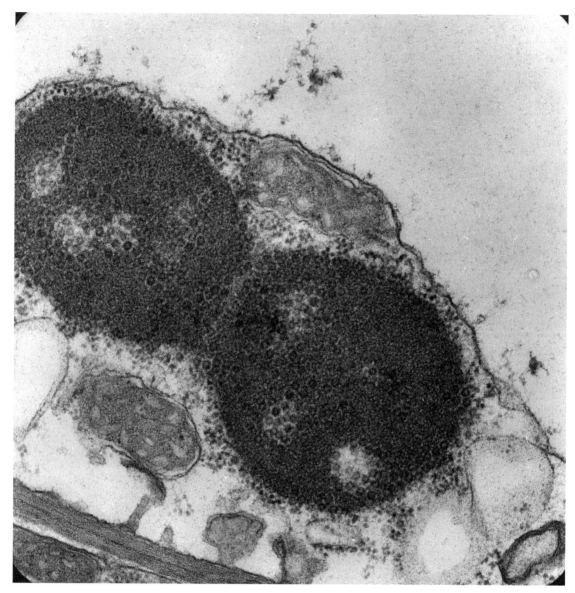

Figure 5. CaMV viroplasm ("inclusion bodies") inside an infected cell. The viroplasm includes virus particles and is surrounded by polysomes.

TAV is a multifunctional protein and as such interacts with a variety of molecules, including the CaMV GAG protein (indicating a role in virus assembly; Himmelbach et al. 1996), single-stranded RNA and DNA (De Tapia et al. 1993), and polysomes (Himmelbach 1995), well in line with the electron microscopy data mentioned in the previous paragraph. The latter interaction might have been due simply to the affinity of TAV for RNA; however, GST pull-down experiments revealed that TAV interacts specifically with eIF3 and with the 60S, but not the 40S ribosome (Park et al. 2001). In accordance with these findings, yeast two-hybrid experiments revealed both eIF3g and 60S ribosome protein L24 as TAV-interaction partners. An additional 60S ribosomal protein, L18, has also been found to interact with TAV using gel overlay technology (Leh et al. 2000). eIF3g has a strong negative effect, and L24 has a strong positive effect on TAV-me-diated polycistronic translation in transfected protoplasts (Park et al. 2001).

As mentioned above, TAV accumulates at a subcellular location, the viroplasm. To get there, an active transport function might be involved. This transport could also target TAV as it is complexed with the translational machinery and CaMV RNA, and this CaMV RNA could be brought to the viroplasm together with TAV. This then finally allows for the seclusion of virus production at and within the viroplasm.

ACKNOWLEDGMENTS

We thank all our past and present colleagues who contributed to the present status of our work through discussions, advice, and constructive criticism. We greatly appreciate the critical reading by Helen Rothnie and the expert technical assistance of Sandra Corsten, Matthias Müller, and David Kirk. We acknowledge financial support from the Novartis Research Foundation, the Roche Research Foundation, Eidgeössische Stipendienaktion für ausländische Studierende, the EMBO, FEBS, and IN-TAS.

REFERENCES

Baughman G. and Howell S.H. 1988. Cauliflower mosaic virus 35 S RNA leader region inhibits translation of downstream genes. *Virology* **167:** 125.

Bonneville J.M., Sanfaçon H., Fütterer J., and Hohn T. 1989. Posttranscriptional *trans*-activation in cauliflower mosaic virus. *Cell* **59:** 1135.

Curran J. and Kolakofsky D. 1989. Scanning independent ribosomal initiation of the Sendai virus Y proteins in vitro and in vivo. *EMBO J.* **8:** 521.

De Tapia M., Himmelbach A., and Hohn T. 1993. Molecular dissection of the cauliflower mosaic virus translation transactivator. *EMBO J.* **12:** 3305.

De Zoeten G.A., Penswick J.R., Horisberger M.A., Ahl P., Schultze M., and Hohn T. 1989. The expression, localization, and effect of a human interferon in plants. *Virology* **172:** 213.

Dixon L.K. and Hohn T. 1984. Initiation of translation of the cauliflower mosaic virus genome from a polycistronic mRNA: Evidence from deletion mutagenesis. *EMBO J.* **3:** 2731.

Dominguez D.I., Ryabova L.A., Pooggin M.M., Schmidt-Puchta W., Fütterer J., and Hohn T. 1998. Ribosome shunting in cauliflower mosaic virus: Identification of an essential and sufficient structural element. *J. Biol. Chem.* **273:** 3669.

Furusawa I., Yamaoka N., Okuno T., Yamamoto M., Kohno M., and Kunoh H. 1980. Infection of turnip protoplasts with cauliflower mosaic virus. *J. Gen. Virol.* **48:** 431.

Fütterer J. and Hohn T. 1991. Translation of a polycistronic mRNA in the presence of the cauliflower mosaic virus transactivator protein. *EMBO J.* **10:** 3887.

———. 1992. Role of an upstream open reading frame in the translation of polycistronic mRNA in plant cells. *Nucleic Acids Res.* **20:** 3851.

Fütterer J., Kiss-Lászlo Z., and Hohn T. 1993. Nonlinear ribosome migration on cauliflower mosaic virus 35S RNA. *Cell* **73:** 789.

Fütterer J., Rothnie H.M., Hohn T., and Potrykus I. 1997. Rice tungro bacilliform virus open reading frames II and III are translated from polycistronic pregenomic RNA by leaky scanning. *J. Virol.* **71:** 7984.

Fütterer J., Gordon K., Sanfaçon H., Bonneville J.M., and Hohn T. 1990. Positive and negative control of translation by the leader sequence of cauliflower mosaic virus pregenomic 35S RNA. *EMBO J.* **9:** 1697.

Fütterer J., Gordon K., Bonneville J.M., Sanfaçon H., Pisan B., Penswick J., and Hohn T. 1988. The leading sequence of caulimovirus large RNA can be folded into a large stem-loop structure. *Nucleic Acids Res.* **16:** 8377.

Fütterer J., Gordon K., Pfeiffer P., Sanfaçon H., Pisan B., Bonneville J.M., and Hohn T. 1989. Differential inhibition of downstream gene expression by the CaMV 35S RNA leader. *Virus Genes* **3:** 45.

Fütterer J., Potrykus I., Bao Y., Li L., Burns T.M., Hull R., and Hohn T. 1996. Position dependent ATT translation initiation in gene expression of the plant pararetrovirus rice tungro bacilliform virus. *J. Virol.* **70:** 2999.

Gronenborn B. 1987. The molecular biology of cauliflower mosaic virus and its application as plant vector. *Plant Gene Res.* **4:** 1.

Gowda S., Wu F.C., Scholthof H.B., and Shepherd R.J. 1989. Gene VI of figwort mosaic virus (caulimovirus group) functions in posttranscriptional expression of genes on the full-length RNA transcript. *Proc. Natl. Acad. Sci.* **86:** 9203.

Guerra-Peraza O., DeTapia M., Hohn T., and Hemmings-Mieszczak M. 2000. Interaction of the cauliflower mosaic virus coat protein with the pregenomic RNA leader. *J. Virol.* **74:** 2067.

Hemmings-Mieszczak M., and Hohn T. 1999. A stable hairpin preceded by a short open reading frame promotes nonlinear ribosome migration on a synthetic mRNA leader. *RNA* **5:** 1149.

Hemmings-Mieszczak M., Hohn T., and Preiss T. 2000. Termination and peptide release at the upstream open reading frame are required for downstream translation on synthetic shunt-competent mRNA leaders. *Mol. Cell. Biol.* **20:** 6212.

Hemmings-Mieszczak M., Steger G., and Hohn T. 1997. Alternative structures of the cauliflower mosaic virus 35 S RNA leader: Implications for viral expression and replication. *J. Mol. Biol.* **267:** 1075.

———. 1998. Regulation of CaMV 35S RNA translation is mediated by a stable hairpin in the leader. *RNA* **4:** 101.

Hill J.R. and Morris D.R. 1992. Cell-specific translation of S-adenosylmethionine decarboxylase mRNA. *J. Biol. Chem.* **267:** 21886.

Himmelbach A. 1995. "Cauliflower mosaic virus ORF VI protein: Interactions with the translational machinery and the ORF IV protein." Ph. D. thesis. University of Basel, Switzerland.

Himmelbach A., Chapdelaine Y., and Hohn T. 1996. Interaction between cauliflower mosaic virus inclusion body protein and capsid protein: Implications for viral assembly. *Virology* **217:** 147.

Hinnebusch A.G. 1997. Translational regulation of yeast GCN4. A window on factors that control initiator-tRNA binding to the ribosome. *J. Biol. Chem.* **272:** 21661.

Jackson R.J. 2000. A comparative view of initiation site selection mechanisms. In *Translational control of gene expression* (ed. N. Sonenberg et al.), p. 127. Cold Spring Harbor Laboratory Press, Cold Spring Harbor, New York.

Kiss-Lászlo Z., Blanc S., and Hohn T. 1995. Splicing of cauliflower mosaic virus 35S RNA is essential for viral infectivity. *EMBO J.* **14:** 3552.

Kitajima E.W., Lauritis J.A., and Swift H. 1969. Fine structure of zinnial leaf tissues infected with dahlia mosaic virus. *Virology* **39:** 240.

Kozak M. 1986. Influence of mRNA secondary structure on initiation by eukaryotic ribosomes. *Proc. Natl. Acad. Sci.* **83:** 2850.

———. 1990. Downstream secondary structure facilitates recognition of initiator codons by eukaryotic ribosomes. *Proc. Natl. Acad. Sci.* **87:** 8301.

Latorre P., Kolakofsky D., and Curran J. 1998. Sendai virus Y proteins are initiated by a ribosomal shunt. *Mol. Cell. Biol.* **18:** 5021.

Lawson R.H. and Hearon S.S. 1974. Ultrastructure of carnation etched ring virus-infected *Saponaria vaccaria* and *Dianthus caryophyllis*. *J. Ultrastruct. Res.* **48:** 201.

Leh V., Yot P., and Keller M. 2000. The cauliflower mosaic virus translational transactivator interacts with the 60S ribosomal subunit protein L18 of *Arabidopsis thaliana*. *Virology* **266:** 1.

Li J. 1996. "Molecular analysis of late gene expression in budgerigar fledgling disease virus." University of Giessen, Germany.

Maiti I.B., Richins R.D., and Shepherd R.J. 1998. Gene expression regulated by gene VI of caulimovirus: Transactivation of downstream genes of transcripts by gene VI of peanut chlorotic streak virus in transgenic tobacco. *Virus Res.* **57:** 113.

O'Connor J.B. and Brian D.A. 2000. Downstream ribosomal entry for translation of coronavirus TGEV gene 3b. *Virology* **269:** 172.

Odell J.T., Dudley R.K., and Howell S.H. 1981. Structure of the 19 S RNA transcript encoded by the cauliflower mosaic virus genome. *Virology* **111:** 377.

Park H.-S., Hohn T., and Ryabova L.A. 2001. A plant viral "Reinitiation" factor interacts with the host transcriptional machinery. *Cell* **106:** 723.

Pfeiffer P. and Hohn T. 1983. Involvement of reverse transcription in the replication of cauliflower mosaic virus: A detailed model and test of some aspects. *Cell* **33:** 781.

Pooggin M.M., Hohn T., and Fütterer J. 1998. Forced evolution reveals the importance of short open reading frame A and secondary structure in the cauliflower mosaic virus 35S RNA leader. *J. Virol.* **72:** 4157.

Pooggin M.M., Fütterer J., Skryabin K.G., and Hohn T. 1999. A short open reading frame terminating in front of a stable hairpin is the conserved feature in pregenomic RNA leaders of plant pararetroviruses. *J. Gen. Virol.* **80:** 2217.

———. 2000. Translation of a short ORF in ribosome shunt on the CaMV RNA leader. *J. Biol. Chem.* **275:** 17288.

———. 2001. Ribosome shunt is essential for infectivity of cauliflower mosac virus. *Proc. Natl. Acad. Sci.* **98:** 886.

Ryabova L.A. and Hohn T. 2000. Ribosome shunting in cauliflower mosaic virus 35S RNA leader is a special case of reinitiation of translation functioning in plant and animal systems. *Genes Dev.* **14:** 817.

Ryabova L.A., Pooggin M.M., Dominguez D.I., and Hohn T. 2000. Continuous and discontinuous ribosome scanning on the cauliflower mosaic virus 35S RNA leader is controlled by sORFs. *J. Biol. Chem.* **275:** 37278.

Schärer-Hernandez N. and Hohn T. 1998. Nonlinear ribosome migration on cauliflower mosaic virus 35S RNA in transgenic tobacco plants. *Virology* **242:** 403.

Schmidt-Puchta W., Dominguez D.I., Lewetag D., and Hohn T.

1997. Plant ribosome shunting in vitro. *Nucleic Acids Res.* **25:** 2854.

Scholthof H.B., Gowda S., Wu F.C., and Shepherd R.J. 1992. The full-length transcript of a caulimovirus is a polycistronic mRNA whose genes are trans activated by the product of gene VI. *J. Virol.* **66:** 3131.

Sha Y., Broglio E.P., Cannon J.F., and Schoelz J.E. 1995. Expression of a plant viral polycistronic mRNA in yeast, *Saccharomyces cerevisiae*, mediated by a plant virus translational transactivator. *Proc. Natl. Acad. Sci.* **92:** 8911.

Shepherd R.J., Richins R.D., and Shalla T.A. 1979. Isolation and properties of the inclusion bodies of cauliflower mosaic virus. *Virology* **102:** 389.

Tarn W.Y. and Steitz J.A. 1996. A novel spliceosome containing U11, U12, and U5 snRNPs excises a minor class (AT-AC) intron in vitro. *Cell* **84:** 801.

Yueh A. and Schneider R.J. 1996. Selective translation initiation by ribosome jumping in adenovirus-infected and heat-shocked cells. *Genes Dev.* **10:** 1557.

———. 2000. Translation by ribosome shunting on adenovirus and hsp70 mRNAs facilitated by complementarity to 18S rRNA. *Genes Dev.* **14:** 414.

Zijlstra C. and Hohn T. 1992. Cauliflower mosaic virus gene VI controls translation from dicistronic expression units in transgenic *Arabidopsis* plants. *Plant Cell* **4:** 1471.

Mechanisms of Internal Ribosome Entry in Translation Initiation

J.S. KIEFT,* A. GRECH, P. ADAMS, AND J.A. DOUDNA*

*Deptartment of Molecular Biophysics and Biochemistry, *Howard Hughes Medical Institute,
Yale University, New Haven, Connecticut 06511*

During eukaryotic translation initiation, the 40S ribosomal subunit must be recruited to a messenger RNA (mRNA) and positioned at the correct initiation codon. In most mRNAs, this is achieved via a series of intermolecular events involving a group of protein factors that assemble on the capped 5′ end of the mRNA. This assembly recruits the 40S ribosomal subunit and enables it to scan to the translational start site (Merrick and Hershey 1996). The mRNA is thought to play a passive role in this process and typically lacks significant secondary structure in the 5′-untranslated region that might interfere with scanning (Fig. 1). In contrast, an alternative mechanism of translation initiation involves active roles of the 5′- and in some cases the 3′-untranslated regions of an mRNA (Jackson 1996; Sachs et al. 1997). In these cases, the untranslated regions are often highly conserved, may extend for several hundred nucleotides, and appear to contain extensive secondary and tertiary structures. Here we describe structural features of the 5′-untranslated region of hepatitis C virus (HCV) that enable its function as an internal ribosome entry site (IRES). We also discuss evidence for IRES-mediated translation in certain cellular genes, and the possible roles of conserved 3′-untranslated regions of HCV mRNA in translational control.

IRES: CONTROL AT THE 5′ END

IRESs occur in the 5′-untranslated regions of numerous viral genomic RNAs as well as some eukaryotic cellular RNAs. In contrast to the 7-methyl-guanosine cap-dependent mechanism used by most eukaryotic messenger RNAs, these RNA structures recruit and activate the translation machinery independent of the 5′ end of the mRNA (Fig. 1) (Sachs et al. 1997). IRES RNAs have varied sequences, proposed secondary structures, and cofactor requirements, but all IRES RNAs induce translation initiation in the absence of the 5′ cap structure and in the absence of a 5′ terminus (they are cap- and end-independent).

IRESs were first identified in the picornaviruses, and these provided the initial paradigm for understanding internal translation initiation (Jackson and Kaminski 1995). Within this family of viruses, several different IRES classes were identified based first on secondary structure and later on cofactor requirements. Site-directed mutagenesis of these IRESs, which are between 300 and 500 nucleotides in length, led to the proposal that IRESs form complex three-dimensional tertiary structures that interact with the translation machinery using contacts located throughout the IRES RNA sequence.

HEPATITIS C VIRUS IRES STRUCTURE AND FUNCTION

Recently, intense interest has focused on the IRES of HCV, a viral pathogen of worldwide health concern. This IRES was first identified by Wang et al. (1993), and since that time, extensive mutation and deletion analysis coupled with translation assays both in vivo and in cell-free extracts have identified the sequences and secondary structures necessary for translation initiation (for review, see Rijnbrand and Lemon 2000). The minimal HCV IRES is ~330 nucleotides in length and is highly conserved both in terms of primary sequence and secondary structure. An important step forward in understanding the mechanism of HCV IRES action (and viral IRESs in general) came when Pestova et al. (1998) demonstrated that

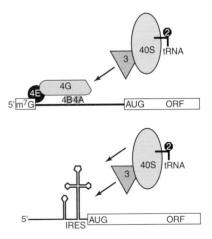

Figure 1. Schematic representation of cap-dependent (*top*) and HCV IRES-driven internal initiation of translation (*bottom*). Cap-dependent initiation requires a set of protein factors that interact with the 5′ cap structure which are not required for 40S subunit recruitment by the HCV IRES RNA.

the HCV IRES binds directly to the 40S ribosomal subunit and eukaryotic initiation factor-3 (eIF3) in the absence of any additional protein factors (Fig. 1). Quantitative biochemical and biophysical analysis of these interactions then identified the presence of a tertiary fold within the HCV IRES RNA and the mechanism by which this fold recruits the translational machinery (Kieft et al. 1999, 2001; Kolupaeva et al. 2000). Binding of the 40S subunit to the HCV IRES RNA involves intermolecular contacts in all three independently folded tertiary domains of the IRES, yet only two of these domains contribute substantially to binding affinity. Within the two domains that provide affinity for the 40S subunit, multiple intermolecular contacts are made. In contrast, binding to eIF3 involves only a single IRES tertiary domain and is weaker than binding to the 40S subunit, and mutations that disrupt the eIF3 IRES interaction but not the 40S–IRES interactions are less detrimental to IRES translation initiation activity. Thus, recruitment of the 43S particle (which contains the 40S subunit, eIF3, eIF2, initiator met-tRNA, and GTP) is driven primarily through multidomain contacts between the 40S subunit and the IRES RNA, suggesting that the strategy of HCV IRES 40S subunit recruitment is fundamentally different from both prokaryotic and canonical eukaryotic mechanisms (Kieft et al. 2001).

The novel nature of HCV IRES translation initiation suggests that behind this mechanism lies an equally unique RNA structure. Detailed, high-resolution studies of the HCV IRES have been hindered by the fact that the IRES RNA is large, multidomained, somewhat flexible, and forms an extended scaffold structure even when folded. These challenges have thus far dictated a "divide and conquer" type of approach to high-resolution structural determination. Recently, nuclear magnetic resonance (NMR) structures of several isolated stem-loops have been solved (Klinck et al. 2000; Lukavsky et al. 2000), and we have recently solved the structure of a folded IRES RNA junction by X-ray crystallography (J.S. Kieft and J.A. Doudna, unpubl.). These structures yield insight into the local folding of individual parts of the IRES, and could be useful in future drug development strategies. However, in order to understand the determinants of IRES affinity and specificity for the translation machinery, structures of IRES RNA bound to its biological targets are needed.

We recently reported the determination of two structures of HCV IRES RNA bound to isolated 40S ribosomal subunit from rabbit, solved by cryo-electron microscopy (cryo-EM) in combination with the single particle approach (Spahn et al. 2001). The two structures were of the full-length IRES RNA and a mutant IRES RNA lacking domain II (Fig. 2). Domain II contacts the 40S subunit and is important for full IRES activity, but contributes little to binding affinity (Kieft et al. 2001). The 20 Å resolution cryo-EM structures reveal that the IRES RNA binds to the head and platform of the 40S subunit in a single extended conformation, making intermolecular contacts that match biochemical predictions. Furthermore, the HCV IRES RNA induces a conforma-

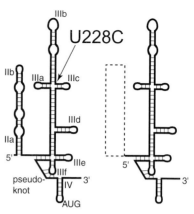

Figure 2. Schematic secondary structures of the HCV IRES constructs used in the study. Mutant U228C has a single point mutation in junction IIIabc (*left*, indicated with *arrow*). The domain II deletion mutant is shown on the right, with a dashed box indicating the location of the deleted domain.

tional change in the 40S subunit that involves fusion of the ribosome density in the head and shoulder regions. This conformational change was not observed in the complex with the domain II deletion mutant, and thus it is the presence of this IRES RNA structural domain (domain II) that induces the conformational change. Domain II contacts the head near or in the E site for tRNA binding and effectively transforms the mRNA decoding channel into a tunnel. In addition, the contact induces fusing of ribosome head to shoulder density at the site where mRNA enters the decoding cleft. Previous biochemical data suggested that domain II affects the way in which the 40S subunit and the IRES RNA interact (Pestova et al. 1998; Kolupaeva et al. 2000); the cryo-EM images provide a plausible structural explanation.

ANALYSIS OF 48S AND 80S PREINITIATION COMPLEX FORMATION WITH THE HCV IRES

Structural and biochemical studies of the HCV IRES provide a foundation for understanding the stepwise mechanism of internal initiation (Fig. 3). In this model, the IRES serves as a binding site for the components of the 43S particle through contacts with eIF3 and the 40S subunit. Once the 43S complex is bound to the IRES (Fig. 3, [4]), the mature 48S complex is formed, with the tRNA anticodon base-paired to the mRNA initiator codon. The 48S complex can then recruit the 60S subunit in a GTP hydrolysis-dependent step, releasing eIF2 and forming the 80S complex. In many ways, this mechanism is very similar to the canonical cap-dependent mechanism, differing only at the step of 43S particle recruitment (Fig. 3, going from [1] to [4]). The realization that HCV IRES RNA binding to the 40S subunit alters the conformation of the ribosomal subunit suggests a role for the IRES RNA as an active manipulator of the translation machin-

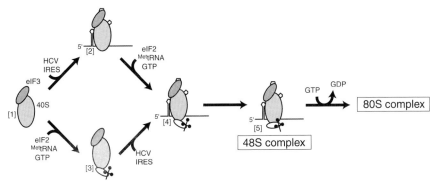

Figure 3. Simplified general mechanism of translation initiation by the HCV IRES. Initiation involves the binding of the components of the 43S particle directly to the IRES RNA ([1] through [4]), formation of the codon–anticodon interaction to form the mature 48S complex ([5]), and subsequent hydrolysis of eIF2-bound GDP, which allows 60S subunit joining. (Adapted from Pestova et al. 1998.)

ery beyond its ability to bind the 40S subunit and eIF3. This observation also raises several new questions regarding the detailed mechanism of HCV IRES-driven internal initiation of translation, including:

- What is the purpose of the 40S subunit conformational change induced by the IRES RNA?

- What are the contributions of other factors on the mechanism of HCV initiation?

- What other dynamic events does the IRES RNA orchestrate within the preinitiation complexes?

To begin a preliminary investigation of these questions, we have taken advantage of the ability of the HCV IRES to assemble preinitiation complexes in rabbit reticulocyte lysate (RRL), and of the ability of small molecule inhibitors to capture these complexes. Specifically, the addition of a nonhydrolyzable GTP analog prevents eIF2-driven GTP hydrolysis and subsequent dissociation of eIF2 from the ribosome, leading to a buildup of 48S complexes. The complexes can then be fractionated via ultracentrifugation through a sucrose gradient, enabling the

quantitation of relative amounts formed under different conditions. Incubation of radiolabeled wild-type HCV IRES RNA in RRL in the absence of any small-molecule inhibitors leads to the sucrose gradient profile shown in Figure 4a. After 15 minutes, both 48S and 80S complexes are present, and there is virtually no free IRES RNA remaining. Thus, assembly of preinitiation complexes by the HCV IRES is quite efficient. The efficiency of assembly drops, however, when mutations are introduced into the IRES RNA (Fig. 4b). Two different mutants are shown, one with a single U to C point mutation at nucleotide 228 and the other with domain II completely deleted (Fig. 2). Of these two mutations, mutant U228C is the most inhibited, with a relatively large amount of free RNA remaining after 15 minutes in RRL. This mutant has been shown to initiate translation at only 5% of wild-type level and has reduced affinity for both the 40S subunit and eIF3 (Kieft et al. 1999; Kieft et al. 2001). In contrast, the domain II deletion mutant appears to incorporate into 48S complexes more readily than U228C, but still below wild-type level. Interestingly, this mutant has

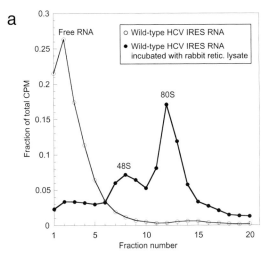

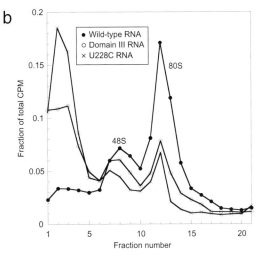

Figure 4. Sucrose gradient profiles of wild-type and mutant IRES RNAs incubated in rabbit reticulocyte lysate for 15 minutes in the absence of small-molecule inhibitors. The locations of free RNA, the 48S complex, and 80S particle are shown.

reduced initiation activity, but has nearly wild-type binding affinity for the 40S subunit and eIF3. This suggests that its reduced ability to form 48S complexes is not due to reduced affinity for the translation machinery. Rather, the fact that domain II induces a conformational change in the 40S subunit through contacts near the tRNA/eIF2-binding site may indicate the presence of additional assembly steps dependent on this IRES structure. It is worth noting that not only do both mutants show a reduced ability to form 48S complexes, but both mutants also display a reduced ratio of 80S:48S when compared to wild-type IRES RNA. Thus, it appears that the assembly pathway may be affected at multiple steps, both before 48S formation and between 48S and 80S formation. Once again, it seems possible that the IRES must do more than just bind the translation machinery; it must somehow interact in very specific ways that signal the assembly of the higher-order complexes.

The above-described analysis reveals the relative populations of free IRES, 48S, and 80S at a single time point, in a fully dynamic system. To focus on the specific step of initial 48S formation, we used a nonhydrolyzable GTP analog to capture 48S complexes and analyzed these IRES RNA-RRL reactions at more than one time point. Figure 5a contains the gradient profiles of wild-type and domain II deletion mutants treated in this manner. Again, after 15 minutes, wild-type RNA is nearly completely incorporated into the 48S complex, while the mutant still displays a large amount of free RNA. Interestingly, after 90 minutes, the mutant RNA has "caught up," now appearing nearly identical to wild-type IRES RNA. Thus, there is a difference in the *rate* of 48S buildup between these two mutants, despite the fact that their *affinities* for the ribosome are virtually the same. Furthermore, this may explain the high amount of variability in reported translation initiation values for mutants involving domain II, as measurements taken at different time points could reveal radically different values. This behavior is in stark contrast to the U228C mutant, shown in Figure 5b. Even after 90 minutes of incubation, this mutant is unable to build up 48S complexes to wild-type levels. Thus, although the two mutants both fail to assemble 48S com-

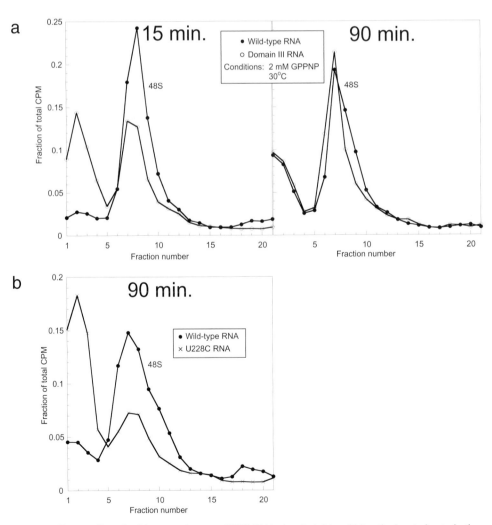

Figure 5. Sucrose gradient profiles of wild-type and mutant IRES RNAs incubated in rabbit reticulocyte lysate in the presence of a nonhydrolyzable GTP analog (GPPNP) for various times.

plexes as well as the wild type, they are failing for different reasons.

ADDITIONAL LAYERS OF COMPLEXITY

From the above-described results it appears that IRES-mediated preinitiation complex formation is an ordered, multistep process. It is possible that factors that do not bind directly to the IRES may transiently contact the IRES RNA or influence IRES-driven assembly. Good candidates for these types of interactions are eIF2 and eIF5. eIF2 should lie near the mRNA-binding groove, close to domain II of the IRES. Indeed, one subunit of eIF2 has been implicated as being important from HCV IRES function in a functional genomics assay (Kruger et al. 2000). eIF5, on the other hand, has been described as a "proofreader" that allows 48S complexes to form 80S particles based on correct formation of the codon–anticodon interaction. Either one of these factors could be receiving or sending signals to the HCV IRES RNA, a possibility that remains to be explored.

IRESs IN CELLULAR mRNAs

In addition to viruses, certain cellular mRNAs appear to utilize an IRES mechanism to facilitate protein expression under conditions where cap-dependent translation may be compromised. Such genes include those encoding human immunoglobin heavy-chain binding protein (BiP) (Macejak and Sarnow 1991), human fibroblast growth factor (FGF-2) (Vagner et al. 1995), insulin-like growth factor (IGF)II (Teerink et al.1995), eIF4G (Gan and Rhoads 1996), platelet-derived growth factor (PDGF)2 (Bernstein et al. 1997), the proto-oncogene c-*myc* (Nanbru et al. 1997), and the vascular endothelial growth factor (VEGF) (Stein et al. 1998). These mRNAs characteristically contain 5′UTRs longer than 200 nucleotides, often with numerous AUGs upstream from the initiation codon. The presence of an IRES in these 5′UTRs was established by the use of dicistronic constructs containing the putative IRES sequence inserted between two protein-encoding cistrons, or by the ability of the 5′UTR in question to initiate synthesis of a reporter protein during picornavirus infection (when cap-dependent translation is compromised) (van der Velden and Thomas 1999).

Although cellular IRESs have not been studied as extensively as the viral IRESs, initial differences between them can be noted. When compared to picornavirus IRESs, which can stimulate expression of the second cistron by 100-fold, most cellular IRESs only modestly increase second cistron expression (2- to 20-fold) (although the c-*myc* IRES stimulates expression ~70-fold) (Stoneley et al. 1998). In addition, some cellular IRESs demonstrate an ability to modulate activity in response to the cell cycle (Cornelis et al. 2000; Pyronnet et al. 2000). Furthermore, no secondary structural homology appears to exist between viral and cellular IRESs, although the structural organization of cellular IRES elements is currently poorly defined. There may be functional differences between the viral and cellular IRESs, as well. For example, deletion of the 3′ end of viral IRESs abolishes

IRES activity, whereas activity is only gradually lost as segments are deleted from the c-*myc* IRES.

An intriguing question is why a certain subset of mRNAs has evolved to be translated by both the "conventional" cap-dependent mechanism and the IRES-mediated cap-independent mechanism. Interestingly, the genes implicated in using an IRES encode factors involved in cell growth, transcription, and translation initiation. Expression of such proteins must be tightly regulated during growth, differentiation, and apoptosis. IRES-mediated translation of these cellular mRNAs may represent a vital regulatory mechanism for the survival and proliferation of cells under acute but transient stress conditions. It remains to be discovered whether cellular IRESs represent a unique strategy for recruiting ribosomes and whether all cellular IRESs use the same mechanism to interact with the translational machinery. In addition, it will be important to identify protein components mediating IRES-dependent translation of cellular mRNAs, and to understand how these IRES complexes assemble and manipulate the translation apparatus.

REGULATORY ELEMENTS IN VIRAL 3′-UNTRANSLATED REGIONS

Like 5′UTRs, the 3′UTR of RNA virus genomes is often conserved and has roles in the translation, replication, and packaging of viral RNA (Chambers et al. 1990; Gale et al. 2000). In HCV, the 3′UTR consists of a variable region (30–40 nucleotides) and a homopolymeric poly(U)/polypyrimidine tract (20–200 nucleotides) followed by the 3X region (98 nucleotides), which is highly conserved across all serotypes (Chambers et al. 1990). Chemical and biochemical studies together with modeling (MFOLD) suggest that the 3X region consists of three stem-loops (SL1, SL2, and SL3, Fig. 6), although chemical probing data indicate that the SL3 and SL2 region is structurally heterogeneous and may have alternate conformations. In chimpanzees, both the poly(U) and the 3X region are necessary for viral infection and replication, whereas the variable region appears to be dispensable (Yanagi et al. 1999; Kolykhalov et al. 2000).

A common aspect of mechanistically distinct cellular and viral translation mechanisms is synergistic interac-

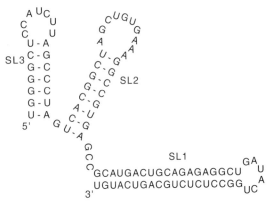

Figure 6. Secondary structure of the HCV 3X region.

tion between the 5′ and 3′ ends of mRNA, which affects translational efficiency. In several cases a physical link between the 5′ and 3′ ends via protein and/or RNA contacts leads to circularization of the mRNA, and this may be a general phenomenon (Wells et al. 1998). Although there are regions of sequence complementarity in the 5′ and 3′ UTRs of HCV, no RNA:RNA interactions have been detected (Blight and Rice 1997). Similarly, studies with a replicon of the related *pestivirus* bovine diarrhea virus (BVDV) showed that there were no RNA:RNA interactions occurring between the 5′ and 3′ UTRs (Yu et al. 1999). This raises the possibility that circularization of the HCV genome, if it does occur, may involve a linkage involving protein:RNA interactions.

Proteins that have been shown to bind elements of the HCV 3′UTR include La (Spangberg et al. 2001), hnRNP C (Gontarek et al. 1999), HuR (Spangberg et al. 2000), glyceraldehyde 3-phosphate dehydrogenase (Petrik et al. 1999), viral NS3 (Banerjee and Dasgupta 2001), ribosomal protein L22 (Wood et al. 2001), and PTB (pyrimidine tract binding protein) (Ito and Lai 1997; Tsuchihara et al. 1997; Chung and Kaplan 1999), but the biological relevance of these interactions remains unclear. Both La and GAPDH recognize the polypyrimidine tract, and La may have a role in maintaining stability of the RNA (Spangberg et al. 2001). PTB and the ribosomal protein L22 are the only proteins identified so far that appear to bind specifically to the 3X region. Interestingly, L22 has also been shown to bind to EBER RNA of the Epstein-Barr virus, an RNA that has no sequence homology with the HCV 3X region (Toczyski et al. 1994). A specific interaction between PTB and the 3X RNA has been demonstrated by a number of investigators (Ito and Lai 1997; Tsuchihara et al. 1997; Chung and Kaplan 1999). PTB has also been shown to play a role in the translation of several RNAs (Hunt and Jackson 1999; Gosert et al. 2000) and to interact with the HCV IRES (Anwar et al. 2000). The region of the 3X RNA that interacts with PTB appears to be limited to stem-loops SL2 and SL3, and there is a requirement for the integrity of the structure and sequence of the stem-loops (Ito and Lai 1997; Chung and Kaplan 1999). However, the sequence immediately prior to stem-loop SL3 is partially homologous to the consensus sequence for cellular polypyrimidine tract targets of PTB recognition, and there are conflicting results as to its requirement for efficient PTB binding of 3X (Tsuchihara et al. 1997). Since L22 and PTB are relatively promiscuous and bind other RNA targets that are dissimilar to 3X, it seems unlikely that these proteins are responsible for the high of degree of conservation observed in the 3X region. Furthermore, no quantitative data are available yet regarding the affinity and specificity of these interactions.

Sequences that interact with a specific protein and enhance translation have been identified in the 3′UTR of a number of cellular mRNAs. In in vitro translation assays, the 3X region enhances the efficiency of both cap-dependent and IRES-mediated translation 3- to 4-fold (Ito et al. 1998; Michel et al. 2001). However, recent work which examined these effects in the context of the complete 3′UTR found that the presence of the 3X region de-

pressed translation (Murakami et al. 2001). Perhaps the use of HCV replicons will enable these conflicting results to be resolved and should facilitate investigations into the role of the HCV 3′UTR in the viral life cycle (Blight et al. 2000).

CONCLUSIONS

Regulation of translation via 5′- and 3′-untranslated regions is ubiquitous, yet the molecular mechanisms of these processes are not well understood. Structural and biochemical analyses of viral IRES elements and 3′ UTRs will reveal details of such pathways that will both illuminate basic virus biology and provide the basis for rational design of antiviral therapeutics. These investigations may also shed light on fundamental mechanisms of translational control present in eukaryotic cells.

ACKNOWLEDGMENTS

The authors thank the National Science Foundation and the National Institutes of Health for training grant support to A.G. and a Program Project grant to J.A.D. J.A.D. is a fellow of the David and Lucile Packard Foundation.

REFERENCES

Anwar A., Ali N., Tanveer R., and Siddiqui A. 2000. Demonstration of functional requirement of polypyrimidine tract-binding protein by SELEX RNA during hepatitis C virus internal ribosome entry site-mediated translation initiation. *J. Biol. Chem.* **275:** 34231.

Banerjee R. and Dasgupta A. 2001. Specific interaction of hepatitis C virus protease/helicase NS3 with the 3′-terminal sequences of viral positive- and negative-strand RNA. *J. Virol.* **75:** 1708.

Bernstein J., Sella O., Le S.Y., and Elroy-Stein O. 1997. PDGF2/c-sis mRNA leader contains a differentiation-linked internal ribosomal entry site (D-IRES). *J. Biol. Chem.* **272:** 9356.

Blight K.J. and Rice C.M. 1997. Secondary structure determination of the conserved 98-base sequence at the 3′ terminus of hepatitis C virus genome RNA. *J. Virol.* **71:** 7345.

Blight K.J., Kolykhalov A.A., and Rice C.M. 2000. Efficient initiation of HCV RNA replication in cell culture. *Science* **290:** 1972.

Chambers T.J., Hahn C.S., Galler R., and Rice C.M. 1990. Flavivirus genome organization, expression, and replication. *Annu. Rev. Microbiol.* **44:** 649.

Chung R.T. and Kaplan L.M. 1999. Heterogeneous nuclear ribonucleoprotein I (hnRNP-I/PTB) selectively binds the conserved 3′ terminus of hepatitis C viral RNA. *Biochem. Biophys. Res. Commun.* **254:** 351.

Cornelis S., Bruynooghe Y., Denecker G., Van Huffel S., Tinton S., and Beyaert R. 2000. Identification and characterization of a novel cell-cycle-regulated internal ribosome entry site. *Mol. Cell* **5:** 597.

Gale M., Jr., Tan S.L., and Katze M.G. 2000. Translational control of viral gene expression in eukaryotes. *Microbiol. Mol. Biol. Rev.* **64:** 239.

Gan W. and Rhoads R.E. 1996. Internal initiation of translation directed by the 5′-untranslated region of the mRNA for eIF4G, a factor involved in the picornavirus-induced switch from cap-dependent to internal initiation. *J. Biol. Chem.* **271:** 623.

Gontarek R.R., Gutshall L.L., Herold K.M., Tsai J., Sathe G.M., Mao J., Prescott C., and Del Vecchio A.M. 1999. hnRNP C and polypyrimidine tract-binding protein specifically interact with the pyrimidine-rich region within the 3′NTR of the HCV RNA genome. *Nucleic. Acids Res.* **27:** 1457.

Gosert R., Chang K.H., Rijnbrand R., Yi M., Sangar D.V., and Lemon S.M. 2000. Transient expression of cellular polypyrimidine-tract binding protein stimulates cap-independent translation directed by both picornaviral and flaviviral internal ribosome entry sites In vivo. *Mol. Cell.Biol* **20:** 1583.

Hunt S.L. and Jackson R.J. 1999. Polypyrimidine-tract binding protein (PTB) is necessary, but not sufficient, for efficient internal initiation of translation of human rhinovirus-2 RNA. *RNA* **5:** 344.

Ito T. and Lai M.M. 1997. Determination of the secondary structure of and cellular protein binding to the 3′-untranslated region of the hepatitis C virus RNA genome. *J. Virol.* **71:** 8698.

Ito T., Tahara S.M., and Lai M.M. 1998. The 3′-untranslated region of hepatitis C virus RNA enhances translation from an internal ribosomal entry site. *J. Virol.* **72:** 8789.

Jackson R.J. 1996. A comparative view of initiation site selection mechanisms. In *Translational Control* (ed. J. W. B. Hershey et al.), p 71. Cold Spring Harbor Laboratory Press, Cold Spring Harbor, New York.

Jackson R.J. and Kaminski A. 1995. Internal initiation of translation in eukaryotes: The picornavirus paradigm and beyond. *RNA* **1:** 985.

Kieft J.S., Zhou K., Jubin R., and Doudna J.A. 2001. Mechanism of ribosome recruitment by hepatitis C IRES RNA. *RNA* **7:** 194.

Kieft J.S., Zhou K., Jubin R., Murray M.G., Lau J.Y., and Doudna J.A. 1999. The hepatitis C virus internal ribosome entry site adopts an ion-dependent tertiary fold. *J. Mol. Biol.* **292:** 513.

Klinck R., Westhof E., Walker S., Afshar M., Collier A., and Aboul-Ela F. 2000. A potential drug target in the hepatitis C virus internal ribosome entry site. *RNA* **6:** 1423.

Kolupaeva V.G., Pestova T.V., and Hellen C.U. 2000. An enzymatic footprinting analysis of the interaction of 40S ribosomal subunits with the internal ribosomal entry site of hepatitis C virus. *J. Virol.* **74:** 6242.

Kolykhalov A.A., Mihalik K., Feinstone S.M., and Rice C.M. 2000. Hepatitis C virus-encoded enzymatic activities and conserved RNA elements in the 3′ nontranslated region are essential for virus replication in vivo. *J. Virol.* **74:** 2046.

Kruger M., Beger C., Li Q.-X., Welch P.J., Tritz R., Leavitt M., Barber J.R., and Wong-Staal F. 2000. Identification of eIF2Bγ and eIF2γ as cofactors of hepatitis C virus internal ribosome entry site-mediated translation using a functional genomics approach. *Proc. Natl. Acad. Sci.* **97:** 8566.

Lukavsky P.J., Otto G.A., Lancaster A.M., Sarnow P., and Puglisi J.D. 2000. Structures of two RNA domains essential for hepatitis C virus internal ribosomal entry site formation. *Nat. Struct. Biol.* **7:** 1105.

Macejak D.G. and Sarnow P. 1991. Internal initiation of translation mediated by the 5′ leader of a cellular mRNA. *Nature* **353:** 90.

Merrick W.C. and Hershey J.W.B. 1996. The pathway and mechanism of eukaryotic protein synthesis. In *Translational Control* (ed. J.W.B. Hershey et al.), p 31. Cold Spring Harbor Laboratory Press, Cold Spring Harbor, New York.

Michel Y.M., Borman A.M., Paulous S., and Kean K.M. 2001. Eukaryotic initiation factor 4g-poly(a) binding protein interaction is required for poly(a) tail-mediated stimulation of picornavirus internal ribosome entry segment-driven translation but not for x-mediated stimulation of hepatitis C virus translation. *Mol. Cell. Biol.* **21:** 4097.

Murakami K., Abe M., Kageyama T., Kamoshita N., and Nomoto A. 2001. Down-regulation of translation driven by hepatitis C virus internal ribosome entry site by the 3′ untranslated region of RNA. *Arch. Virol.* **146:** 729.

Nanbru C., Lafon I., Audigier S., Gensac M.C., Vagner S., Huez G., and Prats A.C. 1997. Alternative translation of the proto-oncogene c-myc by an internal ribosome entry site. *J. Biol. Chem.* **272:** 32061.

Pestova T.V., Shatsky I.N., Fletcher S.P., Jackson R.J., and Hellen C.U.T. 1998. A prokaryotic-like mode of cytoplasmic eukaryotic ribosome binding to the initiation codon during internal translation initiation of hepatitis C and classical swine fever virus RNAs. *Genes Dev.* **12:** 67.

Petrik J., Parker H., and Alexander G.J. 1999. Human hepatic glyceraldehyde-3-phosphate dehydrogenase binds to the poly(U) tract of the 3′ non-coding region of hepatitis C virus genomic RNA. *J. Gen. Virol.* **80:** 3109.

Pyronnet S., Pradayrol L., and Sonenberg N. 2000. A cell-cycle dependent internal ribosome entry site. *Mol. Cell* **5:** 607.

Rijnbrand R.C. and Lemon S.M. 2000. Internal ribosome entry site-mediated translation in hepatitis C virus replication. *Curr. Top. Microbiol. Immunol.* **242:** 85.

Sachs A.B., Sarnow P., and Henze M.W. 1997. Starting at the beginning, middle, and end: Translation initiation in eukaryotes. *Cell* **89:** 831.

Spahn C.M.T., Kieft J.S., Grassucci R.A., Penczek P.A., Zhou K., Doudna J.A., and Frank J. 2001. Hepatitis C virus IRES RNA-induced changes in the conformation of the 40S ribosomal subunit. *Science* **291:** 1959.

Spangberg K., Wiklund L., and Schwartz S. 2000. HuR, a protein implicated in oncogene and growth factor mRNA decay, binds to the 3′ ends of hepatitis C virus RNA of both polarities. *Virology* **274:** 378.

———. 2001. Binding of the La autoantigen to the hepatitis C virus 3′ untranslated region protects the RNA from rapid degradation in vitro. *J. Gen. Virol.* **82:** 113.

Stein I., Itin A., Einat P., Skaliter R., Grossman Z., and Keshet E. 1998. Translation of vascular endothelial growth factor mRNA by internal ribosome entry: Implications for translation under hypoxia. *Mol. Cell. Biol.* **18:** 3112.

Stoneley M., Paulin F.E., Le Quesne J.P., Chappell S.A., and Willis A.E. 1998. C-myc 5′ untranslated region contains an internal ribosome entry segment. *Oncogene* **16:** 423.

Teerink H., Voorma H.O., and Thomas A.A. 1995. The human insulin-like growth factor II leader 1 contains an internal ribosomal entry site. *Biochim. Biophys. Acta* **1264:** 403.

Toczyski D.P., Matera A.G., Ward D.C., and Steitz J.A. 1994. The Epstein-Barr virus (EBV) small RNA EBER1 binds and relocalizes ribosomal protein L22 in EBV-infected human B lymphocytes. *Proc. Natl. Acad. Sci.* **91:** 3463.

Tsuchihara K., Tanaka T., Hijikata M., Kuge S., Toyoda H., Nomoto A., Yamamoto N., and Shimotohno K. 1997. Specific interaction of polypyrimidine tract-binding protein with the extreme 3′-terminal structure of the hepatitis C virus genome, the 3′X. *J. Virol.* **71:** 6720.

Vagner S., Gensac M.C., Maret A., Bayard F., Amalric F., Prats H., and Prats A.C. 1995. Alternative translation of human fibroblast growth factor 2 mRNA occurs by internal entry of ribosomes. *Mol. Cell. Biol.* **15:** 35.

van der Velden A.W. and Thomas A.A. 1999. The role of the 5′ untranslated region of an mRNA in translation regulation during development. *Int. J. Biochem. Cell. Biol.* **31:** 87.

Wang C., Sarnow P., and Siddiqui A. 1993. Translation of human hepatitis C virus RNA in cultured cells is mediated by an internal ribosome-binding mechanism. *J. Virol.* **67:** 3338.

Wells S.E., Hillner P.E., Vale R.D., and Sachs A.B. 1998. Circularization of mRNA by eukaryotic translation initiation factors. *Mol. Cell.* **2:** 135.

Wood J., Frederickson R.M., Fields S., and Patel A.H. 2001. Hepatitis C virus 3′x region interacts with human ribosomal proteins. *J. Virol.* **75:** 1348.

Yanagi M., St. Claire M., Emerson S.U., Purcell R.H., and Bukh J. 1999. In vivo analysis of the 3′ untranslated region of the hepatitis C virus after in vitro mutagenesis of an infectious cDNA clone. *Proc. Natl. Acad. Sci.* **96:** 2291.

Yu H., Grassmann C.W., and Behrens S.E. 1999. Sequence and structural elements at the 3′ terminus of bovine viral diarrhea virus genomic RNA: Functional role during RNA replication. *J. Virol.* **73:** 3638.

Initiator Met-tRNA-independent Translation Mediated by an Internal Ribosome Entry Site Element in Cricket Paralysis Virus-like Insect Viruses

E. JAN,* S.R. THOMPSON,* J.E. WILSON,* T.V. PESTOVA,† C.U.T. HELLEN,† AND P. SARNOW*

*Department of Microbiology and Immunology, Stanford University School of Medicine, Stanford, California 94305; †Department of Microbiology and Immunology, Morse Institute of Molecular Genetics, State University of New York Health Sciences Center at Brooklyn, Brooklyn, New York 11203

In eukaryotic cells, the translation of the majority of messenger RNAs is initiated by a scanning mechanism (Kozak 1989) whereby the 40S ribosomal subunit is recruited to the 5′-terminal end of the capped mRNA as a 43S complex, consisting of the 40S ribosome, the ternary Met-tRNA$_i$/eIF2-GTP complex, and factors eIF1A and eIF3 (Hershey and Merrick 2000). The 43S complex then scans the mRNA in an ATP-dependent manner until it locates an appropriate start AUG where the 60S subunit joins to form an 80S ribosome.

An alternate mechanism for recruiting 40S ribosomal subunits to the mRNA is through an internal ribosome entry site (IRES) (for review, see Hellen and Sarnow 2001). IRES elements can directly recruit 40S ribosomes to the mRNA and bypass, in most cases, the requirement for the cap-binding protein, eIF4E (Hellen and Sarnow 2001). Therefore, RNAs containing IRESs can be preferentially translated during cellular stress, viral infection, or differentiation when overall translation is compromised by modification or sequestration of eIF4E (Hellen and Sarnow 2001). Since the discovery of IRESs in picornaviral RNA genomes (Jang et al. 1988; Pelletier and Sonenberg 1988), several IRES elements have been found in other viruses, including hepatitis C virus (HCV) and classic swine fever virus (CSFV), and in some cellular RNAs (for review, see Hellen and Sarnow 2001). Unlike the picornaviral IRESs, the HCV and CSFV IRESs do not require the RNA helicase eIF4A or initiation factors eIF4B, eIF3, and eIF4F to recruit the 40S subunit to the start AUG codon (Pestova et al. 1998). Thus, the requirement of initiation factors for 40S recruitment can vary among different IRES elements.

Recently, an unusual IRES was uncovered in the family of insect viruses called the cricket paralysis virus-like viruses (for review, see Hellen and Sarnow 2001). Specifically, it was shown that the IRESs in two members of this family, the cricket paralysis virus (CrPV) and the *Plautia stali* intestine virus (PSIV), initiate at a non-methionine codon without Met-tRNA$_i$ (Sasaki and Nakashima 1999; Wilson et al. 2000a,b). Therefore, translation by these IRESs does not require initiator methionine or eIF2. Furthermore, these IRESs can directly recruit the 40S ribosomal subunit and can form 80S complexes in the absence of initiation factors (Wilson et al.

2000a). Remarkably, translation by the IRES is initiated from the A site of the ribosome with the first amino acid of the protein encoded by the codon positioned in the A site. In this paper, we summarize and discuss these recent findings on an unusual IRES-mediated mechanism of translational initiation.

PROPERTIES OF THE CRICKET PARALYSIS VIRUS-LIKE VIRUSES

CrPV was originally isolated in infected Australian field crickets. It causes paralysis in the hind legs of crickets and eventually leads to death (Reinganum et al. 1970). Other insect viruses belonging to this family are *Plautia stali* intestine virus, *Drosophila* C virus, *Rhopalosiphum padi* virus, himetobi P virus, Triatoma virus, and black queen-cell virus (for review, see Moore and Tinsley 1982; Christian and Scotti 1998; Hellen and Sarnow 2001). CrPV-like viruses are positive-strand RNA viruses that infect insect cells and contain properties similar to mammalian picornaviruses. Electron micrograph arrays of sections of CrPV-infected crickets show virus-like particles that are similar to those seen in picornavirus-infected cells (Moore and Tinsley 1982). Recent sequencing of the insect viruses indicates that the genomes encoded proteins similar to mammalian picornaviruses such as the RNA polymerase, helicase, and capsid (Christian and Scotti 1998). Although the insect viruses and the mammalian picornaviruses share similar protein sequences, the genomic organization of the insect viruses (Fig. 1) is vastly different (Christian and Scotti 1998; Hellen and Sarnow 2001). First, the insect viral genomes encode two open reading frames (ORF1 and ORF2), whereas the mammalian picornavirus genomes encode one ORF. The two ORFs are separated by an intergenic region ~180 nucleotides in length. Second, the first ORF of the insect genome encodes nonstructural proteins such as RNA polymerase and helicase, whereas the second ORF encodes the structural proteins such as capsid proteins. This organization is in contrast to the picornaviral genome, where the structural proteins are encoded upstream of the nonstructural proteins. However, as in the picornaviruses, ORF1 is translated by internal initiation (Wilson et al. 2000b). Interestingly, in cells infected by

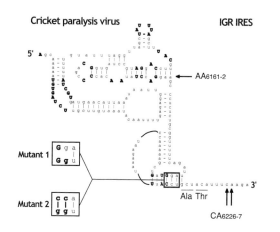

IRES	IRES activity	Toeprints
wt	100%	AA6161-2 + CA6226-7
Mutant 1	<1%	AA6161-2
Mutant 2	50%	AA6161-2 + CA6226-7

Figure 1. Features of the cricket paralysis virus (CrPV) IRES. A diagram displaying the genome organization of CrPV with its two IRES elements is shown on the top. The predicted secondary structure of the CrPV IRES is remodeled after the *Plautia stali* intestine virus IRES (Kanamori and Nakashima 2001). Conserved residues are shown in bold. The locations and genotypes of mutant 1 and mutant 2 are indicated. The bottom panel displays IRES activities and locations of toeprints in the mutated IRES elements. (Reprinted, with permission, from Hellen and Sarnow 2001.)

the insect viruses, protein expression by ORF2 is in supramolar excess over that by ORF1 (Moore et al. 1980; Wilson et al. 2000b). This finding suggests that the regulation of the expression of the two ORFs is independent. Indeed, the intergenic region located between ORF1 and ORF2 has been found to contain an IRES as well (Sasaki and Nakashima 1999; Domier et al. 2000; Wilson et al. 2000b), demonstrating that these viral genomes are naturally dicistronic mRNAs.

A PSEUDOKNOT STRUCTURE DIRECTS IGR–IRES-MEDIATED METHIONINE-INDEPENDENT TRANSLATIONAL INITIATION

Insight into the translational regulation of ORF2 came from work by Nakashima and coworkers (Sasaki and Nakashima 2000). These authors discovered that the PSIV intergenic region (IGR) had IRES activity. Interestingly, they found that the first amino acid of ORF2 is not a methionine, but instead a glutamine encoded by a CAA triplet. In an elegant set of experiments, it was demon-

strated that methionyl peptidases did not remove a putative amino-terminal methionine, thus demonstrating that glutamine was indeed the first amino acid incorporated into the ORF2 product (Sasaki and Nakashima 2000).

Similarly, the CrPV IGR contains an IRES (IGR–IRES) (Wilson et al. 2000b). In this case, the first amino acid of ORF2, as determined by mutagenesis analysis and amino-terminal protein sequencing, is an alanine encoded by a GCU codon (Fig. 1) (Wilson et al. 2000a). Sequence analysis of the other CrPV-like IRES elements indicates that the first amino acid of ORF2 is either an alanine or glutamine. These findings raise the possibility that translation initiation mediated by these insect IRESs may occur through an unusual pathway that does not require eIF2 or Met-tRNA$_i$.

Structural probing, employing site-directed mutagenesis, has shown that both PSIV (Kanamori and Nakashima 2001) and CrPV (Wilson et al. 2000a,b; Hellen and Sarnow 2001) IGR–IRESs contain a pseudoknot structure immediately upstream of the start codon that is essential for IRES function (Fig. 1). In the case of the CrPV IGR-IRES, the pseudoknot structure includes an inverted repeat formed by interaction between a UUACCU motif, located upstream of the alanine GCU start codon, and a AGGUAA motif, located 18 nucleotides upstream of IRES (Fig. 1). Disruption of the pseudoknot by mutating the UUACCU motif eliminated IRES activity, which could be restored by compensatory mutations that restored the predicted pseudoknot (Wilson et al. 2000b). Similar analyses have shown that the start codon-encompassing pseudoknot structure is also essential for the functioning of the PSIV IRES (Kanamori and Nakashima 2001). Furthermore, the same pseudoknot structure can be predicted to form in the other five insect CrPV-like viruses, suggesting that the pseudoknot structure has an important function (Kanamori and Nakashima 2001).

Recently, Nakashima and colleagues have generated an enormous number of mutants and compensatory mutants to deduce a tertiary structure model of the PSIV IRES that can be used to predict the tertiary structures of the CrPV-like viruses (Kanamori and Nakashima 2001). Remarkably, this analysis revealed that the IGR–IRESs in the CrPV-like viruses contain a highly conserved triple pseudoknot structure (Fig. 1). The roles of these structural elements in ribosome recruitment and translational initiation are not yet known.

CrPV IGR–IRES INITIATES TRANSLATION FROM THE A SITE OF THE RIBOSOME

To investigate the mechanism of methionine-independent translation, we first examined the factors that are required for the recruitment of 40S subunit to the IGR–IRES using toeprinting analysis (Wilson et al. 2000a). Briefly, cDNA is produced by reverse transcriptase from a primer oligonucleotide that is complementary to a sequence element downstream of the suspected RNA/40S complex. If the reverse transcriptase encounters a 40S/RNA complex, cDNA synthesis is arrested, re-

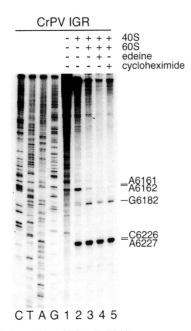

Figure 2. Assembly of IGR–IRES/ribosome complexes from purified subunits. Locations of toeprints on the CrPV IGR–IRES RNA in the presence of 40S or 60S subunits are shown. Edeine (10 μM) or cycloheximide (500 μg/m) was added as indicated. A sequencing ladder is shown at left. (Reprinted, with permission from Elsevier Science, from Wilson et al. 2000a.)

sulting in a stop or toeprint at its leading edge of the ribosome. Similarly, cDNA synthesis may be arrested by a stable secondary structure. Using this assay, it was found that the EMCV IRES requires all initiation factors except eIF4E, eIF1, and eIF1A to recruit the 40S subunit to the proper start codon (Pestova et al. 1996). In contrast, the CrPV IGR–IRES recruited the 40S subunit in the absence of all initiation factors (Fig. 2). Specifically, 40S binding to the CrPV IGR–IRES produced a toeprint at nucleotides CA6226–6227 and AA6161–6162 (Fig. 2). Formation of this toeprint was insensitive to edeine (Wilson et al. 2000a), a compound that is known to interfere with start codon recognition by the ternary 40S complex (Kozak and Shatkin 1978; Odom et al. 1978), further suggesting that Met-tRNA$_i$/eIF2–GTP complexes are not involved in the recruitment of ribosomes to the IGR–IRES.

The nucleotides positioned in the ribosomal P and A sites of the ribosome can be inferred by the location of the toeprint or the arrested primer extension. For example, positioning of the ternary 40S/Met-tRNA$_i$/eIF2–GTP complex at the start AUG codon on the HCV IRES results in primer extension arrest at 15–17 nucleotides downstream from the AUG, if the A of the AUG is designated as +1 (Pestova et al. 1998). In contrast, a binary 40S/HCV IRES complex reveals a 13- to 15-nucleotide toeprint (Pestova et al. 1998), indicating that the reverse transcriptase can penetrate farther into the 40S/IRES complex. This is presumably due to the fact that both the P and A sites of the 40S/HCV IRES complex are empty. Given this assumption, the ribosomal P site in the 40S/CrPV IGR-IRES complex should be occupied by the

CCU triplet (Fig. 1), because it is 13–15 nucleotides from the toeprint at CA 6226–6227. The fact that the alanine is the amino-terminal amino acid of ORF2 (Wilson et al. 2000a) suggests that the proline CCU triplet was not decoded. In support of this argument, a mutant IRES, which contained a UGA stop codon in place of the CCU proline codon, was functional, provided that the upstream pseudoknot structure was maintained (Wilson et al. 2000a). Together, these findings argue that the CCU codon is not decoded and that initiation by the IGR–IRES commences from the ribosomal A site without the requirement of Met-tRNA$_i$ or any charged tRNAs in the P site.

For most mRNAs, a series of steps is required for the large 60S ribosomal subunit to join 40S to form an 80S complex. For example, hydrolysis of GTP by eIF5 and eIF5B are required for 60S joining and proper positioning of the ribosome at the initiation codon. To test the requirements for the formation of 80S/IGR–IRES complexes, radiolabeled IGR–IRES RNAs were incubated with purified 40S and 60S subunits and the reaction products were analyzed after sedimentation on 10–30% sucrose gradients (Fig. 3). Surprisingly, the CrPV IGR–IRES could form 80S complexes in the absence of any initiation factors, GTP, or tRNAs (Fig. 3). Formation

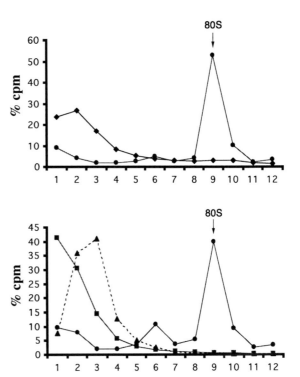

Figure 3. Sucrose gradient analysis of CrPV IGR–IRES RNA/ribosome complexes. Sedimentation profiles of radiolabeled CrPV IGR–IRES RNA incubated with 40S and 60S subunits (*circles*), 60S subunits alone (*diamonds*), or with both 40S and 60S subunits in the presence of 15 mM EDTA (*squares*) are shown. In addition, the EMCV IRES incubated with 40S and 60S subunits is shown (*triangles*). The top of the gradient is at left, and the percentage of cpms in each gradient fraction is displayed.

of 80S/CrPV IGR–IRES complexes was specific because purified 60S subunits did not bind to the IGR–IRES, and the encephalomyocarditis virus (EMCV) IRES could not form 80S complexes in the absence of initiation factors (Fig. 3). Furthermore, the 80S/CrPV IGR–IRES complex was disrupted when EDTA was added to the reactions, indicating that the IGR–IRES was bound to the 80S ribosomes and was not an RNA aggregate that fortuitously cosedimented with the 80S fraction. Furthermore, 80S/CrPV IGR–IRES complexes revealed a toeprint at CA6226–6227 that was identical to the toeprint obtained with 40S/CrPV IGR–IRES complexes (Fig. 2). Curiously, the AA6161–6162 toeprint was reduced in the 80S/CrPV IGR–IRES complexes.

Because two toeprints were revealed by the 80S/CrPV IGR–IRES complex and the 40S/CrPV IGR–IRES complex, we determined whether the presence of either toeprints correlated with IRES activity. Disruption of the pseudoknot structure immediately upstream of the alanine GCU codon destroyed translational activity and led to the loss of the toeprint at CA6226–6227 (Fig. 1) (Wilson et al. 2000b); in contrast, disruption of the pseudoknot did not affect the AA6161–6162 toeprint in 80S/CrPV IGR–IRES and 40S/CrPV IGR–IRES complexes, suggesting that the presence of the CA6226–6227 toeprint correlated with the function of the IRES (data not shown; Wilson et al. 2000a). These data suggest a novel pathway for 80S assembly on the IGR–IRES that does not require initiation factors. It remains to be determined whether preformed 80S complexes can bind to the IGR–IRES or whether 40S subunits bind first, followed by 60S subunits joining.

MODEL FOR CrPV IGR–IRES-MEDIATED TRANSLATION

To further examine the factor requirement for 80S assembly on the CrPV IGR–IRES, various drugs that target and inhibit different aspects of initiation were used to analyze the mechanism by which the IGR–IRES can form 80S complexes in rabbit reticulocyte lysates (RRL). GMP-PNP, which is a nonhydrolyzable GTP analog, blocks the recycling of eIF2α and the joining of 60S subunits on most mRNAs. Edeine inhibits recognition of the AUG by the ternary 40S complexes, and L-methioninol is a substrate analog inhibitor of methionine-tRNA synthetases (for discussion, see Wilson et al. 2000a). It was found that 80S complexes formed on the IGR–IRES in the presence of GMP-PNP, edeine, or L-methioninol; in contrast, 80S complexes did not form on the EMCV IRES in the presence of any of these compounds (Wilson et al. 2000a). These data are consistent with the model that the IGR–IRES can form 80S complexes in the absence of initiation factors and GTP, and does not require eIF2 or initiator tRNA.

Because the CCU triplet in the P site is not decoded and the alanine-encoding GCU is in the A site, this suggests that the first translocation event of the 80S ribosome does not involve peptide-bond formation. To further examine this hypothesis, toeprinting analysis of the IGR–IRES

was performed in the RRL. As seen previously with purified 40S and 60S, toeprints at CA6226–6227 were observed in RRL when treated with high concentrations of edeine (Fig. 4). Edeine, when used at high concentrations, can inhibit the interaction of the aminoacylated-tRNA with the A site of the ribosome (Szer and Kurylo-Burowska 1970; Carrasco et al. 1974). Most likely, the 80S/CrPV IGR–IRES complex in RRL treated with edeine contained an empty A site. When the RRL was treated with cycloheximide, which inhibits elongation, a strong toeprint was observed at UAA6231–6233 and weaker toeprints were observed at G6229 and CA6226–6227 (Fig. 4). The weaker toeprint at CA6226–6227 likely represents 80S/CrPV IGR–IRES complexes with empty P and A sites with the CCU and GCU codons in the P and A sites, respectively, as was seen with toeprints in 80S/CrPV IGR–IRES complexes assembled from purified 40S and 60S subunits (Fig. 2). It would

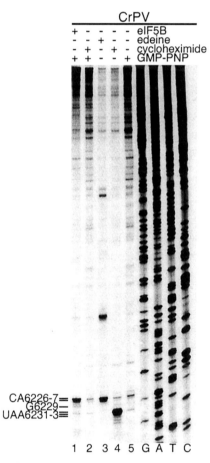

Figure 4. Toeprint analyses of translation initiation complexes formed on IGR–IRES elements in the RRL. Patterns of arrest of primer extension on the IGR–IRES in the presence of 1 mM GMP-PNP plus 3 µg of eIF5B (lane *1*), 1 mM GMP-PNP plus 500 µg/m cycloheximide (lane *2*), 10 µM edeine (lane *3*), 500 µg/m cycloheximide (lane *4*), or 1 mM GMP–PNP (lane *5*). The positions of the toeprints are indicated on the left. A sequencing ladder is shown at right. Autoradiographs of polyacrylamide gels are shown. (Reprinted, with permission, from Elsevier Science, from Wilson et al. 2000a.)

be predicted that 80S/CrPV IGR–IRES complexes that have an Ala-tRNA in the A site should produce a toeprint approximately 3 nucleotides from the CA6226–6227 toeprint. Indeed, the weaker toeprint at G6229 most likely represents such a scenario (Fig. 4). Since it has been observed that the P-site triplet in an 80S/RNA complex produces a 15–17-nucleotide toeprint when a tRNA occupies the P site of the ribosome, the strong toeprint at UAA6231–6233 most likely represents 80S/CrPV IGR–IRES complexes with an alanine-tRNA and the alanine GCU codon in the P site (Fig. 5).

The fact that the 80S/CrPV IGR–IRES complex produced a toeprint at UAA6231–6233 when the RRL was treated with cycloheximide suggests that a pseudotranslocation event occurred, that is, a translocation without bond formation (step IV, Fig. 5). In support of the pseudotranslocation hypothesis, addition of eIF5B and GMP-PNP, which blocks the delivery of aminoacylated tRNAs to the A site (Pestova et al. 2000), produced a

prominent toeprint at CA6226–6227, indicating that pseudotranslocation was inhibited. In summary, after an Ala-tRNA enters the A site, the 80S/CrPV IGR–IRES complex quickly translocates the alanine codon to the P site of the ribosome (Fig. 5). Since cycloheximide inhibits EF2 activity, this pseudotranslocation event does not require EF2 or prior peptide-bond formation.

The pseudotranslocation event is probably rapid as the toeprint at G6229, which represents the transient intermediate of an 80S/CrPV IGR–IRES complex with an Ala-tRNA in the A site, is weak. Addition of GMP-PNP, or GMP-PNP and cycloheximide, resulted in weak toeprints at CA6226–6227, G6229, and UAA6231–6233 (Fig. 4). This result indicates that pseudotranslocation can still occur in the presence of GMP–PNP. Because the toeprint at UAA6231-6233 is much weaker than the corresponding toeprints treated with cycloheximide alone, it suggests that the rate of pseudotranslocation is reduced when GMP-PNP is added. In summary, upon 80S assem-

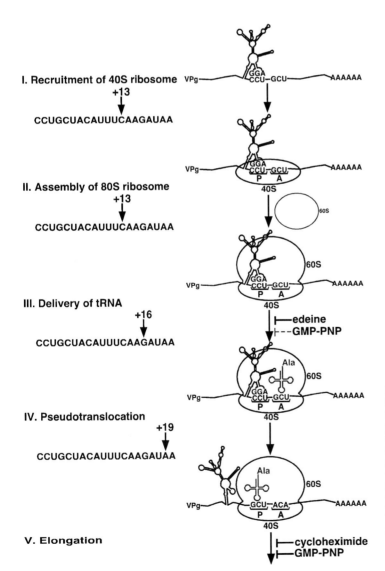

I. Recruitment of 40S ribosome +13

CCUGCUACAUUUCAAGAUAA

II. Assembly of 80S ribosome +13

CCUGCUACAUUUCAAGAUAA

III. Delivery of tRNA +16

CCUGCUACAUUUCAAGAUAA

IV. Pseudotranslocation +19

CCUGCUACAUUUCAAGAUAA

V. Elongation

Figure 5. Model for IGR–IRES-mediated translation initiation. See text for details. The position of the ribosome on the IGR–IRES at each step in the pathway is summarized at left. For simplicity, only nucleotides 6214–6233 of the IGR–IRES are shown (numbering as in Fig. 1), and only the strongest toeprint defining each step in this pathway is indicated, with the number above each arrow showing the position of the toeprint relative to C_{6214}. The steps that are blocked by various inhibitors are also indicated. The viral mRNA with genome-linked protein, VPg, 40S and 60S ribosomal subunits, and ribosomal P and A sites are indicated. (Reprinted, with permission, from Elsevier Science, Wilson et al. 2000a.)

bly, the IGR–IRES triggers the ribosome to perform a pseudotranslocation that does not depend on prior peptide-bond formation.

CrPV IGR–IRES FUNCTIONS IN *SACCHAROMYCES CEREVISIAE* WHEN TERNARY Met-tRNA$_i$/eIF2-GTP COMPLEX IS LOW

IRES elements in viral and mammalian cellular mRNAs do not function in living *S. cerevisiae*. The reasons for this are not entirely clear; however, it has been proposed that a particular yeast-encoded RNA inhibits the poliovirus and hepatitis C virus IRES elements by sequestration of factors that are required for internal initiation of translation (Coward and Dasgupta 1992; Das et al. 1998).

The unusual properties of the CrPV IGR–IRES, specifically the minimal requirement for translation initiation factors, prompted us to examine whether the IGR–IRES functions in yeast. To this end, we constructed a yeast copper promoter-containing plasmid that can direct the synthesis of dicistronic mRNAs containing *Leu2* as the first cistron, followed by the IGR–IRES, which is followed by *Ura3* as the second cistron (Fig. 6A). This plasmid was introduced into wild-type yeast cells (H1402) and grown on selective medium lacking uracil. However, no growth was observed under this condition, suggesting that an insufficient amount of Ura3 protein was produced in the transformed wild-type yeast strain (Fig. 6B).

It is known that the IGR–IRES functions poorly in eukaryotic cells when the rate of overall translation is high. However, lowering the amounts of ternary complexes by induction of eIF2 phosphorylation dramatically raises IGR–IRES efficiency (Wilson et al. 2000a). To test whether the IGR–IRES functions in yeast when overall levels of ternary complexes are low, we monitored IGR–IRES activity in a mutant strain (H1613) expressing a constitutively active *GCN2* (*GCN2^c*) allele, containing a double mutation in *GCN2*, E601K and E1591K, in the protein kinase domain and in the ribosome-binding domain, respectively. It was shown that overexpression of this allele results in a low abundance of ternary complexes (eIF2-GTP-tRNA^{met}) (Ramirez et al. 1992; Hinnebusch 1997). Briefly, GCN2 phosphorylates the α-subunit of eIF2, resulting in irreversible binding of eIF2 to eIF2B, an eIF2-recycling factor. Because eIF2B is limiting in the cell, eIF2 becomes sequestered in inactive complexes that are unavailable to be recycled from eIF2-GDP to eIF2-GTP; as a consequence, intracellular levels of ternary complexes become reduced (Hinnebusch 2000). Western blot analysis using antibodies directed against various forms of eIF2α showed that both wild-type and mutant strains contained equivalent amounts of eIF2α, but the fraction of phosphorylated eIF2α was significantly greater in the *GCN2^c* (H1613) strain (Fig. 6C). The dicistronic plasmid was introduced into the yeast *GCN2^c* strain, and it was noted that the IGR–IRES was capable of expressing sufficient amounts of Ura3 protein to support growth on plates in the absence of uracil (Fig. 6B). In contrast, a mutated IGR–IRES (mut14; Wilson et

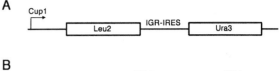

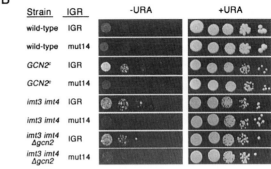

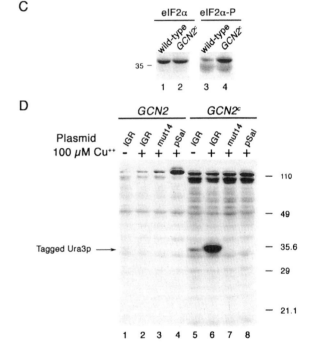

Figure 6. The CrPV IGR–IRES functions in living *S. cerevisiae* when ternary complex is low. (*A*) Diagram of the dicistronic reporter gene expressed from a copper-inducible promoter (Cup1). The upstream *Leu2* cistron and downstream *Ura3* cistron separated by the CrPV IGR–IRES are shown. (*B*) Growth assay for yeast strains H1402 (wild-type), H1613 (*GCN2^c*), H2545 (*imt3 imt4*), and H2546 (*imt3 imt4 Δgcn2*) transformed with the dicistronic reporter containing either the wild-type IGR–IRES (IGR) or a mutant IGR with a CC to GG change (mut14; also see mutant 1 in Fig.1). To induce expression of the reporter gene, transformed yeast strains were grown in the presence of 100 μM copper. See text for details. (*C*) Immunoblot analysis to measure levels of eIF2α, phosphorylated eIF2α, and FLAG-tagged Ura3 protein. Total levels of eIF2α were detected with the CM-217 polyclonal antibody (a gift from Tom Dever) that detects both phosphorylated and non-phosphorylated forms of eIF2α; the phosphorylated form of eIF2α was detected with a polyclonal antibody from Research Genetics. The FLAG-tagged Ura3 protein (Tagged Ura3p) was visualized using the anti-FLAG M2 monoclonal antibody (Sigma).

al. 2000b), which disrupts the pseudoknot in the IGR–IRES, did not promote growth on plates lacking uracil, suggesting that the expression of Ura3 protein is dependent on a functional IGR–IRES. Similar results were obtained in strains expressing a constitutively active *GCN2* mutant allele (H1692) containing a single mutation of E1606G in the ribosome-binding and dimerization domains of GCN2p (data not shown). Overall, these findings suggest that the IGR–IRES functions in yeast when eIF2α is phosphorylated.

Next, IGR–IRES activity was examined in yeast strains in which ternary complexes were lowered by limiting the pool of initiator tRNA molecules. To this end, the dicistronic plasmid was introduced into a yeast strain harboring disruptions in two initiator tRNA^{met} genes, *IMT3* and *IMT4* (H2545) (Dever et al. 1995). As predicted, the IGR–IRES mediated URA3 protein production in the mutant, *imt3 imt4* strain, as demonstrated by growth on plates lacking uracil (Fig. 6B). In addition, the IGR–IRES was active in strain H2546 (*imt3 imt4 Δgcn2*), containing a deletion in the *GCN2* gene in addition to the disruption in the two initiator tRNA^{met} genes, demonstrating that IGR–IRES activity was not dependent on the expression of wild-type GCN2 or constitutively active GCN2. Consistent with previous results, the dicistronic reporter containing the mutant IGR–IRES (mut14) did not support growth in the absence of uracil in either the H2545 or the H2546 strain, suggesting that synthesis of Ura3 protein is dependent on a wild-type IGR–IRES.

To examine whether Ura3 protein expression levels correlated with growth of yeast cells in the absence of uracil, proteins were isolated from various yeast strains grown under conditions in which the copper promoter was induced (+ Cu^{++}) or uninduced (–Cu^{++}). Ura3 protein was detected by an antibody directed against the FLAG-tag epitope located at the carboxyl terminus of URA3. The anti-flag antibody detected a protein that migrated at the predicted size of Ura3 protein in the *GCN2^c* mutant yeast cells, which were expressing dicistronic mRNAs with a wild-type IGR–IRES in the intergenic spacer region, in the presence of copper (Fig. 6D, lane 6). Only small amounts of Ura3p were produced in the absence of copper (Fig. 6D, lane 5). To test the possibility that the IGR–IRES cDNA contained cryptic promoter activity, which would result in the synthesis of Ura3 protein from functionally, capped monocistronic mRNAs, the size and integrity of expressed RNA species was examined by northern analysis. Only full-length dicistronic mRNA of similar amounts was detected in strains expressing wild-type or mutant IGR–IRES elements (data not shown). Taken together, these results suggest that the yeast translational apparatus is capable of performing efficient internal initiation on the CrPV IGR–IRES when the intracellular levels of the ternary complex are reduced.

CONCLUSIONS

IRES elements use canonical eukaryotic initiation factors for the recruitment of ribosomal subunits and for the positioning of the initiator tRNA into the ribosomal P site.

In contrast, the CrPV-like IGR–IRES elements can assemble 80S ribosomes without any canonical factor, including the initiator Met-tRNA. Upon assembly of an 80S ribosome, the IGR–IRES triggers the 80S into a pseudo-translocation reaction that proceeds without peptide-bond formation. Thus, the IGR–IRES may be a divergent RNA element that performs functions which were acquired by the ribosomal P site occupying initiator tRNA. The presence of such elements in eukaryotic host-cell mRNAs would ensure translation initiation of the IRES-containing mRNAs when canonical initiation factors are inactive. Finally, the IGR–IRES seems to be beneficial for virus spread, because it would ensure continued translation when antiviral responses, such as phosphorylation of eIF2α, have been launched in virus-infected cells.

ACKNOWLEDGMENTS

Work performed in the authors' laboratories was supported by grants from the Damon Runyon-Walter Winchell Foundation (E.J.), the National Institutes of Health (S.R.T., T.P., C.U.T.H., and P.S.), and the National Science Foundation (C.U.T.H.).

REFERENCES

Carrasco L., Battaner E., and Vazquez D. 1974. The elongation steps in protein synthesis by eukaryotic ribosomes: Effect of antibiotics. *Methods Enzymol.* **30:** 282.

Christian P.D. and Scotti P.D. 1998. Picornalike viruses of insects. In *The insect viruses* (ed. L.K. Miller and L.A. Ball), p. 301. Plenum Press, New York.

Coward P. and Dasgupta A. 1992. Yeast cells are incapable of translating RNAs containing the poliovirus 5′ untranslated region: Evidence for a translational inhibitor. *J. Virol.* **66:** 286.

Das S., Ott M., Yamane A., Venkatesan A., Gupta S., and Dasgupta A. 1998. Inhibition of internal entry site (IRES)-mediated translation by a small yeast RNA: A novel strategy to block hepatitis C virus protein synthesis. *Front. Biosci.* **3:** D1241.

Dever T.E., Yang W., Astrom S., Bystrom A.S., and Hinnebusch A.G. 1995. Modulation of tRNA(iMet), eIF-2, and eIF-2B expression shows that GCN4 translation is inversely coupled to the level of eIF-2.GTP.Met-tRNA(iMet) ternary complexes. *Mol. Cell. Biol.* **15:** 6351.

Domier L.L., McCoppin N.K., and D'Arcy C.J. 2000. Sequence requirements for translation initiation of *Rhopalosiphum padi* virus ORF2. *Virology* **268:** 264.

Hellen C.U.T. and Sarnow P. 2001. Internal ribosome entry sites in eukaryotic mRNA molecules. *Genes Dev.* **15:** 1593.

Hershey J.W.B. and Merrick W.C. 2000. The pathway and mechanism of initiation of protein synthesis. In *Translational control of gene expression* (ed. N. Sonenberg et al.), p. 33. Cold Spring Harbor Laboratory Press, Cold Spring Harbor, New York.

Hinnebusch A.G. 1997. Translational regulation of yeast GCN4. A window on factors that control initiator-tRNA binding to the ribosome. *J. Biol. Chem.* **272:** 21661.

———. 2000. Mechanism and regulation of initiator methionyl-tRNA binding to ribosomes. In *Translational control of gene expression* (ed. N. Sonenberg et al.), p. 185. Cold Spring Harbor Laboratory Press, Cold Spring Harbor, New York.

Jang S.K., Krausslich H.G., Nicklin M.J., Duke G.M., Palmenberg A.C., and Wimmer E. 1988. A segment of the 5′ nontranslated region of encephalomyocarditis virus RNA directs internal entry of ribosomes during in vitro translation. *J. Virol.* **62:** 2636.

Kanamori Y. and Nakashima N. 2001. A tertiary structure model of the internal ribosome entry site (IRES) for methionine-independent initiation of translation. *RNA* **7:** 266.

Kozak M. 1989. The scannning model for translation: An update. *J. Cell Biol.* **108:** 229.

Kozak M. and Shatkin A.J. 1978. Migration of 40S ribosomal subunits on mRNA in the presence of edeine. *J. Biol. Chem.* **253:** 6568.

Moore N.F. and Tinsley T.W. 1982. The small RNA-viruses of insects. *Arch. Virol.* **72:** 229.

Moore N.F., Kearns A., and Pullin J.S.K. 1980. Characterization of cricket paralysis virus-induced polypeptides in *Drosophila* cells. *J. Virol.* **33:** 1.

Odom O.W., Kramer G., Henderson A.B., Pinphanichakarn P., and Hardesty B. 1978. GTP hydrolysis during methionyl-tRNAf binding to 40 S ribosomal subunits and the site of edeine inhibition. *J. Biol. Chem.* **253:** 1807.

Pelletier J. and Sonenberg N. 1988. Internal initiation of translation of eukaryotic mRNA directed by a sequence derived from poliovirus RNA. *Nature* **334:** 320.

Pestova T.V., Hellen C.U., and Shatsky I.N. 1996. Canonical eukaryotic initiation factors determine initiation of translation by internal ribosomal entry. *Mol. Cell. Biol.* **16:** 6859.

Pestova T.V., Shatsky I.N., Fletcher S.P., Jackson R.J., and Hellen C.U. 1998. A prokaryotic-like mode of cytoplasmic eukaryotic ribosome binding to the initiation codon during internal translation initiation of hepatitis C and classical swine fever virus RNAs. *Genes Dev.* **12:** 67.

Pestova T.V., Lomakin I.B., Lee J.H., Choi S.K., Dever T.E., and Hellen C.U. 2000. The joining of ribosomal subunits in eukaryotes requires eIF5B. *Nature* **403:** 332.

Ramirez M., Wek R.C., Vazquez de Aldana C.R., Jackson B.M., Freeman B., and Hinnebusch A.G. 1992. Mutations activating the yeast eIF-2 alpha kinase GCN2: Isolation of alleles altering the domain related to histidyl-tRNA synthetases. *Mol. Cell. Biol.* **12:** 5801.

Reinganum C., O'Loughlin G.T., and Hogan T.W. 1970. A non-occluded virus of the field crickets *Teleogryllus oceanicus* and *T. commodus* (Orthoptera: Gryllidae). *J. Invertebr. Pathol.* **16:** 214.

Sasaki J. and Nakashima N. 1999. Translation initiation at the CUU codon is mediated by the internal ribosome entry site of an insect picorna-like virus in vitro. *J. Virol.* **73:** 1219.

———. 2000. Methionine-independent initiation of translation in the capsid protein of an insect RNA virus. *Proc. Natl. Acad. Sci.* **97:** 1512.

Szer W. and Kurylo-Borowska Z. 1970. Effect of edeine on aminoacyl-tRNA binding to ribosomes and its relationship to ribosomal binding sites. *Biochim. Biophys. Acta.* **224:** 447.

Wilson J.E., Pestova T.V., Hellen C.U., and Sarnow P. 2000a. Initiation of protein synthesis from the A site of the ribosome. *Cell* **102:** 511.

Wilson J.E., Powell M.J., Hoover S.E., and Sarnow P. 2000b. Naturally occurring dicistronic cricket paralysis virus RNA is regulated by two internal ribosome entry sites. *Mol. Cell. Biol.* **20:** 4990.

The mRNA Closed-loop Model: The Function of PABP and PABP-interacting Proteins in mRNA Translation

A. KAHVEJIAN, G. ROY, AND N. SONENBERG

Department of Biochemistry and McGill Cancer Center, McGill University, Montréal, Québec, Canada

Translational control is an important means by which cells govern gene expression. It provides a rapid response to growth and proliferation stimuli and plays a role in controlling feedback mechanisms in the cell. In systems with little or no transcriptional control (e.g., reticulocytes and oocytes), translation is the predominant mode of regulation of gene expression (for review, see Mathews et al. 2000). Initiation, the rate-limiting step of translation, is often the target of translational control. All nuclear-transcribed eukaryotic mRNAs possess a 5′ cap (m^7GpppN, where m is a methyl group and N is any nucleotide), and most possess a 3′ poly(A) tail. The focus of this paper is the role of the poly(A) tail and the poly(A)-binding protein (PABP) in translation initiation. Such an analysis requires a brief introduction of translation initiation and its control.

Initiation of translation consists of recruitment of the ribosome to mRNA, negotiation of the 5′-untranslated region (5′UTR), and recognition of an initiation codon (for review, see Hershey and Merrick 2000). Eukaryotic translation initiation factors (eIFs) mediate the ribosome recruitment process. The eIF4 family of initiation factors is responsible for the recruitment of the small 40S ribosomal subunit to the 5′ cap structure. The cap is recognized by the eIF4F complex, which consists of eIF4E, eIF4A, and eIF4G. eIF4E binds directly to the mRNA 5′ cap structure, eIF4A is an RNA helicase, and eIF4G is a modular scaffolding protein that binds eIF4E, eIF4A, eIF3, PABP, and Mnk, a Ser/Thr kinase that phosphorylates eIF4E (for review, see Gingras et al. 1999; Pyronnet et al. 1999). eIF3 also interacts with the 40S ribosomal subunit, thus serving as a link between the mRNA/eIF4F complex and the ribosome. eIF4A, in conjunction with eIF4B and eIF4H, is thought to unwind the mRNA 5′ secondary structure to facilitate 40S ribosome binding (for review, see Hershey and Merrick 2000). eIF4A may also be involved in the progression of the 40S ribosomal subunit along the mRNA 5′UTR. Upon recognition of an initiation codon, most of the initiation factors are released, followed by the joining of the large 60S ribosomal subunit (for review, see Hershey and Merrick 2000). Ribosomes can also be recruited to the mRNA via an alternate pathway through an *internal ribosome entry site* (IRES). A number of viral RNAs and cellular mRNAs contain IRESs, in their 5′UTR, which bypass the requirement for a cap structure for translation initiation. Certain IRESs

interact with translation factors (e.g., the encephalomyocarditis virus IRES; Lomakin et al. 2000), whereas others directly interact with the 40S ribosomal subunit (e.g., the hepatitis C virus IRES; Kieft et al. 2001).

There are numerous examples of translational control which are exerted at the ribosome recruitment step, both general (applying to large numbers of mRNAs) and mRNA-specific. Notably, cap-dependent translation is modulated by the *eIF4E binding proteins* (4E-BPs). These small repressor molecules interact with eIF4E to compete with eIF4G binding, and thus inhibit eIF4F complex formation and initiation of translation (for review, see Raught et al. 2000). This inhibition is reversible, as the affinity of 4E-BPs for eIF4E is modulated by 4E-BP phosphorylation. Phosphorylation is regulated by many types of extracellular stimuli, including growth factors, hormones, and mitogens, all of which activate the PI3-kinase signaling pathway (for review, see Gingras et al. 2001). This mechanism highlights the importance of the link between extracellular stimuli and translational control.

Regulation of translation may also be mediated via the mRNA 3′UTR. For instance, short-lived mRNAs, which code for proto-oncogene products, cytokines, and early response gene products, contain AU-rich elements (ARE) in their 3′UTR which, under most circumstances, confer mRNA instability (Chen and Shyu 1994; Stoecklin et al. 1994; Xu et al. 1997). However, for some mRNAs the ARE affects translation. For example, the mRNA encoding tumor necrosis factor-α (TNF-α) contains an ARE that represses translation (Gueydan et al. 1999; Piecyk et al. 2000). TNF-α translation is stimulated in macrophages in response to lipopolysaccharides (LPS), by a mechanism that involves the removal of a repressor protein from the ARE (Piecyk et al. 2000). A large number of mRNAs contain 3′UTR elements that regulate translation during early development; the majority of these studies have been conducted in *Drosophila* (for review, see Wickens et al. 2000). The formation of the caudal protein gradient in the *Drosophila* embryo is dependent on a bicoid-response element (BRE) in the 3′UTR of the caudal mRNA. The bicoid protein interacts with the BRE and suppresses caudal mRNA translation (for review, see Wickens et al. 2000). Another example is the 3′UTR of 15-lipoxygenase (LOX), which contains a differentiation control element (DICE). When bound by heterogeneous nuclear ribonucleoproteins K and E1 (hnRNP-K and hnRNP-E1), this element represses translation (Ostareck et al. 1997). Recent studies have demonstrated that the DICE inhibits 60S subunit joining, rather than 40S recruitment (Ostareck et al. 2001). This mechanism is unexpected, as regulation of translation initiation usually occurs at the 40S ribosome recruitment step.

How mRNA 3′UTR elements may regulate translation is not well understood, but the recent findings demonstrating the circularization of eukaryotic mRNAs (see below) provide a possible framework to explain their mechanism of action. A key player in this mechanism is the poly(A) tail, which is bound by PABP. Here, we describe the mechanism by which PABP stimulates translation, and the various factors involved in this process, including two proteins recently identified in our laboratory.

THE POLY(A) TAIL AND TRANSLATION

The function of the poly(A) tail has been an active area of research for three decades (Kates 1971; Adesnik et al. 1972). Early in vitro experiments suggested a role for the poly(A) tail in translation initiation. The poly(A) tail was demonstrated to confer a translational advantage to the mRNA in reticulocyte lysate, and addition of poly(A) RNA inhibited the translation of poly(A)$^+$ mRNA (Doel and Carey 1976; Jacobson and Favreau 1983; Grossi de Sa et al. 1988). More detailed experiments comparing the translatability, degradation, polysome profile, and assembly into messenger ribonucleoprotein complexes (mRNPs) of synthetic poly(A)$^+$ versus poly(A)$^-$ mRNA were carried out in a reticulocyte lysate translation system by Munroe and Jacobson (1990b). These studies demonstrated a two- to threefold translational stimulation conferred by the poly(A) tail, which was not attributable to its mRNA-stabilizing effect (Munroe and Jacobson 1990b). Furthermore, addition of exogenous poly(A) to these extracts inhibited the translation of poly(A)$^+$ mRNA, while stimulating the translation of poly(A)$^-$ mRNA. Another finding consistent with the poly(A) tail's importance in translation is the positive correlation between the polyadenylation state of an mRNA and its translational activation during development. In many systems (e.g., *Xenopus, Drosophila,* mouse), the translation of maternal mRNAs in the oocyte coincides with their polyadenylation state (for review, see Wickens et al. 2000).

PABP

PABP was discovered in 1973 as the major protein associated with the poly(A) tail of eukaryotic mRNAs (Blobel 1973). Numerous studies in yeast have implicated PABP in mediating the effects of the poly(A) tail on translation initiation. Depletion of PABP in yeast by promoter inactivation or use of a temperature-sensitive mutation reduced translation initiation and cell growth (Sachs and Davis 1989), whereas immunodepletion of PABP from yeast extracts repressed translation of polyadenylated mRNAs (Tarun and Sachs 1995). PABP is an essential protein, because in yeast, deletion of the *PAB1* gene is lethal (Sachs et al. 1987).

PABP is a relatively abundant protein; HeLa cells contain approximately six molecules per ribosome (Görlach et al. 1994). It is a multidomain mRNA-binding protein containing four phylogenetically conserved RNA recognition motifs (RRMs) (Adam et al. 1986; Sachs et al. 1987). The RRM is the most common RNA-binding domain, and exists in more than 200 putative RNA-binding proteins (for review, see Nagai 1996; Varani and Nagai 1998; Perez-Canadillas and Varani 2001). X-Ray crystallography studies have demonstrated that the RRM consists of four antiparallel β sheets flanked by two α helices (Oubridge et al. 1994). Two conserved sequence motifs, RNP1 and RNP2, are located in the central two β strands and make contact with the RNA (Nagai 1996; Deo et al. 1999). PABP fragments containing RRMs 1 and 2 bind poly(A) RNA with an affinity similar to wild-type. RRMs 3 and 4 display a tenfold lower affinity for poly(A) (Burd et al. 1991; Kuhn and Pieler 1996; Deo et al. 1999). The proline-rich, carboxy-terminal one-third of the protein serves as a docking site for a number of proteins, including PABP interacting proteins 1 and 2 (Paip1 and Paip2) (see below; Deo et al. 2001; Khaleghpour et al. 2001b; Kozlov et al. 2001).

Different fragments of PABP can individually stimulate translation in *Xenopus* oocytes, independently of their poly(A)-binding activity (Gray et al. 2000). A fragment containing RRMs 1 and 2 of PABP, which binds eIF4G (see below), is more effective than full-length PABP in stimulating translation (Gray et al. 2000). A fragment consisting of RRMs 3 and 4 or the carboxyl terminus of PABP can also stimulate translation, albeit to a lesser degree (Gray et al. 2000). Otero et al. (1999) demonstrated that exogenous PABP stimulates the translation of capped poly(A)$^+$ and, to a lesser extent, poly(A)$^-$ mRNA in yeast extracts via distinct mechanisms. These findings suggest that the mechanism by which PABP stimulates translation is complex and may involve redundant or alternative pathways.

TRANSLATIONAL SYNERGY BETWEEN THE 5′ CAP AND THE POLY(A) TAIL

Suggestive evidence for the circularization of mRNA existed long before the discovery of the underlying protein interactions responsible for bridging the 5′ and 3′ termini (see below). The closed-loop model for mRNA circularization was proposed almost two decades ago (Jacobson and Favreau 1983; Palatnik et al. 1984) and reiterated in the following years (Sachs and Davis 1989; Munroe and Jacobson 1990a,b; Jacobson 1996). Furthermore, electron micrographs of rat pituitary cell cross-sections revealed a predominance of circular polysomes on the surface of the rough endoplasmic reticulum (Christensen et al. 1987). The synergistic enhancement of translation observed in mRNAs containing both a 5′ cap and 3′ poly(A) tail also suggested a physical interaction between the mRNA extremities (Gallie 1991). Electroporation of mRNAs into cells demonstrated that those possessing both a cap and a poly(A) tail were more efficiently translated than mRNAs possessing only one of the structures, and that this enhancement was synergistic (Gallie 1991). Capped and polyadenylated mRNAs display similar synergistic properties in yeast extracts (Iizuka et al. 1994), indicating that the poly(A) tail plays an important role in stimulating cap-dependent translation initiation.

THE eIF4G/PABP INTERACTION

The synergism between the mRNA 5′ and 3′ ends could be readily explained if they were to physically interact. Proof of a direct interaction between the 5′ and 3′ ends of mRNA proved elusive until the discovery of the physical association between eIF4G and PABP in yeast (Tarun and Sachs 1996) and plant systems (Le et al. 1997). However, at the same time, an interaction between mammalian PABP and eIF4G could not be demonstrated (see e.g., Craig et al. 1998). The subsequent discovery of an extended amino terminus of human eIF4G explains this difficulty, as it revealed the binding site for PABP (Imataka et al. 1998; Piron et al. 1998). A stretch of 29 amino acids in the amino terminus of human eIF4G interacts with RRMs 1 and 2 of human PABP (Imataka et al. 1998), as in yeast PABP. However, despite its high homology with yeast PABP, human PABP does not interact with yeast eIF4G (Otero et al. 1999), likely because the PABP-binding sites in the human and yeast eIF4G proteins are quite divergent. The fact that this interaction has been conserved through evolution (or arose twice), despite the divergence of the protein sequences, underscores its functional significance.

The interaction between PABP and eIF4G results in the circularization of the mRNA. Indeed, atomic force microscopy experiments demonstrated the formation of RNA circles when recombinant yeast eIF4G, eIF4E, and PABP were mixed with an mRNA possessing a 5′ cap and a 3′ poly(A) tail (Wells et al. 1998). The absence of any of the proteins prevented circle formation, as did the use of eIF4G mutants defective in eIF4E or PABP binding (Wells et al. 1998).

mRNA CIRCULARIZATION: BIOLOGICAL SIGNIFICANCE

What is the purpose of mRNA circularization? An interesting idea is that circularization permits efficient translation of intact mRNAs only, thus diminishing the possibility of generating potentially dominant negative forms of proteins from nicked mRNAs. The biological significance of circularization has been studied in several systems. Surprisingly, yeast strains containing an eIF4G mutant defective in PABP binding are viable, unless combined with a mutation in eIF4E (Tarun et al. 1997). This synthetic lethality suggests a redundancy in PABP and eIF4E function (Tarun et al. 1997). In contrast, the eIF4G–PABP interaction appears to play a critical role in *Xenopus* oocytes, because expression of an eIF4G mutant that does not bind PABP repressed translation of polyadenylated mRNAs and inhibited progesterone-

induced oocyte maturation (Wakiyama et al. 2000). Recently, the interaction between eIF4G and PABP was shown to be required for translational control of ceruloplasmin (Cp) (Mazumder et al. 2001). Binding of cytosolic factors to the 3′UTR of Cp mRNA represses translation by a mechanism dependent on circularization; immunodepeletion of PABP or eIF4G, or the absence of a poly(A) tail, abrogated the translational silencing of Cp mRNA (Mazumder et al. 2001).

5′–3′ interactions also occur in viral mRNAs lacking a cap structure or a poly(A) tail. Picornaviral RNAs do not possess a 5′ cap structure, and they initiate translation via ribosome binding to an IRES (Belsham and Jackson 2000). In this case, the poly(A) tail and the IRES synergistically promote translation, which may be explained by the direct binding of eIF4G to both the IRES and PABP (Bergamini et al. 2000; Lomakin et al. 2000). Hepatitis C virus (HCV) RNA also contains an IRES, but no poly(A) tail, and does not require eIF4G for translation (Kieft et al. 2001). In a model proposed by Ito and Lai, the polypyrimidine-tract-binding protein (PTB) serves to circularize HCV RNA and thus enhance its translation by interacting with an element in the 3′UTR and to a protein that interacts with the HCV IRES (Ito and Lai 1999). Strong evidence for the functional requirement of circularization exists for rotavirus mRNAs, which are capped but not polyadenylated. These mRNAs contain a conserved sequence (UGACC) in their 3′UTR that is recognized by the viral NSP3 protein (Poncet et al. 1993). NSP3 interacts with the amino terminus of eIF4G to stimulate viral mRNA translation (Poncet et al. 1993, 1996; Piron et al. 1998; Vende et al. 2000). In addition, NSP3 displaces PABP from eIF4G and thus inhibits host protein synthesis (Piron et al. 1998). These findings ascribe a key role for PABP as a translation factor and stress the importance of circularization for translational activation.

MECHANISM OF PABP STIMULATION OF TRANSLATION INITIATION

Several models have been proposed to explain the mechanism by which circularization promotes translation. First, circularization could promote the recycling of terminating ribosomes on the same mRNA by bringing the two ends of the mRNA together. Another model suggests that PABP increases 60S ribosomal subunit joining. A third model posits that PABP stabilizes the eIF4F–cap interaction and thus stimulates 40S ribosomal subunit binding.

In early experiments, Sachs and Davis observed that mutations in a 60S ribosomal protein or in a helicase required for 60S ribosomal subunit biosynthesis were capable of rescuing the phenotype of the PABP deletion in yeast (Sachs and Davis 1989, 1990). These genetic data are supported by biochemical experiments, as Munroe and Jacobson (1990b) reported that the absence of the poly(A) tail led to a decrease in 60S ribosomal subunit joining. Subsequent genetic studies in yeast are also consistent with such a model: Wickner's group showed that the poly(A) tail derepresses translation (Widner and

Wickner 1993). Poly(A)$^+$ mRNA lost its translational advantage over poly(A)$^-$ mRNA in yeast mutants lacking two Ski RNA helicase proteins, Ski2p and Slh1p. This is due to a dramatic increase in the translation of poly(A)-deficient mRNA, with no effect on poly(A)$^+$ mRNA (Widner and Wickner 1993). Deletion of eIF5B or mutation of eIF5, two initiation factors involved in 60S ribosomal subunit joining, drastically reduced translation of poly(A)$^+$ mRNA, with little effect on poly(A)$^-$ mRNA. Furthermore, the increase in poly(A)$^-$ mRNA translation in the Ski2Δ Slh1Δ strain was lost in an eIF5B null background (Searfoss et al. 2001). The Ski proteins were suggested to act by inhibiting eIF5 and eIF5B, thus impeding 60S ribosomal subunit joining. Importantly, this inhibition is repressed by PABP (Searfoss et al. 2001). Notwithstanding these data, other experiments argue for a role of PABP in recruiting the 40S ribosomal subunit to the mRNA. In extracts immunodepleted of PABP, mRNA translation and 40S ribosomal subunit recruitment were inhibited (Tarun and Sachs 1995). Biochemical data also suggest that the mRNA 5′–3′ interactions may stabilize the formation of the 5′ initiation complex. PABP was reported to increase the affinity of eIF4F for the cap structure (Wei et al. 1998) and to interact with eIF4B in plant and mammalian systems (Le et al. 1997; Bi and Goss 2000; Bushell et al. 2001). One possible explanation for the discrepancy between the models is that when 60S ribosomal subunit joining is inhibited (as in the eIF5 or eIF5B mutants, or by diminishing 60S biogenesis), 60S ribosome joining becomes rate-limiting for initiation, and thus the difference between poly(A)$^+$ and poly(A)$^-$ mRNAs in 40S recruitment would be difficult to detect.

PABP INTERACTING PROTEINS 1 AND 2

Interaction screening using the far western technique with radiolabeled PABP led our laboratory to the cloning of two PABP-binding partners: PABP interacting proteins 1 and 2 (Paip1 and Paip2) (Craig et al. 1998; Khaleghpour et al. 2001a). Paip1 is a 54-kD protein that stimulates translation in vivo. It shares homology with the central region of eIF4G, which contains an eIF4A-binding site, and it interacts with eIF4A (Craig et al. 1998). Paip1 is also involved in mRNA turnover. It is found in a protein complex that stabilizes the c-*fos* proto-oncogene mRNA by binding to the major protein-coding-region determinant of instability (mCRD) (Grosset et al. 2000). Conversely, Paip2 (MW = 14 kD) preferentially inhibits the translation of mRNAs possessing a poly(A) tail (Fig. 1A) (Khaleghpour et al. 2001a). Translation of an mRNA containing the HCV IRES is not inhibited by Paip2 (Fig.1B) (Khaleghpour et al. 2001a), most likely because this IRES translates independently of eIF4G (Pestova et al. 1998). Paip2 inhibits 80S ribosome complex formation in a ribosome-binding assay (Khaleghpour et al. 2001a), but its effect on 40S ribosome complex formation has not been examined. How might Paip2 inhibit ribosome complex formation? Paip2 competes with Paip1 for binding to PABP, and in an RNA-binding as-

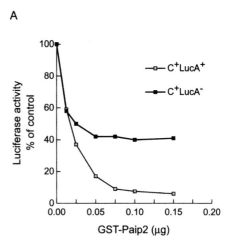

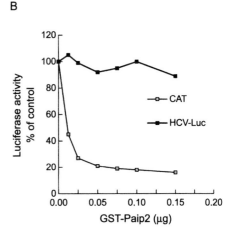

Figure 1. Paip2 inhibits translation in vitro. (*A*) Translation of capped poly(A)$^+$ or poly(A)$^-$ luciferase mRNA in the presence of Paip2 in a Krebs-2 cell-free translation system. Shown as relative luciferase activity (% of control). (*B*) Effect of Paip2 on translation of capped, poly(A)$^+$ bicistronic CAT-HCV IRES-luc mRNA in a Krebs-2 cell-free translation system. Shown as relative luciferase activity (% of control). (Adapted, with permission, from Khaleghpour et al. 2001a.)

say, strongly interdicts the interaction of PABP with poly(A), thus disrupting the poly(A)-organizing activity of PABP (Fig. 2A, lanes 1–4) (Khaleghpour et al. 2001a). The simplest model to depict Paip2 function is that it inhibits translation initiation by disrupting the circularization of mRNAs (Fig. 3). The binding sites for Paip2 and eIF4G on PABP overlap (see below), suggesting that they compete for binding to PABP. A mutant of Paip2, Paip2 (1-42), which does not bind to PABP, and a mutant that binds only to the carboxyl terminus, but not to the RRM region of PABP, Paip2 (76-127), do not affect the PABP/poly(A) interaction (Fig. 2A, lanes 5–8 and 13–16, and Fig. 2B). Furthermore, these fragments of Paip2 do not inhibit translation (Fig. 2C) (Khaleghpour et al. 2001b).

Paip1 interacts with RRMs 1 and 2 and the carboxyl terminus of PABP (G. Roy et al., unpubl.), whereas Paip2 interacts with RRMs 2 and 3, and the carboxyl terminus

A

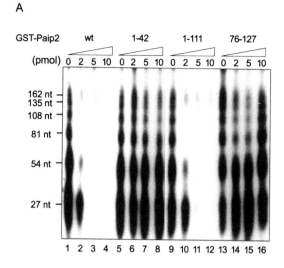

B

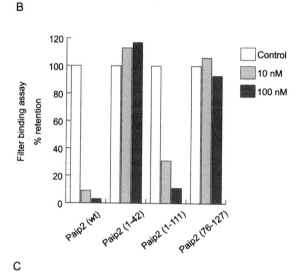

C

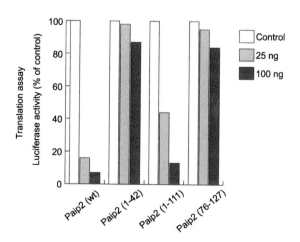

Figure 2. Functional dissection of Paip2. (*A*) Disruption of poly(A)-organizing activity of PABP by Paip2 fragments. (*B*) Inhibition of PABP binding to poly(A) by Paip2 fragments in a filter binding assay. (*C*) Effects of Paip2 fragments on capped poly(A)$^+$ luciferase mRNA translation in a Krebs-2 cell-free translation system. (Adapted, with permission, from Khaleghpour et al. 2001b.)

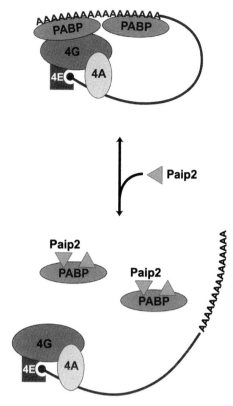

Figure 3. Model for the inhibition of translation by Paip2. Paip2 inhibits PABP binding to the mRNA and may interfere with the PABP/eIF4G interaction on the mRNA, resulting in the disruption of the circular conformation.

of PABP (Khaleghpour et al. 2001b). Paip2 binds to PABP with a 2:1 stoichiometry with two independent K_ds of 0.66 and 74 nM (Khaleghpour et al. 2001b). Paip1 binds to PABP with a K_d of 1.9 nM, and the stoichiometry of this interaction is currently under investigation. Importantly, the interactions of Paip1 and Paip2 with PABP are mutually exclusive (Khaleghpour et al. 2001a). The binding of both Paip1 and Paip2 to PABP requires RRM2 (Khaleghpour et al. 2001b; G. Roy, unpubl.), and both proteins contain a conserved 15-amino-acid stretch necessary for binding the carboxyl terminus of PABP (Fig. 4) (Khaleghpour et al. 2001b).

NEW PABP INTERACTING PARTNERS

The 15-amino-acid stretch, which is shared by Paip1 and Paip2, is also present in a number of other proteins (Fig. 4), suggesting that the carboxyl terminus of PABP has other binding partners and perhaps undiscovered roles unrelated to translation (Deo et al. 2001; Kozlov et al. 2001). Paip1 and Paip2 may therefore compete with the carboxy-terminal binding partners of PABP and modulate other aspects of PABP function. For example, eRF3, the eukaryotic translation termination factor, also contains the conserved 15-amino-acid sequence in its

A

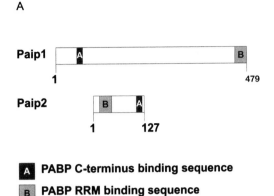

A PABP C-terminus binding sequence

B PABP RRM binding sequence

B

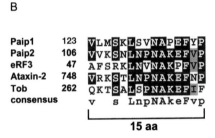

Figure 4. PABP interaction sites. (*A*) Respective PABP-binding sites in Paip1 and Paip2. (*B*) Alignment of the PABP carboxy-terminal binding site in Paip1 and Paip2 with amino acid sequences in eukaryotic release factor 3 (eRF3), Ataxin-2, and Tob. The interactions between the first three proteins in this list and the carboxyl terminus of PABP have been reported.

amino terminus and interacts with the carboxyl terminus of PABP (Hoshino et al. 1999). The large distance between the translation termination codon and the poly(A) tail in mRNAs with long 3′UTRs has served as an argument against the poly(A) tail's role in ribosome recycling. However, the interaction between eRF3 and PABP may provide a physical link between a translation-terminating ribosome and the 3′ poly(A) tail for recruitment to the 5′ end of the mRNA. Another surprising and interesting protein that contains the PABP carboxy-terminal binding motif, Ataxin-2, is implicated in spinocerebellar ataxia (Lorenzetti et al. 1997) and interacts with another RRM-containing protein, A2BP1 (Shibata et al. 2000). The function of Ataxin-2 is unknown, but its possible interaction with PABP may shed new light on spinocerebellar ataxia. The 15-amino-acid motif is also found in the *t*ransducer *of* Erb*B*-2 (Tob) (Deo et al. 2001), which interacts with p185^erbB2, a transmembrane glycoprotein containing an SH2-binding domain that stimulates cell growth (Matsuda et al. 1996). Tob has antiproliferative properties and is homologous to two other suppressors of cell growth, *B*-cell *t*ranslocation *g*ene protein 1 (BTG-1) and the *a*bundant in *n*euroepithelium *a*rea protein (ANA) (Yoshida et al. 1998). Members of this family of proteins also interact with the human homolog of yeast Caf1

(Yoshida et al. 2001), an RNA-binding protein involved in mRNA deadenylation (Tucker et al. 2001). The possible interaction of Tob with the carboxyl terminus of PABP may implicate Tob in mediating signaling to the translational and deadenylation apparatus. Paip1 and Paip2 also interact with the carboxyl terminus of the HYD ubiquitin ligase via the same 15-amino-acid stretch (Deo et al. 2001), potentially targeting these proteins for degradation. It would be of interest to determine whether ubiquitination serves as a regulator of mRNA circularization and translation.

REGULATION OF Paip FUNCTION

Paip2 is a phosphoprotein, the phosphorylation state of which is modulated by serum (A. Kahvejian and B. Raught, unpubl.). The functional significance, the intracellular signaling pathways involved in this regulation, and the locations of the phosphorylation sites are currently unknown. Paip1 is also phosphorylated (B. Raught and A.-C. Gingras, unpubl.). Does phosphorylation of Paip1 and Paip2 serve as a switch to promote or inhibit the circularization of mRNA? Elucidation of the pathways responsible for these modifications and of the biochemical consequences that ensue will help to answer this question. If phosphorylation indeed modulates Paip1 and Paip2 function at the 3′ end of the mRNA, the mechanism would resemble the modulation of 4E-BP function by phosphorylation (Raught et al. 2000).

CONCLUSIONS

Translational control had long been thought to be mediated primarily via the 5′ end of the mRNA. However, over the last decade, observations of the effects of the poly(A) tail and PABP on translation initiation have provided renewed importance to the 3′UTR of mRNA. Yet, the mechanism of poly(A) action is not well understood. The discovery of the interaction of PABP with both eIF4G and the Paips may explain the link between the 5′ and 3′ ends of the mRNA and shed new light on the role of PABP as a translation factor. It is noteworthy that Paip1 and Paip2 exist only in metazoans. They may thus function in a higher-order mechanism of translational control in the metazoa. Further research into the regulation of these proteins and the discovery of new protein–protein or protein–RNA interactions will result in a more in-depth understanding of the importance of mRNA circularization for translation and translational control.

ACKNOWLEDGMENTS

We thank Brian Raught, Francis Poulin, Mathieu Miron, and Rahul C. Deo for critical reading of this manuscript. This work was supported by a grant from the Canadian Institute of Health Research to N.S. N.S. is a distinguished scientist of the Canadian Institute of Health Research and a Howard Hughes International Scholar. A.K. and G.R. were recipients of doctoral studentships from the Canadian Institute of Health Research.

REFERENCES

Adam S.A., Nakagawa T., Swanson M.S., Woodruff T.K., and Dreyfuss G. 1986. mRNA polyadenylate-binding protein: Gene isolation and sequencing and identification of a ribonucleoprotein consensus sequence. *Mol. Cell. Biol.* **6:** 2932.

Adesnik M., Salditt M., Thomas W., and Darnell J.E. 1972. Evidence that all messenger RNA molecules (except histone messenger RNA) contain poly(A) sequences and that the poly(A) has a nuclear function. *J. Mol. Biol.* **71:** 21.

Belsham G.J. and Jackson R.J. 2000. Translation initiation on picornavirus RNA. In *Translational control of gene expression* (ed. N. Sonenberg et al.), p. 869. Cold Spring Harbor Laboratory Press, Cold Spring Harbor, New York.

Bergamini G., Preiss T., and Hentze M.W. 2000. Picornavirus IRESes and the poly(A) tail jointly promote cap-independent translation in a mammalian cell-free system. *RNA* **6:** 1781.

Bi X. and Goss D.J. 2000. Wheat germ poly(A)-binding protein increases the ATPase and the RNA helicase activity of translation initiation factors eIF4A, eIF4B, and eIF-iso4F. *J. Biol. Chem.* **275:** 17740.

Blobel G. 1973. A protein of molecular weight 78,000 bound to the polyadenylate region of eukaryotic messenger RNAs. *Proc. Natl. Acad. Sci.* **70:** 924.

Burd C.G., Matunis E.L., and Dreyfuss G. 1991. The multiple RNA-binding domains of the mRNA poly(A)-binding protein have different RNA-binding activities. *Mol. Cell. Biol.* **11:** 3419.

Bushell M., Wood W., Carpenter G., Pain V.M., Morley S.J., and Clemens M.J. 2001. Disruption of the interaction of mammalian protein synthesis eukaryotic initiation factor 4B with the poly(A)-binding protein by caspase- and viral protease-mediated cleavages. *J. Biol. Chem.* **276:** 23922.

Chen C.Y. and Shyu A.B. 1994. Selective degradation of early-response-gene mRNAs: Functional analyses of sequence features of the AU-rich elements. *Mol. Cell. Biol.* **14:** 8471.

Christensen A.K., Kahn L.E., and Bourne C.M. 1987. Circular polysomes predominate on the rough endoplasmic reticulum of somatotropes and mammotropes in the rat anterior pituitary. *Am. J. Anat.* **178:** 1.

Craig A.W., Haghighat A., Yu A.T., and Sonenberg N. 1998. Interaction of polyadenylate-binding protein with the eIF4G homologue PAIP enhances translation. *Nature* **392:** 520.

Deo R.C., Sonenberg N., and Burley S.K. 2001. X-ray structure of the human hyperplastic discs protein: An ortholog of the C-terminal domain of poly(A)-binding protein. *Proc. Natl. Acad. Sci.* **98:** 4414.

Deo R.C., Bonanno J.B., Sonenberg N., and Burley S.K. 1999. Recognition of polyadenylate RNA by the poly(A)-binding protein. *Cell* **98:** 835.

Doel M.T. and Carey N.H. 1976. The translational capacity of deadenylated ovalbumin messenger RNA. *Cell* **8:** 51.

Gallie D.R. 1991. The cap and poly(A) tail function synergistically to regulate mRNA translational efficiency. *Genes Dev.* **5:** 2108.

Gingras A.-C., Raught B., and Sonenberg N. 1999. eIF4 initiation factors: Effectors of mRNA recruitment to ribosomes and regulators of translation. *Annu. Rev. Biochem.* **68:** 913.

———. 2001. Regulation of translation initiation by FRAP/mTOR. *Genes Dev.* **15:** 807.

Görlach M., Burd C.G., and Dreyfuss G. 1994. The mRNA poly(A)-binding protein: Localization, abundance, and RNA-binding specificity. *Exp. Cell Res.* **211:** 400.

Gray N.K., Coller J.M., Dickson K.S., and Wickens M. 2000. Multiple portions of poly(A)-binding protein stimulate translation *in vivo*. *EMBO J.* **19:** 4723.

Grosset C., Chen C.Y., Xu N., Sonenberg N., Jacquemin-Sablon H., and Shyu A.B. 2000. A mechanism for translationally coupled mRNA turnover: Interaction between the poly(A) tail and a c-fos RNA coding determinant via a protein complex. *Cell* **103:** 29.

Grossi de Sa M.F., Standart N., Martins de Sa C., Akhayat O., Huesca M., and Scherrer K. 1988. The poly(A)-binding protein facilitates *in vitro* translation of poly(A)-rich mRNA. *Eur. J. Biochem.* **176:** 521.

Gueydan C., Droogmans L., Chalon P., Huez G., Caput D., and Kruys V. 1999. Identification of TIAR as a protein binding to the translational regulatory AU-rich element of tumor necrosis factor alpha mRNA. *J. Biol. Chem.* **274:** 2322.

Hershey J.W.B. and Merrick W.C. 2000. Pathway and mechanism of initiation of protein synthesis. In *Translational control of gene expression* (ed. N. Sonenberg et al.), p. 33. Cold Spring Harbor Laboratory Press, Cold Spring Harbor, New York.

Hoshino S., Imai M., Kobayashi T., Uchida N., and Katada T. 1999. The eukaryotic polypeptide chain releasing factor (eRF3/GSPT) carrying the translation termination signal to the 3′-poly(A) tail of mRNA. Direct association of eRF3/GSPT with polyadenylate-binding protein. *J. Biol. Chem.* **274:** 16677.

Iizuka N., Najita L., Franzusoff A., and Sarnow P. 1994. Cap-dependent and cap-independent translation by internal initiation of mRNAs in cell extracts prepared from *Saccharomyces cerevisiae*. *Mol. Cell. Biol.* **14:** 7322.

Imataka H., Gradi A., and Sonenberg N. 1998. A newly identified N-terminal amino acid sequence of human eIF4G binds poly(A)-binding protein and functions in poly(A)-dependent translation. *EMBO J.* **17:** 7480.

Ito T. and Lai M.M. 1999. An internal polypyrimidine-tract-binding protein-binding site in the hepatitis C virus RNA attenuates translation, which is relieved by the 3′-untranslated sequence. *Virology* **254:** 288.

Jacobson A. 1996. Poly(A) metabolism and translation: The closed-loop model. In *Translational control* (ed. J.W.B. Hershey et al.), p. 451. Cold Spring Harbor Laboratory Press, Cold Spring Harbor, New York.

Jacobson A. and Favreau M. 1983. Possible involvement of poly(A) in protein synthesis. *Nucleic Acids Res.* **11:** 6353.

Kates J. 1971. Transcription of the vaccinia virus genome and the occurrence of polyriboadenylic acid sequences in messenger RNA. *Cold Spring Harbor Symp. Quant. Biol.* **35:** 743. Cold Spring Harbor Laboratory Press, Cold Spring Harbor, New York.

Khaleghpour K., Svitkin Y.V., Craig A.W., DeMaria C.T., Deo R.C., Burley S.K., and Sonenberg N. 2001a. Translational repression by a novel partner of human poly(A) binding protein, Paip2. *Mol. Cell.* **7:** 205.

Khaleghpour K., Kahvejian A., De Crescenzo G., Roy G., Svitkin Y.V., Imataka H., O'Connor-McCourt M., and Sonenberg N. 2001b. Dual interactions of the translational repressor Paip2 with poly(A) binding protein. *Mol. Cell. Biol.* **21:** 5200.

Kieft J.S., Zhou K., Jubin R., and Doudna J.A. 2001. Mechanism of ribosome recruitment by hepatitis C IRES RNA. *RNA* **7:** 194.

Kozlov G., Trempe J.F., Khaleghpour K., Kahvejian A., Ekiel I., and Gehring K. 2001. Structure and function of the C-terminal PABC domain of human poly(A)-binding protein. *Proc. Natl. Acad. Sci.* **98:** 4409.

Kuhn U. and Pieler T. 1996. *Xenopus* poly(A) binding protein: Functional domains in RNA binding and protein-protein interaction. *J. Mol. Biol.* **256:** 20.

Le H., Tanguay R.L., Balasta M.L., Wei C.C., Browning K.S., Metz A.M., Goss D.J., and Gallie D.R. 1997. Translation initiation factors eIF-iso4G and eIF-4B interact with the poly(A)-binding protein and increase its RNA binding activity. *J. Biol. Chem.* **272:** 16247.

Lomakin I.B., Hellen C.U., and Pestova T.V. 2000. Physical association of eukaryotic initiation factor 4G (eIF4G) with eIF4A strongly enhances binding of eIF4G to the internal ribosomal entry site of encephalomyocarditis virus and is required for internal initiation of translation. *Mol. Cell. Biol.* **20:** 6019.

Lorenzetti D., Bohlega S., and Zoghbi H.Y. 1997. The expansion of the CAG repeat in ataxin-2 is a frequent cause of autosomal dominant spinocerebellar ataxia. *Neurology.* **49:** 1009.

Mathews M.B., Sonenberg N., and Hershey J.W.B. 2000. Origins and principles of translational control. In *Translational control of gene expression* (ed. N. Sonenberg et al.), p. 1. Cold Spring Harbor Laboratory Press, Cold Spring Harbor, New York.

Matsuda S., Kawamura-Tsuzuku J., Ohsugi M., Yoshida M., Emi

M., Nakamura Y., Onda M., Yoshida Y., Nishiyama A., and Yamamoto T. 1996. Tob, a novel protein that interacts with p185erbB2, is associated with anti-proliferative activity. *Oncogene* **12:** 705.

Mazumder B., Seshadri V., Imataka H., Sonenberg N., and Fox P.L. 2001. Translational silencing of ceruloplasmin requires the essential elements of mRNA circularization: Poly(A) tail, poly(A)-binding protein, and eukaryotic translation initiation factor 4G. *Mol. Cell. Biol.* **21:** 6440.

Munroe D. and Jacobson A. 1990a. Tales of poly(A): A review. *Gene* **91:** 151.

———. 1990b. mRNA poly(A) tail, a 3′ enhancer of translational initiation. *Mol. Cell. Biol.* **10:** 3441.

Nagai K. 1996. RNA-protein complexes. *Curr. Opin. Struct. Biol.* **6:** 53.

Ostareck D.H., Ostareck-Lederer A., Shatsky I.N., and Hentze M.W. 2001. Lipoxygenase mRNA silencing in erythroid differentiation: The 3′UTR regulatory complex controls 60S ribosomal subunit joining. *Cell.* **104:** 281.

Ostareck D.H., Ostareck-Lederer A., Wilm M., Thiele B.J., Mann M., and Hentze M.W. 1997. mRNA silencing in erythroid differentiation: hnRNP K and hnRNP E1 regulate 15-lipoxygenase translation from the 3′ end. *Cell.* **89:** 597.

Otero L.J., Ashe M.P., and Sachs A.B. 1999. The yeast poly(A)-binding protein Pab1p stimulates *in vitro* poly(A)-dependent and cap-dependent translation by distinct mechanisms. *EMBO J.* **18:** 3153.

Oubridge C., Ito N., Evans P.R., Teo C.H., and Nagai K. 1994. Crystal structure at 1.92 Å resolution of the RNA-binding domain of the U1A spliceosomal protein complexed with an RNA hairpin. *Nature* **372:** 432.

Palatnik C.M., Wilkins C., and Jacobson A. 1984. Translational control during early *Dictyostelium* development: Possible involvement of poly(A) sequences. *Cell* **36:** 1017.

Perez-Canadillas J.M. and Varani G. 2001. Recent advances in RNA-protein recognition. *Curr. Opin. Struct. Biol.* **11:** 53.

Pestova T.V., Shatsky I.N., Fletcher S.P., Jackson R.J., and Hellen C.U. 1998. A prokaryotic-like mode of cytoplasmic eukaryotic ribosome binding to the initiation codon during internal translation initiation of hepatitis C and classical swine fever virus RNAs. *Genes Dev.* **12:** 67.

Piecyk M., Wax S., Beck A.R., Kedersha N., Gupta M., Maritim B., Chen S., Gueydan C., Kruys V., Streuli M., and Anderson P. 2000. TIA-1 is a translational silencer that selectively regulates the expression of TNF-alpha. *EMBO J.* **19:** 4154.

Piron M., Vende P., Cohen J., and Poncet D. 1998. Rotavirus RNA-binding protein NSP3 interacts with eIF4GI and evicts the poly(A) binding protein from eIF4F. *EMBO J.* **17:** 5811.

Poncet D., Aponte C., and Cohen J. 1993. Rotavirus protein NSP3 (NS34) is bound to the 3′ end consensus sequence of viral mRNAs in infected cells. *J. Virol.* **67:** 3159.

———. 1996. Structure and function of rotavirus nonstructural protein NSP3. *Arch. Virol. Suppl.* **12:** 29.

Pyronnet S., Imataka H., Gingras A.-C., Fukunaga R., Hunter T., and Sonenberg N. 1999. Human eukaryotic translation initiation factor 4G (eIF4G) recruits Mnk1 to phosphorylate eIF4E. *EMBO J.* **18:** 270.

Raught B., Gingras A.-C., and Sonenberg N. 2000. Regulation of ribosomal recruitment in eukaryotes. In *Translational control of gene expression* (ed. N. Sonenberg et al.), p. 245. Cold Spring Harbor Laboratory Press, Cold Spring Harbor, New York.

Sachs A.B. and Davis R.W. 1989. The poly(A) binding protein is required for poly(A) shortening and 60S ribosomal subunit-dependent translation initiation. *Cell.* **58:** 857.

———. 1990. Translation initiation and ribosomal biogenesis: Involvement of a putative rRNA helicase and RPL46. *Science*

247: 1077.

Sachs A.B., Davis R.W., and Kornberg R.D. 1987. A single domain of yeast poly(A)-binding protein is necessary and sufficient for RNA binding and cell viability. *Mol. Cell. Biol.* **7:** 3268.

Searfoss A., Dever T.E., and Wickner R. 2001. Linking the 3′ poly(A) tail to the subunit joining step of translation initiation: Relations of pab1p, eukaryotic translation initiation factor 5b (Fun12p), and Ski2p-Slh1p. *Mol. Cell. Biol.* **21:** 4900.

Shibata H., Huynh D.P., and Pulst S.M. 2000. A novel protein with RNA-binding motifs interacts with ataxin-2. *Hum. Mol. Genet.* **9:** 1303.

Stoecklin G., Hahn S., and Moroni C. 1994. Functional hierarchy of AUUUA motifs in mediating rapid interleukin-3 mRNA decay. *J. Biol. Chem.* **269:** 28591.

Tarun S.Z. and Sachs A.B. 1995. A common function for mRNA 5′ and 3′ ends in translation initiation in yeast. *Genes Dev.* **9:** 2997.

———. 1996. Association of the yeast poly(A) tail binding protein with translation initiation factor eIF-4G. *EMBO J.* **15:** 7168.

Tarun S.Z., Jr., Wells S.E., Deardorff J.A., and Sachs A.B. 1997. Translation initiation factor eIF4G mediates *in vitro* poly(A) tail-dependent translation. *Proc. Natl. Acad. Sci.* **94:** 9046.

Tucker M., Valencia-Sanchez M.A., Staples S.R., Chen J., Denis C.L., and Parker R. 2001. The transcription factor associated Ccr4 and Caf1 proteins are components of the major cytoplasmic mRNA deadenylase in *Saccharomyces cerevisiae. Cell.* **104:** 377.

Varani G. and Nagai K. 1998. RNA recognition by RNP proteins during RNA processing. *Annu. Rev. Biophys. Biomol. Struct.* **27:** 407.

Vende P., Piron M., Castagne N., and Poncet D. 2000. Efficient translation of rotavirus mRNA requires simultaneous interaction of NSP3 with the eukaryotic translation initiation factor eIF4G and the mRNA 3′ end. *J. Virol.* **74:** 7064.

Wakiyama M., Imataka H., and Sonenberg N. 2000. Interaction of eIF4G with poly(A)-binding protein stimulates translation and is critical for *Xenopus* oocyte maturation. *Curr. Biol.* **10:** 1147.

Wei C.C., Balasta M.L., Ren J., and Goss D.J. 1998. Wheat germ poly(A) binding protein enhances the binding affinity of eukaryotic initiation factor 4F and (iso)4F for cap analogues. *Biochemistry* **37:** 1910.

Wells S.E., Hillner P.E., Vale R.D., and Sachs A.B. 1998. Circularization of mRNA by eukaryotic translation initiation factors. *Mol. Cell.* **2:** 135.

Wickens M., Goodwin E.B., Kimble J., Strickland S., and Hentze M.W. 2000. Translational control of developmental decisions. In *Translational control of gene expression* (ed. N. Sonenberg et al.), p. 295. Cold Spring Harbor Laboratory Press, Cold Spring Harbor, New York.

Widner W.R. and Wickner R.B. 1993. Evidence that the SKI antiviral system of *Saccharomyces cerevisiae* acts by blocking expression of viral mRNA. *Mol. Cell. Biol.* **13:** 4331.

Xu N., Chen C.Y., and Shyu A.B. 1997. Modulation of the fate of cytoplasmic mRNA by AU-rich elements: Key sequence features controlling mRNA deadenylation and decay. *Mol. Cell. Biol.* **17:** 4611.

Yoshida Y., Hosoda E., Nakamura T., and Yamamoto T. 2001. Association of ANA, a member of the antiproliferative Tob family proteins, with a caf1 component of the ccr4 transcriptional regulatory complex. *Jpn. J. Cancer Res.* **92:** 592.

Yoshida Y., Matsuda S., Ikematsu N., Kawamura-Tsuzuku J., Inazawa J., Umemori H., and Yamamoto T. 1998. ANA, a novel member of Tob/BTG1 family, is expressed in the ventricular zone of the developing central nervous system. *Oncogene* **16:** 2687.

The mRNA Capping Apparatus as Drug Target and Guide to Eukaryotic Phylogeny

S. Shuman

Molecular Biology Program, Sloan-Kettering Institute, New York, New York 10021

The 5′ cap is a distinctive feature of eukaryotic cellular and viral messenger RNA. The cap structure consists of 7-methyl guanosine linked via an inverted 5′–5′ triphosphate bridge to the initiating nucleoside of the transcript (Fig. 1). The cap is formed by enzymatic modification of nascent pre-mRNAs as they are being synthesized by cellular or viral RNA polymerases. Capping of cellular RNAs in vivo is effectively restricted to the transcripts made by nuclear RNA polymerase II (i.e., pre-mRNAs and certain small nuclear RNAs). Recent studies have shown that the targeting of cap formation is achieved through direct binding of the cellular capping apparatus to the RNA polymerase II elongation complex (Shuman 1997). Many eukaryotic DNA and RNA viruses, especially those that replicate in the cytoplasm, encode their own RNA polymerase and their own capping apparatus. In such systems, the viral capping enzymes interact with, or are subunits of, the viral transcription complex (Hagler and Shuman 1992; Guarino et al. 1998).

Cellular mRNAs are capped via three enzymatic reactions: (1) The 5′ triphosphate end of the nascent pre-mRNA is hydrolyzed to a diphosphate by RNA 5′ triphosphatase; (2) the diphosphate RNA end is capped with GMP by RNA guanylyltransferase; and (3) the GpppN cap is methylated by RNA (guanine-N7) methyltransferase (Fig. 1). This pathway was originally elucidated via the analysis of vaccinia virus capping enzymes (Ensinger et al. 1975). The same pathway is conserved in all eukaryotic organisms that have been examined. Key points are: (1) The triphosphate bridge of the cap is formed from the α-phosphate of the GTP donor and the 5′ diphosphate of the RNA acceptor; (2) RNA guanylylation precedes cap methylation; (3) ^{m7}GTP is not a cap donor; (4) cap methylation renders the blocking GMP refractory to decapping by the guanylyltransferase; i.e., once methylation occurs, the reaction is irreversible.

Although the three capping reactions are universal in eukaryotes, there is a surprising diversity in the genetic organization of the cap-forming enzymes in different taxa as well as a complete divergence in the structure and catalytic mechanism of the RNA triphosphatase component as one moves from lower to higher eukaryotic species (Shuman 2000). These differences, described herein, can be exploited to develop novel approaches to anti-infective therapy directed against capping of the pathogen's

mRNAs, and they provide new and instructive clues to eukaryotic phylogeny.

THE CAPPING APPARATUS AND THE RIBOSOME

The link between capping and eukaryotic translation initiation in vitro has been appreciated for more than 25 years (Muthukrishnan et al. 1975), and it is generally accepted that specific cap-binding proteins recruit mRNA to the eukaryotic ribosome. The mechanism of translation is fundamentally conserved in all domains of life, and the atomic structures of the bacterial ribosome presented at the Cold Spring Harbor Symposium provided spectacular insights into how a large (2.3×10^6 D) RNA-based ribonucleoprotein machine transforms the information content of mRNA into protein primary structure (Ban et al. 2000; Wimberly et al. 2000; Yusupov et al. 2001). In contrast, the capping apparatus is a relatively small (1.2×10^5 to 2.3×10^5 D) machine, composed only of protein, whose job is to confer "mRNA identity" on a subset of eukaryotic transcripts.

The bacterial ribosome has been a resoundingly successful target for lifesaving antibiotics for the treatment of bacterial infections. Such antibiotics display remarkable selectivity for interaction with the prokaryotic translation apparatus. Recent crystal structures of the bacterial ribosome with bound antibiotics (Brodersen et al. 2000; Carter et al. 2000; Pioletti et al. 2001) will stimulate the design of new compounds that block bacterial protein synthesis. An outstanding challenge is to develop agents that selectively block gene expression in eukaryotic pathogens such as fungi and protozoa. Protozoan diseases, especially malaria, exact a huge toll in human morbidity and mortality. Recent structural and phylogenetic studies described below highlight the capping apparatus, and the RNA triphosphatase component in particular, as a promising target for the discovery of new antifungal, antiprotozoal, and antiviral drugs.

The present conception of the universal phylogenetic tree comprising three evolutionary domains—bacteria, archaea, and eukarya—was inferred from comparative sequence analyses of the rRNA components of the ribosome (Woese et al. 1990; Woese 2000). I propose here a heuristic scheme of eukaryotic phylogeny based on the

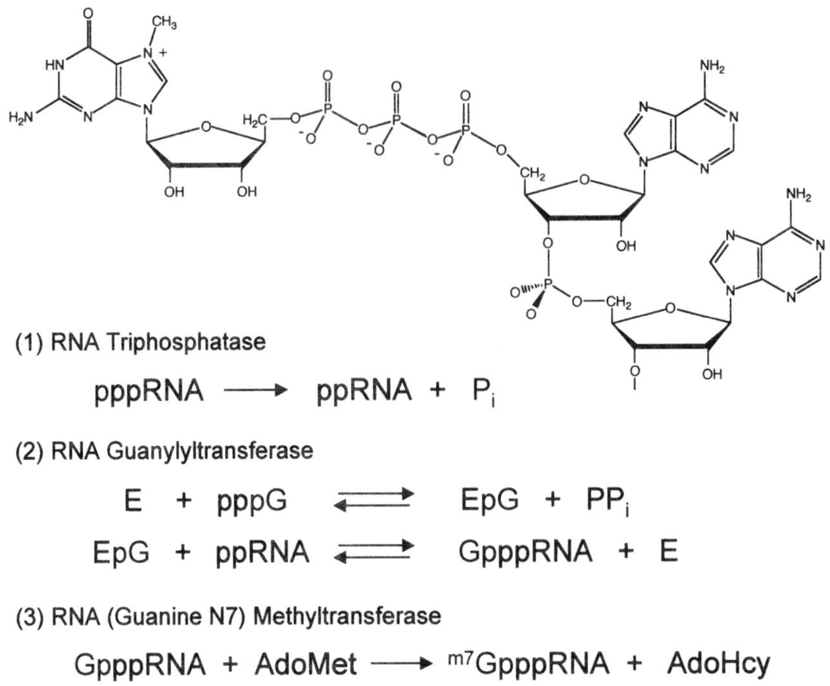

Figure 1. Structure and enzymatic synthesis of the ^{m7}G cap of eukaryotic mRNA.

(1) RNA Triphosphatase

$$pppRNA \longrightarrow ppRNA + P_i$$

(2) RNA Guanylyltransferase

$$E + pppG \rightleftharpoons EpG + PP_i$$

$$EpG + ppRNA \rightleftharpoons GpppRNA + E$$

(3) RNA (Guanine N7) Methyltransferase

$$GpppRNA + AdoMet \longrightarrow {}^{m7}GpppRNA + AdoHcy$$

structures and physical linkage of the triphosphatase and guanylyltransferase components of the mRNA capping apparatus.

EVOLUTIONARY FORCES SHAPING THE APPEARANCE OF THE CAP IN EUKARYOTES

The cap structure is unique to eukaryotic cellular and viral mRNAs. The mRNAs of eubacteria and archaea do not have caps. The reasons are simple: The enzymes that catalyze cap synthesis exist only in eukaryotic cells and certain eukaryotic viruses and are lacking in the proteomes of bacteria and archaea. Genetic analysis of the cap-forming enzymes in fungi demonstrates that capping is essential for cell growth and also highlights the distinct functions of the cap guanylate and the cap N7 methyl group in gene expression in vivo. The cap guanylate is necessary and sufficient to protect mRNA from premature degradation by 5′ to 3′ exoribonucleases, whereas the cap methyl group is required for translation (Schwer et al. 1998, 2000). The fact that the guanylyltransferase component specifically acts on 5′ diphosphate RNA ends, and is incapable of catalyzing GMP transfer to a 5′ monophosphate RNA, ensures that caps are added only to the 5′ end of the primary transcript and not to processed 5′ ends arising from endonucleolytic cleavage.

In considering why the cap appeared in eukarya, I would invoke a confluence of evolutionary forces including (1) the loss of Shine-Dalgarno base-pairing as a means of directing ribosomes to mRNA, (2) the emergence of three RNA polymerases with distinct functional repertoires, and (3) the appearance of 5′ exoribonucleases.

Prokaryotic "mRNA identity" is signaled by a Shine-Dalgarno sequence proximal to the site of translation initiation that can base-pair with 16S rRNA and thereby position the mRNA start codon at the P site on the ribosome. With no hard-wired specification of ribosome binding in their mRNA sequences, eukaryotes employ the cap as an alternative signal to direct the translation apparatus to the 5′ end of protein-encoding RNAs.

Bacteria and archaea have a single DNA-dependent RNA polymerase responsible for the transcription of all classes of cellular RNAs. Eukarya have three RNA polymerases, and the protein-encoding genes are (with rare exceptions) transcribed exclusively by RNA polymerase II. The capping enzymes are targeted to RNA polymerase II elongation complexes at the 5′ end of the transcription unit. The acquisition of the cap is the earliest modification event in mRNA biogenesis and appears to be the critical event in establishing eukaryotic mRNA identity. The consequences of indiscriminate capping of the 5′ ends of RNAs made by polymerase I and III are potentially disastrous, because subsequent endonucleolytic maturation of capped pre-tRNAs and pre-rRNAs would generate a vast cellular pool of noncoding capped molecules that would compete with bona fide mRNAs for binding to the ribosome. By the same reasoning, I would argue eukaryotic cells ought to have a mechanism to defer the capping of pol II transcripts until the RNA polymerase ternary complex has committed to processive elongation—so as to avoid generating short abortive transcripts with ^{m7}G caps.

A major function of the cap guanylate is to protect mRNA from digestion by 5′ exoribonucleases. The existence of 5′ exoribonucleases is confined to eukarya; such enzymes are absent from bacteria (by biochemical and phylogenetic criteria) and are also undetectable in archaeal proteomes (Deutscher and Li 2000). It is conceivable that the 5′ exoribonucleases and cap synthesis evolved in tandem in early eukarya to provide a primitive immunity from RNA viruses or viroids, i.e., whereby the uncapped RNA invader would be degraded while the capped cellular transcripts would be shielded. Of course, RNA viruses have since evolved diverse strategies to contend with the capping problem, including the encoding of their own cap-forming enzymes (reoviruses, alphaviruses), stealing the caps from cellular mRNAs (influenza virus), and the adoption of unique RNA structures that bypass the cap requirement for translation and stability (picornaviruses, hepatitis C virus). Coordinated appearance of cap and 5′ exoribonucleases in eukarya also provides a mechanism to regulate gene expression posttranscriptionally, whereby enzymatic decapping of the mRNA triggers its decay.

THE CAPPING APPARATUS DIFFERS IN HIGHER AND LOWER EUKARYOTES

The genetic organization and quaternary structure of the capping apparatus differ in significant respects in metazoans and fungi. Mammals and other metazoa encode a two-component capping system consisting of a bifunctional triphosphatase-guanylyltransferase polypeptide and a separate methyltransferase polypeptide. The mammalian triphosphatase-guanylyltransferase, Mce1, is a 597-amino-acid protein composed of two functionally autonomous catalytic domains: an amino-terminal triphosphatase domain, Mce1(1-210), and a carboxy-terminal guanylyltransferase domain, Mce1(211-597). Recombinant versions of Mce1 and the component domains have been produced and exploited for biochemical and structural analyses. Mce1 is a monomeric protein, as are the isolated triphosphatase and guanylyltransferase domains (Ho et al. 1998). The mammalian cap methyltransferase (Hcm1; 476 amino acids) is also a monomeric protein (Saha et al. 1999). The capping apparatus is organized identically in the nematode *Caenorhabditis elegans*, the arthropod *Drosophila melanogaster*, and the amphibian *Xenopus laevis* (Takagi et al. 1997; Yokoska et al. 2000).

Fungi encode a three-component system consisting of separate triphosphatase, guanylyltransferase, and methyltransferase gene products. Genetic and biochemical analysis of the triphosphatase (Cet1), guanylyltransferase (Ceg1), and methyltransferase (Abd1) components in budding yeast *Saccharomyces cerevisiae* has resulted in the delineation of their minimal functional domains, the fine-mapping of essential residues at their active sites, and the demonstration that each of the cap-forming activities is essential for cell viability (Wang et al. 1997; Wang and Shuman 1997; Bisaillon and Shuman 2001). The *S. cerevisiae* RNA triphosphatase Cet1 is a homodimeric protein that interacts avidly in *trans* with the guanylyltransferase Ceg1 to form a bifunctional capping enzyme complex. The cap methyltransferase Abd1 is a

monomeric enzyme. The structure of the capping apparatus in the pathogenic fungus *Candida albicans*, consisting of separately encoded triphosphatase (CaCet1), guanylyltransferase (Cgt1), and methyltransferase (Ccm1) enzymes, appears to mimic that of *S. cerevisiae*, including the physical association of the triphosphatase and guanylyltransferase enzymes (Yamada-Okabe et al. 1988; Schwer et al. 2001). The *Schizosaccharomyces pombe* triphosphatase (Pct1), guanylyltransferase (Pce1), and cap methyltransferase (Pcm1) enzymes have also been identified and characterized. The *S. pombe* triphosphatase Pct1 is a homodimeric enzyme (like *S. cerevisiae* Cet1), but remarkably, it does not interact physically with the *S. pombe* guanylyltransferase Pce1 (Pei et al. 2001). The *PCT1* and *PCE1* genes are both essential for growth of *S. pombe* (Y. Pei et al., unpubl.).

The primary structures and biochemical mechanisms of the fungal and mammalian guanylyltransferases are conserved. The primary structures and mechanisms of the fungal and mammalian cap methyltransferases are also highly conserved. In contrast, the atomic structures and catalytic mechanisms of the fungal and mammalian RNA triphosphatases are completely different (Lima et al. 1999; Changela et al. 2001).

CONSERVED STRUCTURE AND MECHANISM OF RNA GUANYLYLTRANSFERASE

Transfer of GMP from GTP to the 5′-diphosphate terminus of RNA is a two-stage ping-pong reaction involving a covalent enzyme–GMP intermediate (Shuman and Hurwitz 1981). In the first step, attack on the α phosphorus of GTP by a lysine side chain on the capping enzyme results in the release of pyrophosphate and the formation of the covalent intermediate, in which GMP is linked via a phosphoamide bond to lysine. In the second step, the GMP is transferred to the 5′ diphosphate RNA end to form the cap (Fig. 1). Both steps require a divalent cation cofactor and are readily reversible.

The capping reaction is chemically similar to the first two nucleotidyl transfer steps in the eukaryotic DNA ligase reaction. The first step entails attack on the α phosphorus of ATP by ligase, resulting in the release of pyrophosphate and the formation of a covalent intermediate (ligase-adenylate) in which AMP is linked via a phosphoamide bond to lysine. In the second step, the AMP is transferred from ligase-adenylate to the 5′ end of a 5′ phosphate-terminated DNA strand to form a DNA-adenylate intermediate (AppN). The inverted 5′–5′ pyrophosphate bridge of DNA-adenylate is chemically similar to the triphosphate bridge in capped GpppRNA.

A common structural basis for covalent nucleotidyl transfer was suggested by the mapping of the covalent nucleotide attachment sites of DNA ligase and capping enzyme sites to a lysine within a conserved motif KxDG (Tomkinson et al. 1991; Cong and Shuman 1993). The KxDG signature (now called motif I) is present in a wide variety of guanylyltransferases and DNA ligases (Fig. 2). That the nucleotide attachment sites of capping enzymes

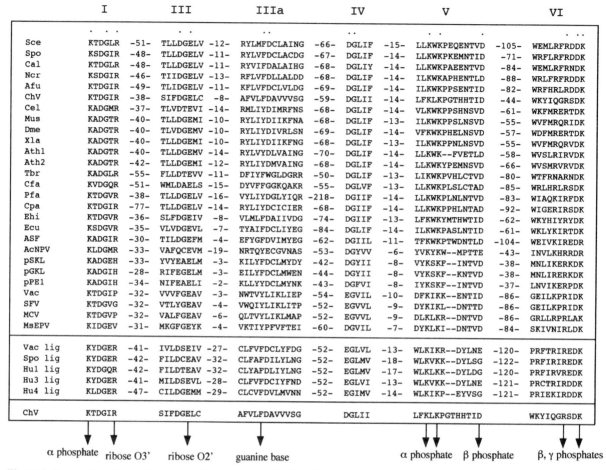

Figure 2. Guanylyltransferase signature motifs are conserved in fungal, protozoal, metazoan, plant, viral, and fungal plasmid capping enzymes. The amino acid sequences of motifs I, III, IIIa, IV, V, and VI are aligned for the capping enzymes of *S. cerevisiae* (Sce), *S. pombe* (Spo), *C. albicans* (Cal), *N. crassa* (Ncr), *A. fumigatus* (Afu), *Chlorella* virus PBCV-1 (ChV), *C. elegans* (Cel), mouse (Mus), *D. melanogaster* (Dme), *X. laevis* (Xla), *A. thaliana* (Ath), *T. brucei gambiense* (Tbr), *Crithidia fasciculata* (Cfa), *P. falciparum* (Pfa), *Cryptosporidium parvum* (Cpa), *Entamoeba histolytica* (Ehi), *Encephalitozoon cuniculi* (Ecu), African swine fever virus (ASF), AcNPV baculovirus (AcNPV), the linear DNA plasmids of *Saccharomyces kluyveri* (pSKL), *Kluyveromyces lactis* (pGKL), and *Pichia etchellsii* (pPE1), vaccinia virus (Vac), Shope fibroma virus (SFV), molluscum contagiosum virus (MCV), and *Melanoplus sanguinipes* entomopoxvirus (MsEPV). Aligned below the capping enzymes are the sequences of the nucleotidyl transferase motifs of the DNA ligases of vaccinia virus (Vac lig), *S. pombe* (Spo lig), and human DNA ligases I, III, and IV. The numbers of amino acid residues separating the motifs are indicated. The amino acids of the *S. cerevisiae* guanylyltransferase motifs that are essential for function in vivo are denoted by dots. Specific contacts between amino acid side chains and the nucleotide substrate in the ChV capping enzyme–GTP co-crystal are indicated by arrowheads.

and ligases are so similar suggested that other structural features might also be conserved. Hence, we scanned "by eye" for local similarities and thereby identified five other motifs (named motifs III, IIIa, IV, V, and VI) that are arrayed in the same order, and with similar spacing, among the guanylyltransferases and ATP-dependent DNA ligases (Fig. 2). Mutational analysis of capping enzymes and DNA ligases has shown that conserved side chains within the six motifs are essential for the capping and ligation activities; essential side chains in *S. cerevisiae* capping enzyme (Wang et al. 1997) are denoted by dots above the aligned sequences in Figure 2. The mutational studies engendered the prediction that the six conserved motifs comprise the nucleotide-binding pockets of capping enzymes and DNA ligases (Shuman and Schwer 1995), and this was proved to be the case by the crystal structures of T7 DNA ligase bound to ATP (Subramanya

et al. 1996), *Chlorella* virus guanylyltransferase bound noncovalently to GTP and covalently to GMP (Håkansson et al. 1997), and *Chlorella* virus DNA ligase bound covalently to AMP (Odell et al. 2000). The atomic contacts between GTP and the conserved amino acids in the *Chlorella* virus guanylyltransferase structure are indicated below the sequence alignment in Figure 2.

Remarkably, X-ray crystallography revealed a shared overall tertiary structure for DNA ligases and capping enzyme, implying their evolution from a common ancestral nucleotidyl transferase that acts via a phosphoramidate intermediate. Thus, nature didn't waste a good idea; having found a nice structural solution to a chemical problem, it recycled it in different biological contexts. It is reasonable to suppose that ATP-dependent polynucleotide ligases antedated the capping enzyme, insofar as ATP-dependent DNA ligases are present in all archaea and eukarya, whereas RNA guanylyltransferases exist only in eukarya. Thus, the capping enzyme may have evolved very early during the emergence of eukarya by ligase gene duplication and subsequent mutation leading to the acquisition of new specificity for GTP as the nucleotide donor and a 5′ diphosphate RNA end as the nucleotide acceptor. Ligase gene duplications must have occurred early in eukaryotic phylogeny, insofar as present fungi, plants, and metazoans encode two or more DNA ligase enzymes per organism.

The sequence conservation among the several guanylyltransferases that have been characterized biochemically permits one to identify, with a high degree of confidence, the capping enzymes from new sources based on searches of genomic and cDNA databases. The fruits of this exercise include the capping enzymes from the fungi *Neurospora crassa* and *Aspergillus fumigatus*, the microsporidian *Encephalitozoon cuniculi*, and the protozoa *Trypanosoma brucei*, *Crithidia fasciculata*, *Plasmodium falciparum*, *Entamoeba histolytica*, and *Cryptosporidium parvum* (Fig. 2). The catalytic activities of the *C. fasciculata*, *T. brucei*, *P. falciparum*, and *E. cuniculi* guanylyltransferases have been demonstrated biochemically and, in the case of the *E. cuniculi* enzyme, by genetic complementation in *S. cerevisiae* (Silva et al. 1998; Ho and Shuman 2001; S. Hausmann et al., unpubl.).

A notable finding is that the multicellular plant *Arabidopsis thaliana* has two capping enzyme genes located on different chromosomes. Both of the plant gene products are putative bifunctional triphosphatase-guanylyltransferase enzymes, and they are both similar in structure and domain organization to the metazoan capping enzymes. *A. thaliana* is thus far unique in having multiple capping enzyme genes. The polypeptides differ slightly in size, primary structure, and spacing between the nucleotidyl transferase motifs (Fig. 2). Whether there are functional differences between the two *A. thaliana* capping enzymes, or perhaps differences in their expression patterns in vivo, has not been addressed.

METAZOAN AND PLANT RNA TRIPHOSPHATASES COMPRISE A DISTINCT BRANCH OF THE CYSTEINE PHOSPHATASE SUPERFAMILY

Fungal RNA triphosphatase and guanylyltransferase activities reside in separate polypeptides, whereas metazoans and *Arabidopsis* contain a bifunctional enzyme composed of an amino-terminal triphosphatase and a carboxy-terminal guanylyltransferase domain. The fungal and metazoan RNA triphosphatases can be further distinguished by their cofactor requirements: The RNA triphosphatases of fungi are strictly dependent on a divalent cation, whereas the RNA triphosphatases of mammals and other metazoa do not require a metal for activity and are instead inhibited by divalent cations.

The triphosphatase domains of metazoan and plant capping enzymes contain a HCxxxxxR(S/T) motif (the P loop) that defines the cysteine phosphatase superfamily (Fig. 3B). The metazoan RNA triphosphatases catalyze a

two-step ping-pong phosphoryl transfer reaction (Fig. 3A). First, the conserved cysteine of the signature motif, which is a cysteine thiolate in the ground state, attacks the γ phosphorus of triphosphate-terminated RNA to form a covalent protein-cysteinyl-*S*-phosphate intermediate and expel the diphosphate RNA product (Changela et al. 2001). Second, the covalent phosphoenzyme intermediate is hydrolyzed to liberate inorganic phosphate. Mutation of the P-loop cysteine to either alanine or serine abrogates RNA triphosphatase activity in vitro and in vivo (Takagi et al. 1997; Changela et al. 2001).

The cysteine phosphatase superfamily also includes protein tyrosine phosphatases, dual-specificity protein phosphatases, and phosphoinositide phosphatases (e.g., the tumor suppressor PTEN). An alignment of the sequences of the metazoan and plant RNA 5′ phosphatases shows that they are highly similar to one another, but display only scant similarity to other branches of the superfamily exclusive of the signature P-loop motif (Fig. 3B). Thus, the specificity for the hydrolysis of phosphoanhydride substrates (as opposed to phosphomonoesters, which are the substrates for protein and lipid phosphatases) is presumably contributed by structural elements unique to and conserved among the RNA phosphatase branch.

The 1.65 Å crystal structure of mammalian RNA triphosphatase shows that the enzyme is a globular monomeric protein with a deep, positively charged active-site pocket that can accommodate a 5′ triphosphate end (Changela et al. 2001). The structure consists of five β strands that form a central twisted sheet flanked by two α helices on one side and four α helices on the other (Fig. 4). The carboxyl terminus of the triphosphatase domain travels back toward the amino terminus where it leads into the guanylyltransferase domain of the bifunctional capping enzyme. The HCxxxxxR(S/T) P-loop motif is highlighted in blue and the active-site cysteine (Cys-126) is shown as a yellow sphere. The overall fold of mammalian RNA triphosphatase is quite similar to that of other cysteine phosphatases, e.g., see the aligned secondary structures of Mce1 and PTEN in Figure 3B. The core elements of the tertiary structure of Mce1 that are conserved in the other cysteine phosphatases are highlighted in brown in Figure 4.

Mutational analysis shows that replacement of P-loop residues Asn-131 or Arg-132 with alanine abolished the in vivo activity of Mce1, whereas alanine substitution for P-loop residue His-129 resulted in temperature-sensitive function in vivo (Changela et al. 2001). Arg-132, which is conserved in every member of the cysteine phosphatase superfamily, is predicted to stabilize the transition state of the γ phosphate during catalysis. His-128 and Asn-131 are unique to the RNA triphosphatase branch of the superfamily and may therefore determine the specificity of these enzymes for the hydrolysis of phosphoanhydrides. The position of Asn-131 in the crystal structure and the orientation of Nδ into the active site suggest that it forms a hydrogen bond to the β phosphate of the substrate. His-128, which hydrogen-bonds to a water in the active-site pocket, may also interact with one of the 5′ phosphates of the substrate.

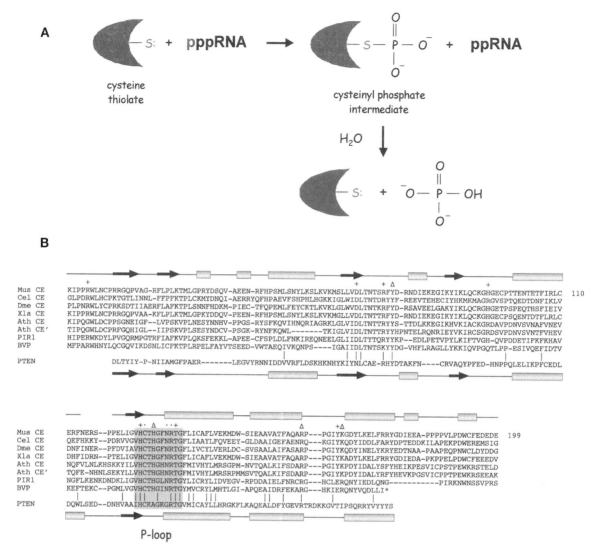

Figure 3. Metazoan RNA triphosphatases. (*A*) Reaction mechanism for phosphoanhydride cleavage involving formation of a cysteinyl phosphate intermediate. (*B*) Structure-based sequence alignment of metazoan RNA triphosphatases. The amino acid sequences of the amino-terminal RNA triphosphatase domains of mouse capping enzyme (Mus CE), *C. elegans* capping enzyme (Cel CE), *D. melanogaster* capping enzyme (Dme CE), *X. laevis* capping enzyme (Xla CE), and two capping enzymes of *A. thaliana* (Ath CE) are aligned to the baculovirus RNA phosphatase BVP, human RNA phosphatase PIR1, and the phosphoinositide phosphatase PTEN. Gaps in the sequences are indicated by dashes. The protein phosphatase signature motif (the P loop) is highlighted in the shaded box. Secondary structure elements of the mammalian RNA triphosphatase (PDB ID code: 1I9S) and PTEN (PDB ID code 1D5R) are shown above and below the respective sequences with α helices depicted as boxes, β strands as arrows, and loops as solid lines. The effects of single alanine substitutions on Mce1 function are denoted above the amino acid sequence as follows: no effect on Mce1 function (+); loss of function (•); *ts* function (Δ).

FUNGAL, VIRAL, AND PROTOZOAN RNA TRIPHOSPHATASES COMPRISE A NEW FAMILY OF METAL-DEPENDENT PHOSPHOHYDROLASES

The RNA triphosphatases of fungal species such as *S. cerevisiae*, *C. albicans*, and *S. pombe* are strictly dependent on a divalent cation. The fungal enzymes belong to a new family of metal-dependent phosphohydrolases that embraces the triphosphatase components of the poxvirus, baculovirus, phycodnavirus, and *P. falciparum* mRNA capping systems. The signature biochemical property of this enzyme family is the ability to hydrolyze nucleoside triphosphates to nucleoside diphosphates and inorganic phosphate in the presence of either manganese or cobalt. The defining structural features of the metal-dependent RNA triphosphatases are two glutamate-containing motifs (β1 and β11 in Fig. 5A) that are required for catalysis by every family member.

The 2.05 Å crystal structure of the *S. cerevisiae* RNA triphosphatase Cet1 illuminates an amazing structural complexity for an enzyme that catalyzes a mundane phos-

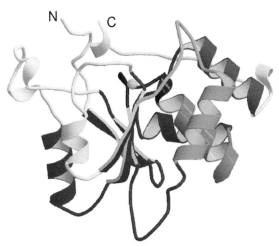

N C

Figure 4. Tertiary structure of the RNA triphosphatase domain of mammalian mRNA capping enzyme. Ribbon diagram of the crystal structure. The P loop is colored blue with the active-site cysteine depicted as a yellow sphere. Other core elements common to cysteine phosphatases are shaded brown.

phohydrolase reaction (Lima et al. 1999) (Fig. 6A). Cet1 crystallized as a homodimer, which is consistent with solution studies of its quaternary structure. The striking feature of the novel tertiary structure is the formation of a topologically closed hydrophilic tunnel composed of eight antiparallel β strands. The β strands comprising the tunnel walls are displayed over the Cet1 protein sequence in Figure 5A. In the dimer, the two tunnels are parallel and oriented in the same direction; i.e., the tunnel "entrances" are on the same face of the dimer (Fig. 6A). The active site of the enzyme is located within the cavity of the "triphosphate tunnel." It can be readily appreciated that the structure of the fungal RNA triphosphatase is completely different from that of the mammalian enzyme.

The tunnel has a distinctive baroque architecture supported by an intricate network of hydrogen bonds and electrostatic interactions within the tunnel cavity, of which a surprisingly high proportion are required for enzyme activity (Fig. 6B). The tunnel contains a single sulfate ion coordinated by multiple basic side chains projecting into the cavity. We proposed that the side-chain interactions of the sulfate reflect contacts made by the enzyme with the γ phosphate of the substrate (Lima et al. 1999). A manganese ion within the tunnel cavity is coordinated with octahedral geometry to a sulfate, to the side-chain carboxylates of the two glutamates in β1, and to a glutamate in β11.

Alanine scanning mutagenesis has identified 15 individual side chains within the tunnel that are important for Cet1 function in vitro and in vivo (denoted by dots in Fig. 5A). Moreover, each of the eight strands of the β barrel contributes at least one functional constituent of the active site. The relevant structural features of the 15 key amino acids have been determined through the analysis of conservative mutational effects (Pei et al. 1999; Bisaillon and Shuman 2001). We have grouped the active-site

residues into three functional classes. Class I residues participate directly in catalysis via coordination of the γ phosphate (Arg-393, Lys-456, Arg-458) or the essential metal (Glu-305, Glu-307, Glu-494). Class II residues make water-mediated contacts with the γ phosphate (Asp-377, Glu-433) or the metal (Asp-471, Glu-496). Class III residues function indirectly in catalysis via their interactions with other essential side chains and/or their stabilization of the tunnel architecture (Lys-409, Arg-454, Arg-469, Thr-473, Glu-492).

On the basis of the structure and the mutational results, we have proposed a one-step in-line mechanism whereby the metal ion (coordinated by acidic residues on the tunnel floor) plus the Arg-393, Arg-458, and Lys-456 side chains (emanating from the walls and roof) activate the γ phosphate for attack by water and stabilize a pentacoordinate phosphorane transition state in which the attacking water is apical to the β phosphate leaving group (Bisaillon and Shuman 2001). We further speculated that the Glu-433 side chain coordinates the nucleophilic water molecule (Fig. 6B) and serves as a general base catalyst.

Alignment of the sequences of RNA triphosphatases from *S. cerevisiae*, *C. albicans*, and *S. pombe* underscores the conservation of the β strands that make up the Cet1 tunnel (Fig. 5A). Mutational analysis of *C. albicans* RNA triphosphatase suggests that its active site is similar, if not identical, to that of Cet1 (Pei et al. 2000). Thus, I would predict that all fungi have a Cet1-like triphosphatase in their mRNA capping apparatus and that the tunnel architecture of the triphosphatase is conserved.

Several new members of the Cet1-like triphosphatase family have recently been identified and characterized. *Chlorella* virus PBCV-1 encodes a metal-dependent RNA triphosphatase that is physically separate from the viral guanylyltransferase (Ho et al. 2001). The *Chlorella* virus triphosphatase appears to contain equivalents of all eight strands of the triphosphate tunnel, and most of the side chains found to be essential for Cet1 activity are conserved in the viral protein. Indeed, extensive mutational analysis of the *Chlorella* virus triphosphatase shows that at least one conserved side chain in each of the putative β strands is essential for catalytic activity (C. Gong and S. Shuman, unpubl.).

The microsporidian *E. cuniculi* is an amitochondrial intracellular parasite that has the smallest genome (~3 Mbp) of any eukaryotic organism characterized to date. *E. cuniculi* encodes a three-component capping system similar to that of fungi. The *E. cuniculi* RNA triphosphatase (EcCet1) has been purified and characterized; it displays the signature property of hydrolyzing NTPs in the presence of manganese or cobalt (S. Hausmann et al., unpubl.). The microsporidian triphosphatase protein sequence includes putative equivalents of the eight β strands of the Cet1 tunnel (Fig. 5A).

PROTOZOA ENCODE A YEAST-LIKE RNA TRIPHOSPHATASE

Protozoan parasites pose a devastating human health problem, especially in the third world. For example, there

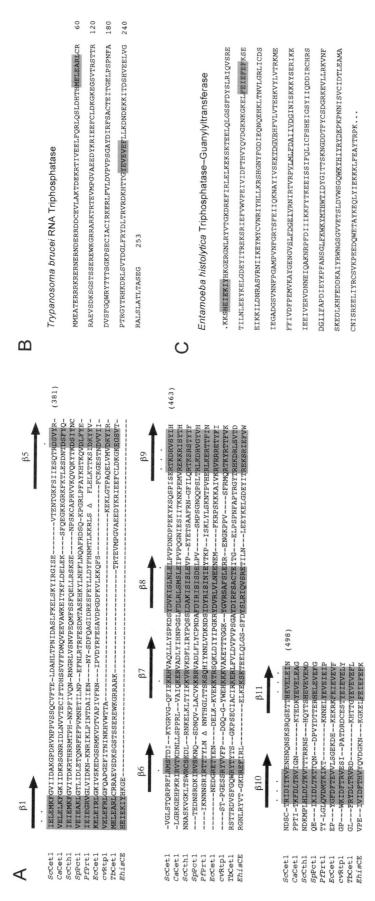

Figure 5. Structural conservation among fungal, viral, microsporidian, and protozoan RNA triphosphatases. (*A*) The amino acid sequence of the catalytic domain of *S. cerevisiae* RNA triphosphatase Cet1 is aligned to the sequences of *C. albicans* CaCet1, *S. pombe* Pct1, *P. falciparum* Prt1, *E. cuniculi* Cet1, and *Chlorella* virus cvRtp1. Also included in the alignment are putative RNA triphosphatases from *T. brucei* (*Tb*Cet1) and the putative RNA triphosphatase domain of a bifunctional capping enzyme from *E. histolytica* (*Ehis*CE). Gaps in the alignment are indicated by dashes. Poly-asparagine inserts in *Pf*Prt1 are omitted from the alignment and are denoted by Δ. The β strands that form the triphosphate tunnel of *Sc*Cet1 are denoted above the sequence. Peptide segments with the highest degree of conservation in all nine proteins are highlighted by the shaded boxes. Hydrophilic amino acids that comprise the active site within the *Sc*Cet1 tunnel are denoted by dots. (*B*) Sequence of the putative *T. brucei* RNA triphosphatase with the metal-binding motifs highlighted in shaded boxes. (*C*) Sequence (partial) of the putative *E. histolytica* RNA triphosphatase domain are highlighted in shaded boxes; the nucleotidyl transferase motifs of the guanylyltransferase domain are underlined.

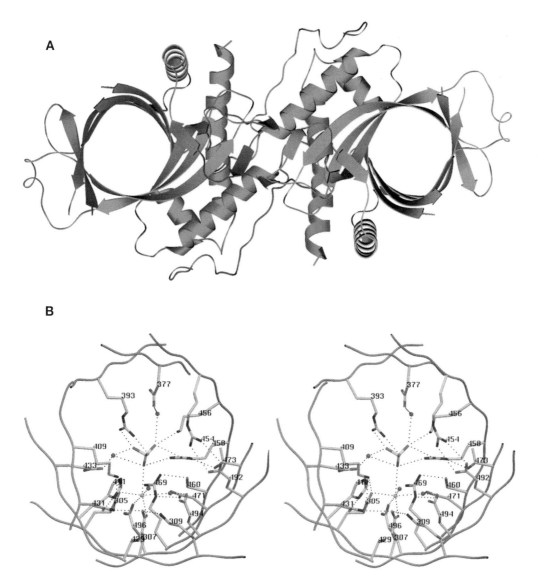

Figure 6. Structure of *S. cerevisiae* RNA triphosphatase Cet1. (*A*) Ribbon diagram of the dimer structure (PDB ID code: 1D8I) looking into the entrances of the parallel triphosphate tunnels. One protomer is colored cyan and the other magenta. (*B*) Stereo view of a cross-section of the triphosphate tunnel of Cet1. The figure highlights the elaborate network of bonding interactions, especially those that coordinate sulfate (γ phosphate) and manganese. The manganese (*blue sphere*) interacts with octahedral geometry with the sulfate, three glutamates, and two waters (*red spheres*). The putative nucleophilic water is coordinated by Glu-433, which is posited to act as a general base catalyst.

are some 400 million new cases of malaria each year, resulting in more than 1 million deaths. Malaria treatment and prevention strategies have been undermined by the spreading resistance of the *Plasmodium* pathogen to erstwhile effective drugs. Protozoan genome projects are undertaken with an explicit goal of identifying new drug targets, the most promising of which would be gene products or metabolic pathways that are essential for the parasite life cycle, but either absent or fundamentally different in the human host.

Given the divergence of the RNA triphosphatase component of the capping apparatus in fungi versus mammals, we have sought to characterize the capping enzymes of protozoan parasites, beginning with the malaria parasite *P. falciparum*. We found that *P. falciparum* en-

codes separate RNA guanylyltransferase (Pgt1) and RNA triphosphatase (Prt1) enzymes and that the triphosphatase component is a member of the fungal/viral family of metal-dependent phosphohydrolases (Ho and Shuman 2001). A separate candidate cap methyltransferase from *P. falciparum* has also been identified (K. Ho and S. Shuman, unpubl.). Thus, *Plasmodium* has a yeast-like capping apparatus. The *Plasmodium* guanylyltransferase and triphosphatase were purified and then shown to possess the capping activities imputed to them on phylogenetic grounds (Ho and Shuman 2001). The *P. falciparum* triphosphatase contains counterparts of the eight β strands of the Cet1 tunnel (Fig. 5A).

The kinetoplastid protozoan *T. brucei* causes sleeping sickness in Africa, and the related species, *Trypanosoma*

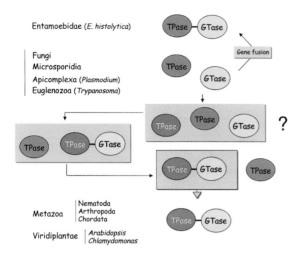

Figure 7. Capping enzyme-based scheme of eukaryotic phylogeny. The ancestral capping system consists of a separately encoded metal-dependent RNA triphosphatase (TPase, *blue*) and guanylyltransferase (GTase, *yellow*) enzymes. The metazoan and plant capping systems, consisting of a cysteine phosphatase-type RNA triphosphatase (TPase, *red*) fused to a guanylyltransferase (Gtase, *yellow*) may have evolved via a series of intermediate steps highlighted in the shaded boxes.

cruzi, causes Chagas' disease in South America. The guanylyltransferase of *Trypanosoma* has been cloned and characterized (Silva et al. 1998), but the triphosphatase component of the capping apparatus has not been reported. I have now identified a candidate RNA triphosphatase (TbCet1) from *T. brucei* (Fig. 5B). TbCet1 is a 253-amino-acid polypeptide (encoded separately from the *T. brucei* guanylyltransferase) that includes the metal-binding motifs characteristic of the fungal/viral triphosphatases, plus putative homologs of all of the other β strands that comprise the active site of Cet1 (Fig. 5A).

Invasive infection by the unicellular parasite *E. histolytica* results in amoebic dysentery, amoebic liver abscess, and extraintestinal lesions. *E. histolytica* infection is widespread and is estimated to cause 50,000 to 100,000 deaths annually. I have identified a candidate *E. histolytica* capping enzyme composed of an amino-terminal triphosphatase domain fused to a guanylyltransferase domain. The true termini of the gene product cannot be surmised from available genome data, but the incomplete *E. histolytica* protein sequence is convincing with respect to the guanylyltransferase domain, which contains the six diagnostic nucleotidyl transferase motifs. The putative amino-terminal triphosphatase domain includes the two metal-binding signature motifs (Fig. 5C), plus likely equivalents of five of the other β strands of Cet1 (Fig. 5A). The *E. histolytica* triphosphatase domain lacks an obvious counterpart of β5 of Cet1. The domain organization of the *E. histolytica* capping enzyme is reminiscent of the vaccinia virus capping enzyme, which also consists of an amino-terminal metal-dependent triphosphatase linked to a downstream guanylyltransferase domain (Yu et al. 1997). It would appear that the *Entamoeba* capping enzyme evolved from a more primitive fungal-type capping apparatus via fusion of the triphosphatase and

guanylyltransferase genes. A similar bifunctional capping enzyme may also exist in *Dictyostelium discoideum* (not shown).

RNA TRIPHOSPHATASE AS AN ANTIFUNGAL AND ANTIPROTOZOAL DRUG TARGET

RNA triphosphatase is an attractive anti-infective drug target because (1) the structure and catalytic mechanism of the fungal/protozoal/viral triphosphatase family are completely different from that of the metazoan capping enzyme and (2) metazoan genomes encode no identifiable homologs of the fungal or protozoal RNA triphosphatases. Thus, a mechanism-based inhibitor of *Plasmodium* RNA triphosphatase Prt1 should be highly selective for the malaria parasite and have minimal effect on the human host (or the mosquito vector). Given the central role of the mRNA cap in eukaryotic gene expression, an antimalarial drug that targets Prt1 would presumably be effective at all stages of the parasite's life cycle. In addition, the structural similarity between fungal and *Plasmodium* RNA triphosphatases raises the exciting possibility of achieving antifungal and antimalarial activity with a single class of mechanism-based inhibitors. Indeed, given that other protozoan parasites such as *Trypanosoma* and *Entamoeba* also encode proteins that resemble the fungal RNA triphosphatases, it is reasonable to think that inhibitors of fungal or *Plasmodium* RNA triphosphatase might be effective against a broad spectrum of protozoa that cause human disease.

HEURISTIC TAXONOMY OF EUKARYOTES BASED ON THEIR CAPPING APPARATUS

Capping enzymes are a good focal point for considering eukaryotic evolution because the mRNA cap structure is ubiquitous in eukaryotic organisms, but absent from the bacterial and archaeal kingdoms. Thus, any differences in the capping apparatus between taxa would reflect events that post-date the emergence of ancestral nucleated cells. The enzymes that catalyze the basic nucleic acid transactions (DNA replication, DNA repair, RNA synthesis, and RNA processing) are generally conserved in lower and higher eukaryotes. Yet, in the case of the capping apparatus, we see a complete divergence of the triphosphatase component and of the physical linkage of the triphosphatase and guanylyltransferase between taxa.

This suggests a heuristic scheme of eukaryotic phylogeny based on two features of the mRNA capping apparatus: the structure and mechanism of the triphosphatase component (metal-dependent "fungal" type versus metal-independent cysteine-phosphatase type) and whether the triphosphatase is physically linked in *cis* to the guanylyltransferase component (Fig. 7). By these simple criteria, relying on "black-and-white" differences in the same metabolic pathway, one arrives at different relationships among taxa than those suggested by comparisons of sequence variations among proteins that are themselves highly conserved in all eukaryotes (Baldauf et al. 2000). For example, the capping-based phylogeny would place metazoans in a common lineage with Viridiplantae (exemplified by the metaphyta *Arabidopsis* and the unicellular alga *Chlamydomomas reinhardtii*) because all of these organisms have a cysteine-phosphatase-type RNA triphosphatase fused in *cis* to their guanylyltransferase (Fig. 7). Fungi, microsporidia, *Plasmodia* (which are classified as Apicomplexa along with other pathogenic parasites *Toxoplasma* and *Cryptosporidia*), and *Trypanosoma* (which are classified as Euglenozoa along with the human parasite *Leishmania*) fall into a different lineage distinguished by a "Cet1-like" RNA triphosphatase that is physically separate from RNA guanylyltransferase. In contrast, the protein sequence variation-based scheme proposed by Baldauf et al. (2000) places fungi in the same supergroup as metazoa and puts the Apicomplexa nearer to plants.

Assuming that multicellular animals evolved from unicellular ancestors, I envision that the three-component capping system with a metal-dependent triphosphatase is the ancestral state from which other eukarya evolved (Fig. 7). We see evidence of evolution in two directions. Certain protozoa (exemplified by *Entamoeba*) acquired a bifunctional capping enzyme by fusion of the ancestral triphosphatase and guanylyltransferase genes. Metazoans and plants have experienced a different gene rearrangement event that transferred a cysteine-phosphatase domain into the same transcription unit as the guanylyltransferase, leading to creation of the triphosphatase-guanylyltransferase fusion protein that we see today. A plausible pathway of evolution could entail the appearance of a new cysteine phosphatase enzyme (e.g., via duplication and mutation of one of the protein phosphatase genes present in lower eukarya) that gained the capacity to hydrolyze an RNA 5′ phosphate instead of, or in addition to, a phosphoprotein (Fig. 7). The model implies a "transition state" wherein an organism contained both a metal-dependent and a cysteine-phosphatase type triphosphatase. The fusion of the cysteine phosphatase to the guanylyltransferase presumably allowed the loss of the Cet1-like enzyme from the genome of the common metazoan/plant ancestor or else the divergence of the protein to a point that it is no longer discernable as Cet1-like. The alternative explanation, which adheres to the sequence-based scheme (Baldauf et al. 2000), would be that plants and metazoans independently experienced this gene fusion in distant branches of the phylogenetic tree, a prospect that seems less appealing to me.

The scheme is useful in that it raises some interesting questions about missing links and the order of events in the progression from fungal and protozoal-type to metazoan and plant-type capping systems. It is undoubtedly oversimplified because it is based on knowledge of only a fraction of eukaryal taxa. As more genomes are sequenced, we may see additional species that have a Cet1-like triphosphatase fused to a guanylyltransferase, others with a cysteine-phosphatase-type RNA triphosphatase that participates in cap formation but is physically separate from the guanylyltransferase, and yet others that encode a novel class of RNA triphosphatase enzyme. Of particular interest will be to characterize the mRNA capping apparatus in the most primitive metazoan organisms.

ACKNOWLEDGMENTS

I am grateful to my colleagues and collaborators Kiong Ho, Beate Schwer, Chris Lima, Yi Pei, Li Kai Wang, Stephane Hausmann, Xana Martins, Martin Bisaillon, Kevin Lehman, Chunling Gong, Alfonso Mondragon, Anita Changela, Dale Wigley, David Bentley, Rob Fisher, and Christian Vivares for their contributions to the work described herein.

REFERENCES

Baldauf S.L., Roger A.J., Wenk-Siefart I., and Doolittle W.F. 2000. A kingdom-level phylogeny of eukaryotes based on combined protein data. *Science* **290:** 972.

Ban N., Nissen P., Hansen J., Moore P.B., and Steitz T.A. 2000. The complete atomic structure of the large ribosomal subunit at 2.4 Å resolution. *Science* **289:** 905.

Bisaillon M. and Shuman S. 2001. Structure-function analysis of the active site tunnel of yeast RNA triphosphatase. *J. Biol. Chem.* **276:** 17261.

Brodersen D.E., Clemons W.M., Jr., Carter A.P., Morgan-Warren R.J., Wimberly B.T., and Ramakrishnan V. 2000. The structural basis for the action of the antibiotics tetracycline, pactamycin, and hygromycin B on the 30S ribosomal subunit. *Cell* **103:** 1143.

Carter A.P., Clemons W.M., Jr., Brodersen D.E., Morgan-Warren R.J., Wimberly B.T., and Ramakrishnan V. 2000. Functional insights from the structure of the 30S ribosomal subunit and its interactions with antibiotics. *Nature* **407:** 340.

Changela A., Ho C.K., Martins A., Shuman S., and Mondragon A. 2001. Structure and mechanism of the RNA triphosphatase component of mammalian mRNA capping enzyme. *EMBO J.* **20:** 2575.

Cong P. and Shuman S. 1993. Covalent catalysis in nucleotidyl transfer: A KTDG motif essential for enzyme-GMP complex formation by mRNA capping enzyme is conserved at the active sites of RNA and DNA ligases. *J. Biol. Chem.* **268:** 7256.

Deutscher M.P. and Li Z. 2000. Exoribonucleases and their multiple roles in RNA metabolism. *Prog. Nucleic Acid Res. Mol. Biol.* **66:** 67.

Ensinger M.J., Martin S.A., Paoletti E., and Moss B. 1975. Modification of the 5′ terminus of mRNA by soluble guanylyl and methyl transferases from vaccinia virus. *Proc. Natl. Acad. Sci.* **72:** 2525.

Guarino L.A., Jin J., and Dong W. 1998. Guanylyltransferase activity of the LEF-4 subunit of baculovirus RNA polymerase. *J. Virol.* **72:** 10003.

Hagler J. and Shuman S. 1992. A freeze-frame view of eukaryotic transcription during elongation and capping of nascent mRNA. *Science* **255:** 983.

Håkansson K., Doherty A.J., Shuman S., and Wigley D.B. 1997. X-ray crystallography reveals a large conformational change during guanyl transfer by mRNA capping enzymes. *Cell* **89:** 545.

Ho C.K. and Shuman S. 2001. A yeast-like mRNA capping apparatus in *Plasmodium falciparum*. *Proc. Natl. Acad. Sci.* **98:** 3050.

Ho C.K., Gong C., and Shuman S. 2001. RNA triphosphatase component of the mRNA capping apparatus of *Paramecium bursaria Chlorella* virus 1. *J. Virol.* **75:** 1744.

Ho C.K., Sriskanda V., McCracken S., Bentley D., Schwer B., and Shuman S. 1998. The guanylyltransferase domain of mammalian mRNA capping enzyme binds to the phosphorylated carboxyl-terminal domain of RNA polymerase II. *J. Biol. Chem.* **273:** 9577.

Lima C.D., Wang L.K., and Shuman S. 1999. Structure and mechanism of yeast RNA triphosphatase: An essential component of the mRNA capping apparatus. *Cell* **99:** 533.

Muthukrishnan S., Both G.W., Furuichi Y., and Shatkin A.J. 1975. 5′-Terminal 7-methylguanosine in eukaryotic mRNA is required for translation. *Nature* **255:** 33.

Odell M., Sriskanda V., Shuman S., and Nikolov D. 2000. Crystal structure of eukaryotic DNA ligase-adenylate illuminates the mechanism of nick sensing and strand joining. *Mol. Cell* **6:** 1183.

Pei Y., Ho C.K., Schwer B., and Shuman S. 1999. Mutational analyses of yeast RNA triphosphatases highlight a common mechanism of metal-dependent NTP hydrolysis and a means of targeting enzymes to pre-mRNAs in vivo by fusion to the guanylyltransferase component of the capping apparatus. *J. Biol. Chem.* **274:** 28865.

Pei Y., Lehman K., Tian L., and Shuman S. 2000. Characterization of *Candida albicans* RNA triphosphatase and mutational analysis of its active site. *Nucleic Acids Res.* **28:** 1885.

Pei Y., Hausmann S., Ho C.K., Schwer B., and Shuman S. 2001. The length, phosphorylation state, and primary structure of the RNA polymerase II carboxyl-terminal domain dictate interactions with mRNA capping enzymes. *J. Biol. Chem.* **276:** 28075.

Pioletti M., Schlünzen F., Harms J., Zarivach R., Glühmann M., Bartels H., Auerbach T., Jacobi C., Hartsch T., Yonath A., and Franceschi F. 2001. Crystal structures of the small ribosomal subunit with tetracycline, edein, and IF3. *EMBO J.* **20:** 1829.

Saha N., Schwer B., and Shuman S. 1999. Characterization of human, *Schizosaccharomyces pombe* and *Candida albicans* mRNA cap methyltransferases and complete replacement of the yeast capping apparatus by mammalian enzymes. *J. Biol. Chem.* **274:** 16553.

Schwer B., Mao X., and Shuman S. 1998. Accelerated mRNA decay in conditional mutants of yeast mRNA capping enzyme. *Nucleic Acids Res.* **26:** 2050.

Schwer B., Lehman K., Saha N., and Shuman S. 2001. Characterization of the mRNA capping apparatus of *Candida albicans*. *J. Biol. Chem.* **276:** 1857.

Schwer B., Saha N., Mao X., Chen H.W., and Shuman S. 2000. Structure-function analysis of yeast mRNA cap methyltransferase and high-copy suppression of conditional mutants by AdoMet synthase and the ubiquitin conjugating enzyme Cdc34p. *Genetics* **155:** 1561.

Shuman S. 1997. Origins of mRNA identity: Capping enzymes bind to the phosphorylated C-terminal domain of RNA polymerase II. *Proc. Natl. Acad. Sci.* **94:** 12758.

———. 2000. Structure, mechanism, and evolution of the mRNA capping apparatus. *Prog. Nucleic Acid Res. Mol. Biol.* **66:** 1.

Shuman S. and Hurwitz J. 1981. Mechanism of mRNA capping by vaccinia virus guanylyltransferase: Characterization of an enzyme-guanylate intermediate. *Proc. Natl. Acad. Sci.* **78:**

187.

Shuman S. and Schwer B. 1995. RNA capping enzyme and DNA ligase—A superfamily of covalent nucleotidyl transferases. *Mol. Microbiol.* **17:** 405.

Silva E., Ullu E., Kobayashi R., and Tschudi C. 1998. Trypanosome capping enzymes display a novel two-domain structure. *Mol. Cell. Biol.* **18:** 4612.

Subramanya H.S., Doherty A.J., Ashford S.R., and Wigley D.B. 1996. Crystal structure of an ATP-dependent DNA ligase from bacteriophage T7. *Cell* **85:** 607.

Takagi T., Moore C.R., Diehn F., and Buratowski S. 1997. An RNA 5′-triphosphatase related to the protein tyrosine phosphatases. *Cell* **89:** 867.

Tomkinson A.E., Totty N.F., Ginsburg M., and Lindahl T. 1991. Location of the active site for enzyme-adenylate formation in DNA ligases. *Proc. Natl. Acad. Sci.* **88:** 400.

Wang S.P. and Shuman S. 1997. Structure-function analysis of the mRNA cap methyltransferase of *Saccharomyces cerevisiae*. *J. Biol. Chem.* **272:** 14683.

Wang S.P., Deng L., Ho C.K., and Shuman S. 1997. Phylogeny of mRNA capping enzymes. *Proc. Natl. Acad. Sci.* **94:** 9573.

Wimberly B.T., Brodersen D.E., Clemons W.M., Morgan-Warren R.J., Carter A.P., Vornhein C., Hartsch T., and Ramakrishnan V. 2000. Structure of the 30S ribosomal subunit. *Nature* **407:** 327.

Woese C.R. 2000. Interpreting the universal phylogenetic tree. *Proc. Natl. Acad. Sci.* **97:** 8392. Woese C.R., Kandler O., and Wheelis M.L. 1990. Towards a natural system of organisms: Proposal for the domains archaea, bacteria, and eucarya. *Proc. Natl. Acad. Sci.* **87:** 4576.

Yamada-Okabe T., Mio T., Matsui M., Kashima Y., Arisawa M., and Yamada-Okabe H. 1998. Isolation and characterization of the *Candida albicans* gene for mRNA 5′ triphosphatase: association of mRNA 5′ triphosphatase and mRNA 5′ guanylyltransferase activities is essential for the function of mRNA 5′ capping enzyme in vivo. *FEBS Lett.* **435:** 49.

Yokoska J., Tsukamoto T., Miura K., Shiokawa K., and Mizumoto K. 2000. Cloning and characterization of mRNA capping enzyme and mRNA (guanine-7-)-methyltransferase cDNAs from *Xenopus laevis*. *Biochem. Biophys. Res. Commun.* **268:** 617.

Yu L., Martins A., Deng L., and Shuman S. 1997. Structure-function analysis of the triphosphatase component of vaccinia virus mRNA capping enzyme. *J. Virol.* **71:** 9837.

Yusupov M.M., Yusupova G.Z., Baucom A., Liberman K., Earnest T.N., Cate J.H.D., and Noller H.F. 2001. Crystal structure of the ribosome at 5.5 Å resolution. *Science* **292:** 883.

Nonsense-mediated mRNA Decay: Insights into Mechanism from the Cellular Abundance of Human Upf1, Upf2, Upf3, and Upf3X Proteins

L.E. Maquat and G. Serin

Department of Biochemistry and Biophysics, School of Medicine and Dentistry,
University of Rochester, Rochester, New York 14642

Nonsense-mediated mRNA decay (NMD), also called mRNA surveillance, is one of several posttranscriptional mechanisms that eukaryotic cells employ to control the quality of mRNA function (for review, see Maquat and Carmichael 2001). NMD degrades aberrant transcripts that prematurely terminate translation in order to eliminate the production of incomplete proteins that could function in dominant-negative or other deleterious ways (for review, see Maquat 1995, 2000; Jacobson and Peltz 1996, 2000; Li and Wilkinson 1998; Culbertson 1999; Frischmeyer and Dietz 1999; Hentze and Kulozik 1999; Hilleren and Parker 1999). Insights into the mechanism of NMD will derive, in part, from an analysis of required factors. Here, we report the cellular abundance of the four human (h) Upf proteins known to function in NMD, and we use this information to extend our understanding of how the premature termination of mRNA translation elicits mRNA decay.

YEAST Upf PROTEINS

The hUpf proteins are named after their counterparts in *Saccharomyces cerevisiae*: Upf1p, Upf2p/Nmd2p, and Upf3p (Leeds et al. 1991, 1992; Atkin et al. 1995, 1997; Cui et al. 1995; He and Jacobson 1995; Lee and Culbertson 1995). Upf1p is an RNA-dependent ATPase and 5′-to-3′ RNA helicase, and mutations that disrupt helicase activity disrupt NMD (Czaplinski et al. 1995; Weng et al. 1996a,b, 1998). Upf2p is rich in acidic residues but otherwise has no distinguishing features (Cui et al. 1995). Upf3p is rich in basic residues (Lee and Culbertson 1995), consists of three sequences that resemble nuclear localization signals and two sequences that resemble leucine-rich nuclear export signals, and shuttles between nuclei and the cytoplasm (Shirley et al. 1998). Since mutation of either Upf2p or Upf3p also disrupts NMD (Cui et al. 1995; He and Jacobson 1995; Lee and Culbertson 1995; He et al. 1996), all three Upf proteins are thought to be integral components of the NMD pathway. Consistent with this idea, Upf1p, Upf2p, and Upf3p colocalize with 80S ribosome particles and polysomes (Peltz et al. 1993; Atkin et al. 1995, 1997), where NMD occurs (Zhang et al. 1997), and Upf2p interacts in two-hybrid assays with both Upf1p and Upf3p (He and Jacobson 1995;

He et al. 1996, 1997; Weng et al. 1996b). Additionally, Upf1p interacts in two-hybrid assays with Dcp2p/Nmd1p (He and Jacobson 1995). Dcp2p/Nmd1p forms a complex with the decapping protein Dcp1p, suggesting a role for Upf1p in the deadenylation-independent decapping that distinguishes NMD from the general decay of nonsense-free mRNA (Muhlrad and Parker 1994; Dunkley and Parker 1999). Consistent with this idea, deletion of Upf1p, Upf2p, or Upf3p was found to inhibit both Dcp1p-mediated decapping and the subsequent Xrn1p-mediated 5′-to-3′ exonucleolytic degradation of many nonsense-containing (and, under certain circumstances, nonsense-free) mRNAs (He and Jacobson 2001).

In addition to functioning in NMD, Upf1p, Upf2p, and Upf3p enhance translation termination (Leeds et al. 1992; Weng et al. 1996a,b, 1998; Maderazo et al. 2000; Wang et al. 2001), and certain mutations within Upf1p or the absence of Upf3p affects programmed –1 frameshifting (Cui et al. 1996; Ruiz-Echavarrìa et al. 1998). Upf1p interacts with release factors eRF1 and eRF3, and eRF3 inhibits Upf1p ATPase/helicase activity and competes with RNA for interaction with Upf1p (Czaplinski et al. 1998). Upf2p and Upf3p interact only with eRF3 (Wang et al. 2001). Upf2p, Upf3p, and eRF1 compete with one another for binding to the essential GTPase domain of eRF3; in contrast, Upf1p binds independently to a more amino-terminal domain (Wang et al. 2001). A series of steps has been proposed to take place following translation termination, be it premature or not, and each step is thought to be required for NMD (Wang et al. 2001): (1) The eRF1–eRF3 complex binds to the ribosome A site as a consequence of 80S ribosome pausing at a termination codon; (2) Upf1p associates with the eRF1–eRF3 complex; (3) the peptidyl-tRNA bond is hydrolyzed; (4) eRF1 dissociates from the ribosome, allowing Upf2p (or Upf3p) to join the complex by binding to eRF3; (5) Upf3p associates with Upf2p (or Upf2p associates with Upf3p), releasing eRF3 and activating the ATPase-helicase activity of Upf1p; and (6) the assembled surveillance complex, consisting minimally of Upf1p, Upf2p, and Upf3p, surveys mRNA sequences located downstream of the termination codon. Provided that a destabilizing element exists downstream (Gonzalez et al. 2000 and references therein), interactions between the surveillance complex and the element trigger NMD.

HUMAN Upf PROTEINS

hUpf1, which consists of 1118 amino acids, was first identified on the basis of similarities to yeast Upf1p (Perlick et al. 1996; Applequist et al. 1997). Conserved domains include a cysteine- and histidine-rich region followed by a group 1 RNA helicase domain. However, hUpf1 differs from yeast Upf1p, which consists of only 853 amino acids, most notably by having 63 additional amino acids at the amino terminus and 83 additional amino acids at the carboxyl terminus. Despite these differences, hUpf1 harboring an arginine-to-cysteine change at position 844 within its putative helicase domain has a dominant-negative effect on NMD in mammalian cells (Sun et al. 1998), as does a similarly mutated Upf1p in yeast (Leeds et al. 1992). Also like yeast Upf1p, hUpf1 manifests RNA-dependent ATPase activity, ATP-sensitive RNA binding, and 5′-to-3′ RNA helicase activity (Bhattacharya et al. 2000). Additionally, in yeast, even though full-length hUpf1 fails to function in either nonsense suppression or NMD, hUpf1 flanked by the extreme amino and carboxyl termini of yeast Upf1p, although inactive in NMD, functions in nonsense suppression (Perlick et al. 1996; Czaplinski et al. 1998). These findings, together with data demonstrating that in-vitro-synthesized hUpf1 interacts with eRF1 and eRF3 and the interaction with eRF3 is inhibited by RNA (Czaplinski et al. 1998), indicate that the activities of Upf1 protein are conserved between yeast and mammalian cells. hUpf1 is essentially equally distributed among polysomal, subpolysomal, and ribosome-free fractions in exponentially growing HeLa cells (Pal et al. 2001), which is compatible with data indicating that most Upf1p in yeast colocalizes with polysomes (Atkin et al. 1995). hUpf1 is a phosphoprotein, as is its SMG-2 ortholog in *Caenorhabditis elegans* (Page et al. 1999). The importance of phosphorylation to NMD in *C. elegans* is illustrated by the finding that *smg-1, -3, -4, -5, -6*, and *-7* mutants, which are defective in NMD, are defective in either the phosphorylation or dephosphorylation of SMG-2 (Page et al. 1999). hUpf1 phosphorylation is sensitive to wortmannin and, to a lesser extent, rapamycin (Pal et al. 2001). The wortmannin sensitivity is attributable, at least in part, to hSMG-1, which is a phosphoinositol kinase-related kinase that is thought to directly phosphorylate hUpf1 (Denning et al. 2001; Yamashita et al. 2001). Although there has been no explicit indication that yeast Upf1p is phosphorylated or that an *S. cerevisiae* ortholog to *C. elegans* SMG-1/hSMG-1 exists, epitope-tagged yeast Upf1p has been observed to migrate as a doublet in acrylamide, suggesting that it is a substrate for post translational modification (Atkin et al. 1997).

Human orthologs to yeast Upf2p and Upf3p, which are also orthologs to *C. elegans* SMG-3 and SMG-4, respectively, have also been identified (Lykke-Andersen et al. 2000; Mendell et al. 2000; Serin et al. 2001; Aronoff et al. 2001). hUpf2 is cytoplasmic but perinuclear, whereas hUpf3 is mostly nuclear and shuttles between nuclei and the cytoplasm (Lykke-Andersen et al. 2000; Serin et al. 2001). Interestingly, there are two genes for hUpf3, and each generates at least two isoforms due to alternative pre-mRNA splicing. Since one gene is X-linked, it has been

named hUpf3X (Serin et al. 2001; or Upf3b by Lykke-Andersen et al. 2000). To date, hUpf3 and hUpf3X are functionally indistinguishable. hUpf1p, hUpf2p, hUpf3, and hUpf3X coimmunoprecipitate (Lykke-Andersen et al. 2000; Serin et al. 2001), and tethering any one of the proteins to the 3′-untranslated region of β-globin mRNA elicits NMD (Lykke-Andersen et al. 2000). The tethered proteins are thought to bypass the general requirement of NMD for a splicing-generated exon–exon junction located more than 50–55 nucleotides downstream of a termination codon (Cheng et al. 1994; Carter et al. 1996; Thermann et al. 1998; Zhang et al. 1998a, b; Sun et al. 2000). Drawing analogy to the situation in yeast, splicing-generated exon–exon junctions in mammalian cells can be thought of as the functional equivalent to destabilizing elements in yeast (for review, see Maquat 2000).

It has been proposed that the requirement of NMD for a splicing-generated exon–exon junction involves the ~335-kD complex of proteins deposited in the nucleus ~20–24 nucleotides upstream of exon–exon junctions as a consequence of splicing (Le Hir et al. 2000a,b; Kim et al. 2001b). According to a prevailing model (Fig.1): (1) One or more components of this complex are thought to recruit hUpf3/3X in the nucleus, (2) hUpf3/3X is then exported in association with mRNA to the cytoplasm where it recruits hUpf2, (3) provided that translation then terminates more than 50–55 nucleotides upstream of a junction, i.e., 25–30 nucleotides upstream of bound hUpf2-hUpf3/3X, hUpf1p is also recruited to form the complex that is normally required for NMD (Lykke-Andersen et al. 2000, 2001; Ishigaki et al. 2001; Kim et al. 2001a; Le Hir et al. 2001). In support of this model, NMD mediated by any tethered hUpf requires translation (Lykke-Andersen et al. 2000), and hUpf2 and hUpf3 are detected in association with cytoplasmic mRNP, but hUpf1 is not (Lykke-Andersen et al. 2000; Ishigaki et al. 2001). In contrast, if translation were to terminate either less than 50–55 nucleotides upstream of a junction or downstream of a junction, it is thought that (1) hUpf1 would not associate with the mRNA-bound complex of hUpf2 and hUpf3 and (2) the complex would be inactivated (possibly by being removed from mRNA) by a translating ribosome. This scenario implies that NMD is restricted to newly synthesized mRNA, consistent with the finding that most, if not essentially all, of NMD takes place on CBP80-bound rather than eIF4E-bound mRNA (Ishigaki et al. 2001; see below). Since studies of hUpf orthologs in *C. elegans* (Page et al. 1999) predict that hUpf2 and hUpf3/3X also function in the dephosphorylation of hUpf1, it is conceivable that interactions between hUpf1 and the mRNA-bound complex of hUpf2-hUpf3/3X could also alter the phosphorylation status of hUpf1.

hUpf PROTEIN ABUNDANCE VARIES FROM AN ESTIMATED 4×10^4 TO 4×10^6 MOLECULES PER HeLa CELL, DEPENDING ON THE PARTICULAR PROTEIN

One way to evaluate the validity of the above-described scenario for hUpf function is to determine hUpf

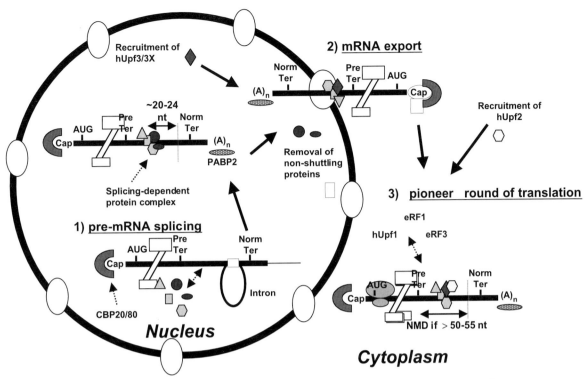

Figure 1. Model for NMD in human cells. In the nucleus, introns are removed from pre-mRNA by splicing and, as a consequence, a ~335-kD complex of proteins binds 20–24 nucleotides upstream of the resulting exon–exon junctions. The mostly nuclear hUpf3/3X is recruited to the complex, and nonshuttling components of the complex are removed. mRNA is then exported to the cytoplasm, where the cytoplasmic but mostly perinuclear hUpf2 associates with hUpf3/3X. During the "pioneer" round of translation, so called because the mRNA substrate is still bound by CBP80/CBP20 at the cap structure and PABP2 at the poly(A) tail, hUpf1, eRF1, and eRF3 become involved at the point of translation termination. If termination takes place more than 50–55 nucleotides upstream of an exon–exon junction that is "marked" by the complex of hUpf2 and hUpf3/3X, then an interaction between hUpf1 and this complex will elicit NMD. In contrast, if termination takes place where there is no exon–exon junction located more than 50–55 nucleotides downstream, then translating ribosomes will remove the complex of hUpf2 and hUpf3/3X, eliminating the possibility for NMD. Either scenario may change the phosphorylation status of hUpf1, given the putative role of hUpf2 and hUpf3/3X in dephosphorylating hUpf1. Dotted vertical lines specify an exon–exon junction of mRNA. Pre Ter specifies a premature termination codon, and Norm Ter specifies the normal termination codon. Although others are shown (with question marks), only two components (RNPS1 and Y14) of the splicing-dependent complex, each of which shuttles to the cytoplasm, appear to remain bound to mRNA during the "pioneer" round of translation (Y. Ishigaki and L.E. Maquat, unpubl.). Notably, depending on the particular mRNA, NMD is cytoplasmic, as shown, or nucleus-associated NMD. Nucleus-associated NMD may take place in the nucleoplasm and involve an uncharacterized translation-like mechanism or during mRNA export and involve cytoplasmic translation as shown here for cytoplasmic NMD.

cellular abundance. Previous quantitations of hUpf1 in human HeLa cells revealed ~3 x 10^6 molecules per exponentially growing cell (Pal et al. 2001). This indicates that there is one molecule of hUpf1 per ~3 ribosomes, based on a calculation of 50 pg of RNA per HeLa cell, 80% of which is ribosomal RNA (Pal et al. 2001). This ratio of hUpf1 to ribosomes is compatible with the putative role of hUpf1 in translation termination.

To supplement this information, the cellular abundance of hUpf2, hUpf3, and hUpf3X was also determined. To this end, expression vectors that produced FLAG-tagged versions of each protein were generated. HeLa cells that had been grown to 80–90% confluency on three 20-cm plates were then transfected using lipofection (Serin et al. 2001) with one of the three expression vectors or, as a control, a vector that expresses FLAG-hUpf1 (Pal et al. 2001). After 48 hours each FLAG-tagged protein was immunopurified from cleared lysates using anti-FLAG M5 antibody and protein A agarose (Serin et al. 2001). Using

western blot analysis and anti-FLAG M5 antibody, the reactivity of serial dilutions of each immunopurified protein was compared to the reactivity of known amounts of commercially available FLAG-bacterial alkaline phosphatase (BAP) (Fig. 2A). Those dilutions of FLAG-hUpf that reacted within the linear range of the FLAG-BAP reactions revealed concentrations of 500 ± 75, 9 ± 1.5, 30 ± 2, and 13 ± 2 fmole/µl for, respectively, FLAG-hUpf1, FLAG-hUpf2, FLAG-hUpf3, and FLAG-hUpf3X. Using western blot analysis and the appropriate anti-hUpf antibody (Ishigaki et al. 2001; Pal et al. 2001), the reactivity of specified amounts of each FLAG-hUpf was compared to the reactivity of extract prepared from a known number of untransfected HeLa cells (Fig. 2B). Those dilutions of cell extract that reacted within the linear range of the appropriate FLAG-hUpf reactions revealed concentrations of 4 ± 2 x 10^6, 2 ± 1 x 10^5, 8 ± 2 x 10^4, and 4 ± 2 x 10^4 molecules per cell for, respectively, hUpf1, hUpf2, hUpf3, and hUpf3X. This determination

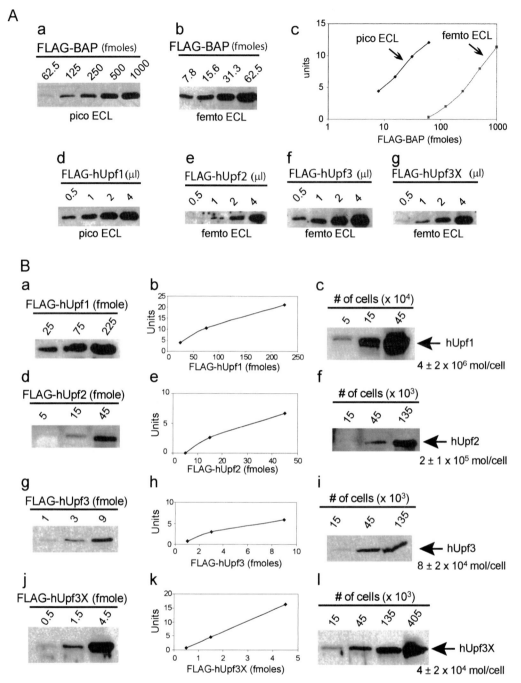

Figure 2. Abundance of human Upf proteins in HeLa cells. (*A*) Quantitation of immunopurified FLAG-hUpf1, FLAG-hUpf2, FLAG-hUpf3, and FLAG-hUpf3X. (*a,b*) pFLAG-hUPF2, pFLAG-hUPF3, and pFLAG-hUPF3X were generated from the corresponding pCI-neo derivative (Serin et al. 2001). pFLAG-hUPF1 (Sun et al. 1998) and each of the newly generated plasmids were transiently expressed in HeLa cells, and FLAG-tagged proteins were immunopurified (Serin et al. 2001). Serial dilutions of FLAG-bacterial alkaline phosphatase (FLAG-BAP; Sigma) were analyzed by western blotting using anti-FLAG M5 antibody (Sigma) and either Pico-ECL (Pierce) or Femto-ECL (Pierce). (*c*) The extent of hybridization was quantitated using autoradiographic densitometry (Image-Quant). (*d–g*) At the same time, serial dilutions of unknown amounts of immunopurified FLAG-hUpf1, FLAG-hUpf2, FLAG-hUpf3, and FLAG-hUpf3X were similarly analyzed. Those dilutions characterized by an extent of reactivity that resided within the linear range of the FLAG-BAP reactivities as measured using either Pico-ECL or Femto-ECL revealed the following concentrations: FLAG-hUpf1, 500 ± 75 fmole/μl; FLAG-hUpf2, 9 ± 1.5 fmole/μl; FLAG-hUpf3, 30 ± 2 fmole/μl; FLAG-hUpf3X, 13 ± 2 fmole/μl. (*B*) Calculation of the number of endogenous hUpf1, hUpf2, hUpf3, and hUpf3X molecules per cell. (*a,d,g,j*) Serial dilutions of each FLAG-hUpf protein and (*c,f,i,l*) serial dilutions of extract from the specified number of exponentially growing HeLa cells were analyzed as described for *A* except that rabbit polyclonal anti-hUpf1 (Pal et al. 2001), anti-Upf2 (Ishigaki et al. 2001), anti-hUpf3X (Ishigaki et al. 2001) and anti-hUpf3 antibodies were used in the western blot analyses. (*b,e,h,k*) Those extract dilutions characterized by an extent of reactivity that resided within the linear range of the corresponding FLAG-hUpf reactivities revealed the specified number of molecules per cell (mol/cell).

for hUpf1 fits well with the previous estimate of 3×10^6 hUpf1 molecules per cell (Pal et al. 2001). Notably, antibody to hUpf3 fails to recognize the isoform that derives from alternatively spliced pre-mRNA in which exonic sequences are removed. However, this isoform is estimated to constitute < 10% of HeLa-cell hUpf3 based on RT-PCR analyses of hUPF3 mRNA abundance (Serin et al. 2001).

MUSINGS ON THE MECHANISM OF MAMMALIAN-CELL NMD

As alluded to above, immunopurifications of RNP from Cos cells using antibody to either the primarily nuclear cap binding protein (CBP) 80 or its primarily cytoplasmic counterpart, eukaryotic initiation factor (eIF) 4E, revealed that NMD characterizes nonsense-containing mRNA bound primarily by CBP80 rather than eIF4E (Ishigaki et al. 2001). Other components of NMD-susceptible mRNP were defined as CBP20, which forms a heterodimer with CBP80; PABP2, which is the mostly nuclear poly(A)-binding protein; the translation initiation factor eIF4G; hUpf2; and hUpf3 (Ishigaki et al. 2001). The relative immunity of eIF4E-bound mRNA to NMD was based on a number of findings: (1) The extent of NMD for CBP80-bound mRNA was comparable to the extent of NMD for eIF4E-bound mRNA (notably, CPB80 binds to the cap co-transcriptionally and at some point after splicing is replaced by eIF4E), (2) hUpf2 and hUpf3 were detected on CBP80-bound but not eIF4E-bound mRNA, and (3) mRNP immunopurified using anti-hUpf3 antibody was associated with CBP80 but not eIF4E (Ishigaki et al. 2001). Consistent with the dependence of NMD on translation termination, the NMD of CBP80-bound mRNA was found to be blocked by either cycloheximide, which freezes elongating ribosomes, or suppressor tRNA, which recognizes a termination codon as a coding codon (Ishigaki et al. 2001).

These data provide the first evidence that translation in mammalian cells can take place in association with CBP80. They also indicate that NMD is confined to a "pioneer" round of translation. The pioneer round would precede mRNP remodeling that includes replacement of CPB80 (and CBP20) by eIF4E, replacement of PABP2 by PABP1, and dissociation of hUpf2 and hUpf3. The pioneer round would also precede the bulk of mRNA translation. Notably, even though CBP80, CBP20, PABP2, and hUpf3 associate with mRNA in the nucleus, they also shuttle to the cytoplasm (presumably while bound to mRNA). This is consistent with the findings that the NMD of some mRNAs takes place during translation by cytoplasmic ribosomes (Sun et al. 2000; Rajavel and Neufeld 2001). Those mRNAs that are subject to "nucleus-associated" NMD could be degraded either in the nucleoplasm or, if translation is restricted to the cytoplasm, during or immediately after transport across the nuclear pore complex but prior to release from nuclei into the cytoplasm (for review, see Maquat 2000).

There are an estimated 30,000 mammalian genes (for review, see Claverie 2001), not all of which are simultaneously transcribed in a cell, and an estimated 75,000 transcripts per cell in the process of elongation by the combination of RNA polymerases II and III (Jackson et al. 1998). Therefore, the number of CBP80-bound nascent pre-mRNAs at any given time is less than 75,000 but probably on the order of 10^4. Primary transcripts are typically 2–6 times longer than the mature mRNAs they encode, which have an average size of ~2000 nucleotides (for review, see Jackson et al. 2000). Recent estimates indicate that there are an average of four to seven exon–exon junctions per mRNA (Venter et al. 2001) or seven to eight exon–exon junctions per mRNA (Lander et al. 2001). Given that internal exons are usually not larger than 250 nucleotides due to splicing-imposed constraints but 3′-most exons appear to be without size constraints (Sterner et al. 1996 and references therein), these estimates are reasonable. Therefore, there are probably no more than ~5 × 10^5 nascent splicing-generated exon–exon junctions per cell, which is on the order of or slightly more than the ~1 × 10^5 molecules of hUpf3 and hUpf3X. This exon–exon junction calculation does not consider those CBP80-bound transcripts, currently of unknown number, that have been released from chromatin and are in transit to translationally active ribosomes. If hUpf3 and hUpf3X have the potential to bind every junction, which is a reasonable supposition, then there would likely not be enough hUpf3/3X for every junction. This shortage may, at least in part, explain why NMD is less than 100% efficient: for example, only 80–85% of nonsense-containing β-globin or triose phosphate isomerase mRNAs is generally subject to NMD (Zhang et al. 1998a,b; Thermann et al. 1998). Nevertheless, the level of hUpf3/3X appears to be sufficient to support the idea that hUpf3/3X binds to most newly synthesized exon–exon junctions.

During the process of mRNA export, junction-bound hUpf3/3X is thought to recruit cytoplasmic hUpf2, possibly near the nuclear periphery where hUpf2 appears to concentrate (Lykke-Andersen et al. 2000; Serin et al. 2001). Considering that hUpf2 is approximately 10-fold more abundant than hUpf3/3X, recruitment could be very efficient.

A role for hUpf1 in translation termination (Czaplinski et al. 1998) is consistent with its abundance of ~4 × 10^6 molecules per cell, which is equivalent to ~0.35 molecules per ribosome (see above). This finding, plus data demonstrating that a dominant-negative hUpf1 mutant impairs NMD when hUpf2 or hUpf3/3X, but not hUpf1, is tethered downstream of a termination codon, indicate that hUpf1 is the last of the hUpf proteins to join the complex required for NMD (Lykke-Andersen et al. 2000). Since active translation is required for tethered hUpf1 to elicit NMD, hUpf1 may interact with the exon–exon junction-bound complex of hUpf2 and hUpf3/3X before termination (Lykke-Andersen et al. 2000). As one of several other possibilities, a post-termination complex that includes translational factors in addition to hUpf1 could be required for NMD.

There is significant difference between the cellular abundance of hUpf proteins and yeast Upf proteins, as would be expected considering that HeLa cells have an estimated 30,000 genes and $\sim 9 \times 10^6$ ribosomes, whereas *S. cerevisiae* has an estimated 6000 genes (for review, see Mewes et al. 1997) and 3×10^5 ribosomes (Waldron and Lacroute 1975). The abundance of Upf1p, Upf2p, and Upf3p in *S. cerevisiae* has been estimated by comparing the intensities of western blot analyses using crude extracts and purified proteins: There are ~ 1600 molecules of Upf1p, ~ 160 molecules of Upf2p, and ~ 80 molecules of Upf3p per cell (Maderazo et al. 2000). It follows that the ratio of Upf1p to ribosomes in yeast is $\sim 1:190$, which is significantly less than the $\sim 1:3$ ratio in mammalian cells. Lending credence to this calculation for yeast, the ratio of Upf1p to ribosomes was independently derived to be 1:120 and 1:90 based on the ratio of, respectively, *Upf1* mRNA to ribosomal protein S3 mRNA and *Upf1* mRNA to ribosomal protein L3 mRNA (Atkin et al. 1995). Despite significant interspecies differences in the ratio of Upf proteins and ribosomes, there is remarkable interspecies similarity in the ratio of Upf1:Upf2:Upf3, which is 20:2:1 in yeast and 40:2:1 in mammalian cells.

CONCLUSIONS

hUpf2p and hUpf3/3X, at $\sim 2 \times 10^5$ and $\sim 1 \times 10^5$ molecules per HeLa cell, respectively, are the least abundant of the hUpf proteins. hUpf2p and hUpf3/3X are thought to be associated with splicing-generated exon–exon junctions and are detected associated with CBP80-bound but not eIF4E-bound mRNA (Ishigaki et al. 2001). Consistent with this finding, NMD in mammalian cells appears to target primarily CBP80-bound rather than eIf4E-bound mRNA (Ishigaki et al. 2001). The abundance of hUpf3/3X, together with projections made from the recently acquired human genome sequence, indicates that many, if not most, exon–exon junctions of CBP80-bound mRNA could be associated with hUpf3/3X. Those junctions that are not associated with hUpf3/3X may contribute to the <100% efficiency that typifies NMD. hUpf1, at $\sim 4 \times 10^6$ molecules per HeLa cell, is the most abundant of the hUpf proteins. Its abundance is on a par with the abundance of ribosomes, consistent with its putative role in translation termination, which is thought to be a prerequisite for its role in NMD. Future studies aim to elucidate interactions between hUpf proteins before, during, and subsequent to translation termination and how hUpf1 phosphorylation influences these interactions.

ACKNOWLEDGMENTS

We thank Ben Blencowe for helpful conversations. This work was supported by U.S. Public Health Service grants DK 33938 and GM 59614 to L.E.M. from the National Institutes of Health.

REFERENCES

Applequist S.E., Selg M., Raman C., and Jäck H.-M. 1997. Cloning and characterization of HUPF1, a human homolog of the *Saccharomyces cerevisiae* nonsense mRNA-reducing UPF1 protein. *Nucleic Acids Res.* **25:** 814.

Aronoff R., Baran R., and Hodgkin J. 2001. Molecular identification of *smg-4*, required for mRNA surveillance in *C. elegans. Gene* **268:** 153.

Atkin A.L., Altamura N., Leeds P., and Culbertson M.R. 1995. The majority of yeast UPF1 colocalizes with polyribosomes in the cytoplasm. *Mol. Biol. Cell* **6:** 611.

Atkin A.L., Schenkman L.R., Eastham M., Dahlseid J.N., Lelivelt M.J., and Culbertson M.R. 1997. Relationship between yeast polyribosomes and Upf proteins required for nonsense mRNA decay. *J. Biol. Chem.* **272:** 22163.

Bhattacharya A., Czaplinski K., Trifillis P., He F., Jacobson A., and Peltz S.W. 2000. Characterization of the biochemical properties of the human Upf1 gene product that is involved in nonsense-mediated mRNA decay. *RNA* **6:** 1226.

Carter M.S., Li S., and Wilkinson M.F. 1996. A splicing-dependent regulatory mechanism that detects translation signals. *EMBO J.* **15:** 5965.

Cheng J., Belgrader P., Zhou X., and Maquat L.E. 1994. Introns are *cis* effectors of the nonsense-codon-mediated reduction in nuclear mRNA abundance. *Mol. Cell. Biol.* **14:** 6317.

Claverie J.-M. 2001. What if there are only 30,000 human genes? *Science* **291:** 1257.

Culbertson M.R. 1999. RNA surveillance: Unforeseen consequences for gene expression, inherited genetic disorders and cancer. *Trends Genet.* **15:** 74.

Cui Y., Dinman J.D., and Peltz S.W. 1996. Mof4-1 is an allele of the UPF1/IFS2 gene which affects both mRNA turnover and -1 ribosomal frameshifting efficiency. *EMBO J.* **15:** 5726.

Cui Y., Hagan K.W., Zhang S., and Peltz S.W. 1995. Identification and characterization of genes that are required for the accelerated degradation of mRNAs containing a premature translational termination codon. *Genes Dev.* **9:** 423.

Czaplinski K., Weng Y., Hagan K.W., and Peltz S.W. 1995. Purification and characterization of the Upf1p: A factor involved in translation and mRNA degradation. *RNA* **1:** 610.

Czaplinski K., Ruiz-Echevarría M.J., Paushkin S.V., Han X., Weng Y., Perlick H.A., Dietz H.C., Ter-Avanesyan M.D., and Peltz S.W. 1998. The surveillance complex interacts with the translation release factors to enhance termination and degrade aberrant mRNAs. *Genes Dev.* **12:** 1665.

Denning G., Jamieson L., Maquat L.E., Thompson E.A., and Fields A.P. 2001. Cloning of a novel phosphoinositide kinase-related kinase: Characterization of the human SMG-1 RNA surveillance protein. *J. Biol. Chem.* **276:** 22709.

Dunkley T. and Parker R. 1999. The DCP2 protein is required for mRNA decapping in *Saccharomyces cerevisiae* and contains a functional MuT motif. *EMBO J.* **18:** 5411.

Frischmeyer P.A. and Dietz H.C. 1999. Nonsense-mediated mRNA decay in health and disease. *Hum. Mol. Genet.* **8:** 1893.

Gonzalez C.I., Ruiz-Echevarría M.J., Vasudevan S., Henry M.F., and Peltz S.W. 2000. The yeast hnRNP-like protein Hrp1/Nab4 marks a transcript for nonsense-mediated mRNA decay. *Mol. Cell* **5:** 1580.

He F. and Jacobson A. 1995. Identification of a novel component of the nonsense-mediated mRNA-decay pathway by use of an interacting-protein screen. *Genes & Dev.* **9:** 437.

———. 2001. Upf1p, Nmd2p, and Upf3p regulate the decapping and exonucleolytic degradation of both nonsense-containing mRNAs and wild-type mRNAs. *Mol. Cell. Biol.* **21:** 1515.

He F., Brown A.H., and Jacobson A. 1996. Interaction between Nmd2p and Upf1p is required for activity but not for dominant-negative inhibition of the nonsense-mediated mRNA decay pathway in yeast. *RNA* **2:** 153.

———. 1997. Upf1p, Nmd2p and Upf3p are interacting components of the yeast nonsense-mediated mRNA decay pathway. *Mol. Cell. Biol.* **17:** 1580.

Hentze M.W. and Kulozik A.E. 1999. A perfect message: RNA surveillance and nonsense-mediated decay. *Cell* **96:** 307.

Hilleren P. and Parker R. 1999. Mechanisms of mRNA surveillance in eukaryotes. *Annu. Rev. Genet.* **33:** 229.

Ishigaki Y., Li X., Serin G., and Maquat L.E. 2001. Evidence for a pioneer round of mRNA translation: mRNAs subject to nonsense-mediated decay in mammalian cells are bound by CBP80 and CBP20. *Cell* **106:** 607.

Jacobson A. and Peltz S.W. 1996. Interrelationships of the pathways of mRNA decay and translation in eukaryotic cells. *Annu. Rev. Biochem.* **65:** 693.

———. 2000. Destabilization of nonsense-containing transcripts in *Saccharomyces cerevisiae*. In *Translational control of gene expression* (eds. N. Sonenberg et al.), p. 849. Cold Spring Harbor Laboratory Press, Cold Spring Harbor, New York.

Jackson D.A., Pombo A., and Iborra F. 2000. The balance sheet for transcription: An analysis of nuclear RNA metabolism in mammalian cells. *FASEB J.* **14:** 242.

Jackson D.A., Iborra F.J., Manders E.M.M., and Cook P.R. 1998. Numbers and organization of RNA polymerase, nascent transcripts, and transcription units in HeLa nuclei. *Mol. Biol. Cell* **9:** 1523.

Kim V.N., Kataoka N., and Dreyfuss G. 2001a. Role of the nonsense-mediated decay factor Upf3 in the splicing-dependent exon-exon junction complex. *Science* **293:** 1832.

Kim V.N., Yong J., Kataoka N., Abel L., Diem M.D., and Dreyfuss G. 2001. The Y14 protein communicates to the cytoplasm the position of exon-exon junctions. *EMBO J.* **20:** 2062.

Lander E.S., Linton L.M., Birren B. et al. 2001. Initial sequencing and analysis of the human genome. *Nature* **409:** 860.

Lee B.S. and Culbertson M.R. 1995. Identification of an additional gene required for eukaryotic nonsense mRNA turnover. *Proc. Natl. Acad. Sci.* **92:** 10354.

Leeds P., Peltz S.W., Jacobson A., and Culbertson M.R. 1991. The product of the yeast UPF1 gene is required for rapid turnover of mRNAs containing a premature translational termination codon. *Genes Dev.* **5:** 2303.

Leeds P., Wood J.M., Lee B.S., and Culbertson M.R. 1992. Gene products that promote mRNA turnover in *Saccharomyces cerevisiae*. *Mol. Cell. Biol.* **12:** 2165.

Le Hir H., Moore M.J., and Maquat L.E. 2000a. Pre-mRNA splicing alters mRNP composition: Evidence for a stable association of proteins at exon-exon junctions. *Genes Dev.* **14:** 1098.

Le Hir H., Izaurralde E., Maquat L.E., and Moore M.J. 2000b. The spliceosome deposits multiple proteins 20-24 nucleotides upstream of mRNA exon-exon junctions. *EMBO J.* **15:** 6860.

Le Hir H., Gatfield D., Izaurralde E., and Moore M.J. 2001. The exon-exon junction complex provides a binding platform for factors involved in mRNA export and nonsense-mediated mRNA decay. *EMBO J.* **20:** 4987.

Li S. and Wilkinson M.F. 1998. Nonsense surveillance in lymphocytes? *Immunity* **8:** 135.

Lykke-Andersen J., Shu M.-D., and Steitz J.A. 2000. Human Upf proteins target an mRNA for nonsense-mediated decay when bound downstream of a termination codon. *Cell* **103:** 1121.

———. 2001. Communication of the position of exon-exon junctions to the mRNA surveillance machinery by the protein RNPS1. *Science* **293:** 1836.

Maderazo A., He F., Mangus D.A., and Jacobson A. 2000. Upf1p control of nonsense mRNA translation is regulated by Nmd2p and Upf3p. *Mol. Cell. Biol.* **20:** 4591.

Maquat L.E. 1995. When cells stop making sense: Effects of nonsense codons on RNA metabolism in vertebrate cells. *RNA* **1:** 453.

———. 2000. Nonsense-mediated RNA decay in mammalian cells: A splicing-dependent means to down-regulate the levels of mRNAs that prematurely terminate translation. In *Translational control of gene expression* (ed. N. Sonenberg et al.), p. 849. Cold Spring Harbor Laboratory Press, Cold Spring Harbor, New York.

Maquat L.E. and Carmichael G.G. 2001. Quality control of mRNA function. *Cell* **104:** 173.

Mendell J.T., Medghalchi S.M., Lake R.G., Noensie E.N., and

Dietz H.C. 2000. Novel Upf2p orthologues suggest a functional link between translation initiation and nonsense surveillance complexes. *Mol. Cell. Biol.* **20:** 8944.

Mewes H.W., Albermann K., Bähr M., Frishman D., Gleissnere A., Hanl J., Heumann K., Kleine K., Maierl A., Oliver S.G., Pfeiffer R., and Zollner A. 1997. Overview of the yeast genome. *Nature* **387:** 7.

Muhlrad D. and Parker R. 1994. Premature translational termination triggers mRNA decapping. *Nature* **370:** 578.

Page M.F., Carr, B., Anders K.R., Grimson A., and Anderson P. 1999. SMG-2 is a phosphorylated protein required for mRNA surveillance in *Caenorhabditis elegans* and related to Upf1p of yeast. *Mol. Cell. Biol.* **19:** 5943.

Pal M., Ishigaki Y., Nagy E., and Maquat L.E. 2001. Evidence that phosphorylation of human Upf1 protein varies with intracellular location and is mediated by a wortmannin-sensitive and rapamycin-sensitive PI 3-kinase-related kinase signaling pathway. *RNA* **5:** 5.

Peltz S.W., Brown A.H., and Jacobson A.J. 1993. mRNA destabilization triggered by premature translational termination depends on at least three *cis*-acting sequence elements and one *trans*-acting factor. *Genes Dev.* **7:** 1737.

Perlick H.A., Medghalchi S.M., Spencer F.A., Kendzior R.J., Jr., and Dietz H.C. 1996. Mammalian orthologues of a yeast regulator of nonsense transcript stability. *Proc. Natl. Acad. Sci.* **93:** 10928.

Rajavel K.S. and Neufeld E.F. 2001. Nonsense mediated decay of the human *HEXA* mRNA. *Mol. Cell. Biol.* **21:** 5512.

Ruiz-Echevarría M.J., Yasenchak J.M., Han X., Dinman J.D., and Peltz S.W. 1998. The Upf3 protein is a component of the surveillance complex that monitors both translation and mRNA turnover and affects viral propagation. *Proc. Natl. Acad. Sci.* **95:** 87216.

Serin G., Gersappe A., Black J.D., Aronoff R., and Maquat L.E. 2001. Identification and characterization of human orthologues to *S. cerevisiae* Upf2 protein and *S. cerevisiae* Upf3 protein (*C. elegans* SMG-4). *Mol. Cell. Biol.* **21:** 209.

Shirley R.L., Lelivelt M.J., Schenkman L.R., Dahlseid J.N., and Culbertson M.R. 1998. A factor required for nonsense-mediated mRNA decay in yeast is exported from the nucleus to the cytoplasm by a nuclear export signal sequence. *J. Cell Sci.* **111:** 3129.

Sterner D.A., Carlo T., and Berget S.M. 1996. Architectural limits on split genes. *Proc. Natl. Acad. Sci.* **93:** 15081.

Sun X., Moriarty P.M., and Maquat L.E. 2000. Nonsense-mediated decay of glutathione peroxidase 1 mRNA in the cytoplasm depends on intron position. *EMBO J.* **19:** 4734.

Sun X., Perlick H.A., Dietz H.C., and Maquat L.E. 1998. A mutated human homologue of yeast Upf1 protein has a dominant-negative effect on the decay of nonsense-containing mRNAs in mammalian cells. *Proc. Natl. Acad. Sci.* **95:** 10009.

Thermann R., Neu-Yilik G., Deters A., Frede U., Wehr K., Hagenmeier C., Hentze M.W., and Kulozik A.E. 1998. Binary specification of nonsense codons by splicing and cytoplasmic translation. *EMBO J.* **12:** 3484.

Venter J.C., Adams M.D., Myers E.W., Li P.W., Mural R.J., Sutton G.G., Smith H.O., Yandell M., Evans C.A., Holt R.A., et al. 2001. The sequence of the human genome. *Science* **291:** 1304.

Waldron C. and Lacroute F. 1975. Effect of growth rate on the amounts of ribosomal and transfer ribonucleic acids in yeast. *J. Bacteriol.* **122:** 855.

Wang W., Czaplinski K., Rao Y., and Peltz S.W. 2001. The role of Upf proteins in modulating the read-through of nonsense-containing transcripts. *EMBO J.* **20:** 880.

Weng Y., Czaplinski K., and Peltz S.W. 1996a. Genetic and biochemical characterization of the mutations in the ATPase and helicase regions of Upf1 protein. *Mol. Cell. Biol.* **16:** 5477.

———. 1996b. Identification and characterization of mutations in the UPF1 gene that affect nonsense suppression and the formation of the Upf protein complex, but not mRNA turnover. *Mol. Cell. Biol.* **16:** 5491.

———. 1998. ATP is a cofactor of the Upf1 protein that modulates its translation termination and RNA binding activities. *RNA* **4:** 205.

Yamashita A., Ohnishi T., Kashima I., Tayua Y., and Ohno S. 2001. Human SMG-1, a novel phosphatidylinositol 3-kinase-related protein kinase, associates with components of the mRNA surveillance complex and is involved in the regulation of nonsense-mediated mRNA decay. *Genes Dev.* **15:** 2215.

Zhang J., Sun X., Qian Y., and Maquat L.E. 1998a. Intron function in the nonsense-mediated decay of β-globin mRNA: Indications that pre-mRNA splicing in the nucleus can influence mRNA translation in the cytoplasm. *RNA* **4:** 801.

Zhang J., Sun X., Qian Y., LaDuca J.P., and Maquat L.E. 1998b. At least one intron is required for the nonsense-mediated decay of triosephosphate isomerase mRNA: A possible link between nuclear splicing and cytoplasmic translation. *Mol. Cell. Biol.* **18:** 5272.

Zhang S., Welch E.M., Hogan K., Brown A.H., Peltz S.W., and Jacobson A. 1997. Polysome-associated mRNAs are substrates for the nonsense-mediated mRNA decay pathway in *Saccharomyces cerevisiae*. *RNA* **3:** 234.

Nonsense-mediated mRNA Decay in *Saccharomyces cerevisiae:* A Quality Control Mechanism That Degrades Transcripts Harboring Premature Termination Codons

C.I. González,* W. Wang,† AND S.W. Peltz†‡¶

*Department of Biology, University of Puerto Rico, San Juan, Puerto Rico 00931; †Department of Molecular Genetics and Microbiology, Robert Wood Johnson Medical School, University of Medicine and Dentistry of New Jersey, Piscataway, New Jersey 08854; ‡PTC Therapeutics, South Plainfield, New Jersey 07080; ¶Cancer Institute of New Jersey, Piscataway, New Jersey 08854

Messenger RNA (mRNA) turnover is a process that plays a major role in the control of gene expression. Although transcription serves as the initial regulatory event, it is clear that mRNA degradation can also influence cellular growth and differentiation. In eukaryotic cells, mRNA decay rates vary extensively. In *S. cerevisiae,* for example, some mRNAs have half-lives of about 60 minutes, and very unstable messages exhibit half-lives as short as 1 minute (Peltz and Jacobson 1993). Several studies have demonstrated that the decay rates of certain mRNAs can be modulated in response to environmental signals (Ross 1995; Gonzalez and Martin 1996; Scheffler et al. 1998). Furthermore, the expression of many highly regulated transcripts involved in cellular proliferation and differentiation is controlled by mRNA stability. Altered control of their mRNA degradation rates can lead to aberrant gene expression and disease (Raymond et al. 1989; Taylor et al. 1996). Therefore, the importance of mRNA turnover to the regulation of gene expression has gained much interest in recent years, and the structures and mechanisms that control mRNA decay rates are currently under investigation (for review, see McCarthy 1999; Jacobson and Peltz 2000).

A large set of observations point to an intimate connection between the processes of mRNA turnover and translation. These findings include: (1) inhibition of translational elongation by the use of inhibitors, as a consequence of tRNA mutations, and by the presence of secondary structure that prevents ribosome scanning, reduces mRNA turnover rates (Peltz et al. 1992; Beelman and Parker 1994; Muhlrad et al. 1995; Zuk and Jacobson 1998; Zuk et al. 1999); (2) the destabilizing function of various instability elements depends on ribosome translocation (Aharon and Schneider 1993; He et al. 1993); and (3) premature translation termination can trigger mRNA turnover (for review, see Jacobson and Peltz 2000). In addition, recent studies suggest that the link between mRNA turnover and translation is controlled by the translation initiation efficiency of the transcript. Collectively, these studies have suggested that the process of decapping is controlled by a competition between translation initiation factors and the decapping enzyme (for review, see Schwartz and Parker 2000). Data from these experiments have demonstrated that the 5′ cap structure is the

site for both decapping and the assembly of the eIF-4F complex (cap-binding complex), which enhances translation initiation. A clear understanding of this relationship will play a dominant role in the elucidation of how cellular mRNA levels are controlled in order to regulate gene expression.

MECHANISMS OF mRNA TURNOVER

Most eukaryotic mRNAs possess a cap structure at the 5′ end and a poly(A) tail at the 3′ end that protect the encoded message from exoribonuclease degradation. Studies in yeast have shown that the majority of wild-type mRNAs are degraded through a pathway that is initiated with the poly(A) shortening to an oligo(A) tail. Recent studies have identified the Pan2/Pan3 complex and the Ccr4 and Caf1 transcription factors as part of a large complex that possess poly(A)-shortening activity (Boeck et al. 1996; Brown and Sachs 1998; Tucker et al. 2001). Following poly(A) shortening, the mRNA is cleaved at its 5′ end by the *DCP1* gene product (Stevens 1988; Beelman et al. 1996; LaGrandeur and Parker 1998). This process is known to delete one or two nucleotides and remove the 5′ cap structure of the transcript. Other factors involved in the decapping process have been recently identified, including Dcp2, Vps16, Pat1, and the Lsm protein complex (Hatfield et al. 1996; Dunckley and Parker 1999; Zhang et al. 1999; Bonnerot et al. 2000; Bouveret et al. 2000; Tharun et al. 2000). The uncapped and deadenylated mRNA then becomes a substrate for the 5′ to 3′ exoribonuclease Xrn1p (Hsu and Stevens 1993; Muhlrad et al. 1994). Alternatively, some yeast transcripts are degraded 3′ to 5′ after poly(A) shortening (Jacobs et al. 1998). Several studies have revealed a third mode of mRNA decay for some vertebrate transcripts which is triggered by a site-specific endonucleolytic cleavage (Brown et al. 1993; Cunningham et al. 2000; Wang and Kiledjian 2000). One mechanism of mRNA turnover that has also been intensively investigated is the nonsense-mediated mRNA decay (NMD) pathway (for review, see Jacobson and Peltz 2000). This process is required for the degradation of transcripts that contain premature translation termination codons. The yeast *S. cerevisiae* has proven very useful in the identification and characterization of *cis*-acting sequences and *trans*-acting factors involved in this pathway, and it is the main focus of this chapter. The reader is referred to recent reviews (Maquat 2000; Lykke-Andersen 2001) for more detailed information on the mammalian nonsense-mediated mRNA decay pathway.

NONSENSE-MEDIATED mRNA DECAY, A QUALITY CONTROL MECHANISM

A clear example of the link between mRNA turnover and translation is the fact that nonsense mutations in a gene can reduce the steady-state levels of the mRNA transcribed from that gene via the NMD pathway (Jacobson and Peltz 2000). The NMD pathway has been observed in all eukaryotic cells examined (Pulak and Anderson 1993; Hentze and Kulozik 1999; Maquat 2000), and it seems to have evolved as a quality control mechanism to ensure that translation termination occurs at the appropriate location within the mRNA. Transcripts harboring premature nonsense codons are degraded rapidly, ensuring that polypeptide fragments synthesized from aberrant mRNAs do not dominantly interfere with the wild-type form of the protein. The phenomenon of nonsense-mediated mRNA decay is shown in Figure 1. The PGK1 mRNA in the yeast *S. cerevisiae* is very stable with a decay rate of more than 45 minutes. Insertion of a premature termination codon, however, reduces its half-life to less than 3 minutes. In contrast to most wild-type mRNAs, the turnover of nonsense-containing mRNAs is deadenylation-independent, entering the predominant 5′ to 3′ decay pathway with an intact poly(A) tail (Muhlrad and Parker 1994; Hagan et al. 1995; Beelman et al. 1996). These nonsense-containing mRNAs are then decapped by Dcp1p and subsequently degraded by the Xrn1p exoribonuclease.

Additional substrates for the NMD pathway include intron-containing pre-mRNAs that enter the cytoplasm (He et al. 1993); mRNAs that are subjected to leaky scanning

(Welch and Jacobson 1999); mRNAs harboring upstream open reading frames (uORFs, Vilela et al. 1998; Ruiz Echevarria and Peltz 2000); and transcripts with extended 3′UTRs (Muhlrad and Parker 1999; Das et al. 2000). Recent results have shown that more than 225 of the approximately 6000 transcripts of *S. cerevisiae* are affected by inactivation of the NMD pathway (Lelivelt and Culbertson 1999). In addition, this cellular process appears to have a role in the regulation of telomeric length (Lew et al. 1998), the levels of kinetochore subunits (Dahlseid et al. 1998), and the maintenance of the double-stranded RNA interference phenotype in *Caenorhabditis elegans* (Domeier et al. 2000).

In *S. cerevisiae*, the NMD pathway is a translation-dependent event. The following results support this conclusion. (1) NMD can be prevented by nonsense-suppressing tRNAs (Gozalbo and Hohmann 1990). (2) NMD is inhibited by drugs and mutations that block translation initiation and elongation (Peltz et al. 1992; Welch and Jacobson 1999; Zuk et al. 1999). (3) Nonsense-containing mRNAs are associated with polysomes whose size reflects the position of the premature termination codon within the ORF (He et al. 1993). (4) NMD resumes immediately after the removal of cycloheximide from the growth medium (Zhang et al. 1997). (5) Proteins essential for NMD interact with the translation termination release factors eRF1 and eRF3 (Czaplinski et al. 1998; Wang et al. 2001).

TRANS-ACTING FACTORS INVOLVED IN NMD PATHWAY

Mutations that inactivate the NMD pathway were initially isolated in a genetic screen to identify allosuppressors of the *his4-38* frameshift mutation (Culbertson et al. 1980). Subsequent analysis of these mutants, as well as mutants isolated from other screens, demonstrated that mutations in the *UPF1*, *UPF2/NMD2*, *UPF3*, *MOF2/SUI1*, *MOF5*, *MOF8*, *PRT1*, and *HRP1/NAB4* genes result in the stabilization of nonsense-containing

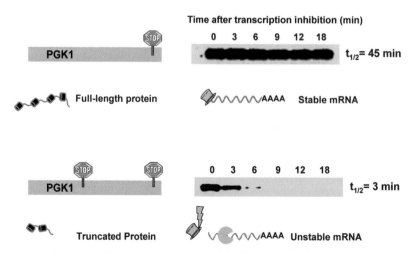

Figure 1. The effect of a premature termination codon on the decay of the PGK1 mRNA. Decay rates for the mRNAs encoded by the wild-type and the nonsense-containing PGK1 alleles were determined ny northern blot analysis of RNAs isolated at different times after transcription was inhibited by a shift from 24°C to 37°C in a yeast strain harboring a temperature-sensitive RNA polymerase II.

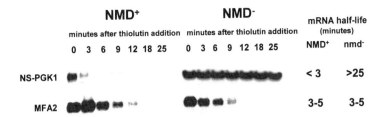

Figure 2. The effects of a mutation in a *trans*-acting factor required for the activation of the NMD pathway. The decay rates of a nonsense-containing PGK1 mRNA and the wild-type MFA2 mRNA were monitored by northern blotting. Transcription was inhibited in wild-type and mutant cell cultures, total RNA was isolated, and mRNA decay rates were determined. (For details, see Gonzalez et al. 2000.)

mRNAs while having no effect on the stability of most wild-type transcripts (for review, see Jacobson and Peltz 2000). An example of the effect of inactivation of the NMD pathway is shown in Figure 2.

The Upf1, Upf2, and Upf3 proteins have been the most extensively investigated components of the NMD pathway in yeast. The Upf1p contains a cysteine- and histidine-rich region near its amino-terminal domain and all the motifs required to be a member of the superfamily group I helicases (Jacobson and Peltz 2000). Upf1p has been purified and demonstrates RNA binding, RNA-dependent ATPase, and RNA helicase activities (Czaplinski et al. 1995, 1999; Weng et al. 1996a,b). *UPF2* encodes a large acidic protein (for review, see Jacobson and Peltz 2000) whereas *UPF3* encodes a basic protein harboring several nuclear localization signals and nuclear export sequences and shuttles from the nucleus to the cytoplasm (Lee and Culbertson 1995; Shirley et al. 1998). The Upf1, Upf2, and Upf3 proteins interact, which suggests that they form a surveillance complex. Single or multiple mutations of the *UPF* genes have the same effect on mRNA turnover, further suggesting that these proteins function in a complex.

In addition to its role in accelerating the turnover of nonsense-containing mRNAs, more recent evidence suggests that Upf1p enhances translation termination at the nonsense codon. Nonsense suppression occurs when a near-cognate tRNA successfully competes with the translation termination factors at a nonsense codon, and amino acid incorporation occurs rather than premature termination. A role for the Upf1 protein in translation termination became evident after the finding that a *upf1* Δ strain not only caused stabilization of nonsense-containing RNAs, but also demonstrated suppression of certain nonsense alleles (Weng et al. 1996a,b; Czaplinski et al. 1998). Consistent with this notion, a set of mutations in the *UPF1* gene was isolated that separated the mRNA decay phenotype from its function in modulating premature termination (Weng et al. 1996a,b). Subsequent studies have shown that Upf1p interacts with the termination release factors eRF1 and eRF3 (Czaplinski et al. 1998). Consistent with the view that the Upf proteins modulate translation termination efficiency at a nonsense codon, recent studies have demonstrated that deletion of either *UPF2* or *UPF3* can also lead to a nonsense suppression phenotype (Maderazo et al. 2000; Wang et al. 2001). In

addition, Upf2p and Upf3p have been shown to interact with release factor eRF3 (Wang et al. 2001). Interestingly, neither Upf2p nor Upf3p interacts with eRF1. In fact, Upf2p, Upf3p, and eRF1 compete with each other for binding to eRF3. As described in more detail below, these results suggest the assembly of various "transient" complexes involving the Upf proteins and the termination release factors at the nonsense codon. This surveillance complex ensures efficient translation termination before it becomes committed to scanning and degrading the aberrant RNA.

CIS-ACTING SEQUENCES INVOLVED IN NMD PATHWAY
Downstream Sequence Elements

Previous results have indicated that a nonsense codon is recognized as aberrant due to downstream sequence elements (DSEs) located 3′ of a premature termination codon (Peltz et al. 1993; Zhang et al. 1995). Transcripts containing a premature termination codon, but lacking a DSE, are not degraded by the NMD pathway; introduction of a DSE 3′ of a nonsense codon in a mRNA, however, activates the NMD pathway (Ruiz-Echevarria and Peltz 1996). Thus, the signal required for activation of NMD is a bipartite element consisting of a premature termination codon and a 3′ DSE. The sequences in the DSE important for activation of the NMD pathway have been extensively characterized (Zhang et al. 1995; Jacobson and Peltz 2000). The DSE is degenerate and can be present in multiple copies in a mRNA (Zhang et al. 1995; Ruiz-Echevarria and Peltz 1996). Deletion analyses within the PGK1 transcript identified a repeated sequence (TGYYGATGYYYY, referred to as the sequence motif) that is essential for the function of the DSE (Zhang et al. 1995). Using this consensus sequence, other DSEs have been identified in many yeast mRNAs.

The importance of the DSE has been further analyzed using the leader region of the GCN4 gene, which encodes a transcriptional activator of amino acid biosynthetic genes. The GCN4 mRNA contains four short uORFs, preceding the main ORF, and translation initiation/re-initiation events occur in this region that regulate the translation of the Gcn4 polypeptide (Hinnebusch 1997). These experiments have demonstrated that (1) DSEs are functional after a translation initiation/termination cycle has been completed; (2) DSEs are functional when located within 150 nucleotides 3′ of the stop codon; and (3) DSEs are not functional if they are scanned by the ribosomes during the normal process of translation (Ruiz-Echevarria and Peltz 1996; Ruiz-Echevarria et al. 1998).

Stabilizer Sequence Elements

Although previous results suggest that the NMD pathway is not activated in some transcripts due to their lack of DSEs (Ruiz-Echevarria and Peltz 1996), recent evidence has shown that there are specific sequences, defined as stabilizer elements (STEs), that inactivate NMD. Two different classes of STEs have been identified. One type is located within the protein coding region and must be translated to inactivate the NMD pathway (Peltz et al. 1993; Ruiz Echevarria et al. 2001). A second class has been identified in the GCN4 leader region (Ruiz-Echevarria et al. 1998). This STE blocks the NMD pathway when positioned downstream of the termination codon and upstream of the DSE. A similar STE has been recently identified in the YAP1 mRNA (Ruiz-Echevarria and Peltz 2000). The poly(U)-binding protein, Pub1p, has been shown to bind both STEs, and these elements fail to exert their inactivating function in *pub1* Δ strains. These observations suggest that Pub1p can bind to the STE and inactivate the NMD pathway.

COMPARISON OF THE NMD PATHWAYS IN YEAST AND HIGHER EUKARYOTES

The NMD pathway has also been observed in higher eukaryotes including fungi, plants, *C. elegans,* and humans (Pulak and Anderson 1993; for review, see Maquat 2000). In *C. elegans,* seven genes (*smg1–smg7*) are required for NMD. Mutations in these genes stabilize nonsense-containing mRNAs (Pulak and Anderson 1993). Several of the *smg* genes have recently been cloned and characterized: *smg-2* encodes a phosphoprotein homologous to the yeast Upf1p (Page et al. 1999), and *smg-4* is homologous to the yeast Upf3p (Serin et al. 2001).

In humans, the medical importance of the NMD pathway has been well reported in recent years (for review, see Maquat 2000). For example, cystic fibrosis and Duchenne muscular dystrophy can be caused by mutations that generate premature termination codons. Human homologs of *UPF1*, *UPF2,* and *UPF3* have also been identified (Perlick et al. 1996; Applequist et al. 1997; Lykke-Andersen et al. 2000; Mendell et al. 2000; Serin et al. 2001). HUpf1p functions in mammalian cells to control the stability of nonsense-containing transcripts and has similar enzymatic characteristics to the yeast counterpart (Sun et al. 1998; Bhattacharya et al. 2000). In addition, like smg-2, it is a phosphoprotein (Pal et al. 2001; Yamashita et al. 2001). Consistent with the view that Upf1p is involved in translation termination, recombinant HUpf1 causes anti-suppression in vitro (A. Bhattacharya and S.W. Peltz, in prep.). Interestingly, HUpf2 and HUpf3 promote rapid degradation similar to that triggered by a premature termination codon when fused to

MS2-coat protein and expressed together with a β-globin mRNA harboring MS2-binding sites in its 3′UTR (Lykke-Andersen et al. 2000).

Many studies have revealed that some mammalian nonsense-containing mRNAs are subjected to NMD in the cytoplasm, and others are degraded in association with the nucleus (Lykke-Andersen et al. 2000; Maquat 2000). Although the cellular location for mammalian NMD has been controversial, numerous experiments have shown that the process of splicing is involved in mammalian NMD pathway. For example, insertion of an intron downstream of a normal stop codon triggers the degradation of this normal transcript lacking a premature termination codon via the NMD pathway, indicating that an intron can function as a DSE (Carter et al. 1996; Thermann et al. 1998). More recently, it has been demonstrated that splice junctions serve as markers to target aberrant mRNAs with premature termination codons to the NMD pathway (Kataoka et al. 2000; Le Hir et al. 2000a,b, 2001; Kim et al. 2001; Lykke-Andersen et al. 2001). These results strongly suggest that, like yeast, the signal for NMD in mammalian cells is also a dual element consisting of a premature termination codon and a 3′ DSE.

THE RNA-BINDING PROTEIN Hrp1/Nab4 IS A SIGNAL THAT MARKS THE TRANSCRIPT FOR NMD

Based on the requirements of a DSE to promote NMD, the current thinking is that recognition of this sequence is a critical event in the NMD pathway. We have recently identified the RNA-binding protein, Hrp1/Nab4, as a factor directly involved in the NMD pathway (Gonzalez et al. 2000). Mutations in the *HRP1/NAB4* gene specifically stabilized nonsense-containing mRNAs. Electrophoretic mobility shift assays revealed that Hrp1p binds specifically to a DSE identified in the PGK1 transcript. A mutation in *HRP1* that resulted in the stabilization of nonsense-containing mRNAs abolished the affinity of the Hrp1p for the DSE. Furthermore, Hrp1p was found to interact with both Upf1p and Upf2p, two important components of the NMD pathway (Gonzalez et al. 2000; W. Wang et al., unpubl.). These results suggest that the DSE 3′ of a premature termination codon is critical to promote NMD because it can promote an interaction with proteins such as Hrp1p/Nab4p and signals to the surveillance complex that a premature termination event has occurred.

A MODEL FOR THE MECHANISM OF NMD

On the basis of the results described above, we proposed the following model for the mechanism of NMD in the yeast *S. cerevisiae*. During or immediately after transcription, the mRNA is processed and packaged into an RNP that is transported from the nucleus to the cytoplasm. Concurrent with or following transport, ribosomes become associated with the RNP and begin translating the message. The process of translation remodels the RNP by displacing the proteins that bound the RNA in the nucleus, such as Hrp1/Nab4. We believe that the successful completion of the first round of translation will trigger a conformational rearrangement in the RNP such that the 5′ and 3′ ends become linked. This step is critical in determining the subsequent translational efficiency and stability of the mRNA.

However, the presence of a nonsense codon causes premature translation termination (Fig. 3). At this point, we think that a termination/pre-surveillance complex, composed of at least the translation termination factors (eRF1 and eRF3) and *UPF1*, is assembled (Czaplinski et al. 1998; Jacobson and Peltz 2000; Wang et al. 2001). An interaction between these proteins facilitates the activity of the termination factors and promotes their dissociation from the complex. We believe that this step enhances the RNA-binding and ATPase activities of Upf1p, which can advance a posttermination surveillance complex (*UPF/NMD* factors) downstream to search 3′ of the nonsense codon. How does the surveillance complex change from a pretermination mode to a posttermination searching mode? Our recent results suggest that phosphorylation of Upf2p might be required for this rearrangement of the surveillance complex (W. Wang et al., unpubl.). According to our data, phosphorylation of Upf2p allows it to interact more efficiently with the Hrp1p/DSE marker complex, is required for the activity of the NMD pathway, and might trigger disassembly of the release factors. Previous results have demonstrated that the *UPF/NMD* factors are present at very different levels in yeast cells (Maderazo et al. 2000). Therefore, the Upf proteins are thought to function at different stages of the NMD pathway and might not be all associated with the termina-

tion/surveillance complex at the same time. Upon assembly, the posttermination mature surveillance complex (*UPF/NMD* factors) can scan downstream of the premature termination event in search of a DSE/Hrp1p marker complex. An interaction between the posttermination surveillance complex and the DSE/Hrp1p marker complex promotes rapid decapping of the mRNA and subsequent degradation of the body of the transcript by a 5′ → 3′ exoribonuclease.

In mammalian cells, a spliceable intron functions like a yeast DSE. This "mark" on the RNA at the exon–exon junction is used by the degradation machinery to recognize premature termination codons. In fact, recent results have demonstrated that the splicing process alters mRNP protein composition (Kataoka et al. 2000; Le Hir et al. 2000a,b, 2001; Kim et al. 2001; LeHir et al. 2001; Lykke-Andersen et al. 2001). Several proteins have been identified that associate with mRNA exon–exon junctions as a consequence of splicing. These include the nuclear matrix-associated splicing factor, SRm160; two general splicing factors, DEK and RNPS1; and two proteins that interact with the mRNA export factor TAP, Y14 and REF. From this postsplicing protein complex, only the interaction of Y14 persists on the mRNA in the cytoplasm at the same position. Therefore, Y14 may serve as a mark for premature termination codons that communicate the information to the cytoplasm and triggers the NMD pathway. However, a direct role for Y14 in the NMD pathway has not yet been identified. Interestingly, hUpf3 shuttles from the nucleus to the cytoplasm, interacts with Y14, and binds to nascent transcripts near the exon–exon junctions following splicing (Lykke-Andersen et al. 2000;

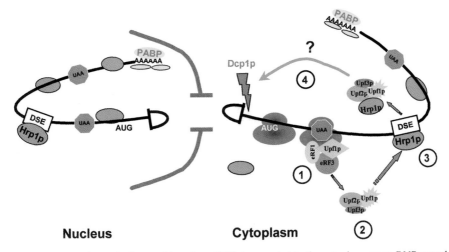

Figure 3. Model for the NMD pathway in *S. cerevisiae*. An mRNA is exported to the cytoplasm as an RNP complex with nuclear RNA-binding proteins attached, such as Hrp1. During the pioneer round of translation, the attached nuclear proteins are displaced by the ribosome, and cytoplasmic RNP remodeling occurs. A premature nonsense codon prevents the remodeling process and activates NMD. The translating ribosome pauses at a premature termination codon and signals to the eRF1–eRF3 complex to bind to the A site. (1) The Upf1p becomes associated with the eRF1–eRF3 termination/pre-surveillance complex during the termination process. After hydrolysis of the peptidyl tRNA bond, eRF1 dissociates from the ribosome (see Wang et al. 2001). Dissociation of eRF1 allows Upf2p and/or Upf3p to bind the eRF3–Upf1p complex. (2) Rearrangement of the complex: Upf2p and/or Upf3p joins the complex and displaces eRF3 to form the mature posttermination surveillance complex. (3) Because the ribosome failed to displace Hrp1 (or other yet-unidentified DSE-marker factors) from the DSE, the mature surveillance complex can recognize the DSE–marker complex. (4) This interaction indicates that the RNP remodeling process was incomplete and promotes rapid decapping of the aberrant transcript and degradation of the body of the mRNA by a 5′ → 3′ exoribonuclease.

Kim et al. 2001). In addition, recent results have demonstrated that RNPS1 also interacts with the human Upf complex and triggers NMD when tethered to the 3′UTR of the β-globin mRNA (Lykke-Andersen et al. 2001). Thus, like Hrp1p, hUpf3 and RNPS1 may function in mammalian cells as marker proteins by associating with nascent transcripts in the nucleus, interacting with other components of the NMD pathway and triggering rapid degradation of aberrant transcripts in the cytoplasm.

Our model for NMD can be applied to all substrates in yeast, except those transcripts with extended 3′UTRs (for review, see Hilleren and Parker 1999; Jacobson and Peltz 2000). Alternatively, the cellular machinery can recognize a DSE as a defective 3′UTR created by the presence of a premature termination codon (Hilleren and Parker 1999; Jacobson and Peltz 2000). Whereas the first model suggests that the translating ribosome fails to displace DSE-binding proteins (i.e., Hrp1p, Upf3p) due to premature termination, the second model suggests that the DSE triggers NMD because it lacks a factor(s) normally present in a 3′UTR of a wild-type mRNA. More recent results, however, suggest that transcripts with extended 3′UTRs may harbor DSE-like sequences in this region that can promote rapid turnover of the aberrant transcript via the NMD pathway (Das et al. 2000).

CONCLUSIONS

Studies of the NMD pathway have revealed an intimate link between multiple cellular processes. Factors involved in the NMD pathway also have important roles in processes such as translation termination, frameshifting, pre-mRNA processing, and nucleocytoplasmic transport. The recent identification of shuttling proteins that move with the mRNA from the nucleus to the cytoplasm will help us understand how nuclear processes influence the fate of the mRNA in the cytoplasm. Future experiments will address the precise mechanism by which these proteins function to trigger decapping of nonsense-containing mRNAs and how they modulate the NMD pathway. Further understanding of the NMD pathway in yeast may hold valuable answers to the study of cellular mechanisms involved in the onset of many human genetic disorders.

ACKNOWLEDGMENTS

This work was supported by a grant from the National Institutes of Health (K01 HL-04355-02) given to C.I.G. S.W.P. is supported by grants from the National Institutes of Health (GM-48631) and an American Heart Association Established Investigator Award.

REFERENCES

Aharon T. and Schneider R.J. 1993. Selective destabilization of short-lived mRNAs with the granulocyte-macrophage colony-stimulating factor AU-rich 3′ noncoding region is mediated by a cotranslational mechanism. *Mol. Cell. Biol.* **13**: 1971.

Applequist S.E., Selg M., Raman C., and Jack H.M. 1997. Cloning and characterization of HUPF1, a human homolog of the *Saccharomyces cerevisiae* nonsense mRNA-reducing UPF1 protein. *Nucleic Acids Res.* **25**: 814.

Beelman C.A. and Parker R. 1994. Differential effects of translational inhibition in *cis* and in *trans* on the decay of the unstable yeast MFA2 mRNA. *J. Biol. Chem.* **269**: 9687.

Beelman C.A., Stevens A., Caponigro G., LaGrandeur T.E., Hatfield L., Fortner D.M., and Parker R. 1996. An essential component of the decapping enzyme required for normal rates of mRNA turnover. *Nature* **382**: 642.

Bhattacharya A., Czaplinski K., Trifillis P., He F., Jacobson A., and Peltz S.W. 2000. Characterization of the biochemical properties of the human Upf1 gene product that is involved in nonsense-mediated mRNA decay. *RNA* **6**: 1226.

Boeck R., Tarun S., Rieger M., Deardorff J.A., Muller-Auer S., and Sachs A.B. 1996. The yeast Pan2 protein is required for poly(A)-binding protein-stimulated poly(A) nuclease activity. *J. Biol. Chem.* **271**: 432.

Bonnerot C., Boeck R., and Lapeyre B. 2000. The two proteins Pat1p (Mrt1p) and Spb8p interact in vivo, are required for mRNA decay and are functionally linked to Pab1p. *Mol. Cell. Biol.* **20**: 5939.

Bouveret E., Rigaut G., Shevchenko A., Wilm M., and Seraphin B. 2000. A Sm-like protein complex that participates in mRNA degradation. *EMBO J.* **19**: 1661.

Brown B.D., Zipkin I.D., and Harland R.M. 1993. Sequence-specific endonucleolytic cleavage and protection of mRNA in *Xenopus* and *Drosophila*. *Genes Dev.* **7**: 1620.

Brown C.E. and Sachs A.B. 1998. Poly(A) tail length control in *Saccharomyces cerevisiae* occurs by message-specific deadenylation. *Mol. Cell. Biol.* **18**: 6548.

Carter M.S., Li S., and Wilkinson M.F. 1996. A splicing-dependent regulatory mechanism that detects translation signals. *EMBO J.* **15**: 5965.

Culbertson M.R., Underbrink K.M., and Fink G.R. 1980. Frameshift suppression *Saccharomyces cerevisiae*. II. Genetic properties of group II suppressors. *Genetics* **95**: 833.

Cunningham K.S., Dodson R.E., Nagel M.A., Shapiro D.J., and Schoenberg D.R. 2000. Vigilin binding selectively inhibits cleavage of the vitellogenin mRNA 3′-untranslated region by the mRNA endonuclease polysomal ribonuclease 1. *Proc. Natl. Acad. Sci.* **97**: 12498.

Cazplinski K., Ruiz-Echevarria M.J., Gonzalez C.I., and Peltz S.W. 1999. Should we kill the messenger? The role of the surveillance complex in translation termination and mRNA turnover. *Bioessays* **21**: 685.

Czaplinski K., Weng Y., Hagan K.W., and Peltz S.W. 1995. Purification and characterization of the Upf1 protein: A factor involved in translation and mRNA degradation. *RNA* **1**: 610.

Czaplinski K., Ruiz-Echevarria M.J., Paushkin S.V., Han X., Perlick H.A., Dietz H.C., Ter-Avanesyan M.D. and Peltz S.W. 1998. The surveillance complex interacts with the translation release factors to enhance termination and degrade aberrant mRNAs. *Genes Dev.* **12**: 1665.

Dahlseid J.N., Puziss J., Shirley R.L., Atkin A.L., Hieter P., and Culbertson M.R. 1998. Accumulation of mRNA coding for the ctf13p kinetochore subunit of *Saccharomyces cerevisiae* depends on the same factors that promote rapid decay of nonsense mRNAs. *Genetics* **150**: 1019.

Das B., Guo Z., Russo P., Chartrand P., and Sherman F. 2000. The role of nuclear cap binding protein Cbc1p of yeast in mRNA termination and degradation. *Mol. Cell. Biol.* **20**: 2827.

Domeier M.E., Morse D.P., Knight S.W., Portereiko M., Bass B.L., and Mango S.E. 2000. A link between RNA interference and nonsense-mediated decay in *Caenorhabditis elegans*. *Science* **289**: 1928.

Dunckley T. and Parker R. 1999. The Dcp2 protein is required for mRNA decapping in *Saccharomyces cerevisiae* and contains a functional MutT motif. *EMBO J.* **18**: 5411.

Gonzalez C.I. and Martin C.E. 1996. Fatty acid-responsive control of mRNA stability: Unsaturated fatty acid-induced degradation of the *Saccharomyces* OLE1 transcript. *J. Biol. Chem.* **271**: 25801.

Gonzalez C.I., Ruiz-Echevarria M.J., Vasudevan S., Henry

M.F., and Peltz S.W. 2000. The yeast hnRNP-like protein Hrp1/Nab4 marks a transcript for nonsense-mediated mRNA decay. *Mol. Cell* 5: 489.

Gozalbo D. and Hohmann S. 1990. Nonsense suppressors partially revert the decrease of the mRNA level of a nonsense mutant allele in yeast. *Curr. Genet.* 17: 77.

Hagan K.W., Ruiz-Echevarría M.J., Quan Y., and Peltz S.W. 1995. Characterization of *cis*-acting sequences and decay intermediates involved in nonsense-mediated mRNA turnover. *Mol. Cell. Biol.* 15: 809.

Hatfield L., Beelman C.A., Stevens A., and Parker R. 1996. Mutations in *trans*-acting factors affecting mRNA decapping in *Saccharomyces cerevisiae. Mol. Cell. Biol.* 16: 5830.

He F., Peltz S.W., Donahue J.L., Rosbash M., and Jacobson A. 1993. Stabilization and ribosome association of unspliced pre-mRNAs in a yeast *upf1-* mutant. *Proc. Natl. Acad. Sci.* 90: 7034.

Hentze M.W. and Kulozik A.E. 1999. A perfect message: RNA surveillance and nonsense-mediated decay. *Cell* 96: 307.

Hilleren P. and Parker R. 1999. Mechanisms of mRNA surveillance in eukaryotes. *Annu. Rev. Genet.* 33: 229.

Hinnebusch A.G. 1997. Translational regulation of yeast GCN4: A window on factors that control initiator-trna binding to the ribosome. *J. Biol. Chem.* 272: 21661.

Hsu C.L. and Stevens A. 1993. Yeast cells lacking 5′→3′ exoribonuclease 1 contain mRNA species that are poly(A) deficient and partially lack the 5′ cap structure. *Mol. Cell Biol.* 13: 4826.

Jacobs J.S., Anderson A.R., and Parker R.P. 1998. The 3′ to 5′ degradation of yeast mRNAs is a general mechanism for mRNA turnover that requires the SKI2 DEVH box protein and 3′ to 5′ exonucleases of the exosome complex. *EMBO J.* 17: 1497.

Jacobson A. and Peltz S.W. 2000. Destabilization of nonsense-containing transcripts in *Saccharomyces cerevisiae*. In: *Translational control of gene expression* (ed. Sonenberg et al.), p. 827. Cold Spring Harbor Laboratory Press, New York.

Kataoka N., Yong J., Kim V.N., Velazquez F., Perkinson R.A., Wang F., and Dreyfuss G. 2000. Pre-mRNA splicing imprints mRNA in the nucleus with a novel RNA-binding protein that persists in the cytoplasm. *Mol. Cell* 6: 673.

Kim V.N., Kataoka N., and Dreyfuss G. 2001. Role of the nonsense-mediated decay factor hUpf3 in the splicing-dependent exon-exon junction complex. *Science* 293: 1832.

LaGrandeur T.E. and Parker R. 1998. Isolation and characterization of Dcp1p, the yeast mRNA decapping enzyme. *EMBO J.* 17: 1487.

Lee B.S. and Culbertson M.R. 1995. Identification of an additional gene required for eukaryotic nonsense mRNA turnover. *Proc. Natl. Acad. Sci.* 92: 10354.

Le Hir H., Moore M.J., and Maquat L.E. 2000a. Pre-mRNA splicing alters mRNP composition: evidence for stable association of proteins at exon-exon junctions. *Genes Dev.* 14: 1098.

Le Hir H., Gatfield D., Izaurralde E., and Moore M.J. 2001. The exon-exon junction complex provides a binding platform for factors involved in mRNA export and nonsense-mediated mRNA decay. *EMBO J.* 20: 4987.

Le Hir H., Izaurralde E., Maquat L.E., and Moore M.J. 2000b. The spliceosome deposits multiple proteins 20-24 nucleotides upstream of mRNA exon-exon junctions. *EMBO J.* 19: 6860.

Lelivelt M.J. and Culbertson M.R. 1999. Yeast Upf proteins required for RNA surveillance affect global expression of the yeast transcriptome. *Mol. Cell Biol.* 19: 6710.

Lew J.E., Enomoto S., and Berman J. 1998. Telomere length regulation and telomeric chromatin require the nonsense-mediated mRNA decay pathway. *Mol. Cell. Biol.* 18: 6121.

Lykke-Andersen J. 2001. mRNA quality control: Marking the message for life or death. *Curr. Biol.* 11: R88.

Lykke-Andersen J., Shu M.D., and Steitz J.A. 2000. Human Upf proteins target an mRNA for nonsense-mediated decay when bound downstream of a termination codon. *Cell* 103: 1121.

———. 2001. Communication of the position of exon-exon junctions to the mRNA surveillance machinery by the protein RNPS1. *Science* 293: 1836.

Maderazo A.B., He F., Mangus D.A., and Jacobson A. 2000. Upf1p control of nonsense mRNA translation is regulated by Nmd2p and Upf3p. *Mol. Cell. Biol.* 20: 4591.

Maquat L.E. 2000. Nonsense-mediated RNA decay in mammalian cells: A splicing-dependent means to down-regulate the levels of mRNAs that prematurely terminate translation. In: *Translational control of gene expression* (ed. Sonenberg et al.), p. 849. Cold Spring Harbor Laboratory Press, Cold Spring Harbor, New York.

McCarthy J.E.G. 1999. Posttranscriptional control of gene expression in yeast. *Microbiol. Mol. Biol. Rev.* 62: 1492.

Mendell J.T., Medghalchi S.M., Lake R.G., Noensie E.N., and Dietz H.C. 2000. Novel upf2p orthologues suggest a functional link between translation initiation and nonsense surveillance complexes. *Mol. Cell Biol.* 20: 8944.

Muhlrad D. and Parker R. 1994. Premature transaltional termination triggers mRNA decapping. *Nature* 370: 578.

Muhlrad D., Decker C.J., and Parker R. 1994. Deadenylation of the unstable mRNA encoded by the yeast MFA2 gene leads to decapping followed by 5′→3′ digestion of the transcript. *Genes Dev.* 8: 855.

Muhlrad D. and Parker R. 1999. Aberrant mRNAs with extended 3′ UTRs are substrates for rapid degradation by mRNA surveillance. *RNA* 5: 1299.

———. 1995. Turnover mechanisms of the stable yeast PGK1 mRNA. *Mol. Cell. Biol.* 15: 2145.

Page M.F., Carr B., Anders K.R., Grimson A., and Anderson P. 1999. SMG-2 is a phosphorylated protein required for mRNA surveillance in *Caenorhabditis elegans* and related to Upf1p of yeast. *Mol. Cell Biol.* 19: 5943.

Pal M., Ishigaki Y., Nagy E., and Maquat L.E. 2001. Evidence that phosphorylation of human Upf1 protein varies with intracellular location and is mediated by a wortmannin-sensitive and rapamycin-sensitive PI 3-kinase-related kinase signalling pathway. *RNA* 7: 5.

Peltz S.W. and Jacobson A. 1993. mRNA turnover in the yeast *Saccharomyces cerevisiae*. In *Regulation of mRNA turnover*. (ed. G. Brawerman and J. Belasco), p. 291, Academic Press, New York. 291.

Peltz S.W., Donahue J.L., and Jacobson A. 1992. A mutation in the tRNA nucleotidyltransferase gene promotes stabilization of mRNAs in *Saccharomyces cerevisiae. Mol. Cell. Biol.* 12: 5778.

Peltz S.W., Brown A.H., and Jacobson A. 1993. mRNA destabilization triggered by premature translational termination depends on three mRNA sequence elements and at least one *trans*-acting factor. *Genes Dev.* 7: 1737.

Perlick H.A., Medghalchi S.M., Spencer F.A., Kendzior R.J., Jr., and Dietz H.C. 1996. Mammalian orthologues of a yeast regulator of nonsense transcript stability. *Proc. Natl. Acad. Sci.* 93: 10928.

Pulak R. and Anderson P. 1993. mRNA surveillance by the *Caenorhabditis elegans smg* genes. *Genes Dev.* 7: 1885.

Raymond V., Atwater J.A., and Verma I.M. 1989. Removal of an mRNA destabilizing element correlates with the increased oncogenicity of proto-oncogene *fos. Oncogene Res.* 5: 1.

Ross J. 1995. mRNA stability in mammalian cells. *Microbiol. Rev.* 59: 423.

Ruiz-Echevarria M.J. and Peltz S.W. 1996. Utilizing the GCN4 leader region to investigate the role of the sequence determinants in nonsense-mediated mRNA decay. *EMBO J.* 15: 2810.

———. 2000. The RNA binding protein Pub1 modulates the stability of transcripts containing upstream open reading frames. *Cell* 101: 741.

Ruiz-Echevarria M.J., Gonzalez C.I., and Peltz S.W. 1998. Identifying the right stop: Determining how the surveillance complex recognizes and degrades an aberrant mRNA. *EMBO J.* 17: 575.

Ruiz-Echevarria M.J., Munshi R., Tomback J., Kinzy T.G., and Peltz S.W. 2001. Characterization of a general stabilizer element that blocks deadenylation-dependent mRNA decay. *J Biol Chem.* 276: 30995.

Scheffler I.E., de la Cruz B.J., and Prieto S. 1998. Control of mRNA turnover as a mechanism of glucose repression in *Saccharomyces cerevisiae*. *Int. J. Biochem. Cell Biol.* **30:** 1175.

Schwartz D.C. and Parker R. 2000. Interaction of mRNA translation and mRNA degradation in *Saccharomyces cerevisiae*. In: *Translational control of gene expression* (ed. N. Sonenberg et al.), p. 807. Cold Spring Harbor Laboratory Press, Cold Spring Harbor, New York.

Serin G., Gersappe A., Black J.D., Aronoff R., and Maquat L.E. 2001. Identification and characterization of human orthologues to *Saccharomyces cerevisiae* Upf2 protein and Upf3 protein. *Mol. Cell. Biol.* **21:** 209.

Shirley R.L., Lelivelt M.J., Schenkman L.R., Dahlseid J.N., and Culbertson M.R. 1998. A factor required for nonsense-mediated mRNA decay in yeast is exported from the nucleus to the cytoplasm by a nuclear export signal sequence. *J. Cell Sci.* **111:** 3129.

Stevens A. 1988. mRNA-decapping enzyme from *Saccharomyces cerevisiae*: Purification and unique specificity for long RNA chains. *Mol. Cell. Biol.* **8:** 2005.

Sun X., Perlick H.A., Dietz H.C., and Maquat L.E. 1998. A mutated human homologue to yeast Upf1 protein has a dominant-negative effect on the decay of nonsense-containing mRNAs in mammalian cells. *Proc. Natl. Acad. Sci.* **95:** 10009.

Taylor G.A., Carballo E., Lee D.M., Lai W.S., Thompson M.J., Patel D.D., Schenkman D.I., Gilkeson G.S., Broxmeyer H.E., Haynes B.F., and Blackshear P.J. 1996. A pathogenetic role for TNF alpha in the syndrome of cachexia, arthritis, and autoimmunity resulting from tristetraprolin (TTP) deficiency. *Immunity* **4:** 445.

Tharun S., He W., Mayes A.E., Lennertz P., Beggs J.D., and Parker R. 2000. Yeast Sm-like proteins function in mRNA decapping and decay. *Nature* **404:** 515.

Thermann R., Neu-Yilik G., Deters A., Frede U., Wehr K., Hagemier C., Hentze M.W., and Kulozik A.E. 1998. Binary specification of nonsense codons by splicing and cytoplasmic translation. *EMBO J.* **17:** 3484.

Tucker M., Valencia-Sanchez M.A., Staples R.R., Chen J., Denis C.L., and Parker R. 2001. The transcription factor associated Ccr4 and Caf1 proteins are components of the major cytoplasmic mRNA deadenylase in *Saccharomyces cerevisiae*. *Cell* **104:** 377.

Vilela C., Linz B., Rodrigues-Pousada C., and McCarthy J.E. 1998. The yeast transcription factor genes *YAP1* and *YAP2* are subject to differential control at the levels of both translation and mRNA stability. *Nucleic Acids Res* **26:** 1150.

Wang W., Czaplinski K., Rao Y., and Peltz S.W. 2001. The role of Upf proteins in modulating the translation read-through of nonsense-containing transcripts. *EMBO J.* **20:** 880.

Wang Z. and Kiledjian M. 2000. Identification of an erythroid-enriched endoribonuclease activity involved in specific mRNA cleavage. *EMBO J.* **19:** 295.

Welch E.M. and Jacobson A. 1999. An internal open reading frame triggers nonsense-mediated decay of the yeast SPT10 mRNA. *EMBO J.* **18:** 6134.

Weng Y., Czaplinski K., and Peltz S.W. 1996a. Genetic and biochemical characterization of mutations in the ATPase and helicase regions of the Upf1 protein. *Mol. Cell. Biol.* **16:** 5477.

———. 1996b. Identification and characterization of mutations in the *UPF1* gene that affect nonsense suppression and the formation of the Upf protein complex but not mRNA turnover. *Mol. Cell. Biol.* **16:** 5491.

Yamashita A., Ohnishi T., Kashima I., Taya Y., and Ohno S. 2001. Human SMG-1, a novel phosphatidylinositol 3-kinase-related protein kinase, associates with components of the mRNA surveillance complex and is involved in the regulation of nonsense-mediated mRNA decay. *Genes Dev.* **15:** 2215.

Zhang S., Ruiz-Echevarria M.J., Quan Y., and Peltz S.W. 1995. Identification and characterization of a sequence motif involved in nonsense-mediated mRNA decay. *Mol. Cell. Biol.* **15:** 2231.

Zhang S., Williams C.J., Hagan K., and Peltz S.W. 1999. Mutations in *VPS16* and *MRT1* stabilize mRNAs by activating an inhibitor of the decapping enzyme. *Mol. Cell. Biol.* **19:** 7568.

Zhang S., Welch E.M., Hagan K., Brown A.H., Peltz S.W., and Jacobson A. 1997. Polysome-associated mRNAs are substrates for the nonsense-mediated mRNA decay pathway in *Saccharomyces cerevisiae*. *RNA* **3:** 234.

Zuk D. and Jacobson A. 1998. A single amino acid substitution in yeast eIF-5A results in mRNA stabilization. *EMBO J.* **17:** 2914.

Zuk D., Belk J.P., and Jacobson A. 1999. Temperature-sensitive mutations in the *Saccharomyces cerevisiae* MRT4, GRC5, SLA2, and THS1 genes result in defects in mRNA turnover. *Genetics* **153:** 35.

Translational Control of 15-Lipoxygenase and msl-2 mRNAs: Single Regulators or Corepressor Assemblies?

F. Gebauer, D.H. Ostareck, A. Ostareck-Lederer, M. Grskovic, and M.W. Hentze
Gene Expression Programme, European Molecular Biology Laboratory, D-69117 Heidelberg, Germany

Meiosis, early development with its rapid first embryonic divisions, and the differentiation of enucleated reticulocytes into red blood cells represent situations where nuclear gene expression is sidelined and regulation occurs translationally. For example, early embryonic development is driven by the regulated, differential translation of maternal mRNAs that were deposited into the egg before fertilization (for review, see Richter 1999; de Moor and Richter 2001). Later during development, numerous biological decisions involved in cell-fate specification, patterning, or dosage compensation are based on translational control circuits (for review, see Wickens et al. 2000). During adult life, aspects of cell differentiation, cell metabolism, or memory and learning are regulated at the level of translation (Rouault 2000; Wells et al. 2000). The regulation of mRNA translation has thus been widely adopted to control gene expression spatially and temporally. With few exceptions, translational control is exerted via regulatory sequences that are located within the untranslated regions of the mRNAs. Most commonly, such regulatory sequences are found within the 3´-untranslated region. This is the case for the 15-lipoxygenase (LOX) and *male-specific lethal-2* (*msl-2*) mRNAs, two examples that we discuss in more detail.

The initiation phase of translation usually represents the rate-limiting step. Regulators most commonly target translation initiation. Although translational regulation could be positive (activation) or negative (inhibition), most of the currently known examples are of the inhibitory type, as if active translation were a default state suitable to be impeded for regulatory purposes. This contrasts with transcriptional regulation, which frequently involves activation from a "default" silent state. Because translation starts from the 5´ end of the mRNA, one of the most direct ways to modulate translation would be to place a regulatory element into the 5´UTR. This mode of translational regulation is exemplified by ferritin mRNA, which harbors an iron-responsive element (IRE) within the first 40 nucleotides downstream of the transcription start site. In iron-deficient cells, the IRE is bound by the iron regulatory protein (IRP)-1 or IRP-2 (Hentze and Kühn 1996), and IRP binding sterically blocks one of the earliest steps of translation initiation: the recruitment of the 43S preinitiation complex, which includes the 40S ribosomal subunit, to the mRNA (Gray and Hentze 1994; Muckenthaler et al. 1998). Despite the attractive simplic-

ity of 5´UTR-mediated translational regulation, most of the translational regulatory elements are found in the 3´UTR. In fact, the 3´UTR appears to serve as a repository for control elements not only in mRNA translation, but also in the regulation of mRNA polyadenylation, stability, and subcellular localization (Richter 2000; Jansen 2001; Macdonald 2001; Wilusz et al. 2001). By comparison to the 5´UTR and to the open reading frame in particular, the composition and nucleotide sequence of the 3´ UTR are evolutionarily less constrained, which may explain this bias. Importantly, translational regulation via 3´UTR regulatory elements highlights the relevance of 5´-3´ communication for mRNA translation.

The recruitment of the 43S translation preinitiation complex is mediated by the ^{7}mGpppN cap structure and the cytoplasmic cap-binding complex eIF4F, which consists of the cap-binding subunit eIF4E, the RNA helicase eIF4A, and the adapter protein eIF4G (Gingras et al. 1999). The poly(A)-binding protein (PABP) can also interact directly with eIF4G (Tarun and Sachs 1996; Le et al. 1997; Imataka et al. 1998). This is important for the synergistic stimulation of translation by the cap structure and the poly(A) tail (Gallie 1991; Iizuka et al. 1994; Tarun and Sachs 1995; Tarun et al. 1997; Preiss and Hentze 1998). The eIF4E–eIF4G–PABP interactions can mediate mRNA circularization (Wells et al. 1998), which represents an important aspect of 5´-3´ communication in translation initiation. Consequently, 3´UTR-mediated translational control can be envisaged to occur via at least two alternative routes: either by affecting the function of the poly(A) tail and the ability of the mRNA to undergo circularization, or independent of it. Regulated cytoplasmic polyadenylation/deadenylation represents a prominent mechanism of translational control particularly during development, and it is controlled by elements located in the 3´UTR (for reviews, see Wickens et al. 2000; Mendez and Richter 2001).

This review focuses on two examples of 3´UTR-mediated translational control that do not appear to target the poly(A) tail. Unlike the case of ferritin mRNA regulation via the 5´UTR, high-affinity binding of proteins to the 3´UTR per se does not affect translation (Casey et al. 1989). This raises the question of how 3´UTR regulatory proteins may act. Considering further examples from the literature (see Table 1), we propose that the binding of 3´UTR regulatory proteins to their sites may provide "nu-

Table 1. Translational Inhibition Via the 3′UTR: Selected Examples

Organism	Biological process	mRNA	cis-element	RNA-binding protein	Other identified factors/genes	Involves poly(A)?
Drosophila	A-P axis formation	oskar	BRE	Bruno	p50, Bic-C Apontic?	no[a]
		nanos	TCE	Smaug		ND
		hb[mat]	NRE	Pum	Nos, Brat	yes[b]
	dosage compensation	msl-2	U$_{7-16}$	SXL		no[c]
C. elegans	sperm production	tra-2	TGE	GLD-1	laf-1	yes[d]
	sperm–oocyte switch	fem-3	PME	FBF-1	Nos-3	ND
Xenopus	oocyte maturation	c-mos	CPE	CPEB	Maskin	(*)
		cyclin B1				
Mammals	erythroid-cell maturation	LOX	DICE	hnRNPs K/E1		no[e]

[a] Lie and Macdonald 1999b; [b]Wreden et al. 1997; see also Note Added in Proof; [c]Gebauer et al. 1999; [d]Thompson et al. 2000; [e]Ostareck-Lederer et al. 1994. *Maskin seems to inhibit translation by interacting with eIF4E (see text). Whether Maskin activity promotes deadenylation or contributes to the maintenance of a short poly(A) tail has not been addressed.

cleation points" to recruit additional corepressors and to establish 3′UTR corepressor assemblies. Such corepressor assemblies may contact the translation initiation apparatus at the 5′ end of the mRNA and control its function. This concept bears parallels to how transcription can be controlled by regulatory promoter and/or enhancer elements.

REGULATION OF ERYTHROID 15-LIPOXYGENASE mRNA VIA THE DIFFERENTIATION CONTROL ELEMENT

The degradation of mitochondria represents a characteristic late step in red blood cell maturation. This step is mediated by the phospholipase 15-lipoxygenase (LOX), which is only expressed in late-stage reticulocytes (Rapoport and Schewe 1986; Höhne et al. 1988). Until this stage, LOX mRNA is maintained in a translationally silent state via a 3′UTR element that consists of 10 consecutive repeats of a CU-rich 19-nucleotide sequence (Ostareck-Lederer et al. 1994). Only two of these repeats are sufficient to confer translational silencing following insertion into the 3′UTR of a heterologous transcript (Ostareck et al. 1997). With reference to its function in the control of LOX mRNA expression during red blood cell differentiation, this element is called the differentiation control element (DICE).

Two KH-domain RNA-binding proteins, hnRNP K and hnRNP E1, bind specifically to the DICE and mediate translational silencing both in transfected HeLa cells and in vitro (Ostareck et al. 1997). The hnRNP E2 protein can substitute for hnRNP E1 in these assays (Ostareck-Lederer et al. 1998). In rabbit reticulocyte lysates, each protein displays repressor function on its own, but optimal inhibition occurs when they act together (Ostareck et al. 1997). Importantly, silenced LOX mRNA is specifically associated with hnRNP K in erythroid precursor cells, supporting the results of the transfection and in vitro experiments (Ostareck et al. 1997). Recent biochemical experiments in rabbit reticulocyte lysates have shed light on the mechanism by which LOX mRNA translation is regulated. The hnRNPs K and E1 bound to the DICE were shown to inhibit the assembly of 80S ribosomes. This inhibition was independent of whether the mRNA was polyadenylated and whether translation was initiated in a cap-independent fashion from the en-

cephalomyocarditis virus (EMCV) or the classical swine fever virus (CSVF) IRES (Ostareck et al. 1997, 2001). Therefore, the regulatory mechanism does not appear to act by affecting the function of either the poly(A) tail or the cap structure.

Linear sucrose gradient density centrifugation and toeprinting assays revealed that LOX mRNA translation is silenced by a novel mechanism that allows 43S preinitiation complex recruitment and scanning to occur, but blocks the joining of the 60S ribosomal subunit at the translation initiation codon (Fig. 1a) (Ostareck et al. 2001). Whether this block is caused by inhibition of one (or several) of the translation initiation factors involved in the 60S joining step or by a direct effect on one of the two ribosomal subunits has not yet been answered conclusively. However, experiments with the cricket paralysis virus (CrPV) IRES argue in favor of the former scenario. CrPV IRES-mediated translation does not require any of the known translation initiation factors and results from the direct joining of the ribosomal subunits at the translation initiation codon (Wilson et al. 2000a,b). Unlike initiation from the EMCV or CSFV IRESs, CrPV-mediated translation bypasses the block that is imposed via the DICE-hnRNP K/E1 complex (Fig. 2a) (Ostareck et al. 2001). This result shows that productive 80S assembly can occur on a DICE-containing mRNA when no translation initiation factors are needed for this assembly.

Interestingly, the DICE can mediate translational repression by hnRNP E1, but an RNA element from human α-globin mRNA that binds hnRNP E1 fails to confer a regulatory effect on translation when inserted into the 3′UTR of a luciferase reporter mRNA in place of the DICE (Fig. 2b) (D.H.Ostareck et al., unpubl). This result suggests that the DICE does not solely act as a binding site for this protein (and hnRNP K). It rather raises the possibility that hnRNP K and E1 may recruit additional factors to the DICE which may play an important role for the silencing mechanism.

COREPRESSOR ASSEMBLIES IN DEVELOPMENTAL REGULATION

Whereas DICE-mediated translational control currently provides a unique example of regulated 60S subunit joining, the developmental regulation of several mRNAs bear-a)

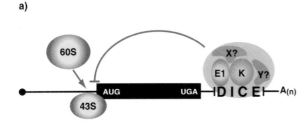

b)

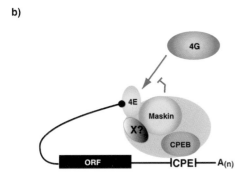

a)

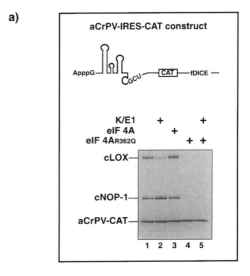

b)

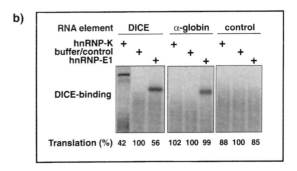

Figure 1. (*a*) Mechanism of inhibition of LOX mRNA translation by the hnRNP K/E1–DICE complex. The KH-domain proteins hnRNP K and E1 bind to the differentiation control element (DICE) in the 3′UTR of LOX mRNA. The hnRNP K/E1–DICE complex inhibits the joining of the 60S ribosomal subunit to the 43S complex at the initiator AUG. The regulatory complex, which may include additional unidentified factors, appears to target translation initiation factors involved in 60S subunit joining. (*b*) Translational repression mediated by the cytoplasmic polyadenylation element (CPE). The U-rich CPE is recognized by the CPE-binding protein (CPEB), which interacts with a protein termed Maskin. This protein contacts the cap-binding protein eIF-4E in resting *Xenopus* oocytes and is thought to prevent translation initiation by competing with eIF-4G for eIF-4E binding. Further factors regulating the interaction of Maskin with the CPEB and/or eIF-4E may be part of the inhibitory complex.

ing 3′UTR "cytoplasmic polyadenylation elements" (CPE) appears to act on an earlier and more commonly targeted step in the translation initiation pathway.

During *Xenopus* oocyte maturation, many mRNAs are translationally activated by cytoplasmic poly(A) tail elongation (for review, see Richter 1999). These mRNAs contain a distinctive feature in their 3′UTRs, the cytoplasmic polyadenylation element (CPE), a U-rich sequence that is recognized by the conserved protein CPEB. The CPE is a bifunctional element: Although it is required for translational activation of mRNAs during oocyte maturation, it is also necessary for translational silencing of the same mRNAs prior to maturation (Stutz et al. 1998; de Moor and Richter 1999; Ralle et al. 1999; Barkoff et al. 2000). Because CPEB binds to the CPE, it was suggested that CPEB was involved in both seemingly opposing functions of this element (de Moor and Richter

Figure 2. (*a*) Translation initiated via the internal ribosome entry site (IRES) in the intergenic region of the cricket paralysis virus (CrPV) (CrPV-IRES) bypasses the silencing mechanism by the hnRNPs K/E1–DICE complex. A schematic representation of the CrPV-IRES-CAT reporter mRNA carrying the functional DICE (fDICE) in the 3′UTR is shown in the upper panel. NOP-1 and LOX mRNAs were used as negative and positive controls, respectively, and were cotranslated with the CrPV-IRES-CAT-fDICE mRNA (lanes *1–5*) in nuclease-treated rabbit reticulocyte lysates (Ostareck et al. 2001). Dialysis buffer (lane *1*), hnRNPs K and E1 (lanes *2* and *5*), eIF4A (lane *3*), or a dominant negative mutant of eIF4A (R362Q) (Pause and Sonenberg 1992) (lanes *4* and *5*) was added to the translation reaction. Whereas LOX mRNA translation was silenced in the presence of the repressor proteins hnRNP K and E1, translation of the CrPV-IRES-CAT-fDICE mRNA remained unchanged in the same reaction (lane *2*). All three mRNAs were efficiently translated in the presence of excess eIF4A (lane *3*). Cap-dependent translation of LOX and NOP-1 mRNA is strongly impaired by the mutant eIF4A (R362Q) (lanes *4* and *5*). In contrast, translation initiation at the CrPV-IRES was not affected, showing that this IRES is indeed responsible for CAT translation. (*b*) ^{32}P-labeled RNA probes consisting of either the DICE element, α-globin mRNA 3′UTR, or the nonrepeated sequence element of the LOX mRNA 3′UTR as negative control were photo-crosslinked to hnRNP K or hnRNP E1 (each 500 ng). The crosslinked products were separated in a denaturing acrylamide gel. The translation rates indicated at the bottom panel represent the efficiency of translation of luciferase reporter constructs bearing the respective 3′UTRs in transfected HeLa cells. (Data from Ostareck et al. 1997.)

1999). Indeed, CPE-bound CPEB has been shown to serve as an interaction platform for factors involved in translational repression as well as in cytoplasmic polyadenylation (for review, see Mendez and Richter 2001). *Xenopus* CPEB interacts with a protein called Maskin, which has been suggested to mediate the translational repressor function of CPEB (Stebbins-Boaz et al. 1999). Maskin can bind to the cap-binding protein eIF4E via a domain that is similar to the region with which eIF4G binds eIF4E. Maskin competes with eIF4G for eIF4E binding, and this competition hinders the formation of a translationally competent eIF4F cap-binding complex, thereby preventing translation (Fig. 1b). Maskin hence acts as a co-repressor that is recruited to the regulated mRNAs via CPEB. This model can account for the finding that CPE/CPEB-mediated repression mainly interferes with cap-dependent translation (de Moor and Richter 1999). However, biochemical experiments aimed at assessing the affinity of Maskin for eIF4E raise the question of whether the affinity of a bimolecular interaction between these two proteins suffices to efficiently out-compete the eIF4E–eIF4G association (Stebbins-Boaz et al. 1999). This suggests that additional factors could be involved, perhaps to stabilize the Maskin–eIF4E interaction. Interestingly, the CPE sequence context can affect the extent of translational repression (Simon et al. 1992; de Moor and Richter 1999; Barkoff et al. 2000), and proteins other than CPEB can be cross-linked to certain CPEs (Wu et al. 1997; Ralle et al. 1999). Since growing oocytes contain no detectable Maskin (Groisman et al. 2000), the translational silencing of CPE-regulated mRNAs such as that encoding cyclin B1 must be achieved at this stage by a different assembly of regulatory factors (Lia de Moor, pers. comm.).

Another illustrative example of 3′UTR-mediated translational control involving multiple regulatory proteins is *oskar* mRNA. One of the earliest events in A-P axis formation during *Drosophila* development is the local translation of Oskar at the posterior pole of the oocyte from localized *oskar* mRNA (for review, see Lipshitz and Smibert 2000). Unlocalized *oskar* mRNA translation is repressed by the interaction of the RNA-binding protein Bruno with specific elements in the *oskar* 3′UTR, referred to as the Bruno response elements (BREs). Bruno binding alone is not sufficient for repression. Another protein, p50, binds to the BREs and contributes to translational inhibition (Gunkel et al. 1998). In addition, mutations in the gene encoding Bicaudal C (Bic-C), a KH domain-containing protein, cause premature translation of *oskar* mRNA in vivo (Saffman et al. 1998). A role for the RNA-binding protein Apontic has been proposed based on its interaction with Bruno (Lie and Macdonald 1999a), although at present no direct evidence for an involvement of Apontic in the translational regulation of *oskar* mRNA has been reported. Collectively, the current evidence suggests that translational repression of unlocalized *oskar* mRNA involves at least Bruno, p50, and BicC, and may include Apontic or others. The mechanism of translational inhibition of *oskar* mRNA is unresolved, but recent experiments using cell-free translation systems from *Drosophila* ovaries may pave the way for further biochemical analysis (Lie and Macdonald 1999b; Castagnetti et al. 2000).

REGULATION OF MALE-SPECIFIC-LETHAL-2 mRNA BY SEX-LETHAL INVOLVES COOPERATION BETWEEN 5′ AND 3′ UTR REPRESSOR ELEMENTS

The translational inhibition of *male-specific-lethal-2 (msl-2)* mRNA in females of *Drosophila melanogaster* requires the functional cooperation between RNA-binding proteins bound to both the 5′ and 3′UTRs of the message. It remains to be seen how frequently such a complex regulatory setting is used in biology, but *msl-2* mRNA regulation provides a prime opportunity to study 5′- to 3′-end communication in translation.

The MSL-2 protein is a component of the *Drosophila* dosage compensation complex, which is expressed only in males and promotes hypertranscription of the single male X chromosome to adjust its output to that of both female X chromosomes. Dosage compensation is restricted to males via the translational repression of *msl-2* mRNA in females by the female-specific RNA-binding protein Sex-lethal (SXL) (Kelley et al. 1995, 1997; Bashaw and Baker 1997; Gebauer et al. 1998). In other words, *msl-2* mRNA is expressed in both sexes, but its translation must be permanently repressed in females. SXL binds to runs of uridines (U_7 to U_{16}) that are present in two copies in the 5′UTR and in four copies in the 3′UTR of *msl-2* mRNA (see Fig. 4A). The SXL-binding sites in both UTRs synergize to inhibit MSL-2 expression in females. In transgenic flies, the expression of reporter mRNAs is moderately inhibited if only the *msl-2* 5′UTR or the 3′UTR is present, whereas the presence of the SXL-binding sites in both UTRs results in tight repression by SXL (Bashaw and Baker 1997; Kelley et al. 1997).

To study the mechanism of SXL action biochemically, a cell-free translation system from *Drosophila* embryos was developed (Gebauer et al. 1999). The translation of luciferase reporter mRNAs bearing both untranslated regions of *msl-2* mRNA is specifically repressed by the addition of recombinant SXL from *D. melanogaster*, but not by unrelated RNA-binding proteins (Fig. 3A). Moreover, the synergism between the two mRNA ends for effective translational inhibition is closely reflected in this in vitro system. Of the two SXL-binding sites present in the 5′UTR, only the downstream one is required for repression, whereas the upstream site can be deleted without effect on translation (Gebauer et al. 1999). In vitro, the translational repression by SXL is equally effective for polyadenylated and nonadenylated mRNAs, suggesting that the poly(A) tail is not a primary target for the effect of SXL (Gebauer et al. 1999).

Recent results have indicated that a *Drosophila*-specific cofactor(s) of SXL could be required to inhibit *msl-2* mRNA. First, the contribution of the 3′UTR to the translational repression by SXL cannot be recapitulated in rabbit reticulocyte lysates, and inhibition via the 5′UTR requires ~10 times higher concentrations of SXL in this system (Gebauer et al. 1999; see also Fig. 3B). In a HeLa cell translation extract, SXL completely fails to

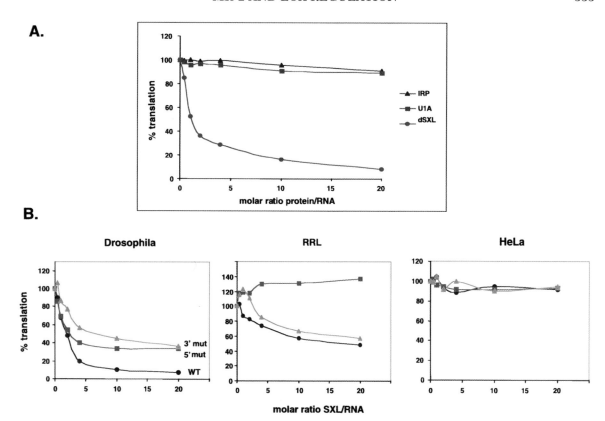

Figure 3. (*A*) Translational repression by *Drosophila* Sex-lethal (dSXL) is specific. A reporter luciferase construct containing the 5′ and 3′ UTRs of *msl-2* mRNA was incubated with the indicated RNA-binding proteins in a cell-free translation system from *Drosophila* embryos. Luciferase activity was subsequently measured as a translation readout. The stability of the mRNAs was not affected by SXL (Gebauer et al. 1999). (*B*) Functional cooperation between the 5′ and 3′ UTRs of *msl-2* mRNA for translational repression is only achieved in *Drosophila* extracts. Translation of a luciferase reporter mRNA containing the wild-type 5′ and 3′ UTRs of *msl-2* mRNA (WT), or mutated versions lacking SXL-binding sites at the 5′ (5′ mut) or the 3′ (3′ mut) UTRs, in the presence of increasing amounts of added recombinant SXL. Translation was assayed in *Drosophila* extracts, HeLa extracts, and rabbit reticulocyte lysates. Typical translational repression was only observed in *Drosophila* extracts, suggesting that a *Drosophila*-specific factor may be required for inhibition.

exert a regulatory effect (Fig. 3B). Second, a derivative of the SXL homolog from the housefly *Musca domestica* binds to *msl-2* mRNA with an affinity similar to that of the *Drosophila* protein but is unable to effectively inhibit *msl-2* translation (F. Gebauer et al., unpubl.). This suggests that the *Drosophila* SXL protein possesses functional domains in addition to the RNA-binding domain, possibly to help recruit a co-repressor(s) (Fig. 4A). How SXL and additional factors may inhibit translation via both UTRs is not yet clear. One possibility is that SXL-driven interactions between the 5′ and 3′ UTRs package the *msl-2* mRNP to render it poorly accessible to the translation machinery. Alternatively, SXL could inhibit different but coupled processes at the 5′ and 3′ UTRs. Answers to these intriguing questions should arise from further biochemical experiments, complemented by genetic experiments in mutant flies.

THE ORDERED ASSEMBLY OF REPRESSOR mRNPs

The translational inhibition of maternal *hunchback* mRNA (*hb*^mat) in the early *Drosophila* embryo may fore-

shadow an aspect of translational control from the 3′UTR that could also apply to numerous other cases, perhaps including those discussed above: the establishment of a silenced mRNP by the sequential and ordered assembly of repressors and co-repressors. *hb*^mat mRNA contains two copies of a 32-nucleotide element just downstream of the translation termination codon that are necessary for the translational repression by the morphogen Nanos (Nos) (Wharton and Struhl 1991). These elements have been referred to as the Nanos response elements or NREs. However, Nos does not bind directly to the NREs, but its interaction requires the previous binding of Pumilio (Pum) to the NRE (Murata and Wharton 1995; Sonoda and Wharton 1999) (Fig. 4B). Formation of the Nos–Pum–NRE ternary complex depends on central nucleotides of the NRE, the RNA-binding domain of Pum, and the carboxy-terminal Cys/His-rich domain of Nos, all of which are essential for translational regulation (Curtis et al. 1997; Wharton et al. 1998; Sonoda and Wharton 1999). The ternary complex subsequently recruits Brain Tumor (Brat), another factor required for the repression of *hb*^mat mRNA (Sonoda and Wharton 2001). Whether Brat interacts directly with the translation machinery at

A

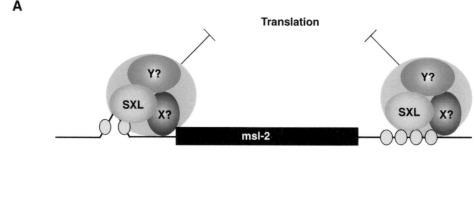

B

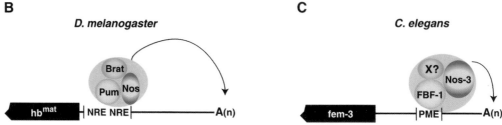

Figure 4. (*A*) Translational repression of *msl-2* mRNA. SXL binds to stretches of uridines, represented as yellow ovals, in the 5′ and 3′ UTRs of *msl-2* mRNA and inhibits translation via a mechanism that may require the interaction with *Drosophila*-specific cofactors, denoted as X and Y. (*B*) Translational repression of maternal *hunchback* (*hb^{mat}*) mRNA. Binding of *Drosophila* Pum to the nanos response elements (NRE) in the 3′ UTR of maternal *hunchback* (*hb^{mat}*) mRNA allows the successive interactions of Nos and Brat. This translation inhibition step is necessary for A-P axis formation. (*C*) Translational inhibition of *C. elegans fem-3* mRNA. The Puf domain-containing protein FBF-1 binds to the point mutation element (PME) in the 3′ UTR of *fem-3* mRNA. FBF-1 interacts with Nos-3 to repress translation of *fem-3*, a necessary step for the sperm–oocyte switch in the hermaphrodite. Both in *Drosophila* and *C. elegans*, the complexes are thought to inhibit translation via shortening of the poly(A) tail.

the 5′ end is an open question, but translational inhibition of *hb^{mat}* mRNA has been suggested to involve deadenylation of the message (Wreden et al. 1997).

The regulation of *hb^{mat}* mRNA in *Drosophila* may have close counterparts in *Caenorhabditis elegans*, because the sperm–oocyte switch in the *C. elegans* hermaphrodite is regulated by a similar set of factors that acts on the *fem-3* mRNA (Fig. 4C). Fem-3 gain-of-function mutations map to a 5-nucleotide element (UCUUG) that is located more or less centrally in the 3′UTR of the message (Ahringer and Kimble 1991). This element has been called the point mutation element (PME). A protein with an RNA-binding domain closely related to that found in *Drosophila* Pum, FBF-1, binds to the PME and is required for the sperm–oocyte switch (Zhang et al. 1997). The RNA-binding domain of FBF-1 and Pum has been named the Puf domain (after *Pu*milio and *F*BF), and it is found in a conserved family of proteins with diverse functions in development and cell differentiation. Three homologs of *Drosophila* Nos have been identified in *C. elegans*. As in *Drosophila*, Nos-3 interacts with FBF-1, but unlike *Drosophila*, this interaction does not appear to require the previous association of FBF-1 with the mRNA (Kraemer et al. 1999). The other two *C. elegans* homologs of Nos, Nos-1 and Nos-2, also function in the sperm–oocyte switch, but they do not seem to interact directly with FBF-1 (Kraemer et al. 1999). Translational repression of *fem-3* mRNA by FBF-1/Nos-3 may also involve deadenylation (Ahringer and Kimble 1991),

although it is not clear whether this is a cause or a consequence of repression. Future work should clarify whether a homolog of the Brat protein is assembled into the PME-bound repressor complex. Interestingly, a family member of the Brat protein in *C. elegans*, LIN41, has been implicated in posttranscriptional regulation in this organism (Slack et al. 2000).

CONCLUSIONS

Translational repression via the 3′UTR is a recurring theme in biology. The central point of this review is the proposal that, in contrast to the 5′UTR, 3′UTR-mediated translational regulation is generally carried out by assemblies of repressor proteins that directly bind to the mRNA, and of co-repressors that must additionally be recruited to these sites. Such assemblies then intersect translation pathways, either in poly(A) tail-dependent or independent ways. This view resembles aspects of transcriptional activation by transcription factors that recruit multiprotein chromatin-remodeling complexes to specific promoter sites on the DNA. What could be the reason for the use of protein assemblies rather than single regulators? One explanation may be simple economy: One given regulator can be used for multiple tasks. In *Drosophila*, single mutations in RNA-binding proteins often cause pleiotrophic effects indicative of the involvement of the same factor in several processes. In addition, the formation of multicomponent assemblies allows more versatile and flexible modes of regulation. Despite the broad biological relevance of translational control, much remains to be discovered about the underlying mechanisms.

Note Added in Proof

While this review was in press, Chagnovich and Lehmann (2001) challenged the idea of the poly(A) tail in hunchback mRNA translational repression by showing that hb^mat RNAs lacking a poly(A) tail or containing a histone stem-loop in place of the poly(A) tail were efficiently repressed.

ACKNOWLEDGMENTS

We thank Cornelia de Moor for communication of results prior to publication, and Nahum Sonenberg, Steve Liebhaber, and Peter Sarnow for generously providing plasmids. A. O-L. was supported by a fellowship from the Deutsche Akademie der Naturforscher Leopoldina, and D.H.O. by grants from the Deutsche Forschungsgemeinschaft and HSFP to M.W.H. M. G. is a Louis-Jeantet Foundation predoctoral fellow of the EMBL International Ph.D. Program.

REFERENCES

Ahringer J. and Kimble J. 1991. Control of the sperm-oocyte switch in *Caenorhabditis elegans* hermaphrodites by the fem-3 3′ untranslated region. *Nature* **349:** 346.

Barkoff A., Dickson K., Gray N., and Wickens M. 2000. Translational control of cyclin B1 mRNA during meiotic maturation: Coordinated repression and cytoplasmic polyadenylation. *Dev. Biol.* **220:** 97.

Bashaw G. and Baker B. 1997. The regulation of the *Drosophila msl-2* gene reveals a function for Sex-lethal in translational control. *Cell* **89:** 789.

Casey J., Koeller D., Ramin V., Klausner R., and Harford J. 1989. Iron regulation of transferrin receptor mRNA levels requires iron-responsive elements and a rapid turnover determinant in the 3′ untranslated region of the mRNA. *EMBO J.* **8:** 3693.

Castagnetti S., Hentze M., Ephrussi A., and Gebauer F. 2000. Control of oskar mRNA translation by Bruno in a novel cell-free system from *Drosophila* ovaries. *Development* **127:** 1063.

Chagnovich D. and Lehmann R. 2001. Poly(A)-independent regulation of maternal hunchback translation in the *Drosophila* embryo. *Proc. Natl. Acad. Sci.* **98:** 11359.

Curtis D., Treiber D., Tao F., Zamore P., Williamson J., and Lehmann R. 1997. A CCHC metal-binding domain in Nanos is essential for translational regulation. *EMBO J.* **16:** 834.

de Moor C. and Richter J. 1999. Cytoplasmic polyadenylation elements mediate masking and unmasking of cyclin B1 mRNA. *EMBO J.* **18:** 2294.

———. 2001. Translational control in vertebrate development. *Int. Rev. Cytol.* **203:** 567.

Gallie D. 1991. The cap and poly(A) tail function synergistically to regulate mRNA translational efficiency. *Genes Dev.* **5:** 2108.

Gebauer F., Merendino L., Hentze M., and Valcárcel J. 1998. The *Drosophila* splicing regulator sex-lethal directly inhibits translation of male-specific-lethal-2 mRNA. *RNA* **4:** 142.

Gebauer F., Corona D., Preiss T., Becker P., and Hentze M. 1999. Translational control of dosage compensation in *Drosophila* by Sex-lethal: Cooperative silencing via the 5′ and 3′ UTRs of msl-2 mRNA is independent of the poly(A) tail. *EMBO J.* **18:** 6146.

Gingras A., Raught B., and Sonenberg N. 1999. eIF4 initiation factors: Effectors of mRNA recruitment to ribosomes and regulators of translation. *Annu. Rev. Biochem.* **68:** 913.

Gray N. and Hentze M. 1994. Iron regulatory protein prevents binding of the 43S translation pre-initiation complex to ferritin and eALAS mRNAs. *EMBO J.* **13:** 3882.

Groisman I., Huang Y.-S., Mendez R., Cao Q., Theurkauf W., and Richter J. 2000. CPEB, maskin and cyclin B1 mRNA at the mitotic apparatus: Implications for local translational control of cell division. *Cell* **103:** 435.

Gunkel N., Yano T., Markussen F.-H., Olsen L., and Ephrussi A. 1998. Localization-dependent translation requires a functional interaction between the 5′ and 3′ ends of oskar mRNA. *Genes Dev.* **12:** 1652.

Hentze M. and Kühn L. 1996. Molecular control of vertebrate iron metabolism: mRNA-based regulatory circuits operated by iron, nitric oxide and oxidative stress. *Proc. Natl. Acad. Sci.* **93:** 8175.

Höhne M., Thiele B., Prehn S., Giessmann E., Nack B., and Rapoport S. 1988. Activation of translationally inactive lipoxygenase mRNP particles from rabbit reticulocytes. *Biomed. Biochim. Acta* **47:** 75.

Iizuka N., Najita L., Franzusoff A., and Sarnow P. 1994. Cap-dependent and cap-independent translation by internal initiation of mRNAs in cell extracts prepared from *Saccharomyces cerevisiae*. *Mol. Cell. Biol.* **14:** 7322.

Imataka H., Gradi A., and Sonenberg N. 1998. A newly identified N-terminal amino acid sequence of human eIF4G binds poly(A)-binding protein and functions in poly(A)-dependent translation. *EMBO J.* **17:** 7480.

Jansen R. 2001. mRNA localization: Message on the move. *Nat. Rev. Mol. Cell Biol.* **2:** 247.

Kelley R., Wang J., Bell L., and Kuroda M. 1997. Sex lethal controls dosage compensation in *Drosophila* by a non-splicing mechanism. *Nature* **387:** 195.

Kelley R., Solovyeva I., Lyman L., Richman R., Solovyev V., and Kuroda M. 1995. Expression of Msl-2 causes assembly of dosage compensation regulators on the X chromosomes and female lethality in *Drosophila*. *Cell* **81:** 867.

Kraemer B., Crittenden S., Gallegos M., Moulder G., Barstead R., Kimble J., and Wickens M. 1999. Nanos-3 and FBF proteins physically interact to control the sperm-oocyte switch in *Caenorhabditis elegans*. *Curr. Biol.* **9:** 1009.

Le H., Tanguay R., Balasta M., Wei C., Browning K., Metz A., Gross J., and Gallie D. 1997. Translation initiation factors eIF-iso4G and eIF-4B interact with the poly(A)-binding protein and increase its RNA-binding activity. *J. Biol. Chem.* **272:** 16247.

Lie Y. and Macdonald P. 1999a. Apontic binds the translational repressor Bruno and is implicated in regulation of oskar mRNA translation. *Development* **126:** 1129.

———. 1999b. Translational regulation of oskar mRNA occurs independent of the cap and poly(A) tail in *Drosophila* ovarian extracts. *Development* **126:** 4989.

Lipshitz H. and Smibert C. 2000. Mechanisms of RNA localization and translational regulation. *Curr. Opin. Genet. Dev.* **10:** 476.

Macdonald P. 2001. Diversity in translational regulation. *Curr. Opin. Cell Biol.* **13:** 326.

Mendez R. and Richter J. 2001. Translational control by CPEB: A means to the end. *Nat. Rev. Mol. Cell Biol.* **2:** 521.

Muckenthaler M., Gray N., and Hentze M. 1998. IRP-1 binding to ferritin mRNA prevents the recruitment of the small ribosomal subunit by the cap-binding complex eIF4F. *Mol. Cell* **2:** 383.

Murata Y. and Wharton R. 1995. Binding of Pumilio to maternal hunchback mRNA is required for posterior patterning in *Drosophila* embryos. *Cell* **80:** 747.

Ostareck D., Ostareck-Lederer A., Shatsky I., and Hentze M. 2001. Lipoxygenase mRNA silencing in erythroid differentiation: The 3′ UTR regulatory complex controls 60S ribosomal subunit joining. *Cell* **104:** 281.

Ostareck D., Ostareck-Lederer A., Wilm M., Thiele B., Mann

M., and Hentze M. 1997. mRNA silencing in erythroid differentiation: hnRNP K and hnRNP E1 regulate 15-lipoxygenase translation from the 3′ end. *Cell* **89:** 597.

Ostareck-Lederer A., Ostareck D., and Hentze M. 1998. Cytoplasmic regulatory functions of the KH-domain proteins hnRNPs K and E1/E2. *Trends Biochem. Sci.* **275:** 409.

Ostareck-Lederer A., Ostareck D., Standart N., and Thiele B. 1994. Translation of 15-lipoxygenase mRNA is inhibited by a protein that binds to a repeated sequence in the 3′ untranslated region. *EMBO J.* **13:** 1476.

Pause A. and Sonenberg N. 1992. Mutational analysis of a DEAD box RNA helicase: The mammalian translation initiation factor eIF-4A. *EMBO J.* **11:** 2643.

Preiss T. and Hentze M. 1998. Dual function of the messenger cap structure in poly(A)-tail-promoted translation in yeast. *Nature* **392:** 516.

Ralle T., Gremmels D., and Stick R. 1999. Translational control of nuclear lamin B1 mRNA during oogenesis and early development of *Xenopus*. *Mech. Dev.* **84:** 89.

Rapoport S. and Schewe T. 1986. The maturational breakdown of mitochondria in reticulocytes. *Biochim. Biophys. Acta* **864:** 471.

Richter J. 1999. Cytoplasmic polyadenylation in development and beyond. *Microbiol. Mol. Biol. Rev.* **63:** 446.

———. Influence of polyadenylation-induced translation on metazoan development and neuronal synaptic function. In *Translational control of gene expression* (ed. N. Sonenberg et al.), p. 785. Cold Spring Harbor Laboratory Press, Cold Spring Harbor, New York.

Rouault T. 2000. Translational control of ferritin synthesis. In *Translational control of gene expression* (ed. N. Sonenberg et al.), p. 655. Cold Spring Harbor Laboratory Press, Cold Spring Harbor, New York.

Saffman E., Styhler S., Rother K., Li W., Richard S., and Lasko P. 1998. Premature translation of oskar in oocytes lacking the RNA-binding protein bicaudal-C. *Mol. Cell. Biol.* **18:** 4855.

Simon R., Tassan J.-P., and Richter J. 1992. Translational control by poly(A) elongation during *Xenopus* development: Differential repression and enhancement by a novel cytoplasmic polyadenylation element. *Genes Dev.* **6:** 2580.

Slack F.J., Basson M., Liu Z., Ambros V., Horvitz H.R., and Ruvkun G. 2000. The LIN-41 RBCC gene acts in the *C. elegans* heterochronic pathway between the let-7 regulatory RNA and the LIN-49 transcription factor. *Mol. Cell* **5:** 659.

Sonoda J. and Wharton R. 1999. Recruitment of Nanos to hunchback mRNA by Pumilio. *Genes Dev.* **13:** 2704.

———. 2001. *Drosophila* Brain Tumor is a translational repressor. *Genes Dev.* **15:** 762.

Stebbins-Boaz B., Cao Q., de Moor C., Mendez R., and Richter J. 1999. Maskin is a CPEB-associated factor that transiently interacts with eIF-4E. *Mol. Cell* **4:** 1017.

Stutz A., Conne B., Huarte J., Gubler P., Völkel V., Flandin P., and Vassalli J.-D. 1998. Masking, unmasking, and regulated polyadenylation cooperate in the translational control of a dormant mRNA in mouse oocytes. *Genes Dev.* **12:** 2535.

Tarun S. and Sachs A. 1995. A common function for mRNA 5′ and 3′ ends in translation initiation in yeast. *Genes Dev.* **9:** 2997.

———. 1996. Association of the yeast poly(A) tail binding protein with translation initiation factor eIF-4G. *EMBO J.* **15:** 7168.

Tarun S., Wells S., Deardorff J., and Sachs A. 1997. Translation initiation factor eIF4G mediates in vitro poly(A) tail-dependent translation. *Proc. Natl. Acad. Sci.* **94:** 9046.

Thompson S.R., Goodwin E.B., and Wickens M. 2000. Rapid deadenylation and poly(A)-dependent translational repression mediated by the *Caenorhabditis elegans* tra-2 3′untranslated region in *Xenopus* embryos. *Mol. Cell. Biol.* **20:** 2129.

Wells D., Richter J., and Fallon J. 2000. Molecular mechanisms for activity-regulated protein synthesis in the synapto-dendritic compartment. *Curr. Opin. Neurobiol.* **10:** 132.

Wells S.E., Hillner P.E., Vale R.D., and Sachs A.B. 1998. Circularization of mRNA by eukaryotic translation initiation factors. *Mol. Cell* **2:** 135.

Wharton R. and Struhl G. 1991. RNA regulatory elements mediate control of Drosophila body pattern by the posterior morphogen nanos. *Cell* **67:** 955.

Wharton R., Sonoda J., Lee T., Patterson M., and Murata Y. 1998. The pumilio RNA-binding domain is also a translational regulator. *Mol. Cell* **1:** 863.

Wickens M., Goodwin E., Kimble J., Strickland S., and Hentze M. 2000. Translational control of developmental decisions. In *Translational control of gene expression* (ed. N. Sonenberg et al.), p. 295. Cold Spring Harbor Laboratory Press, Cold Spring Harbor, New York.

Wilson J., Pestova T., Hellen C., and Sarnow P. 2000a. Initiation of protein synthesis from the A site of the ribosome. *Cell* **102:** 511.

Wilson J., Powell M., Hoover S., and Sarnow P. 2000b. Naturally occuring dicistronic cricket paralysis virus RNA is regulated by two internal ribosome entry sites. *Mol. Cell. Biol.* **20:** 4990.

Wilusz C., Wormington M., and Peltz S. 2001. The cap-to-tail guide to mRNA turnover. *Nat. Rev. Mol. Cell Biol.* **2:** 237.

Wreden C., Verroti A., Schisa J., Lieberfarb M., and Strickland S. 1997. Nanos and pumilio establish embryonic polarity in *Drosophila* by promoting posterior deadenylation of hunchback mRNA. *Development* **124:** 3015.

Wu L., Good P., and Richter J. 1997. The 36-kilodalton embryonic-type cytoplasmic polyadenylation element-binding protein in *Xenopus laevis* is ElrA, a member of the ELAV family of RNA-binding proteins. *Mol. Cell. Biol.* **17:** 6402.

Zhang B., Gallegos M., Puoti A., Durkin E., Fields S., Kimble J., and Wickens M. 1997. A conserved RNA-binding protein that regulates sexual fates in the *C. elegans* hermaphrodite germ line. *Nature* **390:** 477.

PUF Proteins and 3´UTR Regulation in the *Caenorhabditis elegans* Germ Line

M. Wickens,* D. Bernstein,* S. Crittenden,† C. Luitjens,* and J. Kimble*†‡

*Department of Biochemistry, †Howard Hughes Medical Institute, ‡Department of Genetics and Laboratory of Molecular Biology, University of Wisconsin, Madison, Wisconsin 53706

In a simple and too common view, mRNAs are a dull milepost between DNA and protein in the Central Dogma. They acquire middling interest as guides for the ribosome, tRNAs, and translation factors. Yet mRNAs have lives of their own, lives that the cell cares about enormously. mRNAs are born, leave home, mate with ribosomes, produce proteins, and die, all as the cell sees fit. Powerful forces—in the form of specific mRNA-binding proteins and their cohorts—guide individual mRNAs into unique variations on this common path. These proteins govern mRNA stability, localization, and translation.

In eukaryotic cells, the sequence elements responsible can lie anywhere in the mRNA: Indeed, certain instances of regulation mediated by 5´UTRs are among the best-understood examples of translational control (for review, see Sachs et al. 1997; Geballe and Sachs 2000; Hinnebusch 2000; Rouault and Harford 2000). However, the region between the termination codon and poly(A) tail—the 3´ untranslated region, or 3´UTR—has emerged as preeminent (for review, see Wickens et al. 2000). Why?

3´UTRs AS REGULATORY NODES

3´UTRs possess distinctive properties that lend themselves to service in regulation. These properties echo those of promoters, in which the information for transcriptional regulation is embedded.

- 3´UTRs are relatively unconstrained evolutionarily: They diverge more rapidly than coding regions, but are marked by pockets of conserved (and presumably regulatory) sequence (Duret et al. 1993; Duret and Bucher 1997; Spicher et al. 1998; Pesole et al. 2000). They escape the obvious constraints imposed on the coding region, as well as those imposed on the 5´UTR to support eukaryotic initiation.

- 3´UTRs can be long. Mammalian 3´UTRs range up to several kilobases. This provides fertile ground for the birth and evolution of new regulatory elements.

- The machinery that cleaves and polyadenylates mRNAs is inherently inefficient, such that mRNA precursors can extend many kilobases past a potential polyadenylation site. As a result, 3´UTRs of different lengths can be attached to the same ORF, and evolution gains a flexible field in which new circuitry can arise.

Regulatory opportunities also arise because 3´UTRs are modules in mRNA biogenesis. This is suggested even by the organization of 3´UTRs in the genome. For example, 3´UTRs in mammalian cells are almost always contained in a single terminal exon, bounded by the last 3´ splice site and the polyadenylation site (Nagy and Maquat 1998). The positioning of splice sites, termination sites, and the 3´ end defines a 3´UTR domain and must be correct: mRNAs that do not conform to the rules—for example, a mammalian mRNA with an intron downstream of a termination codon—are purged by the nonsense-mediated decay (NMD) system (Sun and Maquat 2000). From this perspective (Hilleren and Parker 1999), NMD is a 3´UTR problem.

In this review, we describe experiments that reveal interacting 3´UTR regulatory proteins that control a wide variety of decisions in a complex tissue. Our story begins with the germ line of *C. elegans* and broadens with the finding that many of the regulators are members of families conserved among eukaryotes. These interacting regulators likely have common, conserved mechanisms and biological roles.

MULTIPLE DECISIONS IN THE *C. ELEGANS* GERM LINE

Figure 1 summarizes the postembryonic development of the *C. elegans* hermaphrodite germ line; hermaphrodites are essentially females that make some sperm and then switch to oogenesis (Schedl 1997). During embryogenesis, two germ-line precursor cells arise from a single germ-line blastomere; after hatching, these two cells begin to proliferate. Initially, all germ-line cells are in the mitotic cell cycle (Fig. 1, yellow); then, as the animal progresses through four larval stages (L1, L2, L3, and L4), the more proximal germ-line cells leave the mitotic cell cycle and enter meiosis (Fig. 1, green). During L4, the most proximal germ-line cells undergo spermatogenesis (Fig. 1, blue); once the proper number of sperm have been generated, the germ line switches to oogenesis (Fig. 1, pink) as the animal enters adulthood. Thus, the germ line grows and differentiates in a spatially and temporally regulated manner. The decisions between mitosis and meiosis and between spermatogenesis and oogenesis must be coordinated to achieve the adult pattern. Remarkably, both decisions rely on many of the same 3´UTR regula-

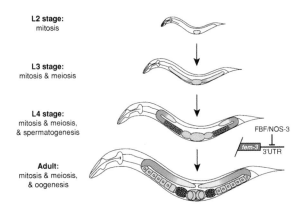

Figure 1. *C. elegans* germ-line development. (*Yellow*), germ-line cells in the mitotic cell cycle; (*green*), germ-line cells that have entered the meiotic cell cycle and are arrested in the pachytene stage of meiotic prophase I; (*pink*) oogenesis; (*blue*) spermatogenesis. L2 to L4 indicate the larval stages.

tors. FBF protein, a plexus in this web of interconnected proteins, was first uncovered through studies of the sperm/oocyte switch (Zhang et al. 1997). We begin there.

FBF AND REGULATION BY THE *fem-3* 3′UTR

The *fem-3* sex-determination gene directs spermatogenesis: Loss-of-function *fem-3 (lf)* mutants make only oocytes and no sperm, and gain-of-function *fem-3(gf)* mutants make only sperm and no oocytes (Hodgkin 1986; Barton et al. 1987). The *fem-3(gf)* mutations identified a regulatory element in the *fem-3* 3′UTR, called the PME (Ahringer and Kimble 1991). More recent studies identified an RNA-binding protein, called FBF (for *fem-3*-binding factor), which binds to the PME and represses *fem-3* activity (Zhang et al. 1997).

C. elegans contains two FBF proteins, FBF-1 and FBF-2, which are 91% identical in amino acid sequence; their functions to date are indistinguishable, and so they are often referred to collectively as FBF. Both FBF-1 and FBF-2 bind the wild-type *fem-3* PME, but not mutant PMEs or other RNA sequences. Animals lacking both *fbf-1* and *fbf-2*, generated by RNA-mediated interference, make only sperm and fail to switch into oogenesis, consistent with a role for FBF-1 and FBF-2 in *fem-3* repression. Intriguingly, FBF-deficient germ lines are small (Zhang et al. 1997), and all germ cells enter meiosis (S. Crittenden et al., in prep.). Because no role for *fem-3* has been deduced in germ-line proliferation, it seems likely that FBF controls some other mRNA to influence this process.

PUF PROTEINS: A FAMILY OF 3′UTR REGULATORS

FBF is a member of a large family of RNA-binding proteins, the PUF (or Pum-HD) family (Fig. 2) (Wickens et al. 2002). The two founding members of this family with demonstrable roles in mRNA function were *Drosophila* Pumilio and *C. elegans* FBF (Barker et al. 1992; Zamore et al. 1997; Zhang et al. 1997), but mem-

PUF protein family

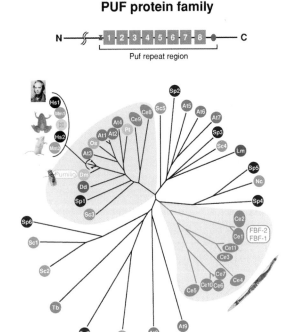

Figure 2. FBF and the PUF protein family. (*Top*) Conserved features of PUF proteins. The eight tandem repeats of ~40 amino acids (Puf repeats) are shown as grey boxes. In many PUF proteins, these are flanked by two short sequences (*anvil* and *oval*) that form Puf-repeat-like structures, named Csp1 and Csp2. The Puf repeat region is typically near the carboxyl terminus of the protein overall; many PUF proteins possess substantial amino-terminal extensions that are not obviously related among family members. For further details (including maps of other PUF proteins), see (Wickens et al. 2002) and supplementary information for that paper at www.biochem.wisc.edu/wickens. (*Bottom*) An unrooted tree was derived by aligning only the Puf repeat regions. Partial ESTs, which suggest many other PUF genes, were not included. Two major groupings of PUF proteins are indicated by shading in gray: the "*C. elegans* cluster" (*lower right*, with image of *C. elegans*) and the "Pumilio cluster" (*upper left*; image of *Drosophila* and vertebrates, including Darwin). The Pumilio cluster includes the vertebrate PUFs uncovered to date. Subfamilies of PUF proteins in *C. elegans* are grouped by related ovoid symbols; FBF-1, FBF-2, and Pumilio, the main PUF proteins discussed here, are specifically marked. PUF proteins are named by their sequence relatedness *within* a species (Wickens et al. 2002); thus PUF-1 (FBF-1) of *C. elegans* is more closely related to PUF-2 (FBF-2) of worms than it is to the other worm PUFs, but is not particularly related to PUF1 of other organisms. Colors indicate different species: (*blue to purple*), fungi (*S. cerevisiae* [Sc], *S. pombe* [Sp], *N. crassa* [Nc], *Dictyostelium discoideum* [Dd]); gray to black, vertebrates (*Homo sapiens* [Hs], *Mus musculus* [Mm], *Xenopus laevis* [Xl]); (*green*) plants (*Arabidopsis thaliana* [At], *Oryza sativa* [Os], *Populus tremula x P. tremuloides* (poplar) [Pt]); brown, trypanosomes (*Leishmania major* [Lm], *Trypanosoma brucei* [Tb]); (*red*), *C. elegans* (Ce); (*orange*), *D. melanogaster* (Dm). For further details, see Wickens et al. (2002) and supplementary information for that paper at www.biochem.wisc.edu/wickens.

bers have now been found in virtually all eukaryotes examined, including vertebrates, plants, yeast, and slime molds (Fig. 2).

Pumilio, like FBF, binds to a 3′UTR regulatory element and represses mRNA activity (for review, see Curtis et al. 1995b). Furthermore, PUF proteins in yeast and slime molds also repress expression of target mRNAs by binding to sequences in the 3′UTR (Souza et al. 1999; Olivas and Parker 2000; Tadauchi et al. 2001). Thus, the PUF family proteins are commonly 3′UTR repressors.

Sequence and Structural Similarity

PUF proteins are characterized by the presence of eight consecutive repeats (Puf repeats) of approximately 40 amino acids (Fig. 2) (Zamore et al. 1997; Zhang et al. 1997; Wharton et al. 1998; Wickens et al. 2002). This Puf repeat region is sufficient to bind to specific RNA sequences (Zamore et al. 1997; Zhang et al. 1997; Wharton et al. 1998) and to provide many of the protein's biological functions (Wharton et al. 1998).

The recently determined structures of human and *Drosophila* Pumilio reveal a striking, extended crescent (Edwards et al. 2001; Wang et al. 2001). All of the individual Puf repeats form nearly identical three-helix triangles; these lie next to one another to form the elongated structure. The inner surface of the crescent likely contacts RNA, and the outer surface contacts proteins. The inner surface carries the conserved aromatic and charged amino acids of the core consensus, exposed to solvent, apparently poised to bind RNA.

Related but Distinct RNA-binding Specificities

PUF proteins bind selectively to their target mRNAs. Individual PUF proteins bind to different RNA sequences and are affected differentially by base changes in a single sequence (Zamore et al. 1997; D. Bernstein et al., unpubl.). Human and fly PUF proteins bind differentially to mutant RNA derivatives of *hunchback*, despite the two proteins being 80% identical (Zamore et al. 1997). Nevertheless, all known PUF-binding sites possess a common tetranucleotide, UGUR (Murata and Wharton 1995; Zamore et al. 1997; Zhang et al. 1997; Wharton et al. 1998; Tadauchi et al. 2001); other nucleotides presumably determine specificity.

C. elegans PUF Proteins

Eleven *C. elegans* PUF proteins are predicted from the complete genome sequence, and all but PUF 4 are represented among cDNAs (Fig. 2) (Wickens et al. 2002). In the nematode PUF proteins, as with PUFs in general, the Puf repeat region is near the carboxyl terminus and the proteins diverge toward the amino terminus. The *C. elegans* PUF proteins fall into subfamilies, in which members of the subfamily are more than 90% identical in amino acid sequence: FBF-1 and FBF-2 comprise one group, PUF-6, PUF-7, and PUF-10 another, and PUF-3 and PUF-11 a third. PUFs 8 and 9 are more closely related to Pumilio than are the other *C. elegans* PUF proteins, although they are not particularly similar to each other (50% identical). Among these proteins, FBF-1 and

FBF-2 bind *fem-3* detectably in a three-hybrid assay; the other PUF proteins tested do not.

PROTEIN PARTNERS

PUF proteins do not act alone to control mRNAs, but instead interact with protein partners. We have used a combination of two-hybrid screens coupled with RNAi functional assays to identify protein partners for FBF. To date, we have found two such partners, both of which belong to families of 3′UTR regulators with homologs widely spread among metazoans (Kraemer et al. 1999; Luitjens et al. 2000).

The Nanos Family

Drosophila Nanos (Nos) possesses two distinctive CCHC zinc fingers that bind RNA nonspecifically (Curtis et al. 1995a, 1997). Nos is required together with Pumilio for repression of *hunchback* mRNA in the *Drosophila* embryo (for review, see Curtis et al. 1995b). *C. elegans* possesses three Nanos homologs. One of these, NOS-3, binds directly to FBF-1 and FBF-2 (Kraemer et al. 1999). The FBF/NOS-3 interaction occurs between the Puf repeat region of FBF and a region of NOS-3 that lies amino-terminal of its zinc fingers. Functional studies indicate that all three *C. elegans* NOS genes affect the sperm/oocyte switch in vivo (Kraemer et al. 1999). *Drosophila* Nanos also binds Pumilio, but this interaction, unlike that between the nematode proteins, is only detected in the ternary complex containing RNA as well (Sonoda and Wharton 1999). Certain *pumilio* mutations, located near the eighth Puf repeat, disrupt both the Pum/Nos interaction and regulation in vivo, emphasizing the importance of the interaction (Sonoda and Wharton 1999). We conclude that PUF/NOS interactions are essential for mRNA repression in both flies and worms. Other PUF/Nanos partnerships are detected in *C. elegans* (see below) and in *Xenopus* oocytes (Nakahata et al. 2001), suggesting that this collaboration of RNA-binding proteins is common.

FBF binds NOS-3, but not NOS-1 or NOS-2 (Kraemer et al. 1999). Rather, as revealed by two-hybrid screens, both NOS-1 and NOS-2 (but not NOS-3) interact with PUF-6, PUF-7, and PUF-10 (D. Bernstein et al., unpubl.). Thus, a set of distinct PUF/NOS partnerships appears to exist. Genetic analysis by RNAi supports the view that all three NOS genes participate in the sperm/oocyte switch: Reduction in the activity of any two *nos* genes yields a substantial switch defect, indicating that the three NOS proteins overlap in function (Kraemer et al. 1999). The RNAi experiments reveal additional complexity as well: RNAi directed separately against interacting PUF or NOS partners does not always yield identical phenotypes. For example, whereas *fbf(RNAi)* animals exhibit a robust sperm/oocyte switch defect, *nos-3 (RNAi)* animals do not; instead, the activity of either *nos-1* or *nos-2* must also be reduced. Similarly, RNAi directed against *nos-1* and *nos-2* yields a significant switch defect, but RNAi directed only against their partners *(puf-6/puf-7/puf-10)* does not.

Although technical limitations of RNAi could underlie these differences, we suspect cross-talk among the proteins; for example, FBF might be required to express the other PUF or NOS proteins.

The CPEB Family

CPEB (cytoplasmic *polyadenylation element binding* protein) binds short U-rich sequences in the 3′UTRs of specific mRNAs (Hake et al. 1998; for review, see Richter 2000). Homologs are found throughout metazoa, and include the *Drosophila* protein *orb*. CPEB family members are characterized by two RRM domains followed by distinctive CCCC and CCHH zinc fingers, both of which are required for RNA binding (Hake et al. 1998). The protein is involved in cytoplasmic polyadenylation reactions that appear to be required to activate certain mRNAs during development (for review, see Richter 2000). To date, CPEB homologs have been implicated in oogenesis in a wide variety of vertebrates and invertebrates, and in mammalian neuronal plasticity (for review, see Richter 2000).

C. elegans possesses four CPEB homologs. One, CPB-1, binds physically to FBF (Luitjens et al. 2000). The requisite regions are the amino-terminal portion of CPB-1, outside its putative RNA-binding region, and the Puf repeat region of FBF (Luitjens et al. 2000). CPEB homologs have been characterized most extensively in the female germ line of vertebrates, where they promote oogenesis and progression through meiosis. Strikingly, *cpb-1* is dispensable for oogenesis in *C. elegans*, but rather is required for spermatogenesis: In *cpb-1* defective animals, gametes enter spermatogenesis, but arrest during first meiosis (Luitjens et al. 2000). Similarly, *fbf*-deficient animals appear also to produce defective sperm, although *fbf* affects a step of spermatogenesis past the *cpb-1* arrest point. We suggest that CPB-1 and FBF collaborate in a subsequent step in spermatogenesis.

A second CPEB homolog, FOG-1, is required for specification of germ-line cells as sperm. Loss-of-function mutants lacking this homolog fail to make sperm; rather, cells that would normally become sperm are transformed into oocytes (Barton and Kimble 1990; Luitjens et al. 2000; Jin et al. 2001). Thus, this putative 3′UTR regulator controls a discrete binary decision of cell fate. In this respect, the FOG-1 translational regulator is akin to transcription regulators, like MyoD, that govern binary cell fate decisions. Neither FOG-1 nor the other two CPEB homologs have been found to interact with a PUF protein. However, the interaction between FBF and CPB-1 and the recent observation that *Xenopus* CPEB coimmunoprecipitates with a *Xenopus* PUF protein suggest that the interaction may be common (Nakahata et al. 2001).

Brat and the NHL Family

Brat, a *Drosophila* protein containing an NHL domain (Slack et al. 2000), binds to the ternary complex of *hunchback* mRNA, Pumilio, and Nanos (Sonoda and Wharton 2001). Again, the Puf repeat region is critical for the interaction. Two well-characterized genes in *C. elegans*, *ncl-1* and *lin-41*, also contain NHL domains. *ncl-1* regulates nucleolar size and rRNA abundance (Frank and Roth 1998). *lin-41* acts during larval development to repress differentiation and promote mitosis of somatic, hypodermal cells (Slack et al. 2000). This role in promoting mitoses echoes PUF functions in stem cells. However, interactions between PUF and NHL proteins have not yet been reported outside *Drosophila*.

To date, these common PUF protein partners have been detected throughout metazoans, but not in unicellular eukaryotes. This is paradoxical, since *Saccharomyces cerevisiae* and *Schizosaccharomyces pombe* contain multiple PUF proteins. Perhaps functional homologs of the interactors exist but have not been uncovered by sequence analysis; alternatively, these partners, or even the existence of PUF partnerships, may be a metazoan invention.

A BRIEF DISCUSSION OF MECHANISM

Mechanistic models of PUF protein action have recently been discussed in some detail (Wickens et al. 2002) and are not recapitulated here. We note that data on FBF suggest that it acts at the level of either translation or transport of mRNAs; in other systems, PUF proteins regulate translation or mRNA stability, most likely through a single underlying event (for review, see Wickens et al. 2000, 2002). Regardless, two points concerning mechanism merit mention.

Links to Poly(A) Length

PUF-mediated repression is correlated with a decrease in poly(A) length, which itself is commonly correlated with reduced translational activity (for review, see Richter 2000; Wickens et al. 2000). For example, PME mutant mRNAs have longer poly(A) tails than their wild-type counterparts (Ahringer and Kimble 1991), and cytoplasmic polyadenylation of *bicoid* mRNA is necessary for its full activity (Salles et al. 1994). Studies of a yeast PUF protein, Puf3p, suggest that poly(A) shortening is a direct effect of PUF action, rather than a secondary consequence of repression (Olivas and Parker 2000). The finding that PUF and CPEB proteins interact is provocative in that context, since CPEB interacts with CPSF, a direct participant in cytoplasmic poly(A) addition (Bilger et al. 1994; Dickson et al. 1999 and in prep.; Mendez et al. 2000). If PUF proteins counteract cytoplasmic polyadenylation, the affected poly(A) tails likely would shorten during early development through a default deadenylation system (Fox and Wickens 1990; Wickens 1992). The well-studied case of U1A autoregulation is striking in this context: U1A binds to the 3′UTR of its own mRNA and inhibits nuclear polyadenylation by directly contacting the poly(A) polymerase (Gunderson et al. 1994). This may presage similar mechanisms in PUF protein action, whether in the nucleus or cytoplasm.

Multiple Mechanisms

PUF proteins may act at multiple levels. For example, yeast PUF3 both accelerates deadenylation and stimulates decapping of the deadenylated mRNA (Olivas and Parker 2000). Similarly, *Drosophila* Pumilio appears to repress *hunchback* mRNA both by stimulating deadenylation and through a poly(A)-independent process (Chagnovich and Lehmann 2001). This has prompted a model in which PUF proteins mediate a change in mRNP structure (Wickens et al. 2002).

COORDINATING COMBINATORIAL REGULATION

We suggest that regulation by FBF, and PUF proteins more generally, is combinatorial in two respects. First, a single regulator controls multiple mRNAs to control distinct processes. This follows, for example, from the fact that FBF-defective animals display a broader range of phenotypes than seen by mutation of the binding site in its one known target, *fem-3* mRNA. The same argument applies to *Drosophila* Pumilio and its target, *hunchback* mRNA, and to yeast Puf5p/Mpt5 and its target, *HO* mRNA (Wickens et al. 2002). Moreover, two distinct Pumilio target mRNAs have been identified: *hb* mRNA, which is controlled in the early embryo to pattern the anterior/posterior axis, and *cyclin B* mRNA, which Pumilio likely represses to slow the cell cycle (Asaoki-Taguchi et al. 1999; Deshpande et al. 1999; for review, see Parisi and Lin 2000; Sonoda and Wharton 2001). Second, the various proteins that interact with Puf proteins are involved in diverse controls. This follows from a comparison of the phenotypes of animals defective in either FBF or FBF's partners, for example: Control of the sperm/oocyte switch depends on both FBF and NOS (Zhang et al. 1997; Kraemer et al. 1999), but control of germ-line proliferation by FBF does not rely on NOS in a similar fashion (Kraemer et al. 1999). Other examples are reviewed elsewhere (Wickens et al. 2000).

The idea that FBF acts combinatorially does not address the biochemical nature of PUF complexes. At the extremes, one can imagine multiple two-member complexes, or a single large complex. Furthermore, we do not know whether the binding of different partners is cooperative or mutually exclusive. In *Drosophila*, at least three proteins—Pumilio, Nanos, and Brat—can form a quaternary complex with *hunchback* mRNA (Sonoda and Wharton 2001). It will be important to learn what other components join this complex in vivo.

Whatever the nature of the complexes and their mode of action, their activities must be controlled in space and time in vivo. In *C. elegans*, mitosis occurs before cells switch into meiosis, and at least some cells remain in mitosis to serve as germ-line stem cells, a control that depends on proximity to the regulatory distal tip cell (Kimble and White 1981). In parallel, cells must be specified as spermatocytes first and then as oocytes. Each of these choices relies on FBF. An important direction of future research is to determine how FBF activity is itself controlled to be active at the right time and place of development to achieve a normally patterned germ line. Regulation of the PUF-interacting proteins, rather than the Puf protein per se, may be critical. In *Drosophila*, spatial regulation of *hunchback* mRNA is controlled by localization of Pumilio's collaborator, Nanos (for review, see Curtis et al. 1995b; Kennedy et al. 1997; Lin and Spradling 1997). It is possible that FBF's collaborators are arranged spatially and temporally to achieve the correct patterns of expression. Indeed, CPB-1 is expressed immediately prior to overt spermatogenesis, and then disappears (Luitjens et al. 2000). In contrast, NOS-3 is distributed uniformly in the developing germ line (Kraemer et al. 1999), suggesting that NOS localization is not a universal solution to this problem.

AN ANCESTRAL FUNCTION: SUSTAINING STEM CELL PROLIFERATION

The original view of PUF protein function highlighted the diversity of regulatory processes that they control. For example, in *Drosophila* embryos, the anterior–posterior axis relies on Pumilio and Nanos, and in the *C. elegans* germ line, the sperm–oocyte switch is mediated by FBF and NOS (for review, see Curtis et al. 1997; Wickens et al. 2000). More recently, a common function has emerged from a comparison of studies in numerous species (see Wickens et al. 2002).

Both in *Drosophila* and in *C. elegans*, germ-line stem cells require PUF function for continued mitoses. In *Drosophila*, germ-line stem cells differentiate prematurely as cystoblasts in Pumilio mutants (Lin and Spradling 1997; Forbes and Lehmann 1998). Similarly, in *C. elegans*, germ-line stem cells enter meiosis prematurely in FBF mutants (S. Crittenden et al., in prep.). Therefore, PUF proteins in these two distantly related metazoans function in the control of continued proliferation of germ-line stem cells.

The role of PUF proteins in vertebrates has not been reported, and therefore the extent to which these ancient proteins control germ-line stem cells is not known. Nonetheless, clues from lower organisms suggest that this may be a broad function. In *Dictyostelium*, PufA is required for continued vegetative divisions; in its absence, cells aggregate and differentiate prematurely into a miniature slug (Souza et al. 1999). In the yeast *S. cerevisiae*, *PUF5* (also known as *MPT5* and *UTH4*) is required for continued mitotic divisions; mutants lacking *puf-5* stop dividing prematurely (Kennedy et al. 1997). The promotion of continued mitotic divisions by PUF proteins in *S. cerevisiae* and *Dictyostelium*, as well as in the germ lines of *Drosophila* and *C. elegans,* suggests an ancient and common function. We speculate that this function is conserved broadly and that PUF proteins are critical for germ-line stem cell divisions throughout the animal kingdom, and perhaps the plant kingdom as well. An extension of this already speculative hypothesis suggests that PUF proteins may be involved in promoting mitotic divisions more generally, including somatic tissues and perhaps somatic stem cells. These ideas will surely be tested within the next few years in organisms from *Arabidopsis* to *Homo sapiens*.

PERSPECTIVE

Ribosomes are the central focus of this volume. In this paper, we have shifted focus to the regulation of eukaryotic mRNAs, and to *trans*-acting proteins that bind sites in the 3′UTR in particular. Eukaryotic mRNAs are surrounded by machines ready to engage them: Ribosomes translate them, other machines move or destroy them. 3′UTR regulators ensure that machines and mRNAs only meet appropriately at the right time and place.

Our studies in *C. elegans* reveal what we suspect are just the beginnings of complex networks of contacts among 3′UTR regulators in its germ line, networks that very likely are conserved in the germ line of other organisms. It is not yet clear to what extent the conservation can be extended to the mRNA targets that are controlled; however, the proposed conservation of function in proliferation of stem cells suggests that common targets exist, along with ones that are idiosyncratic to each species. We suspect that comparable networks exist in somatic cells.

ACKNOWLEDGMENTS

We are grateful for stimulating discussions with numerous colleagues, including in particular the members of the Wickens and Kimble laboratories. Work in the Wickens laboratory is supported by the National Institutes of Health. Judith Kimble is a member of the Howard Hughes Medical Institute.

REFERENCES

Ahringer J. and Kimble J. 1991. Control of the sperm-oocyte switch in *Caenorhabditis elegans* hermaphrodites by the *fem-3* 3′ untranslated region. *Nature* **349:** 346.

Asaoki-Taguchi M., Yamada M., Nakamura A., Hanyu K., and Kobayashi S. 1999. Maternal Pumilio acts together with Nanos in germline development in *Drosophila* embryos. *Nat. Cell Biol.* **1:** 431.

Barker D., Wang C., Moore J., Dickinson L., and Lehmann R. 1992. *Pumilio* is essential for function but for distribution of the *Drosophila* abdominal determinant nanos. *Genes Dev.* **6:** 2312.

Barton M.K. and Kimble J. 1990. *fog-1*, a regulatory gene required for specification of spermatogenesis in the germ line of *Caenorhabditis elegans*. *Genetics* **125:** 29.

Barton M.K., Schedl T.B., and Kimble J. 1987. Gain-of-function mutations of *fem-3*, a sex-determination gene in *Caenorhabditis elegans*. *Genetics* **115:** 107.

Bilger A., Fox C.A., Wahle E., and Wickens M. 1994. Nuclear polyadenylation factors recognize cytoplasmic polyadenylation elements. *Genes Dev.* **8:** 1106.

Chagnovich D. and Lehmann R. 2001. Poly(A)-independent regulation of maternal hunchback translation in the *Drosophila* embryo. *Proc. Natl. Acad. Sci.* **98:** 11359.

Curtis D., Apfeld J., and Lehmann R. 1995a. *nanos* is an evolutionarily conserved organizer of anterior-posterior polarity. *Development* **121:** 1899.

Curtis D., Lehmann R., and Zamore P.D. 1995b. Translational regulation in development. *Cell* **81:** 171.

Curtis D., Treiber D.K., Tao F., Zamore P.D., Williamson J.R., and Lehmann R. 1997. A CCHC metal-binding domain in Nanos is essential for translational regulation. *EMBO J.* **16:** 834.

Deshpande G., Calhoun G., Yanowitz J., and Schedl P. 1999. Novel functions of *nanos* in downregulating mitosis and transcription during the development of the *Drosophila* germline. *Cell* **99:** 271.

Dickson K.S., Bilger A., Ballantyne S., and Wickens M.P. 1999. The cleavage and polyadenylation specificity factor in *Xenopus laevis* oocytes is a cytoplasmic factor invloved in regulated polyadenylation. *Mol. Cell. Biol.* **19:** 5707.

Duret L. and Bucher P. 1997. Searching for regulatory elements in human noncoding sequences. *Curr. Opin. Struct. Biol.* **7:** 399.

Duret L., Dorkeld F., and Gautier C. 1993. Strong conservation of non-coding sequences during vertebrate evolution: Potential involvement in post-transcriptional regulation of gene expression. *Nucleic Acids Res.* **21:** 2315.

Edwards T.A., Pyle S.E., Wharton R.P., and Aggarwal A.K. 2001. Structure of pumilio reveals similarity between RNA and peptide binding motifs. *Cell* **105:** 281.

Forbes A. and Lehmann R. 1998. Nanos and Pumilio have critical roles in the development and function of *Drosophila* germline stem cells. *Development* **125:** 679.

Fox C.A. and Wickens M. 1990. Poly(A) removal during oocyte maturation: A default reaction selectively prevented by specific sequences in the 3′ UTR of certain maternal mRNAs. *Genes Dev.* **4:** 2287.

Frank D.J. and Roth M.B. 1998. *ncl-1* is required for the regulation of cell size and ribosomal RNA synthesis in *Caenorhabditis elegans*. *J. Cell Biol.* **140:** 1321.

Geballe A.P. and Sachs M.S. 2000. Translational control by upstream open reading frames. In *Translational control of gene expression* (ed. N. Sonenberg et al.), p. 595. Cold Spring Harbor Laboratory Press, Cold Spring Harbor, New York.

Gunderson S.I., Beyer K., Martin G., Keller W., Boelens W.C., and Mattaj I.W. 1994. The human U1A snRNP protein regulates polyadenylation via a direct interaction with poly(A) polymerase. *Cell* **76:** 531.

Hake L.E., Mendez R., and Richter J.D. 1998. Specificity of RNA binding by CPEB: Requirement for RNA recognition motifs and a novel zinc finger. *Mol. Cell. Biol.* **18:** 685.

Hilleren P. and Parker R. 1999. mRNA surveillance in eukaryotes: Kinetic proofreading of proper translation termination as assessed by mRNP domain organization. *RNA* **5:** 711.

Hinnebusch AG. 2000. Mechanism and regulation of initiator methionyl-tRNA binding to ribosomes. In *Translational control of gene expression* (N. Sonenberg et al.), p. 185. Cold Spring Harbor Laboratory Press, Cold Spring Harbor, New York.

Hodgkin J. 1986. Sex determination in the nematode *C. elegans*: Analysis of tra-3 suppressors and characterization of fem genes. *Genetics* **114:** 15.

Jin S.W., Kimble J., and Ellis R.E. 2001. Regulation of cell fate in *Caenorhabditis elegans* by a novel cytoplasmic polyadenylation element binding protein. *Dev. Biol.* **229:** 537.

Kennedy B.K., Gotta M., Sinclair D.A., Mills K., McNabb D.S., Murthy M., Pak S.M., Laroche T., Gasser S.M., and Guarente L. 1997. Redistribution of silencing proteins from telomeres to the nucleolus is associated with extension of life span in *S. cerevisiae*. *Cell* **89:** 381.

Kimble J.E. and White J.G. 1981. On the control of germ cell development in *Caenorhabditis elegans*. *Dev. Biol.* **81:** 208.

Kraemer B., Crittenden S., Gallegos M., Moulder G., Barstead R., Kimble J., and Wickens M. 1999. NANOS-3 and FBF proteins physically interact to control the sperm/oocyte switch in *C. elegans*. *Curr. Biol.* **9:** 1009.

Lin H. and Spradling A. 1997. A novel group of *pumilio* mutations affects the asymmetric division of germline stem cells in the *Drosophila* ovary. *Development* **124:** 2463.

Luitjens C., Gallegos M., Kraemer B., Kimble J., and Wickens M. 2000. CPEB proteins control two key steps in spermatogenesis in *C. elegans*. *Genes Dev.* **14:** 2596.

Mendez R., Murthy K.R., Ryan K., Manley J.L., and Richter J.D. 2000. Phosphorylation of CPEB by Eg2 mediates the recruitment of CPSF into an active cytoplasmic polyadenylation complex. *Mol. Cell* **6:** 1253.

Murata Y. and Wharton R. 1995. Binding of pumilio to maternal hunchback mRNA is required for posterior patterning in *Drosophila* embryos. *Cell* **80:** 747.

Nagy E. and Maquat L. 1998. A rule for termination codon position within intron-containing genes: When nonsense affects RNA abundance. *Trends Biochem. Sci.* **23:** 198.

Nakahata S., Katsu Y., Mita K., Inoue K., Nagahama Y., and Yamashita M. 2001. Biochemical identification of *Xenopus* pumilio as a sequence-specific cyclin B1 mRNA-binding protein that physically interacts with a Nanos homolog, Xcat-2, and a cytoplasmic polyadenylation element-binding protein. *J. Biol. Chem.* **276:** 20945.

Olivas W. and Parker R. 2000. The Puf3 protein is a transcript-specific regulator of mRNA degradation in yeast. *EMBO J.* **19:** 6602.

Parisi M. and Lin H. 2000. Translational repression: A duet of Nanos and Pumilio. *Curr. Biol.* **10:** R81.

Pesole G., Liuni S., Grillo G., Licciulli F., Larizza A., Makalowski W., and Saccone C. 2000. UTRdb and UTRsite: Specialized databases of sequences and functional elements of 5´ and 3´ untranslated regions of eukaryotic mRNAs. *Nucleic Acids Res.* **28:** 193.

Richter J.D. 2000. Influence of polyadenylation-induced translation on metazoan development and neuronal synaptic function. In *Translational control of gene expression* (ed. N. Sonenberg et al.), p. 785. Cold Spring Harbor Laboratory Press, Cold Spring Harbor, New York.

Rouault T.A. and Harford J.B. 2000. Translational control of ferritin synthesis. In *Translational control of gene expression* (ed. J.W.B. Hershey, M.B. Mathews, and N. Sonenberg), pp. 655. Cold Spring Harbor Laboratory Press, Cold Spring Harbor, New York.

Sachs A.B., Sarnow P., and Hentze M.W. 1997. Starting at the beginning, middle, and end: Translation initiation in eukaryotes. *Cell* **89:** 831.

Salles F.J., Lieberfarb M.E., Wreden C., Gergen J.P., and Strickland S. 1994. Coordinate initiation of *Drosophila* development by regulated polyadenylation of maternal messenger RNAs. *Science* **266:** 1996.

Schedl T. 1997. Developmental genetics of the germline. In *C. elegans II* (ed. D.L. Riddle et al.), p. 241. Cold Spring Harbor Laboratory Press, Cold Spring Harbor, New York.

Slack F.J., Basson M., Liu Z., Ambros V., Horvitz H.R., and Ruvkun G. 2000. The *lin-41* RBCC gene acts in the *C. elegans* heterochronic pathway between the *let-7* regulatory RNA and the *lin-29* transcription factor. *Mol. Cell* **5:** 659.

Sonoda J. and Wharton R.P. 1999. Recruitment of Nanos to hunchback mRNA by Pumilio. *Genes Dev.* **13:** 2704.

———. 2001. *Drosophila* brain tumor is a translational repressor. *Genes Dev.* **15:** 762.

Souza G.M., da Silva A.M., and Kuspa A. 1999. Starvation promotes *Dictyostelium* development by relieving PufA inhibition of PKA translation through the YakA kinase pathway. *Development* **126:** 3263.

Spicher A., Guicherit O.M., Duret L., Aslanian A., Sanjines E.M., Denko N.C., Giaccia A.J., and Blau H.M. 1998. Highly conserved RNA sequences that are sensors of environmental stress. *Mol. Cell. Biol.* **18:** 7371.

Sun X. and Maquat L.E. 2000. mRNA surveillance in mammalian cells: The relationship between introns and translation termination. *RNA* **6:** 1.

Tadauchi T., Matsumoto K., Herskowitz I., and Irie K. 2001. Post-transcriptional regulation through the HO 3´-UTR by Mpt5, a yeast homolog of Pumilio and FBF. *EMBO J.* **20:** 552.

Wang X., Zamore P.D., and Hall T.M. 2001. Crystal structure of a pumilio homology domain. *Mol. Cell* **7:** 855.

Wharton R.P., Sonoda J., Lee T., Patterson M., and Murata Y. 1998. The Pumilio RNA-binding domain is also a translational regulator. *Mol. Cell* **1:** 863.

Wickens M. 1992. Introduction: RNA and the early embryo. *Semin. Dev. Biol.* **3:** 363.

Wickens M., Bernstein D., Kimble J., and Parker R. 2002. A PUF family portrait: 3´UTR regulation as a way of life. *Trends Genet.* (in press).

Wickens M., Goodwin E.B., Kimble J., Strickland S., and Hentze M. 2000. Translational control of developmental decisions. In *Translational control of gene expression* (ed. N. Sonenberg et al.), p. 295. Cold Spring Harbor Laboratory Press, Cold Spring Harbor, New York.

Zamore P.D., Williamson J.R., and Lehmann R. 1997. The Pumilio protein binds RNA through a conserved domain that defines a new class of RNA-binding proteins. *RNA* **3:** 1421.

Zhang B., Gallegos M., Puoti A., Durkin E., Fields S., Kimble J., and Wickens M. 1997. A conserved RNA binding protein that regulates patterning of sexual fates in the *C. elegans* hermaphrodite germ line. *Nature* **390:** 477.

Translational Control of Embryonic Cell Division by CPEB and Maskin

I. GROISMAN, Y.-S. HUANG, R. MENDEZ, Q. CAO, AND J.D. RICHTER
Program in Molecular Medicine, University of Massachusetts Medical School,
Worcester, Massachusetts 01605

Translational control of specific messenger RNAs mediates such basic processes as cell division, cell differentiation, and cell polarity. Although studies over the past dozen years have illustrated multiple ways in which translation is regulated, there are surprisingly few examples where the mechanism of specific mRNA translation is known in detail (Wickens et al. 2000; Macdonald 2001; Mendez and Richter 2001). One of these mechanisms occurs in early development, and although there are likely to be some major differences among various species, the overall framework appears to be similar in most metazoans. Translational control by cytoplasmic polyadenylation has been observed in the oocytes and/or embryos of mammals, amphibians, insects, and mollusks (McGrew et al. 1989; Vassalli et al. 1989; Salles et al. 1994; Minshall et al. 1999). Polyadenylation-induced translation has also been detected in the mammalian brain, where it may contribute to synaptic plasticity, a phenomenon of neurons that may underlie long-term memory storage (Wu et al. 1998). Finally, recent evidence indicates that cytoplasmic polyadenylation also occurs on spindles and centrosomes, where it controls the local translation of a specific mRNA that modulates cell division (Groisman et al. 2000). Because most of the essential features of cytoplasmic polyadenylation have been defined primarily using the oocytes of *Xenopus laevis*, most of our comments are confined to a description of this process in this species.

Early development in *Xenopus* (and many other animals) requires the temporally and spatially restricted translation of mRNAs inherited by the egg at the time of fertilization. As a group, these transcripts are referred to as maternal mRNA, and although they are often dormant, or masked, in oocytes, they become activated (are unmasked) at certain subsequent developmental times. The masking and unmasking processes are intimately linked and control such important developmental phenomena as germ-layer formation, establishment of the body plan, and programming of the germ-cell progenitors (Richter 2000; Wickens et al. 2000; Richter and Theurkauf 2001). Oocytes are arrested in the diplotene stage at the end of prophase I, in a G_2-like stage of the cell cycle, and can be fertilized only after they re-enter the meiotic divisions (oocyte maturation) and arrest a second time at metaphase II. Following fertilization, the embryonic cell divisions occur very quickly and are composed only of S and M phases. When the embryo consists of 4000 cells (at the mid-blastula transition or MBT), the cell divisions slow and become "somatic like" in that they comprise the normal G_1-S-G_2-M phases. Because there is no transcription from the time of oocyte maturation until the MBT, this phase of early development is directed by maternally inherited mRNAs and proteins (Fig. 1).

Progesterone induces oocyte maturation and acts through a receptor that stimulates cytoplasmic, but not nuclear, signaling events (Bayaa et al. 2000; Tian et al. 2000). One key molecule whose activity is stimulated soon after oocytes are exposed to progesterone is Mos, a serine/threonine kinase that activates the Map kinase cascade (Sagata 1997; Ferrell 1999). This cascade ultimately leads to the activation of M-phase promoting factor (MPF), a heterodimer of cyclin B and the kinase cdc2 (also known as cyclin-dependent kinase 1, cdk1), which phosphorylates a number of substrates that induce many manifestations of oocyte maturation (Mendez and Richter 2001). Although Mos is essential for maturation, oocytes contain none of this protein, but instead they contain dormant *mos* mRNA. Masked *mos* mRNA is appended with an unusually short poly(A) tail, about 20 nucleotides or fewer; in response to progesterone, the poly(A) tail elongates up to about 100 nucleotides, and translation ensues (Sheets et al. 1995; Stebbins-Boaz et al. 1996). In addition to *mos*, oocytes have several mRNAs whose translation is controlled by cytoplasmic polyadenylation, including cyclins A2, B1 and B2, wee1, and cdk2 (Sheets et al. 1994, 1995; Stebbins-Boaz and Richter 1994; de Moor and Richter 1997; Nakajo et al. 2000).

CYTOPLASMIC POLYADENYLATION

Cytoplasmic polyadenylation is controlled by two *cis*-acting elements in the 3´ untranslated region (UTR) of responding mRNAs: the hexanucleotide AAUAAA, which is also necessary for nuclear pre-mRNA cleavage and polyadenylation, and the cytoplasmic polyadenylation element (CPE), whose structure varies somewhat, but the consensus is UUUUUAU (Richter 2000). CPEB, a conserved protein containing a zinc finger and two RNA recognition motifs (RRMs) interacts with the CPE both before and after polyadenylation (Hake and Richter 1994; Hake et al. 1998). The initiation of polyadenylation occurs when CPEB is phosphorylated on Ser-174 by the ki-

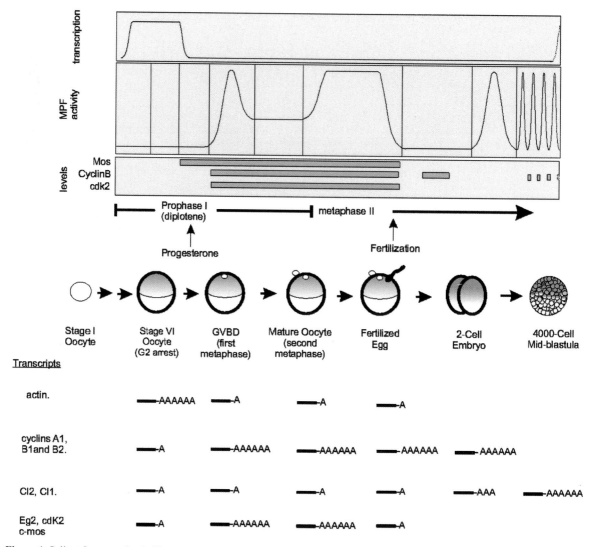

Figure 1. Salient features of early *Xenopus* development. The top panel shows that transcription is ongoing during prophase I of meiosis but ceases when progesterone stimulates oocyte maturation. M-phase promoting factor (MPF) activity, which is tantamount to cdc2 kinase activity, peaks twice during meiosis, at metaphase I and II. MPF also peaks at metaphase in the cleaving embryo. The steady-state levels of Mos, cyclin B, and cdk2, proteins that are involved in cell division, are indicated. The middle panel depicts various stages of oogenesis and embryogenesis. The white spot on the oocytes at GVBD (germinal vesicle, or nuclear envelope, breakdown) is indicative of metaphase I. The small white circle on the outside of the oocytes at second meiosis represents a polar body. Finally, the lower panel shows representative RNAs that gain or lose poly(A) tails during development. For details, see Mendez and Richter (2001).

nase Eg2 (Mendez et al. 2000b), which itself is activated soon after progesterone is applied to the oocytes (Andresson and Ruderman 1998). This phosphorylation event stimulates CPEB to bind and recruit cleavage and polyadenylation specificity factor (CPSF) into an active polyadenylation complex, presumably by stabilizing it on the AAUAAA (Mendez et al. 2000a). By analogy with nuclear pre-mRNA polyadenylation, poly(A) polymerase (PAP) is attracted to the end of the mRNA by CPSF. Oocytes, like somatic cells, contain multiple forms of PAP, and it is not yet clear which isoform of this enzyme is used in cytoplasmic polyadenylation. For example, one PAP (the "long form"), which is both cytoplasmic and nuclear, is phosphorylated, probably by cdc2, as oocytes

enter M phase (Ballantyne et al. 1995); data from mammalian somatic cells have indicated that such phosphorylation events lead to the inactivation of the enzyme (Colgan et al. 1996, 1998; Colgan and Manley 1997). Because cytoplasmic polyadenylation is observed when oocytes enter M phase, it seems plausible that this PAP is not the enzyme that is used for this process. On the other hand, a shorter PAP which lacks the carboxy-terminal portion that contains both the nuclear localization signal and the major cdc2 recognition sites presumably would remain active during M phase, and thus may be the protein that catalyzes cytoplasmic polyadenylation (Gebauer and Richter 1995). Of course, it may be that another PAP altogether is responsible for cytoplasmic polyadenylation.

TRANSLATIONAL CONTROL BY CYTOPLASMIC POLYADENYLATION

The CPE is the only obvious feature that is uniquely common among mRNAs undergoing cytoplasmic polyadenylation. Because these mRNAs are mostly translationally quiescent in immature oocytes, it seems possible that the CPE mediates translational repression in addition to polyadenylation. This possibility is suggested by the fact that endogenous CPE-containing mRNA is unmasked by the injection of multiple copies of the CPE (de Moor and Richter 1999). These results, plus those that demonstrate that reporter RNAs containing a CPE in their 3′UTRs are masked when injected into oocytes, strongly suggest that this is the case (Stutz et al. 1998; de Moor and Richter 1999; Barkoff et al. 2000; Tay et al. 2000). By extension, these data indicate that CPEB is a translation inhibitory factor as well as a polyadenylation-inducing factor. However, a CPEB interacting protein seems to hold the key as to how mRNA translation is regulated. This factor, called maskin, interacts simultaneously with both CPEB and eIF4E (Stebbins-Boaz et al. 1999). The interaction between maskin and eIF4E is mediated by an eIF4E-binding motif that is present (in a closely related form) in all metazoan eIF4Gs as well as other eIF4E-binding proteins (eIF4EBPs). Recall that eIF4G stimulates translation by indirectly positioning the 40S ribosomal subunit on the 5′ end of the mRNA, and accomplishes this task by binding to eIF4E, which in turn binds the cap structure on the 5′ end of mRNAs. The eIF4EBPs are translational inhibitors because they compete with eIF4G for binding to eIF4E (Gingras et al. 1999). Because maskin contains an eIF4E-binding re-

gion, it also competes with eIF4G for binding to eIF4E (Stebbins-Boaz et al. 1999). Thus, when maskin and eIF4E interact, there is no translation because eIF4G is not present to assemble the 40S ribosomal-subunit-containing initiation complex.

The hypothesis that maskin is an inhibitor of translation would indicate that mRNA unmasking must involve the dissociation of maskin from eIF4E. This dissociation does indeed take place, and it occurs at a time that is coincident with cytoplasmic polyadenylation (Stebbins-Boaz et al. 1999). Although these two events may merely be coincidental, it is plausible that polyadenylation induces translation by instigating the dissociation of maskin from eIF4E (Fig. 2). How this could occur might be related to the observations of Tarun and Sachs (1996), who used yeast cells to demonstrate that the 5′ cap and the poly(A) tail act synergistically to stimulate translation. This synergism may be due to a stabilization of the eIF4E–eIF4G complex by poly(A)-binding protein (PABP), which interacts with eIF4G (Tarun and Sachs 1996, 1997; Imataka et al. 1998; Kessler and Sachs 1998). When oocytes mature, the newly elongated mRNA poly(A) tails bind PABP, which assists eIF4G in out-competing maskin for eIF4E, and results in translation initiation. One conundrum with this hypothesis, however, is the observation that classic PABP is very scarce in oocytes and early embryos (Zelus et al. 1989). On the other hand, recent evidence suggests that oocytes contain an unusual form of PABP that, most importantly, contains a putative eIF4G-binding site (Voeltz et al. 2001). Finally, posttranslational modifications of CPEB (Paris et al. 1991; Hake et al. 1998; Mendez et al. 2000b), eIF4G (Morley and Pain 1995), eIF4E (Morley and Pain

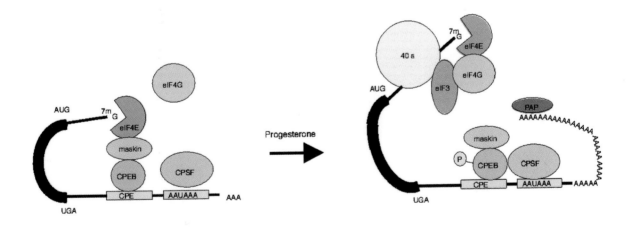

Figure 2. Model for translational masking and unmasking during development. In oocytes, CPE-containing mRNAs are dormant and are bound by CPEB, which in turn is bound by maskin, which in turn is bound by the cap (^{7m}G)-binding protein eIF4E. Cleavage and polyadenylation specificity factor (CPSF) and eIF4G are at best only loosely associated with this mRNA complex. Following the exposure of oocytes to progesterone, CPEB is phosphorylated on Ser-174 by the kinase Eg2, a member of the Aurora family of serine/threonine kinases. Phospho-CPEB then binds to and recruits CPSF to the AAUAAA, which in turn recruits poly(A) polymerase (PAP) to the end of the mRNA. Coincident with polyadenylation is the dissociation of maskin from eIF4E. This dissociation allows eIF4G to bind eIF4E, which through eIF3 positions the 40S ribosomal subunit on the 5′ of the mRNA. For details, see Stebbins-Boaz et al. (1999).

1995), or perhaps maskin could each have an impact on the initiation of translation.

POLYADENYLATION AND THE REGULATION OF CELL DIVISION

Soon after the onset of polyadenylation in maturing oocytes, during the metaphase I to metaphase II transition, as much as 90% of the CPEB is destroyed (Hake and Richter 1994; Reverte et al. 2001; R. Mendez and J.D. Richter, in prep.). The CPEB that remains stable through this period, as well as through fertilization and early embryogenesis, is mainly restricted to blastomeres of the animal pole (i.e., cells of the embryonic blastula that give rise to ectodermal structures such as skin and the nervous system). Most of the maskin that is detected in embryos is also found in animal pole blastomeres (Groisman et al. 2000). Within these cells, both CPEB and maskin are associated with centrosomes and spindles (Fig. 3). This association suggests that CPEB and maskin could interact with microtubules, and in cell extracts they do indeed cofractionate with these structures. At least when tested in vitro with CPEB, this interaction is a direct one, and is mediated by a small PEST (proline, aspartic acid, serine, threonine)-like region that lies within the amino-terminal region of CPEB. The importance of this PEST sequence in vivo was illustrated with chimeric GFP–CPEB proteins; when DNA encoding this fusion protein was transfected into normal rat kidney (NRK) cells, fluorescence was detected at centrosomes. On the other hand, fluorescence resulting from the transfection of a similar construct that lacked the CPEB PEST sequence was observed rather uniformly in cells (Groisman et al. 2000). This

CPEB PEST-less deletion protein, denoted as Δ4, was an important reagent for helping determine local translation on the mitotic apparatus (see below).

When injected into embryos, reagents that are known to disrupt polyadenylation-induced translation (e.g., CPEB antibody, CPEB dominant negative mutant, 3'-deoxyadenosine) inhibit cell division and produce abnormal mitotic structures such as multiple centrosomes, centrosomes detached from spindles, and tripolar spindles (Fig. 4). These results suggest that embryonic cell division requires polyadenylation-induced translation, but they do not indicate where this requirement occurs (i.e., soluble or spindle-associated), or what mRNA(s) might be involved. Four observations pointed to cyclin B1 mRNA as the key molecule involved in local translation: (1) It has a CPE and is regulated by cytoplasmic polyadenylation, at least in maturing oocytes (Stebbins-Boaz et al. 1996). (2) Its translation is necessary for cell division (Murray and Kirschner 1989). (3) It is found on spindles in *Drosophila* embryos (Raff et al. 1990). (4) Cyclin protein is found on spindles in HeLa cells (Hagting et al. 1998). Given these observations, it is perhaps not surprising that both cyclin B1 mRNA and protein were found to be spindle-associated in *Xenopus* embryos (Groisman et al. 2000). These data alone would argue that cell division could require cyclin B1 mRNA translation on spindles. Going further, however, the injection of CPEB Δ4, the mutant protein that lacks the PEST domain and thus is unable to associate with microtubules, has little effect on cyclin B1 mRNA translation, but causes this message to dissociate from spindles. The consequence of this dissociation is the loss of cyclin B1 protein from spindles and, as a result, inhibited cell division (Fig. 5) (Groisman et al. 2000). Therefore, CPEB controls not only cyclin mRNA translation, but its localization to spindles as well.

To begin to decipher the molecular characteristics of CPEB activity during the cell cycle, I. Groisman and J.D. Richter (in prep.) employed a calcium-ionophore-activated cycling egg extract (Murray 1991). These investigators showed that cyclin B1 RNA undergoes polyadenylation during the cell cycle with the peak activity occurring at or near M phase, which corresponds to the zenith of steady-state cyclin levels. Poly(A) removal is also cell-cycle-regulated, but the peak of this activity occurs at S phase, or when cyclin protein is at its nadir. Most importantly, an antibody that neutralizes CPEB activity not only prevents polyadenylation, but also inhibits cyclin protein accumulation and holds the cell cycle in S phase. Thus, these results suggest that CPEB-mediated cytoplasmic polyadenylation controls the embryonic cell cycle by regulating the timing of cyclin B1 mRNA translation.

CONCLUSIONS

The key factor that drives cells into mitosis is M-phase-promoting factor, a heterodimer of cyclin B and the kinase cdc2. Whereas cdc2 must undergo both phosphorylation and dephosphorylation events to be catalytically active, cyclin B is an essential cofactor for this activity, and the modulation of its steady-state levels is critical for cell cycle progression. It is axiomatic that the regulated

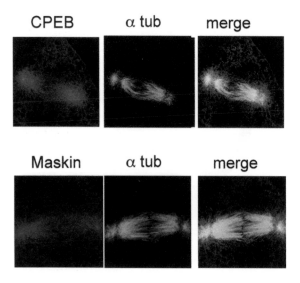

Figure 3. CPEB and maskin on the mitotic apparatus. Blastomeres from *Xenopus* embryos were immunostained for CPEB and maskin (*red*) and costained with antibody for α-tubulin (*green*). The merged images (*yellow*) show colocalization of these two proteins. Confocal images are shown (for details, see Groisman et al. 2000).

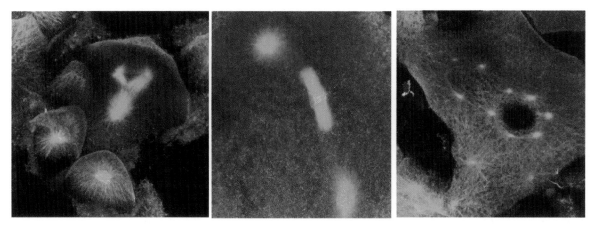

Figure 4. Defects in the mitotic apparatus following disruption of cytoplasmic polyadenylation. *Xenopus* embryos injected with affinity-purified CPEB antibody were cultured for several hours, then fixed and immunostained with antibody against α-tubulin (*green*). Confocal imaging reveals that the inhibition of polyadenylation by CPEB antibody (Stebbins-Boaz et al. 1996) results in multiple centrosomes (*left* panel), centrosomes detached from the spindles (*middle* panel), and tripolar spindles (*right* panel). Toto-3 (orange stain) was used to identify DNA. For details, see Groisman et al. (2000).

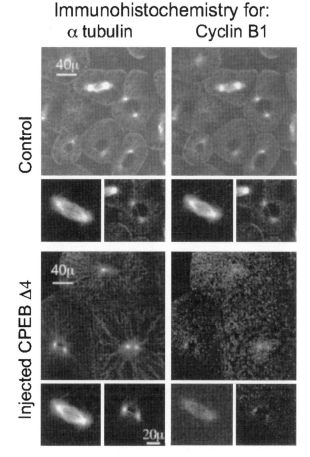

Figure 5. CPEB mediates the accumulation of cyclin B1 protein on spindles. *Xenopus* embryos were injected with purified *E. coli*-expressed CPEB deletion mutant protein Δ4, which lacks an internal PEST region and cannot interact with microtubules. Following several hours of culture, the embryos were fixed and immunostained for α-tubulin and cyclin B1 protein. CPEB Δ4 destroys the anchoring of cyclin B1 mRNA on the mitotic apparatus, which prevents cyclin B1 protein from accumulating on this structure. For details, see text and Groisman et al. (2000).

destruction of cyclin B drives cells into and out of mitosis. That is, as cyclin B begins to accumulate in interphase, it binds to cdc2 and promotes entry into mitosis. In mitosis, when cyclin levels are high, an elaborate protein destruction machine known as the anaphase-promoting complex (APC) is activated. The APC/proteasome then destroys cyclin B, which stimulates the cells to go from M phase into anaphase. When cyclin B protein levels are low, the APC is inactivated, thus allowing cyclin B to again accumulate and stimulate another round of mitosis. Whereas cyclin protein levels oscillate throughout the cell cycle, it has been presumed that cyclin mRNA translation occurs at a more-or-less constant rate. Our studies suggest that at least in the early embryo, cyclin mRNA translation is regulated during the cell cycle by cytoplasmic mRNA polyadenylation. Moreover, our data indicate that local cyclin mRNA translation, on spindles and/or on centrosomes, is important for the integrity of the mitotic apparatus and resulting cell division. Because the early embryonic cell cycle of *Xenopus* is unusual in that it lacks appreciable G_1 and G_2 phases, one might appropriately wonder whether regulated cyclin mRNA translation is important for typical somatic cell cycle progression. One observation suggests that regulated polyadenylation-induced cyclin mRNA translation does not apply to mammals, or at least to the mouse. A CPEB knockout mouse has recently been generated, and it is viable (although sterile) (Tay and Richter 2001). On the other hand, many organisms, including mammals, contain several CPEB-like proteins, and these may stimulate polyadenylation-induced cyclin mRNA translation in somatic cells in a manner similar to the way "classic" CPEB stimulates cyclin mRNA translation in *Xenopus* embryos (Mendez and Richter 2001).

ACKNOWLEDGMENTS

Work in the authors' laboratory was supported by grants from the National Institutes of Health. Y.-S.H. was supported by a fellowship from the Charles A. King Trust, and R.M. was supported by a fellowship from the Leukemia and Lymphoma Society of America.

REFERENCES

Andresson T. and Ruderman J.V. 1998. The kinase Eg2 is a component of the *Xenopus* oocyte progesterone-activated signaling pathway. *EMBO J.* **17:** 5627.

Ballantyne S., Bilger A., Astrom J., Virtanen A., and Wickens M. 1995. Poly(A) polymerases in the nucleus and cytoplasm of frog oocytes: Dynamic changes during oocyte maturation and early development. *RNA* **1:** 64.

Barkoff A.F., Dickson K.S., Gray N.K., and Wickens M. 2000. Translational control of cyclin B1 mRNA during meiotic maturation: Coordinated repression and cytoplasmic polyadenylation. *Dev. Biol.* **220:** 97.

Bayaa M., Booth R.A., Sheng Y., and Liu X.J. 2000. The classical progesterone receptor mediates *Xenopus* oocyte maturation through a nongenomic mechanism. *Proc. Natl. Acad. Sci.* **97:** 12607.

Colgan D.F. and Manley J.L. 1997. Mechanism and regulation of mRNA polyadenylation. *Genes Dev.* **11:** 2755.

Colgan D.F., Murthy K.G., Prives C., and Manley J.L. 1996. Cell-cycle related regulation of poly(A) polymerase by phosphorylation. *Nature* **384:** 282.

Colgan D.F., Murthy K.G., Zhao W., Prives C., and Manley J.L. 1998. Inhibition of poly(A) polymerase requires p34cdc2/cyclin B phosphorylation of multiple consensus and non-consensus sites. *EMBO J.* **17:** 1053.

de Moor C.H. and Richter J.D. 1997. The Mos pathway regulates cytoplasmic polyadenylation in *Xenopus* oocytes. *Mol. Cell. Biol.* **17:** 6419.

———. 1999. Cytoplasmic polyadenylation elements mediate masking and unmasking of cyclin B1 mRNA. *EMBO J.* **18:** 2294.

Ferrell J.E., Jr. 1999. Building a cellular switch: More lessons from a good egg. *Bioessays* **21:** 866.

Gebauer F. and Richter J.D. 1995. Cloning and characterization of a poly(A) polymerase from *Xenopus* oocytes. *Mol. Cell. Biol.* **15:** 1422.

Gingras A.C., Raught B., and Sonenberg N. 1999. eIF4 initiation factors: Effectors of mRNA recruitment to ribosomes and regulators of translation. *Annu. Rev. Biochem.* **68:** 913.

Groisman I., Huang Y.S., Mendez R., Cao Q., Therukauf W., and Richter J.D. 2000. CPEB, maskin, and cyclin B1 mRNA at the mitotic apparatus: Implications for local translational control of cell division. *Cell* **103:** 435.

Hagting A., Karlsson C., Clute P., Jackman M., and Pines J. 1998. MPF localization is controlled by nuclear export. *EMBO J.* **17:** 4127.

Hake L.E. and Richter J.D. 1994. CPEB is a specificity factor that mediates cytoplasmic polyadenylation during *Xenopus* oocyte maturation. *Cell* **79:** 617.

Hake L.E., Mendez R., and Richter J.D. 1998. Specificity of RNA binding by CPEB: Requirement for RNA recognition motifs and a novel zinc finger. *Mol. Cell. Biol.* **18:** 685.

Imataka H., Gradi A., and Sonenberg N. 1998. A newly identified N-terminal amino acid sequence of human eIF4G binds poly(A)-binding protein and functions in poly(A)-dependent translation. *EMBO J.* **17:** 7480.

Kessler S.H. and Sachs A.B. 1998. RNA recognition motif 2 of yeast Pab1p is required for its functional interaction with eukaryotic translation initiation factor 4G. *Mol. Cell. Biol.* **18:** 51.

Macdonald P. 2001. Diversity in translational regulation. *Curr. Opin. Cell Biol.* **13:** 326.

McGrew L.L., Dworkin-Rastl E., Dworkin M.B., and Richter J.D. 1989. Poly(A) elongation during *Xenopus* oocyte maturation is required for translational recruitment and is mediated by a short sequence element. *Genes Dev.* **3:** 803.

Mendez R. and Richter J.D. 2001. CPEB-mediated translation: A means to an end. *Nat. Rev. Mol. Cell Biol.* **2:** 521.

Mendez R., Murthy K.G.K., Manley J.L., and Richter J.D.

2000a. Phosphorylation of CPEB by Eg2 mediates the recruitment of CPSF into an active cytoplasmic polyadenylation complex. *Mol. Cell* **6:** 1253.

Mendez R., Hake L.E., Andresson T., Littlefield L.E., Ruderman J.V., and Richter J.D. 2000b. Phosphorylation of CPE binding factor by Eg2 regulates *c-mos* mRNA translation. *Nature* **404:** 302.

Minshall N., Walker J., Dale M., and Standart N. 1999. Dual roles of p82, the clam CPEB homolog, in cytoplasmic polyadenylation and translational masking. *RNA* **5:** 27.

Morley S.J. and Pain V.M. 1995. Hormone-induced meiotic maturation in *Xenopus* oocytes occurs independently of p70s6k activation and is associated with enhanced initiation factor (eIF)-4F phosphorylation and complex formation. *J. Cell Sci.* **108:** 1751.

Murray A.W. 1991. Cell cycle extracts. *Methods Cell Biol.* **36:** 581.

Murray A.W. and Kirschner M.W. 1989. Cyclin synthesis drives the early embryonic cell cycle. *Nature* **339:** 275.

Nakajo N., Yoshitome S., Iwashita J., Iida M., Uto K., Ueno S., Okamoto K., and Sagata N. 2000. Absence of Wee1 ensures the meiotic cell cycle in *Xenopus* oocytes. *Genes Dev.* **14:** 328.

Paris J., Swenson K., Piwnica-Worms H., and Richter J.D. 1991. Maturation-specific polyadenylation: In vitro activation by p34cdc2 and phosphorylation of a 58-kD CPE-binding protein. *Genes Dev.* **5:** 1697.

Raff J.W., Whitfield W.G., and Glover D.M. 1990. Two distinct mechanisms localise cyclin B transcripts in syncytial *Drosophila* embryos. *Development* **110:** 1249.

Reverte C.G., Ahearn M.D., and Hake L.E. 2001. CPEB degradation during *Xenopus* oocyte maturation requires a PEST domain and the 26S proteasome. *Dev. Biol.* **231:** 447.

Richter J.D. 2000. The influence of polyadenylation-induced translation on metazoan development and neuronal synaptic plasticity. In *Translational control of gene expression* (ed. N. Sonenberg et al.), p. 785. Cold Spring Harbor Laboratory Press, Cold Spring Harbor, New York.

Richter J.D. and Theurkauf W.E. 2001. Development: The message is in the translation. *Science* **293:** 60.

Sagata N. 1997. What does Mos do in oocytes and somatic cells? *Bioessays* **19:** 13.

Salles F.J., Lieberfarb M.E., Wreden C., Gergen J.P., and Strickland S. 1994. Coordinate initiation of *Drosophila* development by regulated polyadenylation of maternal messenger RNAs. *Science* **266:** 1996.

Sheets M.D., Wu M., and Wickens M. 1995. Polyadenylation of c-mos mRNA as a control point in *Xenopus* meiotic maturation. *Nature* **374:** 511.

Sheets M.D., Fox C.A., Hunt T., Vande Woude G., and Wickens M. 1994. The 3'-untranslated regions of c-mos and cyclin mRNAs stimulate translation by regulating cytoplasmic polyadenylation. *Genes Dev.* **8:** 926.

Stebbins-Boaz B. and Richter J.D. 1994. Multiple sequence elements and a maternal mRNA product control cdk2 RNA polyadenylation and translation during early *Xenopus* development. *Mol. Cell. Biol.* **14:** 5870.

Stebbins-Boaz B., Hake L.E., and Richter J.D. 1996. CPEB controls the cytoplasmic polyadenylation of cyclin, cdk2, and c-mos mRNAs and is necessary for oocyte maturation in *Xenopus*. *EMBO J.* **15:** 2582.

Stebbins-Boaz B., Cao Q.P., de Moor C.H., Mendez R., and Richter J.D. 1999. Maskin is a CPEB-associated factor that transiently interacts with eIF-4E. *Mol. Cell* **4:** 1017.

Stutz A., Conne B., Huarte J., Gubler P., Volkel V., Flandin P., and Vassalli J.D. 1998. Masking, unmasking, and regulated polyadenylation cooperate in the translational control of a dormant mRNA in mouse oocytes. *Genes Dev.* **12:** 2535.

Tarun S.Z., Jr. and Sachs A.B. 1996. Association of the yeast poly(A) tail binding protein with translation initiation factor eIF-4G. *EMBO J.* **15:** 7168.

———. 1997. Binding of eukaryotic translation initiation factor 4E (eIF4E) to eIF4G represses translation of uncapped mRNA. *Mol. Cell. Biol.* **17:** 6876.

Tay J. and Richter J.D. 2001. Germ cell differentiation and synaptonemal complex formation are disrupted in CPEB knockout mice. *Dev. Cell* **1:** 201.

Tay J., Hodgman R., and Richter J.D. 2000. Translational control of cyclin B1 mRNA in maturing mouse oocytes. *Dev. Biol.* **221:** 1.

Tian J., Kim S., Heilig E., and Ruderman J.V. 2000. Identification of XPR-1, a progesterone receptor required for *Xenopus* oocyte activation. *Proc. Natl. Acad. Sci.* **97:** 14358.

Vassalli J.D., Huarte J., Belin D., Gubler P., Vassalli A., O'Connell M.L., Parton L.A., Rickles R.J., and Strickland S. 1989. Regulated polyadenylation controls mRNA translation during meiotic maturation of mouse oocytes. *Genes Dev.* **3:** 2163.

Voeltz G.K., Ongkasuwan J., Standart N., and Steitz J.A. 2001. A novel embryonic poly(A) binding protein, ePAB, regulates

mRNA deadenylation in *Xenopus* egg extracts. *Genes Dev.* **15:** 774.

Wickens M., Goodwin E.B., Kimble J., Strickland S., and Hentze M. 2000. Translational control of developmental decisions. In *Translational control of gene expression* (ed. N. Sonenberg et al.). p. 295. Cold Spring Harbor Laboratory Press, Cold Spring Harbor, New York.

Wu L., Welles D., Tay J., Mendis D., Abbot M., Barnitt A., Quinlan E., Heynen A., Fallon J., and Richter J.D. 1998. CPEB-mediated cytoplasmic polyadenylation and the regulation of experience-dependent translation of α-CaMKII mRNA at synapses. *Neuron* **21:** 1129.

Zelus B.D., Giebelhaus D.H., Eib D.W., Kenner K.A., and Moon R.T. 1989. Expression of the poly(A)-binding protein during development of *Xenopus laevis*. *Mol. Cell. Biol.* **9:** 2756.

Small RNA Regulators of Translation: Mechanisms of Action and Approaches for Identifying New Small RNAs

S. GOTTESMAN,* G. STORZ,† C. ROSENOW,‡ N. MAJDALANI,* F. REPOILA,*§
AND K.M. WASSARMAN†¶

*Laboratory of Molecular Biology, National Cancer Institute, Bethesda, Maryland 20892-4264;
†Cell Biology and Metabolism Branch, National Institute of Child Health and Human Development, Bethesda,
Maryland 20892-5430; ‡Affymetrix, Santa Clara, California 95051

In the last few years, it has become increasingly clear that small RNAs (sRNAs) can act as important regulators, transmitting environmental signals to up- or down-regulate the synthesis and activity of proteins. However, sRNA genes are not immediately obvious from genomic sequences. Here we review what has been learned about the activities of the known sRNAs in *Escherichia coli*, with an emphasis on three involved in translational regulation of a major transcriptional regulator, the *rpoS*-encoded sigma factor σ^S. In addition, we discuss the approaches developed to search the *E. coli* genome for additional novel sRNAs, approaches that also can be used to detect mRNAs subject to translational regulation.

sRNAs POSSESS A RANGE OF FUNCTIONS

Table 1 summarizes information about the known sRNAs encoded in the *E. coli* chromosome. A number of these were first identified as relatively abundant discrete sRNA species (4.5S, 6S, tmRNA, RNase P, and Spot 42 RNAs); others were found by their function in screens of multicopy plasmid libraries (DicF, MicF, DsrA, and RprA RNAs) or by their ability to bind a protein of interest (CsrB RNA). Still others were first detected as transcripts during studies of their neighboring genes (OxyS, GcvB, and Crp Tic RNAs). None was identified via mutations, possibly because they represent relatively small mutagenic targets and because sRNA genes frequently have fairly subtle phenotypes when inactivated. For instance, mutations in *ssrS*, which encodes the highly conserved 6S RNA, alter the relative usage of σ^{70}- and σ^S-dependent promoters in stationary phase, but this effect was appreciated only after it was known that 6S RNA binds to RNA polymerase (Wassarman and Storz 2000). Mutations in *dsrA* reduce σ^S expression at low temperature in exponentially growing cells, but have much milder effects at higher temperatures or in stationary-phase cells (Sledjeski et al. 1996; Repoila and Gottesman 2001).

What do sRNAs do? For the set of 13 known sRNAs in Table 1, a variety of functions have been discovered. RNase P RNA acts as a ribozyme in RNA processing. CsrB RNA antagonizes the function of the CsrA repres-

sor, and the 6S RNA modifies RNA polymerase activity. 4.5S RNA affects protein localization in *E. coli*, as its counterpart 7SL does in eukaryotes. tmRNA interacts with the ribosome, directing the addition of an 11-amino-acid tag to stalled polypeptide chains, thus releasing the polypeptide chain from the ribosome and targeting the resulting protein for degradation. Each of these five sRNAs uses or interacts with one or more proteins. The RNase P protein is required for optimal RNase P activity in vivo (Jung et al. 1992). As noted, CsrB RNA and 6S RNA bind specific proteins as targets. The 4.5S RNA interacts with Ffh to form the signal recognition particle (Herskovits et al. 2000), and tmRNA function requires the ribosome-binding protein SmpB (Karzai et al. 1999).

Many of the other sRNAs in Table 1 have been implicated as positive or negative regulators of translation. It is notable that these same sRNAs also bind to the abundant Hfq protein. Thus, for the moment, it seems reasonable to extrapolate that sRNAs binding tightly to Hfq are involved in translational regulation. Below we consider how three of these sRNAs regulate translation of the *rpoS* mRNA.

sRNAs INTEGRATE MULTIPLE ENVIRONMENTAL SIGNALS IN REGULATING *rpoS* mRNA TRANSLATION

The DsrA, RprA, and OxyS RNAs all regulate the translation of a common target, the *rpoS* mRNA. The *rpoS* gene encodes σ^S, a sigma factor found in *E. coli* and other gram-negative bacteria. It closely resembles the major vegetative sigma factor, σ^{70}, in both sequence and promoter site selection; the elements that distinguish a σ^{70}-dependent promoter from a σ^S-dependent promoter are still being defined (Becker and Hengge-Aronis 2001). σ^S was first recognized as important for the expression of a large number of genes when cells enter stationary phase. It is also a central regulator after a variety of stress treatments. In addition, σ^S has been implicated in virulence in a number of organisms (for review, see Hengge-Aronis 2000). Changes in the σ^S levels in the cell are primarily responsible for changes in the expression of σ^S-dependent promoters; σ^S amounts are very low during exponential growth under optimum growth conditions, but rapidly increase as cells enter stationary phase or after stress treatments. The change in σ^S levels reflects changes in the rate of degradation of the protein, some

---

Present addresses: ¶Department of Bacteriology, University of Wisconsin-Madison, Wisconsin 53706; §UMR960 INRA-ENVT, Toulouse, France.

Table 1. Known sRNAs

Gene RNA	Ig Start	RNA size	First identification	Function	Known protein binding[a]	Conservation[b] (%)	Microarray detection[c]
ffs 4.5S	475596	114	abundant RNA	protein localization	Ffh	99	++++
ssrA tmRNA	2753397	363	abundant RNA	truncated protein tagging for degradation	SmpB	97	++++
ssrS 6S	3053959	184	abundant RNA	RNA polymerase regulation	RNA polymerase	98	++++
rnpB RNase P	3267466	377	abundant RNA	RNase P component	RnpA	99	++++
spf Spot 42	4047330	109	abundant RNA	unknown	Hfq	98	++
dicF DicF	1647063	53	multicopy phenotype	anti-*ftsZ* translation		98	++
rprA RprA	1768208	105	multicopy phenotype	stimulate *rpoS* translation	Hfq	97	–
dsrA DsrA	2023233	85	multicopy phenotype	stimulate *rpoS* translation, anti-*hns* translation	Hfq	79	–
micF MicF	2310770	93	multicopy phenotype	anti-*ompF* translation		86	++
csrB CsrB	2922136	360	protein copurification	anti-CsrA	CsrA	84	++++
oxyS OxyS	4155800	109	in vivo transcription	anti-*fhlA* translation, anti-Hfq activation of *rpoS*	Hfq	76	–
gcvB GcvB	2940590	205	in vitro transcription	regulation of *oppA, dppA*		95	++++
crpT Crp Tic	3483456	300	in vitro transcription	regulation of *crp*		76[d]	+++

The references for RprA RNA, GcvB RNA, CsrB RNA, 6S RNA, and Crp Tic RNA are Majdalani et al. (2001), Urbanowski et al. (2000), Romeo (1998), Wassarman and Storz (2000), and Okamoto and Freundlich (1986), respectively. Other information was taken from Wassarman et al. (1999).

[a]Based on immunoprecipitation with Hfq (Wassarman et al. 2001 and unpubl.) or other published information (for reviews, see Karzai et al. 1999; Wassarman et al. 1999; Wassarman and Storz 2000).

[b]Percentage of identical residues in an alignment of the *E. coli* sRNA with the best conserved *Salmonella* species. *dicF* is only homologous to *S. typhi* and is not found in other *Salmonella* or in *Klebsiella*.

[c]Detection on high-density oligonucleotide probe arrays carried out using RNA isolated from MG1655 cells grown to $OD_{600} = 0.8$ at 37°C in LB medium: (–) no signal, (+) low signal, (++) intermediate signal, (+++) high signal, (++++) very high signal.

[d]Small RNA transcript overlaps ORF. The region necessary for activity, upstream of the ORF, is >85% conserved.

changes in transcription, and a major change in the level of translation.

Translational regulation of the *rpoS* mRNA is mediated by a long upstream leader sequence, initiating at promoters within or in front of the upstream *nlpD* gene (Hengge-Aronis 2000). Translation is dependent on the RNA-binding protein Hfq. Mutations in the leader that render *rpoS* translation independent of Hfq affect a secondary structure that would be predicted to occlude the ribosome-binding site (Fig. 1A) (Brown and Elliot 1997). The central elements of this secondary structure are well conserved. Presumably, the stresses that increase translation somehow lead to opening of this structure with the aid of Hfq, thereby allowing ribosome binding and translation initiation.

What was unexpected in this scheme was the critical role played by *trans*-acting sRNAs. At least two sRNAs stimulate *rpoS* mRNA translation. The first identified of these, DsrA RNA, was demonstrated to pair with the up-stream region of the *rpoS* leader (Fig. 1B) (Majdalani et al. 1998). RprA RNA, a second stimulatory sRNA described recently (Majdalani et al. 2001), bears short regions of sequence similarity to DsrA RNA and also acts by pairing to the upstream region (Fig. 1C) (N. Majdalani et al., in prep.). Both DsrA and RprA RNAs also bind to Hfq, and presumably Hfq helps to stimulate the interaction of the sRNAs with the *rpoS* mRNA (Sledjeski et al. 2001; Wassarman et al. 2001).

Osmotic shock induction of RpoS is dependent on both DsrA and RprA RNAs, demonstrating their important role in allowing translation after stress (Majdalani et al. 2001). Possibly osmotic shock increases the sensitivity of the inhibitory *rpoS* mRNA structure to basal levels of the sRNAs. In addition to allowing translation after osmotic shock, these two sRNAs increase RpoS translation when they are overproduced. Thus, conditions that increase their synthesis induce *rpoS* mRNA translation. DsrA RNA synthesis is increased at low temperature, resulting

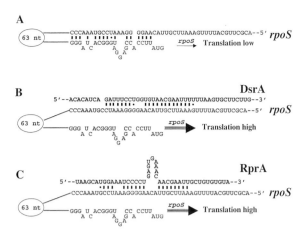

Figure 1. (*A*) Current model for secondary structure of the *rpoS* mRNA leader (Brown and Elliott 1997). Sequences on the top strand pair with the sequence just upstream of the AUG initiation codon for RpoS. (*B*) Model for pairing of DsrA RNA to the top strand of the *rpoS* mRNA leader (Majdalani et al. 1998). (*C*) Model for pairing of RprA RNA to the top strand of the *rpoS* mRNA leader (N. Majdalani et al., in prep.).

in elevated σ^S levels during exponential growth at low temperature (Repoila and Gottesman 2001). The expression of the RprA RNA is not regulated by temperature, but is under the control of a two-component regulatory system that responds to cell-surface perturbations. When this system is activated, RpoS translation is again increased (N. Majdalani et al., in prep.). Therefore, the availability of these two sRNAs, each capable of relieving the translational repression of the *rpoS* mRNA, allows a variety of environmental signals to result in activation of the σ^S-dependent transcription program.

In addition to the two positive sRNA regulators, one sRNA that represses *rpoS* mRNA translation, OxyS RNA, has been identified (Altuvia et al. 1997). This sRNA is induced in response to treatment with hydrogen peroxide and also binds the Hfq protein (Zhang et al. 1998). It is not yet clear how the OxyS RNA decreases σ^S expression. Conceivably, the OxyS RNA base-pairs with the *rpoS* mRNA in a manner similar to the DsrA and RprA RNAs, in this case preventing rather than facilitating the opening of the inhibitory secondary structure. Alternatively, given that OxyS RNA levels are quite high after oxidative stress, the OxyS RNA may shut down *rpoS* mRNA translation by effectively competing for Hfq binding with other sRNA regulators such as DsrA and RprA.

We have more information about the sRNAs regulating *rpoS* mRNA translation than we do for any other target. σ^S may be an especially important cellular target, and therefore subject to particularly complex regulation. However, we expect that many other targets of sRNAs will be discovered. As was found for σ^S, the participation of one or more sRNAs may not be obvious from the nature of the *cis*-acting regulatory signals. Thus, we suggest that sRNA participation in translational regulation should be considered seriously in all cases.

An aspect of sRNA activity that has not been fully explored is the ability of these molecules to act on multiple

targets. When overproduced, DsrA RNA is known to negatively regulate a global transcriptional silencer, H-NS, and has been postulated to regulate a number of other genes as well (Sledjeski and Gottesman 1995; Lease et al. 1998). OxyS RNA has been shown to regulate *fhlA* translation in addition to *rpoS* (Altuvia et al. 1998; Argaman and Altuvia 2000). For both of these examples, different regions of the sRNA probably are required to regulate the different target genes. Additional targets for RprA RNA have not yet been defined, but the experience with the similarly sized DsrA and OxyS RNAs suggests that such targets may exist. The physiological importance of co-regulation of multiple targets or the hierarchy with which they are regulated remains to be investigated (Fig. 2).

STRATEGIES FOR HUNTING FOR sRNAs

The set of 13 sRNAs listed in Table 1 provides a broad set of functions; however, there is no reason to believe that this list is all-inclusive. Given that many of these sRNAs were discovered serendipitously, it seems likely that numerous other sRNAs with equally interesting functions remain to be discovered. Thus, an important question is, How many sRNAs are encoded in a sequenced and well-studied organism like *E. coli*? We and Argaman et al. have recently completed a genome-wide search for sRNAs in *E. coli* (Argaman et al. 2001; Wassarman et al. 2001). The combined results of these searches have been the identification of 22 new sRNAs (Table 2). Several other searches for additional sRNAs are under way (Carter et al. 2001; Rivas et al. 2001; S. Chen et al., pers. comm.). Coupled with those sRNAs already known, the number of sRNAs encoded by the *E. coli* chromosome is probably between 40 and 100. From this expanded family of sRNAs, we can begin to draw some general conclusions about the best strategy for identifying such molecules. Possible approaches and their success in identifying sRNAs are described below. Understanding the functions of the new sRNAs will be a greater challenge but is discussed briefly.

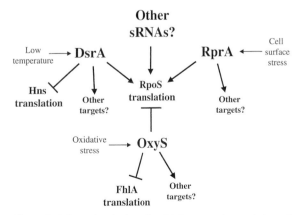

Figure 2. Schematic of roles of multiple sRNAs in regulation of *rpoS* mRNA translation. Arrows indicate positive control; bars indicate negative control.

Table 2. New sRNAs

Gene RNA	Ig start	RNA size	Expression[a]	Hfq binding[b]	Conservation[c] (%)	Microarray detection[d]
rydB	1762411	60	M >> S > E	NT		++++
ryeE	2165049	86	E, S >M	+ (E)	89	+++
ryfA/PAIR3	2651357	320	E, M	NT		+++
ryhA/sraH	3348110	45/120	S >> M > E	+ (S)	95	+++
sraA	457922	120	M,S>E	NT	84	–
sraB	1145858	149–168	S>E	NT	78	++++
sraD	2812755	70	S>>M>E	NT	91	++
sraF	3235936	189	CS	NT	91	+++
sraG	3308879	146–174	HS, CS, E, M > S	NT	92	–
ryhB/sraI	3578437	90	M >> S	+ (M)	79	–
ryiA/sraJ	3983621	210	E > M, S	+ (E)	89	++
ryjA/sraL	4275510	140	S >> M	– (S)	92	+++
rybB	887180	80	S >> M	+ (S)	99	–
ryiB/sraK/csrC	4048313	270	M >S >> E	– (M)	87	+++
rybA	852161	205	S > M > E	– (S)	85	+++
rygA/sraE/PAIR2	2974037	89	S >> M, E	+ (S)	(87)	++
rygB/PAIR2	2974037	83	S, E > M	+ (S)	(87)	++
ryeA/sraC	1920997	275	M > E > S	–/+ (M)	(98)	++
ryeB	1920997	100	S >> M	+ (S)	(96)	+++
ryeC/QUAD1a	2151151	143 / 107	S > M > S / M > E, S	NT		+++
RyeD/QUAD1b	2151151	137 / 102	M > E > S / M > E	NT		+++
rygC/QUAD1c	3054807	139 / 107	S >> M > E / S, M > E	NT		+++

ryx names are from Wassarman et al. (2001). *sra* names are from Argaman et al. (2001). QUAD and PAIR names are from Rudd (1999). CsrC is from T. Romeo, unpublished. Methods of determining sizes are found in those references.

[a]Relative expression in three growth conditions: (E) LB medium, exponential phase; (M) minimal medium, exponential phase; (S) LB medium, stationary phase. (CS) Cold shock; (HS) heat shock.

[b]RNA coimmunoprecipitation with Hfq as detected by northern analysis: (+) strong binding (>30% of RNA bound); (+/–) weak binding (5–10%); (–/+) minimal binding (<5%); (–) no detectable binding. E, M, S refer to cell growth conditions as in *a*. (NT) Not tested.

[c]Conservation as in Table 1 for sRNAs whose promoters and terminators can be predicted or whose 5′ and 3′ ends have been mapped precisely. The 5′ and 3′ ends of the sRNAs are approximate for the numbers in parentheses.

[d]Detection on high-density oligonucleotide probe arrays as in Table 1.

Strategies for detecting sRNAs in a genome-wide manner fall into two general classes, those that depend primarily on sequence characteristics (pattern of conservation between species, presence of promoter and terminator sequences, secondary structure) and those that depend primarily on direct detection (microarrays, isolation and cloning of sRNAs based on size, precipitation with RNA-binding proteins). The approaches discussed below focus on the sequences outside of those encoding ORFs. sRNAs encoded on the nonsense strand of genes or cleaved from within mRNAs will elude most currently available genome-wide search methods. It should be noted that, in addition to detecting the sRNAs, these approaches will identify short mRNAs encoding small peptides. In many cases, the patterns of conservation between species for mRNAs will be characteristic of that for protein coding sequences, distinguishing them from sRNAs.

DETECTING sRNA GENES USING SEQUENCE CHARACTERISTICS

Conservation

High conservation between species has proven to be a very useful hallmark of sRNAs. It is a striking characteristic of almost all of the previously identified sRNAs (see Table 1); conservation of coding sequence for a typical ORF is generally considerably less than that for the sRNAs. Presumably, the high level of conservation observed for the sRNA genes reflects a requirement for sequences to co-evolve with other sequences, either to encode specific structures and/or to interact with specific sequences in target genes. We chose to look for conservation within intergenic (Ig) regions as a first indicator of the presence of novel sRNAs (Wassarman et al. 2001). In our work, we used comparisons primarily between *E. coli,* *Salmonella* species, and *Klebsiella pneumoniae*. These have diverged sufficiently for our purposes. For instance, two ORFs around *rprA* are 78% and 75% conserved at the DNA sequence level between *E. coli* and *Salmonella,* whereas *rprA* itself is 96% conserved (Table 1). Most importantly, the regions between genes not involved in regulation can fall to less than 50% conservation and are characterized by insertions and deletions. Thus, "junk" or spacer DNA has diverged to a significant degree.

The most abundant highly conserved sequences within Ig regions were those that were immediately 5′ to an ORF. These regions of conservation likely correspond to complex promoters or conserved leader sequences. In many cases, either published information or microarray data confirmed the presence of a leader. Since high conservation suggests that leader structure is important, we suggest that many of the *E. coli* genes containing conserved leaders are targets for translational regulation;

some specific examples are discussed below. Distinguishing between such leaders and conservation indicative of a sRNA is one of the challenges in sorting through comparative genome data for novel sRNAs. In our work, we used location relative to the ORF, predicted structure elements, and microarray expression data (see below) as secondary criteria for choosing among highly conserved Ig sequences as candidates for encoding sRNAs. Another approach to searching for small RNAs, using comparative genomics to look for conserved and co-evolving structures (stem-loops), has recently been described (Rivas et al. 2001).

In searching for sRNAs in a given organism, it will be necessary to identify one or two additional organisms at a similarly useful evolutionary distance for which sequence information is available (although not necessarily fully assembled or annotated). The approach of using conservation within Igs to identify regulatory elements has recently been tested with *S. cerevisiae* in a comparison to related *Saccharomyces* species (Cliften et al. 2001). Cliften et al. found using both a closely related species (70% DNA homology) and a less well conserved species (40% homology) most useful for detecting conservation over both short (promoter elements) and long (characteristic of sRNAs) segments. The emerging data from the sequence of the mouse genome suggests that, surprisingly, significant regions of the Ig sequences are conserved with the human genome (Shabalina et al. 2001). Thus, it seems likely that important regulatory information, including sRNAs, will be identified by comparing the Ig regions of the mouse and human genomes.

Structural Motifs

A sRNA must, by definition, start and stop within a relatively short region (50–400 nucleotides). Therefore, if the signals that define starting and stopping, promoters and terminators, can be described sufficiently to allow a computer search, their presence at such distances should be a sign of a sRNA. In addition, many of the known sRNAs are structured and contain stem-loops, characterized at the sequence level by inverted repeat sequences. We used the presence of inverted repeats as one of our secondary criteria in selecting candidates for further analysis. Of the 18 candidate Ig regions we initially rated as containing either an inverted repeat or a possible Rho-independent terminator (an inverted repeat followed by a stretch of T's on one side), 9 proved to encode sRNAs. Others encoded leaders, and secondary structure elements also may exist within these leaders. Argaman et al. (2001) analyzed Ig regions for promoters and rho-independent terminators as a primary criterion, using conservation as a secondary criterion. They tested 23 predicted sRNAs and characterized 14 sRNAs; 2 of these are listed as known RNAs in Table 1, and 7 of these overlapped with those found in our search (Table 2). Because both of these approaches required conservation, we cannot independently assess the usefulness of using structural motifs compared to conservation. Recently, another group has carried out an analysis for promoters and rho-independent terminators without using interspecies conservation

as a criterion; they are able to identify a number of additional small RNAs not found by the Argaman et al. study (S. Chen et al. [Ibis Therapeutics], pers. comm.). It is apparent that searching Ig regions for both conservation and structural motifs is productive. However, expansion of the approach of using structural motifs as predictors of sRNAs to other organisms obviously will require the ability to predict these elements.

DIRECT DETECTION OF sRNAs

Below we discuss two different approaches for directly detecting sRNAs. Inherent in either of these methods is the requirement that a given RNA be expressed under the condition being studied. Many sRNAs are in fact relatively stable and expressed at reasonably high levels. Those that are expressed under very specific growth conditions, however, are unlikely to be revealed by a direct approach. We chose three growth conditions for northern blots to confirm the presence of sRNAs in conserved candidates (growth in rich broth, either exponential or stationary phase, and growth in minimal medium). There were significant differences in expression for different RNAs under these conditions (Table 2) (Wassarman et al. 2001). Argaman et al. (2001) used similar conditions as well as heat shock and cold shock, and one of the sRNAs was only detected at a significant level after cold shock (Table 2). Certainly, other growth conditions might be necessary to discover some sRNAs. One of our top candidates had the conservation and termination signal characteristic of a sRNA but was not detected in our northern blots.

Examination of Whole-genome Expression Patterns

One genome-wide approach for the direct detection of sRNAs is hybridization to microarrays. Optimally, microarrays should have complete genome representations with probes corresponding to Ig regions and the antisense as well as sense strand of ORFs. A sRNA is easier to detect if it yields a hybridization signal separate from and discontinuous with the signals from neighboring ORFs. One limitation is that the small size and structure of the sRNAs may interfere with detection under some conditions. Any method of labeling that puts short, highly structured sequences at a disadvantage will discriminate to some extent against sRNAs. With these caveats in mind, we used Affymetrix high-density oligonucleotide microarrays to assay RNA from a single condition of growth (rich broth, late exponential phase). Eleven of the RNAs found by northern blot also were detected by the microarray. It seems likely that further exploitation of such high-density oligonucleotide microarrays, under multiple growth conditions, coupled with further development of labeling protocols, will make this approach highly useful.

Another recent approach also dependent on expression has been used in a eukaryote (Huttenhofer et al. 2001). First, there was an enrichment for RNAs of a size consistent with sRNAs. Subsequently, the individual separated RNAs were sequenced. Although it is not yet possible to compare this approach with those used above, it has ob-

Table 3. Candidate Genes for Translational Regulation

A. Ribosomal Proteins and Translational Machinery

Gene	Protein[a]	Conservation length (nt)[b]	Promoter position (nt)[c]	Transcript[d]	Comment[e]	Reference[f]
rplU	L21	115				
rpmB	L28	209		rif	complex operon	Gifford and Wallace (1999)
rpsB	S2	106		rif	mutants improve λcI translation	Shean and Gottesman (1992)
rpsF	S6	105	120 98		inverted repeat in leader	Schnier et al. (1986)
rpsJ	S10	184	172	rif	L4 regulates translation	Zengel and Lindahl (1996)
rpsP	S16	247			attenuation	Bystrom et al. (1989)
rpsU	S21	233	134 68		multiple promoters	
leuS	synthetase	106	69			
glyQ	synthetase	79				
thrS	synthetase	107	162		known translational regulation	Nogueria et al. (2001)

B. Exported Proteins: Inner and Outer Membrane Proteins, Periplasmic Proteins

Gene	Protein[a]	Conservation length (nt)[b]	Promoter position (nt)[c]	Transcript[d]	Comment[e]	Reference[f]
btuB	B12 receptor (OM)	109	240	rif	known translational regulation	Nou and Kadner (2000)
corA	Mg transport (IM)	276				
fkpA	prolyl isomerase (P)	141	105		σ^E-dependent promoter	
imp/ostA	affects permeability (OM)	173	226		σ^E-dependent promoter	
lepA	GTP binding (IM)	87 (42-nt gap)	73			
mdoG	oligosaccharide synthesis (glucan)	140	105		σ^E-dependent promoter	
nagE	PTS (IM)	62	103		complex promoter	
nuoA	NADH dehydrogenase subunit (IM)	196	94		#16 (large + 300 nt) complex promoter	
ompA	porin, receptor (OM)	184	360		#1 (large)	
ompX	regulatory (OM)	116	P1: 240; P2: 32		regulates σ^E activity	
rfaQ	Lps synthesis	180	138		jumpstart sequence for RfaH-dependent antitermination	Leeds and Welch (1997)
spr	lipoprotein	230				
tolC	porin (OM)	97				
zipA	cell division (IM)	162				

vious advantages for a host with a large and complex genome for which microarrays representing full coverage of the Ig regions are not yet available.

Enrichment for RNAs Binding to Specific Proteins

We noted that all of the sRNA regulators of *rpoS* mRNA translation, DsrA, RprA, and OxyS RNA, bind the Hfq protein and all require Hfq for function. In eukaryotic cells, RNA-binding proteins have been useful tools for identifying classes of sRNAs (Montzka and Steitz 1988;

Tyc and Steitz 1989). We found that a variety of specific RNA bands coimmunoprecipitate with Hfq. When the selected RNAs were analyzed on northern blots, a significant number of the known and the newly identified sRNAs were detected in the selected RNA samples (summarized in Tables 1 and 2) (Wassarman et al. 2001) .

Using microarrays, coimmunoprecipitation with Hfq can be generalized to sRNAs not previously identified. To improve detection of the small quantity of RNA in the immunoprecipitated samples, a hybrid capture protocol was used (C. Rosenow, unpubl.). Unlabeled RNA was hybridized directly to the microarray and detected using

Table 3. (*Continued*)

C. Energy Metabolism and TCA Cycle Genes

Gene	Protein[a]	Conservation length (nt)[b]	Promoter position (nt)[c]	Transcript[d]	Comment[e]	Reference[f]
aceB	malate synthase	223	83		complex regulation	
acnA	aconitase; regulatory role	319	407 50		mRNA binding (3′UTR); Fe-S switch? complex promoter	Tang and Guest (1999)
acnB	aconitase; regulatory role	148 (30 nt gap)	95		mRNA binding (3′UTR)	Tang and Guest (1999)
acpP	acyl carrier protein	139			#3 (300 + nt)	
aspA	aspartase	260	106		Crp, Fnr regulated	
fumA	fumarase	119	62		aspartase relative	
gapA	glyceraldehyde 3P-dehydrogenase	341	152, others	rif	#13 (large) known leader, 4 promoters	
gpmA	phospho-glycerate mutase	100 (24-nt gap)				
icd	isocitrate dehydrogenase	172	115		#4 (large)	
mdh	malate dehydrogenase	145 (interrupted)	280	rif	ArcA/heme regulated	
pckA	PEP carboxykinase	133	139		CsrA regulated	Sabnis et al. (1995)
pykF	pyruvate kinase	251	240		CsrA regulated	Sabnis et al. (1995)

D. Other Known Genes

Gene	Protein[a]	Conservation length (nt)[b]	Promoter position (nt)[c]	Transcript[d]	Comment[e]
ahpC	hydroxyl damage repair	86	200	rif	
bolA	morphogene	102	63		
clpP	protease	142	110		heat shock promoter
cspD	cold shock family	239	87	rif	growth phase regulated, stationary phase
csrA	glycogen regulator	187		rif	RNA-binding protein; translational regulator
groS	chaperonin	112	72		
hupA	HU subunit	184	104	rif	
hupB	HU subunit	211	119; 57; 15	rif	
moaA	molybdenum cofactor synthesis	262	S1: 131; S2: 218		ModE repressor; complex promoter
tig	trigger factor; prolyl isomerase; protein export	343			#29 (large)
topA	topoisomerase	111	P4: 230; P3: 200; P2: 165; P1: 68		#7 (large) multiple promoters

All genes in the table had well-conserved regions in the Ig region adjacent to or close to their predicted ATG start (http://dir2.nichd.nih.gov/nichd/cbmb/segr/segrPublications.html) and also had detectable RNA signal from high-density oligonucleotide array experiments carried out with RNA isolated under the conditions listed in Table 1 (C. Rosenow, in prep.). A number of genes encoding unknown ORFs (including *ybeB*, *yeeF*, *yobF*, *yejG*, *yfbV*, *yfhK*, *ygiM*, *yjeB*, *yjgP*) fit the criteria of the genes listed in the table.

[a] Protein or function encoded by the gene. For exported proteins, localization is indicated when known (IM: inner membrane; OM: outer membrane; P: periplasm).

[b] Conservation generally extended to within 20 nt of ATG start. Where a larger gap between the ATG and the start of the conservation was present, it is noted.

[c] mRNA transcript start, relative to ATG, from literature or database searches. Where mRNA start was mapped close to start of ORF, gene was deleted from this table. Where two or more promoters have been documented, the length of each start to the ATG is listed on a separate line.

[d] Rif indicates transcript persisting after 20-minute chase with rifampycin.

[e] Translational regulation reported in the literature or possible other explanations for conservation (complex promoters) are listed. Candidate numbers and size of transcript are listed if the Ig region was specifically probed in northern blots (Wassarman et al. 2001). Large indicates RNA band present at top of gel, consistent with mRNA (>350 nt).

[f] Only references demonstrating translational regulation are listed.

antibody to the resulting RNA:DNA hybrid. Enrichment by immunoprecipitation and the novel detection protocol, which should be less sensitive to the size and quantities of the RNAs, allowed detection of some sRNAs not found previously. Although these experiments are still under development, they show promise of providing a sensitive screen for RNAs binding to a given protein. Therefore, screening a variety of expression conditions for RNAs binding a known general RNA-binding protein should be a powerful approach for identifying one class of novel RNAs. Hfq has been implicated in the degradation and poly(A) tailing of some messages (Hajndsorf and Regnier 2000; Vytvytska et al. 2000), and the preliminary immunoprecipitation experiments also suggest that some mRNAs may bind Hfq. Thus, precipitation of an RNA with Hfq is unlikely in itself to be sufficient to identify a sRNA; additional analysis will be needed to decide whether a given precipitated RNA is likely to encode a small regulatory RNA or a message.

Although Hfq is the only *E. coli* protein thus far found to have broad binding to sRNAs, coimmunoprecipitation with other proteins may allow the detection of other classes of sRNAs. If a given RNA-binding protein is associated with a given function, this approach provides the possibility of linking given sRNAs to function.

EXPLORING THE FUNCTIONS OF NEW sRNAs

Once searches such as those described above have succeeded, we rapidly approach the next problem: What do the new sRNAs do? The task of elucidating the new sRNA functions should occupy many investigators for quite a while. A few general approaches are summarized here.

As discussed above, the identification of a binding protein may suggest function. This approach was very successful with 6S RNA, where identifying RNA polymerase as the primary target of this well-conserved and abundant sRNA was a key to understanding its function (Wassarman and Storz 2000). We believe that the sRNAs binding strongly to Hfq are translational regulators that act by base-pairing interactions. If so, their targets might be predicted by computational searches for regions of complementarity to the sRNA. The RyhB RNA shows complementarity to the *sdh* (succinate dehydrogenase) operon and is involved in the regulation of *sdh* synthesis (Wassarman et al. 2001; E. Masse and S. Gottesman, in prep.). Additional targets for DsrA also were suggested by a search for complementarity, at least some of which have been confirmed (Lease et al. 1998). As we identify more targets and become more familiar with the types of complementarity that are sufficient and necessary to lead to regulation, it should be possible to improve computer searches for sRNA antisense targets.

Another possible hint to function may be the pattern of expression of a given sRNA. It is striking that many sRNAs vary greatly in amount between exponential and stationary-phase growth, suggesting that they may play an important role in mediating a shift from one state to the other. Thus, sRNAs in addition to the DsrA, RprA, and

OxyS RNAs may regulate *rpoS* mRNA translation. Consistent with this, we found that a number of the new sRNAs affect expression of σ^S when overproduced (Wassarman et al. 2001).

Assays of phenotypes of cells carrying mutations or overproducing a given sRNA also are an important test of function. However, for many of the sRNAs studied thus far, the phenotypes are subtle or are not detected in the absence of some information on probable function. Whole-genome expression patterns in cells lacking or overexpressing a particular sRNA also may give clues to function, but given that many of the known sRNAs regulate translation rather than transcription, these screens may need to be protein-based rather than solely RNA-based.

EVIDENCE OF TRANSLATIONAL REGULATION: HIGHLY CONSERVED AND EXPRESSED LEADER SEQUENCES

As noted above, in about half of the cases where a significant patch of conservation was present in an Ig region, the pattern of conservation suggested the presence of a complex promoter and/or an untranslated 5′-leader sequence. In many cases, expression of RNA in the region expected for an untranslated leader also was detected in the high-density oligonucleotide arrays. In some cases, data on the mapping of the promoter present in the literature confirm the presence of a leader. Table 3 lists a number of examples from our work in which both conservation and microarray analysis suggest the presence of an expressed leader (http://eclipse.nichd.nih.gov/nichd/cbmb/segr/segrPublications.html, and C. Rosenow et al., in prep.). Several of these genes have previously been shown to be subject to translational regulation, consistent with our suggestion that many of the genes in Table 3 will have interesting regulation dependent on the upstream RNA. Both translational regulation and regulation by antitermination are expected to be found. Exceptions may be cases where the sequence constraints of multiple overlapping promoters lead to conservation and an extended leader, without the message sequence itself playing a regulatory role.

Table 3 is divided into four sections reflecting the general roles the genes play in the cell. We observed conservation and expression of an apparent leader for many ribosomal protein genes and synthetases (Table 3A). In many of these cases, the microarray signal persisted after cells were treated with rifamycin for 20 minutes, suggesting either a particularly strong signal or a stable leader sequence. The *rbsJ* operon was previously shown to be translationally regulated by ribosomal protein L4 binding to sequences within the untranslated leader (Zengel and Lindahl 1996). The leader also has been shown to be required for the translational regulation of *thrS* (Nogueria et al. 2001). We have not found references for translational regulation for other genes in this section. Nonetheless, the existence of the conserved leader strongly suggests that other ribosomal operons and other synthetases will be subject to translational regulation that is dependent on an untranslated leader. A second group of genes that were relatively abundant in our list were those encoding exported proteins (Table 3B). Possibly the untranslated leader plays a role in pausing translation to allow export machinery to associate with the message. Again, evidence for translational regulation has not been reported for most of these genes, and there is generally not much information on the function of the mRNA leaders. One exception is *btuB*, encoding the B12 receptor. Translational regulation for this gene has been proposed to operate by direct binding of adenosylcobalamin to the mRNA (Nou and Kadner 2000). In another case, *rfaQ*, the JUMPSTART sequence found in the upstream region is characteristic of genes regulated by an RfaH-dependent antitermination mechanism (Leeds and Welch 1997). Another notable characteristic of a number of genes in this section is the abundance of genes transcribed by σ^E. An explanation for this remains to be found. An unexpected group of genes associated with conserved, expressed leaders were those of the TCA cycle and related genes of energy metabolism (Table 3C). As described above, the expression of the *sdh* operon, encoding another enzyme of the TCA cycle, is regulated by the *ryhB* sRNA from sites within the operon. It will be interesting to determine the physiological importance of the posttranscriptional regulation of this group of genes. At least two of these genes are regulated by CsrA, whose function is antagonized by the CsrB RNA. Finally, Table 3D lists yet other potentially interesting targets that do not fall into a single well-represented category. We also noted that some unknown ORFs are associated with conserved upstream sequences with detectable expression.

Since the microarrays considered in this work were performed with RNA extracted under a single growth condition, we expect many other untranslated leader sequences will be detected as other conditions are explored. High-density oligonucleotide arrays have been used to examine 14 different growth conditions (C. Rosenow, in prep.). Certainly, if conservation upstream of the 5′ end of the ORF is the only criterion used, many additional genes are candidates for translational regulation. In addition, the criterion used here does not identify any genes for which translational regulation operates with sequences within the ORF-coding region itself or within the mRNA encoding an upstream gene. For instance, the regulatory leader RNA for σ^S is contained within the upstream *nlpD* gene with only 63 nucleotides between the two coding regions. Assuming that conservation within a leader reflects functional importance, we predict that a substantial portion of the genes in *E. coli* are subject to translational regulation. The mechanisms of translational regulation may involve only *cis*-acting sequences, or they may depend on interactions with protein or small molecule regulators. However, given the example of σ^S, it seems likely that in some cases sRNAs will be central mediators of the regulation.

PERSPECTIVES

sRNAs can no longer be considered rare and unusual. In eukaryotes, they are involved in RNA processing and are increasingly being implicated in RNA interference and sRNA-dependent developmental pathways (see, e.g., Grishok et al. 2001). In prokaryotes, they have already been implicated in regulation of translation and transcription, in many cases modulating central developmental switches. Therefore, the development of methods for the detection and study of sRNAs has become a focus of attention. The identification of many *E. coli* genes with highly conserved untranslated leaders also suggests that many regulatory circuits will include a significant contribution by translation regulation, in some cases mediated by sRNAs.

ACKNOWLEDGMENT

We thank S. Chen and S. Holbrook for communicating results prior to publication.

REFERENCES

Altuvia S., Weinstein-Fischer D., Zhang A., Postow L., and Storz G. 1997. A small stable RNA induced by oxidative stress: Role as a pleiotropic regulator and antimutator. *Cell* **90:** 43.

Altuvia S., Zhang A., Argaman L., Tiwari A., and Storz G. 1998. The *Escherichia coli oxyS* regulatory RNA represses *fhlA* translation by blocking ribosome binding. *EMBO J.* **17:** 6069.

Argaman L. and Altuvia S. 2000. *fhlA* repression by OxyS RNA: Kissing complex formation at two sites results in a stable antisense-target RNA complex. *J. Mol. Biol.* **300:** 1101.

Argaman L., Hershberg R., Vogel J., Bejerano G., Wagner E.G.H., Margalit H., and Altuvia S. 2001. Novel small RNA-encoding genes in the intergenic region of *Escherichia coli*. *Curr. Biol.* **11:** 941.

Becker G. and Hengge-Aronis R. 2001. What makes an *Escherichia coli* promoter sigma dependent? Role of the -13/-14 nucleotide promoter positions and region 2.5 of sigma. *Mol. Microbiol.* **39:** 1153.

Brown L. and Elliott T. 1997. Mutations that increase expression of the *rpoS* gene and decrease its dependence on *hfq* function in *Salmonella typhimurium*. *J. Bacteriol.* **179:** 656.

Bystrom A.S., von Gabain A., and Bjork G.R. 1989. Differentially expressed *trmD* ribosomal protein operon of *Escherichia coli* is transcribed as a single polycistronic mRNA species. *J. Mol. Biol.* **208:** 575.

Carter R.J., Dubchak I., and Holbrook S.R. 2001. A computational approach to identify genes for functional RNAs in genomic sequences. *Nucleic Acids Res.* **29:** 3928.

Cliften P.F., Hillier L.W., Fulton L., Graves T., Miner T., Gish W.R., Waterston R.H., and Johnston M. 2001. Surveying *Saccharomyces* genomes to identify functional elements by comparative DNA sequence analysis. *Genome Res.* **11:** 1175.

Gifford C.M. and Wallace S.S. 1999. The genes encoding formamidopyrimidine and MutY DNA glycosylases in *Escherichia coli* are transcribed as part of complex operons. *J. Bacteriol.* **181:** 4223.

Grishok A., Pasquinelli A.E., Conte D., Li N., Parrish S., Ha I., Baillie D.L., Fire A., Ruvkun G., and Mello C.C. 2001. Genes and mechanisms related to RNA interference regulate expression of the small temporal RNAs that control *C. elegans* developmental timing. *Cell* **106:** 23.

Hajnsdorf E. and Regnier P. 2000. Host factor Hfq of *Escherichia coli* stimulates elongation of poly(A) tails by poly(A) polymerase I. *Proc. Natl. Acad. Sci.* **97:** 1501.

Hengge-Aronis R. 2000. The general stress response in *Escherichia coli*. In *Bacterial stress responses* (ed. G. Storz and R. Hengge-Aronis), p. 161. ASM Press, Washington, D.C.

Herskovits A.A., Bochkareva E.S., and Bibi E. 2000. New prospects in studying the bacterial signal recognition particle pathway. *Mol. Microbiol.* **38:** 927.

Huttenhofer A., Kiefmann M., Meier-Ewert S., O'Brien J., Lehrach H., Bachellerie J.-P., and Brosius J. 2001. RNomics: An experimental approach that identifies 201 candidates for novel, small, non-messenger RNAs in mouse. *EMBO J.* **20:** 2943.

Jung Y.H., Park I., and Lee Y. 1992. Alteration of RNA I metabolism in a temperature-sensitive *Escherichia coli rnpA* mutant strain. *Biochem. Biophys. Res. Commun.* **186:** 1463.

Karzai A.W., Susskind M.M., and Sauer R.T. 1999. SmpB, a unique RNA-binding protein essential for the peptide-tagging activity of SsrA (tmRNA). *EMBO J.* **18:** 3793.

Lease R.A., Cusick M., and Belfort M. 1998. Riboregulation in *Escherichia coli:* DsrA RNA acts by RNA:RNA interactions at multiple loci. *Proc. Natl. Acad. Sci.* **95:** 12456.

Leeds J.A. and Welch R.A. 1997. Enhancing transcription through the *Escherichia coli* hemolysin operon, *hlyCABD*: RfaH and upstream JUMPStart DNA sequences function together via a postinitiation mechanism. *J. Bacteriol.* **179:** 3519.

Majdalani N., Chen S., Murrow J., St. John K., and Gottesman S. 2001. Regulation of RpoS by a novel small RNA: The characterization of RprA. *Mol. Microbiol.* **39:** 1382.

Majdalani N., Cunning C., Sledjeski D., Elliott T., and Gottesman S. 1998. DsrA RNA regulates translation of RpoS message by an anti-antisense mechanism, independent of its action as an antisilencer of transcription. *Proc. Natl. Acad. Sci.* **95:** 12462.

Montzka K.A. and Steitz J.A. 1988. Additional low-abundance human small nuclear ribonucleoproteins: U11, U12, etc. *Proc. Natl. Acad. Sci.* **85:** 8885.

Nogueira T., de Smit M., Graffe M., and Springer M. 2001. The relationship between translational control and mRNA degradation for the *Escherichia coli* threonyl-tRNA synthetase gene. *J. Mol. Biol.* **310:** 709.

Nou X. and Kadner R.J. 2000. Adenosylcobalamin inhibits ribosome binding to *btuB* RNA. *Proc. Natl. Acad. Sci.* **97:** 7190.

Okamoto K. and Freundlich M. 1986. Mechanism for the autogenous control of the *crp* operon: Transcriptional inhibition by a divergent RNA transcript. *Proc. Natl. Acad. Sci.* **83:** 5000.

Repoila F. and Gottesman S. 2001. Signal transduction cascade for regulation of RpoS: Temperature regulation of DsrA. *J. Bacteriol.* **183:** 4012.

Rivas E., Klein R.J., Jones T.A., and Eddy S.R. 2001. Computational identification of noncoding RNAs in *E. coli* by comparative genomics. *Curr. Biol.* **11:** 1369.

Romeo T. 1998. Global regulation by the small RNA-binding protein CsrA and the non-coding RNA molecule CsrB. *Mol. Microbiol.* **29:** 1321.

Rudd K.E. 1999. Novel intergenic repeats of *Escherichia coli* K-12. *Res. Microbiol.* **150:** 653.

Sabnis N.A., Yang H., and Romeo T. 1995. Pleiotropic regulation of central carbohydrate metabolism in *Escherichia coli* via the gene *csrA*. *J. Biol. Chem.* **270:** 29096.

Schnier J., Kitakawa M., and Isono K. 1986. The nucleotide sequence of an *Escherichia coli* chromosomal region containing the genes for ribosomal proteins S6, S18, L9 and an open reading frame. *Mol. Gen. Genet.* **204:** 126.

Shabalina S.A., Ogurtsov A.Y., Kondrashov V.A., and Kondrashov A.S. 2001. Selective constraint in intergenic regions of human and mouse genomes. *Trends Genet.* **17:** 373.

Shean C.S. and Gottesman M.E. 1992. Translation of the prophage lambda cI transcript. *Cell* **70:** 513.

Sledjeski D. and Gottesman S. 1995. A small RNA acts as an antisilencer of the H-NS-silenced *rcsA* gene of *Escherichia coli*. *Proc. Natl. Acad. Sci.* **92:** 2003.

Sledjeski D.D., Gupta A., and Gottesman S. 1996. The small RNA, DsrA, is essential for the low temperature expression of RpoS during exponential growth in *Escherichia coli*. *EMBO J.* **15:** 3993.

Sledjeski D.D., Whitman C., and Zhang A. 2001. Hfq is necessary for regulation by the untranslated RNA DsrA. *J. Bacteriol.* **183:** 1997.

Tang Y. and Guest J.R. 1999. Direct evidence for mRNA binding and post-transcriptional regulation by *Escherichia coli* aconitases. *Microbiology* **145:** 3069.

Tyc K. and Steitz J.A. 1989. U3, U8 and U13 comprise a new class of mammalian snRNPs localized in the cell nucleolus. *EMBO J.* **8:** 3113.

Urbanowski M.L., Stauffer L.T., and Stauffer G.V. 2000. The *gcvB* gene encodes a small untranslated RNA involved in expression of the dipeptide and oligopeptide transport systems in *Escherichia coli*. *Mol. Microbiol.* **37:** 856.

Vytvytska O., Moll I., Kaberdin V.R., von Gabain A., and Blasi U. 2000. Hfq (HF1) stimulates *ompA* mRNA decay by interfering with ribosome binding. *Genes Dev.* **14:** 1109.

Wassarman K.M. and Storz G. 2000. 6S RNA regulates *E. coli* RNA polymerase activity. *Cell* **101:** 613.

Wassarman K.M., Zhang A., and Storz G. 1999. Small RNAs in *Escherichia coli*. *Trends Microbiol.* **7:** 37.

Wassarman K.M., Repoila F., Rosenow C., Storz G., and Gottesman S. 2001. Identification of novel small RNAs using comparative genomics and microarrays. *Genes Dev.* **15:** 1637.

Zengel J.M. and Lindahl L. 1996. A hairpin structure upstream of the terminator hairpin required for ribosomal protein L4-mediated attenuation control of the S10 operon of *Escherichia coli*. *J. Bacteriol.* **178:** 2383.

Zhang A., Altuvia S., Tiwari A., Argaman L., Hengge-Aronis R., and Storz G. 1998. The *oxyS* regulatory RNA represses *rpoS* translation by binding Hfq (HF-1) protein. *EMBO J.* **17:** 6061.

Initiation Factors in the Early Events
of mRNA Translation in Bacteria

C.O. Gualerzi, L. Brandi, E. Caserta, C. Garofalo,* M. Lammi,† A. La Teana,*
D. Petrelli, R. Spurio, J. Tomsic, and C.L. Pon
Laboratory of Genetics, Department of Biology, MCA University of Camerino 62032, Camerino (MC), Italy;
*Institute of Biochemistry, University of Ancona, 60131 Ancona, Italy; †Department of Cell Biology,
University of Calabria, 87036 Arcavacata di Rende (CS), Italy

Initiation of mRNA translation in bacteria proceeds through several steps. The early events entail the selection of the mRNA initiation codon by the 30S ribosomal subunit with the help of initiator fMet-tRNA and initiation factors IF1, IF2, and IF3, which all together form a "30S initiation complex." Subsequently, the choice of the start site is made irreversible by the binding of the 50S ribosomal subunit to the 30S initiation complex, which expels IF1 and IF3 and yields a "70S initiation complex." Among the late initiation events are the adjustment of fMet-tRNA in the ribosomal P site, the A-site binding of the EF-Tu ternary complex carrying the elongator aminoacyl-tRNA encoded by the second mRNA codon, and the subsequent formation of the first peptide bond yielding the initiation dipeptide. At this stage, with the first translocation, the ribosome enters the elongation phase.

Before discussing in more detail the mechanistic aspects of these steps, we describe the molecular and structural bases underlying the specific interactions among initiation components, their mutual selective recognition, and their mechanism of action.

mRNA SELECTION AND PROPERTIES OF
TRANSLATIONAL INITIATION REGIONS

The recent advances in our understanding of the 3D structure of the ribosome and of its active sites made possible by cryoelectron microscopy (for review, see Agrawal and Frank 1999) and crystallography (Cate et al. 1999; Schluenzen et al. 2000; Wimberly et al. 2000) allow us to depict the path of the mRNA through the 30S ribosomal subunit with considerable precision; by comparison, this path had been defined with remarkable accuracy in previous studies by EM, chemical probing, and site-directed cross-linking (for review, see Gualerzi and Pon 1996). The recognition and binding of the translational initiation region (TIR) of the mRNA by the 30S ribosomal subunit depends to various degrees on the structural elements of a canonical TIR, which include the initiation triplet (most frequently AUG), the purine-rich Shine-Dalgarno (SD) sequence complementary to the 3′-end region of 16S rRNA, and a spacer, of variable length, separating SD and the initiation triplet. The optimal combinations of these TIR elements that can maximize translation of mRNA have been reviewed previously (Gualerzi and Pon

1990, 1996; McCarthy and Brimacombe 1994, Gualerzi et al. 2000). In addition to the majority of mRNAs having a canonical TIR, bacteria (and archaea) contain a small number of a special kind of mRNA, the "leaderless mRNAs" which begin at their 5′end with an initiation triplet AUG. These mRNAs are particularly interesting insofar as they may represent an ancestral form of template that can be recognized by the translational apparatus of all three kingdoms of life (Grill et al. 2000). Initiation site selection in these mRNAs does not rely on a detectable thermodynamic affinity for the 30S subunit but stringently depends on the decoding of their essential AUG start codon (Wu and Janssen 1996; Van Etten and Janssen 1998) by the initiator tRNA assisted by IF2 (Grill et al. 2001). Unlike IF2, which strongly favors translation of leaderless mRNAs, IF3 favors the dissociation of the 30S initiation complexes formed with leaderless mRNAs and antagonizes the translation of these mRNAs (Tedin et al. 1999), which is also strongly influenced by environmental factors such as temperature (S. Grill et al., in prep.).

A *cis* element that seems to affect positively the translation of leaderless (but also of leadered) mRNAs are multimers of the CA dinucleotide present 3′ of the initiation codon (Martin-Farmer and Janssen 1999), and a controversy has arisen concerning the possibility that a DB sequence located to the 3′ side of the initiation codon (the "downstream box") might participate in the selection of leaderless (Sprengart et al. 1996; Sprengart and Porter 1997), as well as other classes of leadered, mRNAs (Etchegaray and Inouye 1999), by base-pairing with a complementary region of 16S rRNA. This issue seems to have been resolved with the compelling evidence accumulated that such a base-pairing does not and cannot occur, under normal as well as under stress conditions (O'Connor et al. 1999; La Teana et al. 2000; Moll et al. 2001) and that the presence of DB sequences in mRNAs has no statistical significance in either *Escherichia coli* or *Bacillus subtilis* (Rocha et al. 1999).

STRUCTURE AND FUNCTION OF
THE INITIATION FACTORS

For a long time after their discovery in the second half of the 1960s, IF2 was regarded as a carrier of the initiator fMet-tRNA, IF3 was thought to be a factor essential for the binding of natural mRNAs to the ribosome, and IF1 was thought to be a more or less dispensable "helper" of the other two factors. This "thermodynamic" interpretation of the function of these factors resulted in the systematic overlooking of data indicating instead that all factors are primarily kinetic effectors of translation initiation. Discussion of the roles and properties of the initiation factors follows.

IF1

IF1 is a small β barrel RNA-binding protein consisting of five β strands and a loop connecting strands 3 and 4 containing a short 3_{10} helix endowed with high flexibility (Sette et al. 1997). The amino acid residues of IF1 involved in or affected by the binding to the 30S ribosomal subunits have been identified by site-directed mutagenesis and nuclear magnetic resonance (NMR) spectroscopy. More recently, the crystallographic structure of an IF1–30S complex (from *Thermus thermophilus*) has been elucidated at 3.2 Å resolution (Carter et al. 2001). The crystallographic data indicate that IF1 binds in the cleft between protein S12, helix 44, and the 530 loop of 16S rRNA. In agreement with the biochemical (Celano et al. 1988) and NMR (Sette et al. 1997) data, the interaction is mainly electrostatic, involving the positively charged surface of the protein and the phosphate backbone of the two aforementioned rRNA structures. The crystallographic data confirm the interaction of IF1 with the ribosomal A site but do not confirm the proposal that IF1 mimics A-site-bound tRNA (Brock et al. 1998); the localization of IF1 further allows the rationalization of the effects of this factor on the chemical reactivity of specific bases (A1492 and A1493) of 16S rRNA (Moazed et al. 1995) and the inhibition of IF1 binding upon mutagenesis of the same nucleotides (Dahlquist and Puglisi 2000). The contacts between IF1 and 30S emerging from this topographical localization are also in reasonable agreement with at least some of the data obtained by mutagenesis (Gualerzi et al. 1989) and NMR spectroscopy (Sette et al. 1997). Furthermore, the crystallographic data shed new light on the possible mechanism of action of this factor. In fact, the data provide compelling evidence that IF1 binding causes conformational changes of the ribosomal subunit, as earlier predicted from the "titration" of the kinetic effects of IF1 on 30S initiation complex formation (Pon and Gualerzi 1984). The IF1-induced conformational change entails localized changes of helix 44 with A1492 and A1493 being flipped out in a non-stacked configuration (unlike what happens upon paromomycin binding) by the insertion of a loop of IF1 into the minor groove of the helix and a 5 Å lateral displacement of C1411 and C1412. This displacement affects the conformation of helix 44 over a long distance ($\cong$ 70 Å) and causes a rotation of head, platform, and shoulder of the 30S particle toward the A site (Carter et al. 2001).

In addition to its aforementioned effects on 30S initiation complex (Wintermeyer and Gualerzi 1983; Pon and Gualerzi 1984), IF1 affects the association/dissociation kinetics of ribosomal subunits, thereby favoring the ribosome dissociation activity of IF3 (Godefroy-Colburn et al. 1975). In our perception (see also below), a very important function of IF1 is that of modulating the affinity of IF2 for the ribosome by favoring its binding to the 30S subunit and, by being ejected from the 30S upon subunit association, causing a weakening of the interaction of IF2 with the 70S initiation complex (Stringer et al. 1977). The influence of IF1 on the affinity of IF2 for the 30S ribosomal subunit may result from a physical contact between these two factors on the ribosome as suggested by cross-linking studies (Boileau et al. 1983) and phylogenetic comparisons (Choi et al. 2000) or may be indirectly caused by the IF1-induced conformational change described above. In either case, the role of IF1 in the ribosomal recycling of IF2 resembles that of the N-domain of IF3 with respect to the recycling of IF3 (see below and Fig. 6). Finally, the binding of IF1 to the A site of the 30S subunit suggests that this factor may participate in conferring specificity to the formation of the 30S initiation complex by occluding the access of a tRNA to this site until a 70S initiation complex is formed, but direct evidence for this function is so far lacking. In any event, the ejection of IF1 from the 30S ribosomal subunit during 70S initiation complex formation (Celano et al. 1988; Gualerzi et al. 1989) would serve the dual purpose of opening the A site for an incoming aminoacyl tRNA and weakening the interaction of IF2 in preparation for its ejection from the ribosome.

IF2

IF2 is an essential protein (Cole et al. 1987) whose main function is the stimulation of fMet-tRNA binding to the ribosomal P site. Unlike EF-Tu, which functions as an aminoacyl-tRNA carrier, IF2 is presumably already 30S-bound when it interacts with fMet-tRNA and increases the on-rate of its codon–anticodon interaction at the P site and decreases the off-rate of its dissociation from the ribosomes. The affinity of IF2 for the 30S subunit is somewhat increased by GTP and greatly increased by the presence of IF1.

The topography of IF2 on the ribosome has been investigated by cross-linking and, more recently, by chemical probing (for review, see Gualerzi et al. 2000). The results concerning the 30S subunit do not give a clear picture of the IF2 location because several sites spread throughout the particle are protected by the factor. Nevertheless, the data suggest a location within the cleft, the lateral protrusion, and the part of the head facing the protrusion (Wakao et al. 1991). More recently, using *Bacillus stearothermophilus* IF2 and *E. coli* 70S ribosomes, it was possible to demonstrate that IF2 interacts with 23S rRNA and specifically protects from DMS A2476 and A2478 in helix 89 and A2665 in the sarcin-ricin domain while it enhances the reactivity of A2660. In the same domain, IF2 protects from Kethoxal G2655 and G2661 (La Teana et al. 2001). As seen in Figure 1a, where these sites are highlighted in the 3D structure of the 50S subunit (Ban et al. 2000), the IF2-protected region is located just below L11 on the right edge of the subunit interface site

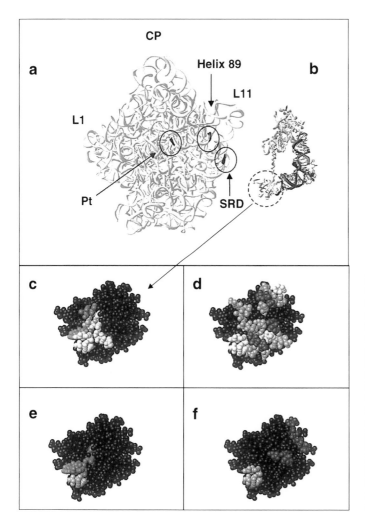

Figure 1. (*a*) The 23S rRNA sites whose chemical reactivity is decreased/increased in the presence of IF2 are shown in the circles indicated as helix 89 and sarcin ricin domain (SRD) within the 3D structure of the 50S ribosomal subunit (Ban et al. 2000) together with some landmarks, such as the binding sites of proteins L1 and L11, the peptidyl transferase center (Pt), and the central protuberance (CP). (*b*) To the right of the 50S we present the 3D structure of eIF5B, the archaeal homolog of IF2 (Roll-Mecak et al. 2000) with a hypothetical docking of its carboxy-terminal domain (within the dotted circle) with the acceptor end of fMet-tRNA. The 3D structure of bacterial IF2C-2 (Meunier et al. 2000) is shown in the remaining four panels, which illustrate the residues involved in the recognition and binding of fMet-tRNA as determined by genetic (*c*) and NMR (*d*) approaches and those involved by 3′ACCAAC (*e*) and by N-formylmethionine (*f*) as also revealed by NMR spectroscopy (Guenneugues et al. 2000).

of the 50S. The distance of these sites from the peptidyl transferase center, identified by Nissen et al. (2000) as A2451, where the acceptor end of initiator tRNA (fMet-ACC) must bind, is quite long, 53.5 Å and 74.7 Å, but the structure of IF2 is probably very elongated (see below). Overall, these results confirm that the IF2-binding site overlaps that of EF-Tu and EF-G (La Teana et al. 2001 and references therein).

The kingdom-specific recognition and binding of fMet-tRNA is one of the main properties of bacterial IF2. Formylation of the αNH_2 group of Met-tRNA$_{Metf}$ is catalyzed by 10-formyltetrahydrofolate:L-methionyl-tRNA$_{Metf}$ N-formyltransferase (the product of *fmt*) and is essential for the interaction of initiator tRNA with IF2. Disruption of *fmt* results in an almost tenfold reduction of the growth rate and in a conditional lethal phenotype (Guillon et al. 1992).

The IF2-fMet-tRNA interaction involves the most carboxy-terminal region of IF2 and the acceptor end of fMet-tRNA (Guenneugues et al. 2000; Spurio et al. 2000). Indeed, fMet-ACCAAC has almost the same affinity for IF2 as the intact fMet-tRNA, whereas a much lower yet detectable affinity is displayed also by the acceptor hexanucleotide ACCAAC, by fMet-adenosine, and by N-

formylmethionine (Guenneugues et al. 2000; Szkaradkiewicz et al. 2000). Concerning IF2, its smallest fragment (IF2C-2) preserving full activity and specificity in the recognition and binding of fMet-tRNA is a 90-amino-acid-long domain consisting of six β strands forming an open β barrel (Meunier et al. 2000). Despite a very low degree of sequence similarity, this domain is structurally similar to domain II of EF-Tu and EF-G (Aevarsson et al. 1994; Nissen et al. 1995). A combination of NMR spectroscopy and genetic approaches allowed the identification of the amino acids of IF2C-2 involved in the interaction with fMet-tRNA. The genetic approach consisted in the selection of mutants capable of suppressing a lethal phenotype caused by a mutation (H301Y) in *B. stearothermophilus* IF2 that inactivates the GTPase activity of the factor. As it turned out, the intragenic suppressors had more or less severe defects in fMet-tRNA binding. The residues identified by this approach and by subsequent site-directed mutagenesis (Fig. 1c) are remarkably coincident with those identified by NMR (monitoring specific chemical shift variations and/or broadening of the peaks in the HSQC spectra) upon titration with fMet-tRNA (Fig.1d) with a hexanucleotide with the same sequence as the acceptor end of tRNA$_{Metf}$ (Fig.1e) and with N-formylmethionine (Fig.1f). Overall, these residues define a continuous patch on the same IF2C-2 surface (Guenneugues et al. 2000). The discrimination operated by IF2 and EF-Tu against aminoacyl-tRNAs with free and blocked αNH_2 groups, respectively, could be rationalized with the different chemical properties (hydrophobic/non-charged versus hydrophilic/positively charged) of the residues localized in corresponding positions within the 3D structures of the two proteins. However, substitution of the tripeptide K_{699}-R_{700}-Y_{701} of IF2 presumably involved in the discrimination with the tripeptide (E-M-H) present in the corresponding positions of EF-Tu did not change the binding specificity of IF2. This result, and the fact that the acceptor ends of initiator and elongator tRNAs seem to have somewhat different orientations on IF2C-2 and on domain II of EF-Tu, respectively (Guenneugues et al. 2000), indicate that the structural resemblances between these two domains might have less straightforward functional implications than superficially expected. This premise is further supported by the fact that although IF2C-2 binds autonomously fMet-tRNA (whereby GTP does not influence this interaction), domain II of EF-Tu constitutes only a portion of the aminoacyl-tRNA-binding region of this elongation factor (Nissen et al. 1995), and the homologous domain II of EF-G does not interact with aminoacyl-tRNAs (Guenneugues et al. 2000).

The 3D structure of *Methanobacterium thermoautotrophicum* eIF5B has been elucidated by X-ray crystallography (Roll-Mecak et al. 2000). Compared to its *B. stearothermophilus* homolog IF2, this protein lacks ~240 amino-terminal residues, contains two additional α-helices at the carboxyl terminus and presumably plays a different role in the archaeal cell. Nevertheless, we give a brief account of eIF5B structure since this is likely homologous to that of bacterial IF2. Archaeal IF2 is a very long (110 Å) molecule with an unusual chalice-shape. The protein consists of a G domain (structurally similar to that of EF-Tu and EF-G), a β barrel, and an α/β/α sandwich that together constitute the cup of the chalice (maximum diameter 66 Å). The stem consists of a long α-helix ending with a second β barrel. Both β barrels are homologous to domain II of EF-Tu, the one located near the carboxyl terminus having the same structure (but obviously not the function) as IF2C-2. The other β barrel is probably implicated in ribosomal binding, as suggested by Roll-Mecak et al. (2000) in the archaeal case, and as indicated by our preliminary data for the bacterial protein. The comparison of the eIF5B structure in the GDP and the GTP forms indicates that a sizable yet not enormous conformational change occurs in the "cup" of the chalice with the G domain being repositioned with respect to the other two domains of the cup. This conformational change is amplified and transmitted to the β barrel at the carboxyl terminus by a swinging of the long α helix constituting the stem. The latter movement may be relevant in bacterial IF2 only insofar as the region homologous to the stem is also an α helix, an assumption that we cannot take for granted. Roll-Mecak et al. (2000) suggest that the function of eIF5B and of its GTPase is to

facilitate the joining of the ribosomal subunits. Bacterial IF2 indeed stimulates 30S joining to 50S, but this activity does not require the presence of GTP (Godefroy-Colburn et al. 1975; La Teana et al. 2001).

As mentioned above, like EF-Tu and EF-G, IF2 is a GTP/GDP-binding protein and a ribosome-dependent GTPase but, unlike the elongation factors, the function of its hydrolytic activity is controversial (Rodnina et al. 2000). In fact, GTP does not influence either IF2-fMet-tRNA interaction or formation and stability of the 30S initiation complex and it is also, as mentioned above, dispensable for the association of the 30S initiation complex with the 50S subunit to yield the 70S initiation complex. This latter process triggers the GTPase activity but the subsequent adjustment of the initiator tRNA in the P site (as evidenced by both puromycin reaction and formation of the initiation dipeptide) occurs also in the absence of GTP or in the presence of GDP if fMet-tRNA is properly coded by a canonical initiation triplet contained within a polyribonucleotide template (La Teana et al. 1996 and references therein). Omission of GTP caused only a slight reduction of the rate of fMet-puromycin formation without a significant change of the activation energy, whereas omission of the template, which resulted in requirement for a higher activation energy, made mandatory the presence of GTP (La Teana et al. 1996). Similar results have also been obtained analyzing the formation of the initiation dipeptide by fast kinetics. These data demonstrate also that the transition from initiation to elongation is not affected by the nucleotide in IF2, because in the absence of GTP in the 30S initiation complex, the fMet-tRNA is correctly positioned in the P site, and the aminoacyl tRNA can bind correctly to the A site to allow formation of the first dipeptide bond (Tomsic et al. 2000). These data indicate that the GTPase activity of IF2 is at best only marginally involved in the adjustment of the initiator tRNA in the P site, one of two functions that traditionally had been attributed to it.

The second role attributed to the IF2 GTPase is to allow release of IF2 from the 70S initiation complex (Hershey 1987). This explanation, recently reiterated by Luchin et al. (1999), rests on the claim that G-domain mutants of *E. coli* IF2 (e.g., H448E), having 40-fold lower GTPase activity than wild type, remain stuck on the ribosome after formation of the 70S initiation complex, thereby causing a dominant lethal phenotype. Being familiar with the same IF2 mutation (whose suppression, as mentioned above, allowed us to screen for mutants affected in fMet-tRNA binding), we could not agree with the data indicating that this IF2 mutant fails to dissociate from the ribosome and with the consequent conclusion reached by Luchin et al. (1999). Thus, we undertook the systematic replacement of H301 with other residues until we found one (H301R) incapable of hydrolyzing GTP yet having no dominant lethal phenotype. Furthermore, we were unable to find a strict correlation between lethality and residual GTPase activity of the mutants. In addition, although we found, in agreement with Luchin et al. (1999), that all H301 mutants tested were active in fMet-tRNA binding but inactive in fMet-puromycin and in initiation dipeptide formation, we also found that the (small) number of IF2 molecules (mutants and wild type) remaining ribosome-associated after 70S initiation complex formation did not correlate with their residual GTPase activity. Thus, considering our data and also in light of the localization of H301 in the critical "Switch 2" region of the molecule (Roll-Mecak et al. 2000), we conclude that substitutions of this residue generate "conformational mutants" whose phenotypes are unrelated to GTP hydrolysis. We further suggest that all these mutants place fMet-tRNA in a wrong orientation, which prevents its acceptor end from reaching the peptidyl transferase center and, at least in some cases, gives rise to dead-end complexes that inhibit translation in vivo as well as in vitro. This premise is supported by at least two additional considerations: (1) the failure of the H301 mutants to support fMet-puromycin (and initiation dipeptide) formation cannot be correlated with their residual GTPase activity, also because GTP hydrolysis is not required for these activities (see above) and (2) the intragenic suppressors of the lethality caused by overproduction of the H301 mutants yielded molecules with a severely impaired fMet-tRNA-binding capacity (Guenneugues et al. 2000).

Finally, in light of the fact that IF2, EF-G, and EF-Tu bind to overlapping ribosomal sites (see also above), one would expect that, in the absence of GTPase activity, inefficient IF2 release from the ribosome would block, or at least interfere with, EF-Tu binding. However, the experimental evidence shows that the absence of GTP or the presence of GDP in the initiation complex does not affect the binding kinetics of the EF-Tu ternary complex to the A site (Tomsic et al. 2000). This can be regarded as decisive evidence that GTP hydrolysis is not involved with the ejection of IF2 from the ribosome.

Since GTP/GDP binding and GTPase activity of IF2 do not seem to be required for any translational functions examined (see also below), it seems legitimate to expect that they must have a functional role which justifies their

evolutionary conservation. A potential GTPase function that has not yet been investigated is proofreading of the initiation complex, and further studies should clarify this point. However, one should also consider the possibility that the conservation of the GTP/GDP-binding site of IF2 serves a regulatory function. In this connection, a plausible role so far overlooked is that the GTP/GDP-binding site could be a sensor of the metabolic state of the cell. Thus, under optimal growth conditions, the G domain of IF2 would be occupied exclusively by GTP, whose cellular concentration is 1–2 mM, and not by GDP whose concentration is very low. Under nutritional stress, however, ppGpp is synthesized and its level increases to millimolar concentrations at the expense of GTP, whose level decreases. Experimental data indicate that under these conditions, ppGpp binds in place of GTP and induces an inactive conformation of IF2, which is now impaired in the formation of both 30S initiation complex (Fig. 2a) and initiation dipeptide (Fig.2b) (J. Tomsic et al., in prep.). Furthermore, there is evidence that the presence of IF2 is required for the translational feedback regulation of stable RNA transcription (Cole et al. 1987) and that both IF2 and its ligands fMet-tRNA and ppGpp can interact with the RNA polymerase and influence its activity at stable RNA promoters. Indeed, the structure and function of the RNA polymerase can be modulated through the interaction with these molecules; promoter selectivity by σ^{70} is altered by ppGpp, and both tRNA$_{Metf}$ and Met-tRNA$_{Metf}$ inhibit RNA polymerase, whereas fMet-tRNA$_{Metf}$ can stimulate transcription from some promoters (Debenham et al. 1980; Nomura et al. 1986) and IF2 can modulate transcriptional specificity (Travers et al. 1980).

Thus, IF2 can be envisaged as a global regulator of the translational rate of the cell as a function of the allowable growth rate as well as serving as the link between translational activity and transcriptional control. It is probably not by chance that the synthesis of fMet-tRNA requires the availability of two one-carbon donors, first to synthe-

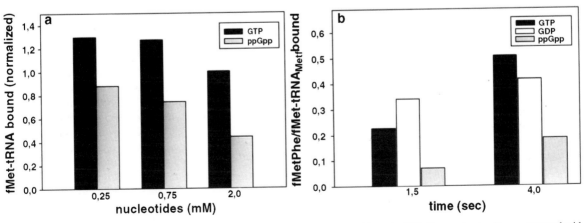

Figure 2. Effect of GTP and ppGpp on the IF2-dependent (*a*) binding of fMet-tRNA to the 30S ribosomal subunits programmed with canonical mRNA and (*b*) formation of initiation dipeptide (fMet-Phe). The data in *a* are obtained at the concentrations of guanosine nucleotides indicated in the abscissa, and the activity values shown in the ordinate are normalized taking the binding in the absence of guanosine nucleotides = 1. The data shown in *b* refer to transpeptidation reactions measured by quenched flow (Tomsic et al. 2000) at the indicated times. The amount of initiation dipeptide formed is normalized for the amount of fMet-tRNA ribosome-bound in each case.

size methionine and, later, to formylate it; therefore, that the cellular concentrations of these molecules can play a key role in metabolic regulation and in the control of the rate of translation initiation is not surprising (Matthews 1996).

IF3

Like the other two factors, IF3, a ≅ 20-kD protein displaying a fairly high degree of sequence conservation in bacteria, is also essential for cell viability (Olsson et al. 1996). For what concerns the function of IF3, the belief that IF3 was the physical intermediate in mRNA–ribosome interaction has dominated the stage for almost two decades since the discovery of this factor. In fact, several early studies have been devoted to the identification of mRNA features presumed to be recognized and bound by IF3 and to the search for IF3 isoforms specialized in the recognition and binding of "different classes" of mRNA to the 30S ribosomal subunit. These deep-rooted convictions were not abandoned with the discovery of the SD sequence, characteristic of the TIR of the majority of bacterial mRNA (see above), but fueled instead the construction of elaborate models in which IF3 could be promoting mRNA–30S interaction through "opening" or "making available" the "anti-SD" sequence of 16S rRNA for the interaction with the complementary region of the mRNA. Against this consolidated interpretational background, the data indicating a kinetic role of IF3 in promoting 30S initiation complex formation (Gualerzi et al.1977; Wintermeyer and Gualerzi 1983) and determining initiation fidelity (Pon and Gualerzi 1974; Risuleo et al. 1976) went mostly unnoticed. However, starting from the late 1980s it became generally accepted that the mRNA, regardless of whether it contains a strong, a weak, or no SD sequence, does not require IF3 to bind to the ribosome, and that although the thermodynamics of 30S–mRNA interaction is only marginally affected by this factor, the actual mRNA initiation site selection is a kinetic process, also involving fMet-tRNA decoding, which is under IF3 (and IF2) control (Calogero et al. 1988). Meanwhile, the notion that IF3 can control initiation fidelity became widely accepted when our earlier results were reproduced by the newly developed "toeprinting" technique (Hartz et al. 1989) and when it was discovered that this activity can occur not only in vitro but also in vivo (Sussman et al. 1996; Haggerty and Lovett 1997). However, since IF3 mutants impaired in the fidelity function do not display drastic reduction of cell viability, it should be clear that this activity, no matter how important, cannot represent the vital function performed by IF3 (Olsson et al. 1996). More likely essential for cell survival are the stimulation of the kinetics of 30S initiation complex formation by IF3 and the ribosome dissociation activity of the factor that supplies the pool of free 30S subunits required for translation initiation by antagonizing 30S–50S subunit association (Hershey 1987 and references therein).

As to the mechanism by which IF3 stimulates mRNA translation while ensuring initiation fidelity, IF3 increases both on- and off-rates of codon–anticodon inter-

action in the P site of the 30S subunit. The on-rate increase is reflected in an overall increased rate of translation, since formation of the 30S initiation complex is one of the slowest steps of the initiation pathway. The translation stimulation by IF3 does not affect all mRNAs to the same extent, and these differences may turn out to have a regulatory significance. Through the increase of the off-rate of codon–anticodon interaction at the P site, IF3 can discriminate between "best-fit" and "awkward" ternary complexes of 30S subunits, template, and aminoacyl tRNA, dissociating the latter type of complex at a much faster rate. A scheme depicting all complexes kinetically discriminated by IF3 is shown in Figure 3.

An additional property of IF3 is that of inducing a repositioning of 30S-bound mRNA, which is shifted from the "standby site" to the "P-decoding site" of the subunit as evidenced by the different pattern of site-directed cross-linking between mRNA and different ribosomal components obtained in the presence and absence of IF3 (La Teana et al. 1995). This shift (illustrated in Fig. 4a), whose functional significance is not entirely clear, has been correlated to the kinetic selection of the initiation triplet by IF3 and probably results from the same factor-induced conformational change of the 30S ribosomal subunit responsible for the different cross-linking pattern induced by IF3 described by Shapkina et al. (2000) and illustrated in Figure 4b. Indeed, an IF3-induced conformational change of the 30S subunit was proposed as a mechanistic explanation for the new properties acquired by the 30S subunit in the presence of IF3 (Pon and Gualerzi 1974). The model presented a few years later to rationalize the large number of IF3 functions with a single mechanism predicted the existence of two active sites in IF3 (one interacting with the head and the other with the platform of the 30S subunit) and the establishment of a fluctuating interaction of the two IF3-binding sites with the two 30S-binding sites. P-site decoding and other 30S functions would be influenced by the effect of IF3 on the conformational dynamics of the subunit envisaged as a "nodding" of its head (Pon et al. 1982).

The subsequent elucidation of the 3D structure of IF3 demonstrated that this protein indeed contains domains (Fortier et al. 1994; Biou et al. 1995; Garcia et al. 1995a,b) separated by a long and flexible linker (Moreau et al. 1997; Hua and Raleigh 1998). The two domains move (Moreau et al. 1997) and bind to the 30S subunit (Sette et al. 1999) independently. The rather unusual structural organization of IF3 suggested that its different functions might be carried out independently by either one of the two domains. In particular, ribosomal binding and ribosome dissociation activity were attributed to the C domain whereas the "initiation fidelity function" was suggested to reside in the N domain (de Cock et al. 1999).

Recent data demonstrated, however, that isolated IF3C not only binds to 30S, as predicted by a large body of biochemical and genetic data, but also is capable of performing all the other known functions of the intact molecule provided that its concentration is high enough to compensate for its reduced affinity for the 30S subunit (Fig. 5); the isolated IF3N, on the other hand, has no autonomous function and under no circumstance, in the ab-

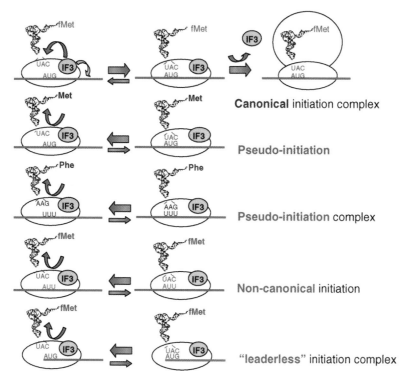

Figure 3. Schematic representation of the complexes recognized by IF3 as "correct" and "incorrect." The "canonical initiation complex" contains 30S subunits programmed with a mRNA containing a canonical initiation triplet (AUG, GUG, or UUG) and the initiator fMet-tRNA. This is the only complex not kinetically discriminated against by IF3. The factor discriminates instead against: (1) "pseudo-initiation complexes" containing noninitiator Met-tRNA with best-fit decoding by an AUG triplet or any correctly coded elongator aminoacyl-tRNA (or tRNA); (2) "noncanonical initiation complex" containing initiator fMet-tRNA bound in response to a noncanonical initiation triplet (e.g., AUU, AUA); and (3) "leaderless initiation complex" in which the real initiator fMet-tRNA is decoded by a canonical AUG triplet placed at the 5′ end of a leaderless mRNA.

sence of physical connection between the two domains, does it improve the activity of isolated IF3C (Petrelli et al. 2001). These findings have a number of important implications. First of all, they demonstrate that the two-domain organization of IF3 has no immediate bearing on the mechanism by which this factor carries out its various

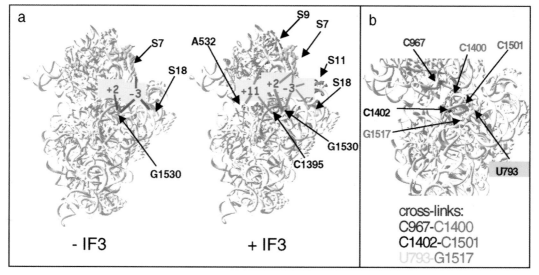

Figure 4. The IF3-dependent mRNA shift is illustrated (*a*) by highlighting the positions, within the 3D structure of the 30S subunit (Wimberly et al. 2000), of the individual components/nucleotides cross-linked to either –3, +2, or +11 of 4-thioU in the mRNA bound in the absence and in the presence of IF3 (La Teana et al. 1995). The locations of the 16S intramolecular cross-linking sites in the P-site decoding region of the 30S subunit influenced by IF3 (Shapkina et al. 2000) are shown in *b*.

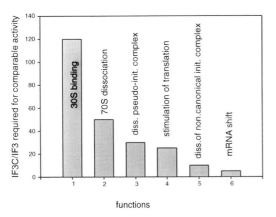

Figure 5. Activities that can be performed in vitro by the isolated carboxy-terminal domain of IF3. Although IF3C is active in all the indicated functions, its efficiency in performing them with respect to native IF3 varies greatly from case to case, decreasing from right to left. The histogram presents a quantitative view of these differences.

functions and suggest a new way to rationalize its two-domain structure. Thus, we suggest that this structure is required to modulate binding and release of IF3 to and from the 30S subunit. This suggestion is based on the notion that the affinity of IF3C for the ribosome is approximately two orders of magnitude lower than that of the intact molecule (Fig. 5), that IF3N does not increase the ribosomal affinity of IF3C in the absence of an intact linker, and that IF3 interacts with the 30S subunits first

through its C domain and then with its N domain (Sette et al. 1999). Accordingly, the model presented in Figure 6a postulates that IF3C establishes an initial somewhat unstable contact with the platform of the free 30S subunit and that the subsequent, intrinsically weaker interaction between IF3N and a second site located on the neck or head of the 30S subunit is sufficient to reduce the dissociation rate of IF3 and to stabilize the 30S–IF3 complex. Upon subunit association, the 50S subunit induces a conformational change of the 30S subunit, pushing platform and body away from each other (Lata et al. 1996; Gabashvili et al. 1999) and widening the gap between the two IF3-binding sites. The loss of the stabilizing interaction established by IF3N causes the dissociation of IF3 from the 30S subunit. In this model, the role of the linker is that of ensuring a physical connection between IF3C and its amino-terminal "anchor." This obviously important function requires the physical integrity of the linker, but does not require a particularly sophisticated sequence, in line with the finding that this structure does not display particularly relevant sequence conservation and can be extensively mutagenized without detectable effect on IF3 function in vivo (Yu and Spremulli 1997; de Cock et al. 1999).

Two alternative mechanisms have been proposed to explain the fidelity function of IF3 in translation initiation. The first entails an "indirect" inspection of the 30S-bound ligands mediated by the factor-induced conformational change of the 30S subunit (Pon and Gualerzi 1974; Pon et al. 1982). The second postulates a "direct" inspec-

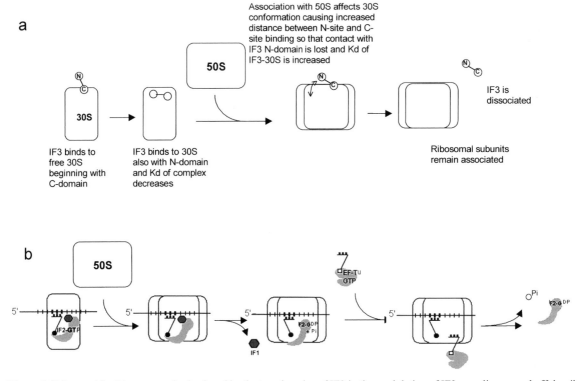

Figure 6. Schemes (*a*) of the proposed role played by the two domains of IF3 in the modulation of IF3 recycling on and off the ribosome and (*b*) of the proposed mechanism by which IF2 is ejected from the ribosomes (more details can be found in the text).

tion by IF3 of the codon–anticodon base-pairing and/ or of the anticodon stem-loop of the P-site-bound tRNA (Hartz et al. 1990; Meinnel et al.1999). However, the finding that IF3C is fully competent in performing the IF3 fidelity function seems to rule out the latter mechanisms, since the localization of IF3C by both cryo-electron microscopy (McCutcheon et al. 1999) and X-ray crystallography (Pioletti et al. 2001), although they contrast with one another, is too far away from P-site decoding and from the anticodon stem-loop of P-site-bound tRNA (Cate et al. 1999; Schluenzen et al. 2000; Wimberly et al. 2000) to allow IF3C to establish a physical contact with either structure.

Furthermore, other considerations seem to clash with the "direct" inspection model. First of all, it is difficult to envisage how 30S-bound IF3 might contact the codon and the anticodon stem-loop in a way to ensure the required discrimination, since both codon–anticodon interaction and the anticodon stem-loop of P-site-bound tRNA are buried within the 30S structure, surrounded and held by six molecular clamps (Cate et al. 1999) and inaccessible to chemical probing (Hüttenhofer and Noller 1992). Second, it is hard to imagine the particular structure/sequence supposedly inspected by IF3 whose discrimination, as seen from the scheme presented in Figure 3, is not restricted to noninitiator tRNA or to non-best-fit codon–anticodon interactions, but also includes fMet-tRNA when bound in response to a 5′AUG start codon or to an internal 5′ (CUG) or 3′ (AUU) wobbling triplet.

If the fidelity function and possibly all other activities of IF3 are due to an IF3-induced conformational change of the 30S subunit (Pon and Gualerzi 1974), it is obvious that, at variance with the earlier proposal which attributed an effect on the conformational dynamics to the fluctuating interaction of two IF3 domains (Pon et al. 1982), isolated IF3C must be sufficient to induce it. Such an occurrence is not unlikely in light of the structural flexibility of the domains of the 30S subunit and of the aforementioned conformational changes induced by IF1 (Carter et al. 2001), a single-domain protein smaller than IF3C, whose binding can indeed affect the conformation of the ribosomal subunit. Furthermore, preliminary data (P. Wollenzien, pers. comm.) indicate that isolated IF3C can promote the same conformational changes induced by native IF3 within the decoding region of the 30S P site evidenced by the cross-links (U793/G1517, C967/C1400, C1402/C1501) shown in Figure 4b.

The finding that IF3C encompasses all IF3 activities poses the question of whether a single mechanism can account for these functions. In an attempt to answer this question, a functional mapping of IF3C was performed by subjecting all eight Arg residues of this domain to a systematic mutagenesis program. The decision to mutagenize the Arg residues was taken after obtaining the evidence (following chemical modification) that at least one of these residues is essential for function and protected by the 30S subunit and by RNA. Thus, each Arg residue was replaced by Lys, Ser, His, or Leu. The resulting 32 variant IF3 molecules were tested for their residual activity in the various IF3 functions. To assess the importance of the

positive charge in each position, the activities of the His mutants were compared in some tests at both low (5.8) and high (7.7) pH. For binding to the 30S subunits, R112, R147, and R168 are of utmost importance and R99 and R116 are also implicated, albeit to a lesser extent (Fig.7a,b) The presence of positive charge was found to be particularly relevant in positions 116, 147, and to a lesser extent, 112, and much less important or totally irrelevant in all other positions. The residues involved in 30S binding roughly define a hemispherical surface, suggesting that IF3C becomes embedded within structural elements (mainly rRNA) of the 30S subunit, leaving exposed the surface of the molecule where R99, R129, R131, and R133 are located.

According to the crystallographic data (Pioletti et al. 2001), specific regions of IF3 make direct contacts with specific structural elements of the 30S subunit. In particular, the following contacts could be accounted for by both our present data and by the crystallographic results: Residue R99 is in direct contact with the bulge of helix 23 (719–723), whereas R112 and R116 contact the 3′-proximal end of helix 45 (1532–1534). On the other hand, residues R147 and R168 do not seem to make any contact with specific regions of the 30S. The region including R129, R131, and R133 from the crystallographic data is in contact with protein S18. This seems to be in contrast with our results which show that these residues are not important for 30S binding; however, R131 and R133 are involved in mRNA shift (see below), and S18 has been shown to cross-link to the mRNA (La Teana et al. 1995 and references therein), suggesting that the two regions are in very close proximity.

Dissociation of the 70S ribosomes, like the 30S binding, is little affected by mutations of R129, R131, and R133 and strongly affected by mutations of R112 and R147. However, compared to ribosomal binding, the dissociation activity is very little affected by mutations at R116 and R168 and substantially affected by R99 (Fig.7a,b). Stimulation and inhibition of translation of mRNA with a canonical AUG or noncanonical AUU start codon display roughly the same high sensitivity to substitutions of R99, R112, and R133 and low sensitivity to substitutions of R116, R133, and R168. Finally, the substitutions of R129 and R147 produce opposite effects on the two activities, the former being important for the stimulation and not for the inhibition, and vice versa for the latter (Fig.7a,b). The dissociation of a pseudo-initiation complex (Fig.7a,b) has similar requirements as the translational inhibition of noncanonical mRNAs (R99, R112, and R147) but is more sensitive than translational inhibition to substitutions of R129 and R168 and less to those of R131. Finally, it should be noted that whenever an mRNA is involved in the IF3 activity, the clustered residues R99, R129, R131, R133, and R168 become important. This is particularly evident for the mRNA shift which, in addition to R112, mainly depends on R131, R133, and R168 (Fig.7a,b).

Overall, these results (summarized in Fig.7c) indicate that the individual amino acid replacements within IF3C affect to different extents the various IF3 functions. The

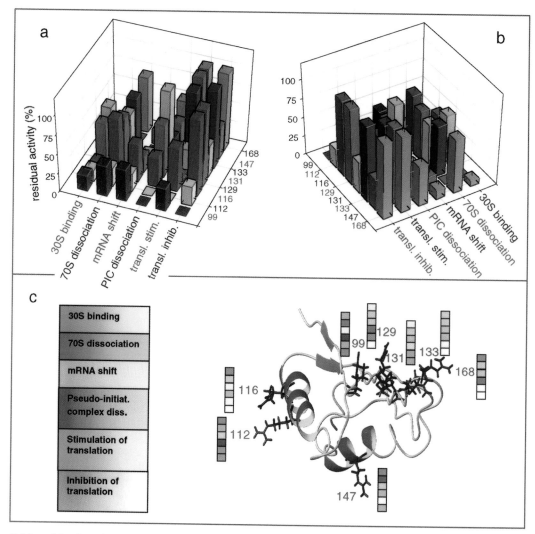

Figure 7. Map of the sites of IF3C active in performing various IF3 functions. The two specular 3D histograms (*a*) and (*b*) show the residual activity (%) in the indicated IF3 functions displayed by IF3 variants, generated by site-directed mutagenesis, in which the eight Arg residues were substituted with Lys, His, Ser, and Leu. Each residual activity value represents the average of the activity of the four mutants for a given position; (*c*) the eight Arg residues of IF3C are shown in the 3D structure of this domain (Biou et al. 1995). Near each residue, the six functions tested in vitro are indicated by different colors whose intensity decreases with decreasing importance of the corresponding residues.

only Arg residue whose mutation has a strong adverse effect on all functions is R112, which is involved in 30S binding being located in the primary RNA-binding site of the protein (which includes also Y107 and K110) as previously defined by selective chemical modification, site-directed mutagenesis, and NMR spectroscopy (for review, see Sette et al. 1999). However, ribosomal binding alone is not sufficient to elicit all IF3 functions, and a strong affinity for the 30S subunit is not a prerequisite to perform them. In fact, mutations increasing more drastically the off-rate of the IF3–30S interaction may affect more drastically those functions (e.g., 70S dissociation and 30S binding) in which the mutant IF3 must compete with high-affinity ligands such as 50S subunits and wild-type IF3. Other functions such as those involving effects of IF3 on 30S ligands (e.g., mRNA and aminoacyl tRNA) not directly competing with IF3 binding may be less affected by these mutations. Thus, the lower residual activ-

ity displayed by some mutants in 30S binding compared to other functions which also obviously require this activity may simply reflect the different rates of the various activities examined in relation to the on- and off-rates that characterize the ribosomal interactions of wild-type and mutant IF3 molecules. In agreement with this premise is the finding that, as seen from the histogram of Figure 5, the efficiency by which IF3C can replace native IF3 is very different for different functions. Overall, the data indicate that IF3 performs its various functions with somewhat different mechanisms, different parts of IF3C being more crucial for some activities than for others.

MECHANISTIC ASPECTS OF THE INITIATION PROCESS

Kinetic analyses of the individual steps of translation initiation are compatible with the pathway shown in Fig-

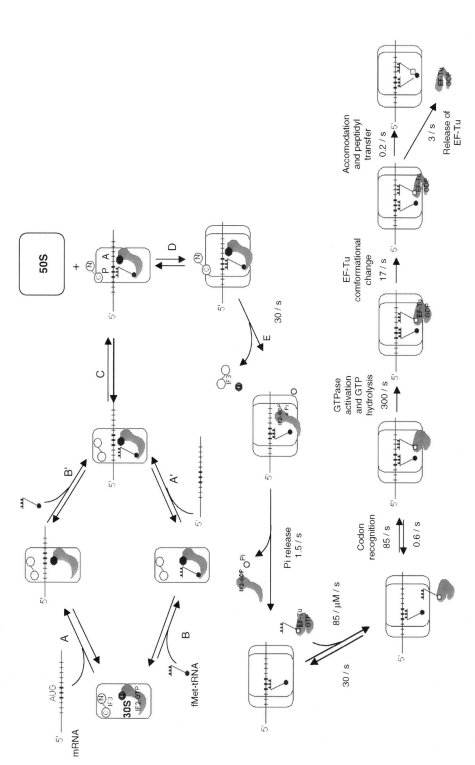

Figure 8. Scheme of the translation initiation events starting from "initiation-competent" 30S subunits and free ligands (mRNA and fMet-tRNA) and ending with the formation of the initiation dipeptide. Further details can be found in the text.

ure 8. As seen from this scheme, the 30S ribosomal sub-unit with a full complement of each initiation factor interacts in stochastic order with the first ligand, either mRNA or fMet-tRNA, to yield two binary complexes which, upon binding the second ligand, form an unstable "pre-ternary complex" consisting of a 30S subunit bearing two noninteracting ligands (mRNA and initiator tRNA). This complex is a kinetic intermediate of the bona fide 30S initiation complex which is formed through a ribosomal conformational rearrangement that induces the mRNA start codon and the anticodon of fMet-tRNA to base-pair in the P site (Gualerzi and Pon 1990; Gualerzi et al. 2000). All the steps preceding the formation of the 30S initiation complex are in rapid equilibrium, and its formation occurs through an isomerization of the pre-ternary complex. This isomerization consists of at least two first-order rearrangements kinetically controlled by the initiation factors: One, revealed by stopped-flow kinetics, is temperature-independent whereas the other, revealed by nitrocellulose filtration assays, is strongly temperature-sensitive. The rate of the slower step is estimated to be between 0.3 and 1.5 s^{-1}. The formation of the 30S initiation complex is reversible in the presence of IF3 and mRNA, and fMet-tRNA can either dissociate or become stabilized on the ribosome by the association of the 50S ribosomal subunit with the 30S initiation complex. With the ejection of IF1 and IF3 from the 30S subunit that accompanies the latter process, the ribosomal interaction of IF2 is weakened, and the fidelity function of IF3 ceases. Joining of the 50S ribosomal subunit with the 30S initiation complex is a very fast process that triggers the rapid (30 s^{-1}) hydrolysis of GTP by IF2 and a slower conformational transition of the ribosome (1.5 s^{-1}) which allows (1) release of Pi from the complex, (2) dissociation of IF2, and (3) EF-Tu•GTP•aminoacyl-tRNA binding to the A-site. Due to the overlapping binding sites of IF2 and EF-Tu, the ribosomes can become competent for EF-Tu•GTP•aminoacyl-tRNA binding only after release of IF2. EF-Tu-assisted binding of the aminoacyl tRNA to the A site is a multistep and overall rapid process which is followed by the peptidyl transfer reaction to yield the initiation dipeptide. The rate of initiation dipeptide formation is fairly slow (0.2 s^{-1}) in contrast to the faster rate (3 s^{-1}) of transpeptidation observed when the donor is an elongator tRNA (Tomsic et al. 2000 and references therein). The kinetic data indicate that after the association of the 50S subunits with the 30S initiation complex, IF2 remains bound to the 70S initiation complex for a considerable time (i.e., at least 35 msec as IF2•GTP followed by 165 msec as IF2•GDP•Pi). Binding of EF-Tu•GTP•aminoacyl-tRNA complex to the A site begins almost instantaneously on preformed 70S initiation complexes but is delayed by ~200 msec when this EF-Tu complex is offered to the 30S initiation complex together with the 50S subunits. Because, as mentioned above, only the release of IF2 can render the A site accessible to the EF-Tu•GTP•aminoacyl-tRNA complex, it can be assumed that 200 msec is also the time required by IF2 to dissociate from its ribosomal site and that the delay corresponds to the time required by the associating ribosomal subunits to undergo the conformational transition

that triggers both Pi and IF2 release. Thus, the measured rate of Pi release from 70S-bound IF2•GDP (1.5 s^{-1}) and the rate of IF2 dissociation from the ribosome are approximately the same. On the other hand, neither GTP hydrolysis nor Pi release is rate-limiting for the dissociation of IF2, since the same time delay and the same kinetics of EF-Tu•GTP•aminoacyl-tRNA binding to the A site was measured when IF2 was present in the 30S initiation complex in the presence of either GTP or GDP as well as in the complete absence of nucleotides.

Since the rate of initiation dipeptide formation was shown to be identical regardless of whether the 30S initiation complex contained IF2•GTP, IF2•GDP, or IF2 alone, it can be concluded that IF2-dependent GTPase is not involved in any of the steps (lasting ~750 msec) preceding and including the first transpeptidation. As mentioned above, these findings are at odds with the traditional views that the function of the IF2-dependent GTPase is that of allowing the final adjustment of fMet-tRNA in the P site and/or the recycling of the factor.

As to the mechanism responsible for the ejection of IF2 from the ribosomes, we can propose the existence of the following model (Fig. 6b). IF2 is initially bound to the 30S subunit, and its relatively low affinity for this particle is considerably increased by the presence of IF1. After losing the support of IF1, which is ejected from the ribosomes upon subunit association, IF2 remains anchored to the 70S ribosomes mainly via a direct, albeit weak, interaction with the 50S subunit and, indirectly, through the binding of its C-2 domain with the acceptor end of the ribosome-bound fMet-tRNA. The conformational transition of the ribosomes induced by subunit association could induce the adjustment of the fMet-tRNA acceptor arm in the peptidyl transferase center and its separation from the C-2 domain of IF2; losing the main (or only) contact between IF2 and fMet-tRNA would cause a further weakening of the IF2–ribosome interaction and thereby favor IF2 recycling.

ACKNOWLEDGMENTS

The following grants are gratefully acknowledged: MURST-PRIN 2000 to C.O.G., C.L.P., and A.L.T.; CNR 95/95 and CNR P.S. Biotecnologia to C.O.G., and CNR P.F. Biotecnologie to C.L.P.

REFERENCES

Aevarsson A., Brazhnikov E., Garber M., Zheltonosova J., Chirgadze Y., Al-Karadaghi S., Svensson L.A., and Liljas A. 1994. Three-dimensional structure of the ribosomal translocase: Elongation factor G from *Thermus thermophilus*. *EMBO J.* **13:** 3669.

Agrawal R.K. and Frank J. 1999. Structural studies of the translational apparatus. *Curr. Opin. Struct. Biol.* **9:** 215.

Ban N., Nissen P., Hansen J., Moore P.B., and Steitz T.A. 2000. The complete atomic structure of the large ribosomal subunit at 2.4 Å resolution. *Science* **289:** 905.

Biou V., Shu F., and Ramakrishnan V. 1995. X-ray crystallography shows that translational initiation factor IF3 consists of two compact alpha/beta domains linked by an alpha-helix. *EMBO J.* **14:** 4056.

Boileau G., Butler P., Hershey J.W., and Traut R.R. 1983. Direct cross-links between initiation factors 1, 2 and 3 and ribosomal proteins promoted by 2-iminothiolane. *Biochemistry* **22:** 3162.

Brock S., Szkaradkiewicz K., and Sprinzl M. 1998. Initiation factors of protein biosynthesis in bacteria and their structural relationship to elongation and termination factors. *Mol. Microbiol.* **29:** 409.

Calogero R.A., Pon C.L., Canonaco M.A., and Gualerzi C.O. 1988. Selection of the mRNA translation initiation region by *Escherichia coli* ribosomes. *Proc. Natl. Acad. Sci.* **85:** 6427.

Carter A.P., Clemons W.M., Jr., Brodersen D.E., Morgan-Warren R.J., Hartsch T., Wimberly B.T., and Ramakrishnan V. 2001. Crystal structure of an initiation factor bound to the 30S ribosomal subunit. *Science* **291:** 498.

Cate J.H., Yusupov M.M., Yusopova G.Z., Earnest T.N., and Noller H.F. 1999. X-ray crystal structures of 70S ribosome functional complexes. *Science* **285:** 2095.

Celano B., Pawlik R.T., and Gualerzi C.O. 1988. Interaction of *Escherichia coli* translation initiation factor IF1 with ribosomes. *Eur. J. Biochem.* **178:** 351.

Choi S.K., Olsen D.S., Roll-Mecak A., Martung A., Remo K.L., Burley S.K., Hinnebusch A.G., and Dever T.E. 2000. Physical and functional interaction between the eukaryotic orthologs of prokaryotic translation initiation factors IF1 and IF2. *Mol. Cell. Biol.* **20:** 7183.

Cole J.R., Olsson C.L., Hershey J.W., Grunberg-Manago M., and Nomura M. 1987. Feedback regulation of rRNA synthesis in *Escherichia coli*. Requirement for initiation factor IF2. *J. Mol. Biol.* **198:** 383.

Dahlquist K.D. and Puglisi J.D. 2000. Interaction of translation initiation factor IF1 with the *E. coli* ribosomal A site. *J. Mol. Biol.* **299:** 1.

Debenham P.G., Pongs O., and Travers A.A. 1980. Formylmethionyl-tRNA alters RNA polymerase specificity. *Proc. Natl. Acad. Sci.* **77:** 870.

de Cock E., Springer M., and Dardel F. 1999. The inter-domain linker of *Escherichia coli* initiation factor IF3: A possible trigger of translation initiation specificity. *Mol. Microbiol.* **32:** 193.

Etchegaray J.P. and Inouye M. 1999. Translational enhancement by an element downstream of the initiation codon in *Escherichia coli*. *J. Biol. Chem.* **274:** 10079.

Fortier P.L., Schmitter J.M., Garcia C., and Dardel F. 1994. The N-terminal half of initiation factor IF3 is folded as a stable independent domain. *Biochimie* **76:** 376.

Gabashvili I.S., Agrawal R.K., Grassucci R., Squires C.L., Dahlberg A.E., and Frank J. 1999. Major rearrangements in the 70S ribosomal 3D structure caused by a conformational switch in 16S ribosomal RNA. *EMBO J.* **18:** 6501.

Garcia C., Fortier P.L., Blanquet S., Lallemand J.Y., and Dardel F. 1995a. ¹H and ¹⁵N resonance assignments and structure of the N-terminal domain of *Escherichia coli* initiation factor 3. *Eur. J. Biochem.* **228:** 395.

———. 1995b. Solution structure of the ribosome-binding domain of *E. coli* translation initiation factor 3. Homology with the U1 A protein of the eukaryotic spliceosome. *J. Mol. Biol.* **254:** 247.

Godefroy-Colburn T., Wolfe A.D., Dondon J., Grunberg-Manago M., Dessen P., and Pantaloni D. 1975. Light-scattering studies showing the effect of initiation factors on the reversible dissociation of *Escherichia coli* ribosomes. *J. Mol. Biol.* **94:** 461.

Grill S., Gualerzi C.O., Londei P., and Bläsi U. 2000. Selective stimulation of translation of leaderless mRNA by initiation factor 2: Evolutionary implications for translation. *EMBO J.* **19:** 4101.

Grill S., Moll I., Hasenohrl D., Gualerzi C.O., and Bläsi U. 2001. Modulation of ribosomal recruitment to 5′-terminal start codons by translation initiation factors IF2 and IF3. *FEBS Lett.* **496:** 167.

Gualerzi C.O. and Pon C.L. 1990. Initiation of mRNA translation in prokaryotes. *Biochemistry* **29:** 5881.

———. 1996. mRNA-ribosome interaction during initiation of protein synthesis. In *Ribosomal RNA: Structure, evolution, processing, and function in protein biosynthesis* (ed. R.A.

Zimmermann and A.E. Dahlberg), p. 259. CRC Press, Boca Raton, Florida.

Gualerzi C.O., Risuleo G., and Pon C.L. 1977. Initial rate kinetic analysis of the mechanism of initiation complex formation and the role of IF3. *Biochemistry* **16:** 1684.

Gualerzi C.O., Spurio R., La Teana A., Calogero R.A., Celano B., and Pon C.L. 1989. Site-directed mutagenesis of *Escherichia coli* translation initiation factor IF1. Identification of the amino acid involved in its ribosomal binding and recycling. *Protein Eng.* **3:** 133.

Gualerzi C.O., Brandi L., Caserta E., La Teana A., Spurio R., Tomšic J., and Pon C.L. 2000. Translation initiation in bacteria. In *The ribosome: Structure, function, antibiotics, and cellular interactions* (ed. R.A Garrett et al.), p. 477. ASM Press, Washington, D.C.

Guenneugues M., Meunier S., Boelens R., Caserta E., Brandi L., Spurio R., Pon C.L., and Gualerzi C.O. 2000. Mapping the fMet-tRNA binding site of initiation factor IF2. *EMBO J.* **19:** 5233.

Guillon J.M., Mechulam Y., Schmitter J.M., Blanquet S., and Fayat G. 1992. Disruption of the gene for Met-tRNA(fMet) formyltransferase severely impairs growth of *Escherichia coli*. *J. Bacteriol.* **174:** 4294.

Haggerty T.J. and Lovett S.T. 1997. IF3-mediated suppression of a GUA initiation codon mutation in the recJ gene of *Escherichia coli*. *J. Bacteriol.* **179:** 6705.

Hartz D., McPheeters D.S., and Gold L. 1989. Selection of the initiator tRNA by *Escherichia coli* initiation factors. *Genes Dev.* **3:** 1899.

Hartz D., Binkley J., Hollingsworth T., and Gold L. 1990. Domains of initiator tRNA and initiation codon crucial for initiator tRNA selection by *Escherichia coli* IF3. *Genes Dev.* **4:** 1790.

Hershey J.W.B. 1987. Protein synthesis. In Escherichia coli *and* Salmonella typhimurium: *Cellular and molecular biology*. (ed. F.C. Neidhardt et al.), p. 613. ASM Press, Washington, D.C.

Hua Y. and Raleigh D.P. 1998. On the global architecture of initiation factor IF3: A comparative study of the linker regions from the *Escherichia coli* protein and the *Bacillus stearothermophilus* protein. *J. Mol. Biol.* **278:** 871.

Huttenhofer A. and Noller H.F. 1992. Hydroxyl radical cleavage of tRNA in the ribosomal P site. *Proc. Natl. Acad. Sci.* **89:** 7851.

Lata K.R., Agrawal R.K., Penczek P., Grassucci R., Zhu J., and Frank J. 1996. Three-dimensional reconstruction of the *Escherichia coli* 30 S ribosomal subunit in ice. *J. Mol. Biol.* **262:** 43.

La Teana A., Gualerzi C.O., and Brimacombe R. 1995. From stand-by to decoding site. Adjustment of the mRNA on the 30S ribosomal subunit under the influence of the initiation factors. *RNA* **1:** 772.

La Teana A., Gualerzi C.O., and Dahlberg A.E. 2001. Initiation factor IF2 binds to the α-sarcin loop and helix 89 of *Escherichia coli* 23S ribosomal RNA. *RNA* **7:** 1173.

La Teana A., Pon C.L., and Gualerzi C.O. 1996. Late events in translation initiation. Adjustment of fMet-tRNA in the ribosomal P-site. *J. Mol. Biol.* **256:** 667.

La Teana A., Brandi A., O'Connor M., Freddi S., and Pon C.L. 2000. Translation during cold adaptation does not involve mRNA-rRNA base pairing through the downstream box. *RNA* **6:** 1393.

Luchin S., Putzer H., Hershey J.W., Cenatiempo Y., Grunberg-Manago M., and Laalami S. 1999. In vitro study of two dominant inhibitory GTPase mutants of *Escherichia coli* translation initiation factor IF2. Direct evidence that GTP hydrolysis is necessary for factor recycling. *J. Biol. Chem.* **274:** 6074.

Martin-Farmer J. and Janssen G.R. 1999. A downstream CA repeat sequence increases translation from leadered and unleadered mRNA in *Escherichia coli*. *Mol. Microbiol.* **31:** 1024.

Matthews R.G. 1996. One-carbon metabolism. In Escherichia coli *and* Salmonella typhimurium: *Cellular and molecular biology*. (ed. F.C. Neidhardt et al.), p. 600. ASM Press, Washington, D.C.

McCarthy J.E. and Brimacombe R. 1994. Prokaryotic translation: The interactive pathway leading to initiation. *Trends Genet.* **10:** 402.

McCutcheon J.P., Agrawal R.K., Philips S.M., Grassucci R.A., Gerchman S.E., Clemons W.M., Jr., Ramakrishnan V., and Frank J. 1999. Location of translational initiation factor IF3 on the small ribosomal subunit. *Proc. Natl. Acad. Sci.* **96:** 4301.

Meinnel T., Sacerdot C., Graffe M., Blanquet S., and Springer M. 1999. Discrimination by *Escherichia coli* initiation factor IF3 against initiation on non-canonical codons relies on complementarity rules. *J. Mol. Biol.* **290:** 825.

Meunier S., Spurio R., Czisch M., Wechselberger R., Guenneugues M., Gualerzi C.O., and Boelens R. 2000. Structure of the fMet-tRNA(fMet)-binding domain of *B. stearothermophilus* initiation factor IF2. *EMBO J.* **19:** 1918.

Moazed D., Samaha R.R., Gualerzi C., and Noller H.F. 1995. Specific protection of 16S rRNA by translational initiation factors. *J. Mol. Biol.* **248:** 207.

Moll I., Huber M., Grill S., Sairafi P., Mueller F., Brimacombe R., Londei P., and Bläsi U. 2001. Evidence against an interaction between the mRNA downstream box and 16S rRNA in translation initiation. *J. Bacteriol.* **183:** 3400.

Moreau M., de Cock E., Fortier P.L., Garcia C., Albaret C., Blanquet S., Lallemand J.Y., and Dardel F. 1997. Heteronuclear NMR studies of *E. coli* translation initiation factor IF3. Evidence that the inter-domain region is disordered in solution. *J. Mol. Biol.* **266:** 15.

Nissen P., Hansen J., Ban N., Moore P.B., and Steitz T.A. 2000. The structural basis of ribosome activity in peptide bond synthesis. *Science* **289:** 920.

Nissen P., Kjeldgaard M., Thirup S., Clark B.F., and Nyborg J. 1995. Crystal structure of the ternary complex of PhetRNAPhe, EF-Tu, and a GTP analog. *Science* **270:** 1464.

Nomura T., Fujita N., and Ishihama A. 1986. Promoter selectivity of *Escherichia coli* RNA polymerase: Alteration by fMet-tRNAfMet. *Nucleic Acids Res.* **14:** 6857.

O'Connor M., Asai T., Squires C.L., and Dahlberg A.E. 1999. Enhancement of translation by the downstream box does not involve mRNA-rRNA base pairing. *Proc. Natl. Acad. Sci.* **96:** 8973.

Olsson C.L., Graffe M., Springer M., and Hershey J.W.B. 1996. Physiological effects of translation initiation factor IF3 and ribosomal protein L20 limitation in *Escherichia coli. Mol. Gen. Genet.* **250:** 705.

Petrelli D., La Teana A., Garofalo C., Spurio R., Pon C.L., and Gualerzi C.O. 2001. Translation initiation factor IF3: Two domains, five functions, one mechanism? *EMBO J.* **20:** 4560.

Pioletti M., Schlünzen F., Harms J., Zarivach R., Glühmann M., Avila H., Bashan A., Bartels H., Auerbach T., Jacobi C., Hartsch T., Yonath A., and Franceschi F. 2001. Crystal structure of complexes of small ribosomal subunit with tetracycline, edeine and IF3. *EMBO J.* **20:** 1829.

Pon C.L. and Gualerzi C. 1974. Effect of initiation factor 3 binding on the 30S ribosomal subunits of *Escherichia coli. Proc. Natl. Acad. Sci.* **71:** 4950.

———. 1984. Mechansim of protein biosynthesis in prokaryotic cells. Effect of IF1 on the initial rate of 30S initiation complex formation. *FEBS Lett.* **175:** 203.

Pon C.L., Pawlik R.T., and Gualerzi C. 1982. The topographical localization of IF3 on *Escherichia coli* 30S ribosomal subunits as a clue to its way of functioning. *FEBS Lett.* **137:** 163.

Risuleo G., Gualerzi C., and Pon C.L. 1976. Specificity and properties of the destabilization, induced by initiation factor IF-3, of ternary complexes of the 30-S ribosomal subunit, aminoacyl-tRNA and polynucleotides. *Eur. J. Biochem.* **67:** 603.

Rocha E.P., Danchin A., and Viari A. 1999. Translation in *Bacillus subtilis:* Roles and trends of initiation and termination, insights from a genome analysis. *Nucleic Acids Res.* **27:** 3567.

Rodnina M.V., Stark H., Savelsbergh A., Wieden H.J., Mohr D., Matassova N.B., Peske F., Daviter T., Gualerzi C.O., and Wintermeyer W. 2000. GTPases mechansims and functions of translation factors on the ribosome. *Biol. Chem.* **381:** 377.

Roll-Mecak A., Cao C., Dever, T.E., and Burley S.K. 2000. X-ray structures of the universal translation initiation factor IF2/eIF5B: Conformational changes on GDP and GTP binding. *Cell* **103:** 781.

Schluenzen F., Tocilj A., Zarivach R., Harms J., Gluehmann M., Janell D., Bashan A., Bartels H., Agmon I., Franceschi F., and Yonath A. 2000. Structure of functionally activated small ribosomal subunit at 3.3 angstroms resolution. *Cell* **102:** 615.

Sette M., Spurio R., van Tilborg P., Gualerzi C.O., and Boelens R. 1999. Identification of the ribosome binding sites of translation initiation factor IF3 by multidimensional heteronuclear NMR spectroscopy. *RNA* **5:** 82.

Sette M., van Tilborg P., Spurio R., Kaptein R., Paci M., Gualerzi C.O., and Boelens R. 1997. The structure of the translational initiation factor IF1 from *E. coli* contains an oligomer-binding motif. *EMBO J.* **16:** 1436.

Shapkina T.G., Dolan M.A., Babin P., and Wollenzien P. 2000. Initiation factor 3-induced structural changes in the 30S ribosomal subunit and in complexes containing tRNA(f)(Met) and mRNA. *J. Mol. Biol.* **299:** 615.

Sprengart M.L. and Porter A.G. 1997. Functional importance of RNA interactions in selection of translation initiation codons. *Mol. Microbiol.* **24:** 19.

Sprengart M.L., Fuchs E., and Porter A.G. 1996. The downstream box: An efficient and independent translation initiation signal in *Escherichia coli. EMBO J.* **15:** 665.

Spurio R., Brandi L., Caserta E., Pon C.L., Gualerzi C.O., Misselwitz R., Krafft C., Welfle K., and Welfle H. 2000. The C-terminal sub-domain (IF2 C-2) contains the entire fMet-tRNA binding site of initiation factor IF2. *J. Biol. Chem.* **275:** 2447.

Stringer E.A., Sarkar P., and Maitra U. 1977. Function of initiation factor 1 in the binding and release of initiation factor 2 from ribosomal initiation complexes in *E. coli. J. Biol. Chem.* **252:** 1739.

Sussman J.K., Simons E., and Simons R.W. 1996. *Escherichia coli* translation initiation factor 3 discriminates the initiation codon in vivo. *Mol. Microbiol.* **21:** 347.

Szkaradkiewicz K., Zuleeg T., Limmer S., and Sprinzl M. 2000. Interaction of fMet-tRNA^fMet and fMet-AMP with the C-terminal domain of *Thermus thermophilus* translation initiation factor 2. *Eur. J. Biochem.* **267:** 4290.

Tedin K., Moll I., Grill S., Resch A., Graschopf A., Gualerzi C.O., and Bläsi U. 1999. Translation initiation factor 3 antagonizes authentic start codon selection on leaderless mRNAs. *Mol. Microbiol.* **31:** 67.

Tomsic J., Vitali L.A., Daviter T., Savelsbergh A., Spurio R., Striebeck P., Wintermeyer W., Rodnina M.V., and Gualerzi C.O. 2000. Late events of translation initiation in bacteria: A kinetic analysis. *EMBO J.* **19:** 2127.

Travers A.A., Debenham P.G., and Pongs O. 1980. Translation initiation factor 2 alters transcriptional selectivity of *Escherichia coli* ribonucleic acid. *Biochemistry* **19:** 1651.

Van Etten W.J. and Janssen G.R. 1998. An AUG initiation codon, not codon-anticodon complementarity, is required for the translation of unleadered mRNA in *Escherichia coli. Mol. Microbiol.* **27:** 987.

Wakao H., Romby P., Ebel J.P., Grunberg-Manago M., Ehresmann C., and Ehresmann B. 1991. Topography of the *Escherichia coli* ribosomal 30S subunit-initiation factor 2 complex. *Biochimie* **73:** 991.

Wimberly B.T., Brodersen D.E., Clemons W.M., Jr., Morgan-Warren R.J., Carter A.P., Vonrhein C., Hartsch T., and Ramakrishnan W. 2000. Structure of the 30S ribosomal subunit. *Nature* **407:** 327.

Wintermeyer W. and Gualerzi C.O. 1983. Effect of *E. coli* initiation factors on the kinetics of N-AcPhe-tRNA^Phe binding to 30S ribosomal subunits: A fluorescence stopped-flow study. *Biochemistry* **22:** 690.

Wu C.J. and Janssen G.R. 1996. Expression of a streptomycete leaderless mRNA encoding chloramphenicol acetyltransferase in *Escherichia coli. J. Bacteriol.* **179:** 6824.

Yu N.J. and Spremulli L.L. 1997. Structural and mechanistic studies on chloroplast translational initiation factor 3 from *Euglena gracilis. Biochemistry* **36:** 14827.

The Translation of Capped mRNAs Has an Absolute Requirement for the Central Domain of eIF4G but Not for the Cap-binding Initiation Factor eIF4E

I.K. ALI AND R.J. JACKSON

Department of Biochemistry, University of Cambridge, Cambridge CB2 1GA, United Kingdom

Current models propose that initiation of mRNA translation in prokaryotes requires just three initiation factors, each of them a single polypeptide chain, with an aggregate mass of 125 kD. In eukaryotes, by contrast, no fewer than eleven separable initiation factors are required, some of which consist of several polypeptide chains, and the aggregate mass of the >26 polypeptides is more than 1600 kD (for review, see Hershey and Merrick 2000), which is somewhat larger than the 40S subunit itself. No other aspect of mRNA translation shows such a large difference in complexity between prokaryotes and eukaryotes.

In contrast to prokaryotic initiation, it is an almost universal rule that washed eukaryotic small (40S) ribosomal subunits cannot bind to mRNA (certainly not at the correct initiation site) in the absence of initiation factors, with the notable exceptions of the internal ribosome entry site/segments (IRESs) of hepatitis C virus (HCV) and the closely related pestiviruses (for review, see Jackson 2000). Coupled with this unique property of direct binding of washed 40S subunits to the correct site, initiation on these IRESs is also unique in that it does not require eIF1 and 1A (probably because there is no ribosome scanning), nor eIF4A, 4B, 4E, or 4G (either as singular entities, or as the eIF4F complex), and does not require ATP hydrolysis (Pestova et al. 1998). How then, do 40S ribosomal subunits get delivered to the mRNA in all other cases apart from these exceptional IRESs? The key players in this process are the eIF4F complex, as is evident from the unique properties of the HCV and pestivirus IRESs, and eIF3, which alone accounts for ~40% of the aggregate mass and ~40% of the ~26 polypeptide chains of the complete set of eukaryotic initiation factors. In view of this complexity, it is not surprising that eIF3 has many roles ascribed to it, and this is reflected in its interactions with many other factors: eIF1, eIF4B, eIF4G, and eIF5 (for review, see Hershey and Merrick 2000). It was originally ascribed the function of an anti-association factor, binding to 40S subunits and preventing their association with 60S subunits until the appropriate step in the initiation pathway. However, eIF3 is still required for initiation complex formation in assays where the anti-association activity is not required because 60S subunits have been omitted, conditions that reveal additional eIF3 functions of stabilizing eIF2/Met-tRNA$_i$/GTP ternary complex binding to 40S subunits, and promoting the binding of these primed 40S subunits to the mRNA. In the case of the HCV and pestivirus IRESs, where eIF3 is not required for binding of the 40S subunits to the correct site on the RNA, it is nevertheless still required for the overall process of initiation, apparently for the ribosomal subunit joining step (Pestova et al. 1998).

The other key player, the eIF4F holoenzyme complex, consists of a backbone or scaffold polypeptide, eIF4G, to which several other initiation factors bind (Fig.1), although in some cases the association may be transitory rather than permanent. There are two eIF4G species in mammalian cells: the originally described form, now known as eIF4GI, and a more recently discovered and less abundant species, eIF4GII (Gradi et al. 1998b). Since the two species seem to share all functions in common, they will be referred to collectively as eIF4G. Mammalian eIF4G is considered to be composed of three domains: an amino-terminal one-third domain defined by the site of cleavage of eIF4G by picornavirus proteases (Fig. 1); a central domain, which is the absolutely indispensable fragment; and a carboxy-terminal one-third domain which has no counterpart in yeast and plant eIF4Gs (for review, see Gingras et al. 1999). The amino-terminal domain interacts with eIF4E (Lamphear et al. 1995), the only initiation factor that binds directly to the 5′-cap structure, and also with poly(A)-binding protein (Imataka et al. 1998). The central domain has RNA-binding properties and interacts with eIF4A (Imataka and Sonenberg 1997; Korneeva et al. 2000, 2001; Lomakin et al. 2000; Morino et al. 2000), which has ATP-dependent RNA helicase activity. It also interacts with eIF3 (Lamphear et al. 1995; Korneeva et al. 2000; Lomakin et al. 2000; Morino et al. 2000), which itself binds 40S ribosomal subunits, and since the two interactions of eIF3 seem to be not mutually exclusive (Hershey and Merrick 2000), there is the potential for a tripartite eIF4G/eIF3/40S subunit interaction relay.

In the translation of capped mRNAs by the scanning ribosome mechanism, eIF4F is thought to interact with the 5′ cap via its eIF4E component, as well as via direct interactions between the RNA and the central domain of eIF4G (Haghighat and Sonenberg 1997), and as a consequence, the helicase activity of the eIF4A associated with the eIF4G central domain is directed toward the 5′-proximal region of the RNA, and the eIF4G/eIF3/40S subunit interaction relay delivers the 40S subunit to the same region of the mRNA, in preparation for scanning (for review, see Jackson 2000). Several, but not all, picor-

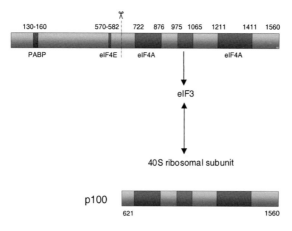

Figure 1. Domain structure of eIF4G(I) and sites of interaction of other proteins with eIF4G. The amino acid numbering of the sites on eIF4GI for interaction with PABP and eIF4E are from Gingras et al. (1999), and the sites for eIF4A and eIF3 interaction are from Korneeva et al. (2000, 2001). Note, however, that the central domain binding site for eIF4A has been defined by others as amino acids 722–969 (Lomakin et al. 2000) or 672–970 (Morino et al. 2000), and the eIF3 interaction site as amino acids 746–1076 (Lomakin et al. 2000) or 672–1065 (Morino et al. 2000). The scissors symbol denotes the site(s) of cleavage of eIF4GI by picornavirus proteases; entero-/rhinovirus 2A proteases and FMDV L-protease actually cleave at different sites, just 7 amino acid residues apart. Below is depicted the p100 fragment of eIF4GI (amino acids 621–1560) used in this work.

navirus species encode a protease that cleaves eIF4G into the amino-terminal one-third domain, and a carboxy-terminal two-thirds fragment, which will be referred to here as p100 (Kirchweger et al. 1994; Lamphear et al. 1995). This cleavage does not compromise translation of the viral RNA, which is invariably by an IRES-dependent mechanism. In the best-studied example, encephalomyocarditis virus (EMCV) RNA, the intact eIF4F complex or p100 binds with high affinity to a site in the IRES near the actual internal initiation codon (Pestova et al. 1996a,b; Kolupaeva et al. 1998; Lomakin et al. 2000), and this binding is thought to lead to 40S subunit delivery to the initiation codon via the eIF4G/eIF3/40S subunit interaction relay.

The cleavage of eIF4G that occurs during infection by many picornavirus species generally coincides with the shutoff of host-cell mRNA translation, and addition of the proteases to cell-free translation assays generally results in an inhibition of capped mRNA translation. These observations have led to the supposition that neither of the two cleavage products can support the translation of capped mRNAs, presumably because the cap-binding function of the eIF4E/eIF4G (amino-terminal domain) complex is now separated from the p100 fragment with the all-important eIF4G central domain. However, we show here that this presumption is incorrect, and that p100, surprisingly, can support quite efficient translation of capped mRNA translation from the 5′-proximal AUG codon. These findings necessitate a re-evaluation of the mechanism of host-cell shutoff by picornaviruses.

RESULTS AND DISCUSSION

Development of an eIF4G-depleted Reticulocyte Lysate System

We initially set out with a quite different aim, namely to test the ability of various eIF4G deletion and point mutants to support translation driven by the EMCV IRES. This aim necessitated the development of an eIF4G-depleted rabbit reticulocyte lysate system for assaying these mutants in the absence of a high background of translation promoted by endogenous intact eIF4G. Several affinity depletion approaches were tried but found wanting, either because depletion was incomplete, or because the recovery on eIF4G add-back was poor: m⁷GDP-Sepharose (to deplete eIF4G indirectly by depleting eIF4E); recombinant wild-type eIF4A or dominant negative R362Q mutant eIF4A (Pause et al. 1994a) covalently linked to CNBr-activated Sepharose; and an RNA affinity column, comprising the binding site of eIF4G on the EMCV IRES, commonly known as the J-K domain (Duke et al. 1992; Pestova et al. 1996b; Kolupaeva et al. 1998; Lomakin et al. 2000), covalently linked to CNBr-activated Sepharose. Success was finally achieved with a method based on a strategy used by Stassinopoulos and Belsham (2001) for depletion but not tested in add-back assays: poly(A)-tailed J-K domain of the EMCV IRES bound to oligo-dT magnetic beads.

A western blot, using an antibody raised against a peptide epitope in the carboxy-terminal one-third domain of eIF4GI (which reacts equally with human and rabbit eIF4GI and has been observed to cross-react with eIF4GII), shows that extensive depletion of eIF4G was achieved (Fig. 2A), and quantitative blotting assays put the degree of depletion at >95% (Ali et al. 2001a). The translation assays show that, as expected, the depleted system is defective in translating both the upstream scanning-dependent cistron of a dicistronic mRNA and the downstream IRES-dependent cistron, which in this particular example is driven by the EMCV IRES (Fig. 2B). Indeed, we find that translation of all mRNAs is severely reduced in the depleted lysate, except translation driven by the HCV IRES (Fig. 2C) or pestivirus IRESs, again as predicted. On supplementing the depleted system with recombinant p100, a very efficient rescue of EMCV IRES activity was observed (Fig. 2B), but there was no recovery of upstream cistron translation. As shown below, this is not due to an intrinsic inability of p100 to support capped mRNA translation, but is the result of competition by the EMCV IRES. Consistent with this explanation, in the case of the RNA with the HCV IRES, which does not require and does not bind eIF4G or p100, and is therefore not a competitor for p100, addition of p100 did rescue upstream cistron translation (Fig. 2C).

What other factors are depleted by this procedure, and to what extent? Although a frequently asked question, this is really not a significant issue, since very good recovery of EMCV IRES activity was observed despite the fact that high, near-saturating RNA concentrations were used (to provide a stringent test for recovery). This shows that whatever other factors may have been depleted, they

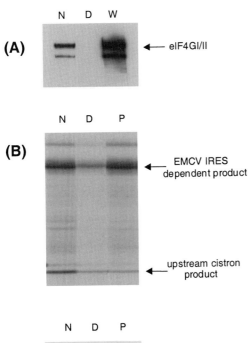

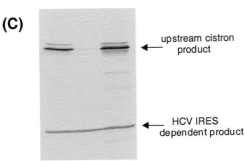

Figure 2. Affinity depletion of eIF4G from reticulocyte lysates. (*A*) Samples (1 μl) of parent lysate (N) and depleted lysate (D) were separated by gel electrophoresis, the gel blotted and probed with anti-eIF4G antiserum. Lane W was loaded with approximately one-fifth of the total SDS-sample buffer eluate of affinity matrix that had been used to deplete 50 μl of lysate. (*B, C*) Translation assays were carried in the parent (non-depleted) lysate (N), in the depleted lysate (D), or in depleted lysate supplemented with 20 μg/ml recombinant p100 (P), with the following RNAs: (*B*) 25 μg/ml capped dicistronic mRNA with an upstream cistron coding for influenza virus NS1, an EMCV IRES, and EMCV coding sequences for L-VP0 (~55 kD) as downstream cistron; (*C*) 25 μg/ml capped dicistronic mRNA with an upstream cistron coding for *Xenopus laevis* cyclin B2, a hepatitis C virus IRES, and a downstream cistron coding for influenza virus NS1. Radiolabeled translation products were examined by gel electrophoresis and autoradiography. The products encoded by the upstream scanning-dependent cistron and the downstream IRES-dependent cistron are indicated. (Reprinted, with permission, from Ali et al. 2001a [copyright Oxford University Press].)

cannot have been depleted to the point where they became limiting. Nevertheless, we have used western blotting to assess the extent of depletion of some other factors, with the following results: About 10% of the eIF3 was depleted, 30–40% of eIF4A, 10–20% of eIF4B, ~25% of

eIF4E, and 50–60% of poly(A)-binding protein (PABP). Is the partial depletion of these factors significant? This question can only be answered by functional tests, i.e., translation assays. We found that in the absence of p100, supplementation of the depleted lysate with any of these other factors, either alone or in combination, caused absolutely no stimulation of translation. With some batches of depleted lysate, the p100-dependent rescue of translation of capped RNAs (discussed below), particularly globin mRNA rather than brome mosaic virus (BMV) RNA, was modestly enhanced by coaddition of eIF4A and eIF4B, but eIF4E and PABP had no influence.

We conclude that this depleted lysate can be regarded as a system that is strictly dependent on at least a fragment of eIF4G but not absolutely dependent on any other factor. The only caveat concerns RNAs with the poliovirus or rhinovirus IRES, whose translation in the depleted lysate is rescued rather poorly by addition of p100 either alone, or together with eIF4A, 4B, 4E, and PABP. Evidently the procedure depletes the lysate of some other, as yet unidentified, component that is required uniquely for initiation on these IRESs.

Translation of Capped mRNAs in the eIF4G-depleted System

We next examined the translation of capped or uncapped dicistronic transcripts with a hepatitis A virus IRES as intercistronic spacer (Ali et al. 2001b) and were very surprised to see that in both cases translation of the upstream (scanning-dependent) cistron was strongly stimulated by addition of p100 to the depleted lysate (Fig. 3). Stimulation of the uncapped transcript was not unexpected, since it has been shown previously that p100 addition to standard (non-depleted) lysates stimulates translation of uncapped versions of normally capped mRNAs (De Gregorio et al. 1998). However, the stimulation of the upstream cistron of the capped species was surprising in view of the prevailing presumption, based on the shutoff of host-cell mRNA translation in poliovirus-infected cells, that the eIF4G cleavage products (which include p100) cannot support capped mRNA translation. Given that capping of in vitro transcripts is far from 100% efficient, in fact only about 70% in our hands (Dasso and Jackson 1989), our initial presumption was that what we had observed in the case of the capped transcript might in fact be p100-dependent stimulation of just the uncapped RNAs in the preparation. However, the fact that the yield of upstream cistron product in the presence of p100 was the same with the two types of mRNAs (Fig. 3) made this seem an implausible explanation, which provoked us into speculating whether p100 might be able to support capped mRNA translation in the depleted system. Because of the problem of incomplete capping of in vitro transcripts, a rigorous test of this question obviously required the use of "natural" mRNAs that are 100% capped, and for this purpose we chose BMV RNA extracted from virions, and also globin mRNA prepared from reticulocyte lysates.

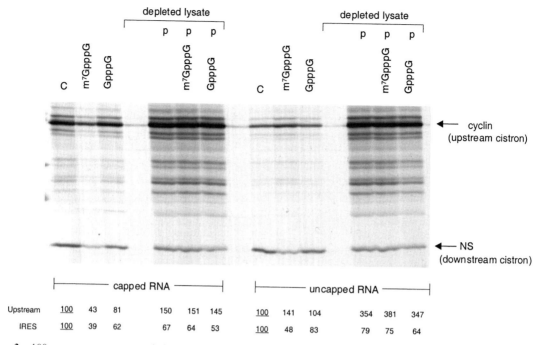

Figure 3. p100 appears to rescue translation of the upstream scanning-dependent cistron of both capped and uncapped dicistronic mRNAs with the hepatitis A virus (HAV) IRES. Capped and uncapped dicistronic mRNAs, as indicated, were translated at 25 μg/ml in either eIF4G-depleted reticulocyte lysate or parent (non-depleted lysate), in the presence of added KCl at 70 mM. Cap analogs, either m⁷GpppG or GpppG, were added at 0.4 mM, where indicated, together with 0.32 mM additional MgCl₂. Where indicated (lanes labeled *p*), recombinant p100 was added at 20 μg/ml. Translation was at 30°C for 60 minutes and the translation products were analyzed by SDS-PAGE followed by autoradiography. The positions of the upstream (cyclin) cistron product, and downstream, HAV IRES-dependent, NS product are shown. The yields of radiolabeled translation products of the upstream and IRES-driven cistrons were determined by scanning densitometry and are expressed relative to the yield in the corresponding control assay which was set at 100. (Reprinted, with permission, from Ali et al. 2001b [copyright ASM Press].)

In fact, the characteristics of BMV RNA and globin mRNA translation were almost identical to those of the upstream cistron of the dicistronic mRNA with the HAV IRES: Translation was severely impaired in the depleted lysate and efficiently rescued by addition of p100. Moreover, just as was the case with the upstream cistron of capped dicistronic mRNAs with the HAV IRES (Fig. 3), the translation of BMV RNA (and globin mRNA) in the depleted lysate supplemented with p100 was completely resistant to inhibition by m⁷GpppG cap analog, whereas translation in the parent (non-depleted) lysate was very sensitive to cap analog (but not to GpppG). Thus, in operational terms, translation of capped mRNAs in the depleted system supplemented with p100 is "cap-independent."

To put these observations into a physiological perspective, we examined the concentration dependence of this rescue by p100, and compared these data with the endogenous concentration of eIF4G present in our reticulocyte lysates (Fig. 4). For the latter purpose, a western blotting analysis was carried out using depleted lysate supplemented with increasing concentrations of p100, and the signals obtained were compared with that given by the parent (non-depleted) lysate after treatment with foot and mouth disease virus (FMDV) L-protease to cleave the endogenous eIF4G to p100, a step that was considered necessary in order to compare "like with like," because the intensity of the western blot signal is

greater after eIF4G cleavage than with the intact factor. The results indicate that the concentration of endogenous eIF4G in the parent lysate is equivalent to ~ 3 μg/ml (Fig. 4). Addition of 2.5 μg/ml p100 to the depleted lysate effected a discernible rescue of BMV RNA translation, and maximum rescue was achieved with just over 10 μg/ml. (Slightly higher concentrations were required for maximum rescue of globin mRNA translation.) Therefore, maximum rescue of BMV RNA translation requires three- to fourfold more p100 than the endogenous eIF4G concentration in our lysates, but a significant rescue can be detected at a concentration equal to the endogenous eIF4G level.

There are two aspects of p100-driven BMV RNA translation in eIF4G-depleted lysates that deserve comment. One is that there are remarkably few incomplete products, indeed no more than when BMV RNAs are translated in the parent lysate (Fig. 4). This implies that p100 is delivering the 40S subunits to a location between the 5′ cap and the first AUG codon with the same degree of accuracy as when it is the intact eIF4F holoenzyme complex that promotes this delivery; p100 is not causing an increased level of initiation at random internal sites. The other point of interest is that the relative yields of the four BMV RNA translation products are different in an eIF4G-depleted lysate supplemented with p100 than in the parent lysate. Translation of RNA-4 is rescued rather inefficiently, perhaps because it has a very short 5′UTR

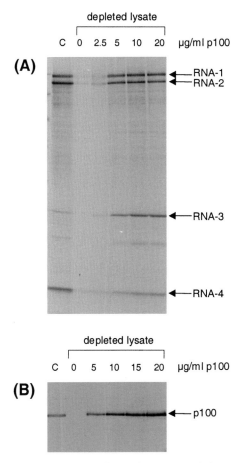

Figure 4. The dependency of capped mRNA translation on p100 concentration, in relation to the concentration of endogenous eIF4G in the parent (non-depleted) lysate. (*A*) BMV RNA was translated at a final concentration of 20 μg/ml in either the control (non-depleted) lysate (C), or the eIF4G-depleted lysate supplemented with 0, 2.5, 5, 10, or 20 μg/ml p100 as indicated. Radiolabeled translation products were visualized by gel electrophoresis and autoradiography. The products encoded by BMV RNA-1, RNA-2, RNA-3, and RNA-4 are indicated. (*B*) Western blot analysis. eIF4G-depleted lysate was supplemented with 0, 5, 10, 15, or 20 μg/ml recombinant p100 as indicated. Parent (non-depleted) lysate (C) was preincubated for 10 minutes at 30°C with in-vitro-expressed FMDV L-protease. Aliquots (equivalent to 1 μl of lysate) were separated by gel electrophoresis, and p100 was detected by western blotting using an anti-eIF4G antiserum raised against a peptide epitope in the carboxy-terminal region of eIF4G. (Reprinted, with permission, from Ali et al. 2001a [copyright Oxford University Press].)

of 9 nucleotides (Dasgupta and Kaesberg 1982). On the other hand, RNA-3 (5′UTR of 91 nucleotides) is rescued especially efficiently, such that it is a better mRNA in the depleted lysate than in the parent lysate. Of the two large RNAs, rescue is rather better for RNA-1 (74-nucleotide 5′UTR) than RNA-2 (103-nucleotide 5′UTR), even though the two 5′UTRs differ in only 3 positions in the first 47 nucleotides (Ahlquist et al. 1984). The efficiency of rescue of globin mRNA was intermediate between the extremes of BMV RNA-3 and -4, and close to that of BMV RNA-2 (Ali et al. 2001a). Therefore, the relative functional affinity of p100 for the four RNAs must be different from that of the eIF4F complex.

Is the Amino-terminal Cleavage Product of eIF4F an Inhibitor of Capped mRNA Translation?

Although the evidence that p100 can support efficient translation of capped mRNAs is unambiguous, the results were surprising in view of the well-known fact that cleavage of the eIF4G component of eIF4F by picornavirus proteases, which generates p100 and an eIF4E/eIF4G (amino-terminal domain) complex, inhibits translation of capped mRNAs. This prompts the question as to what exactly are the differences between the depleted system supplemented with p100 and a system treated with viral proteases (or prepared from infected cells), which might explain why p100 can support capped mRNA translation in the one case, but not, apparently, in the other. There are three main differences: (1) As shown above (Fig. 4), the concentration of p100 in the protease-treated extract will be lower than that which we have generally added to the depleted system, and there can be little doubt that this difference partly explains the inhibition caused by the protease; (2) the depleted system translating BMV RNA has no picornaviral RNA that could act as a competitor, an issue that is examined in the next section; and (3) the depleted system does not contain the amino-terminal cleavage product of eIF4G. The last of these differences raises the question of whether the eIF4E/eIF4G (amino-terminal domain) cleavage product might retain the ability to bind to 5′ caps, and whether this binding might, by steric hindrance, prevent p100 from interacting with the cap-proximal part of the mRNA. It is true that both singular eIF4E and 4E-BP1/eIF4E complex, and presumably also the eIF4E/eIF4G (amino-terminal domain) complex, have been reported to bind to capped RNAs with much lower affinity than that of the intact eIF4F complex (Haghighat and Sonenberg 1997), a difference that was attributed to the RNA-binding properties of the central domain of the eIF4G component of eIF4F. However, unlike the case of singular eIF4E or eIF4E/4E-BP1 complex, the eIF4E/eIF4G (amino-terminal domain) cleavage product generated by picornavirus proteases retains the ability to interact with PABP (Fig. 1) and therefore would be expected to remain tethered to a capped polyadenylated mRNA by interacting with PABP bound at the 3′-poly(A) tail (Imataka et al. 1998). This tethering might allow appreciable binding of the eIF4E/eIF4G (amino-terminal domain) complex to the mRNA cap, despite the lower intrinsic affinity in comparison with eIF4F.

If the eIF4E/eIF4G (amino-terminal domain) cleavage product does bind to the 5′ cap and thereby sterically hinders functional interaction of the p100 cleavage product with the mRNA, the prediction is that inhibition of translation of capped mRNAs caused by addition of picornavirus proteases should be at least partly reversed by addition of m⁷GpppG cap analog, which would prevent the putative inhibitory interaction of the eIF4E/eIF4G (amino-terminal domain) complex with the cap. We have indeed observed that when high concentrations of FMDV L-protease were used to inhibit capped globin mRNA translation, the addition of m⁷GpppG cap analog did, in complete contrast to its usual effect, actually stimulate translation by up to twofold. However, viewed as relief of inhibition caused by the protease, this stimulation is inconsequential; a 97–98% inhibition caused by the protease is decreased to 94–96% inhibition when cap analog was also added. We conclude that any binding of the eIF4E/eIF4G (amino-terminal domain) complex to the cap does not inhibit initiation significantly, and that the absence of this complex from our eIF4G-depleted system cannot be the explanation for differences between this system and one treated with picornavirus proteases, or a cell-free extract prepared from virus-infected cells.

Picornavirus IRESs Outcompete Capped mRNA for p100

This leaves the question of whether competition between viral RNA and capped cellular mRNAs for p100 could provide an explanation for the shutoff of host-cell mRNA translation, notwithstanding the fact that our eIF4G-depleted lysate supplemented with p100 can translate capped mRNA quite efficiently. Ideally, this issue would be tested using poliovirus RNA as the competitor. However, this cannot be done because, as explained above, the translation of such RNAs in the eIF4G-depleted system is not efficiently rescued by p100. Therefore, our choice of competitor is limited to the FMDV IRES, or cardiovirus IRESs, such as EMCV. When a capped dicistronic mRNA with one of these IRESs was tested in the depleted system, IRES-dependent translation was recovered at remarkably low concentrations of p100, but the upstream scanning-dependent cistron was only poorly rescued even at quite high p100 concentrations (Fig. 5A). However, if precisely the same construct was linearized in the early part of the IRES prior to transcription, so that the RNA product was a capped transcript comprising just the upstream cistron with no competing IRES-linked cistron, then addition of p100 resulted in quite efficient recovery of scanning-dependent translation (Fig. 5B).

Similar results were obtained when the test RNA was a mixture of BMV RNAs and a monocistronic mRNA with the EMCV IRES. Again, good recovery of EMCV IRES activity was achieved at low p100 concentrations (Fig. 5C). However, the recovery of BMV RNA translation required much higher concentrations of factor, higher than in the absence of the IRES competitor, when maximum recovery was seen at just over 10 µg/ml p100 (Fig. 4).

It is clear that the EMCV IRES strongly outcompetes capped scanning-dependent mRNAs for functional interaction with p100, and we obtained very similar results with the FMDV IRES. Although we have been unable, for the reasons explained above, to carry out the direct test with the poliovirus and rhinovirus IRESs as potential competitors, the fact that treatment of standard (not eIF4G-depleted) systems with picornavirus proteases generally results in stimulation of the activity of these IRESs (Borman et al. 1995; Ali et al. 2001b) provides strong reasons for supposing that these IRESs also outcompete capped mRNAs for p100. In contrast, the data of

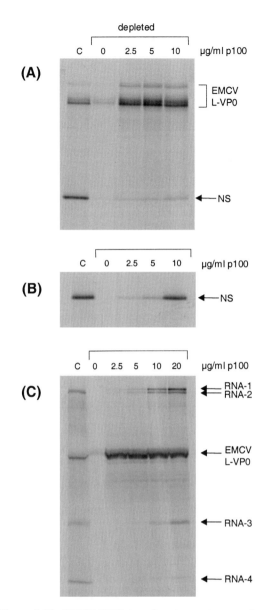

Figure 5. The EMCV IRES strongly outcompetes capped scanning-dependent mRNAs for functional interaction with p100. (*A*) Capped dicistronic mRNA with an upstream cistron coding for influenza virus NS1, an EMCV IRES, and EMCV coding sequences for L-VP0 (~55 kD) as downstream cistron, was translated at a final concentration of 20 µg/ml in the control (nondepleted) lysate (C), and in the eIF4G-depleted lysate supplemented with 0, 2.5, 5, or 10 µg/ml p100, as indicated. Radiolabeled translation products were visualized by gel electrophoresis and autoradiography. The products encoded by the upstream scanning-dependent cistron and the downstream IRES-dependent cistron are indicated. (*B*) A capped monocistronic mRNA encoding NS, transcribed from the same cDNA construct as the dicistronic mRNA in *A* but with linearization in the 5′-proximal part of the IRES, was translated at a final concentration of 8 µg/ml (same molar concentration as the dicistronic mRNA assayed in panel *A*), under the same conditions as in *A*. (*C*) An equimolar mixture of BMV RNAs (13 µg/ml final concentration) and an uncapped monocistronic mRNA (7 µg/ml) with the EMCV IRES linked to viral L-VP0 coding sequences was translated under the same conditions as for *A* and *B*. The EMCV L-VP0 product and products encoded by BMV RNA-1, RNA-2, RNA-3, and RNA-4 are indicated. (Reprinted, with permission, from Ali et al. 2001a [copyright Oxford University Press].)

Figure 3 showing that p100 rescued HAV IRES activity to a lesser extent than upstream cistron translation implies that this IRES (which is atypical of picornavirus IRESs in many respects) competes poorly against capped mRNA for p100, which is entirely consistent with the fact that cleavage of eIF4G by picornaviral proteases inhibits the HAV IRES (Borman et al. 1995; Ali et al. 2001b).

A Reinterpretation of the Cause of Host-cell Shutoff by Poliovirus Infection

These results show that the shutoff of host-cell mRNA translation following picornavirus infection cannot be due to an intrinsic inability of p100 to drive capped host-cell mRNA translation. What, then, is the cause of this shutoff? As shown above, an inhibitory effect of the amino-terminal cleavage product, the eIF4E/eIF4G (amino-terminal domain) complex, is unlikely to make a significant contribution to the shutoff. Rather, the principal explanation for inhibition of host-cell protein synthesis would seem to be the limiting concentration of the p100 cleavage product, coupled with strong competition by the viral RNA for this limiting p100. Interestingly, this revised explanation nicely accounts for some aspects of the shutoff that have long been considered perplexing. Although it is generally the case that in a normal infection the shutoff of host-cell mRNA coincides with the time at which all the eIF4G is cleaved by the viral protease, this temporal correlation breaks down if viral RNA replication is inhibited; for example, by guanidinium chloride, monensin, 3-methyl quercetin, or nigericin (Bonneau and Sonenberg 1987; Irurzon et al. 1995). Under these conditions, when there would be little competing viral RNA, host-cell mRNA translation may persist at 30–50% of the uninfected control cell rate at a time when all eIF4G has apparently been cleaved. The more recent discovery of a second species of eIF4G (eIF4GII) appeared to offer an explanation for this discrepancy, since eIF4GII is cleaved more slowly than eIF4GI during poliovirus infection, and under the special circumstances of inhibited viral RNA replication, about 50% of the endogenous eIF4GII still remains intact at a time when all eIF4GI has been cleaved and host-cell mRNA translation is nevertheless persisting at 30–50% of control rate (Gradi et al. 1998a). However, since eIF4GII is much less abundant than eIF4GI (Gradi et al. 1998b), but appears to be no more active than eIF4GI, there are reasons for questioning whether this residual 50% of eIF4GII could sustain such a high rate of host-cell protein synthesis. It seems more likely that under these special circumstances of inhibited viral RNA replication, and hence little competing viral RNA, the host-cell mRNA translation is being sustained not just by the very low level of residual intact eIF4GII, but also by the relatively abundant p100 cleavage product of eIF4GI and II.

Although a normal poliovirus infection results in a general shutdown of host-cell protein synthesis, the translation of a few mRNAs persists. In a cDNA microarray analysis, some 200 cellular mRNAs, out of 7000

screened, remained polysome-associated in poliovirus-infected cells after the general shutdown had occurred (Johannes et al. 1999). Hitherto, these exceptional mRNAs have been considered to fall into two special classes. Some of them are thought to be translated by an IRES-dependent mechanism, as exemplified by BiP mRNA (Sarnow 1989; Macejak and Sarnow 1991). Others, as illustrated by adenovirus late mRNAs (Dolph et al. 1988), may be translated by the nonlinear scanning mechanism known as ribosome shunting, which seems to be favored over linear scanning when eIF4F activity is low (Schneider 2000; Yueh and Schneider 2000). Our results, showing that the four BMV RNAs differ quite markedly in the extent to which their translation can be supported by p100, suggest that there may be a third class of cellular mRNAs which would be at least partially resistant to the general shutoff caused by poliovirus infection. These would be mRNAs that are translated by the conventional linear scanning mechanism, but which, like BMV RNA-3, have an unusually high functional affinity for the p100 cleavage product of eIF4G.

p100 Can Reverse the Inhibition of Capped mRNA Translation Caused by Cap Analog, 4E-BP1, or FMDV L-Protease

We reasoned that if p100 can support capped mRNA translation in the eIF4G-depleted lysate, albeit in a cap-independent manner, it might also be able to drive BMV RNA translation in normal (non-depleted) systems in which eIF4F function had been inhibited, either by m^{7}GpppG cap analog, or by picornavirus proteases, or by 4E-BP1, which binds eIF4E and effectively strips it from its previous association with eIF4G (Pause et al. 1994b). Figure 6 shows that in all three cases this expectation was fulfilled. Significantly, the rescue of BMV RNA translation in these three circumstances resembles the p100-dependent recovery in the depleted lysate both in its dose-response to varying p100 concentration, and in the fact that rescue of BMV RNA-3 translation is much better than that of RNA-4. Thus, the underlying mechanism is certainly the same, which confirms that the ability of p100 to support capped mRNA translation is not a peculiarity of the depleted lysate system. Also of interest is the fact that addition of p100 to the normal, uninhibited lysate had little effect on overall translation, but there was a slight increase in BMV RNA-3 translation at the expense of a small decrease in RNA-4 translation (Fig. 6).

What Is the Minimum Fragment of eIF4G Required to Support IRES-dependent Translation and Capped or Uncapped mRNA Translation by the Scanning Mechanism?

As stated previously, the original motivation behind our efforts to prepare an eIF4G-depleted lysate was the wish to test various eIF4G deletion and point mutants for their ability to support translation dependent on the EMCV IRES. Clearly, we are now in a position to test

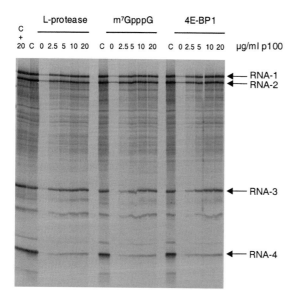

Figure 6. p100 can reverse the inhibition of capped mRNA translation caused by either FMDV L-protease, m⁷GpppG cap analog, or 4E-BP1. BMV RNA was translated at a final concentration of 20 μg/ml in either control (untreated) lysate (C), or control lysate supplemented with 20 μg/ml recombinant p100 (C + 20), or lysate subjected to one of the following regimes, and then supplemented with 0, 2.5, 5, 10, or 20 μg/ml recombinant p100 as indicated: preincubation for 10 minutes at 30°C with 5 μg/ml recombinant FMDV L-protease; supplementation with 0.4 mM m⁷GpppG (with 0.32 mM additional MgCl₂); or preincubation for 10 minutes at 30°C with 10 μg/ml recombinant 4E-BP1. Radiolabeled translation products were examined by gel electrophoresis and autoradiography. (Reprinted, with permission, from Ali et al. 2001a [copyright Oxford University Press].)

these mutants not only on translation driven by the EMCV IRES, but also in other assays: the translation of capped mRNAs in the eIF4G-depleted lysate, and the translation of capped mRNAs in a normal (non-depleted) lysate supplemented with either m⁷GpppG cap analog, or FMDV L-protease, or 4E-BP1. An additional assay is the strong stimulation of translation of uncapped versions of normally capped mRNAs that is observed when p100 is added to standard non-depleted lysates (De Gregorio et al. 1998). Figure 7A summarizes the data obtained with a number of eIF4GI deletion mutants in these various assays. The signal obtained with p100 in each assay has been scored as ++++, and the activity of the other mutants is rated in proportion to the effect of p100.

In all assays with capped BMV RNA or globin mRNA, what can be regarded as the central fragment of eIF4GI (amino acids 643–1076) exhibited not more than half the activity of p100, and deletions that intrude just a short distance into the amino-terminal region of this central fragment completely inactivated it (Fig. 7A). This reduction of activity by ~50% on deleting the carboxy-terminal one-third of eIF4G parallels the results obtained by Morino et al. (2000), who examined capped mRNA translation driven by an eIF4G fragment representing p100 that extended 63 amino acids toward the amino terminus of eIF4G and thus included the eIF4E interaction site (Fig. 1). In their system, too, deletion of the carboxy-terminal tail reduced the activity of the eIF4G fragment by half.

In the case of stimulation of translation (by the scanning mechanism) of uncapped versions of normally capped mRNAs, deletion of the carboxy-terminal tail of p100 again reduced activity, but by a significantly smaller margin than in the case of capped mRNA translation (Fig. 7A). The ratio of the activity of the central domain to the activity of p100 was consistently higher for uncapped mRNA as opposed to capped, and even deletions into the amino-terminal part of the central domain retained a little activity on the uncapped mRNA. Thus, there must be a subtle difference between the interaction of the eIF4G central domain with uncapped as opposed to capped mRNA.

Deletion of the carboxy-terminal tail of p100 also reduced translation dependent on the EMCV IRES, but, as with uncapped RNA, by a smaller margin than was the case with capped mRNAs. Moreover, EMCV IRES activity is still maintained when quite substantial deletions are made into the central domain. Gratifyingly, the relative activity of these deletion mutants in our translation assay is very similar to their relative binding affinity in assays of IRES/eIF4A/eIF4G fragment ternary complex formation, and their relative potency in supporting 48S initiation complex formation on the EMCV IRES, as reported by Lomakin et al. (2000). Particularly noteworthy is the fact that the 697–969 truncation mutant retains some activity in our translation assay and in the two assays of Lomakin et al. (2000), despite the fact that it does not seem to interact directly with eIF3. However, as Lomakin et al. (2000) have pointed out, the possibility of an indirect interaction between eIF3 and this eIF4G deletion mutant via some bridging factor, for example, eIF4B, cannot yet be ruled out.

We also tested these deletion mutants in the eIF4G-depleted lysate programmed with an RNA consisting of the IRES and the first part of viral polyprotein coding sequences of FMDV. There are two initiation sites in the FMDV genome, and as the first part of the polyprotein codes for a self-excising protease, the translation products are two polypeptides of different size, one initiated at the upstream L$_{ab}$ initiation site, which is equivalent to the single initiation site of the closely related EMCV IRES, and the other initiated at the L$_b$ site 84 nucleotides downstream (Sangar et al. 1987). It has been suggested that ribosomes access the L$_b$ initiation site by first binding at, or very near to, the upstream L$_{ab}$ site, and then scanning to the downstream initiation site (Belsham 1992). However, more recent publications have questioned whether all the ribosomes that initiate at the L$_b$ site necessarily follow this route, or whether there may be an alternative mode of accessing the L$_b$ site (Lopez de Quinto and Martined-Salas 1999; Pöyry et al. 2001).

As compared with the relative utilization of the two sites in the parent lysate, addition of p100 to the depleted lysate rescued initiation at the downstream L$_b$ site much more effectively than at the upstream site (Fig. 7B). This differential, as well as the overall efficiency of rescue, was reduced when p100 was replaced by the eIF4G central domain. Curiously, when truncated forms of the central domain were tested, the upper band split into a doublet, suggesting that some initiation occurred slightly

(A)

	EMCV IRES	Uncapped mRNA	BMV RNA	BMV RNA	Globin mRNA
	Depleted system	*Normal system*	*Depleted system*	*Normal + m7GpppG*	*Normal + m7GpppG*
p100	++++	++++	++++	++++	++++
643-1076	++	+++	++	++	++
697-1076	+++	+	0	(+)	0
734-1076	0	0	0	0	0
697-969	+(+)	0	0	0	0

(B)

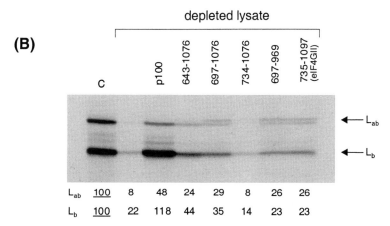

depleted lysate

	c	p100	643-1076	697-1076	734-1076	697-969	735-1097 (eIF4GII)
L$_{ab}$	100	8	48	24	29	8	26	26
L$_b$	100	22	118	44	35	14	23	23

Figure 7. The activity of eIF4G deletion mutants in supporting translation dependent on the EMCV IRES, and translation of capped or uncapped mRNAs by the scanning ribosome mechanism. (*A*) Translation assays were carried out with the designated RNAs in the eIF4G-depleted lysate or in normal lysate with or without 0.4 mM m7GpppG, as indicated. The activity of p100 in each assay is scored as ++++, and the activity of the other eIF4GI deletion mutants (defined in terms of amino acid numbering) was scored in relation to p100 activity. (*B*) Uncapped RNA corresponding to the IRES and the first part of the viral polyprotein coding sequence of FMDV (Asia 4 strain) was translated at 20 µg/ml in eIF4G-depleted lysate supplemented with p100 or other eIF4GI deletion mutants as indicated. The product designated L$_{ab}$ is the self-excised protease translated from the upstream (L$_{ab}$) initiation site, and the smaller product is the self-excised protease initiated at the downstream (L$_b$) site.

upstream of the L$_{ab}$ site (Fig. 7B), which would have to be initiation at a non-AUG codon.

CONCLUDING REMARKS

The work described here has led to a reevaluation of the mechanism by which picornavirus infection results in shutoff of host-cell protein synthesis, and the revised explanation we have proposed can account for many observations that were very puzzling in the context of the previous presumption, prevalent over the past ~15 years, that p100 cannot support capped mRNA translation. Our results also provide an explanation for other observations previously considered puzzling; for example, the fact that injection of coxsackie 2A protease into *Xenopus* oocytes

resulted in cleavage of all the eIF4G, yet the overall translation rate decreased by only 35% (Keiper and Rhoads 1997), with some mRNAs more severely inhibited and others hardly affected at all, much like the difference between BMV RNAs-3 and -4 in our eIF4G-depleted system (Fig. 4).

Viewed from a different perspective from that of picornavirus-induced shutoff, the ability of p100 to support capped mRNA translation is not so surprising after all, given the well-established fact that p100 stimulates the translation of uncapped versions of normally capped mRNAs (De Gregorio et al. 1998). Since p100 is clearly able to drive translation of an mRNA with a pppG... end, would it not be rather extraordinary if it were totally inactive on an RNA that had an m7GpppG... end, but was otherwise identical? The results in Figure 3 suggest that p100 drives the translation of capped and uncapped mRNAs with similar efficiency, in contrast to the large difference in efficiency observed when translation is promoted by the intact eIF4F complex. p100 is less efficient than the intact eIF4F complex on capped mRNAs, in the sense that more p100 than eIF4F is required to give the same translational yield (Fig. 4), whereas the converse seems to be the case with uncapped RNAs that are translated by the scanning mechanism.

Our results serve to (re)emphasize the paramount importance of the central domain of eIF4G in initiation, which was first highlighted by studies of initiation on the EMCV IRES (Pestova et al. 1996b) and later confirmed by the ability of a tethered eIF4G central domain to direct internal initiation even in the absence of a specific IRES (De Gregorio et al. 1999). The role of eIF4E and the eIF4E/4G interaction in capped mRNA translation by the scanning mechanism seems to be largely just an increase in the functional affinity of the factor for capped mRNAs, such that ~fourfold lower concentrations of eIF4F complex than of p100 suffice to give the same yield of translation product (Fig. 4). It is also possible that eIF4E and the eIF4E/4G interaction help to ensure accurate delivery of the 40S subunit to the 5′-proximal region of the capped mRNA and hence increase the accuracy of selection of the cap-proximal AUG as initiation site. However, in a system in which translation is driven by p100, initiation still occurs overwhelmingly at the 5′-proximal AUG, implying that the functional interaction of p100 with the mRNA occurs selectively at a 5′-proximal site, so that eIF4E and the eIF4E/4G interaction only assist this selectivity rather than being absolutely necessary for it.

The fact that p100 interacts with an internal site on the EMCV IRES, coupled with the ability of eIF4G central domain deletion mutants to support EMCV IRES activity but not translation of capped or uncapped mRNAs by the scanning mechanism (Fig. 7A), suggests that there is something fundamentally different about the nature of the interactions of p100 with the IRES. It is almost as though the viral IRES has evolved to function as a high-affinity aptamer for the central domain of eIF4G, an aptamer that is likely to be targeted at a different region of the eIF4G central domain from that which interacts with the 5′-proximal region of capped or uncapped mRNAs.

ACKNOWLEDGMENTS

We thank Simon Morley and Linda McKendrick (University of Sussex) for gifts of antibodies and constructs for expression of p100 and 4E-BP1; Ivan Lomakin, Tatyana Pestova, and Christopher Hellen (SUNY Health Center, Brooklyn, New York) for constructs for expression of eIF4G central domain mutants; Graham Belsham for advice on the eIF4G depletion method; and Rosemary Farrell for providing technical assistance and infrastructure support. This work was supported by grants from the Wellcome Trust (051424 and 062348). I.K.A. was supported by a Medical Research Council postgraduate research studentship and gratefully acknowledges additional travel grants from Gonville and Caius College, Cambridge.

REFERENCES

Ahlquist P., Dasgupta R., and Kaesberg P. 1984. Nucleotide sequence of the brome mosaic virus genome and its implications for viral replication. *J. Mol. Biol.* **172:** 369.

Ali I.K., McKendrick L., Morley S.J., and Jackson R.J. 2001a. Truncated initiation factor eIF4G lacking an eIF4E binding site can support capped mRNA translation. *EMBO J.* **20:** 4233.

———. 2001b. The activity of the hepatitis A virus IRES requires association between the cap-binding translation initiation factor (eIF4E) and eIF4G. *J. Virol.* **75:** 7854.

Belsham G.J. 1992. Dual initiation sites of protein synthesis on foot-and mouth disease virus RNA are selected following internal entry and scanning of ribosomes in vivo. *EMBO J.* **11:** 1105.

Bonneau A.-M. and Sonenberg N. 1987. Proteolysis of the p220 component of the cap-binding protein complex is not sufficient for complete inhibition of host cell protein synthesis after poliovirus infection. *J. Virol.* **61:** 986.

Borman A.M., Bailly J.-L., Girard M., and Kean K.M. 1995. Picornavirus internal ribosome entry segments: Comparison of translation efficiency and the requirements for optimal internal initiation in vitro. *Nucleic Acids Res.* **23:** 3656.

Dasgupta R. and Kaesberg P. 1982. Complete nucleotide sequence of the coat protein messenger RNAs of brome mosaic virus and cowpea chlorotic mottle virus. *Nucleic Acids Res.* **10:** 703.

Dasso M.C. and Jackson R.J. 1989. On the fidelity of mRNA translation in the nuclease-treated rabbit reticulocyte lysate system. *Nucleic Acids Res.* **17:** 3129.

De Gregorio E., Preiss T., and Hentze M.W. 1998. Translational activation of uncapped mRNAs by the central part of human eIF4G is 5′ end dependent. *RNA* **4:** 828.

———. 1999. Translation driven by an eIF4G core domain in vivo. *EMBO J.* **18:** 4865.

Dolph P.J., Racaniello V., Villamarin A., Palladion F., and Schneider R.J. 1988. The adenovirus tripartite leader eliminates the requirement for cap binding protein during translation initiation. *J. Virol.* **62:** 2059.

Duke G.M., Hoffman M.A., and Palmenberg A.C. 1992. Sequence and structure elements that contribute to efficient encephalomyocarditis virus RNA translation. *J. Virol.* **66:** 1602.

Gingras A.C., Raught B., and Sonenberg N. 1999. eIF4 initiation factors: Effectors of mRNA recruitment to ribosomes and regulators of translation. *Annu. Rev. Biochem.* **68:** 913.

Gradi A., Svitkin Y.V., Imataka H., and Sonenberg N. 1998a. Proteolysis of human eukaryotic initiation factor eIF4GII, but not eIF4GI, coincides with the shutoff of host protein synthesis after poliovirus infection. *Proc. Natl. Acad. Sci.* **95:** 11089.

Gradi A., Imataka H., Svitkin Y.V., Rom E., Raught B., Morino S., and Sonenberg N. 1998b. A novel functional human eukaryotic translation initiation factor eIF4G. *Mol. Cell. Biol.* **18:** 334.

Haghighat A. and Sonenberg N. 1997. eIF4G dramatically enhances the binding of eIF4E to the mRNA 5′-cap structure. *J. Biol. Chem.* **272:** 21677.

Hershey J.W.B. and Merrick W.C. 2000. The pathway and mechanism of initiation of protein synthesis. In *Translational control of gene expression* (ed. N. Sonenberg et al.), p 33. Cold Spring Harbor Laboratory Press, Cold Spring Harbor, New York.

Imataka H. and Sonenberg N. 1997. Human eukaryotic translation initiation factor 4G (eIF4G) possesses two separate and independent binding sites for eIF4A. *Mol. Cell. Biol.* **17:** 6940.

Imataka H., Gradi A., and Sonenberg N. 1998. A newly identified N-terminal amino acid sequence of human eIF4G binds poly(A)-binding protein and functions in poly(A)-dependent translation. *EMBO J.* **17:** 7480.

Irurzon A., Sanchez-Palomino S., Nonoa I., and Carrasco L. 1995. Monensin and nigericin prevent the inhibition of host translation by poliovirus without affecting p220 cleavage. *J. Virol.* **69:** 7453.

Jackson R.J. 2000. A comparative view of initiation site selection mechanisms. In *Translational control of gene expression* (ed. N. Sonenberg et al.), p. 127. Cold Spring Harbor Laboratory Press, Cold Spring Harbor, New York.

Johannes G., Carter M.S., Eisen M.B., Brown P.O., and Sarnow P. 1999. Identification of eukaryotic mRNAs that are translated at reduced cap-binding complex eIF4F concentrations using a cDNA microarray. *Proc. Natl. Acad. Sci.* **96:** 13118.

Keiper B.D. and Rhoads R.E. 1997. Cap-independent translation initiation in *Xenopus* oocytes. *Nucleic Acids Res.* **25:** 395.

Kirchweger R., Ziegler E., Lamphear B.J., Waters D., Liebig H.D., Sommergruber W., Sobrino F., Hohenadl C., Blaas D., Rhoads R.E., and Skern T. 1994. Foot-and-mouth disease virus leader proteinase: Purification of the Lb form and determination of its cleavage site on eIF-4 gamma. *J. Virol.* **61:** 2711.

Kolupaeva V.G., Pestova T.V., Hellen C.U.T., and Shatsky I.N. 1998. Translation eukaryotic initiation factor 4G recognizes a specific structural element within the internal ribosome entry site of encephalomyocarditis virus RNA. *J. Biol. Chem.* **273:** 18599.

Korneeva N.L., Lamphear B.J., Hennigan F.L.C., and Rhoads R.E. 2000. Mutually cooperative binding of eukaryotic translation initiation factor (eIF) 3 and eIF4A to human eIF4G-1. *J. Biol. Chem.* **275:** 41369.

Korneeva N.L., Lamphear B.J., Hennigan F.L.C., Merrick W.C., and Rhoads R.E. 2001. Characterization of the two eIF4A-binding sites on human eIF4G-1. *J. Biol. Chem.* **276:** 2872.

Lamphear B.J., Kirchweger R., Skern T., and Rhoads R.E. 1995. Mapping of functional domains in eukaryotic protein synthesis initiation factor 4G (eIF4G) with picornaviral proteases. *J. Biol. Chem.* **270:** 21975.

Lomakin I.B., Hellen C.U.T., and Pestova T.V. 2000. Physical association of eukaryotic initiation factor 4G (eIF4G) with eIF4A strongly enhances binding of eIF4G to the internal ribosome entry site of encephalomyocarditis virus and is required for internal initiation of translation. *Mol. Cell. Biol.* **20:** 6019.

Lopez de Quinto S. and Martinez-Salas E. 1999. Involvement of the aphthovirus RNA region located between the two functional AUGs in start codon selection. *Virology* **255:** 324.

Macejak D.G. and Sarnow P. 1991. Internal initiation of translation mediated by the 5′ leader of a cellular mRNA. *Nature* **353:** 90.

Morino S., Imataka H., Svitkin Y.V., Pestova T.V., and Sonenberg N. 2000. Eukaryotic translation initiation factor 4E (eIF4E) binding site and the middle one-third of eIF4GI constitute the core domain for cap-dependent translation, and the C-terminal one-third functions as a modulatory region. *Mol. Cell. Biol.* **20:** 468.

Pause A., Méthot N., Svitkin Y., Merrick W.C., and Sonenberg N. 1994a. Dominant negative mutants of mammalian translation initiation factor eIF4A define a critical role for eIF4F in cap-dependent and cap-independent initiation of translation. *EMBO J.* **13:** 1205.

Pause A., Belsham G.J., Gingras A.C., Donze O., Lin T.A., Lawrence J.C., and Sonenberg N. 1994b. Insulin-dependent stimulation of protein synthesis by phosphorylation of a regulator of 5′ cap function. *Nature* **371:** 762.

Pestova T.V., Hellen C.U.T., and Shatsky I.N. 1996a. Canonical eukaryotic initiation factors determine translation by internal ribosome entry. *Mol. Cell. Biol.* **16:** 6859.

Pestova T.V., Shatsky I.N., and Hellen C.U.T. 1996b. Functional dissection of eukaryotic initiation factor eIF4F: The 4A subunit and the central domain of the 4G subunit are sufficient to mediate internal entry of 43S preinitiation complexes. *Mol. Cell. Biol.* **16:** 6870.

Pestova, T.V., Shatsky I.N., Fletcher S.P., Jackson R.J., and Hellen C.U.T. 1998. A prokaryotic-like mode of cytoplasmic eukaryotic ribosome binding to the initiation codon during internal initiation of hepatitis C and classical swine fever virus RNAs. *Genes Dev.* **12:** 67.

Pöyry T.A.A., Hentze M.W., and Jackson R.J. 2001. Construction of regulatable picornavirus IRESes as a test of current models of the mechanism of internal translation initiation. *RNA* **5:** 647.

Sangar D.V., Newton S.E., Rowlands D.J., and Clarke B.E. 1987. All foot-and-mouth disease virus serotypes initiate protein synthesis at 2 separate AUGs. *Nucleic Acids Res.* **15:** 3305.

Sarnow P. 1989. Translation of glucose regulated protein 78/immunoglobulin heavy chain binding protein mRNA is increased in poliovirus-infected cells at a time when cap-dependent translation of cellular mRNAs is inhibited. *Proc. Natl. Acad. Sci.* **86:** 5795.

Schneider R.J. 2000. Adenovirus inhibition of cellular protein synthesis and preferential translation of viral mRNAs. In *Translational control of gene expression* (ed. N. Sonenberg et al.), p. 901. Cold Spring Harbor Laboratory Press, Cold Spring Harbor, New York.

Stassinopoulos I.A. and Belsham G.J. 2001. A novel protein-RNA binding assay: Functional interactions of the foot-and-mouth disease virus internal ribosome entry site with cellular proteins. *RNA* **7:** 114.

Yueh A. and Schneider R.J. 2000. Translation by ribosome shunting on adenovirus and Hsp70 mRNAs facilitated by complementarity to 18S rRNA. *Genes Dev.* **14:** 414.

Functions of Eukaryotic Factors in Initiation of Translation

T.V. Pestova†*‡ and C.U.T. Hellen*‡

*Department of Microbiology and Immunology, State University of New York Health Science Center
at Brooklyn, New York 11203; †A.N. Belozersky Institute of Physico-chemical Biology,
Moscow State University, Moscow 119899, Russia; ‡Gene Expression Programme, European Molecular Biology
Laboratory, D-69117, Heidelberg, Germany

Initiation of protein synthesis in both eukaryotes and prokaryotes involves the assembly by initiation factors of separated large and small ribosome subunits into a ribosomal initiation complex in which initiator tRNA is positioned in the ribosomal P site so that it is base-paired with the initiation codon of the mRNA. In prokaryotes, the small (30S) ribosomal subunit binds directly to the initiation codon region, and initiation involves only three single-subunit initiation factors termed IF1, IF2, and IF3. The initiation process in eukaryotes is significantly more complex than this; it may involve more than ten eukaryotic initiation factors (eIFs), and on most mRNAs, the small (40S) ribosomal subunit first binds to the 5′ end of an mRNA and then scans downstream searching for the initiation codon.

Translation initiation on the majority of eukaryotic mRNAs can be divided into a series of interrelated stages (Figs. 1, 2). These stages are:

1. Recycling by eIF2B of inactive eIF2–GDP to eIF2–GTP (active in binding Met-tRNA$_i^{Met}$)

2. Selection of aminoacylated initiator Met-tRNA$_i^{Met}$ by eIF2–GTP from the cytoplasmic pool of aminoacylated and deacylated initiator and elongator tRNAs.

3. Binding of eIF2–GTP/Met-tRNA$_i^{Met}$ and other eIFs to the 40S subunit to form a 43S complex, which may occur concomitantly with dissociation of an 80S ribosome into separate 40S and 60S subunits.

4. Attachment of the 43S complex to the capped 5′ end of an mRNA, which is enhanced by binding of the eIF4E (cap-binding) subunit of eIF4F to the 5′-terminal m^{7}G cap.

5. Movement of the ribosomal complex from this initial binding site along the 5′-untranslated region (5′UTR) until it locates the initiation codon.

6. Displacement of initiation factors from the resulting 48S complex in which the anticodon of Met-tRNA$_i^{Met}$ is base-paired to the initiation codon.

7. Joining of a large (60S) ribosomal subunit to form an 80S ribosome, leaving Met-tRNA$_i^{Met}$ in the ribosomal P site. Stages 6 and 7 may occur simultaneously.

8. Displacement of joining factors from the 80S ribosome, leaving it in an active state to enter the elongation phase of translation and to begin polypeptide synthesis.

We have used in vitro biochemical reconstitution with purified components to identify the minimum set of factors that is required for initiation and are now beginning to elucidate the mechanism of each stage in this process. Native β-globin mRNA was used in initial experiments because it is efficiently translated and because it is a representative mRNA, having a 5′-terminal m^{7}G cap, a 3′ poly(A) tail and a moderately structured 5′UTR of average length upstream of the initiation codon, which is the first AUG triplet from the 5′ terminus.

FACTOR REQUIREMENTS FOR RIBOSOMAL ATTACHMENT AND SCANNING ON β-GLOBIN mRNA

A 43S complex comprising a 40S subunit, eIF3, and the eIF2–GTP/Met-tRNA$_i^{Met}$ complex together with ATP and the factors eIF4A, eIF4B, and eIF4F was not sufficient for assembly of a 48S complex at the initiation codon of β-globin mRNA but instead resulted in formation of a ribosomal "complex I" in a cap-proximal position on the mRNA (Pestova et al. 1998). Toeprinting showed that this complex was arrested 16–23 nucleotides from the 5′ end of the mRNA; its proximity to the mRNA's 5′ end suggests that it may not even have begun scanning.

Two additional activities present in rabbit reticulocyte lysate (RRL) were required for 43S complexes to scan to the initiation codon to form a 48S complex ("complex II") without being arrested at the initial binding site. These two factors were purified on the basis of activity in the reconstituted initiation assay and were identified by sequencing as eIF1 and eIF1A. They synergistically promoted formation of a 48S complex at the correct initiation codon without any trace of complex I. eIF1 in the absence of eIF1A was able to recognize and destabilize the aberrantly assembled complex I, reducing its prominence. eIF1 is also active in destabilizing aberrantly assembled ribosomal complexes on other mRNAs (Pestova et al. 1998), and this activity represents an important function for eIF1 that is consistent with previous reports that it acts as a monitor of translational fidelity (Yoon and Donahue 1992; Cui et al. 1998). Inclusion of eIF1 in assembly reactions without eIF1A also yielded small amounts of complex II but with only two (+16 and 17) rather than three (+15–17) toeprints. eIF1A in the absence of eIF1 increased formation of complex I without promoting for-

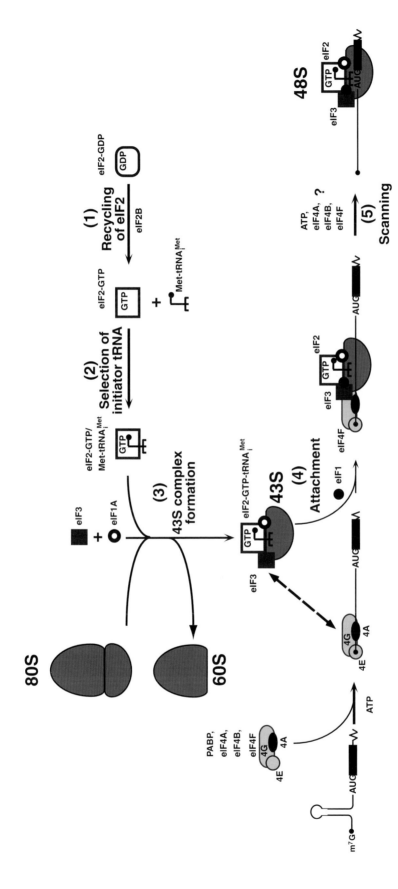

Figure 1. Schematic representation of the pathway of 48S initiation complex formation on a capped eukaryotic mRNA. This model depicts the initial stage of action of initiation factors that participate in 48S complex formation. A ternary complex comprising initiator tRNA, GTP, and eIF2 is bound with eIF1A and eIF3 to a 40S subunit to form a 43S preinitiation complex. At least four factors, eIF4A, eIF4B, eIF4F, and the poly(A)-binding protein PABP, may cooperate in ATP-dependent binding of the 43S complex to mRNA, probably by creating an unstructured cap-proximal ribosomal binding site. PABP and eIF4B are nonessential for this process. Loading of the 43S complex onto this binding site may be mediated by interaction between the eIF4G subunit of cap-bound eIF4F and the eIF3 component of the 43S complex (indicated by a dashed, double-headed arrow) as well as by interactions of eIF2, eIF3, and the 40S subunit with the mRNA. The bound 43S complex requires eIF1 to scan the 5′UTR in a 5′–3′ direction until it recognizes the initiation codon. The other factor requirements for scanning have not been determined. The initiation codon and the anticodon of initiator tRNA are base-paired to the initiation codon in the resulting stable 48S complex.

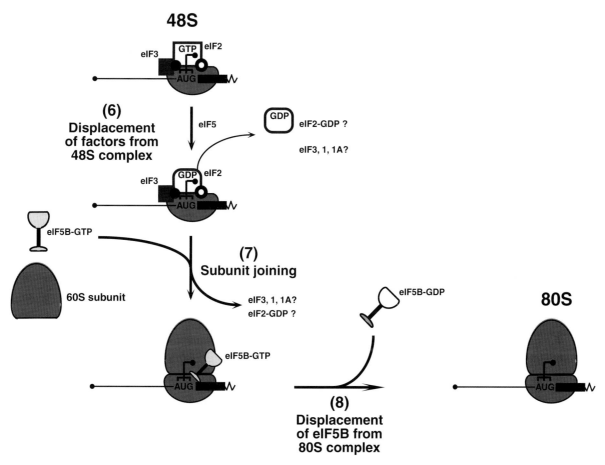

Figure 2. Schematic representation of the pathway of 80S complex formation on a eukaryotic mRNA. Hydrolysis of eIF2-bound GTP in the 48S complex is triggered by eIF5, possibly leading to release of eIF2–GDP and allowing eIF5B–GTP to mediate joining of the resulting complex to a 60S ribosomal subunit. The stage at which eIF1, eIF1A, and eIF3 are released is not known. Ribosome-activated hydrolysis of eIF5B-bound GTP leads to release of eIF5B–GDP to form a 80S ribosome that is competent to begin protein synthesis.

mation of complex II. eIF1A therefore enhances the ability of 43S complexes to bind mRNA and most likely the processivity of scanning 43S/eIF1/mRNA complexes, whereas eIF1 plays a role in ensuring the fidelity of initiation codon selection.

Complex I is intrinsically unstable and cannot be "chased" directly into complex II by eIF1 and eIF1A. Instead, eIF1 and eIF1A together promote dissociation of complex I and enable the released ribosomal complexes to rebind mRNA in a competent state to scan to the initiation codon. These factors may contribute to the interaction of 43S complexes with mRNA that enable them to enter the scanning mode, for example by contributing to formation of a channel through which the mRNA moves during scanning, directly, by inducing conformational changes in the 40S subunit or both.

MECHANISM OF 5′-END-DEPENDENT RIBOSOMAL ATTACHMENT TO mRNA AND OF RIBOSOMAL SCANNING

The ability to reconstitute the process of 48S complex formation on a model native capped mRNA enables us to address specific and detailed questions about this process and to investigate the functions of individual factors in it. For example, the molecular interactions and conformational changes that lead to end-dependent binding of the 43S complex to mRNA are incompletely characterized. Several lines of evidence suggest that the m^7G cap and the cap-binding protein eIF4E strongly enhance the efficiency of 5′-end-dependent ribosomal attachment to mRNA but are not essential for this process. Attachment of mammalian ribosomes to uncapped mRNA occurs inefficiently but is strictly 5′-end-dependent; after attachment, scanning ribosomes initiate translation at AUG triplets inserted upstream of the initiation codon of an uncapped mRNA with an efficiency that depends on their context just as for capped mRNAs (Kozak 1989; Gunnery et al. 1997). 48S complex formation on capped β-globin mRNA in an in-vitro-reconstituted reaction was substantially reduced but not abolished by substituting eIF4F by eIF4A and eIF4G$_{613-1560}$ or even eIF4G$_{613-1090}$ (Morino et al. 2000). In a similar reaction done using highly purified initiator tRNA (instead of a crude mixture of total tRNA) and therefore lacking competitor RNA, eIF4F could be substituted by eIF4A and eIF4G$_{697-949}$ (T.

Pestova, in prep.). This domain of eIF4G consists of five pairs of α helices (Marcotrigiano et al. 2001) and binds both RNA and eIF4A but does not form a binary complex with eIF3 (Lomakin et al. 2000).

The eIF4A component of eIF4G/4A and eIF4F complexes likely mediates localized unwinding of 5′-proximal residues on β-globin mRNA to create a single-stranded region that is potentially a ribosome-binding site (Rozen et al. 1990). In this model, the primary function of eIF4G/4A is to generate an unstructured region to which the 43S complex can bind, and eIF4E functions primarily to enhance binding of this complex to a defined site. A prediction of this model is that an mRNA with a wholly unstructured 5′UTR might be able to recruit 43S complexes without any requirement for eIF4G/4A. Experiments done with mRNAs of this type have confirmed that 48S complex formation can occur efficiently in the absence of ATP and eIF4G/4A (T. Pestova, in prep.).

Current models for recruitment of 43S complexes to the 5′ end of an mRNA have stressed the importance of the interaction between eIF3 (as part of the 43S complex) and eIF4G (as part of the cap-bound eIF4F complex) (Hershey and Merrick 2000). The observations discussed above that eIF4A and a domain of eIF4G that does not bind eIF3 can mediate end-dependent ribosomal attachment on globin mRNA and that 43S complexes can bind directly to unstructured 5′UTRs in the absence of eIF4G indicate that a reappraisal of this model is warranted. We discussed above that an unstructured region of RNA is important for ribosomal binding and can either be "prepared" for binding and possibly "presented" to the 43S complex by eIF4G/4A or can occur naturally. A role for the eIF3–eIF4G interaction is therefore not ruled out, but it may be partially redundant, and interactions of components of the 43S complex with the prepared mRNA-binding site may be more important than has so far been appreciated. These components could include eIF2 and eIF3, as well as the 40S subunit itself. Moreover, it is likely that conformational changes in these components induced by their assembly into the 43S complex are required to permit correct interactions with the mRNA.

The ribosomal scanning model has been proposed to account for the mechanism of ribosomal start-site selection (Kozak 1989). In its simplest form, the 40S subunit carrying initiator tRNA and associated with factors migrates linearly in a 5′ to 3′ direction from the 5′ end of an mRNA to the first AUG triplet, which is recognized as the start site for initiation. This process can be considered as a combination of two linked processes, which are ribosomal movement (scanning) and initiation codon selection. Significant progress has been made using genetic suppressor analysis in *Saccharomyces cerevisiae* to identify components of the ribosomal preinitiation complex that determine start-site selection (for review, see Donahue 2000). Factors implicated in AUG recognition include eIF1, all three subunits of eIF2, and eIF5. On the other hand, the process of scanning is almost wholly uncharacterized (Pestova and Hellen 1999). Understanding the role of individual initiation factors in the scanning process may be facilitated by considering it as comprising three distinct processes: detachment of the 40S sub-

unit and associated factors from the binding site at the 5′ end of the mRNA, unwinding of structured RNA in the 5′UTR, and ribosomal movement.

Ribosomal scanning may reflect an intrinsic ability of the 40S subunit to move on mRNA by linear diffusion or by transient dissociation–reassociation. Alternatively, the 40S subunit may be able to move on mRNA only by helicase-mediated translocation. The observation that 48S complexes can assemble on mRNA with an unstructured 5′UTR in the absence of ATP, eIF4A, and eIF4F (T. Pestova, in prep.) suggests that 43S complexes are intrinsically capable of movement on unstructured mRNA. However, no 48S complexes were assembled on mRNAs with unstructured 5′UTRs interrupted by defined internal stems in the absence of ATP and factors associated with ATP hydrolysis (T. Pestova, in prep.). That means that ribosomal scanning on native 5′UTRs with even modest secondary structure requires unwinding of these structures by eIFs 4A, 4B, and 4F. We do not know whether and how ribosomal scanning is coupled to unwinding of secondary structure by these factors. If eIF4F is involved in this process, we do not know whether it is the same molecule that mediated attachment of the 43S complex to the 5′ end of the mRNA. If the same molecule of eIF4F assists the 43S complex in scanning and also participates in the earlier stage of ribosomal attachment, we do not know whether the eIF4E–cap interaction is broken when the ribosome begins to scan, or whether eIF4F remains attached to the cap throughout the scanning process. We have noted previously that if eIF4F remains bound to the cap and to the scanning complex, it could help this complex to rebind the same mRNA if it dissociates prematurely before reaching the initiation codon (Pestova and Hellen 2000).

INITIATION CODON RECOGNITION DURING RIBOSOMAL SCANNING

The critical interaction that leads to ribosomal recognition of a start site for translation is the base-pairing between the AUG codon and the CAU anticodon of initiator tRNA. Initiation can occur at an AGG triplet if a corresponding change is made in the anticodon of initiator tRNA to maintain base-pairing (Cigan et al. 1988). Initiation efficiency is reduced if the sequence "context" flanking an AUG triplet deviates from the experimentally determined optimum GCC(**A/G**)CC**AUG**G (in which the residues in bold have the strongest effects and the initiation codon is underlined) (Kozak 1991). The components of the scanning ribosomal complex that are responsible for context recognition are not known, nor is it known whether poor context results in decreased recognition of the initiation codon or whether there is a mechanism that discriminates against assembly of 48S complexes at initiation codons with poor context. The observation that leaky scanning past an initiation codon that lies in an unfavorable context is suppressed by insertion of downstream secondary structure has led to the suggestion that circumstances that slow scanning ribosomes can enhance initiation codon recognition (Kozak 1991). We have noted similar reductions in leaky scanning on mRNAs containing tandem-spaced initiation codons in favorable context caused by omission from reconstituted initiation reactions of factors required for processive ribosomal scanning (T. Pestova, in prep.).

Changes in the context of an initiation codon can modulate initiation efficiency at least tenfold (Kozak 1991). The function for context nucleotides would necessarily imply that they interact with a component of the scanning ribosomal context. To account for such significant effects, we can speculate further that a failure to establish such interactions could be sensed by a component of the translation apparatus, leading to discrimination against assembly of 48S complexes at initiation codons with poor context. Another circumstance in which initiation codons are bypassed is if they occur close to the 5′ end of a mRNA (Kozak 1991). The molecular basis for the inability of 5′-proximal AUG triplets to promote efficient initiation is unclear: The anticodon loop of initiator tRNA may be unable to inspect the extreme 5′-terminal nucleotides of an mRNA, or the interaction of these nucleotides with the mRNA-binding cleft of the 40S subunit may be unstable.

STRUCTURE AND FUNCTION OF eIF1A

eIF1A is one of the most conserved initiation factors in eukaryotes, and there is significant sequence homology between eIF1A and the prokaryotic initiation factor IF1 (Kyrpides and Woese 1998). This region of sequence homology corresponds to a structurally homologous β barrel oligomer-binding (OB) fold (Sette et al. 1997; Battiste et al. 2000) that in eIF1A comprises amino acid residues 33–94. In addition, eIF1A contains a small domain with two helices (residues 95–114) and extended amino-terminal and carboxy-terminal strands. The protein surface of eIF1A is very highly charged, and a large positive electrostatic surface on it matches quite well to an extensive RNA-binding surface identified by nuclear magnetic resonance (NMR) titration experiments. The RNA ligand for eIF1A is not known but, by analogy with IF1, could well be 18S ribosomal RNA. Like eIF1A, IF1 also has a highly asymmetric charge distribution that is probably important in stabilizing its binding to the 30S subunit. The binding

site for IF1 on the 30S subunit has been localized by X-ray crystallography to a cleft formed between helix 44, the 530 rRNA loop, and ribosomal protein S12 (Carter et al. 2001). As noted by Carter et al., as a result, bound IF1 occludes the ribosomal A site and "covers, but doesn't block a channel at the base of the A site through which mRNA could pass." IF1 binding to the 30S subunit also causes conformational changes in ribosomal RNA and in the relative positions of the domains of the 30S subunit.

eIF1A increases the ability of 43S complexes to bind mRNA, and we found that it increases the processivity of scanning ribosomal complexes (Pestova et al. 1998). eIF1A also stabilizes binding of the ternary complex to the 40S subunit in the absence of mRNA (Chaudhuri et al. 1999; Battiste et al. 2000), possibly by an allosteric mechanism. Our preliminary mutational analysis of eIF1A (Battiste et al. 2000) was guided by considerations such as the location of the RNA-binding surface and of conserved amino acid residues. Coincidentally, we mutated exactly those residues in eIF1A whose equivalents in IF1 are involved in contacts with the 30S subunit (Fig. 3). Thus, mutations H59D, R65D, and K67D are located in the region of eIF1A which is equivalent to the IF1 region that interacts with the 530 loop. Residues in IF1 equivalent to the substitutions R45D and W69A are in the region that interacts with helix 44, and the substitution Y83A in eIF1A is of a residue equivalent to Y60 of IF1, which interacts with S12. The close structural homology of the OB domains of eIF1A and IF1 leads us to cautiously assume that equivalent regions of eIF1A interact with the 40S subunit. Qualitatively, these mutants all had the same phenotype in initiation. They acted as efficiently as wild-type eIF1A in eIF1-dependent destabilizing of complex I on β-globin mRNA, but all also behaved aberrantly, strongly enhancing toeprints at positions +8 to +9 relative to the A of the initiation codon, indicative of a stalled or incorrectly assembled ribosomal complex. These complexes could be converted into correctly assembled 48S complexes by delayed addition of wild-type eIF1A, a result which indicates that mutant forms of eIF1A do not remain stably associated with the 40S subunit during scanning. We do not know whether wild-type eIF1A is stably associated with ribosomes during the scanning process or

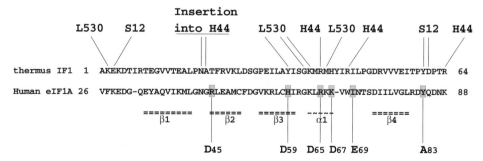

Figure 3. Sequence alignment of the OB-like domains of human eIF1A and *Thermus thermophilus* IF1 (Kyrpides and Woese 1998; A. Carter, pers. comm.) annotated to show secondary structure elements in human eIF1A (labeled β for β strand and α for 3_{10} helix) and amino acid substitutions made in human eIF1A mutants (Battiste et al. 2000), and interactions of individual amino acid residues in *T. thermophilus* IF1 with helix 44 (H44), the 530 loop (L530), and ribosomal protein S12 of the 30S ribosomal subunit (Carter et al. 2001).

cycles on and off. By analogy with IF1, we suggest that eIF1A might participate in formation of the channel through which mRNA passes during scanning. However, we cannot exclude the possibility that its influence on scanning is indirect; for example, by altering the conformation of rRNA or the relative positions of domains in the 40S subunit in a similar manner to changes in the 30S subunit induced by IF1.

STRUCTURE AND FUNCTION OF eIF1

The known functions of eIF1 are that its acts weakly alone and synergistically with eIF1A to promote ribosomal scanning to the initiation codon, and that it acts as a monitor of translational accuracy. eIF1 recognizes and destabilizes ribosomal complexes that have stalled at the 5′ end of the 5′UTR during end-dependent initiation and those that have assembled aberrantly by internal ribosomal entry at an incorrect AUG codon in encephalomyocarditis virus mRNA (Pestova et al. 1998). A series of mutations in yeast eIF1 that allow initiation at a UUG codon and that therefore have a defect in preventing initiation at non-AUG codons have been identified (Yoon and Donahue 1992; Cui et al. 1998). eIF1 has a similar length and conserved sequence in different eukaryotes, and human eIF1 can replace yeast eIF1 in vivo. The solution structure of human eIF1 has been determined by NMR spectroscopy (Fletcher et al. 1999). Significantly, all mutated residues that altered the accuracy of initiation codon selection in yeast mapped close together on a surface that is almost perfectly conserved in all eIF1 homologs, likely forming a binding site for an as-yet-unidentified component of the translation apparatus.

IDENTIFICATION OF THE REQUIREMENT FOR eIF5B IN RIBOSOMAL SUBUNIT JOINING

The final stage in initiation is joining of a 60S subunit to the 48S complex assembled at the initiation codon to form an 80S ribosome. The 48S complex is associated with a number of factors (minimally eIFs 1, 1A, 2, and 3) that must be displaced before subunit joining can occur. Substitution of GTP by a nonhydrolyzable analog such as guanosine 5′-[β,γ-imido]-triphosphate (GMPPNP) does not prevent binding of initiator tRNA to the 40S subunit by eIF2, but arrests initiation at the stage of 48S complex formation. Displacement of factors and joining of a 60S subunit therefore both depend on hydrolysis of GTP bound to eIF2 in 48S complexes.

Hydrolysis of GTP by the eIF2 γ subunit is activated by the 49-kD eIF5, reportedly by an "arginine finger" mechanism (Paulin et al. 2001). Although eIF5 interacts strongly and specifically with eIF3 and with the eIF2 β subunit off the ribosome (Das et al. 1997; Asano et al. 2000), it induces eIF2's GTPase activity only when eIF2 is incorporated into 43S complexes. The factor interactions and conformational changes necessary for activation of eIF2 by eIF5 are not known; GTP hydrolysis by eIF2 is thought to be induced by eIF5 only once base-pairing between the initiation codon and the anticodon of initiator tRNA has been established (Huang et al. 1997).

Initiator tRNA remains associated with the 40S subunit after GTP hydrolysis, whereas eIF2–GDP dissociates. This step is irreversible, so the activity of eIF5 is thought to ensure stringent selection of the initiation codon during ribosomal scanning.

Until recently, eIF5 was considered to be the only factor required for subunit joining (for review, see Pestova et al. 2000a). Although eIF5 is necessary for joining of 60S subunits to 48S complexes assembled on native β-globin mRNA, it is not sufficient. We found that a second factor that we purified as a single homogeneous 175-kD polypeptide from an active mouse ribosomal salt-wash fraction was also required (Pestova et al. 2000b). We named this factor eIF5B because it acts together with eIF5 in mediating subunit joining, and we identified it by amino-terminal sequencing as a murine homolog of prokaryotic initiation factor IF2. A function for a eukaryotic homolog of IF2 in translation was first discovered by studies in yeast; deletion of the yeast IF2 homolog resulted in a reduction in polysomes and in accumulation of 80S ribosomal particles (Choi et al. 1998; Lee et al. 1999). Native eIF5 and eIF5B could be replaced in reconstituted initiation reactions by recombinant eIF5 and eIF5B$_{587–1220}$ (truncated to remove most of the nonconserved amino-terminal domain) (Pestova et al. 2000b).

eIF5B sequences are highly conserved, and these proteins can be divided into three regions, corresponding approximately to amino acid residues 1–628, 629–850, and 851–1220 of human eIF5B. The first of these regions is very basic and highly variable in sequence between different organisms, whereas the second and third are strongly conserved. The central domain of eIF5B/IF2 contains sequence motifs characteristic of GTP-binding proteins (Lee et al. 1999). Prokaryotic IF2 binds four major ligands (GTP, initiator tRNA, 30S subunits, and 50S subunits) and is involved in two essential steps in translation initiation. It binds initiator tRNA and promotes its entry into the ribosomal P site, and it promotes ribosomal subunit joining. IF2 is a GTPase, and its activity is strongly induced by binding to both ribosomal subunits. We tested the influence of purified native mouse eIF5B in 48S complex formation on β-globin mRNA using a toeprinting assay and found that it could not substitute for eIF2 in this process, nor did it influence 48S complex formation in the presence of eIF2. eIF5B was also unable to substitute for eIF2 in assembly of 80S ribosomes that were active in catalyzing methionylpuromycin synthesis. Thus, despite the sequence homology with IF2 and their common role in promoting subunit joining, eIF5B differs from IF2 in not promoting recruitment of initiator tRNA to the small ribosomal subunit.

eIF1 AND eIF3 INFLUENCE THE REQUIREMENT FOR eIF5B IN RIBOSOMAL SUBUNIT JOINING

The physical and biochemical properties of eIF5B suggest that it is probably a protein that had previously been implicated in subunit joining and then (erroneously) discounted as an inactive contaminant of the 49-kD eIF5 (for review, see Pestova et al. 2000a). Earlier reports indicated that eIF5 and eIF5B had overlapping and even redundant functions in promoting hydrolysis of eIF2-bound GTP on 48S complexes and in joining of 60S subunits to them, whereas we had found that both factors were essential for assembly of 80S ribosomes on native mRNA. These apparently contradictory conclusions are probably due to the fact that whereas 48S complexes assembled on native mRNA contain eIFs 1, 1A, 2, and 3, many earlier analyses were done using 48S complexes assembled on AUG triplets using only 40S subunits and the eIF2/GTP/initiator tRNA ternary complex. This difference suggested that eIFs 1, 1A, and 3 might influence the factor requirements for subunit joining. To investigate this, we first assembled 48S complexes on AUG triplets using different combinations of these three factors together with 40S subunits and the ternary complex.

Assembly of active 80S ribosomes was assayed by their ability to catalyze synthesis of methionylpuromycin, a reaction that mimics formation of the first peptide bond. Individually, eIF5 and eIF5B were equally active in promoting 80S ribosome formation from 48S complexes that contained only eIF1A and eIF2, and they did not act synergistically or even additively. Additional inclusion of eIF1 and eIF3 in assembly reactions reduced the individual activities of both eIF5B, and to a greater extent, eIF5. However, in the presence of these four factors, eIF5 and eIF5B acted synergistically, indicating that they have complementary functions in subunit joining. eIF5B acted catalytically in methionylpuromycin synthesis, promoting several cycles of subunit joining.

Hydrolysis of eIF2-bound GTP was induced equally by eIF5 and by eIF5B when 48S complexes contained only eIF1A and eIF2. The activity of eIF5B in inducing GTP hydrolysis was inhibited by eIF1 and eIF3, whereas the activity of eIF5 was unaffected. This difference and the lack of sequence homology between eIF5 and eIF5B suggests that these two factors may induce eIF2's GTPase activity in different ways. The activity of eIF5B$_{587-1220}$ was reduced significantly more than the activity of full-length eIF5B, which indicates that the highly charged amino-terminal domain of eIF5B plays a role in mediating interactions with a component(s) of the 48S complex.

Taken together, these results indicate that in the presence of the full set of factors that are normally associated with 48S complexes, eIF5 induces hydrolysis of eIF2-bound GTP and that eIF5B is required to play another essential role. These experiments do not indicate whether the roles of eIF5 and eIF5B in subunit joining are limited to merely preparing the 48S complex for joining to a 60S subunit or whether one or both participate directly in the joining process itself.

TWO ESSENTIAL GTP HYDROLYSIS STEPS IN EUKARYOTIC INITIATION OF TRANSLATION

The central domain of eIF5B contains sequence motifs characteristic of GTP-binding proteins. eIF5B bound readily to GTP to form a binary complex independently of ribosomes, but bound GTP was readily displaced by unlabeled GTP, GMP-PNP, or GDP. eIF5B did not have a detectable intrinsic GTPase activity, but its ability to

hydrolyze GTP was induced by 60S subunits and even more so by 40S and 60S subunits together. 40S subunits alone had no effect. Ribosomal activation of eIF5B's GTPase activity can most readily be accounted for by direct physical interaction, which we have observed by centrifugation of complexes through a sucrose cushion (T. Pestova, unpubl.).

The GTPase activity of IF2 is also specifically activated by large and small ribosomal subunits together (Kolakofsky et al. 1968). This interesting similarity between eIF5B and IF2 may indicate that activation of their GTPase activities may occur by a common mechanism, possibly involving conserved or functionally related ribosomal elements. The binding site of eIF5B on the ribosome is not known, although some evidence suggests that binding of eIF5B prevents binding of the elongation factor 1/GTP/tRNA ternary complex to the ribosome (Wilson et al. 2000) and that bound eIF5B occludes the ribosomal A site (see below).

To investigate whether eIF5B needs to bind GTP to adopt an active conformation, 48S complexes assembled in the presence of GTP were purified by gel filtration (to remove unincorporated GTP) and were then assayed for the ability to form 80S ribosomes in reactions that also contained 60S subunits, eIF5, either full-length native eIF5B or recombinant eIF5B$_{587-1220}$, and different nucleotides. The activity of eIF5B$_{587-1220}$ was completely GTP-dependent. Although native full-length eIF5B was active in the absence of GTP, its activity was stimulated about threefold in its presence (T. Pestova, unpubl.). Subunit joining was also promoted by eIF5B$_{587-1220}$/ GMP-PNP, a result which indicates that GTP hydrolysis is not required for this process. However, in the presence of GMP-PNP, eIF5B acted stoichiometrically rather than catalytically. The ability of eIF5B to catalyze multiple rounds of ribosomal subunit joining in the presence of GTP (but not GMP-PNP) strongly suggests that after hydrolysis of GTP, the resulting GDP product readily exchanges for GTP without the need for a guanine nucleotide exchange factor.

These results lead to the conclusion that translation initiation involves two successive GTP hydrolysis steps. The first is catalyzed by eIF2 in the 48S complex and is induced by eIF5, and the second is catalyzed by eIF5B. The requirement for GTP hydrolysis by eIF5B could be because GTP hydrolysis is required for the displacement of factors from the 48S complex or for the release of eIF5B itself from assembled ribosomes. Analysis of ribosomal complexes resolved by sucrose density gradient centrifugation indicated that eIF1, eIF2, and eIF3 were displaced from 48S complexes on assembly into 80S ribosomes irrespective of whether GTP or GMP-PNP was included in the reaction. However, eIF5B$_{587-1220}$ remained stably associated with 80S ribosomes, and to a lesser extent with 60S subunits in the presence of GMP-PNP, whereas in the presence of GTP, eIF5B was not bound to ribosomal subunits or to any ribosomal complex (Pestova et al. 2000b).

The requirement for hydrolysis of eIF5B-bound GTP for this factor to dissociate from 80S ribosomes can explain the inability of 80S ribosomes assembled using eIF5B/GMPPNP to catalyze methionylpuromycin synthesis (Pestova et al. 2000b). Strikingly, addition of eIF5B$_{587-1220}$/GMP-PNP to active 80S ribosomes also inhibited their ability to catalyze this reaction, whereas the addition of eIF5B$_{587-1220}$, GMP-PNP, or eIF5B$_{587-1220}$/GTP had no such effect. The specificity of inhibition suggests that eIF5B binds to the ribosomal A site. This remains to be confirmed, but it is consistent with the physical interaction of eIF5B with eIF1A (Choi et al. 2000), which has a binding site that has been mapped to the vicinity of the A site by enzymatic footprinting (V. Kolupaeva and T. Pestova, unpubl.).

ACKNOWLEDGMENTS

We acknowledge our colleagues V. Kolupaeva and I. Lomakin at SUNY HSC Brooklyn who contributed to the work reviewed in this manuscript. Research done in our laboratories was supported by grant GM-59660 from the National Institutes of Health (to C.U.T.H.), grant GM-63940 from the National Institutes of Health (to T.V.P.), and by grants MCB-9726958 and MCB-0110834 from the National Science Foundation (to C.U.T.H. and T.V.P.).

REFERENCES

Asano K., Clayton J., Shalev A., and Hinnebusch A.G. 2000. A multifactor complex of eukaryotic initiation factors, eIF1, eIF2, eIF3, eIF5 and initiator tRNA(Met) is an important translation initiation intermediate in vivo. *Genes Dev.* **14:** 2534.

Battiste J.L., Pestova T.V., Hellen C.U.T., and Wagner G. 2000. The eIF1A solution structure reveals a large RNA-binding surface important for scanning function. *Mol. Cell* **5:** 109.

Carter A.P., Clemons W.M., Jr., Brodersen D.E., Morgan-Warren R.J., Hartsch T., Wimberly B.T., and Ramakrishnan V. 2001. Crystal structure of an initiation factor bound to the 30S ribosomal subunit. *Science* **291:** 498.

Chaudhuri J., Chowdhury D., and Maitra U. 1999. Distinct functions of eukaryotic translation initiation factors eIF1A and eIF3 in the formation of the 40 S ribosomal preinitiation complex. *J. Biol. Chem.* **274:** 17975.

Cigan A.M., Feng L., and Donahue T.F. 1988. tRNA$_i$^{Met} functions in directing the scanning ribosome to the start site of translation. *Science* **242:** 93.

Choi S.K., Lee J.H., Zoll W.L., Merrick W.C., and Dever T.E. 1998. Promotion of binding of Met-tRNA$_i$^{Met} to ribosomes by yIF2, a bacterial IF2 homolog in yeast. *Science* **280:** 1757.

Choi S.K., Olsen D.S., Roll-Mecak A., Martung A., Remo K.L., Burley S.K., Hinnebusch A.G., and Dever T.E. 2000. Physical and functional interaction between the eukaryotic orthologs of prokaryotic translation initiation factors IF1 and IF2. *Mol. Cell. Biol.* **20:** 7183.

Cui Y., Dinman J.D., Kinzy T.G., and Peltz S.W. 1998. The mof2/Sui1 protein is a general monitor of translational accuracy. *Mol. Cell. Biol.* **12:** 248.

Das S., Maiti T., Das K., and Maitra U. 1997. Specific interaction of eukaryotic translation initiation factor 5 (eIF5) with the β-subunit of eIF2. *J. Biol. Chem.* **272:** 31712.

Donahue T.F. 2000. Genetic approaches to translation in *Saccharomyces cerevisiae*. In *Translational control of gene expression.* (ed. N. Sonenberg et al.), p. 487. Cold Spring Harbor Laboratory Press, Cold Spring Harbor, New York.

Fletcher C.M., Pestova T.V., Hellen C.U.T., and Wagner G. 1999. Structure and interactions of the translation initiation factor eIF1. *EMBO J.* **18:** 2631.

Gunnery S., Mälvali Ü., and Mathews M.B. 1997. Translation of

an uncapped mRNA involves scanning. *J. Biol. Chem.* **272:** 21642.

Hershey J.W.B. and Merrick W.C. 2000. The pathway and mechanism of initiation of protein synthesis. In *Translational control of gene expression* (ed. N. Sonenberg et al.), p. 33. Cold Spring Harbor Laboratory Press, Cold Spring Harbor, New York.

Huang H.-K., Yoon H., Hannig E.M., and Donahue T.F. 1997. GTP hydrolysis controls stringent selection of the AUG start codon during translation initiation in *Saccharomyces cerevisiae. Genes Dev.* **11:** 2396.

Kolakofsky D., Ohta T., and Thach R.E. 1968. Junction of the 50S ribosomal subunit with the 30S initiation complex. *Nature* **220:** 244.

Kozak M. 1989. The scanning model for translation: An update. *J. Cell Biol.* **108:** 229.

———. 1991. Structural features in eukaryotic mRNAs that modulate the initiation of translation. *J. Biol. Chem.* **266:** 19867.

Kyrpides N.C. and Woese C.R. 1998. Universally conserved translation initiation factors. *Proc. Natl. Acad. Sci.* **95:** 224.

Lee J.H., Choi S.K., Roll-Mecak A., Burley S.K., and Dever T.E. 1999. Universal conservation in translation initiation revealed by human and archaeal homologs of bacterial translation factor IF2. *Proc. Natl. Acad. Sci.* **96:** 1066.

Lomakin I.B., Hellen C.U.T., and Pestova T.V. 2000. Physical association of eukaryotic initiation factor 4G (eIF4G) with eIF4A strongly enhances binding of eIF4G to the internal ribosomal entry site of encephalomyocarditis virus and is required for internal initiation of translation. *Mol. Cell. Biol.* **20:** 6019.

Marcotrigiano J., Lomakin I.B., Sonenberg N., Pestova T.V., Hellen C.U.T., and Burley S.K. 2001. A conserved HEAT domain within eIF4G directs assembly of the translation initiation machinery. *Mol. Cell* **7:** 193.

Morino S., Imataka H., Svitkin Y., Pestova T.V., and Sonenberg N. 2000. Eukaryotic translation initiation factor (eIF) 4E binding site and the middle one-third of eIF4GI constitute the core domain for cap-dependent translation, and the C-terminal one-third of eIF4GI functions as a modulatory region. *Mol. Cell. Biol.* **20:** 468.

Paulin F.E., Campbell L.E., O'Brien K., Loughlin J., and Proud C.G. 2001. Eukaryotic translation initiation factor 5 (eIF5) acts as a classical GTPase activator protein. *Curr. Biol.* **11:** 55.

Pestova T.V. and Hellen C.U.T. 1999. Ribosome recruitment and scanning: What's new? *Trends Biochem. Sci.* **24:** 85.

———. 2000. The structure and function of initiation factors in eukaryotic protein synthesis. *Cell. Mol. Life Sci.* **57:** 651.

Pestova T.V., Borukhov S.I., and Hellen C.U.T. 1998. Eukaryotic ribosomes require initiation factors 1 and 1A to locate initiation codons. *Nature* **394:** 854.

Pestova T.V., Dever T.E., and Hellen C.U.T. 2000a. Ribosomal subunit joining. In *Translational control of gene expression* (ed. N. Sonenberg et al.), p. 425. Cold Spring Harbor Laboratory Press, Cold Spring Harbor, New York.

Pestova T.V., Lomakin I.B., Lee J.H., Choi S.K., Dever T.E., and Hellen C.U.T. 2000b. The joining of ribosomal subunits in eukaryotes requires eIF5B. *Nature* **403:** 332.

Rozen F., Edery I., Meerovitch K., Dever T.E., Merrick W.C., and Sonenberg N. 1990. Bidirectional RNA helicase activity of eukaryotic initiation factors 4A and 4F. *Mol. Cell. Biol.* **10:** 1134.

Sette M., van Tilborg P., Spurio R., Kaptein R., Paci M., Gualerzi C.O., and Boelens R. 1997. The structure of the translational initiation factor IF1 from *E. coli* contains an oligomer-binding motif. *EMBO J.* **16:** 1436.

Wilson J., Pestova T.V., Hellen C.U.T., and Sarnow P. 2000. Initiation of protein synthesis from the A site of the ribosome. *Cell* **102:** 511.

Yoon H. and Donahue T.F. 1992. The sui1 suppressor locus in *Saccharomyces cerevisiae* encodes a translation factor that functions during tRNA$_i$^{Met} recognition of the start codon. *Mol. Cell. Biol.* **12:** 248.

Interactions of the Eukaryotic Translation Initiation Factor eIF4E

J.D. Gross,* H. Matsuo,* M. Fletcher,* A.B. Sachs,†
and G. Wagner*

*Department of Biological Chemistry and Molecular Pharmacology, Harvard Medical School,
Boston, Massachusetts 02115; †Department of Molecular and Cell Biology,
University of California at Berkeley, Berkeley, California 94720

Initiation of translation in eukaryotes is a highly regulated process by which a cell convinces the ribosome to start protein synthesis. This involves a multitude of initiation factors that interact with the small ribosomal subunit, recruit the initiator tRNA, and enable binding of the 43S particle to the messenger RNA (Sonenberg et al. 2000). Recruitment of the ribosome to mRNA is thought to happen via binding of the eIF3 component of the 43S ribosomal complex to eIF4G. The latter protein binds to the cap-binding protein eIF4E, recruiting the ribosome to the 5′ end of the mRNA. Ribosome recruitment to mRNA is enhanced by the poly(A)-binding protein, Pabp, and it has been shown in yeast that Pabp can mediate translation initiation if the eIF4E/eIF4G interaction is impaired (Tarun et al. 1997). Using a third recruitment mechanism, some mRNAs have internal ribosome entry sites for direct recruitment of the small ribosomal subunit to the mRNA, primarily in cooperation with eIF4G. Thus, the protein eIF4G plays a central role in the initiation of eukaryotic translation. Mammalian and yeast forms of eIF4G consist of about 1550 and 950 residues, respectively. A highly conserved segment consisting of YxxxxLΦ (Φ stands for a hydrophobic residue) responsible for eIF4E binding has been identified. In addition, yeast and human eIF4G share related functionalities, such as binding eIF4A, eIF3, and Pabp.

It has been shown in the past that the interaction of eIF4E and eIF4G is controlled by the regulatory 4E-binding protein, 4E-BP, for which several isoforms have been found in man, rat, and yeast (Haghighat et al. 1995; Altmann et al. 1997; Poulin et al. 1998). The 4E-BPs share the short consensus binding sequence YxxxxLΦ with eIF4G and compete with eIF4G for binding to eIF4E inhibiting translation. Hyperphosphorylation of 4E-BP mediated by hormone-stimulated signaling events causes dissociation of 4E-BP from eIF4E so that eIF4G can bind and translation is initiated (Pause et al. 1994; Mader et al. 1995; Gingras et al. 1999). The goal of our work was to investigate structural details of the interactions between eIF4E, the mRNA cap, eIF4G, and 4EBP, and to understand the mechanisms of this molecular switch. The recurring theme of the research was that domains interacting with eIF4E were unstructured by themselves but folded up at least partially when binding eIF4E. The purpose of this feature is still not fully understood.

eIF4E HAS TWO DISTINCT BINDING SITES FOR THE CAP AND eIF4G/4E-BP

To understand the mechanisms of eIF4E function, we have solved the structure of yeast eIF4E in complex with a cap analog using nuclear magnetic resonance (NMR) spectroscopy (Fig. 1) (Matsuo et al. 1997). Independently, Burley and coworkers have solved a crystal structure of mouse eIF4E (Marcotrigiano et al. 1997). Due to the low solubility and rapid irreversible precipitation of yeast eIF4E at concentrations needed for NMR spectroscopy, we placed the protein in a CHAPS micelle. This had the benefit of increasing the solubility from about 0.3 mM to almost 1 mM, and the time course of irreversible precipitation was slowed from a few days to several months. However, this approach increased the molecular mass of the system to about 40 kD. The large molecular mass and the presence of the detergent represented a challenge for NMR spectroscopy.

The structure of the 213-residue yeast eIF4E exhibits a folded core that consists of the highly conserved residues 36–213. The amino-terminal 35 residues are unstructured, including a conserved His-32 and hydrophobic residues in positions 33 and 35 (Matsuo et al. 1997; McGuire et al. 1998). The folded portion of the eIF4E structure consists of an eight-stranded central β sheet that is decorated on one side with three long α helices. The other face of the sheet exhibits a deep cleft that forms the binding site for the cap structure. The binding pocket is formed by three short helices located in the loops connecting strands b1 and b2, b3 and b4, and b7 and b8 (Matsuo et al. 1997). The specificity of the binding pocket for the m^{7}G base is due to the enhanced stacking interaction of the base between two tryptophanes located at helices a1 and a3. The electron-deficient π orbitals of the methylated G (formal charge +1) interact favorably with the filled π orbitals of the Trp side chains (Matsuo et al. 1997). Despite the fact that the cap-binding pocket is deep and the m^{7}G group fits snugly into the pocket, the equi-

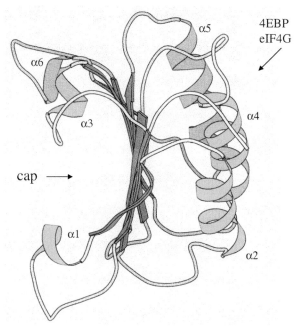

Figure 1. Ribbon diagram of the yeast eIF4E structure (Matsuo et al. 1997). The figure was produced with Molscript (Kraulis 1991).

librium dissociation constant for a small cap analog m⁷Gpp is rather high in the low micromolar range (Cai et al. 1999).

4E-BP IS PRIMARILY UNSTRUCTURED BUT CONTAINS TWO SHORT SEGMENTS THAT ARE PARTIALLY FOLDED

To study the mechanism by which 4E-BP inhibits translation initiation, we pursued structural studies of this

class of molecules. With 118 residues, these molecules seemed ideal for NMR structure determination. We expressed 4E-BP1 and 4E-BP2 and recorded a ^{1}H-^{15}N correlated spectrum. These spectra exhibited rather poor spectral dispersion and indicated that 4EBPs are unstructured (Fig. 2A). Adding mouse eIF4E to the sample caused the appearance of a few new cross-peaks. Thus, binding of eIF4E induces structure but only in a small part of 4E-BP1; the majority of the resonances are still in a state characteristic of an unfolded protein (Fig. 2B) (Fletcher et al. 1998). Comparison of the spectra of free and bound 4E-BP1 indicated that ~20 residues become structured upon interaction with eIF4E. Subsequently, we synthesized a peptide consisting of residues 49–68 that contained the consensus binding sequence. The peptide produced similar but not identical changes in the spectrum of mouse eIF4E as full-length 4E-BP1. In addition, the peptide inhibited translation of capped CAT mRNA in reticulocyte lysate (Fletcher et al. 1998). Subsequently, crystal structures of complexes between two related 17-residue peptides and mouse eIF4E were solved and revealed an ordered strand-helix arrangement of the bound peptide (Marcotrigiano et al. 1999).

To explore whether free 4E-BP is entirely without structure, we pursued sequential assignments for the peptide (Fig. 3). By doing this, we realized that a number of signals are absent in the spectra due to extensive line broadening. Signals are missing for two short stretches of the amino acid sequence, residues 43–46 and 56–65 (Fletcher and Wagner 1998). Extensive line broadening of signals in an NMR spectrum is usually due to slow exchange between different folded conformations. Here it indicates the presence of partially folded structure. The sequence 56–65 overlaps in part with the consensus binding sequence YxxxxLΦ, indicating that this segment may be partially folded even prior to binding eIF4E. The sequence 43–46 is amino-terminal to the consensus binding sequence, and its significance for eIF4E binding was not

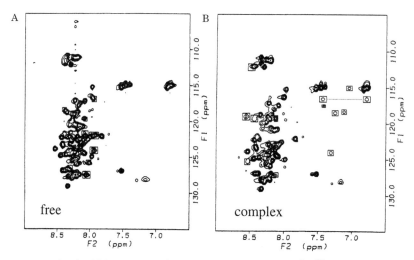

Figure 2. 4EBP is unstructured and exhibits only minor changes upon binding eIF4E. ^{1}H-^{15}N correlated spectra of 4E-BP1 alone (*A*) and in complex with mouse eIF4E (*B*).

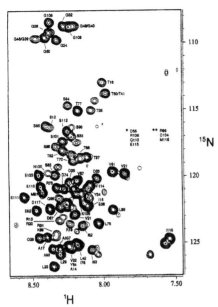

Figure 3. High-resolution ^{1}H-^{15}N correlated spectrum of free 4E-BP1. The sequence-specific assignments are given in the figure.

4E-BP, we measured heteronuclear NOEs of the amide nitrogens. In this experiment, amide proton resonances are saturated by radio-frequency irradiation, and the effect on the intensity of the amide nitrogens is recorded. In a folded protein the relative change is rather small, typically –0.2, whereas in an unfolded peptide it is on the order of –1.0 or even lower. The data shown in Figure 4 indicate that all the signals which could be observed are from completely disordered regions of the peptide. Thus, outside the two partially folded segments, 4E-BP is completely unstructured and mobile.

IDENTIFICATION OF A FRAGMENT OF YEAST EIF4G THAT IS UNSTRUCTURED BY ITSELF BUT PROTEOLYSIS-PROTECTED IN COMPLEX WITH EIF4E

Since 4E-BP competes with eIF4G for binding eIF4E, we wondered whether the 4E-binding domain of eIF4IG behaves similarly. Hershey et al. (1999) performed limited proteolysis experiments on various fragments of yeast eIF4G1 in the presence and absence of eIF4E. This work identified a fragment that is proteolysis resistant when forming a complex with eIF4E but is rapidly proteolyzed in the absence of the cap-binding protein. Thus, binding to eIF4E seemed to protect proteolytic sites or even induce the formation of a folded structure. This was confirmed with NMR experiments. eIF4G $_{393–490}$ exhibits

obvious. This segment was not included in the X-ray structure of the complex with eIF4E (Fig. 4) (Marcotrigiano et al. 1999). To further characterize the state of free

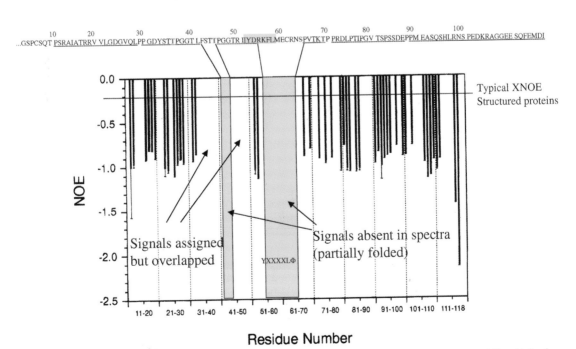

Figure 4. Position of disordered and partially ordered segments in 4E-BP1. Residues for which assignments could be obtained are underlined. The consensus binding segment YxxxxLΦ is colored. The lower panel reports the heteronuclear NOEs, which are a measure of mobility/disorder. Typical values for a structured protein are about –0.2, indicated with the horizontal line. All the residues that can be observed and are well enough dispersed to be analyzed have NOE values of about –1.0, consistent with complete absence of structure (Fletcher and Wagner 1998).

eIF4G1₃₉₃₋₄₉₀ folds upon binding eIF4E

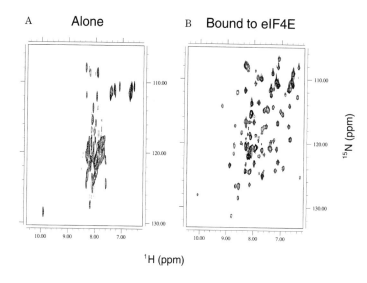

A   Alone   B   Bound to eIF4E

Figure 5. The 4E-binding domain of eIF4G1 is unstructured by itself (*A*) but adopts a folded structure when bound to eIF4E (*B*). The ^{1}H-^{15}N correlated spectrum of free eIF4G1 393–490 is typical for an unstructured protein but is well dispersed in the complex with eIF4E (Hershey et al. 1999).

a ^{1}H-^{15}N correlated spectrum characteristic of an unfolded polypeptide, similar to that of free 4E-BP1 (Fig. 5A). However, addition of eIF4E causes a dramatic dispersion of the signals, revealing that the domain folds upon binding eIF4E (Fig. 5B). This is entirely different from the 4E-BP case. Probing of eIF4G1 with tandem alanine-screening mutagenesis identified three sites in eIF4G ₃₉₃₋₄₉₀ that were crucial for eIF4E function in vivo: E427D428, K446K447, and L459L460 (Hershey et al. 1999). The latter sequence is the LΦ segment of the consensus binding sequence, but the former two are not.

STRUCTURAL STUDIES OF THE COMPLEX BETWEEN eIF4E AND eIF4G₃₉₃₋₄₉₀

Presently, we are pursuing a solution structure analysis of the complex between eIF4E and eIF4G₃₉₃₋₄₉₀ to understand the differences between the eIF4G and 4E-BP

complexes with eIF4E and thus to understand the mechanism of the translational control achieved by the action of 4E-BP. Starting with the complex characterized with the spectrum shown in Figure 5B, we obtained sequence-specific assignments for eIF4G₃₉₃₋₄₉₀ in the complex with eIF4E. We identified the locations of regular secondary structure and determined a preliminary structure of the 4E-binding domain of eIF4G1. The domain contains four helices and two extended strands (Fig. 6). The 4E-binding face is shown in Figure 7. The consensus sequence is oriented vertically on the left-hand side. Y454, L459, and L460 of the YxxxxLΦ sequence are labeled in the figure. There is a cavity in the middle of the displayed structure, which accommodates a large portion of the amino terminus of eIF4E. The locations of the three tandem mutations identified as crucial for the eIF4E/eIF4G interaction (Hershey et al. 1999) are indicated with the short horizontal bars in Figure 6. E427 and D428 are labeled in the

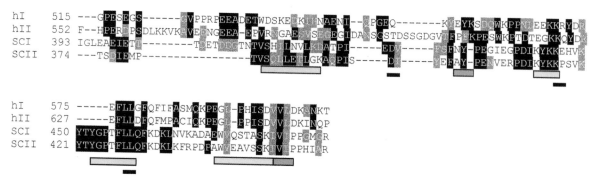

Figure 6. Sequence comparison of the 4E-binding domains of human eIF4G1 (hI), human eIF4G2 (hII), yeast eIF4G1 (SCI), and yeast eIF4G2 (SCII). The locations in yeast eIF4G1 of helices (*yellow*) and ordered extended segments (*orange*) are indicated. The positions of the three tandem mutations that affected eIF4E binding in vivo are marked with black horizontal bars. The two segments of 4E-BP1 that have partial structure are in register with part of the consensus binding sequence and a short helix containing the K446K447 dipeptide.

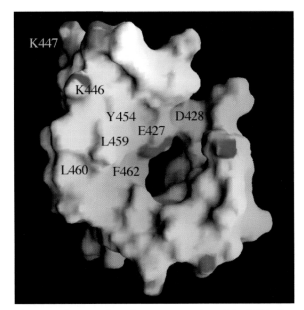

Figure 7. Electrostatic surface representation of the 4E-binding domain of yeast eIF4G1 determined in complex with eIF4E. The residues Y454, L459, and L460 of the YxxxxLΦ sequence are labeled in the figure. E427 and D428 are close to the consensus binding sequence. The figure was produced with GRASP (Nicholls et al. 1991).

surface representation of Figure 7. They are in close proximity to the residues of the YxxxxLΦ sequence and make numerous contacts with eIF4E, explaining their importance for the interaction. K446 and K447 are on an adjacent face of the protein, and contact an acidic loop connecting a strand and a helix in eIF4E.

It is interesting to compare where the regions of partially folded structure in 4E-BP1 are located relative to the secondary structure elements of eIF4G1. The segment of residues 56–65 (Fig. 4) is located at the helix containing the YxxxxLΦ sequence. The other partially structured segment of 4EBP1 (residues 43–46) is located amino-terminal to the consensus binding sequence and is in register with the short helix containing the segment KYKK in eIF4G1. Interestingly, in eIF4G, this contains the dipeptide K446-K447 that was identified by alanine screening as important for the function of eIF4G (Hershey et al. 1999). Thus, it may be that 4E-BP adopts a similar helix in complex with eIF4E, and formation of the helix may be crucial for eIF4E interaction.

CONCLUSION

The interaction of the translation initiation factor eIF4E with eIF4G is regulated by the 4E-binding proteins. Both eIF4G and 4E-BP are primarily unstructured in their free form. A more detailed inspection of free 4E-BP1 revealed two small segments of partially folded structure located at the consensus binding sequence YxxxxLΦ and a number of residues amino-terminal to this segment, respectively. When binding eIF4E, 4E-BP folds up partially, but the majority of the protein remains unstructured. In contrast, almost the entire 4E-binding

domain of yeast eIF4G, eIF4G$_{393–490}$, adopts a globular structure in complex with eIF4E. Thus, unfolded-to-folded transitions appear to be a common element of translation initiation. Why binding of a partially folded 4E-BP to eIF4E can disrupt the interaction of the folded eIF4G with eIF4E remains to be determined.

ACKNOWLEDGMENTS

This research was supported by grants from the National Institutes of Health and the Human Frontier Science Program Organization. J.D.G. acknowledges a postdoctoral fellowship from the American Cancer Society.

REFERENCES

Altmann M., Schmitz N., Berset C., and Trachsel H. 1997. A novel inhibitor of cap-dependent translation initiation in yeast: p20 competes with eIF4G for binding to eIF4E. *EMBO J.* **16:** 1114.

Cai A., Jankowska-Anyszka M., Centers A., Chlebicka L., Stepinski J., Stolarski R., Darzynkiewicz E., and Rhoads R.E. 1999. Quantitative assessment of mRNA cap analogues as inhibitors of in vitro translation. *Biochemistry* **38:** 8538.

Fletcher C.M. and Wagner G. 1998. The interaction of eIF4E with 4E-BP1 is an induced fit to a completely disordered protein. *Protein Sci.* **7:** 1639.

Fletcher C.M., McGuire A.M., Gingras A.C., Li H., Matsuo H., Sonenberg N., and Wagner G. 1998. 4E binding proteins inhibit the translation factor eIF4E without folded structure. *Biochemistry* **37:** 9.

Gingras A.C., Raught B., and Sonenberg N. 1999. eIF4 initiation factors: Effectors of mRNA recruitment to ribosomes and regulators of translation. *Annu. Rev. Biochem.* **68:** 913.

Haghighat A., Mader S., Pause A., and Sonenberg N. 1995. Repression of cap-dependent translation by 4E-binding protein 1: Competition for binding to eukaryotic initiation factor 4E. *EMBO J.* **14:** 5701.

Hershey P.E., McWhirter S.M., Gross J.D., Wagner G., Alber T., and Sachs A.B. 1999. The Cap-binding protein eIF4E promotes folding of a functional domain of yeast translation initiation factor eIF4G1. *J. Biol. Chem.* **274:** 21297.

Kraulis P.J. 1991. Molscript—A program to produce both detailed and schematic plots of protein structures. *J. Appl. Crystallogr.* **24:** 946.

Mader S., Lee H., Pause N., and Sonenberg N. 1995. The translation initiation factor eIF-4E binds to a common motif shared by the translation factor eIF-4γ and the translational repressors 4E binding proteins. *Mol. Cell. Biol.* **15:** 4990.

Marcotrigiano J., Gingras A.C., Sonenberg N., and Burley S.K. 1997. X-ray studies of the messenger RNA 5′ cap-binding protein (eIF4E) bound to 7-methyl-GDP. *Nucleic Acids Symp. Ser.* **36:** 8.

———. 1999. Cap-dependent translation initiation in eukaryotes is regulated by a molecular mimic of eIF4G. *Mol. Cell.* **3:** 707.

Matsuo H., Li H., McGuire A.M., Fletcher C.M., Gingras A.C., Sonenberg N., and Wagner G. 1997. Structure of translation factor eIF4E bound to m^{7}GDP and interaction with 4E-binding protein. *Nat. Struct. Biol.* **4:** 717.

McGuire A.M., Matsuo H., and Wagner G. 1998. Internal and overall motions of the translation factor eIF4E: Cap binding and insertion in a CHAPS detergent micelle. *J. Biomol. NMR* **12:** 73.

Nicholls A., Sharp K.A., and Honig B. 1991. Protein folding and association: Insights from the interfacial and thermodynamic properties of hydrocarbons. *Proteins* **11:** 281.

Pause A., Belsham G.J., Gingras A.C., Donze O., Lin T.A., Lawrence J.C., Jr., and Sonenberg N. 1994. Insulin-dependent stimulation of protein synthesis by phosphorylation of a regulator of 5´-cap function. *Nature* **371:** 762.

Poulin F., Gingras A.C., Olsen H., Chevalier S., and Sonenberg N. 1998. 4E-BP3, a new member of the eukaryotic initiation factor 4E-binding protein family. *J. Biol. Chem.* **273:** 14002.

Sonenberg N., Hershey J.W.B., and Mathews M.B., Eds. 2000. *Translational control of gene expression.* Cold Spring Harbor Laboratory Press, Cold Spring Harbor, New York.

Tarun S.Z., Jr., Wells S.E., Deardorff J.A., and Sachs A.B. 1997. Translation initiation factor eIF4G mediates in vitro poly(A) tail-dependent translation. *Proc. Natl. Acad. Sci.* **94:** 9046.

A Multifactor Complex of eIF1, eIF2, eIF3, eIF5, and tRNA$_i$^Met Promotes Initiation Complex Assembly and Couples GTP Hydrolysis to AUG Recognition

K. Asano,* L. Phan,* L. Valásek,* L.W. Schoenfeld,* A. Shalev,* J. Clayton,*
K. Nielsen,* T.F. Donahue,† and A.G. Hinnebusch*

*Laboratory of Gene Regulation and Development, National Institute of Child Health and Human Development,
National Institutes of Health, Bethesda, Maryland 20892; †Department of Biology, Indiana University,
Bloomington, Indiana 47405

Translation initiation in mammalian cells requires the formation of an 80S initiation complex consisting of the 80S ribosome with methionyl initiator tRNA (Met-tRNA$_i$^Met) bound to the P site and base-paired with the AUG start codon in mRNA. Formation of this complex is catalyzed by ~30 polypeptides comprising 11 different eukaryotic initiation factors (eIFs). According to current models, the first step in the initiation pathway involves dissociation of 80S ribosomes into free 40S and 60S subunits, and is stimulated by eIF1A and the multisubunit eIF3. The eIF3 remains bound to the 40S subunit and participates in subsequent reactions. The second step involves the transfer of Met-tRNA$_i$^Met to the 40S ribosome in a ternary complex (TC) with the heterotrimeric factor eIF2 and GTP, producing the 43S complex. This step is stimulated by eIFs 1, 1A, and 3 (Fig. 1). Binding of mRNA to the 43S complex is catalyzed by eIF4F, containing the m^{7}G cap-binding protein (eIF4E), RNA helicase eIF4A, and the scaffolding subunit eIF4G. The eIF4F binds at the cap, and the helicase activity of eIF4A removes secondary structure from the mRNA leader, dependent on the accessory factors eIF4B and eIF4H. Interaction of the 43S complex with the unstructured mRNA leader, producing the 48S complex, is facilitated by physical interaction between eIF4G and eIF3. It is also stimulated by the mRNA poly(A) tail and poly(A)-binding protein (PABP), which can additionally interact with eIF4G. The 48S complex scans the mRNA for an AUG triplet in a reaction facilitated by eIFs 1 and 1A. On base-pairing between AUG and the anticodon of Met-tRNA$_i$^Met, the GTP in the TC is hydrolyzed and most (if not all) eIFs are released, yielding the 40S initiation complex. The hydrolysis of GTP is dependent on eIF5, functioning as a GTPase activating protein (GAP). In the final step, the 60S subunit joins the 40S complex in a reaction stimulated by eIF5B and requiring hydrolysis of a second molecule of GTP, yielding the 80S initiation complex (Fig. 1) (for review, see Hershey and Merrick 2000).

The eIF2–GDP released from the initiation complex is inactive for binding Met-tRNA$_i$^Met and must be converted to eIF2–GTP in order to regenerate the TC. This recycling reaction is facilitated by the heteropentameric guanine nucleotide exchange factor (GEF) eIF2B and is a major target of translational control by a conserved mechanism involving phosphorylation of eIF2. The eIF2 phosphorylated on Ser-51 of its α subunit (eIF2[αP]) is functional for transferring Met-tRNA$_i$^Met to the ribosome; but eIF2(αP)–GDP is a competitive inhibitor, not a substrate, of eIF2B. Because eIF2 occurs in excess of eIF2B, the recycling of eIF2 can be strongly inhibited by phosphorylation of only a fraction of eIF2 (Fig. 1) (for review, see Asano et al. 1999).

The broad outlines of the initiation pathway were elucidated more than 20 years ago by reconstituting different steps in the pathway with purified ribosomes, mRNA (or AUG triplet), Met-tRNA$_i$^Met, and the eIFs. Until recently, little was known about molecular interactions among the different factors that participate in this pathway. Fractions of eIF2 can be found stably associated with its GAP (eIF5) (Chaudhuri et al. 1994) or its GEF (eIF2B) (Proud 1992) in cell extracts. However, there is no evidence for a physical link between eIF2 and eIF3 or eIF1A, despite the stimulatory roles of the latter factors in recruiting TC to the ribosome. Direct interaction between TC and eIFs 1A or 3 may occur only when they are bound to the 40S ribosome; alternatively, another factor could serve as an adapter between TC and eIFs 1A or 3. The eIFs 1A and 3 might also produce an allosteric alteration of the 40S ribosome that stimulates TC binding indirectly. As indicated above, eIF4G can interact with both ends of the mRNA by binding to eIF4E at the cap or to PABP at the poly(A) tail, and mammalian eIF4G additionally interacts with eIF3. Hence, eIF4G serves as an adapter between mRNA and eIF3, and since eIF3 binds directly to the ribosome, these interactions should promote mRNA binding to the 43S complex (Hentze 1997; Sachs et al. 1997). However, the interaction between eIF4G and eIF3 is difficult to detect in budding yeast (Phan et al. 1998; Asano et al. 2001). Even in mammals, the binding partner for eIF4G among the eIF3 subunits is unknown, and it is unclear whether eIF4G–eIF3 interaction is crucial for mRNA binding to 40S ribosomes in vivo. The eIF2 interacts directly with mRNA (Laurino et al. 1999; Hinnebusch 2000), and this interaction may also contribute to mRNA binding by the 40S ribosome.

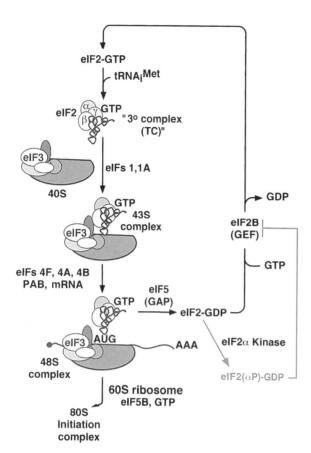

Figure 1. The eukaryotic translation initiation pathway. A conventional depiction of the pathway involving sequential binding of eIF3 and the TC, promoted by eIFs 1 and 1A to form the 43S preinitiation complex. The mRNA joins, facilitated by the eIF4 factors and PABP, to produce the 48S complex, which scans the mRNA for an AUG start codon. On AUG recognition (promoted by eIFs 1 and 1A), eIF5 binds and stimulates GTP hydrolysis, followed by release of factors and joining of the 60S subunit to produce the 80S initiation complex. Recycling of eIF2–GDP to eIF2–GTP is catalyzed by the GEF eIF2B. This reaction is inhibited by phosphorylation of eIF2–GDP on the α subunit, converting eIF2–GDP to a competitive inhibitor of eIF2B.

Recent work from our laboratory has revealed a network of physical interactions among eIFs that participate in the binding of Met-tRNA$_i^{Met}$ and mRNA to 40S subunits during assembly of the 48S complex. Unexpectedly, the carboxy-terminal domain (CTD) of eIF5, a GAP for the TC, plays a central role in these interactions, bridging association of eIF3 with the TC, eIF1, and eIF4G. We have identified a multifactor complex (MFC) containing eIFs 1, 2, 3, 5, and Met-tRNA$_i^{Met}$ that can exist free of 40S ribosomes, suggesting that constituents of the MFC bind to the 40S ribosome as a preformed unit. Results from in vitro experiments indicate that formation of the MFC promotes binding of TC and mRNA to the 40S ribosome. In vivo analysis indicates that stable incorporation of eIFs 1 and 5 into the 48S complex, via the MFC, is required for efficient coupling of eIF5-dependent GTP hydrolysis with start codon recognition during the scanning process.

SACCHAROMYCES CEREVISIAE CONTAINS AN EVOLUTIONARILY CONSERVED CORE eIF3 COMPLEX THAT IS TIGHTLY ASSOCIATED WITH eIF5 IN VIVO

Considering the central role of eIF3 in assembly of the 48S complex, it was imperative to define the subunit composition of eIF3 in budding yeast. Human eIF3 is a complex factor containing 11 subunits: p170, p116, p110, p66, p48, p47, p44, p40, p36, p35, and p28, designated a–k, respectively (Hershey and Merrick 2000; Browning et al. 2001). *S. cerevisiae* contains orthologs of human eIF3 subunits a, b, c, g, i, and j, known as TIF32/RPG1, PRT1, NIP1, TIF35, TIF34, and HCR1, respectively, but there are no obvious counterparts for the other human eIF3 subunits encoded in the yeast genome. Thus, it was expected that budding yeast would contain a simpler eIF3 compared to the mammalian factor.

Multisubunit complexes containing PRT1 (ortholog of human eIF3b) have been purified from yeast by several laboratories. A Ts⁻ lethal mutation in PRT1 leads to dramatic loss of polyribosomes at 36°C in vivo, indicating a severe initiation defect (Hartwell and McLaughlin 1969). Heat-treated *prt1-1* extracts are defective for ternary complex binding to 40S subunits (Feinberg et al. 1982), and this activity was rescued with a five-subunit complex containing PRT1 purified from wild-type cells (Danaie et al. 1995). We purified a similar complex by nickel-affinity chromatography directed against polyhistidine-tagged PRT1 (His₈-PRT1), followed by gel filtration. The proteins copurifying with His₈-PRT1 were identified by mass spectrometry as orthologs of human eIF3 subunits a, c, i, and g (TIF32, NIP1, TIF34, and TIF35, respectively). Interestingly, our purified complex also contained nearly stoichiometric amounts of eIF5, the GAP for eIF2 (Fig. 2A). The purified eIF3–eIF5 complex complemented the defects in binding radiolabeled Met-tRNA$_i^{Met}$ (TC) to 40S subunits (Fig. 2B), and also the translation of a LUC mRNA reporter, in a heat-treated *prt1-1* extract (Phan et al. 1998). Recently, we showed that the binding to 40S subunits of radiolabeled *MFA2* mRNA (Fig. 2C) and of endogenous eIF5 also are defective in *prt1-1* extracts, and both activities can be rescued with purified eIF3 (Phan et al. 2001). Our results imply that these critical functions of eIF3 can be carried out by a core complex containing only eIF3 subunits a, b, c, g, and i. Consistently, all five genes encoding these core subunits are essential in *S. cerevisiae*, and it was shown that depletion of the wild-type factors TIF32, TIF34, and TIF35, or Ts⁻ lethal mutations in *TIF32* and *TIF34*, all lead to inhibition of translation initiation in vivo (for review, see Hinnebusch 2000).

Yeast eIF3 also has been purified by its ability to substitute for human eIF3 in promoting 80S initiation complex formation and methionyl-puromycin (Met-puromycin) synthesis (Naranda et al. 1994). The purified complex contained PRT1, TIF34, TIF35, and a proteolytic fragment of TIF32, but lacked NIP1 (Naranda et al. 1994, 1997; Hershey et al. 1996). However, we found that

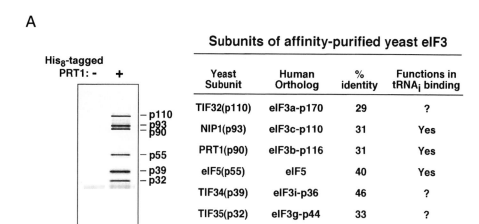

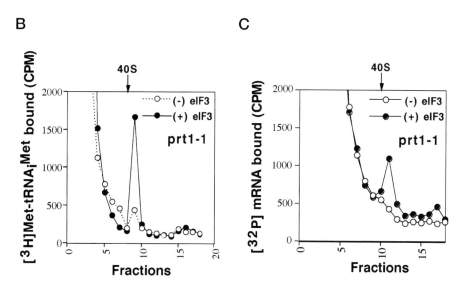

Figure 2. Affinity-purified yeast eIF3 contains eIF5 and stimulates binding of Met-tRNA$_i^{Met}$ and mRNA to 40S ribosomes in vitro. (*A*) In the panel on the left in the lane labeled "+", subunits of a purified yeast eIF3–eIF5 complex were resolved by SDS-PAGE and stained with silver. The complex was isolated from a ribosomal salt wash (RSW) by nickel chelation chromatography directed against the His$_8$-tagged PRT1 subunit of eIF3, followed by gel filtration on a Superose-6 FPLC column. The lane labeled "–" is derived from a mock purification conducted using RSW from a strain expressing untagged PRT1. The identities of all the copurifying proteins were determined by mass measurements of tryptic peptides by mass spectrometry. As shown in the table, they correspond to orthologs of known human eIF3 subunits or eIF5. The last column summarizes results of experiments in which extracts were prepared from strains with a mutation in PRT1 (*prt1-1*) or eIF5 (*tif5-7A*), or depleted of NIP1, and were shown to be defective for binding of exogenous radiolabeled Met-tRNA$_i^{Met}$ to 40S subunits, in a manner that was rescued with purified eIF3 or eIF5 (Phan et al. 1998; Asano et al. 2001; see text for further details). (*B*) After incubating [^{3}H]Met-tRNA$_i^{Met}$ with a heat-inactivated *prt1-1* extract, with or without added purified eIF3 as indicated, the reaction was resolved by sedimentation through a sucrose density gradient and the fractions (numbered from the top) were assayed for radioactivity. The arrow marks the position of 40S ribosomes. (*C*) The same experiment described in *B* was carried out using [^{32}P]-*MFA2* mRNA. (*A, B,* Reprinted, with permission, from Phan et al. 1998; *C,* reprinted from Phan et al. 2001.)

yeast extracts depleted of NIP1 were defective for translation and ternary complex binding to 40S subunits and that both defects could be partially rescued with affinity-purified eIF3–eIF5 complex (Phan et al. 1998). Moreover, the translation rate and polysome content were severely reduced following NIP1 depletion in vivo, confirming the importance of this subunit for translation initiation (Greenberg et al. 1998).

When purified by its ability to stimulate Met-puromycin synthesis (Naranda et al. 1994), yeast eIF3

contained three polypeptides of 135 kD, 62 kD, and 16 kD in addition to the core eIF3 subunits described above. Two of these proteins, TIF31 (Vornlocher et al. 1999) and GCD10 (Garcia-Barrio et al. 1995), are not related to subunits of human eIF3, and the third is eIF1 (encoded by *SUI1* in yeast) (Naranda et al. 1996). We confirmed the interaction of eIF1 with affinity-purified eIF3 (Phan et al. 1998) but found that the association is very salt-sensitive (Asano et al. 1998; Phan et al. 1998). The physical interaction of TIF31 with eIF3 seems firmly established

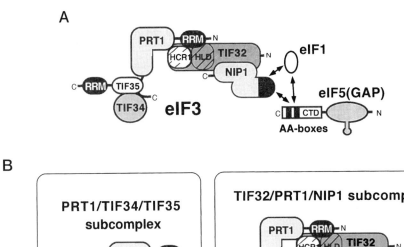

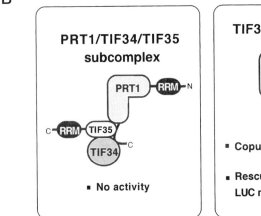

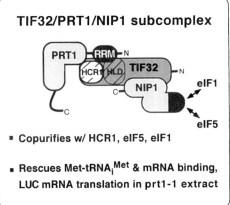

Figure 3. A complex of the three largest subunits of eIF3 can bind HCR1, eIF5, and eIF1 in vivo and stimulate binding of Met-tRNA$_i$^Met and mRNA to 40S ribosomes in vitro. (*A*) Model showing interactions among the five core subunits of yeast eIF3 (TIF32, PRT1, NIP1, TIF34, TIF35), the eIF3-associated protein HCR1, eIF5, and eIF1, identified using yeast two-hybrid assays and in vitro GST-pulldown experiments (Asano et al. 1998, 2000; Phan et al. 1998; Valásek et al. 2001). PRT1 and TIF35 contain RNA recognition motifs (RRM), and TIF32 contains an HCR1-like domain (HLD). Interactions are depicted schematically as points of contact between the representative shapes. The carboxy-terminal domain of eIF5 (CTD) contains a conserved bipartite motif (AA boxes) required for interaction with eIF3–NIP1. NIP1 additionally interacts with eIF1, and eIF1 interacts weakly with the eIF5–CTD. Both eIF1 and eIF5–CTD interact with the amino-terminal one-fifth of NIP1 (shaded black). The amino-terminal portion of eIF5 contains a zinc-finger motif depicted as a prong. (*B*) Summary of experiments showing that two different trimeric eIF3 subcomplexes can be overexpressed in yeast and purified by nickel-chelation chromatography directed against His$_8$-tagged PRT1. The TIF32/His$_8$-PRT1/NIP1 trimeric complex copurifies with HCR1, eIF5, and eIF1, in accordance with the subunit interaction model in *A*. It also has the biochemical activities listed in the right-hand panel. In contrast, the His$_8$-PRT1/TIF34/TIF35 subcomplex depicted on the left has little or no activity (Phan et al. 2001; see text for details).

(Hanachi et al. 1999; Vornlocher et al. 1999), although deletion of *TIF31* has no effect on cell growth or translation (Vornlocher et al. 1999). *GCD10* is essential (Garcia-Barrio et al. 1995), but our analysis revealed that GCD10 and GCD14 are subunits of a nuclear tRNA methyltransferase that is required for synthesis of 1-methyladenosine at position 58 (m^{1}A58) in all tRNAs containing this modification (Anderson et al. 1998, 2000; Cuesta et al. 1998). It remains unclear whether GCD10 has a cytoplasmic function in the eIF3 complex.

Yeast *HCR1* encodes an ortholog of human eIF3j that was first identified as a dosage suppressor of the Ts$^-$ phenotype of the *rpg1-1* allele of *TIF32* and was found associated with TIF32 in cell extracts. Deletion of *HCR1* leads to a slow-growth phenotype (Slg$^-$) and exacerbates the Ts$^-$ phenotype of *rpg1-1*, indicating functional interaction between HCR1 and TIF32 (Valásek et al. 1999). We found that high-copy-number *HCR1* also suppresses

certain Ts$^-$ alleles of *PRT1*. Additionally, we reported that a fraction of HCR1 is associated with all of the eIF3 core subunits, and that HCR1 is a component of 43–48S initiation complexes in vivo (Valásek et al. 2001).

A STABLE SUBCOMPLEX OF THE THREE LARGEST eIF3 SUBUNITS INTERACTS WITH eIF1, eIF5, AND HCR1 AND PROMOTES Met-tRNA$_i$^Met AND mRNA BINDING TO 40S RIBOSOMES IN VITRO

In an effort to map subunit interactions within eIF3 and to identify interactions between eIF3 and other factors, we conducted a large-scale yeast two-hybrid analysis to detect interactions among eIFs 1 and 5, and each of the subunits of eIFs 2 and 3. This analysis was complemented by in vitro binding experiments using GST fusions purified from *Escherichia coli* and proteins translated in vitro

(GST pulldown assays). The results of these studies suggested that eIF3 consists of two distinct domains organized at opposite ends of the PRT1 subunit. TIF34 and TIF35 bind to the carboxy-terminal end of PRT1, whereas TIF32 and NIP1 are tethered to the amino terminus (summarized in Fig. 3A). The carboxy-terminal portion of TIF32 interacts directly with the RNA recognition motif (RRM) at the amino terminus of PRT1, but no direct interactions between NIP1 and PRT1 were detected. Hence, NIP1 is tethered to eIF3 through the amino-terminal portion of TIF32, and this latter interaction maps to the carboxy-terminal half of NIP1 (Fig. 3A) (Asano et al. 1998; Valásek et al. 2001).

Interestingly, eIFs 1 and 5 both bound to the amino-terminal 20% of NIP1 (Phan et al. 1998; Asano et al. 1999, 2000). This interaction was mapped to the carboxy-terminal 40% of eIF5 (eIF5–CTD). eIF1 binds weakly to the eIF5–CTD, but a very stable trimeric complex containing eIF1, the amino-terminal 20% of NIP1, and the eIF5–CTD could be formed in vitro with recombinant polypeptides (summarized in Fig. 3A) (Asano et al. 2000). Yeast eIF1 was originally identified by the isolation of mutations that relax the stringency of AUG recognition on $HIS4$ mRNA (Sui$^-$ phenotype) (Yoon and Donahue 1992). Sui$^-$ mutations also have been isolated in eIF5, and the rate of GTP hydrolysis in the TC stimulated by eIF5 is a major determinant of the accuracy of selecting AUG triplets as start codons (Huang et al. 1997). The fact that both eIF1 and eIF5 are tethered to the amino-terminal segment of NIP1 (Fig. 3) suggests that eIF3 plays an important structural role in coordinating the functions of these factors in AUG recognition during mRNA scanning.

The results of GST pulldown assays indicate that HCR1 interacts with PRT1 via the RRM domain and also with TIF32, with which it makes multiple contacts. Interestingly, TIF32 contains an internal domain related to HCR1 that is part of its binding domain for the PRT1 RRM. Thus, HCR1 and the HCR1-like domain (HLD) in TIF32 both interact with the PRT1 RRM domain (Fig. 3A). We showed that these interactions can occur simultaneously by purifying a trimeric complex containing HCR1, the HLD of TIF32, and the PRT1 RRM (Valásek et al. 2001). Given that overexpression of HCR1 can suppress certain $prt1$ and $tif32$ mutations, it appears that direct binding of HCR1 to these two core subunits promotes one or more essential functions of the eIF3 complex. Because HCR1 is expressed at lower levels than are the core eIF3 subunits, its overexpression should increase the proportion of eIF3 containing HCR1 (Valásek et al. 2001).

We found recently that two trimeric eIF3 subcomplexes, His$_8$-PRT1/NIP1/TIF32 (PN2) and His$_8$-PRT1/TIF34/TIF35 (P45), can be overexpressed in yeast and affinity-purified by nickel chelation chromatography. Although a binary His$_8$-PRT1/TIF32 (P2) complex was also obtained, a stable His$_8$-PRT1/NIP1 binary complex was not recovered. Together, these findings provide in vivo support for the conclusion that TIF32 bridges interaction between PRT1 and NIP1 in eIF3 (summarized in Fig. 3B). Additionally, we found that eIF5, eIF1, and HCR1 copurified with the PN2 subcomplex but not with the P2 or P45 subcomplexes. These last results support

our conclusion that NIP1 provides a critical binding domain in eIF3 for eIFs 1 and 5 (Fig. 3B) (Phan et al. 2001). As mentioned above, HCR1 bound individually to TIF32 and PRT1, but not to NIP1, in GST pulldown assays. The fact that HCR1 copurified with the PN2 subcomplex, but not with P2, suggests that NIP1 is required for the interactions of HCR1 with TIF32 and PRT1 in the eIF3 complex at physiological concentrations of these proteins. NIP1 may provide an additional contact with HCR1 or promote the correct conformations of the HCR1-binding domains in TIF32 and PRT1.

The purified eIF3 subcomplexes described above also were tested for the ability to rescue Met-tRNA$_i^{Met}$ and mRNA binding to 40S subunits, and the translation of luciferase (LUC) mRNA, in a heat-inactivated $prt1-1$ extract. Surprisingly, the PN2 trimeric subcomplex, and to a lesser extent, the binary P2 subcomplex, showed substantial activity in all three assays compared to the P45 subcomplex or His$_8$-PRT1 alone (summarized in Fig. 3B). Thus, it appears that all of the critical functions of eIF3 can be carried out effectively by a subcomplex containing only the three largest core eIF3 subunits (Phan et al. 2001). The five-subunit eIF3 complex had a higher specific activity than did the PN2 subcomplex, suggesting that TIF34 and TIF35 augment the functions of the three largest subunits in eIF3. Because TIF34 and TIF35 are essential proteins in vivo (Naranda et al. 1997; Hanachi et al. 1999), their stimulatory activities may be crucial for maximal translation of a subset of mRNAs, including some that encode essential proteins.

Additional support for our eIF3/eIF5/eIF1 protein linkage map (Fig. 3A) was provided by purifying a mutant complex containing an amino-terminally truncated His$_8$-PRT1 subunit (His$_8$-prt1-Δ100) lacking the RRM domain. As predicted, this complex contained TIF34 and TIF35, which bind to the carboxyl terminus of PRT1, but lacked TIF32, NIP1, HCR1, and eIF5, whose interactions with eIF3 depend on the PRT1 RRM. We confirmed previous findings that prt1-Δ100 forms a defective eIF3 complex that cannot associate with 40S ribosomes in vivo, and that its expression has a dominant-negative Slg$^-$ phenotype (Evans et al. 1995). On the basis of the model in Figure 3A, we predicted that this phenotype results from sequestration of TIF34 and TIF35 in defective subcomplexes containing prt1-Δ100 that cannot bind to 40S ribosomes. In accordance with this idea, expression of $prt1-\Delta100$ increased the nonribosomal fractions of TIF34 and TIF35, but not of TIF32 or NIP1, in cell extracts. Additionally, we found that the Slg$^-$ phenotype of $prt1-\Delta100$ was suppressed specifically by overexpressing TIF34 and TIF35 simultaneously (Valásek et al. 2001).

THE CTD OF eIF5 BRIDGES eIF3 AND eIF2 IN A MULTIFACTOR COMPLEX THAT IS AN IMPORTANT INTERMEDIATE IN TRANSLATION INITIATION IN VIVO

We showed that binding of the eIF5–CTD to eIF3c/NIP1 is dependent on a sequence motif conserved between yeast and mammalian eIF5, consisting of two segments rich in acidic and aromatic residues, that we termed AA boxes 1 and 2 (Fig. 3A). A cluster of 12 alanine substitutions in AA box 1 (*tif5-12A*) or 7 alanine substitutions in AA box 2 (*tif5-7A*) destroyed the binding of eIF5 to NIP1 in vitro, and the latter mutation abolished coimmunoprecipitation from cell extracts of native eIF3 with a functional FLAG-tagged form of eIF5 (FL-eIF5) expressed in vivo. The *tif5-7A* mutation confers Slg⁻ and Ts⁻ phenotypes, whereas *tif5-12A* mutants are inviable (Asano et al. 1999).

As mentioned above, eIF5 forms a stable complex with eIF2, the substrate for its GAP activity. Interestingly, this interaction also was localized to the eIF5–CTD and involves the β subunit of eIF2 (Fig. 4A). The *tif5-7A* mutation in the CTD destroyed binding of eIF5 to both recombinant eIF2β and purified eIF2 holoprotein in vitro, and it abolished coimmunoprecipitation from cell extracts of native eIF2 with eIF5-FL expressed in vivo. Mutational analysis of eIF2β (SUI3) showed that the amino-terminal domain of this protein is necessary and sufficient for interaction with eIF5, and that three lysine stretches (K boxes) in this region make redundant contributions to binding of recombinant eIF2β or eIF2 holoprotein to eIF5 in vitro (Fig. 4A) (Asano et al. 1999). Although mutation of all 3 K boxes in eIF2β/SUI3 is lethal, mutants lacking only K boxes 1-2 or 2-3 are viable and show weakened association of the native eIF5 and eIF2 proteins in vivo. Interest-ingly, the Slg⁻ and Ts⁻ phenotypes of *tif5-7A* were partially suppressed by overexpressing all three subunits of eIF2 along with tRNAᵢ^Met. These findings suggest that the AA boxes in eIF5 mediate an important interaction between eIF5 and the ternary complex in vivo (Asano et al. 1999).

Given that the eIF5–CTD interacts with both eIF3c/NIP1 and eIF2β, it was possible that these interactions occur simultaneously and that eIF5–CTD bridges interaction between eIF3 and eIF2 in vivo. Consistent with this hypothesis, we showed that the eIF5–CTD can bind simultaneously to the amino-terminal segments of eIF3c/NIP1 and eIF2β in vitro. Furthermore, we could coimmunoprecipitate native eIF2 with eIF3 and eIF5 from cell extracts using antibodies against an HA-tagged form of eIF3i/TIF34. Importantly, the coimmunoprecipitation of eIF2 with HA-TIF34 was destroyed by the *tif5-7A* mutation in the eIF5–CTD (Fig. 4B,C). Similarly, coimmunoprecipitation of eIF3 with FL-eIF2β was abolished by mutations in K boxes 1-2 or 2-3 that weakened eIF2–eIF5 interactions in vitro. We also observed cosedimentation of eIFs 1, 2, 3, and 5 in a high-molecular-weight complex (~15S) during velocity sedimentation through sucrose gradients, and this complex was disrupted by the *tif5-7A* mutation (Fig. 5A). Together, these findings indicate that portions of eIFs 1, 2, 3, and 5 are present in a multifactor complex (MFC) that can exist

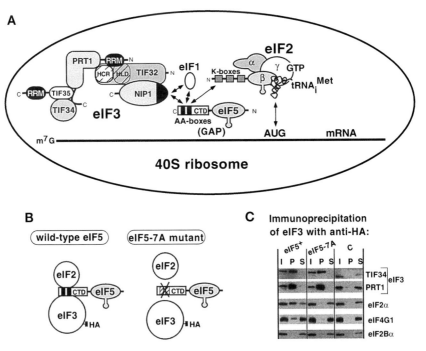

Figure 4. The eIF5-CTD bridges interactions between eIF3, eIF1, and eIF2. (*A*) Model summarizing interactions between the eIF5–CTD and the amino-terminal domain of eIF3–NIP1, eIF1, and the amino-terminal half of eIF2β containing three lysine stretches (K boxes), identified using the yeast two-hybrid assay and in vitro GST-pulldown experiments. Because the eIF5–CTD can interact simultaneously with its binding domains in eIF3–NIP1 and eIF2β in vitro (Asano et al. 2000), eIF5 is shown bridging eIF3 and the eIF2/GTP/Met-tRNAᵢ^Met ternary complex on the 40S ribosome in a 48S preinitiation complex. (*B*) Coimmunoprecipitation experiment showing that native eIF2 and eIF3 are associated in cell extracts, dependent on the AA boxes in the eIF5–CTD. The schematics on the left describe the design of the experiment shown on the right. Extracts were prepared from two yeast strains containing HA₃-tagged eIF3–TIF34 and either wild-type eIF5 (*TIF5*) or eIF5-7A (*tif5-7A*), and from a third strain containing untagged TIF34 (*C*). After immunoprecipitating with anti-HA antibodies, the immune complexes were probed for the five proteins listed on the right by western blotting. (I) 20% of the input extracts, (P) immunoprecipitated fractions, (S) supernatant fractions. (Reprinted, with permission, from Asano et al. 2000.)

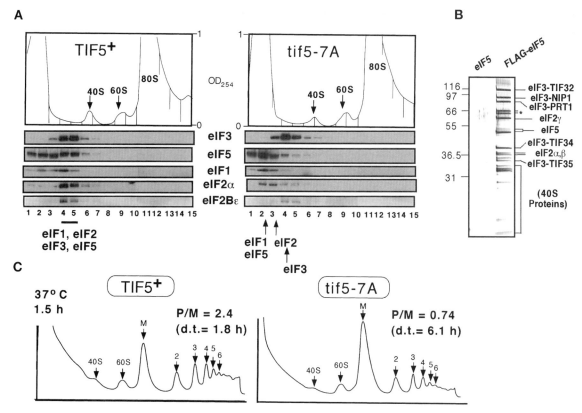

Figure 5. The mulifactor complex containing eIF1, eIF2, eIF3, and eIF5 can be isolated free of 40S ribosomes, and its disruption impairs the rate of translation initiation in vivo. (*A*) Cells from isogenic *TIF5* (*left*) and *tif5-7A* (*right*) strains were grown at 30°C and treated with cycloheximide for 5 minutes before harvesting. WCEs were prepared (in the absence of heparin to destabilize 43S-48S complexes) and resolved by centrifugation on sucrose gradients. The gradient was scanned continuously at 254 nm while collecting fractions, yielding the A_{254} profiles shown on top, with the positions of the ribosomal species indicated. In the panels beneath the A_{254} profiles are the results of western blot analysis of the gradient fractions using antibodies against the indicated initiation factors. (*B*) Silver staining of the multifactor complex purified by anti-FLAG antibody resin from a strain expressing FLAG-tagged eIF5 and resolved by SDS-PAGE. A mock-purified preparation obtained from a strain containing untagged eIF5 was analyzed in parallel. The identities of the initiation factor polypeptides were established by western analysis with the appropriate antibodies (data not shown). * indicates degradation products of TIF32. The presence of 40S ribosomal proteins was deduced by comparing the staining pattern with purified 40S ribosomes. (*C*) Wild-type and *tif5-FL-7A* strains growing at 30°C were shifted to 37°C for 1.5 hours, and treated with cycloheximide as above. WCEs were prepared and resolved by sedimentation on sucrose gradients. Fractions were collected while scanning continuously at 254 nm. The positions of 40S and 60S subunits, 80S ribosomes, and polysomes containing 1 to 6 translating ribosomes are indicated. (P/M) Ratio of A_{254} in the combined polysome fractions to that in the 80S peak. (d.t.) Cell doubling time. (Reprinted, with permission, from Asano et al. 2000.)

free of the ribosome in vivo and is dependent on the eIF5–CTD for its integrity (Asano et al. 2000).

We went on to purify the MFC by affinity chromatography (Fig. 5B) followed by centrifugation through a sucrose gradient (to remove contaminating 40S ribosomes) and found that the eIF2 could still be coimmunoprecipitated with eIF3 from the final preparation. Furthermore, the affinity-purified MFC contained stoichiometric amounts of eIF2 and tRNA$_i^{Met}$, implying that the eIF2 in the MFC exists in the form of TC (Asano et al. 2000). The ability of eIF1, eIF3, eIF5, and TC to associate independently of the 40S ribosome raises the possibility that these factors bind to the ribosome as a preformed unit (Fig. 4A).

The *tif5-7A* mutation in AA box 2 of the eIF5–CTD destabilized the MFC in cell extracts and, importantly, produced a decrease in polysome content and an accumulation of vacant 80S couples in vivo (Fig. 5C) (Asano et

al. 2000). Thus, integrity of the MFC is important for high-level translation initiation in vivo. The *tif5-7A* mutation strongly inhibits growth at elevated temperatures but is not lethal (Asano et al. 1999). Because this mutation does not completely eliminate the MFC in vivo (Asano et al. 2000), a more disruptive mutation in the eIF5–CTD that abolishes MFC formation could be lethal.

Interestingly, deletion of *HCR1* destabilized a fraction of the MFC, with a particularly strong effect on the association between eIFs 1 and 3; however, it did not disrupt the eIF3 complex itself. Thus, HCR1 seems to stabilize interactions between eIF3 and the other components of the MFC (Valásek et al. 2001). Given that HCR1 specifically interacted with the PRT1/NIP1/TIF32 subcomplex in eIF3 (Fig. 3B), perhaps HCR1 stabilizes the association of NIP1 with eIFs 1 and 5. Interestingly, in the *hcr1Δ* mutant, we observed accumulation of the MFC constituents on 40S ribosomes in addition to reduced amounts of free MFC. This may indicate that removal of HCR1 reduces the activities of eIF1 or eIF5 in converting 48S to 80S complexes as the rate-limiting defect in the *hcr1Δ* mutant. This would be consistent with the known function of eIF1 in locating the start codon, and of eIF5 in triggering GTP hydrolysis on AUG recognition.

Unexpectedly, deletion of *HCR1* decreased the steady-state level of 40S subunits and produced a concomitant increase in free 60S subunits. This reduction in 40S subunits may contribute to the diminished polysome size and slow-growth phenotype of the *hcr1Δ* strain. However, it is unlikely to account for the decreased amounts of MFC or accumulation of 43S or 48S complexes observed in *hcr1Δ* cells, as these defects were not observed in an *rp51aΔ* mutant that suffers from an even greater depletion of 40S subunits. Hence, it appears that HCR1 has a dual function in translation initiation, enhancing the activity of eIF3 and associated factors in the MFC, and promoting the biosynthesis or stability of 40S subunits. An attractive hypothesis is that both functions depend on the ability of HCR1 to bind to a specific site on the 40S ribo-

some, interacting with factors involved in 40S biogenesis and also promoting the correct conformation of the MFC in 43-48S initiation complexes (Valásek et al. 2001).

THE eIF5 CTD ENHANCES 40S BINDING OF Met-tRNA$_i$^{Met} AND mRNA IN VITRO

Given that the eIF5–CTD mediates stable association between eIF3 and TC in the MFC, and the fact that eIF3 (at least the mammalian factor) can bind directly to 40S subunits (Hinnebusch 2000), it seemed likely that formation of the MFC would enhance binding of TC to the 40S ribosome. Consistent with this prediction, we found that binding of exogenous [^{35}S]Met-tRNA$_i$^{Met} to the 40S ribosome was defective in a *tif5-7A* cell extract, and that this defect was complemented by addition of purified wild-type eIF5 but not the *tif5-7A* product (eIF5-7A). Interestingly, binding of exogenous [^{32}P]*MFA2* mRNA to the 40S ribosome also was defective in the *tif5-7A* extract and could be rescued with purified eIF5 but not with eIF5-7A (Fig. 6A,B) (Asano et al. 2001). The mRNA-

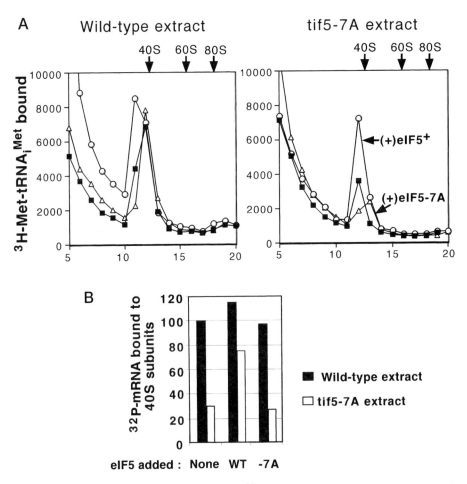

Figure 6. The *tif5-7A* mutation in the eIF5–CTD reduces Met-tRNA$_i$^{Met} and mRNA binding to 40S ribosomes in vitro. (*A*) After incubating [^{3}H]Met-tRNA$_i$^{Met} with WCEs from *TIF5* and *tif5-7A* strains, along with purified eIF5 (*circles*), purified eIF5-7A (*triangles*), or with no addition (*filled squares*), the reaction was resolved by sedimentation through a sucrose density gradient and the fractions (numbered from the top) were assayed for radioactivity. (*B*) The same experiment described in *A* was carried out using [^{32}P]-*MFA2* mRNA, and the amounts of radioactivity in the 40S region were quantified and plotted. (Reprinted, with permission, from Asano et al. 2001.)

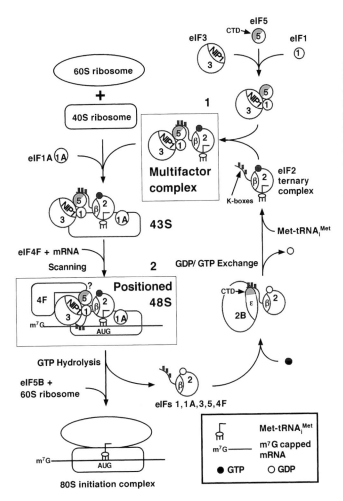

Figure 7. Model for the role of the eIF5–CTD in assembling the translation initiation complex in *S. cerevisiae*. The eIF5–CTD (gray sector in circle labeled 5), containing the conserved AA boxes, bridges interaction between eIF3 (3) and eIF2 (2) and mediates formation of the multifactor complex (MFC) (box 1) by interacting simultaneously with eIF3–NIP1 and the K-box domain at the amino terminus of eIF2β (*wavy line with filled boxes*). As the MFC occurs free of the ribosomes, it may carry out the binding of TC, eIFs 1, 3, and 5 to the 40S subunit in a single step, to form the 43S complex. The eIF1A (1A) may bind directly to the 40S ribosome (Hershey and Merrick 2000). Capped poly(A) mRNA bound to eIF4F is recruited to the 43S complex by interactions between eIF3 and the eIF4G subunit of eIF4F. Interaction between eIF4G and the eIF5–CTD (marked with ?) may also enhance mRNA binding. The eIF2β–eIF5 interaction in the MFC may be replaced by the eIF4G–eIF5 interaction in the 48S complex, when eIF2β interacts with mRNA. The 48S complex scans to the AUG start codon. Base-pairing between AUG and Met-tRNA$_i^{Met}$ triggers hydrolysis of GTP bound to eIF2, dependent on the eIF5 amino-terminal domain (*white sector*), followed by ejection of eIF2–GDP and other eIFs. Joining of the 60S subunit is stimulated by eIF5B. The GDP bound to eIF2 is subsequently replaced with GTP by the guanine-nucleotide exchange factor eIF2B (2B). This last interaction is mediated, at least partly, by a second AA-box-containing motif (shown as *gray shading*) in the catalytic (ε) subunit of eIF2B (Asano et al. 1999). Our results indicate that formation of the MFC is required for binding of Met-tRNA$_i^{Met}$ and mRNA to 40S ribosomes in vitro. It is also required for stable incorporation of eIF5 into 48S complexes and conversion of 48S to 80S complexes. In vivo, the latter appears to be the rate-limiting defect in initiation produced by the *tif5-7A* mutation in the eIF5–CTD, which destabilizes the MFC. (Reprinted, with permission, from Asano et al. 2001.)

binding defect in the *tif5-7A* extract could result indirectly from its reduced TC-binding activity, as TC binding is a prerequisite for mRNA binding to 40S subunits in mammalian extracts (Hinnebusch 2000). Nevertheless, this finding led us to investigate whether the eIF5–CTD can promote an interaction between eIF3 and eIF4G, the adapter subunit of the eIF4F complex.

Consistent with this possibility, we found that a small fraction of native eIF4G coimmunoprecipitated with eIF3 from cell extracts, and that this association was reduced by the *tif5-7A* mutation in the eIF5–CTD (Fig. 4B,C) (Asano et al. 2001). Using purified recombinant proteins, we found that the carboxy-terminal half of eIF4G can bind directly to the eIF5–CTD, in a manner stimulated by the AA boxes. Furthermore, this interaction can occur simultaneously with the eIF5–CTD/NIP1 interaction in vitro, leading to formation of a ternary complex containing eIF5–CTD, the amino terminus of NIP1, and the carboxyl terminus of eIF4G (Asano et al. 2001). It is thought that a direct interaction between eIF3 and eIF4G promotes mRNA binding to ribosomes in mammalian cells (Hentze 1997; Sachs et al. 1997). The ability of eIF5–CTD to interact simultaneously with eIF3–NIP1 and eIF4G may help to stabilize eIF3–eIF4G interaction and enhance mRNA binding to 40S ribosomes in yeast.

It is intriguing that the eIF5–CTD/eIF4G interaction was found to be mutually exclusive with the eIF5–CTD/eIF2β interaction in vitro (Asano et al. 2001), suggesting that the two interactions occur at different stages of the initiation pathway. The eIF5–CTD/eIF2β interaction clearly takes place in the MFC free of the ribosome (Asano et al. 2000) and promotes the 40S binding of these factors as a preformed unit (see box 1 in Fig. 7). This interaction may give way to the eIF5–CTD/eIF4G interaction in the 48S complex to enhance mRNA binding or facilitate scanning (box 2 in Fig. 7). It was shown that the K boxes in eIF2β mediate mRNA binding by eIF2 (Laurino et al. 1999); thus, the mRNA–eIF2β interaction in the 48S complex might displace eIF5–CTD from the K boxes in eIF2β and allow the eIF5–eIF4G interaction to proceed (box 2 in Fig. 7) (Asano et al. 2001).

THE eIF5 CTD IS REQUIRED IN VIVO FOR CONVERSION OF 48S TO 80S COMPLEXES

Although the *tif5-7A* mutation reduced Met-tRNA$_i^{Met}$ binding to 40S ribosomes in extracts (Fig. 6A), in vivo we observed accumulation of eIF2 on free 40S subunits and a reduction in the nonribosomal pool of eIF2 in *tif5-7A* cells (Fig. 8B, *TIF5*[+] versus *tif5-7A*). Additionally, free 40S

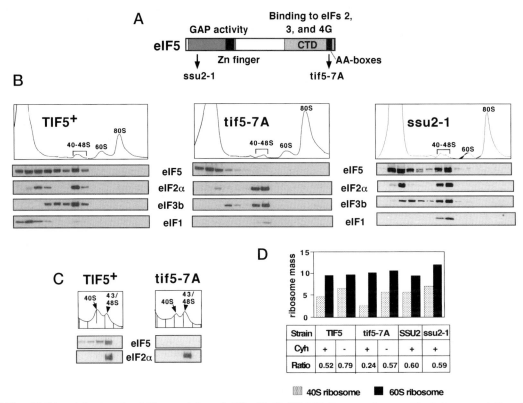

Figure 8. The *tif5-7A* mutation impairs stable association of eIF5 with 43-48S initiation complexes and leads to accumulation of 48S complexes in vivo. (*A*) The rectangle depicts the primary structure of yeast eIF5, with shaded areas denoting regions highly conserved in its eukaryotic homologs. Filled squares denote the zinc-finger motif or AA boxes, as indicated. Arrows indicate positions of *ssu2-1* (G62S) or *tif5-7A* mutations. The deduced roles of the amino-terminal domain and CTD in GAP activity or initiation complex assembly, respectively, are indicated (see text for further details). (*B*) Cells from *TIF5*, *tif5-7A*, and *ssu2-1* strains were grown at 30°C and treated with cycloheximide, as above. WCEs (prepared in the presence of heparin to preserve 43S-48S complexes) were resolved by centrifugation on 15–40% sucrose gradients. The fractions were scanned for A_{254} and analyzed by western blotting using antibodies against the indicated factors. (*C*) The same extracts analyzed in *B* were fractionated on 7.5–30% sucrose gradients and analyzed as above. (*D*) Histogram showing the free 40S and 60S subunit masses, quantitated by the area under A_{254} profiles of the type shown in panel *B*, plus a similar profile not shown for the wild-type strain isogenic to the *ssu2-1* mutant (*SSU2*). The results of several independent experiments using extracts prepared in the presence or absence of cycloheximide were quantified. The bottom panel indicates the strains used, the presence or absence of cycloheximide (Cyh), and the calculated 40S/60S ribosome mass ratios. The histogram shows that, in the presence of cyh (which freezes polysomes), the *tif5-7A* extract, but not the *ssu2-1* extract, shows reduced free 40S subunits compared to that seen in isogenic wild-type extracts (*TIF5* or *SSU2*, respectively). The depletion of free 40S subunits in the *tif5-7A* extract was dimished when polysomes were dissociated by omission of cyh, as expected if 40S subunits are sequestered in polysomes as 48S complexes. (Reprinted, with permission, from Asano et al. 2001.)

subunits devoid of eIFs 2 and 3 were depleted in this mutant (Fig. 8C,D), and an increased proportion of 40S subunits was sequestered in polysomes (48S complexes). Together, these observations suggest that the initiation pathway is blocked at the stage of 48S to 80S conversion, and not at the level of TC binding to 40S subunits, in *tif5-7A* cells. We presume that the rate of 43S complex formation is reduced in the *tif5-7A* mutant, as seen in vitro (Fig. 6A), but because the defect in 48S to 80S conversion is more severe, we observe the accumulation of 48S complexes containing the TC. Importantly, eIF5 was lost from 43-48S complexes in *tif5-7A* cells (Fig. 8B, *TIF5⁺* versus *tif5-7A*), showing that the eIF5–CTD is crucial for stable incorporation of eIF5 into the preinitiation complex. Hence, the absence of eIF5 from 48S complexes impedes or delays their conversion to 80S complexes in vivo (Asano et al. 2001).

The simplest way to explain the defect in 48S to 80S conversion in *tif5-7A* cells is to propose that the loss of eIF5 from 48S complexes impairs activation of GTP hydrolysis by the TC on recognition of the start codon. Ostensibly at odds with this idea, we observed no effect of *tif5-7A* or *tif5-12A* on the ability of eIF5 to stimulate GTP hydrolysis by model 48S complexes containing the TC and AUG triplets bound to 40S subunits (Asano et al. 2001). These in vitro assay conditions are nonphysiological in several respects, containing only eIFs 2 and 5, an AUG triplet versus mRNA, and high Mg⁺⁺ concentration; nevertheless, our data suggest that the AA boxes in eIF5–CTD are not essential for the catalytic activity of eIF5. In the same assay, the *ssu2-1* mutation in the amino terminus of eIF5 (G62S) (Fig. 8A) led to a substantial reduction in eIF5 GAP activity. Consistently, *ssu2-1* reduced the rate of translation initiation but did not diminish the amount of eIF5 associated with the 43-48S complexes in vivo (Fig. 8B, *TIF5*[+] versus *ssu2-1*) (Asano et al. 2001). Accordingly, we propose that *ssu2-1* impairs the GAP activity of eIF5, lodged in the amino terminus of the protein, but not its interactions with other eIFs in 43S-48S initiation complexes. In contrast, *tif5-7A* leaves the catalytic activity of eIF5 largely intact but eliminates eIF5 as an integral component of 48S complexes. This model is consistent with the previous finding that the Sui⁻ mutation in the amino terminus of eIF5 altering Gly-31 to arginine (*SUI5*[G31R]) increased GAP activity in vitro by about twofold (Huang et al. 1997).

Even if the CTD is not required for eIF5 catalytic function per se, the loss of eIF5 from 48S complexes in the *tif5-7A* mutant could still reduce the rate of GTP hydrolysis in vivo by decreasing the concentration of eIF5 in the vicinity of 48S complexes paired with start codons. It is also possible that eIF5 must be anchored precisely in native 48S complexes in order to trigger GTP hydrolysis effectively. The eIF5 cannot stimulate GTP hydrolysis by the TC unless both factors are bound to the 40S subunit and the Met-tRNA$_i$[Met] is base-paired with AUG (Chakrabarti and Maitra 1991). Hence, GTP hydrolysis may require conformational changes on recognition of the start codon that depend on precise juxtaposition of eIF5 with other components of the MFC or with eIF4G. This hypothetical function of the eIF5-CTD could be bypassed in the model 48S complexes containing only TC, or at the nonphysiological Mg⁺⁺ concentrations, used for the in vitro GAP assays (Asano et al. 2001).

A less conventional explanation for the delayed conversion of 48S to 80S complexes produced by *tif5-7A* is prompted by our finding that 48S complexes accumulated in the *tif5-7A* mutant but not in *ssu2-1* cells (Fig. 8B, D). This distinction suggests that different steps in the maturation of 48S complexes are disrupted by these two mutations in eIF5. If *ssu2-1* impairs only the GAP activity of eIF5, then, in this mutant, 48S complexes should scan the mRNA effectively but fail to hydrolyze GTP on reaching the start codon. Because 48S complexes do not accumulate in *ssu2-1* cells, it appears that positioned 48S complexes decay to free 40S subunits if GTP hydrolysis does not occur in a prescribed period following AUG recognition by the TC (Asano et al. 2001). If so, then the accumulation of stable 48S complexes seen in *tif5-7A* cells may signify a delay in reaching the start codon during the scanning process. Considering that eIF1 interacts with eIF5-CTD (Asano et al. 2000), and that eIF1 is implicated in AUG recognition (Pestova et al. 1998; Donahue 2000), *tif5-7A* might disrupt a function of eIF1 in scanning. It is also conceivable that the weakened eIF5-CTD/eIF4G interaction produced by *tif5-7A* reduces the rate of scanning.

In summary, we found that *tif5-7A* greatly reduced the association of eIF5 with native 43-48S complexes in vivo. Moreover, 48S complexes containing all other components of the MFC accumulated in the *tif5-7A* mutant, indicating a block late in the pathway where eIF5 GAP function is required. Thus, whereas the eIF5-CTD promotes Met-tRNA$_i$[Met] and mRNA binding to 40S subunits

in vitro, its rate-limiting function in vivo is the incorporation of eIF5 into preinitiation complexes in a manner required for efficient recognition of the start codon. The fact that 48S complexes accumulated in *tif5-7A* cells (assembly defect) but not in *ssu2-1* cells (GAP catalytic defect) suggests that the absence of eIF5 as a stable constituent of the 48S complex may impede the scanning process. It remains to be determined whether the eIF5–eIF4G interaction we detected is crucial for mRNA binding per se or for AUG recognition during scanning.

THE CONSERVED BIPARTITE MOTIF CONTRIBUTES TO BINDING OF eIF2 BY ITS GAP (eIF5) AND GEF (eIF2B)

Interestingly, the bipartite motif in the CTD of eIF5 (AA boxes 1 and 2) also occurs at the extreme carboxyl terminus of the ε subunit of eIF2B, the GEF for eIF2. There are five subunits in native eIF2B, which can be divided into two stable subcomplexes. The γ and ε subunits (encoded by *GCD1* and *GCD6* in yeast) comprise the catalytic subcomplex, which has wild-type levels of GEF activity in vitro (Pavitt et al. 1998). As first shown for rat eIF2B (Fabian et al. 1997), the ε subunit alone has GEF activity, although its specific activity is low compared to the γ/ε subcomplex (Pavitt et al. 1998). The CTD of eIF2Bε/GCD6 (residues 518–712) is required for GEF activity and for interaction with eIF2, whereas the amino-terminal portion of the protein mediates complex formation with other subunits of eIF2B and the stimulatory effect of eIF2Bγ/GCD1 on catalytic activity (Gomez and Pavitt 2000).

We found that *12Ala* and *7Ala* mutations in AA boxes 1 and 2, respectively, of the CTD in recombinant eIF2Bε/GCD6 reduced its binding to eIF2β and eIF2 holoprotein in vitro, just as observed for the corresponding mutations in the eIF5–CTD. Additionally, the *gcd6-7A* mutation reduced complex formation between native eIF2B and eIF2 holoproteins in vivo, as judged by coimmunoprecipitation assays. It also led to constitutively derepressed *GCN4* translation (Gcd⁻ phenotype), consistent with a reduction in the GEF activity of eIF2B in vivo (Asano et al. 1999). Similarly, the K boxes in eIF2β were required for interaction of recombinant eIF2β and eIF2 holoprotein with eIF2Bε in vitro, and the K-box mutations conferred a Gcd⁻ phenotype in vivo. Thus, the AA boxes in the eIF2Bε–CTD are required for tight binding of eIF2B to eIF2, in a manner dependent on the K boxes in eIF2β (Fig. 7). Intriguingly, the K-box domain in eIF2β mediates stable interaction of eIF2 with both its GAP (eIF5) and GEF (eIF2Bε) through homologous CTDs containing conserved AA boxes. These interactions are mutually exclusive, as eIF2 was not found simultaneously associated with eIF5 and eIF2B in vivo (Asano et al. 1999).

Interestingly, archaea contain all three subunits of eIF2 but lack obvious orthologs of eIF5 and eIF2B. Archaeal eIF2β consistently lacks the K-box domain (Asano et al. 1999). It is tempting to speculate that during eukaryotic evolution the primordial eIF5 and eIF2Bε acquired domains containing the AA boxes, and the K-box domain was added to the β subunit of eIF2, their common substrate. This would provide a high-affinity binding site on eIF2 for its GAP and GEF without compromising basic functions of eIF2 in transferring tRNA$_i^{Met}$ to the small ribosomal subunit and in AUG selection. The dependence on eIF5 for GTP hydrolysis by the TC seems to provide a proofreading capability, as Sui⁻ mutations in eIF5 increase the probability of initiation at UUG triplets (Huang et al. 1997). The dependence on eIF2B for GDP–GTP exchange confers the ability to regulate the concentration of active GTP-bound eIF2. Inhibition of eIF2B by phosphorylation of eIF2α is a mechanism employed from yeast to man for down-regulating general translation, and gene-specific translational induction, in response to starvation or stress (Harding et al. 2000; Hinnebusch 2000). Thus, the appearance of eIF5 and eIF2B in eukaryotic evolution increased the accuracy of start codon selection and provided a means to regulate translation at the tRNA$_i^{Met}$-binding step of initiation.

ACKNOWLEDGMENTS

We thank Tom Dever and other members of the LGRD for many helpful suggestions during the course of this work.

REFERENCES

Anderson J., Phan L., and Hinnebusch A.G. 2000. The Gcd10p/Gcd14p complex is the essential two-subunit tRNA(1-methyladenosine) methyltransferase of Saccharomyces cerevisiae. *Proc. Natl. Acad. Sci.* **97:** 5173.

Anderson J., Phan L., Cuesta R., Carlson B.A., Pak M., Asano K., Bjork G.R., Tamame M., and Hinnebusch A.G. 1998. The essential Gcd10p-Gcd14p nuclear complex is required for 1-methyladenosine modification and maturation of initiator methionyl-tRNA. *Genes Dev.* **12:** 3650.

Asano K., Clayton J., Shalev A., and Hinnebusch A.G. 2000. A multifactor complex of eukaryotic initiation factors eIF1, eIF2, eIF3, eIF5, and initiator tRNA^{Met} is an important translation initiation intermediate in vivo. *Genes Dev.* **14:** 2534.

Asano K., Phan L., Anderson J., and Hinnebusch A.G. 1998. Complex formation by all five homologues of mammalian translation initiation factor 3 subunits from yeast *Saccharomyces cerevisiae*. *J. Biol. Chem.* **273:** 18573.

Asano K., Krishnamoorthy T., Phan L., Pavitt G.D., and Hinnebusch A.G. 1999. Conserved bipartite motifs in yeast eIF5 and eIF2Bε, GTPase-activating and GDP-GTP exchange factors in translation initiation, mediate binding to their common substrate eIF2. *EMBO J.* **18:** 1673.

Asano K., Shalev A., Phan L., Nielsen K., Clayton J., Valasek L., Donahue T.F., and Hinnebusch A.G. 2001. Multiple roles for the carboxyl terminal domain of eIF5 in initiation complex assembly and GTPase activation. *EMBO J.:* **20:** 2326.

Browning K.S., Galle D.R., Hershey J.W.B., Hinnebush A.G., Maitra U., Merrick W.C., and Norbury C. 2001. Unified nomenclature for the subunits of eukaryotic initiation factor 3′. *Trends Biochem. Sci.* **26:** 284.

Chakrabarti A. and Maitra U. 1991. Function of eukaryotic initiation factor 5 in the formation of an 80 S ribosomal polypeptide chain initiation complex. *J. Biol. Chem.* **21:** 14039.

Chaudhuri J., Das K., and Maitra U. 1994. Purification and characterization of bacterially expressed mammalian translation initiation factor (eIF-5): Demonstration that eIF-5 forms a specific complex with eIF-2. *Biochemistry* **33:** 4794.

Cuesta R., Hinnebusch A.G., and Tamame M. 1998. Identifica-

tion of *GCD14* and *GCD15*, novel genes required for translational repression of *GCN4* mRNA in *Saccharomyces cerevisiae*. *Genetics* **148:** 1007.

Danaie P., Wittmer B., Altmann M., and Trachsel H. 1995. Isolation of a protein complex containing translation initiation factor Prt1 from *Saccharomyces cerevisiae*. *J. Biol. Chem.* **270:** 4288.

Donahue T. 2000. Genetic approaches to translation initiation in *Saccharomyces cerevisiae*. In *Translational control of gene expression* (ed. N. Sonenberg et al.), p. 487. Cold Spring Harbor Laboratory Press, Cold Spring Harbor, New York.

Evans D.R.H., Rasmussen C., Hanic-Joyce P.J., Johnston G.C., Singer R.A., and Barnes C.A. 1995. Mutational analysis of the Prt1 protein subunit of yeast translation initiation factor 3. *Mol. Cell. Biol.* **15:** 4525.

Fabian J.R., Kimball S.R., Heinzinger N.K., and Jefferson L.S. 1997. Subunit assembly and guanine nucleotide exchange activity of eukaryotic initiation factor-2B expressed in Sf9 cells. *J. Biol. Chem.* **272:** 12359.

Feinberg B., McLaughlin C.S., and Moldave K. 1982. Analysis of temperature-sensitive mutant ts187 of *Saccharomyces cerevisiae* altered in a component required for the initiation of protein synthesis. *J. Biol. Chem.* **257:** 10846.

Garcia-Barrio M.T., Naranda T., Cuesta R., Hinnebusch A.G., Hershey J.W.B., and Tamame M. 1995. GCD10, a translational repressor of *GCN4*, is the RNA-binding subunit of eukaryotic translation initiation factor-3. *Genes Dev.* **9:** 1781.

Gomez E. and Pavitt G.D. 2000. Identification of domains and residues within the epsilon subunit of eukaryotic translation initiation factor 2B (eIF2β) required for guanine nucleotide exchange reveals a novel activation function promoted by eIF2B complex formation. *Mol. Cell. Biol.* **20:** 3965.

Greenberg J.R., Phan L., Gu Z., deSilva A., Apolito C., Sherman F., Hinnebusch A.G., and Goldfarb D.S. 1998. Nip1p associates with 40S ribosomes and the Prt1p subunit of eIF3 and is required for efficient translation initiation. *J. Biol. Chem.* **273:** 23485.

Hanachi P., Hershey J.W.B., and Vornlocher H.P. 1999. Characterization of the p33 subunit of eukaryotic translation initiation factor-3 from *Saccharomyces cerevisiae*. *J. Biol. Chem.* **274:** 8546.

Harding H.P., Novoa I., Zhang Y., Zeng H., Wek R., Schapira M., and Ron D. 2000. Regulated translation initiation controls stress-induced gene expression in mammalian cells. *Mol. Cell* **6:** 1099.

Hartwell L.H. and McLaughlin C.S. 1969. A mutant of yeast apparently defective in the initiation of protein synthesis. *Proc. Natl. Acad. Sci.* **62:** 468.

Hentze M.W. 1997. eIF4G: A multipurpose ribosome adapter. *Science* **275:** 500.

Hershey J.W.B. and Merrick W.C. 2000. Pathway and mechanism of initiation of protein synthesis. In *Translational control of gene expression* (ed. N. Sonenberg et al.), pp. 33. Cold Spring Harbor Laboratory Press, Cold Spring Harbor, New York.

Hershey J.W.B., Asano K., Naranda T., Vornlocher H.P., Hanachi P., and Merrick W.C. 1996. Conservation and diversity in the structure of translation initiation factor eIF3 from humans and yeast. *Biochimie* **78:** 903.

Hinnebusch A.G. 2000. Mechanism and regulation of initiator methionyl-tRNA binding to ribosomes. In *Translational control of gene expression* (ed. N. Sonenberg et al.), p. 185. Cold Spring Harbor Laboratory Press, Cold Spring Harbor, New York.

Huang H., Yoon H., Hannig E.M., and Donahue T.F. 1997. GTP hydrolysis controls stringent selection of the AUG start codon during translation initiation in *Saccharomyces cerevisiae*. *Genes Dev.* **11:** 2396.

Laurino J.P., Thompson G.M., Pacheco E., and Castilho B.A. 1999. The β subunit of eukaryotic translation initiation factor 2 binds mRNA through the lysine repeats and a region comprising the C_2-C_2 motif. *Mol. Cell. Biol.* **19:** 173.

Naranda T., MacMillan S.E., and Hershey J.W.B. 1994. Purified yeast translational initiation factor eIF-3 is an RNA-binding protein complex that contains the PRT1 protein. *J. Biol. Chem.* **269:** 32286.

Naranda T., Kainuma M., McMillan S.E., and Hershey J.W.B. 1997. The 39-kilodalton subunit of eukaryotic translation initiation factor 3 is essential for the complex's integrity and for cell viability in *Saccharomyces cerevisiae. Mol. Cell. Biol.* **17:** 145.

Naranda T., MacMillan S.E., Donahue T.F., and Hershey J.W. 1996. SUI1/p16 is required for the activity of eukaryotic translation initiation factor 3 in Saccharomyces cerevisiae. *Mol. Cell. Biol.* **16:** 2307.

Pavitt G.D., Ramaiah K.V.A., Kimball S.R., and Hinnebusch A.G. 1998. eIF2 independently binds two distinct eIF2B subcomplexes that catalyze and regulate guanine-nucleotide exchange. *Genes Dev.* **12:** 514.

Pestova T.V., Borukhov S.I., and Hellen C.U.T. 1998. Eukaryotic ribosomes require initiation factors 1 and 1A to locate initiation codons. *Nature* **394:** 854.

Phan L., Schoenfeld L., Valasek L., and Hinnebusch A.G. 2001. The TIF32/NIP1/PRT1 subcomplex of yeast eIF3 binds HCR1, eIF5, and EIF1, and stimulates mRNA and Met-tRNA$_i^{Met}$ binding to 40S ribosomes. *EMBO J.* **20:** 2954..

Phan L., Zhang X., Asano K., Anderson J., Vornlocher H.P., Greenberg J.R., Qin J., and Hinnebusch A.G. 1998. Identification of a translation initiation factor 3 (eIF3) core complex, conserved in yeast and mammals, that interacts with eIF5.

Mol. Cell. Biol. **18:** 4935.

Proud C.G. 1992. Protein phosphorylation in translational control. *Curr. Top. Cell Regul.* **32:** 243.

Sachs A.B., Sarnow P., and Hentze M.W. 1997. Starting at the beginning, middle, and end: Translation initiation in eukaryotes. *Cell* **89:** 831.

Valásek L., Hasek J., Trachsel H., Imre E.M., and Ruis H. 1999. The *Saccharomyces cerevisiae HCRI* gene encoding a homologue of the p35 subunit of human translation eukaryotic initiation factor 3 (eIF3) is a high copy suppressor of a temperature-sensitive mutation in the Rpg1p subunit of yeast eIF3. *J. Biol. Chem.* **274:** 27567.

Valásek L., Phan L., Schoenfeld L.W., Valásková V., and Hinnebusch A.G. 2001. Related eIF3 subunits TIF32 and HCR1 interact with an RNA recognition motif in PRT1 required for eIF3 integrity and ribosome binding. *EMBO J.* **20:** 891.

Vornlocher H.P., Hanachi P., Ribeiro S., and Hershey J.W.B. 1999. a 110-kilodalton subunit of translation initiation factor eIF3 and an associated 135-kilodalton protein are encoded by the *Saccharomyces cerevisiae TIF32* and *TIF31* genes. *J. Biol. Chem.* **274:** 16802.

Yoon H.J. and Donahue T.F. 1992. The *sui1* suppressor locus in *Saccharomyces cerevisiae* encodes a translation factor that functions during tRNA$_i^{Met}$ recognition of the start codon. *Mol. Cell. Biol.* **12:** 248.

Universal Translation Initiation Factor IF2/eIF5B

T.E. DEVER,*§ A. ROLL-MECAK,‡ S.K. CHOI,* J.H. LEE,* C. CAO,*
B.-S. SHIN,* AND S.K. BURLEY†‡

*Laboratory of Gene Regulation and Development, National Institute of Child Health and Human Development,
National Institutes of Health, Bethesda, Maryland 20892-2716; †Laboratories of Molecular Biophysics and
‡Howard Hughes Medical Institute, The Rockefeller University, New York, New York 10021

Translation factors are thought to accelerate the rate of protein synthesis and/or increase the fidelity of the process. In addition, the positive contributions of translation factors to cellular protein synthesis provide a means to regulate this process in response to cellular or environmental cues. Along with the requirement for factors to facilitate translation elongation, a distinct set of factors have been identified that promote assembly of a functional ribosome•mRNA•initiator Met-tRNA$_i^{Met}$ complex in which the anticodon of the Met-tRNA$_i^{Met}$ and the AUG codon of the mRNA base-pair within the ribosomal P site. Whereas three translation initiation factors (IF) have been identified in prokaryotes, translation initiation in eukaryotes requires at least 12 independent factors (Fig. 1) (for review, see Hershey and Merrick 2000). In addition, GTP is an essential requirement for translation in both prokaryotes and eukaryotes. Although biochemical and genetic analyses have provided insights into the roles of the translation initiation factors, the precise molecular function for most of these factors has not been resolved.

The recent deciphering of the genomes of a large number of organisms has revealed that a subset of translation initiation factors has been conserved among prokaryotes, archaea, and eukaryotes. As indicated in Figure 1, the factors IF1/eIF1A and IF2/eIF5B are conserved in all three kingdoms. The factor EF-P/eIF5A has also been conserved; however, whereas eIF5A stimulates model assays of translation initiation, in vivo studies have not confirmed a role for the protein in translation (Kang and Hershey 1994). The homology of the factors IF1 and eIF1A has been revealed by both amino acid sequence conservation (Kyrpides and Woese 1998) and similarity of factors' three-dimensional structures (Sette et al. 1997; Battiste et al. 2000). The conservation between IF2 and eIF5B is discussed in detail below. In addition to these universally conserved factors, it is interesting to note that factors eIF1 and eIF2 have been conserved between archaea and eukaryotes. Genetic studies in yeast have implicated both of these factors in AUG start codon recognition by scanning 40S ribosomal complexes (for review, see Donahue 2000). It is intriguing to speculate that eIF2 may functionally substitute for the mRNA Shine-Dalgarno sequence/16S rRNA interaction that promotes AUG start codon recognition in bacterial cells. As the structures of eIF1 (Fletcher et al. 1999) and the prokary-

otic factor IF3 (Biou et al. 1995) show some resemblance, and a role of IF3 is to ensure AUG start-site specificity, it is also intriguing to speculate that the function of IF3 may have been assumed by eIF1 during evolution. Finally, the function of many of the eukaryotic-specific translation factors, notably in the eIF4 family and eIF3, is to facilitate binding of the 40S ribosomal complexes to the 5′-capped end of the eukaryotic mRNAs.

CONSERVATION OF TRANSLATION FACTOR IF2/eIF5B

Analyses of the amino acid sequences of IF2 and eIF5B from prokaryotes and eukaryotes reveal that the protein can be subdivided into three regions (Fig. 2): an amino-terminal region with an abundance of charged residues, a highly conserved consensus GTP-binding domain, and a carboxy-terminal region that is fairly well-conserved. The eIF5B from archaea is highly similar to the prokaryotic and eukaryotic factors; however, it lacks the amino-terminal region (Fig. 2). Interestingly, the amino-terminal region of yeast eIF5B can be removed with no effect on yeast cell growth or on the ability of eIF5B to promote translation in vitro (Choi et al. 2000). The amino-terminal region of bacterial IF2 has been reported to bind ribosomes (Moreno et al. 1999), and like eIF5B, this region of IF2 is nonessential in vivo; however, deletion of the amino-terminal region of IF2 severely cripples bacterial cell growth (Laalami et al. 1991).

In the yeast Saccharomyces cerevisiae, eIF5B is encoded by the gene FUN12. Deletion of the FUN12 gene results in a severe slow-growth phenotype, and polyribosome profiles from isogenic wild-type and fun12Δ strains reveal a translation initiation defect in the strains lacking eIF5B (Choi et al. 1998). This translation defect is also observed in vitro. When programmed with a luciferase reporter mRNA, whole-cell translation extracts prepared from wild-type strains produce ~15- to 25-fold more luciferase than extracts prepared from strains lacking eIF5B (Fig. 3) (Choi et al. 1998). Importantly, addition of recombinant GST-yeast-eIF5B to extracts prepared from strains lacking eIF5B fully restored translational activity (Fig. 3). This in vitro result demonstrates that eIF5B directly stimulates translation and indicates that eIF5B is a translation initiation factor.

Human and archaeal eIF5B fully or partially, respectively, substituted for yeast eIF5B and restored high-level

---

§Corresponding author.

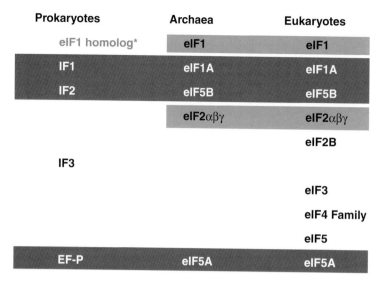

Figure 1. A core set of translation initiation factors is conserved through evolution. The factors in archaea have been identified on the basis of amino acid sequence similarity to the factors in prokaryotes and eukaryotes. As indicated, an ortholog of eukaryotic translation factor eIF1 has been identified in some, but not all, of the prokaryotic genomes that have been sequenced.

translational activity in extracts prepared from the eIF5B-deficient yeast strain (Fig. 3) (Lee et al. 1999). Consistent with their ability to restore translational activity in vitro, human and archaeal eIF5B substituted for yeast eIF5B in vivo and partially suppressed the slow-growth phenotype of strains lacking eIF5B (Lee et al. 1999). These in vivo and in vitro results demonstrate that the function of eIF5B has been conserved between yeast and mammals, and likely reflects a conservation in eIF5B function among all archaea and eukaryotes. To date, in vivo cross-complementation studies between archaeal or yeast eIF5B and *Escherichia coli* IF2 have been unsuccessful (J.H. Lee et al., unpubl.), suggesting that IF2 and eIF5B make critical contacts with other components of the translational machinery or ribosomes that are specific for bacterial versus archaeal/eukaryotic organisms. This lack of complementation between IF2 and eIF5B may reflect the fact that we have been unable to detect binding of eIF5B to initiator

Met-tRNA$_i^{Met}$ (T.V. Pestova et al., unpubl.), whereas IF2 has been reported to specifically bind fMet-tRNA$_f^{Met}$ (see Wu and RajBhandary 1997; Guenneugues et al. 2000; Szkaradkiewicz et al. 2000). To summarize, archaeal and eukaryotic eIF5B appear to be functional homologs, and their function may be slightly diverged from that of bacterial IF2.

REQUIREMENT FOR AN ADDITIONAL MOLECULE OF GTP IN EUKARYOTIC VERSUS PROKARYOTIC TRANSLATION INITIATION

Role for eIF5B in Ribosomal Subunit Joining

As described elsewhere in this volume, Tatyana Pestova in a biochemical study identified eIF5B as a fac-

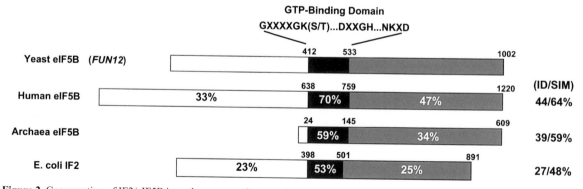

Figure 2. Conservation of IF2/eIF5B in prokaryotes, archaea, and eukaryotes. The schematics representing the full-length IF2/eIF5B from the yeast *S. cerevisiae* (encoded by the gene *FUN12*), humans, the archaea *Methanococcus jannaschii*, and *E. coli* are aligned through their highly conserved GTP-binding domains (containing the indicated consensus sequence motifs). Numbers above the schematics indicate the amino acid residue. Numbers within the schematics of the human, archaea, and *E. coli* factors are the percentages of amino acid sequence identities in the amino-terminal, GTP-binding, and carboxy-terminal domains relative to yeast eIF5B. At the right is indicated the percentage of amino acid sequence identities (ID) and similarities (SIM) of the full-length proteins compared to yeast eIF5B.

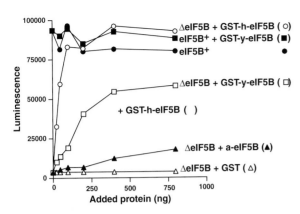

Figure 3. Restoration of translational activity in extracts from *fun12Δ* yeast strains lacking eIF5B by addition of recombinant yeast, human, or archaeal (*M. jannaschii*) eIF5B. In vitro translation extracts were prepared from isogenic wild-type (eIF5B⁺) and eIF5B-deletion (ΔeIF5B) yeast strains. The extracts were supplemented with the indicated amounts of recombinant proteins and programmed with an in-vitro-transcribed luciferase reporter mRNA. Translational activity was determined after 15 minutes' incubation by measuring luminescence. (GST-y-eIF5B) Yeast eIF5B fusion protein; (GST-h-eIF5B) human eIF5B fusion protein; (a-eIF5B) *M. jannaschii* eIF5B (no GST).

tor required for the ribosomal subunit joining step of protein synthesis. Following scanning of the small ribosomal complex containing Met-tRNAᵢᴹᵉᵗ and most likely the factors eIF2, eIF3, eIF5, eIF1, and eIF1A from the 5′ end of the mRNA to the AUG start codon, eIF5 is thought to promote GTP hydrolysis by eIF2, resulting in release of the factors from the ribosome (for review, see Hershey and Merrick 2000). Pestova et al. (2000) found that an additional factor, eIF5B, was required for 60S subunit joining. Recombinant human eIF5B substituted for the native protein and stimulated 60S subunit joining as well as methionyl-puromycin synthesis, a model assay for first peptide-bond formation. Consistent with the presence of the GTP-binding domain in eIF5B, Pestova et al. (2000) found that eIF5B bound GTP and that the factor hydrolyzed GTP in a ribosome-dependent reaction. Finally, eIF5B acted catalytically to promote subunit joining, and blocking the GTPase activity of eIF5B using the nonhydrolyzable GTP analog GDPNP permitted subunit joining but prevented the release of eIF5B from the 80S ribosomes following subunit joining (Pestova et al. 2000). Interestingly, the requirement for eIF5B for subunit joining is consistent with the observation that bacterial IF2 promotes subunit association (Godefroy-Colburn et al. 1975) and that the GTPase activity of IF2 is triggered upon subunit association (Kolakofsky et al. 1968; Tomsic et al. 2000).

Kinetic Studies Reveal GTP Requirement in a Late Step of Translation Initiation

Two recent kinetic studies have provided new insights into the roles of GTP and IF2 in translation initiation. Lorsch and Herschlag (1999) studying translation initiation in a mammalian in vitro system obtained evidence

for a new GTP-dependent step late in the translation initiation pathway. This GTP requirement followed subunit joining and resulted in a 30-fold activation of the 80S complex, generating what they referred to as the 80S* complex (Lorsch and Herschlag 1999). This GTP-dependent conversion of 80S to 80S* was attributed to a soluble factor, as opposed to a ribosomal constituent. Finally, conversion of 80S to 80S* was inhibited by inclusion of GMP-PNP, resulting in formation of dead 80S complexes. These attributes are strikingly similar to the properties reported by Pestova et al. (2000) for eIF5B, and it will be very interesting to see whether recombinant human eIF5B can promote the conversion of 80S to 80S* in this system.

In a related study using a reconstituted bacterial translation initiation system, Tomsic et al. (2000) reported that GTP hydrolysis by IF2 occurred with fast kinetics following subunit joining; however, Pᵢ release from the IF2•ribosomal complex was slow and rate-limiting for subsequent binding of the first elongating aminoacyl tRNA in complex with EF1A. It was proposed that a conformational rearrangement of IF2 or the ribosome is required subsequent to subunit joining to allow for IF2 release and first peptide-bond formation (Rodnina et al. 2000). This model is strikingly similar to the 80S to 80S* conversion proposed by Lorsch and Herschlag (1999). Surprisingly, the results of the kinetic study on bacterial translation suggest that GTP is not required for subunit joining and that GTP hydrolysis by IF2 is not linked to release of IF2 from the 70S ribosome. However, a point mutation in the IF2 GTP-binding domain that blocks GTP hydrolysis causes a dominant-lethal phenotype in vivo and results in retention of IF2 on the 70S ribosome following subunit joining (Luchin et al. 1999). Interestingly, this latter result is consistent with results of Pestova et al. which indicated that GTP hydrolysis by eIF5B is required for release of the factor following 80S complex formation (Pestova et al. 2000). The source of the discrepancy between the results of the kinetic and mutational studies on IF2 is unclear at present, and additional work is necessary to resolve the role of GTP hydrolysis by IF2 in translation initiation.

Model for the Roles of GTP in Translation Initiation

The discovery of eIF5B together with the previous identification of eIF2 and the work of Lorsch and Herschlag (1999), indicates that there are at least two GTP-dependent steps in eukaryotic translation initiation. In contrast, IF2 is the only GTPase required for prokaryotic translation initiation. A model to account for the additional GTP requirement in eukaryotic translation initiation in comparison to prokaryotic translation is presented in Figure 4. In prokaryotes, IF2 facilitates fMet-tRNAf ᴹᵉᵗ binding to the 30S ribosomal subunit, and in addition, IF2 promotes subunit joining. Hydrolysis of GTP by IF2 occurs rapidly following subunit joining and may be necessary for IF2 release from the 70S ribosome following subunit joining. In eukaryotes, the IF2 homolog eIF5B is

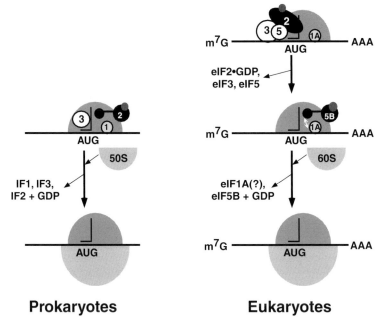

Prokaryotes   **Eukaryotes**

Figure 4. Model depicting two GTP-dependent steps in eukaryotic translation initiation versus only a single GTP requirement in prokaryotic translation initiation. (Left) In prokaryotic translation initiation, IF2 is the lone GTPase. The preinitiation complex with bound initiator fMet-tRNAfMet and factors localizes to the AUG start codon via interactions with the Shine-Dalgarno sequence on the mRNA. Following binding of the large 50S ribosomal subunit, hydrolysis of GTP by IF2 is coupled to release of the factors from the 70S ribosome. (Right) In eukaryotic translation initiation, the preinitiation complex containing factors eIF2, eIF3, eIF5, eIF1 (not depicted), and eIF1A, and initiator Met-tRNAiMet scans on the mRNA to the AUG codon. Proper base-pairing between the anticodon of the Met-tRNAiMet and the AUG start codon triggers eIF5-dependent GTP hydrolysis by eIF2 and release of the factors and deposition of Met-tRNAiMet in the ribosomal P site. The factor eIF5B then binds and promotes binding of the large 60S ribosomal subunit. Assembly of the 80S ribosome triggers the GTPase activity of eIF5B, leading to release of the remaining initiation factors from the ribosome.

required for efficient subunit joining, and GTP hydrolysis by eIF5B is required for release of the factor from the 80S ribosome following subunit joining. Accordingly, IF2 and eIF5B are performing complementary roles to promote subunit joining in prokaryotes and eukaryotes, respectively.

The presence of a second GTPase, eIF2, is restricted to archaea and eukaryotes (Fig. 1). The factor eIF2 binds the Met-tRNA$_i^{Met}$ to the ribosome in eukaryotes (for review, see Hershey and Merrick 2000). Thus, together eIF2 and eIF5B perform the functions attributed solely to IF2 in bacteria. The eIF2 forms a ternary complex with GTP and Met-tRNA$_i^{Met}$, and this ternary complex then binds to a 40S ribosomal subunit together with the translation factors eIF1, eIF1A, eIF3, and eIF5 to form a 43S preinitiation complex. This complex associates with the mRNA and scans to the AUG start codon. Recognition of the AUG start codon is accomplished by base-pairing with the anticodon loop of the tRNA$_i^{Met}$. This base-pairing is thought to trigger eIF5 to activate GTP hydrolysis by eIF2. In this way, eIF5, together with eIF2 and the Met-tRNA$_i^{Met}$, stringently ensures the fidelity of the initiation process (Huang et al. 1997; Donahue 2000). In prokaryotes, the selection of the AUG start codon is specified by interactions between the ribosome-binding site (Shine-Dalgarno sequence) on the mRNA and the 3′ end of the 16S rRNA in the 30S subunit. Thus, the novel GTP-dependent step in eukaryotic translation initiation is a fidelity checkpoint that replaces the simple base-pairing in-

teraction used by prokaryotes to specify the translational start site. In addition, it should be noted that introduction of eIF2 in eukaryotic translation initiation provides a new opportunity to regulate protein synthesis. Like EF1A, eIF2 requires a guanine nucleotide exchange factor (eIF2B) to regenerate the active eIF2•GTP complex following each round of translation initiation. Phosphorylation of eIF2 on Ser-51 of its α subunit by specific stress-responsive kinases converts eIF2 from a substrate to a competitive inhibitor of eIF2B (Dever 1999). The phosphorylation of eIF2 is a common mechanism employed by eukaryotic cells to regulate both general and gene-specific mRNA translation (Dever 1999).

Finally, a question that arises concerns the function of eIF2 in archaea. Several studies have reported the presence of Shine-Dalgarno-like sequences, complementary to the 3′ end of 16S rRNA, near the AUG start codons of open reading frames in various archaea. These findings suggest that archaea use a prokaryotic-like mechanism to locate translational start sites and, as such, archaea should not require eIF2. However, a recent report suggests that in the archaeon *Sulfolobus solfataricus* the first open reading frame in a number of operons lacks Shine-Dalgarno ribosome-binding sequences, whereas subsequent open reading frames in these operons are preceded by a Shine-Dalgarno sequence (Tolstrup et al. 2000). Thus, it can be proposed that archaea employ a eukaryotic scanning-type mechanism involving eIF2 to locate the first open reading frame on a polycistronic mRNA, whereas the subsequent open reading frames are translated via a prokaryotic-like mechanism which may or may not require eIF2 to bind the Met-tRNA$_i^{Met}$ to the ribosome.

STRUCTURAL ANALYSIS OF IF2/eIF5B

Recently, we determined the structure of eIF5B from the archaeon *Methanobacterium thermoautotrophicum* (Roll-Mecak et al. 2000). The protein consists of four domains and resembles a molecular chalice. The GTP-binding domain and domains II and III form the cup of the chalice, and they are connected via a long α helix (the stem of the chalice) to domain IV, which forms the base of the chalice (Fig. 5A). The GTP-binding domain resembles the GTP-binding domains of the translation factors EF1A and EF2, as well as that of Ras and the heterotrimeric GTP-binding proteins. Domains II and IV are antiparallel β barrels, whereas domain III is a novel α/β/α-sandwich.

Domain IV of eIF5B, as expected, resembles the carboxyl terminus of IF2 from *Bacillus stearothermophilus* (Meunier et al. 2000). This carboxy-terminal domain of IF2 is responsible for binding fMet-tRNA$_f^{Met}$, and mutagenesis studies and spectroscopic methods have implicated several residues of IF2 in fMet-tRNA$_f^{Met}$ binding (Misselwitz et al. 1999; Guenneugues et al. 2000). Interestingly, these residues are not conserved in archaeal and eukaryotic eIF5B, consistent with the inability to detect specific binding of eIF5B to Met-tRNA$_i^{Met}$. In contrast to the binding of aminoacyl tRNA to EF1A, the binding of fMet-tRNA$_f^{Met}$ to IF2 appears to only require fMet and

the 3′ end of the tRNA. Competitive binding studies revealed that fMet-AMP can effectively compete the binding of full-length fMet-tRNA$_f^{Met}$ to IF2 (Szkaradkiewicz et al. 2000). In addition, deacylation protection and spectroscopic titration experiments demonstrated that fMet linked to the last six residues of the acceptor end of tRNA$_f^{Met}$ interacted with similar affinity as the full-length fMet-tRNA$_f^{Met}$ to IF2 (Guenneugues et al. 2000). Thus, it can be proposed that a primary function of the fMet-tRNA$_f^{Met}$ binding property of prokaryotic IF2 is to ensure that translation initiates with a formylated amino acid.

Comparison of the active and inactive structures of IF2/eIF5B revealed a concerted movement of domains II, III, and IV triggered by small conformational changes in the active site of the G domain induced by Mg^{++}/GTP binding (Fig. 5B) (Roll-Mecak et al. 2000). eIF5B seems to employ an articulated lever mechanism to amplify the small conformational changes in the active site over a distance of 90 Å to the carboxy-terminal domain IV. We presume that these conformational changes induced by Mg^{++}/GTP binding allow the proper interaction of the factor with the ribosome.

Similarity of IF2/eIF5B to Translation Elongation GTPases Suggests a Model for eIF5B Function in Translation Initiation

As mentioned above, the GTP-binding domain of eIF5B is structurally similar to the GTP-binding domains of translation factors EF1A and EF2. However, the structural similarity between eIF5B, EF1A, and EF2 extends to the domain II β barrel (see Fig. 6) (Roll-Mecak et al. 2000). Not only are the first two domains of eIF5B•GTP, EF2•GDP, and EF1A•GTP structurally similar, they also display the same relative orientation, suggesting that these domains form a common ribosome-binding platform. However, domains III and IV in IF2/eIF5B are not structurally similar to the corresponding domains in EF2. Comparison of the structures of EF1A•GTP•Phe-tRNA^{Phe} and EF2•GDP revealed that domains III–V of EF2 appear to mimic the shape of the tRNA moiety of the EF1A ternary complex, with domain III acting as the acceptor stem, domain V as the T stem, and domain IV as the anticodon helix (Nyborg et al. 1996). The surface charge distribution on domain IV of EF2 is similar to the anticodon helix of the tRNA, and electron microscopic studies have shown that this domain inserts into the ribosomal A site (Agrawal et al. 1998; Stark et al. 2000), presumably to promote ribosomal translocation during translation elongation. When the GTP-binding domains of eIF5B and EF1A are superimposed, domain IV of eIF5B is positioned at roughly 90° relative to the anticodon helix of the tRNA in the EF1A ternary complex (Fig. 6). Thus, as opposed to domain IV of EF2, which inserts into the decoding center on the small subunit, we predict that domain IV of IF2/eIF5B traverses across the top of the A site where it can be in close proximity to the 3′ end of the P-site tRNA, as discussed below.

A model that takes into account the structure of eIF5B and its resemblance to the elongation factors EF1A and

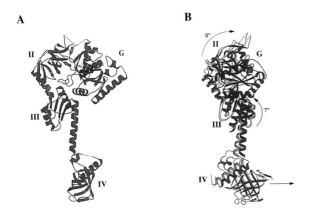

Figure 5. Three-dimensional structure of IF2/eIF5B. (*A*) Ribbon diagram showing the nucleotide-binding face of IF2/eIF5B from the archaeon *M. thermoautotrophicum* in complex with the non-hydrolyzable GTP analog GDPNP (Roll-Mecak et al. 2000). The locations of the GTP-binding domain (G domain) and domains II–IV are indicated. (*B*) Domain movements in IF2/eIF5B induced by Mg^{++}/GTP binding. Ribbon diagrams of IF2/eIF5B•GDPNP (*black*) and IF2/eIF5B•GDP (*gray*) superpositioned based on conserved elements in their GTP-binding domains (Roll-Mecak et al. 2000). The view of eIF5B is rotated 90° about a vertical axis relative to the image in *A*. The arrows denote the relative domain movements induced by GTP binding to eIF5B. (Reprinted, with permission, from Roll-Mecak et al. 2000 [copyright Elsevier Science].)

IF2/eIF5B   **EF1A ternary complex**

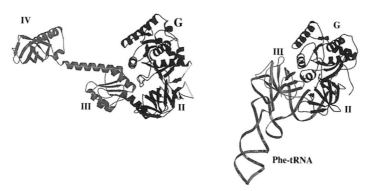

Figure 6. Comparison of the structures of the translational GTPases IF2/eIF5B•GDP and EF1A•GDPNP•Phe-tRNA^{Phe}. Ribbon diagrams of IF2/eIF5B•GDP (*left*, Roll-Mecak et al. 2000) and EF1A•GDPNP•Phe-tRNA^{Phe} (*right*, Nissen et al. 1995) oriented based on the alignment of the conserved P loop and α-helix H1 in the GTP-binding domains of the factors. The conserved GTP-binding (G domain) and β-barrel (II) domains of the factors are depicted in black; domains III and IV of IF2/eIF5B, domain III of EF1A, and Phe-tRNA^{Phe} are depicted in gray. The dorsal face of the factors is shown from the perspective of the small 40S ribosomal subunit. (Adapted and reprinted, with permission, from Roll-Mecak et al. 2000 [copyright Elsevier Science].)

EF2 is presented in Figure 7. Prior to describing the model, it is important to note that domain IV of yeast eIF5B directly interacts with the translation factor eIF1A. As mentioned earlier, eIF1A structurally resembles IF1 and, together with IF2/eIF5B, these proteins are the only universally conserved translation initiation factors (Fig. 1). Protein–protein interaction assays revealed that the carboxyl terminus of eIF5B directly interacts with eIF1A, and the two proteins form a complex independent of the ribosome (Choi et al. 2000). Similarly, the factors IF1 and IF2 have been reported to be cross-linked when bound to the ribosome (Boileau et al. 1983). Thus, these homologous factors likely physically interact on the ribosome to promote translation initiation in all organisms. Previ-

ously, the IF1-binding site on the ribosome was mapped to the A site (Moazed et al. 1995), and more recently, the X-ray structure of IF1 bound to the 30S subunit confirmed that the factor binds to the base of the A site adjacent to the mRNA (Carter et al. 2001). In our proposed model for eukaryotic translation initiation (Fig. 7), the 40S ribosomal subunit scans to the AUG codon where eIF2 hydrolyzes its GTP and many of the factors are released. The resulting complex contains Met-tRNA$_i^{Met}$ base-paired to the AUG codon of the mRNA in the P site and eIF1A in the A site of the 40S subunit. We propose that this complex is unstable in the absence of eIF5B, and that this instability at least partly accounts for the slow-growth phenotype of yeast strains lacking eIF5B. Inter-

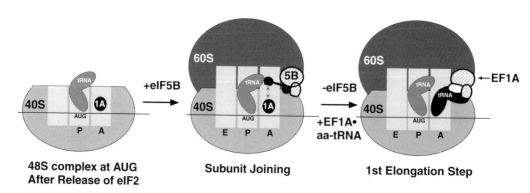

Figure 7. Structure-based model for eIF5B function in promoting the final steps of translation initiation. (*Left*) Following scanning of the 40S ribosomal complex to the AUG start codon, GTP hydrolysis by eIF2 is coupled to release of factors from the ribosome. The Met-tRNA$_i^{Met}$ remains bound in the P site, and we propose that eIF1A remains in the A site. (*Middle*) eIF5B binds to the complex mediated by interactions between domain IV and eIF1A. Domain IV of eIF5B is also in a position to interact with the aminoacyl end of the initiator Met-tRNA$_i^{Met}$ or perhaps with ribosomal constituents of the P site. The binding of eIF5B to the 40S complex facilitates 60S subunit binding. Hydrolysis of GTP by eIF5B following subunit joining is presumably accompanied by conformational changes that could reposition Met-tRNA$_i^{Met}$ in the P site and/or trigger release of eIF1A and eIF5B. (*Right*) The release of eIF5B and eIF1A presents a vacant A site on the 80S ribosome that is filled by the first elongating tRNA as part of the ternary complex EF1A•GTP•aminoacyl-tRNA.

estingly, this slow-growth phenotype can be partially rescued by increasing the dosage of tRNA$_i$^Met genes in the cell (Choi et al. 1998). We propose that the increased abundance of Met-tRNA$_i$^Met in these strains alters the equilibrium of the binding reaction, resulting in more Met-tRNA$_i$^Met bound to the ribosome, and thus increasing the efficiency of protein synthesis.

According to our model, the eIF1A in the A site serves as a docking point for eIF5B. Domain IV of eIF5B will also be in position to interact with the aminoacyl end of Met-tRNA$_i$^Met bound in the ribosomal P site (Fig. 7). This latter interaction may be more significant in bacterial translation, where domain IV of IF2 specifically recognizes fMet as described above. The binding of eIF5B to the 40S subunit facilitates joining of the large ribosomal subunit. Upon subunit joining, eIF5B hydrolyzes its GTP and presumably undergoes a conformational change that could dislodge eIF1A, weaken the binding of eIF5B to the ribosome, and/or alter the conformation of the ribosome or the Met-tRNA$_i$^Met in the P site to facilitate the transition to translation elongation. (These changes may reflect the conversion of 80S to 80S* as described above.) Finally, the domain movements of eIF5B may result in formation of a functional A site such that following release of eIF5B, a molecular imprint is left for binding the EF1A ternary complex in the first step of translation elongation.

CONCLUDING REMARKS

Two translation initiation factors have been conserved through evolution. These factors, IF1/eIF1A and IF2/eIF5B, physically and functionally interact to promote assembly of a translationally competent ribosome with Met-tRNA$_i$^Met in the P site base-paired with the AUG start codon of the mRNA. Interestingly, the third prokaryotic translation initiation factor IF3 and the eukaryotic factor eIF1 appear to have somewhat overlapping roles to ensure fidelity in the initiation process and specify initiation at an AUG codon. The identification of eIF5B reveals that there are at least two GTP-dependent steps in eukaryotic translation initiation as opposed to the single GTP requirement in prokaryotes. In eukaryotes and archaea, the GTPases eIF2 and eIF5B together perform the functions attributed to IF2 in prokaryotes. Whereas eIF5B is required for subunit joining, the factor eIF2 is needed for AUG start-codon selection and appears to functionally substitute for the ribosome-binding site (Shine-Dalgarno sequence) in prokaryotic mRNAs.

Key questions to address concerning the function of eIF5B in translation initiation include: (1) Which translation factors are released from the 40S subunit upon GTP hydrolysis by eIF2? (2) When does eIF5B bind to the ribosome and what factors are present on the 40S subunit when eIF5B binds? (3) What ribosomal constituents are required to stimulate the GTPase activity of eIF5B? (4) What is the conformational transition of eIF5B upon binding to the ribosome and after GTP hydrolysis? In addition, based on the critical role of eIF5B to promote subunit joining, it is surprising that eIF5B is not essential for viability in *S. cerevisiae*. In *Drosophila melanogaster*, eIF5B appears to be required for proper development and viability of the organism (Carrera et al. 2000); however, it remains to be determined whether eIF5B is required for cellular viability in *Drosophila* or other higher eukaryotes. Finally, electron microscopic and/or X-ray crystallographic images of IF2/eIF5B bound to the ribosome will provide important insights to further reveal the function of this universally conserved translation initiation factor.

ACKNOWLEDGMENTS

We thank Tatyana Pestova and Alan Hinnebusch for insights and useful discussions, and members of our labs for advice during the course of our studies. S.K.B. is an investigator in the Howard Hughes Medical Institute. A.R.-M. was supported by a National Science Foundation graduate fellowship and a Burroughs-Welcome Fund Interfaces training grant to The Rockefeller University.

REFERENCES

Agrawal R., Penczek P., Grassucci R., and Frank J. 1998. Visualization of elongation factor G on the *Escherichia coli* 70S ribosome: The mechanism of translocation. *Proc. Natl. Acad. Sci.* **95:** 6134.

Battiste J.B., Pestova T.V., Hellen C.U.T., and Wagner G. 2000. The eIF1A solution structure reveals a large RNA-binding surface important for scanning function. *Mol. Cell* **5:** 109.

Biou V., Shu F., and Ramakrishnan V. 1995. X-ray crystallography shows that translational initiation factor IF3 consists of two compact α/β domains linked by an α-helix. *EMBO J.* **14:** 4056.

Boileau G., Butler P., Hershey J.W., and Traut R.R. 1983. Direct cross-links between initiation factors 1, 2, and 3 and ribosomal proteins promoted by 2-iminothiolane. *Biochemistry* **22:** 3162.

Carrera P., Johnstone O., Nakamura A., Casanova J., Jackle H., and Lasko P. 2000. VASA mediates translation through interaction with a *Drosophila* yIF2 homolog. *Mol. Cell* **5:** 181.

Carter A., Clemons W., Brodersen D., Morgan-Warren R., Hartsch T., Wimberly B., and Ramakrishnan V. 2001. Crystal structure of an initiation factor bound to the 30S ribosomal subunit. *Science* **291:** 498.

Choi S.K., Lee J.H., Zoll W.L., Merrick W.C., and Dever T.E. 1998. Promotion of Met-tRNA$_i$^Met binding to ribosomes by yIF2, a bacterial IF2 homolog in yeast. *Science* **280:** 1757.

Choi S.K., Olsen D., Roll-Mecak A., Martung A., Remo K., Burley S., Hinnebusch A., and Dever T. 2000. Physical and functional interaction between eukaryotic orthologs of prokaryotic translation initiation factors IF1 and IF2. *Mol. Cell. Biol.* **20:** 7183.

Dever T.E. 1999. Translation initiation: Adept at adapting. *Trends Biochem. Sci.* **24:** 398.

Donahue T. 2000. Genetic approaches to translation initiation in *Saccharomyces cerevisiae*. In *Translational control of gene expression* (ed. N. Sonenberg et al.), p. 487. Cold Spring Harbor Laboratory Press, Cold Spring Harbor, New York.

Fletcher C.M., Pestova T.V., Hellen C.U.T., and Wagner G. 1999. Structure and interactions of the translation initiation factor eIF1. *EMBO J.* **18:** 2631.

Godefroy-Colburn T., Wolfe A.D., Dondon J., Grunberg-Manago M., Dessen P., and Pantaloni D. 1975. Light-scattering studies showing the effect of initiation factors on the reversible dissociation of *Escherichia coli* ribosomes. *J. Mol. Biol.* **94:** 461.

Guenneugues M., Caserta E., Brandi L., Spurio R., Meunier S., Pon C., Boelens R., and Gualerzi C. 2000. Mapping the fMet-tRNA$_f$^Met binding site of initiation factor IF2. *EMBO J.* **19:** 5233.

Hershey J.W.B. and Merrick W.C. 2000. Pathway and mechanism of initiation of protein synthesis. In *Translational control of gene expression* (ed. N. Sonenberg et al.), p. 33. Cold Spring Harbor Laboratory Press, Cold Spring Harbor, New York.

Huang H., Yoon H., Hannig E.M., and Donahue T.F. 1997. GTP hydrolysis controls stringent selection of the AUG start codon during translation initiation in *Saccharomyces cerevisiae*. *Genes Dev.* **11:** 2396.

Kang H.A. and Hershey J.W. 1994. Effect of initiation factor eIF-5A depletion on protein synthesis and proliferation of *Saccharomyces cerevisiae*. *J. Biol. Chem.* **269:** 3934.

Kolakofsky D., Dewey K.F., Hershey J.W., and Thach R.E. 1968. Guanosine 5′-triphosphatase activity of initiation factor f2. *Proc. Natl. Acad. Sci.* **61:** 1066.

Kyrpides N.C. and Woese C.R. 1998. Universally conserved translation initiation factors. *Proc. Natl. Acad. Sci.* **95:** 224.

Laalami S., Putzer H., Plumbridge J.A., and Grunberg-Manago M. 1991. A severely truncated form of translation initiation factor 2 supports growth of *Escherichia coli*. *J. Mol. Biol.* **220:** 335.

Lee J.H., Choi S.K., Roll-Mecak A., Burley S.K., and Dever T.E. 1999. Universal conservation in translation initiation revealed by human and archaeal homologs of bacterial translation initiation factor IF2. *Proc. Natl. Acad. Sci.* **96:** 4342.

Lorsch J.R. and Herschlag D. 1999. Kinetic dissection of fundamental processes of eukaryotic translation initiation in vitro. *EMBO J.* **18:** 6705.

Luchin S., Putzer H., Hershey J.W., Cenatiempo Y., Grunberg-Manago M., and Laalami S. 1999. In vitro study of two dominant inhibitory GTPase mutants of *Escherichia coli* translation initiation factor IF2. Direct evidence that GTP hydrolysis is necessary for factor recycling. *J. Biol. Chem.* **274:** 6074.

Meunier S., Spurio R., Czisch M., Wechselberger R., Guenneugues M., Gualerzi C., and Boelens R. 2000. Structure of the fMet-tRNA^fMet-binding domain of *B. stearothermophilus* initiation factor IF2. *EMBO J.* **19:** 1918.

Misselwitz R., Welfle K., Krafft C., Welfle H., Brandi L., Caserta E., and Gualerzi C. 1999. The fMet-tRNA binding domain of translational initiation factor IF2: Role and environment of its two Cys residues. *FEBS Lett.* **459:** 332.

Moazed D., Samaha R.R., Gualerzi C., and Noller H.F. 1995. Specific protection of 16 S rRNA by translational initiation factors. *J. Mol. Biol.* **248:** 207.

Moreno J.M., Drskjotersen L., Kristensen J.E., Mortensen K.K., and Sperling-Petersen H.U. 1999. Characterization of the domains of *E. coli* initiation factor IF2 responsible for recognition of the ribosome. *FEBS Lett* **455:** 130.

Nissen P., Kjeldgaard M., Thirup S., Polekhina G., Reshetnikova L., Clark B.F., and Nyborg J. 1995. Crystal structure of the ternary complex of Phe-tRNA^Phe, EF-Tu, and a GTP analog. *Science* **270:** 1464.

Nyborg J., Nissen P., Kjeldgaard M., Thirup S., Polekhina G., and Clark B.F.C. 1996. Structure of the ternary complex of EF-Tu: Macromolecular mimicry in translation. *Trends Biochem. Sci.* **21:** 81.

Pestova T.V., Lomakin I.B., Lee J.H., Choi S.K., Dever T.E., and Hellen C.U.T. 2000. The joining of ribosomal subunits in eukaryotes requires eIF5B. *Nature* **403:** 332.

Rodnina M., Stark H., Savelsbergh A., Wieden H.-J., Mohr D., Matassova N., Peske F., Daviter T., Gualerzi C., and Wintermeyer W. 2000. GTPase mechanisms and functions of translation factors on the ribosome. *Biol. Chem.* **381:** 377.

Roll-Mecak A., Cao C., Dever T.E., and Burley S.K. 2000. X-ray structures of the universal translation initiation factor IF2/eIF5B: Conformational changes on GDP and GTP binding. *Cell* **103:** 781.

Sette M., van Tilborg P., Spurio R., Kaptein R., Paci M., Gualerzi C.O., and Boelens R. 1997. The structure of the translational initiation factor IF1 from *E. coli* contains an oligomer-binding motif. *EMBO J.* **16:** 1436.

Stark H., Rodnina M., Wieden H.-J., van Heel M., and Wintermeyer W. 2000. Large-scale movement of elongation factor G and extensive conformational change of the ribosome during translocation. *Cell* **100:** 301.

Szkaradkiewicz K., Zuleeg T., Limmer S., and Sprinzl M. 2000. Interaction of fMet-tRNA^fMet and fMet-AMP with the C-terminal domain of *Thermus thermophilus* translation initiation factor 2. *Eur. J. Biochem.* **267:** 4290.

Tolstrup N., Sensen C., Garrett R., and Clausen I. 2000. Two different and highly organized mechanisms of translation initiation in the archaeon *Sulfolobus solfataricus*. *Extremophiles* **4:** 175.

Tomsic J., Vitali L., Daviter T., Savelsbergh A., Spurio R., Striebeck P., Wintermeyer W., Rodnina M., and Gualerzi C. 2000. Late events of translation initiation in bacteria: A kinetic analysis. *EMBO J.* **19:** 2127.

Wu X.-Q. and RajBhandary U. 1997. Effect of the amino acid attached to *Escherichia coli* initiator tRNA on its affinity for the initiation factor IF2 and on the IF2 dependence of its binding to the ribosome. *J. Biol. Chem.* **272:** 1891.

Structural Studies of Eukaryotic Elongation Factors

G.R. Andersen and J. Nyborg

Department of Molecular and Structural Biology, University of Aarhus, Denmark

Translation of the genetic information in the genes of DNA is performed by the peptide-bond-synthesizing machinery of the ribosome. The chemical reaction producing polypeptides takes place at the peptidyl-transfer center of the ribosome and is an RNA-catalyzed process (Nissen et al. 2000b). However, the transitions between several different functional states of the ribosome are catalyzed by a number of protein factors only loosely associated with the ribosome (Garrett et al. 2000; Sonenberg et al. 2000). The complete translation of genetic information into a given protein brings the ribosome and its factors through four different phases: (1) The initiation phase governed by start codons on mRNA and initiation factors (IFs) where ribosomal subunits are assembled and where initiator tRNAs bind to the ribosome, (2) the elongation phase where codons on mRNA are matched with anticodons on tRNAs under the strict control of elongation factors (EFs), (3) the termination phase during which stop codons are recognized by release factors (RFs) and fully synthesized polypeptides are released from the ribosome, and (4) the recycling phase where ribosomal subunits are dissociated from each other under the influence of a ribosome recycling factor (RRF).

The elongation phase is remarkably conserved throughout evolution (Merrick and Nyborg 2000), whereas the eukaryotic initiation phase is complex and controlled by a large number of eukaryotic IFs (eIFs) which are again the focal points of regulation of protein biosynthesis (Hershey and Merrick 2000), in contrast to prokaryotic initiation, which is influenced by only three IFs. On the other hand, eukaryotic termination (Kisselev and Buckingham 2000) has only two RFs (of which the eRF1 recognizes all three stop codons) and is simpler than prokaryotic termination, where three factors are involved and where factors RF1 and RF2 recognize the stop codons.

The elongation phase works in a cyclic manner and adds one amino acid at a time to the growing polypeptide using the sequence of codons on mRNA. (For an overview of the elongation cycle of protein biosynthesis, see Figure 2 in Merrick and Nyborg [2000].) The elongation cycle is controlled by three factors: EF1A, EF1B, and EF2 (formerly called EF-Tu, EF-Ts, and EF-G) in prokaryotes and in organelles, and eEF1A, eEF1B, and eEF2 (formerly called EF-1α, EF-1$\beta\gamma\delta$, and EF2) in eukaryotes. In fungi an additional elongation factor, eEF3, is present. When the next codon to be translated is exposed in the A site (aminoacyl site) of the small subunit of the ribosome, aminoacylated tRNAs (aa-tRNAs) in complex with EF1A:GTP in the so-called ternary complex test their anticodons against the exposed codon. This interac-

tion is very rapid. When a match between codon and anticodon occurs, the EF1A:GTP of the ternary complex is brought into contact with the GTPase activating center of the large ribosomal subunit, and EF1A:GDP is released from the ribosome. The amino acid attached to the CCA end of tRNA is brought into the peptidyl-transferase center where the new peptide bond is produced. The growing peptide is thereby attached to the incoming tRNA while the tRNA in the peptidyl site (P site) of the ribosome becomes deacylated. EF1A:GDP is recycled into its active state EF1A:GTP by the nucleotide exchange factor EF1B. The last part of the elongation cycle involves EF2:GTP, which controls the advance of the ribosome by exactly three nucleotides (one codon) of the mRNA, such that the incoming tRNA is moved into the P site, the deacylated tRNA is moved into the exit site (E site), and the next codon is exposed in the A site. During this translocation, EF2:GTP is in contact with the GTPase activating center of the ribosome, and EF2:GDP is released. There is so far no known nucleotide exchange factor for EF2.

The structural studies of protein biosynthesis have resulted in many remarkable structural models of the ribosome and of translation factors within the past few years. The recent model at 2.4 Å resolution of the large ribosomal subunit from *Haloarcula marismortui* (Ban et al. 2000) revealed that the enzymatic reaction of the peptidyl-transferase center is based on RNA (Nissen et al. 2000b). The models at the same time of the small ribosomal subunit from *Thermus thermophilus* both at 3.0 Å resolution (Wimberley et al. 2000) and at 3.3 Å resolution (Schluenzen et al. 2000), and the subsequent models of binding of IF1 (Carter et al. 2001), of the anticodon stem-loop of cognate tRNA (Ogle et al. 2001), of IF3 (Pioletti et al. 2001), and of antibiotics (Carter et al. 2000; Pioletti et al. 2001) have revealed how cognate codon–anticodon recognition is associated with specific local rearrangements of the small ribosomal subunit, in such a way that this recognition can be transduced to other parts of the ribosome, and that antibiotics known to inhibit the function of the small subunit interfere with these rearrangements. At the same time, the impressive work of pushing the cryo-electron microscopy (cryo-EM) technique forward (Agrawal and Frank 1999; van Heel 2000) has resulted in image reconstructions of ribosomes in complex with tRNAs (Stark et al. 1997b; Agrawal et al. 2000), with the ternary complex of EF1A (Stark et al. 1997a), and with EF2 (Agrawal et al. 1998). Cryo-EM reconstructions have been published of prokaryotic ribosomes or subunits at a resolution of about 10 Å (Matadeen et al. 1999; Gabashvili et al. 2000). Recently, the crystal structure of the whole ribosome from *T. thermophilus* at 5.5 Å resolution has also been published (Yusupov et al. 2001). The combination of cryo-EM and crystallography will undoubtedly reveal many new pictures of various functional states of the ribosome in the very near future.

Most of the structural studies of elongation factors have been performed on bacterial systems as described in more detail below. However, structural studies of eukaryotic elongation factors have now been initiated. At about the same time, the first cryo-EM image reconstructions of eukaryotic ribosomes (Dube et al. 1998a,b), and recently also of the localization of yeast eEF2 on the yeast ribosome (Gomez-Lorenzo et al. 2000), have been published. Structures of some eukaryotic initiation factors are known, some of which are reviewed in Hershey and Merrick (2000). More recently determined structures are for eIF4G, PABP, and eIF5B (Deo et al. 1999; Roll-Mecak et al. 2000; Marcotrigiano et al. 2001). It is very likely that the next decade will witness a large increase in the structural knowledge of eukaryotic translation.

STUDIES OF PROKARYOTIC FACTORS

EF1A:GDP and EF1A:GTP

Structural studies of elongation factors of bacterial protein biosynthesis give an almost complete picture of the various functional states of these factors. The inactive state of EF1A in complex with GDP was earlier determined from *Escherichia coli* (Kjeldgaard and Nyborg 1992; Abel et al. 1996) and from *Thermus aquaticus* (Polekhina et al. 1996), but also recently to high resolution from *E. coli* (Song et al. 1999) and from bovine mitochondria (Andersen et al. 2000a). The last two papers use a standard notation for secondary structural elements of EF1A that is also used in this paper. Furthermore, the structure of the active state of EF1A:GDPNP, where GDPNP is a nonhydrolyzable analog, was determined from *T. aquaticus* (Kjeldgaard et al. 1993) and from *T. thermophilus* (Berchtold et al. 1993). These studies show that EF1A has three structural domains. Domain 1 is a G domain of about 200 residues with a fold as found in all G proteins (Kjeldgaard et al. 1996) and contains a central β sheet that consists of the parallel strands d, a, e, f, and g, with strand c on the edge of the structure antiparallel to strand d. The β sheet is surrounded by α helices A, A′, E, and F on one side and by helices B, C, and D on the other side. The domain has consensus sequence motifs characteristic of G proteins (Bourne et al. 1991). Domains 2 and 3 of about 100 residues each are both β-barrel structures, and are together seen as one unit in all known structures. When the structures of EF1A:GDP and EF1A:GDPNP are compared, a large conformational change is observed (Berchtold et al. 1993; Kjeldgaard et al. 1993).

The EF1A:GDP structure is open with a short peptide linker of 10 residues between domain 1 and domain 2, and with major contacts between domains 1 and 3 only. Upon activation, EF1A:GDPNP becomes much more closed, with extensive contacts between all three domains, and with the contacts between domains 1 and 3 changed to such a degree that one has to assume a tem-

porary dissociation of these two domains when the structure is changed between the two forms. When domains 2 and 3 of the two structures are superimposed, domain 1 is seen to rotate by about 90° on the surface of domain 3 (Kjeldgaard et al. 1993). The large conformational change progresses through the molecule initiated by the fact that GTP has an extra γ phosphate in the nucleotide-binding pocket. The site for the γ phosphate is surrounded by three of the five consensus sequences. Using residue numbering from *T. aquaticus*, these sequences are: 18-GXXXXGKT-25, T-62, and 81-DXXG-84. The first of these consensus sequences is found between strand a and helix A and creates a so-called P loop, as the loop between the two conserved glycine residues holds the β phosphate in place by hydrogen bonds from main-chain NH groups, and the conserved lysine residue makes a salt bridge with the terminal phosphates. In the GDP form, a Mg^{++} ion in this site has as ligands one oxygen from the β phosphate, one side-chain O from the conserved Thr-25, and four oxygens from water molecules. In the GTP form, the Mg^{++} ion receives an extra oxygen ligand from the γ phosphate. The γ phosphate also engages in a hydrogen bond with NH of the peptide bond at the conserved Gly-84 of the so-called switch II region, which consists of the loop from strand d, helix B, and the loop to strand e. This is achieved by a flip of about 150° in the peptide bond between Pro-83 and Gly-84, which goes through backbone conformations only possible for a glycine residue. This local change in the backbone creates a structural change of the switch II region, which results in a shift of 4 residues along the sequence of helix B, such that this is rotated by about 45°. This is exactly the helix involved in the major part of the contacts between domains 1 and 3. Furthermore, the new position of the side chain of Pro-83 creates steric clashes with Thr-62 in the switch I region, which is between helix A′ and strand c. This region is therefore changed from a β-hairpin structure of strands b′ and b in the GDP form into the short helix A″ in the GTP form. This change brings the side-chain O of Thr-62 in switch I into a position as a ligand to Mg^{++}. The structural changes of both switch I and switch II force the dramatic overall change of the spatial orientation of domain 1 relative to domains 2 and 3. In summary, the introduction of the extra γ phosphate into the nucleotide-binding site induces progressively larger and larger conformational changes: first a peptide flip at Gly-84, then changes in the secondary structures of switch regions I and II, and finally, the large reorientation of domain 1 relative to domains 2 and 3. The Mg^{++} ion and its various ligands in the two forms seem to play a central structural role in this conformational change.

EF1B and Nucleotide Exchange

Structures of the guanine nucleotide exchange factor (GEF) EF1B in complex with EF1A have been determined both from *E. coli* (Kawashima et al. 1996) and from *T. thermophilus* (Wang et al. 1997). Although there are very interesting differences between the two structures, the main features involved in nucleotide exchange

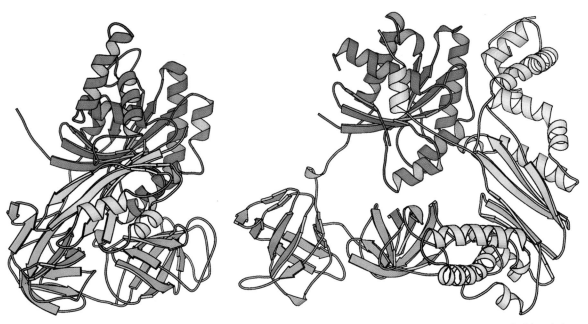

Figure 1. Structures of eEF1A:eEF1Bα and EF1A:EF1B. Color codes in this and all other figures are: (*domain 1*) red with switch regions in magenta; (*domain 2*) green; (*domain 3*) blue; (*exchange factor*) yellow. Yeast eEF1A:eEF1Bα is shown to the left, *E. coli* EF1A:EF1B to the right. Domains 1 of both structures are in similar orientations. This and all other figures are produced using the program Molscript (Kraulis 1991).

are the same, and only the *E. coli* complex is dealt with here. EF1B contacts domain 1 of EF1A close to the nucleotide-binding site, but it also contacts domain 3 such that EF1B effectively separates the two domains of EF1A and thus catalyzes the temporary dissociation of domains 1 and 3 of EF1A (Fig. 1). This separation then allows the

local change of the secondary structures of switch regions I and II.

The conserved Phe-81 of EF1B intrudes between His-84 (His-85 in *T. aquaticus*) and His-118 of EF1A (Fig. 2). His-84 is in switch II, and this interaction separates the switch II helix from the rest of the structure. Another im-

Figure 2. Interactions of nucleotide exchange factors with nucleotide-binding sites. (*Left*) eEF1Bα interacting with its carboxy-terminal residues with eEF1A. In the middle of the figure, Lys-205, and below it pointing to the left, Gln-204 are seen. The switch I region of eEF1A is above and the switch II region below Gln-204. Ser-21 of eEF1A forms a hydrogen bond with Lys-205. At the top right is Glu-122 in helix C* of eEF1A interacting with Lys-156, which has contact with Asp-21 of the P loop. (*Right*) Shows how Phe-81 of EF1B intrudes between His-84 and His-118 of EF1A. Asp-80 of EF1B also makes contacts with helix B of EF1A. At the top is Lys-51 of EF1B forming a hydrogen bond with Asp-17 of the P loop of EF1A.

portant interaction is that of Asp-80 of EF1B with the backbone of the switch II helix, which also adds to this separation. The net result is that 80-DCPG-83 of EF1A moves away from a position where the residues make hydrogen bonds to a number of water molecules in or close to the ligand sphere of the Mg^{++} ion, thereby destabilizing its binding. Asp-80 of EF1A is also forced to break a hydrogen bond to the Mg^{++} ion ligand Thr-25. Furthermore, Lys-51 of EF1B interacts with the P loop in such a way that a peptide bond is flipped, a hydrogen bond to the β phosphate is broken, and the structure of the P loop is slightly altered, with the result that the β phosphate is expelled. Finally, EF1B interacts with the pocket that holds the G base. Thus, many interactions result in the destabilization of the binding of the nucleotide and of the Mg^{++} ion.

TERNARY COMPLEXES OF EF1A AND MACROMOLECULAR MIMICRY

The structures of ternary complexes of *T. aquaticus* EF1A:GDPNP with yeast Phe-tRNA (Nissen et al. 1995) and with *E. coli* Cys-tRNA (Nissen et al. 1999) have been determined. The structures are very similar, and the structure of the bacterial EF1A with yeast tRNA indicates that ternary complexes have the same overall conformation in all organisms. The complex is very elongated, with the anticodon of tRNA pointing away from EF1A. Interactions between protein and RNA are such that EF1A recognizes all elongator tRNAs. There is a broad unspecific interaction between the surface of domain 3 of EF1A with one side of the T-stem helix of tRNA. This involves nonconserved residues of EF1A and only backbone features of tRNA. However, specific interactions occur at the 5′ phosphate and at the terminal A base together with the aminoacyl bond in a cleft between domains 1 and 2. This cleft seems to be large enough to accommodate all amino acid side chains.

The 5′ phosphate is bound at the junction of the three domains of EF1A, where the conserved residue Arg-300 makes a salt bridge with the phosphate, and the conserved Lys-90 and Asn-91 interact with the ribose. The A76 base of tRNA is bound in a pocket on the surface of domain 2. This pocket has conserved or semi-conserved hydrophobic residues Ile-231, Val-237, and Leu-289 at one wall, and the conserved residue Glu-271 which stacks on the base plane of A76 on the other. Residue Glu-271 also forms a hydrogen bond with the 2′ OH of the ribose. The aminoester only interacts with main-chain atoms of EF1A. The carbonyl oxygen makes a hydrogen bond with the main-chain NH of Arg-274, which has its side chain close to the phosphate of A76. The amino group forms hydrogen bonds to main-chain CO of Asn-285, main-chain NH of His-273, and possibly also main-chain CO of Glu-271. Because these groups represent two hydrogen-bond acceptors and one hydrogen donor, it is very likely that the amino group of the aminoester is de-protonated in the complex.

The overall shape of aa-tRNA:EF1A:GDPNP is very similar to that of EF2:GDP (Nissen et al. 1995). The do-

mains 1 and 2 of EF2 and EF1A have the same fold, apart from an insert in domain 1 of EF2 (Czworkowski et al. 1994; Al-Karadaghi et al. 1996), and in EF2:GDP they also have the same relative orientation as in EF1A:GDPNP. This overall similarity points to the fact that they act in similar ribosomal environments in the elongation cycle. The EF2:GTP interacts with the ribosome in the pretranslocation state, induces translocation, and leaves the ribosome as EF2:GDP while the ribosome has changed to the posttranslocation state. Whatever the structural differences are between the pre- and posttranslocation states of the ribosome, they are likely to be sufficiently different so that EF2:GTP and the ternary complex will recognize such differences. In this sense, it can be said that EF2:GDP dissociates from the ribosome in a state which can be recognized by the ternary complex and that EF2:GDP leaves an imprint of the ternary complex on the ribosome by mimicking its shape (Liljas 1996). Thus, the domains 3, 4, and 5 of EF2 structurally mimic the shape of a tRNA (Nissen et al. 1995). Especially the elongated domain 4 mimics the anticodon stem-loop of tRNA. This gave rise to the so-called macromolecular mimicry hypothesis of protein factors in translation that are likely to interact with the A and P sites of the ribosome (Nissen et al. 1995, 2000b). This hypothesis was later confirmed by the structures of ribosome recycling factors (Selmer et al. 1999; Kim et al. 2000; Yoshida et al. 2001) and of eukaryotic release factor eRF1 (Song et al. 2000). Macromolecular mimicry further suggests that all G proteins in translation interact with the same GTPase stimulating center of the ribosome.

STUDIES OF EUKARYOTIC FACTORS

eEF1A and a Fragment of eEF1Bα

Recently, the structure of yeast eEF1A was determined in complex with the carboxy-terminal half of its nucleotide exchange factor eEF1Bα (Andersen et al. 2000b). This structure is shown in Figure 1, along with the *E. coli* EF1A:EF1B complex. In yeast the eEF1B has two subunits: eEF1Bα (formerly called EF-1β) has exchange activity mostly residing in its carboxy-terminal half, whereas the amino-terminal half interacts with the amino terminus of eEF1Bγ (formerly EF-1γ) of unknown function. Higher eukaryotes have an additional subunit eEF1Bβ (formerly EF-1δ) with a carboxyl terminus homologous to that of eEF1Bα (Janssen et al. 1994), and this subunit has exchange activity as well (van Damme et al. 1990). The carboxy-terminal half of eEF1Bα is sufficient for viability of yeast (Carr-Schmid et al. 1999b) and has for the human factor been shown by nuclear magnetic resonance (NMR) to form a structural unit, and to retain 50% of its activity (Pérez et al. 1999). Furthermore, archaea have versions of aEF1Bα that are half the size of the eukaryotic counterpart and have sequences that compare to the carboxy-terminal half of yeast and human eEF1Bα.

The amino acid sequences of eEF1A show high homology with those of prokaryotic EF1A, although it is obvious that eEF1A has inserts in some of its loops as seen from Figure 1 and Figure 6 in Merrick and Nyborg (2000). On the other hand, eEF1Bα shows no detectable sequence homology with its prokaryotic counterpart EF1B. It is obvious from the comparison of the structure of the prokaryotic EF1A:EF1B with that of eukaryotic eEF1A:eEF1Bα (Fig. 1) that the two complexes are different. The prokaryotic EF1B contacts at one end domain 1 of EF1A near all the loops involved in the nucleotide-binding site and, at the other end, contacts the surface of domain 3, whereas eukaryotic eEF1Bα at one end interacts with domain 1 of eEF1A only close to switch regions I and II and at the other with the surface of domain 2.

As expected from the high sequence homology, eEF1A has three domains with the same basic topology as found in EF1A. Domain 1 of eEF1A thus has a similar central β sheet of strands a–g, surrounded by α helices A, A′, B, C, D, and F. Helix E is missing, helix A′ is much longer and shifted in position, and helix F is shorter than in EF1A. The switch I region, which is the most variable region of all G proteins, and which is found between helix A and strand c, has a large insert between helices A and A′ forming the long helix A*. A new helix C* is formed by a large insert in the loop between strand e and helix C. Finally, an insert of two antiparallel β strands is found just before the shortened helix F. Domain 2 is very similar in eEF1A and EF1A and so is domain 3, except for an insert at 375–388 of eEF1A.

The structure of the carboxy-terminal half of eEF1Bα starts at residue 117 and ends at residue 206, which is the wild-type carboxyl terminus of eEF1Bα. It contains a four-stranded antiparallel β sheet with two α helices packed on one side of the sheet. It has a fold similar to the ones found in many ribosomal proteins and, in fact, has two copies of the ribosome RNA-binding motif (the RRM motif), also called the split β-α-β motif (Liljas and al-Karadaghi 1997; Ramakrishnan and White 1998). The shortened version of archaeal aEF1Bα starts at a position corresponding to residue 118. The structure of the fragment from yeast (Andersen et al. 2000b) is very similar to the structure of the fragment from human eEF1Bα determined by NMR (Pérez et al. 1999). The largest difference is found at a semi-conserved loop at 161-IGFG-164 in yeast and 179-VGYG-182 in human eEF1Bα, which is flexible or disordered in the NMR structure (Pérez et al. 1999).

Eukaryotic Nucleotide Exchange

One end of the elongated eEF1Bα fragment contacts domain 1 of eEF1A near the expected Mg^{++} ion and γ-phosphate-binding sites. The Lys-205 residue of eEF1Bα forms a salt bridge with the carboxy-terminal carboxylate. Its side chain is positioned across the expected site for the γ phosphate, and its NH$_3^+$ group is positioned at the expected Mg^{++} ion site. It therefore also forms a hydrogen bond with the expected Mg^{++} ion ligand Ser-21 of eEF1A (corresponding to Thr-25 of EF1A). Its role is essential for activity, as it has been shown that a Lys205Ala mutant was unable to function in vivo, although the protein was produced and folded correctly (Andersen et al. 2001). This direct introduction of Lys-205 into the Mg^{++} ion and γ-

phosphate-binding sites of eEF1A must contribute significantly to the destabilization of nucleotide binding to eEF1A, and is very different from the indirect way that prokaryotic EF1B influences these two binding sites, as seen in Figure 2 (see above). In addition, the way that Gln-204 of eEF1Bα is squeezed between the strands d and e of eEF1A with influence on the positions of switch regions I and II is a much more direct interaction than for the EF1A:EF1B complex. By breaking a hydrogen bond of the β sheet and creating a new hydrogen bond with the main chain of Asp-91 in the consensus sequence 91-DGPG-94, this part is moved away from the Mg^{++} ion and γ-phosphate-binding pockets. Although the side chain of Asp-91 retains its hydrogen bond with the Mg^{++} ion ligand Ser-21, it now also forms a salt bridge with Lys-20 of the consensus sequence 14-GXXXXGKS-21, which is expected to interact with the β and γ phosphates of the nucleotides. These changes further destabilize nucleotide binding. There are some similarities between prokaryotic and eukaryotic nucleotide exchange, in that they both result in displacement of the consensus sequence DXXG (see above). However, in eukaryotes the interaction of the nucleotide exchange factor is much more directly aimed at residues involved in the binding sites.

There is no interaction between eEF1Bα and the remaining loops of eEF1A involved in nucleotide binding. This can possibly be explained by the fact that the amino-terminal part of eEF1B and residues not seen in the structure do contribute to nucleotide exchange by contacting these loops. The first residue in the structure of the fragment of eEF1Bα is indeed not far from the nucleotide-binding site. That such an explanation is not very likely comes from two facts. One is that in archaea the nucleotide exchange factor is not longer than seen in the structure, although it will later be argued that the exchange mechanisms in archaea and in eukarya are possibly not similar (see below). Another is that the alteration of the structure of the P loop as seen in prokaryotes (see above) is simulated by internal residues of eEF1A. The side chain of Glu-122 in helix C* forms a salt bridge with Lys-154 of the consensus sequence 153-NKMD-156 involved in direct recognition of the G base. Lys-154 in turn forms a hydrogen bond with the main chain at Asp-17 of the P loop. These interactions then provoke a peptide flip in the P loop at the same position as in prokaryotes (see above) and thus create a similar effect on nucleotide binding. One has to assume that this interaction of Glu-122 is changed in both of the nucleotide-binding forms of eEF1A. Mutations of Glu-122 or of Asp-156 show that in vivo eEF1A becomes less dependent on the exchange activity of eEF1Bα, and this confirms the functional importance of Glu-122 (Kinzy and Woolford 1995; Carr-Schmid et al. 1999a).

Recently, the first structures of nucleotide exchange intermediates of any G protein were determined for the yeast eEF1A:eEF1Bα complex (Andersen et al. 2001), as shown in Figure 3. Crystals were obtained of eEF1A:eEF1Bα:GDP in the presence of Mg^{++} and of eEF1A:eEF1Bα:GDPNP in the absence of Mg^{++}. Crystals of the GDP form were soaked in EDTA with a re-

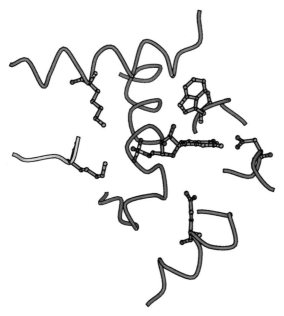

Figure 3. Structure of an intermediate of nucleotide exchange of eEF1A:eEF1Bα:GDPNP. In the middle is seen GDPNP for which the β and γ phosphates are disordered. On top of the G base is Trp-194, forming part of the hydrophobic pocket. To the right is Asp-156 recognizing the rim of the G base. Below is Glu-122 of helix C*. To the right is Lys-205 of eEF1Bα, and above that Lys-64 of eEF1A.

duction in a low-density peak, therefore proposed to be a Mg^{++} site. Diffracting crystals of the GDPNP form in the presence of Mg^{++} were not obtained. Furthermore, the crystals of the GDPNP form obtained in the absence of Mg^{++}, although stable for days in a cryo-protectant buffer, were converted into poorly diffracting crystals by soaking in Mg^{++}-containing buffer for 30 minutes. This strongly indicates some structural rearrangement of eEF1A:eEF1Bα induced by GDPNP:Mg^{++}. Both structures show interpretable electron density for the G base, the ribose, and the α phosphate bound in the way expected from the prokaryotic EF1A:GDP and EF1A:GDPNP structures. Only weak density for a possible Mg^{++} is seen in the GDP form, and none for the β phosphate. In the GDPNP form, only weak noninterpretable density is seen for the β phosphate and none for the γ phosphate.

It was earlier proposed that a nucleotide first binds the G base and the ribose, and thereafter positions the phosphates and Mg^{++} correctly into their binding sites of the nucleotide-free G protein, and accordingly, that the reverse order takes place when a nucleotide leaves its binding site (Cherfils and Chardin 1999). The structures of nucleotides in complex with eEF1A:eEF1Bα for the first time directly confirm this hypothesis by visualizing intermediates in the nucleotide exchange mechanism (Andersen et al. 2001). The possible position of the Mg^{++} near the α phosphate in the GDP form only has one observable ligand, which is one of the oxygens of the α phosphate. It cannot be ruled out that this is very weak density for the

β phosphate. In any case, the structures show that the phosphates beyond the α phosphate are disordered. What is interesting is that whatever their positions are, they will be close to Lys-205 of eEF1Bα occupying the Mg^{++} ion site and to Lys-64 of eEF1A. This points to a possible mechanism by which the γ phosphate of GTP can interact with these residues to alter the P-loop structure and to expel Lys-205 of eEF1Bα from the Mg^{++} ion site. Again, Lys-205 is seen to be central in the nucleotide exchange mechanism for eukaryotic eEF1A.

Recently, the structure of archaeal aEF1A:GDP from *Sulfolobus solfataricus* has been determined (A. Zagari and L. Vitagliano, pers. comm.). The overall folds of the three domains of aEF1A are the same as described here for yeast eEF1A, although the orientation of domain 1 relative to domains 2 and 3 is altered (see Fig. 4). Surprisingly, the structure around the GDP-binding site is not comparable to that found in prokaryotic EF1A:GDP. Especially, no Mg^{++} ion is seen in its expected binding site, nor are some of the ligands known from the EF1A:GDP structure found in equivalent positions in aEF1A:GDP. This is especially true for the local structures of the consensus sequence 90-DAPG-93 and of switch II.

This structural result seems to be in stark contrast to the mechanism of nucleotide exchange as deduced above from the structures of the yeast elongation factors, because this mechanism as described is strongly based on the suggestion that destabilization of the Mg^{++} binding in eEF1A:GDP is one of the first steps in the nucleotide exchange. However, the possibility exists that the mechanisms are slightly different in archaea and in eukarya. The fact that archaeal aEF1Bα is only half the size of eukaryal eEF1Bα could indicate this. Also, and perhaps more importantly, the residues corresponding to 204-QK-205 are not conserved in aEF1Bα. Thus, for the archaeal nucleotide exchange factor the destabilization of Mg^{++} binding might not be as important as it seems to be for the eukaryal one.

Is EF1B Simulating tRNA?

At the other end of the structure of eEF1Bα is a semiconserved loop of sequence 161-IGFG-164. Residue 161 is always hydrophobic V/L/I, residue 162 is G/A, residue 163 is aromatic, and residue 164 is strictly conserved as glycine. The glycine residues make this loop flexible, as seen in the structure of human eEF1Bα (Pérez et al. 1999). The aromatic residue, Phe-163 in the yeast elongation factor, is found in the pocket expected to hold the terminal A76 base of aa-tRNA in the ternary complex of eEF1A. Thus, the hydrophobic residues I254, I257, and V260 are found on one side of the aromatic side chain, whereas Glu-291 is stacked onto the aromatic ring of Phe-163 on the other side (see Fig. 5). The Glu-291 furthermore makes a hydrogen bond with the NH of the backbone at Phe-163. This simulates an important hydrogen bond in the ternary complex of the prokaryotic ternary complex (see above) and thereby stabilizes the structure of the otherwise flexible loop. On top of this, Arg-320 of

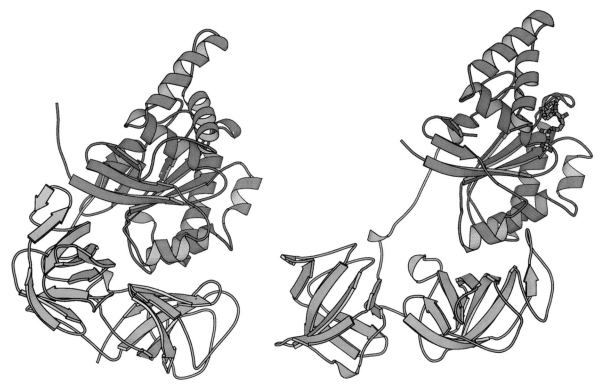

Figure 4. Structures of eukaryal eEF1A and archaeal aEF1A:GDP. (*Left*) Yeast eEF1A as seen in the complex with eEF1Bα. (*Right*) *S. solfataricus* aEF1A:GDP. The nucleotide of aEF1A:GDP is shown in a ball-and-stick diagram. Notice the differences in the overall structures and of the switch II regions in the middle of the molecules.

eEF1A, equivalent to the Arg-300 of EF1A that makes a salt bridge with the 5′ phosphate of aa-tRNA in the ternary complex (see above), interacts with the strictly conserved Asp-199 of eEF1Bα (Fig. 5). Thus, in several aspects this end of eEF1Bα simulates the interaction of aa-tRNA with eEF1A.

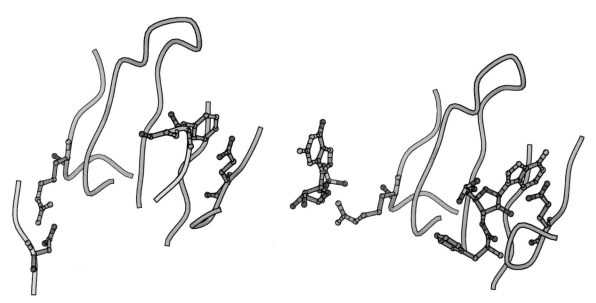

Figure 5. Interaction of eEF1Bα with domain 2 of eEF1A simulates aa-tRNA binding. (*Left*) In the middle of the figure is seen Phe-163 of eEF1Bα stacking with Glu-291 of eEF1A, which forms a hydrogen bond with the backbone NH of Phe-163. To the left is Asp-120 of eEF1Bα forming a salt bridge with Arg-320 of eEF1A. (*Right*) In the middle of the figure is seen the terminal A base and the Phe-aminoester of the ternary complex of EF1A. The A base is stacking on Glu-271, which forms a hydrogen bond with 2′ OH of the ribose. To the left is Arg-300 of EF1A forming a salt bridge with the 5′ phosphate of tRNA.

This simulation of the interaction of aa-tRNA by eEF1Bα is interesting, considering the hypothesis of channeling of tRNA in eukaryotic protein biosynthesis (Negrutskii et al. 1996). The channeling hypothesis claims that aa-tRNAs are transferred directly from tRNA synthetases, possibly in a multi-synthetase complex, onto elongation factor eEF1A without dissociation into the cytosol. The hypothesis is supported by the fact that endogenous aa-tRNAs in the cell are protected from degradation by RNases, in contrast to exogenous aa-tRNA introduced into permeabilized cells (Negrutskii and Deutscher 1992). Furthermore, in contrast to the prokaryotic system, eEF1A is known to stimulate the activity of valyl and phenylalanyl tRNA synthetase in mammalian cells (Negrutskii et al. 1996), where valyl tRNA synthetase is also known to associate with EF1 through interaction with eEF1Bβ (Bec et al. 1994). All these observations indicate an intimate coupling in eukaryotes between tRNA synthetases and elongation factors to protect aa-tRNAs and to transfer them efficiently between components of the protein synthesis machinery.

The observed overlap of the binding of eEF1Bα with the expected aa-tRNA recognition site on the surface of domain 2 of eEF1A in the crystal structure of the eEF1A:eEF1Bα complex supports such a relationship between tRNA synthetases and elongation factors. The flexibility of the interacting loop of eEF1Bα observed in the NMR structure suggests a mechanism by which this loop and the CCA-aa end of tRNA can compete for the same binding site. If the flexible loop of eEF1Bα flips away from its binding to the pocket on domain 2 to the mean conformation seen in the NMR structure, then the binding site for the terminal A base and ribose of aa-tRNA becomes accessible. The CCA-aa end of tRNA can then zip into its binding site, which would result in weaker binding and release of eEF1Bα from domain 2 of eEF1A. The remaining interaction of aa-tRNA with eEF1A can be formed after complete dissociation of eEF1Bα. Thus, eEF1A:GTP is not necessarily released into solution after nucleotide exchange, but an intermediate complex eEF1A:eEF1Bα:GTP may react directly with aa-tRNA (Andersen et al. 2000b). The formation of a complex between *Artemia* eEF1A:eEF1Bα and GDPCP that can be dissociated by aa-tRNA supports such a mechanism (Janssen and Möller 1988). The observation that the Phe163Ala mutant cannot be suppressed by overexpression of eEF1A (T.G. Kinzy, pers. comm.) further indicates that the interaction of the flexible loop is important for the correct formation of a ternary complex of eEF1A, otherwise the increased amount of eEF1A:GTP in the cytosol would on its own support protein biosynthesis.

INTERACTION OF G-PROTEIN TRANSLATION FACTORS WITH THE RIBOSOME

Although there is extensive structural information on the interaction of eukaryotic elongation factors with the eukaryotic ribosome, much can be inferred from structural studies of prokaryotic translation, and the fact that

the elongation cycle is well conserved in all organisms. The interactions of the ternary complex of EF1A and of EF2:GTP with the ribosome have been studied by cryo-EM image reconstructions. The antibiotic kirromycin stalls the ternary complex on the ribosome after GTP hydrolysis. Such a formation of a stable ribosomal state has been used in the cryo-EM reconstruction of the ternary complex on the *E. coli* ribosome (Stark et al. 1997a). Similarly, the antibiotic fusidic acid stalls EF2 on the ribosome and has been the basis of another cryo-EM reconstruction (Agrawal et al. 1998). These results show that the ternary complex and EF2 interact with the ribosome in very similar ways, but also that there are some alterations of their structures relative to the ones known from crystal structures, either because they are altered by their interaction with the antibiotic or because they are altered when they interact with the ribosome. That the latter could be true is indicated by a reconstruction of the ribosome with EF2:GTP (Agrawal et al. 1999), which shows similar features as the reconstruction to the antibiotic stalled one. More detailed models of how EF1A and EF2 interact with the 50S ribosomal subunit have been proposed (Ban et al. 1999). They all show that domain 1 is in contact with the 50S subunit, most likely close to the switch 1 region, and that domain 2 is in contact with the 30S subunit, probably close to the pocket binding the A76 base. Similar interactions are likely to occur also for IF2 and for RF3, which are both G proteins and which were predicted to have domains 1 and 2 with folds as in EF1A and EF2 (Aevarsson 1995).

Two of the translation initiation factors are known to be universal. These are in eubacteria and are called IF1 and IF2. IF1 interacts with the ribosomal A site to prevent interaction of any tRNA during initiation, as can be seen in the recent structure determination of IF1 in complex with the 30S subunit (Carter et al. 2001). IF2 forms a ternary complex with initiator tRNA and assists in assembly of the ribosome particle. However, the prokaryotic initiator tRNA cannot form a stable ternary complex with IF2 outside the ribosome, in contrast to the eukaryotic counterpart eIF2. In eukaryotes the initiation factors having homology with the prokaryotic factors are IF1/eIF1A and IF2/eIF5B. It is not clear whether they interact with eukaryotic initiator tRNA$_i$. Recently, the structure of IF2/eIF5B from *Methanobacterium thermoautotrophicum* was determined (Roll-Mecak et al. 2000), and this confirms not only that IF2/eIF5B has a domain 1 and 2 with folds similar to the ones found in EF1A and EF2, but also that in both IF2/eIF5B:GTP and IF2/eIF5B:GDP the relative orientation of the two domains is similar to that in EF2:GDP and in EF1A:GTP. This indicates that IF2/eIF5B, like EF2, leaves the ribosome in a state ready to accept the ternary complex of EF1A (Roll-Mecak et al. 2000).

One puzzling observation, when the fold of domain 2 of EF1A is compared to those of EF2, eEF1A, and IF2/eIF5B, is that all four apparently have the pocket where the terminal A base is found in the ternary complex of EF1A (see Fig. 6). Neither EF2 nor IF2/eIF5B will bind a tRNA in this position. In fact, IF2 is known to bind initiator tRNA at its domain 4, which also has a fold like domain 2 of EF1A (Guenneugues et al. 2000; Meunier et

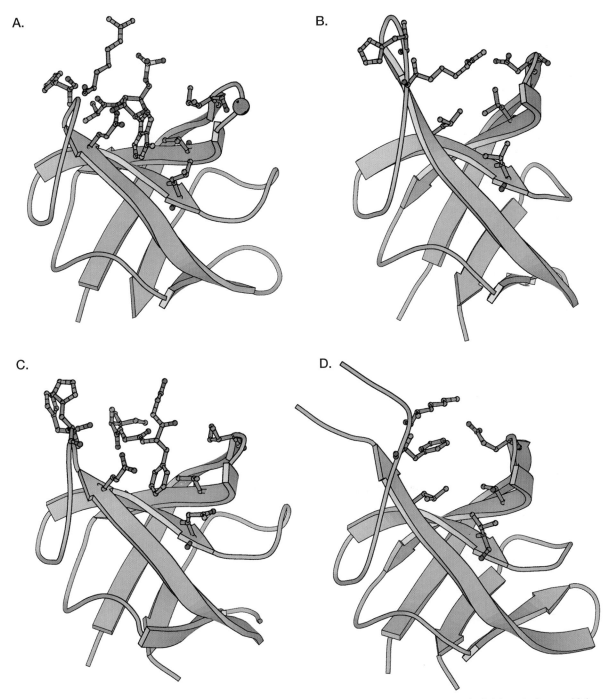

Figure 6. The pocket on domain 2 of translation factors. (*A*) The ternary complex of EF1A. The terminal A base is shown with hydrophobic residues on the right and Glu-271 and Arg-274 on the left (see text). The position of Gly-233 is shown by a gray sphere. (*B*) Elongation factor EF2. Notice that the pocket has somewhat larger hydrophobic residues and that a salt bridge is formed at the entrance to the pocket. (*C*) Complex of eEF1A and eEF1Bα. Notice similarity with the ternary complex. Glu-291 of eEF1A stacks on Phe-163 of eEF1Bα (see text). (*D*) Initiation factor IF2/eIF5B. Notice the large phenylalanine residue and a salt bridge at the entrance of the pocket. A loop on one side of the pocket is long and partly disordered.

al. 2000). So why do EF2 and IF2/eIF5B have this pocket? On closer inspection, the pockets of EF2 and IF2/eIF5B are filled with side chains of residues from both walls of the pocket (Fig. 6), and especially IF2/eIF5B has a much longer loop between the β strands at one side of the pocket (Roll-Mecak et al. 2000). However, the other side, which has a β-hairpin structure with

a β turn, is very well conserved in all three domains. These are residues 225-VEDVFTI**TG**RGTVATGRI-242 of *T. aquaticus* EF1A, residues 313-AFKIMAD**PY**V-GRLTFIRV-330 of *T. thermophilus* EF2, residues 248-LQDVYKI**GG**IGTVPVGRV-265 of yeast eEF1A, and residues 236-ILEVKEE**TG**LGMTIDAVI-253 of *M. thermoautotrophicum* IF2/eIF5B, where residues in the β turn are shown in bold. It is striking that Gly-233 in the loop (Gly-222 in *E. coli*) of EF1A when mutated to aspartic acid becomes the so-called B₀ mutant, which has been known for a long time to be defective in its interaction with the ribosome (Swart et al. 1987). One possible explanation is a purely structural one, as the loop structure could be distorted by this mutation. A much more interesting explanation would be that this is a very important and therefore conserved point of interaction between domain 2 of all G proteins in translation and the ribosomal 30S subunit (Nyborg and Liljas 1998). Perhaps in the case of EF1A the 30S subunit by such an interaction could alter the affinity of the A76 base of tRNA for its binding pocket.

STUDIES OF MITOCHONDRIAL FACTORS

Bovine mtEF1A

The crystal structure of bovine mitochondrial mtEF1A:GDP has been determined to high resolution (Andersen et al. 2000a). The overall structure is very similar to those known from prokaryotes (Kjeldgaard and Nyborg 1992; Abel et al. 1996; Polekhina et al. 1996; Song et al. 1999), although the orientation of domain 1 relative to domains 2 and 3 is slightly altered (see Fig. 7). Comparisons of all determined structures of EF1A:GDP do indicate that this relative orientation can undergo some variation. The large variation in affinities of the nucleotides to EF1A, and especially the fact that GDP has an affinity for *E. coli* EF1A that is higher by two orders of magnitude

Figure 7. Structure of bovine mitochondrial EF1A:GDP. The nucleotide is shown in a ball-and-stick diagram, and the Mg⁺⁺ ion as a gray sphere. The carboxy-terminal extension is shown in orange.

than that of GTP, cannot easily be reconciled with structural information. The immediate sphere of interactions between nucleotides and proteins is very well conserved in all G proteins. Attempts to find an explanation by looking at structural differences in the next layer of the protein, i.e., variations in the residues that are close to the very conserved ones, are seemingly not very successful. An indication of a possible distant influence from domains 2 and 3 comes from the observation that for the isolated domain 1 of *E. coli* EF1A, the affinities for GDP and GTP are the same and both slightly lower than the affinity of GTP for the wild-type protein (Jensen et al. 1989).

The only significant difference between the bacterial and the mitochondrial EF1A is a 10-residue helical extension at the carboxyl terminus. A longer extension of 18 residues is found in yeast eEF1A but is not ordered in the crystals of eEF1A:eEF1Bα. The short helix of mtEF1A has structural and sequence similarities with the zinc fingers that are known to specifically recognize DNA or RNA (Nolte et al. 1998). This strongly indicates that this carboxy-terminal helix does interact with RNA. From a model of a possible ternary complex of mtEF1A, it is not obvious that this interaction is with tRNA. Much more likely is a sequence-specific interaction with RNA of the mitochondrial ribosome.

Nematode mtEF1A

Mitochondrial protein biosynthesis is characterized by minimal tRNAs and minimal ribosomes (Watanabe and Osawa 1995). Some mitochondrial tRNAs are very short, such as tRNA$_{GCU}^{Ser}$, which in most mitochondria is missing the D stem. More interestingly, for nematode mitochondria, many tRNAs are shortened and, for example, the tRNA^{Met} is missing the T stem. From the knowledge of the structure of bacterial ternary complexes of EF1A (Nissen et al. 1995, 1999), this is indeed surprising, as the T stem is part of the interaction between tRNA and EF1A (see above). Recently it has been shown that the nematode *Caenorhabditis elegans* has two different versions of mtEF1A. One of these, mtEF1A2, is homologous with bovine mtEF1A and has been shown to interact with tRNAs with a normal T stem, whereas the other, mtEF1A1, interacts with tRNAs missing the T stem (Ohtsuki et al. 2001). It was further shown that this mtEF1A1 has a 50-residue carboxy-terminal extension with the possibility that this extension forms a completely new extra domain in this elongation factor. An obvious explanation for this extra domain would then be that it somehow substitutes for the missing T stem of the tRNA and perhaps stabilizes the structure of this minimal tRNA (Ohtsuki et al. 2001).

CONCLUSION

There is no doubt that the next few years will see a growing amount of structural information on eukaryotic translation. One can foresee many interesting and surprising results, especially on factors of the initiation phase, which is so complex and so tightly controlled. Eukaryotic elongation factors are also slightly more complex than the prokaryotic factors, not least in the fact that the nucleotide exchange factor eEF1B is more complicated and has a varying number of subunits in eukaryotes. The eEF1A also seems to be involved in many other cellular functions, such as its well-established interaction with actin and with other components of the cellular cytoskeleton.

Some part of the functions of elongation factors remains to be elucidated more closely by structural studies. One of these is the nucleotide exchange mechanism. The work described here contributes to a first glimpse of this mechanism in eukaryotes. However, much work must still be performed. It is not known whether the exchange factor is actively involved in the large conformational change of EF1A or eEF1A. Indeed, it is not known why there is such a large change of structure during the activation of this elongation factor. Nothing similar seems to be happening for the other translation factors, for other G proteins involved in translation, or indeed for any other G protein. Of course, it is fair to say that at the moment it is not known whether this is indeed only the case for prokaryotic EF1A! Nevertheless, one cannot help but speculate that this could be a useful feature of helping the aa-tRNA pass from a successful state of testing the codon–anticodon interaction into the state of catalysis at the peptidyl transferase center of the ribosome.

Another puzzling feature of the nucleotide exchange factor is that it is involved in initiation of viral gene replication as in the Qβ replicase complex of bacteriophage Qβ. This complex, apart from a replicase subunit from the viral genome, consists of EF1A, EF1B, and ribosomal protein S1 from the host. On top of this, the 3´ ends of some plant viral genomes have a tRNA mimicking structure. Why is this so? Do these bacteriophages and viruses use some intrinsic activity of the host cell that involves these proteins? Why should they together help in recognizing a tRNA-like structure?

Another function that is poorly understood is the GTPase activity of EF1A. It is known that EF1A has very low intrinsic GTPase activity, and that this is highly stimulated by the ribosome. This in itself is understandable, as EF1A has to provide a pool of ternary complexes with various aa-tRNAs, and should only leave an aa-tRNA on the ribosome after cognate recognition of the codon exposed in the A site. But how is the GTPase activity stimulated by the ribosome? Structural information is available on transition-state analogs of GTP hydrolysis for heterotrimeric G proteins, and it is known that the GAP proteins of the small p21-like G proteins provide an "Argfinger" to stabilize the transition state. The AlF$_3$ which has been so useful in these investigations does not work on isolated EF1A, but most likely only on its complex with the ribosome. An Arg finger of the ribosome does not seem too likely, as arginine residues of L7/L12 have been tested with no clear result. Does the ribosome alter the known structure of EF1A in the ternary complex to such a degree that it can use its internal arginine residue of the switch I region for transition-state stabilization in analogy with the heterotrimeric G proteins?

Needless to say, there are still numerous research projects involving structural studies of elongation factors

that must be undertaken in the future. Of these, the ones involving eukaryotic elongation factors should be very interesting and should provide answers to some of the questions raised above.

ACKNOWLEDGMENTS

We thank our technician Lan Bich Van for excellent help at various stages of the work described. We also thank Adriana Zagari and Luigi Vitagliano for access to their structure of *S. solfataricus* aEF1A:GDP prior to publication. This work has been supported by the Danish Natural Science Research Council through its program for Biotechnological Research, by the Danish center for Synchrotron Radiation Research, The Carlsberg Foundation, and the EU Research Project BIO4-97-2188. We thank Terri G. Kinzy and Linda L. Spremulli for stimulating cooperation on yeast elongation factors and on bovine mtEF1A, respectively.

REFERENCES

Abel K., Yoder M.D., Hilgenfeld R., and Jurnak F. 1996. An α to β conformational switch in EF-Tu. *Structure* **4:** 1153.

Aevarsson A. 1995. Structure-based sequence alignment of elongation factors Tu and G with related GTPases involved in translation. *J. Mol. Evol.* **41:** 1096.

Agrawal R.K. and Frank J. 1999. Structural studies of the translational apparatus. *Curr. Opin. Struct. Biol.* **9:** 215.

Agrawal R.K., Penczek P., Grassucci R.A., and Frank J. 1998. Visualization of the elongation factor G on the *Escherichia coli* 70S ribosome: The mechanism of translocation. *Proc. Natl. Acad. Sci.* **95:** 6134.

Agrawal R.K., Heagle A.B., Penczek P., Grassucci R.A., and Frank J. 1999. EF-G-dependent GTP hydrolysis induces translocation accompanied by large conformational changes in the 70S ribosome. *Nat. Struct. Biol.* **6:** 643.

Agrawal R.K., Spahn C.M., Penczek P., Grassucci R.A., Nierhaus K.H., and Frank J. 2000. Visualization of tRNA movements on the *Escherichia coli* 70S ribosome during the elongation cycle. *J. Cell Biol.* **150:** 447.

Al-Karadaghi S., Aevarsson A., Garber M., Zheltonosova J., and Liljas A. 1996. The structure of elongation factor G in complex with GDP: Conformational flexibility and nucleotide exchange. *Structure* **4:** 555.

Andersen G.R., Thirup S., Spremulli L.L., and Nyborg J. 2000a. High resolution crystal structure of bovine mitochondrial EF-Tu in complex with GDP. *J. Mol. Biol.* **297:** 421.

Andersen G.R., Valente L., Pedersen L., Kinzy T.G., and Nyborg J. 2001. Crystal structures of nucleotide exchange intermediates in the eEF1A:eEF1Bα complex. *Nat. Struct. Biol.* **8:** 531.

Andersen G.R., Pedersen L., Valente L., Chatterjee I., Kinzy T.G., Kjeldgaard M., and Nyborg J. 2000b. Structural basis for the nucleotide exchange and competition with tRNA in the yeast elongation factor complex eEF1A:eEF1Bα. *Mol. Cell* **6:** 1261.

Ban N., Nissen P., Hansen J., Moore P.B., and Steitz T.A. 2000. The complete atomic structure of the large ribosomal subunit at 2.4 Å resolution. *Science* **289:** 905.

Ban N., Nissen P., Hansen J., Capel M., Moore P.B., and Steitz T.A. 1999. Placement of protein and RNA structures into a 5 Å-resolution map of the 50S ribosomal subunit. *Nature* **400:** 841.

Bec G., Kerjan P., and Waller J.P. 1994. Reconstitution in vitro of valyl-tRNA synthetase-elongation factor (EF) 1βγδ complex. Essential role of the NH$_2$-terminal extension of valyl-tRNA synthetase and of the EF-1δ subunit in complex formation. *J. Biol. Chem.* **269:** 2086.

Berchtold H., Reshetnikova L., Reiser C.O.A., Schirmer N.K.,

Sprinzl M., and Hilgenfeld R. 1993. Crystal structure of active elongation factor Tu reveals major domain rearrangements. *Nature* **365:** 126.

Bourne H.R., Sanders D.A., and McCormick F. 1991. The GTPase superfamily: Conserved structure and molecular mechanism. *Nature* **349:** 117.

Carr-Schmid A., Durko N., Cavallius J., Merrick W.C., and Kinzy T.G. 1999a. Mutations in a GTP-binding motif of eukaryotic elongation factor 1A reduce both translational fidelity and the requirement for nucleotide exchange. *J. Biol. Chem.* **274:** 30297.

Carr-Schmid A., Valente L., Loik V.I., Williams T., Starita L.M., and Kinzy T.G. 1999b. Mutations in elongation factor 1β, a guanine nucleotide exchange factor, enhance translational fidelity. *Mol. Cell. Biol.* **19:** 5257.

Carter A.P., Clemons W.M., Brodersen D.E., Morgan-Warren R.J., Wimberley B.T., and Ramakrishnan V. 2000. Functional insights from the structure of the 30S ribosomal subunit and its interactions with antibiotics. *Nature* **407:** 340.

Carter A.P., Clemons W.M., Brodersen D.E., Morgan-Warren R.J., Hartsch T., Wimberley B.T., and Ramakrishnan V. 2001. Crystal structure of an initiation factor bound to the 30S ribosomal subunit. *Science* **291:** 498.

Cherfils J. and Chardin P. 1999. GEFs: Structural basis for their activation of small GTP-binding proteins. *Trends Biochem. Sci.* **24:** 306.

Czworkowski J., Wang J., Steitz T.A., and Moore P.B. 1994. The crystal structure of elongation factor G complexed with GDP, at 2.7Å resolution. *EMBO J.* **13:** 3661.

Deo R.C., Bonanno J.B., Sonenberg N., and Burley S.K. 1999. Recognition of polyadenylate RNA by the poly(A)-binding protein. *Cell* **98:** 835.

Dube P., Bacher G., Stark H., Mueller F., Zemlin F., van Heel M., and Brimacombe R. 1998a. Correlation of the expansion segments in mammalian rRNA with the fine structure of the 80 S ribosome; a cryoelectron microscopic reconstruction of the rabbit reticulocyte ribosome at 21 Å resolution. *J. Mol. Biol.* **279:** 403.

Dube P., Wieske M., Stark H., Schatz M., Stahl J., Zemlin F., Lutsch G., and van Heel M. 1998b. The 80S rat liver ribosome at 25 Å resolution by electron cryomicroscopy and angular reconstitution. *Structure* **6:** 389.

Gabashvili I.S., Agrawal R.K., Spahn C.M., Grassucci R.A., Svergun D.I., Frank J., and Penczek P. 2000. Solution structure of the E. coli 70S ribosome at 11.5 Å resolution. *Cell* **100:** 537.

Garrett R.A., Douthwaite S.R., Liljas A., Matheson A.T., Moore P.B., and Noller H.F., Eds. 2000. *The ribosome: Structure, function, antibiotics, and cellular interactions*, ASM Press, Washington D.C.

Gomez-Lorenzo M.G., Spahn C.M., Agrawal R.K., Grassucci R.A., Penczek P., Chakraburtty K., Ballesta J.P., Lavandera J.L., Garcia-Bustos J.F., and Frank J. 2000. Three-dimensional cryo-electron microscopy localization of EF2 in the *Saccharomyces cerevisiae* 80S ribosome at 17.5 Å resolution. *EMBO J.* **19:** 2710.

Guenneugues M., Caserta E., Brandi L., Spurio R., Meunier S., Pon C.L., Boelens R., and Gualerzi C.O. 2000. Mapping the fMet-tRNA$_f^{Met}$ binding site of initiation factor IF2. *EMBO J.* **19:** 5233.

Hershey J.W.B. and Merrick W.C. 2000. Pathway and mechanism of initiation of protein synthesis. In *Translational control of gene expression* (ed. N. Sonenberg et al.), p. 33. Cold Spring Harbor Laboratory Press, Cold Spring Harbor, New York.

Janssen G.M. and Möller W. 1988. Kinetic studies on the role of elongation factors 1β and 1γ in protein synthesis. *J. Biol. Chem.* **263:** 1773.

Janssen G.M.C., van Damme H.T.F., Kriek J., Amons R., and Möller W. 1994. The subunit structure of elongation factor 1 from *Artemia:* Why two α-chains in this complex? *J. Biol. Chem.* **269:** 31410.

Jensen M., Cool R.H., Mortensen K.K., Clark B.F.C., and Parmeggiani A. 1989. Structure-function relationships of elongation factor Tu. Isolation and activity of the guanine-nucleotide-binding domain. *Eur. J. Biochem.* **182:** 247.

Kawashima T., Berthet-Colominas C., Wulff M., Cusack S., and Leberman R. 1996. The structure of the *Escherichia coli* EF-Tu:EF-Ts complex at 2.5 Å resolution. *Nature* **379:** 511.

Kim K.K., Min K., and Suh S.W. 2000. Crystal structure of the ribosome recycling factor from *Escherichia coli. EMBO J.* **19:** 2362.

Kinzy T.G. and Woolford J.L., Jr. 1995. Increased expression of *Saccharomyces cerevisiae* translation elongation factor 1α bypasses the lethality of a TEF5 null allele encoding elongation factor 1β. *Genetics* **141:** 481.

Kisselev L.L. and Buckingham R.H. 2000. Translational termination comes of age. *Trends Biochem. Sci.* **25:** 561.

Kjeldgaard M. and Nyborg J. 1992. Refined structure of elongation factor EF-Tu from *Escherichia coli. J. Mol. Biol.* **223:** 721.

Kjeldgaard M., Nyborg J., and Clark B.F.C. 1996. The GTP-binding motif: Variations on a theme. *FASEB J.* **10:** 1347.

Kjeldgaard M., Nissen P., Thirup S., and Nyborg J. 1993. The crystal structure of elongation factor EF-Tu from *Thermus aquaticus* in the GTP conformation. *Structure* **1:** 35.

Kraulis P.J. 1991. MOLSCRIPT: A program to produce both detailed and schematic plots of protein structures. *J. Appl. Cryst.* **24:** 946.

Liljas A. 1996. Protein synthesis: Imprinting through molecular mimicry. *Curr. Biol.* **6:** 247.

Liljas A. and Al-Karadaghi S. 1997. Structural aspects of protein synthesis. *Nat. Struct. Biol.* **4:** 767.

Marcotrigiano J., Lomakin I.B., Sonenberg N., Pestova T.V., Hellen C.U., and Burley S.K. 2001. A conserved HEAT domain within eIF4G directs assembly of the translation initiation machinery. *Mol. Cell* **7:** 193.

Matadeen R., Patwardhan A., Gowen B., Orlova E.V., Pape T., Cuff M., Mueller F., Brimacombe R., and van Heel M. 1999. The *Escherichia coli* large ribosomal subunit at 7.5 Å resolution. *Structure* **7:** 1575.

Merrick W.C. and Nyborg J. 2000. The protein biosynthesis elongation cycle. In *Translational control of gene expression* (ed. N. Sonenberg et al.,), p. 89. Cold Spring Harbor Laboratory Press, Cold Spring Harbor, New York.

Meunier S., Spurio R., Czisch M., Wechselberger R., Guenneugues M., Gualerzi C.O., and Boelens R. 2000. Structure of the fMet-tRNA^{fMet}-binding domain of *B. stearothermophilus* initiation factor IF2. *EMBO J.* **19:** 1918.

Negrutskii B.S. and Deutscher M.P. 1992. A sequestered pool of aminoacyl-tRNA in mammalian cells. *Proc. Natl. Acad. Sci.* **89:** 3601.

Negrutskii B.S., Budkevich T.V., Shalak V.F., Turkovskaya G.V., and El'Skaya A.V. 1996. Rabbit translation factor 1α stimulates the activity of homologous aminoacyl-tRNA synthetase. *FEBS Lett.* **382:** 18.

Nissen P., Kjeldgaard M., and Nyborg J. 2000a. Macromolecular mimicry. *EMBO J.* **19:** 489.

Nissen P., Thirup S., Kjeldgaard M., and Nyborg J. 1999. The crystal structure of Cys-tRNA^{Cys}:EF-Tu:GDPNP reveals general and specific features in the ternary complex and in tRNA. *Structure* **7:** 143.

Nissen P., Hansen J., Ban N., Moore P.B., and Steitz T.A. 2000b. The structural basis of ribosome activity in peptide bond synthesis. *Science* **289:** 920.

Nissen P., Kjeldgaard M., Thirup S., Polekhina G., Reshetnikova L., Clark B.F.C., and Nyborg J. 1995. Crystal structure of the ternary complex of Phe-tRNA^{Phe}, EF-Tu, and a GTP analog. *Science* **270:** 1464.

Nolte R.T., Conlin R.M., Harrison S.C., and Brown R.S. 1998. Differing roles for zinc fingers in DNA recognition: Structure of a six-finger transcription factor IIIA complex. *Proc. Natl. Acad. Sci.* **95:** 2938.

Nyborg J. and Liljas A. 1998. Protein biosynthesis: Structural studies of the elongation cycle. *FEBS Lett.* **430:** 95.

Ogle J.M., Brodersen D.E., Clemons Jr. W.M., Tarry M.J., Carter A.P., and Ramakrishnan V. 2001. Recognition of cognate transfer RNA by the 30S ribosomal subunit. *Science* **292:** 897.

Ohtsuki T., Watanabe Y.Y., Takemoto C., Kawai G., Ueda T., Kita K., Kojima S., Kaziro Y., Nyborg J., and Watanabe K. 2001. An "elongated" translation elongation factor Tu for truncated tRNAs in nematode mitochondria. *J. Biol. Chem.* **276:** 21571.

Pérez J.M.J., Siegal G., Kriek J., Hård K., Dijk J., Canters G.W., and Möller W. 1999. The solution structure of the guanine nucleotide exchange domain of elongation factor 1β reveals a striking resemblance to that of EF-Ts from *Escherichia coli*. *Structure* **7:** 217.

Pioletti M., Schluenzen F., Harms J., Zarivach R., Gluhmann M., Avila H., Bashan A., Bartels H., Auerbach T., Jacobi C., Hartsch T., Yonath A., and Franceschi F. 2001. Crystal structures of complexes of the small ribosomal subunit with tetracycline, edeine and IF3. *EMBO J.* **20:** 1829.

Polekhina G., Thirup S., Kjeldgaard M., Nissen P., Lippmann C., and Nyborg J. 1996. Helix unwinding in the effector region of elongation factor EF-Tu:GDP. *Structure* **4:** 1141.

Ramakrishnan V. and White S.W. 1998. Ribosomal protein structures: Insights into the architecture, machinery and evolution of the ribosome. *Trends Biochem. Sci.* **23:** 208.

Roll-Mecak A., Cao C., Dever T.E., and Burley S.K. 2000. X-ray structures of the universal translation initiation factor IF2/eIF5B: Conformational changes on GDP and GTP binding. *Cell* **103:** 781.

Schluenzen F., Tocilj A., Zarivach R., Harms J., Gluehmann M., Janell D., Bashan A., Bartels H., Agmon I., Franceschi F., and Yonath A. 2000. Structure of functionally activated small ribosomal subunit at 3.3 Å resolution. *Cell* **102:** 615.

Selmer M., Al-Karadaghi S., Hirokawa G., Kaji A., and Liljas A. 1999. Crystal structure of *Thermotoga maritima* ribosome recycling factor: A tRNA mimic. *Science* **286:** 2349.

Sonenberg N., Hershey J.W.B., and Mathews M.B., Eds. 2000. *Translational control of gene expression*, Cold Spring Harbor Laboratory Press, Cold Spring Harbor, New York.

Song H., Parsons M.R., Rowsell S., Leonard G., and Philips S.E.V. 1999. Crystal structure of intact elongation factor EF-Tu from *Escherichia coli* in GDP conformation at 2.05 Å resolution. *J. Mol. Biol.* **285:** 1245.

Song H., Mugnier P., Das A.K., Webb H.M., Evans D.R., Tuite M.F., Hemmings B.A., and Barford D. 2000. The crystal structure of human eukaryotic release factor eRF1: Mechanism of stop codon recognition and peptidyl-tRNA hydrolysis. *Cell* **100:** 311.

Stark H., Rodnina M.V., Rinke-Appel J., Brimacombe R., Wintermeyer W., and van Heel M. 1997a. Visualization of elongation factor Tu on the *Escherichia coli* ribosome. *Nature* **389:** 403.

Stark H., Orlova E.V., Rinke-Appel J., Jünke N., Mueller F., Rodnina M., Wintermeyer W., Brimacombe R., and van Heel M. 1997b. Arrangement of tRNAs in pre- and posttranslational ribosomes revealed by electron cryomicroscopy. *Cell* **88:** 19.

Swart G.W., Parmeggiani A., Kraal B., and Bosch L. 1987. Effects of the mutation glycine-222-aspartic acid on the functions of elongation factor Tu. *Biochemistry* **26:** 2047.

van Damme H.T., Amons R., Karssies R., Timmers C.J., Janssen G.M., and Möller W. 1990. Elongation factor 1β of *Artemia*: Localization of functional sites and homology to elongation factor 1δ. *Biochim. Biophys. Acta* **1050:** 241.

van Heel M. 2000. Unveiling ribosomal structures: The final phases. *Curr. Opin. Struct. Biol.* **10:** 259.

Wang Y., Jiang Y., Meyering-Voss M., Sprinzl M., and Sigler P.B. 1997. Crystal structure of the EF-Tu:EF-Ts complex from *Thermus thermophilus*. *Nat. Struct. Biol.* **4:** 650.

Watanabe K. and Osawa S. 1995. tRNA sequences and variations in the genetic code. In *tRNA: Structure, biosynthesis and function* (ed. D. Söll and U. RajBhandary), p. 225. ASM Press, Washington, D.C.

Wimberley B.T., Brodersen D.E., Clemens W.M., Morgan-Warren R.J., Carter A.P., Vonrhein C., Hartsch T., and Ramakrishnan V. 2000. Structure of the 30S ribosomal subunit. *Nature* **407:** 327.

Yoshida T., Uchiyama S., Nakano H., Kashimori H., Kijima H., Ohshima T., Saihara Y., Ishino T., Shimahara H., Yoshida T., Yokose K., Ohkubo T., Kaji A., and Kobayashi Y. 2001. Solution structure of the ribosome recycling factor from *Aquifex aeolicus*. *Biochemistry* **40:** 2387.

Yusupov M.M., Yusupova G.Z., Baucom A., Lieberman K., Earnest T.N., Cate J.H.D., and Noller H.F. 2001. Crystal structure of the ribosome at 5.5 Å resolution. *Science* **292:** 883.

Translation Elongation Factor 1 Functions in the Yeast *Saccharomyces cerevisiae*

M. ANAND, L. VALENTE, A. CARR-SCHMID, R. MUNSHI,
O. OLAREWAJU, P.A. ORTIZ, AND T.G. KINZY*

*Department of Molecular Genetics and Microbiology, UMDNJ Robert Wood Johnson Medical School and
The Cancer Institute of New Jersey, Piscataway, New Jersey 08854

During the process of protein synthesis, aminoacyl-tRNAs (aa-tRNA) are delivered to the translating ribosome by soluble protein synthesis factors. Whereas the initiator met-tRNA is delivered by a specialized initiation factor, all other standard aa-tRNAs are delivered by one protein, the G-protein translation elongation factor 1A. In prokaryotes and eukaryotes this protein, EF1A (formerly EF-Tu) or eEF1A (formerly EF-1α), respectively, binds aa-tRNA in a GTP-specific manner and delivers it to the ribosome (Fig. 1). The kinetics of the delivery step have been elegantly dissected for the prokaryotic homolog EF-Tu (for review, see Rodnina and Wintermeyer 2001). The yeast *Saccharomyces cerevisiae*, on the other hand, has proven an excellent system to address elongation in eukaryotic organisms with a particular focus on the fidelity of reading frame maintenance, misreading, and nonsense suppression. The availability of the cloned genes for the

subunits of eEF1 (Table 1) and the genetic approaches available in yeast have opened a window on this process in vivo. Furthermore, as the first eukaryotic eEF1 protein whose crystal structure has been solved (Andersen et al. 2000), an integrated structure/function approach has allowed development of hypotheses on the role of the eEF1 complex in translational fidelity. Based on the role of eEF1A in elongation, multiple types of fidelity are affected, allowing analysis of numerous events occurring at the ribosome. Further work in yeast has shown that the other subunits of the eEF1 complex can affect translational fidelity.

The eEF1B complex performs guanine nucleotide exchange for eEF1A, and consists of two subunits in *S. cerevisiae* (Fig. 1). eEF1Bα (formerly EF-1β) performs the catalytic function of the complex, and mutations in this protein affect nonsense suppression. eEF1Bγ (for-

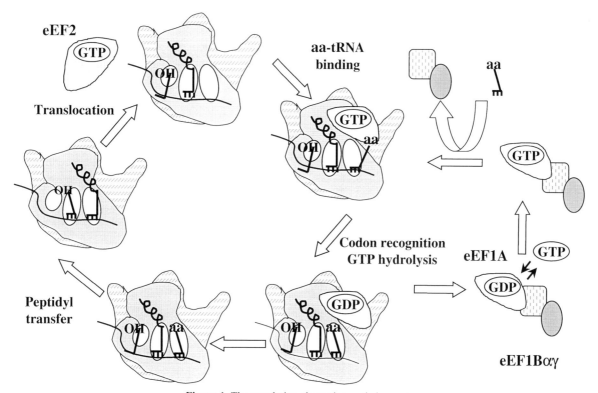

Figure 1. The translation elongation cycle in yeast.

Table 1. Genes Encoding the Translation Elongation Factor 1 Subunits in *S. cerevisiae* and Conservation with the Human Homologs

eEF1	Genes	ORF name	Identity (similarity) to human	Reference
A	*TEF1*	*YPR080W*	81% (89%)	Schirmaier and Phillipson (1984)
	TEF2	*YBR118W*		
Bα	*EFB1 (TEF5)*	*YAL003W*	55% (77%)	Hiraga et al. (1993)
Bγ	*CAM1 (TEF3)*	*YPL048W*	33% (51%)	Kambouris et al. (1993); Kinzy et al. (1994)
	TEF4	*YKL081W*	33% (51%)	
Bγ-like		*YGR201C*	27% (44%)	Guerreiro et al. (1996)

merly EF-1γ), although associated with eEF1Bα in all organisms studied to date, is not essential in yeast (Kinzy et al. 1994). There are several isoforms of eEF1Bγ in yeast, and altered expression of at least one also affects nonsense suppression. Thus, all three subunits in yeast affect accurate gene expression.

eEF1A also provides the basis for analysis of several other important cellular events. eEF1A is frequently identified in association with complexes performing a variety of cellular functions aside from protein synthesis, as well as events in viral processes. Perhaps the best characterized in vitro is the unique bundling of actin (for review, see Condeelis 1995). This function is conserved from fungi through plants and metazoans, and effects of eEF1A on actin can be seen in vivo using yeast. The study of this association may provide new insight into the effect of this interaction on the cell and the biological significance of the link between the cytoskeleton and the protein synthesis machinery beyond localization of mRNA. Last, elegant structural examples for molecular mimicry (Nissen et al. 1995) make the number of eEF1A-like proteins apparent in organisms such as *S. cerevisiae* an excellent starting place for understanding the structural and functional conservation of these G proteins.

METHODS

Substitution of eEF1A residues E286 and E291 to alanine were prepared in plasmid pTKB351 (*TEF2 TRP1*) and E286 to lysine on pTKB164 (*TEF1 TRP1*) using the QuickChange Kit (Stratagene). The oligonucleotides used were (5′-GCTGGTGTTACCACCGCGGTCAAGTCGT-TGAAATGC-3′) for E286A, (5′-CCACTGAAGTCAA GTCCGTTGCCATGCATCACGAA-3′) for E291A, and (5′-CTTTTGCCCCAGCCGGTGTTACCACTAAA-GT-CAAGTCCGTTG-3′) for E286K, producing plasmids pTKB536, pTKB542, and pTKB585, respectively. These plasmids were transformed into MC213 (*Matα, leu2-3, 112, his4-713, ura3-52, lys2-20, met2-1, trp1Δ, tef1::LEU2, tef2Δ2, pTEF2:URA3*) and loss of the helper plasmid monitored by growth on 5-fluoro-orotic acid to produce strains TKY550 (E286A), TKY562 (E291A), and TKY588 (E286K). Growth on YPD at different temperatures was assayed as described previously (Carr-Schmid et al. 1999b). Reading-frame maintenance was monitored by the ability to grow on complete synthetic media lacking methionine (C-met), indicating suppression of the *met2-1* nonprogrammed +1 frameshift allele.

GENETIC ANALYSIS OF THE ELONGATION FACTORS IN YEAST

Translational Fidelity during Elongation: Multiple Effects and Assays

Translational fidelity is monitored at several points during protein synthesis. During initiation, fidelity can refer to either the selection of the appropriate AUG for initiation or the use of an AUG as opposed to other codons. During elongation, an even wider range of events can occur. First, after start-site selection, the correct three-base reading frame must be maintained. A shift in reading frame of one or two bases in either the 5′ or 3′ direction will lead to a quick encounter with a stop codon and termination to produce an incomplete polypeptide. Furthermore, many viruses utilize programmed frameshifting sites as a mechanism to express the proper proportions of the viral proteins, exemplified in yeast by the –1 signal in the LA virus (Dinman 1995) and the +1 signal in the Ty retrotransposons (Farabaugh 1995). Work in yeast has identified not only the *cis*-acting sequences of the viral mRNA required, but also cellular proteins that function as *trans*-acting factors critical for efficient viral maintenance. Second, during elongation, insertion of each aa-tRNA is monitored to assure that the correct amino acid is incorporated into the protein. Misincorporation of the wrong amino acid can happen by mischarging of a tRNA; however, it can also occur if the ribosome fails to recognize an improper codon–anticodon interaction, allowing the deposition of an incorrect aa-tRNA in the A site. Last, the termination process requires that only the three stop codons signal the end of protein synthesis. Nonsense suppression, or readthrough of a stop codon, allows incorporation of an amino acid. Based on the role of the eEF1 complex in protein synthesis, its subunits have the potential to affect all of the processes occurring during elongation.

Elongation Factor 1A: Effects on Multiple Types of Translational Fidelity

As a central component in the delivery of aa-tRNA, not surprisingly, eEF1A can affect reading-frame maintenance, misincorporation, and nonsense suppression. Sandbaken and Culbertson (1988) isolated nine mutant alleles in the *TEF2* gene encoding eEF1A that function as dominant suppressors of a nonprogrammed +1 frameshift in yeast, measured by readthrough of a metabolic reporter gene (i.e., *met2-1*) containing a one-base insertion and monitoring growth on media lacking that amino acid. Analysis of the recessive phenotypes of these genes indicated that not all are functional as the only form of the protein in vivo; however, the functional mutants demonstrate altered sensitivity to paromomycin, a phenotype linked to translational fidelity in yeast (Palmer et al. 1979; Singh et al. 1979). The recessive effects of these mutations on programmed ribosomal frameshifting in the yeast viral systems have shown that not all mutants that read through a nonprogrammed frameshift signal affect the programmed signals. The *TEF2-4* allele, which alters E122 to K, results in a significant reduction in –1 programmed frameshifting, with lesser effects seen for *TEF2-9* (E295K), *TEF2-7* (T142I), and *TEF2-10* (E122Q) (Dinman and Kinzy 1997). Recessive allele-specific effects are also seen for a programmed +1 frameshift signal for alleles *TEF2-2* (E317K) and *TEF2-3* (E40K). Additionally, the dominant programmed +1 frameshift phenotypes of these mutants are also different between alleles, which may be linked to eEF1A functions in error correction (Farabaugh and Vimaladithan 1998). Thus, unique alleles that affect +1 versus –1 programmed frameshifting may indicate specific functions of eEF1A required for the two very different *cis*-acting signals used in these two viral systems.

Alterations in the NXKD GTP-binding consensus element of eEF1A that alter basic functions of the protein related to nucleotide binding affect misreading as monitored in vitro by misincorporation of leucine during a poly(U)-based phenylalanine polymerization assay (Cavallius and Merrick 1998). Substitution of N153T increases misincorporation 2.3-fold, and several others show 1.5- to 1.9-fold effects. As both decoding and termination occur at a similar state of the A site, a functional link between these two processes is also possible. Initial studies have indicated that a reduction in the gene dosage of eEF1A from the normal complement of two genes to one results in enhanced fidelity as monitored by readthrough of a metabolic reporter containing a nonsense mutation (Song et al. 1989). In vivo, alterations of N153T, D156N, and N153T/D156E result in omnipotent nonsense suppression, increasing readthrough at all three stop codons by as much as 4-fold measured quantitatively with a *lacZ* reporter construct with a premature stop codon (Carr-Schmid et al. 1999a). It remains to be dissected how much of this is due to misreading versus direct effects on the termination process, in particular since both misreading and nonsense suppression are affected in the NKXD mutations.

It is now possible to integrate a structural interpretation of these mutants and their in vivo effects based on the recent co-crystal structure of yeast eEF1A and the catalytic fragment of eEF1Bα (Fig. 2) (Andersen et al. 2000). Mutations in the GTP-binding domain that affect nonsense suppression (yellow circles) and +1 reading frame maintenance (gray circles) are perhaps easiest to interpret based on the proposed role of nucleotide hydrolysis on fidelity. In vitro, the N153T mutant shows a 4.6-fold in-

crease in intrinsic GTPase activity. The predicted model of aa-tRNA binding is based on the structure (Andersen et al. 2000) and aa-tRNA cross-linking data (Kinzy et al. 1992). This model proposes a channeling of aa-tRNA directly to eEF1A; however, it also potentially links domain II mutants in perhaps sensing signals from the tRNA or in maintaining tRNA presentation for reading frame maintenance (Fig. 3). The original E291K mutation

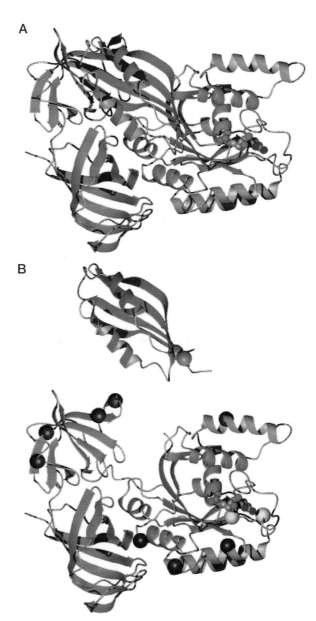

Figure 2. (*A*) X-ray crystal structure of *S. cerevisiae* eEF1A domains I (*magenta*), II (*blue*), and III (*red*) complexed with the catalytic fragment of eEF1Bα (*green*) (Andersen et al. 2000). (*B*) Mutations in *S. cerevisiae* eEF1A and the catalytic fragment of eEF1Bα that affect the fidelity of protein synthesis. Gray balls indicate sites of mutations that result in dominant suppression of a nonprogrammed +1 frameshift (Sandbaken and Culbertson 1988), yellow balls indicate increased nonsense suppression (Carr-Schmid et al. 1999a), and red balls indicate reduced nonsense suppression (Carr-Schmid et al. 1999b).

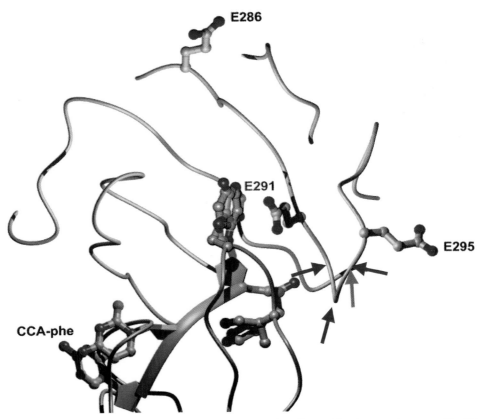

Figure 3 Mutations in eEF1A near the proposed aminoacyl-tRNA-binding site affect reading frame maintenance. The CCA-phe end of the phe-tRNA$_{phe}$ from the ternary complex of the prokaryotic homolog EF1A (Nissen et al. 1995) in gold is overlaid with eEF1Bα (*gray*). Side chains are indicated for the *TEF2-1* (E286K), *TEF2-16* (E291K), and *TEF2-9* (E295K) alleles (Sandbaken and Culbertson 1988). Green arrows indicate the three residues homologous to those in rabbit eEF1A, one of which cross-linked to four different aa-tRNAs (Kinzy et al. 1992). The blue arrow indicates the residue homologous to that in rabbit eEF1A which cross-linked to an activated lysine tRNA (Kinzy et al. 1992).

(*TEF2-16*) confers dominant +1 frameshift suppression but is inviable as the only form of the protein whereas E286K (*TEF2-1*) is dominant and viable but slow growing (Sandbaken and Culbertson 1988). The location of E291 at the eEF1Bα-binding site and the proposed aa-tRNA-binding site further supports a critical role for this residue. Previously using a *los- arc-* tRNA export mutant, eEF1A overexpression was determined to suppress tRNA mislocalization (Grosshans et al. 2000). Overexpression of the mutant forms encoded by *TEF2-1* and *TEF2-16* were unable to suppress the defect, perhaps due to reduced aa-tRNA-binding and thus consistent with the aa-tRNA-binding model. These authors propose channeling by eEF1A from the earliest times as aa-tRNA is made available in the cell during export.

To further define the roles of residues E286 and in particular E291, alanine substitutions were prepared and analyzed in vivo. Comparison of the growth of wild-type yeast versus strains expressing mutant forms of eEF1A indicate that only the E286K mutant shows slow growth at all temperatures (Fig. 4). Both E286K and E286A show nonprogrammed +1 frameshift suppression of the *met2-1* allele as monitored by growth on C-met. This growth assay may underrepresent the difference in frameshift suppression, as the E286K mutant strain grows dramatically

more slowly at the temperatures shown, yet displays greater growth on C-met. The E291A mutation is viable and shows no effect on growth or frameshift suppression. Thus, the charge reversal from negative to positive is likely to be important for the dramatic fidelity phenotype, but the side chain itself is not essential in vivo. Future genetic approaches utilizing mutants in eEF1A that alter fidelity can lead to other cellular factors such as modifying proteins or ribosomal components that affect the ability of eEF1A to aid in maintaining translational accuracy.

Guanine Nucleotide Exchange during Elongation: Mechanisms and Effects on Translation Fidelity

As a guanine nucleotide exchange factor, eEF1Bα's main role in the cell is to regulate the activity of eEF1A by catalyzing the release of GDP in exchange for GTP. eEF1Bα has been shown to interact with eEF1A through its carboxyl terminus and was demonstrated to catalyze the exchange in vitro (van Damme et al. 1990). Utilizing the co-crystal structure of yeast eEF1A and the catalytic fragment of eEF1Bα (Andersen et al. 2000), it is clear that mutations in eEF1Bα behave differently dependent on their location on either side of the protein. At one face

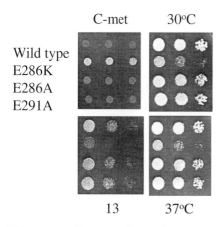

C-met   30°C

Wild type
E286K
E286A
E291A

13   37°C

Figure 4. Yeast expressing mutant forms of eEF1A located near the proposed aa-tRNA-binding site of eEF1A alter reading frame maintenance. Isogenic yeast strains deleted for both chromosomal genes encoding eEF1A and supported for growth by a plasmid-encoded gene expressing wild-type eEF1A or E286K, E286A, or E291A (*top row* to *bottom*) forms of eEF1A were grown on C-met at 30°C or on YEPD at 13°C, 30°C, or 37°C. The E291K mutation does not support growth as the only form of eEF1A (Sandbaken and Culbertson 1988).

of eEF1Bα, lysine residues 120 and 205 are positioned at the guanine nucleotide-binding site on eEF1A in a region proposed to be important for guanine nucleotide exchange. Mutational analysis determined that K205 is critical for eEF1Bα's function, because a change to an alanine is not functional as the only form of the protein but is still able to bind eEF1A in vitro (Andersen et al. 2001). The essential K205 is buried in domain I of eEF1A and destroys the binding site for the Mg^{++} ion associated with the bound guanine nucleotide, which then allows the release of the GDP molecule. Random PCR mutagenesis targeted to K120 and S121 of a highly conserved region within the carboxyl terminus of eEF1Bα yielded 21 different alleles with varying effects on the cell (Carr-Schmid et al. 1999b). Analysis of a subset of mutant strains showed a decreased overall rate of total protein synthesis and sensitivity to translation elongation inhibitors. In addition, these strains demonstrate reduced nonsense suppression of all three stop codons based on *lacZ* reporter constructs and the inability to misread the *lys2-801* UAG allele. This may indicate that mutations in residues of eEF1Bα that are interacting with the G domain of eEF1A could regulate eEF1Bα's activity in the traditional exchange role and thus indirectly affect fidelity based on the pool of active eEF1A. The crystal structure indicates that the F163 residue of eEF1Bα makes important interactions with a hydrophobic pocket in domain II of eEF1A that is hypothesized to bind the aminoacylated CCA end of the tRNA (Fig. 3). This residue, when mutated to an alanine, behaves similarly in some aspects to mutations in K120 and S121, such as a reduction in overall translation rates and sensitivity to translation elongation inhibitors. Interestingly, the F163A mutant strain is temperature sensitive, as opposed to the K120 S121 mutants that are predominantly cold sensitive, and shows no significant effect on the level of

nonsense suppression. It was expected that the F163A mutation would display effects similar to the K120 S121 mutations if the enhanced fidelity is an indirect effect based on the level of active eEF1A and translation rate. However, the ability of mutations on different faces of eEF1Bα to confer different effects suggests that the fidelity effect is not just a result of slowed elongation. Further analysis is needed to determine how eEF1Bα affects fidelity, what specific aspect of its activity is compromised in the mutant proteins and guanine nucleotide exchange, and the consequences for efficient and accurate gene expression.

Altered eEF1Bγ Levels and Fidelity, Through or Around eEF1Bα

The eEF1Bγ subunit is found associated with eEF1Bα in a variety of eukaryotic organisms. Whereas eEF1Bα performs the catalytic function in vitro, the role of eEF1Bγ is less clear. Prior studies in *Artemia salina* have demonstrated that eEF1Bγ can stimulate eEF1Bα's activity (Janssen and Moller 1988). However, neither of the two full-length forms of eEF1Bγ, encoded by the *TEF3* and *TEF4* genes in *S. cerevisiae*, is essential (Kinzy et al. 1994). The presence of two isoforms may be unique to *S. cerevisiae*, as *Caenorhabditis elegans*, for example, appears to have one form (Chervitz et al. 1998). Like eEF1A and eEF1Bα, it appears that eEF1Bγ levels can affect the accuracy of the translation process. *TEF4* was identified as a gene that, when overexpressed from a *GAL* promoter, results in antisuppression, or reduced readthrough of a stop codon (Benko et al. 2000), although the potential effect of *TEF3* has not been addressed. It is of interest that this phenotype is similar to the effect of mutations in eEF1Bα that affect the catalytic face of the proteins. Thus, although eEF1Bγ is not essential in vivo, it may help modulate the activity of eEF1Bα and, as a result, similarly affect nonsense suppression. The use of yeast strains viable in the absence of the normally essential eEF1Bα protein by the presence of an extra gene encoding eEF1A (Kinzy and Woolford 1995) can determine whether this effect requires the complete eEF1B complex or is a separate function of eEF1Bγ.

eEF1Bγ-like Proteins and the Link to Glutathione *S*-transferases

The Tef3p and Tef4p isoforms of eEF1Bγ are dissimilar from the other examples of duplicate genes encoding components of the translational apparatus in yeast. The genes encoding eEF1A, for example, have identical coding regions, and many pairs of ribosomal proteins differ by only a few amino acids (for review, see Woolford and Warner 1991). Tef3p and Tef4p, however, are only 60% identical, most of which is due to the 90% identity of the carboxyl termini. The amino terminus of several eukaryotic eEF1Bγ proteins was identified to contain similar motifs to the glutathione-binding site of θ class glutathione *S*-transferases (Koonin et al. 1994). The GST-like motif, proposed to be involved in glutathione binding, is shown as an alignment in Figure 5A compared to *Issatchenkia orientalis* GST and is conserved in all three yeast eEF1Bγ-like genes. This motif may link the well-known role of GSTs in detoxification of radicals with posttranscriptional control, or as previously proposed, it may function in dimerization or formation of multisubunit complexes including synthetases (Koonin et al. 1994). Furthermore, yeast have an additional gene, YGR201C (Guerreiro et al. 1996), encoding an ORF that is similar to only the amino terminus of eEF1Bγ (Fig. 5A). Links between the stress response and the eEF1B complex are also supported by the identification of the *PPZ1*-encoded protein as a phosphatase targeting Ser-86 of eEF1Bα (de Nadal et al. 2001). The Ppz1 protein is a phosphatase involved in a variety of processes, including salt tolerance. Cells lacking *PPZ1* show an increase in a nonphosphorylated form of eEF1Bα and paromomycin resistance. Furthermore, cells expressing excess wild-type but not a S86A mutant form of eEF1Bα are resistant to the negative effects of *PPZ1* overexpression. Gene chip experiments may begin to identify cellular conditions that require the eEF1Bγ family of proteins. In this case, whereas the mRNAs encoding the full-length eEF1Bγs (encoded by *TEF3* and *TEF4*) appear to change

expression coordinately and in a similar way to other components of the translational apparatus, YGR201C is unique. For example, YGR201C is up-regulated in response to diauxic shift (DeRisi et al. 1997) and histone depletion (Wyrick et al. 1999). In these studies, the mRNAs encoded by *TEF3* and *TEF4* are down-regulated similarly to other components of the elongation pathway such as eEF1A, eEF1Bα, eEF2, and eEF3. YGR201C exhibits induced expression following an osmotic shift, whereas the mRNAs of eEF1Bα and eEF2 are diminished (Rep et al. 2000). These experiments show different patterns of gene expression between YGR201C and the other proteins of the translation elongation machinery. Last, it is possible that the YGR201C protein is related to eEF1Bγ yet functioning in a different cellular compartment. The original identification placed the initiation codon at a position which is actually 50 amino acids downstream of the initiation codon predicted by the annotated yeast genome. It is of interest that this leader is similar in length to mitochondrial leader sequences and has the predicted mitochondrial intermediate peptidase site RX↓(F/L/I)XX(T/S/G)XXXX↓ (Branda and Isaya 1995). The sequence **RK↓LQMSDGTL↓** in YGR201C would result in removal of the first 55 amino acids of the ORF. The biological function of both the GST-like motif and the eEF1Bγ-like protein may expand our understanding of this protein class.

ELONGATION FACTOR HOMOLOGS IN YEAST

The conservation of domains and the low but significant homology found in the family of eEF1Bγ-related proteins may indicate nonoverlapping but related functions. eEF1A is also one of a family of related proteins in *S. cerevisiae*. The G-protein consensus elements found in eEF1A are conserved in a broad range of G proteins such as Ras, and in the heterotrimeric G proteins (Dever et al. 1987). The elongation factor class of G proteins contains bona fide translation factors, including eEF1A and eEF2. In yeast there are several other related gene products. The genes encoding those proteins most similar to eEF1A include not only mitochondrial EF-Tu (*TUF1*), but also *SKI7* and *HBS1*. This is not unique to eEF1A, as there are several genes related to eEF2 including not only mitochondrial EF-G (*MEF2*), but also *SNU114*, *GUF1*, and *MEF1*. Those closest to eEF1A are also most similar to eRF3 (Fig. 6). Both Ski7p and Hbs1p have an amino-terminal extension prior to the most highly conserved region in the G domain. However, phylogenetic analysis utilizing 16 eRF3 and 7 Hbs1p species indicates that no obvious sequence similarity between the amino-terminal portions of Hbs1p and eRF3 exists (Inagaki and Doolittle 2000). There is little to no divergence in the length of the carboxy-terminal domain. It remains to be seen whether there is a close link between the proteins in this class and their roles in vivo. Interestingly, both *SKI7* and *HBS1* are linked to mRNA metabolism events. *SKI7* was originally isolated by the *super killer* phenotype, an increased ability to kill sensitive cells (Ridley et al. 1984). Further work

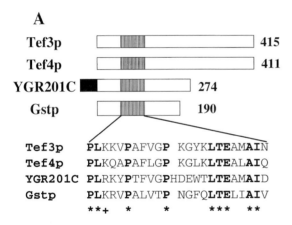

Figure 5. Multiple eEF1Bγ-like proteins are encoded in the *S. cerevisiae* genome. (*A*) Cartoons indicate the different isoforms of eEF1Bγ compared to GST from *Issatchenkia orientalis*, where the solid box indicates the unique amino-terminal extension of the YGR201C ORF and the striped box indicates the glutathione-binding motif region shown in detail below, where * indicates identical residues between all 4 proteins and + indicates a conserved positively charged side chain. (*B*) The box indicates pairwise identity between the Tef3p (3), Tef4p (4), YGR201C (201C), and *I. orientalis* GST (GST).

		Identity/Similarity to eEF1A
eEF1A	458	100/100%
eRF3	685	37/57
Hbs1p	611	31/50
Ski7p	747	26/46

Figure 6. *S. cerevisiae* contains multiple elongation factor class G proteins related to eEF1A. Cartoons of eEF1A, eRF3, Hbs1p, and Ski7p are shown indicating the varying amino-terminal extensions (*gray boxes*), conserved G-domain (*dotted box*), and carboxy-terminal regions (*empty box*).

has demonstrated a role for this protein in mRNA degradation (van Hoof et al. 2000). The *HBS1* gene was determined to suppress the cold-sensitive growth defect of a strain lacking the two Hsp70 subfamily B genes *SSB1* and *SSB2*, ribosome-associated proteins (Nelson et al. 1992). The identification of putative *HBS1* orthologs from mouse and human (Wallrapp et al. 1998), and more recently from *Drosophila melanogaster, Schizosaccharomyces pombe,* and *Candida albicans* (Inagaki and Doolittle 2000), suggests *HBS1* is evolutionarily conserved. The examples of both the GST-like domain in three eEF1Bγ-like proteins and the large number of proteins related to the elongation factor class of G proteins may indicate either recruitment of domains for multiple purposes or the adaptation of one motif to multiple but related functions.

eEF1A AND THE ACTIN CYTOSKELETON: A LINK TO PROTEIN SYNTHESIS OR A NONTRANSLATIONAL FUNCTION

Actin Alterations Affect Protein Synthesis in Metazoans

A wealth of prior work in metazoan organisms has demonstrated a link between the cytoskeleton and the protein synthetic apparatus. These links occur at several levels. Studies in permeabilized cells demonstrate that the presence of the cellular organization results in protein synthesis rates that are higher than in cell-free extracts and are closer to intact cell levels (Negrutskii et al. 1994). Dissociation of the actin cytoskeleton, but not microtubules, by drugs or cold shock results in a significant reduction in total protein synthesis (Stapulionis et al. 1997). Second, mRNAs are frequently localized within the cell, and an interaction with the cytoskeleton can affect the translation of the mRNA (for review, see Hesketh 1994; Bassell and Singer 1997; Jansen 1999). Last, in some cases, there is an association of components of the translational apparatus. Ribosomes themselves have been linked to the cytoskeleton (Hesketh and Pryme 1991). For example, treating cells with extracellular matrix-coated beads results in the localization of ribosomes to the focal adhesion complexes and thus may result in expression of certain gene products (Chicurel et al. 1998). Additionally,

some of the soluble protein synthesis factors interact with cytoskeletal components. This includes eEF2 (Shestakova et al. 1991; Bektas et al. 1994), eEF1Bγ (Janssen and Moller 1988), and eEF1A (Yang et al. 1990; Moore et al. 1998).

eEF1A Binds and Bundles Actin in a Unique Conformation and Excludes aa-tRNA

The eEF1A–actin interaction has been well characterized in vitro (for review, see Condeelis 1995). This interaction, which occurs with both globular and filamentous actin (Dharmawardhane et al. 1991), is independent of the nucleotide-bound state of eEF1A for F-actin, whereas GTP inhibits binding to G-actin (Dharmawardhane et al. 1991). This results in unique bundles of actin that are predicted to exclude other actin-bundling proteins (Owen et al. 1992). Most importantly, the eEF1A–actin interaction is mutually exclusive to aa-tRNA binding (Liu et al. 1996). One question that remains is whether the localization of these components is required for the organization of the cytoskeleton, mRNA localization, regulated gene expression, or all.

Multiple Findings Link Actin and Translation/mRNA Metabolism in Yeast

The yeast *S. cerevisiae* has been used to study both basic aspects of cell biology including actin (for review, see Botstein et al. 1997) as well as protein synthesis and mRNA metabolism. In several cases, genetic and biochemical studies show points where these processes overlap. Several proteins with genetic links to actin function interact with components of the translational apparatus. Bailleul et al. identified Sla1p, a protein linked to cortical actin cytoskeleton organization and localization of the Rho1p GTPase mentioned below (Holtzman et al. 1993; Ayscough et al. 1999), via a two-hybrid interaction with the amino-terminal 113 amino acids of eRF3 (Bailleul et al. 1999). These residues of eRF3 are not required for the protein's function and instead are involved in [PSI] prion formation. It was hypothesized that this interaction may be involved in regulation of cellular structures. Additionally, treatment of yeast with low levels of the actin-filament-destabilizing drug Latrunculin A can cure the [PSI] phenotype (Bailleul-Winslett et al. 2000). Two-hybrid analysis has also identified an interaction between a subunit of the translation initiation factor 3 (eIF3) and Sla2p, a protein linked to cytoskeletal membrane assembly and Sla1p function (Palecek et al. 2001). Several cytoskeletal proteins are also proposed to confer phenotypes related to protein synthesis or RNA metabolism. A *sla1* null mutant, for example, is sensitive to translation inhibitors (Bailleul et al. 1999). Temperature-sensitive mutations in the *SLA2* gene mentioned above also affect mRNA metabolism by stabilizing a series of different mRNAs, including mRNA with short and long half-lives or a premature nonsense codon (Zuk et al. 1999). Thus, yeast, like other eukaroyotes, shows links between RNA metabolism and the cytoskeleton.

Actin and eEF1A Are Genetically Linked in Yeast

Recent work has produced bridges between actin and eEF1A through known actin-associated factors. In yeast, eEF1A interacts with Bni1, a downstream target of the Rho1p that helps regulate actin cytoskeleton reorganization through an interaction with profilin (Umikawa et al. 1998). This interaction inhibits the binding of eEF1A to globular actin, consistent with the role of profilin in binding actin monomers. eEF1A also interacts with the yeast-adenylyl-cyclase-associated protein CAP, which is associated with actin monomers and sequesters actin (Yanagihara et al. 1997). Additionally, eEF1A is a target for the Rho-associated-kinase (Rho-kinase) and binds to the myosin-binding subunit of myosin phosphatase (Izawa et al. 2000). Both of these proteins are linked to cellular structures. Last, eEF1A interacts with the zinc finger protein Zpr1p, and the loss of this interaction affects morphology and growth and results in an accumulation of cells at G_2/M, an effect perhaps linked to localization (Gangwani et al. 1998).

Work in the fission yeast *S. pombe* (Suda et al. 1999) and *S. cerevisiae* (Munshi et al. 2001) has demonstrated that the phenotypes associated with eEF1A overexpression are clearly related to the actin cytoskeleton. In *S. pombe,* the eEF1A effect was identified in a genetic screen for cDNAs that affect cell morphology when overexpressed. This effect was suppressed by excess actin, and destabilization of the actin filaments did not effect eEF1A localization within the cell. One possible explanation consistent with the cellular concentration of both proteins and the K_d of eEF1A for actin is that eEF1A is binding the G-form of actin, and thus affecting its ability to enter actin filaments. Here it is of note that most of the actin-associated proteins that interact with eEF1A affect G-actin as well. In *S. cerevisiae*, the effect of excess eEF1A was further shown to not involve total translation or translational fidelity, but a subset of actin mutants strains was particularly sensitive to eEF1A overexpression. Overall, these results support a biological consequence of eEF1A on the actin cytoskeleton. Further work assessing the question of whether this links eEF1A-localized translation or merely the recruitment of eEF1A to an additional function will now benefit from a genetic approach in yeast to complement the well-characterized in vitro interaction.

CONCLUSIONS

The yeast system has proven an excellent model system for studies of both the mechanism and regulation of protein synthesis and the basic principles of cell biology. With the emerging detailed molecular view of the ribosome, the genetic systems of yeast, mutant forms of key players in translational fidelity such as eEF1A, and systems to manipulate not only ribosomal protein genes but also the rRNA itself, new work is likely to shed light on the important effects of elongation factors on ribosome-associated functions. Furthermore, the link between the cytoskeleton and protein synthesis, both structurally and functionally, offers the potential to develop a new understanding of another area of posttranscriptional control of gene expression.

ACKNOWLEDGMENTS

We acknowledge Kimberly Kandl, Ishita Chatterjee, Lea Starita, Gregers Andersen, and Valerie Loik for contributions to the work, and Gregers Andersen for assistance with figures. T.G.K. is supported by National Institutes of Health grant GM-57483 and National Science Foundation grant MCB-9983565; L.V. by National Institutes of Health grant F31 GM-20445; and P.O. by National Institutes of Health grant R25 GM-55145.

REFERENCES

Andersen G.R., Valente L., Pedersen L., Kinzy T.G., and Nyborg J. 2001. Crystal structures of nucleotide exchange intermediates in the eEF1A-eEF1Bα complex. *Nat. Struct. Biol.* **8:** 531.

Andersen G.R., Pedersen L., Valente L., Chatterjee I., Kinzy T.G., Kjeldgaard M., and Nyborg J. 2000. Structural basis for nucleotide exchange and competition with tRNA in the yeast elongation factor complex eEF1A:eEF1Bα. *Mol. Cell* **6:** 1261.

Ayscough K.R., Eby J.J., Lila T., Dewar H., Kozminski K.G., and Drubin D.G. 1999. Sla1p is a functionally modular component of the yeast cortical actin cytoskeleton required for correct localization of both Rho1p-GTPase and Sla2p, a protein with talin homology. *Mol. Cell. Biol.* **10:** 1061.

Bailleul P.A., Newnam G.P., Steenbergen J.N., and Chernoff Y.O. 1999. Genetic study of interactions between the cytoskeletal assembly protein Sla1 and prion-forming domain of the release factor Sup35 (eRF3) in *Saccharomyces cerevisiae*. *Genetics* **153:** 81.

Bailleul-Winslett P.A., Newnam G.P., Wegrzyn R.D., and Chernoff Y.O. 2000. An antiprion effect of the anticytoskeletal drug latrunculin A in yeast. *Gene Expr.* **9:** 145.

Bassell G. and Singer R.H. 1997. mRNA and cytoskeletal filaments. *Curr. Opin. Cell Biol.* **9:** 109.

Bektas M., Nurten R., Gurel Z., Sayers Z., and Bermek E. 1994. Interactions of eukaryotic elongation factor 2 with actin: A possible link between protein synthetic machinery and cytoskeleton. *FEBS Lett.* **356:** 89.

Benko A.L., Vaduva G., Martin N.C., and Hopper A.K. 2000. Competition between a sterol biosynthetic enzyme and tRNA modification in addition to changes in the protein synthesis machinery causes altered nonsense suppression. *Proc. Natl. Acad. Sci.* **97:** 61.

Botstein D., Amberg D., Mulholland J., Huffaker T., Adams A., Drubin D., and Stearns T. 1997. The yeast cytoskeleton. In *The molecular and cellular biology of the yeast Saccharomyces,* vol. 3 (ed. J.R. Pringle et al.), p. 1. Cold Spring Harbor Laboratory Press, Cold Spring Harbor, New York.

Branda S.S. and Isaya G. 1995. Prediction and identification of new natural substrates of the yeast mitochondrial intermediate peptidase. *J. Biol. Chem.* **270:** 27366.

Carr-Schmid A., Durko N., Cavallius J., Merrick W.C., and Kinzy T.G. 1999a. Mutations in a GTP-binding motif of eEF1A reduce both translational fidelity and the requirement for nucleotide exchange. *J. Biol. Chem.* **274:** 30297.

Carr-Schmid A., Valente L., Loik V.I., Williams T., Starita L.M., and Kinzy T.G. 1999b. Mutations in elongation factor 1β, a guanine nucleotide exchange factor, enhance translational fidelity. *Mol. Cell. Biol.* **19:** 5257.

Cavallius J. and Merrick W.C. 1998. Site-directed mutagenesis of yeast eEF1A: Viable mutants with altered nucleotide specificity. *J. Biol. Chem.* **273:** 28752.

Chervitz S.A., Aravind L., Sherlock G., Ball C.A., Koonin E.V.,

Dwight S.S., Harris M.A., Dolinski K., Mohr S., Smith T., Weng S., Cherry J.M., and Botstein D. 1998. Comparison of the complete protein sets of worm and yeast: Orthology and divergence. *Science* **282:** 2022.

Chicurel M.E., Singer R.H., Meyer C.J., and Ingber D.E. 1998. Integrin binding and mechanical tension induce movement of mRNA and ribosomes to focal adhesions. *Nature* **392:** 730.

Condeelis J. 1995. Elongation factor 1α, translation and the cytoskeleton. *Trends. Biochem. Sci.* **20:** 169.

de Nadal E., Fadden R.P., Ruiz A., Haystead T., and Arino J. 2001. A role for the ppz ser/thr protein phosphatases in the regulation of translation elongation factor $1B\alpha$. *J. Biol. Chem.* **276:** 14829.

DeRisi J.L., Iyer V.R., and Brown P.O. 1997. Exploring the metabolic and genetic control of gene expression on a genomic scale. *Science* **278:** 680.

Dever T.E., Glynias M.J., and Merrick W.C. 1987. GTP-binding domain: Three consensus sequence elements with distinct spacing. *Proc. Natl. Acad. Sci.* **84:** 1814.

Dharmawardhane S., Demma M., Yang F., and Condeelis J. 1991. Compartmentalization and actin binding properties of ABP50: The elongation factor-1 alpha of *Dictyostelium*. *Cell. Motil. Cytoskelet.* **20:** 279.

Dinman J.D. 1995. Ribosomal frameshifting in yeast viruses. *Yeast* **11:** 1115.

Dinman J.D. and Kinzy T.G. 1997. Translational misreading: mutations in translation elongation factor 1α differentially affect programmed ribosomal frameshifting and drug sensitivity. *RNA* **3:** 870.

Farabaugh P.J. 1995. Post-transcriptional regulation of transposition by Ty retrotransposons of *Saccharomyces cerevisiae*. *J. Biol. Chem.* **270:** 10361.

Farabaugh P.J. and Vimaladithan A. 1998. Effect of frameshift-inducing mutants of elongation factor 1α on programmed +1 frameshifting in yeast. *RNA* **4:** 38.

Gangwani L., Mikrut M., Galcheva-Gargova Z., and Davis R.J. 1998. Interaction of ZPR1 with translation elongation factor-1α in proliferating cells. *J. Cell. Biol.* **143:** 1471.

Grosshans H., Hurt E., and Simos G. 2000. An aminoacylation-dependent nuclear tRNA export pathway in yeast. *Genes Devel.* **14:** 830.

Guerreiro P., Barreiros T., Soares H., Cyrne L., Silva A.M., and Rodrigues-Pousada C. 1996. Sequencing of a 17.6 kb segment on the right arm of yeast chromosome VII reveals 12 ORFs, including *CCT, ADE3* and *TR-1* genes, homologues of the yeast *PMT* and *EF1G* genes, of the human and bacterial electron-transferring flavoproteins (β-chain) and of the *Escherichia coli* phosphoserine phosphorylase, and five new ORFs. *Yeast* **12:** 273.

Hesketh J. 1994. Translation and the cytoskeleton: A mechanism for targeted protein synthesis. *Mol. Biol. Rep.* **19:** 233.

Hesketh J.E. and Pryme I.F. 1991. Interaction between mRNA, ribosomes and the cytoskeleton. *Biochem. J.* **277:** 1.

Hiraga K., Suzuki K., Tsuchiya E., and Miyakawa T. 1993. Cloning and characterization of the elongation factor EF-1β homologue of *Saccharomyces cerevisiae*. EF-1β is essential for growth. *FEBS Lett.* **316:** 165.

Holtzman D.A., Yang S., and Drubin D.G. 1993. Synthetic-lethal interactions identify two novel genes, *SLA1* and *SLA2*, that control membrane cytoskeleton assembly in *Saccharomyces cerevisiae*. *J. Cell. Biol.* **122:** 635.

Inagaki Y. and Doolittle W.F. 2000. Evolution of the eukaryotic translation termination system: Origins of release factors. *Mol. Biol. Evol.* **17:** 882.

Izawa T., Fukata Y., Kimura T., Iwamatsu A., Dohi K., and Kaibuchi K. 2000. Elongation factor-1 alpha is a novel substrate of rho-associated kinase. *Biochem. Biophys. Res. Commun.* **278:** 72.

Jansen R.P. 1999. RNA-cytoskeletal associations. *FASEB J.* **13:** 455.

Janssen G.M.C. and Moller W. 1988. Elongation factor $1\beta\gamma$ from *Artemia*: Purification and properties of its subunits. *Eur. J. Biochem.* **171:** 119.

Kambouris N.G., Burke D.J., and Creutz C.E. 1993. Cloning and genetic characterization of a calcium and phospholipid binding protein from *Saccharomyces cerevisiae* that is homologous to translation elongation factor 1-γ. *Yeast* **9:** 151.

Kinzy T.G. and Woolford J.L., Jr. 1995. Increased expression of *Saccharomyces cerevisiae* translation elongation factor EF-1α bypasses the lethality of a *TEF5* null allele encoding EF-1β. *Genetics* **141:** 481.

Kinzy T.G., Freeman J.P., Johnson A.E., and Merrick W.C. 1992. A model of the aminoacyl-tRNA binding site of eukaryotic elongation factor 1α. *J. Biol. Chem.* **267:** 1623.

Kinzy T.G., Ripmaster T.R., and Woolford J.L., Jr. 1994. Multiple genes encode the translation elongation factor EF-1γ in *Saccharomyces cerevisiae*. *Nucleic Acids Res.* **22:** 2703.

Koonin E.V., Mushegian A.R., Tatusov R.L., Altschul S.F., Bryant S.H., Bork P., and Valencia A. 1994. Eukaryotic translation elongation factor 1γ contains a glutathione transferase domain—Study of a diverse, ancient protein superfamily using motif search and structural modeling. *Protein Sci.* **3:** 2045.

Liu G., Tang J., Edmonds B.T., Murray J., Levin S., and Condeelis J. 1996. F-actin sequesters elongation factor 1α from interaction with aminoacyl-tRNA in a pH-dependent reaction. *J. Cell Biol.* **135:** 953.

Moore R.C., Durso N.A., and Cyr R.J. 1998. Elongation factor-1alpha stabilizes microtubules in a calcium/calmodulin-dependent manner. *Cell. Motil. Cytoskeleton* **41:** 168.

Munshi R., Kandl K.A., Carr-Schmid A., Whitacre J.L., Adams A.E., and Kinzy T.G. 2001. Overexpression of translation elongation factor 1alpha affects the organization and function of the actin cytoskeleton in yeast. *Genetics* **157:** 1425.

Negrutskii B.S., Stapulionis R., and Deutscher M.P. 1994. Supramolecular organization of the mammalian translation system. *Proc. Natl. Acad. Sci.* **91:** 964.

Nelson R.J., Ziegelhoffer T., Nicolet C., Werner-Washburne M., and Craig E.A. 1992. The translation machinery and 70 kd heat shock protein cooperate in protein synthesis. *Cell* **71:** 97.

Nissen P., Kjeldgaard M., Thirup S., Polekhina G., Reshetnikova L., Clark B.F.C., and Nyborg J. 1995. Crystal structure of the ternary complex of Phe-tRNAPhe, EF-Tu, and a GTP analog. *Science* **270:** 1464.

Owen C.H., DeRosier D.J., and Condeelis J. 1992. Actin crosslinking protein EF-1α of *Dictyostelium discoideum* has a unique bonding rule that allows square-packed bundles. *J. Struct. Biol.* **109:** 248.

Palecek J., Hasek J., and Ruis H. 2001. Rpg1/Tif32p, a subunit of translation initiation factor 3, interacts with actin-associated protein Sla2p. *Biochem. Biophys. Res. Commun.* **282:** 1244.

Palmer E., Wilhelm J.M., and Sherman F. 1979. Phenotypic suppression of nonsense mutants in yeast by aminoglycoside antibiotics. *Nature* **277:** 148.

Rep M., Krantz M., Thevelein J.M., and Hohmann S. 2000. The transcription response of *Saccharomyces cerevisiae* to osmotic shock. *J. Biol. Chem.* **275:** 8290.

Ridley S.P., Sommer S.S., and Wickner R.B. 1984. Superkiller mutations in *Saccharomyces cerevisiae* suppress exclusion of M2 double-stranded RNA by L-A-HN and confer cold sensitivity in the presence of M and L-A-HN. *Mol. Cell. Biol.* **4:** 761.

Rodnina M.V. and Wintermeyer W. 2001. Fidelity of aminoacyl-tRNA selection on the ribosome: Kinetic and structural mechanisms. *Annu. Rev. Biochem.* **70:** 415.

Sandbaken M.G. and Culbertson M.R. 1988. Mutations in elongation factor EF-1α affect the frequency of frameshifting and amino acid misincorporation in *Saccharomyces cerevisiae*. *Genetics* **120:** 923.

Schirmaier F. and Phillipson P. 1984. Identification of two genes coding for the translation elongation factor EF-1α of *S. cerevisiae*. *EMBO J.* **3:** 3311.

Shestakova E.A., Motuz L.P., Minin A.A., Gelfand V.I., and Gavrilova L.P. 1991. Some of eukaryotic elongation factor 2 is colocalized with actin microfilament bundles in mouse embryo fibroblasts. *Cell Biol. Int. Rep.* **15:** 75.

Singh A., Ursic D., and Davies J. 1979. Phenotypic suppression and misreading in *Saccharomyces cerevisiae. Nature* **277:** 146.

Song J.M., Picologlou S., Grant C.M., Firoozan M., Tuite M.F., and Liebman S. 1989. Elongation factor EF-1α gene dosage alters translational fidelity in *Saccharomyces cerevisiae. Mol. Cell. Biol.* **9:** 4571.

Stapulionis R., Kolli S., and Deutscher M.P. 1997. Efficient mammalian protein synthesis requires an intact F-actin system. *J. Biol. Chem.* **272:** 24980.

Suda M., Fukui M., Sogabe Y., Sato K., Morimatsu A., Arai R., Motegi F., Miyakawa T., Mabuchi I., and Hirata D. 1999. Overproduction of elongation factor 1α, an essential translational component, causes aberrant cell morphology by affecting the control of growth polarity in fission yeast. *Genes Cells* **4:** 517.

Umikawa M., Tanaka K., Kamei T., Shimizu K., Imamura H., Sasaki T., and Takai Y. 1998. Interaction of Pho1p target Bni1p with F-actin-binding elongation factor 1α: Implications in Rho1p-regulated reorganization of the actin cytoskeleton in *Saccharomyces cerevisiae. Oncogene* **16:** 2011.

van Damme H.T.F., Karssies R., Timmers C.J., Janssen G.M.C., and Moller W. 1990. Elongation factor 1β of artemia: Localization of functional sites and homology to elongation factor 1δ. *Biochim. Biophys. Acta* **1050:** 241.

van Hoof A., Staples R.R., Baker R.E., and Parker R. 2000. Function of the ski4p (Csl4p) and Ski7p proteins in 3′-to-5′ degradation of mRNA. *Mol. Cell. Biol.* **20:** 8230.

Wallrapp C., Verrier S., Zhouravleva G., Philippe H., Philippe M., Gress T.M., and Jean-Jean O. 1998. The product of the mammalian orthologue of the *Saccharomyces cerevisiae HBS1* gene is phylogenetically related to eukaryotic release factor 3 (eRF3) but does not carry eRF3-like activity. *FEBS Lett.* **440:** 387.

Woolford J.L., Jr. and Warner J.R. 1991. The ribosome and its synthesis. In *The molecular and cellular biology of the yeast Saccharomyces: Genome dynamics, protein synthesis, and energetics* (ed. J.R. Broach et al.), vol. 1, p. 587. Cold Spring Harbor Laboratory Press, Cold Spring Harbor, New York.

Wyrick J.J., Holstege F.C.P., Jennings E.G., Causton H.C., Shore D., Grunstein M., Lander E.S., and Young R.A. 1999. Chromosomal landscape of nucleosome-dependent gene expression and silencing in yeast. *Nature* **402:** 418.

Yanagihara C., Shinkai M., Kariya K., Yamawaki-Kataoka Y., Hu C.D., Masuda T., and Kataoka T. 1997. Association of elongation factor 1 alpha and ribosomal protein L3 with the proline-rich region of yeast adenylyl cyclase-associated protein CAP. *Biochem. Biophys. Res. Commun.* **232:** 503.

Yang F., Demma M., Warren V., Dharmawardhane S., and Condeelis J. 1990. Identification of an actin-binding protein from *Dictyostelium* as elongation factor 1α. *Nature* **347:** 494.

Zuk D., Belk J.P., and Jacobson A. 1999. Temperature-sensitive mutations in the *Saccharomyces cerevisiae MRT4, GRC5, SLA2* and *THS1* genes result in defects in mRNA turnover. *Genetics* **153:** 35.

Mechanism of Elongation Factor G Function in tRNA Translocation on the Ribosome

W. WINTERMEYER,* A. SAVELSBERGH,* Y.P. SEMENKOV,‡ V.I. KATUNIN,‡ AND M.V. RODNINA†

*Institutes of *Molecular Biology and †Physical Biochemistry, University of Witten/Herdecke,
58448 Witten, Germany; ‡Sankt Petersburg Nuclear Physics Institute,
Russian Academy of Sciences, 188350 Gatchina, Russia*

The elongation cycle of ribosomal protein synthesis is completed by translocation, that is, the movement of the tRNA•mRNA complex on the ribosome. Translocation is catalyzed by elongation factor G (EF-G) at the expense of GTP. Early models of EF-G function and of the role of GTP hydrolysis in translocation were based on two key results. One was that there is slow, spontaneous translocation, indicating that the reaction thermodynamically is exergonic without an additional energy source, such as GTP hydrolysis, and that the basic structural mechanism resides in the ribosome (Pestka 1969; Gavrilova and Spirin 1972, 1974). Another was that the extent of EF-G-catalyzed translocation remained unchanged when GTP was replaced with nonhydrolyzable analogs, whereas the dissociation of EF-G from the ribosome was inhibited (Modolell et al. 1975; Belitsina et al. 1976). Thus, by analogy with the canonical GTPase switch mechanism, it was concluded that translocation is brought about by the binding of EF-G•GTP to the ribosome, and that subsequent GTP hydrolysis is required for the dissociation of EF-G in the GDP-bound form (Kaziro 1978; Spirin 1985).

Recent pre-steady-state kinetic results show, however, that GTP hydrolysis initiates, rather than completes, the functional cycle of EF-G on the ribosome (Rodnina et al. 1997), suggesting a revised model of EF-G function in translocation (Rodnina et al. 2000a,b). A number of functional states of EF-G on the ribosome have been defined by kinetic analysis and, in part, characterized structurally by cryoelectron microscopy (cryo-EM) (Agrawal et al. 1998, 1999; Frank and Agrawal 2000; Stark et al. 2000). Furthermore, the contributions to translocation catalysis of tRNA hybrid-state formation, EF-G binding, and GTP hydrolysis have been determined by quantitative kinetic analysis, providing a consistent model of the molecular mechanism of translocation. The recent results of kinetic and structural analysis of translocation are reviewed in this paper.

KINETIC MECHANISM OF TRANSLOCATION

To study translocation by rapid kinetic methods, a number of observables were used to monitor different partial reactions (Rodnina et al. 1997, 1999). Binding of EF-G to the ribosome, tRNA movement, and conformational rearrangements in EF-G result in fluorescence changes of reporter groups in tRNA (proflavin) or GTP (mant-dGTP), such that kinetics of these reactions can be followed by fluorescence stopped-flow; GTP hydrolysis can be measured by quench flow; finally, the release of inorganic phosphate, Pi, from EF-G after GTP hydrolysis can be measured by stopped-flow monitoring the fluorescence increase of coumarin-labeled phosphate-binding protein upon binding of Pi (Brune et al. 1994). Kinetic measurements were performed at varying concentrations, and from the rates measured at saturating concentration of EF-G, rate constants were determined for most of the steps.

The kinetic model of translocation derived from the kinetic analysis is depicted in Figure 1. In the initial pretranslocation state, deacylated tRNA resides in the P site (P/E hybrid state) and peptidyl tRNA in the A site (A/P hybrid state). Immediately following the binding of EF-G•GTP to form a readily reversible initial complex, the GTPase activity of EF-G is induced and GTP is hydrolyzed rapidly. Subsequent release of Pi is much slower and is immediately followed (or accompanied) by tRNA movement. Of the following rearrangements, at least one involves a conformational change of EF-G. At 5 s⁻¹, this rearrangement is probably the rate-limiting step of the reaction sequence and mainly determines the rate of the turnover reaction, which is around 3 s⁻¹ (M.V. Rodnina, unpubl.). Translocation is completed by the rapid dissociation of EF-G•GDP (and deacylated tRNA) to reach the final posttranslocation state of the ribosome with peptidyl tRNA in the P site and a free A site.

COUPLING OF TRANSLOCATION AND Pi RELEASE

Regarding the mechanism of how EF-G may promote translocation, two features of the kinetic mechanism (Fig. 1) are most relevant. First, the dissociation of Pi is much slower than GTP hydrolysis, indicating that Pi release is limited by a slow rearrangement around the binding site of the γ-phosphate. Second, Pi release and tRNA movement take place at similar rates (Fig. 2a), suggesting that Pi release precedes, and possibly drives, tRNA movement or a rearrangement that determines tRNA movement. To more clearly establish the latter point, experiments were performed in which translocation was inhibited either by mutations in EF-G or by adding antibiotics known to interfere with translocation. EF-G•GTP binding to the ribo-

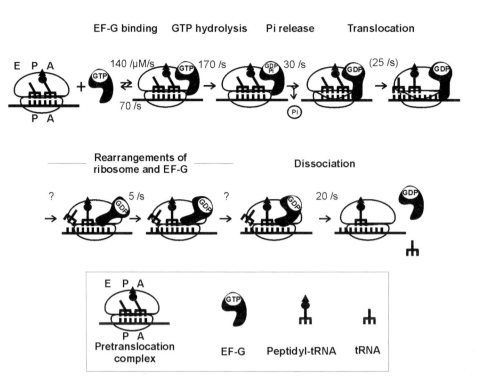

Figure 1. Kinetic mechanism of translocation. Rate constants are derived from rapid kinetic measurements performed at 7 mM Mg^{++}, 37°C, and at saturating concentration of EF-G (2 μM) (Rodnina et al. 1997 and unpubl.). The value for translocation is an apparent rate constant, because the observed rate of translocation is probably limited by the preceding step; the rate constant of the dissociation of EF-G·GDP was measured separately (B. Wilden, unpubl.). Note that EF-G and ribosomes are depicted in varying shapes to indicate their presence in different conformations.

some and rapid hydrolysis of GTP were unaffected in all cases studied.

When translocation was inhibited by single amino acid exchanges at position 583 (H583K/R) at the tip of domain 4 (>100-fold inhibition), the release of Pi was unaffected (Fig. 2b) (Savelsbergh et al. 2000a). The same result was obtained with mutant EF-G in which domains 1 and 5 were cross-linked by an engineered disulfide bridge (~10^4-fold inhibition) (Peske et al. 2000). Finally, adding the antibiotic viomycin, which is known to block translocation (Modolell and Vazquez 1977) by binding to the 30S decoding region (Moazed and Noller 1990), also did not affect Pi release. These results show that Pi release can be uncoupled from translocation, i.e., precedes translocation.

The mechanism of uncoupling seems to be different in the three cases studied. Mutations at the 583 position may interfere with a contact between domain 4 of EF-G and the 30S subunit (see below, Fig. 6) which is important for translocation, indicating that this contact is an important element of the coupling mechanism. Introducing the cross-link between domains 1 and 5 reduces the conformational flexibility of EF-G, possibly inhibiting a conformational change that is necessary for the coupling. Viomycin stabilizes peptidyl tRNA in the A site (Modolell and Vazquez 1977) by several orders of magnitude (Y.P. Semenkov, unpubl.), most likely by stabilizing the binding conformation of 16S rRNA in the decoding center (Yusupov et al. 2001), and this effect probably ex-

plains the inhibition of translocation. Thus, we speculate that viomycin, by binding to 16S rRNA, blocks a conformational transition of the ribosome that is required for translocation, thereby uncoupling Pi release from translocation and allowing EF-G turnover, which is only slightly inhibited by the antibiotic.

Complementary information was obtained by experiments with a deletion mutant of EF-G that lacked domains 4 and 5. EF-GΔ4/5, as EF-GΔ4 (Rodnina et al. 1997), binds to ribosomes and hydrolyzes GTP at the same rate as wild-type EF-G, but has >1000-fold reduced activity in translocation (Savelsbergh et al. 2000a). With the deletion mutant, Pi release is inhibited to the same extent as translocation, suggesting that Pi release and translocation are coupled as in the wild-type situation, albeit in an entirely different time domain (Fig. 2c). A similar inhibition of Pi release and tRNA movement, and no inhibition of GTP hydrolysis, were observed when experiments were performed with wild-type EF-G in the presence of thiostrepton (Rodnina et al. 1999). The inhibition by thiostrepton of Pi release was the same with vacant ribosomes, i.e., it was not due to the inhibition of tRNA movement.

Thiostrepton binds to the L11-RNA subdomain of 23S rRNA and is thought to restrict the structural flexibility of the RNA in that region. The strong inhibition by thiostrepton of Pi release from ribosome-bound EF-G therefore suggests that a conformational change within the G domain of EF-G, required for Pi dissociation, is in-

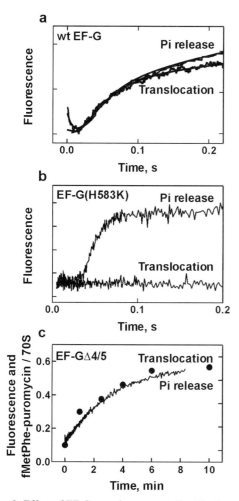

Figure 2. Effect of EF-G mutations on coupling Pi release and tRNA movement. (*a*) Wild-type EF-G. (*b*) EF-G(H583K). (*c*) EF-GΔ4/5. Pi release was monitored by the fluorescence of coumarin-labeled phosphate-binding protein (Brune et al. 1994; Savelsbergh et al. 2000a), tRNA movement by the fluorescence of proflavin-labeled fMetPhe-tRNA^Phe (Rodnina et al. 1997; Savelsbergh et al. 2000a). In *c*, translocated fMetPhe-tRNA^Phe was determined by reaction with puromycin (*filled circles*).

These results may be rationalized by assuming that an important element of translocation catalysis by EF-G is mutual structural adaptation of factor and ribosome to rearrange from an initial, readily reversible complex to a more tightly bound complex that then rearranges further on the pathway to the transition state. One way to perturb the rearrangement is to block the respective conformational transition of the ribosome in a state where Pi release is still possible (viomycin); another is to interfere with structural coupling between ribosome and factor (583 mutations). In these cases, Pi release is unaffected, but tRNA movement is strongly impaired, presumably due to inhibition of subsequent rearrangement(s) of the complex. At a later stage, after the formation of the productive complex, Pi release may be inhibited by interfering with subsequent rearrangements. Again, inhibition can be exerted either through the ribosome (thiostrepton) or through EF-G (domain 4/5 deletion). In both cases, tRNA movement is strongly inhibited as well, supporting the idea that tRNA movement is coupled to rearrangement(s) of the ribosome–EF-G complex induced by Pi release.

The formation of a tightly bound ribosome–EF-G complex, inferred above from slow Pi release, is also supported by direct evidence, because in several cases of slow translocation, the dissociation of EF-G from the ribosome was inhibited as well. This is demonstrated by the inhibition of EF-G turnover under conditions of ribosome excess where EF-G has to cycle several times to produce measurable amounts of products, either of GTP hydrolysis or of translocation. Extremely slow turnover ($<10^{-4}$ s^{-1}) was observed in the presence of thiostrepton (Rodnina et al. 1999), with truncated EF-G lacking domains 4/5 (Rodnina et al. 1997; Savelsbergh et al. 2000a), or with EF-G that was conformationally restricted by a disulfide cross-link between domains 1 and 5 (Peske et al. 2000). Slow turnover of EF-G (and slow translocation) was observed in translocation experiments with nonhydrolyzable GTP analogs or GDP (see below, Fig. 5). Taken together, these results indicate that, after translocation, further rearrangement(s) of the ribosome–EF-G complex has to take place for EF-G to complete its functional cycle and to dissociate from the ribosome.

DIFFERENT TRANSITION STATES IN TRANSLOCATION CATALYSIS BY EF-G WITH AND WITHOUT GTP HYDROLYSIS

From the early work on translocation, it is known that EF-G efficiently promotes translocation without GTP hydrolysis (for review, see Kaziro 1978; Spirin 1985, 1999). However, compared to the reaction with GTP hydrolysis, translocation is much slower with EF-G and nonhydrolyzable GTP analogs or GDP (Rodnina et al. 1997), or when a GTPase-inactive mutant, EF-G(R29M), is used with GTP (Fig. 3a) (Mohr et al. 2000). At the same time, the turnover of EF-G is strongly inhibited in all these cases. The kinetic analysis with caged GTP or GDP, using both fluorescence stopped-flow and a newly developed time-resolved puromycin assay (Fig. 3b), yields the

hibited indirectly by restricting the structural mobility of the ribosome; i.e., that the ribosome and EF-G are conformationally coupled. Since thiostrepton does not influence initial complex formation and GTP hydrolysis (Rodnina et al. 1997), it is likely that coupling is established immediately after GTP hydrolysis, but before Pi release. Based on the low-resolution cryo-EM structure of the complex prior to tRNA movement (Fig. 6) (Stark et al. 2000), the predominant interactions are those between domain 1 (G domain) of EF-G and the L7/12 stalk of the 50S subunit and between domain 4 and the shoulder of the 30S subunit. The L7/12 interaction is important for GTPase stimulation (Savelsbergh et al. 2000b) and may be responsible for impeding Pi (and nucleotide) release either by obstructing the exit from the nucleotide-binding pocket or by inducing a structural change within the G domain that interferes with dissociation.

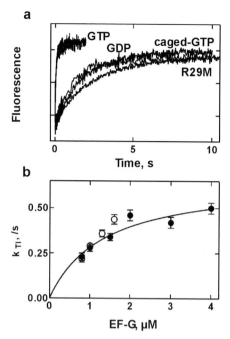

Figure 3. Influence of GTP hydrolysis on the rate of tRNA movement. (*a*) Time courses of tRNA movement monitored by fluorescence stopped-flow. Exponential fitting yields apparent rate constants of translocation. (*b*) Concentration dependence of k_{TL} for GDP. Translocation was monitored either by the fluorescence of proflavin-labeled fMetPhe-tRNA^{Phe} (*open circles*) or by time-resolved puromycin reaction (*filled circles*). Using the Michaelis-Menten formalism for fitting yields $k_{TL}= 0.5$ s^{-1} at saturation. Similar values for k_{TL} were obtained for caged-GTP and EF-G(R29M) (both 0.7 s^{-1}); the value for GTP was 25 s^{-1} (Rodnina et al. 1997).

Table 1. Transition-state Parameters of EF-G-dependent Translocation (37°C)

Nucleotide	$\Delta G^{\neq}$	$\Delta H^{\neq}$ (kJ/mole)	$T\Delta S^{\neq}$
GTP	67 ± 7	57 ± 6	-10 ± 2
Caged-GTP	77 ± 8	94 ± 5	17 ± 2
GDP	79 ± 8	91 ± 4	12 ± 1

entropic term, $T*\Delta S^{\neq}$, which is favorable in the absence of GTP hydrolysis, becomes unfavorable when GTP is hydrolyzed, the difference amounting to about 25 kJ/mole. These differences indicate that translocation catalysis by EF-G proceeds by different pathways with and without GTP hydrolysis. A considerable fraction of the free energy of GTP hydrolysis is coupled to the reaction to lower the activation enthalpy, indicating that additional interactions between EF-G and the ribosome are established in the transition state. In keeping with the formation of additional interactions, the activation entropy is also lowered, suggesting that the transition-state complex formed with GTP hydrolysis is characterized by fewer degrees of freedom, compared to the one formed without GTP hydrolysis.

Another argument supporting the view that translocation with and without GTP hydrolysis proceeds by different pathways comes from a comparison of the rates of EF-G turnover measured at conditions of ribosome excess. In the presence of GTP, the turnover rate is about 3 s^{-1} (M.V. Rodnina, unpubl.) and determined by a rate-limiting rearrangement of the ribosome–EF-G complex taking place after translocation (Fig. 1), whereas with GDP the turnover reaction is >100 times slower, 0.02 s^{-1} (Fig. 4). Because in both cases it is EF-G•GDP that eventually dissociates from the ribosome ($k_{off} = 20$ s^{-1}), the large kinetic effect of GTP hydrolysis on turnover indicates that GTP hydrolysis accelerates a rearrangement of the ribosome–EF-G complex preceding dissociation, rather than the dissociation step itself. It is not known how the system stores the free energy of GTP hydrolysis, which is the first step of the reaction sequence (Fig. 1), up to the stage after tRNA movement. However, the observed behavior is consistent with the formation of a ribosome–EF-G complex that undergoes coupled conformational transitions throughout the reaction sequence.

rate constants summarized in the kinetic scheme (Fig. 4). Compared to the uncatalyzed reaction, which is extremely slow (5•10^{-4} s^{-1}, 37°C; Semenkov et al. 1992), translocation without GTP hydrolysis has a 20 kJ/mole-lower free energy of activation, $\Delta G^{\neq}$, and GTP hydrolysis accelerates the reaction further (25 s^{-1}, Rodnina et al. 1997) by lowering $\Delta G^{\neq}$ by another 10 kJ/mole (Fig. 5).

Transition-state parameters of translocation (Table 1) were determined by kinetic analysis at different temperatures between 15°C and 37°C (V.I. Katunin, unpubl.). The activation enthalpy, $\Delta H^{\neq}$, is lowered considerably (by about 35 kJ/mole) by GTP hydrolysis, whereas the

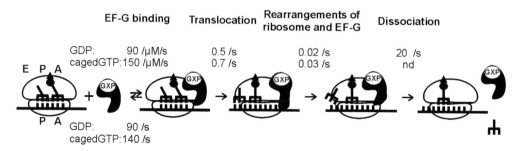

Figure 4. Kinetic scheme of translocation without GTP hydrolysis. Rate constants of binding and translocation were determined from single-round kinetic experiments, monitoring the fluorescence of proflavin-labeled fMetPhe-tRNA^{Phe} (cf. Figs. 2, 3); the rates of subsequent rearrangements represent the rate of EF-G turnover, measured at conditions of ribosome excess. The rate constant of EF-G–GDP dissociation was measured separately (B. Wilden, unpubl.). Symbols as in Fig. 1. (GXP), GDP or caged-GTP.

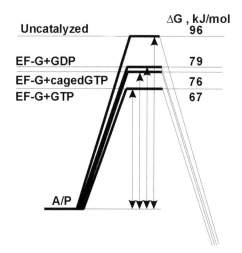

Figure 5. Free energies of activation of translocation. Values of $\Delta G^{\neq}$ (37°C) were derived from rate constants of translocation (Figs. 3, 4) according to the equation: $\Delta G^{\neq} = -RT \cdot \ln(k \cdot h/k_B/T)$ (Dixon and Webb 1979). The $\Delta G^{\neq}$ value for the uncatalyzed reaction is calculated from the rate measured under conditions comparable to those used for the reactions with EF-G (Semenkov et al. 1992).

tRNA MOVEMENTS DURING TRANSLOCATION

In the pretranslocation state, A- and P-site-bound tRNAs are arranged such that the anticodons and acceptor ends are close to one another, as shown by cryo-EM (Stark et al. 1997) and crystallography (Yusupov et al. 2001). The detailed positions of the acceptor ends on the 50S subunit depend on the functional state of the tRNA; upon forming peptidyl tRNA in the A site, the acceptor end moves toward, or into, the 50S P site and the acceptor end of deacylated tRNA toward the E site to form A/P and P/E "hybrid states," respectively (Moazed and Noller 1989a,b). Peptidyl tRNA in the A/P state hardly reacts with puromycin (Semenkov et al. 1992), indicating that the structural arrangement of peptidyl tRNA with respect to the peptidyl transferase is different in A/P and P/P states, and that the fully reactive arrangement is only established by translocation.

The formation of the A/P hybrid state destabilizes peptidyl-tRNA binding in the A site by 19 kJ/mole (37°C) (Semenkov et al. 2000). The group transfer potential of the ester linkage in aminoacyl tRNA (aa-tRNA), and probably in peptidyl tRNA, amounts to about –30 kJ/mole (25°C). Thus, the observed destabilization suggests that a substantial fraction of the free energy of transpeptidylation is utilized for conformational work, most likely the formation of the hybrid A/P state. The destabilization of tRNA binding in the A site is correlated with an acceleration of tRNA movement in translocation (Semenkov et al. 2000). Peptidyl tRNA (fMetPhe-tRNA[Phe]) is translocated about 130 times faster than aa-tRNA (Phe-tRNA[Phe]), corresponding to a 13 kJ/mole difference in the free energies of activation, $\Delta\Delta G^{\neq}$, of translocation from the A/A and A/P states. The fact that $\Delta\Delta G^{\neq}$ amounts to only about two-thirds of the free energy

difference between A/A and A/P states may indicate that the transition states of translocation are not the same for aa-tRNA and peptidyl tRNA.

The ability of P-site-bound deacylated tRNA to interact with the E site, possibly by entering the P/E hybrid state, is important for translocation as well, since modifications at the 3′ end of the tRNA that impair the interaction with the E site (Lill et al. 1986) strongly reduce the rate of translocation (Lill et al. 1989), and a tRNA fragment comprising only the anticodon arm bound to the 30S P site inhibits translocation (Joseph and Noller 1998). From the P/E hybrid state, the tRNA moves toward the E site to establish an intermediate E-site-bound state, E′, with the anticodon still bound to the mRNA. The codon–anticodon interaction in the E′ state is weak (Lill and Wintermeyer 1987) and is probably disrupted during the rearrangement of the tRNA to another state, E, which is probably the one characterized structurally by cryo-EM (Stark et al. 1997; Agrawal et al. 1999). The arrangement of the E-site tRNA in the recent crystallographic structure of the 70S ribosome probably represents a state close to the E′ state with an interaction of the anticodon with the (noncomplementary) codon present in the 30S E site (Yusupov et al. 2001).

The overall affinity of tRNA binding in the E site and the distribution of the tRNA between E′ and E states strongly depend on the concentrations of Mg[++] (Robertson and Wintermeyer 1987) and polyamines (Gnirke et al. 1989; Semenkov et al. 1996). At ionic conditions that are optimized for rapid and accurate protein synthesis, the tRNA is bound to the E site in a kinetically labile fashion (Semenkov et al. 1996), and its dissociation is fast and does not limit the rates of translocation (V.I. Katunin, unpubl.) or of the EF-Tu-tRNA complex to the A site (Bilgin et al. 1992; Rodnina et al. 1994; Semenkov et al. 1996).

STRUCTURAL REARRANGEMENTS OF THE RIBOSOME–EF-G COMPLEX DURING TRANSLOCATION

Structural changes of the ribosome during translocation were proposed very early (Bretscher 1968; Spirin 1968, 1970). In these models, tRNA movement is coupled to movements of one ribosomal subunit relative to the other. Initial attempts to characterize structural changes of the ribosome related to translocation were restricted to the comparison of pre- and posttranslocation ribosomes. Scattering methods revealed only slight differences (Spirin et al. 1987; Serdyuk et al. 1992), as did the direct visual comparison of the pre- and the posttranslocation state of the ribosome (Stark et al. 1997), most likely because EF-G was not present in these complexes and the structure of the transition state does not prevail in the posttranslocation state.

Structures of ribosome–EF-G complexes were studied by a variety of methods which invariably required that the ribosome–EF-G complex was stabilized. Thus, many studies were performed with fusidic acid-stabilized complexes, i.e., in a state after translocation. Similarly, ribosome complexes formed with EF-G and nonhydrolyzable

GTP analogs do perform translocation quite rapidly (Rodnina et al. 1997) and therefore represent the post-translocation state as well. Chemical probing of both types of posttranslocation complexes has revealed EF-G footprints in two regions of 23S rRNA, in the L11 region (helices 43 and 44) in domain II (residues 1067 and 1069), and in the sarcin–ricin stem-loop (SRL; helix 95) in domain VI (residues 2655, 2660, 2661, and 2665) (Moazed et al. 1988). In the three-dimensional structure of the ribosome, these parts of 23S rRNA are situated close to each other on the 50S subunit, the L11-RNA forming the base of the stalk, which comprises proteins L10-(L7/L12)$_2$, and the SRL being situated below the stalk base (Ban et al. 1999; Yusupov et al. 2001). As suggested by hydroxyl radical cleavage of rRNA directed from Fe-EDTA tethered to various surface positions of EF-G (Wilson and Noller 1998), in the fusidic acid-stabilized complex domain 5 of EF-G faces the L11-RNA, whereas domain 1 contacts the SRL. Domain 4 in this complex reaches into the intersubunit space and comes close to the decoding region of 16S rRNA on the 30S side and residues of domain IV of 23S rRNA on the 50S side.

Low-resolution cryo-EM reconstructions of the post-translocation complex are consistent with the arrangement derived from the chemical data (post 2 complex, Fig. 6) (Agrawal et al. 1998, 1999; Stark et al. 2000). In another complex, stabilized by replacing GTP with GDPCP (and lacking tRNA), EF-G is arranged in the same way, i.e., in the posttranslocation position, although structural details at the stalk base and the stalk appear different, compared to the complex stabilized by fusidic acid; the differences were attributed to the fact that GDPCP, rather than GDP and fusidic acid, was bound to EF-G (Agrawal et al. 1999). When the complex after translocation was stabilized by thiostrepton, which binds to the L11-RNA, the cryo-EM reconstruction (Stark et al. 2000) revealed a similar gross orientation, but different arrangement of EF-G relative to the subunits (post 1, Fig. 6). In particular, the close proximity

with the stalk base and the SRL was not seen in the latter complex, in keeping with the lack of DMS footprints on the SRL (Rodnina et al. 1999).

Studying the arrangement of EF-G on the ribosome prior to translocation is less straightforward, because it is difficult to inhibit translocation. For instance, replacing GTP with GDPCP yielded a heterogeneous mixture of ribosome–tRNA–EF-G complexes that were partially translocated and partially occupied with EF-G, and in the cryo-EM reconstruction fragmented density was observed that was interpreted to coincide with the density obtained for EF-G in the posttranslocation position (Agrawal et al. 1999). From the comparison of the reconstructions (obtained at different resolutions) of a ribosome with fMet-tRNA in the P site, but no EF-G, and of a ribosome without tRNA and with EF-G–GDPCP bound in the posttranslocation position (see above), the same authors deduced a slight "ratchet-like" rotational movement of the 30S subunit relative to 50S that they related to translocation (Agrawal et al. 1999; Frank and Agrawal 2000). However, if the proposed arrangement of EF-G (cf. Fig. 1c–f in Agrawal et al. 1999) were pretranslocational, it would create a steric clash between domain 4 reaching into the decoding region and the anticodon stem-loop of A-site tRNA, unless an extensive distortion of the EF-G molecule is assumed, which is unlikely. Thus, for both biochemical and structural reasons, in the complexes studied by Agrawal and colleagues, EF-G almost certainly was in the posttranslocation position, and the reported structural differences do not pertain to structural changes of the ribosome related to translocation.

To study the pretranslocation ribosome–EF-G complex in a biochemically well-defined state, we have resorted to stabilization by antibiotic binding, taking advantage of the inhibition of translocation by thiostrepton. It was possible to prepare a complex that was fully occupied with two tRNAs in their respective pretranslocation positions and had EF-G bound at high occupancy (>80%). The

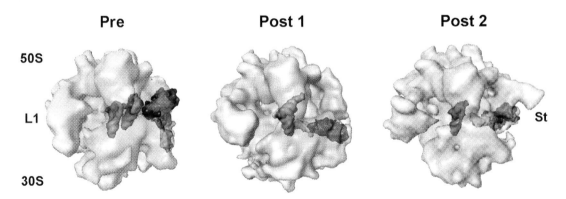

Figure 6. Three-dimensional cryo-EM reconstructions of ribosome–EF-G complexes before and after translocation. Ribosomes are displayed in semitransparent manner, viewed by looking onto the 50S central protuberance; landmarks: (L1) protein L1; (St) stalk. (Pre) Ribosomes carrying tRNA^{fMet} in the P site, fMetPhe-tRNA^{Phe} in the A site, and EF-G in the pretranslocation position stabilized by thiostrepton. (Post 1) Same complex after translocation. (Post 2) Ribosome–EF-G–GDP complex formed in the presence of fusidic acid; the P-site tRNA is modeled as taken from the pre complex. tRNA, green. The domains of EF-G are color-coded: domain 1, magenta; domain 2, blue; domain 3, green; domain 4, yellow; domain 5, red. In the three complexes, EF-G is depicted in different conformations, as obtained by moving domains 3–5 relative to domains 1–2 for optimum fits of the density. (Data from Stark et al. 2000.)

cryo-EM structure (Stark et al. 2000) shows EF-G arranged across the subunit interface (pre, Fig. 6), suggesting that, in the pretranslocation position, domain 1 (G domain) of EF-G makes extensive contact with the L7/L12 part of the stalk, whereas domain 4 contacts the shoulder of the 30S subunit. In this complex, the structure of the 30S subunit is changed substantially, in particular in the head and neck regions, compared to the EF-G-free pretranslocation complex and to the two complexes (post 1, post 2; Fig. 6) after translocation.

A structural change in the 30S head, induced by EF-G binding in the pretranslocation state stabilized by thiostrepton, is also suggested by chemical probing data which show strongly decreased accessibility for DMS modification and hydroxyl radical cleavage of several residues in helix 34 (N.B. Matassova, unpubl.). Protections are seen with wild-type EF-G, both with GTP and GDPNP, i.e., when there is translocation, whereas no protections are observed with mutants of EF-G that are strongly impaired (583 mutations in domain 4) or inactive (cross-linked EF-G) in translocation. There are earlier reports suggesting that helix 34 has an important role in translation. Mutations at different positions in helix 34 severely impair cell growth by diverse effects on translation, including reduced incorporation of ribosomes into polysomes, erroneous termination, decreased fidelity of protein synthesis by increased readthrough of stop codons, and increased frameshifting (Prescott and Kornau 1992; Moine and Dahlberg 1994). Helix 34 comprises the binding site of spectinomycin, an antibiotic that inhibits translocation by binding to helix 34 (Bollen et al. 1968; Sigmund et al. 1984; Carter et al. 2000). Thus, it is likely that the structural characteristics of the pre complex, as revealed by cryo-EM and chemical probing, represent salient features of the transition-state complex.

There are probably more structural changes required for rapid translocation, including EF-G-induced changes in the 50S subunit, which may be inhibited by thiostrepton binding and, therefore, not be observed by cryo-EM or chemical probing. This limitation pertains particularly to the L11 region of 23S rRNA, which contains the binding site for thiostrepton. Furthermore, there is no interaction of EF-G with the SRL in the pre (cryo-EM) or post (cryo-EM, DMS probing) complexes stabilized by thiostrepton, whereas it is present in the posttranslocation complex stabilized by fusidic acid. Thus, the chemical data support the cryo-EM results and suggest that thiostrepton stabilizes the ribosome–EF-G complex in a posttranslocation state (post 1, Fig. 6) prior to that stabilized by fusidic acid (post 2, Fig. 6) (Rodnina et al. 1999). The inhibition by thiostrepton of translocation and EF-G turnover, therefore, is attributed to the inhibition of structural transitions in the 50S subunit that, in the absence of antibiotic, are induced by EF-G binding and GTP hydrolysis.

The structures of tRNA–ribosome complexes, as determined by cryo-EM (Frank 2000; van Heel 2000) and, in particular, by crystallography (Yusupov et al. 2001), reveal a number of connections (bridges) between the subunits, some of which appear to block the pathway of tRNA movement and will probably have to move to al-

low, or induce, translocation (Yusupov et al. 2001). Particularly prominent are the bridges B1a (also referred to as A-site finger), connecting the 30S head to the central protuberance of 50S, and B2a (helix 69 of 23S rRNA), which is situated between the two tRNAs and appears to make a contact with the A-site tRNA. These, and a number of other intersubunit bridges, form a network of interactions that provide a structural basis for conformational coupling between the subunits and for signaling the conformational effects of EF-G binding, GTP hydrolysis, and Pi release from the EF-G-binding site to the tRNA-binding sites, as well as between the subunits.

STRUCTURAL CHANGES OF EF-G DURING TRANSLOCATION

The structural changes of EF-G that are induced by binding to the ribosome and GTP hydrolysis are not known at high resolution. By analogy with the well-characterized structural changes in other G proteins, it is likely that the loss of the γ-phosphate of GTP introduces conformational strain in the G domain (domain 1) of EF-G. Rearrangements in the G domain as such may affect the conformation of the 50S subunit and/or induce a change of the overall architecture of the EF-G molecule which, in turn, may drive a ribosome rearrangement leading to translocation. Cryo-EM structures of EF-G–ribosome complexes suggest that a movement of domain 4 is involved (Agrawal et al. 1998; Stark et al. 2000). It may be induced by a change of the interactions of domain 1 with domain 5, which is probably rigidly connected to domain 4. In addition, domain 3 of EF-G may be involved in sensing structural changes in domain 1 (Laurberg et al. 2000). The interactions of domains 1/2 with domains 3/5 indeed seem to be functionally important, since numerous fusidic-acid-resistance mutations and revertants have been found at the respective domain interfaces (Johanson and Hughes 1994; Johanson et al. 1996; Laurberg et al. 2000).

An idea about the conformations assumed by EF-G in different intermediate states of translocation is provided by comparing the EF-G density in the cryo-EM reconstructions (Fig. 6) with the crystal structure of EF-G–GDP (Aevarsson et al. 1994; Czworkowski et al. 1994; al-Karadaghi et al. 1996). In the pre state, EF-G appears to assume a conformation that is slightly more open compared to the crystal structure, the largest change entailing a movement of the tip of domain 4. In the posttranslocation state stabilized by thiostrepton, density due to EF-G appears in a different position on the ribosome and in yet another conformation in which domain 5 is displaced from domain 1 toward domain 2 in a way that results in a large (>40 Å) displacement of the tip region of domain 4. The fit suggests an arrangement where the body of EF-G is involved in several contacts with the 30S subunit close to the head-to-body junction and domain 4 reaches into the decoding center and approaches the anticodon of the P-site-bound peptidyl tRNA. In the complex blocked by the antibiotic fusidic acid in a state preceding the dissociation from the ribosome, domain 5 of EF-G is moved toward domain 2, again resulting in a large displacement of the tip of domain 4 (Agrawal et al. 1998; Stark et al. 2000).

Conformational changes of EF-G are essential for translocation catalysis. When the relative mobility of domains 1 and 5 was restricted by introducing an interdomain disulfide bridge, tRNA movement was blocked practically completely, while the steps up to and including Pi release were not affected (Peske et al. 2000). The interesting implication is that the acceleration of translocation brought about by EF-G binding to the pretranslocation complex, regardless of whether or not there is GTP hydrolysis, depends on the structural flexibility of the factor, suggesting that a relative movement of domains 1 and 5, and probably the concomitant movement of domain 4, are required for EF-G to promote translocation.

Following tRNA movement, further rearrangements of the ribosome–EF-G complex have to take place in order to enable the factor to dissociate from the ribosome. These appear to be coupled rearrangements of the factor and the ribosome as well, as indicated by the finding that the dissociation of EF-G is inhibited by certain mutations in EF-G (Mohr et al. 2000; Savelsbergh et al. 2000a), by binding of antibiotics to either EF-G (fusidic acid) or the ribosome (thiostrepton) (Rodnina et al. 1999), or by replacing GTP with nonhydrolyzable GTP analogs (Modolell et al. 1975; Belitsina et al. 1976; Kaziro 1978; Rodnina et al. 1997). Furthermore, the dissociation of EF-G is completely blocked by introducing a disulfide bridge between domains 1 and 5 (Peske et al. 2000), indicating that the mobility of the structural unit formed of domains 5 and 4 is essential for dissociation, in keeping with the finding that the deletion of domain 4 blocks the dissociation (Rodnina et al. 1997).

In conclusion, the available data indicate that EF-G assumes different conformations at different stages of translocation, mainly by a movement of domains 3/4/5 relative to domains 1/2. Although only relatively small movements of domain 5 relative to domains 1/2 are found, the resulting movements of the tip region of domain 4 are quite extensive, due to the extended structure of the arm formed of domains 5 and 4. These conformational changes of EF-G are probably responsible for the coupling of GTP hydrolysis and Pi release to the rearrangements of the ribosome and are therefore crucial for translocation catalysis.

CONCLUSIONS

Results from kinetic and structural studies suggest that translocation entails conformational transitions of the ribosome that, in the transition state, result in the concerted movement of the two tRNAs together with the mRNA. These rearrangements can take place spontaneously, albeit with very low probability, and are accelerated by EF-G binding to the ribosome. Conformational flexibility of EF-G is essential, suggesting that by mutual structural adaptation, binding interactions are established that lead to a conformationally coupled complex in which conformational fluctuations are biased in favor of the productive state, thereby lowering the energy level of the transition

state. Following tRNA movement, further rearrangements of the ribosome–EF-G complex are required to resolve these interactions and allow the dissociation of the factor. GTP hydrolysis accelerates both tRNA movement and EF-G turnover, indicating that conformational changes of EF-G that are induced by GTP hydrolysis or, more likely, by the release of the γ-phosphate, due to conformational coupling with the ribosome, promote rearrangements that both precede and follow the movement of the tRNAs.

These features of the functional cycle of EF-G in translocation bear strong resemblance to those of motor ATPases, such as myosin. Characteristic features of these systems are the delay of Pi release relative to ATP hydrolysis and, during the delay phase, the formation of a tightly coupled complex, e.g., between myosin and actin, as a basis for exerting force and creating directed movement. Both features are characteristic for the function of EF-G as well. Based on this analogy, it appears that EF-G in the complete system, i.e., when there is GTP hydrolysis and undisturbed conformational coupling, actively promotes conformational transitions on the productive pathway. On the other hand, when there is no GTP hydrolysis, EF-G still promotes translocation in a way that the productive conformational state of the ribosome is stabilized by mutual structural adaptation, albeit less effectively and with the consequence of turnover inhibition.

According to the model, the ribosome–EF-G complex, from GTP hydrolysis on, behaves as a tightly coupled structural unit, and subsequent conformational changes of EF-G induced by GTP hydrolysis and Pi release exert structural effects on both ribosomal subunits. The model provides a consistent explanation for various kinds of translocation inhibition. Antibiotic binding to either subunit strongly inhibits tRNA movement and, to variable extents, the turnover of EF-G, most likely by interfering with rearrangements of the ribosome. Furthermore, both steps are inhibited when there is no GTP hydrolysis. Finally, there are mutations in EF-G that lead to inhibition of translocation by either uncoupling Pi release from subsequent rearrangements or by inhibiting rearrangements of the ribosome–EF-G complex in a concerted fashion.

Although recent work has revealed many mechanistic and structural features of translocation, important questions are still unresolved and have to be addressed in future work. The structural states of the ribosome that allow, and control, the movement of the tRNA–mRNA complex as well as the conformations of EF-G in intermediate states of translocation have to be determined at high resolution; the molecular mechanism of how the GTPase activity of EF-G is stimulated by the ribosome has to be defined; the function of EF-G as a motor GTPase remains to be fully established, including the elucidation of the structural basis for the inhibition of nucleotide and phosphate release from EF-G while it is bound to the ribosome. Although there are powerful methods available to study translocation both kinetically and structurally, no doubt additional new methods, e.g., single-molecule techniques, will have to be applied in order to reach a comprehensive appreciation of the molecular mechanism of translocation.

ACKNOWLEDGMENTS

Work in our laboratories was supported by the Deutsche Forschungsgemeinschaft, the European Commission, the Volkswagen-Stiftung, the Alfried Krupp von Bohlen und Halbach-Stiftung, and the Fonds der Chemischen Industrie.

REFERENCES

Aevarsson A., Brazhnikov E., Garber M., Zheltonosova J., Chirgadze, al-Karadaghi S., Svensson L.A., and Liljas A. 1994. Three-dimensional structure of the ribosomal translocase: Elongation factor G from *Thermus thermophilus*. *EMBO J.* **13:** 3669.

Agrawal R.K., Penczek P., Grassucci R.A., and Frank J. 1998. Visualization of elongation factor G on the *Escherichia coli* 70S ribosome: The mechanism of translocation. *Proc. Natl. Acad. Sci.* **95:** 6134.

Agrawal R.K., Heagle A.B., Penczek P., Grassucci R.A., and Frank J. 1999. EF-G-dependent GTP hydrolysis induces translocation accompanied by large conformational changes in the 70S ribosome. *Nat. Struct. Biol.* **6:** 643.

al-Karadaghi S., Aevarsson A., Garber M., Zheltonosova J., and Liljas A. 1996. The structure of elongation factor G in complex with GDP: Conformational flexibility and nucleotide exchange. *Structure* **4:** 555.

Ban N., Nissen P., Hansen J., Capel M., Moore P.B., and Steitz T.A. 1999. Placement of protein and RNA structures into a 5 Å-resolution map of the 50S ribosomal subunit. *Nature* **400:** 841.

Belitsina N.V., Glukhova M.A., and Spirin A.S. 1976. Stepwise elongation factor G-promoted elongation of polypeptides on the ribosome without GTP cleavage. *J. Mol. Biol.* **108:** 609.

Bilgin N., Claesens F., Pahverk H., and Ehrenberg M. 1992. Kinetic properties of *Escherichia coli* ribosomes with altered forms of S12. *J. Mol. Biol.* **224:** 1011.

Bollen A., Davies J., Ozaki M., and Mizushima S. 1968. Ribosomal protein conferring sensitivity to the antibiotic spectinomycin in *Escherichia coli*. *Science* **165:** 85.

Bretscher M.S. 1968. Translocation in protein synthesis: A hybrid structure model. *Nature* **218:** 675.

Brune M., Hunter J.L., Corrie J.E., and Webb M.R. 1994. Direct, real-time measurement of rapid inorganic phosphate release using a novel fluorescent probe and its application to actomyosin subfragment 1 ATPase. *Biochemistry* **33:** 8262.

Carter A.P., Clemons W.M., Brodersen D.E., Morgan-Warren R.J., Wimberly B.T., and Ramakrishnan V. 2000. Functional insights from the structure of the 30S ribosomal subunit and its interactions with antibiotics. *Nature* **407:** 340.

Czworkowski J., Wang J., Steitz T.A., and Moore P.B. 1994. The crystal structure of elongation factor G complexed with GDP, at 2.7 Å resolution. *EMBO J.* **13:** 3661.

Dixon M. and Webb E.C. 1979. *Enzymes*. Academic Press, New York.

Frank J. 2000. The ribosome—A macromolecular machine par excellence. *Chem. Biol.* **7:** R133.

Frank J. and Agrawal R.K. 2000. A ratchet-like inter-subunit reorganization of the ribosome during translocation. *Nature* **406:** 318.

Gavrilova L.P. and Spirin A.S. 1972. Mechanism of translocation in ribosomes. II. Activation of spontaneous (nonenzymic) translocation in ribosomes of *Escherichia coli* by p-chloromercuribenzoate. *Mol. Biol.* **6:** 248.

———. 1974. "Nonenzymatic" translation. *Methods Enzymol.* **30:** 452.

Gnirke A., Geigenmuller U., Rheinberger H.J., and Nierhaus K.H. 1989. The allosteric three-site model for the ribosomal elongation cycle. Analysis with a heteropolymeric mRNA. *J. Biol. Chem.* **264:** 7291.

Johanson U. and Hughes D. 1994. Fusidic acid-resistant mutants define three regions in elongation factor G of *Salmonella typhimurium*. *Gene* **143:** 55.

Johanson U., Aevarsson A., Liljas A., and Hughes D. 1996. The dynamic structure of EF-G studied by fusidic acid resistance and internal revertants. *J. Mol. Biol.* **258:** 420.

Joseph S. and Noller H.F. 1998. EF-G-catalyzed translocation of anticodon stem-loop analogs of transfer RNA in the ribosome. *EMBO J.* **17:** 3478.

Kaziro Y. 1978. The role of guanosine 5′-triphosphate in polypeptide chain elongation. *Biochim. Biophys. Acta* **505:** 95.

Laurberg M., Kristensen O., Martemyanov K., Gudkov A.T., Nagaev I., Hughes D., and Liljas A. 2000. Structure of a mutant EF-G reveals domain III and possibly the fusidic acid binding site. *J. Mol. Biol.* **303:** 593.

Lill R. and Wintermeyer W. 1987. Destabilization of codon-anticodon interaction in the ribosomal exit site. *J. Mol. Biol.* **196:** 137.

Lill R., Robertson J.M., and Wintermeyer W. 1986. Affinities of tRNA binding sites of ribosomes from *Escherichia coli*. *Biochemistry* **25:** 3245.

———. 1989. Binding of the 3′ terminus of tRNA to 23S rRNA in the ribosomal exit site actively promotes translocation. *EMBO J.* **8:** 3933.

Moazed D. and Noller H.F. 1989a. Interaction of tRNA with 23S rRNA in the ribosomal A, P, and E sites. *Cell* **57:** 585.

———. 1989b. Intermediate states in the movement of transfer RNA in the ribosome. *Nature* **342:** 142.

———. 1990. Binding of tRNA to the ribosomal A and P sites protects two distinct sets of nucleotides in 16 S rRNA. *J. Mol. Biol.* **211:** 135.

Moazed D., Robertson J.M., and Noller H.F. 1988. Interaction of elongation factors EF-G and EF-Tu with a conserved loop in 23S RNA. *Nature* **334:** 362.

Modolell J. and Vazquez D. 1977. The inhibition of ribosomal translocation by viomycin. *Eur. J. Biochem.* **81:** 491.

Modolell J., Girbes T., and Vazquez D. 1975. Ribosomal translocation promoted by guanylylimido diphosphate and guanylyl-methylene diphosphonate. *FEBS Lett.* **60:** 109.

Mohr D., Wintermeyer W., and Rodnina M.V. 2000. Arginines 29 and 59 of elongation factor G are important for GTP hydrolysis or translocation on the ribosome. *EMBO J.* **19:** 3458.

Moine H. and Dahlberg A.E. 1994. Mutations in helix 34 of *Escherichia coli* 16 S ribosomal RNA have multiple effects on ribosome function and synthesis. *J. Mol. Biol.* **243:** 402.

Peske F., Matassova N.B., Savelsbergh A., Rodnina M.V., and Wintermeyer W. 2000. Conformationally restricted elongation factor G retains GTPase activity but is inactive in translocation on the ribosome. *Mol. Cell* **6:** 501.

Pestka S. 1969. Studies on the formation of transfer ribonucleic acid-ribosome complexes. VI. Oligopeptide synthesis and translocation on ribosomes in the presence and absence of soluble transfer factors. *J. Biol. Chem.* **244:** 1533.

Prescott C.D. and Kornau H.C. 1992. Mutations in *E. coli* 16S rRNA that enhance and decrease the activity of a suppressor tRNA. *Nucleic. Acids Res.* **20:** 1567.

Robertson J.M. and Wintermeyer W. 1987. Mechanism of ribosomal translocation. tRNA binds transiently to an exit site before leaving the ribosome during translocation. *J. Mol. Biol.* **196:** 525.

Rodnina M.V., Fricke R., and Wintermeyer W. 1994. Transient conformational states of aminoacyl-tRNA during ribosome binding catalyzed by elongation factor Tu. *Biochemistry* **33:** 12267.

Rodnina M.V., Savelsbergh A., Katunin V.I., and Wintermeyer W. 1997. Hydrolysis of GTP by elongation factor G drives tRNA movement on the ribosome. *Nature* **385:** 37.

Rodnina M.V., Pape T., Savelsbergh A., Mohr D., Matassova N.B., and Wintermeyer W. 2000a. Mechanism of partial reactions of the elongation cycle catalyzed by elongation factors Tu and G. In *The ribosome: Structure, function, antibiotics, and cellular interactions* (ed. R.A. Garrett et al.), p. 301. ASM Press, Washington, D.C.

Rodnina M.V., Savelsbergh A., Matassova N.B., Katunin V.I., Semenkov Y.P., and Wintermeyer W. 1999. Thiostrepton inhibits turnover but not GTP hydrolysis by elongation factor G on the ribosome. *Proc. Natl. Acad. Sci.* **96:** 9586.

Rodnina M.V., Stark H., Savelsbergh A., Wieden H.-J., Mohr D., Matassova N.B., Peske F., Daviter T., Gualerzi C.O., and Wintermeyer W. 2000b. GTPase mechanisms and functions of translation factors on the ribosome. *Biol. Chem.* **381:** 377.

Savelsbergh A., Matassova N.B., Rodnina M.V., and Wintermeyer W. 2000a. Role of domains 4 and 5 in elongation factor G functions on the ribosome. *J. Mol. Biol.* **300:** 951.

Savelsbergh A., Mohr D., Wilden B., Wintermeyer W., and Rodnina M.V. 2000b. Stimulation of the GTPase activity of translation elongation factor G by ribosomal protein L7/12. *J. Biol. Chem.* **275:** 890.

Semenkov Y.P., Rodnina M.V., and Wintermeyer W. 1996. The "allosteric three-site model" of elongation cannot be confirmed in a well-defined ribosome system from *Escherichia coli. Proc. Natl. Acad. Sci.* **93:** 12183.

———. 2000. Energetic contribution of tRNA hybrid state formation to translocation catalysis on the ribosome. *Nat. Struct. Biol.* **7:** 1027.

Semenkov Y.P., Shapkina T.G., and Kirillov S.V. 1992. Puromycin reaction of the A-site bound peptidyl-tRNA. *Biochimie* **74:** 411.

Serdyuk I., Baranov V., Tsalkova T., Gulyamova D., Pavlov M., Spirin A., and May R. 1992. Structural dynamics of translating ribosomes. *Biochimie* **74:** 299.

Sigmund C.D., Ettayebi M., and Morgan E.A. 1984. Antibiotic resistance mutations in 16S and 23S ribosomal RNA genes of *Escherichia coli. Nucleic. Acids Res.* **12:** 4653.

Spirin A.S. 1968. On the mechanism of ribosome function. The hypothesis of locking-unlocking of subparticles. *Dokl. Akad. Nauk. SSSR* **179:** 1467.

———. 1970. A model of the functioning ribosome: Locking and unlocking of the ribosome subparticles. *Cold Spring Harbor Symp. Quant. Biol.* **34:** 197.

———. 1985. Ribosomal translocation: Facts and models. *Prog. Nucleic. Acid Res. Mol. Biol.* **32:** 75.

———. 1999. *Ribosomes.* Kluwer Academic / Plenum, New York.

Spirin A.S., Baranov V.I., Polubesov G.S., Serdyuk I.N., and May R.P. 1987. Translocation makes the ribosome less compact. *J. Mol. Biol.* **194:** 119.

Stark H., Rodnina M.V., Wieden H.-J., van Heel M., and Wintermeyer W. 2000. Large-scale movement of elongation factor G and extensive conformational change of the ribosome during translocation. *Cell* **100:** 301.

Stark H., Orlova E.V., Rinke-Appel J., Jünke N., Mueller F., Rodnina M.V., Wintermeyer W., Brimacombe R., and van Heel M. 1997. Arrangement of tRNAs in pre- and post-translocational ribosomes revealed by electron cryomicroscopy. *Cell* **88:** 19.

van Heel M. 2000. Unveiling ribosomal structures: The final phases. *Curr. Opin. Struct. Biol.* **10:** 259.

Wilson K.S. and Noller H.F. 1998. Mapping the position of translational elongation factor EF-G in the ribosome by directed hydroxyl radical probing. *Cell* **92:** 131.

Yusupov M.M., Yusupova G.Z., Baucom A., Lieberman K., Earnest T.N., Cate J.H., and Noller H.F. 2001. Crystal structure of the ribosome at 5.5 Å resolution. *Science* **292:** 883.

Modulation of Translation Termination Mechanisms by *cis*- and *trans*-Acting Factors

D.M. Janzen and A.P. Geballe

Divisions of Human Biology and Clinical Research, Fred Hutchinson Cancer Research Center,
Seattle, Washington 98109; Departments of Medicine and Microbiology,
University of Washington, Seattle, Washington 98115

Translation termination is a ubiquitous but incompletely characterized step in gene expression. Recent structural and functional studies are providing an increasingly refined view of this complex series of reactions (Fig. 1). For example, the identity of the termination codons in most organisms has been known for decades, but the reassignment of these codons for use as sense codons in some settings and the influence of the surrounding nucleotides and codons on termination efficiency have been discovered only more recently (Tate and Mannering 1996; Mottagui-Tabar et al. 1998; Bertram et al. 2001; Cassan and Rousset 2001; Lehman 2001). Two classes of proteins are necessary for efficient termination. Class I release factors (RF1 and RF2 in prokaryotes, eRF1 in eukaryotes) recognize the termination codon and trigger peptidyl-tRNA hydrolysis in a reaction catalyzed by the peptidyl-transferase center of the ribosome. The recent solution of the eRF1 crystal structure supports the concept of mimicry of tRNAs by proteins, such as eRF1, that target the ribosomal A site (Ito et al. 1996; Nakamura et al. 2000; Song et al. 2000). Class II release factors (RF3 in prokaryotes, eRF3 in eukaryotes) interact with class I factors, hydrolyze GTP, and enhance the termination activity, at least in part by stimulating release of class I factors from the A site and thereby enabling their reuse (Zhouravleva et al. 1995; Le Goff et al. 1997; Crawford et al. 1999; Karimi et al. 1999). Despite their similar activities, eRF3, unlike RF3, associates with its class I partner independent of ribosomes and is encoded by an essential gene (Stansfield et al. 1995; Zhouravleva et al. 1995; Ito et al. 1996; Paushkin et al. 1997).

Along with discoveries that support the standard model of the termination reaction are other findings that require consideration of modified and new models. For example, the non-Mendelian heritable trait Psi (which increases nonsense suppression efficiency) appears to result from prion-like conformational alterations in the yeast eRF3 protein (Wickner et al. 2000). Whether this is a general property of eukaryotic class II release factors is not yet known (Jean-Jean et al. 1996). The simple view that class I and class II release factors alone constitute the complete termination machinery needs to be reconsidered due to the discovery of interactions between these factors and other proteins. Upf proteins essential for nonsense-mediated decay in *Saccharomyces cerevisiae* interact with eRF1 and eRF3, thus providing a biochemical link between termination and subsequent steps in RNA metabolism (Welch et al. 2000).

Another phenomenon in which termination appears to be critical is the translational control of eukaryotic genes by short upstream open reading frames (uORFs) contained in transcript leader regions (Geballe and Sachs 2000; Morris and Geballe 2000). In at least some cases, synthesis of the major gene product requires that ribosomes translate the uORF, terminate, and then reinitiate downstream. The factors and mechanisms that determine whether a ribosome will remain associated with an mRNA after having completed translation of a uORF and whether it will be able to reinitiate downstream are not well understood.

Here, we report progress in understanding the regulatory mechanism mediated by the unusual class of uORFs that act in a peptide-sequence-dependent manner (Geballe and Sachs 2000; Morris and Geballe 2000). Although few such uORFs have been characterized thus far, their presence in divergent biological systems, ranging from yeast to mammals, suggests that they are likely to be widespread. In fact, analogous mechanisms in which the nascent peptide affects ribosome function occur in prokaryotes as well (Lovett and Rogers 1996). Increasing evidence implicates termination as the key regulatory step in these cases. Dissection of the mechanism of these *cis*-acting regulators of termination, along with the functions of the participating *trans*-acting factors, should enhance our understanding of the basic termination mechanism and of its role in regulating gene expression.

INHIBITION OF TRANSLATION TERMINATION BY HUMAN CYTOMEGALOVIRUS uORF2

The potential of eukaryotic uORFs to inhibit translation in *cis* is consistent with the model that 43S ribosomal subunits load on an mRNA near the 5′ end and scan in a 3′ direction in search of an initiation codon (Kozak 1999). In some cases, an upstream AUG codon and its associated uORF obstruct the ribosome from reaching the downstream AUG codon. Alternatively, a ribosome may leak past an upstream AUG codon that is surrounded by a suboptimal context of nucleotides or may reinitiate after having translated the uORF and thus translate the down-

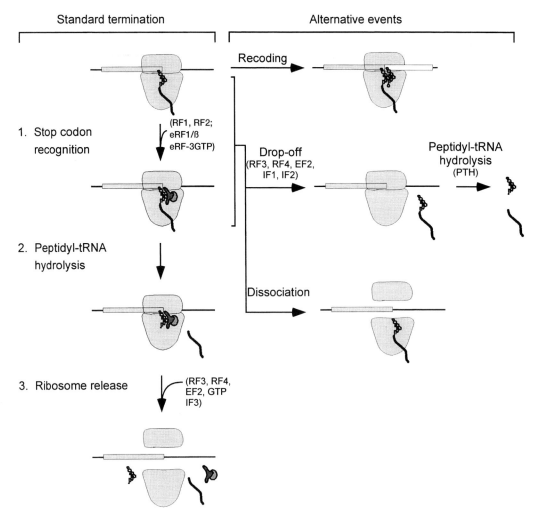

Figure 1. Standard and alternative translation termination mechanisms. Factors involved in the various steps are indicated in parentheses.

stream cistron efficiently despite the uORF. An unexpected feature of a few well-characterized uORFs is that their peptide-coding sequences are critical determinants of their inhibitory activity, suggesting that the nascent peptides participate in the regulatory mechanism.

In the case of the cytomegalovirus (CMV) UL4 gene, a 22-codon uORF (uORF2) inhibits downstream translation (Fig. 2). Mutation of the uORF2 initiation codon in the virus or in transgenes expressed in cells or cell-free extracts increases expression of the downstream cistron (Schleiss et al. 1991; Cao and Geballe 1996a; Alderete et al. 2001). Importantly, missense mutations at each of several positions within the uORF allow moderate (arrowheads in Fig. 2) or efficient (arrows) downstream translation, whereas synonymous mutations consistently retain the inhibitory properties of the wild-type uORF2 (Degnin et al. 1993; Alderete et al. 1999). The codons essential for inhibition are dispersed along ORF2, suggesting that more than one region of the nascent peptide might be required for its activity.

Several observations suggest that uORF2 inhibits downstream translation by repressing termination at its own stop codon. Mutation of the final proline codon to alanine or threonine, or of the termination codon to a sense codon, eliminates the inhibitory effect (Degnin et al. 1993). Thus, translation of the entire uORF is necessary for inhibition, and carboxy-terminal extensions, even by only a single amino acid, allow efficient downstream translation. Although the uORF2 AUG codon has a suboptimal context and captures only a small fraction of ribosomes that load on the mRNA, it exerts a potent inhibitory effect (Cao and Geballe 1995). A model to reconcile this paradox proposes that ribosomes that translate uORF2 stall at the termination site. The resulting roadblock obstructs subsequent ribosomes that otherwise would have been able to bypass uORF2 by leaky scanning.

Two sets of experiments provide direct support for this model. First, toeprint assays, in which the positions of ribosomes on an mRNA are detected by inhibition of a reverse transcription reaction, show that ribosomes stall at the uORF2 termination site (Fig. 2) (Cao and Geballe 1996a, 1998; Alderete et al. 2001). Second, the nascent uORF2 peptide remains covalently linked to the tRNA[Pro]

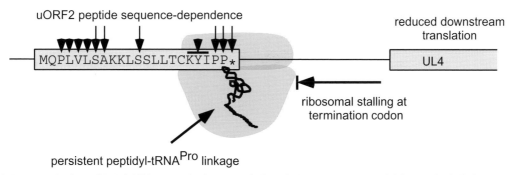

Figure 2. Cytomegalovirus uORF2 inhibitory mechanism. Translation of uORF2 represses peptidyl-tRNA hydrolysis, causes ribosomal stalling, and inhibits downstream translation by a peptide-sequence-dependent mechanism. Missense mutations of individual codons (or jointly of the indicated KYI triplet) alleviate most (*vertical arrows*) or part (*arrowheads*) of the inhibitory effect of uORF2 on downstream translation.

that decodes the last sense codon of uORF2 (Cao and Geballe 1996b). Both of these properties depend on the same uORF2 codons as does inhibition of downstream translation, consistent with the hypothesis that the translational inhibition is caused by the failure of the peptidyl-tRNA hydrolysis reaction and the resulting ribosomal stalling.

Analyses of the kinetics of the ribosomal blockade and of peptidyl-tRNA hydrolysis suggest that termination of uORF2 translation follows an alternative sequence of events compared to conventional translation (Fig. 1). The half-life of ribosomal stalling at the uORF2 termination site in reticulocyte extracts is less than 15 minutes, whereas linkage of the uORF2 peptide and tRNA is quite stable for more than 30 minutes (Cao and Geballe 1998). Thus, ribosomes vacate the termination site on the mRNA without prior hydrolysis of the peptidyl tRNA. The fact that the ribosome dissociates from the mRNA does not clarify whether the peptidyl tRNA dissociates from the ribosome. In fact, a substantial portion of the peptidyl tRNA remains associated with ribosomes even when the ribosomes are no longer stalled at the termination site (Cao and Geballe 1998). Thus, dissociation of the ribosome after translation of uORF2 appears to differ from the process of "drop-off" (Fig. 1) described in prokaryotes, in which intact peptidyl tRNAs are released from the ribosome (Menninger 1976).

Our current model of the uORF2 mechanism suggests that uORF2 inhibits a step prior to peptidyl-tRNA hydrolysis during translation termination (step 2 in Fig. 1). As a result, ribosomes persist for a prolonged but not indefinite period at the termination codon and block other ribosomes from gaining access to the downstream cistron. Eventually, ribosomes dissociate from the mRNA, enabling the downstream ORF to be translated by a leaky scanning mechanism until uORF2 is again translated and the cycle repeats (Cao and Geballe 1995; Morris and Geballe 2000). This mechanism seems likely to be mediated by an interaction of the nascent peptidyl tRNA with a component of the ribosome or with a translation factor, resulting in either a block to eRF1 entry into the ribosome or inhibition of the activity of eRF1 or another factor needed for peptidyl-tRNA hydrolysis.

INHIBITION OF GENERAL TRANSLATION BY WILD-TYPE AND MUTANT eRF1 IN CELL-FREE EXTRACTS

Before exploring the role of eRF1 in the uORF2 inhibitory mechanism, we first measured the effects of eRF1 on general translation in cell-free extracts. We programmed rabbit reticulocyte lysates, a system in which all the hallmarks of the uORF2 mechanism have been observed (Cao and Geballe 1996a,b), with *lacZ* transcripts containing various transcript leader sequences. Addition of human eRF1 inhibited β-gal expression in a dose-dependent manner (Fig. 3). For these experiments, eRF1 containing an amino-terminal histidine tag was expressed in *Escherichia coli* and purified on Ni-NTA, followed by MonoQ column chromatography (Frolova et al. 1998). The addition of 0.2 μM eRF1, an amount approximately equal to the level of the endogenous protein (data not shown), reduced translation by ~25%, whereas 0.8 μM added eRF1 inhibited β-gal expression by ~10-fold. Nonspecific protein effects were not likely to explain the reduced translation since no inhibition was detected when the eRF1 was replaced with bovine serum albumin at the same molar concentrations or when the eRF1 was heat-denatured (data not shown). Similar results were obtained with independent preparations of the eRF1 and in assays in which translation was measured by autoradiography of ^{35}S-labeled protein products. The data shown in Figure 3 were obtained by translation of a transcript containing a P22A mutant version of uORF2 (made from pEQ439 [described in Cao and Geballe 1996a]) that does not inhibit downstream translation or cause ribosomal stalling or persistent linkage of the peptidyl-tRNA bond. Similar results were obtained using monocistronic mRNAs containing only *lacZ*, without any uORF2 sequences (data not shown, and see Fig. 5). In another report, addition of eRF1 to cell-free extracts reduced synthesis of a protein expressed by nonsense suppression, consistent with eRF1 acting as an antisuppressor (Drugeon et al. 1997). However, the shorter protein product was not consistently increased in proportion to the decrease in the longer product, suggesting that overall protein synthesis was inhibited by eRF1 in that system, too.

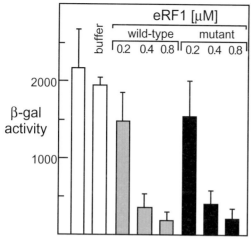

Figure 3. Inhibition of translation by wild-type and mutant eRF1. *LacZ* transcripts containing the P22A mutant uORF2 were translated in reticulocyte lysates with no additions (*left bar*), with added eRF1 storage buffer (*buffer*), or with addition of the indicated amounts of purified wild-type or mutant histidine-tagged eRF1.

To determine whether the inhibitory effect of eRF1 requires fully functional protein, we constructed and purified a histidine-tagged mutant form of eRF1 in which the two glycines (codons 183 and 184) within the GGQ minidomain were changed to alanines. The GGQ domain is conserved among all known class I release factors of prokaryotes, eukaryotes, and archaea (Frolova et al. 1999). Mutations of these amino acids inactivate the termination activity of eRF1 but appear not to preclude its binding to the ribosomal A site or interaction with eRF3 (Frolova et al. 1999; Merkulova et al. 1999; Song et al. 2000). In our assays, this mutant form of eRF1 inhibited translation to a similar extent as did the wild-type protein (Fig. 3). Thus, the inhibitory effect of eRF1 on general translation does not require catalytic activity mediated by the GGQ motif.

In principle, eRF1 might inhibit translation at any of several steps, including ribosomal association with the mRNA, initiation, elongation, or even, paradoxically, termination. To differentiate among these mechanisms, we conducted toeprint assays (Cao and Geballe 1996a) using transcripts containing the uORF2 P22A mutant, along with varying amounts of wild-type or mutant eRF1 (Fig. 4). Addition of even a high concentration of wild-type or mutant eRF1 did not result in any new toeprint products. Thus, the inhibitory effect of eRF1 on translation does not appear to induce ribosomal stalling at any unique position, including the uORF2 initiation or termination sites. In the presence of hygromycin, which blocks elongation, wild-type eRF1 did not alter the ability of ribosomes to initiate at the uORF2 AUG codon. Thus, eRF1 appears not to interfere with, or prior to, initiation. A plausible explanation for these results is that eRF1 slows elongation, resulting in ribosomes being dispersed over multiple codons of uORF2. Alternatively, eRF1 may cause the ribosomes to dissociate from the transcript at some point after initiation.

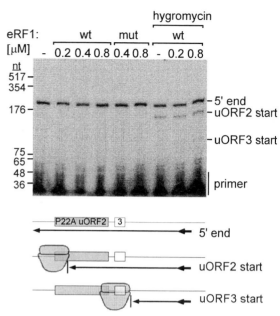

Figure 4. Excess eRF1 does not prevent translation initiation or cause ribosomal stalling at a unique position on uORF2. Toeprint assays were performed (Cao and Geballe 1996a) after translation of transcripts containing the P22A mutant uORF2 with addition of wild-type or mutant eRF1. Where indicated, hygromycin (16 μg/ml) was added before translation. The interpretation of the extension products is depicted (*bottom*).

Addition of eRF1 could inhibit elongation by one of several mechanisms. High levels of wild-type eRF1 might stimulate premature termination at sense codons. In cell-free termination assays using prokaryotic extracts, RF1 and RF2 have measurable termination activity at certain sense codons (Freistroffer et al. 2000). However, mutant eRF1 does not catalyze termination at stop codons (Frolova et al. 1999) and would not be expected to do so at sense codons. Alternatively, eRF1 could simply compete with aminoacyl tRNAs for A-site binding and thus retard elongation, as has been suggested for bacterial RF1 (Short et al. 1999). Since the wild-type and mutant forms of eRF1 are likely to have similar affinities for the ribosomal A site, both would be expected to inhibit elongation by this mechanism with similar dose-response profiles as was observed (Fig. 3). A third possibility is that noncognate binding of eRF1 to a sense codon in the A site might stimulate drop-off of the peptidyl tRNA from the ribosome–mRNA complex or provoke dissociation of the ribosome–peptidyl-tRNA from the mRNA (Fig. 1). In prokaryotes, drop-off is stimulated by some factors that bind the A site, including IF1, IF2, RF3, and RF4 (formerly RRF) (Heurgue-Hamard et al. 1998; Karimi et al. 1998). The interaction of eRF1 and eRF3 (Stansfield et al. 1995; Zhouravleva et al. 1995), coupled with the speculation that eRF3 might have RF4-like activity (Nakamura and Ito 1998; Kisselev and Buckingham 2000), provides a potential mechanism by which added eRF1 could trigger drop-off. Finally, our studies do not exclude the possibility that bacterially expressed histidine-tagged eRF1 differs from the natural protein. Posttranslational modification of the glutamine in the GGQ motif is necessary for full activity of RF2 and is incomplete when this protein is overexpressed in bacteria (Dincbas-Renqvist et al. 2000). However, histidine-tagged eRF1 expressed in bacteria has been shown to be active (Drugeon et al. 1997; Frolova et al. 1999, 2000; Merkulova et al. 1999).

If high levels of eRF1 inhibit translation, we might expect that overexpression of eRF1 would not be tolerated in vivo. Overexpression of *E. coli* RF2 but not *Salmonella* RF2 is toxic (Uno et al. 1996). Although several reports have described eRF1 overexpression, we are not aware of data that clarify whether overexpression of wild-type eRF1 inhibits translation in cells. For example, Le Goff et al. (1997) documented eRF1 overexpression in transient transfection assays, but the viability and translational activity in the particular cells that expressed high levels of eRF1 were not ascertained. Genetic experiments have revealed phenotypes attributable to overexpression of eRF1 (SUP45) in yeast, but the actual levels of eRF1 were not examined (Stansfield et al. 1995; Derkatch et al. 1998). Thus, defining the settings in which excess eRF1 inhibits translation and clarifying the mechanism will require further study.

THE eRF1 GGQ DOMAIN IS REQUIRED FOR FULL INHIBITION BY uORF2

One possible mechanism by which a nascent peptide might inhibit termination is by blocking entry of eRF1 to the A site. If the uORF2 peptidyl tRNA competes with eRF1 for the A site, then excess eRF1 might abrogate uORF2 inhibitory activities. Therefore, we next measured the effects of the wild-type and mutant eRF1 on expression of transcripts containing the wild-type uORF2 (pEQ438 in Cao and Geballe [1996a]). Unexpectedly, compared to the effect of the wild-type eRF1, mutant eRF1 caused a relative increase in translation downstream from wild-type uORF2 (gray bars in Fig. 5). In Figure 5, β-gal activities are depicted as the ratio of activities obtained with mutant compared to wild-type eRF1 to correct for uORF2-independent translational inhibition by eRF1 (Fig. 3). Consistent with previous results, the wild-type and mutant forms of eRF1 inhibited β-gal activity to a similar extent when control transcripts containing no uORF2 sequences (white bars) or the P22A uORF2 mutant (black bars) were used.

To elucidate the mechanism underlying this puzzling result, we measured the effect of added eRF1 on the duration of ribosomal stalling at the uORF2 termination site using toeprint assays (Fig. 6). After 10 minutes of translation with no additions or with addition of 0.2 μM wild-type or mutant eRF1, 16 μg/ml hygromycin or 16 μg/ml anisomycin was added to prevent new ribosomes from translating uORF2 and reaching the termination codon after the stalled ribosomes vacate the site (Cao and Geballe 1998). Aliquots were removed at subsequent times and analyzed by toeprint assay. The intensity of the termination site toeprint bands was quantified by PhosphorImager analysis, with each time point normalized to the total RNA abundance estimated from the intensity of

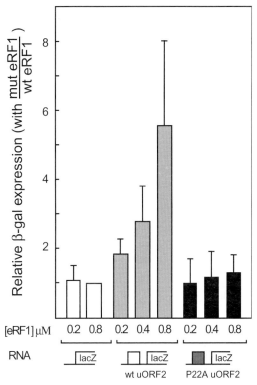

Figure 5. Mutant eRF1 causes a relative increase in translation downstream from wt uORF2. β-gal expression was measured after translation of transcripts containing no uORF2, wild-type uORF2, or the P22A uORF2 mutant in the presence of wild-type or mutant eRF1. The ratios of β-gal expression in the presence of mutant eRF1 compared to wild-type eRF1 are shown.

the 5'-end signal. Consistent with previous results (Cao and Geballe 1998), the $t_{1/2}$ of the termination toeprint in the absence of added eRF1 was ~6 minutes (white bars in Fig. 6). Addition of wild-type eRF1 had little effect of the stalling kinetics (gray bars). In contrast, addition of the mutant eRF1 reduced the duration of stalling by a factor of 2 to 3 (black bars). The mean $t_{1/2}$ (+S.E.) values shown in Figure 6 are based on three to six independent experiments (except for the $t_{1/2}$ for wild-type eRF1 using anisomycin, which was measured only once). These results suggest that the mutant eRF1 augments downstream translation by reducing the duration of stalling at the uORF2 termination codon.

Several potential alternatives to the conventional termination reaction can be envisioned to explain these results (Fig. 1). First, mutant eRF1 might promote nonsense suppression of the uORF2 termination codon, resulting in the stalled ribosome translating to the next in-frame stop codon. At that point, the ribosome would presumably terminate efficiently and then either reinitiate downstream or vacate the transcript and allow other 43S ribosomal subunits that have leaked past the uORF2 initiation codon to scan to the downstream AUG codon. Consistent with this explanation, eRF1 mutants that have alterations in the GGQ motif have been shown to act as dominant negative inhibitors of termination and to augment nonsense suppression (Frolova et al. 1999; Song et al. 2000). Al-

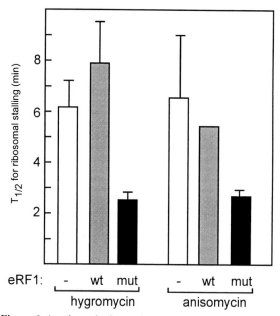

Figure 6. Accelerated release of the ribosome from the uORF2 termination site by mutant eRF1. The half-life of ribosomal stalling at the end of the wild-type uORF2 was measured using toeprint assays as described in the text.

ternatively, mutant eRF1 might act by causing a relative increase in the peptidyl-tRNA drop-off from the stalled ribosome, followed by disengagement of the ribosome from the transcript. Finally, mutant eRF1 might directly promote disengagement of the ribosomes from the mRNA, even without prior drop-off of the peptidyl tRNA from the ribosome.

Nonsense suppression is an unlikely explanation for our results. First, the toeprint assay used to measure the duration of ribosomal stalling employed elongation inhibitors that would preclude nonsense suppression and elongation to the next stop codon. Thus, nonsense suppression cannot easily explain the accelerated release of the ribosome from the uORF2 termination site. As well, the uORF2 peptide and peptidyl tRNA detected after addition of mutant eRF1 were the same apparent sizes as when wild-type eRF1 or no eRF1 was added, rather than larger as would be predicted by the nonsense suppression model (our unpublished data). Finally, notwithstanding data demonstrating an increase in expression of a nonsense-containing gene in yeast that express such an eRF1 GGQ mutant (Song et al. 2000), it is difficult to envision how such a mutant could augment nonsense suppression. Mutant eRF1 would be expected to compete with wild-type eRF1 for binding to the termination codon and thus would reduce termination efficiency, but it would also compete with suppressor tRNAs and thereby diminish nonsense suppression as well. Together, these arguments make the nonsense suppression hypothesis of the mutant eRF1 effects improbable.

The fact that wild-type and mutant eRF1s have differing effects on translation of the wild-type uORF2 but not the P22A mutant suggests that a specific interaction in-

volving the GGQ motif in the wild-type eRF1 and the proline at codon 22 of the wild-type uORF2 peptidyl tRNA might be critical for the uORF2 inhibitory mechanism. Mutation of either partner could disrupt such an interaction. Thus far, we have been unable to demonstrate an interaction between these two factors using yeast two-hybrid assays or GST-pulldown assays (data not shown). However, the postulated interaction might well be a weak one, except in the natural setting when the two components are held in close juxtaposition in the ribosomal A and P sites. In essence, this model proposes that an interaction of the wild-type eRF1 with the wild-type uORF2 peptidyl tRNA stabilizes an intermediate in the termination reaction. Our results do not distinguish whether mutant eRF1 actively accelerates the dissociation of the ribosome and the mRNA or simply allows it to occur, unimpeded by any stabilizing effect resulting from the interaction between the nascent peptidyl tRNA and eRF1. Regardless, these results suggest that the uORF2 peptidyl tRNA blocks termination after eRF1 has entered the ribosomal A site. Exactly how this interaction blocks hydrolysis activity will require further study. Since several amino acids of uORF2 that are quite distal from the terminal proline and that have presumably already entered the peptide exit channel of the 60S ribosomal subunit are also required for inhibitory activity (Fig. 2), other interactions between the uORF2 peptide and components of the termination machinery may also contribute to the inhibitory mechanism.

CONCLUSIONS

Recent results from a variety of systems provide increasing evidence that translation termination entails a series of more intricate and complex steps than has been appreciated in the past. In addition to recoding that results in readthrough past a termination codon, release of the nascent peptide while still attached to a tRNA and disassembly of the ribosome–mRNA complex without prior peptidyl-tRNA hydrolysis illustrate alternatives to the conventional termination reaction (Fig. 1). Beyond its importance as a ubiquitous step in the expression of all proteins, the likelihood that termination may be regulated and rate-limiting for expression of certain genes heightens the importance of understanding these mechanisms in more detail.

Results presented here suggest that eRF1 can inhibit translation, most likely during elongation. Several translation factors that target the ribosomal A site, including eRF1, mimic the structure of tRNA (Selmer et al. 1999; Kim et al. 2000; Nakamura et al. 2000; Song et al. 2000). Thus, it seems plausible that excess eRF1 and tRNAs might compete for A-site binding during elongation and at termination (Short et al. 1999). The ability of suppressor tRNAs to compete with class I release factors during termination is well established. Conversely, the release factors might enter the A site during elongation and thereby block access of the cognate tRNA and might even accidentally trigger peptide release. If such premature termination occurs, it is likely to be more frequent at near-cognate, compared to non-cognate, codons. In fact, UAU and UGG were found to be hot spots for erroneous in vitro termination mediated by RF1 and RF2, respectively (Freistroffer et al. 2000). Whether such premature termination contributes significantly to translational errors and whether it is ever used as a mechanism for producing a functional truncated protein is not clear. Regardless, the potential for interference suggests a need for balanced and regulated expression of class I release factors and other factors that compete for binding to the ribosomal A site.

Translation is generally considered to be a highly processive and accurate process. However, this view is challenged by several studies in *E. coli* (Menninger 1976; Manley 1978; Jorgensen and Kurland 1990; Freistroffer et al. 2000). For example, the drop-off of peptidyl tRNAs may be as high as 10^{-2} to 10^{-3} per elongation event, based on measurements made in peptidyl-tRNA hydrolase mutants (Menninger 1976). These released peptidyl tRNAs are then cleaved extraribosomally by peptidyl-tRNA hydrolase, an essential protein presumably needed to free the tRNAs for reuse (Menninger 1976; Hernandez-Sanchez et al. 1998; Tenson et al. 1999; Heurgue-Hamard et al. 2000; Menez et al. 2000). Drop-off may be an accidental event or a manifestation of a proofreading system that reduces synthesis of proteins in which a translational error has occurred. Whether peptidyl-tRNA drop-off occurs in eukaryotes is uncertain, although eukaryotes do have peptidyl-tRNA hydrolases (Kossel and Raj-Bhandary 1968; Jost and Bock 1969; Gross et al. 1992; Ouzounis et al. 1995), suggesting that they too must sometimes deal with the peptidyl-tRNA drop-off products.

Despite the occurrence of drop-off events, the ribosome typically continues translation until it reaches the termination codon. After peptidyl-tRNA hydrolysis and release of the nascent peptide, the ribosome is thought to disengage from the mRNA. Recent studies in prokaryotes suggest a model in which, after peptidyl-tRNA hydrolysis, RF3 removes RF1 from the A site (Karimi et al. 1999). Subsequently, RF4 and EF2 (formerly called EF-G) act to disengage the 50S subunit. Like eRF1, RF4 structurally mimics tRNA, binds to the A site, and may have activities during elongation as well as at termination (Janosi et al. 1996; Selmer et al. 1999; Kim et al. 2000). Finally, IF3 causes release of the uncharged tRNA from the P site of the 30S ribosome.

In contrast to this sequence of reactions and to drop-off of peptidyl tRNAs from ribosomes, our studies of uORF2 suggest that dissociation of a eukaryotic ribosome from an mRNA can occur prior to hydrolysis and even while the peptidyl tRNA remains associated with the ribosome (Fig. 1). Such a pathway might be advantageous by preventing sequestration of ribosomes or mRNAs in an inactive state.

Another mechanism for preventing trapping of ribosomes on mRNAs that lack stop codons has recently been elucidated in prokaryotes (Karzai et al. 2000). In such cases, there is no mRNA in the A site. An RNA molecule (known as SsrA, tmRNA, or 10Sa RNA) that has both tRNA and mRNA functions, permits continued translation with addition of a carboxy-terminal protein tag that targets the aberrant protein for rapid degradation. This system is essential in some but not all prokaryotes and has not yet been described in eukaryotes. Regardless, this mechanism further illustrates the diversity of events that can occur during and after translation termination.

A critical unanswered question raised by the study of uORFs is what determines the fate of the ribosome after peptidyl-tRNA hydrolysis. RF4 participates in the release of the ribosome from the mRNA in prokaryotes, but no RF4-like activity has been found in the eukaryotic translation machinery. The essentiality of eRF3, like RF4 but unlike RF3, and its extraribosomal association with eRF1 led to the speculation the eRF3 might have an RF4-like activity (Nakamura and Ito 1998; Kisselev and Buckingham 2000). Regardless of the underlying mechanism, the determination of whether a ribosome remains associated with or dissociates from an mRNA after termination can be consequential. Ribosomes clearly remain associated with the yeast *GCN4* mRNA after translating the first uORF, but after translation of the fourth uORF, they are unable to reinitiate, possibly because they dissociate (Hinnebusch 1996). The penultimate codon and the ten nucleotides downstream from the stop codon of the fourth uORF are key determinants of reinitiation efficiency. Studies in other systems point to intercistronic sequences and spacing as major determinants in the ability of the ribosomes to reinitiate (Kozak 1987; Lincoln et al. 1998; Wang and Wessler 1998; Child et al. 1999). Termination efficiency is mediated in part by the context defined by codons near the end of the reading frame and by nucleotides flanking the termination codon (Tate and Mannering 1996; Mottagui-Tabar et al. 1998; Bertram et al. 2001; Cassan and Rousset 2001; Lehman 2001). Whether these parameters influence the posttermination fate of the ribosome is an important unanswered question.

Events occurring during and after termination may be critical for translation of conventional monocistronic mRNAs as well as those with uORFs. The emerging hypothesis that mRNAs exist as functional circles, with their ends tethered by the interaction of the poly(A)-binding and cap-binding proteins, suggests that ribosomes may remain associated with the mRNA after termination and preferentially reinitiate translation on the same transcript (Sachs 2000). If this view is correct, then determinants that modulate the association of ribosome and mRNA after termination might have substantial effects on the synthesis of many proteins.

Termination and post-terminaton steps in translation are complex and incompletely characterized, yet they have potentially profound regulatory effects on gene expression. In addition to the benefits of dissecting this ubiquitous step for our fundamental understanding of gene expression, studies of termination and its regulation have potential practical applications. For example, modulation of termination may be useful as a therapy for genetic diseases in which nonsense mutations interrupt the critical gene-coding regions (Bedwell et al. 1997; Barton-Davis et al. 1999; Keeling et al. 2001). Although such mutations typically account for only a minority of cases of any particular genetic disease, they occur in many different diseases and thus such a therapeutic strategy could benefit a large number of individuals. As well, manipulation of termination efficiency has the potential to be useful in cancer therapy (Cazzola and Skoda 2000). In fact, the drug girodazole has antitumor activity and is thought to inhibit translation termination (Lavelle et al. 1991; Colson et al. 1992). Modulating termination efficiency may provide a way to selectively control expression of oncogenes and other regulatory genes, many of which are encoded by mRNAs with uORFs (Kozak 1991). Integration of advances in understanding the structures of the ribosome and translation factors, along with insights derived from functional analyses utilizing genetic and biochemical methods, and even possibly from clinical studies, portends an exciting future for research on this fundamental step in gene expression.

ACKNOWLEDGMENTS

We thank Jianhong Cao (Fred Hutchinson Cancer Research Center), Michael Torgov (University of Washington), and the Biotechnology Resource of the Fred Hutchinson Cancer Research Center for technical assistance. We also thank Dr. Gerald Fuller (University of Alabama) for providing a human eRF1 plasmid. This work was supported by grant AI26672 from the National Institutes of Health.

REFERENCES

Alderete J.P., Child S.J., and Geballe A.P. 2001. Abundant early expression of gpUL4 from a human cytomegalovirus mutant lacking a repressive upstream open reading frame. *J. Virol.* **75:** 7188.

Alderete J.P., Jarrahian S., and Geballe A.P. 1999. Translational effects of mutations and polymorphisms in a repressive upstream open reading frame of the human cytomegalovirus UL4 gene. *J. Virol.* **73:** 8330.

Barton-Davis E.R., Cordier L., Shoturma D.I., Leland S.E., and Sweeney H.L. 1999. Aminoglycoside antibiotics restore dystrophin function to skeletal muscles of mdx mice. *J. Clin. Invest.* **104:** 375.

Bedwell D.M., Kaenjak A., Benos D.J., Bebok Z., Bubien J.K., Hong J., Tousson A., Clancy J.P., and Sorscher E. J. 1997. Suppression of a CFTR premature stop mutation in a bronchial epithelial cell line. *Nat. Med.* **3:** 1280.

Bertram G., Innes S., Minella O., Richardson J., and Stansfield I. 2001. Endless possibilities: Translation termination and stop codon recognition. *Microbiology* **147:** 255.

Cao J. and Geballe A.P. 1995. Translational inhibition by a human cytomegalovirus upstream open reading frame despite inefficient utilization of its AUG codon. *J. Virol.* **69:** 1030.

———. 1996a. Coding sequence-dependent ribosomal arrest at termination of translation. *Mol. Cell. Biol.* **16:** 603.

———. 1996b. Inhibition of nascent-peptide release at translation termination. *Mol. Cell. Biol.* **16:** 7109.

———. 1998. Ribosomal release without peptidyl tRNA hydrolysis at translation termination in a eukaryotic system. *RNA* **4:** 181.

Cassan M. and Rousset J.P. 2001. UAG readthrough in mammalian cells: Effect of upstream and downstream stop codon contexts reveal different signals. *BMC Mol. Biol.* **2:** 3.

Cazzola M. and Skoda R. C. 2000. Translational pathophysiology: A novel molecular mechanism of human disease. *Blood*

95: 3280.

Child S.J., Miller M.K., and Geballe A.P. 1999. Translational control by an upstream open reading frame in the HER-2/neu transcript. *J. Biol. Chem.* **274:** 24335.

Colson G., Rabault B., Lavelle F., and Zerial A. 1992. Mode of action of the antitumor compound girodazole (RP 49532A, NSC 627434). *Biochem. Pharmacol.* **43:** 1717.

Crawford D.J., Ito K., Nakamura Y., and Tate W.P. 1999. Indirect regulation of translational termination efficiency at highly expressed genes and recoding sites by the factor recycling function of *Escherichia coli* release factor RF3. *EMBO J.* **18:** 727.

Degnin C.R., Schleiss M.R., Cao J., and Geballe A.P. 1993. Translational inhibition mediated by a short upstream open reading frame in the human cytomegalovirus gpUL4 (gp48) transcript. *J. Virol.* **67:** 5514.

Derkatch I.L., Bradley M.E., and Liebman S.W. 1998. Overexpression of the SUP45 gene encoding a Sup35p-binding protein inhibits the induction of the de novo appearance of the [PSI+] prion. *Proc. Natl. Acad. Sci.* **95:** 2400.

Dincbas-Renqvist V., Engstrom A., Mora L., Heurgue-Hamard V., Buckingham R., and Ehrenberg M. 2000. A post-translational modification in the GGQ motif of RF2 from *Escherichia coli* stimulates termination of translation. *EMBO J.* **19:** 6900.

Drugeon G., Jean-Jean O., Frolova L., Le Goff X., Philippe M., Kisselev L., and Haenni A.L. 1997. Eukaryotic release factor 1 (eRF1) abolishes readthrough and competes with suppressor tRNAs at all three termination codons in messenger RNA. *Nucleic Acids Res.* **25:** 2254.

Freistroffer D.V., Kwiatkowski M., Buckingham R.H., and Ehrenberg M. 2000. The accuracy of codon recognition by polypeptide release factors. *Proc. Natl. Acad. Sci.* **97:** 2046.

Frolova L.Y., Merkulova T.I., and Kisselev L.L. 2000. Translation termination in eukaryotes: Polypeptide release factor eRF1 is composed of functionally and structurally distinct domains. *RNA* **6:** 381.

Frolova L.Y., Tsivkovskii R.Y., Sivolobova G.F., Oparina N.Y., Serpinsky O.I., Blinov V.M., Tatkov S.I., and Kisselev L.L. 1999. Mutations in the highly conserved GGQ motif of class 1 polypeptide release factors abolish ability of human eRF1 to trigger peptidyl-tRNA hydrolysis. *RNA* **5:** 1014.

Frolova L.Y., Simonsen J.L., Merkulova T.I., Litvinov D.Y., Martensen P.M., Rechinsky V.O., Camonis J.H., Kisselev L.L., and Justesen J. 1998. Functional expression of eukaryotic polypeptide chain release factors 1 and 3 by means of baculovirus/insect cells and complex formation between the factors. *Eur. J. Biochem.* **256:** 36.

Geballe A.P. and Sachs M.S. 2000. Translational control by upstream open reading frames. In *Translational control of gene expression* (ed. N. Sonenberg et al.), p. 595. Cold Spring Harbor Laboratory Press, Cold Spring Harbor, New York.

Gross M., Crow P., and White J. 1992. The site of hydrolysis by rabbit reticulocyte peptidyl-tRNA hydrolase is the 3′-AMP terminus of susceptible tRNA substrates. *J. Biol. Chem.* **267:** 2080.

Hernandez-Sanchez J., Valadez J.G., Herrera J.V., Ontiveros C., and Guarneros G. 1998. Lambda bar minigene-mediated inhibition of protein synthesis involves accumulation of peptidyl-tRNA and starvation for tRNA. *EMBO J.* **17:** 3758.

Heurgue-Hamard V., Dincbas V., Buckingham R.H., and Ehrenberg M. 2000. Origins of minigene-dependent growth inhibition in bacterial cells. *EMBO J.* **19:** 2701.

Heurgue-Hamard V., Karimi R., Mora L., MacDougall J., Leboeuf C., Grentzmann G., Ehrenberg M., and Buckingham R.H. 1998. Ribosome release factor RF4 and termination factor RF3 are involved in dissociation of peptidyl-tRNA from the ribosome. *EMBO J.* **17:** 808.

Hinnebusch A.G. 1996 Translational control of *GCN4*: Gene specific regulation by phosphorylation of eIF2. In *Translational control* (ed. J.W.B. Hershey et al.), p. 199. Cold Spring Harbor Laboratory Press, Cold Spring Harbor, New York.

Ito K., Ebihara K., Uno M., and Nakamura Y. 1996. Conserved motifs in prokaryotic and eukaryotic polypeptide release factors: tRNA-protein mimicry hypothesis. *Proc. Natl. Acad. Sci.* **93:** 5443.

Janosi L., Hara H., Zhang S., and Kaji A. 1996. Ribosome recycling by ribosome recycling factor (RRF)—An important but overlooked step of protein biosynthesis. *Adv. Biophys.* **32:** 121.

Jean-Jean O., Le Goff X., and Philippe M. 1996. Is there a human [psi]? *C.R. Acad. Sci. III* **319:** 487.

Jorgensen F. and Kurland C.G. 1990. Processivity errors of gene expression in *Escherichia coli. J. Mol. Biol.* **215:** 511.

Jost J.P. and Bock R.M. 1969. Enzymatic hydrolysis of N-substituted aminoacyl transfer ribonucleic acid in yeast. *J. Biol. Chem.* **244:** 5866.

Karimi R., Pavlov M.Y., Buckingham R.H., and Ehrenberg M. 1999. Novel roles for classical factors at the interface between translation termination and initiation. *Mol. Cell* **3:** 601.

Karimi R., Pavlov M.Y., Heurgue-Hamard V., Buckingham R.H., and Ehrenberg M. 1998. Initiation factors IF1 and IF2 synergistically remove peptidyl-tRNAs with short polypeptides from the P-site of translating *Escherichia coli* ribosomes. *J. Mol. Biol.* **281:** 241.

Karzai A.W., Roche E.D., and Sauer R.T. 2000. The SsrA-SmpB system for protein tagging, directed degradation and ribosome rescue. *Nat. Struct. Biol.* **7:** 449.

Keeling K.M., Brooks D.A., Hopwood J.J., Li P., Thompson J.N., and Bedwell D.M. 2001. Gentamicin-mediated suppression of Hurler syndrome stop mutations restores a low level of alpha-L-iduronidase activity and reduces lysosomal glycosaminoglycan accumulation. *Hum. Mol. Genet.* **10:** 291.

Kim K.K., Min K., and Suh S.W. 2000. Crystal structure of the ribosome recycling factor from *Escherichia coli. EMBO J.* **19:** 2362.

Kisselev L.L. and Buckingham R.H. 2000. Translational termination comes of age. *Trends Biochem. Sci.* **25:** 561.

Kossel H. and RajBhandary U.L. 1968. Studies on polynucleotides. LXXXVI. Enzymic hydrolysis of N-acylaminoacyl-transfer RNA. *J. Mol. Biol.* **35:** 539.

Kozak M. 1987. Effects of intercistronic length on the efficiency of reinitiation by eucaryotic ribosomes. *Mol. Cell. Biol.* **7:** 3438.

———. 1991. An analysis of vertebrate mRNA sequences: Intimations of translational control. *J. Cell Biol.* **115:** 887.

———. 1999. Initiation of translation in prokaryotes and eukaryotes. *Gene* **234:** 187.

Lavelle F., Zerial A., Fizames C., Rabault B., and Curaudeau A. 1991. Antitumor activity and mechanism of action of the marine compound girodazole. *Investig. New Drugs* **9:** 233.

Le Goff X., Philippe M., and Jean-Jean O. 1997. Overexpression of human release factor 1 alone has an antisuppressor effect in human cells. *Mol. Cell. Biol.* **17:** 3164.

Lehman N. 2001. Molecular evolution: Please release me, genetic code. *Curr. Biol.* **11:** R63.

Lincoln A.J., Monczak Y., Williams S.C., and Johnson P.F. 1998. Inhibition of CCAAT/enhancer-binding protein alpha and beta translation by upstream open reading frames. *J. Biol. Chem.* **273:** 9552.

Lovett P.S. and Rogers E.J. 1996. Ribosome regulation by the nascent peptide. *Microbiol. Rev.* **60:** 366.

Manley J.L. 1978. Synthesis and degradation of termination and premature-termination fragments of beta-galactosidase in vitro and in vivo. *J. Mol. Biol.* **125:** 407.

Menez J., Heurgue-Hamard V., and Buckingham R.H. 2000. Sequestration of specific tRNA species cognate to the last sense codon of an overproduced gratuitous protein. *Nucleic Acids Res.* **28:** 4725.

Menninger J.R. 1976. Peptidyl transfer RNA dissociates during protein synthesis from ribosomes of *Escherichia coli. J. Biol. Chem.* **251:** 3392.

Merkulova T.I., Frolova L.Y., Lazar M., Camonis J., and Kisse-

lev L.L. 1999. C-terminal domains of human translation termination factors eRF1 and eRF3 mediate their in vivo interaction. *FEBS Lett.* **443:** 41.

Morris D.R. and Geballe A.P. 2000. Upstream open reading frames as regulators of mRNA translation. *Mol. Cell. Biol.* **20:** 8635.

Mottagui-Tabar S., Tuite M.F., and Isaksson L.A. 1998. The influence of 5′ codon context on translation termination in *Saccharomyces cerevisiae. Eur. J. Biochem.* **257:** 249.

Nakamura Y. and Ito K. 1998. How protein reads the stop codon and terminates translation. *Genes Cells* **3:** 265.

Nakamura Y., Ito K., and Ehrenberg M. 2000. Mimicry grasps reality in translation termination. *Cell* **101:** 349.

Ouzounis C., Bork P., Casari G., and Sander C. 1995. New protein functions in yeast chromosome VIII. *Protein Sci.* **4:** 2424.

Paushkin S.V., Kushnirov V.V., Smirnov V.N., and Ter-Avanesyan M.D. 1997. Interaction between yeast Sup45p (eRF1) and Sup35p (eRF3) polypeptide chain release factors: Implications for prion-dependent regulation. *Mol. Cell. Biol.* **17:** 2798.

Sachs A. 2000. Physical and functional interactions between the mRNA cap structure and the poly(A) tail. In *Translational control of gene expression* (ed. N. Sonenberg et al.), p. 447. Cold Spring Harbor Laboratory Press, Cold Spring Harbor, New York.

Schleiss M.R., Degnin C.R., and Geballe A.P. 1991. Translational control of human cytomegalovirus gp48 expression. *J. Virol.* **65:** 6782.

Selmer M., Al-Karadaghi S., Hirokawa G., Kaji A., and Liljas A. 1999. Crystal structure of *Thermotoga maritima* ribosome recycling factor: A tRNA mimic. *Science* **286:** 2349.

Short G.F., Golovine S.Y., and Hecht S.M. 1999. Effects of release factor 1 on in vitro protein translation and the elaboration of proteins containing unnatural amino acids. *Biochemistry* **38:** 8808.

Song H., Mugnier P., Das A.K., Webb H.M., Evans D.R., Tuite M.F., Hemmings B.A., and Barford D. 2000. The crystal structure of human eukaryotic release factor eRF1 — Mechanism of stop codon recognition and peptidyl-tRNA hydrolysis. *Cell* **100:** 311.

Stansfield I., Jones K.M., Kushnirov V.V., Dagkesamanskaya A.R., Poznyakovski A.I., Paushkin S.V., Nierras C.R., Cox B.S., Ter-Avanesyan M.D., and Tuite M.F. 1995. The products of the SUP45 (eRF1) and SUP35 genes interact to mediate translation termination in *Saccharomyces cerevisiae. EMBO J.* **14:** 4365.

Tate W.P. and Mannering S.A. 1996. Three, four or more: The translational stop signal at length. *Mol. Microbiol.* **21:** 213.

Tenson T., Herrera J.V., Kloss P., Guarneros G., and Mankin A.S. 1999. Inhibition of translation and cell growth by minigene expression. *J. Bacteriol.* **181:** 1617.

Uno M., Ito K., and Nakamura Y. 1996. Functional specificity of amino acid at position 246 in the tRNA mimicry domain of bacterial release factor 2. *Biochimie* **78:** 935.

Wang L. and Wessler S.R. 1998. Ineffeicient reinitiation is responsible for upstream open reading frame-mediated translational repression of the maize R gene. *Plant Cell* **10:** 1733.

Welch E.M., Wang W., and Peltz S.W. 2000. Translation termination: It's not the end of the story. In *Translational control of gene expression* (ed. N. Sonenberg et al.), p. 467. Cold Spring Harbor Laboratory Press, Cold Spring Harbor, New York.

Wickner R.B., Taylor K.L., Edskes H.K., and Maddelein M.L. 2000. Prions: Portable prion domains. *Curr. Biol.* **10:** R335.

Zhouravleva G., Frolova L., Le Goff X., Le Guellec R., Inge-Vechtomov S., Kisselev L., and Philippe M. 1995. Termination of translation in eukaryotes is governed by two interacting polypeptide chain release factors, eRF1 and eRF3. *EMBO J.* **14:** 4065.

Protein tRNA Mimicry in Translation Termination

Y. Nakamura, M. Uno,* T. Toyoda, T. Fujiwara, and K. Ito

*Department of Basic Medical Sciences, Institute of Medical Science, University of Tokyo,
Minato-ku, Tokyo 108-8639, Japan*

Termination of protein synthesis takes place on the ribosomes as a response to a stop, rather than a sense, codon in the "decoding" site (A site). Translation termination requires two classes of polypeptide release factors (RFs): a class I factor, codon-specific RFs (RF1 and RF2 in prokaryotes; eRF1 in eukaryotes), and a class II factor, nonspecific RFs (RF3 in prokaryotes; eRF3 in eukaryotes) that bind guanine nucleotides and stimulate class I RF activity. Although the termination process and the RF activity were discovered in vitro in the late 1960s, much of the mechanism and the apparatus have remained obscure. The underlying mechanism for stop codon decoding represents a longstanding coding problem of considerable interest, since it entails protein–RNA recognition instead of the well-understood codon–anticodon pairing during the mRNA–tRNA interaction. After four decades of investigation, we are now aware that molecular mimicry between protein and tRNA provides a clue to this problem.

Molecular mimicry between protein and nucleic acid is a novel concept in biology, proposed in 1995 from three crystallographic discoveries, one, on protein–RNA mimicry, and the other two, on protein–DNA mimicry. Nyborg, Clark, and colleagues first described this concept when they solved the crystal structure of elongation factor EF-Tu:GTP:aminoacyl-tRNA ternary complex and found its overall structural similarity with another elongation factor EF-G including the resemblance of part of EF-G to the anticodon stem of tRNA (Nissen et al. 1995). Protein mimicry of DNA has been shown in the crystal structure of the uracil-DNA glycosylase-uracil glycosylase inhibitor protein complex (Mol et al. 1995; Savva and Pear 1995), as well as in the nuclear magnetic resonance (NMR) structure of transcription factor TBP–TAF$_{II}$230 complex (Liu et al. 1998). Consistent with this discovery, functional mimicry of a major autoantigenic epitope of the human insulin receptor by RNA has been suggested (Doudna et al. 1995), but its nature of mimicry is still largely unknown. The milestone of functional mimicry between protein and nucleic acid has been achieved by the discovery of a "peptide anticodon" that deciphers stop codons in mRNA (Ito et al. 2000). It is surprising that it took four decades after the discovery of the genetic code to figure out the basic mechanisms behind the deciphering of its 64 codons.

tRNA MIMICRY BY ELONGATION FACTOR EF-G

In 1994, two crystallography groups solved the three-dimensional (3D) structure of *Thermus thermophilus* EF-G, a translocase protein that forwards peptidyl tRNA from the A site to the P site on the ribosome (Ævarsson et al. 1994; Czworkowski et al. 1994). They proposed five subdomains of EF-G: G; G´; II–V, whose carboxy-terminal part, domains III–V, appears to mimic the shapes of the acceptor stem, the anticodon helix, and the T stem of tRNA, respectively (Fig. 1A). This resemblance between part of EF-G and tRNA was first described when the 3D structure of the ternary complex of Phe-tRNA, *Thermus aquaticus* elongation factor EF-Tu, and the nonhydrolyzable GTP analog, GDPNP, was solved (Nissen et al. 1995). The 3D structure of the ternary complex is almost completely superimposable with the EF-G:GDP complex, and domain IV of EF-G forms a protruding "rod" conformation, which is similar to the shape of the anticodon arm of tRNA (Fig. 1A). This mimic was also noticed independently by Ito et al. (1996); when the Phe-tRNA structure per se was aligned with the carboxy-terminal part of EF-G, using the Cα coordinates from domains III–V, the two structures were superimposable except for minor differences.

Another structural test of EF-G tRNA mimicry is to map the position of EF-G in the ribosome by directed hydroxyl radical probing. Wilson and Noller (1998) have mapped the location and orientation of EF-G in the ribosome using Fe(II) tethered to 18 different (single cysteine) positions on the surface of EF-G bound to the ribosome. The data provide convincing evidence for the ribosomal A-site occupation by domain IV of EF-G, as well as the proximity of the tip of domain IV to the 30S-decoding site. Not only tethered radical footprinting, but also cryo-electron microscopy analysis of the EF-G ribosome complex clearly point out the ribosomal A-site occupation by domain IV of EF-G as well as the proximity of the tip of domain IV to the 30S decoding site (Agrawal et al. 1998). Thus, the structural mimicry of domain IV, inferred from the crystallographic comparison, extends to its position in or near the tRNA-binding region of the ribosome, suggesting common requirements for the structure and function on the ribosome.

tRNA MIMICRY BY RELEASE FACTORS

A tRNA-like property has been speculated for RFs because they are involved in reading stop codons, instead of

*Present address: Biological Research Laboratories, Sankyo Co. LTD., 2-58-1, Hiromachi, Shinagawa-ku, Tokyo 140-8710, Japan.

tRNA, during the termination of protein synthesis (see Fig. 2A) (Nakamura et al. 1996; Tate et al. 1996). The fact that two RFs from prokaryotes exhibit codon specificity led many researchers to think that they interact directly with their codons. However, evidence has been lacking for such direct contact until recently. A clue to this problem emerged, prematurely but promisingly, when we cloned and sequenced the structural gene for RF3 (Mikuni et al. 1994). Its primary protein sequence resembled the amino-terminal part of EF-G, rather than EF-Tu, suggesting that RF1 and RF2 might be equivalent to the carboxy-terminal part of EF-G (for review, see Nakamura et al. 2000b). Although the structural data greatly facilitated the mimicry hypothesis (see below), this initial idea urged us to test the primary sequence comparisons of RFs from different organisms and other translation factors. Thus, a major breakthrough was achieved via findings of universally conserved seven-domain structures, domains A through G, in prokaryotic and eukaryotic class I RFs. This led to a novel hypothesis of "molecular mimicry" between RF and tRNA (Ito et al. 1996; Nakamura et al. 1996). Domains D and E of RFs appear to share primary (and secondary) sequence homology with the carboxy-terminal portion, domain IV, of EF-G that mimics the shape of the anticodon helix of tRNA. Therefore, it seemed that domains D/E of RF constitute a tRNA-mimicry domain necessary for RF binding to the ribosomal A site, and encode an anticodon-mimicry element to recognize the stop codons.

This prediction was confirmed recently by the discovery of a peptide determinant in RFs equivalent to the anticodon of tRNA (Ito et al. 2000). Genetic selection combined with biochemical studies showed that the tripeptides Pro-Ala-Thr in RF1 and Ser-Pro-Phe in RF2 determine the RF identity and that the first and third amino acids independently discriminate between the second and third purine bases, respectively (Fig. 2B). Thus, at the first position, Pro is restrictive to A (RF1), whereas Ser is permissive to both A and G (RF2). At the third position, Thr is permissive to A and G (RF1), whereas Phe is restrictive to A (RF2). These two discrimination switches operate separately, since the Pro-Pro-Phe variant recognizes only UAA, whereas the Ser-Pro-Thr variant recognizes three stop codons and UGG as well. Hence, they were referred to as a tripeptide "anticodon" that deciphers stop codons in mRNA (for review, see Nakamura et al. 2000a). Impressively, the experimentally defined tripeptide anticodon locus perfectly fits to the one that we have predicted by our RF-tRNA mimicry hypothesis (Ito et al. 1996).

Functional sites of interaction between RF1 and the ribosome have been assigned by directed hydroxyl radical

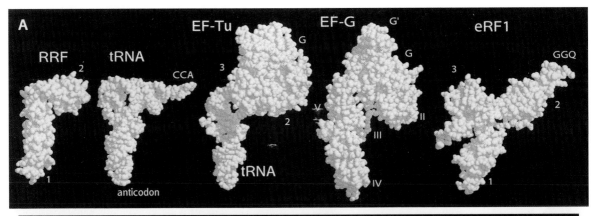

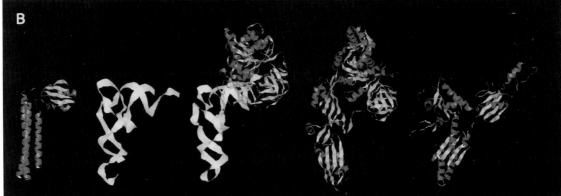

Figure 1. Crystal structures of translation factors that mimic tRNA. (*A*) Space-filling models. (*B*) Ribbon diagram models. (*Left to right*): *T. thermophilus* RRF (Protein Data Bank accession code 1EH1), yeast Phe-tRNA[Phe], *T. aquaticus* EF-Tu:GDPNP:Phe-tRNA[Phe] (1TTT), *T. thermophilus* EF-G:GDP (1DAR), and human eRF1 (1DT9). Protein domains are shown with the relevant numbers.

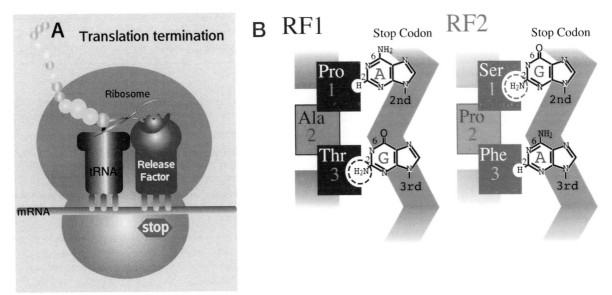

Figure 2. Schematic presentation of translation termination. (*A*) Dual function of polypeptide release factors: to decipher stop codons in the 30S decoding site and to activate peptidyl-tRNA hydrolysis in the 50S peptidyltransferase site. (Courtesy of Mitsuko Kudo, Biohistory Research Hall.) (*B*) The tripeptide anticodon of bacterial release factors. The first and third amino acids discriminate the second and third purine bases. The C-2 amino group of G is a primary target for discrimination by Pro and Phe, whereas Thr and Ser permit both C-2 amino group and proton of purine (*dotted circles* suggest a contribution of hydrogen bonding). (Adapted, with permission, from Nakamura et al. 2000 [copyright Elsevier Science].)

probing (Wilson et al. 2000). Interestingly, the site-directed radical cleavages from positions 187 and 192 of RF1 in 16S rRNA are very similar to the corresponding positions on EF-G, when bound to the ribosome in the posttranslocational state. In particular, Fe(II) tethered to EF-G at the tip of its anticodon-mimicking domain IV results in identical rRNA cleavages in the head and platform of the small subunit surrounding the decoding site. RF1 positions 187 and 192 flank the tripeptide anticodon (188–190). Thus, the anticodon mimic inferred from the functional study extends to its position in or near the tRNA-binding region of the ribosome, providing a strong support for the RF-tRNA mimicry model.

The crystal structure of human eRF1 to 2.8 Å has been published by Barford and colleagues (see Fig. 1) (Song et al. 2000). They pointed out that the overall shape and dimensions of eRF1 resemble a tRNA molecule with domains 1, 2, and 3 of eRF1 corresponding to the anticodon stem, aminoacyl acceptor stem, and T stem of a tRNA molecule. This domain assignment relies on the assumptions that the universal GGQ motif (Frolova et al. 1999) located at the tip of domain 2 is assumed to be a structural counterpart of the tRNA aminoacyl group on the CCA-3′ acceptor stem and that domain 1, in which a codon-specific discrimination defect can be created (Bertram et al. 2000), may be equivalent to the anticodon of tRNA (Song et al. 2000). Nevertheless, if we extend the bacterial tripeptide anticodon analogy to eRF1s, it can be speculated that a Thr-Ala-Ser tripeptide adjacent to the helical hairpin might play the role of an omnipotent discriminator tripeptide (Nakamura et al. 2000a). A simple omnipotent discriminator tripeptide of *Escherichia coli* type, however, could not account for the exclusive recognition

of all three stop codons, since it would recognize UGG as well (Ito et al. 2000). Therefore, this putative discriminator, if real, must be designed to exclude UGG by any means.

ACCURACY OF STOP CODON DECODING

The error rate of polypeptide termination is less than 10^{-5}, showing that the recognition of stop codons is highly accurate. The molecular basis of this accuracy, however, is unknown. The main mechanism of discrimination of second/third purine bases by the peptide anticodons is steric exclusion of the C-2 amino group of G by either of the two bulky amino acids, whereas "wobble" recognition of both purines by the two hydrophilic amino acids is likely to involve hydrogen bonding (Ito et al. 2000). In addition to the recognition of two stop codons each, RF1 and RF2 must discriminate against the 61 sense codons of the genetic code. Freistroffer et al. (2000) have reported that GTP energy-driven proofreading is not involved in enhancing the accuracy of termination, which is contrary to the elongation process, suggesting that protein–RNA interactions can be so precise. This remarkable selectivity of RFs must be based on a very careful design of the A site to hold both protein and mRNA in sterically well defined positions via specific positional interactions between the RF domain(s) around the peptide anticodon and the docking site(s) of the ribosome.

An intriguing RF2 variant affected in the accuracy of stop codon decoding and acquired in the omnipotent decoding activity has been isolated as a plasmid-borne RF2 mutant that restored the growth of a temperature-sensitive RF1 strain of *E. coli* (Ito et al. 1998). This omnipotent activity is caused by a single Glu→Lys mutation at position 167 of RF2, and we referred to this variant as RF2*. In both in vivo and in vitro polypeptide termination assays, RF2* catalyzed UAG/UAA termination, as does RF1, as well as UGA termination, showing that the mutation did not switch stop codon selectivity from RF2 to RF1, but conferred on RF2 the omnipotent recognition activity. Although Glu-167 is not part of the peptide anti-codon, RF2* was the first protein that is altered in stop codon selectivity. Therefore, we speculated that Glu-167 constitutes an effector domain required for correct inter-action between the codon and the peptide anticodon on the ribosome, and that E167K might block this process.

Further analysis revealed that the accuracy of stop codon recognition is severely impaired by multiple Glu→Lys (negative-to-positive) charge-flip alterations in domain C, adjacent to the peptide anticodon domain, of RF, and this impairment is not to reduce the activity of peptidyl-tRNA hydrolysis, but to enhance or (newly) in-duce the hydrolysis not only at cognate stop codons but also at noncognate codons including sense codons (M. Uno et al., in prep.). Table 1 represents part of the pheno-types of such single or double mutants at positions 149 and 170 of E. coli RF2, showing that these charge-flip RF2 variants accelerate peptide release at cognate stop codons, UGA and UAA, and are able to trigger peptide release not only at noncognate UAG stop codons, but also at sense codons such as UGG, UAC, and UUA. Consis-tently, the double flip variant was able to complement the RF1 and RF2 knockout strains. All these charge-flip changes in domain C, including E167K, appear to reduce the accuracy in a quite general fashion. Therefore, what we found here seemed to be a bypassing of any codons by RF at the A site of the ribosome, and this bypassing is suf-ficient to trigger peptidyl-tRNA hydrolysis in the pep-tidyl transferase center of the 50S subunit. How codon bypassing triggers peptidyl-tRNA hydrolysis, however, is puzzling. Nevertheless, these findings showed that the electrostatic interaction involving negative charges in do-main C controls the accurate docking of RF in the ribo-some and that the charge-flipping creates a novel pheno-type of translation termination by codon bypassing via relaxed positioning of the peptide anticodon in the de-coding pocket of the ribosome. Further studies will pro-vide us with a clue to the mechanisms underlying the ac-curacy of stop codon recognition and how the cognate stop codon recognition by RF activates peptidyl-tRNA hydrolysis.

tRNA MIMICRY BY RIBOSOME RECYCLING FACTOR

After release of nascent polypeptides by RFs, the post-termination complex composed of the ribosome, deacy-lated tRNA (in P site), a class I RF (in A site), and mRNA needs to be dissociated for the next round of protein syn-thesis. As a first step, in bacteria, RF3 accelerates the dis-sociation of RF1 and RF2 from the ribosome in a GTP-de-pendent manner, thereby playing as a class-I-RF-recycling factor in vitro (Freistroffer et al. 1997). Upon GTP hydrol-ysis, RF3 is also released from the ribosome, thereby leav-ing behind the posttermination complex with mRNA, de-acylated tRNA in the P site, and the empty A site, which is believed to be a substrate for another factor, ribosome re-cycling factor (RRF) (Fig. 3). Early studies by Kaji and colleagues in the 1970s revealed that RRF is required for the dissociation of the posttermination ribosomal complex in bacteria in concert with EF-G (for review, see Janosi et al. 1996). More recent studies by Ehrenberg and colleagues have found that fast recycling of ribosomes requires both RF3 and RRF in vitro (Freistroffer et al. 1997; Pavlov et al. 1997). Nevertheless, how RRF dissociates the posttermi-nation complex remains unknown.

The crystal structure of RRF has recently been solved to 2.55, 2.3, and 2.6 Å resolution by three groups using RRF proteins from Thermotoga maritima (Selmer et al. 1999), E. coli (Kim et al. 2000), and T. thermophilus (see Fig. 1) (Toyoda et al. 2000). These three molecules are composed of two domains, a long three-helix bundle (do-main 1) and a three-layer β/α/β sandwich (domain 2), and superimpose with tRNA[Phe] except for the amino-acid-binding 3′ end. Selmer et al. (1999) have proposed that RRF is a near-perfect tRNA mimic to explain the mecha-nistic disassembly of the posttermination ribosomal com-plex. They speculate that RRF binds to the A site of the ribosome and that EF-G translocates RRF from the A to the P site and deacylated tRNA from the P to the E site of the ribosome in a GTP-dependent manner, where it

Table 1. Codon Bypassing Release Activity of Domain C Charge-Flip RF2 Variants

Domain C RF2 variants	In vivo complementation[a]		In vitro peptide release at codons[b]					
	RF1 knockout	RF2 knockout	UAG	UAA	UGA	UGG	UAC	UUA
Wild-type	−	+	−	+	+	−	−	−
E149K	−	+	+	+*	+*	n.d.	n.d.	n.d.
E167K	+	+	+	+*	+*	n.d.	n.d.	n.d.
E170K	−	+	+	+*	+*	n.d.	n.d.	n.d.
E149K E170K	+	+	+	+*	+*	+	+	+

[a]RM789A (prfA::Km[R]) and RM789B (prfB::Cm[R]) cells carrying the maintainer plasmid pSUIQT-RF2* (Ito et al. 1998) were trans-formed with pBR322-RF2 derivatives containing relevant single or double mutations in domain C. Growth of these transformants was monitored in the absence of IPTG: (+) growth (complementation); (−) no growth (no complementation).
[b]f[³H]Met release from the (f[³H]Met-tRNA[f]•mini-messenger•ribosome) complex upon addition of RFs was determined (Ito et al. 2000). Mini-messenger RNAs used in these reactions were 9-mer sequences consisting of 5′-UUC AUG-3′ followed by stop (or sense) triplets. (+) Release activity; (+*) enhanced release activity compared with wild-type RF2; (−) no activity. n.d. indicates no data.

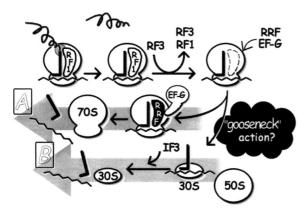

Figure 3. Two models describing how RRF is involved in recycling of posttermination complexes after release of polypeptides by release factors.

would dissociate rapidly (RRF translocation model, Fig. 3A). However, the model by Selmer et al. (1999) is not consistent with the biochemical findings of Karimi et al. (1999), which show, first, that RRF and EF-G split the ribosome into subunits in a reaction that requires GTP hydrolysis (subunit disassembly model, Fig. 3B) and, second, that the initiation factor IF3 is required for the removal of deacylated tRNA from the P site of the 30S particle. The near-perfect structural similarity between tRNA and RRF as proposed by Selmer et al. (1999) may not be necessarily true, since all the three proteins possess the markedly distinct interdomain angles (see below).

RIBOSOME RECYCLING FACTOR BEYOND A tRNA MIMIC

Regardless of the resemblance between the shapes of RRF and tRNA, very little is known about the structure-and-function relationship of RRF. RRF is a tRNA-like L-shape molecule consisting of two domains that are bridged by two loops that function as a flexible hinge. Although the individual domain structures are similar, the interdomain angle is potentially variable, and the hinge flexibility is vital for the function of RRF (gooseneck model; Toyoda et al. 2000). *T. thermophilus* RRF is normally inactive in *E. coli* and fails to complement the *E. coli* RRF defect. However, it becomes fully active upon truncation of the carboxy-terminal 5 amino acids or substitutions of the interdomain residues. Most of these alterations were interpreted as increasing the hinge flexibility of RRF, making the *T. thermophilus* RRF function in the heterologous condition (Fig. 4). Since the specific interaction between RRF and EF-G is crucial for the action of RRF (Rao and Varshney 2001; K. Ito et al., unpubl.), we assume that the increased flexibility of the hinge may allow functional positioning of heterologous RRF and EF-G proteins in the *E. coli* ribosome, although other possibilities cannot be excluded at present.

The truncation of the carboxy-terminal 9 amino acids, which are in direct contact with the hinge, impairs the active conformation of the hinge and/or the three-helix bundle in domain 1, so that RRF no longer binds to the ribosome (Fujiwara et al. 2001). This RRF variant defective in ribosome binding regains the binding capacity through

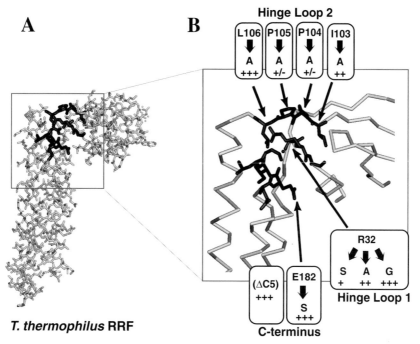

Figure 4. Molecular mobility of *T. thermophilus* RRF. (*A*) Three-dimensional structure of *T. thermophilus* RRF, in which hinge loops and carboxyl terminus are shown in bold. (*B*) Amino acid substitutions or deletions that affect the hinge flexibility. *T. thermophilus* RRF derivatives carrying the relevant alterations were expressed in *E. coli* and examined to see whether they can restore the growth of *E. coli* null RRF mutant of Fujiwara et al. (1999). The complementation capacity was scored by colony size and growth: (+++) normal growth (large colony); (++) fair growth (medium colony); (+) weak growth (small colony); and (+/-) sick growth (tiny colony).

multiple secondary changes occurring in three topologically distinct regions of RRF, two of which are equivalent to the tip of the anticodon stem and the upper surface of the acceptor stem of tRNA (Fujiwara et al. 2001). These findings suggest that RRF interacts with the ribosome in a way similar to tRNA, spanning 30S and 50S subunits, for exerting its action. If RRF binds to the A site of the ribosome in a manner similar to tRNA and splits the ribosome into subunits, RRF must exert its action within the A site in concert with EF-G. EF-G can generate a post- to pre-peptidyl transfer transition state of the ribosome coupled with GTP hydrolysis. This energy-driven transition may involve distortion of the interface between 30S and 50S ribosome particles. Therefore, it is tempting to speculate that either of the domains connected by the flexible gooseneck of RRF may penetrate into a distorted interface and interfere with post- to pre-peptidyl transfer transition, shifting the equilibrium toward a direct uncoupling of 30S and 50S.

Recently, Ishino et al. (2000) have monitored the formation of ribosome complexes on the surface-coupled RRF of *E. coli* in real time with a BIACORE 2000 instrument based on the surface plasmon resonance technique. They have found that RRF tends to interact more efficiently with 50S subunits, which seems to be of biological significance through mutational and antibiotic analysis (Ishino et al. 2000). This might be interpreted as indicating that part of the RRF-binding site in the 50S subunit is sequestered in the 70S ribosome state and exposed in the free 50S subunit state. Hence, it is tempting to speculate that such RRF-binding sites in 50S subunits are located at or near the interface between 50S and 30S subunits, which are hardly accessible in the 70S ribosome. Of two models proposed for RRF, the preferential and stable binding of RRF to 50S subunits cannot be easily accounted for by the RRF translocation model, but rather favors the subunit disassembly model. The polysome-to-monosome breakdown assay has been used as a conventional assay for the RRF activity (Hirashima and Kaji 1972), but this does not necessarily reflect the primary action of RRF because the assay uses crude polysome fractions under conditions that favor the reassembly of 30S and 50S subunits to 70S ribosomes. We assume that nature may not have created such protein of a tRNA mimic to simply substitute for tRNA unless protein is required to pursue some function(s) that tRNA cannot do. Further mechanistic analysis will be needed.

PERSPECTIVES

What is most remarkable in molecular mimicry is the fact that the three proteins, EF-G, RRF, and eRF1, structurally known as a tRNA mimic, possess completely different protein folds with unrelated primary and secondary structures of protein. How did these unrelated protein architectures evolve to mimic a tRNA shape? These distinct protein folds are interpreted as indicating that a mimic of the shape of a tRNA works as an entrance pass to sit in the cockpit (A site) in a ribosome "machine"; however, the action once sitting there is different for different translation factors, which should require the different protein folds for the action. This is, in some sense, equivalent to the animal or plant mimicry where the mimic itself is not the purpose, but the purpose is to cheat objects by imitating a shape or a color for diverse purposes such as to prey, evade, lure, pollinate, threaten, etc. Nature must have evolved this art of molecular mimicry using different protein architectures for the diverse actions, still keeping a similar shape to fulfill the requirement of the ribosome.

Given the RNA World hypothesis, one might speculate that most or many proteins might have evolved to substitute for the RNA ancestors during evolution. We assume that molecular mimicry of RNA by protein may have played a central role in the evolutionary process. What RNA ancestors could have been replaced by protein and what could not during the process of the world transition? Are there any ancestors for tRNA mimics? How did different architectures for tRNA mimics evolve independently? In the modern DNA/protein world, most such RNA ancestors might have disappeared and only a few RNA molecules might have survived as living molecular fossils, as they could not have been replaced by protein. tRNA should be one such molecular fossil, whose molecular mimicry we became aware of in the translational apparatus, such as EF-G, eRF1, and RRF.

CONCLUSIONS

Recent advances in structural and molecular biology uncovered that a set of translation factors resembles a tRNA shape and, in one case, even mimics a tRNA function for deciphering the genetic code. Nature must have evolved this art of molecular mimicry between protein and ribonucleic acid using different protein architectures to fulfill the requirement of a ribosome machine.

ACKNOWLEDGMENTS

We are grateful to Elsevier Science for allowing us to reproduce previously published material. This work was supported by grants from The Ministry of Education, Sports, Culture, Science and Technology, Japan; the Human Frontier Science Program; and the Basic Research for Innovation Biosciences Program of Bio-oriented Technology Research Advancement Institution (BRAIN).

REFERENCES

Agrawal R.K., Penczek P., Grassucci R.A., and Frank J. 1998. Visualization of the elongation factor G on the *Escherichia coli* 70S ribosome: The mechanism of translation. *Proc. Natl. Acad. Sci.* **95:** 6134.

Ævarsson A., Brazhnikov E., Garber M., Zheltonosova J., Chirgadze Y., al-Karadaghi S., Svensson L.A., and Liljas A. 1994. Three-dimensional structure of the ribosomal translocase: elongation factor G from *Thermus thermophilus. EMBO J.* **13:** 3669.

Bertram G., Bell H.A., Ritchie D.W., Fullerton G., and Stansfield I. 2000. Terminating eukaryote translation: Domain 1 of release factor eRF1 functions in stop codon recognition. *RNA* **6:** 1236.

Czworkowski J., Wang J., Steitz T.A., and Moore P.B. 1994. The crystal structure of elongation factor G complexed with GDP, at 2.7 Å resolution. *EMBO J.* **13:** 3661.

Doudna J.A., Cech T.R., and Sullenger B.A. 1995. Selection of an RNA molecule that mimics a major autoantigenic epitope of human insulin receptor. *Proc. Natl. Acad. Sci.* **92:** 2355.

Freistroffer D.V., Kwaitkowski M., Buckingham R.H., and Ehrenberg M. 2000. The accuracy of codon recognition by polypeptide release factors. *Proc. Natl. Acad. Sci.* **97:** 2046.

Freistroffer D.V., Pavlov M.Y., MacDougall J., Buckingham R.H., and Ehrenberg M. 1997. Release factor RF3 in *E. coli* accelerates the dissociation of release factors RF1 and RF2 from the ribosome in a GTP-dependent manner. *EMBO J.* **16:** 4126.

Frolova L.Y., Tsivkovskii R.Y., Sivolobova G.F., Oparina N.Y., Serpinsky O.I., Blinov V.M., Tatkov S.I., and Kisselev L.L. 1999. Mutations in the highly conserved GGQ motif of class 1 polypeptide release factor abolish ability of human eRF1 to trigger peptidyl-tRNA hydrolysis. *RNA* **5:** 1014.

Fujiwara T., Ito K., and Nakamura Y. 2001. Functional mapping of the ribosome contact site in the ribosome recycling factor. *RNA* **7:** 64.

Fujiwara T., Ito K., Nakayashiki T., and Nakamura Y. 1999. Amber mutations in ribosome recycling factors of *Escherichia coli* and *Thermus thermophilus:* Evidence for C-terminal modulator element. *FEBS Lett.* **447:** 297.

Hirashima A. and Kaji A. 1972. Factor-dependent release of ribosomes from messenger RNA: Requirement for two heat-stable factors. *J. Mol. Biol.* **65:** 43.

Ishino T., Atarashi K., Uchiyama S., Yamami T., Yoshida T., Hara H., Yokose K., Kobayashi Y., and Nakamura Y. 2000. Interaction of ribosome recycling factor and elongation factor EF-G with *E. coli* ribosomes studied by surface plasmon resonance technique. *Genes Cells* **5:** 953.

Ito K., Uno M., and Nakamura Y. 1998. Single amino acid substitution in prokaryote polypeptide release factor 2 permits it to terminate translation at all three stop codons. *Proc. Natl. Acad. Sci.* **95:** 8165.

———. 2000. A tripeptide 'anticodon' deciphers stop codons in messenger RNA. *Nature* **403:** 680.

Ito K., Ebihara K., Uno M., and Nakamura Y. 1996. Conserved motifs of prokaryotic and eukaryotic polypeptide release factors: tRNA-protein mimicry hypothesis. *Proc. Natl. Acad. Sci.* **93:** 5443.

Janosi L., Hara H., Zhang S., and Kaji A. 1996. Ribosome recycling by ribosome recycling factor (RRF) —An important but overlooked step of protein biosynthesis. *Adv. Biophys.* **32:** 121.

Karimi R., Pavlov M., Buckingham R., and Ehrenberg M. 1999. Novel roles for classical factors at the interface between translation termination and initiation. *Mol. Cell* **3:** 601.

Kim K.K., Min K., and Suh S.W. 2000. Crystal structure of the ribosome recycling factor from *Escherichia coli. EMBO J.* **19:** 2362.

Liu D., Ishima R., Tong K.I., Bagby S., Kokubo T., Muhandiram D.R., Kay L.E. Nakatani Y., and Ikura M. 1998. Solution structure of a TBP-ATF$_{II}$230 complex: Protein mimicry of the minor groove surface of the TATA box unwound by TBP. *Cell* **94:** 573.

Mikuni O., Ito, K., Moffat J., Matsumura K., McCaughan K., Nobukuni T., Tate W., and Nakamura Y. 1994. Identification of the *prfC* gene, which encodes peptide-chain-release factor 3 of *Escherichia coli. Proc. Natl. Acad. Sci.* **91:** 5798.

Mol C.D., Arvai A.S., Sanderson R.J., Slupphaug G., Kavli B., Krokan H.E., Mosbaigh D.W., and Tainer J.A. 1995. Crystal structure of human uracil-DNA glycosylase in complex with a protein inhibitor: Protein mimicry of DNA. *Cell* **82:** 701.

Nakamura Y., Ito K., and Ehrenberg M. 2000a. Mimicry grasps reality in translation termination. *Cell* **101:** 349.

Nakamura Y., Ito K., and Isaksson L.A. 1996. Emerging understanding of translation termination. *Cell* **87:** 147.

Nakamura Y., Kawazu Y., Uno M., Yoshimura, K., and Ito K. 2000b. Genetic probes to bacterial release factors: tRNA mimicry hypothesis and beyond. In *The ribosome: Structure, function, antibiotics and cellular interactions* (ed. R.A. Garrett et al.), p.519. ASM Press, Washington, D.C.

Nissen P., Kjeldgaard M., Thirup S., Polekhina G., Reshetnikova L., Clark B.F.C., and Nyborg J. 1995. Crystal structure of the ternary complex of Phe-tRNA^{Phe}, EF-Tu, and a GTP analog. *Science* **270:** 1464.

Pavlov M.Y., Freistroffer D.V., MacDougall J., Buckingham R.H., and Ehrenberg M. 1997. Fast recycling of *Escherichia coli* ribosomes requires both ribosome recycling factor (RRF) and release factor RF3. *EMBO J.* **16:** 4134.

Rao A.R. and Varshney U. 2001. Specific interaction between the ribosome recycling factor and the elongation factor G from *Mycobacterium tuberculosis* mediates peptidyl-tRNA release and ribosome recycling in *Escherichia coli. EMBO J.* **20:** 2977.

Savva R. and Pear L.H. 1995. Nucleotide mimicry in the crystal structure of the uracil-DNA glycosylase-uracil glycosylase inhibitor protein complex. *Nat. Struct. Biol.* **2:** 752.

Selmer M., al-Karadaghi S., Hirokawa G., Kaji A., and Liljas A. 1999. Crystal structure of *Thermotoga maritima* ribosome recycling factor: A tRNA mimic. *Science* **286:** 2349.

Song H., Mugnier P., Das A.K., Webb H.M., Evans D.R., Tuite M.F., Hemmings B.A., and Barford D. 2000. The crystal structure of human eukaryotic release factor eRF1—Mechanism of stop codon recognition and peptidyl-tRNA hydrolysis. *Cell* **100:** 311.

Tate W.P., Poole E.S., and Mannering S.A. 1996. Hidden infidelities of the translational stop signal. *Prog. Nucleic Acid Res.* **52:** 293.

Toyoda T., Tin O.F., Ito K., Fujiwara T., Kumasaka T., Yamamoto M., Garber M.B., and Nakamura Y. 2000. Crystal structure combined with genetic analysis of the *Thermus thermophilus* ribosome recycling factor shows that a flexible hinge may act as a functional switch. *RNA* **6:** 1432.

Wilson K., Ito K., Noller H., and Nakamura Y. 2000. Functional sites of interaction between release factor RF1 and the ribosome. *Nat. Struct. Biol.* **7:** 866.

Wilson K.S. and Noller H.F. 1998. Mapping the position of translational elongation factor EF-G in the ribosome by directed hydroxyl radical probing. *Cell* **92:** 131.

Mitogenic and Nutritional Signals Are Transduced into Translational Efficiency of TOP mRNAs

E. HORNSTEIN, H. TANG, AND O. MEYUHAS

Department of Biochemistry, Hebrew University-Hadassah Medical School, Jerusalem 91120, Israel

Modulation of the abundance of the translational apparatus appears to enable eukaryotic cells to cope with changing requirements for protein synthesis during transitions between extreme growth and nutritional states. Thus, rRNA synthesis is down-regulated when cells cease to proliferate or are deprived of amino acids and is up-regulated upon reversal of such conditions (Grumm 1999 and references therein). Likewise, the translational efficiency of mRNAs encoding many protein components of the translational machinery is similarly regulated. These include ribosomal proteins (Meyuhas et al. 1996a; Meyuhas and Hornstein 2000 and references therein); elongation factor 1A (Rao and Slobin 1987; Avni et al. 1994; Jefferies et al. 1994a) and elongation factor 2 (Terada et al. 1994; Avni et al. 1997); poly(A)-binding protein (Hornstein et al. 1999), which has been implicated in both translation initiation and ribosome assembly (Sachs 2000); and a few other proteins that have been implicated in ribosome assembly or nuclear-cytoplasmic transport of RNA (for review, see Meyuhas 2000). The corresponding mRNAs are characterized by the presence of a 5′ *t*erminal *o*ligo*p*yrimidine tract (5′TOP) and therefore are referred to as TOP mRNAs. This structural motif comprises the core of the translational *cis*-regulatory element of these mRNAs, and its features are summarized elsewhere (Meyuhas and Hornstein 2000).

The apparent advantages in regulating the synthesis of the protein components of the translational apparatus at the translational level are the rate and the readily reversible nature of the response to altering physiological conditions. According to one estimate, most of the energy, consumed during cellular growth, is utilized for generating components of protein synthesis machinery (Schmidt 1999). Hence, the ability of cells to rapidly repress the biosynthesis of the translational machinery upon shortage of amino acids or growth arrest ensures prompt blocking of unnecessary energy wastage. Likewise, when amino acids are replenished or mitogenic stimulation is applied, cells can then rapidly respond in resuming the costly biosynthesis of the translational apparatus.

This paper focuses on the mechanism underlying the translational control of mammalian TOP mRNAs upon mitogenic and nutritional stimuli, with special emphasis on the characterization of the respective signal transduc-

tion pathways. It should be noted that these mRNAs and their unique mode of regulation are not confined to mammals, as TOP mRNAs have been identified in other vertebrates and even in insects (Meyuhas et al. 1996a and references therein). Furthermore, a large body of information regarding this class of mRNAs resulted from studies with *Xenopus laevis* TOP mRNA (Amaldi and Pierandrei-Amaldi 1997), whose translational control mechanism utilizes the same factors as the mammalian counterparts (Avni et al. 1994). Finally, several aspects of the translational control of TOP mRNAs, which are only briefly mentioned here, have been discussed thoroughly in recent reviews (Fumagalli and Thomas 2000; Meyuhas 2000; Meyuhas and Hornstein 2000).

THE TRANSLATIONAL EFFICIENCY OF TOP mRNAs CORRELATES WITH THE CELLULAR MITOGENIC ACTIVITY

Previous reports have repeatedly shown that cessation of proliferation leads to selective repression of the translation of TOP mRNAs, as judged by their shift from polysomes into mRNP particles (subpolysomal fraction) (for review, see Meyuhas and Hornstein 2000). This behavior has been demonstrated when cells are arrested at G_0 by a wide variety of treatments, including dexamethasone treatment of P1798 lymphosarcoma cells, induction for terminal differentiation of murine erytholeukemia (MEL), or contact-inhibition of NIH-3T3 cells (Fig. 1). Translational repression of TOP mRNAs has also been observed when NIH-3T3 cells were arrested at S phase by aphidicolin, an inhibitor of DNA polymerase, or even when arrested at M phase by inhibiting the assembly of the mitotic spindle by nocodazole (Fig. 1). A direct relationship between mitogenic activity and translational efficiency of TOP mRNAs is applicable also for whole animals, as translation of these mRNAs is repressed upon transition from the rapidly growing state in fetal liver to the quiescent state in adult liver. Likewise, resumption of translation can be observed in the regenerating liver (Fig. 1) (Aloni et al. 1992; Hornstein et al. 1999).

TOP mRNAs are inefficiently translated even in growing cells, as about 30% of these mRNAs are sequestered in mRNP. However, the proportion of translationally inactive TOP mRNAs is at least doubled in quiescent cells (Fig. 1). Interestingly, the translation of TOP mRNAs ex-

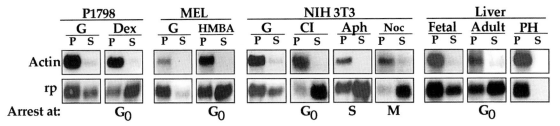

Figure. 1. TOP mRNAs are translationally repressed in mitogenically inactive cells or tissues. Cytoplasmic extracts were prepared from proliferating (G) or resting cells whose growth was arrested by 0.1 μM dexamethasone (Dex) treatment for 24 hr (P1798 cell); induction for differentiation by 5 μM hexamethylene bisacetamide (HMBA) treatment for 96 hr (MEL cells); growth to confluence (CI), inhibition of DNA polymerase by 5 μg/ml aphidicolin treatment (Aph) for 24 hr or inhibition of mitosis by 25 μM nocodazole (Noc) treatment for 24 hr (NIH-3T3 cells). Similarly, cytoplasmic extracts were prepared from fetal (dividing), adult (nondividing), or mitogenically active regenerating (PH) rat liver. These extracts were centrifuged through sucrose gradients and separated into polysomal (P) and subpolysomal (S) fractions. RNA from equivalent aliquots of these fractions was analyzed by northern blot hybridization with the probes indicated at the left. rp refers to the probes used to detect rpS4 mRNA in P1798, rpL30 mRNA in MEL, rpS16 mRNA in NIH-3T3, and rpL7 in the liver (for further experimental details, see Meyuhas et al. 1996b).

hibits an "all-or-none" phenomenon, i.e., these mRNAs alternate between repressed and active states, and, when in the active state, they are translated at near maximum efficiency (Agrawal and Bowman 1987; Meyuhas et al. 1987; Loreni and Amaldi 1992; Shima et al. 1998; Hornstein et al. 1999). This bimodal distribution clearly indicates that the translational repression results from a blockage at the translational initiation step.

TOP mRNAs ARE TRANSLATIONALLY CONTROLLED BY AMINO ACID SUFFICIENCY

The supply of amino acids to tissues and organs has an important role in protein homeostasis in the whole animal. Protein anabolic response after a protein-rich meal was described in humans and rodents and was related to postprandial increase in concentrations of plasma amino acids (for review, see Fafournoux et al. 2000; Kimball and Jefferson 2000). A tight regulatory mechanism, by which amino acid sufficiency regulates protein homeostasis, is performed at multiple levels. These include (1) inhibition of autophagy, a lysosomal proteolytic process (Rabkin et al. 1991; Chua 1994; Dennis et al. 1999), which is regulated by amino-acid-induced secretion of anabolic hormones (for review, see Fafournoux et al. 2000), and (2) through controlling initiation of translation at several steps (for review, see Kimball and Jefferson 2000): recycling of eIF2 in its active form, eIF2-GTP (Kimball et al. 1998); phosphorylation of the eIF4E-binding protein, 4E-BP1, which is followed by its dissociation from the cap-binding protein, eIF4E (Raught et al. 2000); and phosphorylation of eIF4E.

Through these multiple effects, amino acids increase the affinity of phosphorylated eIF4E to the cap structure and to eIF4G, thus inducing more frequent formation of 43S preinitiation complex (Kimball and Jefferson 2000; Raught et al. 2000).

The apparent down-regulation of ribosomal RNA synthesis upon amino acid withdrawal (Grummt 1999) has led us to reason that the synthesis of the protein components of

the translational apparatus might be similarly regulated. Indeed, Figure 2 demonstrates that TOP mRNAs encoding ribosomal protein L32 (rpL32) and elongation factor 1A are much more sensitive to this cue than actin mRNA, a typical non-TOP mRNA. Thus, TOP mRNAs are translationally repressed shortly after amino acid withdrawal and are rapidly derepressed after amino acid re-addition. Notably, translational control of TOP mRNAs by both mitogenic and nutritional stimuli is strictly dependent on the integrity of the 5′TOP motif (Levy et al. 1991; Avni et al. 1994; Biberman and Meyuhas 1997; Hornstein et al. 1999; Tang et al. 2001).

TRANSDUCTION OF MITOGENIC SIGNALS INTO S6 KINASE ACTIVITY AND TRANSLATIONAL EFFICIENCY OF TOP mRNAs

Shortly after mitogenic stimuli in cultured cells (Thomas et al. 1980) or the whole organism (Gressner and Wool 1974), ribosomal protein S6 (rpS6) is phosphorylated at five residues (Ser-235, Ser-236, Ser-240, Ser-244, and Ser-247) clustered at its carboxyl terminus (Krieg et al. 1988). This phosphorylation has attracted much attention due to its temporal correlation with the

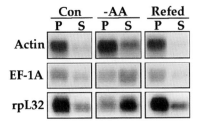

Figure 2. Amino acid deficiency induces reversible translational repression of TOP mRNAs. Human embryonic kidney 293 cells were either untreated (Con), amino-acid-starved for 2 hr (–AA), or amino-acid-refed for 0.5 hr (Refed). Analysis of the polysomal/subpolysomal distribution of TOP mRNAs encoding rpL32 and elongation factor 1A (EF-1A), as well as actin (a non-TOP mRNA), was carried out as described in the legend to Fig. 1.

initiation of protein synthesis and the suggestion that ribosomes with the highest proportion of phosphorylated rpS6 have a selective advantage in mobilization into polysomes (Jefferies and Thomas 1996 and references therein). Furthermore, rpS6 is located near the mRNA/tRNA-binding site at the interface between the small and large ribosomal subunits, and potentially, this location enables rpS6 to alter translation efficiency of at least a subset of mRNAs (Nygard and Nika 1982). Phosphorylation of rpS6 is carried out by two closely related kinases, S6K1 (also known as p70[S6K]) and S6K2 (Gout et al. 1998; Saitoh et al. 1998; Shima et al. 1998). Several studies have shown that mitogenic stimulation of quiescent cells induces activation of S6K and increases translational efficiency of TOP mRNAs (for review, see Meyuhas 2000). These studies led to the hypothesis that rpS6 phosphorylation, following S6K activation, increases the affinity of ribosomes for TOP mRNAs and thus facilitates their initiation (Thomas and Thomas 1986; Jefferies et al. 1994a,b). It should be noted, however, that this model, although mechanistically attractive, has remained purely speculative. Thus, neither the involvement of rpS6 in the translational control of TOP mRNAs nor its being the only physiological substrate of S6K has been experimentally proven.

Mitogenic activation of S6K1 requires multiple phosphorylation events (Fumagalli and Thomas 2000) and seems to involve at least three upstream regulatory pathways: (1) through *phosphatidylinositol 3-kinase* (PI3K) (Chung et al. 1994; Ming et al. 1994), which signals through a ras/MAPK-independent pathway (Ballou et al. 1991; Ming et al. 1994); (2) through *phosphoinositide-dependent kinase 1* (PDK1), which signals to S6K1 in a phosphatidylinositol-3-phosphate-independent fashion (Meyuhas 2000); and (3) through *mammalian target of rapamycin* (mTOR; also known as FRAP of RAFT) (for review, see Dufner and Thomas 1999; Meyuhas 2000). The immunosuppressant rapamycin selectively inhibits mTOR and consequently its downstream effectors S6K1/2 (Chung et al. 1992; Price et al. 1992; Shima et al. 1998) and 4E-BP1 (for review, see Gingras et al. 2001). The mechanism by which mTOR activates S6K is ill-defined, yet the prevailing model indicates that it might do so by restraining a phosphatase (for review, see Meyuhas 2000).

The initial strategy to examine the causal relationship between S6K activity and translational efficiency of TOP mRNAs involved the assessment of the repressive effect of rapamycin on the recruitment of TOP mRNAs into polysomes, upon mitogenic stimulation (Table 1). Such studies in multiple experimental systems have yielded controversial results. Thus, in some cases, rapamycin exerted no (or only a minor) repressive effect on the translational efficiency of TOP mRNAs (Table 1, lines 1 to 5), whereas in other cases, it inhibited the translational activation of these mRNAs (Table 1, lines 6 to 10). It should be noted that the assessment of the percentage of an mRNA associated with polysomes strictly depends on the recovery of the RNA extracted from the polysomal and subpolysomal fractions. Hence, assessment of the data presented in Table 1 should take into account two critical considerations: whether they were statistically significant (two or more measurements) and whether the behavior of a non-TOP mRNA was comonitored as a proof for the selectivity of the effect. Moreover, results derived using rapamycin should be interpreted cautiously since (1) it has been shown to block cell cycle progression and to inhibit the proliferation of a variety of cell types (for review, see Sehgal 1998) and (2) it might reflect, at least partly, an inhibition of eIF4E, through dephosphorylation of 4E-BP1, in an S6K1-independent manner (Gingras et al. 2001).

An alternative approach, to more directly address the causal relationships between S6K activity and translational efficiency of TOP mRNAs, was based on transfection of cells with expression vectors encoding a mutant version of S6K1, p70[S6K]A229. This mutant appears to selectively mistarget upstream signals and to render S6K1 inactive (i.e., it functions as a dominant-interfering mutant). Nevertheless, the overexpression of this mutant in one repetition experiment exerted a modest inhibitory effect on the translational activation of a chimeric TOP mRNA following mitogenic stimulation (Jefferies et al. 1997). The apparent inability of p70[S6K]A229 to efficiently repress the translation of the examined TOP

Table 1. Effect of Rapamycin on Translational Activation of TOP mRNAs in Mitogenically Stimulated Cells

Cells	TOP mRNA	% mRNA in polysomes −rapamycin	% mRNA in polysomes +rapamycin	Effect	Non-TOP mRNAs as internal control	Reference
1. Swiss-3T3	various	82±2 (3)	77±4 (3)	no effect	none	Jefferies et al. (1994b)
2. ES cells (p70[S6K−/−])	eEF1A	51, 57	60	no effect	GAPDH	Kawasome et al. (1998)
3. NIH-3T3	S16-GH	87	69	minor effect	none	Jefferies et al. (1997)
4. 293 + wt p70[S6K]	eEF1A	78	60	minor effect	none	Jefferies et al. (1997)
5. 293 + p70[S6K] E389	eEF1A	85	77	minor effect	none	Jefferies et al. (1997)
6. ES cells (p70[S6K+/+])	eEF1A	63, 63	27	repression	GAPDH	Kawasome et al. (1998)
7. MEF (p70[S6K+/+])	eEF1A	67	32	repression	actin	Shima et al. (1998)
8. MEF (p70[S6K−/−])	eEF1A	64	18	repression	actin	Shima et al. (1998)
9. Lymphoblastoids	various	44±3 (6)	22±3 (6)	repression	6 different	Terada et al. (1994)
10. T cells	various	27±2 (6)	15±1 (6)	repression	actin, GAPDH, IL-2	Terada et al. (1995)

Cells were induced to proliferate by serum refeeding (lines 1 to 8), hemagglutinin (line 9), or phorbol 12, 13-dibutyrate/ionomycin (line 10). In all cases, the results represent a single measurement per individual mRNA. The term "Effect" refers to a decrease in polysomal association and "repression" represents a decrease of the percentage of polysomal association below 50% or by a factor of at least 2. In all cases, with the exception of line 10, the percentage of mRNA in polysomes was calculated by extrapolating the numerical values from the published figures.

mRNA is consistent with the failure of p70^{S6K}E389, a constitutively active mutant of S6K1, to recruit TOP mRNA into polysomes in quiescent cells (Jefferies et al. 1997). Finally, dominant-interfering mutants of S6K1, like rapamycin, blocked an upstream signal directed at the activation of 4E-BP1 (von Manteuffel et al. 1997), thus inhibiting general translation initiation.

AMINO ACID-INDUCED TRANSLATION OF TOP mRNAs REQUIRES PI3 KINASE BUT IS INDEPENDENT OF mTOR AND S6K1

Study of rpS6 phosphorylation in rat liver revealed that this protein is phosphorylated not only following mitogenic stimulation, but also upon refeeding of starved animals (Kozma et al. 1989). The importance of amino acids in this nutritional stimulation has been demonstrated in hepatocytes isolated from starved rats. Thus, supplementing these cells with a complete mixture of amino acids led to phosphorylation of rpS6, which could be abolished by rapamycin (Blommaart et al. 1995). This observation has demonstrated that amino acids, independently of insulin or growth factors, can regulate the activity of S6K1 through an mTOR-mediated mechanism. Indeed, recent studies with various cell lines have indicated that S6K1 activity is rapidly modulated by amino acid deprivation or by their reintroduction (Fox et al. 1998; Hara et al. 1998; Patti et al. 1998; Wang et al. 1998; Xu et al. 1998). Concomitantly, the phosphorylation of rpS6 is regulated with similar kinetics (Tang et al. 2001). Notably, the amino-acid-induced activation of S6K1 has been observed following a postprandial increase in plasma amino acid concentration (for review, see Fafournoux et al. 2000; Kimball and Jefferson 2000). Detailed analysis of the involvement of individual amino acids in this mode of regulation has established a critical role for branched amino acid (Xu et al. 1998) or even leucine alone (Kimball et al. 1999). Regulation of S6K1 by amino acid sufficiency is mediated in a cell-type-specific manner by the loss of aminoacylated tRNA (Iiboshi et al. 1999; Pham et al. 2000).

The apparent correlation between the amino-acid-induced activation of S6K1 and the translation of TOP mRNAs has prompted us to study the role of rpS6 phosphorylation, as well as the activity of S6K, in the translational control of these mRNAs. Initially, we examined the involvement of phosphorylated rpS6 in the translational repression of TOP mRNA in amino-acid-starved cells. To this end, we induced phosphorylation of this protein in an S6K-independent fashion by transfecting cells with p90 ribosomal protein S6 kinase (RSK2). This protein was originally isolated due to its ability to phosphorylate rpS6 (Erikson and Maller 1985), with the same specificity of S6K1 (Sturgill and Wu 1991). Indeed, overexpression of this kinase led to efficient phosphorylation of rpS6, yet it failed to relieve the translational repression of a TOP mRNA (Tang et al. 2001). Therefore, it appears that rpS6 phosphorylation per se is insufficient for translational activation of TOP mRNAs.

Next, we examined the translational efficiency of TOP mRNAs in cells lacking S6K1 activity, and consequently, their rpS6 was permanently hypo- or unphosphorylated. Such cells were obtained either by overexpression of dominant-negative S6K1 mutants or by targeted disruption of both S6K1 alleles (Kawasome et al. 1998). In all of these cells, TOP mRNAs were efficiently translated when untreated or upon amino acid refeeding. Hence, it appears that translational control of TOP mRNAs by amino acid sufficiency does not involve S6K activity or rpS6 phosphorylation (Tang et al. 2001).

The involvement of mTOR in the amino-acid-mediated regulation of S6K1 has been demonstrated by studies using rapamycin, which blocks the activation of S6K1 by amino acid refeeding (for review, see Gingras et al. 2001). Nevertheless, whether mTOR directly senses and is regulated by amino acid sufficiency is not clear. No effect of amino acid withdrawal and readdition on the kinase activities associated with mTOR has been demonstrated as yet, although dephosphorylation of mTOR in vivo in response to amino acid withdrawal has been reported (Westphal et al. 1999). The current prevailing model suggests that amino acid starvation, like rapamycin, leads to inhibition of mTOR kinase activity, resulting in the disinhibition of one or more protein phosphatases, which inactivate S6K1 (Hara et al. 1998; Peterson et al. 1999).

To examine the possible role of mTOR in the translational activation of TOP mRNAs, cells were treated by rapamycin upon amino acid refeeding. The rpS6 phosphorylation was completely blocked under these circumstances, yet rapamycin failed to significantly inhibit the amino-acid-induced translational activation of TOP mRNAs, indicating that the latter response occurs essentially in an mTOR-independent fashion (Tang et al. 2001).

The minor, if any, inhibitory effect of rapamycin on the amino-acid-induced activation of TOP mRNA translation led us to search for the involvement of an alternative signal transduction pathway. Previous reports have shown that the activity of PI3K and one of its downstream targets, PKB, is refractory to alterations in amino acid sufficiency (Hara et al. 1998; Patti et al. 1998; Wang et al. 1998; Iiboshi et al. 1999; Kimball et al. 1999). Hence, it has been argued that the effect of amino acids on S6K1 activity, unlike that of mitogenic stimulations, is not transduced through a PI3K-mediated pathway. Other experiments, however, demonstrated that S6K1 activity failed to respond to nutrient stimuli in the presence of inhibitors of PI3K (Fox et al. 1998; Patti et al. 1998; Shigemitsu et al. 1999; Tang et al. 2001), thus attesting to the involvement of a PI3K-mediated signaling pathway in the amino-acid-induced activation of S6K1.

Two complementary experimental approaches were used to examine the possible involvement of PI3K in the nutritional activation of TOP mRNAs translation: (1) Addition of LY294002, a specific inhibitor of PI3K, to amino acid refed cells completely abolished the recruitment of TOP mRNAs into polysomes (Tang et al. 2001) and (2) overexpression of PTEN, a tumor suppressor with a lipid phosphatase activity that antagonizes the activity of PI3K in vivo (Cantley and Neel 1999), or of a dominant-negative mutant of p85, the regulatory subunit of PI3K, suppressed the amino-acid-induced translational activation of TOP mRNAs (Tang et al. 2001). It therefore appears that PI3K activity is required for this mode of regulation.

A tentative model depicting the signaling pathways leading to the translational activation of TOP mRNAs by amino acid sufficiency is presented in Figure 3. According to this model, the signaling from amino acids bifurcates upstream of mTOR, as inferred from the ability of rapamycin to discern between the activity of mTOR and S6K, on the one hand, and the translational efficiency of TOP mRNAs, on the other hand. Moreover, amino acids signal into TOP mRNAs through an unknown target (denoted as X) in a PI3K-dependent fashion.

CANDIDATE *TRANS*-ACTING FACTORS

The discrete translational behavior of TOP mRNAs suggests that the 5′TOP motif is recognized by a specific translational *trans*-acting factor. This contention is supported by circumstantial evidence based on (1) the ability of a high-salt wash of mRNP from growth-arrested cells to selectively repress in vitro the translation of elongation factor 1A mRNA (Slobin and Rao 1993) and (2) the ability of short synthetic RNA oligonucleotides containing the first 16 nucleotides of rpS16 mRNA to specifically and completely relieve the translational repression of various chimeric TOP mRNAs in cell-free translation systems (Biberman and Meyuhas 1999). Hence, it appears that translational repression of TOP mRNAs is caused in vitro by the accumulation of a titratable repressor.

A search for proteins that interact with the 5′TOP motif has yielded a cytoplasmic protein of about 56 kD from mouse T lymphocytes, p56[L32], which specifically binds DNA or RNA sequences containing the first 34 nucleotides of mouse rpL32 mRNA (Kaspar et al. 1992; Severson et al. 1995). However, no experimental data are currently available to show that it can affect the translation of TOP mRNA in vitro or in vivo.

A similar study with *X. laevis* cytoplasmic extracts has established that La autoantigen and cellular nucleic-acid-binding protein (CNBP) bind the 5′TOP and the CG-rich sequence immediately downstream within the 5′UTR of ribosomal protein mRNAs, respectively (Pellizzoni et al. 1996, 1997). The binding of both these proteins requires the assistance of a third protein, Ro60 autoantigen (Pellizzoni et al. 1998). Induction of La expression in resting *Xenopus* cells led to a small elevation in the translational efficiency of endogenous TOP mRNAs (from ~40% to ~50% in polysomes), which, although statistically significant, was far less than that characteristic for growing cells (~80%) (Crosio et al. 2000). Furthermore, establishing the physiological significance of the small positive effect exerted by induced overexpression of La requires demonstration that the abundance or activity of La is upregulated by mitogenic stimulation.

CONCLUDING REMARKS

Resting or amino-acid-starved cells maintain a large excess of inefficiently translated mRNAs, which enables them to respond rapidly to mitogenic and nutritional stimulations by having immediately available a large protein synthesis capacity. Evidently, significant progress has been made in recent years toward understanding the mechanism of the translational control of vertebrate TOP mRNAs. Nevertheless, two fundamental questions remain unsatisfactorily answered: (1) What is the number and the nature of the *trans*-acting factors involved in the

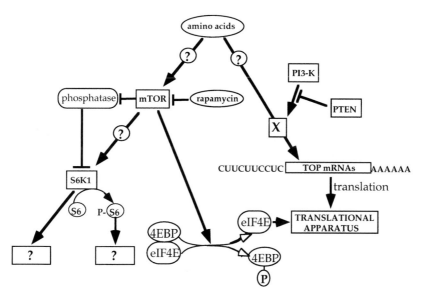

Figure 3. Schematic representation of signal transduction pathways involved in activation of rpS6 phosphorylation and translational control of TOP mRNAs in amino-acid-stimulated cells. Arrows delineate the flow of information. Circled and boxed question marks represent putative links and unknown targets, respectively. See text for details.

regulatory apparatus? and (2) Is the mTOR/S6K pathway involved in transducing mitogenic signals into the translational efficiency of TOP mRNAs?

The apparent nutritional control of TOP mRNAs in an mTOR/S6K-independent fashion might reflect the fact that the respective signaling pathway differs from that transducing mitogenic signals. However, several lines of evidence have cast doubts regarding the involvement of the mTOR/S6K pathway also in the translational activation of TOP mRNAs by mitogenic stimuli. Thus, TOP mRNAs are efficiently translated in dividing embryonic stem (ES) cells, in which both alleles of S6K1 were knocked out and rpS6 was constitutively unphosphorylated (Tang et al. 2001). Similarly, when the activity of S6K1 and rpS6 phosphorylation was blocked in proliferating cells by overexpressing of dominant-negative S6K1 mutants, the translation of cotransfected chimeric TOP mRNA was as efficient as in cells expressing an empty vector (Tang et al. 2001). Likewise, the marked discrepancy between the complete inhibition of S6K1 activation by rapamycin, on the one hand, and the minor effect on translational efficiency of TOP mRNAs upon mitogenic stimulation, on the other hand (Jefferies et al. 1994b, 1997), is inconsistent with causal relationships between these variables.

Verifying the relative contribution of the mTOR/S6K pathway to the mitogenically stimulated translation of TOP mRNAs might benefit from the analysis of the translational behavior of these mRNAs in already available relevant mutant organisms, including liver S6 knockout mice (Volarevic et al. 2000) and *Drosophila* lacking their single S6K gene, *dS6K* (Montagne et al. 1999). Notably, *Drosophila* cells not only contain TOP mRNAs, but the latter are also translationally regulated (for review, see Meyuhas et al. 1996a). In addition, establishment of an S6K1/S6K2 double-knockout mouse is crucial for studying the role of S6K activity in the translational control of TOP mRNAs in cells other than ES cells. If indeed these kinases are shown to be essential for this mode of regulation, disruption of the phosphorylation sites in rpS6 will still be necessary to elucidate whether the effect of S6K1/S6K2 is exerted solely through phosphorylation of rpS6 or other, yet unknown, substrates.

ACKNOWLEDGMENTS

This work was supported by grants to O. M. from the Israel Science Foundation founded by The Academy of Sciences and Humanities and by the United States–Israel Binational Science Foundation (BSF 97-00055).

REFERENCES

Agrawal A.G. and Bowman L.H. 1987. Transcriptional and translational regulation of ribosomal protein formation during mouse myoblast differentiation. *J. Biol. Chem.* **262:** 4868.

Aloni R., Peleg D., and Meyuhas O. 1992. Selective translational control and nonspecific posttranscriptional regulation of ribosomal protein gene expression during development and regeneration of rat liver. *Mol. Cell. Biol.* **12:** 2203.

Amaldi F. and Pierandrei-Amaldi P. 1997. TOP genes: A translationally controlled class of genes including those coding for

ribosomal proteins. *Prog. Mol. Subcell. Biol.* **18:** 1.

Avni D., Biberman Y., and Meyuhas O. 1997. The 5′ terminal oligopyrimidine tract confers translational control on TOP mRNAs in a cell type- and sequence context-dependent manner. *Nucleic Acids Res.* **25:** 995.

Avni D., Shama S., Loreni F., and Meyuhas O. 1994. Vertebrate mRNAs with a 5′-terminal pyrimidine tract are candidates for translational repression in quiescent cells: Characterization of the translational *cis*-regulatory element. *Mol. Cell. Biol.* **14:** 3822.

Ballou L., Luther H., and Thomas G. 1991. MAP2 kinase and 70K S6 kinase lie on distinct signalling pathways. *Nature* **349:** 348.

Biberman Y. and Meyuhas O. 1997. Substitution of just five nucleotides at and around the transcription start site of rat β-actin promoter is sufficient to render the resulting transcript a subject for translational control. *FEBS Lett.* **405:** 333.

———. 1999. TOP mRNAs are translationally inhibited by a titratable repressor in both wheat germ extract and reticulocyte lysate. *FEBS Letters* **456:** 357.

Blommaart E.F.C., Luiken J.J., Blommaart P.J., van Woerkom G.M., and Meijer A.J. 1995. Phosphorylation of ribosomal protein S6 is inhibitory for autophagy in isolated rat hepatocytes. *J. Biol. Chem.* **270:** 2320.

Cantley L. and Neel B. 1999. New insights into tumor suppression: PTEN suppresses tumor formation by restraining the phosphoinositide 3-kinase/AKT pathway. *Proc. Natl. Acad. Sci.* **96:** 4240.

Chua B.H. 1994. Specificity of leucine effect on protein degradation in perfused rat heart. *J. Mol. Cell. Cardiol.* **26:** 743.

Chung J., Kuo C.J., Crabtree G.R., and Blenis J. 1992. Rapamycin-FKBP specifically blocks growth-dependent activation of and signaling by the 70 kd S6 kinases. *Cell* **69:** 1227.

Chung J., Grammer T., Lemon K., Kaziauskas A., and Blenis J. 1994. PDGF- and insulin-dependent pp70S6k activation mediated by phosphatidylinositol-3-OH kinase. *Nature* **370:** 71.

Crosio C., Boyl P., Loreni F., Pierandrei-Amaldi P., and Amaldi F. 2000. La protein has a positive effect on the translation of TOP mRNAs in vivo. *Nucleic Acids Res.* **28:** 2927.

Dennis P., Fumagalli S., and Thomas G. 1999. Target of rapamycin (TOR): Balancing the opposing forces of protein synthesis and degradation. *Curr. Opin. Genet. Dev.* **9:** 49.

Dufner A. and Thomas G. 1999. Ribosomal S6 kinase signaling and the control of translation. *Exp. Cell Res.* **253:** 100.

Erikson E. and Maller J. 1985. A protein kinase from *Xenopus* eggs specific for ribosomal protein S6. *Proc. Natl. Acad. Sci.* **82:** 742.

Fafournoux P., Bruhat A., and Jousse C. 2000. Amino acid regulation of gene expression. *Biochem. J.* **351:** 1.

Fox H.L., Kimball S.R., Jefferson L.S., and Lynch C.J. 1998. Amino acids stimulate phosphorylation of p70S6k and organization of rat adipocytes into multicellular clusters. *Am. J. Physiol.* **274:** C206.

Fumagalli S. and Thomas G. 2000. S6 phosphorylation and signal transduction. In *Translational control of gene expression* (ed. N. Sonenberg et al.). p. 695. Cold Spring Harbor Laboratory Press, Cold Spring Harbor, New York.

Gingras A.C., Raught B., and Sonenberg N. 2001. Regulation of translation initiation by FRAP/mTOR. *Genes Dev.* **15:** 807.

Gout I., Minami T., Hara K., Tsujishita Y., Filonenko V., Waterfield M., and Yonezawa K. 1998. Molecular cloning and characterization of a novel p70 S6 kinase, p70 S6 kinase beta containing a proline-rich region. *J. Biol. Chem.* **273:** 30061.

Gressner A.M. and Wool I.G. 1974. The phosphorylation of liver ribosomal proteins in vivo . Evidence that only a single small subunit protein (S6) is phosphorylated. *J. Biol. Chem.* **249:** 6917.

Grummt I. 1999. Regulation of mammalian ribosomal gene transcription by RNA polymerase I. *Prog. Nucleic Acid Res. Mol. Biol.* **62:** 109.

Hara K., Yonezawa K., Weng Q.-P., Kozlowski M.T., Belham C., and Avruch J. 1998. Amino acid sufficiency and mTOR regulate p70 S6 kinase and eIF-4E BP1 through a common effector mechanism. *J. Biol. Chem.* **273:** 14484.

Hornstein E., Git A., Braunstein I., Avni D., and Meyuhas O.

1999. The expression of poly (A)-binding protein gene is translationally regulated in a growth dependent fashion through a 5′-terminal oligopyrimidine tract motif. *J. Biol. Chem.* **274:** 1708.

Iiboshi Y., Papst P.J., Kawasome H., Hosoi H., Abraham R.T., Houghton P.J., and Terada N. 1999. Amino acid-dependent control of p70(s6k). Involvement of tRNA aminoacylation in the regulation. *J. Biol. Chem.* **274:** 1092.

Jefferies H.B.J. and Thomas G. 1996. Ribosomal protein S6 phosphorylation and signal transduction. In *Translational control* (ed. J.W.B. Hershey et al.), p. 389. Cold Spring Harbor Laboratory Press, Cold Spring Harbor, New York.

Jefferies H.B.J., Thomas G., and Thomas G. 1994a. Elongation factor-1 α mRNA is selectively translated following mitogenic stimulation. *J. Biol. Chem.* **269:** 4367.

Jefferies H.B.J., Reinhard C., Kozma S.C., and Thomas G. 1994b. Rapamycin selectively represses translation of the 'polypyrimidine tract' mRNA family. *Proc. Natl. Acad. Sci.* **91:** 4441.

Jefferies H.B., Fumagalli S., Dennis P., Reinhard C., Pearson R., and Thomas G. 1997. Rapamycin suppresses 5′TOP mRNA translation through inhibition of p70^{s6k}. *EMBO J.* **16:** 3693.

Kaspar R.L., Kakegawa T., Cranston H., Morris D.R., and White M.W. 1992. A regulatory *cis* element and a specific binding factor involved in the mitogenic control of murine ribosomal protein L32 translation. *J. Biol. Chem.* **267:** 508.

Kaspar R.L., Morris D.R., and White M. 1993. Control of ribosomal protein synthesi in eukaryotic cells. in *Translational Regulation of Gene Expression* (ed. J. Ilan). pp. 335-348. Plenum Press, New York.

Kawasome H., Papst P., Webb S., Keller G.M., Johnson G.L., Gelfand E.W., and Terada N. 1998. Targeted disruption of p70$^{(s6k)}$ defines its role in protein synthesis and rapamycin sensitivity. *Proc. Natl. Acad. Sci.* **95:** 5033.

Kimball S.R. and Jefferson L.S. 2000. Regulation of translation initiation in mammalian cells by amino acids. In *Translational control of gene expression* (ed. N. Sonenberg et al.), p. 561. Cold Spring Harbor Laboratory Press, Cold Spring Harbor, New York.

Kimball S.R., Shantz L.M., Horetsky R.L., and Jefferson L.S. 1999. Leucine regulates translation of specific mRNAs in L6 myoblasts through mTOR-mediated changes in availability of eIF4E and phosphorylation of ribosomal protein S6. *J. Biol. Chem.* **274:** 11647.

Kimball S.R., Fabian J.R., Pavitt G.D., Hinnebusch A.G., and Jefferson L.S. 1998. Regulation of guanine nucleotide exchange through phosphorylation of eukaryotic initiation factor eIF2alpha. Role of the alpha- and delta-subunits of eIF2b. *J. Biol. Chem.* **273:** 12841.

Kozma S.C., Lane H.A., Ferrari S., Luther H., Siegmann M., and Thomas G. 1989. A stimulated S6 kinase from rat liver: Identity with the mitogen-activated S6 kinase from 3T3 cells. *EMBO J.* **8:** 4125.

Krieg J., Hofsteenge J., and Thomas G. 1988. Identification of the 40 S ribosomal protein S6 phosphorylation sites induced by cycloheximide. *J. Biol. Chem.* **263:** 11473.

Levy S., Avni D., Hariharan N., Perry R.P., and Meyuhas O. 1991. Oligopyrimidine tract at the 5′ end of mammalian ribosomal protein mRNAs is required for their translational control. *Proc. Natl. Acad. Sci.* **88:** 3319.

Loreni F. and Amaldi F. 1992. Translational regulation of ribosomal protein synthesis in *Xenopus* cultured cells: mRNA relocation between polysomes and RNP during nutritional shifts. *Eur. J. Biochem.* **205:** 1027.

Meyuhas O. 2000. Synthesis of the translational apparatus is regulated at the translational level. *Eur. J. Biochem.* **267:** 6321.

Meyuhas O. and Hornstein E. 2000. Translational control of TOP mRNAs. In *Translational control of gene expression* (ed. N. Sonenberg et al.). p. 671. Cold Spring Harbor Laboratory Press, Cold Spring Harbor, New York.

Meyuhas O., Avni D., and Shama S. 1996a. Translational control of ribosomal protein mRNAs in eukaryotes. In *Translational control* (ed. J.W.B. Hershey et al.), p. 363. Cold Spring Harbor Laboratory Press, Cold Spring Harbor, New York.

Meyuhas O., Thompson A.E., and Perry R.P. 1987. Glucocorticoids selectively inhibit the translation of ribosomal protein mRNAs in P1798 lymphosarcoma cells. *Mol. Cell. Biol.* **7:** 2691.

Meyuhas O., Biberman Y., Pierandrei-Amaldi P., and Amaldi F. 1996b. Analysis of polysomal RNA. In *A laboratory guide to RNA: Isolation, analysis, and synthesis* (ed. P. Krieg), p. 65. Wiley-Liss, New York.

Ming X., Burgering B., Wennstrom S., Claesson-Welsh L., Heldin C., Bos J., Kozma S., and Thomas G. 1994. Activation of p70/p85 S6 kinase by a pathway independent of p21ras. *Nature* **371:** 426.

Montagne J., Stewart M., Stocker H., Hafen E., Kozma S., and Thomas. 1999. *Drosophila* S6 kinase: A regulator of cell size. *Science* **285:** 2126.

Nygard O. and Nika H. 1982. Identification by RNA-protein cross-linking of ribosomal proteins located at the interface between the small and the large subunits of mammalian ribosomes. *EMBO J.* **1:** 357.

Patti M.E., Brambilla E., Luzi L., Landaker E.J., and Kahn C.R. 1998. Bidirectional modulation of insulin action by amino acids. *J. Clin. Invest.* **101:** 1519.

Pellizzoni L., Lotti F., Maras B., and Pierandrei-Amaldi P. 1997. Cellular nucleic acid binding protein binds a conserved region of the 5′ UTR of *Xenopus laevis* ribosomal protein mRNAs. *J. Mol. Biol.* **267:** 264.

Pellizzoni L., Lotti F., Rutjes S., and Pierandrei-Amaldi P. 1998. Involvement of the *Xenopus laevis* Ro60 autoantigen in the alternative interaction of La and CNBP proteins with the 5′UTR of L4 ribosomal protein mRNA. *J. Mol. Biol.* **281:** 593.

Pellizzoni L., Cardinali B., Lin-Marq N., Mercanti D., and Pierandrei-Amaldi P. 1996. A *Xenopus laevis* homologue of the La autoantigen binds the pyrimidine tract of the 5′UTR of ribosomal protein mRNAs in vitro: Implication of a protein factor in complex formation. *J. Mol. Biol.* **259:** 904.

Peterson R.T., Desai B.N., Hardwick J.S., and Schreiber S.L. 1999. Protein phosphatase 2A interacts with the 70-kDa S6 kinase and is activated by inhibition of FKBP12-rapamycin-associated protein. *Proc. Natl. Acad. Sci.* **96:** 4438.

Pham P.T., Heydrick S.J., Fox H.L., Kimball S.R., Jefferson L.S., Jr., and Lynch C.J. 2000. Assessment of cell-signaling pathways in the regulation of mammalian target of rapamycin (mTOR) by amino acids in rat adipocytes. *J. Cell. Biochem.* **79:** 427.

Price D.J., Grove J.R., Calvo V., Avruch J., and Bierer B.E. 1992. Rapamycin-induced inhibition of the 70-kilodalton protein kinase. *Science* **257:** 973.

Rabkin R., Tsao T., Shi J., and Mortimore G. 1991. Amino acids regulate kidney cell protein breakdown. *J. Lab. Clin. Med.* **117:** 505.

Rao T.R. and Slobin L.I. 1987. Regulation of the utilization of mRNA for eucaryotic elongation factor Tu in Friend erythroleukemia cells. *Mol. Cell. Biol.* **7:** 687.

Raught B., Gingras A.C., and Sonenberg N. 2000. Regulation of ribosomal recruitment in eukaryotes. In *Translational control of gene expression* (ed. N. Sonenberg et al.), p. 245. Cold Spring Harbor Laboratory Press, Cold Spring Harbor, New York.

Sachs A. 2000. Physical and functional interactions between the mRNA cap structure and the poly(A) tail. In *Translational control of gene expression* (ed. N. Sonenberg et al.), p. 447. Cold Spring Harbor Laboratory Press, Cold Spring Harbor, New York.

Saitoh M., ten Dijke P., Miyazono K., and Ichijo H. 1998. Cloning and characterization of p70^{s6k} β defines a novel family of p70 S6 kinases. *Biochem. Biophys. Res. Commun.* **253:** 470.

Schmidt E.V. 1999. The role of c-*myc* in cellular growth control. *Oncogene* **18:** 2988.

Sehgal S.N. 1998. Rapamune (RAPA, rapamycin, sirolimus): Mechanism of action immunosuppressive effect results from blockade of signal transduction and inhibition of cell cycle progression. *Clin. Biochem.* **31:** 335.

Severson W.E., Mascolo P.L., and White M.W. 1995. Lymphocyte p56^{L32} is a RNA/DNA-binding protein which interacts with conserved elements of the murine L32 ribosomal protein mRNA. *Eur. J. Biochem.* **229:** 426.

Shigemitsu K., Tsujishita Y., Hara K., Nanahoshi M., Avruch J., and Yonezawa K. 1999. Regulation of translational effectors by amino acid and mammalian target of rapamycin signaling pathways. Possible involvement of autophagy in cultured hepatoma cells. *J. Biol. Chem.* **274:** 1058.

Shima H., Pende M., Chen Y., Fumagalli S., Thomas G., and Kozma S. 1998. Disruption of the p70(s6k)/p85(s6k) gene reveals a small mouse phenotype and a new functional S6 kinase. *EMBO J.* **17:** 6649.

Slobin L.I. and Rao M.N. 1993. Translational repression of EF-1α mRNA in vitro. *Eur. J. Biochem.* **213:** 919.

Sturgill T.W. and Wu J. 1991. Recent progress in characterization of protein kinase cascades for phosphorylation of ribosomal protein S6. *Biochim. Biophys. Acta.* **1092:** 350.

Tang H., Hornstein E., Stolovich M., Levy G., Livingstone M., Templeton D., Avruch J., and Meyuhas O. 2001. Amino acid-induced translation of TOP mRNAs is fully dependent on PI3-kinase-mediated signaling, is partially inhibited by rapamycin, and is independent of S6K1 and rpS6 phosphorylation. *Mol. Cell. Biol.* **21:** 8671.

Terada N., Takase K., Papst P., Nairn A.C., and Gelfand E.W. 1995. Rapamycin inhibits ribosomal protein synthesis and induces G1 prolongation in mitogen-activated T lymphocytes. *J. Immunol.* **155:** 3418.

Terada N., Patel H.R., Takase K., Kohno K., Nairn A.C., and Gelfand E.W. 1994. Rapamycin selectively inhibits translation of mRNAs encoding elongation factors and ribosomal proteins.

Proc. Natl. Acad. Sci. **91:** 11477.

Thomas G. and Thomas G. 1986. Translational control of mRNA expression during the early mitogenic response in Swiss mouse 3T3 cells: Identification of specific proteins. *J. Cell Biol.* **103:** 2137.

Thomas G., Siegmann M., Kubler A., Gordon J., and Jimenez de Asua L. 1980. Regulation of 40S ribosomal protein S6 phosphorylation in Swiss mouse 3T3 cells. *Cell* **19:** 1015.

Volarevic S., Stewart M., Ledermann B., Zilberman F., Terracciano L., Montini E., Grompe M., Kozma S., and Thomas G. 2000. Proliferation, but not growth, blocked by conditional deletion of 40S ribosomal protein S6. *Science* **288:** 2045.

von Manteuffel S., Dennis P., Pullen N., Gingras A., Sonenberg N., and Thomas G. 1997. The insulin-induced signalling pathway leading to S6 and initiation factor 4E binding protein 1 phosphorylation bifurcates at a rapamycin-sensitive point immediately upstream of p70s6k. *Mol. Cell. Biol.* **17:** 5426.

Wang X., Campbell L.E., Miller C.M., and Proud C.G. 1998. Amino acid availability regulates p70 S6 kinase and multiple translation factors. *Biochem. J.* **334:** 261.

Westphal R., Coffee R.J., Marotta A., Pelech S., and Wadzinski B. 1999. Identification of kinase-phosphatase signaling modules composed of p70 S6 kinase-protein phosphatase 2A (PP2A) and p21-activated kinase-PP2A. *J. Biol. Chem.* **274:** 687.

Xu G., Kwon G., Marshall C.A., Lin T.A., Lawrence J.C.J., and McDaniel M.L. 1998. Branched-chain amino acids are essential in the regulation of PHAS-I and p70 S6 kinase by pancreatic beta-cells. A possible role in protein translation and mitogenic signaling. *J. Biol. Chem.* **273:** 28178.

Double-stranded RNA-binding Proteins and the Control of Protein Synthesis and Cell Growth

L.M. Parker,* I. Fierro-Monti,*† T.W. Reichman,‡ S. Gunnery, and M.B. Mathews†‡¶

Department of Biochemistry and Molecular Biology and ‡Graduate School of Biomedical Sciences, New Jersey Medical School, UMDNJ, Newark, New Jersey 07103-2714

RNA-binding proteins play central roles in cellular metabolism. They are responsible for the transcription, processing, localization, transport, and translation of RNA; accordingly, they are essential for controlled cell growth and proper organismal development. Several families of RNA-binding proteins have been identified that share common RNA-binding motifs. These include the RRM (RNA recognition motif) proteins that primarily bind to single-stranded RNA (ssRNA) (Shamoo et al. 1995), the zinc finger motif proteins that bind to duplexed DNA as well as double-stranded RNA (dsRNA) (Finerty and Bass 1997, 1999; Yang et al. 1999), and the dsRNA-binding motif (dsRBM) proteins that bind to duplexed and highly structured RNA (Fierro-Monti and Mathews 2000). The dsRBM is characterized by an α–β–β–β–α structure that is involved in binding to dsRNA and also serves as a protein–protein interaction motif.

Sequence homologies lying outside their dsRBMs divide the dsRBM family of RNA-binding proteins into at least nine subfamilies (Fierro-Monti and Mathews 2000). Enzymatic functions of these proteins vary from the unwinding activity of RNA helicase A (RHA) (Lee and Hurwitz 1993), to the hydrolytic and processing activity of RNase III (Nicholson 1996), and the RNA-editing activities of the deaminases ADAR1 and ADAR2 (Gerber and Keller 2001). These proteins all cause structural changes in RNA via covalent or noncovalent modifications, and play key roles in cell function. For example, RHA acts as a transcriptional cofactor (Nakajima et al. 1997; Aratani et al. 2001) and is required for early embryonic development (Lee et al. 1998), whereas its *Drosophila* homolog *maleless* is essential for dosage compensation in the fruit fly (Lee et al. 1997). The dsRNA-dependent protein kinase PKR represents a second category of dsRBM-containing enzymes. This enzyme does not modify RNA; rather, its activity is modulated by structured and dsRNAs. PKR is widespread in vertebrate cells in a latent (inactive) form, and its synthesis is induced transcriptionally by the cytokines interferon α and β as part of the cellular antiviral response. PKR is a serine/threonine kinase that gains enzymatic activity as a result of autophosphorylation. Once activated, PKR phosphorylates the eukaryotic initiation factor eIF2 on its α subunit, leading to inhibition of protein synthesis at the level of initiation as a result of impaired recycling of eIF2 by the guanosine nucleotide exchange factor eIF2B (for review, see Kaufman 2000). The significance of PKR is underlined by the number and variety of mechanisms that viruses have evolved to block or otherwise evade the consequences of its activation, evidently in order to maintain protein synthesis and productive infection (for review, see Pe'ery and Mathews 2000).

Other dsRBM-containing proteins have roles in cell growth control, gene expression, and RNA localization, although they exert no known enzymatic activities. The *Drosophila* protein Staufen, for example, plays an important part in the developing embryo and oocyte; it is responsible for the proper localization of *bicoid* and *oskar* mRNAs at the anterior pole of the embryo and the posterior pole of the oocyte, respectively, by binding to regions within the 3´UTRs of these mRNAs (St Johnston et al. 1991; Ferrandon et al. 1994). The human homolog of this protein is also involved in mRNA localization and transport (Roegiers and Jan 2000). Finally, a fourth group of dsRBM-containing proteins includes those that modulate PKR function. In the case of the poxvirus proteins exemplified by vaccinia E3L (Romano et al. 1998) and the cellular protein PACT (Patel and Sen 1998), this role is well-established; in other cases, such as the TAR RNA-binding protein TRBP (Park et al. 1994; Cosentino et al. 1995) and nuclear factor 90 (NF90), which is the focus of the present study, there is reason to question whether this is the proteins' sole or principal role.

PKR REGULATION

PKR is able to bind to dsRNA that is 11 base pairs (bp) in length, although such short RNA molecules cannot activate PKR and instead act as competitive inhibitors of the enzyme. The minimal RNA duplex needed for PKR activation approaches 30 bp, and efficiency increases steadily with chain length up to a maximum at ~85 bp (Manche et al. 1992; Schmedt et al. 1995). It is thought that this requirement for longer RNA reflects an activation mechanism in which two monomers of PKR are brought together on a single dsRNA molecule, allowing for intermolecular autophosphorylation (Kostura and

---

*These authors made equivalent contributions to this paper.

†Present address: Department of Biochemistry, University of Cambridge, Cambridge, United Kingdom.

¶Corresponding author: mathews@umdnj.edu

Mathews 1989). Correspondingly, long viral dsRNAs generated during infection are thought to serve as natural PKR activators, whereas highly structured viral RNAs such as adenovirus VA RNA$_I$ that contain short duplexed regions inhibit the function of PKR (Kostura and Mathews 1989; Mellits et al. 1990). Other viral PKR inhibitors include the vaccinia virus proteins E3L and K3L. K3L has homology with eIF2α in the region of its phosphorylation site and functions as a pseudosubstrate (Davies et al. 1993; Kawagishi-Kobayashi et al. 1997; Sharp et al. 1997), whereas E3L sequesters the dsRNA activator via its dsRBM (Davies et al. 1993; Romano et al. 1998; Sharp et al. 1998).

Cellular regulators of PKR function have also been identified. One of these, P58, appears to be a viral target. Influenza virus infection converts this normally inactive inhibitor of PKR into an inhibitory form that can downregulate PKR function (for review, see Tan et al. 2000). The cellular proteins, PACT and its rat homolog RAX, are exceptional in that they activate PKR. PACT and RAX contain three dsRBMs and can directly interact with PKR, activating it in the absence of dsRNA in response to cell stress (Patel et al. 2000; Ruvolo et al. 2001). The cellular *Alu* RNAs, also produced under cell stress conditions (Chu et al. 1998), activate PKR kinase activity, although they inhibit the function of PKR at high concentrations. Thus, PKR shuts down translational initiation in response to cellular stresses and viral infections.

Although eIF2 is the best-known substrate for PKR, other substrates are beginning to emerge, such as the regulatory subunit of PP2A (Xu and Williams 2000) and p53 (Cuddihy et al. 1999a), although the functional consequences of their phosphorylation by PKR are not well understood. PKR also phosphorylates RHA, NF90, and its binding partner nuclear factor 45 (NF45), at least in vitro (Parker et al. 2001). The enzyme has also been implicated in several pathways and functions other than protein synthesis, ranging from apoptosis and splicing (Balachandran et al. 1998; Gil et al. 1999; Osman et al. 1999) to differentiation and cancer (Meurs et al. 1993; Barber et al. 1995; Petryshyn et al. 1997; Abraham et al. 1998; Kim et al. 2000). Its presence in complexes with a variety of cellular proteins is suggestive of roles distinct from translation. For example, PKR has been found in the IκB kinase (IKK) complex and is thought to mediate the activation of NFκB through this pathway (Bonnet et al. 2000; Gil et al. 2000; Zamanian-Daryoush et al. 2000). It is associated with the STAT1 and STAT3 transcription factors (Wong et al. 1997; Deb et al. 2001) and coimmunoprecipitates with NF90 (Parker et al. 2001) and RHA (I. Fierro-Monti and M.B. Mathews, unpubl.).

Taken together, these data provide *prima facie* evidence for the involvement of PKR in cell growth regulation distinct from its translational function. In support of this idea, work on related eIF2 kinases GCN2 (Hinnebusch 2000) and PERK (Harding et al. 2000) has shown that partial inhibition of protein synthesis can be an important mechanism to up-regulate critical genes, as is the case for the transcription activators GCN4 and ATF4. Furthermore, recent work has demonstrated that PKR

participates in the PDGF signaling pathway, important for cell proliferation (Deb et al. 2001). On the other hand, the role of PKR in malignancy has proved elusive. Although several reports showed that PKR can function as a tumor suppressor, two independent strains of PKR knock-out mice failed to show evidence of increased susceptibility to malignancy (Yang et al. 1995; Abraham et al. 1999). In fact, fibroblasts from PKR null mice grow more slowly than their wild-type counterparts, implying that PKR plays a positive role in cell growth (Zamanian-Daryoush et al. 1999).

NF90 AND NF45

NF90 was originally purified and characterized by virtue of its ability to bind, in complex with NF45, to the antigen response recognition element ARRE-2, a DNA element in the IL-2 promoter (Corthésy and Kao 1994; Kao et al. 1994). Subsequently, we identified NF90 (together with NF45) as one of a small number of proteins that bind to dsRNA and to a second small adenoviral RNA, VA RNA$_{II}$ (Liao et al. 1998). Other labs have also identified NF90, as well as the related proteins DRBP76, ILF3, and NFAR1 and 2, on the basis of their ability to interact with PKR (Buaas et al. 1999; Patel et al. 1999; Saunders et al. 2001). Another closely related protein, MPP4, is phosphorylated during the M phase of the cell cycle (Matsumoto-Taniura et al. 1996). Both NF90 and NF45 are found in a complex containing eIF2 and the catalytic subunit of the DNA-dependent protein kinase DNA-PK (Ting et al. 1998).

Recent work is supportive of activities related to both transcription (Saunders et al. 2001; Reichman et al. 2002) and translation (Parker et al. 2001). NF90 interacts with PKR in RNA-dependent and RNA-independent fashions through regions in its amino and carboxyl termini, respectively (Parker et al. 2001). Furthermore, NF90 serves as a substrate for PKR in vitro: PKR phosphorylation of NF90 occurs in the RNA-binding region of NF90, and the carboxy-terminal half of NF90 can inhibit this phosphorylation *in trans*. Translational effects have also been reported by others. TCP80, thought to be an alternatively spliced variant of NF90, has been implicated in the translational regulation of acid β-glucosidase (Xu et al. 2000). The rat homolog of NF90, p74, interacts with PKR physically as well as functionally in yeast, giving a phenotype suggesting that PKR forms a higher-order complex with p74 and eIF2 that may be important for PKR function (Coolidge and Patton 2000).

NF90 can act as a transcriptional regulator, serving to either up- or down-regulate promoter activity. Its ability to activate transcription is enhanced by NF45 in vivo, apparently through a conformational change in NF90. NF90 binds to NF45 primarily through its central region, which shares homology with NF45 (Reichman et al. 2002). The homology between NF90 and NF45 is the only recognizable sequence motif within NF45, and its function remains unknown.

From this summary, it is evident that the biological role of NF90 is poorly understood and that of NF45 is even more nebulous. Here we investigate the translational roles of these proteins. Our data demonstrate that NF90 can serve as a positive activator of PKR function, whereas NF45 down-regulates the function of PKR. We speculate on the integration of these findings into an overall scheme of PKR's role as a pivotal regulator of gene expression.

NF90 FUNCTION IN YEAST

Expressed on its own, NF90 did not perturb yeast growth (data not shown), but functional interactions were observed when it was coexpressed with PKR (Fig. 1). Like the *Saccharomyces cerevisiae* kinase GCN2, PKR phosphorylates yeast eIF2α on Ser-51 (Dever et al. 1992). Consequently, high levels of PKR inhibit translation initiation and yeast cell growth (Dever et al. 1993).

These inhibitory effects can be overcome by the vaccinia virus E3L or K3L proteins that block eIF2 phosphorylation (Romano et al. 1998; Sharp et al. 1998), whereas the PKR activator PACT enhances the slow-growth phenotype caused by PKR (Patel and Sen 1998).

Figure 1A displays the relative growth rates of yeast cells expressing PKR together with E3L, PACT or NF90, in comparison with cells containing PKR alone. Western blots showed that full-length NF90 is stably expressed in yeast (data not shown). Coexpression of NF90 with PKR slowed growth to about the same extent as seen with PACT, whereas E3L led to a slightly increased growth rate. This observation is consistent with NF90 functioning as a PKR activator. To verify that the growth effect is mediated by the kinase activity of PKR, we used the catalytically inactive mutant PKR[K296R], which does not exert an inhibitory effect on growth in *S. cerevisiae*. Co-

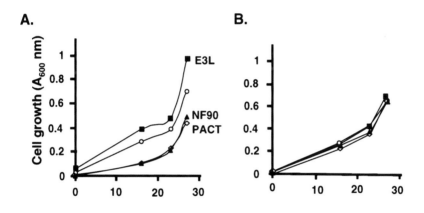

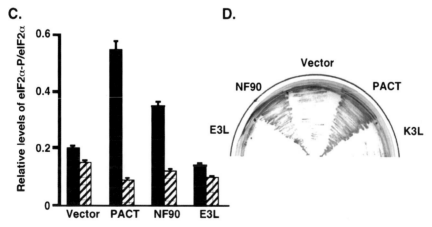

Figure 1. NF90 exacerbates the PKR slow-growth phenotype in yeast. (*A, B*) Growth curves of INVsc1 yeast cells (Invitrogen; *MAT his31 leu2 trp1-289 ura 3-52/MAT his31 leu2, trp1-289, ura3-52* [diploid]) expressing wild-type PKR (*A*) or catalytically inactive PKR[K296R] (*B*), together with the empty vector pYES2 (*open circles*), PKR inhibitor vaccinia virus E3L (*closed squares*), PKR activator PACT (*open diamonds*), or NF90 (*closed triangles*). (*C*) Cell extracts from INVsc1 yeast strains (as in *A* and *B*) were analyzed by immunoblotting with antibodies directed against eIF2α and eIF2α-P. Signals were quantified by densitometry. The bar graph displays the ratio of eIF2α-P:eIF2α in extracts from cells expressing PKR (*solid bars*) or PKR[K296R] (*hatched bars*) together with PACT, NF90, E3L, or containing the empty vector as indicated. (*D*) The empty vector pYES2, and plasmids expressing NF90, PACT, E3L, or K3L, all under the control of the galactose inducible (GAL) promoter, were introduced into the yeast strain RY1-1. This gcn2Δ strain contains two integrated copies of the wild-type PKR gene. Transformants were streaked from single colonies on SD medium in the absence of histidine and the presence of 30 mM 3-AT and were incubated for 5–6 days at 30°C.

expression of NF90, E3L, or PACT with PKR[K296R] did not cause any significant change in growth rate (Fig. 1B).

These genetic data indicate that NF90 activates PKR in vivo, and implicate PKR's kinase activity in the response to NF90. Because eIF2α is not the only substrate for PKR, at least in higher cells, we measured eIF2α phosphorylation levels in the yeast used for growth assays. As shown in Figure 1C, the expression of NF90 or PACT in the presence of PKR led to increased eIF2α phosphorylation. Conversely, E3L reduced the phosphorylation of eIF2α observed in the presence of PKR. As expected, no such changes were observed in the presence of PKR[K296R]. These biochemical results are consistent with the view that the growth effects of NF90 are mediated by eIF2 phosphorylation.

For a further genetic test of NF90's function, we used the gcn2-deleted S. cerevisiae strain RY1-1, which carries the PKR gene integrated into the yeast genome (Romano et al. 1995). Growth of yeast in the absence of exogenous histidine is dependent on induction of the transcriptional activator GCN4. This induction takes place at the translational level and requires a modest elevation in eIF2 phosphorylation that is usually brought about by the GCN2 kinase (Hinnebusch 2000). When grown in glucose-containing medium, the RY1-1 strain produces wild-type PKR at low levels that can induce the expression of GCN4 via limited phosphorylation of eIF2α. GCN4 induction results in the synthesis of several enzymes involved in amino acid biosynthesis. Thus, expression of PKR allows RY1-1 cells to grow on medium lacking histidine and supplemented with 3-aminotriazol (3-AT), an inhibitor of the histidine biosynthetic pathway. Under these conditions, the PKR inhibitors K3L and E3L inhibited growth, whereas NF90 and the PKR activator PACT enhanced growth (Fig. 1D). The opposing phenotypic effects of NF90 in the experiments of Figure 1A (growth inhibition) and 1D (growth stimulation) provide strong evidence that it functions by increasing the phosphorylation of eIF2α in the presence of PKR.

NF90 INHIBITS CELL-FREE TRANSLATION

On the basis of the genetic experiments described above, we surmised that if NF90 were activating PKR, it would also inhibit protein synthesis in the reticulocyte lysate translation system since PKR is present in reticulocytes (Farrell et al. 1977). To address this possibility, we examined the effect of a GST-NF90 fusion protein on the translation of mRNAs in the rabbit reticulocyte translation system. Globin mRNA or luciferase mRNA was incubated with the reticulocyte lysate in the presence of [35S]methionine, and the products were quantified after autoradiography. Figure 2 displays the levels of globin and luciferase synthesized relative to a buffer-only control. The GST-NF90 fusion protein was inhibitory (60%) toward the translation of the capped and polyadenylated globin mRNA, whereas no effect was seen with the GST control. Similar translational inhibition was obtained

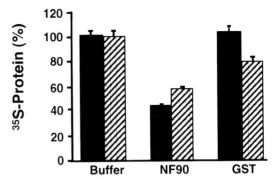

Figure 2. Inhibition of cell-free translation by NF90. The effect of NF90 on translation of reporter mRNAs was tested in vitro using the rabbit reticulocyte lysate system with capped and polyadenylated globin mRNA (*solid bars*) or polyadenylated luciferase mRNA (*hatched bars*). GST-NF90 (NF90; 150 ng) or GST (150 ng) in PBS buffer, or PBS buffer alone (Buffer), was included in the translation assays. Radiolabeled protein products were separated by SDS-polyacrylamide gel electrophoresis and quantified using a Packard Instant Imager. Protein synthesis is expressed as percentage of the control (Buffer) [35S]methionine incorporation after correcting for background. Error bars indicate standard deviation.

with luciferase mRNA, although GST alone gave some inhibition (~20%) of luciferase synthesis. Whether this is due to sequence differences between the mRNAs or to the fact that the luciferase mRNA was uncapped (though polyadenylated) is unclear at present.

DOES NF90 ACTIVATE PKR IN VITRO?

One possible inference from the foregoing results (Figs. 1 and 2) is that NF90 activates PKR function. This deduction would run counter to our recent demonstration that NF90, at relatively high concentrations, inhibits PKR function in vitro through competitive binding to dsRNA via its dsRBMs (Parker et al. 2001). On the other hand, the finding that the amino-terminal region of NF90 interacts with PKR at a second site in an RNA-independent manner raises the possibility of additional NF90 effects on PKR. We therefore wanted to determine directly whether NF90 up-regulates PKR autophosphorylation activity, or its ability to phosphorylate eIF2, in vitro. Kinase assays were performed at concentrations of NF90 that do not inhibit PKR function through sequestration of dsRNA. Full-length NF90, or a truncated version of the protein containing only its amino-terminal 334 amino acids, was added to PKR kinase assays in which the protein components were preassembled and held on ice for 30 minutes.

PKR failed to phosphorylate either itself or eIF2 in the absence of dsRNA (Fig. 3A, lane 1) when full-length NF90 or its amino-terminal half was present (lanes 2 and 3). Thus, NF90 has no effect on PKR in the absence of dsRNA when preincubated with the kinase and eIF2; in this respect, it differs from PACT, which activates PKR in the absence of dsRNA (Patel and Sen 1998). When dsRNA was added to the kinase assay mixture in which protein components had been preincubated on ice (Fig.

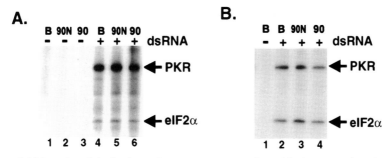

Figure 3. NF90 does not inhibit PKR activity in vitro. Kinase assays were performed in the presence or absence of dsRNA as indicated. The protein components of the assay were preincubated on ice for 30 min in the absence (*A*) or presence (*B*) of dsRNA. PKR, eIF2, and full-length NF90 (90), or an amino-terminal truncation of NF90 (residues 1–334; 90N), were assayed in kinase assays as described previously (Manche et al. 1992). Following preincubation, [γ-^{32}P]ATP was added, with reovirus dsRNA where indicated, and reaction mixtures were incubated at 30°C for 30 min. Shown are autoradiograms of SDS-polyacrylamide gels with the positions of PKR and eIF2α indicated by arrows.

3A, lanes 4–6), neither NF90 nor its amino-terminal region had any discernible effect on PKR autophosphorylation or eIF2 phosphorylation activity. When the proteins were preincubated on ice together with dsRNA, no effect of NF90 or its amino-terminal domain was observed (Fig. 3B). Although a variety of experimental conditions were tested, including different concentrations of NF90, a positive effect of NF90 on PKR kinase activity was not seen in vitro (data not shown). These data suggest that the translational inhibition effect of NF90 observed in yeast and reticulocyte lysates requires additional components that are not present in the purified kinase assay.

IS NF90 SPECIFIC FOR PKR?

Reticulocytes also contain a second kinase, the heme-regulated inhibitor HRI, which can phosphorylate eIF2α and inhibit cellular protein synthesis (for review, see Chen 2000). Therefore, HRI activation could potentially contribute to, or even be responsible for, the translational inhibition seen in Figure 2. Since HRI causes a slow-growth phenotype when expressed in yeast (Dever et al. 1993), we were led to ask whether the coexpression of NF90 can exacerbate this effect. Figure 4A shows that, as expected, the yeast strain INVsc1 expressing HRI grew more slowly than a strain expressing NF90. Yeast coexpressing HRI and NF90 grew even more slowly than yeast expressing HRI alone (Fig. 4B). Thus, NF90 exacerbates the slow-growth phenotype caused by either PKR or HRI in yeast. Conceivably, NF90 might be a general activator of eIF2 kinases; arguing against this possibility, however, there was no sign that NF90 activates the endogenous GCN2 kinase present in the INVsc1 strain (data not shown).

NF90 ASSOCIATION WITH eIF2

The experiments shown in Figures 3 and 4 do not support the view that NF90 is a specific activator of PKR; rather, they raise the possibility that NF90's site of action lies downstream from eIF2 phosphorylation, possibly at the level of eIF2 dephosphorylation or mediated via an association with eIF2. Consistent with this idea, all three

subunits of eIF2 copurified in a complex containing NF90, NF45, and DNA-PK (Ting et al. 1998). We therefore determined whether NF90 associates with eIF2α using the "^{35}S-coIP" protocol described previously (Parker et al. 2001). Briefly, this coimmunoprecipitation protocol assesses the ability of a radiolabeled protein to interact with an unlabeled protein. Both proteins are synthesized in vitro, and complex formation is registered by coprecipitation of the labeled protein by antibody directed against the *unlabeled* protein.

The experiment shown in Figure 5A compared NF90's ability to interact with eIF2 (lanes 1–8) against its ability to bind its partner, NF45 (lanes 9–16). Antibody against NF45 immunoprecipitated NF45 (lane 9) and brought down NF90 when unlabeled NF45 was present in the mixture but not when NF45 was absent (compare lanes 11 and 10 in Fig. 5). A labeled control protein, luciferase, was not coimmunoprecipitated (lane 12). Reciprocally, NF90 antibody only immunoprecipitated labeled NF45 in the presence of unlabeled NF90 (lanes 14 and 15); again, this antibody did not immunoprecipitate luciferase (lane 16). These data provide evidence for a direct interaction between NF90 and NF45.

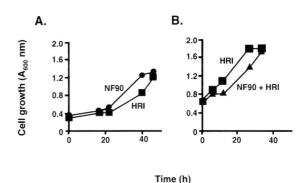

Figure 4. Effect of NF90 and HRI on yeast growth. (*A, B*) INVsc1 yeast strain expressed NF90 (*solid circles*), HRI (*solid squares*), or both NF90 and HRI (*solid triangles*). Growth in liquid SD medium was monitored by measuring A$_{600}$ nm. Points represent averages of four independent clones.

Similarly, antibody against NF90 immunoprecipitated eIF2α only when unlabeled NF90 was present in the mixture (compare lanes 2 and 3 in Fig. 5). Reciprocally, antibody directed against eIF2α brought down NF90 only when extract containing unlabeled eIF2α was mixed with NF90-containing extract (lanes 6 and 7). None of the antibodies immunoprecipitated luciferase even when it was mixed with other unlabeled proteins, and preimmune serum did not immunoprecipitate any of the labeled proteins (data not shown). The interactions between NF90 and eIF2α, like those between NF90 and NF45, were resistant to treatment with RNase A and to salt washes up to 800 mM KCl (data not shown), suggesting that these interactions are stable and RNA-independent.

For confirmation, we conducted pull-down experiments with GST fusion proteins. GST-NF90 brought down eIF2α as well as PKR (Fig. 5B, lanes 3 and 5), whereas GST alone did not bring down either of these labeled proteins (lanes 4 and 6). Similarly, NF90 and eIF2α were pulled down by a GST-PKR fusion but not by GST alone (lanes 10 and 12). We conclude that NF90, PKR, and eIF2α are able to interact, suggestive of functional interplay between them.

DOES NF45 INTERACT WITH PKR IN VITRO?

Since NF45 interacts with NF90 and NF90 interacts with PKR in vitro (Parker et al. 2001), we set out to determine whether NF45 could interact directly with PKR. Using a variety of GST fusion proteins, we conducted pull-down experiments with labeled NF45, PKR, or lu-

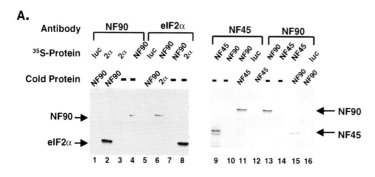

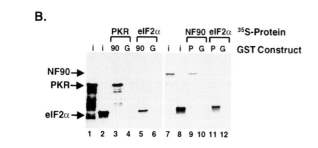

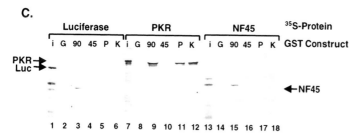

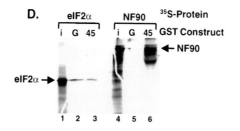

Figure 5. Interactions between NF90 and NF45 and with eIF2α and PKR. (*A*) Coimmunoprecipitation analysis with NF90. NF90, NF45, eIF2α (2α), and luciferase (luc) were synthesized in wheat-germ extracts either unlabeled (Cold Protein) or radiolabeled (^{35}S-Protein). Extracts containing radiolabeled protein, or mixtures of radiolabeled and unlabeled proteins, were subjected to immunoprecipitation with antibodies to NF90 (lanes *1–4*, *13–16*), eIF2α (lanes *5–8*), or NF45 (lanes *9–12*) as indicated. Immunoprecipitates were analyzed by SDS-polyacrylamide gel electrophoresis and autoradiography. Arrows mark positions of NF90, NF45, and eIF2α.

(*B*) GST pull-down analysis of NF90 interactions. Aliquots of protein extract containing 10 μg of GST (G), GST-NF90 (90), or GST-PKR (P) fusion proteins, as indicated, were combined with ^{35}S-labeled PKR (lanes 3 and 4), NF90 (lanes 9 and 10), or eIF2α (lanes 5, 6, 11, and 12) synthesized in wheat-germ extract. Proteins were precipitated using glutathione-Sepharose 4B beads, separated by SDS-polyacrylamide gel electrophoresis, and visualized by autoradiography. (Lanes *1*, *2*, *7*, and *8*) 20% of the input radiolabeled proteins (i) used in the binding assay.

(*C*) GST pull-down analysis of NF45 and PKR interactions. Aliquots of protein extracts containing 10 μg of GST (G), GST-NF90 (90), GST-NF45 (45), GST-PKR (P), or GST-PKR[K296R] (K) fusion proteins, as indicated, were combined with ^{35}S-labeled luciferase (lanes *2–6*), PKR (lanes *8–12*), or NF45 (lanes *14–18*) synthesized in wheat-germ extract, and precipitated, separated and visualized as above. (Lanes *1*, *7*, and *13* [i]) 10% of the input radiolabeled proteins used in the binding assay.

(*D*) GST pull-down analysis of interactions between NF45 and eIF2α. Protein extracts containing GST (G) or GST-NF45 (45) fusion protein were incubated with ^{35}S-labeled eIF2α (lanes *2* and *3*) or NF90 (lanes *5* and *6*). (Lanes *1* and *4* [i]) 10% of the input radiolabeled proteins used in the binding assay.

ciferase as a control (Fig. 5C). PKR formed heterodimers with GST-NF90 and homodimers with GST-PKR and GST-PKR[K296R] (lanes 9, 11, and 12), as expected, but no interaction was detected with GST-NF45 (lane 10). Similarly, NF45 did not interact detectably with GST-PKR or GST-PKR[K296R] (lanes 11 and 12), although it formed the expected heterodimers with GST-NF90 (lane 15). Parenthetically, NF45:GST-NF45 interactions were not detected, suggesting that NF45 does not form homodimers under these conditions (lane 16). We also failed to observe significant binding of eIF2α to the GST-NF45 fusion construct (Fig. 5D). These experiments gave no evidence of stable interactions between NF45 and PKR or eIF2α, implying that the NF45:PKR and NF45:eIF2 interactions observed in coimmunoprecipitation and copurification experiments from human cells and tissue (Ting et al. 1998; Parker et al. 2001) are likely bridged by their interactions with NF90 or other proteins.

NF45 INHIBITS PKR IN YEAST

Although we were unable to find evidence for a direct interaction between NF45 and PKR, the ability of NF45 to form a complex with NF90 in vivo (Corthésy and Kao 1994; Ting et al. 1998; T.W. Reichman et al., in prep.) and in vitro (Fig. 5A) prompted us to determine whether NF45 can modulate the phenotypic effects of NF90 in the yeast growth assay (see Fig. 1). First, we transformed the INVsc1 strain of yeast with plasmids expressing PKR or PKR[K296R], or an empty vector control. The resulting strains carried a low-copy plasmid that produced PKR under the control of the GAL promoter. These strains were then transformed with a plasmid expressing NF90 (or empty vector) also under the control of the GAL promoter (as in the experiments of Fig. 1). Finally, the yeast strains were transformed with a plasmid expressing NF45 fused to the GAL4 DNA-binding domain (or the GAL4 DNA-binding domain alone) under the control of the ADH promoter.

Under GAL-inducing conditions, PKR is overexpressed, resulting in the failure of the yeast to grow (Fig. 6A), presumably because the high level of eIF2 phosphorylation leads to the cessation of protein synthesis. The presence of the NF45 fusion protein restored growth to PKR-containing yeast strains, in the presence of both NF90 and its vector control. These data imply that NF45 acts as an inhibitor of PKR, independent of NF90. The NF45 fusion protein had no effect on the growth of yeast containing PKR[K296R] or the empty vector control.

We also examined the effect of NF45 on PKR using the yeast strain RY1-1. This strain's integrated copy of PKR under the control of the GAL promoter is induced to express high levels of PKR by growth on galactose-containing medium (in contrast to the low levels produced in Fig. 1D). This high level of PKR expression inhibits growth of yeast (Romano et al. 1995). We transformed the yeast strain RY1-1 with several different modulators of PKR, all under the control of the GAL promoter. These modulators include E3L, K3L, PACT, NF90, the carboxy-terminal half of NF90 (containing its dsRBMs), and NF45, which in this experiment was expressed as a His-

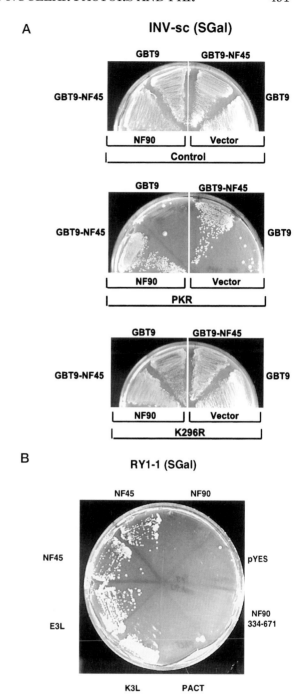

Figure 6. Effect of NF45 on growth of yeast. (*A*) *S. cerevisiae* strain INVSc1 expressing GBT9 or GBT9-NF45, Y-NF90, or pYES2 (Vector), and p415 (Control), pRS415-PKR, or pRS415-K296R, as indicated, was grown for 3–5 days on medium containing 10% galactose and 1% raffinose. (*B*) The RY1-1 gcn2Δ strain, carrying two copies of the human *PKR* integrated into the *LEU2* locus under the control of the galactose-inducible *GAL1-CYC1* promoter (Romano et al. 1995) and expressing pYES2, pY-NF90, pY-NF90 334-671, pY-NF45, pY-PACT, p2245 (expressing E3L), or pCL140 (expressing K3L), was grown as above.

tagged protein (not fused to the GAL DNA-binding domain). Like the strains expressing E3L and K3L, NF45 restored the growth of yeast expressing PKR when expressed under inducing conditions (Fig. 6B).

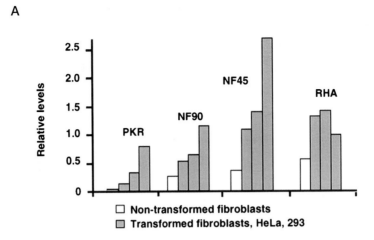

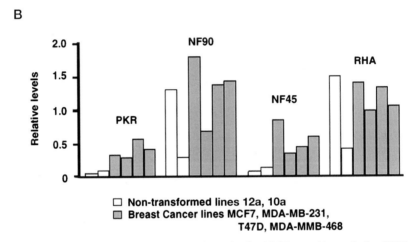

Figure 7. Levels of dsRNA-binding proteins in normal and transformed cells. (*A*) Western blot analysis of PKR, NF90, NF45, RHA, and actin was performed in nontransformed human embryo fibroblasts (*open bars*) and SV40-transformed human embryo fibroblasts, HeLa, and 293 cell lines (*shaded bars*). Western blots were quantified by densitometry, and protein levels were plotted relative to the actin control. (*B*) Western blot analysis, as above, of nontransformed breast cancer cell lines 12a and 10a (*open bars*) and transformed breast cancer cell lines MCF7, MDA-MB-231, T47D, and MDA-MMB-468 (*shaded bars*).

These data indicate that NF45 acts as an inhibitor of PKR in yeast. Western blot analysis confirmed that eIF2α phosphorylation levels are elevated in yeast expressing PKR alone and are reduced to control levels in yeast expressing either GBT9-NF45 or NF45 alone (data not shown). Since NF45 serves as a substrate for PKR (Parker et al. 2001), it is possible that it may reduce the phosphorylation of eIF2 by competition. Alternatively, NF45 may be enhancing the function of a yeast phosphatase such as GLC7 (Wek et al. 1992), which acts in opposition to the GCN2 kinase to affect eIF2 phosphorylation levels.

dsRNA-BINDING PROTEINS IN TRANSFORMED CELLS

The finding that both NF90 and NF45 modulate yeast growth via PKR (see Figs. 1 and 6), together with the earlier observations that PKR acts as a tumor suppressor, led us to compare the expression levels of these proteins in

normal and transformed cells. Using western blot analysis, we examined the relative levels of PKR, NF90, NF45, and RHA, another dsRBM-containing protein, normalizing each to the level of actin.

First, we compared a series of human cell lines including a matched pair of human fibroblasts (SV40-transformed SV.RNS/HF-1 and nontransformed HS74; Banga et al. 1997), together with HeLa and 293 cells (derived from a cervical carcinoma and by transformation of human embryo kidney cells with adenovirus-5, respectively). The levels of PKR were higher in all transformed cell lines as compared to the nontransformed cell lines (Fig. 7A). Similarly, NF90, NF45, and RHA levels were also elevated in the transformed cell lines as compared to the nontransformed fibroblasts. With the exception of RHA in 293 cells, the expression levels of the four proteins were roughly proportional to each other.

A similar pattern of expression was seen in a set of breast cancer cell lines that were compared to two different nontransformed mammary ductal epithelial cell lines

(Fig. 7B). With the exception of the normal cell line 12a, which displayed high levels of NF90 and RHA, the four proteins were all expressed at higher levels in the transformed cells than in the normal cells. Thus, the expression profile of NF45 more closely mirrored that of PKR than that of its binding partner NF90. The NF90 profile resembled that of RHA. It was surprising that NF45 and NF90 expression did not always correlate, since they copurify in a variety of systems (Corthésy and Kao 1994; Liao et al. 1998; Ting et al. 1998).

dsRNA-BINDING PROTEINS IN THE CELL CYCLE

These findings, together with the observation that PKR levels fluctuate during the cell cycle (Zamanian-Daryoush et al. 1999), prompted us to examine the relative levels of PKR, NF90, and RHA during the cell cycle. To this end, progression of HeLa cells through the cell cycle was blocked in S phase with hydroxyurea, then the cells were washed and released to continue cycling. Cells were harvested every 3 hours and processed for analysis of protein levels by immunoblotting (Fig. 8A) and for flow cytometric analysis to give the distribution of cell cycle stages (Fig. 8B). The results show that PKR exhibited a shallow

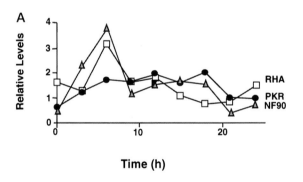

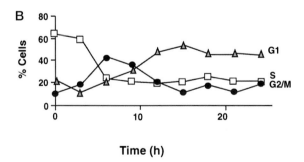

Figure 8. Levels of dsRNA-binding proteins through the cell cycle. HeLa cells were blocked in S-phase with hydroxyurea, washed, and released to progress through the cycle. Cell samples were harvested every 3 hr. (*A*) Cell extracts were separated by SDS-polyacrylamide gel electrophoresis, immunoblotted, and probed with anti-RHA, anti-NF90, anti-PKR, and anti-actin antibodies. Protein levels, in arbitrary units, are displayed relative to the levels of actin. (*B*) Cells were processed for single-color flow cytometric analysis of DNA content. Distribution of the cell cycle stages was calculated based on DNA histograms using Cell Fit software.

peak during G_1. The levels of NF90 and RHA varied in concert, with lowest expression during S-phase. These proteins both reached peak accumulation during G_2 and M phases, possibly coincident with the phosphorylation of NF90 during mitosis (Matsumoto-Taniura et al. 1996).

CONCLUSIONS AND SPECULATION

As shown here (Fig. 5) and elsewhere, NF90 and NF45 are predominantly nuclear proteins that physically interact with one another, both alone and in higher-order complexes (Corthésy and Kao 1994; Ting et al. 1998; Satoh et al. 1999; Parker et al. 2001; Reichman et al. 2002). NF90 also interacts directly with the predominantly cytoplasmic proteins PKR and eIF2 (Fig. 5), as well as with other proteins and dsRNA (Liao et al. 1998; Ting et al. 1998; Langland et al. 1999; Coolidge and Patton 2000; Parker et al. 2001). Strikingly, both NF90 and NF45 also interact functionally with PKR in yeast, but their actions are exerted in opposite directions. Whereas NF90 cooperates with PKR in slowing yeast growth, NF45 counters this action of PKR (see Figs. 1 and 6).

What do we know about the mechanisms of action of NF90 and NF45 in causing these phenotypes? In both cases, the effect correlates with changes in the phosphorylation status of eIF2α. Moreover, neither NF90 nor NF45 affects yeast growth in the absence of PKR activity, implying a link to PKR itself or to the elevated level of eIF2 phosphorylation that results from its expression. NF90 also inhibits translation in reticulocyte lysates (Fig. 2) and exerts a cooperative effect with HRI in yeast (Fig. 4), suggesting that it may be a vertebrate-specific modulator of this subset of kinases. Surprisingly, NF90 does not appear to affect PKR activity directly since it does not activate the kinase in an in vitro kinase assay (Fig. 3). In the case of NF45, direct physical interactions with PKR were not detected (Fig. 5), although the presence of higher-order complexes containing PKR, NF90, and NF45 is evidenced by their coimmunoprecipitation from cell lysates (Parker et al. 2001). These observations imply that additional factors—either RNAs or proteins—may be involved in the modulation of PKR activity by NF90 and NF45.

A core complex of NF90 and NF45 has been purified (Corthésy and Kao 1994), as well as a larger complex that also contains all three subunits of eIF2 and the catalytic subunit of DNA-PK (Ting et al. 1998). In addition, as documented in our laboratory and others, a series of other components interacts both with NF90 and with NF45 in vivo as well as in vitro. Although discrete complexes containing all of these components have not been characterized to date, we envision the existence of a large complex or set of complexes that integrate cues from diverse signal transduction pathways in controlling translation (Fig. 9).

Protein complexes composed of many subunits allow sensitivity and flexibility in the control of several critical enzymatic processes. It is by organizing and reorganizing multiple protein complexes that complexity is elaborated within relatively simple systems. This theme has been clearly established in transcription, where combinatorial arrays of transcriptional enhancers, coactivators, and enzymes that modify the basal transcription machinery allow the efficient control of transcriptional initiation and elongation of target genes. We speculate that analogous complexes may regulate PKR in response to diverse cellular stimuli.

PKR interacts with the ribosome as well as with components involved in a variety of signal transduction pathways including p53, IKK-β, and STAT1 (Wong et al. 1997; Cuddihy et al. 1999a,b; Zamanian-Daryoush et al. 2000). PKR also interacts with a variety of viral and cellular RNAs that regulate its function. NF90 and NF45 join PACT and P58 as members of a class of proteins that regulate the activity of PKR. These proteins are thought to be part of higher-order complexes that serve to integrate various stimuli and to orchestrate appropriate responses by regulating PKR. This enzyme is found in the nucleus (Jiménez-García et al. 1993; Jeffrey et al. 1995), where its function remains to be elucidated, as well as in the cytoplasm, where it is largely ribosome-associated and regulates translational initiation. NF90 and NF45, which are primarily nuclear proteins (Corthésy and Kao 1994; Reichman et al. 2002) but are also found in association with the ribosomes (Liao et al. 1998), may function to regulate PKR in both locations.

The variations observed in the abundancy ratio of NF90 to NF45 (Fig. 7) imply that they are not always coordinately regulated and that they are not invariably present in heterodimeric complexes (of fixed stoichiometry, at least). We demonstrate here (Fig. 8) that NF90 levels vary throughout the cell cycle, peaking at G_2/M. Thus, one means of regulation of PKR may be through modulation of the levels of NF90 and NF45 within a single large complex. Nonequivalent levels of the proteins associated with PKR may determine whether the complex is activated (by NF90) or inhibited (by NF45). A second possi-

ble means of regulating PKR is via the posttranslational modification of NF90 or NF45. Circumstantial support for this notion comes from the observation that MPP4, a close relative of NF90, is phosphorylated in mitosis (Matsumoto-Taniura et al. 1996). Whether NF45 is regulated during the cell cycle is not known, although, like NF90, it is a substrate for phosphorylation by PKR in vitro (Parker et al. 2001). Moreover, the regulation of PKR by PACT is thought to be modulated through posttranslational modification, as phosphorylation of PACT increases its affinity for PKR (Patel et al. 2000).

In summary, we propose that NF90 and NF45 play a part in the control of PKR in response to different cellular cues and in its integration with other cellular responses. On this view, PKR is coordinately activated and inhibited through a series of different complexes in order to allow a rapid cessation of protein synthesis during cell stress, such as that induced by DNA damage. It will be of interest to determine which cellular stimuli activate NF90 and NF45 to modulate PKR. The observations that NF90 and NF45 copurify with the catalytic subunit of DNA-PK, and serve as substrates for DNA-PK in vitro (Ting et al. 1998), suggest a way in which DNA damage could signal to PKR and hence to the protein synthesis and signal transduction pathways that it regulates.

ACKNOWLEDGMENTS

We thank John Hershey, Tom Dever, Alan Hinnebusch, Charles Thornton, Patrick Romano, Fatah Kashanshi, and Harvey Ozer for advice, reagents, and cell lines. We also thank members of the Mathews lab for helpful discussions. This work was supported by grant AI-34552 from the National Institutes of Health.

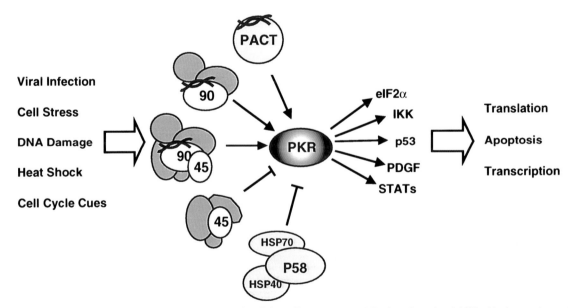

Figure 9. Model for regulation of PKR. In response to stimuli from diverse sources, PKR is activated or inhibited by interactions with a series of macromolecular complexes. These include complexes containing NF90 and/or NF45. Once activated, PKR can interact with and phosphorylate eIF2 and additional regulatory proteins, leading to effects on gene expression, cell growth, and apoptosis. Wavy lines symbolize duplexed or structured RNAs.

REFERENCES

Abraham N., Jaramillo M.L., Duncan P.I., Methot N., Icely P.L., Stojdl D.F., Barber G.N., and Bell J.C. 1998. The murine PKR tumor suppressor gene is rearranged in a lymphocytic leukemia. *Exp. Cell Res.* **244:** 394.

Abraham N., Stojdl D.F., Duncan P.I., Methot N., Ishii T., Dube M., Vanderhyden B.C., Atkins H.L., Gray D.A., McBurney M.W., Koromilas A.E., Brown E.G., Sonenberg N., and Bell J.C. 1999. Characterization of transgenic mice with targeted disruption of the catalytic domain of the double-stranded RNA-dependent protein kinase, PKR. *J. Biol. Chem.* **274:** 5953.

Aratani S., Fujii R., Oishi T., Fujita H., Amano T., Ohshima T., Hagiwara M., Fukamizu A., and Nakajima T. 2001. Dual roles of RNA helicase a in CREB-dependent transcription. *Mol. Cell. Biol.* **21:** 4460.

Balachandran S., Kim C.N., Yeh W.C., Mak T.W., Bhalla K., and Barber G.N. 1998. Activation of the dsRNA-dependent protein kinase, PKR, induces apoptosis through FADD-mediated death signaling. *EMBO J.* **17:** 6888.

Banga S.S., Kim S., Hubbard K., Dasgupta T., Jha K.K., Patsalis P., Hauptschein R., Gamberi B., Dalla-Favera R., Kraemer P., and Ozer H.L. 1997. SEN6, a locus for SV40-mediated immortalization of human cells, maps to 6q26-27. *Oncogene* **14:** 313.

Barber G.N., Wambach M., Thompson S., Jagus R., and Katze M.G. 1995. Mutants of the RNA-dependent protein kinase (PKR) lacking double-stranded RNA binding domain I can act as transdominant inhibitors and induce malignant transformation. *Mol. Cell. Biol.* **15:** 3138.

Bonnet M.C., Weil R., Dam E., Hovanessian A.G., and Meurs E.F. 2000. PKR stimulates NF-κB irrespective of its kinase function by interacting with the IκB kinase complex. *Mol. Cell. Biol.* **20:** 4532.

Buaas F.W., Lee K., Edelhoff S., Disteche C., and Braun R.E. 1999. Cloning and characterization of the mouse inter*leukin* enhancer binding factor 3 (Ilf3) homolog in a screen for RNA binding proteins. *Mamm. Genome* **10:** 451.

Chen J.-J. 2000. Heme-regulated eIF2α kinase. In *Translational control of gene expression* (ed. N. Sonenberg et al.), p. 529. Cold Spring Harbor Laboratory Press, Cold Spring Harbor, New York.

Chu W.M., Ballard R., Carpick B.W., Williams B.R., and Schmid C.W. 1998. Potential Alu function: Regulation of the activity of double-stranded RNA-activated kinase PKR. *Mol. Cell. Biol.* **18:** 58.

Coolidge C.J. and Patton J.G. 2000. A new double-stranded RNA-binding protein that interacts with PKR. *Nucleic Acids Res.* **28:** 1407.

Corthésy B. and Kao P.N. 1994. Purification by DNA affinity chromatography of two polypeptides that contact the NF-AT DNA binding site in the interleukin 2 promoter. *J. Biol. Chem.* **269:** 20682.

Cosentino G.P., Venkatesan S., Serluca F.C., Green S.R., Mathews M.B., and Sonenberg N. 1995. Double-stranded-RNA-dependent protein kinase and TAR RNA-binding protein form homo- and heterodimers in vivo. *Proc. Natl. Acad. Sci.* **92:** 9445.

Cuddihy A.R., Wong A.H., Tam N.W., Li S., and Koromilas A.E. 1999a. The double-stranded RNA activated protein kinase PKR physically associates with the tumor suppressor p53 protein and phosphorylates human p53 on serine 392 in vitro. *Oncogene* **18:** 2690.

Cuddihy A.R., Li S., Tam N.W., Wong A.H., Taya Y., Abraham N., Bell J.C., and Koromilas A.E. 1999b. Double-stranded-RNA-activated protein kinase PKR enhances transcriptional activation by tumor suppressor p53. *Mol. Cell. Biol.* **19:** 2475.

Davies M.V., Chang H.W., Jacobs B.L., and Kaufman R.J. 1993. The E3L and K3L vaccinia virus gene products stimulate translation through inhibition of the double-stranded RNA-dependent protein kinase by different mechanisms. *J. Virol.* **67:** 1688.

Deb A., Zamanian-Daryoush M., Xu Z., Kadereit S., and Williams B.R. 2001. Protein kinase PKR is required for platelet-derived growth factor signaling of c-fos gene expression via Erks and Stat3. *EMBO J.* **20:** 2487.

Dever T.E., Feng L., Wek R.C., Cigan A.M., Donahue T.F., and Hinnebusch A.G. 1992. Phosphorylation of initiation factor 2 alpha by protein kinase GCN2 mediates gene-specific translational control of GCN4 in yeast. *Cell* **68:** 585.

Dever T.E., Chen J.-J., Barber G.N., Cigan A.M., Feng L., Donahue T.F., London I.M., Katze M.G., and Hinnebusch A.G. 1993. Mammalian eukaryotic initiation factor 2 alpha kinases functionally substitute for GCN2 protein kinase in the GCN4 translational control mechanism of yeast. *Proc. Natl. Acad. Sci.* **90:** 4616.

Farrell P.J., Balkow K., Hunt T., Jackson R.J., and Trachsel H. 1977. Phosphorylation of initiation factor elF-2 and the control of reticulocyte protein synthesis. *Cell* **11:** 187.

Ferrandon D., Elphick L., Nüsslein-Volhard C., and St Johnston D. 1994. Staufen protein associates with the 3′UTR of bicoid mRNA to form particles that move in a microtubule-dependent manner. *Cell* **79:** 1221.

Fierro-Monti I. and Mathews M.B. 2000. Proteins binding to duplexed RNA: One motif, multiple functions. *Trends Biochem. Sci.* **25:** 241.

Finerty P.J., Jr. and Bass B.L. 1997. A *Xenopus* zinc finger protein that specifically binds dsRNA and RNA-DNA hybrids. *J. Mol. Biol.* **271:** 195.

———. 1999. Subsets of the zinc finger motifs in dsRBP-ZFa can bind double-stranded RNA. *Biochemistry* **38:** 4001.

Gerber A.P. and Keller W. 2001. RNA editing by base deamination: More enzymes, more targets, new mysteries. *Trends Biochem. Sci.* **26:** 376.

Gil J., Alcami J., and Esteban M. 1999. Induction of apoptosis by double-stranded-RNA-dependent protein kinase (PKR) involves the alpha subunit of eukaryotic translation initiation factor 2 and NF-κB. *Mol. Cell. Biol.* **19:** 4653.

———. 2000. Activation of NF-κB by the dsRNA-dependent protein kinase, PKR involves the I κB kinase complex. *Oncogene* **19:** 1369.

Harding H.P., Novoa I.I., Zhang Y., Zeng H., Wek R., Schapira M., and Ron D. 2000. Regulated translation initiation controls stress-induced gene expression in mammalian cells. *Mol. Cell* **6:** 1099.

Hinnebusch A.G. 2000. Mechanism and regulation of initiator methionyl-tRNA binding to ribosomes. In *Translational control of gene expression* (ed. N. Sonenberg et al.), p. 185. Cold Spring Harbor Laboratory Press, Cold Spring Harbor, New York.

Jeffrey I.W., Kadereit S., Meurs E.F., Metzger T., Bachmann M., Schwemmle M., Hovanessian A.G., and Clemens M.J. 1995. Nuclear localization of the interferon-inducible protein kinase PKR in human cells and transfected mouse cells. *Exp. Cell Res.* **218:** 17.

Jiménez-García L.F., Green S.R., Mathews M.B., and Spector D.L. 1993. Organization of the double-stranded RNA-activated protein kinase DAI and virus-associated VA RNA₁ in adenovirus-2 infected HeLa cells. *J. Cell Sci.* **106:** 11.

Kao P.N., Chen L., Brock G., Ng J., Kenny J., Smith A.J., and Corthésy B. 1994. Cloning and expression of cyclosporin A- and FK506-sensitive nuclear factor of activated T-cells: NF45 and NF90. *J. Biol. Chem.* **269:** 20691.

Kaufman R. 2000. The double-stranded RNA-activated protein kinase, PKR. In *Translational control of gene expression* (ed. N. Sonenberg et al.), p. 503. Cold Spring Harbor Laboratory Press, Cold Spring Harbor, New York.

Kawagishi-Kobayashi M., Silverman J.B., Ung T.L., and Dever T.E. 1997. Regulation of the protein kinase PKR by the vaccinia virus pseudosubstrate inhibitor K3L is dependent on residues conserved between the K3L protein and the PKR substrate eIF2α. *Mol. Cell. Biol.* **17:** 4146.

Kim S.H., Forman A.P., Mathews M.B., and Gunnery S. 2000. Human breast cancer cells contain elevated levels and activity of the protein kinase, PKR. *Oncogene* **19:** 3086.

Kostura M. and Mathews M.B. 1989. Purification and activation of the double-stranded RNA-dependent eIF-2 kinase DAI. *Mol. Cell. Biol.* **9:** 1576.

Langland J.O., Kao P.N., and Jacobs B.L. 1999. Nuclear factor-90 of activated T-cells: A double-stranded RNA-binding protein and substrate for the double-stranded RNA-dependent protein kinase, PKR. *Biochemistry* **38:** 6361.

Lee C.G. and Hurwitz J. 1993. Human RNA helicase A is homologous to the maleless protein of *Drosophila*. *J. Biol. Chem.* **268:** 16822.

Lee C.G., Chang K.A., Kuroda M.I., and Hurwitz J. 1997. The NTPase/helicase activities of *Drosophila maleless*, an essential factor in dosage compensation. *EMBO J.* **16:** 2671.

Lee C.G., da Costa Soares V., Newberger C., Manova K., Lacy E., and Hurwitz J. 1998. RNA helicase A is essential for normal gastrulation. *Proc. Natl. Acad. Sci.* **95:** 13709.

Liao H.J., Kobayashi R., and Mathews M.B. 1998. Activities of adenovirus virus-associated RNAs: Purification and characterization of RNA binding proteins. *Proc. Natl. Acad. Sci.* **95:** 8514.

Manche L., Green S.R., Schmedt C., and Mathews M.B. 1992. Interactions between double-stranded RNA regulators and the protein kinase DAI. *Mol. Cell. Biol.* **12:** 5238.

Matsumoto-Taniura N., Pirollet F., Monroe R., Gerace L., and Westendorf J.M. 1996. Identification of novel M phase phosphoproteins by expression cloning. *Mol. Biol. Cell* **7:** 1455.

Mellits K.H., Kostura M., and Mathews M.B. 1990. Interaction of adenovirus VA RNAI with the protein kinase DAI: Nonequivalence of binding and function. *Cell* **61:** 843.

Meurs E.F., Galabru J., Barber G.N., Katze M.G., and Hovanessian A.G. 1993. Tumor suppressor function of the interferon-induced double-stranded RNA-activated protein kinase. *Proc. Natl. Acad. Sci.* **90:** 232.

Nakajima T., Uchida C., Anderson S.F., Lee C.G., Hurwitz J., Parvin J.D., and Montminy M. 1997. RNA helicase A mediates association of CBP with RNA polymerase II. *Cell* **90:** 1107.

Nicholson A.W. 1996. Structure, reactivity, and biology of double-stranded RNA. *Prog. Nucleic Acid Res. Mol. Biol.* **52:** 1.

Osman F., Jarrous N., Ben-Asouli Y., and Kaempfer R. 1999. A *cis*-acting element in the 3'-untranslated region of human TNF-alpha mRNA renders splicing dependent on the activation of protein kinase PKR. *Genes Dev.* **13:** 3280.

Park H., Davies M.V., Langland J.O., Chang H.W., Nam Y.S., Tartaglia J., Paoletti E., Jacobs B.L., Kaufman R.J., and Venkatesan S. 1994. TAR RNA-binding protein is an inhibitor of the interferon-induced protein kinase PKR. *Proc. Natl. Acad. Sci.* **91:** 4713.

Parker L.M., Fierro-Monti I., and Mathews M.B. 2001. Nuclear factor 90 is a substrate and regulator of the eukaryotic initiation factor 2 kinase double-stranded RNA-activated protein kinase. *J. Biol. Chem.* **276:** 32522.

Patel C.V., Handy I., Goldsmith T., and Patel R.C. 2000. PACT, a stress-modulated cellular activator of interferon-induced double-stranded RNA-activated protein kinase, PKR. *J. Biol. Chem.* **275:** 37993.

Patel R.C. and Sen G.C. 1998. PACT, a protein activator of the interferon-induced protein kinase, PKR. *EMBO J.* **17:** 4379.

Patel R.C., Vestal D.J., Xu Z., Bandyopadhyay S., Guo W., Erme S.M., Williams B.R., and Sen G.C. 1999. DRBP76, a double-stranded RNA-binding nuclear protein, is phosphorylated by the interferon-induced protein kinase, PKR. *J. Biol. Chem.* **274:** 20432.

Pe'ery T. and Mathews M.B. 2000. Viral translation strategies and host defense mechanisms. In *Translational control of gene expression* (ed. N. Sonenberg et al.), p. 371. Cold Spring Harbor Laboratory Press, Cold Spring Harbor, New York.

Petryshyn R.A., Ferrenz A.G., and Li J. 1997. Characterization and mapping of the double-stranded regions involved in activation of PKR within a cellular RNA from 3T3-F442A cells. *Nucleic Acids Res.* **25:** 2672.

Reichman T.W., Muñiz L.C., and Mathews M.B. 2002. The RNA binding protein nuclear factor 90 functions as both a positive

and negative regulator of gene expression in mammalian cells. *Mol. Cell Biol.* **22:** 343.

Roegiers F. and Jan Y.N. 2000. Staufen: A common component of mRNA transport in oocytes and neurons? *Trends Cell Biol.* **10:** 220.

Romano P.R., Green S.R., Barber G.N., Mathews M.B., and Hinnebusch A.G. 1995. Structural requirements for double-stranded RNA binding, dimerization, and activation of the human eIF-2 alpha kinase DAI in *Saccharomyces cerevisiae*. *Mol. Cell. Biol.* **15:** 365.

Romano P.R., Zhang F., Tan S.L., Garcia-Barrio M.T., Katze M.G., Dever T.E., and Hinnebusch A.G. 1998. Inhibition of double-stranded RNA-dependent protein kinase PKR by vaccinia virus E3: Role of complex formation and the E3 N-terminal domain. *Mol. Cell. Biol.* **18:** 7304.

Ruvolo P.P., Gao F., Blalock W.L., Deng X., and May W.S. 2001. Ceramide regulates protein synthesis by a novel mechanism involving the cellular PKR activator RAX. *J. Biol. Chem.* **276:** 11754.

Satoh M., Shaheen V.M., Kao P.N., Okano T., Shaw M., Yoshida H., Richards H.B., and Reeves W.H. 1999. Autoantibodies define a family of proteins with conserved double-stranded RNA-binding domains as well as DNA binding activity. *J. Biol. Chem.* **274:** 34598.

Saunders L.R., Perkins D.J., Balachandran S., Michaels R., Ford R., Mayeda A., and Barber G.N. 2001. Characterization of two evolutionarily conserved, alternatively spliced nuclear phosphoproteins, NFAR-1 and 2, that function in mRNA processing and interact with the dsRNA-dependent protein kinase, PKR. *J. Biol. Chem.* **276:** 32300.

Schmedt C., Green S.R., Manche L., Taylor D.R., Ma Y., and Mathews M.B. 1995. Functional characterization of the RNA-binding domain and motif of the double-stranded RNA-dependent protein kinase DAI (PKR). *J. Mol. Biol.* **249:** 29.

Shamoo Y., Abdul-Manan N., and Williams K.R. 1995. Multiple RNA binding domains (RBDs) just don't add up. *Nucleic Acids Res.* **23:** 725.

Sharp T.V., Witzel J.E., and Jagus R. 1997. Homologous regions of the alpha subunit of eukaryotic translational initiation factor 2 (eIF2α) and the vaccinia virus K3L gene product interact with the same domain within the dsRNA-activated protein kinase (PKR). *Eur. J. Biochem.* **250:** 85.

Sharp T.V., Moonan F., Romashko A., Joshi B., Barber G.N., and Jagus R. 1998. The vaccinia virus E3L gene product interacts with both the regulatory and the substrate binding regions of PKR: Implications for PKR autoregulation. *Virology* **250:** 302.

St Johnston D., Beuchle D., and Nüsslein-Volhard C. 1991. Staufen, a gene required to localize maternal RNAs in the *Drosophila* egg. *Cell* **66:** 51.

Tan S.-L., Gale M., Jr., and Katze M.G. 2000. Translational reprogramming during influenza virus infection. In *Translational control of gene expression* (ed. N. Sonenberg et al.), p. 933. Cold Spring Harbor Laboratory Press, Cold Spring Harbor, New York.

Ting N.S., Kao P.N., Chan D.W., Lintott L.G., and Lees-Miller S.P. 1998. DNA-dependent protein kinase interacts with antigen receptor response element binding proteins NF90 and NF45. *J. Biol. Chem.* **273:** 2136.

Wek R.C., Cannon J.F., Dever T.E., and Hinnebusch A.G. 1992. Truncated protein phosphatase GLC7 restores translational activation of GCN4 expression in yeast mutants defective for the eIF-2 alpha kinase GCN2. *Mol. Cell. Biol.* **12:** 5700.

Wong A.H., Tam N.W., Yang Y.L., Cuddihy A.R., Li S., Kirchhoff S., Hauser H., Decker T., and Koromilas A.E. 1997. Physical association between STAT1 and the interferon-inducible protein kinase PKR and implications for interferon and double-stranded RNA signaling pathways. *EMBO J.* **16:** 1291.

Xu Y.H., Busald C., and Grabowski G.A. 2000. Reconstitution of TCP80/NF90 translation inhibition activity in insect cells. *Mol. Genet. Metab.* **70:** 106.

Xu Z. and Williams B.R. 2000. The B56alpha regulatory subunit of protein phosphatase 2A is a target for regulation by double-stranded RNA-dependent protein kinase PKR. *Mol. Cell. Biol.* **20:** 5285.

Yang M., May W.S., and Ito T. 1999. JAZ requires the double-stranded RNA-binding zinc finger motifs for nuclear localization. *J. Biol. Chem.* **274:** 27399.

Yang Y.L., Reis L.F., Pavlovic J., Aguzzi A., Schafer R., Kumar A., Williams B.R., Aguet M., and Weissmann C. 1995. Deficient signaling in mice devoid of double-stranded RNA-de-

pendent protein kinase. *EMBO J.* **14:** 6095.

Zamanian-Daryoush M., Der S.D., and Williams B.R. 1999. Cell cycle regulation of the double stranded RNA activated protein kinase, PKR. *Oncogene* **18:** 315.

Zamanian-Daryoush M., Mogensen T.H., DiDonato J.A., and Williams B.R. 2000. NF-κB activation by double-stranded-RNA-activated protein kinase (PKR) is mediated through NF-κB-inducing kinase and IκB kinase. *Mol. Cell. Biol.* **20:** 1278.

Translational Regulation in the Cellular Response to Biosynthetic Load on the Endoplasmic Reticulum

H.P. HARDING, I. NOVOA, A. BERTOLOTTI, H. ZENG, Y. ZHANG, F. URANO, C. JOUSSE, AND D. RON

Skirball Institute of Biomolecular Medicine, Departments of Medicine, Cell Biology, and the Kaplan Cancer Center, New York University School of Medicine, New York, New York 10016

Proteins destined for secretion, membrane insertion, and retention within the lumen of the exocytic compartment are synthesized on ribosomes bound to the endoplasmic reticulum (ER). The nascent peptide is translocated across the ER membrane and encounters the luminal environment, where it undergoes specific post-translational modifications and folding reactions. The flux of such client proteins through the ER varies considerably between different cell types and is influenced substantially by physiological conditions. For example, the load on the ER of a pancreatic acinar cell that secretes large quantities of digestive enzymes will be much greater than that placed on a maturing erythroblast whose ribosomes are active mainly in the synthesis of cytoplasmic proteins such as globins.

Cells have evolved specific signaling pathways to ensure a match between the load on their ER and the capacity of the organelle to meet that load. A mismatch between demand and supply is defined as ER stress, and the signaling pathways triggered have been referred to as constituting an unfolded protein response (or UPR). The latter term originated because of the methods used to elicit ER stress experimentally. These consist of the application of toxins that impair the folding of client proteins in the ER and thus presumably lead to the accumulation of unfolded/malfolded proteins in the lumen of the organelle. It has not been determined yet whether exposure to these toxins truly mimics the state of the ER under physiological conditions of mismatch between demand and supply. However, the downstream responses elicited by the pharmacological interventions and physiological states of ER stress are very similar.

The cell's response to ER stress is tripartite and consists of the up-regulation of genes involved in the biosynthetic functions of the ER (translocation, modification, folding, and secretion of client proteins), up-regulation of genes involved in the degradation of malfolded/unfolded client proteins (so-called ER-associated degradation or ERAD), and translational regulation of client protein biosynthesis. The first two components of the UPR reduce ER stress by increasing the capacity of the organelle to fold and degrade client proteins, whereas the third component is adaptive because it reduces the load on the organelle by decreasing the production of client proteins (Kaufman 1999; Mori 2000). This paper focuses on the molecular mechanisms involved in this third translational arm of the UPR.

ER STRESS PROMOTES eIF2α PHOSPHORYLATION AND INHIBITS PROTEIN BIOSYNTHESIS

The first experimental clues that perturbation of ER function leads to reduced polypeptide biosynthesis came from the Brostrom lab. These authors found that adding calcium back to the media of calcium-depleted cells stimulated protein synthesis. Initially, this finding was thought to indicate a requirement for cytosolic calcium in translation (Brostrom et al. 1983). However, subsequent studies from that lab and from Randy Kaufman's group focused attention of the role of the ER in mediating this phenomenon. This hypothesis was further supported by the observation that toxins such as tunicamycin and reducing agents such as dithiothreitol that perturb protein folding in the ER without producing significant changes in calcium dynamics also reduced polypeptide biosynthesis (Prostko et al. 1992; Wong et al. 1993).

These studies also provided the first molecular clues as to how translation might be repressed by perturbed ER function. Reduced polypeptide biosynthesis in thapsigargin-treated cells correlated with dissociation of polyribosomes, suggesting that the initiation step of mRNA translation was affected. This correlated with diminished activity of eIF2B and phosphorylation of its substrate eIF2 on the α subunit (Prostko et al. 1992, 1993). Phosphorylation of eIF2α is a major step in regulating protein biosynthesis in eukaryotes. The trimeric eIF2 complex is required for recruitment of the charged initiator methionyl tRNA (Met-tRNA$_i^{Met}$) to the small ribosomal subunit. However, eIF2 is active in this recruitment step only when bound to GTP forming a ternary complex of Met-tRNA$_i^{Met}$ + eIF2 + GTP. This ternary complex is incorporated into the small ribosomal subunit and disassembled upon recognition of an AUG start codon. Hydrolysis of the γ-phosphate on GTP is an integral part of this disassembly process, liberating eIF2 in a GDP-bound form. To participate in another round of translation initiation, the GDP on eIF2 must be exchanged for GTP, and this is effected by the exchange factor eIF2B. eIF2 trimers phosphorylated on Ser-51 of their α subunit bind to and inhibit eIF2B, blocking this essential exchange reaction. As a result, formation of ternary complexes is inhibited and translation initiation is impeded (Hinnebusch 2000).

This powerful means of regulating translation initiation is used by cells to reduce polypeptide biosynthesis in response to specific signals. Several eIF2α kinases have evolved to couple specific stress signals to phosphorylation of eIF2α on Ser-51. For example, the interferon-induced PKR is activated by double-stranded RNA in virally infected cells and participates in a host adaptation that deprives the invading virus of use of the cell's protein synthesizing machinery (Kaufman 2000), and HRI, a heme-repressed eIF2α kinase, ensures the matching of globin biosynthesis in erythroid precursors to the production of the heme prosthetic group (Chen 2000).

PERK IS AN eIF2α KINASE THAT RESPONDS SPECIFICALLY TO ER STRESS

The well-established role of eIF2α kinases in regulating translation suggested the existence of a similar kinase that may regulate translation in response to ER stress. Clues to its identification were provided by the analysis of the gene expression arm of UPR. In yeast, activation of gene expression by the UPR is mediated by Ire1p, a type I transmembrane ER resident protein whose luminal domain responds to the primary stress signal(s) and whose carboxy-terminal cytoplasmic protein kinase domain is an effector of downstream signaling (Cox et al. 1993; Mori et al. 1993). Homologs of Ire1p have been identified in other species, and alignment of their stress-sensing luminal domains led to the identification of conserved features in the primary amino acid sequence (Sidrauski and Walter 1997; Tirasophon et al. 1998; Wang et al. 1998). This enabled our lab to conduct a search for other proteins that might use such a domain to sense luminal ER stress signals. We identified a type I transmembrane ER resident protein with a luminal domain resembling that of IRE1 and a cytoplasmic domain that most closely resembled other known eIF2α kinases. Homologs of that protein, which we named *PKR-like ER kinase* (or PERK), were found in different vertebrate and invertebrate species but not in yeast.

PERK (also known as PEK) is highly active as an eIF2α kinase, phosphorylating its substrate on Ser-51 and effecting strong inhibition of protein biosynthesis (Shi et al. 1998; Harding et al. 1999). PERK is activated specifically by ER luminal signal(s) and does not respond to stress signals generated in the cytoplasm (Harding et al. 1999; Bertolotti et al. 2000). Gene targeting experiments in mice indicate that PERK is essential to the phosphorylation of eIF2α and the inhibition of protein biosynthesis observed in cells experiencing ER stress. Exposure of *PERK–/–* cells to agents that perturb protein folding in the ER leads to no measurable decrease in protein synthesis rates and not to any increase in eIF2α phosphorylation. Under the same conditions, wild-type cells experience profound decrease in total protein synthesis and significant elevation of eIF2α(P) (Harding et al. 2000a). Although it is possible that in other cell types or under specific physiological conditions (such as viral infection), other eIF2α kinases (such as PKR) may also contribute to translational control by ER stress (Prostko et al. 1995; Srivastava et al. 1995), our results indicate that PERK is the major regulator.

MECHANISMS OF PERK ACTIVATION

PERK and IRE1 are both long-lived type I ER membrane proteins. Their activation by ER stress is, for the most part, posttranslational, although PERK mRNA levels increase slightly (H. Harding, unpubl.). Yeast Ire1p oligomerizes when overexpressed, and oligomerization, which is mediated by the stress-sensing luminal domain, correlates with activation of the kinase resident in the cytoplasmic effector domain of the protein. These findings suggested a model whereby IRE1 is activated by oligomerization and trans-autophosphorylation, like classic cell surface receptors. However, oligomerization of the overexpressed yeast Ire1p occurs constitutively and is not subject to regulation by ER stress (Shamu and Walter 1996). In mammalian cells, too, overexpressed IRE1 and PERK are constitutively active, suggesting that overexpression can override control mechanisms that normally repress these kinases (Tirasophon et al. 1998; Wang et al. 1998; Harding et al. 1999).

Oligomerization in response to ligand binding is believed to have a role in activation of the related eIF2α kinase PKR (Kostura and Mathews 1989; Langland and Jacobs 1992; Wu and Kaufman 1997). This is believed to trigger a trans-autophosphorylation event that enhances kinase activity (Thomis and Samuel 1993). PERK can certainly be activated by forced oligomerization. This has been demonstrated experimentally by expressing the kinase domain of PERK ectopically on the plasma membrane as a fusion protein with the extracellular and transmembrane domains of human CD4 (Fig. 1) (Bertolotti et al. 2000). Similar results were obtained in the Kaufman lab by replacing the luminal domains of IRE1 or PERK with a leucine zipper dimerization domain (Liu et al. 2000).

In unstressed cells, PERK exists as a complex with the major HSP70-like ER chaperone, BiP (Bertolotti et al. 2000). The size of this complex, estimated by glycerol gradient sedimentation, is consistent with the estimated size of a heterodimer composed of one PERK and one BiP molecule. Topological considerations and the finding that IRE1 also forms a similar complex indicate that binding of BiP takes place at the luminal domain. Within minutes after treating cells with agents that perturb protein folding in the ER, BiP dissociates from PERK, PERK is incorporated in a high-molecular-weight complex, consistent with a PERK oligomer, and becomes phosphorylated; its kinase activity toward eIF2α increases significantly. BiP binding is rapidly restored upon resolution of ER stress. Activation of IRE1 is similarly correlated with loss of BiP binding.

These observations suggest a model in which PERK (and IRE1) are held in an inactive state by BiP. Dissociation of BiPs from their luminal domains serves as a trigger for oligomerization (Fig. 2) (Bertolotti et al. 2000). This model is consistent with the tendency of PERK and IRE1 to be activated when overexpressed and with the observation that BiP overexpression represses signaling in the yeast (Kohno et al. 1993) and mammalian UPR (Dorner et al. 1992) and represses PERK and IRE1 activation (Bertolotti et al. 2000). If BiP binding to PERK and IRE1 measures the surplus capacity of the ER to fold

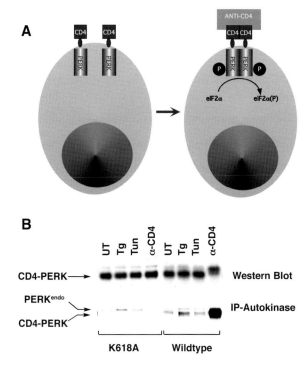

Figure 1. PERK activation by dimerization. (*A*) Cartoon depicting the extracellular and transmembrane domains of the lymphocyte coreceptor CD4 fused to the kinase domain of PERK. The chimera is expressed on the plasma membrane of transfected cells. Ligation with an antibody to CD4 forces dimerization of the kinase and leads to trans-autophosphorylation and activation. (*B*) CD4-PERK chimeric protein detected by immunoblot from cells exposed to thapsigargin (Tg) and tunicamycin (Tun) or ligated with the anti-CD4 antibody (α-CD4). The shift to lower mobility in the α-CD4-treated cells (last lane) reflects trans-autophosphorylation and activation of CD4-PERK, which has been uncoupled from the normal activation by ER stress (*upper panel*). The increase in autokinase activity of CD4-PERK is revealed by an immunoprecipitation in vitro kinase assay-incorporated labeled ^{32}P (*lower panel*). K618A refers to a chimera between CD4 and mutant *PERK* that lacks kinase activity. (Adapted, with permission, from Bertolotti et al. 2000 [copyright Macmillan].)

client proteins, then the UPR is triggered not by the accumulation of malfolded proteins, but by a decline in the functional reserve of the organelle. In this case, the UPR is not an unfolded protein response, but rather a response to the threat of unfolded proteins. The mechanism by which ER stress leads to BiP dissociation from PERK is not known, nor is it understood how BiP dissociation might allow PERK and IRE1 to oligomerize.

Regardless of how it is effected, oligomerization leads to PERK phosphorylation and an increase in kinase activity toward its substrate, eIF2α. The basis for this activation by trans-autophosphorylation is not fully understood, but it is a feature shared by many other protein kinases, including other eIF2α kinases (see above). In the context of its activation, PERK is autophosphorylated on many serine and threonine residues (Ma et al. 2001). Some, like Thr-980 in the predicted activation loop of PERK's kinase domain, are conserved in other kinases and have an important role in PERK's kinase activity. But

the significance of PERK phosphorylation on other residues is not known. One consequence of PERK phosphorylation may be to increase its affinity toward its substrate. This is suggested by in vitro pull-down experiments in which the amount of eIF2α that associates with wild-type phosphorylated PERK is much greater than that associated with a mutant protein that fails to autophosphorylate because it lacks kinase activity (Fig. 3).

THE SIGNIFICANCE OF TRANSLATIONAL CONTROL BY ER FUNCTIONAL RESERVE

The essential role of PERK in repressing translation in ER-stressed cells provides an opportunity to examine the functional importance of this regulatory pathway. Wild-type cells are surprisingly resistant to the effects of agents that perturb protein folding in the ER. *PERK–/–* cells, on the other hand, exhibit marked sensitivity to toxins that perturb ER function (Harding et al. 2000a). The magnitude of the hypersensitivity of *PERK–/–* cells to one such toxin, tunicamycin, was such that it could be used in a selection scheme for *PERK–/–* ES cells that had been rescued with a *PERK* transgene (Harding et al. 2000a).

It seems likely that at least part of the protective role of PERK against ER stress is played out at the level of repression of protein synthesis. This repression would serve to reduce the amount of client proteins translocated into the ER, relieving some of the load placed on the organelle and increasing its functional reserve. This notion regarding PERK's role in adaptation to ER stress is consistent with the observation that *PERK–/–* cells have more activated IRE1 than wild-type cells when exposed to agents that cause ER stress. IRE1 activation reports on the level of ER stress. This enhancement is most readily observed experimentally as sustained activation of IRE1 in *PERK–/–* cells after removal of the toxin causing ER stress. Presumably, translational repression in wild-type cells limits the amount of malfolded proteins that accumulate in their ER and accounts for the more rapid return of wild-type cells to basal condition (Fig. 4) (Harding et al. 2000a).

In several tissues of *PERK–/–* mice, we found lower levels of phosphorylated eIF2α and higher levels of activated IRE1 than in wild-type littermates. This finding suggests that the proposed role for PERK in moderating load on the ER is played out in the cells of intact mammals whose ER is subject to physiological demands to process client proteins (Fig. 4) (Harding et al. 2001). These differences in the level of phosphorylated eIF2α and activated IRE1 between wild-type and mutant tissues were most conspicuous in the pancreas, an organ whose cells are actively engaged in protein secretion.

The *PERK–/–* animals are indistinguishable at birth from their wild-type littermates, but they fail to thrive postnatally and develop both endocrine and exocrine pancreatic insufficiency. A similar phenotype is exhibited by humans homozygous for loss-of-function mutations in the *EIF2AK3* gene encoding PERK, a cause of the rare Wolcott-Rallison syndrome of infantile diabetes mellitus (Wolcott and Rallison 1972; Delepine et al. 2000). In mice, the onset of diabetes mellitus and malabsorption

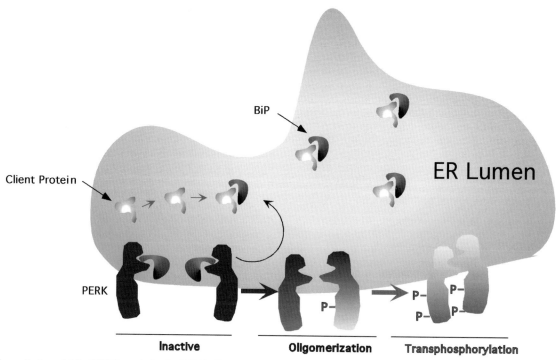

Figure 2. A model for PERK regulation by BiP binding. In unstressed cells (*left*), BiP binding to the PERK luminal domain maintains PERK in a monomeric and inactive form. Increased demands by client proteins lead to dissociation of BiP (which engages the client proteins) and allow PERK to oligomerize and trans-autophosphorylate. Trans-autophosphorylation increases PERK's activity and initiates downstream signaling.

correlate with death of the insulin-secreting β cells of the islets of Langerhans and the digestive enzyme-secreting cells of the pancreatic acinus. Interestingly, remaining viable β cells in islets of *PERK–/–* mice are not measurably impaired in insulin biosynthesis or processing. Instead, these cells were found to be impaired in their ability to moderate insulin biosynthesis. Insulin biosynthesis is

controlled by glucose levels. A switch to hyperglycemic media increases translation of the insulin mRNA (Howell and Taylor 1966; Itoh and Okamoto 1980). This increase was more pronounced in *PERK–/–* than in wild-type islets, suggesting that under physiological conditions, the burst of insulin biosynthesis that accompanies a rise in plasma glucose levels leads to some ER stress. PERK ac-

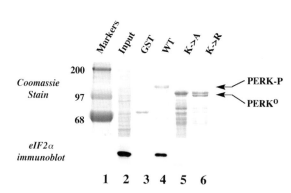

Figure 3. Phosphorylated PERK has a greater affinity for its substrate. Cell lysates were incubated with immobilized glutathione *S*-transferase fusion proteins of the wild-type PERK cytoplasmic domain or the indicated point mutants that impair kinase activity. The bound eIF2α that remained associated with GST-PERK after extensive washing was revealed by immunoblot (*lower panel*). The bacterially expressed wild-type GST-PERK is an active kinase, whereas the mutants lack kinase activity (Harding et al. 1999). These features are reflected in the lower mobility of the wild-type GST-PERK in the Coomassie-stained gel of the bait proteins used in this pull-down experiment (*upper panel*).

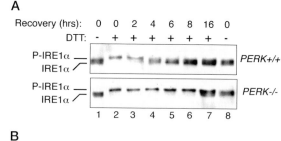

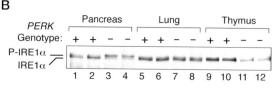

Figure 4. Increased IRE1α activation in the *PERK–/–* cells. (*A*) IRE1α activation detected by immunoblot of immunoprecipitated protein from wild-type (+/+) and *PERK–/–* cells after treatment with an agent that causes ER stress (DTT) and wash-out of the agent. Note the delay in return of IRE1α to the inactive higher mobility state in the mutant cells. (*A*, Reprinted, with permission, from Harding et al. 2000a, [copyright Elsevier Science].) (*B*) Increased IRE1α activation in the pancreas of *PERK–/–* mice, revealed by the same assay (*B*, Reprinted, with permission, from Harding et al. 2001 [copyright Elsevier Science].)

tivation, under these circumstances, likely moderates the increase in insulin biosynthesis and protects the β cell. Loss of this protective function leads to the destruction of the β cells (Harding et al. 2001).

Most insulin-producing cells remaining at any one time in the islets of mutant animals appear normal ultrastructurally and functionally. Rarely, however, we encountered an insulin-producing cell whose ER was grossly distended with electron-dense material. It is tempting to speculate that this material reflects client proteins that were delivered into the lumen of the ER in excess of the organelle's processing capacity. We do not observe transitional stages between the majority of cells with normal-appearing ER and the rare, grossly abnormal cells, and we do not understand why this accumulation would be catastrophic and not gradual (Harding et al. 2001). It is therefore possible that the morphological abnormalities of the ER are a late and indirect consequence of the defect in PERK function. Regardless of whether the distended ER reflects an early or late event in the process that kills the PERK–/– cells, the normal appearance and function of the majority of β cells present at any one time in the mutant endocrine pancreas suggest that death of individual cells is by a stochastic catastrophic process and not by gradual malfunction. The nature of that process remains unknown.

The fate of the PERK–/– pancreas bears an interesting relationship to that of erythropoietic cells in HRI–/– mice. If the supply of heme in erythropoietic cells is limited, the eIF2α kinase HRI is activated and translation is repressed. This regulatory process matches the synthesis of globin chains to the availability of their heme prosthetic group. In wild-type mice, iron deficiency, which limits heme availability, leads to a mild and well-tolerated anemia. Iron deficiency in HRI–/– mice causes a profound hemolytic anemia, with destruction of erythropoietic precursors and mature red blood cells. The cause of cellular destruction is likely to be toxicity from continuously synthesized globin chains that cannot be complexed with heme (Han et al. 2001). Both eIF2α kinases thus appear to protect cells against toxicity caused by unrestrained synthesis of their own proteins; the toxins are the proteins themselves.

SPECULATION OF THE POSSIBILITY OF SELECTIVE CONTROL OF TRANSLATION OF SPECIFIC mRNAs BY PERK

The experiments described above emphasize the global nature of translational repression by eIF2α kinases. This feature is easy to accommodate in the case of PKR, which operates in the context of virally infected cells. It is more difficult to rationalize in the case of kinases such as PERK, which responds to a compartment-specific stress signal or HRI, which responds to the availability of a building block specific to one biosynthetic pathway (hemoglobin). The apparent lack of selectivity in control of mRNA translation by these latter two kinases is surprising. It is possible that lack of selectivity in the attenuation of translation revealed in our experiments may be due to the use of agents that promote levels of PERK activation that are never attained physiologically. Perhaps

in the physiological context, eIF2α phosphorylation by PERK exerts some selectivity toward inhibiting translation of membrane-bound ribosomes.

A hint that this might be the case has been provided by the PERK–/– islets of Langerhans, where we find that the modulatory effect of PERK on glucose-stimulated protein biosynthesis appears to have some selectivity toward insulin. In β cells, glucose stimulates general protein biosynthesis as well as insulin biosynthesis. We find that in glucose-stimulated PERK–/– islets, insulin biosynthesis is proportionately increased over total protein biosynthesis (H. Harding, unpubl.), suggesting that PERK may selectively affect translation on membrane-bound ribosomes that translate insulin.

A prerequisite for any local control of translation is that exchange of GTP on eIF2 should also be regulated locally. In other words, the "signal" generated by phosphorylated eIF2α (i.e., the inhibition of eIF2B) should not diffuse throughout the cell. This is not an unreasonable idea. Recent findings of a physical link between the 3′ and 5′ ends of mRNAs imply that reinitiation of translation can take place locally (Sachs 2000). Guanine nucleotide exchange on eIF2 and ternary complex re-formation may thus also take place locally. The translating ribosomes engaged in biosynthesis of ER client proteins are physically associated with the ER membrane through the translocation complex, and PERK itself is associated with the same membrane. It is therefore easy to imagine that PERK might control the phosphorylation levels of a pool of eIF2α localized to the ER membrane and influence a likewise localized pool of eIF2B.

We have uncovered circumstantial evidence that the eIF2α(P) signal does not equilibrate freely in the cell. When activated at the plasma membrane, the CD4-PERK chimera (see Fig. 1) phosphorylates substantial amounts of eIF2α, yet, the effect on protein biosynthesis is considerably less than that obtained with similar levels of eIF2α phosphorylation effected by activating endogenous PERK through the induction of ER stress (H. Harding, unpubl.). It is possible that these differences between endogenous and ectopic activation of PERK reflect the activities of other signals generated by ER stress, signals that synergize with eIF2α(P) to control translation. However, given that PERK is both necessary and sufficient to control translation when activated at the ER membrane (Harding et al. 1999, 2000a), we regard it as more likely that the eIF2α(P) generated at the plasma membrane is simply less effective at inhibiting the relevant pool of eIF2B than similar quantities of eIF2α(P) generated at the ER membrane.

Taken one step further, these speculations on specificity of translational control by PERK might lead us to imagine a scenario in which translation initiation/reinitiation was controlled regionally on the ER membrane. This model of the ER would envision it as being compartmentalized into functional domains in which translation rates on the cytoplasmic side were responsive to changes in the local folding environment on the luminal side. Such an arrangement would ensure that the functional reserves of the ER were maintained locally. This static view of the translational component of the ER receives some support from recent studies that measured a very low diffusion rate for the oligosaccharide transferase complex (OST) by fluorescent recovery after photo-bleaching (G. Kreibich, pers. comm.). The OST complex is stably associated with the ribosome and translocation pore, and thus, its rate of diffusion in the plane of the ER membrane likely reports on the diffusion rate of the translation and translocation unit. There is, of course, no need to evoke such a model of local control if the client proteins of the ER were allowed to diffuse rapidly and equilibrate throughout the compartment. There is evidence that for some client proteins and under some conditions, this is indeed the case (Nehls et al. 2000). However, when folding is perturbed by tunicamycin, mobility of a client protein such as VSV-G is severely curtailed (Nehls et al. 2000), indicating that under other circumstances, the folding environment in the ER may be functionally segmented.

PERK ACTIVATES AN eIF2α(P)-DEPENDENT GENE EXPRESSION PROGRAM: THE INTEGRATED STRESS RESPONSE

So far, we have discussed PERK's role in decreasing the amount of polypeptides synthesized in cells experiencing ER stress. However, eIF2α phosphorylation in mammalian cells has an additional consequence, the activation of a gene expression pathway that has features in common with the yeast general control response. In yeast, amino acid deprivation activates the eIF2α kinase Gcn2p; this has a relatively modest inhibitory effect on global protein synthesis but markedly enhances the translation of the GCN4 mRNA. Gcn4p, the encoded protein, is a transcription factor that activates genes involved in amino acid biosynthesis. Regulation of GCN4 mRNA translation by eIF2α phosphorylation requires several short upstream open reading frames (uORFs) in the 5′ region of the mRNA. Positioned between the cap and the Gcn4p initiating AUG, these short uORFs are translated by scanning ribosomes. In the repressed state, when ternary complexes are abundant, scanning ribosomes reinitiate at the downstream uORFs and fail to initiate at the Gcn4p AUG. When ternary complexes are limiting, because of eIF2α phosphorylation (or other causes such as mutations that reduce eIF2B function), the scanning ribosomes bypass the downstream uORF and initiate instead at the Gcn4p AUG, increasing its translation (Hinnebusch 1997).

We noted that the CHOP gene, which is positively regulated in the UPR, is not induced in PERK–/– cells. It was also known that CHOP is transcriptionally activated by amino acid deprivation of mammalian cells (Bruhat et al. 1997; Jousse et al. 1999), conditions under which the mammalian homolog of GCN2 is active. Furthermore, we found that CHOP induction by amino acid limitations was impaired in mammalian cells lacking GCN2. These findings suggested the existence of a signaling pathway responsive to eIF2α(P) levels in mammalian cells. If this pathway were organized along the lines of the yeast general control response, it might imply the existence of a transcription factor(s) that would be regulated translationally and would operate at the level of Gcn4p.

Several lines of evidence suggested that the bZIP transcription factor ATF4 might be a candidate for this role: Niki Holbrook and colleagues found that overexpressed ATF4 could activate the CHOP promoter (Fawcett et al. 1999). Years ago, we had noted that although ATF4 mRNA was abundantly expressed in many cell types, the protein was very difficult to detect (Vallejo et al. 1993). ATF4 protein accumulates in stressed hypoxic cells, and this appears to occur without a major change in mRNA abundance (Estes et al. 1995). Finally, we noted the presence of several uORFs in the ATF4 mRNA that are conserved from aplysia to mammals.

In unstressed cells, ATF4 mRNA was found in a relatively light polysomal fraction, consisting with poor translatability of the mRNA. In response to agents that cause ER stress and promote eIF2α phosphorylation, the ATF4 mRNA shifted to heavier polysomal fractions, and its rate of translation, measured by pulse labeling, increased markedly. These paradoxical stress-induced changes in ATF4 mRNA association with ribosomes and translation rates were clearly dependent on the phosphorylation of eIF2α, as they were not observed in PERK–/– cells. A parallel dependence of ATF4 expression on GCN2 was observed in amino-acid-starved cells, further supporting the essential role of eIF2α(P) in up-regulating ATF4 translation (Harding et al. 2000b).

The uORFs of the ATF4 mRNA have an important role in its translation. They repress basal translation of the mRNA and impose on it positive regulation by eIF2α phosphorylation. These features bear great similarity to the regulation of GCN4 mRNA in yeast, although some details may be different. GCN4 mRNA has four upstream uORFs, and deletion of the 5′-most of these abolishes positive regulation by eIF2α(P) without increasing basal translation. This is explained by a model in which the 3′ uORFs are able to impose basal repression of mRNA translation, but only ribosomes that initiate at uORF1 can bypass the 3′ uORFs and reinitiate at the Gcn2p AUG (Hinnebusch 1997). ATF4 has only two conserved uORFs, and deletion of either results in increased basal translation and abolishes regulation by eIF2α(P). The basis for these important differences in the regulation of ATF4 and yeast GCN4 translation is currently not understood.

The regulation of ATF4 mRNA translation by eIF2α(P) provides important insight into how PERK and other eIF2α kinases may regulate gene expression in mammalian cells. At this point, it is not known whether ATF4 is the only mRNA thus regulated or if there are other effectors of PERK function that are similarly translationally regulated. Preliminary experiments with ATF4–/– cells suggest that they are hypersensitive to the effects of certain forms of stress that are associated with eIF2α phosphorylation and are defective in CHOP induction (H. Harding, unpubl.). Mice bearing the knockout allele were kindly provided by Thore Hettman and Jeff Leiden (Hettmann et al. 2000). These findings suggest that ATF4 is an important effector of signaling by eIF2α kinases such as PERK and further indicate that the phenotype of PERK–/– mice may be due not only to unregulated protein biosynthesis, but also to impaired activation of a downstream gene expression program.

In yeast, the general control response mediated by GCN2, eIF2α(P), and GCN4 coordinately activates many genes involved in amino acid biosynthesis. Our analysis of the gene expression aspect of PERK signaling has focused so far on a single marker gene, *CHOP*. It is clear, however, that in mammalian cells as well, signaling through eIF2α kinases affects many other genes. This was revealed by analysis of mice in which the wild-type eIF2α was replaced by a Ser-51 to alanine mutation. This mutation blocks signaling that might be mediated through reduction in ternary complex formation by eIF2α(P). Homozygous mutant mice die shortly after birth. They exhibit severe defects in the activation of genes involved in gluconeogenesis, a process essential for survival of postnatal mammals (Scheuner et al. 2001). Predictably, cells from these mice also fail to activate *CHOP* in response to ER stress, providing formal evidence that signaling from PERK to CHOP requires eIF2α(P).

The extent of the gene-expression program activated by eIF2α kinases in mammalian cells remains to be discovered. It is likely that this genomic response will integrate disparate stress signals, given that the different eIF2α kinases respond to specific upstream physiological regulators. We therefore propose that this signaling pathway be tentatively named the integrated stress response (ISR). Figure 5 compares the known components of the ISR with its yeast ancestor, the general control response. The availability of cells with mutations in defined components of this ISR (*PERK, GCN2, eIF2α, ATF4*) renders it highly amenable to study using expression arrays.

TERMINATING SIGNALING DOWNSTREAM FROM PERK

The advantages cells gain by activating PERK are readily demonstrated by the hypersensitivity of the knock-out cells to ER stress. However, it is clear that inhibition of protein synthesis cannot be maintained indefinitely, and cells must have evolved some mechanism to terminate signaling by PERK. PERK itself is rapidly inactivated when the folding environment in the ER is restored to normal, for example, after washing away the perturbing toxin (Bertolotti et al. 2000). Termination of PERK signaling correlates with re-formation of the PERK-BiP complex and dephosphorylation of PERK. Although the phosphatase(s) responsible for PERK inactivation has not been identified, recent experiments have provided insight into the mechanism of eIF2α(P) dephosphorylation and termination of signaling downstream from PERK, and these are reviewed below.

We recently conducted a genetic screen to uncover new genes that might be implicated in signaling in the integrated stress response. We sought to identify cDNA fragments that when introduced into cells as highly expressed retroviral transgenes would function as genetic suppressor elements for *CHOP* induction by ER stress (genetic suppressor elements are reviewed in Gudkov and Roninson [1997]). A genetic suppressor element fragment isolated by this screen encoded the carboxyl terminus of a GADD34, a known stress-induced protein. Bernard Roizman and coworkers previously called attention to the homology between the carboxyl terminus of GADD34 and a herpes simplex virus protein γ₁34.5 (He et al. 1996). The latter had been found to block the antiviral activity of cellular PKR, an eIF2α kinase. Roizman's lab went on to show that γ₁34.5 interferes with PKR function by recruiting the catalytic subunit of protein phosphatase 1 (PP1c) to eIF2α, promoting its dephosphorylation on Ser-51 (He et al. 1997, 1998). Viral protein synthesis can thus continue unabated despite activation of PKR.

We confirmed that GADD34 reduced eIF2α(P) levels and blocked translational and transcriptional control in stressed cells, this despite normal activity of the upstream kinases PERK and GCN2. Inhibition of *CHOP* induction correlated with the ability of GADD34 to bind PP1c and promote eIF2α dephosphorylation, both in vivo and in vitro. Most interestingly, we observed that *GADD34* itself is induced transcriptionally by ER stress or amino

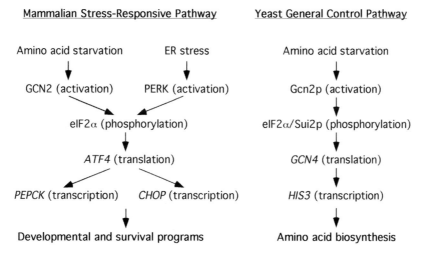

Figure 5. Comparison of the genes active in the mammalian integrated stress response with those of the yeast general control response. The mode by which each gene is activated is indicated in parentheses next to the gene's name. (Reprinted, with permission, from Harding et al. 2000b [copyright Elsevier Science].)

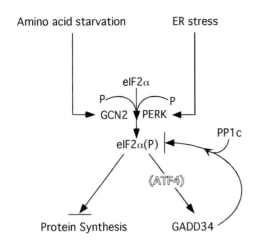

Figure 6. Role of GADD34 in terminating signaling in the integrated stress response. Activation of the upstream eIF2α kinases and eIF2α phosphorylation attenuate protein biosynthesis and activate gene expression. One of the genes activated, *GADD34*, encodes a regulatory subunit of protein phosphatase 1, which recruits the catalytic subunit (PP1c) to effect eIF2α(P) dephosphorylation. (Reprinted, with permission, from Novoa et al. 2001 [copyright Rockefeller University Press].)

acid deprivation, and this induction requires the activity of the eIF2α kinases PERK and GCN2, respectively. Thus, *GADD34* is a target gene of the ISR which it serves to negatively regulate (Fig. 6) (Novoa et al. 2001).

We speculate that activation of *GADD34* may be important in limiting translational repression to a transient early phase of the ER stress response (Wong et al. 1993). Early recovery from translation repression ensures that mRNAs which accumulated as part of the gene expression program associated with the UPR will have an opportunity to be translated before they decay. According to this model, translational repression is the first line of defense in the UPR, preventing the introduction of client proteins into an ER that lacks functional reserve. The induction of genes whose products increase the functional capacity of the ER has a greater latency and provides a more lasting solution, but its implementation requires translation. The latter point is revealed by examining the translation of BiP mRNA as reflected by its incorporation into polysomes.

BiP mRNA and protein are readily detectable basally and increase severalfold under conditions of ER stress. In unstressed cells, BiP mRNA is found in the heaviest fractions of a sucrose gradient, indicating that it is associated with polyribosomes and is actively translated. This correlates with high levels of incorporation of [^{35}S]methionine into BiP protein. Early in the ER stress response, when eIF2α is heavily phosphorylated, BiP mRNA shifts to lighter fractions in the sucrose gradient, and incorporation of [^{35}S]methionine into BiP protein is markedly attenuated (Fig. 7) (Harding et al. 2000b). These results suggest that BiP mRNA is unable to avoid the consequences of translational repression prevailing in the early phases of

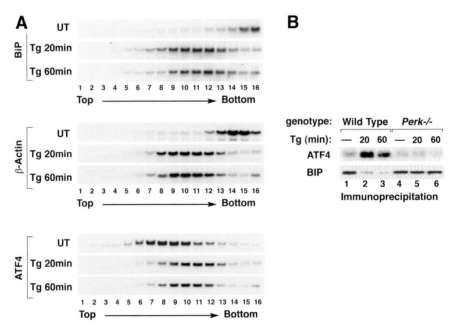

Figure 7. BiP translation is severely attenuated early in the ER stress response. (*A*) The three panels reveal the distribution of the BiP, β-actin, and ATF4 mRNAs in the same fractions of a polysome profile obtained by sucrose-gradient centrifugation from untreated cells (UT) and cells treated with thapsigargin (Tg) for 20 or 60 min. Note that both BiP and β-actin mRNA are found in the heaviest ribosome-rich fractions in unstressed cells, indicating that both proteins are well translated under such conditions. In stressed cells, both mRNAs shift to the lighter fractions, consistent with a block in translation initiation. ATF4, in contrast, shifts to heavier fractions in the stressed cells. (*B*) The translation of ATF4 and BiP is compared by pulse-labeling with radiolabeled methionine and immunoprecipitation of the radiolabeled protein from untreated and thapsigargin-treated (Tg) wild-type and *PERK–/–* mutant cells. Note that at these early points of severe ER stress, BiP translation is profoundly attenuated, although there is a hint toward recovery at the 60-min timepoint. (Portions of this figure were adapted, with permission, from Harding et al. 2000b [copyright Elsevier Science].)

an experimental ER stress response, this despite the presence of an IRES that functions to preserve BiP translation under other stressful circumstances (Macejak and Sarnow 1990). Ultimately, translational recovery allows BiP mRNA molecules that had accumulated earlier in the UPR to be translated, resulting in higher levels of BiP protein.

CONCLUSIONS

Identification of PERK and study of the phenotype of *PERK–/–* cells and animals have brought to our attention the importance of translational control by ER stress. The endocrine and exocrine pancreas, whose cells have specialized in regulated production and secretion of large quantities of polypeptides, are especially dependent on PERK function. Their functional attrition in *PERK* mutant mice and humans is likely caused by death of the secretory cells. The processes leading to death of the *PERK–/–* cells are poorly understood. The insertion of polypeptides into the ER under conditions in which the functional reserve of the organelle has been exceeded may cause these polypeptides to assume configurations that are toxic to the cells. This may come about because of inadequate reserve of chaperones that normally promote folding or because the capacity of the apparatus that carries out ER-associated degradation is exceeded. This hypothesis of death at the hands of endogenous proteotoxins emphasizes the importance of control of polypeptide biosynthesis and fits the emerging concept whereby abnormal protein conformations may be pathogenic (Hightower 1991; Carrell and Lomas 1997).

The proteotoxin hypothesis may also explain why PERK apparently evolved in metazoans and is absent from yeast. Unicellular organisms divide continuously and might be able to dilute proteotoxins in this manner. Metazoan life, with the emergence of terminally differentiated cells, confronted the organism for the first time with the challenge of loss of this dilutional defense mechanism. This challenge became more acute with development of long-lived lineages such as brain and endocrine pancreas. PERK, according to this idea, evolved to limit the production of potential proteotoxins.

Translational control is also linked to a gene expression program, and we propose that this response be named the integrated stress response because it integrates translation and transcription under different stressful circumstances. Details of signal transduction in this pathway have been worked out in yeast and more recently in mammalian cells. The known components include a stress-activated eIF2α kinase (e.g., PERK and GCN2); the eIF2 complex and its guanine nucleotide exchange factor, eIF2B, a transcription factor that is translationally up-regulated (e.g., ATF4); and the downstream genes. Few of the latter have been discovered to date. Surprisingly, however, those that have include some of the canonical ER stress-induced genes such as *CHOP* and *BiP*. The ISR and its more-linear and less-diversified yeast precursor (the general control response) are therefore of considerable consequence to the organism. Unraveling its workings is an important challenge for the future.

ACKNOWLEDGMENTS

This work was supported by National Institutes of Health grants DK47119 and ES08681. D.R. is an Ellison Medical Foundation Senior Scholar. H.H. was supported in part by a National Research Service Award from the National Cancer Institute, I.N. was supported in part by a Basque Government Fellowship, A.B. was supported by a Human Frontier Science Program award, F.U. was supported in part by an award from the Japan Society for the Promotion of Science and by the Uehara memorial foundation, and C.J. was supported in part by a postdoctoral fellowship from the Institut Français de Nutrition.

REFERENCES

Bertolotti A., Zhang Y., Hendershot L., Harding H., and Ron D. 2000. Dynamic interaction of BiP and the ER stress transducers in the unfolded protein response. *Nat. Cell. Biol.* **2:** 326.

Brostrom C.O., Bocckino S.B., and Brostrom M.A. 1983. Identification of a Ca^{2+} requirement for protein synthesis in eukaryotic cells. *J. Biol. Chem.* **258:** 14390.

Bruhat A., Jousse C., Wang X.-Z., Ron D., Ferrara M., and Fafournoux F. 1997. Amino acid limitation induces expression of *chop*, a CCAAT/enhancer binding protein related gene at both transcriptional and post-transcriptional levels. *J. Biol. Chem.* **272:** 17588.

Carrell R.W. and Lomas D.A. 1997. Conformational disease. *Lancet* **350:** 134.

Chen J.-J. 2000. Heme-regulated eIF2α kinase. In *Translational control of gene expression* (ed. N. Sonenberg et al.), pp. 529. Cold Spring Harbor Laboratory Press, Cold Spring Harbor, New York.

Cox J.S., Shamu C.E., and Walter P. 1993. Transcriptional induction of genes encoding endoplasmic reticulum resident proteins requires a transmembrane protein kinase. *Cell* **73:** 1197.

Delepine M., Nicolino M., Barrett T., Golamaully M., Lathrop G.M., and Julier C. 2000. EIF2AK3, encoding translation initiation factor 2-alpha kinase 3, is mutated in patients with Wolcott-Rallison syndrome. *Nat. Genet.* **25:** 406.

Dorner A., Wasley L., and Kaufman R. 1992. Overexpression of GRP78 mitigates stress induction of glucose regulated proteins and blocks secretion of selective proteins in Chinese hamster ovary cells. *EMBO J.* **11:** 1563.

Estes S.D., Stoler D.L., and Anderson G.R. 1995. Normal fibroblasts induce the C/EBP beta and ATF-4 bZIP transcription factors in response to anoxia. *Exp. Cell. Res.* **220:** 47.

Fawcett T.W., Martindale J.L., Guyton K.Z., Hai T., and Holbrook N.J. 1999. Complexes containing activating transcription factor (ATF)/cAMP-responsive-element-binding protein (CREB) interact with the CCAAT/enhancer-binding protein (C/EBP)-ATF composite site to regulate Gadd153 expression during the stress response. *Biochem. J.* **339:** 135.

Gudkov A.V. and Roninson I.B. 1997. Isolation of genetic suppressor elements (GSEs) from random fragment cDNA libraries in retroviral vectors. *Methods Mol. Biol.* **69:** 221.

Han A.P., Yu C., Lu L., Fujiwara Y., Browne C., Chin G., Fleming M., Lebouich P., Orkin S.H, and Chen J.J. 2001. Heme-regulated eIF2alpha kinase (HRI) is required for translational control and survival of erythroid precursors in iron deficiency. *EMBO J.* **20:** 6909.

Harding H., Zhang Y., and Ron D. 1999. Translation and protein folding are coupled by an endoplasmic reticulum resident kinase. *Nature* **397:** 271.

Harding H., Zhang Y., Bertolotti A., Zeng H., and Ron D. 2000a. *Perk* is essential for translational regulation and cell survival during the unfolded protein response. *Mol. Cell* **5:** 897.

Harding H., Novoa I., Zhang Y., Zeng H., Wek R.C., Schapira M., and Ron D. 2000b. Regulated translation initiation controls stress-induced gene expression in mammalian cells. *Mol. Cell* **6:** 1099.

Harding H., Zeng H., Zhang Y., Jungreis R., Chung P., Plesken H., Sabatini D., and Ron D. 2001. Diabetes mellitus and excocrine pancreatic dysfunction in Perk–/– mice reveals a role for translational control in survival of secretory cells. *Mol. Cell* **7:** 1153.

He B., Gross M., and Roizman B. 1997. The gamma(1)34.5 protein of herpes simplex virus 1 complexes with protein phosphatase 1alpha to dephosphorylate the alpha subunit of the eukaryotic translation initiation factor 2 and preclude the shutoff of protein synthesis by double-stranded RNA-activated protein kinase. *Proc. Natl. Acad. Sci.* **94:** 843.

———. 1998. The gamma134.5 protein of herpes simplex virus 1 has the structural and functional attributes of a protein phosphatase 1 regulatory subunit and is present in a high molecular weight complex with the enzyme in infected cells. *J. Biol. Chem.* **273:** 20737.

He B., Chou J., Liebermann D.A., Hoffman B., and Roizman B. 1996. The carboxyl terminus of the murine MyD116 gene substitutes for the corresponding domain of the gamma(1)34.5 gene of herpes simplex virus to preclude the premature shutoff of total protein synthesis in infected human cells. *J. Virol.* **70:** 84.

Hettmann T., Barton K., and Leiden J.M. 2000. Microphthalmia due to p53-mediated apoptosis of anterior lens epithelial cells in mice lacking the CREB-2 transcription factor. *Dev. Biol.* **222:** 110.

Hightower L.E. 1991. Heat shock, stress proteins, chaperones, and proteotoxicity. *Cell* **66:** 191.

Hinnebusch A.G. 1997. Translational regulation of yeast GCN4. A window on factors that control initiator-tRNA binding to the ribosome. *J. Biol. Chem.* **272:** 21661.

———. 2000. Mechanism and regulation of initiator methionyl-tRNA binding to ribosomes. In *Translational control of gene expression* (ed. N. Sonenberg et al.), pp. 185. Cold Spring Harbor Laboratory Press, Cold Spring Harbor, New York.

Howell S.L. and Taylor K.W. 1966. Effects of glucose concentration on incorporation of [3H]leucine into insulin using isolated mammalian islets of Langerhans. *Biochim. Biophys. Acta.* **130:** 519.

Itoh N. and Okamoto H. 1980. Translational control of proinsulin synthesis by glucose. *Nature* **283:** 100.

Jousse C., Bruhat A., Harding H.P., Ferrara M., Ron D., and Fafournoux P. 1999. Amino acid limitation regulates CHOP expression through a specific pathway independent of the unfolded protein response. *FEBS Lett.* **448:** 211.

Kaufman R.J. 1999. Stress signaling from the lumen of the endoplasmic reticulum: Coordination of gene transcriptional and translational controls. *Genes Dev.* **13:** 1211.

———. 2000. The double-stranded RNA-activated protein kinase PKR. In *Translational control of gene expression* (ed. N. Sonenberg et al.), p. 503. Cold Spring Harbor Laboratory Press, Cold Spring Harbor, New York.

Kohno K., Normington K., Sambrook J., Gething M.J., and Mori K. 1993. The promoter region of the yeast KAR2 (BiP) gene contains a regulatory domain that responds to the presence of unfolded proteins in the endoplasmic reticulum. *Mol. Cell. Biol.* **13:** 877.

Kostura M. and Mathews M.B. 1989. Purification and activation of the double-stranded RNA-dependent eIF-2 kinase DAI. *Mol. Cell. Biol.* **9:** 1576.

Langland J.O. and Jacobs B.L. 1992. Cytosolic double-stranded RNA-dependent protein kinase is likely a dimer of partially phosphorylated Mr = 66,000 subunits. *J. Biol. Chem.* **267:** 10729.

Liu C.Y., Schroder M., and Kaufman R.J. 2000. Ligand-independent dimerization activates the stress-response kinases IRE1 and PERK in the lumen of the endoplasmic reticulum. *J. Biol. Chem.* **275:** 44881.

Ma Y., Lu Y., Zeng H., Ron D., Mo W., and Neubert T.A. 2001. Characterization of phosphopeptides from protein digests using matrix-assisted laser desorption/ionization time-of-flight mass spectrometry and nanoelectrospray quadrupole time-of-flight mass spectrometry. *Rapid Commun. Mass Spectrom.* **15:** 1693.

Macejak D.G. and Sarnow P. 1990. Translational regulation of the immunoglobulin heavy-chain binding protein mRNA. *Enzyme* **44:** 310.

Mori K. 2000. Tripartite management of unfolded proteins in the endoplasmic reticulum. *Cell* **101:** 451.

Mori K., Ma W., Gething M.J., and Sambrook J. 1993. A transmembrane protein with a cdc2+/CDC28-related kinase activity is required for signaling from the ER to the nucleus. *Cell* **74:** 743.

Nehls S., Snapp E.L., Cole N.B., Zaal K.J., Kenworthy A.K., Roberts T.H., Ellenberg J., Presley J.F., Siggia E., and Lippincott-Schwartz J. 2000. Dynamics and retention of misfolded proteins in native ER membranes. *Nat. Cell Biol.* **2:** 288.

Novoa I., Zeng H., Harding H., and Ron D. 2001. Feedback inhibition of the unfolded protein response by GADD34-mediated dephosphorylation of eIF2α. *J. Cell Biol.* **153:** 1011.

Prostko C.R., Brostrom M.A., and Brostrom C.O. 1993. Reversible phosphorylation of eukaryotic initiation factor 2 alpha in response to endoplasmic reticular signaling. *Mol. Cell. Biochem.* **127-128:** 255.

Prostko C.R., Brostrom M.A., Malara E.M., and Brostrom C.O. 1992. Phosphorylation of eukaryotic initiation factor (eIF) 2 alpha and inhibition of eIF-2B in GH3 pituitary cells by perturbants of early protein processing that induce GRP78. *J. Biol. Chem.* **267:** 16751.

Prostko C.R., Dholakia J.N., Brostrom M.A., and Brostrom C.O. 1995. Activation of the double-stranded RNA-regulated protein kinase by depletion of endoplasmic reticular calcium stores. *J. Biol. Chem.* **270:** 6211.

Sachs A. 2000. Physical and functional interactions between the mRNA cap structures and the poly(A) tail. In *Translational control of gene expression* (ed. N. Sonenberg et al.), p. 447. Cold Spring Harbor Laboratory Press, Cold Spring Harbor, New York.

Scheuner D., Song B., McEwen E., Liu C., Laybutt R., Gillespie P., Saunders T., Bonner-Weir S., and Kaufman R.J. 2001. Translational control is required for the unfolded protein response and in vivo glucose homeostasis. *Mol. Cell* **7:** 1165.

Shamu C.E. and Walter P. 1996. Oligomerization and phosphorylation of the Ire1p kinase during intracellular signaling from the endoplasmic reticulum to the nucleus. *EMBO J.* **15:** 3028.

Shi Y., Vattem K.M., Sood R., An J., Liang J., Stramm L., and Wek R.C. 1998. Identification and characterization of pancreatic eukaryotic initiation factor 2 alpha-subunit kinase, PEK, involved in translational control. *Mol. Cell. Biol.* **18:** 7499.

Sidrauski C. and Walter P. 1997. The transmembrane kinase Ire1p is a site-specific endonuclease that initiates mRNA splicing in the unfolded protein response. *Cell* **90:** 1031.

Srivastava S.P., Davies M.V., and Kaufman R.J. 1995. Calcium depletion from the endoplasmic reticulum activates the double-stranded RNA-dependent protein kinase (PKR) to inhibit protein synthesis. *J. Biol. Chem.* **270:** 16619.

Thomis D.C. and Samuel C.E. 1993. Mechanism of interferon action: Evidence for intermolecular autophosphorylation and autoactivation of the interferon-induced, RNA-dependent protein kinase PKR. *J. Virol.* **67:** 7695.

Tirasophon W., Welihinda A.A., and Kaufman R.J. 1998. A stress response pathway from the endoplasmic reticulum to the nucleus requires a novel bifunctional protein kinase/endoribonuclease (Ire1p) in mammalian cells. *Genes Dev.* **12:** 1812.

Vallejo M., Ron D., Miller C.P., and Habener J.F. 1993. C/ATF, a member of the activating transcription factor family of DNA-binding proteins, dimerizes with CAAT/enhancer-binding proteins and directs their binding to cAMP response elements. *Proc. Natl. Acad. Sci.* **90:** 4679.

Wang X.Z., Harding H.P., Zhang Y., Jolicoeur E.M., Kuroda M., and Ron D. 1998. Cloning of mammalian Ire1 reveals diversity in the ER stress responses. *EMBO J.* **17:** 5708.

Wolcott C. and Rallison M. 1972. Infancy-onset diabetes mellitus and multiple epiphyseal dysplasia. *J. Pediatr.* **80:** 292.

Wong W.L., Brostrom M.A., Kuznetsov G., Gmitter-Yellen D., and Brostrom C.O. 1993. Inhibition of protein synthesis and early protein processing by thapsigargin in cultured cells. *Biochem. J.* **289:** 71.

Wu S. and Kaufman R.J. 1997. A model for the double-stranded RNA (dsRNA)-dependent dimerization and activation of the dsRNA-activated protein kinase PKR. *J. Biol. Chem.* **272:** 1291.

A Novel Stress-response Protein That Binds at the Ribosomal Subunit Interface and Arrests Translation

D.E. Agafonov, V.A. Kolb, and A.S. Spirin

Institute of Protein Research, Russian Academy of Sciences, Pushchino, Moscow Region, Russia 142290

One of the most direct approaches for studies of protein topography on the ribosome surface is the so-called hot tritium bombardment technique (Yusupov and Spirin 1988). The technique is based on the replacement of hydrogen by tritium in covalent bonds of macromolecules in their thin surface layer (Shishkov et al. 1976; Goldanskii et al. 1988). Studying proteins of the ribosome surface with this methodology has led to the finding that an unidentified minor component of *Escherichia coli* ribosomes corresponding to spot Y on the two-dimensional electrophoretic map of ribosomal proteins becomes highly exposed upon dissociation of the 70S ribosome into subunits (Yusupov and Spirin 1986, 1988; Agafonov et al. 1997).

Later, the protein Y was isolated and its binding with ribosomes and ribosomal subunits was studied (Agafonov et al. 1999). Recently, the protein was shown to appear on ribosomes under some stress conditions and to inhibit translation (Agafonov et al. 2001). It has been proposed to name the protein *Ribosome-associated inhibitor A* (RaiA).

PROTEIN Y, OR RaiA, APPEARS ON RIBOSOMES IN COLD-SHOCKED AND AGED CELLS

The presence of the RaiA in *E. coli* ribosomes was analyzed depending on conditions of the culture growth and cell harvesting. The protein was not detected in ribosomes isolated from *E. coli* cells grown at 37°C and harvested without cooling (Fig. 1a). It appeared in ribosomes in response to cold shock when the growing cells were chilled and incubated under lowered temperatures (Fig. 1b). The presence of the protein in ribosomes from the cells grown at 37°C (Agafonov et al. 1999) was the result of the cooling of cell culture during harvesting of the bacteria from a large volume of fermentation medium.

The protein became detectable in ribosomes within the first hour of incubation under cold shock conditions, e.g., at 15°C, and remained associated with ribosomes during the subsequent four hours of the growth arrest phase. When growth was resumed, the protein disappeared from the ribosomes. The ribosomes from the cells that attained the stationary phase at physiological temperature (37°C) also contained RaiA. Hence, expression of the protein or its binding to ribosomes was induced by environmental stress, such as temperature downshift or excessive cell

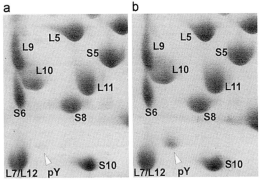

Figure 1. Cold-induced synthesis of RaiA in *E. coli* cells. Shown are fragments of 2D electrophoretic gel slabs stained with Coomassie Brilliant Blue G 250. (*a*) Proteins from ribosomes of the cells grown at 37°C and harvested without cooling. (*b*) Proteins from ribosomes of the cells subjected to temperature downshift (gradually cooling down to 4°C during 1 hour).

culture density. From the intensities of protein spots in electrophoretic slabs, it follows that the molar ratio of RaiA to ribosomes varied from 1:10 to 1:3, depending on conditions of cell cultivation and harvesting.

THE 12.7-kD PROTEIN (pY, OR RaiA) IS ENCODED BY *yfia*, AN OPEN READING FRAME LOCATED UPSTREAM OF THE *phe* OPERON

Determination of amino-terminal amino acid sequence was made with RaiA samples extracted from gels of two-dimensional electrophoresis of total ribosomal protein (Agafonov et al. 1999). The determined part of the protein sequence was found to coincide with the sequence deduced from the *E. coli* ORF *yfia*: (MTMNITSKQME-ITPAIRQHV...). It was concluded that the polypeptide Y (protein RaiA) is encoded by the ORF *yfia*. Previously, *yfia* was referred to as *unidentified reading frame 1* (URF1) (Hudson and Davidson 1984) and located between the 16S rRNA gene and the *phe* operon in the *E. coli* chromosome. The ORF corresponds to a protein of 113 amino acid residues with a molecular mass of 12,785 D.

Earlier, the protein encoded by *yfia* was identified on a two-dimensional electrophoretic map of total *E. coli* protein (Link et al. 1997), but its localization in the cell was not determined. A 30S-subunit-associated protein homologous to the *E. coli* protein Y (RaiA) was isolated from total ribosomal protein of spinach chloroplasts (Johnson et al. 1990); the authors, however, claimed that the protein is unique for chloroplasts.

Taking into account the fact that the spot Y corresponds to the protein named RaiA and is encoded by ORF *yfia*, the gene is proposed to be renamed *rai*A.

THE PROTEIN CAN BE ISOLATED FROM DISSOCIATED RIBOSOMES

The procedure for preparative isolation of protein RaiA was developed based on the assumption that the protein is localized on the ribosomal interface (Agafonov et al. 1997). Thus, it could be extracted from ribosomes by high-salt treatment inducing ribosome dissociation. Crude 70S ribosomes were first washed with a buffer containing high magnesium ion concentration (20 mM), and then treated with 400 mM NaCl, resulting in the dissociation of the 70S ribosomes and the washing-off of RaiA protein. Two steps of column chromatography were used to prepare the protein with more than 90% purity. Two-dimensional co-electrophoresis of the purified product with ribosomal proteins has demonstrated its identity with the protein migrating as spot Y (Agafonov et al. 1999).

Using the mass spectrometry technique, a single component with molecular mass of 12,640 D was detected in the purified protein preparation. Taking into account the absence of amino-terminal methionine residue, the deduced molecular mass of the protein should be 12,654 D, which is in good agreement with the mass spectrometry result.

THERE ARE HOMOLOGS OF THE PROTEIN IN OTHER SPECIES OF EUBACTERIA

Among complete eubacterial genomes available at Gen-Bank, sequences having significant homology with *yfia* were found in the following organisms: *Aquifex aeolicus* (ORF aq1603), *Bacillus subtilis* (gene *yvyD*), *Borrelia burgdorferi* (ORF bb0449), *Chlamidophyla pneumoniae* (gene *yvyD*), *Haemophilus influenzae* (ORF hi0257), *Mycobacterium tuberculosis* (ORF Rv3241c), *Rickettsia prowazekii* (gene *yhbH*), *Synechocystis sp.* (gene *lrt*A), *Thermotoga maritima* (ORF tmTM1607), and *E. coli*

(genes *yfia* and RP5M). Other homologs found among proteins from the SwissProt database were RP5M_AZOVI from *Azotobacter vinelandii*, RP5M_KLEPN from *Klebsiella pneumoniae*, RP5M_SALTY from *Salmonella typhimurium*, RP5M_PSEPU from *Pseudomonas putida*, RP5M_ALCEU from *Alcaligenes eutrophus*, RP5M_RHIME from *Rhizobium meliloti*, RP5M_BRAJA from *Bradyrhizobium japonicum*, RP5M_ACICA from *Acinetobacter calcoaceticus*, RP5M_THIFE from *Thiobacillus ferrooxidans*, YSEA_STACA from *Staphylococcus carnosus,* and RR30_SPIOL from chloroplasts of *Spinacia oleracea*.

Functions of most of them are unknown. The sequences called RP5M were supposed to be transcription modulator protein of transcription cofactor sigma 54. The gene *lrt*A from *Synechocystis sp.* determines a light-repressed transcript (Tan et al. 1994). The homolog of protein Y (RaiA) found in chloroplasts of *S. oleracea* was reported to be a ribosomal protein of the 30S subunit (Johnson et al. 1990).

RaiA thus belongs to a protein family present in a number of bacterial species. No homologs, however, are seen in the genomes of *Chlamydia trachomatis, Helicobacter pylori, Treponema pallidum, Mycoplasma pneumoniae,* and *Mycoplasma genitalium,* nor in archaebacteria. Hence, proteins of this family seem to be much less conservative as compared with the canonical ribosomal proteins.

THE PROTEIN RESIDES ON THE SMALL RIBOSOMAL SUBUNIT AT THE SUBUNIT INTERFACE

To determine which of the two ribosomal subunits is mainly responsible for the binding of protein RaiA, an ultrafiltration binding test was used. Suspensions of 70S ribosomes or isolated ribosomal subunits were filtered through the 100-kD cut-off membrane of Centricon-100 concentrator. The membrane retains ribosomal subunits but allows free protein RaiA to pass through. Analysis of filtrates by SDS-PAGE demonstrated that the protein has an affinity to the 30S subunit and the 70S ribosome but not the 50S subunit (Fig. 2a). The test also showed that the protein RaiA binding with the 30S subunit depends on Mg^{++} concentration: When the binding reaction was carried out in the presence of 1 mM $MgCl_2$, no retention of the protein was detected (Fig. 2b).

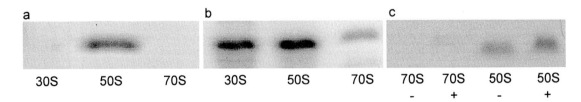

Figure 2. Binding of RaiA to ribosomal particles. SDS-PAGE of filtrates passed through 100-kD cut-off membranes. The binding test was performed in buffer containing 10 mM $MgCl_2$ (*a*), 1 mM $MgCl_2$ (*b*), and 50 µM 7-dimethylamino-6-demethyl-6-deoxytetracycline (minocycline) at 10 mM $MgCl_2$ (*c*). Types of the ribosomal particles used for binding are indicated below corresponding gel lanes. The presence or absence of minocycline (*c*) is indicated by plus and minus, respectively. Photographs of Coomassie-stained gels.

The complex of protein RaiA with the 30S subunit in 10 mM Mg++ was subjected to hot tritium bombardment as such and after addition of the 50S subunit. Analysis of tritium incorporation into proteins showed a significant shielding effect of the 50S subunit on RaiA at the 30S subunit surface. The association of the 30S subunit with the 50S subunit resulted in a 70% decrease of the protein accessibility, whereas the accessibility of other proteins of the small subunit did not change or changed insignificantly (Fig. 3). This result proves that RaiA is localized at the subunit interface.

THE PROTEIN STABILIZES THE ASSOCIATION BETWEEN RIBOSOMAL SUBUNITS

The localization of RaiA at the 30S subunit surface contacting the 50S subunit suggested that the protein might affect association of the ribosomal subunits. To check this, the Mg++ dependence of ribosome dissociation was recorded by light scattering at 400 nm (Fig. 4). It is seen that in the range of Mg++ concentrations from 7 to 4 mM, the intensity of light scattering remained unchanged, being equal to that at 10 mM Mg++, where ribosomes are in the associated (70S) state. At Mg++ concentrations below 1 mM, when only ribosomal subunits were present in the assay mixture, the light-scattering intensity was at the lower level that corresponded to the dissociated (50S + 30S) state of the ribosome. As follows from the intensity value, about half of the ribosomes were dissociated at 2.5 mM Mg++. The addition of RaiA in equimolar ratio to ribosomes resulted in virtually complete association of ribosomal subunits at this Mg++ concentration; about 50% of ribosomes remained associated even at 1.5 mM Mg++.

A similar result was obtained by sedimentation analysis of ribosomes in an analytical ultracentrifuge in the absence and presence of RaiA. Twofold molar excess of the protein over ribosomes resulted in virtually complete as-

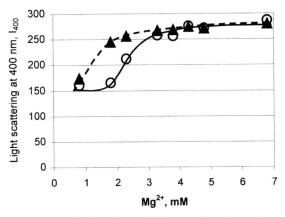

Figure 4. Mg++ dependence of ribosome dissociation determined by light scattering at 400 nm. Scattering light intensity (I_{400}) is shown in arbitrary units. Samples contained ribosomes without RaiA (*open circles, solid line*), or with an equimolar amount of RaiA (*closed triangles, dashed line*).

sociation of ribosomal subunits at 1.8 mM Mg++ (Agafonov et al. 1999).

THE PROTEIN INHIBITS TRANSLATION AT THE ELONGATION STAGE

To test functional consequences of the physical interaction of RaiA with the ribosome, cell-free translation experiments in the presence of RaiA were performed. Figure 5 demonstrates that RaiA strongly inhibited cell-free translation of green fluorescent protein (GFP) mRNA in bacterial (*E. coli*) extract. The inhibition was proportional to the amount of RaiA added, being around 75–80% at the equimolar RaiA-to-ribosome ratio.

Translation of a natural mRNA includes three stages: initiation, elongation, and termination. On the other hand, translation of a synthetic polynucleotide, such as polyuridylic acid, does not involve the mechanisms of

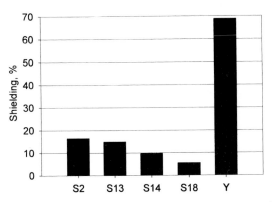

Figure 3. Change in the ribosome protein accessibility to ^{3}H bombardment as a result of the association of 30S and 50S subunits. The complexes of the 30S subunit with RaiA were labeled with the use of the hot tritium bombardment technique prior to and after the association with the large subunit. Extent of the shielding that resulted from the association is shown in percentage for RaiA (pY) and the most shielded ribosomal proteins (S2, S13, S14, and S18).

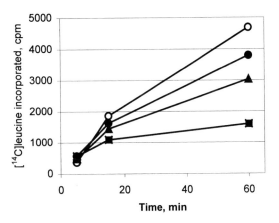

Figure 5. Inhibition of cell-free translation of GFP mRNA by RaiA. Molar ratios of RaiA to ribosomes were 0.2 (*closed circles*), 0.4 (*closed triangles*), and 1 (*closed squares*), or the incubation mixture contained no RaiA (*open circles*). Amount of newly synthesized protein was determined from radioactivity of hot 10% trichloroacetic acid precipitate.

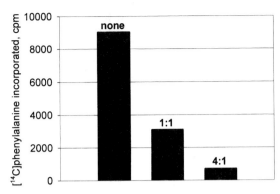

Figure 6. Inhibition of poly(U)-directed polyphenylalanine elongation by RaiA. Molar ratio of RaiA to ribosomes is indicated above bars. Amount of newly synthesized polyphenylalanine was determined from radioactivity of hot 10% trichloroacetic acid precipitate.

initiation and termination and is regarded as the case of elongation only. Figure 6 shows the effect of RaiA on poly(U)-directed elongation of polyphenylalanine in a cell-free translation system. It is seen that the protein effectively inhibited elongation. About 65% inhibition was observed at equimolar amounts of RaiA and ribosomes, and the fourfold excess of RaiA over ribosomes in the translation system led to more than 90% inhibition. Hence, we can conclude that RaiA inhibits translation at the elongation stage.

THE PROTEIN BLOCKS THE RIBOSOMAL A SITE

The elongation stage of translation is a sequence of elongation cycles, each of which consists of three consecutive steps: aminoacyl-tRNA binding, transpeptidation, and translocation. Thus, each elongation cycle of the ribosome results in the selection of one aminoacyl tRNA from the milieu, the formation of one peptide bond, and the readout of one codon of a message. The question arises as to which of the partial steps of the elongation cycle is inhibited by RaiA.

The poly(U)-directed cell-free translation system can be made in such a way that either the catalyst of aminoacyl-tRNA binding, elongation factor Tu (EF-Tu); or the catalyst of translocation, elongation factor G (EF-G); is absent from the translation mixture (Gavrilova et al. 1976; Spirin et al. 1976). The result is that either aminoacyl-tRNA binding or translocation, respectively, passes via the nonenzymatic (factor-free) pathways and, hence, becomes the rate-limiting step of the entire elongation process. In our experiments with the EF-Tu-promoted system where translocation was the rate-limiting step, the addition of RaiA had no influence on the elongation rate. At the same time, the presence of RaiA in the EF-G-catalyzed system, where the aminoacyl-tRNA-binding step limited the rate of translation, inhibited elongation to the same extent as in the case of the full translation system (Fig. 7). Thus, the experiments with the one-factor-dependent systems have indicated that RaiA affects specif-

ically the aminoacyl-tRNA-binding step, rather than other steps of the elongation cycle.

Table 1 shows the results of directly testing the codon-cognate aminoacyl-tRNA binding with poly(U)-programmed 70S ribosomes in the absence and presence of RaiA. Both EF-Tu-promoted ("enzymatic") and "nonenzymatic" binding of phenylalanyl tRNA were tested. The preformed ternary complex [^{14}C]phenylalanyl tRNA^{Phe}•EF-Tu•GTP was incubated with poly(U)-programmed 70S ribosomes, and the resultant enzymatic binding of [^{14}C]phenylalanyl tRNA^{Phe} to the ribosome was measured. As shown in the table, the presence of RaiA strongly inhibited the enzymatic binding of the codon-cognate aminoacyl tRNA to the A site of the 70S ribosome. The table shows that RaiA also blocks the nonenzymatic binding of [^{14}C]phenylalanyl tRNA^{Phe} with the 70S ribosome.

Aminoacyl-tRNA-binding experiments with isolated poly(U)-programmed 30S ribosomal subunits were also conducted. Table 1 shows that RaiA inhibited the codon-dependent binding of [^{14}C]phenylalanyl tRNA^{Phe} with the isolated 30S subunit as well. This indicates that the block-

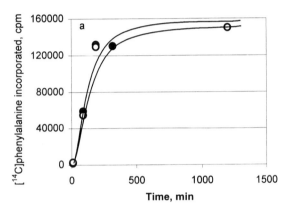

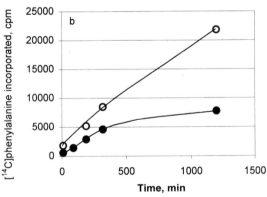

Figure 7. Effect of RaiA on poly(U)-directed one-factor-dependent translation systems. (*a*) Translation of poly(U) in EF-Tu-dependent system. The incubation mixture contained *E. coli* ribosomes, poly(U), [^{14}C]Phe-tRNA, EF-Tu, and GTP in 10 mM MgCl$_2$, 100 mM NH$_4$Cl, at 37°C. EF-G omitted. (*b*) Translation of poly(U) in EF-G-dependent system. The incubation mixture contained *E. coli* ribosomes, poly(U), [^{14}C]Phe-tRNA, EF-Tu, and GTP in 10 mM MgCl$_2$, 100 mM NH$_4$Cl, at 37°C. EF-Tu omitted. Closed circles are translation in the presence of RaiA; open circles, in its absence.

Table 1. Blockade of Aminoacyl-tRNA Binding to the Ribosomal A Site by RaiA

	EF-Tu-promoted binding with 70S ribosomes, cpm	Nonenzymatic binding with 70S ribosomes, cpm	Binding with 30S subunits, cpm
– RaiA	7119	4064	794
+ RaiA	618	221	329

Binding of [^{14}C]phenylalanyl tRNA with poly(U)-programmed 70S ribosomes (Agafonov et al. 2001) and with isolated poly(U)-programmed 30S ribosomal subunits (according to Glukhova et al. 1975) was estimated from the radioactivity retained on nitrocellulose filters.

ade of aminoacyl-tRNA binding by RaiA is due to the direct effect of the protein on the aminoacyl-tRNA-binding site area at the 30S subunit, rather than the result of some changes in the character of the two-subunit association.

At the same time, the experiments with radioactive tetracycline, a specific inhibitor of the ribosomal A site, demonstrated that the antibiotic and RaiA do not compete in their binding to the ribosome. As shown in Figure 8, tetracycline binding to the ribosome remained unaffected in the presence of RaiA. On the other hand, the ultrafiltration test showed that the interaction of the protein with the ribosome was resistant against tetracycline (Fig. 2c). Hence, both the protein and tetracycline do not interfere with each other when they bind to the ribosome, thus implying their nonoverlapping binding within the A site.

The A-site-blocking function of RaiA is consistent with the localization of RaiA on the ribosomal interface. The intersubunit position of the ribosome-bound aminoacyl tRNA and the A site is clearly revealed by recent crystallographic studies of ribosomal functional complexes (Cate et al. 1999; Yusupov et al. 2001).

THE RIBOSOME-ASSOCIATED INHIBITOR PROTEIN MAY BE INVOLVED IN SURVIVAL UNDER STRESS CONDITIONS

The involvement of ribosomes in adaptation of bacteria to low temperature was implied a long time ago (Das and Goldstein 1968; Friedman et al. 1969). It was found that an unknown component (or components) of the ribosomal fraction of the bacterial cell is responsible for the arrest of translation at low temperatures (Das and Goldstein 1968). A number of antibiotics specifically affecting bacterial ribosomes were shown to mimic cold shock or heat shock response, which gave cause for claiming ribosomes to be sensors of cold and heat shock in bacteria (VanBogelen and Neidhardt 1990). More recently, a ribosome-binding factor RbfA was discovered (Dammel and Noller 1995) that proved to be a cold shock protein (Jones and Inouye 1996). It was shown that RbfA binds with the 30S ribosomal subunit and suppresses a cold-sensitive mutation in 16S ribosomal RNA (Dammel and Noller 1995). Another cold shock protein, CsdA, was also characterized as a ribosome-associated protein, and its capability to unwind double-stranded RNA was described (Jones et al. 1996). The major cold shock protein of *E. coli*, CspA (Goldstein et al. 1990), was reported to be an RNA-binding protein and qualified as an RNA chaperone (Jiang et al. 1997). None of the cold shock proteins, however, was shown to be the component responsible for the arrest of translation during the first stage of cold shock response. Discovery of a number of other cold shock proteins in bacteria (Jones and Inouye 1994; Graumann and Marahiel 1996) did not clarify the mechanisms of protein synthesis and cell growth halt either.

The observation that RaiA, a novel ribosome-associated stress protein, inhibits protein synthesis at the elongation stage of translation in cell-free systems is in agreement with the pattern of cold shock response, involving the decline of protein synthesis and the arrest of bacterial growth (Das and Goldstein 1968; Friedman et al. 1969). This also correlates well with the growth inhibition observed at the stationary phase of cell culture. The protein appears in ribosomes at the early stage of the cold shock response and thus can mediate the overall adaptation to lower temperature. Moreover, the protein is present in the ribosomal fraction only when cell growth is arrested, either during cold shock response or at the stationary phase of cell culture. When growth is resumed, the protein disappears from ribosomes. It should be pointed out that RaiA, being a naturally occurring inhibitor of translation, causes no damage to the ribosome and manifests a reversible mode of translational inhibition. It seems plausible that the main function of the protein is to arrest translation in response to an environmental stress.

The reversible inhibition of translation in response to excessive cell density (at the stationary phase of cell growth) was reported also for another protein, the so-called ribosome modulation factor of 6.5 kD (Wada et al. 1990, 1995). The protein was shown to cause dimerization of ribosomes with the formation of 100S particles and, like RaiA, put down the binding of aminoacyl tRNA with ribosomes.

The results reported here give evidence of the direct functional interplay between the ribosome and the protein involved in the stress response. Moreover, the mechanism of RaiA action via the blockade of the ribosomal A site corroborates the assumption that the occupation of the A site triggers cold and heat response in bacteria (VanBogelen and Neidhardt 1990).

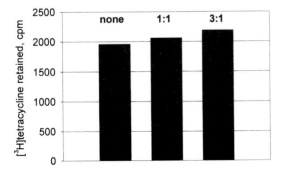

Figure 8. Binding of [^{3}H]tetracycline to 70S ribosomes in the presence of RaiA. Molar ratios of the protein to ribosomes are indicated above bars.

As mentioned above, genes homologous to *yfia (raiA)* of *E. coli* can be found in various species of bacteria. The wide occurrence of *RaiA* homologs suggests that the protein may be generally important for survival under stress conditions. In any case, the discovered protein seems to represent a direct link between ribosomes and the adaptation to environmental stress.

ACKNOWLEDGMENT

This work was supported by grants from the Russian Foundation for Basic Research.

REFERENCES

Agafonov D.E., Kolb V.A., and Spirin A.S. 1997. Proteins on ribosome surface: Measurements of protein exposure by hot tritium bombardment technique. *Proc. Natl. Acad. Sci.* **94:** 12892.

———. 2001. Ribosome-associated protein that inhibits translation at the aminoacyl-tRNA binding stage. *EMBO Rep.* **2:** 399.

Agafonov D.E., Kolb V.A., Nazimov I.V., and Spirin A.S. 1999. A protein residing at the subunit interface of the bacterial ribosome. *Proc. Natl. Acad. Sci.* **96:** 12345.

Cate J.H., Yusupov M.M., Yusupova G.Z., Earnest T.N., and Noller H.F. 1999. X-ray crystal structures of 70S ribosome functional complexes. *Science* **285:** 2095.

Dammel C.S. and Noller H.F. 1995. Suppression of a cold-sensitive mutation in 16S rRNA by overexpression of a novel ribosome-binding factor, RbfA. *Genes Dev.* **9:** 626.

Das H.K. and Goldstein A. 1968. Limited capacity for protein synthesis at zero degrees centigrade in *Escherichia coli. J. Mol. Biol.* **31:** 209.

Friedman H., Lu. P., and Rich A. 1969. Ribosomal subunits produced by cold sensitive initiation of protein synthesis. *Nature* **223:** 909.

Gavrilova L.P., Kostiashkina O.E., Koteliansky V.E., Rutkevitch N.M., and Spirin A.S. 1976. Factor-free ("non-enzymic") and factor-dependent systems of translation of polyuridylic acid by *Escherichia coli* ribosomes. *J. Mol. Biol.* **101:** 537.

Glukhova M.A., Belitsina N.V., and Spirin A.S. 1975. A study of codon-dependent binding of aminoacyl-tRNA with the ribosomal 30S subparticle of *Escherichia coli. Eur. J. Biochem.* **52:** 197.

Goldanskii V.I., Kashirin I.A., Shishkov A.V., Baratova L.A., and Grebenshchikov N.I. 1988. The use of thermally activated tritium atoms for structural-biological investigations: The topography of the TMV protein-accessible surface of the virus. *J. Mol. Biol.* **201:** 567.

Goldstein J., Pollitt N.S., and Inouye M. 1990. Major cold shock protein of *Escherichia coli. Proc. Natl. Acad. Sci.* **87:** 283.

Graumann P. and Marahiel M.A. 1996. Some like it cold: Response of microorganisms to cold shock. *Arch. Microbiol.* **166:** 293.

Hudson G.S. and Davidson B.E. 1984. Nucleotide sequence and transcription of the phenylalanine and tyrosine operons of *Escherichia coli* K12. *J. Mol. Biol.* **180:** 1023.

Jiang W., Hou Y., and Inouye M. 1997. CspA, the major cold-shock protein of *Escherichia coli*, is an RNA chaperone. *J. Biol. Chem.* **272:** 196.

Johnson C.H., Kruft V., and Subramanian A.R. 1990. Identification of a plastid-specific ribosomal protein in the 30 S subunit of chloroplast ribosomes and isolation of the cDNA clone encoding its cytoplasmic precursor. *J. Biol. Chem.* **265:** 12790.

Jones P.G. and Inouye M. 1994. The cold-shock response—A hot topic. *Mol. Microbiol.* **11:** 811.

———. 1996. RbfA, a 30S ribosomal binding factor, is a cold-shock protein whose absence triggers the cold-shock response. *Mol. Microbiol.* **21:** 1207.

Jones P.G., Mitta M., Kim Y., Jiang W., and Inouye M. 1996. Cold shock induces a major ribosomal-associated protein that unwinds double-stranded RNA in *Escherichia coli. Proc. Natl. Acad. Sci.* **93:** 76.

Link A.J., Robison K., and Church G.M. 1997. Comparing the predicted and observed properties of proteins encoded in the genome of *Escherichia coli* K-12. *Electrophoresis* **18:** 1259.

Shishkov A.V., Filatov E.S., Simonov E.F., Unukovich M.S., Goldanskii V.I., and Nesmeyanov A.N. 1976. Production of tritium-labeled biologically active compounds. *Dokl. Akad. Nauk SSSR* **228:** 1237.

Spirin A.S., Kostiashkina O.E., and Jonák J. 1976. Contribution of the elongation factors to resistance of ribosomes against inhibitors: Comparison of the inhibitor effects on the factor-dependent and factor-free translation systems. *J. Mol. Biol.* **101:** 553.

Tan X., Varughese M., and Widger W.R. 1994. A light-repressed transcript found in *Synechococcus* PCC 7002 is similar to a chloroplast-specific small subunit ribosomal protein and to a transcription modulator protein associated with sigma 54. *J. Biol. Chem.* **269:** 20905.

VanBogelen R.A. and Neidhardt F.C. 1990. Ribosomes as sensors of heat and cold shock in *Escherichia coli. Proc. Natl. Acad. Sci.* **87:** 5589.

Wada A., Yamazaki Y., Fujita N., and Ishihama A. 1990. Structure and probable genetic location of a "ribosome modulation factor" associated with 100S ribosomes in stationary-phase *Escherichia coli* cells. *Proc. Natl. Acad. Sci.* **87:** 2657.

Wada A., Igarashi K., Yoshimura S., Aimoto S., and Ishihama A. 1995. Ribosome modulation factor: Stationary growth phase-specific inhibitor of ribosome functions from *Escherichia coli. Biochem. Biophys. Res. Commun.* **214:** 410.

Yusupov M.M. and Spirin A.S. 1986. Are there proteins between the ribosomal subunits? Hot tritium bombardment experiments. *FEBS Lett.* **197:** 229.

———. 1988. Hot tritium bombardment technique for ribosome surface topography. *Methods Enzymol.* **164:** 426.

Yusupov M.M., Yusupova G.Z., Baucom A., Lieberman K., Earnest T.N., Cate J.H., and Noller H.F. 2001. Crystal structure of the ribosome at 5.5 Å resolution. *Science* **292:** 883.

The Fourth Step of Protein Synthesis: Disassembly of the Posttermination Complex Is Catalyzed by Elongation Factor G and Ribosome Recycling Factor, a Near-perfect Mimic of tRNA

A. KAJI,* M.C. KIEL,* G. HIROKAWA,*† A.R. MUTO,* Y. INOKUCHI,‡ AND H. KAJI§

*Microbiology Department, Medical School, University of Pennsylvania, Philadelphia, Pennsylvania 19104;
†Faculty of Pharmaceutical Sciences, Chiba University, Chiba, Japan; ‡Department of Bioscience, Teikyo
University, Utsunomiya, Japan; §Department of Biochemistry and Molecular Pharmacology, Jefferson
Medical College, Thomas Jefferson University, Philadelphia, Pennsylvania 19107

CHARACTERISTICS OF THE DISASSEMBLY OF THE POSTTERMINATION COMPLEX BY RRF

History and Basic Characteristics of RRF

The ribosome recycling factor (RRF) has been slowly recognized as an essential factor for prokaryotic protein synthesis, but it is one of the least-known soluble factors involved in bacterial protein synthesis. Therefore, in this section, we briefly describe the historical background and summarize the general information about this factor. For details, readers are referred to recent reviews written on this subject (Janosi et al. 1996b; Kaji et al. 1998; Kaji and Hirokawa 2000).

Polypeptide synthesis is generally known as consisting of three steps: initiation, elongation, and termination (Gualerzi et al. 2000; Nierhaus et al. 2000; Rodnina et al. 2000; Wilson et al. 2000). Much less known is the fourth step: disassembly of the posttermination ribosomal complex and recycling of the machinery necessary for the next round of translation. There are three reasons why this step has been overlooked for a long time. First, in the pioneering work on the third step of protein synthesis, the Nirenberg group (Scolnick et al. 1968) and others used two triplets, AUG and UAA or other termination codons, to bind formylmethionyl (fMet) tRNA and release factors (RF-1 or RF-2) to the ribosome. After hydrolysis of the bound fMet-tRNA by RF-1 or RF-2 in the presence of RF-3, these triplets dissociated from the ribosomes, giving an impression that the disassembly may take place without any enzymatic help. Second, in most of the early work elucidating the basic steps of protein synthesis, researchers used a large excess of Mg^{++} ions, ribosomes, and artificial synthetic polynucleotides such as polyuridylic acid (polyU) or polyadenylic acid (polyA) (for review, see Nirenberg and Leder 1964). Under these experimental conditions, one does not or cannot observe the fourth step, disassembly of the posttermination complex. Third, as discussed later, in vivo studies created a false impression that *all* ribosomes or their subunits may stay on the termination codon despite the fact that the au-

thors of these papers indicated that only a fraction of the ribosomes would stay on the mRNA. Complete disassembly was thought to occur spontaneously during ribosome wandering on the mRNA after the release of the peptidyl group from the last tRNA (Sarabhai and Brenner 1967; Adhin and van Duin 1990). Thus, the fourth step of protein synthesis remained unrecognized until the factor that catalyzes this step was shown to be essential for bacterial life (Janosi et al. 1994). The importance of this step was further realized by the fact that the genome project revealed that all living matter, except for archaebacteria, possesses the gene encoding RRF (*frr*). *Escherichia coli* RRF consists of 185 amino acids, and most other bacterial RRFs have significant sequence and size homology with *E. coli* RRF.

Although there are a number of ways to examine the RRF reaction, most of the early and present work on this step in our laboratory used a model system, puromycin-treated polysome, as substrate. Each ribosome on the polysome has two deacylated tRNAs at the P and E sites (Remme et al. 1989; Stark et al. 1997) and mRNA bound to it. This configuration is assumed to be very similar to that of the natural posttermination complex, except that the A site is not occupied with the termination codon. Treatment of this polysome with RRF and elongation factor G (EF-G) resulted in conversion of the polysome to monosomes. The time course of this conversion is shown in Figure 1. As seen from this figure, the kinetics of this reaction (40% complete in 5 minutes) is on the same order with that of the recently developed system which measures release of tRNA from the posttermination complex with synthetic oligonucleotides (70% complete in 2 minutes; see Fig. 1 of Karimi et al. [1999]). This system described in Figure 1 was used to identify and purify RRF (Hirashima and Kaji 1972b). A cloned gene encoding RRF (*frr*) (Ichikawa and Kaji 1989) was confirmed by checking the product with this system. Most of the findings obtained with this model system have been confirmed with a more natural system using a R17 RNA coat cistron having an amber codon at the seventh triplet (Ogawa and Kaji 1975), as well as with a simple synthetic

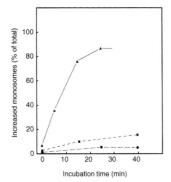

Figure 1. Time course of the conversion of puromycin-treated polysomes into single ribosomes by RRF and EF-G. (*Closed triangle*) The reaction mixture contained both RRF and EF-G; (*closed square*) the reaction mixture contained only puromycin-treated polysomes; (*closed circle*) the reaction mixture contained only polysomes not treated with puromycin. The amount of increased single ribosomes, expressed as the percentage of total ribosomes, was plotted. (Reprinted, with permission, from Hirashima and Kaji 1972a.)

oligonucleotide, such as UUCAUGUAA (Grentzmann et al. 1998).

As described above, disassembly of the posttermination complex depends not only on RRF, but also on EF-G. Figure 2 demonstrates that an optimum condition is met when these factors are present in equal molar quantities. In this experiment, the RRF reaction was measured in the presence of a constant amount of RRF and various amounts of EF-G. When the molar ratio of these two proteins reached 1 to 1, the extent of the reaction reached maximum. This finding is important as discussed in the section dealing with the possible mechanism of the RRF reaction.

Serious Consequences of Depletion of RRF

RRF was originally discovered using in vitro conditions. Therefore, it was very important to establish the role of this factor in vivo before more work in vitro was

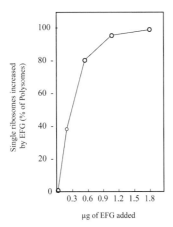

Figure 2. Optimal release of ribosomes from mRNA occurs when the molar ratio of RRF/EF-G equals 1. The release of ribosomes from polysomes was measured with 0.18 μg of purified RRF and the indicated amounts of purified EF-G. (Reprinted, with permission, from Hirashima and Kaji 1972b [copyright American Chemical Society].)

performed. In agreement with the essential nature of this factor for bacterial life, 12 different temperature-sensitive mutants of RRF have been isolated (Janosi et al. 1998). Using one of the mutants, LJ14 (Val117Asp), it was found that in vivo inactivation of RRF resulted in a bactericidal effect during the lag phase, whereas the in vivo inactivation was bacteriostatic during growing phase.

Early in vitro studies with the coat cistron of R17 phage with an amber codon at the seventh triplet demonstrated that coat protein beginning at the eighth amino-terminal amino acid (phenylalanine) was synthesized in the absence of RRF. In this system, ~60% of the ribosome, which completed the synthesis of the short amino-terminal portion of the coat cistron, reinitiated the translation from the codon next to the termination codon (Ryoji et al. 1981). We call this reinitiation due to the lack of RRF "unscheduled downstream translation." With the use of LJ14 described above, we demonstrated that unscheduled downstream translation indeed takes place in vivo. With a reporter gene, β-galactosidase, placed downstream from a short open reading frame (ORF), in vivo inactivation of RRF induced the expression of the reporter gene. An interesting observation is that the unscheduled downstream translation occurs regardless of whether the reporter gene is in-frame with the upstream ORF or not. This means that while the ribosome stays on the posttermination complex due to the inactivation of RRF, it can undergo frameshifting. In the experiment shown in Figure 3, we demonstrate that this frameshifting is temperature-dependent. As shown in the figure, the out-of-frame reporter gene translation took place best at 39°C, whereas

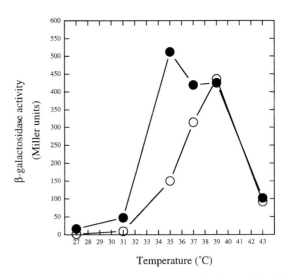

Figure 3. Evidence for temperature-dependent frameshift of ribosome at the posttermination complex in the absence of RRF. *E. coli* LJ14 (with temperature-sensitive RRF) harboring either pPEN2363 (in-frame β-galactosidase reporter placed downstream from short upstream ORF) (*closed circle*) or pPEN2369 (out-of-frame β-galactosidase reporter) (*open circle*) was grown at 31°C. The cultures were shifted to the indicated temperatures and grown for another 4 hr. The induced β-galactosidase activity was then measured and plotted against the temperature. (Reprinted, with permission, from Janosi et al. 1998.)

Table 1. In the Absence of RRF, Ribosomes Reinitiate Randomly without Any Initiation Signal Downstream of the Termination Codon

Reporter vector	Partial nucleotide sequence	Detected amino terminal sequence	Distance of reinitiation point from termination codon (nt)

pPEN2363 in-frame *lacZ'-'lacZ*	5´...**AGGCTCTA**GCTAGAGGATCCGTCG
	         L E D P S
	                 +
	..... ACCTGCAGCCAAGCTTGCGATCCCGTC...3´
	  T C S Q A C D P V
	  + + +
		DPSTCS
		TCSQAC
		SQACDP
			7
			16
			22

pPEN2369 out-of-frame *lacZ'-'lacZ*	5´...**AGGCTCTAG**CTAGAGGATCCGTCGA
	         R I R R
	CCTGCAGCCAAGCTTGCTCCCGTC...3´
	  P A A K L A P V
	  + + +
		PAAKLA
		AAKLAP
		AKLAPV
			17
			20
			23

^a The junction region between the promoter proximal open reading frame (bold typeface characters) and the fused nucleotide sequence (plain typeface characters) is shown. Beneath the nucleotide sequence, the translation of codons which are in-frame with the *lacZ*-coding sequence is also shown (in one-letter codes for the amino acids). + underneath the amino acid indicates a translational reinitiation site. (Adapted, with permission, from Janosi et al. 1998.)

in-frame downstream translation took place best at 34°C. We interpret this to indicate that the ribosome staying at the posttermination complex is subject to thermal agitation that promotes frameshifting. This is perhaps the first example of frameshifting dependent on temperature.

As shown in Table 1, the amino-terminal analysis of the products of unscheduled downstream translation in the absence of RRF revealed that the initiation occurs randomly without any need of the canonical initiation signal. It is clear from this table that the initiation of unscheduled downstream translation in vivo takes place not always from the next codon to the termination triplet, but from points significantly far away from the termination codon. This is in contrast to the in vitro system where the unscheduled initiation starts mostly from the triplet next to the termination codon (Ryoji et al. 1981). It should be emphasized that we never observed initiation upstream from the termination codon. In other words, there was no evidence of ribosome scanning in this system. In a separate experiment, upon inactivation of RRF, we found that almost 100% of the ribosomes that completed the upstream translation reinitiated translation downstream from the stop codon (see Fig. 7 and Table 4 in Janosi et al. 1998).

The random reinitiation of unscheduled downstream translation in the absence of RRF described above has a serious consequence in translationally coupled systems such as the coat protein gene and the lysis gene of RNA phage GA. In this system, these two cistrons are connected by UAAUG, indicating that the one nucleotide (A) is used for both the termination and the initiation signal. For active lysis protein synthesis, the ribosome must translate precisely from the AUG initiation codon because the intact amino terminus is required for the activity of the lysis protein. In this system, we demonstrated that inactivation of RRF in vivo produced a heterogeneous downstream translation product with various truncated amino termini (Inokuchi et al. 2000). This means that for the synthesis of active lysis protein, normal RRF

activity is essential. This system is described in more detail later in this chapter, where mechanism of the RRF action is discussed.

Another important but not very well recognized role of RRF is its role in maintaining translational fidelity during chain elongation. It was known that occasional incorporation of amino acids other than phenylalanine took place in the in vitro polypeptide synthesis system programmed by poly(U). This incorporation, although at low levels, was regarded as a sign of translational error. This translational error increased upon addition of streptomycin (Davies and Davis 1968). The miscoding effect of streptomycin was demonstrated to be at the step of binding of aminoacyl tRNA to the complex of mRNA and ribosomes (Kaji and Kaji 1965). When RRF was removed from this system, a significant increase (three- to tenfold) of misincorporation of leucine, isoleucine, or mixture of amino acids other than phenylalanine was observed. Interestingly, RRF did not influence the miscoding effect of streptomycin (Janosi et al. 1996a).

To examine the role of RRF in maintaining translational fidelity in vivo, we examined β-galactosidase synthesized in the presence of faulty RRF in LJ14. When LJ14 with temperature-sensitive RRF was kept at the semipermissive temperature, a low but significant amount of protein was synthesized. Under these conditions, β-galactosidase was synthesized, isolated, and purified. We then compared the heat stability of this β-galactosidase with that of the enzyme synthesized under the conditions where normal RRF is present. It was found that the β-galactosidase synthesized under faulty RRF (synthesized at semipermissive temperature) was much more heat-labile (A. Kaji et al., unpubl.). It has been accepted that this is an indication of in vivo translational error (Santos et al. 1996).Therefore, our observation discussed above is consistent with the notion that RRF maintains the fidelity of translation in vivo, and inactivation of RRF results in an unstable protein product due to translational error introduced by nonfunctional RRF.

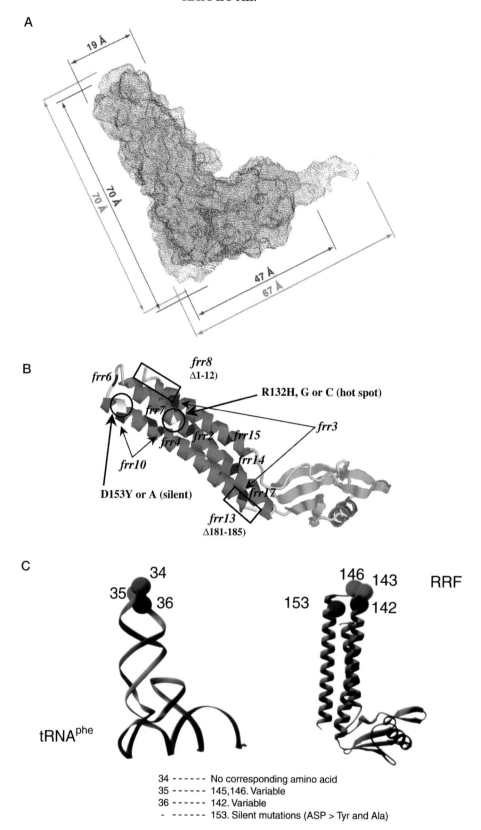

Figure 4. (*A*) Grasp surface representation of the superposition of RRF (*blue*) and yeast tRNA^Phe (*red*). (Reprinted, with permission, from Selmer et al. 1999.) (*B*) Molecular model of the *E. coli* RRF structure (Kim et al. 2000) (PDB ID number 1EK8) with indicated temperature-sensitive mutation spots and hot spot R132 (Janosi et al. 2000) as described in the text. (*C*) Ribbon structure of tRNA (*left*) and RRF (*right*) demonstrating that nonconserved and silent mutations of RRF correspond to the anticodon region of tRNA. Anticodon nucleotide 34 has no corresponding amino acid in RRF. Silent mutation position 153 is shown near the anticodon region.

STRUCTURAL BIOLOGY OF RRF AND ITS POSSIBLE IMPLICATIONS TO THE ACTION OF RRF

In 1999, in collaboration with the Anders Liljas group of Lund University, Sweden, we were able to determine the crystal structure of *Thermotoga maritima* RRF at 2.55 Å resolution by multiwavelength anomalous dispersion (MAD) as shown in Figure 4A (blue structure) (Selmer et al. 1999). Since there are altogether six helices in this RRF, they were named helices 1 to 6. RRF consists of two domains. Domain 1 is longer than domain 2 and consists of a three-helix bundle containing residues 2–30 (H1), 105–145 (H5), and 150–183 (H6). They are tightly packed, suggesting that the relative thermostability of RRF (Hirashima and Kaji 1972a) may be due to this structure. Domain 2 is a three-layer β/α/β sandwich, and it consists of residues 31–104. A two-stranded antiparallel β ribbon and a four-stranded antiparallel β sheet are placed on the top and bottom of three helices H2, H3 (short 3_{10} helices), and H4. One important feature revealed from this structure is that the interaction of these two domains is unusually loose. Thus, the water-accessible surface area lost in the domain–domain interaction is only 8.2%. This is lower than normal. This point becomes important when we discuss the mode of action of RRF in light of its solution structure.

In Figure 4A, the crystal structure of *T. maritima* RRF (blue) is superimposed with that of tRNA (red). As is clear from the figure, RRF is a near-perfect mimic of tRNA. The only part that differs from tRNA is the missing CCA end and the slightly indented anticodon region. Shortly after we solved the crystal structure of *T. maritima* RRF, the structure of *E. coli* RRF with detergent (decyl-β-D-maltopyranoside) was solved (Kim et al. 2000), confirming our overall structure of *T. maritima* RRF. The notable difference is the relative position of domains 1 and 2 because the detergent was inserted at the hinge region between these two domains. In addition, the total number of helices was four in *E. coli* RRF in contrast to six in *T. maritima* RRF. Therefore, domain I of *E. coli* RRF consists of helices 1, 3, and 4.

Prior to the structural information described above, we obtained 52 null, 6 reversion, and 5 silent mutations of *E. coli frr* (Janosi et al. 2000). In addition, as described in the preceding section, we obtained 12 different temperature-sensitive alleles of RRF (Janosi et al. 1998). Since the mutated RRF was that of *E. coli*, we discuss here the possible significance of mutations on the structure and function of RRF using *E. coli* RRF structure. It was first noted that all of our 12 temperature-sensitive mutations are located in domain 1 (see Fig. 4B, all but *frr16* are shown). From the three-helix structure, domain 1 appears to be a relatively stable structure, and mutations in this region give temperature instability perhaps by disturbing this stable structure. The stability of domain 1 must depend on the length of the helices because carboxy-terminal truncation resulting in shortening of helix 4 resulted in temperature-sensitive RRF as can be seen from *frr13* whose stop codon is at position 181 (Δ181–185). This no-

tion was further supported by the fact that slightly longer truncations result in a complete loss of activity (for example, *frr142* [Δ172–185]).

Null mutations can be classified in two categories: those that are distorting the structure of the protein and those that are at the active site and probably interfering with the binding to the substrate, ribosome. Examples of the former are Leu163Pro (*frr4*), Leu175Pro (*frr124 and frr149*), and Leu182Pro (*frr17*). Due to the change to proline, the direction of helix 4 must be dramatically changed to destroy the entire structure of domain 1, resulting in null mutations in some cases. It should be pointed out that Leu182Pro and Δ181–185 are temperature-sensitive, indicating that some of the drastic changes due to proline still maintain the function at the permissive temperature. Since the carboxy-terminal region is close to the hinge region of domains 1 and 2, it is possible that some of these null mutations are influencing the hinge region, resulting in a nonflexible (therefore, nonfunctional) RRF. In this connection, the double mutation of *frr116* (Leu36Gln and Thr106Lys) may interfere with the relative movement of domains 1 and 2 because they are located around both loops of the hinge region.

The group of mutations that appears to belong to the second category is represented by the hot-spot mutation at Arg-132 as exemplified by Arg132Gly (*frr114*), Arg132Cys (*frr132 and frr141*), and Arg132His (*frr133*), which are all lethal. Arg-132 is a highly conserved residue in all prokaryotic RRF so far sequenced, and no alteration of computer-predicted structure was observed by Arg132His substitution. Other possible examples of mutations belonging to this category are Arg110His (*frr106*) and Arg129Cys (*frr119*). These arginines are well-conserved in most eubacterial RRF. These and other conserved basic amino acid residues are located on the same side of helices 3 and 4, suggesting that this region may seek the negatively charged phosphate backbone of ribosomal RNA (Fig. 4B).

Silent mutations are often as informative as null mutations. Three revertants of the temperature-sensitive mutant LJ14 (*frr14* and Val117Asp) were found to have a silent mutation at position 153 (Asp153Tyr and Asp153Ala). The Asp-153 is located on helix 3 close to the tip of the whole molecule (Fig. 4B). Drastic changes such as hydrophilic to aromatic amino acid at position 153 near the anticodon region of tRNA did not influence the function of RRF. As shown in Figure 4C, no corresponding amino acid to the anticodon nucleotide position 34 exists. Furthermore, the amino acid corresponding to anticodon nucleotides 35 and 36 (positions 142, 145, and 146) are not well conserved. It therefore appears that the anticodon region of RRF does not have an important functional significance.

It should be borne in mind that the influence of each mutation on the structure and function of protein as discussed above has serious drawbacks because of the lack of structural data on each mutated protein. We hope to examine our above hypothesis by determining the crystal structure of each of the key mutant proteins in the near future. Perhaps solving the crystal structure of ribosome-

A

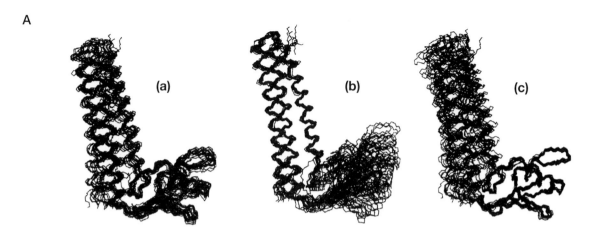

B

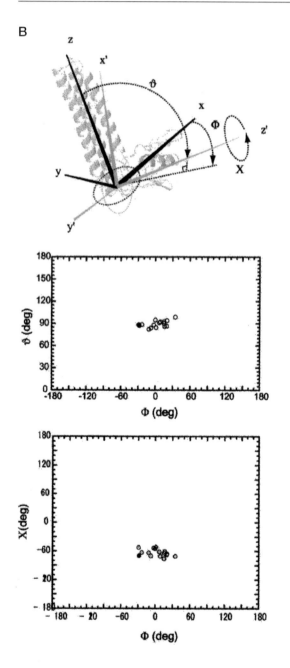

C

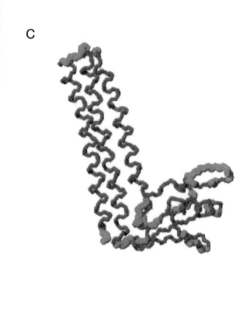

Figure 5. (*A*) Best-fit superpositions of the backbone atoms of whole molecules (*a*), domain 1 (*b*), and domain 2 (*c*) of the 15 NMR-derived structures of *A. aeolicus* RRF. (*B*) Distributions of interdomain angles for the ensemble of the 15 NMR-derived structures of *A. aeolicus* RRF (*open circles*), and for the X-ray structure of *T. maritima* RRF (*closed circle*). The interdomain angles are represented by the set of three spherical polar angles. The definitions for the angles are shown schematically (*top*). (*C*) Rapid internal motion on the subnanosecond time scale for the backbone of *A. aeolicus* RRF. The trace is colored in red where the value of ^{15}N-{^{1}H} is smaller than 0.65. (Reprinted, with permission, from Yoshida et al. 2001 [copyright American Chemical Society].)

bound RRF would tell whether what is described above is true or not.

Since crystal structures of molecules are subject to the physical stress of crystal lattice formation, this force may influence relative positions of loosely connected portions such as domains 1 and 2. We therefore determined the solution structure of *Aquifex aeolicus* RRF by heteronuclear multidimensional nuclear magnetic resonance (NMR) spectroscopy (Yoshida et al. 2001). The structure turned out to be almost identical to what we first reported with *T. maritima* RRF crystallography, confirming the near-perfect mimicry of tRNA in solution; 15 structures were calculated and were superimposed as shown in Figure 5. The superpositions of each domain are very good, indicating the reliability of the structure of each domain. On the other hand, the relative positions of each of the 15 structures varied extensively. It should be emphasized that the relative positions of domains 1 and 2 were determined by methods that have been recently developed for defining the long-range order in NMR structure determination (Tjandra and Bax 1997; Tjandra et al. 1997). These approaches utilize the information from the relaxation time dependence on rotational diffusion anisotropy or the residual dipolar coupling of weakly aligned molecules. Therefore, our solution structure was determined not only with the usual NMR method that relies on short-range distance restraints, but also with these new methods that determine the relative positions of two domains.

The differences between the structure of *A. aeolicus* RRF determined by NMR and the crystal structure of *T. maritima* RRF are found in the rotational direction of domain 2 around the long axis of domain 1 (Φ of Fig. 5B). Figure 5B indicates the quantitative relationship of the possible domain movement. It is important to note that the angle Φ representing the horizontal movement of domain 1 relative to domain 2 was as large as 60°. This is consistent with the relatively loose connection between these domains suggested by small contact surface area of the crystal structure of *T. maritima* RRF. In addition, measurement of $^{15}N-\{^1H\}$ NOE values shows that the residues in the corner of the L-shaped molecule are undergoing fast internal motion as shown in Figure 5C. Although rapid motions of the ends of molecules are expected, a rapid motion of the middle of the molecule corresponding to the hinge region of the molecule suggests an intramolecular movement of RRF. The bending angle of the joint between the two domains υ of RRF in solution was 89.7°, confirming the crystal structure of *T. maritima* RRF. The variation of υ in 15 structures was very small; the rotating motion expressed as X around the axis of domain 2 is also very small. These observations suggest that the molecule maintains the tRNA shape and dimension in solution but that domain 1 may rotate as much as 60° relative to domain 2.

On the basis of this structure, we propose that the horizontal motion represented by Φ is important for the function of RRF. In support of this notion are three observations. First, the *E. coli* RRF crystal was obtained only in the presence of the detergent that is inserted in the hinge region. It appears that the detergent fixed the relative position of two domains at the position indicated in Figure 4B, making it easier to crystallize (Kim et al. 2000). Second, we were able to crystallize *T. maritima* RRF. *T. maritima* is a thermophilic bacterium. RRF from bacteria growing at high temperature must be designed to give appropriate flexibility of Φ at high (90°C) temperature. Therefore, *T. maritima* RRF must be "frozen" at room temperature in the angle of Φ as indicated in Figure 5. This would make it easier to crystallize because of the homogeneous population of this RRF at room temperature. Third, the "frozen" nature of *T. maritima* RRF at 37°C would make it difficult to function at this temperature. This is exactly what we observed, as shown in Table 2 (Atarashi and Kaji 2000). *T. maritima* RRF did not function in *E. coli*, and it was a competitive inhibitor of the *E. coli* reaction. This is consistent with the notion that "frozen RRF" can still bind to the ribosome because of its tRNA mimicry. However, it cannot function because the frozen RRF lacks the flexibility of the hinge region. In support of this notion, expression of *T. maritima* RRF is bactericidal (Atarashi and Kaji 2000). As described below, the intramolecular movement of these two domains probably occurs prior to or simultaneous with the release of RRF from the ribosome (step IV, as described in the next section).

MODE OF ACTION OF RRF ON THE POSTTERMINATION COMPLEX VARIES DEPENDING ON THE mRNA SEQUENCE SURROUNDING THE TERMINATION CODON

Figure 6A describes our hypothesis on how RRF and EF-G catalyze recycling of a typical posttermination complex. The "typical posttermination complex" should satisfy the following conditions. The termination codon should not be close to (1) the initiation codon of the downstream cistron, (2) a strong Shine-Dalgarno (SD) sequence either by nucleotide distance or by hairpin formation, or (3) polyadenylic or polyuridylic sequences that strengthen the affinity of the mRNA to the ribosome. Although the exact configuration of the posttermination complex remains elusive, we believe that the P and E sites are occupied with deacylated tRNA in a fashion similar to that of the polysome we use for routine assay of RRF. As long as these conditions are satisfied, RRF should disas-

Table 2. Inhibition of the *E. coli* RRF Reaction by *T. maritima* RRF

Amount (μg) of RRF		% Conversion of polysome to monosome	% Inhibition
E. coli	*T. maritima*		
5	0	19.8	0.0
5	2.5	19.3	2.6
5	5	18.3	7.2
5	50	6.9	65.3
0	50	-8.8	n.a.[a]

The *E. coli* RRF reaction was followed by the conversion of polysome to monosome. (Adapted from Atarashi and Kaji 2000.)

[a]n.a. indicates not applicable.

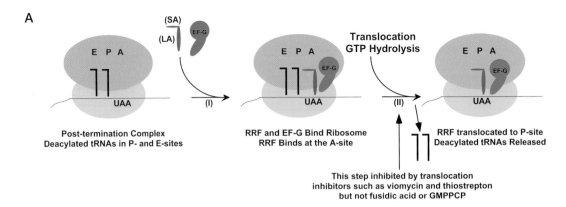

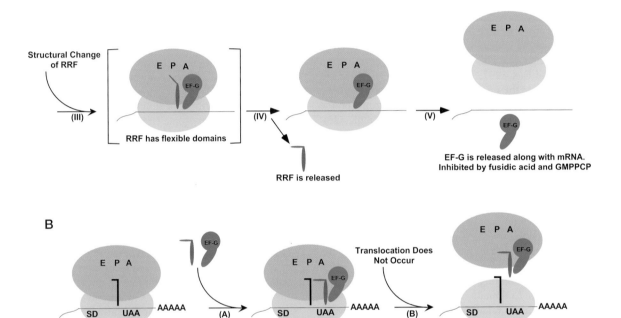

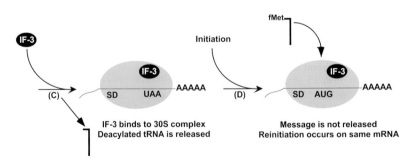

Figure 6. Mechanisms for the action of RRF and EF-G. (*A*) Our model consists of five steps, I–V. In step I, RRF (*orange*) binds to the A site of the ribosome, perhaps simultaneously with the binding of EF-G (*green*). In step II, EF-G translocates RRF from the A site to the P site. In step III, RRF changes its configuration and is subsequently released, in the presence of EF-G (step IV). Final step V is the release of EF-G and mRNA from the ribosome. (*B*) An alternative model as proposed by Karimi et al. (1999). In step A, RRF and EF-G bind the ribosome. In step B, EF-G and RRF do not translocate but rather split the 70S ribosome into subunits, leaving message and tRNA bound to the 30S subunit while RRF and EF-G remain with the 50S subunit. In step C, IF-3 removes the tRNA from the 30S-mRNA complex. Reinitiation of protein synthesis then occurs on the same 30S-mRNA complex, as depicted by step D.

semble the posttermination complex in conjunction with the action of EF-G as shown in Figure 6A. We define "disassemble" to mean the release of mRNA and tRNA from the ribosome.

The disassembly of posttermination complex consists of five steps, and we describe the evidence available to suggest each step. The first step (I) is the binding of RRF to the A site. RRF is shown as an orange, L-shaped figure. The long arm (LA) and short arm (SA) represent domains 1 and 2, respectively. Since domain 2 is similar to the portion represented by the CCA end of tRNA, it is shown to be in the 50S subunit. As indicated in Figure 6A, the binding of RRF to the A site may take place together with or independent of binding of EF-G to the factor-binding site (Wilson and Noller 1998). The first experimental evidence for the A-site binding of RRF is that the RRF reaction is inhibited by tetracycline, although a high concentration was required (Hirashima and Kaji 1973). Competition of RRF with RF-1, as reported by the Ehrenberg-Buckingham group (Pavlov et al. 1997a), is consistent with this concept. With the well-washed ribosome, under the conditions where N-Ac-Phe-tRNA binds both A and P sites, N-Ac-Phe-tRNA competes with binding of RRF to the ribosome, whereas under conditions where N-Ac-Phe-tRNA binds only to the P site, only 50% of RRF is competed (G. Hirokawa et al., unpubl.). This suggests that under these conditions, RRF binds to both A and P sites of naked nonprogrammed ribosome.

The next step (II) is the translocation of RRF from the A site to the P site, probably simultaneously with the release of two deacylated tRNAs (Hirashima and Kaji 1973) from the P and E sites. This step should be sensitive to translocation inhibitors such as viomycin and thiostrepton that inhibit even one round of translocation (Rodnina et al. 1997, 1999). On the other hand, translocation inhibitors such as fusidic acid and nonhydrolyzable GTP analogs (e.g., GMPPCP) that allow for one round of translocation (Rodnina et al. 1997) should not inhibit this step. Our experimental evidence indeed shows that the release of tRNA from the model posttermination complex was observed exactly as expected in the presence of the translocation inhibitors described above (G. Hirokawa et al., unpubl.). This step leaves RRF on the P site of the partially disassembled posttermination complex, as shown in Figure 6. This intermediate has empty A and E sites. The steps following the translocation depicted in this figure may take place simultaneously. For illustration purposes, they are shown as three separate steps.

We propose that RRF undergoes "molecular gymnastics" on the ribosome as shown in step III of Figure 6A. In this figure, the angle between domains 1 and 2 of RRF is changed for illustration purpose only. Our proposal is that the angle Φ changes in this intramolecular movement keeping the angle υ as 90° as discussed in the preceding section (see Fig. 5) (Yoshida et al. 2001). This movement may be dependent on the specific interaction of RRF with EF-G. We propose that, due to this specific intramolecular movement of RRF on the P site, RRF is removed from the P site (step IV). This removal of RRF is dependent on the action of EF-G. We have been able to show that re-

moval of RRF from the well-washed ribosome is dependent on the addition of EF-G. This action of EF-G is different from the translocation activity of EF-G (step II) because none of the translocation inhibitors described above inhibited this part of the disassembly reaction (M. Kiel et al., unpubl.). This action of EF-G is perhaps related to the ribosomal configurational change caused by EF-G in the presence of GMPPCP (Frank and Agrawal 2000). We believe that the last step of the disassembly reaction is the release of the ribosome from mRNA as shown in Figure 6 (step V). This is supported by the observation that the translocation inhibitors that keep EF-G on the ribosome (such as fusidic acid) inhibit the release of ribosome from mRNA (Hirashima and Kaji 1973). It should be noted that steps III, IV, and V may occur simultaneously.

Figure 6A is consistent with the second role of RRF, as discussed in the introduction, keeping the translational fidelity during the chain elongation (Janosi et al. 1996a). For this to occur, we propose that peptidyl tRNA is released in step II if the tRNA of the peptidyl tRNA is noncognate or near-cognate. This should not happen often with cognate long peptidyl tRNA at the P site. If RRF releases cognate peptidyl tRNA often as suggested with short peptidyl tRNA (Heurgue-Hamard et al. 1998), this would be a complete waste of energy and would not survive phylogenetic and evolutionary selection. In support of this concept, RRF releases N-acetyl-Phe-tRNA bound to the nonprogrammed ribosome, whereas RRF does not release N-Ac-Phe-tRNA bound to the ribosome programmed with poly(U) (M. Kiel et al., unpubl.). In further support of the concept that RRF is responsible for the release of peptidyl tRNA, it has been shown that impaired expression of RRF reduced peptidyl tRNA drop-off rates in vivo (Heurgue-Hamard et al. 1998). It would be of interest to examine the amino acid sequence of the released peptidyl tRNA in this system to explore the possibility that at least some of them are noncognate or near-cognate peptidyl tRNA.

In Figure 6A, the released ribosome is pictured as a 70S ribosome because we observe the released ribosome as a 70S ribosome in our routine assay using a model posttermination complex (Hirashima and Kaji 1970, 1973). It is possible that under certain conditions, subunit dissociation may take place together with the disassembly as suggested by Kaempfer (1970, 1972). On the other hand, our experimental evidence is consistent with release of the 70S ribosome from mRNA (Kohler et al. 1968; Davis 1971; Subramanian and Davis 1971, 1973). We consider that this question is rather trivial because everyone agrees that dissociation of the 70S ribosome into subunits is prerequisite to the next round of translation in most cases. Whether or not this dissociation is simultaneous with the disassembly is not an important question. As pointed out previously, this would depend on various factors such as the sequence of mRNA around the termination codon, concentration of IF-3 that dissociates the 70S ribosome into the subunits, local magnesium ion concentration, and presence and absence of other factors, which influence subunit dissociation. For further discussion, see Janosi et

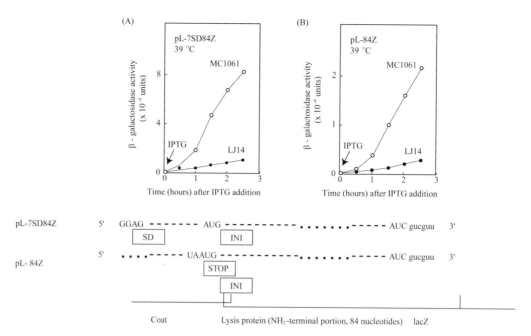

Figure 7. RRF does not release ribosomes from the posttermination complex formed at UAAUG. *E. coli* MC1061 (with the wild-type RRF, *open circle*) or LJ14 (with a temperature-sensitive RRF, *closed circle*) harboring various plasmids as indicated below were grown at 27°C. They were then transferred to 39°C (nonpermissive temperature for temperature-sensitive RRF) in the presence of IPTG for the induction of reporter gene transcription (β-galactosidase). (*A*) pL-7SD84Z, representing β-galactosidase synthesis with an added SD sequence without control by the coupling mechanism. (*B*) pL84Z, representing lysis gene synthesis through coupling. (Reprinted, with permission, from Inokuchi et al. 2000.)

al. (1996b). We discuss Figure 6B after the discussion of Figure 7.

Figure 6A applies only to the typical posttermination complex. The behavior of ribosomes in response to RRF drastically changes depending on the sequence surrounding the termination codon. We describe here one such example (Inokuchi et al. 2000). The coat and lysis proteins of RNA phage GA are translationally coupled, and the intercistronic nucleotide sequence is UAAUG. In this system, about 75% of the ribosomes completing the translation of the upstream open reading frame (ORF) (coat cistron) are released (Fig. 2 in Inokuchi et al. 2000). The remaining 25% are solely responsible for the translation of the lysis gene. To investigate the possible role of RRF in this system, two plasmid constructs, pL-7SD84Z and pL84Z, were made as shown in the lower panel of Figure 7. The former is a typical ORF coding for the β-galactosidase reporter gene. The latter construct is almost identical to what the GA phage has for the coat and the lysis genes except that the reporter gene β-galactosidase is attached downstream from the UAAUG. The amount of reporter gene expression from this construct represents lysis gene expression by those ribosomes that completed the upstream translation (hereafter, called upstream ribosomes). We reasoned that β-galactosidase expression by pL-7SD84Z represents translation of normal ORF, whereas that by pL84Z represents the translation of the reporter gene by the ribosomes surviving the sequence UAAUG.

If RRF were involved in the release of ribosomes from UAAUG, inactivation of RRF would increase the amount of expression of β-galactosidase because more ribosomes would stay on the mRNA and begin the downstream unscheduled translation, as has been shown previously (Janosi et al. 1998) (discussed in Fig. 3). Since RRF is an essential protein for protein synthesis, partial inactivation of RRF would result in a general decrease of β-galactosidase synthesis. However, the ratio of β-galactosidase expression by pL84Z to that by pL-7SD84Z is expected to go up upon inactivation of temperature-sensitive RRF if RRF is releasing ribosomes from UAAUG. This is because, as shown in Figure 3, ribosomes remaining on mRNA due to inactive RRF would translate the reporter gene, β-galactosidase, downstream from UAAUG. In Figure 7A, pL7SD84Z was expressed in wild type (open circle) and LJ14 with temperature-sensitive RRF (closed circle) at the nonpermissive temperature. In Figure 7B, the identical experiment was performed with pL84Z. It is noted that the ratio of the open circle and closed circle was identical in Figures 7 A and B. This means that the relative amount of the gene expression downstream from UAAUG did not increase upon inactivation of RRF. This suggests that RRF is not involved in the release of ribosomes from UAAUG despite the fact that 75% of the upstream ribosomes are released.

Although RRF does not appear to release ribosome from UAAUG, it has an important role to place the upstream ribosome in the correct reading frame starting from AUG of UAAUG (Fig. 4 in Inokuchi et al. 2000). We have not formulated the exact mechanism through which RRF makes the ribosome read correctly from the AUG of UAAUG without releasing the ribosome from the complex.

Another example of the behavior of the posttermination complex different from Figure 6A has been given by the Ehrenberg-Buckingham group in a system with a synthetic messenger, containing the sequence AUG UUU ACG AUU UAA coding for the peptide MFTI (Karimi et al. 1999). The mechanism proposed based on the data obtained from this system is shown in Figure 6B. Two important features of this synthetic mRNA are (1) it has a strong SD sequence 17 nucleotides upstream of UAA and (2) it has a long (21 nucleotides) poly(A) sequence 5 nucleotides downstream from the UAA triplet. These two sequences have strong affinity to ribosomes and are well within the range of the nucleotides, which interact with the ribosome of the posttermination complex (Steitz 1970). Since it is not known when the SD interaction is broken in the regular initiation process (Kozak 1999), the posttermination complex of this system can still be under the influence of this strong SD sequence. Indeed, it has been shown that the ribosome tends to stay on the mRNA under these conditions (see, e.g., Fig. 4 in Pavlov et al. 1997b; Fig. 4 in Dincbas et al. 1999). Furthermore, in Table 2 of Dincbas et al. (1999), it was shown that the stronger the SD sequence, the more likely it became that a cell expressing a particular minigene would die. This suggests that the ribosome may stay on this short mRNA with a strong SD sequence and continue producing short peptidyl tRNA that would be toxic to the cell with reduced peptidyl tRNA hydrolase.

Two important observations described in Karimi et al. (1999) that differ from the scheme shown in Figure 6A are indicated in Figure 6B: (1) The 30S subunit of the posttermination complex remains on the mRNA together with tRNA and (2) the deacylated tRNA remaining on the 30S–mRNA complex is released only with IF-3 (initiation factor 3). We postulate that the strong affinity of mRNA due to nearby strong SD sequence and polyadenylic acid sequence would make it difficult for RRF to catalyze the translocation step described in step II of Figure 6A.

In early studies on RRF, we noted that disassembly of the complex of polyuridylic acid and ribosome with tRNA^{phe} as described in Kaji and Kaji (1963) did not take place in the presence of RRF and EF-G (A. Kaji et al., unpubl.). This is the reason we had to use puromycin-treated polysomes as a substrate of RRF for identification and purification of RRF (Hirashima and Kaji 1972b). We attributed this failure of RRF to disassemble the poly(U)–ribosome-tRNA^{phe} complex to the strong affinity of ribosomes for poly(U). Our recent work using various constructs with or without the strong SD sequence near the termination complex supports this notion (Y. Inokuchi et al., in prep.).

These observations point to the fact that translocation of RRF is different from translocation of tRNA. Indeed, normal translocation of tRNA takes place in the presence of the nearby SD sequence in the typical initiation process, whereas RRF translocation does not. Recent work (Rao and Varshney 2001) also supports the notion that the RRF translocation is different from tRNA translocation. Assuming that translocation of RRF by EF-G does not

take place due to the strong affinity of the ribosome to the message, one can imagine that the intramolecular movement of RRF (as shown in step III of Fig. 6A) could cause dissociation of the 70S ribosome into its subunits, leaving the 30S subunit on the mRNA.

Recent work by the Wintermeyer group has demonstrated that EF-G hydrolyzes GTP before translocation (Rodnina et al. 1997). Additionally, Agrawal et al. (1999) reported large-scale ribosomal conformational changes by the binding of EF-G and GTP hydrolysis. With a tightly bound message and RRF, the hydrolysis of GTP by EF-G may cause the release of 50S subunits, rather than translocation of RRF. If we assume strong affinity of mRNA to ribosome, it is quite understandable that the mRNA remains on the 30S subunit. In the translational coupling of tryptophan synthesis enzymes, the SD sequence is near the termination codon UAA of the upstream ORF (Das and Yanofsky 1984). Therefore, the model shown in Figure 6B may have an important implication under special circumstances such as translational coupling.

Although these two seemingly different mechanisms for the disassembly of the posttermination complex are not mutually exclusive, there are problems with each of the mechanisms proposed above. We describe here the shortcomings of each. We begin with the mechanism described in Figure 6A. First, most of the experiments leading to the mechanism were performed with the model posttermination complex, puromycin-treated polysome, derived from growing bacteria. It is possible that other factors may be present in this substrate. Counterargument for this criticism is that the findings with the polysomes have been confirmed with a semipurified system using the natural posttermination complex involving the amber mutant of phage R17 (Ogawa and Kaji 1975) and a completely synthetic system with nine nucleotides, UUCAU-GUAA (Grentzmann et al. 1998). However, one can still argue that the former is not a completely pure system and the latter is too artificial and does not represent the real situation. Second, part of the scheme was also derived from experiments involving nonprogrammed well-washed ribosomes. In the physiological situation where cells are actively synthesizing protein, RRF does not encounter such a thing as well-washed nonprogrammed ribosomes. Therefore, the finding derived from such studies may or may not have biological significance. Third, most importantly, none of the intermediate steps depicted in Figure 6A have been isolated and proven to exist. They are all inferred from experiments with washed ribosomes and the model posttermination complex. Clearly, much more work is required to support the hypothesis.

We now discuss the problems of the mechanism depicted in Figure 6B. The first problem is the absence of tRNA release during the disassembly reaction. For releasing tRNA, IF-3 is necessary. This is difficult to understand because, with the same synthetic mRNA, the same laboratory clearly demonstrated that the cooperative action of RRF, RF-3, and EF-G releases peptidyl (MFTI)-tRNA from the ribosomal P site (Figs. 3 and 4 in Heurgue-Hamard et al. 1998). No IF-3 was added in this experiment. There was significant release of peptidyl tRNA even without RF-3. If the three factors RRF, EF-G, and RF-3 are sufficient to release peptidyl tRNA from the posttermination complex, it is hard to understand why IF-3 is required to release tRNA after the peptidyl group has been released by release factors RF-1 or RF-2. It is generally accepted that the affinity of peptidyl tRNA to ribosome is higher than that of deacylated tRNA because the nascent polypeptide threads through the tunnel of the 50S subunit (Ban et al. 2000; Yusupov et al. 2001). It is well known that binding of peptidyl tRNA to the 50S subunit is very strong. Indeed, it was for this reason that in the early days, the 50S subunit, but not the 30S subunit, was thought to be the tRNA-binding subunit. This erroneous concept was corrected only after two laboratories independently showed that tRNA specific to phenylalanine binds to the complex of 30S subunit and poly(U), whereas the 50S subunit alone does not bind aminoacyl tRNA even in the presence of mRNA (Kaji et al. 1966; Pestka and Nirenberg 1966). Thus, the failure of RRF and EF-G to release tRNA from the posttermination complex is difficult to understand in view of the release of peptidyl tRNA under almost identical conditions.

One piece of evidence for involvement of IF-3 in the disassembly of the posttermination complex is that IF-3 is required for the maximum synthesis of the peptide fMFTI (Fig. 4 in Karimi et al. 1999). Although this interpretation is perfectly valid, the data may be interpreted to indicate that IF-3 is needed for dissociation of 70S into subunits at the termination codon. This action of IF-3 has long been established (Subramanian et al. 1970). Although the authors claim that dissociation of the 70S ribosome took place in the absence of IF-3, no physical evidence was presented. It is therefore possible that the requirement for IF-3 in this system may not necessarily mean that it is required for the release of tRNA. Furthermore, IF-3 is reported to ensure efficiency and fidelity in the formation of initiation complexes (Gualerzi et al. 2000). Therefore, one would expect that the amount of the synthesized peptide would increase upon addition of IF-3 regardless of whether or not it functions at the disassembly step.

We should emphasize that it has been established that IF-3 releases aminoacyl tRNA from the complex of the 30S subunit and synthetic mRNA, poly(U), and Phe-tRNA (Gualerzi et al. 1971). This reaction probably has general physiological significance in that IF-3 disassembles the unwanted complex of the 30S subunit and mRNA. It is therefore understandable that the complex of short synthetic mRNA, 30S subunit, and tRNA[Phe] releases tRNA[Phe] upon addition of IF-3 (Fig. 5 in Karimi et al. 1999). If RRF and EF-G indeed dissociate the 70S ribosome into subunit without releasing tRNA as shown in Figure 6B, the involvement of IF-3 for the release of tRNA is expected. However, dissociation of 70S ribosomes into subunits by RRF and EF-G is inferred from the effect of added 50S subunits to the system. In addition, there was no direct evidence to show that the stimulatory effect of IF-3 on the short peptide synthesis is due to the removal of tRNA from the complex. Further experiments are required to establish that the release of

tRNA by IF-3 indeed has an important role in the overall recycling of the ribosome in this system where mRNA has a strong affinity to ribosome.

If one assumes the mechanism depicted in Figure 6B to be the general mechanism of the disassembly of all the posttermination complexes, one faces the problem of collision of the upstream 70S ribosome with the 30S subunit remaining on the mRNA. With the short mRNA used by the Ehrenberg-Buckingham group, this will not be a problem because the bound ribosome never leaves the mRNA due to the proximity of the strong SD sequence to the termination codon. A survey of the available papers on scanning by the 30S subunits (Sarabhai and Brenner 1967; Draper 1996) failed to provide an explanation on how bacteria cope with the possible collision between the scanning 30S subunits and 70S ribosomes actively translating the upstream ORF. Even in the translationally coupled system where the upstream ribosome has an important role, most of the upstream ribosomes are released (Fig. 2 in Adhin and van Duin 1990; Table 4 and Fig. 1 of Grodzicker and Zipser 1968; page 153, section d in Sarabhai and Brenner 1967). In fact, only 1% of the upstream ribosomes engaged in the reinitiation (for discussion, see p. 159 in Sarabhai and Brenner 1967). The dissociation of ribosome from mRNA at the end of coat cistron of f2 phage is 20 times more probable than continual association (Fig. 8 in Beremand and Blumenthal 1979). In our routine RRF assay, we observe the release of mRNA from ribosomes. Ribosomes were released from the R17 phage RNA at the coat cistron amber codon (Fig. 1 in Ogawa and Kaji 1975). Even with short synthetic mRNAs, we observe mRNA release by EF-G and RRF (Figs. 4 and 5 and Table 2 in Grentzmann et al. 1998). We therefore propose that Figure 6B applies only to the case where ribosomes have strong affinity to mRNA.

RRF IN EUKARYOTES

In this section, the eukaryotic RRF homolog is briefly discussed. What is responsible for the disassembly of the posttermination complex in eukaryotic cytoplasmic protein synthesis? We have no information to answer this question. However, as described below, it is clear that the RRF homolog in eukaryotes is not fulfilling this task because it does not exist in the cytoplasm. In collaboration with Dr. Joyard's group (Grenoble, France), we showed that the RRF homolog of spinach (*Spinacia oleracea* L) exists in chloroplasts (Rolland et al. 1999). It has 46% sequence identity and 66% sequence homology with *E. coli* RRF. This is called mature RRFHCP (RRF homolog of chloroplast). From the amino-terminal analysis of RRFHCP, and isolated precursor of RRFHCP (called pRRFH), we found that the precursor protein has the 78-amino-acid chloroplast targeting sequence.

As shown in Figure 8A, western blot analysis performed on leaf extract revealed that RRFHCP is in chloroplasts and mitochondria. Upon incubation of in vitro-synthesized precursor protein (pRRFH) with

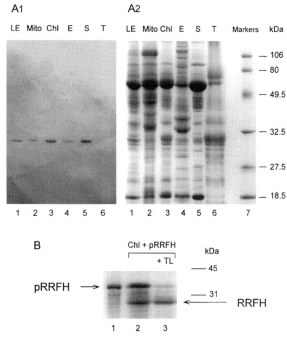

Figure 8. Subcellular localization of a plant RRF homolog (RRFHCP) (Rolland et al. 1999). (*A*) Analysis of polypeptides from different spinach leaf subfractions. (*A1*) Western blot analysis of polypeptides from spinach leaf subfractions with antibody against RRFHCP. (*A2*) Coomassie blue staining of the corresponding gel. (LE) Total leaves; (Mito) mitochondria; (Chl) chloroplast; (E) chloroplast envelope membranes; (S) stroma; (T) thylakoid membranes. (*B*) Import of the [³⁵S]methionine-labeled RRFHCP into isolated pea chloroplasts. Lanes: (*1*) in vitro [³⁵S]methionine-labeled translation products of RRFHCP cDNA; (*2*) the ³⁵S-labeled-RRFHCP was incubated with pea chloroplasts and chloroplasts were isolated after the import reaction; (*3*) the same as lane 2 except that the chloroplasts were treated further with thermolysin (TL) to remove proteins bound to the cytosolic face of the chloroplast outer envelope. The mature and precursor proteins are indicated as RRFH and pRRFH, respectively.

chloroplasts, we were able to show that pRRFH was taken up into the chloroplasts and processed to RRFHCP (Fig. 8B). The processing of pRRFH must take place in chloroplasts because we found the processed product (mature RRF homolog) only in chloroplasts that were inaccessible to the externally added proteolytic enzyme, thermolysin. The expression of plant RRF caused a bactericidal effect on *E. coli* with temperature-sensitive RRF even at the permissive temperature. On the other hand, the plant RRF did not exert any effect on the wild-type *E. coli* (Fig. 4 in Rolland et al. 1999). These observations strongly suggest that plant RRF homologs are localized in organelles such as mitochondria and chloroplasts. They must perform functions in the organelles similar to the functions they perform in prokaryotes.

In yeast, a unique ORF YHR038W codes for a protein whose size is similar to that of *E. coli* RRF (184 vs. 185 residues of *E. coli* RRF). This gene was first isolated by complementation of a mutant affected in the expression of isocitrate lyase. For this reason, it was named FIL1 (factor for isocitrate lyase expression) (Kanai et al. 1998).

We call this gene *FRR* and its product Frrp according to the yeast gene nomenclature convention because it is clearly a yeast homolog of the *E. coli* gene coding for RRF (*frr*). The *FRR* sequence indicates that its product contains a 46-amino-acid-long amino-terminal extension that is a mitochondria targeting sequence. Deletion of *FRR* in haploid strains leads to a respiratory incompetent phenotype (inability to grow on nonfermentable carbohydrates) and to instability of the mitochondrial genome. In our recent studies, it was found that the RRF homolog in yeast is localized in the mitochondria (A. Kaji et al., unpubl.). This protein, Frrp, is therefore a mitochondrial protein and is essential for mitochondrial maintenance and expression.

We created a temperature-sensitive yeast mutant expressing a modified allele of *FRR*. To obtain this temperature-sensitive mutant, we changed a unique amino acid residue of the Frrp. This amino acid that we changed is conserved between the bacterial and the yeast RRF sequence, and this change in *E. coli* RRF also leads to a temperature-sensitive *E. coli* RRF (Janosi et al. 1998). These results suggest the functional homology between *E. coli* RRF and yeast Frrp. At the nonpermissive temperature of this mutant, mitochondrial protein synthesis was severely inhibited. To our knowledge, this is the first example of a temperature-sensitive mutation in a soluble factor of the mitochondrial translation machinery. The fact that this mutation was brought about by the same amino acid change that conferred temperature-sensitivity on *E. coli* RRF strongly suggests that yeast RRF has a structure similar to that of *E. coli* RRF. In fact, computer modeling of the predicted yeast RRF showed it to be a near-perfect mimic of tRNA just like *E. coli* RRF.

We propose that the eukaryotic RRF homolog was originated from prokaryotes, which moved into eukaryotes and became organelles during phylogenetic development. The fact that archaebacteria are the only free-living organisms that do not have an RRF homolog is consistent with this concept. The archaebacteria are phylogenetically between eukaryotes and prokaryotes, but they do not have organelles such as mitochondria and chloroplasts. Therefore, archeal translational machinery is very much like that of eukaryotes. Since no RRF homolog is involved in the cytoplasmic protein synthesis of eukaryotes, it is understandable that archaebacteria do not use RRF for the disassembly of the posttermination complex. These considerations support the concept that eukaryotic RRF homologs originated from the prokaryotic RRF. RRF must perform the function to maintain the organelles just like the prokaryotic RRF does for maintenance of prokaryotes.

CONCLUDING REMARKS AND FUTURE DIRECTIONS

RRF and the fourth step of protein synthesis, disassembly of the posttermination complex, remained obscure for almost 30 years since its discovery. The biological significance of RRF has slowly been recognized after it was established that it is an essential factor for most organisms. Even the smallest free-living organism such as

Mycoplasma genitalium with only 500 genes has an *frr* homolog, while it manages to survive with only one termination factor (release factor) (Fraser et al. 1995). In contrast, *E. coli* has three termination factors RF-1, 2, and 3. The molecular mechanism of the RRF action began to be understood only after it was found that RRF is a near-perfect mimic of tRNA.

There are many aspects of RRF to be elucidated. In addition to the exact mechanism of the disassembly reaction, the problems to be elucidated include the mechanism of error prevention, the role in translational coupling, and interactions with other biological components. Interactions of RRF with other factors are suggested by isolation of many intergenic suppressor mutations of LJ14 (with a temperature-sensitive RRF) (Janosi et al. 2000). It is not known why loss of RRF is bactericidal during the lag phase, but it is bacteriostatic in growing phase (Janosi et al. 1998).

Although the importance of RRF as a possible target of new antimicrobial agents has repeatedly been pointed out (see, e.g., Kaji et al. 1998), no specific inhibitor of RRF has been reported. This aspect of RRF has to be pursued vigorously to take advantage of the discovery of RRF for contribution to health science. In this connection, the exact role of the RRF homologs in organelles of eukaryotes must be elucidated before we can design specific inhibitors of bacterial RRF. Last but not least, the answer to the important question of how eukaryotes manage the fourth step of protein synthesis remains elusive. This step, which is not catalyzed by RRF, should be thoroughly studied in light of the known mechanism of the RRF reaction in prokaryotes.

REFERENCES

Adhin M.R. and van Duin J. 1990. Scanning model for translational reinitiation in eubacteria. *J. Mol. Biol.* **213:** 811.

Agrawal R.K., Heagle A.B., Penczek P., Grassucci R.A., and Frank J. 1999. EF-G-dependent GTP hydrolysis induces translocation accompanied by large conformational changes in the 70S ribosome. *Nat. Struct. Biol.* **6:** 643.

Atarashi K. and Kaji A. 2000. Inhibitory effect of heterologous ribosome recycling factor on growth of *Escherichia coli*. *J. Bacteriol.* **182:** 6154.

Ban N., Nissen P., Hansen J., Moore P.B., and Steitz T.A. 2000. The complete atomic structure of the large ribosomal subunit at 2.4 Å resolution. *Science* **289:** 905.

Beremand M.N. and Blumenthal T. 1979. Overlapping genes in RNA phage: A new protein implicated in lysis. *Cell* **18:** 257.

Das A. and Yanofsky C. 1984. A ribosome binding site sequence is necessary for efficient expression of the distal gene of a translationally-coupled gene pair. *Nucleic Acids Res.* **12:** 4757.

Davies J. and Davis B.D. 1968. Misreading of ribonucleic acid code words induced by aminoglycoside antibiotics: The effect of drug concentration. *J. Biol. Chem.* **243:** 3312.

Davis B.D. 1971. Role of subunits in the ribosome cycle. *Nature* **321:** 153.

Dincbas V., Heurgue-Hamard V., Buckingham R.H., Karimi R., and Ehrenberg M. 1999. Shutdown in protein synthesis due to the expression of mini-genes in bacteria. *J. Mol. Biol.* **291:** 745.

Draper D.E. 1996. Translational initiation. In *Escherichia coli and Salmonella: Cellular and molecular biology* (ed. F.C. Neidhart et al.), p. 902. ASM Press, Washington, D.C.

Frank J. and Agrawal R.K. 2000. A ratchet-like inter-subunit re-

organization of the ribosome during translocation. *Nature* **406:** 318.

Fraser C.M., Gocayne J.D., White O., Adams M.D., Clayton R.A., Fleischmann R.D., Bult C.J., Kerlavage A.R., Sutton G., Kelley J.M., Fritchman J.L., Weidman J.F., Small K.V., Sandusky M., Fuhrmann J., Nguyen D., Utterback T.R., Saudek D.M., Phillips C.A., Merrick J.M., Tomb J.-F., Dougherty B.A., Bott K.F., Hu P.-C., Lucier T.S., Peterson S.N., Smith H.O., Hutchison C.A., III, and Venter J.C. 1995. The minimal gene complement of *Mycoplasma genitalium*. *Science* **270:** 397.

Grentzmann G., Kelly P.J., Laalami S., Shuda M., Firpo M.A., Cenatiempo Y., and Kaji A. 1998. Release factor RF-3 GTPase activity acts in disassembly of the ribosome termination complex. *RNA* **4:** 973.

Grodzicker T. and Zipser D. 1968. A mutation which creates a new site for the re-initiation of polypeptide synthesis in the z gene of the lac operon of *Escherichia coli*. *J. Mol. Biol.* **38:** 305.

Gualerzi C., Pon C.L., and Kaji A. 1971. Initiation factor dependent release of aminoacyl-tRNAs from complexes of 30S ribosomal subunits, synthetic polynucleotide and aminoacyl tRNA. *Biochem. Biophys. Res. Commun.* **45:** 1312.

Gualerzi C.O., Brandi L., Caserta E., Teana A.L., Spurio R., Tomsic J., and Pon C.L. 2000. Translation initiation in bacteria. In *The ribosome: Structure, function, antibiotics, and cellular interactions* (ed. R.A. Garret et al.), p. 477. ASM Press, Washington, D.C.

Heurgue-Hamard V., Karimi R., Mora L., MacDougall J., Leboeuf C., Grentzmann G., Ehrenberg M., and Buckingham R.H. 1998. Ribosome release factor RF4 and termination factor RF3 are involved in dissociation of peptidyl-tRNA from the ribosome. *EMBO J.* **17:** 808.

Hirashima A. and Kaji A. 1970. Factor dependent breakdown of polysomes. *Biochem. Biophys. Res. Commun.* **41:** 877.

————. 1972a. Factor-dependent release of ribosomes from messenger RNA—Requirement for two heat-stable factors. *J. Mol. Biol.* **65:** 43.

————. 1972b. Purification and properties of ribosome releasing factor. *Biochemistry* **11:** 4037.

————. 1973. Role of elongation factor G and a protein factor on the release of ribosomes from messenger ribonucleic acid. *J. Biol. Chem.* **248:** 7580.

Ichikawa S. and Kaji A. 1989. Molecular cloning and expression of ribosome releasing factor. *J. Biol. Chem.* **264:** 20054.

Inokuchi Y., Hirashima A., Sekine Y., Janosi L., and Kaji A. 2000. Role of ribosome recycling factor (RRF) in translational coupling. *EMBO J.* **19:** 3788.

Janosi L., Ricker R., and Kaji A. 1996a. Dual functions of ribosome recycling factor in protein biosynthesis: Disassembling the termination complex and preventing translational errors. *Biochimie* **78:** 959.

Janosi L., Shimizu I., and Kaji A. 1994. Ribosome recycling factor (ribosome releasing factor) is essential for bacterial growth. *Proc. Natl. Acad. Sci.* **91:** 4249.

Janosi L., Hara H., Zhang S., and Kaji A. 1996b. Ribosome recycling by ribosome recycling factor (RRF)—An important but overlooked step of protein biosynthesis. *Adv. Biophys.* **32:** 121.

Janosi L., Mori H., Sekine Y., Abragan J., Janosi R., Hirokawa G., and Kaji A. 2000. Mutations influencing the *frr* gene coding for ribosome recycling factor (RRF). *J. Mol. Biol.* **295:** 815.

Janosi L., Mottagui-Tabar S., Isaksson L.A., Sekine Y., Ohtsubo E., Zhang S., Goon S., Nelken S., Shuda M., and Kaji A. 1998. Evidence for in vivo ribosome recycling, the fourth step in protein biosynthesis. *EMBO J.* **17:** 1141.

Kaempfer R. 1970. Dissociation of ribosomes on polypeptide chain termination and origin of single ribosomes. *Nature* **228:** 534.

————. 1972. Initiation factor IF-3: Specific inhibitor of ribosomal subunit association. *J. Mol. Biol.* **71:** 583.

Kaji A. and Hirokawa G. 2000. Ribosome recycling factor: An essential factor for protein synthesis. In *The ribosome: Structure, function, antibiotics and cellular interactions* (ed. R.A.

Garret et al.), p. 527. ASM Press, Washington, D.C.

Kaji A. and Kaji H. 1963. Specific interaction of soluble RNA with polyribonucleic acid induced polysomes. *Biochem. Biophys. Res. Commun.* **13**: 186.

Kaji A., Teyssier E., and Hirokawa G. 1998. Disassembly of the post-termination complex and reduction of translational error by ribosome recycling factor (RRF)—A possible new target for antibacterial agents. *Biochem. Biophys. Res. Commun.* **250**: 1.

Kaji H. and Kaji A. 1965. Specific binding of sRNA to ribosomes: Effect of streptomycin. *Proc. Natl. Acad. Sci.* **54**: 213.

Kaji H., Suzuka I., and Kaji A. 1966. Binding of specific soluble ribonucleic acid to ribosomes. Binding of soluble ribonucleic acid to the template-30 S subunits complex. *J. Biol. Chem.* **241**: 1251.

Kanai T., Takeshita S., Atomi H., Umemura K., Ueda M., and Tanaka A. 1998. A regulatory factor, Fil1p, involved in derepression of the isocitrate lyase gene in *Saccharomyces cerevisiae*. A possible mitochondrial protein necessary for protein synthesis in mitochondria. *Eur. J. Biochem.* **256**: 212.

Karimi R., Pavlov M.Y., Buckingham R.H., and Ehrenberg M. 1999. Novel roles for classical factors at the interface between translation termination and initiation. *Mol. Cell* **3**: 601.

Kim K.K., Min K., and Suh S.W. 2000. Crystal structure of the ribosome recycling factor from *Escherichia coli*. *EMBO J.* **19**: 2362.

Kohler R.E., Ron E.Z., and Davis B.D. 1968. Significance of the free 70S ribosomes in *Escherichia coli* extracts. *J. Mol. Biol.* **36**: 71.

Kozak M. 1999. Initiation of translation in prokaryotes and eukaryotes. *Gene* **234**: 187.

Nierhaus K.H., Spahn C., Burkhardt N., Dabrowski M., Diedrich G., Einfeldt E., Kamp D., Marquez V., Patzke S., Schafer M.A., Stelzl U., Blaha G., Willumeit R., and Stuhrmann H.B. 2000. Ribosome elongation cycle. In *The ribosome: Structure, function, antibiotics and cellular interactions* (ed. R.A. Garret et al.), p. 319. ASM Press, Washington, D.C.

Nirenberg M. and Leder P. 1964. RNA codewords and protein synthesis: The effect of trinucleotides upon the binding of sRNA to ribosomes. *Science* **145**: 1399.

Ogawa K. and Kaji A. 1975. Requirement for ribosome-releasing factor for the release of ribosomes at the termination codon. *Eur. J. Biochem.* **58**: 411.

Pavlov M.Y., Freistroffer D.V., Heurgue-Hamard V., Buckingham R.H., and Ehrenberg M. 1997a. Release factor RF3 abolishes competition between release factor RF1 and ribosome recycling factor (RRF) for a ribosome binding site. *J. Mol. Biol.* **273**: 389.

Pavlov M.Y., Freistroffer D.V., MacDougall J., Buckingham R.H., and Ehrenberg M. 1997b. Fast recycling of *Escherichia coli* ribosomes requires both ribosome recycling factor (RRF) and release factor RF3. *EMBO J.* **16**: 4134.

Pestka S. and Nirenberg M. 1966. Regulatory mechanisms and protein synthesis. X. Codon recognition on 30 S ribosomes. *J. Mol. Biol.* **21**: 145.

Rao A.R. and Varshney U. 2001. Specific interaction between the ribosome recycling factor and the elongation factor G from Mycobacterium tuberculosis mediates peptidyl-tRNA release and ribosome recycling in *Escherichia coli*. *EMBO J.* **20**: 2977.

Remme J., Margus T., Villems R., and Nierhaus K.H. 1989. The third ribosomal tRNA-binding site, the E site, is occupied in native polysomes. *Eur. J. Biochem.* **183**: 281.

Rodnina M.V., Savelsbergh A., Katunin V.I., and Wintermeyer W. 1997. Hydrolysis of GTP by elongation factor G drives tRNA movement on the ribosome. *Nature* **385**: 37.

Rodnina M.V., Pape T., Savelsbergh A., Mohr D., Matassova N.B., and Wintermeyer W. 2000. Mechanisms of partial reactions of the elongation cycle catalyzed by elongation factors

Tu and G. In *The ribosome: Structure, function, antibiotics and cellular interactions* (ed. R.A. Garret et al.), p. 301. ASM Press, Washington, D.C.

Rodnina M.V., Savelsbergh A., Matassova N.B., Katunin V.I., Semenkov Y.P., and Wintermeyer W. 1999. Thiostrepton inhibits the turnover but not the GTPase of elongation factor G on the ribosome. *Proc. Natl. Acad. Sci.* **96**: 9586.

Rolland N., Janosi L., Block M.A., Shuda A., Teyssier E., Miege C., Cheniclet C., Carde J., Kaji A., and Joyard J. 1999. Plant ribosome recycling factor homolog is a chloroplastic protein and is bactericidal in *Escherichia coli* carrying temperature-sensitive ribosome recycling factor. *Proc. Natl. Acad. Sci.* **96**: 5464.

Ryoji M., Berland R., and Kaji A. 1981. Reinitiation of translation from the triplet next to the amber termination codon in the absence of ribosome-releasing factor. *Proc. Natl. Acad. Sci.* **78**: 5973.

Santos M.A.S., Perreau V.M., and Tuite M.F. 1996. Transfer RNA structural change is a key element in the reassignment of the CUG codon in *Candida albicans*. *EMBO J.* **15**: 5060.

Sarabhai A. and Brenner S. 1967. A mutant which reinitiates the polypeptide chain after chain termination. *J. Mol. Biol.* **27**: 145.

Scolnick E., Tompkins R., Caskey T., and Nirenberg M. 1968. Release factors differing in specificity for terminator codons. *Proc. Natl. Acad. Sci.* **61**: 768.

Selmer M., Al-Karadaghi S., Hirokawa G., Kaji A., and Liljas A. 1999. Crystal structure of *Thermotoga maritima* ribosome recycling factor: A tRNA mimic. *Science* **286**: 2349.

Stark H., Orlova E.V., Rinke-Appel J., Jünke N., Mueller F., Rodnina M., Wintermeyer W., Brimacombe R., and van Heel M. 1997. Arrangement of tRNAs in pre- and posttranslocational ribosomes revealed by electron cryomicroscopy. *Cell* **88**: 19.

Steitz J.A. 1970. Nucleotide sequenes of the ribosomal binding sites of bacteriophage R17 RNA. *Cold Spring Harbor Symp. Quant. Biol.* **34**: 621.

Subramanian A.R. and Davis B.D. 1971. Rapid exchange of subunits between free ribosomes in extracts of *Escherichia coli*. *Proc. Natl. Acad. Sci.* **68**: 2453.

———. 1973. Release of 70S ribosomes from polysomes in *Escherichia coli*. *J. Mol. Biol.* **74**: 45.

Subramanian A.R., Davis B.D., and Beller R.J. 1970. The ribosome dissociation factor and the ribosome-polysome cycle. *Cold Spring Harbor Symp. Quant. Biol.* **34**: 223.

Tjandra N. and Bax A. 1997. Direct measurement of distances and angles in biomolecules by NMR in a dilute liquid crystalline medium. *Science* **278**: 1111.

Tjandra N., Garrett D.S., Gronenborn A.M., Bax A., and Clore G.M. 1997. Defining long range order in NMR structure determination from the dependence of heteronuclear relaxation times on rotational diffusion anisotropy. *Nat. Struct. Biol.* **4**: 443.

Wilson D.N., Dalphin M.E., Pel H.J., Major L.L., Mansell J.B., and Tate W.P. 2000. Factor-mediated termination of protein synthesis: A welcome return to the mainstream of translation. In *The ribosome: Structure, function, antibiotics, and cellular interactions* (ed. R.A. Garret et al.), p. 495. ASM Press, Washington, D.C.

Wilson K.S. and Noller H.F. 1998. Mapping the position of translational elongation factor EF-G in the ribosome by directed hydroxyl radical probing. *Cell* **92**: 131.

Yoshida T., Uchiyama S., Nakano H., Kashimori H., Kijima H., Ohshima T., Saihara Y., Ishino T., Shimahara H., Yoshida T., Yokose K., Ohkubo T., Kaji A., and Kobayashi Y. 2001. Solution structure of the ribosome recycling factor from *Aquifex aeolicus*. *Biochemistry* **40**: 2387.

Yusupov M.M., Yusupova G.Z., Baucom A., Lieberman K., Earnest T.N., Cate J.H., and Noller H.F. 2001. Crystal structure of the ribosome at 5.5 Å resolution. *Science* **29**: 29.

Structure, Function, and Regulation of Free and Membrane-bound Ribosomes: The View from Their Substrates and Products

A.E. Johnson,*†‡ J.-C. Chen,* J.J. Flanagan,† Y. Miao,* Y. Shao,*
J. Lin,§ and P.E. Bock¶

*Department of Medical Biochemistry and Genetics, Texas A&M University System Health Science Center,
College Station, Texas 77843-1114; Departments of ‡Chemistry and of †Biochemistry and Biophysics, Texas A&M
University, College Station, Texas 77843; §Department of Biochemistry and Molecular Biology,
University of Oklahoma Health Sciences Center, Oklahoma City, Oklahoma 73104; and
¶Department of Pathology, Vanderbilt University School of Medicine, Nashville, Tennessee 37232

A myriad of experimental approaches and techniques has been used during the past 45 years to examine the ribosome, and this research has yielded a tremendous amount of information. Yet despite the enormous effort and ingenuity that have been devoted to achieving an understanding of the ribosome, many aspects of its structure, function, assembly, and regulation still remain largely or completely obscure. During the past 25 years, we have chosen to focus on the substrates and products of the ribosome, specifically the aminoacyl tRNAs (aa-tRNAs) that bind to the ribosome in response to specific mRNA codons and the nascent protein chains that are synthesized by the ribosome. This approach has provided us with an unusual perspective of the ribosome and has yielded some unique insights into ribosome structure, function, and regulation. Furthermore, based on past success in identifying and characterizing important aspects of the ribosome, this experimental approach promises to continue to yield information that is essential for a full understanding of the ribosome.

WHY FOCUS ON THE SUBSTRATES AND PRODUCTS?

The wide range of topics covered in this volume and the great diversity of ribosomal topics in the scientific literature as a whole demonstrate that one can focus on any of an extremely large number of structural or functional elements of the ribosome. Each of these elements can be revealing in terms of understanding one or more aspects of the ribosome. Yet, because of the size and varied structural and functional states of the ribosome, each of these structural or functional elements is important for, or is involved in, only a portion of the ribosome's life cycle (e.g., during initiation of protein synthesis, during mRNA selection, during ribosome binding to the membrane). Thus, if one were to position a reporter group at a particular site on a ribosomal protein or RNA to monitor changes at that site, those changes would be observed only when that site on the ribosome participated, either directly or indirectly, in the transition of the ribosome from one functional state to another. One therefore might

expect the reporter group or probe to be informative (i.e., where the action is) only at certain stages of the ribosomal life cycle.

In contrast, the substrates and products of the ribosome are always where the action is when the ribosome is functioning. Since ribosomal operations are designed to select and move aa-tRNA substrates through the ribosome and to create a protein product, positioning a probe in an aa-tRNA or in the nascent chain ensures that that probe will always be located in a functionally important ribosomal site. Furthermore, this probe placement provides a very different perspective on the process of protein biosynthesis. Instead of monitoring the changes undergone by the ribosome, this approach allows one to detect and characterize what the ribosome is doing to the substrate and product. Although the perspectives of the ribosome, the product, and the substrate are not the same, they are complementary. Therefore, each of these perspectives must be examined to achieve a complete understanding of ribosomal function.

Another important reason for focusing on the ribosomal product is that a nascent protein chain may interact with other cellular components while it is still bound to the ribosome. In such a case, processing of the nascent chain could occur co-translationally, a possibility that raises an important question: To what extent does the ribosome participate in and/or regulate processes that involve the nascent chain? Any such involvement of the ribosome would constitute an expansion of its functional role beyond that of simply synthesizing protein, and hence would need to be fully characterized if we are to understand ribosome structure and function in toto.

As it happens, there are already numerous well-documented examples of functionally important interactions of ribosomes with other cellular components during translation that are initiated and/or mediated by the nascent chain. For example, ribosomes synthesizing nascent secretory or membrane proteins in mammalian cells are directed to the membrane of the endoplasmic reticulum (ER), whereupon the nascent chains are translocated across or integrated into the ER membrane as the ribosome interacts with ER membrane components and completes translation (for review, see Johnson and van Waes 1999). It has also become clear in recent years that the folding of the nascent chain commences prior to termination of translation, and that chaperones located in either the cytoplasm or the ER lumen interact co-translationally with ribosome-bound nascent chains to facilitate folding (for review, see Frydman 2001; McCallum et al. 2000). As another example, it is now known that nascent apolipoprotein B (and perhaps other proteins) is targeted for degradation under some metabolic conditions and is proteolyzed by the proteasome before translation is completed (for review, see Fisher et al. 1997; Yao et al. 1997; Pariyarath et al. 2001). To assess the extent of ribosome involvement in such processes, it is necessary to have a probe in the nascent chain that can monitor the processes as a function of the translational state of the ribosome.

THE SUBSTRATE'S POINT OF VIEW

The location of the tRNA substrate at the ribosome can be assessed by determining the covalent status of the substrate (an aa-tRNA, a peptidyl tRNA, or an unacylated tRNA) and its sensitivity to puromycin. One can get further information on the substrate's interactions and environment in the ribosome by using an extrinsic reporter group or probe that is attached to either the amino acid or the tRNA portion of the aa-tRNA. The data obtained from such probes can then be correlated with traditional measures of tRNA location in the ribosome to obtain additional insight into the mechanisms involved in substrate processing by the ribosome.

In one example of such a study, we were able to use fluorescence resonance energy transfer (FRET) to measure the distance moved by the inside corner of the tRNA as it is translocated from the A site to the P site (Johnson et al. 1982). This measurement not only determined the extent of the tRNA conformational change in the ribosomal complex that accompanied tRNA translocation, but also established, for the first time, the relative orientation of the two tRNAs bound in the A and P sites of the ribosome. A fluorescent dye (the donor) was covalently attached to the s^4U-8 in fMet-tRNA, and a different dye (the acceptor) was attached to the s^4U-8 of Phe-tRNA. After binding the donor-labeled fMet-tRNA to the P site and the acceptor-labeled Phe-tRNA to the A site, the efficiency of FRET between the donor and acceptor dyes showed that the dyes on the two tRNAs were separated by 26 ± 4 Å, and hence that the tRNAs in the A and P sites were immediately adjacent to each other, separated by less than 10 Å (Johnson et al. 1982). This result was quite unexpected, because in 1982, the corners of the tRNAs were assumed to be separated as far as possible to minimize electrostatic repulsion between the ribosome-bound tRNAs whose aminoacyl ends had to be adjacent during peptidyl transfer and whose anticodons are bound to adjacent codons in the mRNA. Yet this side-by-side tRNA arrangement has been confirmed by recent cryoelectron microscopic (cryo-EM) data (Agrawal et al. 2000) and crystal structure data (Yusupov et al. 2001).

Fluorescent probes attached to the s^4U-8 in several different tRNAs also allowed us to identify a (the?) primary

mechanism by which elongation factor Tu (EF-Tu) facilitates protein synthesis (Janiak et al. 1990). Although the same probe (fluorescein) was positioned in the same location in each of these aa-tRNAs, the fluorescence intensities differed greatly because the conformations of the individual aa-tRNAs differed near the s^4U-8. However, after EF-Tu•GTP had been added to the samples to form aa-tRNA•EF-Tu•GTP ternary complexes, each ternary complex was found to have the same fluorescence intensity. Thus, the different aa-tRNAs all had the same or very similar conformation in the ternary complex. The binding of EF-Tu•GTP to different aa-tRNAs therefore causes each to adopt a common structure before it begins the mRNA decoding process at the ribosome. Since the ribosome has only a single binding site for aa-tRNA recognition and selection, such a homogenization of the otherwise very different aa-tRNA conformations is required to treat all aa-tRNAs uniformly (Janiak et al. 1990). The conformation-sensitive probe on various aa-tRNA substrates therefore revealed how the system is able to accommodate multiple different substrates at a single recognition site on the ribosome: by using the elongation factor to ensure a uniform aa-tRNA conformation during codon–anticodon recognition.

A probe attached to the amino acid of the aa-tRNA allows one to investigate the aminoacyl ends of the tRNA-binding sites on the ribosome. In the late 1970s and early 1980s, many peptidyl-tRNA analogs were used to position chemically reactive or photoreactive probes in the ribosomal P or A sites to identify the ribosomal proteins and rRNA located adjacent to the peptidyl transferase center. The peptidyl-tRNA analogs were convenient because probes could be easily attached to the α-amino group of the aa-tRNA. However, these analogs could not be used to examine elongation-factor-dependent binding because modification of the α-amino group blocks aa-tRNA association with EF-Tu•GTP (see, e.g., Janiak et al. 1990). Thus, using a Lys-tRNA with a probe covalently attached to the ε-amino group of the lysine side chain rather than the lysine α-amino group (see below), we were able to carry out elongation-factor-dependent affinity labeling studies of the ribosomal A and P sites (Johnson and Cantor 1980). By using elongation factors to control the positioning of tRNAs in the ribosome, the observed cross-links were shown to have originated from a particular tRNA-binding site, thereby identifying ribosomal components adjacent to the aminoacyl ends of the A and P sites. It should also be noted that this approach, using elongation factors to bind and position the aa-tRNA in the ribosome, ensured that only functional ribosomes were being examined.

THE PRODUCT'S POINT OF VIEW

The ribosome differs from many enzymes in that the product of the catalyzed reaction does not leave the enzyme (the ribosome) at the end of each catalytic cycle. The ribosome must therefore have mechanisms for maintaining control of the product and directing it to a specific location within or on the surface of the ribosome. Since the ribosome protects 35–40 amino acids of the nascent chain from proteolytic digestion (Malkin and Rich 1967; Blobel and Sabatini 1970), the nascent chain appears to be sequestered inside the ribosome in some manner. Yet there was considerable disagreement about the location of the nascent chain well into the 1990s, largely because of conflicting electron microscopy data (see, e.g., Bernabeu and Lake 1982; Bernabeu et al. 1983; Ryabova et al. 1988; Wagenknecht et al. 1989) and a reluctance to accept the existence of a tunnel in the large ribosomal subunit proposed on the basis of early diffraction data (Yonath et al. 1987). Of course, the most direct way to examine the location of the nascent chain in the ribosomal complex is to position a probe in the nascent chain and thereby determine its location and surroundings. Such an approach has the obvious additional advantage of allowing the investigator to determine the environment of a nascent chain in a functioning and structurally intact ribosome.

Additionally, as noted above, the polypeptide synthesized by the ribosome often interacts with non-ribosomal complexes while it is still a nascent chain and is bound to the ribosome as a peptidyl tRNA, as is exemplified by the signal-sequence-dependent targeting of ribosomes to the ER membrane. It is reasonable to expect that these interactions with the nascent chain are regulated, and that such regulation may involve the ribosome. To investigate these possibilities, the most direct method is again to examine nascent chain interactions with these complexes and their potential influence on ribosome structure and function (or vice versa) by positioning a probe in the nascent chain and correlating changes in its environment and interactions with the state of the ribosome at different stages of the process being investigated.

PROBE PLACEMENT IN THE NASCENT CHAIN

A direct examination of the environment and interactions of the nascent protein chain can only be accomplished by a reporter group that is covalently attached to the nascent chain. However, standard chemical modification techniques cannot be used to attach the probe to a ribosome-bound nascent chain because ribosomal proteins and, in the case of membrane-bound ribosomes, ER membrane proteins will also be labeled and create a very large background signal (the nascent chain constitutes less than 1% of the polypeptide in a typical sample of ribosomes). Instead, the only way to position a probe solely in the nascent chain is to use the ribosome to incorporate a probe at a specific site in the nascent chain, an approach that requires a functional aa-tRNA analog with a probe covalently attached to its amino acid side chain. We originated and developed this approach more than 25 years ago, and it is now widely used to position probes in nascent chains or full-length proteins.

The first successful incorporation of an amino acid with a nonnatural side chain in vitro was obtained using a chemically modified Lys-tRNA (Johnson et al. 1976). Subsequent studies employed lysine analogs with chemically reactive (see, e.g., Johnson et al. 1978; Johnson and Cantor 1980), photoreactive (see, e.g., Krieg et al. 1986; Do et al. 1996), or fluorescent (see, e.g., Crowley et al.

1993, 1994) probes. A variation of this approach emerged in the late 1980s and early 1990s in which the nonnatural amino acids were esterified to an amber suppressor tRNA, a tRNA that recognizes and translates the amber stop codon (see, e.g., High et al. 1993; Cornish et al. 1994). For proteins that contain multiple lysine codons, the availability of modified amber suppressor tRNAs allowed researchers to position a single probe at a specific site in a nascent chain simply by engineering a UAG stop codon into the coding sequence at the desired location using standard molecular biological techniques.

TRANSLATION INTERMEDIATES

To examine nascent chain environment and interactions, not only does a probe need to be incorporated into the nascent chain, but the probe also has to be positioned at a specific location relative to the ribosome to facilitate interpretation of the data generated by the probe. Furthermore, all of the probes in the sample must be at the same location to allow a clear interpretation of the data. Protein translation in vitro cannot be synchronized because it is impossible to control the time at which a ribosome initiates translation on a particular mRNA or the rate of ribosome movement along the mRNA. However, a homogeneous sample of ribosome•nascent chain complexes (RNCs), each with the same length of nascent chain, can be prepared by using mRNAs that are truncated at a specific site within the coding region of the gene of interest. A ribosome will initiate translation normally on such a truncated mRNA and will elongate until the end of the mRNA is reached. At that point, the ribosome will stall because of the absence of a codon in the A site. But the ribosome and peptidyl tRNA will not dissociate from the truncated mRNA because there is no stop codon. Thus, every ribosome will translate the same length of nascent chain as long as (1) each ribosome is allowed enough time to translate to the end of the truncated mRNA and (2) an excess of truncated mRNA is used to ensure that only one ribosome initiates translation per mRNA. This approach generates a homogeneous sample of translation intermediates (RNCs) in which the length of the nascent chain is dictated by the length of the mRNA added to the translation.

MEMBRANE-BOUND RIBOSOMES

Membrane-bound ribosomes, which constitute ~50% of the total ribosomes in some mammalian cells, have been largely ignored by those investigating ribosome structure and function, as evidenced by the small number of papers in this volume that even mention membrane-bound ribosomes. This lack of interest in membrane-bound ribosomes is usually justified by the assertion that free and membrane-bound ribosomes most likely function in the same way. However, the reality is that membrane-bound ribosomes are much more difficult to work with experimentally, and hence investigators have preferred free ribosomes for their studies. Yet, as noted below, our studies with membrane-bound ribosomes have provided important insights into ribosome structure, function, and regulation that were unobtainable with free ribosomes. Hence, it is our belief that more studies with membrane-bound ribosomes should be undertaken to further our understanding of ribosomes in general. In fact, we believe that membrane-bound ribosomes constitute one of the most intriguing and mechanistically important frontiers in ribosome research.

In eukaryotic cells, protein synthesis is initiated on ribosomes that are free in the cytosol. Those ribosomes synthesizing proteins destined to be secreted from the cell or to be integrated into a membrane are identified by the presence of a signal sequence in the nascent chain, usually at its amino-terminal end (for review, see Walter and Johnson 1994). When a signal sequence emerges from the ribosome, it is recognized and bound by a cytoplasmic ribonucleoprotein, the signal recognition particle (SRP). The binding of SRP to the RNC causes a temporary cessation of protein synthesis that has been termed an "elongation arrest." The elongation-arrested SRP•RNC complex then diffuses in the cytosol until it collides with a protein found only on the membrane of the rough ER, the SRP receptor. The interaction between SRP and the SRP receptor initiates a series of events that includes the binding of the ribosome to the translocon (the site of secretory protein translocation across and membrane protein integration into the ER membrane; for review, see Johnson and van Waes 1999), the release of the signal sequence from SRP, the insertion of the signal sequence into the translocon, the dissociation of SRP and SRP receptor from the ribosome and translocon, and the resumption of protein synthesis (Walter and Johnson 1994). This targeting process is GTP-dependent, and both SRP and SRP receptor bind and hydrolyze GTP (Walter and Johnson 1994).

Thus, targeting of ribosomes to membranes in vitro requires only the addition of SRP and ER microsomes to a cell-free translation system that is synthesizing a polypeptide with a signal sequence. The resulting membrane-bound ribosomes can then be separated from residual free ribosomes and other components by gel filtration chromatography or by sedimentation through a sucrose cushion. Since inactive or nontranslating ribosomes are removed by this purification step (only actively translating ribosomes will have a nascent chain with an exposed signal sequence that can target the ribosomes to the ER microsomes), studies of purified membrane-bound ribosomes are not complicated by the presence of an equal or greater number of nonfunctioning ribosomes, as is the case in studies of free ribosomes.

NASCENT CHAIN ENVIRONMENT AND LOCATION

As noted above, there was no consensus in the early 1990s about whether the nascent chain extended into the cytosol directly from the peptidyl transferase center during translation or first moved through a tunnel in the large ribosomal subunit before emerging into the cytosol, simply because no experimental technique had been able to distinguish unambiguously between these two possibilities. However, by examining membrane-bound ribosomes with fluorescent-labeled nascent chains, we were able to resolve this issue.

First, by incorporating water-sensitive 6-(7-nitrobenz-2-oxa-1,3-diazol-4-yl)aminohexanoyl (NBD) dyes into the nascent chain using N^ϵ-NBD-Lys-tRNA (ϵNBD-Lys-tRNA) and then determining the extent of their exposure to water from the NBD fluorescence lifetimes, we were able to show that the nascent chain is in an aqueous environment inside the ribosome (i.e., where the nascent chain is protected from proteases) (Crowley et al. 1993, 1994). Second, we observed that the aqueous space containing the NBD-labeled nascent chain inside free ribosomes was contiguous with the cytosol (i.e., the probes were accessible from the cytosol) because NBD fluorescence intensity was reduced by the addition of iodide ions to the sample. Iodide ions efficiently quench NBD fluorescence upon colliding with the NBD dye, so this experimental approach directly measures iodide ion accessibility to the probe. Yet when these experiments were repeated using membrane-bound ribosomes, we discovered that the nascent chain, although still in an aqueous environment, was not exposed to the cytosol because cytoplasmic iodide ions did not quench the NBD fluorescence (Crowley et al. 1993; 1994). Since no probes within the protease-protected portion of the nascent chain were accessible to iodide ions in the cytoplasm, the ribosome had to completely surround this portion of the nascent chain. Furthermore, since these nascent chain probes were accessible to cytoplasmic iodide ions when bound to free ribosomes, but not when bound to membrane-bound ribosomes, binding to the ER membrane completely sealed off the nascent chain from the cytosol. Thus, the nascent chain did not enter the cytosol directly from the peptidyl transferase center, and the nascent chain did not move along the surface of the ribosome in a protease-protected groove.

Instead, the fluorescence data demonstrated that the nascent chain moves through a tunnel in the ribosome (Fig. 1). The existence of a nascent chain tunnel through a bacterial ribosome has since been confirmed by the high-resolution crystal structure of its large subunit (Nissen et al. 2000). Cryo-EM studies have also indicated the presence of a nascent chain tunnel in the larger eukaryotic ri-

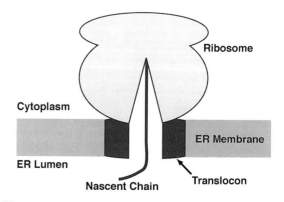

Figure 1. A schematic view of a membrane-bound ribosome. The nascent chain tunnel in a ribosome bound to the ER membrane is aligned with the large aqueous pore formed by the translocon to facilitate the translocation of a secretory protein into the lumen of the ER (Crowley et al. 1994; Hamman et al. 1997).

bosomes (see, e.g., Beckmann et al. 1997). However, because the mammalian ribosome, but not the bacterial ribosome, binds to the membrane and forms a tight seal with the translocon (Johnson and van Waes 1999), the size and conformation of the mammalian nascent chain tunnel may change upon binding to the translocon. This possibility is suggested by experiments in which large collisional quenching agents were able to quench probes located partway up the nascent chain tunnel inside a ribosome that was bound to an intact and functional translocon in the ER membrane (Hamman et al. 1997). At present, no cryo-EM or crystallographic data are available for ribosomes bound in a functional manner to translocons that are intact (i.e., contain all of the translocon components) and are embedded in a phospholipid bilayer. The fluorescence data also showed that the mammalian nascent chain tunnel is coaxially aligned with an aqueous pore in the translocon (Fig. 1) (Crowley et al. 1993, 1994). As expected, this alignment is mediated by direct ribosome–translocon interactions since cryo-EM images show the ribosomal tunnel aligned with, but not joined to, the translocon pore in the absence of the ER membrane and some translocon components (Beckmann et al. 1997).

Placing fluorescent probes in the nascent chains of membrane-bound ribosomes revealed several important structural and mechanistic aspects of the ribosome. First, the nascent chain product is in an aqueous environment while associated with the ribosome. Second, the carboxy-terminal residues of the nascent chain are localized inside the ribosome in an aqueous space that is contiguous with the cytosol in free ribosomes, but not membrane-bound ribosomes. Third, upon binding to a translocon in the ER membrane, the ribosome forms a tight seal that both prevents nascent secretory proteins from moving into the cytosol (the incorrect destination) and also maintains the permeability barrier of the ER membrane despite the presence of a large aqueous pore in the translocon. Fourth, these structural properties of the ribosome and translocon are sufficient to explain the mechanism by which secretory proteins are translocated across the ER membrane. Once the ribosome binds tightly to the translocon, a growing nascent secretory protein has no option but to move through the ribosomal tunnel and translocon pore into the ER lumen because the tight ribosome–translocon seal does not allow any alternative pathways for the nascent chain (Fig. 1). Thus, nascent chain translocation across the membrane is dictated by the topography of the ribosome and translocon (Crowley et al. 1994), a fact that highlights the active and critical involvement of the ribosome in a process other than protein biosynthesis.

NEW RIBOSOMAL FUNCTIONS REVEALED BY MEMBRANE-BOUND RIBOSOMES

Membrane-bound ribosomes in mammals facilitate both the complete translocation of secretory and some organellar proteins through the ER membrane and also the insertion of membrane proteins into the ER bilayer. Each of these cotranslational processes must be carried out

without disrupting the permeability barrier of the ER membrane because calcium ions are stored in the ER and the unregulated release of these potent second messengers would have severe metabolic consequences for the cell. As noted in the previous section, maintaining the permeability barrier during cotranslational translocation is straightforward: The ribosome simply binds to the translocon and maintains a tight seal throughout translocation (Fig. 1) (Crowley et al. 1994). However, membrane protein integration is much more complicated. Not only must transmembrane (TM) sequences in the nascent chain be inserted into the bilayer and moved laterally at some point during integration, but one or more domains of each protein will have to end up on the cytoplasmic side of the ER membrane. If the tight ribosome–translocon junction is responsible for maintaining the permeability barrier, how does a cytoplasmic domain of a nascent membrane protein enter the cytosol without disrupting this seal?

By placing a fluorescent probe in a nascent membrane protein and examining its accessibility to cytoplasmic and lumenal iodide ions at different stages of the integration process (i.e., at different lengths of the nascent chain), we were able to identify the mechanisms by which the system maintains membrane integrity during integration. In short, the ribosome–translocon seal is broken to allow the cytoplasmic domain of a nascent membrane protein to move into the cytosol, but only after the other end of the aqueous translocon pore is first closed (Fig. 2) (Liao et al. 1997). Surprisingly, the lumenal end of the pore is closed while the TM sequence is still inside the ribosome! Thus, it is the *ribosome* that first detects the TM sequence in the nascent chain, and it is the *ribosome* that initiates the conversion of the operational mode of the

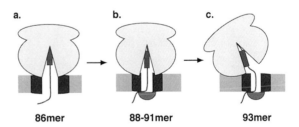

Figure 2. Ribosome-initiated changes at the ER membrane during co-translational protein integration. A newly synthesized TM segment (indicated by the dark gray rectangle in the nascent chain) in a 111p nascent chain elicits changes at the ER membrane from well inside the nascent chain tunnel. (*a*) The ribosome–translocon seal is intact, and the lumenal end of the translocon pore is open when the 111p nascent chain is 86 amino acids in length. (*b*) A 111p nascent chain of 88 to 91 amino acids is sealed off from both the cytosol and the ER lumen. The mechanism by which the lumenal end of the pore is closed is still unknown, but is arbitrarily shown here as resulting from the binding of a soluble lumenal protein that plugs the pore. (*c*) The ribosome–translocon seal is broken when the 111p nascent chain reaches a length of 93 amino acids, but the lumenal seal remains closed. Although the ribosome is depicted as rotating relative to the translocon to expose the nascent chain tunnel to the cytosol, the actual nature and extent of the conformational changes are unknown (see Liao et al. 1997).

molecular machinery at the ER membrane from translocation to integration. Since the TM sequence is very near the peptidyl transferase center when changes occur at the membrane surface, it is clear that there is a very long signal transduction pathway within the ribosome that is able to communicate from near the peptidyl transferase center to the other side of the ER membrane, a distance in excess of 100 Å (Liao et al. 1997).

The functions of the ribosome thus extend far beyond protein synthesis. Besides mediating the targeted delivery of secretory and membrane proteins to the ER membrane, the ribosome is also a critical component of the cellular machinery that maintains the permeability of the ER membrane. Furthermore, it is evident from the above studies that translocation and integration are achieved by the highly coupled and coordinated actions of the ribosome and translocon. In effect, the membrane-bound ribosome and the translocon must function as a single unit both structurally and functionally.

NASCENT CHAIN INTERACTIONS AFTER EMERGING FROM THE RIBOSOME

The use of probes incorporated into the nascent chain to examine the interactions of nascent chains with non-ribosomal molecules and complexes has proven extremely beneficial. Such probes have identified the SRP component to which the signal sequence binds (Krieg et al. 1986; Kurzchalia et al. 1986), as well as the ER membrane proteins that form the translocon pore (for review, see Johnson and van Waes 1999). Nascent chain probes have also revealed many aspects of the structure and function of the molecular machinery that accomplishes protein translocation and integration at the ER membrane (Johnson and van Waes 1999), such as demonstrating that secretory proteins pass through the ER membrane via an aqueous pore in the translocon (Crowley et al. 1994). In addition, such probes have shown, among other things, that chaperonins interact much sooner with ribosome-bound nascent chains than had been thought previously (McCallum et al. 2000). This approach therefore promises to provide a wealth of information about nascent chain interactions and environment subsequent to its departure from the ribosome, as is evidenced by the following.

SRP CONTROL OF RIBOSOME FUNCTION AND MEMBRANE BINDING

SRP regulates ribosome function via a signal-sequence-dependent conversion of a free ribosome to a membrane-bound ribosome (Walter and Johnson 1994). To effect this transition, SRP selects a ribosome that is synthesizing a nascent secretory or membrane protein, arrests elongation, and then targets the ribosome to the ER membrane to carry out cotranslational translocation or integration at the translocon. Thus, as far as the ribosome is concerned, SRP has a critical regulatory role within the cytoplasm of the cell.

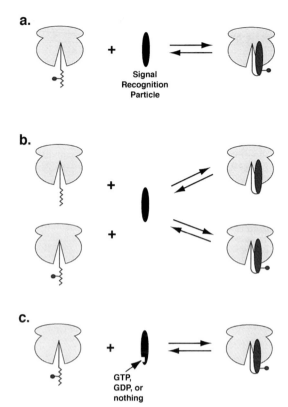

Figure 3. Fluorescence-detected binding of SRP to RNCs. (*a*) Direct titration. The binding of SRP (*black oval*) to NBD-labeled RNCs in a sample reaches equilibrium after each addition of SRP to an RNC-containing cuvette. The signal sequence is indicated by the sawtooth representation at the amino-terminal end of the nascent chain, and the fluorescent probe covalently attached to the signal sequence is indicated by the small black circle. (*b*) Competition experiment. When SRP is titrated into a cuvette containing both NBD-labeled RNCs and nonfluorescent RNCs, SRP binding to each RNC species reaches equilibrium soon after each SRP addition. Since both equilibria must be satisfied simultaneously in the cuvette, the extents of SRP association with NBD-RNC and RNC when SRP is limiting will indicate the relative affinity of SRP for each RNC species. (*c*) Nucleotide effect. The nucleotide dependence of SRP-RNC association is determined by carrying out titrations in the absence of nucleotide or in the presence of excess GTP or GDP.

A ribosome is selected for binding to the membrane when its nascent chain contains a cytosol-exposed signal sequence that is recognized and bound by the SRP. Since the accuracy of the selection process depends on the accuracy of the interaction between the signal sequence and the SRP, it is interesting that the SRP is somewhat promiscuous in its selection of signal sequences. There is no consensus for the sequences that can function as signal sequences (von Heijne 1985). The discrimination between acceptable and unacceptable sequences is presumably based on the binding affinity between the SRP and the putative signal sequence (Fig. 3a), and one would therefore expect that legitimate signal sequences in RNCs would bind much more tightly to SRP than would random nascent chain sequences. An early study by Walter et al. (1981) estimated SRP affinity for RNCs programmed with different mRNAs and concluded that the signal sequence greatly increased SRP affinity for the RNC. Although the reported affinities were recognized to be somewhat lower than the actual affinities because of the non-equilibrium sedimentation technique used to quantify complex formation, the data were consistent with expectations.

Subsequent studies revealed that the selection process was far more complex than originally envisioned. For example, it was discovered that targeting of ribosomes to the ER membrane was blocked when the length of the nascent chain exceeded 140 amino acids (Siegel and Walter 1988). Since addition of excess SRP relieved this blockage, it was concluded that the affinity of SRP for the signal sequence was dependent on the length of nascent chain. However, no direct measurements of SRP affinity for RNCs with different nascent chain lengths have yet been made.

The SRP-dependent selection process became even more complicated when additional information about SRP structure and function became available. Photocrosslinking experiments, using probes incorporated into the signal sequence, identified SRP54, the 54-kD protein subunit of SRP, as the binding site for the signal sequence in SRP (Krieg et al. 1986; Kurzchalia et al. 1986). It was later shown that the SRP54 also contained the sole GTP-binding site in SRP (Bernstein et al. 1989), thereby raising the intriguing possibility that signal sequence binding by SRP may be regulated by the occupancy of the GTP-binding site. Indeed, several laboratories began to investigate this possibility, and the resulting data led to conflicting models. One model proposed that SRP binds to the signal sequence only when the GTP-binding site in SRP54 is empty (Miller et al. 1993; Rapiejko and Gilmore 1997), whereas other models proposed that the RNC signal sequence binds only to SRP54•GDP (Bacher et al. 1996) or, alternatively, to SRP54•GTP (Althoff et al. 1994). Thus, the mechanism by which SRP selects RNCs for targeting to the ER membrane is both complex and controversial.

SPECTROSCOPIC DETECTION AND QUANTIFICATION OF SRP BINDING TO RNCs

Rationale

The most direct way to examine SRP–signal sequence interactions is to position a spectroscopic probe in the signal sequence and monitor its association with SRP by any spectral changes that are elicited as SRP is titrated into a cuvette containing probe-labeled RNCs. The primary advantage of using spectroscopy to detect complex formation is that most spectroscopic techniques are nondestructive. Thus, one can evaluate the extent of complex formation continuously, either as a function of time or under equilibrium conditions. By monitoring the extent of the spectral change (i.e., the extent of SRP•RNC complex formation) as a function of SRP concentration, one can determine the K_d for the complex (Fig. 3a). Furthermore,

if sufficient time is allowed for the molecules in the cuvette to reach equilibrium, the measured K_d will be an actual equilibrium constant, not an estimated or approximate K_d such as those obtained using non-equilibrium techniques for determining the amount of complex in a sample.

Fluorescence spectroscopy is the most sensitive spectroscopic approach available because the number of molecules required to obtain a measurable signal is much smaller than with any other technique. This property is especially important when one wants to measure the K_d for a high-affinity interaction between two ligands. Accurate K_d measurements require the determination of free ligand concentrations, and these are best measured when the ligand concentrations are near or below the K_d. Since SRP binds tightly to signal-sequence-containing RNCs (sedimentation data indicated a $K_d < 8$ nM; Walter et al. 1981), and since fluorescence is the only spectroscopic technique that can measure concentrations at the nanomolar level, SRP binding to RNCs must be detected and quantified using fluorescence. Fluorescence spectroscopy has the additional advantage that signal changes can be productively detected and analyzed in many different ways, including fluorescence lifetime, intensity, energy (excitation and emission wavelengths), polarization, FRET, and quenching. In principle, the extent of SRP binding to NBD-labeled RNCs could be monitored by changes in any one of these properties of NBD.

A major concern, of course, is whether the placement of a fluorescent dye in the signal sequence to monitor SRP binding will interfere with the binding of SRP to the signal sequence. If such interference does occur, then the K_d measured by this approach will be inaccurate. However, the spectroscopic approach also allows us to determine, directly and easily, the K_d for SRP binding to the same RNC lacking the fluorescent probe. The K_d for the nonfluorescent RNC is determined by its ability to compete with the fluorescent RNC (NBD-RNC) for binding to a limited amount of SRP (Fig. 3b). Since the amount of SRP•NBD-RNC in a sample is given directly by the magnitude of the observed spectral change, the extent of competition by the nonfluorescent RNC for binding to a limited amount of SRP is given by the extent to which the RNC reduces SRP•NBD-RNC complex formation (i.e., lowers the extent of the spectral change). The K_d for the nonfluorescent RNC can then be determined from these spectral data because the SRP titration is carried out at equilibrium.

At every concentration of SRP, the distribution of SRP (i.e., free, bound to RNC, or bound to NBD-RNC) is thus dictated by the simultaneous equilibria that reflect SRP affinity for each of the RNC species. By simultaneously solving the equations for the two equilibria (the equations can be combined because the free SRP concentration is the same for both), it is easy to show algebraically that one can determine the K_d for the SRP•RNC complex from the known parameters (the spectroscopically determined concentration of SRP•NBD-RNC in the sample, the previously or simultaneously determined K_d for SRP•NBD-RNC, and the total concentrations of SRP, RNC, and

NBD-RNC). Thus, it is irrelevant whether or not the fluorescent probe interferes, either positively or negatively, with SRP binding to the signal sequence containing RNC, because competition experiments will determine the actual K_d values for any natural species that lacks a probe. In effect, the NBD-RNC species is simply used as a spectral probe to assay, independently, free SRP concentrations in the titration.

By determining the K_d of the SRP•RNC complex under different conditions, the effect, if any, of a particular variable on SRP affinity can be quantified. For example, SRP titrations of fluorescent-labeled RNCs carried out in the presence of excess GTP, excess GDP, or no nucleotide will reveal to what extent occupancy of the nucleotide-binding site on SRP54 affects SRP binding to the ribosome-bound signal sequence (Fig. 3c). The nucleotide dependence of SRP selection of ribosomes for binding to the ER membrane can therefore be determined directly using this approach.

As noted above, studies with free ribosomes are complicated by the fact that not all ribosomes in a sample will synthesize a nascent chain. The reason(s) for this heterogeneity is (are) not understood, but typically only 30–50% of the ribosomes in an in vitro sample function in translation. Since there is no easy way to separate functioning ribosomes from inactive ribosomes (aside from membrane binding, as discussed above), it is conceivable that these inactive ribosomes could interfere with any titration by SRP if the inactive ribosomes have some affinity for SRP. To assess this possibility, one can directly measure the K_d for SRP binding to nontranslating ribosomes by a simple competition experiment as described above. Such experiments reveal that SRP does bind to nontranslating ribosomes with a K_d that is ~200-fold higher than that for RNCs. Thus, we routinely include the association between SRP and nontranslating ribosomes as one of the simultaneous equilibria in our determinations of SRP binding to RNCs.

Preparation of Translation Intermediates

Homogeneous samples of RNCs were prepared by translation of truncated preprolactin (pPL) mRNAs in wheat germ extract as described elsewhere (Crowley et al. 1993). NBD probes were incorporated into the nascent chain wherever there was a lysine codon in the mRNA by doing the translation in the presence of εNBD-[14C]Lys-tRNA as before (Crowley et al. 1993). For example, for pPL nascent chains that were 65 amino acids in length (pPL$_{65}$), εNBD-Lys probes were present either in position 4 or position 9, the locations of lysine in the natural pPL sequence. In some experiments, VW202p, a chimera formed by the fusion of a pPL signal sequence to a long Bcl-2-based polypeptide sequence lacking lysines, was used to ensure that probes were incorporated only into the pPL signal sequence and not into the mature region of the protein (Crowley et al. 1994). (Since the added εNBD-Lys-tRNA must compete with the normal complement of Lys-tRNA in the wheat germ extract, only 25% of the

lysines incorporated into protein are εNBD-Lys [Krieg et al. 1989].) RNCs were then purified by gel filtration to remove residual εNBD-Lys-tRNAs and deacylated εNBD-Lys (Liao et al. 1997). In some experiments, truncated pPL mRNAs with an amber codon inserted into the signal sequence were translated in the presence of the modified amber suppressor tRNA, εNBD-[14C]Lys-tRNA^{amb}, to obtain NBD–RNC complexes in which every nascent chain of the desired length had a single probe. In every experiment, parallel samples were prepared with unmodified [14C]Lys-tRNA or [14C]Lys-tRNA^{amb} to correct for light scattering and background fluorescence (see, e.g., Crowley et al. 1993). The total number of ribosomes in a sample was determined by absorbance at 260 nm (Sperrazza et al. 1980). The total number of probes in a sample was determined by its radioactivity content. The total number of nascent chains was equal to the number of incorporated probes in the amber suppressor experiments, and was twice the number of incorporated probes whenever the probes were incorporated into the pPL or VW202p signal sequence using εNBD-[14C]Lys-tRNA (see above).

Fluorescence Spectroscopy

Fluorescence experiments were done at 4°C in 50 mM HEPES (pH 7.5), 140 mM KOAc, 5 mM Mg(OAc)$_2$, 1 mM DTT, using the instruments, microcuvettes, and other procedures (e.g., mixing, sample-blank matching) described earlier (Crowley et al. 1993). SRP was titrated into samples by sequential addition of known amounts of SRP (determined by absorbance at 280 nm) to the sample volume. After each addition of SRP and mixing, a constant fluorescence intensity signal was obtained (i.e., equilibrium was reached) within 5 minutes. After blank subtraction and dilution correction (in that order), the emission intensity of the sample at any point in the titration (F) was compared to its initial intensity in the absence of any SRP (F$_0$). In all experiments, F/F$_0$ reached a maximum value as SRP was increased, as expected for SRP saturation of NBD-RNC-binding sites. All titration data were analyzed as competition experiments (since every sample contained both NBD–RNCs and inactive ribosomes) using the equations and analysis routines detailed in Bock et al. (1997).

Titration Results

Upon addition of SRP to a sample of NBD-VW202$_{65}$-RNCs formed using εNBD-Lys-tRNA, the fluorescence intensity increases and reaches a maximal intensity increase of 65% (Fig. 4). This spectral change requires functional and intact SRP because neither NEM-inactivated SRP nor SRP lacking SRP54 is able to elicit a fluorescence change in NBD-RNCs (data not shown). Thus, SRP binding to the signal sequence in a ribosome-bound nascent chain can indeed be detected spectroscopically.

Analysis of the extent of the fluorescence change as a function of SRP concentration revealed that the K_d for the

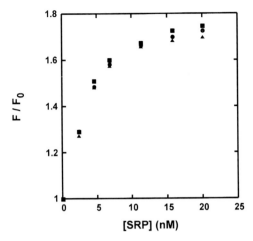

Figure 4. Fluorescence-detected binding of SRP to NBD-RNC and its dependence on GTP or GDP. SRP was titrated into samples containing 12.5 nM NBD-RNCs with a 65-residue-long VW202p nascent chain and either no nucleotide (▲), 0.2 mM GTP (●), or 0.2 mM GDP (■). εNBD-Lys was incorporated into the VW202p nascent chains at positions 4 and/or 9 in the natural pPL signal sequence. Since the signal sequence in this 65-residue nascent chain has not completely emerged from the ribosome, the maximum SRP-dependent change in fluorescence intensity observed here is larger than in samples containing an 84-residue nascent chain (see Fig. 5).

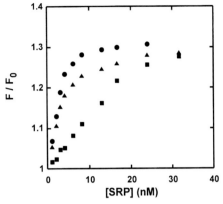

Figure 5. Competition experiments to assess the effects of the NBD probe and of the nascent chain on SRP binding to RNCs. SRP was titrated into a sample containing 4.4 nM NBD-RNCs with an 84-residue-long VW202p nascent chain (▲). Parallel samples also contained, in addition to 4.4 nM NBD-RNC, either 94 nM of nontranslating ribosomes (●) or 17 nM RNCs that lacked any εNBD-Lys residues (■).

SRP•NBD-RNC was sub-nanomolar. For example, the average K_d obtained for the SRP•NBD-pPL$_{86}$-RNC complex prepared with an amber suppressor tRNA was 0.2 ± 0.1 nM ($n = 12$). Furthermore, when competition experiments were done using native pPL$_{86}$-RNCs, it was found that the presence of the probe did not detectably alter the binding affinity of the SRP for the signal sequence (data not shown). Thus, consistent with the idea that nascent chain-SRP affinity dictates which ribosomes are selected for targeting to the ER membrane, the SRP binds with very high affinity to the ribosome-bound signal sequence.

The influence of GTP, if any, on SRP selection of RNCs for binding to the ER membrane was examined by determining the effect of GTP on the K_d of the SRP•NBD-RNC complex. In contrast to the proposed models, the association of SRP with RNCs was found to be independent of GTP or GDP. Not only are parallel titrations unaffected by the presence or absence of GTP or GDP (Fig. 4), but the calculated K_d values for samples containing GTP, GDP, or no nucleotide were equivalent. Thus, the initial binding of SRP to a ribosome-bound signal sequence is nucleotide-independent.

As for SRP affinity for ribosomes lacking signal sequences, competition experiments show that even nontranslating ribosomes bind SRP. However, their affinity for SRP is much lower than that of an RNC with a signal sequence, as is evident by comparing the extents of competition for SRP by a 4-fold excess of RNCs over NBD-RNCs and by a 20-fold excess of nontranslating ribosomes over NBD-RNCs (Fig. 5). The low, but measurable, affinity of SRP for empty ribosomes suggests that SRP will bind to ribosomes lacking exposed signal

sequences and thereby scan the emerging nascent chain for the presence of a signal sequnce.

CONCLUSIONS

Examination of ribosome structure, function, and regulation from the point of view of either the substrate or the product is an extremely valuable and necessary approach for obtaining a complete understanding of the ribosome. The value of this approach is exemplified both by the many previously published studies, some of which are cited in this manuscript, and by the early data we have obtained using fluorescent-labeled nascent chains to characterize SRP binding to signal sequences. When completed, the latter experiments promise to reveal the mechanisms involved in SRP selection of ribosomes for targeting to the membrane, complete with the thermodynamics and kinetics of this process. For example, are RNCs with nascent chains longer than 140 amino acids unable to target to the ER membrane because SRP is unable to bind to nascent chains longer than 140 amino acids? By focusing on the nascent chain product or the aa-tRNA substrate, one can directly test the validity of many hypotheses about the ribosome, including those that have already become dogma.

ACKNOWLEDGMENTS

Work in the authors' laboratories was supported by National Institutes of Health grants GM-26494 (A.E.J.) and H.L.-38779 (P.E.B.), and by the Robert A. Welch Foundation (A.E.J.).

REFERENCES

Agrawal R.K., Spahn C.M.T., Penczek P., Grassucci R.A., Nierhaus K.H., and Frank J. 2000. Visualizaion of tRNA movements on the *Escherichia coli* 70S ribosome during the elongation cycle. *J. Cell Biol.* **150:** 447.

Althoff S.M., Stevens S.W., and Wise J.A. 1994. The Srp54 GTPase is essential for protein export in the fission yeast *Schizosaccharomyces pombe. Mol. Cell. Biol.* **14:** 7839.

Bacher G., Lütcke H., Jungnickel B., Rapoport T.A., and Dobberstein B. 1996. Regulation by the ribosome of the GTPase of the signal-recognition particle during protein targeting. *Nature* **381:** 248.

Beckmann R., Bubeck D., Grassucci R., Penczek P., Verschoor A., Blobel G., and Frank J. 1997. Alignment of conduits for the nascent polypeptide chain in the ribosome-Sec61 complex. *Science* **278:** 2123.

Bernabeu C. and Lake J.A. 1982. Nascent polypeptide chains emerge from the exit domain of the large ribosomal subunit: Immune mapping of the nascent chain. *Proc. Natl. Acad. Sci.* **79:** 3111.

Bernabeu C., Tobin E.M., Fowler A., Zabin I., and Lake J.A. 1983. Nascent polypeptide chains exit the ribosome in the same relative position in both eucaryotes and procaryotes. *J. Cell Biol.* **96:** 1471.

Bernstein H.D., Poritz M.A., Strub K., Hoben P.J., Brenner S., and Walter P. 1989. Model for signal sequence recognition from amino-acid sequence of 54K subunit of signal recognition particle. *Nature* **340:** 482.

Blobel G. and Sabatini D.D. 1970. Controlled proteolysis of nascent polypeptides in rat liver cell fractions. I. Location of the polypeptides within ribosomes. *J. Cell Biol.* **45:** 130.

Bock P.E., Olson S.T., and Björk I. 1997. Inactivation of thrombin antithrombin is accompanied by inactivation of regulatory exosite I. *J. Biol. Chem.* **272:** 19837.

Cornish V.W., Benson D.R., Altenbach C.A., Hideg K., Hubbell W.L., and Schultz P.G. 1994. Site-specific incorporation of biophysical probes into proteins. *Proc. Natl. Acad. Sci.* **91:** 2910.

Crowley K., Reinhart G.D., and Johnson A.E. 1993. The signal sequence moves through a ribosomal tunnel into a noncytoplasmic aqueous environment at the ER membrane early in translocation. *Cell* **73:** 1101.

Crowley K.S., Liao S., Worrell V.E., Reinhart G.D., and Johnson A.E. 1994. Secretory proteins move through the endoplasmic reticulum membrane via an aqueous, gated pore. *Cell* **78:** 461.

Do H., Falcone D., Lin J., Andrews D.W., and Johnson A.E. 1996. The cotranslational integration of membrane proteins into the phospholipid bilayer is a multistep process. *Cell* **85:** 369.

Fisher E.A., Zhou M., Mitchell D.M., Wu X., Omura S., Wang H., Goldberg A.L., and Ginsberg H.N. 1997. The degradation of apolipoprotein B100 is mediated by the ubiquitin-proteasome pathway and involves heat shock protein 70. *J. Biol. Chem.* **272:** 20427.

Frydman J. 2001. Folding of newly translated proteins in vivo: The role of molecular chaperones. *Annu. Rev. Biochem.* **70:** 603.

Hamman B.D., Chen J.-C., Johnson E.E., and Johnson A.E. 1997. The aqueous pore through the translocon has a diameter of 40-60 Å during cotranslational protein translocation at the ER membrane. *Cell* **89:** 535.

High S., Martoglio B., Görlich D., Andersen S.S.L., Ashford A.J., Giner A., Hartmann E., Prehn S., Rapoport T.A., Dobberstein B., and Brunner J. 1993. Site-specific photocrosslinking reveals that Sec61p and TRAM contact different regions of a membrane-inserted signal sequence. *J. Biol. Chem.* **268:** 26745.

Janiak F., Dell V.A., Abrahamson J.K., Watson B.S., Miller D.L., and Johnson A.E. 1990. Fluorescence characterization of the interaction of various transfer RNA species with elongation factor Tu•GTP: Evidence for a new functional role for elongation factor Tu in protein biosynthesis. *Biochemistry* **29:** 4268.

Johnson A.E. and Cantor C.R. 1980. Elongation factor-dependent affinity labelling of *Escherichia coli* ribosomes. *J. Mol. Biol.* **138:** 273.

Johnson A.E. and Van Waes M.A. 1999. The translocon: A dy-

namic gateway at the ER membrane. *Annu. Rev. Cell Dev. Biol.* **15:** 799.

Johnson A.E., Miller D.L., and Cantor C.R. 1978. Functional covalent complex between elongation factor Tu and an analog of lysyl-tRNA. *Proc. Natl. Acad. Sci.* **75:** 3075.

Johnson A.E., Adkins H.J., Matthews E.A., and Cantor C.R. 1982. Distance moved by transfer RNA during translocation from the A site to the P site on the ribosome. *J. Mol. Biol.* **156:** 113.

Johnson A.E., Woodward W.R., Herbert E., and Menninger J.R. 1976. N$^\epsilon$-acetyllysine transfer ribonucleic acid: A biologically active analogue of aminoacyl transfer ribonucleic acids. *Biochemistry* **15:** 569.

Krieg U.C., Johnson A.E., and Walter P. 1989. Protein translocation across the endoplasmic reticulum membrane: Identification by photocross-linking of a 39 kD integral membrane glycoprotein as part of a putative translocation tunnel. *J. Cell Biol.* **109:** 2033.

Krieg U.C., Walter P., and Johnson A.E. 1986. Photocrosslinking of the signal sequence of nascent preprolactin to the 54-kilodalton polypeptide of the signal recognition particle. *Proc. Natl. Acad. Sci.* **83:** 8604.

Kurzchalia T.V., Wiedmann M., Girshovich A.S., Bochkareva E.S., Bielka H., and Rapoport T.A. 1986. The signal sequence of nascent preprolactin interacts with the 54K polypeptide of the signal recognition particle. *Nature* **320:** 634.

Liao S., Lin J., Do H., and Johnson A.E. 1997. Both lumenal and cytosolic gating of the aqueous ER translocon pore is regulated from inside the ribosome during membrane protein integration. *Cell* **90:** 31.

Malkin L.I. and Rich A. 1967. Partial resistance of nascent polypeptide chains to proteolytic digestion due to ribosomal shielding. *J. Mol. Biol.* **26:** 329.

McCallum C.D., Do H., Johnson A.E., and Frydman J. 2000. The interaction of the chaperonin tailless complex polypeptide 1 (TCP1) ring complex (TRiC) with ribosome-bound nascent chains examined using photo-cross-linking. *J. Cell Biol.* **149:** 591.

Miller J.D., Wilhelm H., Gierasch L., Gilmore R., and Walter P. 1993. GTP binding and hydrolysis by the signal recognition particle during initiation of protein translocation. *Nature* **366:** 351.

Nissen P., Hansen J., Ban N., Moore P.B., and Steitz T.A. 2000. The structural basis of ribosome activity in peptide bond synthesis. *Science* **289:** 920.

Pariyarath R., Wang H., Aitchison J.D., Ginsberg H.N., Johnson A.E., and Fisher E.A. 2001. Co-translational interactions of apoprotein B with the ribosome and translocon during its assembly with lipids in the endoplasmic reticulum or its targeting to proteasomal degradation in the cytosol. *J. Biol. Chem.* **276:** 541.

Rapiejko P.J. and Gilmore R. 1997. Empty site forms of the SRP54 and SR$^\alpha$ GTPases mediate targeting of ribosome-nascent chain complexes to the endoplasmic reticulum. *Cell* **89:** 703.

Ryabova L.A., Selivanova O.M., Baranov V.I., Vasiliev V.D., and Spirin A.S. 1988. Does the channel for nascent peptide exist inside the ribosome? *FEBS Lett.* **226:** 255.

Siegel V. and Walter P. 1988. The affinity of signal recognition particle for presecretory proteins is dependent on nascent chain length. *EMBO J.* **7:** 1769.

Sperrazza J.M., Russell D.W., and Spremulli L.L. 1980. Reversible dissociation of wheat germ ribosomal subunits: Cation-dependent equilibria and thermodynamic parameters. *Biochemistry* **19:** 1053.

von Heijne G. 1985. Signal sequences. The limits of variation. *J. Mol. Biol.* **184:** 99.

Wagenknecht T., Carazo J.M., Radermacher M., and Frank J. 1989. Three-dimensional reconstruction of the ribosome from *Escherichia coli. Biophys. J.* **55:** 455.

Walter P. and Johnson A.E. 1994. Signal sequence recognition and protein targeting to the endoplasmic reticulum membrane. *Annu. Rev. Cell Biol.* **10:** 87.

Walter P., Ibrahimi I., and Blobel G. 1981. Translocation of proteins across the endoplasmic reticulum. I. Signal recognition protein (SRP) binds to in-vitro-assembled polysomes synthesizing secretory protein. *J. Cell Biol.* **91:** 545.

Yao Z., Tran K., and Mcleod R.S. 1997. Intracellular degradation of newly synthesized apolipoprotein B. *J. Lipid Res.* **38:** 1937.

Yonath A., Leonard K.R., and Wittmann H.G. 1987. A tunnel in the large ribosomal subunit revealed by three-dimensional image reconstruction. *Science* **236:** 813.

Yusupov M.M., Yusupova G.Z., Baucom A., Lieberman K., Earnest T.N., Cate J.H., and Noller H.F. 2001. Crystal structure of the ribosome at 5.5 Å resolution. *Science* **292:** 883.

The Active 80S Ribosome–Sec61 Complex

R. Beckmann,*† C.M.T. Spahn,‡ J. Frank,‡§ and G. Blobel*
*Laboratory of Cell Biology, Howard Hughes Medical Institute, The Rockefeller University, New York,
New York 10021; ‡Howard Hughes Medical Institute, Health Research Inc. at the Wadsworth Center,
Empire State Plaza, Albany, New York 12201-0509; §Department of Biomedical Science,
State University of New York at Albany, New York 12201

The atomic models of the ribosomal subunits, together with structural investigations of functional ribosomal complexes by X-ray crystallography and 3D cryo-EM, have led to a basic understanding of the structure and dynamics of protein biosynthesis in prokaryotic systems. In contrast, our knowledge about eukaryotic ribosomes is sparse. The evolutionary conservation of rRNA and ribosomal proteins suggests that the ribosomal subunits share a similar spatial arrangement and that the fundamental mechanism of protein biosynthesis is the same. However, the actual degree of similarity is not known, and significant differences are known to exist. Due to insertion elements in the rRNA molecules (Gerbi 1996) and the presence of 20 to 30 additional proteins (Wittmann-Liebold 1986; Wool et al. 1990; Planta and Mager 1998), eukaryotic ribosomes are larger than their prokaryotic counterparts. There are functional differences, e.g., in the mechanism of initiation (Sachs et al. 1997), or in the mechanism of elongation, as demonstrated by EF3, a third elongation factor in yeast (Triana-Alonso et al. 1995). Additionally, several antibiotics are specific for one class of ribosomes or the other (Spahn and Prescott 1996). Furthermore, the ribosome has to play a role in other biological processes in the cell such as protein folding (Hardesty et al. 2000) and protein transport (Beckmann et al. 1997, 2001; Menetret et al. 2000). Binding and recognition events in these processes do not necessarily take place at those ribosomal sites that are highly conserved among bacterial and eukaryotic ribosomes. These and other differences in function are likely to be reflected by differences in structure.

Translocation of nascent secretory proteins and insertion of membrane proteins is one of these prominent processes in the eukaryotic cell involving the ribosome. During cotranslational protein translocation, the translating ribosome is targeted in a signal sequence-dependent manner to the membrane of the eukaryotic endoplasmic reticulum (ER) and associates with the protein-conducting channel (PCC) (Matlack et al. 1998). The PCC forms the so-called translocon with additional membrane protein complexes, which interact with the translocating polypeptide, such as the signal peptidase complex (SPC) and the oligosaccharidyl transferase (OST) (Johnson and van Waes 1999). Specific binding and insertion of the signal sequence by the PCC leads to a high-salt-resistant interaction with the ribosome (Gorlich and Rapoport 1993; Kalies et al. 1994; Jungnickel and Rapoport 1995; Mothes et al. 1998) and to accessibility of an aqueous environment for the translocating peptide in the channel (Gilmore and Blobel 1985; Simon and Blobel 1991; Crowley et al. 1994). Hydrophobic domains of translocating polypeptides have access to the hydrophobic core of the lipid bilayer, and partitioning from the channel into the membrane by lateral opening of the PCC can take place (Martoglio et al. 1995; Borel and Simon 1996; Heinrich et al. 2000). The cytosolic domains of membrane proteins are translated and released from the ribosome–channel complex without dissociation of the RNC–channel complex (Mothes et al. 1997). Notably, during all these processes the ion permeability barrier of the membrane must be maintained. Fluorescence quenching studies suggest that a seal is provided by a tight junction between the translocon and the ribosome (Crowley et al. 1994). On the same basis, a diameter of 15 Å has been estimated for the aqueous pore in the inactive channel (which has been suggested to be plugged on the lumenal side by a Hsp70-like protein called BiP; Hamman et al. 1998), and of up to 60 Å for the pore of the active channel (Hamman et al. 1997). Release of cytosolic loops or transmembrane domains of nascent integral membrane proteins requires the seal between ribosome and channel to be opened, which would lead to ion conductance. However, communication between the ribosome and the PCC has been suggested to coordinate the release of cytosolic loops with gating of the channel such that an ion barrier is maintained at all times (Liao et al. 1997).

In eukaryotic cells, the PCC is formed by the trimeric Sec61 complex, consisting of the Sec61α, β, and γ subunits (Sec61p, Sbh1p, and Sss1p in yeast) with altogether 12 transmembrane helices (Gorlich and Rapoport 1993; Wilkinson et al. 1996). Two-dimensional (2D) electron microscopy (EM) maps revealed that three to four Sec61 complexes can form ring-like structures of 100 Å diameter, with a central indentation or pore (Hanein et al. 1996). The SecYEG complex, the bacterial homolog, forms similar oligomeric structures (Meyer et al. 1999; Manting et al. 2000). The first cryo-EM three-dimensional (3D) structure of the ribosome–Sec61 complex showed a donut-shaped Sec61 complex associated with the large ribosomal subunit (Beckmann et al. 1997). The position of

---

†Present address: Institute for Biochemistry, Charité, Humboldt University of Berlin, Monbijoustr. 2, 10117 Berlin, Germany.

the channel aligns its central pore with the exit of the tunnel in the large ribosomal subunit representing the conduit for the nascent chain. A gap between the channel and the ribosome was observed, which was attributed to the absence of a signal sequence (Beckmann et al. 1997). A cryo-EM study of mammalian ribosome–Sec61 complexes reported a similar spatial arrangement with several connections forming the ribosome–channel junction, but still leaving a gap (Menetret et al. 2000). The gap was observed even under conditions where a signal sequence was assumed to be present (however, these maps of presumably translating ribosomes show no tRNA density). This led to a challenge of the current view of seal formation between ribosome and channel (Menetret et al. 2000). Taken together, it is not clear which structural arrangement allows the ribosome–PCC complex to fulfill its different functions in cotranslational translocation.

A recent cryo-EM study of a complex containing the translating yeast 80S ribosome, a peptidyl tRNA, a nascent polypeptide chain including a signal sequence, and the Sec61 complex provided new structural insights (Beckmann et al. 2001; Spahn et al. 2001a). The P-site tRNA in this map demonstrated that the complex indeed represented a translating ribosome. The combination of the cryo-EM map at approximately 15 Å resolution with molecular rRNA and protein models allowed a molecular analysis of this complex.

PURIFICATION OF RNC AND RECONSTITUTION OF RNC–SEC61 COMPLEXES

A homogeneous population of ribosome–nascent chain complexes (RNCs) carrying a signal sequence was the prerequisite for a structural analysis of a reconstituted RNC–Sec61 complex by cryo-EM and 3D reconstruction. To that end, the first 120 amino acids of the yeast vacuolar type II membrane protein dipeptidylaminopeptidase B (DAP2) were used as a nascent chain. This protein is translocated cotranslationally in vivo (Ng et al. 1996) with the transmembrane domain in position 30 to 45 serving as an uncleaved signal sequence, which inserts into the membrane in a loop. The ribosomes were programmed in a yeast cell-free translation system with a truncated synthetic mRNA lacking a stop codon resulting in stalled nascent chains still bound to tRNA in the P site of the ribosome. To immunopurify RNCs based on the presence of the nascent chain, the mRNA encoded an additional amino-terminal HA tag.

The translation of the truncated mRNA resulted in the appearance of two major bands on the autoradiography after SDS-PAGE (Fig. 1a). The lower band represents the completed polypeptide, and the larger product migrating at about 40 kD represents the polypeptide still bound to tRNA (peptidyl tRNA). The smear in between is a result of ongoing hydrolysis of the peptidyl-tRNA bond in the basic environment of the gel during the run (Fig. 1a,b). A first fraction of crude ribosomes was isolated by spinning the translation reaction through a high-salt/high-sucrose cushion. This also resulted in dissociation of nonribosomal proteins from RNCs. To selectively bind the RNCs, the mixture was then incubated with biotinylated anti-HA antibodies and streptavidin-coupled magnetic beads. The RNCs were eluted by incubation with an excess of corresponding HA peptide and spun through a sucrose cushion again. This procedure results in a highly enriched RNC fraction (Fig. 1a), which, most importantly, is virtually free of empty ribosomes and provides sufficient amounts (0.25 OD_{260}, ~5 pmole) for further studies. The isolated RNCs are stable for at least 9 hours on ice without significant hydrolysis or dissociation of the peptidyl tRNA. Thus, a method has been provided for the isolation of a stable and homogeneous fraction of programmed (80S) ribosomes carrying the same kind of nascent chain kept in place by the P-site tRNA (Beckmann et al. 2001).

For the structural study, the active, i.e., translating and translocating, ribosome–channel complex was reconstituted using the RNC preparation and purified Sec61 complex in a membrane-free system. The nascent chain (HA-DP120) was chosen long enough to allow the signal sequence and an additional 40 amino acids to be fully emerged from the tunnel of the large ribosomal subunit (Blobel and Sabatini 1970). This configuration is known to facilitate a productive interaction of the signal sequence with the channel leading to the insertion of the nascent chain as a loop (Shaw et al. 1988; Jungnickel and Rapoport 1995; Mothes et al. 1998). Successful nascent chain insertion after incubation of the RNCs with solubilized Sec61 complex was monitored by protease protection assays (Fig. 1b). With RNCs alone, the nascent chain signal shifted completely to a lower molecular weight, reflecting that only a small part (30–35 amino acids), buried in the ribosomal tunnel, is protected (Fig. 1b). Addition of the Sec61 complex, however, results in protection of the full-length chain with different degrees of efficiency: Most efficient is an excess of Sec61 complex in the detergent deoxyBigChaps (DBC) with approximately 75–85% protected chains (Fig. 1b). Thus, reconstitution of an active RNC–channel complex with fully inserted nascent chain set the stage for the structural study.

CRYO-EM STRUCTURE OF THE TRANSLATING 80S RIBOSOME–SEC61 COMPLEX

Cryo-EM and single-particle 3D reconstruction resulted in the structure of the active RNC–channel complex determined at a resolution of 15.4 Å based on the Fourier shell correlation criterion using a cutoff at 0.5 (corresponding to 10.9 Å using a cutoff at 3*sigma). It was further processed to separate rRNA from protein densities (Spahn et al. 2000) and interpreted by docking of rRNA and protein structure models (Beckmann et al. 2001; Spahn et al. 2001a). The structure shows the typical overall appearance of the yeast 80S ribosome, with the small (40S) and large (60S) ribosomal subunits clearly recognizable (Fig. 2a) (Beckmann et al. 1997; Gomez-Lorenzo et al. 2000; Morgan et al. 2000). At the exit site of the large ribosomal subunit tunnel (Beckmann et al. 1997; Nissen et al. 2000), extra density, represent-

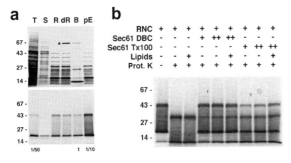

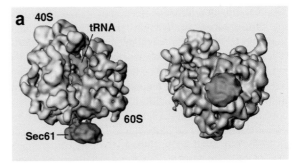

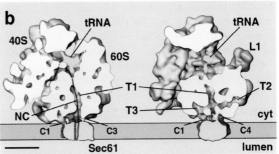

Figure 1. Immunopurification of RNCs and reconstitution of the RNC-Sec61 complex. (*a*) Truncated mRNA coding for the first 120 amino acids of dipeptidylpeptidase B with amino-terminal HA tag (HA-DP120) was translated in a yeast cell-free system in the presence of ^{35}S-labeled methionine. The translation reaction was subjected to an immunopurification procedure using a monoclonal anti-HA antibody. Aliquots of fractions were applied to SDS-PAGE, blotted onto nitrocellulose, amido black-stained (Protein) and autoradiographed to detect nascent polypeptide chains (^{35}S-Met). (T) Total, (S) supernatant, (r) crude ribosomes, (dR) depleted crude ribosomes after antibody and magnetic bead incubation, (pE) eluate using excess HA peptide, (B) magnetic beads after elution. Note the enrichment of HA-DP120 RNCs (pE). Asterisk marks bovine serum albumin, which is part of the antibody preparation. (*b*) Immunopurified RNCs carrying the HA-DP120 chain were reconstituted with different amounts of solubilized Sec61 complex in Deoxy-BigChap (DBC) or Triton X-100 detergent in the presence or absence of phospholipids. Subsequently the samples were subjected to protease digestion by incubation with proteinase K and analyzed by SDS-PAGE followed by autoradiography.

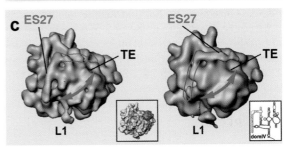

Figure 2. Cryo-EM structures of the translating 80S ribosome–Sec61 complex. (*a*) Reconstruction of the translating and translocating RNC–Sec61 complex in DBC at 15.4 Å resolution. Note the presence of tRNA density in the P site and the Sec61 channel with the shape of a compact disk. Color coding: yellow, small ribosomal subunit (40S); blue, large ribosomal subunit (60S); red, Sec61 complex; green, P-site tRNA. (*b*) The 15.4 Å reconstruction as shown in *a* (*left*) and rotated by 60° around the z axis (*right*) but cut perpendicularly to the plane of the membrane. Color code as in *a*. C1–C4 mark the connections between the ribosome and the channel. T1 marks the vertical tunnel, representing the expected path of the nascent chain, which is indicated (NC). T2 and T3 indicate the positions of two horizontal tunnels connecting the main vertical tunnel (T1) with the cytosolic environment. Note the gap between the ribosome and the channel. (Cyt) Cytosolic side, (lumen) ER lumen. Bar, 100 Å. (*c*) The reconstruction of a ribosome–Sec61 complex in Triton X-100 (Beckmann et al. 1997) shown without the channel density (*left*). TE marks the tunnel exit and L1 the L1 protrusion. Note the expansion segment 27 (ES27) in the L1 position, distant from the tunnel exit. The pink contour shows the exit position of ES27 as observed in another reconstruction of an empty ribosome (*right*). *Left inset:* The 15.4 Å reconstruction in the same orientation with the color code used in *a* and ES27 in the L1 position colored pink. *Right inset:* Secondary structure diagram of domain IV of *S. cerevisiae* 25S rRNA with colored ES27 (*green* and *pink*) and the moving helix labeled pink (diagram from R.R. Gutell, URL: http://www.rna.icmb.utexas.edu).

ing the Sec61 complex, is visible. The shape of the PCC in its active and inactive states (not shown) is similar to the shape observed previously (Beckmann et al. 1997), with a gap of at least 15 Å between the channel surface and the ribosome, regardless of the presence of a signal sequence. However, additional connections between the PCC and the ribosome are visible. Surprisingly, the PCC now appears as a compact disk with a central indentation rather than a toroidal mass with a central pore. Furthermore, additional density is visible in the intersubunit space of the programmed ribosome, unambiguously recognizable as tRNA (Fig. 2a,b).

The P-site tRNA (density shown with the same contour level as the ribosome) corresponds to occupancy of at least 80–90% (Figs. 2 and 6a). Thus, the translation of truncated mRNA indeed leads to a stalled peptidyl tRNA, with the vast majority tightly bound to the P site (resistant to high salt concentrations). The high occupancy of ribosomes with peptidyl tRNA is a good indication for the presence of the signal sequence-containing nascent chain, which is small (~14 kD) and too elongated to be visualized at this resolution. Therefore, the presence of the tRNA density in this map served as a prerequisite to permit the interpretation of the map as a truly active complex.

The connection between ribosome and channel is not circumferentially sealed but leaves a lateral opening of ~15 Å (Figs. 2 and 7). This finding is similar to what has been observed at lower resolution in several ribosome–channel reconstructions from yeast as well as from mammalian complexes (Beckmann et al. 1997;

Menetret et al. 2000). There are no significant differences in the dimensions of these gaps when comparing active and inactive complexes (Beckmann et al. 2001), which agrees well with protease protection experiments involving nascent chains of different lengths (Jungnickel and Rapoport 1995). With a lateral opening as large as 15 Å, it is unlikely that the ribosome–channel connection can function as an ion-tight seal. Moreover, a seal formed by the ribosome–membrane junction may be ineffective in light of the existence of additional tunnels in the large ribosomal subunit (T2 and T3, Fig. 2b), which are also present in bacterial and archaebacterial ribosomes (Frank et al. 1995; Ban et al. 2000; Gabashvili et al. 2001). Since these tunnels connect the conduit for the nascent chain with the cytosolic environment, they may allow ion flow between the cytosol and the ER lumen, even if a seal-forming junction between ribosome and membrane is present (Fig. 2b).

Interestingly, the large subunit bears a rod-like density 150 Å in length close to the channel attachment site (Fig. 2c). It was identified as the main helix of expansion segment 27 (ES27) of 25S rRNA (see Fig. 5), one of the rRNA insertions characteristic for 80S ribosomes (Gerbi 1996). In several reconstructions of different non-translating yeast ribosome complexes (Beckmann et al. 1997; Gomez-Lorenzo et al. 2000; Morgan et al. 2000) this helix can be found close to the L1 arm (L1 position) but also, following a counterclockwise rotation by 90 degrees, reaching over the tunnel exit site (exit position). Therefore, the ES27 helix is a highly dynamic structure with two main conformations in functional equilibrium (Fig. 2c). The ES27 helix was located exclusively in the L1 position in ribosome–channel complexes, thereby not interfering with the channel in the exit position. The ES27 helix may play a role in coordinating access of nonribosomal factors, such as chaperones, modifying enzymes, SRP, or the translocon to the tunnel exit site and to the emerging nascent chain. The function of ES27 is essential and conserved, because in *Tetrahymena*, deletion of this insertion is lethal but can be complemented with a corresponding insert from other species (Sweeney et al. 1994). The ES27 helix is an example of an extremely dynamic rRNA segment in the periphery of the 80S ribosome comparable to the L1 protuberance (Gomez-Lorenzo et al. 2000).

THE 80S RIBOSOME VERSUS THE 70S RIBOSOME

Compared to the 70S ribosome from *Escherichia coli* (Gabashvili et al. 2000), the 80S ribosome from *Saccharomyces cerevisiae* appears to be enlarged in certain areas of the periphery (Fig. 3a,b). The 40S ribosomal subunit shows the classic division into head, body, and platform (Dube et al. 1998; Verschoor et al. 1998). Due to extra density in the 40S subunit and a change in orientation of some elements, the 40S subunit is longer than its bacterial counterpart and shows a strong subdivision at the bottom part of the body into a "left foot" and a "right foot" (Fig. 3c). Extra density is mainly present (1) at the bottom of the 40S subunit, (2) below the platform, forming the "left foot" and the "back lobe" of the 40S subunit, (3) at the "beak", and (4) at the solvent side of the head, building the "head lobe." The 40S ribosomal subunit of *S. cere-*

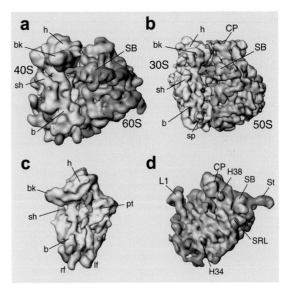

Figure 3. The 80S ribosome from *S. cerevisiae* compared with the *E. coli* 70S ribosome. (*a*) The cryo-EM map of the yeast 80S ribosome is shown together with the cryo-EM map of the *E. coli* 70S ribosome (*b*) (Gabashvili et al. 2000). The ribosomes are shown from the L7/L12 side (*a,b*). The isolated small subunit (*yellow, c*) and large subunit (*blue, d*) are shown from the interface sides. The P-site-bound tRNA is shown in green. Additional parts of the eukaryotic 80S ribosome that are due to expansion segments in the rRNAs and nonhomologous proteins are shown in gold (40S subunit) and purple (60S subunit). Landmarks for the 40S subunit: (b) body; (bk) beak; (h) head; (lf) left foot; (rf) right foot; (pt) platform; (sh) shoulder; (sp) spur. Landmarks for the 60S subunit: (CP) central protuberance; (L1) L1 protuberance; (SB) stalk base; (St) L7/L12 stalk; (H34) helix 34; (H38) helix 38, (SRL) sarcin-ricin loop.

visiae is composed of the 1798-nucleotide-long 18S rRNA and 32 ribosomal proteins. The 18S rRNA is 256 nucleotides longer than the 16S rRNA of *E. coli*, and the yeast 40S subunit contains 11 more proteins than the 30S subunit from *E. coli*. Consequently, both insertion elements in the 18S rRNA and additional ribosomal proteins are responsible for the extra mass of the small subunit in eukaryotes (Spahn et al. 2001a).

The 60S ribosomal subunit is enlarged by expansion segments and additional proteins roughly at two opposite sides (Fig. 3d). One region is behind the central protuberance and the stalk region, when looking at the 60S subunit in the crown view; the other region runs from the bottom of the 60S to the base of the L1 protuberance. The 60S ribosomal subunit of *S. cerevisiae* is built from 25S rRNA (3392 nucleotides), 5.8S rRNA (158 nucleotides), 5S rRNA (121 nucleotides), and 45 ribosomal proteins. Since the 5.8S rRNA is homologous to the 5′ region of the 23S rRNA, the LSU rRNA in yeast is 646 nucleotides longer than the bacterial 23S rRNA from *E. coli* (2904) and 505 nucleotides longer than the archaebacterial 23S rRNA from *Haloarcula marismortui* (3045 nucleotides). The yeast 60S subunit has 12 proteins more than the *E. coli* 50S subunit (33 different proteins) and 14 proteins more than the *H. marismortui* 50S subunit (31 proteins), which, together with the expansion segments in the rRNA, account for the extra mass (Spahn et al. 2001a).

DOCKING OF ATOMIC MODELS INTO SEPARATED DENSITIES OF THE CRYO-EM MAP

Molecular level information can be obtained by docking of atomic models of components into a lower-resolution cryo-EM map of a macromolecular complex. It is estimated that atomic models of known substructures can be positioned in a cryo-EM map with accuracy exceeding the resolution of the map severalfold (Rossmann 2000). To facilitate the docking, the map of the 80S ribosome was computationally separated (Spahn et al. 2000) into rRNA and protein maps first (Figs. 4–7). rRNA models of the recently solved crystal structures of the 30S subunit from *Thermus thermophilus* (Wimberly et al. 2000) and the 50S subunit from *H. marismortui* (Ban et al. 2000) were fitted into the resulting maps for the small subunit (SSU) rRNA (Fig. 4) and large subunit (LSU) rRNA of yeast (Fig. 5), respectively. Where necessary, the X-ray

models were modified by moving nonfitting parts (e.g., helices) as rigid bodies relative to the rest of the model. Homology models were calculated for those yeast ribosomal proteins that were found to be related to a bacterial or an archaebacterial protein of known 3D structure. In total, homology models were obtained for 43 yeast ribosomal proteins (15 for the 40S subunit [Fig. 4], 28 for the 60S subunit [Fig. 5]) and positioned in the protein map of the 80S ribosome (Spahn et al. 2001a).

Although the accuracy of the RNA–protein separation method is not absolute, the overall agreement of the separated maps with the atomic models is very good and allows an interpretation of the model on the molecular level. Excellent agreement between the shape of the protein densities and the homology models shows that the observed sequence similarities are indeed reflected in conserved tertiary structures (Figs. 4 and 5).

THE 40S RIBOSOMAL SUBUNIT

Of the SSU proteins of yeast, 15 have homologous proteins in bacteria whereas 17 have no homologous counterparts. Density in the 40S-protein map that is not accounted for by homology models predicts the locations of these proteins or those of additional domains in the homologous proteins. Most of the regions with additional protein density are present at the solvent side of the 40S subunit (Fig. 4a). Two such clusters of protein density are in the head of the 40S subunit, where they build the head lobe, located above protein rpS0 (S2p), and part of the beak, replacing helix 33a of the 16S rRNA. The protein density at the top of the platform surrounding rpS14 (S11p) is enlarged, leading to a stronger contact with rpS5 (S7p). This head–platform interaction defines part of the mRNA exit channel. Additional clusters are in the region around protein rpS0 (S2p), at the branch site of the helix 21 insertion (ES6), below helix 21, and at the 60S side of the 40S subunit at the bottom of the left and right foot (Fig. 4). There is no protein density corresponding to proteins S6p and S18p, leaving the 18S rRNA (the junction between helices 22 and 23; Fig. 4b) accessible at the solvent side of the platform, which might have functional implications for initiation (Spahn et al. 2001b).

The 18S rRNA map in the region that corresponds to the common core of the SSU rRNA is in good agreement with the X-ray model of the 16S rRNA, showing that the phylogenetic conservation of secondary structure is indeed reflected in the conservation of the tertiary structure (Fig. 4b). However, some differences due to conformational changes are apparent, e.g., in the arrangement of helix 16 and 17. Helix 16 in a mammalian 40S subunit has the same conformation as in the yeast 40S; it becomes broadly fused with the head domain upon binding of the HCV IRES (Spahn et al. 2001b). Therefore, the difference in the shoulder region between bacteria and eukaryotes might be important for the distinct mechanisms of translation initiation. Another difference between prokaryotic and eukaryotic 40S subunits is in the position and shape of the beak. Helix 33 is slightly longer in yeast than helix 33 of 16S rRNA, but it does not contain helix 33a, so that the overall rRNA element is shorter (Fig. 4b).

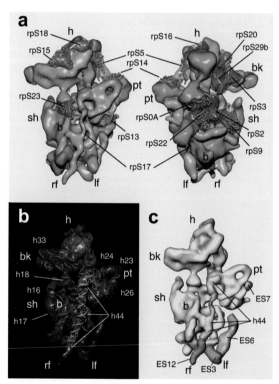

Figure 4. 40S subunit: RNA versus protein separation and docked structure models. (*a*) The 40S subunit is shown from the intersubunit side (*left*) and the solvent side (*right*). The RNA partition is painted in yellow at high contour level, the protein partition in green with transparency wherever homology models could be docked into the map. Solid green parts predict the positions of additional proteins without homologous counterpart in prokaryotes. The fitted homologous proteins are shown as ribbon models and are designated with their name. (*b*) The RNA map is shown transparently superposed with the ribbons model for the common conserved core of the 18S rRNA. The domains of the ribbons model for the SSU rRNA are color-coded: domain I, red; domain II, blue; domain III major, green; domain III minor, yellow. (*c*) The common core of the 18S rRNA is shown in yellow with the expansion segments highlighted in different colors. Several helices of the 18S rRNA are annotated with their numbers. Landmarks of the 40S subunit are as in Fig. 3.

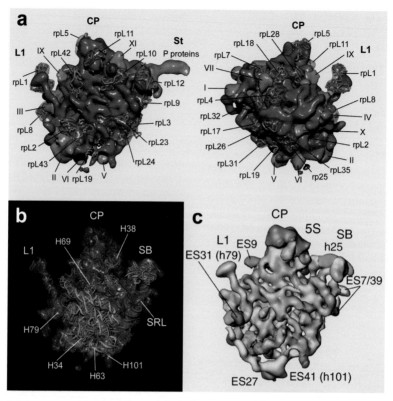

Figure 5. 60S subunit: RNA versus protein separation and docked structure models. (*a*) The 60S subunit is shown from the intersubunit side, in the classic crown view (*left*) and from the bottom of the L1 side (*right*). The RNA partition is shown in blue at a high contour level, the protein partition in orange and transparently wherever homology models could be docked into the map. Therefore, solid parts of the protein partition predict the position of additional proteins with no homologous counterpart in prokaryotes. These clusters of density are labeled with a roman number. The fitted homologous proteins are shown as ribbon models and are designated with their name. (*b*) The RNA map is shown transparently superposed with the ribbons model for the common conserved core of the 5.8S/25S and 5S rRNAs. The domains of the ribbons model are color-coded: domain I, blue; domain II, cyan; domain III, green; domain IV, yellow; domain V, red; domain VI, magenta; 5S rRNA, blue. Several helices are annotated with their numbers. (*c*) The common core of the 60S subunit rRNA is shown from the intersubunit side in blue with the expansion segments highlighted in different colors. Landmarks are as in Fig. 3.

Nevertheless, the length of the beak is increased due to the presence of an additional protein (Fig. 4a).

The localization of insertion element has been possible wherever unaccounted density in the rRNA map corresponds to insertions revealed in the secondary structure diagrams (Fig. 4c). Helix 44 was clearly resolved and its extension (ES12) builds the right foot of the 40S subunit (Dube et al. 1998). ES6, the helix 21 insertion, emerges at the solvent side of the platform, appearing to branch into two irregular helices, the larger branch of which constitutes the left foot. At its lower end, it appears to interact with expansion segment ES3. Only in eukaryotic ribosomes this structure creates a network of interactions connecting the platform of the 40S subunit with the lower part of the body.

THE 60S RIBOSOMAL SUBUNIT

For 28 yeast LSU proteins, homology models were calculated and docked into the cryo-EM map. For the 17 proteins for which the exact position is still unknown, as well as for additional domains in the homologous proteins, the remaining density predicts the positions (Fig. 4). Twelve regions with additional protein density at the surface of

the 60S subunit are present, most of them at the solvent side. The large protein cluster building the extended stalk can be identified as the P0/P1/P2 complex, the homolog of the bacterial L10(L7/L12)$_4$ complex. The extended stalk in yeast has a different orientation compared to the stalk in bacteria with a side lobe that is located close to the amino-terminal domain of rpL12 (L11p), where a contact occurs between the stalk–base region and elongation factor EF2 (Gomez-Lorenzo et al. 2000). This contact has never been observed in 70S·EF-G complexes from *E. coli* (Frank and Agrawal 2000) and might explain the ability of 80S ribosomes to discriminate between EF2 and EF-G (Uchiumi et al. 2001).

The atomic model of the *H. marismortui* 23S rRNA (Ban et al. 2000) could be fitted well into the RNA map of the 60S subunit (Fig. 5b). In addition, the 25S rRNA model was supplemented with the L1 region of domain IV of 23S rRNA from *T. thermophilus* (Yusupov et al. 2001) and L11-binding RNA from *Thermotoga maritima* (Wimberly et al. 1999), respectively. The position of some elements, however, had to be slightly readjusted: (1) domain VI of 25S rRNA, (2) helices 7, 18, 19, 20, and 24 of domain I around the exit site of the nascent peptide, (3) helix 25, (4) the sarcin-ricin-loop (SRL) with helices

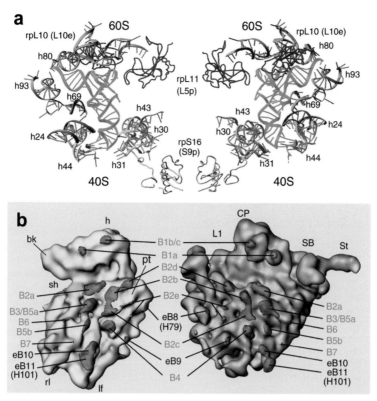

Figure 6. Environment of the P-site-bound tRNA and position of the intersubunit bridges. (*a*) Atomic models of the P-site-bound tRNA (*green*) together with the surrounding environment of the 60S subunit (*blue*) and 40S subunit (*yellow/gold*) are shown in the ribbons representation from the L1 side (*left*) and the L7/L12 side (*right*). The ribosomal components are designated. (*b*) The 40S subunit (*yellow*) and the 60S subunit (*blue*) are shown from their intersubunit sides. The intersubunit bridges are shown in red and annotated with their numbers. Bridges also observed in prokaryotic ribosomes are numbered B1–B7. Additional intersubunit connections in the yeast ribosome are named eB8–eB11.

89 and 91, (5) helix 42, which leads to the stalk base, and (6) 5S rRNA. Compared to the recent X-ray structure of the *T. thermophilus* 70S ribosome (Yusupov et al. 2001), large differences can be seen at the L1- and L11-bearing protuberances. Both structures are highly dynamic, in line with their emerging role in the entering and exiting of the tRNAs (Spahn et al. 2001a).

The yeast LSU rRNA contains several expansion segments (Gerbi 1996b) that are distributed over the whole 5.8S/25S rRNA, located roughly at two opposite sides of the surface of the 60S subunit (Fig. 5c). The expansion segments are not mere solvent-exposed protuberances, but are frequently involved in tertiary and quaternary contacts, providing additional interdomain interactions. Two expansion segments even form additional bridges with the 40S subunit (Fig. 6b). Expansion segments in 25S rRNA domains II and VI are responsible for the additional RNA density behind the stalk region. Two smaller insertions into domain II of 25S rRNA, ES9 and ES12, are located at the back of the central protuberance. Insertions into domains I, II, and V of 5.8S/25S rRNA appear to form a new network of strands below the L1 stalk. The third major insertion, ES27 in helix 63, is located at the L1 side, toward the bottom of the 60S subunit (Fig. 5, see above). Two more expansion segments are located at the bottom of the 60S subunit. ES41, an extension of he-

lix 101 of 25S rRNA, makes an intersubunit bridge with the 40S subunit (Fig. 6, see below). ES24, an extension of helix 59 of 25S rRNA, appears to be involved in the binding of Sec61 (Beckmann et al. 2001).

tRNA–RIBOSOME INTERACTION AT THE P SITE

The translating 80S ribosome clearly displays an L-shaped density in the intersubunit space that corresponds to the P-site-bound peptidyl tRNA in its entirety (Fig. 2). Only the single-stranded 3′ CCA end is not resolved. The density shows indications of major and minor grooves of the tRNA molecule and allows an accurate docking of the atomic structure of the tRNA derived from a tRNA•70S complex from *T. thermophilus* (Cate et al. 1999; Yusupov et al. 2001). The tRNA density is fused with the ribosome density at several discrete positions. These apparent contacts involve all four domains of the tRNA (Fig. 6). Only a smaller part of the tRNA interacts with the 40S subunit of yeast (Nierhaus et al. 2000). The anticodon stem-loop is positioned between head and body/platform of the 40S subunit, where the P-site codon appears to interact with the top of helix 44 of 18S rRNA and tRNA with the apical loop of helix 24 in the platform. From the opposite site, the tRNA is clamped down by 40S components belonging to the head. Helices 30, 31, and 43 of 18S rRNA, as well as an extended part of rpS16 (S9p), are likely to be involved in these interactions (Fig. 6), which in yeast appear to be very similar to those described for a bacterial system (Carter et al. 2000; Yusupov et al. 2001). The long tail of S13 that interacts with the tRNA in *T. thermophilus* does not have a corresponding sequence for rpS18, which has a somewhat changed position compared to the position of S13. Thus, in contrast to the tRNA contact with the S13p protein family, the tRNA contact with the S9p protein family is evolutionarily conserved. Three of the four tRNA domains interact with the large ribosomal subunit. The D-loop makes a contact with helix 69 of 25S rRNA and the T-loop interacts with rpL11 (L5p). The same interactions have been proposed for the *E. coli* ribosome (Spahn et al. 2000) and the *T. thermophilus* ribosome (Yusupov et al. 2001). The end of the tRNA acceptor stem appears to contact the 60S subunit involving helices 80 and 93 of 25S rRNA, as well as protein rpL10 (L10e). An additonal contact may be present that involves the TΨC stem of the tRNA and protein rpL10 (L10e) (Spahn et al. 2001a).

BRIDGES BETWEEN THE 40S AND THE 60S RIBOSOMAL SUBUNITS

The model of the 80S structure allowed us to determine the molecular identity of most of the ribosomal components involved in subunit–subunit interaction (Fig. 6b), as it has been done in the recent 5.5 Å resolution X-ray map of the *T. thermophilus* 70S ribosome (Yusupov et al. 2001). All the bridges that have been discovered in the bacterial 70S ribosomes have a corresponding bridge in the yeast 80S ribosome. This remarkable evolutionary conservation demonstrates the importance of these bridges. Most of the intersubunit contacts involve RNA–RNA interactions, although various RNA–protein interactions and one protein–protein contact can be identified. Several of the RNA interactions clearly involve minor groove sites.

Interestingly, four new bridges have been identified in the yeast 80S ribosome that might be specific for eukaryotic ribosomes (Fig. 6). Thus, the current B1–B7 nomenclature for bacterial ribosomes has been extended by naming the additional intersubunit contacts eB8–eB11. eB8, located below the L1 protuberance, is formed by an unknown 40S protein and ES31, an insertion into helix 79 of 25S rRNA. Since helix 79 is stacked onto the L1-bearing helix 76, bridge eB8 could be important in controlling the large conformational change that affects the position of the L1 protuberance in the yeast 80S ribosome. eB9 is built by components that are not present in bacterial ribosomes. It appears to make use of a protein–protein contact between unknown proteins, or an RNA–protein contact between an unknown LSU protein and the helix 21 insertion in 18S rRNA. Adding to the bridges connecting the main part of the 60S subunit with the body/platform domains of the 40S subunit are the additional bridges eB10 and eB11, which are among the strongest intersubunit connections in the yeast 80S ribosome. eB10 involves an

RNA–RNA contact between 18S rRNA helix 11 and 25S rRNA helix 63, as well as an unknown protein. eB11 is formed by an RNA–RNA interaction between 18S rRNA helix 9 and the expansion of helix 101 of 25S rRNA, and an unknown protein of the 40S subunit (Spahn et al. 2001a).

CONNECTIONS BETWEEN THE RNC AND THE CHANNEL

Docking of rRNA and protein models into separated rRNA and protein densities allowed the analysis of the PCC attachment sites at the molecular level (Fig. 7a–c) (Beckmann et al. 2001). Connection 1 is formed by a protrusion consisting of the helix 59 of 25S rRNA domain III and the amino-terminal domain of the ribosomal protein rpL19 (L19e). Connection 2 most likely also involves rRNA domain III (helix 53 and helix 50) and the amino-terminal domain of rpL19 as well as rpL25 (L23p). Connection 3 is in closest proximity to the tunnel exit engaging rRNA helix 24 of domain I and, in addition, an extended loop of rpL26 (L24p). Connection 4 is the most substantial one, involving two ribosomal proteins, rpL25 and rpL35 (L29p), and the tip of rRNA helix 7 of domain I. Since isolated rRNA of the large ribosomal subunit was shown to be sufficient to bind to the channel with high affinity (Prinz et al. 2000), it is notable that ribosomal proteins are most likely involved in all ribosome–channel connections. In contrast to other parts of the ribosomal periphery, the entire region near the tunnel exit, including rRNA as well as proteins, appears to be well conserved. Four ribosome–channel connections have been observed in similar positions for the mammalian ribosome (Menetret et al. 2000) with a very similar appearance of the exit site region (Morgan et al. 2000). When comparing this region with that of the archaebacterial 50S subunit, only a shorter rRNA helix 59 is evident (Nissen et al. 2000). This helix in the bacterial *E. coli* ribosome (Gabashvili et al. 2000) is comparable in length to the yeast ribosome, but a rpL19-like protein is missing (L19e in *Archaea*). Thus, with minor differences, the overall spatial arrangement of the ribosome–channel interaction is conserved in bacterial, archaebacterial, and eukaryotic cells and involves both large subunit RNA and proteins.

THE PROTEIN-CONDUCTING CHANNEL

The dimensions of the PCC in the recent reconstruction (Beckmann et al. 2001) are similar to those found in earlier EM studies (Hanein et al. 1996; Beckmann et al. 1997; Menetret et al. 2000). The diameter of the translocating channel is between 85 and 95 Å and the thickness is approximately 48 Å, hence, easily spanning the membrane (Fig. 7a,d). The channel associated with the empty ribosome appears slightly flattened and elongated with a diameter of approximately 100 Å (Fig. 7d). A slightly pentagonal shape, as observed previously for the Sec61 (Hanein et al. 1996; Beckmann et al. 1997) and the SecYE complex (Meyer et al. 1999), is not clearly recognizable.

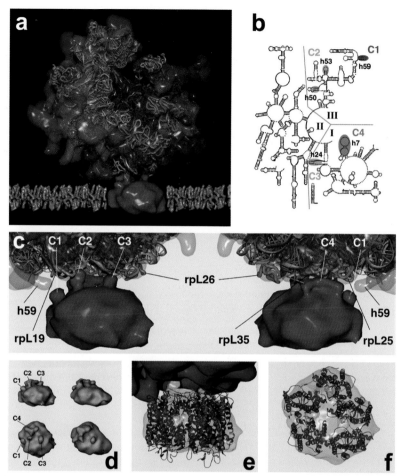

Figure 7. Structure of the Sec61 complex and connections between RNC and the channel. (*a*) Structure of the RNC–Sec61 complex after separating rRNA and protein densities, and docking of homology protein and RNA models (represented as ribbons) into their densities. Color code: red, Sec61 complex; red ribbon in green density, P-site tRNA; blue, 25S/5S rRNA; transparent orange, 60S protein density; orange, backbone of 60S protein models; yellow, 18S rRNA; transparent turquoise, 40S protein density; turquoise, backbone of 40S protein models; gray, lipid bilayer density to indicate position and dimension of the ER membrane. Note the gap between the ribosome and the channel, and the presence of rRNA as well as protein near the channel connections. (*b*) Secondary structure diagram of the yeast 25S rRNA domains I, II, and III (diagram according to R.R. Gutell, URL: http://www.rna.icmb.utexas.edu). The regions involved in connection are labeled accordingly. (*c*) Close-up of the RNC–Sec61 junction in a similar orientation as in *a* and rotated 180° around the z axis (*right*). The 25S rRNA and the 60S protein densities are shown in transparent blue and transparent orange, respectively, with models of the 25S rRNA domain I (*turquoise*), domain III (*green*), and proteins (*orange*) shown as ribbons. The positions of the four connections (C1–C4) are indicated. Density for helix 59 of rRNA domain III is marked by h59. Participation in connection 1, rRNA domain III and rpL19 (L19e); connection 2, rRNA domain III and rpL19 (L19e); connection 3, rRNA domain I and rpL26 (L24p), connection 4, rRNA domain I, rpL25 (L23p) and rpL35 (L29p). (*d*) Isolated density of the Sec61 complex in the translocating state (*left*) and the inactive state (*right*). The orientation on top is the same as in *c*, below is rotated 90° toward the viewer. The connections are marked. The channel attached to the empty ribosome appears elongated and flattened compared to the one engaged in translocation. (*e*) Close-up of the RNC–Sec61 junction in the same orientation as in *a*, with the large ribosomal subunit shown in blue and the Sec61 complex in transparent. In order to estimate the stoichiometry of the Sec61 oligomer, the transmembrane helices of five rhodopsin molecules (*red* ribbons) were fitted into the channel density. (*f*) Top view of *e*, but only the channel density with fitted membrane domains. The number of ~35 helices suggests a trimer of Sec61 trimers (36 helices).

Assuming membrane-specific helix packing, we determined the number of helices that fit into the channel density. The outline of the channel offers space for 35 helices with less dense packing in the central region, for example, provided by five rhodopsin molecules (Fig. 7e,f) (Palczewski et al. 2000). This helix packing corresponds to an average area between 180 and 190 Å^2/helix, which is in good agreement with experimental data (for review, see Collinson et al. 2001). This result argues strongly in favor of three Sec61 trimers forming the active PCC. For the homologous SecYEG complex, it was reported that a slightly higher average area of 199 Å^2 per helix is occupied in the plane of the membrane (Collinson et al. 2001).

Remarkably, in both functional states, the channel appears compact and has only a small indentation instead of a central pore (Figs. 2 and 7d). This indicates that the interaction with the signal sequence does not necessarily lead to a widely open channel conformation. The shape of the inactive channel could indicate that an iris-like movement of the membrane helices is the consequence of gat-

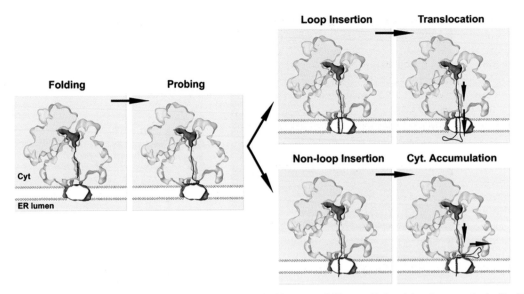

Figure 8. Binary model of cotranslational protein translocation. The binary model has to distinguish only two principally different functional states of the RNC–channel complex: After folding of α-helical segments, the emerging signal sequence is probed by the PCC. Insertion can occur in two different orientations with the channel opening just wide enough to accommodate the inserted nascent polypeptide. In case of *loop insertion*, the following nascent chain is guided through the membrane and translocation is possible. In case of *non-loop insertion*, the following nascent chain cannot translocate, accumulating on the cytosolic side of the membrane. Such cytosolic domains can easily exit into the cytosol at any time without compromising the ion permeability barrier. Nascent chain translocation after loop insertion and cytosolic accumulation after non-loop insertion are the only two principally different functional states of the nascent chain translocating RNC–PCC complex. For a polytopic membrane protein, the states would alternate with every new hydrophobic transmembrane domain. (Reprinted, with permission, from Beckmann et al. 2001 [copyright Cell Press].)

ing by the signal sequence, similar to the conformational change suggested for the gating of gap junctions (Unwin and Zampighi 1980). These results do not support conclusions from fluorescence quenching experiments with native mammalian ER membranes suggesting a pore of 9–15 Å diameter in the inactive translocon (Hamman et al. 1998), and of 40–60 Å diameter in the translocating one (Hamman et al. 1997). However, the channels in native membranes are known to be associated with additional proteins in different oligomeric states (Wang and Dobberstein 1999; Menetret et al. 2000). In conclusion, in all studies of the eukaryotic PCC, using either 2D (Hanein et al. 1996) or 3D reconstructions (Beckmann et al. 1997, 2001; Menetret et al. 2000), a central pore, or at least an indentation, has been observed, arguing strongly that the central region in the channel is very flexible in accordance with its function in protein conductance across the membrane.

BINARY MODEL OF COTRANSLATIONAL TRANSLOCATION

The compact conformation of the channel and the gap between the channel and the ribosome indicates that gating of the channel by the signal sequence can lead to an opening providing enough space for the inserted nascent polypeptide chain without allowing ion conductance. These findings led to a binary model (Beckmann et al. 2001) describing the functioning of the ribosome–PCC complex in cotranslational translocation (Fig. 8): The conserved tunnel in the large ribosomal subunit allows exclusively the folding of α-helical secondary structure.

Probing of the nascent chain for hydrophobic helices by the channel leads to the insertion of hydrophobic segments in an orientation exposing the carboxyl terminus either to the cytosolic or to the lumenal side, in the latter case resulting in loop formation of the remaining nascent chain in the channel (Fig. 8). Importantly, insertion does not lead to further opening of the pore to any predefined size. Therefore, the following nascent chain will either accumulate on the cytosolic side or, only if it has been co-inserted in a loop, translocate across the membrane. In this model, the channel itself has the property of a seal, preventing ion flow across the membrane independent of the presence of the ribosome. Therefore, the model needs to distinguish only two principally different functional states of the RNC–PCC complex (hence, binary) to describe the translocation of secretory proteins or insertion of membrane proteins without allowing ion permeation: (1) translation preceded by a loop insertion resulting in translocation across the membrane (e.g., secretory proteins) and (2) translation preceded by a channel insertion without loop resulting in accumulation in the cytosol (e.g., cytosolic domains of membrane proteins or signal anchor proteins). In the case of polytopic membrane proteins, these two states would alternate.

CONCLUSIONS

On the basis of a new protocol for the immunopurification of RNCs, cryo-EM was used in combination with a homology modeling approach. Hence, an interpretation of the cryo-EM map of an active eukaryotic 80S ribosome–channel complex in terms of atomic RNA and protein models was provided. It was found that the common core of the eukaryotic ribosome agrees very well with X-ray structures of the bacterial and archaebacterial subunits, providing the basis for the common mechanism of protein translation. The intersubunit bridges and the ribosome–tRNA interactions are highly conserved, as well. However, a few unexpected differences exist even in this highly conserved region, and dramatic differences are present in the periphery of the 80S ribosome due to expansion segments and additional proteins. The model of the yeast 80S ribosome provides the basis for molecular-level interpretations of the interactions of the eukaryotic ribosomes within the context of the eukaryotic environment such as the protein-conducting channel. Conserved rRNA segments, as well as several ribosomal proteins, were found to be involved in forming the ribosome–channel junctions. The seal forming compact conformation of the channel and the gap between the channel and the ribosome led to the suggestion of a binary model of co-translational protein translocation, which significantly reduces the regulatory and spatial requirements for the functioning RNC–PCC complex.

ACKNOWLEDGMENTS

We thank Sean Campbell, Bob Grassucci, Mike Lewis, and David Stokes for technical assistance and support with the electron microscopy; Narayanan Eswar, Andrej Sali, Jürgen Helmers, and Pawel A. Penczek for fruitful collaboration; and Jan Giesebrecht for generation of ray-traced illustrations. This work was supported by a grant of the VolkswagenStiftung (to R.B.) and grants from the National Institutes of Health (to J.F.) and the National Science Foundation (to J.F.).

REFERENCES

Ban N., Nissen P., Hansen J., Moore P.B., and Steitz T.A. 2000. The complete atomic structure of the large ribosomal subunit at 2.4 Å resolution. *Science* **289:** 905.

Beckmann R., Bubeck D., Grassucci R., Penczek P., Verschoor A., Blobel G., and Frank J. 1997. Alignment of conduits for the nascent polypeptide chain in the ribosome-Sec61 complex. *Science* **278:** 2123.

Beckmann R., Spahn C.M.T., Narayanan E., Helmers J., Penczek P.A., Sali A., Frank J., and Blobel G. 2001. Architecture of the protein-conducting channel associated with the translating 80S ribosome. *Cell* **107:** 361.

Blobel G. and Sabatini D.D. 1970. Controlled proteolysis of nascent polypeptides in rat liver cell fractions. I. Location of the polypeptides within ribosomes. *J. Cell Biol.* **45:** 130.

Borel A.C. and Simon S.M. 1996. Biogenesis of polytopic membrane proteins: Membrane segments assemble within translocation channels prior to membrane integration. *Cell* **85:** 379.

Carter A.P., Clemons W.M., Jr., Brodersen D.E., Morgan-Warren R.J., Wimberly B.T., and Ramakrishnan V. 2000. Functional insights from the structure of the 30S ribosomal subunit and its interactions with antibiotics. *Nature* **407:** 340.

Cate J.H., Yusupov M.M., Yusupova G.Z., Earnest T.N., and Noller H.F. 1999. X-ray crystal structures of 70S ribosome functional complexes. *Science* **285:** 2095.

Collinson I., Breyton C., Duong F., Tziatzios C., Schubert D., Or E., Rapoport T., and Kuhlbrandt W. 2001. Projection structure and oligomeric properties of a bacterial core protein translocase. *EMBO J.* **20:** 2462.

Crowley K.S., Liao S., Worrell V.E., Reinhart G.D., and John-son A.E. 1994. Secretory proteins move through the endoplasmic reticulum membrane via an aqueous, gated pore. *Cell* **78:** 461.

Dube P., Bacher G., Stark H., Müller F., Zemlin F., van Heel M., and Brimacombe R. 1998. Correlation of the expansion segments in mammalian rRNA with the fine structure of the 80 S ribosome; a cryoelectron microscopic reconstruction of the rabbit reticulocyte ribosome at 21 Å resolution. *J. Mol. Biol.* **279:** 403.

Frank J. and Agrawal R.K. 2000. A ratchet-like inter-subunit reorganization of the ribosome during translocation. *Nature* **406:** 318.

Frank J., Zhu J., Penczek P., Li Y., Srivastava S., Verschoor A., Radermacher M., Grassucci R., Lata R.K., and Agrawal R.K. 1995. A model of protein synthesis based on cryo-electron microscopy of the *E. coli* ribosome. *Nature* **376:** 441.

Gabashvili I.S., Agrawal R.K., Spahn C.M., Grassucci R.A., Svergun D.I., Frank J., and Penczek P. 2000. Solution structure of the *E. coli* 70S ribosome at 11.5 Å resolution. *Cell* **100:** 537.

Gabashvili I.S., Gregory S.T., Valle M., Grassucci R., Worbs M., Wahl M.C., Dahlberg A.E., and Frank J. 2001. Role of ribosomal proteins L4 and L22 in gating mechanisms of the polypeptide tunnel system. *Mol. Cell* **8:** 181.

Gerbi S.A. 1996. Expansion segments: Regions of variable size that interrupt the universal core secondary structure of ribosomal RNA. In *Ribosomal RNA: Structure, evolution, processing, and function in protein biosynthesis* (ed. A.E. Dahlberg and R.A. Zimmermann), p. 71. CRC Press, Boca Raton, Florida.

Gilmore R. and Blobel G. 1985. Translocation of secretory proteins across the microsomal membrane occurs through an environment accessible to aqueous perturbants. *Cell* **42:** 497.

Gomez-Lorenzo M.G., Spahn C.M., Agrawal R.K., Grassucci R.A., Penczek P., Chakraburtty K., Ballesta J.P., Lavandera J.L., Garcia-Bustos J.F., and Frank J. 2000. Three-dimensional cryo-electron microscopy localization of EF2 in the *Saccharomyces cerevisiae* 80S ribosome at 17.5 Å resolution. *EMBO J.* **19:** 2710.

Gorlich D. and Rapoport T.A. 1993. Protein translocation into proteoliposomes reconstituted from purified components of the endoplasmic reticulum membrane. *Cell* **75:** 615.

Hamman B.D., Hendershot L.M., and Johnson A.E. 1998. BiP maintains the permeability barrier of the ER membrane by sealing the lumenal end of the translocon pore before and early in translocation. *Cell* **92:** 747.

Hamman B.D., Chen J.C., Johnson E.E., and Johnson A.E. 1997. The aqueous pore through the translocon has a diameter of 40-60 A during cotranslational protein translocation at the ER membrane. *Cell* **89:** 535.

Hanein D., Matlack K.E., Jungnickel B., Plath K., Kalies K.U., Miller K.R., Rapoport T.A., and Akey C.W. 1996. Oligomeric rings of the Sec61p complex induced by ligands required for protein translocation. *Cell* **87:** 721.

Hardesty B., Kramer G., Tsalkova T., Ramachandiran V., McIntosh B., and Brod D. 2000. Folding of nascent peptides on ribosomes. In *The ribosome: Structure, function, antibiotics, and cellular interactions* (ed. R.A. Garrett et al.), p.287. ASM Press, Washington, D.C.

Heinrich S.U., Mothes W., Brunner J., and Rapoport T.A. 2000. The Sec61p complex mediates the integration of a membrane protein by allowing lipid partitioning of the transmembrane domain. *Cell* **102:** 233.

Johnson A.E. and van Waes M.A. 1999. The translocon: A dynamic gateway at the ER membrane. *Annu. Rev. Cell Dev. Biol.* **15:** 799.

Jungnickel B. and Rapoport T.A. 1995. A posttargeting signal sequence recognition event in the endoplasmic reticulum membrane. *Cell* **82:** 261.

Kalies K.U., Gorlich D., and Rapoport T.A. 1994. Binding of ribosomes to the rough endoplasmic reticulum mediated by the Sec61p-complex. *J. Cell Biol.* **126:** 925.

Liao S., Lin J., Do H., and Johnson A.E. 1997. Both lumenal and cytosolic gating of the aqueous ER translocon pore are regulated from inside the ribosome during membrane protein integration. *Cell* **90:** 31.

Manting E.H., van Der Does C., Remigy H., Engel A., and Driessen A.J. 2000. SecYEG assembles into a tetramer to form the active protein translocation channel. *EMBO J.* **19:** 852.

Martoglio B., Hofmann M.W., Brunner J., and Dobberstein B. 1995. The protein-conducting channel in the membrane of the endoplasmic reticulum is open laterally toward the lipid bilayer. *Cell* **81:** 207.

Matlack K.E., Mothes W., and Rapoport T.A. 1998. Protein translocation: Tunnel vision. *Cell* **92:** 381.

Menetret J., Neuhof A., Morgan D.G., Plath K., Radermacher M., Rapoport T.A., and Akey C.W. 2000. The structure of ribosome-channel complexes engaged in protein translocation. *Mol. Cell* **6:** 1219.

Meyer T.H., Menetret J.F., Breitling R., Miller K.R., Akey C.W., and Rapoport T.A. 1999. The bacterial SecY/E translocation complex forms channel-like structures similar to those of the eukaryotic Sec61p complex. *J. Mol. Biol.* **285:** 1789.

Morgan D.G., Menetret J.F., Radermacher M., Neuhof A., Akey I.V., Rapoport T.A., and Akey C.W. 2000. A comparison of the yeast and rabbit 80 S ribosome reveals the topology of the nascent chain exit tunnel, inter-subunit bridges and mammalian rRNA expansion segments. *J. Mol. Biol.* **301:** 301.

Mothes W., Jungnickel B., Brunner J., and Rapoport T.A. 1998. Signal sequence recognition in cotranslational translocation by protein components of the endoplasmic reticulum membrane. *J. Cell Biol.* **142:** 355.

Mothes W., Heinrich S.U., Graf R., Nilsson I., von Heijne G., Brunner J., and Rapoport T.A. 1997. Molecular mechanism of membrane protein integration into the endoplasmic reticulum. *Cell* **89:** 523.

Ng D.T., Brown J.D., and Walter P. 1996. Signal sequences specify the targeting route to the endoplasmic reticulum membrane. *J. Cell Biol.* **134:** 269.

Nierhaus K.H., Spahn C., Burkhardt N., Dabrowski M., Diedrich G., Einfeldt E., Kamp D., Marquez V., Patzke S., Schafer M.A., Stelzl U., Blaha G., Willumeit R., and Stuhrmann H.B. 2000. Ribosomal elongation cycle. In *The ribosome: Structure, function, antibiotics, and cellular interactions* (ed. R.A. Garrett et al.), p. 319. ASM Press, Washington, D.C.

Nissen P., Hansen J., Ban N., Moore P.B., and Steitz T.A. 2000. The structural basis of ribosome activity in peptide bond synthesis. *Science* **289:** 920.

Palczewski K., Kumasaka T., Hori T., Behnke C.A., Motoshima H., Fox B.A., Le Trong I., Teller D.C., Okada T., Stenkamp R.E., Yamamoto M., and Miyano M. 2000. Crystal structure of rhodopsin: A G protein-coupled receptor. *Science* **289:** 739.

Planta R.J., and Mager W.H. 1998. The list of cytoplasmic ribosomal proteins of *Saccharomyces cerevisiae*. *Yeast* **14:** 471.

Prinz A., Behrens C., Rapoport T.A., Hartmann E., and Kalies K.U. 2000. Evolutionarily conserved binding of ribosomes to the translocation channel via the large ribosomal RNA. *EMBO J.* **19:** 1900.

Rossmann M.G. 2000. Fitting atomic models into electron-microscopy maps. *Acta Crystallogr. D Biol. Crystallogr.* **56:** 1341.

Sachs A.B., Sarnow P., and Hentze M.W. 1997. Starting at the beginning, middle, and end: Translation initiation in eukaryotes. *Cell* **89:** 831.

Shaw A.S., Rottier P.J., and Rose J.K. 1988. Evidence for the loop model of signal-sequence insertion into the endoplasmic reticulum. *Proc. Natl. Acad. Sci.* **85:** 7592.

Simon S.M. and Blobel G. 1991. A protein-conducting channel in the endoplasmic reticulum. *Cell* **65:** 371.

Spahn C.M.T. and Prescott C.D. 1996. Throwing a spanner in the works: Antibiotics and the translation apparatus. *J. Mol. Med.* **74:** 423.

Spahn C.M.T., Penczek P.A., Leith A., and Frank J. 2000. A method for differentiating proteins from nucleic acids in intermediate-resolution density maps: Cryo-electron microscopy defines the quaternary structure of the *Escherichia coli* 70S ribosome. *Struct. Fold. Des.* **8:** 937.

Spahn C.M.T., Beckmann R., Narayanan E., Penczek P.A., Sali A., Blobel G., and Frank J. 2001a. Structure of the 80S ribosome from *Saccharomyces cerevisiae*—tRNA-ribosome and subunit-subunit interactions. *Cell* **107:** 373.

Spahn C.M.T., Kieft J.S., Grassucci R.A., Penczek P., Zhou K., Doudna J.A., and Frank J. 2001b. Hepatitis C virus IRES RNA-induced changes in the conformation of the 40S ribosomal subunit. *Science* **291:** 1962.

Sweeney R., Chen L., and Yao M.C. 1994. An rRNA variable region has an evolutionarily conserved essential role despite sequence divergence. *Mol. Cell. Biol.* **14:** 4203.

Triana-Alonso F.J., Chakraburtty K., and Nierhaus K.H. 1995. The elongation factor 3 unique in higher fungi and essential for protein biosynthesis is an E site factor. *J. Biol. Chem.* **270:** 20473.

Uchiumi T., Hori K., Nomura T., and Hachimori A. 2001. Replacement of L7/L12·L10 protein complex in *Escherichia coli* ribosomes with the eukaryotic counterpart changes the specificity of elongation factor binding. *J. Biol. Chem.* **274:** 27578.

Unwin P.N. and Zampighi G. 1980. Structure of the junction between communicating cells. *Nature* **283:** 545.

Verschoor A., Warner J.R., Srivastava S., Grassucci R.A., and Frank J. 1998. Three-dimensional structure of the yeast ribosome. *Nucleic Acids Res.* **26:** 655.

Wang L. and Dobberstein B. 1999. Oligomeric complexes involved in translocation of proteins across the membrane of the endoplasmic reticulum. *FEBS Lett.* **457:** 316.

Wilkinson B.M., Critchley A.J., and Stirling C.J. 1996. Determination of the transmembrane topology of yeast Sec61p, an essential component of the endoplasmic reticulum translocation complex. *J. Biol. Chem.* **271:** 25590.

Wimberly B.T., Guymon R., McCutcheon J.P., White S.W., and Ramakrishnan V.R. 1999. A detailed view of a ribosomal active site: The structure of the L11-RNA complex. *Cell* **97:** 491.

Wimberly B.T., Brodersen D.E., Clemons W.M., Jr., Morgan-Warren R.J., Carter A.P., von Rhein C., Hartsch T., and Ramakrishnan V. 2000. Structure of the 30S ribosomal subunit. *Nature* **407:** 327.

Wittmann-Liebold B. 1986. Ribosomal proteins: Their structure and evolution. In *Structure, function, and genetics of ribosomes* (ed. B. Hardesty and G. Kramer), p. 326. Springer-Verlag, New York.

Wool I.G., Endo Y., Chan Y.L., and Glück A. 1990. Structure, function, and evolution of mammalian ribosomes. In *The ribosome: Structure, function, and evolution* (ed. W.E.Hill et al.), p. 203. ASM Press, Washington, D.C.

Yusupov M.M., Yusupova G.Z., Baucom A., Lieberman K., Earnest T.N., Cate J.H., and Noller H.F. 2001. Crystal structure of the ribosome at 5.5 Å resolution. *Science* **292:** 883.

Ribosomal RNA Genes, RNA Polymerases, Nucleolar Structures, and Synthesis of rRNA in the Yeast *Saccharomyces cerevisiae*

M. Nomura

Department of Biological Chemistry, University of California, Irvine, California 92697-1700

When the structural complexity of ribosomes became evident in the early 1960s, two obvious questions were asked. The first concerns how the large and small rRNAs, as well as many ribosomal protein molecules, are organized within ribosomal particles and how the ribosome machine makes proteins; that is, the question of structure and function. The second question is how these molecular components are synthesized and then assembled into such complex and yet specific structures, and how such synthetic processes are regulated; that is, the question of biosynthesis and its regulation.

Regarding the first question, recent impressive progress in structural studies of ribosomes using cryo-electron microscopy and X-ray crystallography has culminated in the nearly complete elucidation of the atomic structure of both large and small subunits of ribosomes as reported last year (Ban et al. 2000; Schluenzen et al. 2000; Wimberly et al. 2000). The structure of the complete 70S ribosome containing bound mRNA and tRNA has also been published very recently (Yusupov et al. 2001). Consequently, intensive research efforts on the structure and function of ribosomes have been initiated based on the newly elucidated structural information. Although this volume deals mostly with this first question, the second question, ribosome synthesis, is also important in connection with cellular growth and its regulation. This paper deals with some aspects of this second question, mostly based on our work in the past decade using the yeast *Saccharomyces cerevisiae* as a model eukaryotic organism.

As in the case of earlier studies on ribosome structure and function, intensive studies on the synthesis of ribosomes were carried out using *Escherichia coli* as a model system in earlier days. Thus, we now know a great deal about the main regulatory features as well as actual molecular mechanisms involved in the synthesis of ribosomal components and their assembly (for reviews, see Condon et al. 1995; Gourse et al. 1996; Keener and Nomura 1996). One of the main conclusions of these studies is that the most important step in making ribosomes is to make rRNA, and this step determines the overall rate of ribosome biosynthesis under most conditions of cell growth. Synthesis of ribosomal proteins (r-proteins) and subsequent assembly follows rRNA production. Similarly, although regulation of the synthesis of rRNA and r-proteins is apparently carried out independently in many

instances in *S. cerevisiae* and perhaps most other eukaryotic cells, the major regulatory mechanisms of ribosome synthesis appear to act directly on the synthesis of rRNA, and coordinated accumulation of rRNA and r-proteins appears to be achieved by degradation of r-proteins produced in some excess (for discussion, see Nomura 1999). Thus, our efforts have mainly focused on understanding the molecular mechanisms involved in the synthesis of rRNA. Compared to bacterial systems, the machinery for the synthesis of rRNA in eukaryotes is much more complex, including several specific transcription factors in addition to the specialized enzyme, RNA polymerase I (Pol I), and tandemly repeated rDNA template structures, elaborated below. In this paper, we summarize the recent progress in our studies using *S. cerevisiae* and discuss the results in connection with some unique features of rRNA synthesis in eukaryotes.

SOME UNIQUE FEATURES OF rRNA SYNTHESIS IN EUKARYOTES

There are three features of rRNA synthesis in most eukaryotes that distinguish it clearly from rRNA synthesis in prokaryotes. First, there are three different kinds of nuclear RNA polymerases with different functional assignments in eukaryotes, and Pol I is dedicated to transcription of large rRNAs. In contrast, a single RNA polymerase carries out transcription of all the genes in prokaryotes, including both eubacteria and archaebacteria. Second, genes for rRNAs are tandemly repeated at one or a few chromosomal loci in eukaryotes, whereas rRNA genes in bacteria are not. In *E. coli* there are seven rRNA operons, but they are scattered on the chromosome. In the yeast *S. cerevisiae*, there are approximately 150 tandemly repeated copies of the rRNA gene ("*35S rDNA*" or simply "rDNA" in this paper) encoding 35S precursor rRNA on chromosome XII. The primary transcript, 35S rRNA, is processed into mature 18S, 5.8S, and 25S rRNAs. Among eukaryotes, *S. cerevisiae* (and some other organisms; see Drouin and de Sá 1995) is different from other well-studied higher eukaryotes, such as mammals, in that the gene for 5S RNA, which is transcribed by RNA polymerase III (Pol III), is also present as part of the rDNA repeats (Fig. 1). Each rDNA repeat unit contains some other *cis* elements formally unrelated to transcription, such as the origin of DNA replication (Ori) and the replication fork block (RFB) site (see the legend to Fig. 1 for RFB). Except for the presence of the 5S RNA gene, most eukaryotes have similar repeat structures, but with various sizes for a repeat unit and with various repeat numbers, the latter ranging from less than 100 to more than 10,000 (Long and Dawid 1980). Third, eukaryotic cells have a clearly recognized subnuclear structure, the nucleolus, where rRNA synthesis and ribosome assembly take place. In addition to these unique features, eukaryotic cells obviously have a nuclear membrane separating the nucleus from the cytoplasm. This fact poses the questions of r-protein import and ribosome export, which might somehow influence rDNA transcription, but we limit our discussion mainly to features directly related to transcription of the 35S rRNA gene by Pol I.

THE USE OF Pol I OR TANDEMLY REPEATED STRUCTURES FOR DNA TEMPLATE IS NOT ABSOLUTELY REQUIRED FOR rRNA SYNTHESIS IN VIVO

The fact that the above three features are present in almost all eukaryotic organisms suggests that they must be functionally important. Yet, the first two features, the use of Pol I and the use of a tandemly repeated rDNA template, are clearly not absolute requirements for rRNA synthesis or cell growth. This conclusion was first obtained by constructing a yeast strain that carries a fusion gene, the 35S rRNA coding region fused to the *GAL7* promoter (*GAL7-35S rDNA*, Fig. 1), on a multicopy plasmid, and by demonstrating that such a strain can grow in galactose medium in the absence of a functional Pol I (Nogi et al. 1991b). In this strain, the gene (*RPA135*) encoding an essential subunit (A135) of Pol I is deleted and rRNA is synthesized by transcription of the (artificial) *GAL7-35S rDNA* fusion gene by RNA polymerase II (Pol II). Thus, Pol I is not absolutely essential for ribosomal synthesis or cell growth in this strain. Furthermore, chromosomal rDNA repeats are not used in such yeast strains and are therefore dispensable. In fact, yeast strains were subsequently constructed in which the chromosomal rDNA repeats are completely deleted and a single unit of rDNA (or the *GAL7-35S rDNA* fusion gene together with the 5S RNA gene) is carried by a plasmid ("*rdnΔΔ*" strains; Oakes et al. 1998; Wai et al. 2000; see also the related study of Nierras et al. 1997 and below). Of course, these strains use artificially created fusion genes or a native rDNA on an artificial plasmid as template, and their growth rates are also significantly reduced compared to the wild type. Nevertheless, they clearly demonstrate that neither the use of Pol I nor the use of rDNA template with a tandem-repeat structure is required for the synthesis of rRNA and ribosomes in vivo. In addition, as described below, yeast cells have the ability to transcribe the (natural) chromosomal rDNA using Pol II and to grow in the absence of Pol I when a mutation inactivates certain proteins that normally function in silencing this Pol II transcription activity.

GENETIC AND BIOCHEMICAL APPROACHES TO IDENTIFY Pol-I-SPECIFIC TRANSCRIPTION FACTORS

Specific in vitro transcription of eukaryotic rDNA was first achieved in mouse cell extracts (Grummt 1981), and purification and identification of Pol I transcription factors in mammalian cells have been done biochemically by fractionation of extracts (for reviews, see Grummt 1998; Zomerdijk and Tjian 1998). However, identification of some factors present only in very small amounts was not easy in the past. In addition, as is well recognized generally, observations made in vitro may not necessarily be relevant in vivo. For this reason, genetic approaches in parallel to biochemical ones are essential, as has been demonstrated in studies of various other biological activities. Surprisingly, genetic approaches were limited and were not very successful in the earlier studies of rDNA transcription. The reason may have been difficulty in finding effective ways to screen for relevant mutants. There are many mutations that prevent accumulation of rRNA. Thus, earlier screening of many temperature-sensitive (ts) mutants of *S. cerevisiae* for preferential reduction of RNA accumulation yielded a series of mutants defining ten genes required for rRNA accumulation (Hartwell et al. 1970). However, most of these genes have turned out to be involved in mRNA splicing; the genes for many r-proteins have introns, and mutational defects in splicing caused defects in the synthesis of many r-proteins, leading to defects in ribosome assembly and degra-

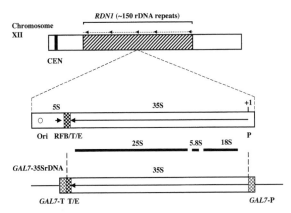

Figure 1. The structures of the *S. cerevisiae RDN1* locus, the rDNA repeat unit, and the *GAL7-35S rDNA* fusion gene. The *RDN1* locus is located on the right arm of chromosome XII and contains ~150 tandem repeats of rRNA genes. A single repeat unit, which is 9.1 kb in length, is expanded, showing the 35S rRNA coding region (*long arrow*) and the 5S RNA coding region (*short arrow*, not to scale) as well as the location of the 35S rRNA gene promoter (P), the 35S rRNA transcription terminator (T), the enhancer (E), replication fork blocking element (RFB), and the origin of replication (Ori, shown as a *circle*). RFB is the DNA element that allows progression of the DNA replication fork in the direction of 35S rRNA transcription, but not in the opposite direction (Brewer et al. 1992; Kobayashi et al. 1992). The *GAL7-35S rDNA* fusion gene is shown at the bottom of the figure. The 35S rRNA gene promoter is replaced by the *GAL7* promoter, and the *GAL7* terminator is added at the end of the enhancer defined originally by Elion and Warner (1984).

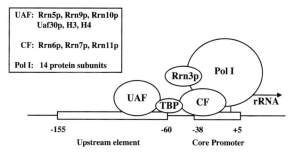

Figure 2. Model for Pol I transcription initiation from the 35S rRNA gene promoter. TBP is placed to indicate contacts with both UAF and CF (Lin et al. 1996; Steffan et al. 1996, 1998), although there are no data to indicate that TBP interacts with DNA directly. UAF may also have a contact with CF (Steffan et al. 1996). Pol I contains 14 polypeptides. Seven of them (A190, A135, A49, A43, A34.5, A14, and A12) are unique to Pol I, two of them (AC40 and AC19) are shared with Pol III, and the remaining five (ABC27, ABC23, ABC14.5, ABC10α, and ABC10β) are shared with both Pol II and Pol III (for review, see Carles and Riva 1998). Among the seven Pol-I-specific subunits, deletion of the genes for A34.5 or A14 does not cause any significant growth defects. Genes for the remaining five are essential or nearly essential and were found among *rrn* genes identified by the genetic screen using the *GAL7-35S rDNA* fusion gene described in the text (for review, see Nomura 1998).

dation of rRNA, and hence, defects in rRNA accumulation (Rosbash et al. 1981).

In designing a method to isolate mutants in Pol I transcriptional machinery, we took advantage of the presence of three nuclear RNA polymerases in eukaryotes and used the *GAL7-35S rDNA* fusion gene described in the previous section. Mutants were isolated from a yeast strain carrying the *GAL7-35S rDNA* fusion gene on a plasmid as those which can form colonies on galactose but not on glucose (Nogi et al. 1991a). It should be noted that mutations defined by such a screen must be inactivating the Pol I machinery directly rather than affecting ribosome synthesis at other steps or influencing it indirectly, e.g., by inhibiting energy production, because the mutations did not affect ribosome synthesis achieved through Pol II transcription of the *GAL7-35S rDNA* gene. In this way, we were able to isolate many specific mutants defective in rRNA synthesis by the Pol I system, classify these mutants, and identify 12 different genes. Five of these genes were shown to encode essential or nearly essential subunit proteins of Pol I not shared by Pol II or Pol III (see Fig. 2). The remaining 7 genes encoded proteins required specifically for transcription of the native chromosomal rRNA gene, that is, presumed transcription factor proteins. All of these 7 genes were subsequently cloned and characterized. Strains were then constructed which carry one of these genes replaced with an engineered version encoding an epitope-tagged gene product, and these strains helped greatly in the purification of transcription factors containing the encoded protein. In this way, three transcription factors, UAF (upstream activation factor), CF (core factor), and Rrn3p, were identified and characterized (Fig. 2) (Keys et al. 1994, 1996; Lalo et al. 1996; Yamamoto et al. 1996; see also the related work of Lin et al. 1996).

Like most other eukaryotic rDNA promoters, the rDNA promoter in *S. cerevisiae* consists of two elements: a core promoter and an upstream element (Fig. 2). For in vitro transcription of rDNA by Pol I, the core element is essential and the upstream element is not essential, but is required for high-level transcription. From in vitro experiments using fractionated cell extracts (Keys et al. 1996; Steffan et al. 1996), as well as a system containing all purified components (Keener et al. 1998), it was concluded that transcription using a DNA template containing only the core promoter ("basal transcription") requires only CF and Rrn3p in addition to Pol I. CF consists of three proteins, Rrn6p, Rrn7p, and Rrn11p (Keys et al. 1994; Lalo et al. 1996; Lin et al. 1996). Rrn3p interacts directly with Pol I (Yamamoto et al. 1996; Keener et al. 1998; Milkereit and Tschochner 1998) and stimulates initiation of transcription (Yamamoto et al. 1996). This stimulation is presumably mediated through its interactions with the A43 subunit of Pol I and with the Rrn6p subunit of CF, facilitating binding of Pol I to the promoter (Peyroche et al. 2000).

UAF was purified and found to contain Rrn5p, Rrn9p, Rrn10p, and three additional proteins, histones H3 and H4 and a protein originally called p30 (Keys et al. 1996; Keener et al. 1997). The gene (*UAF30*) for p30 was recently identified, and the protein is now called Uaf30p (Siddiqi et al. 2001). It has been demonstrated that high-level in vitro transcription from the complete rDNA promoter requires UAF and TBP in addition to Pol I, Rrn3p, and CF. UAF does not stimulate transcription when the DNA template lacks the upstream element. UAF binds to the upstream element tightly and functions to recruit CF with the help of TBP (and possibly other factors). This is probably the basis for stimulation of transcription (Keys et al. 1996; Steffan et al. 1996; Keener et al. 1998). TBP interacts with both UAF and CF in vitro. In addition, interactions of TBP with the Rrn9p subunit of UAF and the Rrn6p subunit of CF were demonstrated both in vitro and in the yeast two-hybrid system (Lalo et al. 1996; Lin et al. 1996; Steffan et al. 1996). Furthermore, genetic experiments demonstrated that the interaction of TBP with Rrn9p detected in the yeast two-hybrid system is essential for the TBP function in Pol I transcription of rDNA in vivo (Steffan et al. 1998).

In the in vitro experiments, basal transcription was observed in the absence of UAF or TBP, or with a DNA template lacking the upstream element as described above. It was therefore expected that *uaf* mutants, carrying a helper plasmid with the *GAL7-35S rDNA* fusion gene and growing in galactose medium, would reveal such a basal transcription after repression of the *GAL7* promoter by shifting to glucose medium. As described below, we did not observe the expected basal transcription by Pol I. Instead, we found that weak transcription by Pol II takes place on the rDNA, using different transcription initiation sites (Oakes et al. 1999). Thus, events taking place in vivo are often different from those expected from observations made in vitro.

It should also be emphasized that the requirement of transcription factors described above for Pol I transcription in the yeast system has been convincingly demonstrated both in vitro and in vivo. We believe that most of the basic components in the Pol I transcription machinery have been discovered. Nevertheless, it is highly likely that there are other (unidentified) components which are not essential for Pol I function, but which modulate Pol I activity in vivo (and perhaps also in vitro) and have not been detected by the genetic screen used. Future studies may uncover the presence of such modulating factors.

CONSTRUCTION OF rDNA DELETION STRAINS AND FUNCTIONAL ANALYSIS OF PROMOTER/ENHANCER ELEMENTS IN VIVO

The redundancy of rDNA genes has been an obstacle to the use of a mutational approach to study the structure–function relationship of rRNAs in connection with ribosomal functions in protein synthesis. Similarly, mutational analysis of *cis* elements, such as promoter and enhancer elements involved in the regulation of rRNA synthesis, was also difficult to perform in vivo; only some artificial reporter systems have been used for the analysis of *cis* elements without assaying real transcription products, functional rRNAs. To overcome these problems, we have constructed yeast mutants in which chromosomal rDNA repeats were deleted completely and rRNA synthesis was achieved by a plasmid ("Pol I helper plasmid") carrying a single rDNA repeat transcribed by Pol I, or a plasmid ("Pol II helper plasmid") carrying, in addition to the 5S RNA gene, the above-described *GAL7-35S rDNA* fusion gene transcribed by Pol II. (Note that a method to delete most of the rDNA repeats using a hygromycin-selection method was previously devised, and a strain with such a deletion was used for mutational analysis of rRNAs [Chernoff et al. 1994]. However, the deletion in the strain was not well characterized. We combined the previous method with a standard gene replacement strategy and constructed yeast strains [called *rdnΔΔ* strains] carrying a complete deletion of rDNA repeats with deletion ends precisely defined [see also Fig. 3] [Oakes et al. 1998; Wai et al. 2000].) Using this system, deletion analysis of the rDNA promoter confirmed the presence of two elements, the upstream element and the core promoter, and showed that basal transcription from the core promoter, if it takes place in vivo as was observed in vitro, is not sufficient to allow cell growth (Wai et al. 2000; see also below).

A surprising discovery made using this system concerns the function of the so-called yeast Pol I enhancer. At the end of the 35S rRNA gene in *S. cerevisiae*, there exists a DNA element called an enhancer (Fig. 1) that has been shown to greatly stimulate rDNA transcription in ectopic reporter systems (Elion and Warner 1984, 1986). In *rdnΔΔ* strains, however, we found that deletion of the enhancer from the Pol I helper plasmid does not cause any decrease in rRNA synthesis or cell growth (Wai et al. 2001). Employing a plasmid reporter system used in the original experiments (Elion and Warner 1984, 1986), we found that the apparent stimulatory effect of the enhancer that was observed originally using ordinary strains (with

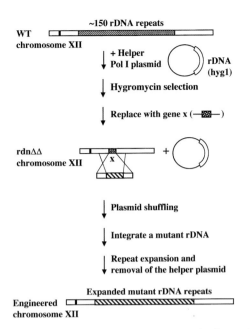

Figure 3. Complete deletion of rDNA repeats and re-integration and expansion of a new rDNA repeat at the original locus. Complete deletion was achieved in two steps, the hygromycin selection step and a standard gene replacement using *HIS3* (gene X in the figure) (Oakes et al. 1998; Wai et al. 2000). For re-integration, *HIS3* was first replaced with *URA3*, and a DNA fragment containing a new rDNA repeat was introduced by homologous recombination, counterselecting Ura⁺ using 5-fluoroorotic acid, as described in Wai et al. (2000).

the intact chromosomal rDNA repeats) could not be observed in *rdnΔΔ* cells, which have disrupted nucleolar structures (Oakes et al. 1998; see below). In addition, the expression of the reporter genes with or without the enhancer was higher than that seen in the ordinary strains. These results suggest that ectopic reporter genes are in general poorly accessible to the Pol I machinery in the nucleolus, and the enhancer somehow improves accessibility (Wai et al. 2001). Interestingly, we also found that a *fob1* mutation abolishes transcription from the enhancer-dependent rDNA promoter integrated at the *HIS4* locus without any effect on transcription from chromosomal rDNA (Wai et al. 2001). *FOB1* is known to be required for recombination hot-spot (*HOT1*) activity (Kobayashi and Horiuchi 1996), which also requires the enhancer region (Voelkel-Meiman et al. 1987), and for recombination within rDNA repeats (Kobayashi et al. 1998). Although the precise function of the *FOB1* gene is unknown, Fob1 protein together with the enhancer apparently stimulates ectopic rDNA promoters to recruit the Pol I machinery, presumably by increasing the probability that the reporter gene is near the nucleolus, perhaps through an interaction of the reporter enhancer with endogenous rDNA. These experiments emphasize that the nucleolus is functionally compartmentalized from the extranucleolar regions and that Pol I and its transcription factors are not readily available to genes outside the nucleolus.

USE OF YEAST MUTANTS TO STUDY NUCLEOLAR STRUCTURES

The presence of a visually recognizable subnuclear structure, the nucleolus, as the site of rDNA transcription by Pol I is one of the unique features of rRNA synthesis in eukaryotes, as mentioned earlier in this paper. Historically, there have been two different ways to think about the nucleolus. One is to consider that the nucleolus is simply an aggregate formed as a result of ribosome biosynthesis at the site of rRNA genes, rDNA. Active rRNA genes are transcribed by the Pol I machinery. The transcripts, nascent precursor rRNAs, then interact with many r-proteins and other components involved in rRNA processing and ribosome assembly, including small nucleolar ribonucloprotein particles (snoRNPs) (for review, see Venema and Tollervey 1999), thus forming an aggregate structure, the nucleolus, at the site of rDNA transcription. According to this model, a single rRNA gene inserted at any chromosomal site would be able to form a mininucleolus. Analyses of P-element transformants carrying a single rRNA gene in *Drosophila melanogaster* integrated outside the nucleolar organizer regions were consistent with this expectation (Karpen et al. 1988; see below for further comments). The second way is to consider the presence of some specific structures, which are present independently of the act of rDNA transcription and ribosome assembly and organize many nucleolar components for efficient rDNA transcription and ribosome assembly. Recent discoveries that the nucleolus has functions other than synthesizing ribosomes (for reviews, see Pederson 1998; Garcia and Pillus 1999; Olson et al. 2000) now tend to support the second viewpoint. We have mentioned in the previous section evidence that the nucleolus in *S. cerevisiae* is functionally compartmentalized, and ectopic rRNA reporter genes inserted at other chromosomal regions or present on a plasmid are not readily accessible to the Pol I machinery (Wai et al. 2001). Such observations are also consistent with the second view. (It should be noted that a single rRNA gene inserted into ectopic chromosomal sites was apparently able to function in the P-element transformation experiments in *Drosophila* mentioned above. This is in apparent conflict with the results obtained in the yeast system. However, the possibility exists that the mininucleoli at ectopic sites were not completely separated from the native nucleolus in the quaternary intranuclear folding of the interphase chromosomes in these flies in vivo, and that transcription of ectopic rRNA genes was achieved through an association of the ectopic template with the native nucleolar structure.)

Observed by immunofluorescence microscopy using antibodies against various known nucleolar proteins, as well as by electron microscopy, the nucleolus in *S. cerevisiae* is a crescent-shaped structure that makes extensive contact with the nuclear envelope (Fig. 4a and Fig. 5B, WT). How are such a structure and its localization determined? What kinds of molecules are involved in specifying the structure and localization? Some insights into these questions have been obtained in the course of our research using various mutant yeast strains described above.

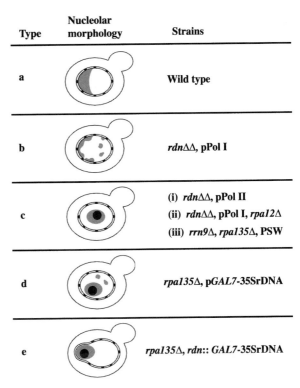

Type	Nucleolar morphology	Strains
a		**Wild type**
b		*rdn*ΔΔ, pPol I
c		(i) *rdn*ΔΔ, pPol II (ii) *rdn*ΔΔ, pPol I, *rpa12*Δ (iii) *rrn9*Δ, *rpa135*Δ, PSW
d		*rpa135*Δ, pGAL7-35SrDNA
e		*rpa135*Δ, *rdn*:: GAL7-35SrDNA

Figure 4. Schematic representation of nucleolar morphologies in *S. cerevisiae* mutants and PSW variants. The morphologies are based on EM images described in Oakes et al. (1993, 1998, 1999). (*a*) Crescent-shaped nucleolus observed in wild-type cells. (*b*) Dispersed nucleolus observed in *rdn*ΔΔ cells bearing rDNA on a multicopy plasmid (pPol I). (*c*) Condensed nucleolus observed in *rdn*ΔΔ cells bearing a multicopy plasmid (pPol II) expressing rDNA from a Pol II promoter. The nucleolus in *rdn*ΔΔ cells bearing pPol I, and with an additional Pol I ts mutation (*rpa12*Δ), also shows this morphology, as does the nucleolus in PSW variants without helper plasmid (*rrn9*Δ, *rpa135*Δ, PSW). (*d*) Nucleolus observed in an *rpa135*Δ mutant that lacks functional Pol I and carries a multicopy plasmid that expresses rDNA from a Pol II promoter. (*e*) Nuclear-envelope-associated nucleolus observed in an *rpa135*Δ mutant containing 20–25 repeats at the rDNA locus that express rDNA from a Pol II promoter. For further explanation, see the text.

As originally demonstrated using metazoan systems, such as *Xenopus laevis* (Brown and Gurdon 1964; Wallace and Birnstiel 1966) and *D. melanogaster* (Ritossa and Spiegelman 1965), rDNA is the nucleolar organizer, and it evidently also plays a crucial role in organizing the nucleolar structure in *S. cerevisiae*. Construction of *rdn*ΔΔ strains with complete chromosomal rDNA deletions as described above has provided more specific information on the role of *cis* elements within rDNA.

We found that *rdn*ΔΔ strains carrying a Pol I helper plasmid contained many small nucleoli distributed throughout the nucleus, primarily localized at the nuclear periphery (Fig. 4b). In contrast, *rdn*ΔΔ strains carrying a Pol II helper plasmid contained a single (and occasionally two, rarely more) round nucleolus that often lacked contact with the nuclear envelope (Fig. 4c, i) (Oakes et al. 1998). The difference in the degree of nucleolar organization and localization is striking. There are two main differences between the two types of strains that may be re-

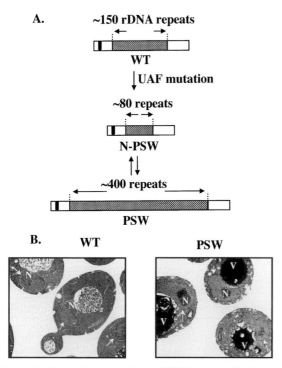

A. ~150 rDNA repeats

WT

↓ UAF mutation

~80 repeats

N-PSW

~400 repeats

PSW

B. WT   PSW

Figure 5. Generation and analysis of PSW variant cells that are able to grow by transcribing chromosomal rDNA repeats using Pol II. (*A*) Two steps that lead to the PSW (*polymerase switched* for growth) state. The first step is a mutation affecting the Pol II silencing function of UAF, e.g., deletion of a UAF subunit gene, *RRN5*, *RRN9*, or *RRN10*. Such a mutant carrying the *GAL7-35S rDNA* on a multicopy plasmid (helper plasmid) can grow in galactose medium. The number of chromosomal rDNA repeats in such a mutant is reduced to ~80 copies, and transcription of these repeats by Pol II is observed. However, without rDNA repeat expansion, cells are unable to form a colony (and hence, cells are in the state defined as N-PSW for *no* growth, but *polymerase switched*). The second step is an expansion of rDNA repeats to ~400, which allows Pol II transcription of rDNA at a high level so that cells are able to grow without a helper plasmid. (*B*) Electron micrographs of the nucleolus in a PSW strain (PSW) and a control strain (WT). The nuclear envelope is marked with an arrow, and the nucleolus (electron-dense area within the nucleus) is indicated as N. (Micrographs reprinted, with permission, from Oakes et al. 1999.)

sponsible for the difference in nucleolar morphology and localization. First, the transcription machinery is different. Second, the *cis* elements, specifically the promoter, on rDNA carried by the plasmids are different. However, the rRNA coding region and the rRNA transcript are identical (except for a few extra nucleotides at the 5′-end of the primary rRNA transcript from the *GAL7* promoter carried on the Pol II helper plasmid). Therefore, models invoking some specific affinity between the rDNA or the rRNA transcript and some structures at the nuclear periphery cannot explain the positioning of the nucleolus at the nuclear periphery in normal yeast cells. In addition, we observed that a *rdnΔΔ* strain with a Pol I ts mutation (*rpa12* deletion) and carrying a Pol I helper plasmid showed a round nucleolus localized away from the nuclear periphery; that is, the same nucleolar morphology and localization seen for the *rdnΔΔ* strains carrying a Pol II helper plasmid, even under permissive growth conditions (Fig. 4c, ii). Thus, the presence of the intact rDNA

with intact *cis* elements alone is not sufficient for predominant localization of the nucleolus at the nuclear periphery, and the presence of an intact Pol I structure is essential for the nucleolar morphology and localization (Oakes et al. 1998).

The importance of the Pol I machinery in maintaining the normal nucleolar structure and localization was also demonstrated using Pol I mutant strains that carry the normal chromosomal rDNA repeats. Yeast strains carrying a ts mutation in the gene encoding one of the essential Pol I subunits showed a rapid fragmentation of the nucleolus upon temperature up-shift, but such fragmentation does not appear to be a consequence of cessation of rRNA synthesis, because other conditions that stop rRNA synthesis, such as the presence of inhibitors, did not cause a fragmentation of the nucleolus (Oakes et al. 1993). Yeast strains with the intact rDNA repeats, but carrying lethal mutations in Pol I machinery and growing by transcribing the *GAL7-35S rDNA* by Pol II also do not have the normal nucleolus, but contain several small nucleoli localized away from the nuclear periphery (Fig. 4d) (Oakes et al. 1993; Trumtel et al. 2000). These mininucleolar bodies contain nucleolar proteins, such as Nop1p and Ssb1p, involved in rRNA processing and ribosome assembly. They are formed presumably around the plasmid rDNA template, as is also the case for the round interior nucleolus present in *rdnΔΔ* strains with a Pol II helper plasmid. In the latter strains, Pol I or Pol-I-specific transcription factor proteins were found to be spread throughout the nucleus (Oakes et al. 1998). Thus, it is evident that there are at least two different kinds of nucleolar proteins. Nucleolar proteins such as those involved in specific transcription, Pol I, UAF, CF, and Rrn3p, may play a primary role in determining the nucleolar localization, whereas others, such as those involved in rRNA processing, apparently do not.

In addition to the promoter elements, there appear to be other chromosomal features that influence nucleolar organization. We found two clearly distinguishable types of nucleolar morphology/localization in Pol I mutant strains, in which Pol II transcribes rDNA template present in chromosomal rDNA repeats. The first type was observed (Oakes et al. 1998) in a *rpa135* deletion strain, in which the template *GAL7-35S rDNA* fusion gene was integrated into chromosomal rDNA repeats as multiple copies (Venema et al. 1995). The copy number of the integrated hybrid gene was estimated to be 20 to 25, and about an equal number of intact rDNA repeats were also present (Venema et al. 1995), so that the total repeat number must be significantly less than the number (~150) present in normal strains. The second type was observed in strains in which the same Pol I subunit is deleted together with the gene (*RRN9*) for a UAF subunit, and Pol II transcribes natural chromosomal rDNA repeats using cryptic Pol II promoter(s) (polymerase switched or PSW strains, see below). In these second type strains, the rDNA repeats are expanded to approximately 400. Both types of nucleoli exhibited a round morphology, but the nucleolus in the first type was more closely associated with the nuclear envelope in the form of a bulge (Fig. 4e) (Oakes et al. 1998), whereas the second type observed in PSW strains did not show such a bulged structure (Fig. 4c, iii) (Oakes et al. 1999). Perhaps, chromosomal elements outside the rDNA repeats constrain the ability of rDNA copies (and the integrated *GAL7-35S rDNA* gene) to move to interior Pol II transcription sites, and the higher repeat numbers in PSW strains might somehow relax this constraint. In summary, it appears that both the type of RNA polymerase transcription machinery involved in rDNA transcription and the organization of the rDNA genes (perhaps including associated non-rDNA elements) strongly influence the intranuclear organization and localization of the gene expression site—the nucleolus.

From all these observations, it is clear that the nucleolar morphology can be altered in response to genetic and epigenetic alterations of the state of yeast cells, with possible alterations in the efficiency of rDNA transcription and/or ribosome assembly (and in some other nucleolar functions such as regulation of cell cycle). Understanding the significance of these alterations in the nucleolar morphology and localization must await more detailed molecular studies of nucleolar organization and its regulatory linkage to overall cell physiology, perhaps extending beyond ribosome biosynthesis (for review, see Pederson 1998; Garcia and Pillus 1999; Olson et al. 2000). Starting from *rdn*ΔΔ strains carrying a helper Pol II plasmid, we have recently succeeded in integrating a cloned rDNA repeat at the original chromosomal locus and its subsequent copy number expansion (Fig. 3) (Wai et al. 2000; see also Kobayashi et al. 2001). It may also be possible to integrate a rDNA repeat at some other chromosomal loci. Such systems will allow mutational analysis not only of DNA elements within rDNA repeats involved in rRNA transcription and nucleolar structures, but also of those in rDNA replication and recombination within a repeated structure. In addition, the possible significance of some other (hypothetical) chromosomal elements outside rDNA repeats discussed above could also be studied in this way.

SILENCING OF Pol II TRANSCRIPTION OF rDNA BY UAF

In the course of studies of mutants defective in transcription factor UAF, we have discovered a phenomenon we call RNA polymerase switching (Oakes et al. 1999; Vu et al. 1999). Mutants defective in UAF subunit genes, *RRN5, RRN9* or *RRN10,* were originally isolated as conditional mutants whose growth depends on the presence of a helper plasmid carrying the *GAL7-35S rDNA* fusion gene and on the presence of galactose as the carbon source, as mentioned earlier in this paper. We have subsequently found that *rrn9* (or *rrn5* or *rrn10*) deletion strains grow extremely slowly on glucose medium (or in the absence of a helper plasmid), but give rise to faster-growing variants. Such variants were found to grow without intact Pol I, synthesizing rRNA by transcribing chromosomal rDNA repeats with Pol II (Vu et al. 1999). We have demonstrated that the switch to growth using the Pol II system (called the PSW state) consists of two steps: a mutational alteration in UAF and an expansion of chromosomal rDNA repeats (Fig. 5A) (Oakes et al. 1999). To

our initial surprise, the first step, a single mutation in a UAF subunit gene, *RRN5, RRN9,* or *RRN10*, was found to be sufficient to allow Pol II transcription of rDNA. No "basal transcription" by Pol I was observed, as mentioned above. In contrast to UAF mutations, mutations in Pol I or other Pol I transcription factors such as CF and Rrn3p do not independently lead to Pol II transcription of rDNA, suggesting a specific role of UAF in preventing the polymerase switching. The second step, expansion of chromosomal rDNA repeats to levels (~400 copies) several-fold higher than the wild type, is required for efficient cell growth. Original UAF mutant cells without repeat expansion are unable to sustain growth, i.e., they do not form colonies (called the N-PSW state for *no* growth, but *po*lymerase *sw*itched). These N-PSW mutant cells have chromosomal rDNA repeats at approximately half (~80) of the normal level (~150) as in the case of Pol I mutant strains with a helper plasmid and growing on galactose. We also found that, as mentioned above, cells in the PSW state have a round nucleolus localized away from the nuclear periphery, which is similar to the nucleolus seen in *rdn*ΔΔ strains carrying a plasmid with the *GAL7-35S rDNA* gene, i.e., a Pol II helper plasmid (Fig. 4c, iii; Fig. 5B, PSW). The increase in rDNA repeat number from ~80 to ~400 is apparently responsible for the switch to the PSW state, that is, the ability to grow continuously without Pol I machinery (and to form a round nucleolus at the interior location). We found that disruption of the *FOB1* gene, which inhibits rDNA repeat expansion and contraction (Kobayashi et al. 1998), decreases the frequency of switching to the PSW state, indicating that the increase in rDNA copy number is a cause rather than a consequence of the switch to the PSW state (Oakes et al. 1999).

There are two possibilities to explain the requirement of rDNA repeat expansion for growth using Pol II: Expansion of the repeats is necessary (1) to increase gene copy numbers, each of which is transcribed only very weakly, or (2) to move rRNA genes from the nuclear periphery to inside the nucleus to a site where Pol II transcription of rDNA takes place relatively more efficiently. We have not yet had an opportunity to distinguish between these two possibilities. Nevertheless, Pol II transcription of chromosomal rDNA does take place at interior Pol II transcription sites in PSW strains. Perhaps, weak Pol II transcription of chromosomal rDNA repeats observed by biochemical assays in N-PSW cells may also be taking place at interior Pol II transcription sites at least transiently, forming a structure resembling the round bulged nucleolus seen in the Pol I deletion strain carrying the *GAL7-rDNA* fusion gene integrated into chromosomal rDNA repeats, as described in the previous section (Fig. 4e). Because of a constraint on movement of rDNA genes to optimal Pol II transcription sites, which is also mentioned above in this connection, we suspect that the efficiency of rDNA transcription by Pol II may not be fully achieved for rDNA copies operating in the N-PSW state. According to this scenario, expansion is required both to relax such a constraint, allowing more efficient Pol II transcription for most rDNA copies, and to increase gene copy numbers, thus increasing the overall rate of rRNA synthesis to attain the rate required for continued cell growth.

EVOLUTION OF THE Pol I MACHINERY, rDNA TANDEM REPEAT STRUCTURES, AND THE NUCLEOLUS

As described above, yeast cells have an inherent ability to transcribe rDNA using Pol II, yielding functional rRNAs and ribosomes. However, UAF normally silences the Pol II transcription of rDNA, and the presence of this ability was only revealed when we undertook studies using mutants that affect UAF components. Thus, this Pol II system could serve as a backup system under conditions where mutations or some other environmental conditions might impair the UAF-mediated Pol I activation system that may also normally function in silencing the Pol II system. There are several reports describing transcription of rDNA by Pol II. In *S. cerevisiae*, Conrad-Webb and Butow (1995) reported Pol II transcription of rDNA in respiratory-deficient mitochondrial [rho°] strains. Although it is not clear how the rho° mutation leads to activation of the silenced Pol II system, this report may encourage further experiments to elucidate the physiological significance of the presence of the silenced Pol II rDNA transcription system. Transcription of rDNA by Pol II has also been reported in other organisms, including mammals. For example, transfection of monkey cells with a plasmid carrying a human rDNA reporter system has demonstrated that deletion of the upstream control element leads to a switch from Pol I to Pol II transcription (Smale and Tjian 1985). The upstream control element of the human rDNA promoter functionally corresponds to the yeast upstream element where UAF binds to achieve both a high-level transcription by Pol I and silencing of Pol II transcription. Although not clearly identified, a factor corresponding to the yeast UAF is likely to exist in mammalian and other metazoan cells, organizing a Pol-I-specific chromatin structure and silencing a cryptic Pol II transcription system. It seems very likely that a silenced Pol-II-mediated rDNA transcription system is present in many or possibly most eukaryotic organisms. If such evolutionary conservation indeed exists, it is likely that the cryptic Pol-II-mediated transcription system may have a physiologically significant role.

It is interesting to note that the transcription machinery in archaebacteria consists of a single RNA polymerase, which is similar to eukaryotic RNA polymerase, and two (or more likely three) transcription factors, which are homologs of TBP and transcription factor IIB (and probably the α subunit of IIE), respectively (Langer et al. 1995; Bell and Jackson 1998; Soppa 1999; Hanzelka et al. 2001). Thus, the transcription machinery in archaea resembles that of eukaryotes, Pol II in particular, rather than that of bacteria. Yet, this single RNA polymerase system in archaea transcribes all RNAs, including rRNA. As discussed by Langer et al. (1995), an ancestor common to both eukaryotic and archaeal organisms may have had a single Pol-II-type RNA polymerase as do present-day archaebacteria. Thus, we can speculate that the emergence of the Pol I system during evolution of eukaryotes was not accompanied by an elimination of the previous rDNA transcription system, and that this ancient system was simply silenced and retained, leading to the present silenced Pol II system for rDNA transcription.

The site of Pol I transcription, i.e., the location of the nucleolus within the nucleus, is clearly separated from the site of Pol II transcription. As mentioned earlier, the dramatic alteration in the localization of the nucleolus in PSW strains confirms this well-recognized fact. Our discovery of the relative inaccessibility of ectopic Pol I reporter genes to the Pol I machinery, especially in the absence of the so-called enhancer, also underscores the compartmentalized nature of the nucleolus. We imagine that the Pol I machinery has evolved for both efficiency and its regulation, and that physical separation of its site from that of the Pol II machinery may have been advantageous. We also imagine that the tandem repeat structure of rDNA genes may have been the simplest way to increase gene dosage and to keep all the rDNA genes in the same region separated from the Pol II transcription machinery. In addition, tandemly repeated rDNA structures, once evolved, might have been useful to alter gene dosage as well as nucleolar structure for adaptation during evolution. It is well known that, even though the frequency of meiotic recombination within rDNA repeats is greatly reduced compared to other chromosomal regions (Petes and Botstein 1977), efficient unequal crossing-over and gene conversion operate during mitotic growth (Petes 1980; Szostak and Wu 1980), and rDNA repeat numbers fluctuate continuously (see, e.g., Kobayashi et al. 1998). Such rDNA repeat expansion and contraction may serve to maintain the well-recognized rRNA sequence homogeneity. In addition, this may also serve to help the cell adapt to environmental changes. There have been reports describing heritable changes in rDNA copy numbers in flux induced by environmental changes (see, e.g., Cullis 1982). Similar heritable changes of rDNA repeat numbers in response to changes of temperature have also been reported for *S. cerevisiae* (Rustchenko et al. 1993). Although we do not know how general such adaptation is, a dramatic increase in rDNA copy numbers accompanied by a switch to the PSW state described above might be considered an extreme example of plasticity of rDNA repeat numbers.

The number of rDNA repeats varies greatly among organisms, ranging from less than 100 to more than 10,000 per haploid genome (Long and Dawid 1980). It was often assumed that the presence of large numbers of rDNA copies reflects a demand for high levels of rRNA transcription for rapid growth. However, there does not appear to be a good correlation between cellular growth rate and numbers of rRNA genes. Because deletion of the *FOB1* gene inhibits rDNA repeat expansion/contraction and stabilizes repeat numbers, we were able to construct isogenic yeast strains in which repeat numbers varied from ~40 to ~150. A strain with ~40 repeats showed rRNA synthesis rates and growth rates identical to the wild type with ~150 repeats under optimal growth conditions (Kobayashi et al. 1998; C.A. Josaitis et al., unpubl.). Yet, introduction of the missing *FOB1* gene into such a strain with ~40 rDNA repeats induced a gradual repeat expansion which ceased at ~150 repeats and did not increase further (Kobayashi et al.

1998). Thus, although there is a considerable plasticity in rDNA repeat numbers and nucleolar structures, ordinary yeast cells maintain ~150 rDNA repeats and use only a fraction of these repeats for rRNA synthesis. (The presence of two different forms of rDNA repeats in growing yeast cells, an active [or potentially active] form and an inactive form, has been demonstrated earlier by psoralen cross-linking experiments [Dammann et al. 1993].) The results of *FOB1*-dependent rDNA repeat expansion experiments suggest that the number of rDNA repeats (~150 for wild-type *S. cerevisiae*) is maintained not by selection to achieve a higher growth rate, but for stability of nucleolar structures that may have evolved as an optimal one not only for rDNA transcription and ribosome assembly, but also for additional (known and unknown) nucleolar functions. The discovery of nucleolar functions such as sequestering of Cdc14p phosphatase through a rDNA chromatin protein, Net1p (or Cfi1p), for regulation of mitotic exit was unexpected (Shou et al. 1999; Visintin et al. 1999), and it is likely that other unidentified nucleolar functions exist (for review, see Pederson 1998; Garcia and Pillus 1999; Olson et al. 2000). We speculate that a primitive nucleolus consisting of the Pol I machinery, repeated rDNA templates, and other factors required for ribosome assembly may have formed a compartmentalized structure separating it from other transcription machineries; and that, once such a structure was formed, its use as a convenient place to sequester a protein (e.g., Cdc14p) for various purposes (e.g., cell cycle regulation) may have evolved. Thus, unique features of rRNA synthesis in eukaryotes discussed earlier, namely the presence of Pol I machinery dedicated to rDNA transcription, tandemly repeated rDNA structures, and the nucleolus with a complex structure as the site of rDNA transcription, are all interrelated—presumably as a result of co-evolution, as discussed above. Events taking place in the nucleolus, which include rDNA transcription and ribosome assembly, their regulation and other nucleolar functions, such as those related to cell cycle regulation, are complex, but represent challenging and likely rewarding subjects for future studies.

ACKNOWLEDGMENTS

I thank previous as well as current members of the laboratory for their contribution to the work summarized in this paper. I also thank Drs. S. M. Arfin, T. Pederson, and R. Steele for critical reading of the manuscript, and M. Oakes and S. VanAmburg for help in preparation of the manuscript. The work in my laboratory described in this article was supported by U.S. Public Health Service grant GM-35949 from the National Institutes of Health.

REFERENCES

Ban N., Nissen P., Hansen J., Moore P.B., and Steitz T.A. 2000. The complete atomic structure of the large ribosomal subunit at 2.4 Å resolution. *Science* **289:** 905.

Bell S.D. and Jackson S.P. 1998. Transcription in Archaea. *Cold Spring Harbor Symp. Quant. Biol.* **63:** 41.

Brewer B.J., Lockshon D., and Fangman W.L. 1992. The arrest of replication forks in the rDNA of yeast occurs independently of transcription. *Cell* **71:** 267.

Brown D.D. and Gurdon J.B. 1964. Absence of ribosomal RNA synthesis in the anucleolate mutant of *Xenopus laevis*. *Proc. Natl. Acad. Sci.* **51:** 139.

Carles C. and Riva M. 1998. Yeast RNA polymerase I subunits and genes. In *Transcription of ribosomal RNA genes by eukaryotic RNA polymerase I* (ed. M.R. Paule), p. 9. Springer-Verlag, Berlin and R.G. Landes, Austin, Texas.

Chernoff Y.O., Vincent A., and Liebman S.W. 1994. Mutations in eukaryotic 18S ribosomal RNA affect translational fidelity and resistance to aminoglycoside antibiotics. *EMBO J.* **13:** 906.

Condon C., Squires C., and Squires C.L. 1995. Control of rRNA transcription in *Escherichia coli*. *Microbiol. Rev.* **59:** 623.

Conrad-Webb H. and Butow R.A. 1995. A polymerase switch in the synthesis of rRNA in *Saccharomyces cerevisiae*. *Mol. Cell. Biol.* **15:** 2420.

Cullis C.A. 1982. Quantitative variation of the ribosomal RNA genes. In *The nucleolus* (ed. E.G. Jordan and C.A. Cullis), p. 103. Cambridge University Press, New York.

Dammann R., Lucchini R., Koller T., and Sogo J.M. 1993. Chromatin structures and transcription of rDNA in yeast *Saccharomyces cerevisiae*. *Nucleic Acids Res.* **10:** 2331.

Drouin G. and de Sá M.M. 1995. The concerted evolution of 5S ribosomal genes linked to the repeat units of other multigene families. *Mol. Biol. Evol.* **12:** 481.

Elion E.A. and Warner J.R. 1984. The major promoter element of rRNA transcription in yeast lies 2 kb upstream. *Cell* **39:** 663.

———. 1986. An RNA polymerase I enhancer in *Saccharomyces cerevisiae*. *Mol. Cell. Biol.* **6:** 2089.

Garcia S.N. and Pillus L. 1999. Net results of nucleolar dynamics. *Cell* **97:** 825.

Gourse R.L., Gaal T., Bartlett M.S., Appleman J.A., and Ross W. 1996. rRNA transcription and growth-dependent regulation of ribosome synthesis in *Escherichia coli*. *Annu. Rev. Microbiol.* **50:** 645.

Grummt I. 1981. Specific transcription of mouse ribosomal DNA in a cell-free system that mimics control *in vivo*. *Proc. Natl. Acad. Sci.* **78:** 727.

———. 1998. Initiation of murine rDNA transcription. In *Transcription of ribosomal RNA genes by eukaryotic RNA polymerase I* (ed. M.R. Paule), p. 135. Springer-Verlag, Berlin and R.G. Landes, Austin, Texas.

Hanzelka B.L., Darcy T.J., and Reeve J.N. 2001. TFE, an archaeal transcription factor in *Methanobacterium thermoautotrophicum* related to eucaryal transcription factor TFIIEα. *J. Bacteriol.* **183:** 1813.

Hartwell L.H., McLaughlin C.S., and Warner J.R. 1970. Identification of ten genes that control ribosome formation in yeast. *Mol. Gen. Genet.* **109:** 42.

Karpen G.H., Schaefer J.E., and Laird C.D. 1988. A *Drosophila* rRNA gene located in euchromatin is active in transcription and nucleolus formation. *Genes Dev.* **2:** 1745.

Keener J. and Nomura M. 1996. Regulation of ribosome synthesis. In *Escherichia coli and Salmonella: Cellular and molecular biology*, 2nd edition (ed. F.C. Neidhardt et al.), p. 1417. American Society for Microbiology, Washington, D.C.

Keener J., Dodd J.A., Lalo D., and Nomura M. 1997. Histones H3 and H4 are components of upstream activation factor required for the high-level transcription of yeast rDNA by RNA polymerase I. *Proc. Natl. Acad. Sci.* **94:** 13458.

Keener J., Josaitis C.A., Dodd J.A., and Nomura M. 1998. Reconstitution of yeast RNA polymerase I transcription in vitro from purified components. *J. Biol. Chem.* **273:** 33795.

Keys D.A., Vu L., Steffan J.S., Dodd J.A., Yamamoto R., Nogi Y., and Nomura M. 1994. *RRN6* and *RRN7* encode subunits of a multiprotein complex essential for the initiation of rDNA transcription by RNA polymerase I in *Saccharomyces cerevisiae*. *Genes Dev.* **8:** 2349.

Keys D.A., Lee B.-S., Dodd J.A., Nguyen T.T., Vu L., Fantino E., Burson L.M., Nogi Y., and Nomura M. 1996. Multiprotein transcription factor UAF interacts with the upstream element of the yeast RNA polymerase I promoter and forms a stable preinitiation complex. *Genes Dev.* **10:** 887.

Kobayashi T. and Horiuchi T. 1996. A yeast gene product, Fob1 protein, required for both replication fork blocking and recombinational hotspot activities. *Genes Cells* **1:** 465.

Kobayashi T., Nomura M., and Horiuchi T. 2001. Identification of DNA *cis*-elements essential for expansion of ribosomal DNA repeats in *Saccharomyces cerevisiae*. *Mol. Cell. Biol.* **21:** 136.

Kobayashi T., Heck D.J., Nomura M., and Horiuchi T. 1998. Expansion and contraction of ribosomal DNA repeats in *Saccharomyces cerevisiae*: Requirement of replication fork blocking (Fob1) protein and the role of RNA polymerase I. *Genes Dev.* **12:** 3821.

Kobayashi T., Hidaka M., Nishizawa M., and Horiuchi T. 1992. Identification of a site required for DNA replication fork blocking activity in the rRNA gene cluster in *Saccharomyces cerevisiae*. *Mol. Gen. Genet.* **233:** 355.

Lalo D., Steffan J.S., Dodd J.A., and Nomura M. 1996. *RRN11* encodes the third subunit of the complex containing Rrn6p and Rrn7p that is essential for the initiation of rDNA transcription by yeast RNA polymerase I. *J. Biol. Chem.* **271:** 21062.

Langer D., Hain J., Thuriaux P., and Zillig W. 1995. Transcription in Archaea: Similarity to that in eucarya. *Proc. Natl. Acad. Sci.* **92:** 5768.

Lin C.W., Moorefield B., Payne J., Aprikian P., Mitomo K., and Reeder R. H. 1996. A novel 66-kilodalton protein complexes with Rrn6, Rrn7, and TATA-binding protein to promote polymerase I transcription initiation in *Saccharomyces cerevisiae*. *Mol. Cell. Biol.* **16:** 6436.

Long E.O. and Dawid I.B. 1980. Repeated genes in eukaryotes. *Annu. Rev. Biochem.* **49:** 727.

Milkereit P. and Tschochner H. 1998. A specialized form of RNA polymerase I, essential for initiation and growth-dependent regulation of rRNA synthesis, is disrupted during transcription. *EMBO J.* **17:** 3692.

Nierras C.R., Liebman S.W., and Warner J.R. 1997. Does *Saccharomyces* need an organized nucleolus? *Chromosoma* **105:** 444.

Nogi Y., Vu L., and Nomura M. 1991a. An approach for isolation of mutants defective in 35S ribosomal RNA synthesis in *Saccharomyces cerevisiae*. *Proc. Natl. Acad. Sci.* **88:** 7026.

Nogi Y., Yano R., and Nomura M. 1991b. Synthesis of large rRNAs by RNA polymerase II in mutants of *Saccharomyces cerevisiae* defective in RNA polymerase I. *Proc. Natl. Acad. Sci.* **88:** 3962.

Nomura M. 1998. Transcriptional factors used by *Saccharomyces cerevisiae* RNA polymerase I and the mechanism of initiation. In *Transcription of ribosomal RNA genes by eukaryotic RNA polymerase I* (ed. M.R. Paule), p. 155. Springer-Verlag, Berlin and R.G. Landes, Austin, Texas

———. 1999. Regulation of ribosome biosynthesis in *Escherichia coli* and *Saccharomyces cerevisiae*: Diversity and common principles. *J. Bacteriol.* **181:** 6857.

Oakes M., Nogi Y., Clark M.W., and Nomura M. 1993. Structural alterations of the nucleolus in mutants of *Saccharomyces cerevisiae* defective in RNA polymerase I. *Mol. Cell. Biol.* **13:** 2441.

Oakes M., Siddiqi I., Vu L., Aris J., and Nomura M. 1999. Transcription factor UAF, expansion and contraction of ribosomal DNA (rDNA) repeats, and RNA polymerase switch in transcription of yeast rDNA. *Mol. Cell. Biol.* **19:** 8559.

Oakes M., Aris J.P., Brockenbrough J.S., Wai H., Vu L., and Nomura M. 1998. Mutational analysis of the structure and localization of the nucleolus in the yeast *Saccharomyces cerevisiae*. *J. Cell Biol.* **143:** 23.

Olson M.O., Dundr M., and Szebeni A. 2000. The nucleolus: An old factory with unexpected capabilities. *Trends Cell Biol.* **10:** 189.

Pederson T. 1998. Survey and summary: The plurifunctional nucleolus. *Nucleic Acids Res.* **26:** 3871.

Petes T. 1980. Unequal meiotic recombination within tandem arrays of yeast ribosomal DNA genes. *Cell* **19:** 765.

Petes T.D. and Botstein D. 1977. Simple Mendelian inheritance of the reiterated ribosomal DNA of yeast. *Proc. Natl. Acad. Sci.* **74:** 5091.

Peyroche G., Milkereit P., Bischler N., Tschochner H., Schultz P., Sentenac A., Carles C., and Riva M. 2000. The recruitment of RNA polymerase I on rDNA is mediated by the interaction of the A43 subunit with Rrn3. *EMBO J.* **19:** 5473.

Ritossa F.M. and Spiegelman S. 1965. Localization of DNA complementary to ribosomal RNA in the nucleolus organizer region of *Drosophila melanogaster*. *Proc. Natl. Acad. Sci.* **53:** 737.

Rosbash M., Harris P.K., Woolford J.L., Jr., and Teem J.L. 1981. The effect of temperature-sensitive RNA mutants on the transcription products from cloned ribosomal protein genes of yeast. *Cell* **24:** 679.

Rustchenko E.P., Curran T.M., and Sherman F. 1993. Variations in the number of ribosomal DNA units in morphological mutants and normal strains of *Candida albicans* and in normal strains of *Saccharomyces cerevisiae*. *J. Bacteriol.* **175:** 7189.

Schluenzen F., Tocilj A., Zarivach R., Harms J., Gluehmann M., Janell D., Bashan A., Bartels H., Agmon I., Franceschi F., and Yonath A. 2000. Structure of functionally activated small ribosomal subunit at 3.3 Å resolution. *Cell* **102:** 615.

Shou W., Seol J.H., Shevchenko A., Baskerville C., Moazed D., Chen Z.W., Jang J., Shevchenko A., Charbonneau H., and Deshaies R.J. 1999. Exit from mitosis is triggered by Tem1-dependent release of the protein phosphatase Cdc14 from nucleolar RENT complex. *Cell* **97:** 233.

Siddiqi I., Dodd J.A., Vu L., Eliason K., Oakes M.L., Keener J., Moore R., Young M.K., and Nomura M. 2001. Transcription of chromosomal rRNA genes by both RNA polymerase I and II in yeast *uaf30* mutants lacking the 30 kDa subunit of transcription factor UAF. *EMBO J.* **20:** 4512.

Smale S.T. and Tjian R. 1985. Transcription of herpes simplex virus *tk* sequences under the control of wild-type and mutant human RNA polymerase I promoters. *Mol. Cell. Biol.* **5:** 352.

Soppa J. 1999. Transcription initiation in Archaea: Facts, factors and future aspects. *Mol. Microbiol.* **31:** 1295.

Steffan J.S., Keys D.A., Dodd J.A., and Nomura M. 1996. The role of TBP in rDNA transcription by RNA polymerase I in *Saccharomyces cerevisiae:* TBP is required for upstream-activation-factor-dependent recruitment of core factor. *Genes Dev.* **10:** 2551.

Steffan J.S., Keys D.A., Vu L., and Nomura M. 1998. Interaction of TATA-binding protein with upstream activation factor is required for activated transcription of ribosomal DNA by RNA polymerase I in *Saccharomyces cerevisiae* in vivo. *Mol. Cell. Biol.* **18:** 3752.

Szostak J.W. and Wu R. 1980. Unequal crossing over in the ribosomal DNA of *Saccharomyces cerevisiae*. *Nature* **284:** 426.

Trumtel S., Léger-Silvestre I., Gleizes P.-E., Teulières F., and Gas N. 2000. Assembly and functional organization of the nucleolus: Ultrastructural analysis of *Saccharomyces cerevisiae* mutants. *Mol. Biol. Cell* **11:** 2175.

Venema J. and Tollervey D. 1999. Ribosome synthesis in *Saccharomyces cerevisiae*. *Annu. Rev. Genet.* **33:** 261.

Venema J., Dirks-Mulder A., Faber A.W., and Raué H.A. 1995. Development and application of an *in vivo* system to study yeast ribosomal RNA biogenesis and function. *Yeast* **11:** 145.

Visintin R., Hwang E.S., and Amon A. 1999. Cfi1 prevents premature exit from mitosis by anchoring Cdc14 phosphatase in the nucleolus. *Nature* **398:** 818.

Voelkel-Meiman K., Keil R.L., and Roeder G.S. 1987. Recombination-stimulating sequences in yeast ribosomal DNA correspond to sequences regulating transcription by RNA polymerase I. *Cell* **48:** 1071.

Vu L., Siddiqi I., Lee B.-S., Josaitis C.A., and Nomura M. 1999. RNA polymerase switch in transcription of yeast rDNA: Role of transcription factor UAF (upstream activation factor) in silencing rDNA transcription by RNA polymerase II. *Proc. Natl. Acad. Sci.* **96:** 4390.

Wai H.H., Vu L., Oakes M., and Nomura M. 2000. Complete deletion of yeast chromosomal rDNA repeats and integration of a new rDNA repeat: Use of rDNA deletion strains for functional analysis of rDNA promoter elements *in vivo*. *Nucleic*

Acids Res. **28:** 3524.

Wai H., Johzuka K., Vu L., Eliason K., Kobayashi T., Horiuchi T., and Nomura M. 2001. Yeast RNA polymerase I enhancer is dispensable for transcription of the chromosomal rRNA gene and cell growth, and its apparent transcription enhancement from ectopic promoters required Fob1 protein. *Mol. Cell. Biol.* **21:** 5541.

Wallace H. and Birnstiel M.L. 1996. Ribosomal cistrons and the nucleolar organizer. *Biochim. Biophys. Acta* **114:** 296.

Wimberly B.T., Brodersen D.E., Clemons W.M., Jr., Morgan-Warren R.J., Carter A.P., Vonrhein C., Hartsch T., and Ramakrishnan V. 2000. Structure of the 30S ribosomal subunit. *Nature* **407:** 327.

Yamamoto R.T., Nogi Y., Dodd J.A., and Nomura M. 1996. *RRN3* gene of *Saccharomyces cerevisiae* encodes an essential RNA polymerase I transcription factor which interacts with the polymerase independently of DNA template. *EMBO J.* **15:** 3964.

Yusupov M.M., Yusupova G.Z., Baucom A., Lieberman K., Earnest T.N., Cate J.H., and Noller H.F. 2001. Crystal structure of the ribosome at 5.5 Å resolution. *Science* **292:** 883.

Zomerdijk J.C.B.M. and Tjian R. 1998. Initiation of transcription on human rRNA genes. In *Transcription of ribosomal RNA genes by eukaryotic RNA polymerase I* (ed. M.R. Paule), p. 121. Springer-Verlag, Berlin and R.G. Landes, Austin, Texas.

Economics of Ribosome Biosynthesis

J.R. WARNER, J. VILARDELL, AND J.H. SOHN*

Department of Cell Biology, Albert Einstein College of Medicine, Bronx, New York 10461

The growing yeast cell is an intricate factory whose major product is ribosomes. The synthesis of ribosomes involves the coordinate activity of three RNA polymerases transcribing more than 300 genes. Analysis of how the cell has dealt with the enormous evolutionary pressure to effect such coordination leads to new appreciation of the integration of the cell's pathways of macromolecular synthesis and poses new questions—bold questions such as how the cell integrates intra- and extracellular signals, and subtle questions such as how the cell titrates accurately the production of a single protein.

On the basis of the size of the yeast genome, 1.3×10^7 bp; the RNA content of a ribosome, 5469 nucleotides; and the approximation that rRNA is ~80% of total yeast RNA; we calculate that there are about 200,000 ribosomes in a yeast cell. Thus, during each generation, of <100 minutes during active log-phase growth on glucose, the cell must synthesize more than 2000 ribosomes per minute. This process consumes a prodigious amount of the cell's resources (for review, see Warner 1999). On the basis of both theoretical calculations and labeling data, we estimate that the transcription of rRNA by RNA polymerase I represents ~60% of total transcription. Direct measurement of the yeast transcriptome showed that the 137 genes encoding the 78 yeast ribosomal proteins represent >4500 of the estimated 15,000 mRNA molecules in the cell, or ~30% of total mRNA (Holstege et al. 1998). Because mRNAs for ribosomal proteins have a relatively short $T_{1/2}$ (Kim and Warner 1983; Li et al. 1999), we estimate that ribosomal protein genes represent ~50% of the total RNA polymerase II initiation events.

Introns are rare in *Saccharomyces cerevisiae*, but ribosomal protein genes account for 101 of the cell's 234 introns (Spingola et al. 1999). That, coupled with the high level of transcription of ribosomal protein genes, means that >90% of all mRNA splicing takes place on transcripts of ribosomal protein genes. Finally, the 150 nuclear pores (Winey et al. 1997) must each transport 78 × 2000/150 = ~1000 ribosomal proteins per minute from the cytoplasm to the nucleus, and 25 ribosomal subunits per minute from the nucleus to the cytoplasm.

To effectively carry out ribosome synthesis, the cell has mastered three levels of coordination: the coordination of ribosome synthesis with the cell's state, the coordination of rRNA transcription with ribosomal protein synthesis, and the coordination of the synthesis of the 78 ribosomal proteins with each other.

COORDINATION OF RIBOSOME SYNTHESIS WITH CELL STATE

Nutritional Regulation of Ribosome Biosynthesis

This prodigious use of resources for ribosome synthesis promotes the vigorous protein synthesis that enables the cell to grow rapidly under favorable environmental conditions, as befits the evolution of a free-living microorganism. However, because the supply of nutrients for such an organism is extremely variable, there is strong evolutionary pressure to restrict the use of resources devoted to rapid growth when conditions may turn unfavorable. An example of the response of *S. cerevisiae* to such conditions is shown in Figure 1, which records the level of mRNAs encoding two ribosomal proteins, L30 and S6, during an ordinary growth cycle of a yeast culture. It is apparent that there is an abrupt repression of ribosome synthesis at OD_{600} of ~3, even though growth continues and the culture will eventually reach a density three to four times as great. This repression extends to the transcription of rRNA as well (Ju and Warner 1994). Interestingly, one can often observe a slight rebound of ribosome synthesis after the cells complete diauxie and begin oxidative growth (lanes 7 and 8).

That ribosome biosynthesis is such a prescient harbinger of future growth of the cell suggests that the cell has sensitive regulatory elements to control it. Indeed, they do. Thus, shifting a culture from a poor C source to a good one causes an immediate stimulation of transcription of both rRNA and ribosomal protein genes (Kief and Warner 1981; Herruer et al. 1987), due to a brief spike of cAMP production leading to activation of the ras-cAMP-PKA pathway (Jiang et al. 1998). Similarly, activation of the TOR pathway couples N availability to ribosome synthesis (Zaragoza et al. 1998; Powers and Walter 1999). Thus, an extensive network of both positive and negative regulatory pathways controls ribosome biosynthesis in response to nutritional signals.

*Present address: Korea Research Institute of Bioscience and Biotechnology (KRIBB), Yusong, P.O. Box 115, Taejon 305-600, Korea.

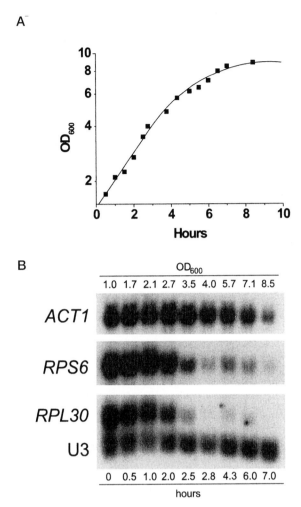

A

B

OD$_{600}$

1.0 1.7 2.1 2.7 3.5 4.0 5.7 7.1 8.5

ACT1

RPS6

RPL30

U3

0 0.5 1.0 2.0 2.5 2.8 4.3 6.0 7.0

hours

Figure 1. Early repression of ribosome biosynthesis: (*A*) The growth of a culture of wild-type *S. cerevisiae* in YPD (broth) medium was monitored; at 23 hours the OD$_{600}$ was 11.6. (*B*) RNA was prepared from cells recovered from aliquots of this culture harvested at intervals. The RNA was subjected to northern analysis and probed for sequences encoding two ribosomal proteins, L30 and S6, as well as for actin (*ACT1*), and for a stable snoRNA, U3. The numbers above represent the OD$_{600}$ at the time of harvest. It is evident that ribosome synthesis almost ceases when the cells are at less than one-third of their final density.

Macromolecular Regulation of Ribosome Biosynthesis

The secretory pathway. Nevertheless, nutritional regulation of ribosome biosynthesis is not sufficient to optimize the production of ribosomes. The cell also has means to control ribosome synthesis by quite a different class of regulatory signals, signals that arise from other pathways of macromolecular synthesis within the cell. The key conclusion is that the cell has evolved mechanisms to permit cross-talk between different elements of its systems for macromolecular synthesis. Although in retrospect this seems obvious, it was not predicted.

The first evidence for the role of other macromolecular systems came from attempts to find mutants that would lead to identification of the components that integrate the incoming signals to determine the level of ribosome syn-

thesis, and that lead to the coordinate production of rRNA and ribosomal proteins. A number of ts mutants were identified in which transcription of both rRNA and ribosomal protein genes was strongly repressed at the nonpermissive temperature. To our surprise, all were identified as mutant in the secretory pathway (Mizuta and Warner 1994). Indeed, examination of many mutants, involving all steps of the secretory pathway, from *SEC61*, at the ER-ribosome junction, to *SEC1*, involved in the fusion of vesicles with the plasma membrane (Fig. 2), led to the conclusion that ANY defect in the secretory pathway leads to a strong repression of ribosome biosynthesis. Inhibitors targeting the secretory pathway, such as tunicamycin and Brefeldin A, had the same effect. Note that the inherently leaky *sec14* mutant elicits less repression, showing that the connection between the two pathways is not an on–off switch, but a variable rheostat. On the other hand, mutants in the branch of the secretory pathway that leads to the vacuole (the *VPS* genes) have no such effect (J. Warner, unpubl.).

What is the connection between the secretory pathway and ribosome biosynthesis? Numerous possibilities, such as the unfolded protein response and the ras-cAMP-PKA pathway, were ruled out experimentally (Nierras and Warner 1999). The key insight was the realization that in

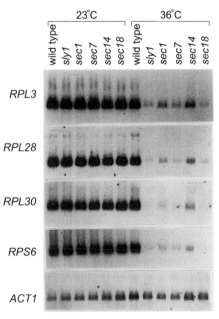

Figure 2. Mutants in the secretory pathway repress ribosome biosynthesis. RNA was prepared from wild-type cells or from cells carrying *ts* mutations in one of several genes representing different steps in the secretory pathway, e.g., *SLY1* in ER-Golgi transport; *SEC1* in the fusion of vesicles with the plasma membrane. Cells were harvested either at permissive temperature or after 90 minutes at the nonpermissive temperature. Their RNA was analyzed on a northern blot and probed with sequences representing genes encoding three different ribosomal proteins and actin (*ACT1*). Not shown in this figure is a mutant in *SEC61*, encoding the pore of the endoplasmic reticulum, at the initial step in the secretory pathway, which has the same effect (J.R. Warner, unpubl.). (Adapted, with permission, from Mizuta and Warner 1994.)

S. cerevisiae the secretory pathway is devoted largely to the synthesis of components of cell membrane and cell wall. In a *sec* mutant, protein synthesis usually continues unabated for nearly a generation, while no new membrane can be formed. The accumulation of new proteins without a concomitant increase in cell surface will lead to the development of a turgor pressure within the cell that in turn triggers a stress response from the stretch sensors within the cell membrane/wall. The predominant stretch sensor of *S. cerevisiae* is Wsc1p (Gray et al. 1997; Verna et al. 1997), which, through intermediates, activates protein kinase C (PKC) (for review, see Delley and Hall 1999). Indeed, deletion of either *WSC1* or *PKC1* abrogates the repression of ribosome synthesis due to a defect in the secretory pathway, suggesting that ribosome synthesis is responsive to stress at the cell surface (Nierras and Warner 1999; Li et al. 2000).

Genetic evidence suggests that below PKC lies a branched pathway (Errede and Levin 1993). One branch is a MAP kinase pathway that leads to increased transcription of genes encoding components of the cell membrane and the cell wall; the other has never been identified. Since deletion of members of the MAP kinase pathway has no effect on the repression of ribosome synthesis (Nierras and Warner 1999; Li et al. 2000), we suggest that this unidentified branch is responsible.

Superficially, one can consider the coupling of the secretory pathway with ribosome synthesis as nothing more than another example of a stress response. At a more subtle level, however, this relationship is clearly an example of the integration of two of the major pathways of macromolecular synthesis in the cell. These observations should, we believe, lead us to reexamine our frequently parochial approach to individual compartments of the cell, realizing that evolution will have demanded a global integration that has so far escaped the cell biologists.

Feedback regulation from 60S subunits. A genetic screen designed to identify elements of the signal transduction pathway between the secretory system and ribosome biosynthesis led to a mutant allele of *RPL1B*, one of two genes encoding ribosomal protein L1 (J. Sohn and J. Warner, in prep.). The nonsense mutation resulted in a nonfunctional protein, leaving the cell with insufficient L1 to assemble a full complement of ribosomes. As a result, the cells became partially starved for 60S subunits, leading to slower growth and accumulation of "halfmers," polyribosomes with an extra 40S subunit awaiting the scarce 60S subunits (Moritz et al. 1990).

This result led us to propose that an imbalance of ribosomal subunits might send a feedback signal to increase ribosome synthesis. A test of that model is shown in Figure 3. Cells were treated with tunicamycin to induce the repression of ribosome synthesis that results from a defect in the secretory pathway. In wild-type cells, tunicamycin causes a decline in mRNA encoding ribosomal protein L30, while mRNA encoding actin remains relatively constant. In cells missing one of the two genes encoding ribosomal protein L11, however, the effect of tunicamycin on *RPL30* mRNA levels, and indeed on ribosome synthesis in general, including rRNA transcrip-

tion, is entirely abrogated (J. Sohn and J. Warner, in prep.). Interestingly, the effect is specific for a deficiency of 60S subunits. Deletion of one of the two genes encoding ribosomal protein S6, resulting in a deficiency of 40S subunits, does not suppress the effect of tunicamycin on ribosome synthesis. In both cases, the cells missing one of the ribosomal protein genes are growing with a doubling time about 40% greater than the wild-type cells. Although only depletion of the 60S subunits has this effect, the transcription of both 60S and 40S protein genes is coordinated.

We have carried out parallel experiments using not only depletion of a different 60S protein, but also mutants of nucleolar proteins that block 60S synthesis. The results are the same. A deficiency of 60S subunits overcomes the repression of ribosome synthesis due to a defect in the secretory pathway.

The mechanism by which a deficiency of 60S subunits can influence the rate of ribosome synthesis remains obscure. Nevertheless, this is just one more example of the way that the expensive process of ribosome synthesis is woven into the web of controls that maintain balanced synthesis of the various components needed for cell growth and maintenance.

COORDINATION OF rRNA AND RIBOSOMAL PROTEIN SYNTHESIS

Under almost all physiological conditions studied, the transcription of rRNA and of ribosomal protein genes is tightly coupled, as would be expected for a process in which synthesis of one without the other is futile. This is true of response both to nutritional signals and to intracellular signals such as those described above. One exception is the effect of heat shock, which brings about a sudden, transient repression of ribosomal protein gene transcription, while having a relatively minor effect on rRNA transcription (Warner and Udem 1972).

Yet, it must be confessed that we know essentially nothing about how this coordination, which involves two different RNA polymerases and some 70–80% of the tran-

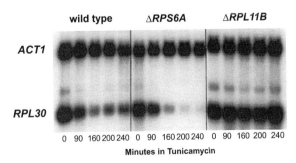

Figure 3. A deficiency of 60S subunits trumps the secretory defect. Cultures of wild-type cells, as well as of cells carrying a deletion of one of the two genes encoding either ribosomal protein S6 or L11, were treated with 2.5 µg/ml tunicamycin, a drug that inhibits the secretory pathway and induces the unfolded protein response. At intervals, aliquots of the cultures were harvested, and RNA was prepared and analyzed on a northern and probed with sequences representing actin and ribosomal protein L30.

scription of the cell, is effected. A number of proteins have been implicated in the transcription of ribosomal protein genes. Rap1p is the most evident, serving as the major transcription factor for most of the ribosomal protein genes (Lascaris et al. 1999). TAF145, a component of TFIID, has been suggested to have a major influence on ribosomal protein gene transcription (Shen and Green 1997), although depletion of this protein affects the transcription of many other genes as well, casting some doubt on its specificity (Holstege et al. 1998). Recently, chromatin remodeling factors such as Esa1p (Reid et al. 2000) and Rsc3p and Rsc30p (Angus-Hill et al. 2001) have been implicated in ribosomal protein gene transcription as well.

In contrast, rRNA transcription depends on an entirely different set of proteins (see Nomura, this volume), with recent data suggesting that regulation is provided by Rrn3p, one of the obligate cofactors of RNA polymerase I (Peyroche et al. 2000). The intriguing observation that Rap1p can also influence the regulation of rRNA transcription (Miyoshi et al. 2001) may point to a connecting link, although no mechanism is evident as yet. There is some evidence that deficiency of many ribosomal proteins leads to reduced rRNA transcription (Warner and Udem 1972). (That paper reported that mutants, later shown to be in the splicing apparatus necessary for production of most ribosomal protein mRNAs, had a collateral effect on rRNA transcription.) On the other hand, reduction of rRNA transcription through a mutant in RNA polymerase I had relatively little effect on the level of ribosomal protein mRNA (Wittekind et al. 1990).

These examples, however, based on defective transcription or splicing, are hardly physiological. The overwhelming evidence leads us to the concept of a core regulatory element that integrates the positive and negative signals to provide a single output that controls transcription of both rRNA and ribosomal protein genes. Identification of that core element is the holy grail of the regulation of ribosome synthesis and would provide the foundation stone to our understanding of the economics of growth of *S. cerevisiae*.

COORDINATION BETWEEN RIBOSOMAL PROTEIN GENES

From the recent crystal structures (Ban et al. 2000; Wimberly et al. 2000) it is evident that each protein in a ribosome is present as a single copy, except perhaps for the acidic proteins that make up the stalk of the large subunit. Thus, there is strong selective pressure for the yeast cell to synthesize equimolar amounts of its 78 individual ribosomal proteins (Fig. 4). This is accomplished through at least three levels of regulation.

The upstream activating sequences of most ribosomal protein genes are constructed in a similar way, with two binding sites for Rap1p, responsible for ~90% of the transcription, followed by T-rich regions that seem to provide the rest (Rotenberg and Woolford, Jr. 1986; Woudt et al. 1986; Schwindinger and Warner 1987). However, several ribosomal protein genes have a single site for Abf1p instead of the Rap1p sites (Hamil et al. 1988; Herruer et al.

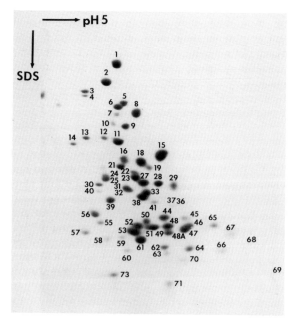

Figure 4. Ribosomal proteins of *S. cerevisiae*. This visual representation emphasizes the challenge of producing equimolar amounts of 78 different proteins. Proteins were prepared from ribosomes isolated from cells of *S. cerevisiae*, and separated on a two-dimensional polyacrylamide gel, at pH 5 in 8 M urea in the first dimension, and in SDS to separate by size in the second (Warner and Gorenstein 1978). The nomenclature is antique, but it can be brought up to date by reference to http://www.mips.biochem.mpg.de/proj/yeast/reviews/rib_nome ncl.html (Mager et al. 1997). Note that some spots, e.g., #15 and #61, have more than a single protein.

1989). It is presumably these common features that lead to the coordinated transcription described above. For convenience, the promoter and activator regions of ribosomal protein genes, as of most other genes, have been studied as fragments in various types of reporter constructs. Reality is undoubtedly more complex; we propose that there has been strong evolutionary pressure on the entire activator/promoter sequence to provide transcription at the appropriate level. The problem is particularly acute, since for some proteins there are two active genes, whereas for others there is only one.

Direct measurements indicate that there is not equimolar expression of the mRNAs for ribosomal proteins. One transcriptome analysis, based on Affymetrix chips, reports an average of about 50 mRNAs per cell for each ribosomal protein (summing over the duplicated genes), but with a fivefold difference between the least and the most abundant (Holstege et al. 1998). The SAGE analysis is somewhat less complete because many of the short ribosomal protein mRNAs do not have a SAGE tag. Nevertheless, for those that can be counted, there is more than a threefold difference between the number of mRNAs for individual ribosomal proteins (Velculescu et al. 1997). The differential is relatively independent of whether the protein is encoded by one gene or two.

If these measurements are correct, then mRNA abundance, the balance between transcription and mRNA turnover, is insufficient to effect equimolar synthesis of the ribosomal proteins. We have no direct measurements of the efficiency of translation of the ribosomal protein mRNAs. However, quantitative analysis indicates that newly formed ribosomal proteins are efficiently incorporated into ribosomes (Gorenstein and Warner 1976), suggesting that there is not wholesale wastage. Although these experiments are subject to uncertainties due to the rapid turnover of unused ribosomal proteins (see below), we suggest, nevertheless, that the survival value of producing equimolar amounts of the ribosomal proteins has led not only to balanced transcription, but also to modulation of the nucleotide sequence that determines for each transcript the efficiency of translation.

A second way in which cells maintain equimolar amounts of ribosomal proteins is by rapid degradation of those that are not assembled into ribosomes. Thus, several measurements of the $T_{1/2}$ of excess ribosomal proteins have yielded values of from 0.5 to 3 minutes (Warner et al. 1985; El-Baradi et al. 1986; Maicas et al. 1988; Tsay et al. 1988; Moritz et al. 1990). The rapid turnover of unused ribosomal proteins is also a characteristic of human cells (Warner 1977). A possible biochemical cause for this rapid turnover is apparent from the X-ray structures, which suggest that until they are assembled into the ribosome, most ribosomal proteins are likely to be at least partially unstructured. Such proteins may trigger the ubiquitin–proteosome pathway, although that has never been demonstrated directly. A reasonable biological cause for the rapid turnover is that it would be dangerous to have abundant rogue RNA-binding proteins at loose in the nucleus.

A third way in which cells of *S. cerevisiae* maintain equimolar amounts of ribosomal proteins is through specific feedback mechanisms. A prime example is the case of L30. *RPL30* and *RPL24A* are two linked ribosomal protein genes that are divergently transcribed. Both carry introns. As shown in Figure 5, when this locus is carried on a multicopy plasmid, the two genes behave quite differently. The mRNA of RPL24A accumulates as expected; however, it is the unspliced transcripts of *RPL30* that accumulate far in excess of the copy number of the plasmid. In analogy with the models of *E. coli*, in which certain ribosomal proteins bind to the mRNA derived from a polycistronic mRNA to regulate the translation and stability of the entire mRNA (Nomura et al. 1984), we suggested that excess L30 binds to its own transcript to prevent its splicing (Dabeva et al. 1986). Indeed, such is the case. L30 binds to the structure that had been predicted from mutational analysis of the nucleotides needed to regulate splicing in vivo (Eng and Warner 1991) shown as a 2D approximation in Figure 6A (Vilardell and Warner 1994). This arrangement, with extensive interactions between the purines in the bulge, has been confirmed by the atomic resolution structure of L30 bound to its transcript, the first such structure of a regulated splicing event (Mao et al. 1999). Interestingly, the key nucleotides are almost entirely within exon I; after splicing, when exon I and exon II have been joined, the nucleotides downstream of the splice site are GCA.... rather than

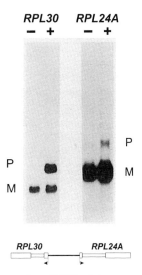

Figure 5. Overexpression of L30 leads to accumulation of the unspliced transcript of RPL30. A fragment of yeast DNA carrying the divergently transcribed pair of ribosomal protein genes, *RPL30* and *RPL24A* (as diagrammed below), was incorporated into a 2μ plasmid and transformed into wild type cells of *S. cerevisiae*. The copy number of the plasmid was measured as about four. RNA was prepared from cells with (+) and without (–) the plasmid, subjected to northern analysis, and probed consecutively for transcripts from the two genes. (Adapted, with permission, from Warner et al. 1985 [copyright ASM, Washington, D.C.].)

GUA.... Thus, it is not surprising that L30 also binds its spliced mRNA, albeit with less affinity, and exerts some inhibition of translation (Dabeva and Warner 1993; Vilardell et al. 2000b).

The structure of the bound complex is interesting in that three distinct segments of the 104-amino-acid protein make contact with the RNA (Mao et al. 1999). This observation, together with analysis of the evolution of the L30 orthologs, led us to propose that the binding of L30 to the *RPL30* transcript mimics the binding of L30 to 25S rRNA. As a test of this hypothesis, we searched for a distant ortholog of L30; L30 is found in all eukaryotes examined and in one branch of the archaea, but in no prokaryotes. The L30 ortholog of the archaeon *Sulfolobus acidocaldarius* has 33% identity with L30 of *S. cerevisiae*. We cloned the *Sulfolobus* gene and expressed it both in *E. coli* and in *S. cerevisiae*. The protein from *E. coli* binds to the *RPL30* transcript and inhibits splicing in vitro. However, the *Sulfolobus* protein will not replace the *Saccharomyces* L30. Indeed, overexpression of the *Sulfolobus* L30 actually prevents growth of the yeast cells (Vilardell et al. 2000a), due to its suppression of splicing of the transcript of the endogenous *RPL30* gene. This was shown not only biochemically, by the accumulation of unspliced transcript, but also genetically, since mutation of the binding site on the *RPL30* transcript permits the cells to grow despite the production of the *Sulfolobus* L30 ortholog (Fig. 6B). Thus, alteration of a single nucleotide, C9>U, some 50 nucleotides from the ORF, protects the cells from the lethal effects of a protein it has not seen for 2 billion years! These observations are a testament both to the evolutionary conservation of ribosome structure,

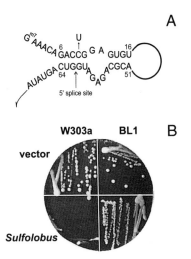

Figure 6. (*A*) Representation of the structure of the *RPL30* transcript that binds L30. The 5′ CAP structure and the 5′ splice site immediately following the initiator AUG are indicated. The loop of 34 nucleotides (from positions 17–50) that is not necessary for binding of L30 is omitted. The mutation of C9 to U, which abolishes binding of L30, and consequently the regulation of splicing, is indicated. (*B*) Expression of *Sulfolobus* L30 inhibits yeast growth. Strains W303a and BL1 differ in a single nucleotide; BL1 carries a C9 to T mutation of the *RPL30* transcription unit (see *A*). Each strain was transformed with a CEN *URA3* vector, or with the vector carrying the *Sulfolobus* L30 ORF under the control of the *GAL1* promoter, and assayed on plates lacking uracil and with 2% galactose as the sole carbon source. It is evident that *Sulfolobus* L30 prevents the growth of the wild-type strain, W303a, but not of strain BL1, whose *RPL30* transcript is not bound by L30. All four strains grow normally in glucose. (Adapted, with permission, from Elsevier Science, from Vilardell et al. 2000a.)

and to the ingenuity of evolution in using this structure not only in the ribosome but also in regulation.

Is the regulation of splicing of the *RPL30* transcript physiologically relevant? After all, normal cells do not have extra copies of the gene, nor are they subjected to an invasion by the *Sulfolobus* gene. The answer is yes! A competition growth experiment between wild-type cells and cells with the same C9>U mutation that permitted growth in the presence of the *Sulfolobus* L30 led to a resounding victory for the wild type. Within 4–5 days, the mutant cells had almost vanished from the culture (Li et al. 1996).

Are all of the ribosomal proteins subject to feedback regulation in the manner of L30? Is that the reason that introns have persisted in ribosomal protein genes? We and other workers have looked for regulation of splicing in other genes; only for *RPS14B* is there strong evidence in its favor (Fewell and Woolford 1999). Our experience suggests that binding of L30 stabilizes the unspliced transcript. However, it is possible that regulation of splicing of transcripts of other genes might occur without the accumulation of apparent intermediates, if the unspliced transcripts were rapidly degraded, perhaps even before their transcription was complete (Elliott and Rosbash 1996). In that case, the regulation might never be detected.

Why might it be important for the cell to regulate the level of L30 especially rigorously? One possibility has

emerged as recent structures have been determined. L30 defines a new family of RNA-binding proteins that bind to purine bulges. Members of this family include another ribosomal protein L7ae (Ban et al. 2000); as well as Snu13p (p15.5 of mammals), which is found in the Box CD class of snoRPs and in the U4snRNP (Watkins et al. 2000); and also Nhp2p, which is found in the H/ACA class of snoRPs (Henras et al. 1998). Indeed, L30 binds to some box CD RNAs in vitro (J. Vilardell and D.A. Samarsky, unpubl.). Perhaps the presence of even trace mounts of L30, interfering with the proper assembly of these RNPs, is sufficient to provide a strong evolutionary pressure for such tight regulation.

CONCLUSIONS

One can hardly consider the ribosome without being deeply influenced by evolutionary considerations. This is true not only because of its long history and its truly remarkable conservation over billions of years, but also because as a central macromolecular complex of cells, whose assembly requires great coordination and whose construction requires substantial resources, these processes have been under continuous, intense evolutionary pressure for optimization. In particular, the three coordinates, of ribosome synthesis with cell growth, of rRNA and ribosomal protein synthesis, and of the synthesis of the 78 ribosomal proteins, must be effected by natural selection.

An understanding of how the three coordinates have been established in organisms with various life-styles would provide fundamental insight into the evolution of a cell's economy. For example, why is the control of ribosomal protein synthesis largely at the level of translation in enteric bacteria, at the level of transcription in *Saccharomyces*, and at the level of translation in mammalian cells (Meyuhas 2000)?

ACKNOWLEDGMENTS

Research from this laboratory has been supported in part by grants from the National Institute of German Medical Sciences GM25532 to J.R.W. and from the National Cancer Institute CA-13330 to the Albert Einstein Cancer Center.

REFERENCES

Angus-Hill M.L., Schlichter A., Roberts D., Erdjument-Bromage H., Tempst P., and Cairns B.R. 2001. A Rsc3/Rsc30 zinc cluster dimer reveals novel roles for the chromatin remodeler RSC in gene expression and cell cycle control. *Mol. Cell* **7:** 741.

Ban N., Nissen P., Hansen J., Moore P.B., and Steitz T.A. 2000. The complete atomic structure of the large ribosomal subunit at 2.4 Å resolution. *Science* **289:** 905.

Dabeva M.D. and Warner J.R. 1993. Ribosomal protein L32 of *Saccharomyces cerevisiae* regulates both splicing and translation of its own transcript. *J. Biol. Chem.* **268:** 19669.

Dabeva M.D., Post-Beittenmiller M.A., and Warner J.R. 1986. Autogenous regulation of splicing of the transcript of a yeast ribosomal protein gene. *Proc. Natl. Acad. Sci.* **83:** 5854.

Delley P.A. and Hall M.N. 1999. Cell wall stress depolarizes cell growth via hyperactivation of RHO1. *J. Cell Biol.* **147:** 163.

El-Baradi T.T., van der Sande A.F.M., Mager W.H., Raue H.A., and Planta R.J. 1986. The cellular level of yeast ribosomal protein L25 is controlled principally by rapid degradation of excess protein. *Curr. Genet.* **10:** 733.

Elliott D.J. and Rosbash M. 1996. Yeast pre-mRNA is composed of two populations with distinct kinetic properties. *Exp. Cell Res.* **229:** 181.

Eng F.J. and Warner J.R. 1991. Structural basis for the regulation of splicing of a yeast messenger RNA. *Cell* **65:** 797.

Errede B. and Levin D.E. 1993. A conserved kinase cascade for MAP kinase activation in yeast. *Curr. Opin. Cell Biol.* **5:** 254.

Fewell S.W. and Woolford J.L.J. 1999. Ribosomal protein S14 of *Saccharomyces cerevisiae* regulates its expression by binding to RPS14B pre-mRNA and to 18S rRNA. *Mol. Cell. Biol.* **19:** 826.

Gorenstein C. and Warner J.R. 1976. Coordinate regulation of the synthesis of eukaryotic ribosomal proteins. *Proc. Natl. Acad. Sci.* **73:** 1547.

Gray J.V., Ogas J.P., Kamada Y., Stone M., Levin D.E., and Herskowitz I. (1997). A role for the Pkc1 MAP kinase pathway of *Saccharomyces cerevisiae* in bud emergence and identification of a putative upstream regulator. *EMBO J.* **16:** 4924.

Hamil K.G., Nam H.G., and Fried H.M. 1988. Constitutive transcription of yeast ribosomal protein gene *TCM1* is promoted by uncommon *cis*- and *trans*-acting elements. *Mol. Cell. Biol.* **8:** 4328.

Henras A., Henry Y., Bousquet-Antonelli C., Nouillac-Depeyre J., Gelugne J.-P., and Caizergues-Ferrer M. 1998. Nhp2p and Nop10p are essential for the function of H/ACA snoRNPs. *EMBO J.* **17:** 7078.

Herruer M.H., Mager W.H., Doorenbosch T.M., Wessels P.L., Wassenaar T.M., and Planta R.J. 1989. The extended promoter of the gene encoding ribosomal protein S33 in yeast consists of multiple protein binding elements. *Nucleic Acids Res.* **17:** 7427.

Herruer M.H., Mager W.H., Woudt L.P., Nieuwint R.T., Wassenaar G.M., Groeneveld P., and Planta R.J. 1987. Transcriptional control of yeast ribosomal protein synthesis during carbon-source upshift. *Nucleic Acids Res.* **15:** 10133.

Holstege F.C.P., Jennings E.G., Wyrick J.J., Lee T.I., Hengartner C.J., Green M.R., Golub T.R., Lander E.S., and Young R.A. 1998. Dissecting the regulatory circuitry of a eukaryotic genome. *Cell* **95:** 717.

Jiang Y., Davis C., and Broach J.R. 1998. Efficient transition to growth on fermentable carbon sources in *Saccharomyces cerevisiae* requires signaling through the Ras pathway. *EMBO J.* **17:** 6942.

Ju Q. and Warner J.R. 1994. Ribosome synthesis during the growth cycle of *Saccharomyces cerevisiae*. *Yeast* **10:** 151.

Kief D.R. and Warner J.R. 1981. Coordinate control of syntheses of ribosomal ribonucleic acid and ribosomal proteins during nutritional shift-up in *Saccharomyces cerevisiae*. *Mol. Cell. Biol.* **1:** 1007.

Kim C.H. and Warner J.R. 1983. Messenger RNA for ribosomal proteins in yeast. *J. Mol. Biol.* **165:** 79.

Lascaris R.F., Mager W.H., and Planta R.J. 1999. DNA-binding requirementrs of the yeast protein Rap1p as selected *in silico* from ribosomal gene promoter sequences. *Bioinformatics* **15:** 267.

Li B., Nierras C.R., and Warner J.R. 1999. Transcriptional elements involved in the repression of transcription of ribosomal protein synthesis. *Mol. Cell. Biol.* **19:** 5393.

Li B., Vilardell J., and Warner J.R. 1996. An RNA structure involved in feedback regulation of splicing and of translation is critical for biological fitness. *Proc. Natl. Acad. Sci.* **93:** 1596.

Mager W.H., Planta R.J., Ballesta J.P., Lee J.C., Mizuta K., Suzuki K., Warner J.R., and Woolford J.L., Jr. 1997. A new nomenclature for the cytoplasmic ribosomal proteins of *Saccharomyces cerevisiae*. *Nucleic Acids Res.* **25:** 4872.

Maicas E., Pluthero F.G., and Friesen J.D. 1988. The accumulation of three yeast ribosomal proteins under conditions of excess mRNA is determined primarily by fast protein decay. *Mol. Cell. Biol.* **8:** 169.

Mao H., White S.A., and Williamson J.R. 1999. Structure of the

yeast RPL30-autoregulatory RNA complex revealing a novel loop-loop recognition motif. *Nat. Struct. Biol.* **6:** 1139.

Meyuhas O. 2000. Synthesis of the translational apparatus is regulated at the translational level. *Eur. J. Biochem.* **267:** 6321.

Miyoshi K., Miyakawa T., and Mizuta K. 2001. Repression of rRNA synthesis due to a secretory defect requires the C-terminal silencing domain of Rap1p in *Saccharomyces cerevisiae*. *Nucleic Acid Res.* **29:** 3297.

Mizuta K. and Warner J.R. 1994. Continued functioning of the secretory pathway is essential for ribosome synthesis. *Mol. Cell. Biol.* **14:** 2493.

Moritz M., Paulovich A.G., Tsay Y.F., and Woolford J.L., Jr. 1990. Depletion of yeast ribosomal proteins L16 or rp59 disrupts ribosome assembly. *J. Cell Biol.* **111:** 2261.

Nierras C.R. and Warner J.R. 1999. Protein kinase C enables the regulatory circuit that connects membrane synthesis to ribosome synthesis in *S. cerevisiae*. *J. Biol. Chem.* **274:** 13235.

Nomura M., Gourse R., and Baughman G. 1984. Regulation of the synthesis of ribosomes and ribosomal components. *Annu. Rev. Biochem.* **53:** 75.

Peyroche G., Milkereit P., Bischler N., Tschochner H., Schultz P., Sentenac A., Carles C., and Riva M. 2000. The recruitment of RNA polymerase I on rDNA is mediated by the interaction of the A43 subunit with Rrn3. *EMBO J.* **19:** 5473.

Powers T. and Walter P. 1999. Regulation of ribosome biogenesis by the rapamycin-sensitive TOR-signalling pathway in *Saccharomyces cerevisiae*. *Mol. Biol. Cell* **10:** 987.

Reid J.L., Iyer V.R., Brown P.O., and Struhl K. 2000. Coordinate regulation of yeast ribosomal protein genes is associated with targeted recruitment of Esa1 histone acetylase. *Mol. Cell* **6:** 1297.

Rotenberg M.O. and Woolford J.L., Jr. 1986. Tripartite upstream promoter element essential for expression of *Saccharomyces cerevisiae* ribosomal protein genes. *Mol. Cell. Biol.* **6:** 674.

Schwindinger W.F. and Warner J.R. 1987. Transcriptional elements of the yeast ribosomal protein gene *CYH2*. *J. Biol. Chem.* **262:** 5690.

Shen W.-C. and Green M.R. 1997. Yeast TAF$_{II}$145 functions as a core promoter selectivity factor, not a general coactivator. *Cell* **90:** 615.

Spingola M., Grate L., Haussler D., and Ares M., Jr. 1999. Genome-wide bioinformatic and molecular analysis of introns in *Saccharomyces cerevisiae*. *RNA* **5:** 221.

Tsay Y.F., Thompson J.R., Rotenberg M.O., Larkin J.C., and Woolford J.L., Jr. 1988. Ribosomal protein synthesis is not regulated at the translational level in *Saccharomyces cerevisiae*: Balanced accumulation of ribosomal proteins L16 and rp59 is mediated by turnover of excess protein. *Genes Dev.* **2:** 664.

Velculescu V.E., Zhang L., Zhou W., Vogelstein J., Basrai M.A., Bassett D.E., Jr., Hieter P., Vogelstein B., and Kinzler K.W. 1997. Characterization of the yeast transcriptome. *Cell* **88:** 243.

Verna J., Lodder A., Lee K., Vagts A., and Ballester R. 1997. A family of genes required for the maintenance of cell wall integrity and for the stress response in *Saccharomyces cerevisiae*. *Proc. Natl. Acad. Sci.* **94:** 13804.

Vilardell J. and Warner J.R. 1994. Regulation of splicing at an intermediate step in the formation of the spliceosome. *Genes Dev.* **8:** 211.

Vilardell J., Yu S.J., and Warner J.R. 2000b. Multiple functions of an evolutionarily conserved RNA binding domain. *Mol. Cell* **5:** 761.

Vilardell J., Chartrand P., Singer R.A., and Warner J.R. 2000a. The odyssey of a regulated transcript. *RNA* **6:** 1773.

Warner J.R. 1977. In the absence of ribosomal RNA synthesis, the ribosomal proteins of HeLa cells are synthesized normally and degraded rapidly. *J. Mol. Biol.* **115:** 315.

Warner J.R. 1999. The economics of ribosome biosynthesis in yeast. *Trends Biochem. Sci.* **24:** 437.

Warner J.R. and Gorenstein C. 1978. The ribosomal proteins of *Saccharomyces cerevisiae*. *Methods Cell Biol.* **20:** 45.

Warner J.R. and Udem S.A. 1972. Temperature sensitive mutations affecting ribosome synthesis in *Saccharomyces cerevisiae*. *J. Mol. Biol.* **65:** 243.

Warner J.R., Mitra G., Schwindinger W.F., Studeny M., and Fried H.M. 1985. *Saccharomyces cerevisiae* coordinates accumulation of yeast ribosomal proteins by modulating mRNA splicing, translational initiation, and protein turnover. *Mol. Cell. Biol.* **5:** 1512.

Watkins N.J., Segault V., Charpentier B., Nottrott S., Fabrizio P., Bachi A., Wilm M., Rosbash M., Branlant C., and Luhrmann R. 2000. A common core RNP structure shared between the small nucleoar box C/D RNPs and the spliceosomal U4 snRNP. *Cell* **103:** 457.

Wimberly B.T., Brodersen D.E., Clemons W.M., Jr., Morgan-Warren R.J., Carter A.P., Vonrhein C., Hartsch T., and Ramakrishnan V. 2000. Structure of the 30S ribosomal subunit. *Nature* **407:** 327.

Winey M., Yarar D., Giddings T.H.J., and Mastronarde D.N. 1997. Nuclear pore complex number and distribution throughout the *Saccharomyces cerevisiae* cell cycle by three-dimensional reconstruction from electron micrographs of nuclear envelopes. *Mol. Biol. Cell* **8:** 2119.

Wittekind M., Kolb J.M., Dodd J., Yamagishi M., Memet S., Buhler J.M., and Nomura M. 1990. Conditional expression of RPA190, the gene encoding the largest subunit of yeast RNA polymerase I: Effects of decreased rRNA synthesis on ribosomal protein synthesis. *Mol. Cell Biol.* **10:** 2049.

Woudt L.P., Smit A.B., Mager W.H., and Planta R.J. 1986. Conserved sequence elements upstream of the gene encoding yeast ribosomal protein L25 are involved in transcription activation. *EMBO J.* **5:** 1037.

Zaragoza D., Ghavidel A., Heitman J., and Schultz M.C. 1998. Rapamycin induces the G0 program of transcriptional repression in yeast by interfering with the TOR signaling pathway. *Mol. Cell. Biol.* **18:** 4463.

Ribosome Biogenesis: Role of Small Nucleolar RNA in Maturation of Eukaryotic rRNA

S.A. Gerbi, A.V. Borovjagin, M. Ezrokhi, and T.S. Lange

Division of Biology and Medicine, Department of Molecular Biology, Cell Biology and Biochemistry,
Brown University, Providence, Rhode Island 02912

Much excitement has been generated recently by the elucidation of the structure of the ribosome by X-ray crystallography and cryo-electron microscopy, providing insights on how this molecular machine works (see other papers in this volume). The ribosome appears to be a ribozyme with catalytic activity for protein synthesis in its RNA moiety (Noller et al. 1992; Green and Noller 1997; Khaitovich et al. 1999; Muth et al. 2000; Nissen et al. 2000). Evidence is beginning to mount that there are conformational changes in the ribosome during translation, reflecting dynamic alterations in the underlying ribosomal RNA (rRNA) structure (Lodmell and Dahlberg 1997; Gabashvili et al. 1999; Frank and Agrawal 2000; Pape et al. 2000; Stark et al. 2000; VanLoock et al. 2000; Ogle et al. 2001). The structure of the ribosome is key to its function, and therefore it is absolutely essential that rRNA be folded correctly from among the myriad possibilities. How does the cell accomplish this awesome task to form the ribosome?

Biogenesis of the ribosome occurs within the nucleolus (a non-membrane-bounded structure in the nucleus) of eukaryotic cells. A large precursor rRNA (pre-rRNA) is synthesized by RNA polymerase I (see Nomura, this volume), and ribosomal proteins begin to associate with the pre-rRNA while transcription is still ongoing (Chooi and Leiby 1981). Presumably, this is a highly ordered process of assembly, following the pathways of protein addition deduced from in vitro reconstitution studies in eubacteria (Nomura 1987, 1997). The structure of the pre-rRNA must still be fairly open to allow the long finger-like projections of the ribosomal proteins to penetrate deep into what will become the interior of the ribosome. The ribosomal proteins constrain the conformation of rRNA, and conformational changes occur as the ribosomal proteins associate with the rRNA (see Recht and Williamson [this volume] and references therein; see also Stelzl and Nierhaus 2001).

Early in ribosome biogenesis, small nucleolar RNAs (snoRNAs) associate with the rRNA precursor. The snoRNAs play several roles in the maturation of eukaryotic rRNA. They serve to guide the modifications in rRNA, are required for the cleavages that liberate mature rRNA from its precursor, and probably serve as chaperones to ensure the correct folding of rRNA. The timing of association of the modification guide snoRNAs is unknown relative to the addition of ribosomal proteins, but

it is clear that the initial pre-rRNA that is released from the template DNA already has virtually all of its modifications (Maden et al. 1977; Maden 1990) and over two-thirds of the 60S and about half of the 40S ribosomal proteins in place (Kuter and Rodgers 1976; Auger-Buendia and Longuet 1978; Augier-Buendia et al. 1979; Fujisawa et al. 1979; Lastick 1980; Todorov et al. 1983). The addition of the remaining ribosomal proteins seems to occur in the nucleolus after rRNA processing has been completed. Subsequently, the 60S and 40S ribosomal subunits are exported to the cytoplasm (see A.W. Johnson et al., this volume) for function in translation. The focus of this paper is the snoRNAs and their roles in ribosome maturation.

snoRNA FORMATION

There are ~200 snoRNAs in vertebrates. Generally, they are rich in uridine and thus are named U followed by a number. There are two major families: the Box C/D and the Box H/ACA snoRNAs. They contain family-specific conserved sequences: Box C (RUGAUGA) and Box D (CUGA) (Tyc and Steitz 1989) or Box H (ANANNA) and Box ACA (ACA) (Balakin et al. 1996; Ganot et al. 1997b). Recently a "Siamese twin" snoRNA was described with characteristics of both major families (Jády and Kiss 2001). A third and minor family of snoRNAs comprises MRP and RNase P.

snoRNAs are synthesized in the nucleoplasm (Gao et al. 1997; Samarsky et al. 1998) from free-standing genes, from polycistronic genes, or from introns of other genes (Maxwell and Fournier 1995). It was surprising to find that in certain cases the exons of the parent gene might not encode an mRNA, and their major function seems to reside in their intron-encoded snoRNA sequences (Tycowski et al. 1996a; Bortolin and Kiss 1998). snoRNAs that are transcribed by RNA polymerase II from a free-standing gene have a trimethylguanosine cap, which is absent from the intron-encoded snoRNAs whose 5′ terminus is a 5′ phosphate. Capping activity is found in the nucleoplasm (Terns and Dahlberg 1994; Terns et al. 1995). For snoRNAs of the Box C/D family, the 5′ cap trimethylation is directed by the Box C/D motif (Box C, Box D, and the terminal stem adjacent to these boxes) (Baserga et al. 1991; Peculis and Steitz 1994; Terns et al. 1995; Speckmann et al. 2000).

A recent study on yeast U3 snoRNA (Kufel et al. 2000) indicates that several events precede the addition of the trimethylguanosine cap at the 5′ end. Yeast U3 snoRNA is made as a precursor with a 3′ extension that can form a hairpin stem which is cleaved by the endonuclease Rnt1p. Subsequently, there is 3′ exonuclease activity (probably the exosome; Allmang et al. 1999; van Hoof et al. 2000) that trims inward until it is further inhibited by Lhp1p (La homologous protein) which is bound to poly(U) tracts in the 3′ tail of pre-U3 snoRNA. Yeast U3 snoRNA contains an intron near its 5′ end (open arrow in Fig. 1) (Myslinski et al. 1990), and this is removed by splicing. Next, core proteins (Nop1p, Nop56p, Nop58p, and presumably Snu13p) bind the Box C/D motif at the 3′ end, and their binding may cause the displacement of Lhp1p. After this has occurred, there is further exonucleolytic trimming by

the exosome until the mature 3′ end is reached. Probably the terminal stem and associated proteins prevent further digestion by the exosome. Only after all these events have occurred is the m^{7}G cap converted to the m$_3^{2,2,7}$G cap at the 5′ end of the U3 molecule.

STRUCTURE OF U3 snoRNP

Sequence alignment of U3 snoRNA sequences indicated regions of evolutionary conservation named Boxes A, B, and C (Wise and Weiner 1980; Hughes et al. 1987; Jeppesen et al. 1988). Shortly thereafter, Box D was identified (Tyc and Steitz 1989) and subsequently Box C′ (Tycowski et al. 1993; also called Box A^0 by Marshallsay et al. 1992), Box A′ (Myslinski et al. 1990), and the GAC

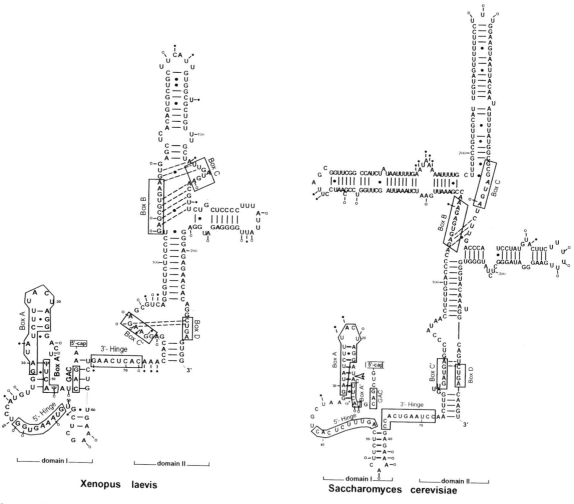

Figure 1. Universal secondary structure model of U3 snoRNA when not bound to pre-rRNA. The secondary structure of domain II (3′ portion) of U3 snoRNA is supported by compensatory base changes seen in phylogenetic comparisons (Jeppesen et al. 1988); the structure of domain I has been redrawn as in Borovjagin and Gerbi (2000). Sites of strong (*filled circles*) or moderate (*open circles*) chemical modification are shown for *Xenopus laevis* U3 snoRNA (Jeppesen et al. 1988) and *Saccharomyces cerevisiae* (Méreau et al. 1997). The base-pairing in the Box B/C and Box C′/D motifs suggested by Watkins et al. (2000) is indicated by dashed lines, but is not consistent with the chemical modification data shown for *Xenopus* U3. The sequences of the U3 conserved boxes and the single-stranded 5′ and 3′ hinge regions are enclosed in boxes. The location of Ψ at nucleotides 8 and 12 in *Xenopus* U3 is inferred from studies in mammalian cells (Reddy and Busch 1983). The position of an intron in yeast U3 (Myslinski et al. 1990) is denoted by an open arrow adjacent to Box A′ of yeast U3.

element (Samarsky and Fournier 1998) were also found. In most Box C/D snoRNAs, the terminal stem is flanked by Box C and Box D. However, in U3 snoRNA, Box C′ replaces Box C in the terminal stem motif and Box C is found elsewhere in the molecule. The structure of U3 (Fig. 1) has been studied by chemical modification and nuclease digestion (Parker and Steitz 1987; Jeppesen et al. 1988; Ségault et al. 1992; Hartshorne and Agabian 1994; Mougin et al. 1996; Méreau et al. 1997; Fournier et al. 1998; Antal et al. 2000). The data are in good agreement that domain II (3′ portion of U3 snoRNA) is highly base-paired; the stems above and adjacent to Boxes B and C are not essential (Samarsky and Fournier 1998; A.V. Borovjagin and S.A. Gerbi, unpubl.). The structure of domain I (5′ portion of U3 snoRNA) was more problematic. The single base-paired stem drawn for vertebrate U3 domain I (Parker and Steitz 1987) contained nucleotides that could be chemically modified in vivo (Jeppesen et al. 1988), suggesting that they must be single-stranded. Moreover, it was not possible to draw one long stem in all organisms; instead, domain I of U3 from yeasts (Porter et al. 1988; Ségault et al. 1992; Mougin et al. 1996; Méreau et al. 1997; Fournier et al. 1998) and plants (Kiss and Solymosy 1990; Marshallsay et al. 1990) could be drawn as two shorter stems. Recently, we suggested a universal model for the structure of U3 snoRNA that fits the sequences and experimental data from yeast to vertebrates (Fig. 1) (Borovjagin and Gerbi 2000). In this new model, the 5′ and 3′ hinge regions (nonconserved sequences known to be highly accessible to chemical modification) are single-stranded, whereas the GAC element and Boxes A′ and A are base-paired—perhaps to mask them from premature action on pre-rRNA. It is believed that domain I of U3 becomes single-stranded once it docks on pre-rRNA, allowing its 5′ end to base-pair with sequences within 18S rRNA (described more fully below) (Hughes 1996; Méreau et al. 1997; Borovjagin and Gerbi 2000, 2001).

Mature snoRNAs occur as ribonucleoproteins. The proteins that associate with the various snoRNA complexes are still being identified. It is likely that several or all of the proteins assemble with a given snoRNA in the nucleoplasm prior to transport to the nucleolus. In the case of U3 snoRNP, most of the proteins described so far associate with the 3′ portion of the RNA (see Table 1). As for all members of the Box C/D snoRNA family, these include Nop1p = fibrillarin, Nop56p, Nop58p/Nop5p on the terminal stem motif. In addition, there are U3-specific proteins such as those listed in Table 1 and several others just being identified now (S. Baserga, pers. comm.). Recently it has been found that Snu13p (15.5 kD), which associates with U4 snRNA, also associates with Box C/D snoRNAs (Watkins et al. 2000). Snu13p associates with Boxes B+C and less strongly with Boxes C′+D of yeast U3 snoRNA (C. Branlant, pers. comm.). Similarities in the Box B+C and Box C′+D motifs with the region in U4 snRNA that binds Snu13p suggested base pairs in the conserved motif (Watkins et al. 2000); although these are indicated in Figure 1 (dashed lines), accessibility of some of these nucleotides to chemical modification calls their base-pairing into question. There are fewer proteins asso-

Table 1. Proteins Associated with U3 snoRNA

Proteins	References
Proteins of Box C/D snoRNAs	
Nop1p (= fibrillarin; 34–40 kD)	Tyc and Steitz (1989); Baserga et al. (1991); Tollervey et al. (1991, 1993); Speckmann et al. (1999); Wormsley et al. (2001)
Nop56p (= 65–68 kD)	Gautier et al. (1997); Lafontaine and Tollervey (2000); Newman et al. (2000)
Nop58p/Nop5p (= 60–62 kD)	Gautier et al. (1997); Wu et al. (1998); Lafontaine and Tollervey (1999); Lyman et al. (1999); Newman et al. (2000)
Snu13p	Lübben et al. (1993); Watkins et al. (2000)
U3 snoRNA-specific proteins	
Dhr1p	Colley et al. (2000)
Imp3p	Lee and Baserga (1999)
Imp4p	Lee and Baserga (1999)
Lcp5p	Wiederkehr et al. (1998)
Mpp10p	Dunbar et al. (1997); Lee and Baserga (1997); Westendorf et al. (1998); Wormsley et al. 2001
Rcl1p (not a core protein)	Billy et al. (2000)
Rrp9p	Lübben et al. (1993); Pluk et al. (1998); Lukowiak et al. (2000); Venema et al. (2000)
Sof1p	Jansen et al. (1993)

ciated with the 5′ portion of U3 snoRNA, but these include transiently associated Rcl1p (Billy et al. 2000) and the protein complex of Mpp10p, Imp3p, and Imp4p where the stem between the 5′ and 3′ hinge regions (Fig. 1) was found to be needed for Mpp10p association (Wormsley et al. 2001).

A different set of proteins associate with snoRNAs of the Box H/ACA family, and include the core proteins Nop10p (10 kD) and Nhp2p (22–25 kD; Henras et al. 1998; Dragon et al. 2000; Pogačić et al. 2000), Gar1p (25 kD; Girard et al. 1992; Bousquet-Antonelli et al. 1997), and Cbf5p (= NAP57/dyskerin; Lafontaine et al. 1998).

Yet other proteins are associated with MRP. These are Pop1p (Lygerou et al. 1994), Pop3p (Dichtl and Tollervey 1997), Pop4p-Pop8p and Rpp1p (Chu et al. 1997; Stolc and Altman 1997; Chamberlain et al. 1998), and

Snm1p (Schmitt and Clayton 1994). Many of these proteins are shared with RNase P (Lygerou et al. 1994; Chu et al. 1997; Dichtl and Tollervey 1997; Stolc and Altman 1997; Chamberlain et al. 1998).

snoRNA TRAFFIC TO
THE NUCLEOLUS

snoRNAs must travel from their site of synthesis in the nucleoplasm to the nucleolus, where they will function in ribosome biogenesis. Sequences needed for function in rRNA maturation are not required for nucleolar localization. Instead, mutagenesis studies revealed that the signature motifs of each family are important for nucleolar localization. For the Box C/D snoRNAs, Box C and Box D are essential and sufficient as *nucleolar localization* elements (NoLEs) (Lange et al. 1998a,b,c; Samarsky et al. 1998; Narayanan et al. 1999a). For U3 snoRNA, we have found that Boxes C and D are essential and Box C′ might also play some role as a NoLE (Lange et al. 1998c). In contrast, other investigators reported that Box C′ is more important than Box C as a NoLE of U3 (Narayanan et al. 1999a). It is not yet settled whether the terminal stem is needed for nucleolar localization (Narayanan et al. 1999a) or not (Lange et al. 1998b,c). We deduced from our mutation and competition studies that Box C and Box D act together, perhaps with a helping role by Box B, for nucleolar localization of U3 snoRNA (Lange et al. 1998c). It is probable that the action of Boxes C and D is mediated by proteins bound to these sequences. In fact, recent evidence supports this idea, since all four proteins associated with the Box C/D motif (Nop1p = fibrillarin, Nop56p, Nop58p/Nop5p, and Snu13p) are needed for localization of U14 constructs to the nucleolus (Verheggen et al. 2001). However, it is not clear whether their role is direct or indirect. It could be that other proteins transport snoRNAs from the nucleoplasm to the nucleolus; possible candidates for this are the protein pair of p50 and p55 (distinct from the U3-specific 55-kD protein) that associate with a U14 construct but are primarily found in the nucleoplasm (Watkins et al. 1998a; Newman et al. 2000). It is likely that the Box C/D snoRNAs traffic from the nucleoplasm through the Cajal (coiled) body (= nucleolar body in yeast) (Samarsky et al. 1998; Narayanan et al. 1999a; Speckmann et al. 1999; Verheggen et al. 2001), where they may interact with Srp40, and then are delivered by Nsr1p to the nucleolus (Verheggen et al. 2001).

For Box H/ACA snoRNAs, Box H and Box ACA are the NoLEs (Narayanan et al. 1999b; Lange et al. 1999; Ruhl et al. 2000). The 3′-terminal domain of mammalian telomerase RNA is similar to a Box H/ACA snoRNP, conferring on it nucleolar localization (Mitchell et al. 1999a,b; Narayanan et al. 1999b; Dragon et al. 2000; Pogačić et al. 2000). Sequences needed for nucleolar localization of MRP (Jacobson et al. 1995) and RNase P (Jacobson et al. 1997) have also been identified. Generally, the NoLEs correspond to sequences that bind proteins common to the family. In addition, these regions are important for the stability of the snoRNA. Interestingly,

"Siamese twin" snoRNAs bearing both Box C/D and Box H/ACA sequence motifs were found to localize exclusively to Cajal bodies and not to nucleoli (T. Kiss, pers. comm.), where one would have predicted. Why the combined NoLEs lead to a different subcellular localization than that of the NoLEs from each snoRNA family is not yet clear.

snoRNAs may traffic through various compartments of the nucleolus once they have arrived at this cellular destination. snoRNAs that are injected into *Xenopus* oocyte nuclei are found in the dense fibrillar component (DFC) of nucleoli where they may be stored until recruited to act on pre-rRNA. However, endogenous U3 snoRNA is also present in the granular component (GC) of nucleoli (for review, see Gerbi and Borovjagin 1996). It has been proposed that U3 travels in association with pre-rRNA from the DFC to the GC where the later steps in rRNA processing occur; when processing is completed and the ribosomal subunits are exported, U3 may recycle back to the DFC for use in another round of rRNA processing (Puvion-Dutilleul et al. 1992; Gerbi and Borovjagin 1996). An interesting corollary of this model is the prediction that U3 snoRNA should dissociate from fibrillarin as it traffics through the nucleolar compartments (Gerbi and Borovjagin 1996), since fibrillarin is found in the DFC but not in the GC.

What serves to retain U3 snoRNA in the nucleolus? The nucleolar binding sites for U3 can be saturated by an excess of U3, in which case some U3 snoRNA is also found in the cytoplasm (Rivera-León 1995; Terns et al. 1995). Depletion of early rRNA precursors by actinomycin D (Rivera-León and Gerbi 1996) or by growth conditions (Glibetic et al. 1992; Sienna et al. 1996) causes a significant fraction of U3 to be lost from the nucleolus, suggesting that at least some of it was anchored through its association with pre-rRNA. In addition, either Boxes B and C or Boxes C′ and D are sufficient to retain U3 in the nucleus (Speckmann et al. 1999), presumably mediated by proteins that bind to these motifs.

MOST snoRNAs GUIDE RNA
MODIFICATION

An early step in the biogenesis of ribosomes is the modification of its rRNA moiety. Vertebrate rRNAs contain about 115 2′-O-methyl ribose (2′-O-Me) moieties, 95 pseudouridines (Ψ), and 10 methylated bases (Maden and Hughes 1997). Yeast contain only about half this number of modifications, namely about 65 2′-O-Me and 45 Ψ (Maden and Hughes 1997), and there are yet fewer in eubacteria. The function of these modifications is not clear. The fact that they reside in evolutionarily conserved sequences of rRNA suggests that they are important for ribosome function, as supported by some recent experiments where the growth rate of yeast was adversely affected when formation of several Ψ in the peptidyl transferase center was prevented (T. King and M.J. Fournier, pers. comm.).

How are specific nucleotides selected for modification during ribosome biogenesis? The mechanism is now being elucidated (see Ofengand et al., this volume) and involves snoRNAs both in eukaryotes (see below) and in archaebacteria (Gaspin et al. 2000; Dennis et al. 2001). It has been possible to identify the Box C/D snoRNAs in sequenced genomes by a computer algorithm (Lowe and Eddy 1999). As depicted in Figure 2, the snoRNA base-pairs with a specific region of rRNA to target the site of modification. Box C/D snoRNAs act as guides for 2′-O-Me (Cavaillé et al. 1996; Kiss-László et al. 1996, 1998; Nicoloso et al. 1996; Tycowski et al. 1996b; Maden and Hughes 1997; Cavaillé and Bachellerie 1998) with the fifth nucleotide in rRNA in the stretch of base-pairing following Box D or Box D′ (an internal Box D sequence) being targeted for modification. Box H/ACA snoRNAs guide Ψ formation (Ganot et al. 1997a; Ni et al. 1997) with the nucleotide targeted for modification being unpaired in the middle of base-pairing between rRNA and an internal loop ("pseudouridylation pocket") in a stem ~14 nucleotides upstream of Box H or Box ACA of the snoRNA. For both kinds of modifications, the guide snoRNA brings with it in "piggyback" style the modification enzyme. Fibrillarin (Nop1p) appears to be the methyltransferase (Tollervey et al. 1993; Niewmierzycka and Clarke 1999; Wang et al. 2000; I. Bozzoni, pers. comm.), and Cbf5p is the pseudouridine synthase (Koonin 1996; Lafontaine et al. 1998; Watkins et al. 1998b; Zebarjadian et al. 1999). This may serve as a paradigm for the few snoRNAs used instead for rRNA processing, which might also piggyback in the proteins needed for cleavage of pre-rRNA (or, alternatively, serve as landing pads for the cleavage proteins). Thus, both for the snoRNAs used to guide modifications and for the snoRNAs used for rRNA processing (see below), base-pairing between the snoRNA and the pre-rRNA seems to define the site for modification and possibly cleavage.

Base-pairing between the ~200 snoRNAs and pre-rRNA coats almost half of the 18S, 5.8S, and 28S coding regions (Smith and Steitz 1997), perhaps serving as chaperones to keep the rRNA from folding prematurely. It is unknown whether there is a prescribed order for snoRNA binding and modification of its targets, nor is it known how many of the guide snoRNAs are bound simultaneously on the pre-rRNA. In some instances, the rRNA sequences for base-pairing with snoRNAs overlap, and in those cases, the snoRNAs must bind sequentially. The initial rRNA precursor is already fully modified (Maden et al. 1977; Maden 1990), but it remains to be elucidated whether snoRNA binding to pre-rRNA is cotranscriptional or immediately after the nascent pre-rRNA is released from the DNA.

Recently, it has been found that guide snoRNAs can modify other classes of RNA besides rRNA. For example, the 2′-O-Me targets of U83 and U85 snoRNAs are not rRNA (Jády and Kiss 2000). In addition, U6 snRNA is transiently localized in nucleoli (Lange and Gerbi 2000), where it is modified by several guide snoRNAs (Tycowski et al. 1998; Ganot et al. 1999). Similarly, U2 snRNA is modified by guide snoRNAs in the nucleolus, probably after re-import to the nucleus from the cytoplasm (Yu et al. 2001). Both 2′-O-Me and Ψ formation in U5 snRNA are directed by the Box C/D–Box H/ACA chimeric U85 snoRNA (Jády and Kiss 2001). Guide snoRNAs have also been identified that have U1 and U4 snRNAs as likely targets (T. Kiss, pers. comm.) and that might guide 2′-O-Me of U2 and U4 and Ψ formation in U2 and U6 snRNAs (Hüttenhofer et al. 2001). Thus, all spliceosomal snRNAs appear to be modified at least to some extent by guide snoRNAs. In addition, snoRNAs such as U3 are modified (Reddy and Busch 1983; Ganot et al. 1999), leading one to wonder whether there are snoRNAs to modify snoRNA, which in turn modify other snoRNAs, etc.?! Instead of having small RNA as a target, a Box C/D snoRNA, HBII-52, expressed specifically in human brain, has an intriguing complementarity to serotonin 2C receptor mRNA (Cavaillé et al. 2000), but its function remains to be elucidated.

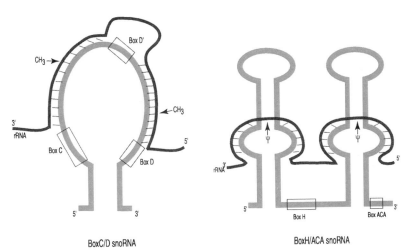

Figure 2. Pre-rRNA is modified by snoRNAs. The black line represents pre-rRNA that is modified by guide snoRNAs of the Box C/D family (*left panel*) for 2-O-Me and the Box H/ACA family (*right panel*) for Ψ. Drawing is modified from Smith and Steitz (1997). See text for further details.

GERBI ET AL.

SOME snoRNAs ARE ESSENTIAL FOR rRNA PROCESSING

Although the vast majority of snoRNAs are guides for modification, genetic depletion suggests that they are not essential for cell viability. In contrast, the few snoRNAs needed for rRNA processing (see Table 2) are essential. There are many similarities in the steps of rRNA processing between yeast and vertebrates to remove the external transcribed spacers (ETS) and internal transcribed spacers (ITS), but there are also some differences (Fig. 3). All the cleavages occur in the nucleolus except for cleavage at site D (3′ end of 18S rRNA) in yeast, which occurs in the cytoplasm (Udem and Warner 1973), unlike vertebrates where cleavage at site 2 occurs in the nucleolus. Two cleavages occur within the ITS1 of yeast at sites A2 and A3, whereas just one cleavage has been reported at site 3 of *Xenopus* (~100 nucleotides upstream of the 5′ end of 5.8S RNA; Peculis and Steitz 1993; Tycowski et al. 1994; Borovjagin and Gerbi 2001), and it remains to be determined whether it might be a composite of yeast sites A2 and A3. The *cis*-acting signals in pre-rRNA needed for several of the cleavages have been elucidated (Venema and Tollervey 1999). The snoRNAs are *trans*-acting factors needed for rRNA processing.

U8 is the only snoRNA known to be needed for 28S rRNA formation (Peculis and Steitz 1993), and a counterpart for it has not been found in yeast. It has been analyzed in *Xenopus* oocytes by injection of mutated transcripts (Peculis and Steitz 1994; Peculis 1997), and the evidence indicates that it links cleavages at the 5′ end of 5.8S RNA and 3′ end of 28S rRNA. This linkage seems to exist in yeast, even though it lacks a U8 snoRNA counterpart (Allmang and Tollervey 1998). In addition, U8 snoRNA appears to act as a chaperone through putative base-pairing with the 5′ end of 28S rRNA to prevent premature pairing of the latter with 5.8S RNA (Peculis 1997).

5.8S RNA of eukaryotes is homologous to the 5′ end of bacterial 23S rRNA (Nazar 1980; Jacq 1981; Clark and Gerbi 1982) and is tethered to the 5′ region of eukaryotic 28S rRNA by hydrogen bonding at both of its ends (Pace et al. 1977; Nazar and Sitz 1980). U8 snoRNA is required for removal of the ITS2 in vertebrates (Peculis and Steitz 1993), and may fold the ITS2 for proper cleavage (Michot et al. 1999). Apparently, structural features rather than a sequence are recognized for ITS2 removal (Cote and Peculis 2001). Both in yeast and in vertebrates, cleavage within the ITS2 occurs as an intermediate step to its removal (Michot et al. 1999). In yeast, generation of the 5′ end of 5.8S RNA involves cleavage at site A3 mediated by the snoRNA MRP (Shuai and Warner 1991; Lindahl et al. 1992; Schmitt and Clayton 1993; Chu et al. 1994; Lindahl and Zengel 1995; Lygerou et al. 1996; Shadel et al. 2000) followed by exonuclease trimming to generate a short form called 5.8S$_S$. An alternate, less common pathway bypasses MRP (Henry et al. 1994) and instead involves endonucleolytic cleavage at B1$_L$ to produce a long form of 5.8S (5.8S$_L$). MRP also exists in vertebrates (Bennett et al. 1992), but its function in rRNA processing has not yet been tested for higher organisms.

Table 2. snoRNAs Required for rRNA Processing

snoRNA	pre-rRNA cleavage site	References
Yeast		
U3	A0, A1, A2	Hughes and Ares (1991)
U14	A1, A2	Li et al. (1990)
MRP	A3	Shuai and Warner (1991); Lindahl et al. (1992); Schmitt and Clayton (1993); Chu et al. (1994); Lygerou et al. (1996); Lindahl and Zengel (1995); Shadel et al. (2000)
snR10	A1, A2	Tollervey (1987)
snR30	A1, A2	Morrissey and Tollervey (1993)
RNase P	3′ end of 5.8S RNA	Chamberlain et al. (1996)
Vertebrates		
U3	A′, A0, 1, 2, 3	Kass et al. (1990); Savino and Gerbi (1990); Hughes and Ares (1991); Savino and Gerbi (1991); Enright et al. (1996); Borovjagin and Gerbi (1999, 2000, 2001)
U8	3, 4, 5, T1	Peculis and Steitz (1993, 1994); Peculis (1997)
U14	A′, 1, 2	Enright et al. (1996); Dunbar and Baserga (1998); Lange et al. (1998b)
U22	A′, 1, 2	Tycowski et al. (1994);
E1 (= U17)	A′, 1	Enright et al. (1996); Mishra and Eliceiri (1997)
E2	2	Mishra and Eliceiri (1997)
E3	3	Mishra and Eliceiri (1997)

Formation of 18S rRNA requires several snoRNAs (Table 2). Since individual depletion often results in the same processing defects, the idea arose that they might work together as a large complex dubbed the "processome" (Fournier and Maxwell 1993) by analogy with the spliceosome, but there is not yet direct evidence for its existence (it has been inferred from indirect evidence that the terminal balls seen on nascent rRNA might be processomes; Mougey et al. 1993b). Mutations of snoRNAs that affect 18S formation do not alter 28S rRNA production, indicating that the pathway to form 18S rRNA is independent of the 28S rRNA pathway. However, both pathways are linked via the 193-kD yeast nucleolar protein Rrp5p (Venema and Tollervey 1996), whose amino-terminal and central regions are essential for 5.8S RNA formation and whose carboxy-terminal portion is re-

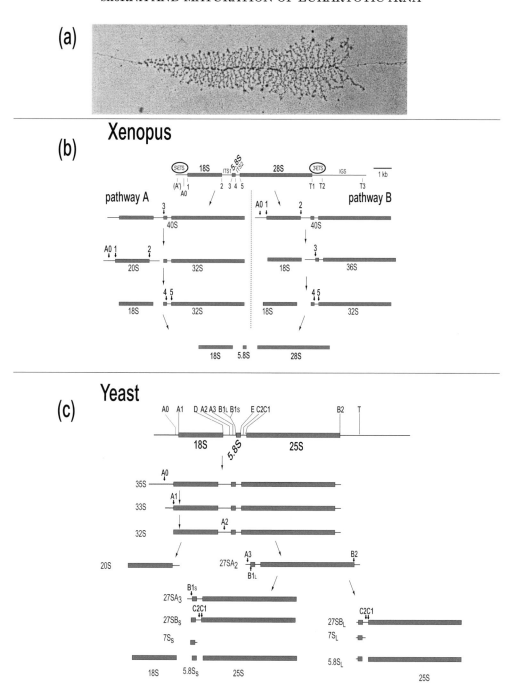

Figure 3. rRNA transcription and processing. (*a*) Electron micrograph of a Miller spread of *Xenopus* rDNA with many nascent pre-rRNA fibrils (courtesy of U. Scheer). Note the electron-dense terminal balls at the 5′ end (where the ETS is situated) of the nascent pre-rRNA. See text for further discussion. rRNA processing pathways are compared for *Xenopus* (*b*) (Savino and Gerbi 1990) and yeast (*c*) (modified from Venema and Tollervey 1999). Both pathways A and B can coexist in a single frog oocyte (Savino and Gerbi 1990), and U3 snoRNA appears to influence which processing pathway is taken (Borovjagin and Gerbi 1999).

quired to produce 18S rRNA (Torchet et al. 1998; Eppens et al. 1999). Moreover, in yeast, mutations at site A2 affect the efficiency of cleavage at site A3 and vice versa (Allmang et al. 1996).

Production of 18S rRNA begins with 5′ ETS cleavage. In mammals, the initial cleavage is at site A′ (near the 5′ end of the 5′ ETS; previously site A′ was called site 0; Gerbi 1995) and utilizes U3, U14, U22, and E1=U17

snoRNAs (Kass et al. 1990; Enright et al. 1996). A prerequisite for cleavage at site A′ is the presence of nucleolin, which appears to bind to this region as well and is posited to help load U3 snoRNA onto the 5′ ETS (Ginisty et al. 1998, 2000). However, cleavage at site A′ is barely or not detectable in *Xenopus* oocytes (Savino and Gerbi 1991; Mougey et al. 1993a,b; Tycowski et al. 1994) and does not seem to occur in yeast (Venema and Tollervey

1999). Moreover, A′ cleavage is not an obligatory early event in processing in *Xenopus* oocytes (Labhart and Reeder 1986; Savino and Gerbi 1991), trypanosomes (White et al. 1986; Hartshorne and Toyofuku 1999), or yeast (Venema and Tollervey 1999). Even if cleavage does not occur at site A′ in some organisms, it is still possible that U3 snoRNA associates with the 5′ ETS in a region nearby (discussed below).

Another U3-mediated cleavage occurs at site A0, which is close to the 3′ end of the 5′ ETS. A0 cleavage is probably followed very rapidly by subsequent cleavages at sites 1 and 2 in *Xenopus* or A1 and A2 in yeast. A0 was first discovered in yeast (Hughes and Ares 1991), and subsequently in trypanosomes (Hartshorne and Toyofuku 1999) and *Xenopus* oocytes (Borovjagin and Gerbi 2001). Despite some confusion in the literature, site A0 is distinct and different from site A′. Cleavage at both sites A′ and A0 occurs in trypanosomes (Hartshorne and Toyofuku 1999; Schnare et al. 2000); similarly, two cleavages occur within the 5′ ETS of an archaebacterium (Russell et al. 1999). Since site A0 is found in a base-paired stem, it was originally thought that RNase III was responsible for its cleavage (Abou Elela et al. 1996), but this turned out later not to be the case (Kufel et al. 1999).

Cleavage at site A0 is closely coupled with cleavages at sites A1 and A2 in yeast or sites 1 and 2 in *Xenopus*. However, it has been possible to uncouple these cleavages by mutation in U3, thus giving insight into the mechanism of rRNA processing. In *Xenopus*, mutation of Box A′ inhibits cleavage at sites 1 and 2, and a novel 19S rRNA intermediate appears with A0 at its 5′ end and site 3 at its 3′ end (Borovjagin and Gerbi 2001). This is comparable to 22S pre-rRNA bounded by A0 and A3 in yeast (Venema and Tollervey 1999) when cleavage at sites A1 and A2 is inhibited by (1) mutation of U3 Box A (Hughes 1996), (2) deletion of sequences between Box A and the 5′ hinge of U3 or point mutation in U3 Box A′ (C. Branlant, pers. comm.), (3) truncation of U3-associated protein Mpp10p (Lee and Baserga 1997), or (4) depletion of the U3-associated helicase Dhr1p (Colley et al. 2000). Similarly, A0 cleavage is unaffected when cleavage at site A1 is inhibited by mutation near the 5′ end of 18S rRNA (Sharma and Tollervey 1999). In yeast, cleavage at site A0 seems to be nonessential for the rest of rRNA processing, which continues when site A0 cleavage is blocked by a mutation in *cis* (Venema et al. 1995).

The data suggest that cleavage at site A0 precedes cleavage at sites 1 and 2. Cleavages at sites 1 and 2 are usually coupled, but recently we have acquired the first evidence that they can be uncoupled; mutation of nucleotides 11–13 (at the 3′ boundary of Box A′) in *Xenopus* U3 snoRNA allows cleavages at sites A0 and 1 but not at site 2 (Borovjagin and Gerbi 2001). This gives rise to a novel pre-rRNA that is 18.5S in size, bounded by site 1 at its 5′ end and site 3 at its 3′ end (Borovjagin and Gerbi 2001). In yeast, cleavage at sites A1 and A2 can be uncoupled by mutation in U3 snoRNA (C. Branlant, pers. comm.) or by mutation of the nucleolar protein Rrp8p (Bousquet-Antonelli et al. 2000).

The mode of action of the other snoRNAs needed for 18S rRNA production (Table 2) has not yet been studied

by mutation as extensively as U3 snoRNA. For all snoRNAs required for rRNA processing, the question remains whether any of them act as ribozymes with catalytic activity in the RNA moiety, or whether they help to target the proteins needed for cleavage. Identification of the putative endonucleases is under way; in addition, exonucleases (e.g., Xrn1p; Rat1p, the exosome) are used for several processing steps (Venema and Tollervey 1999). Since the functional end of U3 snoRNA is its 5′ portion, it will be especially interesting to characterize any proteins bound there (e.g., Rcl1p; Billy et al. 2000, and the complex of Imp3p, Imp4p, and Mpp10p; Wormsley et al. 2001) that could play a direct role in the cleavages.

ROLE OF U3 snoRNA IN 18S rRNA PRODUCTION

Our studies with antisense oligonucleotide-mediated depletion of U3 snoRNA in *Xenopus* oocytes, followed by injection of mutant U3 transcripts, have allowed us to dissect the functional regions of U3. We propose that U3 snoRNA docks on the pre-rRNA substrate by base-pairing between the hinge regions of U3 snoRNA with complementary sequences in the 5′ ETS (Fig. 4) (Borovjagin and Gerbi 2000). Notice that the 5′ and 3′ hinge regions are proposed to be single-stranded in unbound U3 snoRNA (Fig. 1) (Borovjagin and Gerbi 2000) and therefore would be readily available for interaction with pre-rRNA. The sequences of the U3 hinge regions are not conserved between species, and phylogenetic comparisons suggest that there has been covariation to maintain complementarity between the hinge regions and regions in the 5′ ETS (Borovjagin and Gerbi 2000). Experiments with compensatory base changes support the interaction between the 5′ hinge region and the 5′ ETS in yeast (Beltrame and Tollervey 1995) and are in progress to test the proposed interactions in *Xenopus*. Earlier experiments with psoralen cross-linking and mutagenesis in yeast also support the validity of the 5′ hinge base-pairing with the 5′ ETS (Beltrame and Tollervey 1992; Beltrame et al. 1994), as do experiments with in vivo chemical modification (Méreau et al. 1997). In yeast the interaction between the 5′ hinge of U3 and the 5′ ETS is essential for 18S rRNA formation; in contrast, in *Xenopus* it is important but not essential (Borovjagin and Gerbi 2000). Similarly, 18S rRNA is still produced when mutations in trypanosome pre-rRNA disrupt the proposed interaction (Hartshorne and Toyofuku 1999). However, the 3′ hinge region of U3 is critical for 18S rRNA formation in *Xenopus* (Borovjagin and Gerbi 2000) and for viability in yeast (C. Branlant, pers. comm.). In trypanosomes, psoralen cross-linking supports the proposed base-pairing between the 3′ hinge of U3 and the 5′ ETS (Hartshorne 1998), and mutation of the latter obliterates 18S rRNA production (Hartshorne and Toyofuku 1999).

Electron microscopy of Miller spreads suggests that U3 snoRNA associates with the 5′ ETS of pre-rRNA while it is still being transcribed, and it has been speculated that the terminal balls seen at the end of the nascent RNA fibrils (panel A of Fig. 3) may contain U3 (Mougey et al. 1993b). Similarly, inhibitor studies suggest that U8

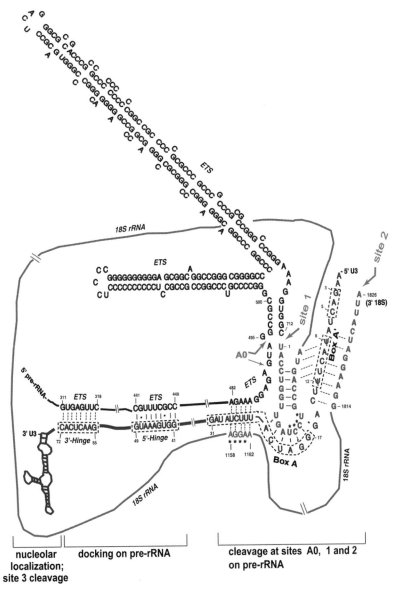

Figure 4. Model of base-pairing interactions between U3 snoRNA and pre-rRNA in *Xenopus*. The model of base-pairing between the 5′ end of U3 snoRNA and 18S rRNA first proposed by Hughes (1996) for yeast has been extended to *Xenopus* and changed to show potential pairing by Box A with ETS sequences near site A0, as an alternate to its pairing with nucleotides 1158–1162 which was originally proposed by Hughes (1996) (the two base-pairings would not occur simultaneously). Asterisks show two areas of 18S rRNA that form a pseudoknot in the mature ribosome; U3 is proposed to prevent premature pseudoknot formation in the precursor rRNA (Hughes 1996). Base-pairing between the 5′ hinge and 3′ hinge of U3 and sequences in the ETS are indicated. Box A′ might switch pairing partners from the 5′ end of 18S to the 3′ end of 18S rRNA (both pairing possibilities are drawn (but would not occur simultaneously). U3 snoRNA is depicted in red and its functional regions are indicated: Domain II is required for nucleolar localization (Lange et al. 1998c) and cleavage of site 3 in pre-RNA (Borovjagin and Gerbi 1999), the 5′ hinge and 3′ hinge dock U3 snoRNA on the ETS of pre-rRNA (Borovjagin and Gerbi 2000), and sequences in U3 domain I are required for cleavage at sites A0, 1, and 2 (Borovjagin and Gerbi 2001). ETS sequences are drawn in dark blue, 18S rRNA in light blue, and the cleavage sites are shown with green arrows.

and U22 snoRNAs may also associate with the nascent transcript of pre-rRNA, since their function in rRNA processing requires their presence during rRNA transcription (Peculis 2000), possibly to dock on the nascent pre-rRNA. Even though U3 snoRNA may be present on the nascent transcript, rRNA processing does not commence until after transcription is completed. We hypothesize that Box A of U3 may base-pair with sequences in the 5′ ETS, thereby preventing it from acting prematurely in

rRNA processing. The interaction between U3 Box A and the 5′ ETS is supported by psoralen cross-links in yeast, trypanosomes, and mammals (Maser and Calvet 1989; Stroke and Weiner 1989; Beltrame and Tollervey 1992; Tyc and Steitz 1992; Hartshorne 1998). These cross-links did not form if mutation prevented the U3 hinge interaction with the 5′ ETS (Tyc and Steitz 1992), suggesting that U3 hinge–5′ ETS base-pairing is a prerequisite for U3 Box A interaction with the 5′ ETS. The observation that 18S rRNA formation can occur in trypanosomes even after mutation of the area of the 5′ ETS proposed to base-pair with U3 Box A (Hartshorne and Toyofuku 1999) is consistent with the notion that the Box A–5′ ETS interaction is not essential for rRNA processing and may actually serve to inhibit premature rRNA processing. When transcription has been completed, nucleolin dissociates from the pre-rRNA; it is conceivable that this could destabilize the U3 Box A–5′ ETS interaction, and thus Box A would be available to base-pair with 18S rRNA sequences in the pre-rRNA (see below).

It has been hypothesized that the Box A′ and Box A sequences in U3 snoRNA base-pair with sequences in 18S rRNA which ultimately will be part of the central pseudoknot (Hughes 1996). Thus, U3 snoRNA would act as an RNA chaperone to prevent premature pseudoknot formation. Phylogenetic comparisons (Hughes 1996) and in vivo chemical modification (Méreau et al. 1997) support the proposed base-pairing with 18S rRNA sequences. Direct evidence by compensatory base changes validates the base-pairing between the 5′ end of U3 Box A and the apical loop of the 5′-most stem in 18S rRNA (Sharma and Tollervey 1999); this interaction is depicted in Figure 4 for *Xenopus*. Based on our finding that Box A is needed for cleavage at site A0 (Borovjagin and Gerbi 2001), we have modified the original model (Hughes 1996) and suggest that the 3′ end of Box A may base-pair with sequences in the 5′ ETS just upstream of the A0 cleavage site (Fig. 4) (Borovjagin and Gerbi 2001). The Hughes (1996) suggestion of base-pairing between U3 Box A′ and sequences of the 5′-most stem of 18S rRNA (Fig. 4) would explain our data that Box A′ is required for site 1 cleavage which is situated nearby in this model.

However, we also found that nucleotides 11–13 at the 3′ boundary of Box A′ are needed for site 2 cleavage. To accommodate this finding, we propose that a molecular switch occurs such that Box A′ changes its pairing partner from the 5′ end of 18S rRNA to the 3′ end of 18S rRNA (Fig. 4), positioned close to site 2. Insertions and deletions imply that the several U3 base-pairing interactions with pre-rRNA occur simultaneously (Borovjagin and Gerbi 2000, 2001). Thus, U3 snoRNA could act as a bridge to draw together the 5′ and 3′ ends of 18S rRNA in the pre-rRNA (Figs. 4 and 5) so that cleavages at sites A0, 1, and 2 may be coordinated. Interestingly, base-pairing between U3 Box A′ and the 3′ end of 18S rRNA cannot be found in yeast, but recall that site D cleavage (3′ end of 18S rRNA in yeast) occurs in the cytoplasm (Udem and Warner 1973), presumably in a U3-independent manner.

In the model just presented, U3 snoRNA acts as a chaperone to prevent premature folding of the central pseudoknot of 18S rRNA, and also U3 draws together the 5′ and 3′ ends of 18S rRNA (Fig. 4). In addition to its role as a putative chaperone, U3 is also important for cleavages in pre-rRNA, and the model places U3 close to cleavage sites A0, 1, and 2 for which it is known to be required.

The proposed dynamic base-pairing between U3 snoRNA and pre-rRNA suggests that helicases may be needed. More than a dozen helicases that affect rRNA processing have been identified (for review, see Venema

and Tollervey 1999). The nucleolar proteins Dhr1p and Dhr2p are of particular interest; they are needed for cleavage at sites A1 and A2 in yeast to form 18S rRNA, and Dhr1p is also needed for cleavage at site A0 (Colley et al. 2000). In addition, Dhr1p (but not Dhr2p) is stably associated with U3 snoRNA, suggesting that U3 snoRNA carries with it a helicase needed for U3-mediated rRNA processing.

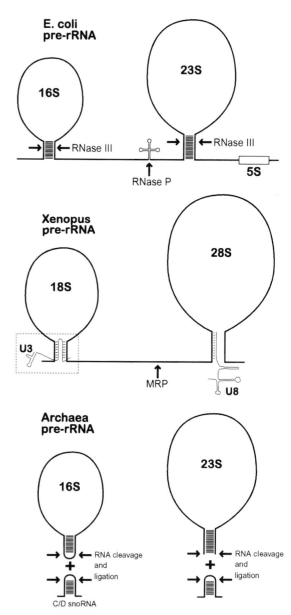

Figure 5. Comparison of pre-rRNA from the three kingdoms of life. Pre-rRNA with two long stems is shown for the eubacterium *E. coli* (Young and Steitz 1978; Bram et al. 1980). In the eukaryote *Xenopus*, these two stems cannot be drawn in pre-rRNA, but the functional equivalent might exist due to base-pairing with snoRNAs (Gerbi 1995; Morrissey and Tollervey 1995; Borovjagin and Gerbi 2001). The dotted box indicates the interactions between U3 and 18S rRNA shown in Fig. 4. U8 can base-pair with the 5′ end of 28S rRNA (Peculis 1997). In archaebacteria, the two stems exist in pre-rRNA but are subject to cleavage and ligation which generates a Box C/D-like snoRNA from the 16S rRNA stem (A. Hüttenhofer et al., pers. comm.).

CONCLUSIONS: EVOLUTIONARY PERSPECTIVES

Eubacterial rRNA is also produced from a large precursor, but the mechanism for rRNA processing differs from eukaryotes. As shown in Figure 5, *E. coli* 16S and 23S rRNA each reside at the top of long base-paired stems that harbor RNase III cleavage sites where the processing reaction begins (Young and Steitz 1978; Bram et al. 1980). In addition, the structure of two stems and loops is also found in archaebacteria (Fig. 5) (Potter et al. 1995), but the stems are considerably shorter in yeast (Veldman et al. 1981) and do not exist in *Xenopus* pre-rRNA. It has been hypothesized that the base-pairing between sequences in *cis* in bacterial pre-rRNA has been replaced in eukaryotes by base-pairing in *trans* between snoRNAs and the termini of the rRNA regions within the pre-rRNA (Morrissey and Tollervey 1995). Our data suggest that U3 snoRNA may play the role of a molecular bridge to bring together the ends of 18S rRNA (Borovjagin and Gerbi 2001). The analogy between bacterial and eukaryotic pre-rRNA continues with the observation that RNase P is used to remove the intergenic tRNA from pre-rRNA in bacteria, and its close relative, the snoRNA MRP, is used for cleavage at site A3 to produce 5.8S RNA in eukaryotes such as yeast (Morrissey and Tollervey 1995). snoRNAs have been identified not only in eukaryotes, but also in archaebacteria, suggesting that they arose before these two kingdoms split (Lafontaine and Tollervey 1998; Omer et al. 2000). Moreover, the Archaea have homologs for fibrillarin (Amiri 1994) and the Nop56p/Nop58p and p50/p55 protein pairs (Lafontaine and Tollervey 1998; Newman et al. 2000), indicating that the proteins bound to the snoRNAs are similar to those found in eukaryotes. An intriguing recent discovery in archaebacteria is that cleavage and ligation of the stem below 16S rRNA results in the formation of a Box C/D-like snoRNA from the base of the stem and a covalently closed circular 16S rRNA (Fig. 5) (A. Hüttenhofer et al., pers. comm.). A small RNA, which however lacks snoRNA motifs, seems to be similarly produced from the base of the stem leading to 23S RNA; whether the termini of 23S rRNA are ligated has not yet been explored (Fig. 5) (A. Hüttenhofer et al., pers. comm.). Thus, in two kingdoms of life, Box C/D snoRNAs are implicated in bringing together the 5′ and 3′ ends of rRNA in the precursor—in eukaryotes where U3 snoRNA is complementary to the terminal regions of 18S rRNA and in the Archaea where a Box C/D snoRNA forms the base of the stem leading to 16S rRNA.

Why do all kingdoms of life bother to have a precursor rRNA, rather than just transcribing the mature forms of rRNA? Clearly, the spacers that are removed by processing must play some role. Indeed, genetic removal of ITS2 is lethal for yeast (Musters et al. 1990). We suggest that the spacers are needed to fold rRNA properly. In eukaryotes they serve as landing sites for snoRNAs that fulfill a chaperone function. In archaebacteria, spacer sequences of pre-rRNA are used instead of U3 snoRNA to base-pair with 16S rRNA; they also prevent premature pseudoknot formation (Dennis et al. 1997). Therefore, the elaborate series of cleavage reactions in rRNA processing and its associated machinery of snoRNAs may have evolved as a way to ensure that mature rRNA will be folded properly. Similarly, the multitude of snoRNAs used to guide modifications in rRNA may also play a chaperone role. A challenge ahead is determination of the probable effect of each of the snoRNAs on the conformation of rRNA, and elucidation of the order of addition of the snoRNAs that might impart a framework for folding rRNA.

ACKNOWLEDGMENT

We are grateful for funding from the National Institutes of Health: GM-20261 (previously) and GM-61945 (currently).

REFERENCES

Abou Elela S., Igel H., and Ares M., Jr. 1996. RNase III cleaves eukaryotic preribosomal RNA at a U3 snoRNP-dependent site. *Cell* **85:** 115.

Allmang C. and Tollervey D. 1998. The role of the 3′ external transcribed spacer in yeast pre-rRNA processing. *J. Mol. Biol.* **278:** 67.

Allmang C., Henry Y., Morrissey J.P., Wood H., Petfalski E., and Tollervey D. 1996. Processing of the yeast pre-rRNA at sites A2 and A3 is linked. *RNA* **2:** 63.

Allmang C., Kufel J., Chanfreau G., Mitchell P., Petfalski E., and Tollervey D. 1999. Functions of the exosome in rRNA, snoRNA and snRNA synthesis. *EMBO J.* **18:** 5399.

Amiri K.A. 1994. Fibrillarin-like proteins occur in the domain Archaea. *J. Bacteriol.* **176:** 2124.

Antal M., Mougin A., Kis M., Boros E., Steger G., Jakab G., Solymosy F., and Branlant C. 2000. Molecular characterization at the RNA and gene levels of U3 snoRNA from a unicellular green alga, *Chlamydomonas reinhardtii*. *Nucleic Acids Res.* **28:** 2959.

Auger-Buendia M.-A. and Longuet M. 1978. Characterization of proteins from nucleolar preribosomes of mouse leukaemia cells by two-dimensional polyacrylamide gel electrophoresis. *Eur. J. Biochem.* **85:** 105.

Auger-Buendia M.-A., Longuet M., and Tavitian A. 1979. Kinetic studies on ribosomal protein assembly in preribosomal particles and ribosomal subunits of mammalian cells. *Biochim. Biophys. Acta* **563:** 113.

Balakin A.G., Smith L., and Fournier M.J. 1996. The RNA world of the nucleolus: Two major families of small RNAs defined by different box elements with related functions. *Cell* **86:** 823.

Baserga S.J., Yang X.W., and Steitz J.A. 1991. An intact box C sequence is required for binding of fibrillarin, the protein common to the major family of nucleolar snRNPs. *EMBO J.* **10:** 2645.

Beltrame M. and Tollervey D. 1992. Identification and functional analysis of two U3 binding sites on yeast pre-ribosomal RNA. *EMBO J.* **11:** 1531.

———. 1995. Base-pairing between U3 and the pre-ribosomal RNA is required for 18S rRNA synthesis. *EMBO J.* **14:** 4350.

Beltrame M., Henry Y., and Tollervey D. 1994. Mutational analysis of an essential binding site for the U3 snoRNA in the 5′ external transcribed spacer of yeast pre-rRNA. *Nucleic Acids Res.* **22:** 5139.

Bennett J.L., Jeong-Yu S., and Clayton D.A. 1992. Characterization of a *Xenopus laevis* ribonucleoprotein endoribonuclease. *J. Biol. Chem.* **267:** 21765.

Billy E., Wegierski T., Nasr F., and Filipowicz W. 2000. Rcl1p, the yeast protein similar to the RNA 3′-phosphate cyclase, associates with U3 snoRNP and is required for 18S rRNA biogenesis. *EMBO J.* **19:** 2115.

Borovjagin A.V., and Gerbi S.A. 1999. U3 small nucleolar RNA is essential for cleavage at sites 1, 2 and 3 in pre-rRNA and determines which rRNA pathway is taken in *Xenopus* oocytes. *J. Mol. Biol.* **286:** 1347.

———. 2000. The spacing between functional *cis*-elements of U3 snoRNA is critical for rRNA processing. *J. Mol. Biol.* **300:** 57.

———. 2001. *Xenopus* U3 snoRNA GAC-Box A′ and Box A sequences play distinct functional roles in rRNA processing. *Mol. Cell. Biol.* **21:** 6210.

Bortolin M.L. and Kiss T. 1998. Human U19 intron-encoded snoRNA is processed from a long primary transcript that possesses little potential for protein coding. *RNA* **4:** 445.

Bousquet-Antonelli C., Henry Y., Gélugne J.-P., Caizergues-Ferrer M., and Kiss T. 1997. A small nucleolar RNP protein is required for pseudouridylation of eukaryotic ribosomal RNAs. *EMBO J.* **15:** 4770.

Bousquet-Antonelli C., Vanrobays E., Gélugne J.P., Caizergues-Ferrer M., and Henry Y. 2000. Rrp8p is a yeast nucleolar protein functionally linked to Gar1p and involved in pre-rRNA cleavage at site A2. *RNA* **6:** 826.

Bram R.J., Young R.A., and Steitz J.A. 1980. The ribonuclease III site flanking 23S sequences in the 30S ribosomal precursor RNA of *E. coli. Cell* **19:** 393.

Cavaillé J. and Bachellerie J.-P. 1998. SnoRNA-guided ribose methylation of rRNA: Structural features of the guide RNA duplex influencing the extent of the reaction. *Nucleic Acids Res.* **26:** 1576.

Cavaillé J., Nicoloso M., and Bachellerie J.-P. 1996. Targeted ribose methylation of RNA *in vivo* directed by tailored antisense RNA guides. *Nature* **383:** 732.

Cavaillé J., Buiting K., Kiefmann M., Lalande M., Brannan C.I., Horsthemke B., Bachellerie J.-P., Brosius J., and Hüttenhofer A. 2000. Identification of brain-specific and imprinted small nucleolar RNA genes exhibiting an unusual genomic organization. *Proc. Natl. Acad. Sci.* **97:** 14311.

Chamberlain J.R., Lee Y., Lane W.S., and Engelke D.R. 1998. Purification and characterization of the nuclear RNase P holoenzyme complex reveals extensive subunit overlap with RNase MRP. *Genes Dev.* **12:** 1678.

Chamberlain J.R., Pagán-Ramos E., Kindelberger D.W., and Engelke D.R. 1996. An RNase P RNA subunit mutation affects ribosomal RNA processing. *Nucleic Acids Res.* **24:** 3158.

Chooi W.Y. and Leiby K.R. 1981. An electron microscopic method for localization of ribosomal proteins during transcription of ribosomal DNA: A method for studying protein assembly. *Proc. Natl. Acad. Sci.* **78:** 4823.

Chu S., Zengel J.M., and Lindahl L. 1997. A novel protein shared by RNase MRP and RNase P. *RNA* **3:** 382.

Chu S., Archer R.H., Zengal J.M., and Lindahl L. 1994. The RNA of RNase MRP is required for normal processing of ribosomal RNA. *Proc. Natl. Acad. Sci.* **91:** 659.

Clark C.G. and Gerbi S.A. 1982. Ribosomal RNA evolution by fragmentation of the 23S progenitor: Maturation pathway parallels evolutionary emergence. *J. Mol. Evol.* **18:** 329.

Colley A., Beggs J.D., Tollervey D., and Lafontaine D.L. 2000. Dhr1p, a putative DEAH-box RNA helicase, is associated with box C+D snoRNP U3. *Mol. Cell. Biol.* **20:** 7238.

Cote C.A. and Peculis B.A. 2001. Role of the ITS2-proximal stem and evidence for indirect recognition of processing sites in pre-rRNA processing in yeast. *Nucleic Acids Res.* **29:** 2106.

Dennis P.P., Omer A., and Lowe T. 2001. A guided tour: Small RNA function in Archaea. *Mol. Microbiol.* **40:** 509.

Dennis P.P., Russell A.G., and Moniz de Sá M. 1997. Formation of the 5′ end pseudoknot in small subunit ribosomal RNA: Involvement of U3-like sequences. *RNA* **3:** 337.

Dichtl B. and Tollervey D. 1997. Pop3p is essential for the activity of the RNase MRP and RNase P ribonucleoproteins *in vivo. EMBO J.* **16:** 417.

Dragon F., Pogačić V., and Filipowicz W. 2000. *In vitro* assembly of human H/ACA small nucleolar RNPs reveals unique features of U17 and telomerase RNAs. *Mol. Cell. Biol.* **20:** 3037.

Dunbar D.A. and Baserga S.J. 1998. The U14 snoRNA is required for 2′-O-methylation of the pre-18S rRNA in *Xenopus* oocytes. *RNA* **4:** 195.

Dunbar D.A., Wormsley S., Agentis T.M., and Baserga S.J. 1997. Mpp10p, a U3 small nucleolar ribonucleoprotein component required for pre-18S rRNA processing in yeast. *Mol. Cell. Biol.* **17:** 5803.

Enright C.A., Maxwell E.S., Eliceiri G.L., and Sollner-Webb B. 1996. ETS rRNA processing facilitated by four small RNAs: U14, E3, U17 and U3 (erratum in *RNA* **2:** 1318 [1996].) *RNA* **2:** 1094.

Eppens N.A., Rensen S., Granneman S., Raué H.A., and Venema J. 1999. The roles of Rrp5p in the synthesis of yeast 18S and 5.8S rRNA can be functionally and physically separated. *RNA* **5:** 779.

Fournier M.J. and Maxwell E.S. 1993. The nucleolar snRNAs: Catching up with the spliceosomal snRNAs. *Trends Biochem. Sci.* **18:** 131.

Fournier R., Brulé F., Ségault V., Mougin A., and Branlant C. 1998. U3 snoRNA genes with and without intron in the *Kluyveromyces* genus. Yeasts can accommodate great variations of the U3 snoRNA 3′-terminal domain. *RNA* **4:** 285.

Frank J. and Agrawal R.K. 2000. A ratchet-like inter-subunit reorganization of the ribosome during translocation. *Nature* **406:** 309.

Fujisawa T., Imai K., Tanaka V., and Ogata K. 1979. Studies on the protein components of 110S and total ribonucleoprotein particles of rat liver. *J. Biochem.* (Tokyo) **85:** 277.

Gabashvili I.S., Agrawal R.K., Grassucci R., Squires C.L., Dahlberg A.E., and Frank J. 1999. Major rearrangements in the 70S ribosomal 3D structure caused by a conformational switch in 16S ribosomal RNA. *EMBO J.* **18:** 6501.

Ganot P., Bortolin M.-L., and Kiss T. 1997a. Site-specific pseudouridine formation in preribosomal RNA is guided by small nucleolar RNAs. *Cell* **89:** 799.

Ganot P., Caizergues-Ferrer M., and Kiss T. 1997b. The family of box ACA small nucleolar RNAs is defined by an evolutionarily conserved secondary structure and ubiquitous sequence elements essential for RNA accumulation. *Genes Dev.* **11:** 941.

Ganot P., Jády B.E., Bortolin M.L., Darzacq X., and Kiss T. 1999. Nucleolar factors direct the 2′-O-ribose methylation and pseudouridylation of U6 spliceosomal RNA. *Mol. Cell. Biol.* **19:** 6906.

Gao L., Frey M.R., and Matera A.G. 1997. Human genes encoding U3 snRNA associate with coiled bodies in interphase cells and are clustered on chromosome 17p11.2 in a complex inverted repeat structure. *Nucleic Acids Res.* **25:** 4740.

Gaspin C., Cavaille J., Erauso G., and Bachellerie J.P. 2000. Archaeal homologs of eukaryotic methylation guide small nucleolar RNAs: Lessons from the *Pyrococcus* genome. *J. Mol. Biol.* **297:** 895 (erratum in *J. Mol. Biol.*[2000] **300:** 1017).

Gautier T., Berges T., Tollervey D., and Hurt E. 1997. Nucleolar KKE/D repeat proteins Nop56p and Nop58p interact and are required for ribosome biogenesis. *Mol. Cell. Biol.* **17:** 7088.

Gerbi S.A. 1995. Small nucleolar RNA. *Biochem. Cell Biol.* **73:** 845.

Gerbi S.A. and Borovjagin A. 1996. U3 snoRNA may recycle through different compartments of the nucleolus. *Chromosoma* **105:** 401.

Ginisty H., Amalric F., and Bouvet P. 1998. Nucleolin functions in the first step of ribosomal RNA processing. *EMBO J.* **17:** 1476.

Ginisty H., Serin G., Ghisolfi-Nieto L., Roger B., Libante V., Amalric F., and Bouvet P. 2000. Interaction of nucleolin with an evolutionarily conserved pre-ribosomal RNA sequence is required for the assembly of the primary processing complex. *J. Biol. Chem.* **275:** 18845.

Girard J.P., Lehtonen H., Caizergues-Ferrer M., Amalric F., Tollervey D., and Lapeyre B. 1992. GAR1 is an essential small nucleolar RNP protein required for pre-rRNA processing in yeast. *EMBO J.* **11:** 673.

Glibetic M., Larson D.E., Sienna N., Bachellerie J.P., and Sells

B.H. 1992. Regulation of U3 snRNA expression during myoblast differentiation. *Exp. Cell Res.* **202:** 183.

Green R. and Noller H.F. 1997. Ribosomes and translation. *Annu. Rev. Biochem.* **66:** 679.

Hartshorne T. 1998. Distinct regions of U3 snoRNA interact at two sites within the 5′ external transcribed spacer of pre-rRNAs in *Trypanosoma brucei* cells. *Nucleic Acids Res.* **26:** 2541.

Hartshorne T. and Agabian N. 1994. A common core structure for U3 small nucleolar RNAs. *Nucleic Acids Res.* **22:** 3354.

Hartshorne T. and Toyofuku W. 1999. Two 5′ ETS regions implicated in interactions with U3 snoRNA are required for small subunit rRNA maturation in *Trypanosoma brucei. Nucleic Acids Res.* **27:** 3300.

Henras A., Henry Y., Bousquet-Antonelli C., Noaillac-Depeyre J., Gélugne J.-P., and Caizergues-Ferrer M. 1998. Nhp2p and Nop10p are essential for the function of H/ACA snoRNPs. *EMBO J.* **17:** 7078.

Henry Y., Wood H., Morrissey J.P., Petfalski E., Kearsay S., and Tollervey D. 1994. The 5′ end of yeast 5.8S rRNA is generated by exonucleases from an upstream cleavage site. *EMBO J.* **13:** 2452.

Hughes J.M.X. 1996. Functional base-pairing interaction between highly conserved elements of U3 small nucleolar RNA and the small ribosomal subunit. *J. Mol. Biol.* **259:** 645.

Hughes J.M.X. and Ares M., Jr. 1991. Depletion of U3 small nucleolar RNA inhibits cleavage in the 5′ external transcribed spacer of yeast pre-ribosomal RNA and prevents formation of 18S ribosomal RNA. *EMBO J.* **10:** 4231.

Hughes J.M.X., Konings D.A.M., and Cesarini G. 1987. The yeast homologue of U3 snRNA. *EMBO J.* **6:** 2145.

Hüttenhofer A., Kiefmann M., Meier-Ewert S., O′Brien J., Lehrach H., Bachellerie J.-P., and Brosius J. 2001. RNomics: An experimental approach that identifies 201 expressed RNA sequences in mouse, candidates for novel, small, non-messenger RNAs. *EMBO J.* **20:** 2943.

Jacobson M.R., Cao L.-G., Wang Y.-L., and Pederson T. 1995. Dynamic localization of RNase MRP RNA in the nucleolus observed by fluorescent RNA cytochemistry in living cells. *J. Cell Biol.* **131:** 1649.

Jacobson M.R., Cao L.-G., Taneja K., Singer R.H., Wang Y.-L., and Pederson T. 1997. Nuclear domains of the RNA subunit of RNase P. *J. Cell Sci.* **110:** 829.

Jacq B. 1981. Sequence homologies between eukaryotic 5.8S rRNA and the 5′ end of prokaryotic 23S rRNA: Evidence for a common evolutionary origin. *Nucleic Acids Res.* **9:** 2913.

Jády B.E. and Kiss T. 2000. Characterization of the U83 and U84 small nucleolar RNAs: Two novel 2′-O-ribose methylation guide RNAs that lack complementarities to ribosomal RNAs. *Nucleic Acids Res.* **28:** 1348.

———. 2001. A small nucleolar guide RNA functions both in 2′-O-ribose methylation and pseudouridylation of the U5 spliceosomal RNA. *EMBO J.* **20:** 541.

Jansen R., Tollervey D., and Hurt E. 1993. A U3 snoRNP protein with homology to splicing factor PRP4 and G beta domains is required for ribosomal RNA processing. *EMBO J.* **12:** 2549.

Jeppesen C., Stebbins-Boaz B., and Gerbi S.A. 1988. Nucleotide sequence determination and secondary structure of *Xenopus* U3 snRNA. *Nucleic Acids Res.* **16:** 2127.

Kass S., Tyc K., Steitz J.A., and Sollner-Webb B. 1990. The U3 small nucleolar ribonucleoprotein functions in the first step of pre-ribosomal RNA processing. *Cell* **60:** 897.

Khaitovich P., Mankin A.S., Green R., Lancaster L., and Noller H.F. 1999. Characterization of functionally active subribosomal particles from *Thermus aquaticus. Proc. Natl. Acad. Sci.* **96:** 85.

Kiss T. and Solymosy F. 1990. Molecular analysis of a U3 gene locus in tomato: Transcription signals, the coding region, expression in transgenic tobacco plants and tandemly repeated pseudogenes. *Nucleic Acids Res.* **18:** 1941.

Kiss-László Z., Henry Y., and Kiss T. 1998. Sequence and structural elements of methylation guide snoRNAs essential for site-specific ribose methylation of pre-rRNA. *EMBO J.* **17:** 797.

Kiss-László Z., Henry Y., Bachellerie J.-P., Caizergues-Ferrer M., and Kiss T. 1996. Site-specific ribose methylation of preribosomal RNA: A novel function for small nucleolar RNAs. *Cell* **85:** 1077.

Koonin E.V. 1996. Pseudouridine synthases: Four families of enzymes containing a putative uridine-binding motif also conserved in dUTPases and dCTP deaminases. *Nucleic Acids Res.* **24:** 2411.

Kufel J., Dichtl B., and Tollervey D. 1999. Yeast Rnt1p is required for cleavage of the pre-ribosomal RNA 3′ ETS but not the 5′ ETS. *RNA* **5:** 909.

Kufel J., Allmang C., Chanfreau G., Petfalski E., Lafontaine D.L.J., and Tollervey D. 2000. Precursors to the U3 small nucleolar RNA lack small nucleolar RNP proteins but are stabilized by La binding. *Mol. Cell. Biol.* **20:** 5415.

Kuter D.J. and Rodgers A. 1976. The protein composiiton of HeLa ribosomal subunits and nucleolar precursor particles. *Exp. Cell Res.* **102:** 205.

Labhart P. and Reeder R.H. 1986. Characterization of three sites of RNA 3′ end formation in the *Xenopus* ribosomal gene spacer. *Cell* **45:** 431.

Lafontaine D.L. and Tollervey D. 1998. Birth of the snoRNPs: The evolution of the modification-guide snoRNAs. *Trends Biochem. Sci.* **23:** 383.

———. 1999. Nop58p is a common component of the box C+D snoRNPs that is required for snoRNA stability. *RNA* **5:** 455.

———. 2000. Synthesis and assembly of the box C+D small nucleolar RNPs. *Mol. Cell. Biol.* **20:** 2650.

Lafontaine D.L.J., Bousquet-Antonelli C., Henry Y., Caizergues-Ferrer M., and Tollervey D. 1998. The box H+ACA snoRNAs carry Cbf5p, the putative rRNA pseudouridine synthase. *Genes Dev.* **12:** 527.

Lange T.S. and Gerbi S.A. 2000. Transient nucleolar localization of U6 small nuclear RNA in *Xenopus laevis* oocytes. *Mol. Biol. Cell* **11:** 2419.

Lange T.S., Borovjagin A.V., and Gerbi S.A. 1998a. Nucleolar localization elements (NoLEs) in U8 snoRNA differ from sequences required for rRNA processing. *RNA* **4:** 789.

Lange T.S., Borovjagin A.V., Maxwell E.S., and Gerbi S.A. 1998b. Conserved boxes C and D are essential nucleolar localization elements of U8 and U14 snoRNAs. *EMBO J.* **17:** 3176.

Lange T.S., Ezrokhi M., Amaldi F., and Gerbi S.A. 1999. Box H and box ACA are nucleolar localization elements of U17 small nucleolar RNA. *Mol. Biol. Cell* **10:** 3877.

Lange T.S., Ezrokhi M., Borovjagin A.V., Rivera-León R., North M.T., and Gerbi S.A. 1998c. Nucleolar localization elements of *Xenopus laevis* U3 small nucleolar RNA. *Mol. Biol. Cell* **9:** 2973.

Lastick S.M. 1980. The assembly of ribosomes in HeLa cell nucleoli. *Eur. J. Biochem.* **113:** 175.

Lee S.J. and Baserga S.J. 1997. Functional separation of pre-rRNA processing steps revealed by truncation of the U3 small nucleolar ribonucleoprotein component, Mpp10. *Proc. Natl. Acad. Sci.* **94:** 13536.

———. 1999. Imp3p and Imp4p: Two specific components of the U3 small nucleolar ribonucleoprotein that are essential for pre-18S rRNA processing. *Mol. Cell. Biol.* **19:** 5441.

Li H.V., Zagorski J., and Fournier M.J. 1990. Depletion of U14 small nuclear RNA (snR128) disrupts production of 18S rRNA in *Saccharomyces cerevisiae. Mol. Cell. Biol.* **10:** 1145.

Lindahl L. and Zengel J.M. 1995. RNase MRP and rRNA processing. *Mol. Biol. Rep.* **22:** 69.

Lindahl L., Archer R.H., and Zengel J.M. 1992. A new rRNA processing mutant of *Saccharomyces cerevisiae. Nucleic Acids Res.* **20:** 295.

Lodmell J.S. and Dahlberg A.E. 1997. A conformational switch in *Escherichia coli* 16S ribosomal RNA during decoding of messenger RNA. *Science* **277:** 1262.

Lübben B., Marshallsay C., Rottman N., and Lührmann R. 1993. Isolation of U3 snoRNP from CHO cells: A novel 55 kDa protein binds to the central part of U3 snoRNA. *Nucleic Acids*

Res. **21:** 5377.

Lukowiak A.A., Granneman S., Mattox S.A., Speckmann W.A., Jones K., Pluk H., Venrooij W.J., Terns R.M. and Terns M.P. 2000. Interaction of the U3-55k protein with U3 snoRNA is mediated by the box B/C motif of U3 and the WD repeats of U3-55k. *Nucleic Acids Res.* **28:** 3462.

Lowe T.M., and Eddy S.R. 1999. A computational screen for methylation guide snoRNAs in yeast. *Science* **283:** 1168.

Lygerou Z., Allmang C., Tollervey D., and Séraphin B. 1996. Accurate processing of a eukaryotic precursor ribosomal RNA by ribonuclease MRP in vitro. *Science* **272:** 268.

Lygerou Z., Mitchell P., Petfalski E., Séraphin B., and Tollervey D. 1994. The *POP1* gene encodes a protein component common to the RNase MRP and RNase P ribonucleoproteins. *Genes Dev.* **8:** 1423.

Lyman S.,K., Gerace L., and Baserga S.J. 1999. Human Nop5/Nop58 is a component common to the box C/D small nucleolar ribonucleoproteins. *RNA* **5:** 1597.

Maden B.E.H. 1990. The numerous modified nucleotides in eukaryotic ribosomal RNA. *Prog. Nucleic Acid Res.* **39:** 241.

Maden B.E.H. and Hughes J.M.X. 1997. Eukaryotic ribosomal RNA: The recent excitement in the nucleotide modification problem. *Chromosoma* **105:** 391.

Maden B.E.H., Khan M.S.N., Hughes D.G., and Goddard J.P. 1977. Inside 45S ribonucleic acid. *Biochem. Soc. Symp.* **42:** 165.

Marshallsay C., Connelly S., and Filipowicz W. 1992. Characterization of the U3 and U6 snRNA genes from wheat: U3 snRNA genes in monocot plants are transcribed by RNA polymerase III. *Plant Mol. Biol.* **19:** 973.

Marshallsay C., Kiss T., and Filipowicz W. 1990. Amplification of plant U3 and U6 snRNA gene sequences using primers specific for an upstream promoter element and conserved intragenic regions. *Nucleic Acids Res.* **18:** 3459.

Maser R.L. and Calvet J.P. 1989. U3 snRNA can be psoralen cross-linked *in vivo* to the 5′ external transcribed spacer of pre-ribosomal RNA. *Proc. Natl. Acad. Sci.* **86:** 6523.

Maxwell E.S. and Fournier M.J. 1995. The small nucleolar RNAs. *Annu. Rev. Biochem.* **35:** 897.

Méreau A., Fournier A., Grégoire A., Mougin P., Fabrizio R., Lührmann R., and Branlant C. 1997. An *in vivo* and *in vitro* structure-function analysis of the *Saccharomyces cerevisiae* U3A snoRNA:protein-RNA contacts and base-pair interactions with pre-ribosomal RNA. *J. Mol. Biol.* **273:** 552.

Michot B., Joseph N., Mazan S., and Bachellerie J.P. 1999. Evolutionarily conserved structural features in the ITS2 of mammalian pre-rRNAs and potential interactions with the snoRNA U8 detected by comparative analysis of new mouse sequences. *Nucleic Acids Res.* **27:** 2271.

Mishra R.K., and Eliceiri G.L. 1997. Three small nucleolar RNAs that are involved in ribosomal RNA precursor processing. *Proc. Natl. Acad. Sci.* **94:** 4972.

Mitchell J.R., Cheng J., and Collins K. 1999a. A box H/ACA small nucleolar RNA-like domain at the human telomerase RNA 3′ end. *Mol. Cell. Biol.* **19:** 567.

Mitchell J.R., Wood E., and Collins K. 1999b. A telomerase component is defective in the human disease dyskeratosis congenita. *Nature* **402:** 551.

Morrissey J.P. and Tollervey D. 1993. Yeast snR30 is a small nucleolar RNA required for 18S rRNA synthesis. *Mol. Cell. Biol.* **13:** 2469.

———. 1995. Birth of the snoRNPs—The evolution of RNase MRP and the eukaryotic pre-rRNA processing system. *Trends Biochem. Sci.* **20:** 78.

Mougey E.B., Pape L.K., and Sollner-Webb B. 1993a. A U3 small nuclear ribonucleoprotein-requiring processing event in the 5′ external transcribed spacer of *Xenopus* precursor rRNA. *Mol. Cell. Biol.* **13:** 5990.

Mougey E.B., O′Reilly M., Oshein Y., Miller O.L., Jr., Beyer A., and Sollner-Webb B. 1993b. The terminal balls characteristic of eukaryotic rRNA transcription units in chromatin spreads are rRNA processing complexes. *Genes Dev.* **7:** 1609.

Mougin A., Grégoire A., Banroques J., Ségault V., Fournier R., Brulé F., Chevrier-Miller M., and Branlant C. 1996. Secondary structure of the yeast *Saccharomyces cerevisiae* pre-U3A snoRNA and its implication for splicing efficiency. *RNA* **2:** 1079.

Musters W., Planta R.J., Van Heerikhuizen H., and Raué H.A. 1990. Functional analysis of the transcribed spacers of *Saccharomyces cerevisiae* ribosomal DNA: It takes a precursor to form a ribosome. In *The ribosome: Structure, function and evolution* (ed. W.E. Hill et al.), p. 435. ASM Press, Washington D.C.

Muth G.W., Ortoleva-Donnely L., and Strobel S. 2000. A single adenine with a neutral pK_a in the ribosomal peptidyl transferase center. *Science* **289:** 947.

Myslinski E., Ségault V., and Branlant C. 1990. An intron in the genes for U3 small nucleolar RNA of the yeast *Saccharomyces cerevisiae*. *Science* **247:** 1213.

Narayanan A., Speckmann W., Terns R., and Terns M.P. 1999a. Role of the box C/D motif in localization of small nucleolar RNAs to coiled bodies and nucleoli. *Mol. Biol. Cell* **10:** 2131.

Narayanan A., Lukowiak A., Jády B.E., Dragon F., Kiss T., Terns R.M., and Terns M.P. 1999b. Nucleolar localization signals of box H/ACA small nucleolar RNAs. *EMBO J.* **18:** 5120.

Nazar R.N. 1980. A 5.8S rRNA-like sequence in prokaryotic 23S rRNA. *FEBS Lett.* **119:** 212.

Nazar R.N. and Sitz T.O. 1980.Role of the 5′-terminal sequence in the RNA binding site of yeast 5.8S rRNA. *FEBS Lett.* **115:** 71.

Newman D.R., Kuhn J.F., Shanab G.M., and Maxwell E.S. 2000. Box C/D snoRNA-associated proteins: Two pairs of evolutionarily ancient proteins and possible links to replication and transcription. *RNA* **6:** 861.

Ni J., Tien A.L., and Fournier M.J. 1997. Small nucleolar RNAs direct site-specific synthesis of pseudouridine in ribosomal RNA. *Cell* **89:** 565.

Nicoloso M., Qu L.H., Michot B., and Bachellerie J.P. 1996. Intron-encoded, antisense small nucleolar RNAs: The characterization of nine novel species points to their direct role as guides for the 2′-O-ribose methylation of rRNAs. *J. Mol. Biol.* **260:** 178.

Niewmierzycka A. and Clarke S. 1999. S-adenosylmethionine-dependent methylation in *Saccharomyces cerevisiae*. Identification of a novel protein arginine methyltransferase. *J. Biol. Chem.* **274:** 814.

Nissen P., Hansen J., Ban N., Moore P.B., and Steitz T.A. 2000. The structural basis of ribosome activity in peptide bond synthesis. *Science* **289:** 920.

Noller H.F., Hoffarth V., and Zimniak. 1992. Unusual resistance of peptidyl transferase to protein extraction procedures. *Science* **256:** 1416.

Nomura M. 1987. The role of RNA and protein in ribosome function: A review of early reconstitution studies and prospects for future studies. *Cold Spring Harbor Symp. Quant. Biol.* **52:** 653.

———. 1997. Reflection on the days of ribosome reconstitution research. *Trends Biochem. Sci.* **22:** 275.

Ogle J.M., Brodersen D.E., Clemons W.M., Jr., Tarry M.J., Carter A.P., and Ramakrishnan V. 2001. Recognition of cognate transfer RNA by the 30S ribosomal subunit. *Science* **292:** 897.

Omer A.D., Lowe T.M., Russell A.G., Ebhardt H., Eddy S.R., and Dennis P.P. 2000. Homologs of small nucleolar RNAs in Archaea. *Science* **288:** 517.

Pace N.R., Walker T.A., and Schroeder E. 1977. Structure of the 5.8S-28S ribosomal RNA junction complex. *Biochemistry* **16:** 5321.

Pape T., Wintermeyer W., and Rodnina M.V. 2000. Conformational switch in the decoding region of 16S rRNA during aminoacyl-tRNA selection on the ribosome. *Nat. Struct. Biol.* **7:** 104.

Parker K.A. and Steitz J.A. 1987. Structural analysis of the human U3 ribonucleoprotein particle reveals a conserved sequence available for base-pairing with pre-rRNA. *Mol. Cell. Biol.* **7:** 2899.

Peculis B.A. 1997. The sequence of the 5′ end of the U8 small nucleolar RNA is critical for 5.8S and 28S rRNA maturation. *Mol. Cell. Biol.* **17:** 3702.

———. 2000. snoRNA nuclear import and potential for co-transcriptional function in pre-rRNA processing. *RNA* **7:** 207.

Peculis B.A. and Steitz J.A. 1993. Disruption of U8 nucleolar snRNA inhibits 5.8S and 28S rRNA processing in the *Xenopus* oocyte. *Cell* **73:** 1233.

———. 1994. Sequence and structural elements critical for U8 snRNP function in *Xenopus* oocytes are evolutionarily conserved. *Genes Dev.* **8:** 2241.

Pluk H., Soffner J., Lührmann R., and van Venrooij W.J. 1998. cDNA cloning and characterization of the human U3 small nucleolar ribonucleoprotein complex-associated 55-kilodalton protein. *Mol. Cell. Biol.* **18:** 488.

Pogačić V., Dragon F., and Filipowicz W. 2000. Human H/ACA small nucleolar RNPs and telomerase share evolutionarily conserved proteins NHP2 and NOP10. *Mol. Cell. Biol.* **20:** 9028.

Porter G.L., Brennwald P.J., Holm K.A., and Wise J.A. 1988. The sequence of U3 from *Schizosaccharomyces pombe* suggests structural divergence of this snRNA between metazoans and unicellular eukaryotes. (erratum in *Nucleic Acids Res.* [1991] **19:** 3484) *Nucleic Acids Res.* **16:** 10131.

Potter S., Durovic P., and Dennis P.P. 1995. Ribosomal RNA precursor processing by a eukaryotic U3 small nucleolar RNA-like molecule in an archeon (retracted in *Science* [1997] **277:** 1189). *Science* **268:** 1056.

Puvion-Dutilleul F., Mazan S., Nicoloso M., Pichard E., Bachellerie J.-P., and Puvion E. 1992. Alterations of nucleolar ultrastructure and ribosome biogenesis by actinomycin D. Implications for U3 snRNP function. *Eur. J. Cell Biol.* **58:** 149.

Reddy R. and Busch H. 1983. Small nuclear RNA and RNA processing. *Prog. Nucleic Acid Res. Mol. Biol.* **30:** 127.

Rivera-León R. 1995. "Nucleolar localization and maintenance of U3 small nucleolar RNA in *Xenopus laevis*." Ph.D. thesis, Brown University, Providence, Rhode Island.

Rivera-León R. and Gerbi S.A. 1996. Delocalization of some small nucleolar RNPs after actinomycin D treatment to deplete early pre-rRNAs. *Chromosoma* **105:** 506.

Ruhl D.D., Pusateri M.E., and Eliceiri G.L. 2000. Multiple conserved segments of E1 small nucleolar RNA are involved in the formation of a ribonucleoproteim particle in frog oocytes. *Biochem. J.* **348:** 517.

Russell A.G., Ebhardt H., and Dennis P.P. 1999. Substrate requirements for a novel archaeal endonuclease that cleaves within the 5′ external transcribed spacer of *Sulfolobus acidocaldarius* precursor rRNA. *Genetics* **152:** 1373.

Samarsky D.A. and Fournier M.J. 1998. Functional mapping of the U3 small nucleolar RNA from the yeast *Saccharomyces cerevisiae*. *Mol. Cell. Biol.* **18:** 3431.

Samarsky D.A., Fournier M.J., Singer R.H., and Bertrand E. 1998. The snoRNA box C/D motif directs nucleolar targeting and also couples snoRNA synthesis and localization. *EMBO J.* **17:** 3747.

Savino R. and Gerbi S.A. 1990. *In vivo* disruption of *Xenopus* U3 snRNA affects ribosomal RNA processing. *EMBO J.* **9:** 2299.

———. 1991. Preribosomal RNA processing in *Xenopus* oocytes does not include cleavage within the external transcribed spacer as an early step (erratum in *Biochimie* [1996] **78:** 295) *Biochimie* **73:** 805.

Schmitt M.E. and Clayton D.A. 1993. Nuclear RNase MRP is required for correct processing of pre-5.8S rRNA in *Saccharomyces cerevisiae*. *Mol. Cell. Biol.* **13:** 7935.

———. 1994. Characterization of a unique protein component of yeast RNase MRP: An RNA-binding protein with a zinc-cluster domain. *Genes Dev.* **8:** 2617.

Schnare M.N., Collings J.C., Spencer D.F., and Gray M.W. 2000. The 28S-18S rDNA intergenic spacer from *Crithidia fasciculata*: Repeated sequences, length heterogeneity, putative processing sites and potential interactions between U3 small nucleolar RNA and the ribosomal RNA precursor. *Nucleic Acids Res.* **28:** 3452.

Ségault V., Mougin A., Grégoire A., Banroques J., and Branlant C. 1992. An experimental study of *Saccharomyces cerevisiae*

U3 snRNA conformation in solution. *Nucleic Acids Res.* **20:** 3443.

Shadel G.S., Buckenmeyer G.A., Clayton D.A., and Schmitt M.E. 2000. Mutational analysis of the RNA component of *Saccharomyces cerevisiae* RNase MRP reveals distinct nuclear phenotypes. *Gene* **245:** 175.

Sharma K. and Tollervey D. 1999. Base pairing between U3 small nucleolar RNA and the 5′ end of 18S rRNA is required for pre-rRNA processing. *Mol. Cell. Biol.* **19:** 6012.

Shuai K. and Warner J.W. 1991. A temperature sensitive mutant of *S. cerevisiae* defective in pre-rRNA processing. *Nucleic Acids Res.* **19:** 5059.

Sienna N., Larson D.E., and Sells B.H. 1996. Altered subcellular distribution of U3 snRNA in response to serum in mouse fibroblasts. *Exp. Cell Res.* **227:** 98.

Smith C.M. and Steitz J.A. 1997. Sno storm in the nucleolus: New roles for myriad small RNPs. *Cell* **89:** 669.

Speckmann W.A., Terns R.M., and Terns M.P. 2000. The Box C/D motif directs snoRNA 5′-cap hypermethylation. *Nucleic Acids Res.* **28:** 4467.

Speckmann W., Narayanan A., Terns R., and Terns M.P. 1999. Nuclear retention elements of U3 small nucleolar RNA. *Mol. Cell. Biol.* **19:** 8412.

Stark H., Rodnina M.V., Wieden H.J., van Heel M., and Wintermeyer W. 2000. Large-scale movement of elongation factor G and extensive conformational change of the ribosome during translocation. *Cell* **100:** 301.

Stelzl U. and Nierhaus K.H. 2001. A short fragment of 23S rRNA containing the binding sites for two ribosomal proteins, L24 and L4, is a key element for rRNA folding during early assembly. *RNA* **7:** 598.

Stolc V. and Altman S. 1997. Rpp1, an essential protein subunit of nuclear RNase P required for processing of precursor tRNA and 35S precursor rRNA in *Saccharomyces cerevisiae*. *Genes Dev.* **11:** 2926.

Stroke I.L. and Weiner A.M. 1989. The 5′ end of U3 snRNA can be cross-linked in vivo to the external transcribed spacer of rat ribosomal RNA precursors. *J. Mol. Biol.* **210:** 497.

Terns M.P. and Dahlberg J.E. 1994. Retention and 5′ cap trimethylation of U3 RNA in the nucleus. *Science* **264:** 959.

Terns M.P., Grimm C., Lund E., and Dahlberg J.E. 1995. A common maturation pathway for small nucleolar RNAs. *EMBO J.* **14:** 4860.

Todorov I., Noll F., and Hadjiolov A.A. 1983. The sequential addition of ribosomal proteins during the formation of the small ribosomal subunit in Friend erythroleukemia cells. *Eur. J. Biochem.* **131:** 271.

Tollervey D. 1987. A yeast small nuclear RNA is required for normal processing of pre-ribosomal RNA. *EMBO J.* **6:** 4169.

Tollervey D., Lehtonen H., Carmo-Fonseca M., and Hurt E. 1991. The small nucleolar RNP protein NOP1 (fibrillarin) is required for pre-rRNA processing in yeast. *EMBO J.* **10:** 573.

Tollervey D., Lehtonen H., Jansen R., Kern H., and Hurt E.C. 1993. Temperature-sensitive mutations demonstrate roles for yeast fibrillarin in pre-rRNA processing, pre-rRNA methylation, and ribosome assembly. *Cell* **72:** 443.

Torchet C., Jacq C., and Hermann-Le Denmat S. 1998. Two mutant forms of the S1/TPR-containing protein Rrp5p affect the 18S rRNA synthesis in *Saccharomyces cerevisiae*. *RNA* **4:** 1636.

Tyc K. and Steitz J.A. 1989. U3, U8 and U13 comprise a new class of mammalian snRNPs localized in the nucleolus. *EMBO J.* **8:** 3113.

———. 1992. A new interaction between the mouse 5′ external transcribed spacer of pre-rRNA and U3 snRNA detected by psoralen crosslinking. *Nucleic Acids Res.* **20:** 5375.

Tycowski K.T., Shu M.-D., and Steitz J.A. 1993. A small nucleolar RNA is processed from an intron of the human gene encoding ribosomal protein S3. *Genes Dev.* **7:** 1176.

———. 1994. Requirement for intron-encoded U22 small nucleolar RNA in 18S ribosomal RNA maturation. *Science* **266:** 1558.

———. 1996a. A mammalian gene with introns instead of exons generating stable RNA products. *Nature* **379:** 464.

Tycowski K.T., Smith C.M., Shu M.-D., and Steitz J.A. 1996b. A small nucleolar RNA requirement for site-specific ribose methylation of rRNA in *Xenopus. Proc. Natl. Acad. Sci.* **93:** 14480.

Tycowski K.T., You Z.-H., Graham P.J., and Steitz J.A. 1998. Modification of U6 spliceosomal RNA is guided by other small RNAs. *Mol. Cell* **2:** 629.

Udem S.A. and Warner J. 1973. The cytoplasmic maturation of a ribosomal precursor ribonucleic acid in yeast. *J. Biol. Chem.* **248:** 1412.

van Hoof A., Lennertz P., and Parker R. 2000. Yeast exosome mutants accumulate 3′-extended polyadenylated forms of U4 small nuclear RNA and small nucleolar RNAs. *Mol. Cell. Biol.* **20:** 441.

VanLoock M.S., Agrawal R.K., Gabashvili I.S., Qi L., Frank J., and Harvey S.C. 2000. Movement of the decoding region of the 16S ribosomal RNNA accompanies tRNA translocation. *J. Mol. Biol.* **304:** 507.

Veldman G.M., Klootwijk J., de Regt V.C., Plants R.J., Branlant C., Krol A., and Ebel J.-P. 1981. The primary and secondary structure of yeast 26S rRNA. *Nucleic Acids Res.* **9:** 6935.

Venema J. and Tollervey D. 1996. RRP5 is required for formation of both 18S and 5.8S rRNA in yeast. *EMBO J.* **15:** 5701.
———. 1999. Ribosome synthesis in *Saccharomyces cerevisiae. Annu. Rev. Genet.* **33:** 261.

Venema J., Henry Y., and Tollervey D. 1995. Two distinct recognition signals define the site of endonucleolytic cleavage at the 5′ end of yeast 18S rRNA. *EMBO J.* **14:** 4883.

Venema J., Vos H.R., Faber A.W., van Venrooij W.J., and Raué H.A. 2000. Yeast Rrp9p is an evolutionarily conserved U3 snoRNP protein essential for the early pre-rRNA processing cleavages and requires box C for its association. *RNA* **6:** 1660.

Verheggen C., Mouaikel J., Thiry M., Blanchard J.M., Tollervey D., Bordonne R., Lafontaine D.L., and Bertrand E. 2001. Box C/D small nucleolar RNA trafficking involves small nucleolar RNP proteins, nucleolar factors and a novel nuclear domain. *EMBO J.* **20:** 5480.

Wang H., Boivert D., Kim K.K., Kim R., and Kim S.H. 2000. Crystal structure of a fibrillarin homologue from *Methanococcus jannaschii*, a hyperthermile, at 1.6 Å resolution. *EMBO J.* **19:** 317.

Watkins N.J., Newman D.R., Kuhn J.F., and Maxwell E.S. 1998a. In vitro assembly of the mouse U14 snoRNP core complex and identification of a 65-kDa box C/D-binding protein. *RNA* **4:** 582.

Watkins N.J., Gottschalk A, Neubauer G., Kastner B., Fabrizio P., Mann M., and Lührmann R. 1998b. Cbf5p, a potential pseudouridine synthase, and Nhp2p, a putative RNA-binding protein, are present together with Gar1p in all H BOX/ACA-motif snoRNPs and constitute a common bipartite structure. *RNA* **4:** 1549.

Watkins N.J., Ségault V., Charpentier B., Nottrott S., Fabrizio P., Bachi A., Wilm M., Rosbash M., Branlant C., and Lührmann R. 2000. A common core RNP structure shared between the small nucleolar Box C/D RNPs and the spliceosomal U4 snRNP. *Cell* **103:** 457.

Westendorf J.M., Konstantinov K.N., Wormsley S., Shu M.D., Matsumoto-Taniura N., Pirollet F., Klier F.G., Gerace L., and Baserga S.J. 1998. M phase phosphoprotein 10 is a human U3 small nucleolar ribonucleoprotein component. *Mol. Biol. Cell* **9:** 437.

White T.C., Rudenko G., and Borst P. 1986. Three small RNAs within the 10 kb trypanosome rRNA transcription unit are analogous to domain VII of other eukaryotic 28S rRNAs. *Nucleic Acids Res.* **14:** 9471.

Wiederkehr T., Pretot R.F., and Minvielle-Sebastia L. 1998. Synthetic lethal interactions with conditional poly(A) polymerase alleles identify *LCP5*, a gene involved in 18S rRNA maturation. *RNA* **4:** 1357.

Wise J.A. and Weiner A.M. 1980. *Dictyostelium* small nuclear RNA D2 is homologous to rat nucleolar RNA U3 and is encoded by a dispersed multigene family. *Cell* **22:** 109.

Wormsley S., Samarsky D.A., Fournier M.J., and Baserga S.J. 2001. An unexpected, conserved element of the U3 snoRNA is required for Mpp10p association. *RNA* **7:** 904.

Wu P., Brockenbrough J.S., Metcalfe A.C., Chen S., and Aris J.P. 1998. Nop5p is a small nucleolar ribonucleoprotein component required for pre-18S rRNA processing in yeast. *J. Biol. Chem.* **273:** 16453.

Young R.A. and Steitz J.A. 1978. Complementary sequences 1700 nucleotides apart form a RNAse III cleavage site in *E. coli* ribosomal precursor RNA. *Proc. Natl. Acad. Sci.* **75:** 3593.

Yu Y.-T., Shu M.-D., Narayanan A., Terns R.M., Terns M.P., and Steitz J.A. 2001. Internal modification of U2 small nuclear (sn)RNA occurs in nucleoli of *Xenopus* oocytes. *J. Cell Biol.* **152:** 1279.

Zebarjadian Y., King T., Fournier M.J., Clarke L., and Carbon J. 1999. Point mutations in yeast CBF5 can abolish in vivo pseudouridylation of rRNA. *Mol. Cell. Biol.* **19:** 7461.

Thermodynamics and Kinetics of Central Domain Assembly

M.I. RECHT AND J.R. WILLIAMSON

Department of Molecular Biology and the Skaggs Institute for Chemical Biology,
The Scripps Research Institute, La Jolla, California 92037

The in vitro reconstitution of the 30S ribosomal subunit from *Escherichia coli* has served as the paradigm for studies of the assembly of large ribonucleoprotein complexes ever since Nomura and colleagues demonstrated that purified ribosomal proteins and 16S rRNA could spontaneously form a functional 30S subunit. Initial studies with partially fractionated sets of ribosomal proteins (Traub and Nomura 1968, 1970), and later, experiments monitoring the binding of individually purified proteins or specific mixtures of proteins to 16S rRNA, allowed the development of an assembly map for the in vitro reconstitution of the *E. coli* 30S subunit (Fig. 1) (Mizushima and Nomura 1970; Nomura et al. 1970; Held et al. 1973, 1974). The assembly map reveals a hierarchy to the binding of ribosomal proteins to form the 30S subunit. The binding of the ribosomal proteins can be divided into three classes: primary, secondary, and tertiary binding proteins. The primary binding proteins (S4, S7, S8, S15, S17, and S20) can bind directly to 16S rRNA in the absence of any other ribosomal proteins. Secondary binding proteins (S5, S6, S9, S12, S13, S16, S18, and S19) require the prior binding of one of the primary binding proteins. Tertiary binding proteins, including S2, S3, S10, S11, S14, and S21, require the prior binding of one or more secondary binding proteins.

The secondary structure of 16S rRNA has been determined using phylogenetic sequence analysis (Noller and Woese 1981; Gutell et al. 1985). 16S rRNA can be divided into four domains: the 5′, central, 3′ major, and 3′ minor domains. In the fully assembled 30S subunit, the 5′, central, and the 3′ major domains form the body, platform, and head, respectively.

The binding sites of the small subunit ribosomal proteins on 16S rRNA have been mapped by a variety of methods. Zimmermann and coworkers used nuclease mapping experiments to determine that S6, S8, S15, and S18 bind to the central domain of 16S rRNA (Zimmermann et al. 1975; Gregory et al. 1984). RNP complexes that contained 16S rRNA bound to S8, S8 + S15, or S8 + S15 + S6 + S18 were isolated by sedimentation in a sucrose gradient or by separation on polyacrylamide gels. Noller and coworkers localized the binding sites of all the ribosomal proteins on 16S rRNA using both base-specific chemical and solution hydroxyl radical footprinting experiments (Svensson et al. 1988; Stern et al. 1989; Powers and Noller 1995). Binding of S8 to 16S rRNA protects

nucleotides in helices 20, 21, 22, and 25 from modification by hydroxyl radicals, and S15 protects nucleotides in helices 20 and 22 from modification by hydroxyl radicals. Binding of S6 and S18 to 16S rRNA caused protection of nucleotides in helices 22, 23, and 26 (Fig. 2).

From the reconstitution assembly map and the information about ribosomal protein-binding sites obtained biochemically, formation of the platform of the 30S subunit from 16S rRNA and ribosomal proteins proceeds as follows: S15 binds directly to 16S rRNA, nucleating formation of the central domain, followed by the cooperative binding of S6 and S18. Binding of S11 and S21 is dependent on the prior binding of S15, S6, and S18.

Measurement of the rate of reconstitution at various temperatures demonstrated that the assembly process follows first-order kinetics (Traub and Nomura 1969, 1970; Held and Nomura 1973). Since the assembly process involves more than 20 components, this result was surprising. It was postulated that although there are several steps in the assembly pathway, there must be a unimolecular rate-limiting step late in assembly. Reconstitution experiments performed under conditions in which the riboso-

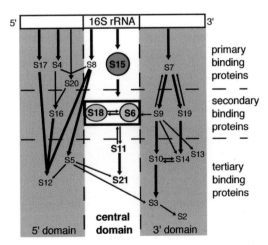

Figure 1. Assembly map for the in vitro reconstitution of the *E. coli* 30S subunit (Mizushima and Nomura 1970; Held et al. 1974). The map has been drawn to correspond to the 16S rRNA domains to which each ribosomal protein binds (Stern et al. 1989; Powers and Noller 1995).

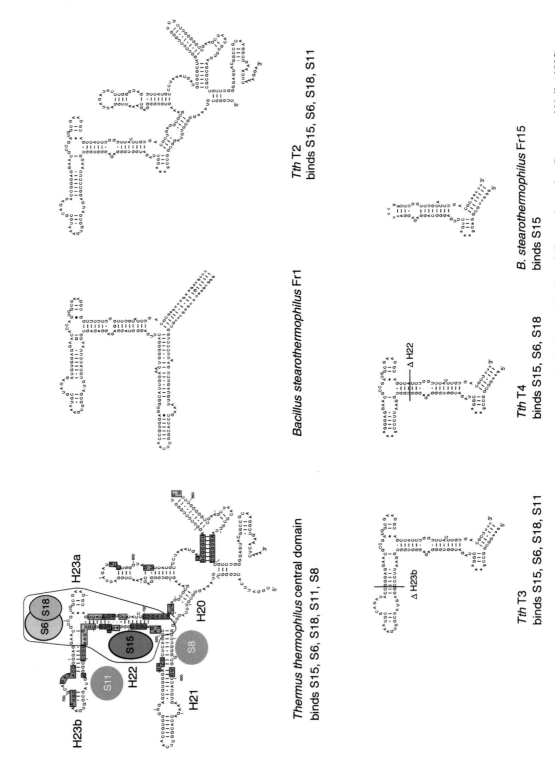

Figure 2. Secondary structures of central domain RNAs. The nucleotides protected from hydroxyl radicals upon binding of ribosomal proteins (Powers and Noller 1995) are mapped on the secondary structure of the *T. thermophilus* 16S rRNA sequence (Gutell 1994). The *B. stearothermophilus* Fr1 and Fr15 fragments are capable of binding S15 (Batey and Williamson 1996a). *T. thermophilus* RNA fragments T2 through T4 form RNP complexes with the ribosomal proteins indicated (Agalarov and Williamson 2000).

mal proteins were limiting demonstrated that the assembly process was not completely cooperative (Nomura et al. 1969). There appear to be at least two nucleation sites on 16S rRNA for the assembly of the 30S subunit. It has been demonstrated that two nucleation sites, one associated with the binding of S4 and the other associated with the binding of S7, are present in 16S rRNA (Nowotny and Nierhaus 1988). This places one site of assembly nucleation in the 5′ domain and the other in the 3′ domain, but provides no nucleation site in the central domain. Although these studies assayed the functional properties of the reconstituted particles, a later study in which the physical state of the reconstituted particles was examined instead also supported the lack of complete cooperativity in 30S subunit assembly (Dodd et al. 1991). The 5′, central, and 3′ domains can be independently reconstituted into RNPs containing the complete set of associated proteins (Weitzmann et al. 1993; Samaha et al. 1994; Agalarov et al. 1998), suggesting that each domain contains a nucleation site for assembly.

The kinetic sequence of in vitro 30S subunit assembly was analyzed by adding an assembly inhibitor (cibacron blue F3GA) at various times during the reconstitution (Datta et al. 1986). In these experiments, S4, S5, S6, S8, and S15 were shown to bind early in the reconstitution, with S7 binding in a subsequent step. Although these results are not in complete agreement with the Nomura assembly map (S6 binds without S18), they are consistent with previous results on the general order of protein binding and they indicate a role for the binding of S15 (and/or S8) in the nucleation of assembly within the independently folding central domain.

The kinetics of 30S assembly have also been measured by monitoring the rate at which protein-specific footprints appear on the 16S rRNA (Powers et al. 1993). This method was used to identify four kinetic classes for the in vitro assembly of 30S subunits: protections that were seen in the absence of any heating step (S4, S16, S17, S20), those that appeared (at 30°C) with a $t_{1/2}<5$ minutes (S15, S6 + S18, S13, S11), a set with a $t_{1/2} = 15–30$ minutes (S5, S7, S8, S9, S12, S19), and a group with $t_{1/2} = 30–60$ minutes (S14, S10, S2, S3, S21) (Powers et al. 1993). Measurements of 30S subunit assembly in vivo following a heat shock that selectively destroys only the 30S subunit indicate that subunit assembly in vivo takes on the order of 15–30 minutes following recovery of RNA synthesis (Tal et al. 1977). In all cases, the kinetics of 30S subunit assembly are slow.

Although the assembly map describes the interdependencies of protein binding and reveals the hierarchical nature of 30S subunit assembly, it does not describe the mechanism by which this hierarchy of protein binding is achieved. We have undertaken detailed studies of smaller systems in an attempt to shed light on the mechanism of central domain assembly.

BINDING OF S15 TO 16S rRNA

The first step in the assembly of the platform is the binding of S15 to 16S rRNA. Although S8 is also a primary binding protein that interacts with the central domain of 16S rRNA, its binding does not potentiate the binding of any other proteins to the central domain during assembly of the platform. To aid in performing detailed kinetic, thermodynamic, and structural studies of the binding of S15 to rRNA, a minimal binding site for S15 on 16S rRNA was defined using RNA and protein components from *Bacillus stearothermophilus*. Using information from both the base-specific (Svensson et al. 1988; Stern et al. 1989) and, more importantly, the solution hydroxyl radical footprinting (Powers and Noller 1995) of *E. coli* S15 on 16S rRNA, a series of RNA fragments were tested for their ability to bind ribosomal protein S15 from *B. stearothermophilus* (BS15). The minimal RNA fragment able to bind BS15 with high affinity and specificity was a 61-nucleotide RNA containing a three-helix junction between helices 20, 21, and 22 (Fr15, Fig. 2) (Batey and Williamson 1996a). Parallel experiments using a *Thermus thermophilus* system identified a similar minimal rRNA-binding site for ribosomal protein S15 (Serganov et al. 1996).

THERMODYNAMICS OF S15 BINDING TO 16S rRNA

The affinity of BS15 for a series of rRNA constructs including full-length 16S rRNA, a large portion of the central domain (Fr1, Fig. 2), and the minimal BS15 binding fragment (Fr15) were determined using a gel mobility shift assay. In each instance, the equilibrium dissociation constant with no divalent cations present and low monovalent salt concentration was approximately 30 nM, indicating that all elements required for the binding of S15 were contained within the minimal binding site (Batey and Williamson 1996a). Although S15 protects nucleotides in the GAAG tetraloop at the end of helix 23a from modification with base-specific chemical probes (Svensson et al. 1988), deletion of this interaction has no thermodynamic effect on the binding of S15 to rRNA (Batey and Williamson 1996a; Serganov et al. 1996). These S15–helix 23a interactions appear to be important for the further assembly of the central domain, as discussed below.

KINETICS OF S15 BINDING

The bimolecular association rate constants for S15 binding to the Fr1 fragment and the Fr15 fragment were determined, and, depending on the buffer conditions, the k_{on} for either RNA was between 10^4 $M^{-1}s^{-1}$ and 10^5 $M^{-1}s^{-1}$ (Batey and Williamson 1996a, 1998). In either case, the binding of BS15 to rRNA is three orders of magnitude slower than expected for a diffusion-controlled process. The unimolecular dissociation rate was also determined for Fr1 and Fr15 and was between 6.7×10^{-4} s^{-1} and 2.4×10^{-4} s^{-1}. These dissociation rates indicate that S15 forms a stable complex with 16S rRNA ($t_{1/2}$ of 18 minutes at 25°C), consistent with the ability to recover S15–rRNA complexes via centrifugation in a sucrose density gradient.

CONFORMATIONAL CHANGE IN 16S rRNA UPON BINDING OF S15

During the course of defining the minimal binding site for BS15 using the gel shift assay, it was observed that the binding of BS15 to RNA constructs containing a full-length helix 21 caused an acceleration, rather than a retardation, in the mobility of the RNA in the gel (Batey and Williamson 1996a,b). This acceleration was attributed to a conformational change in the RNA upon (or required for) binding of S15. A change in the mobility of a central domain RNA fragment from *T. thermophilus* 16S rRNA in the presence of Mg^{++} has also been observed (Serganov et al. 1996). The binding of BS15 to RNA constructs in which helix 21 is truncated shows a retardation, rather than an acceleration, in the gel shift assay, indicating a conformational reorganization of the three-helix junction between helices 20, 21, and 22.

In the experiments used to define the minimal binding site for BS15, a buffer containing 50 mM monovalent and no divalent cations was used. It was later determined that the presence of polyvalent cations, such as Mg^{++} or spermidine, caused an acceleration of the RNA in a gel mobility assay much the same as the binding of BS15 to the RNA (Batey and Williamson 1998).

The quantitative details of cation- and S15-induced conformational change in the three-helix junction between helices 20, 21, and 22 in 16S rRNA were determined by transient electric birefringence (TEB) (Orr et al. 1998). To measure the angles subtended by the three helices in both the absence and presence of BS15 or Mg^{++} by TEB, a set of three RNA constructs was used in which each possible pair of helices was extended by 70 base pairs relative to the third helix. The TEB experiments

demonstrate that in the absence of S15 or divalent cations, the three-helix junction adopts an apparently planar conformation in which all three interhelical angles are ~120° (Fig. 3, RNA in R$_0$ state). Upon addition of Mg^{++} or BS15, helices 21 and 22 become colinear and helix 20 forms a 60° angle with respect to the other two helices (Fig. 3, R$_1$•Mg^{++} or R$_1$•S15). The conformational change induced by binding of Mg^{++} or S15 was also monitored by single-molecule fluorescence resonance energy transfer (Ha et al. 1999), yielding results that are consistent with the gel shift and TEB measurements.

Both the gel mobility shift and TEB experiments demonstrate that the three-helix junction can adopt a conformation similar to that of the S15-bound form of the RNA in the presence of divalent cations. In addition, the rate of BS15 association (k_{on}) with the RNA is increased 100-fold in the presence of divalent cations. From these data, a structure-capture mechanism for the binding of BS15 to the three-way junction was proposed. In this mechanism, the RNA transiently adopts a conformation (R$_1$) that is uniquely competent for BS15 binding.

SUBDOMAINS FOR BINDING OF CENTRAL DOMAIN PROTEINS

According to the Nomura assembly map, the second step in assembly of the platform is binding of S6 and S18 to the S15–rRNA complex. To study these interactions in a minimal system, a series of RNA constructs derived from the central domain of *T. thermophilus* 16S rRNA were made and assayed for their ability to bind proteins of the central domain by purification of the RNPs in a sucrose density gradient (Agalarov and Williamson 2000).

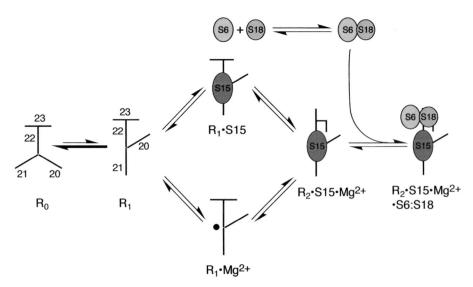

Figure 3. Summary of current mechanism for first two steps of central domain assembly based on biochemical and structural data. In the absence of ribosomal proteins, the three-helix junction between helices 20, 21, and 22 of 16S rRNA exists in an equilibrium between two conformations, R$_0$ and R$_1$. In R$_0$, helices 20, 21, and 22 are planar with an ~120° angle between the three helices. Addition of S15 or divalent cations stabilizes the R$_1$ conformation, in which helices 21 and 22 are colinear and there is an ~60° angle between helix 20 and helix 22. The binding of S15 stabilizes a new conformation of the RNA, R$_2$, in which helix 23a is parallel to helix 22. The S6:S18 heterodimer recognizes the R$_2$ conformation and binds across the upper three-helix junction between helices 22, 23, and 23a.

The bound proteins were identified by SDS-PAGE. As shown in Figure 2, specific RNA fragments were identified that could bind S15 alone, S15+S6+S18, or S15+S6+S18+S11. These particles reflect the hierarchical nature of primary, secondary, and tertiary protein binding in subunit assembly, and it is possible that these minimal RNPs represent intermediates in the assembly pathway of the central domain. An important conclusion from these experiments is that although the chemical footprinting experiments had identified interactions of the ribosomal proteins throughout the central domain RNA, only half of the RNA (helices 20–23) is required for tight binding of all the central domain proteins.

STRUCTURE OF AN ASSEMBLY INTERMEDIATE RNP COMPLEX

During the past few years, enormous advances in the structural biology of ribosomes have been made. High-resolution structures of the 50S subunit (Ban et al. 2000), the 30S subunit (Schluenzen et al. 2000; Wimberly et al. 2000), the T4 RNP (Agalarov et al. 2000), and an S15–RNA complex (Nikulin et al. 2000) have been solved by X-ray crystallography. We focus our discussion on the T4 RNP due to its implications for the mechanism of S6:S18 binding during assembly of the central domain. The T4 RNP (Fig. 4) was crystallized and its structure solved to 2.6 Å by X-ray crystallography (Agalarov et al. 2000). The RNP contains ribosomal proteins S15, S6, and S18, and the minimal RNA fragment able to bind these three proteins. S15 binds two elements of minor groove RNA, one located at the three-helix junction between helices 20, 21, and 22 and another adjacent to a purine-rich bulge in helix 22. In addition, S15 interacts with the tetraloop of helix 23a, possibly stabilizing the coaxial stacking of helix 23 on helix 22 and creating the binding site for S6 and S18.

S6 and S18 bind to the S15–RNA complex as a heterodimer to the upper three-helix junction between helices 22, 23, and 23a. S6 interacts primarily with the minor groove of helices 22 and 23, and S18 makes extensive

contacts that span the entire upper three-helix junction. Although electron density was incomplete for the distal end of helix 23 and for portions of S18, most of these two entities could be placed in the structure. The contacts between S15, S6, and S18 with the rRNA seen in the smaller RNPs are also observed in the structures of the complete 30S subunit. There are extensive protein–protein contacts between S6 and S18, but there are no protein–protein contacts between S15 and either S6 or S18. This implies that the hierarchical nature of S6 and S18 binding only after the binding of S15 must be mediated by S15-induced conformational changes in the 16S rRNA.

BINDING OF S6 AND S18 TO S15–rRNA COMPLEX

Previous studies on the binding of ribosomal proteins to 16S rRNA have used RNA and protein components from a variety of organisms, including *E. coli*, *B. stearothermophilus*, and *T. thermophilus*. For various technical reasons, we chose to study the mechanism of S6 and S18 binding to the S15–rRNA complex using RNA and proteins from the hyperthermophilic bacterium *Aquifex aeolicus*. For a detailed study of the thermodynamics and kinetics of S6 and S18 binding, a minimal RNA fragment corresponding to the T4 RNA (A4 RNA) and ribosomal proteins S15 (AS15), S6 (AS6), and S18 (AS18) were used.

There is evidence for cooperative binding of S6 and S18 to an S15–rRNA complex from the assembly map and other biochemical studies. The structural data demonstrate that S6 and S18 interact upon binding to the S15–rRNA complex, but the mechanism of the cooperative binding of S6 and S18 has not been addressed. S6 and S18 could bind as a preformed heterodimer to the S15–rRNA complex, or they could bind sequentially with one of the two proteins interacting weakly with the complex, followed by cooperative binding of the second protein to create a tightly bound complex.

To begin the study of assembly in the *A. aeolicus* system, the binding affinity of AS15 to 16S rRNA fragments

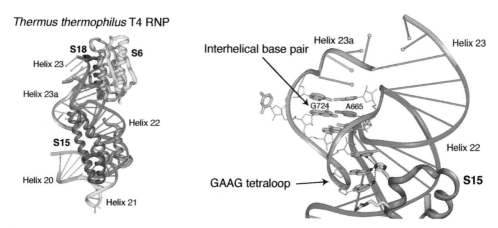

Figure 4. Structure of T4 RNP (Agalarov et al. 2000). An overall view of the structure from the solvent face of the 30S subunit is shown on the left. The interhelical base pair between A665 and G724 and the contacts between S15 and the GAAG tetraloop of helix 23a are shown on the right.

was measured and compared to other systems. Consistent with previous studies of S15 binding to rRNA in other organisms, an equilibrium dissociation constant of 6 nM was measured for binding of AS15 to the A4 RNA. In contrast to the significant increase in binding affinity of BS15 to Fr1 RNA in the presence of Mg^{++}, divalent cations had no effect on the affinity of AS15 binding to A4 RNA. It is possible that these differences in the effects of divalent cations are due to the absence of helix 21 in the A4 construct compared to the Fr1 RNA. The effects of Mg^{++} on the affinity and kinetics of BS15 binding to smaller RNA fragments has not been thoroughly investigated. In addition, the relative K_d for S15 binding in the *T. thermophilus* system in the presence of Mg^{++} was not reported, so the degree of conservation of Mg^{++} effects on the S15-binding-site structure is not clear.

THERMODYNAMICS OF S6 AND S18 BINDING

The binding of S6 and S18 to the S15–RNA complex is cooperative, and there are extensive contacts between the two proteins in the structure of the T4 RNP (and 30S subunit). In addition, S6 has been cross-linked to S18 in the 30S subunit using either methyl-4-mercaptobutyrimidate (Sommer and Traut 1976), various bisimidoesters (Expert-Bezancon et al. 1977), or 2-iminothiolane (Lambert et al. 1983). None of these results, however, describes the mechanism by which S6 and S18 bind to the central domain RNA. An interaction of two proteins once bound to an RNP or the 30S subunit does not prove that this interaction exists when the proteins are not bound to the RNP. Cooperative binding could be due either to sequential binding of the two proteins or to the binding of an S6:S18 heterodimer. There is no evidence in the literature of formation of an S6:S18 heterodimer in solution.

Using isothermal titration calorimetry, we have shown that a 1:1 complex between AS6 and AS18 forms in solution using the standard 30S reconstitution buffer conditions (Fig. 5). The K_d for the formation of the AS6:AS18 heterodimer is 8.5 nM.

The formation of an S6:S18 heterodimer is consistent with, but does not prove, a mechanism in which these proteins bind as a heterodimer. Titration of an equimolar mixture of S6 and S18 into a preformed S15–RNA complex yields an equilibrium dissociation constant for the binding of S6 and S18 to the S15–RNA complex of about 6 nM. This result, along with other biochemical and biophysical data, indicates that S6 and S18 bind as a heterodimer to the S15–RNA complex (Recht and Williamson 2001).

KINETICS OF S6:S18 BINDING

The association and dissociation rates of S6:S18 with the S15–RNA complex have been determined. As was observed for the BS15–rRNA interaction, the bimolecular association rate is slow, 8.0×10^4 $M^{-1}s^{-1}$. The unimolecular dissociation rate, 1.6×10^{-4} s^{-1}, is also slow, indicating formation of a stable complex (Recht and

Williamson 2001). The slow association rate may indicate that S6:S18 only binds to a transiently formed RNA tertiary structure, similar to the structure-capture mechanism proposed for S15-16S rRNA binding. Instead of a cation-stabilized RNA conformation observed in the binding of S15, the binding of S6:S18 requires a unique S15 stabilized RNA conformation.

MECHANISM OF CENTRAL DOMAIN ASSEMBLY

A possible mechanism for the assembly of the central domain has been developed using information from the 30S assembly map, chemical probing experiments, structural studies of central domain RNPs and the 30S subunit, and the thermodynamic and kinetic studies presented above (Fig. 3). The three-helix junction between helices 20, 21, and 22 adopts a particular conformation that is competent for S15 binding (R_1). Interactions between S15 and the GAAG tetraloop at the end of helix 23a apparently stabilize the active conformation of the upper three-helix junction between helices 22, 23, and 23a (R_2), as does an interhelical base pair between A665 in helix 22 and G724 in helix 23a (Fig. 4). This interhelical base pair is critical for binding of S6:S18, as deletion of A665 yields an RNA that can bind S15 but not the S6:S18 heterodimer (Agalarov et al. 2000). This conformation of the upper three-helix junction is recognized and bound by the S6:S18 heterodimer. It is likely to follow that the binding

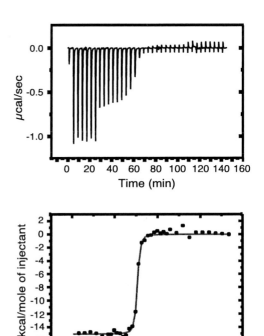

Figure 5. Formation of S6:S18 heterodimer as monitored by isothermal titration calorimetry. Aliquots of a solution of *A. aeolicus* S18 are injected to a solution of *A. aeolicus* S6.

of the S6:S18 heterodimer stabilizes the conformation of helix 23b to which S11 binds. The binding of S11 to the S6:S18:S15–rRNA complex could be stabilized further through direct protein–protein interactions between S18 and S11, which have been observed in the structure of the 30S subunit.

CONCLUSIONS

Now that assembly of the central domain from a variety of organisms has been studied, it appears the hierarchical binding of ribosomal proteins in the assembly map developed by the Nomura lab for reconstitution of the *E. coli* 30S subunit is essentially conserved. It also seems likely that the binding of S15 nucleates assembly within the central domain of 16S rRNA. Our kinetic studies demonstrate that binding of S15 and the S6:S18 heterodimer is slow. The Nomura assembly map was determined under equilibrium conditions, whereas the studies on the possible nucleation sites for ribosome assembly were under non-equilibrium conditions. It remains to be determined whether ribosome assembly in vivo is under kinetic or thermodynamic control.

ACKNOWLEDGMENTS

This paper reviews work performed by Robert T. Batey, Jeffrey W. Orr, Sultan C. Agalarov, Peter M. Funke, and Michael I. Recht. We thank Peter Funke for critical reading of the manuscript. This work was supported by a grant to J.R.W. from the National Institutes of Health (GM-53757).

REFERENCES

Agalarov S.C. and Williamson J.R. 2000. A hierarchy of RNA subdomains in assembly of the central domain of the 30 S ribosomal subunit. *RNA* **6**: 402.
Agalarov S.C., Sridhar Prasad G., Funke P.M., Stout C.D., and Williamson J.R. 2000. Structure of the S15,S6,S18-rRNA complex: Assembly of the 30S ribosome central domain. *Science* **288**: 107.
Agalarov S.C., Zheleznyakova E.N., Selivanova O.M., Zheleznaya L.A., Matvienko N.I., Vasiliev V.D., and Spirin A.S. 1998. In vitro assembly of a ribonucleoprotein particle corresponding to the platform domain of the 30S ribosomal subunit. *Proc. Natl. Acad. Sci.* **95**: 999.
Ban N., Nissen P., Hansen J., Moore P. B., and Steitz T. A. 2000. The complete atomic structure of the large ribosomal subunit at 2.4 Å resolution. *Science* **289**: 905.
Batey R.T. and Williamson J.R. 1998. Effects of polyvalent cations on the folding of an rRNA three-way junction and binding of ribosomal protein S15. *RNA* **4**: 984.
———. 1996a. Interaction of the *Bacillus stearothermophilus* ribosomal protein S15 with 16 S rRNA. I. Defining the minimal RNA site. *J. Mol. Biol.* **261**: 536.
———. 1996b. Interaction of the *Bacillus stearothermophilus* ribosomal protein S15 with 16 S rRNA: II. Specificity determinants of RNA-protein recognition. *J. Mol. Biol.* **261**: 550.
Datta D., Changchien L.M., and Craven G.R. 1986. Studies on the kinetic sequence of in vitro ribosome assembly using cibacron blue F3GA as a general assembly inhibitor. *Nucleic Acids Res.* **14**: 4095.
Dodd J., Kolb J.M., and Nomura M. 1991. Lack of complete cooperativity of ribosome assembly in vitro and its possible relevance to in vivo ribosome assembly and the regulation of ribosomal gene expression. *Biochimie* **73**: 757.
Expert-Bezancon A., Barritault D., Milet M., Guerin M.F., and Hayes D.H. 1977. Identification of neighbouring proteins in *Escherichia coli* 30 S ribosome subunits. *J. Mol. Biol.* **112**: 603.
Gregory R.J., Zeller M.L., Thurlow D.L., Gourse R.L., Stark M.J.R., Dahlberg A.E., and Zimmermann R.A. 1984. Interaction of ribosomal proteins S6, S8, S15 and S18 with the central domain of 16 S ribosomal RNA from *Escherichia coli*. *J. Mol. Biol.* **178**: 287.
Gutell R.R. 1994. Collection of small subunit (16S- and 16S-like) ribosomal RNA structures: 1994. *Nucleic Acids Res.* **22**: 3502.
Gutell R.R., Weiser B., Woese C.R., and Noller H.F. 1985. Comparative anatomy of 16-S-like ribosomal RNA. *Prog. Nucleic Acid Res. Mol. Biol.* **32**: 155.
Ha T., Zhuang X., Kim H.D., Orr J.W., Williamson J.R., and Chu S. 1999. Ligand-induced conformational changes observed in single RNA molecules. *Proc. Natl. Acad. Sci.* **96**: 9077.
Held W.A. and Nomura M. 1973. Rate determining step in the reconstitution of *Escherichia coli* 30S ribosomal subunits. *Biochemistry* **12**: 3273.
Held W.A., Mizushima S., and Nomura M. 1973. Reconstitution of *Escherichia coli* 30 S ribosomal subunits from purified molecular components. *J. Biol. Chem.* **248**: 5720.
Held W.A., Ballou B., Mizushima S., and Nomura M. 1974. Assembly mapping of 30 S ribosomal proteins from *Escherichia coli*. Further studies. *J. Biol. Chem.* **249**: 3103.
Lambert J.M., Boileau G., Cover J.A., and Traut R.R. 1983. Cross-links between ribosomal proteins of 30S subunits in 70S tight couples and in 30S subunits. *Biochemistry* **22**: 3913.
Mizushima S. and Nomura M. 1970. Assembly mapping of 30S ribosomal proteins from *E. coli*. *Nature* **226**: 1214.
Nikulin A., Serganov A., Ennifar E., Tishchenko S., Nevskaya N., Shepard W., Portier C., Garber M., Ehresmann B., Ehresmann C., Nikonov S., and Dumas P. 2000. Crystal structure of the S15-rRNA complex. *Nat. Struct. Biol.* **7**: 273.
Noller H.F. and Woese C.R. 1981. Secondary structure of 16S ribosomal RNA. *Science* **212**: 403.
Nomura M., Traub P., Guthrie C., and Nashimoto H. 1969. The assembly of ribosomes. *J. Cell Physiol* (suppl. 1) **74**: 241.
Nomura M., Mizushima S., Ozaki M., Traub P., and Lowry C.V. 1970. Structure and function of ribosomes and their molecular components. *Cold Spring Harbor Symp. Quant. Biol.* **34**: 49.
Nowotny V. and Nierhaus K.H. 1988. Assembly of the 30S subunit from *Escherichia coli* ribosomes occurs via two assembly domains which are initiated by S4 and S7. *Biochemistry* **27**: 7051.
Orr J.W., Hagerman P.J., and Williamson J.R. 1998. Protein and Mg(2+)-induced conformational changes in the S15 binding site of 16 S ribosomal RNA. *J. Mol. Biol.* **275**: 453.
Powers T. and Noller H.F. 1995. Hydroxyl radical footprinting of ribosomal proteins on 16S rRNA. *RNA* **1**: 194.
Powers T., Daubresse G., and Noller H.F. 1993. Dynamics of in vitro assembly of 16S rRNA into 30S ribosomal subunits. *J. Mol. Biol.* **232**: 362.
Recht M.I. and Williamson J.R. 2001. Central domain assembly: Thermodynamics and kinetics of S6 and S18 binding to an S15-RNA complex. *J. Mol. Biol.* **313**: 35.
Samaha R.R., O'Brien B., O'Brien T.W., and Noller H.F. 1994. Independent in vitro assembly of a ribonucleoprotein particle containing the 3′ domain of 16S rRNA. *Proc. Natl. Acad. Sci.* **91**: 7884.
Schluenzen F., Tocilj A., Zarivach R., Harms J., Gluehmann M., Janell D., Bashan A., Bartels H., Agmon I., Franceschi F., and Yonath A. 2000. Structure of functionally activated small ribosomal subunit at 3.3 angstroms resolution. *Cell* **102**: 615.
Serganov A.A., Masquida B., Westhof E., Cachia C., Portier C., Garber M., Ehresmann B., and Ehresmann C. 1996. The 16S rRNA binding site of *Thermus thermophilus* ribosomal protein S15: Comparison with *Escherichia coli* S15, minimum site and structure. *RNA* **2**: 1124.
Sommer A. and Traut R.R. 1976. Identification of neighboring

protein pairs in the *Escherichia coli* 30 S ribosomal subunit by crosslinking with methyl-4-mercaptobutyrimidate. *J. Mol. Biol.* **106:** 995.

Stern S., Powers T., Changchien L.-M., and Noller H.F. 1989. RNA-protein interactions in 30S ribosomal subunits: Folding and function of 16S rRNA. *Science* **244:** 783.

Svensson P., Changchien L.-M., Craven G.R., and Noller H.F. 1988. Interaction of ribosomal proteins, S6, S8, S15 and S18 with the central domain of 16 S ribosomal RNA. *J. Mol. Biol.* **200:** 301.

Tal M., Silberstein A., and Møyner K. 1977. In vivo reassembly of 30-S ribosomal subunits following their specific destruction by thermal shock. *Biochim. Biophys. Acta.* **479:** 479.

Traub P. and Nomura M. 1968. Structure and function of *E. coli* ribosomes. V. Reconstitution of functionally active 30S ribosomal particles from RNA and proteins. *Proc. Natl. Acad. Sci.* **59:** 777.

―――. 1969. Structure and function of *Escherichia coli* ribosomes. VI. Mechanism of assembly of 30 s ribosomes studied in vitro. *J. Mol. Biol.* **40:** 391.

―――. 1970. Studies on the assembly of ribosomes in vitro. *Cold Spring Harbor Symp. Quant. Biol.* **34:** 63.

Weitzmann C.J., Cunningham P.R., Nurse K., and Ofengand J. 1993. Chemical evidence for domain assembly of the *Escherichia coli* 30S ribosome. *FASEB J.* **7:** 177.

Wimberly B.T., Brodersen D.E., Clemons W.M., Jr., Morgan-Warren R.J., Carter A.P., Vonrhein C., Hartsch T., and Ramakrishnan V. 2000. Structure of the 30S ribosomal subunit. *Nature* **407:** 327.

Zimmermann R.A., Mackie G.A., Muto A., Garrett R.A., Ungewickell E., Ehresmann C., Stiegler P., Ebel J.P., and Fellner P. 1975. Location and characteristics of ribosomal protein binding sites in the 16S RNA of *Escherichia coli. Nucleic Acids Res.* **2:** 279.

Nuclear Export of the Large Ribosomal Subunit

A.W. Johnson,* J.H.-N. Ho,* G. Kallstrom,* C. Trotta,† E. Lund,†
L. Kahan,† J. Dahlberg,† and J. Hedges*

*Section of Molecular Genetics and Microbiology and Institute for Cellular and †Molecular Biology,
University of Texas at Austin, Austin, Texas 78712; †Department of Biomolecular Chemistry,
University of Wisconsin-Madison, Madison, Wisconsin 53706

The nuclear envelope of eukaryotic cells separates the site of ribosome synthesis, in the nucleus, from the site of function, in the cytoplasm. Ribosome biogenesis requires the transport of cargo both into and out of the nucleus through the nuclear pore complex (NPC). Ribosomal proteins, synthesized in the cytoplasm, are imported into the nucleus where their assembly into ribosomal subunits in the nucleolus is coordinated with transcription and extensive processing of the ribosomal RNAs (rRNAs) (Kressler et al. 1999; Venema and Tollervey 1999). The assembled subunits then are exported for translation in the cytoplasm (Fig. 1). The export of ribosomal subunits is a major activity in rapidly growing cells. For example, in the budding yeast *Saccharomyces cerevisiae*, ribosomes are produced and exported at a rate of approximately 2,000 per minute, requiring the nuclear import of over 150,000 proteins per minute (Warner 1999). Thus, the majority of transport across the nuclear envelope in yeast may well be devoted to ribosome biogenesis.

Most nuclear export events depend on the small GTPase Ran and specific transport receptors of the importin β family that recognize nuclear export signals (NESs) on the cargo molecules (for reviews, see Rout et al. 1997; Dahlberg and Lund 1998; Mattaj and Englmeier 1998; Pemberton et al. 1998; Görlich and Kutay 1999; Nakielny and Dreyfuss 1999). Importin β-like export receptors require RanGTP (present in the nucleus) for tight binding to their cargos. The export complex is then translocated through the NPC by a mechanism that appears to involve sequential binding to nucleoporin receptor-binding sites with a gradient of binding affinities (Rout et al. 2000). Upon delivery to the cytoplasm, the intrinsic GTPase activity of Ran is activated by a cytoplasmic GTPase activating protein, releasing the export receptor and cargo.

NESs can be integral protein components of the cargo molecules or they can be provided in *trans* by adapter proteins that bridge the interaction between cargo and receptors (Mattaj and Englmeier 1998). The classic example of an adapter protein is HIV-1 Rev protein, which provides the NES for export of unspliced viral mRNAs. The best-characterized class of NES is the leucine-rich NES, initially identified in HIV-1 Rev and c-AMP-dependent protein kinase inhibitor (PKI) (Fischer et al. 1995; Wen et al. 1995). Subsequently, numerous other shuttling proteins have been shown to utilize a leucine-

rich NES. The importin β-like receptor for this class of NES is CRM1 (Crm1p in yeast) (Fornerod et al. 1997; Fukuda et al. 1997; Ossareh-Nazari et al. 1997). In most eukaryotic cells, the interaction of CRM1 with its NES ligand is inhibited by the drug leptomycin B (LMB), which interacts with a highly conserved cysteine residue (Kudo et al. 1999). Wild-type *S. cerevisiae* is resistant to LMB, but a single point mutation changing threonine at position 539 to cysteine renders yeast sensitive to the drug (Neville and Rosbash 1999).

Although nucleocytoplasmic transport has been a field of great interest and activity, until very recently, little was known about nuclear export of ribosomal subunits. Microinjection experiments in *Xenopus* oocytes showed that subunit export is energy-dependent and receptor-mediated (Bataille et al. 1990). In the yeast *S. cerevisiae*, mutations in Gsp1p (Ran) or in regulators of Ran inhibit ribosomal subunit export (Hutchison et al. 1969; Hurt et al. 1999; Moy and Silver 1999). Mutations in several nucleoporins, the structural proteins of the nuclear pore complex, as well as overexpression of the tRNA export factor Los1p, also affect ribosomal subunit export (Sarkar and Hopper 1998; Hurt et al. 1999; Moy and Silver 1999; Stage-Zimmermann et al. 2000). A conditional mutation in Crm1p, the receptor for NES-containing proteins, inhibits small (40S) subunit export (Moy and Silver 1999),

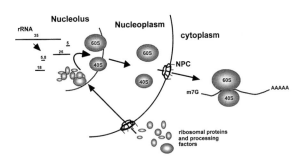

Figure 1. Overview of ribosome assembly in eukaryotic cells. In eukaryotic cells, ribosomal proteins and all *trans*-acting processing proteins are imported from their site of translation in the cytoplasm to the nucleus. rRNA is transcribed, processed, and assembled with ribosomal proteins into nascent subunits in the nucleolus. The subunits are then exported through the nuclear pore complexes (NPC) to the cytoplasm by the receptor protein CRM1. The nuclear export signal for the 60S subunits is provided by NMD3.

but its effect on large (60S) subunit export is less clear (Hurt et al. 1999; Stage-Zimmermann et al. 2000). In addition to the uncertainty of the receptors for ribosomal subunit export, the export signals for the ribosomal subunits had not been identified. Furthermore, it was not known whether the export signals of the subunits are provided in *cis* by proteins intrinsic to the subunits or in *trans* by adapter proteins that transiently associate with the subunits during nuclear export. Recent work from our laboratories has shown that the export signal for the 60S subunit is provided by the adapter protein NMD3 (Nmd3p in yeast). NMD3 contains a leucine-rich NES and requires CRM1 for nuclear export of the NMD3–60S subunit complex. This export pathway for the 60S subunit appears to be conserved throughout eukaryotes.

60S SUBUNIT EXPORT IN YEAST

We had been studying the role of the essential protein Nmd3p. Nmd3p is highly conserved, and orthologs are found in both Archaea and Eukarya. It is required for maturation of the 60S subunits and binds to free 60S subunits. Temperature-sensitive *nmd3* mutants fail to accumulate 60S subunits at nonpermissive temperatures, even though rRNA processing is not significantly perturbed (Ho and Johnson 1999). In these cells, mature subunits made before shift to nonpermissive temperature are stable. Since Nmd3p is predominantly cytoplasmic, we interpreted these results to indicate that Nmd3p acted late in the biogenesis pathway, after assembly of the subunit in the nucleolus. In support of this conclusion, pulse-chase labeling has shown that nascent Nmd3p-bound 60S subunits appear to have a full protein complement (Ho et al. 2000a; G. Kallstrom and A.W. Johnson, unpubl.).

The amino-terminal 40 kD of Nmd3p is conserved in all archaeal and eukaryotic Nmd3p-like proteins (Fig. 2a). This region contains four Cys-X-X-Cys motifs similar to zinc ribbons found in some ribosomal proteins. Analysis of deletion mutants suggests that this region of the protein is primarily responsible for binding to the large ribosomal subunit (J. Hedges and A.W. Johnson, unpubl.). Amino acids 279–336 define a site of genetic interaction with the ribosomal protein Rpl10p (Karl et al. 1999) and may be responsible for direct binding to Rpl10p (Gadal et al. 2001). The eukaryotic members of this protein family differ from their archaeal orthologs by the presence of an additional carboxy-terminal domain (Fig. 2a,b).

Although we initially reported that Nmd3p was cytoplasmic, this is the steady-state distribution of the protein that actively shuttles between the nucleus and cytoplasm. The sequences that direct shuttling are located in the carboxy-terminal domain which is unique to the eukaryotic proteins (Fig. 2b). The highly basic region (amino acids 387–435 in yeast Nmd3p) contains a nuclear localization signal (NLS) that can direct a GFP-fusion protein into the nucleus. The extreme carboxy-terminal 50 amino acids harbor a strong leucine-rich NES (Ho et al. 2000b). Deletion mapping has localized the NES to the last 33 amino acids, and we have shown that amino acids 485–505 (DEDAPQINIDELLDELDEMTL) are sufficient to direct

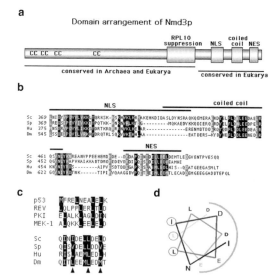

Figure 2. Architecture of Nmd3p. (*a*) A large amino-terminal domain, conserved throughout Archaea and Eukarya, contains four C-x-x-C motifs and a region that is responsible for genetic interaction with RPL10. The carboxyl terminus of the eukaryotic proteins contains the shuttling sequences. (*b*) Alignment of the carboxyl terminus of Nmd3p and related proteins (Sc, *S. cerevisiae*; Sp, *Schizosaccharomyces pombe*; Hu, human; Dm, *Drosophila melanogaster*). The extent of the NLS and NES that is sufficient for function when fused to a reporter protein is indicated by bars. The predicted coiled-coil domain is also indicated. (*c*) An alignment of the hydrophobic core of the NES of yeast Nmd3p (amino acids 490–501) is compared with other NESs from other proteins. (*d*) Projection of an amphipathic helix containing the three conserved residues (*circled*) that form a hydrophobic face critical for function of the NES. The acidic face is indicated by a gray semicircle.

export of a nuclear localized reporter protein (J. Hedges and A.W. Johnson, unpubl.). Mutation of three hydrophobic residues (underlined in the sequence) abolishes NES function and yields a phenotype similar to Nmd3Δ50. These residues are predicted to lie in the hydrophobic face of an α helix spanning the NES (Fig. 2d). Finally, amino acids 426–465 are predicted to form a coiled coil, a structure that could mediate intra- or intermolecular interactions. This region has been proposed to contain a second NES (Gadal et al. 2001). Alternatively, it may regulate the activity of the primary NES or the NLS through an intramolecular interaction or by dimerization.

We constructed a series of deletions within the carboxyl terminus of Nmd3p that lacked the NES (Nmd3Δ50), the coiled-coil domain (Nmd3ΔCC), the NES plus the coiled-coil domain (Nmd3Δ100), the NLS (Nmd3ΔNLS), and the entire carboxyl terminus containing all three domains (Nmd3Δ120). All mutants were carboxy-terminally tagged with 13 tandem copies of the c-myc epitope (Longtine et al. 1998). Mutant proteins lacking the NES but retaining the NLS (Nmd3Δ50 and Nmd3Δ100) accumulated in the nucleus (Fig. 3b) (Ho et al. 2000b) and inhibited growth when expressed at low levels in *NMD3* wild-type cells. This was consistent with an earlier observation that overexpression of Nmd3p

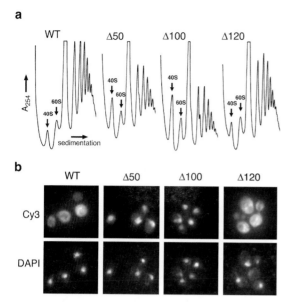

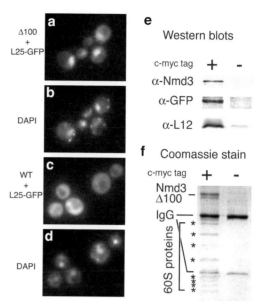

Figure 3. Deletion of the NES of Nmd3p leads to nuclear accumulation of the protein and inhibition of 60S subunit biogenesis. (*a*) Extracts were prepared from wild-type cells expressing the indicated carboxy-terminally truncated proteins and fractionated by ultracentrifugation in 7% to 47% sucrose gradients. The positions of free 40S and 60S subunits are indicated. The relative peak heights of free subunits are a sensitive indicator of total subunit levels and, hence, biogenesis defects. (*b*) Indirect immunofluorescence of the c-myc-tagged mutant proteins in yeast. (Cy3) Rhodamine channel for c-myc-tagged proteins; (DAPI) UV channel indicating the position of nuclei. (Reprinted, with permission, from Ho et al. 2000b [copyright Rockefeller University Press].)

Figure 4. Nuclear-trapped Nmd3p sequesters 60S subunits in the nucleus. L25-GFP and Nmd3Δ100 (*a,b*) or full-length Nmd3p (*c,d*) was coexpressed from plasmids in wild-type cells. DAPI was added and the localization of L25-GFP (*a, c*) or DAPI (*b, d*) was visualized by fluorescence microscopy. (*e*) Extracts were prepared from cells coexpressing L25-GFP and Nmd3Δ100 either with (+) or without (–) 13-myc tag. Proteins were immunoprecipitated with anti-c-myc antibodies and analyzed by SDS-PAGE and western blotting using antibodies to c-myc (Nmd3Δ100), GFP, or wild-type 60S subunit protein L12. (*f*) Gel-separated proteins from *e* were stained with Coomassie Blue. Nmd3Δ100, immunoglobulin heavy and light chains (IgG), and ribosomal proteins (*asterisks*) are indicated. (Reprinted, with permission, from Ho et al. 2000b [copyright Rockefeller University Press].)

lacking the carboxy-terminal 100 amino acids inhibited growth (Belk et al. 1999). Nmd3Δ120, which lacked the NLS, was retained in the cytoplasm. Deletion of the NLS or coiled-coil domain alone was not lethal but gave rise to a slow-growth phenotype when these mutant proteins replaced the endogenous protein (not shown). Since Nmd3p is essential, the viability of the Nmd3ΔNLS mutant suggests that another region of the protein contributes to nuclear localization as well. This may be through direct interaction with a nuclear import receptor or indirectly through a ribosomal protein which in turn contains an NLS.

The dominant negative phenotype of the NES-deficient proteins correlated with nuclear accumulation and inhibition of 60S subunit biogenesis (Fig. 3). Since Nmd3p binds directly to 60S subunits (Ho et al. 2000a) and is required for a late step in their synthesis (Ho and Johnson 1999), we surmised that the dominant negative proteins were binding to the nascent 60S subunits in the nucleus and preventing their export because of the absence of an NES. To follow directly the fate of 60S ribosomal subunits, we monitored the localization of a functional GFP-tagged 60S subunit protein, Rpl25p (Hurt et al. 1999). Coexpression of L25-GFP with Nmd3Δ100 resulted in nuclear retention of the L25-GFP (Fig. 4) (Ho et al. 2000b). Furthermore, the GFP-tagged Rpl25p was in subunits bound by Nmd3p, as these subunits could be coim-

munoprecipitated with myc-tagged Nmd3Δ100 (Fig. 4c, d). In these experiments, the mutant Nmd3 proteins were expressed at levels similar to endogenous Nmd3p. Nevertheless, they displayed a strong dominant negative phenotype. We believe this is due to the accumulation of high levels of the mutant protein in the nucleus compared to wild-type Nmd3p, which is in the nucleus only transiently and thus present at low levels. Although nascent subunits containing L25-GFP accumulated in the nucleus in the presence of Nmd3Δ100, pulse-chase labeling of rRNA under these conditions revealed a substantial amount of rRNA turnover (J. Hedges and A.W. Johnson, unpubl.). Thus, it appears that only a fraction of the nascent 60S subunits that are produced is stable in the nucleus. It is likely that this is the population that is bound by Nmd3p.

Our work suggested that export of 60S subunits depends on the NES of Nmd3p (Ho et al. 2000b). To test this idea, we fused a heterologous NES to the amino terminus of Nmd3Δ100. We chose the well-characterized NES from PKI which previously had been demonstrated to function in yeast (Stade et al. 1997). As a negative control, we used the same NES bearing a point mutation that blocks function. The addition of wild-type but not mutant

PKI NES to Nmd3Δ100 restored complementation of an *nmd3* deletion mutant (Ho et al. 2000b). It also restored cytoplasmic export of Nmd3p as well as 60S subunit production. From this we conclude that Nmd3p is the primary adapter that provides the export signal for the 60S subunit.

In this heterologous NES experiment, Nmd3p and 60S subunit export was made dependent on Crm1p, the importin β-like receptor for the leucine-rich PKI NES. This showed that the Crm1p-dependent pathway was capable of exporting the large subunit. Although there were conflicting data on the possible involvement of Crm1p in the export of the 60S subunit in wild-type yeast cells (Hurt et al. 1999; Stage-Zimmermann et al. 2000), the NES of Nmd3p (amino acids 485–505) is predicted to form a short leucine-rich amphipathic α helix, typical of CRM1 ligands (Fig. 2c,d). Consequently, we reexamined the potential role of Crm1p in the export of 60S subunits and Nmd3p. The amino acids of Nmd3p responsible for its shuttling were fused to GFP (making GFP-NLS-NES) and the localization of the fusion protein was monitored by fluorescence microscopy. Indeed, in the presence of the CRM1 inhibitor LMB, GFP-NLS-NES and full-length Nmd3p accumulated in nuclei of LMB-sensitive but not -resistant cells (Fig. 5 and data not shown). Importantly, L25-GFP accumulated in the nucleus upon treatment with LMB (Fig. 5), indicating that export of 60S subunits was blocked. Similar results have recently been reported independently (Gadal et al. 2001). Thus, Crm1p is an essential transport receptor in the export of 60S subunits in yeast.

60S SUBUNIT EXPORT IN METAZOANS

Since Nmd3p is conserved in eukaryotes, we were interested in the possible conservation of the 60S subunit transport pathway in vertebrate cells. The human ortholog of Nmd3p (hNMD3) (Genbank accession number AF132941) was cloned from a HeLa cell cDNA library and found to complement an *nmd3* null mutant of yeast. The open reading frame was fused 3′ to GFP, and plasmid DNAs encoding GFP-hNMD3 fusion proteins were transiently transfected into HeLa cells. Fluorescence microscopy showed that the protein was distributed throughout the nucleoplasm and cytoplasm (Fig. 6a), but it was retained primarily in the nucleus upon treatment with LMB (Fig. 6b) (C. Trotta et al., in prep.). The arrangement of the sequences of the shuttling signals in the yeast and human proteins is conserved (Fig. 2). A triple leucine-to-alanine mutation in the region corresponding to the NES of the yeast protein resulted in localization in both the nucleoplasm and the nucleolus (Fig. 6c). The nucleolar localization was more apparent when cells were treated with LMB (Fig. 6d) (C. Trotta et al., in prep.).

Deletion of the cluster of basic amino acids (405–417) within the NLS region led to cytoplasmic localization of GFP-hNMD3 (Fig. 7a). However, upon treatment with LMB (Fig. 7b), or when combined with an NES mutant (not shown), the NLS mutant proteins accumulated in the nucleoplasm but not in the nucleolus. These results suggest that the cluster of basic amino acids directs nucleoplasmic localization and that hNMD3, like the yeast protein, shuttles between the nucleus and cytoplasm.

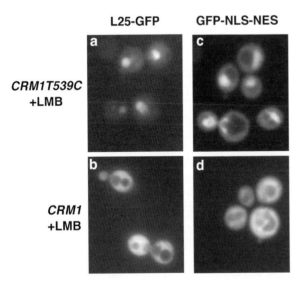

Figure 5. Nmd3p and 60S subunit export depends on Crm1p. (*a,b*) Localization of L25-GFP expressed in (*a*) LMB-sensitive (*CRM1T539C*) or (*b*) wild-type (*CRM1*) yeast cells. (*c,d*) Localization of GFP containing the NLS and NES of Nmd3p (GFP-NLS-NES). Upon treatment with LMB, L25-GFP and GFP-NLS-NES were retained in the nucleus in the sensitive strain (*c*). Because LMB is specific for CRM1-ligand interaction, these results indicate that Nmd3p and 60S subunit export is mediated by Crm1p. (Reprinted, with permission, from Ho et al. 2000b [copyright Rockefeller University Press].)

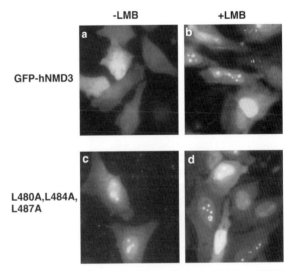

Figure 6. hNMD3 export in HeLa cells is dependent on CRM. (*a,b*) HeLa cells were transiently transfected with plasmid encoding GFP-hNMD3. (*a*) Distribution of GFP-hNMD3 with a wild-type NES, in the absence of LMB; fluorescence was distributed throughout the nucleoplasm and cytoplasm (–LMB). (*b*) Distribution of the same protein, but in cells treated with LMB, to block export (+LMB); GFP fluorescence accumulated in the nucleoplasm and nucleolus. (*c,d*) Distribution of GFP-hNMD3 in which three conserved leucines in the NES (L480A, L484A, L487A) were mutated to alanines; HeLa cells were transfected in the absence (*c*) or presence (*d*) of LMB.

Δ405-417

-LMB   +LMB

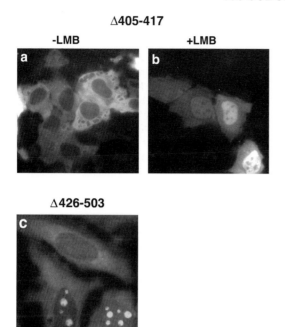

Δ426-503

Figure 7. Nucleolar and nuclear localization of hNMD3. (*a,b*) Distribution of GFP-hNMD3 lacking the basic NLS (amino acids 405–417) expressed in transfected HeLa cells. (*a*) Distribution in the absence of LMB. (*b*) Distribution in the presence of LMB. Treatment of these HeLa cells with LMB resulted in re-localization of the mutant protein to the nucleoplasm with exclusion from the nucleolus; thus, the basic cluster is important for localization in both the nucleoplasm and nucleolus. (*c*) Distribution of GFP-hNMD3 lacking the carboxy-terminal 78 amino acids (Δ426–503). Deletion of the NES plus coiled-coil domain from this protein (Δ426–503) led to accumulation in the nucleolus in both the presence and absence of LMB (data not shown), suggesting that in addition to NES activity, amino acids 426–503 contain a signal for exit from the nucleolus.

export of the large subunit from oocyte nuclei depends on CRM1 and RanGTP.

To determine whether 60S subunit export in oocytes is mediated by NMD3, in-vitro-transcribed mRNAs encoding wild-type or mutant hNMD3 proteins were injected into prelabeled oocytes. Expression of wild-type hNMD3 had little effect on 28S rRNA export, but expression of hNMD3 lacking its NES or containing point mutations within the NES significantly reduced the amounts of 28S rRNA exported to the cytoplasm and increased the amount of 28S rRNA that accumulated in the nucleus (C. Trotta et al., in prep.). These results mirrored the dominant negative phenotype of the corresponding mutations in yeast. Although this work relied on the use of heterologous dominant negative mutant proteins, the results are entirely consistent with those from yeast, suggesting that export of 60S subunits in eukaryotic cells depends on a highly conserved pathway that uses NMD3 as an adapter protein and CRM1 as the Ran-dependent receptor (see Fig. 8).

FUTURE DIRECTIONS

Can It Be So Simple?

The work described here sketches out the rough outline of nuclear export of the 60S subunit. Although NMD3 appears to be the principal contributor of the nuclear export signal, it is likely that additional factors will also be important in regulating this pathway. Furthermore, the export pathway will likely be composed of multiple discrete steps. As noted for the localization of mutant hNMD3 in HeLa cells, the partitioning of NMD3 between the nucle-

However, the accumulation of the deletion mutant protein lacking the basic region indicates that another sequence can serve as a signal for nuclear import. Interestingly, hNMD3Δ78, lacking the NES and the coiled-coil domain, localized almost exclusively to the nucleolus, with little or no protein being evident in the nucleoplasm (Fig. 7c). Thus, the carboxy-terminal 78 amino acids appear to contain both a nuclear export sequence and a novel nucleolar exit sequence and appear to control partitioning of the protein between the nucleoplasm and the nucleolus.

We used *Xenopus laevis* oocytes to test the role of CRM1 in 60S subunit export, since it was likely that hNMD3 would function in *Xenopus*, given the high degree of sequence conservation of NMD3 among eukaryotes and the ability of hNMD3 to complement a yeast deletion mutant. When a dominant negative mutant form of Ran (T24N) was injected into oocyte nuclei, to deplete them of RanGTP, the appearance of newly made rRNAs in the cytoplasm was dramatically reduced, indicating that Ran is required for export of 60S subunits (C. Trotta et al., in prep.). Similarly, saturation of the CRM1-dependent export pathway by injection of NES peptide conjugates blocked the accumulation of labeled 28S rRNA in the cytoplasm (C. Trotta et al., in prep.). Thus, as in yeast,

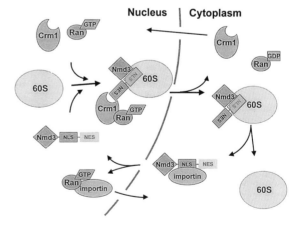

Figure 8. Model for nuclear export of 60S subunits and shuttling of NMD3. NMD3 contains both an NLS and an NES. In the nucleus, it binds 60S subunits to provide the export signal that is recognized by CRM1 in the presence of RanGTP. Once in the cytoplasm, GTP hydrolysis releases CRM1 and RanGDP. NMD3 is released from the newly exported 60S subunit in a separate step, but the factors controlling its release are not known. After release, NMD3 re-enters the nucleus, allowing it to bring out another 60S subunit. We suggest that when bound to the 60S subunit, the NLS of NMD3 is masked (indicated by lighter shading) to prevent retrograde transport. Similarly, efficient recognition of the NES may depend on binding to a 60S subunit.

olus and the nucleoplasm is separate from its localization between the nucleus and the cytoplasm. The nucleolar localization of the "dominant negative" hNMD3 mutant suggests that the protein binds nascent subunits in the nucleolus. Considering that the nascent subunit must be pre-assembled in the nucleolus, we envision that NMD3 must load at a very late step in assembly to prevent precocious export of incompletely assembled particles. Thus, the control of NMD3 loading is of particular interest. Our preliminary results suggest that the binding of Nmd3p to the 60S subunit is mutivalent, leading to the suggestion that Nmd3p binding requires the assembly of multiple proteins to signal the completion of the subunit. Of particular interest are the genetic and physical interactions of Nmd3p with Rpl10p (of the L10e family of ribosomal proteins) (Karl et al. 1999; Gadal et al. 2001). Rpl10p is required for subunit joining (Eisinger et al. 1997) and binds the 60S subunit late in its maturation (Kruiswijk et al. 1978), making Rpl10p an attractive candidate for part of the binding site for Nmd3p. However, it remains to be determined whether Nmd3p interacts with Rpl10p on the ribosome; preliminary results indicate that Rpl10p is not required for Nmd3p binding to the subunit in vitro (G. Kallstrom and A.W. Johnson, unpubl.).

As the subunit matures from a 66S precursor in the nucleolus to a mature 60S subunit, many assembly factors must be released from the pre-complex. Several additional factors have recently been reported to be required for nucleolar exit, although their functions are not yet known (Milkereit et al. 2001). For example, proteins required for the release of assembly factors would be likely to affect the exit of nascent subunits from the nucleolus. The high affinity of Nmd3p for 60S subunits and the nuclear retention of nascent 60S subunit by NES-deficient Nmd3p should provide a powerful technique for examining the protein composition of the nascent subunit during export. We have found that free 60S subunits that coimmunoprecipitated with Nmd3p contain several additional proteins including two putative GTP-binding proteins Nog1p and Kre35p (G. Kallstrom and A.W. Johnson, unpubl.). Whereas Nog1p is nucleolar and nucleoplasmic, Kre35p is cytoplasmic. Both of these proteins are required for 60S subunit maturation. Thus, GTP-binding proteins may play a role in regulating the assembly or disassembly of proteins on the 60S subunit at discrete steps in transport.

Transport of the NMD3–60S subunit complex through the NPC requires, at a minimum, CRM1 and Ran-GTP. The order of protein loading and the site of assembly of the export complex are not yet known. Because exogenous NMD3 can be exported free of ribosomes, its NES must be accessible even when the protein is not bound to ribosomes. However, the efficiency in NMD3 export may be increased upon binding to 60S subunits. It is unclear whether association with CRM1 occurs in the nucleolus or nucleoplasm. Translocation through the NPC is not well understood, but considering that the diameter of a yeast ribosome is similar to the channel diameter of the NPC, both approximately 25–30 nm (Verschoor et al. 1998; Yang et al. 1998), the large subunit may represent the upper size limit of cargo that can pass through the NPC.

Export complexes rapidly disassemble upon presentation to the cytoplasm when the intrinsic GTPase activity of Ran is activated by a cytoplasmic GTPase activating protein (for review, see Koepp and Silver 1996). However, the release of Nmd3p from the 60S subunit is distinct from the release of Crm1p and RanGDP. Evidence from our lab suggests that Nmd3p remains associated with free 60S subunits in the cytoplasm (Ho et al. 2000a). However, since Nmd3p is not abundant in polysomes, it must be released before translation initiation. Upon release, Nmd3p would be free either to re-enter the nucleus for another round of 60S subunit recruitment or to rebind a mature free 60S subunit in the cytoplasm. The binding of Nmd3p to recycling subunits in the cytoplasm could reduce the pool of free Nmd3p able to enter the nucleus, thereby providing a means of regulating the rate of nuclear export of 60S subunits.

But Archaea Lack Nuclei

As noted earlier, archaea contain an NMD3 ortholog. This suggests that NMD3 has a function that predates the evolution of the nucleus. Eukaryotic cells appear to have adapted this protein for nucleocytoplasmic transport by the simple addition of shuttling sequences. However, the ortholog from *Methanococcus janischii* does not complement a yeast *nmd3* null mutant, even when provided with the shuttling sequences of Nmd3p (A.W. Johnson, unpubl.). Thus, NMD3 is likely to play another, more ancient, role in the maturation or maintenance of 60 subunits. This function might be in the maturation and assembly of 60S subunits or in their recruitment during translation initiation. Also, it may monitor preexisting subunits for their structural integrity. Several conditional mutations in yeast *NMD3* have severe defects in 60S subunit production that reduce the levels of 60S subunits but do not affect translation per se, lending support to the idea that Nmd3p has a role in biogenesis that is distinct from transport.

40S Export

The work reviewed here focuses on export of 60S ribosomal subunits. The export pathway of the 40S subunit remains to be delineated. Although the two subunits appear to be exported independently of each other, export of both depends on CRM1 and RanGTP in *Xenopus* oocytes (C. Trotta et al., in prep.) and yeast (Moy and Silver 1999). However, the protein that provides the NES for 40S subunit export has not been identified, and it is unknown whether such a signal is provided by an adapter protein like NMD3 or by an integral ribosomal protein.

CONCLUSION

The evolution of the eukaryotic nucleus allows separation of RNA processing and translation, but this barrier necessitated the evolution of transport mechanisms to selectively move macromolecules across the nuclear envelope. Although this has been a field of active research in the last decade, it is only within the last year that the factors responsible for transporting a ribosomal subunit have been identified. Now, the identification of proteins that bind to the ribosomal subunits at discrete steps in biogenesis and transport, the ability to affinity-purify these complexes, and recent advances in proteomics promise explosive growth in our understanding of the dynamics of ribosomal subunit maturation and export.

ACKNOWLEDGMENTS

This work was supported by National Institutes of Health grants GM-53655 to A.W.J. and GM-30220 to J.E.D. and a Damon Runyon-Walter Winchell Fellowship to C.T.

REFERENCES

Bataille N., Helser T., and Fried H.M. 1990. Cytoplasmic transport of ribosomal subunits microinjected into the *Xenopus laevis* oocyte nucleus: A generalized, facilitated process. *J. Cell Biol.* **111:** 1571.

Belk J.P., He F., and Jacobson A. 1999. Overexpression of truncated Nmd3p inhibits protein synthesis in yeast. *RNA* **5:** 1055.

Dahlberg J.E. and Lund E. 1998. Functions of the GTPase Ran in RNA export from the nucleus. *Curr. Opin. Cell Biol.* **10:** 400.

Eisinger D.P., Dick F.A., and Trumpower B.L. 1997. Qsr1p, a 60S ribosomal subunit protein, is required for joining of 40S and 60S subunits. *Mol. Cell. Biol.* **17:** 5136.

Fischer U., Huber J., Boelens W.C., Mattaj I.W., and Luhrmann R. 1995. The HIV-1 Rev activation domain is a nuclear export signal that accesses an export pathway used by specific cellular RNAs. *Cell* **82:** 475.

Fornerod M., Ohno M., Yoshida M., and Mattaj I.W. 1997. CRM1 is an export receptor for leucine-rich nuclear export signals. *Cell* **90:** 1051.

Fukuda M., Asano S., Nakamura T., Adachi M., Yoshida M., Yanagida M., and Nishida E. 1997. CRM1 is responsible for intracellular transport mediated by the nuclear export signal. *Nature* **390:** 308.

Gadal O., Strauss D., Kessl J., Trumpower B., Tollervey D., and Hurt E. 2001. Nuclear export of 60S ribosomal subunits depends on Xpo1p and requires a nuclear export sequence-containing factor, Nmd3p that associates with the large subunit protein Rpl10p. *Mol. Cell. Biol.* **21:** 3405.

Görlich D. and Kutay U. 1999. Transport between the cell nucleus and the cytoplasm. *Annu. Rev. Cell Dev. Biol.* **15:** 607.

Ho J.H.-N. and Johnson A.W. 1999. *NMD3* encodes an essential cytoplasmic protein required for stable 60S ribosomal subunits in *Saccharomyces cerevisiae*. *Mol. Cell. Biol.* **19:** 2389.

Ho J.H.-N., Kallstrom G., and Johnson A.W. 2000a. Nascent 60S subunits enter the free pool bound by Nmd3p. 2000 *RNA* **11:** 1625.

———. 2000b. Nmd3p is a Crm1p-dependent adapter protein for nuclear export of the large ribosomal subunit. *J. Cell Biol.* **151:** 1057.

Hurt E., Hannus S., Schmelzl B., Lau D., Tollervey D., and Simos G. 1999. A novel in vivo assay reveals inhibition of ribosomal nuclear export in Ran-cycle and nucleoporin mutants. *J. Cell Biol.* **144:** 389.

Hutchison H.T., Hartwell L.H., and McLaughlin C.S. 1969. Temperature-sensitive yeast mutant defective in ribonucleic acid production. *J. Bacteriol.* **99:** 807.

Karl T., Önder K., Kodzius R., Pichová A., Wimmer H., Thür A., Hundsberger H., Löffler M., Klade T., Beyer A., Breitenbach M., and Koller L. 1999. *GRC5* and *NMD3* function in translational control of gene expression and interact genetically. *Curr. Genet.* **34:** 419.

Koepp D.M. and Silver P.A. 1996. A GTPase controlling nuclear trafficking: Running the right way or walking RANdomly? *Cell* **87:** 1.

Kressler D., Linder P., and de la Cruz J. 1999. Protein trans-acting factors involved in ribosome biogenesis in *Saccharomyces cerevisiae*. *Mol. Cell. Biol.* **19:** 7897.

Kruiswijk T., Planta R.J., and Krop J.M. 1978. The course of assembly of ribosomal subunits in yeast. *Biochim. Biophys. Acta* **517:** 378.

Kudo N., Matsumori N., Taoka H., Fujiwara D., Schreiner E.P., Wolff B., Yoshida M., and Horinouchi S. 1999. Leptomycin B inactivates CRM1/exportin 1 by covalent modification at a cysteine residue in the central conserved region. *Proc. Natl. Acad. Sci.* **96:** 9112.

Longtine M.S., McKenzie A.R., Demarini D.J., Shah N.G., Wach A., Brachat A., Philippsen P., and Pringle J.R. 1998. Additional modules for versatile and economical PCR-based gene deletion and modification in *Saccharomyces cerevisiae*. *Yeast* **14:** 953.

Mattaj I.W. and Englmeier L. 1998. Nucleocytoplasmic transport: The soluble phase. *Annu. Rev. Biochem.* **67:** 265.

Milkereit P., Gadal O., Podtelejnikov A., Trumtel S., Gas N., Petfalski E., Tollervey D., Mann M., Hurt E., and Tschochner H. 2001. Maturation and intranuclear transport of pre-ribosomes requires Noc proteins. *Cell* **105:** 499.

Moy T.I. and Silver P.A. 1999. Nuclear export of the small ribosomal subunit requires the Ran-GTPase cycle and certain nucleoporins. *Genes Dev.* **13:** 2118.

Nakielny S. and Dreyfuss G. 1999. Transport of proteins and RNAs in and out of the nucleus. *Cell* **99:** 677.

Neville M. and Rosbash M. 1999. The NES-Crm1p export pathway is not a major mRNA export route in *Saccharomyces cerevisiae*. *EMBO J.* **18:** 3746.

Ossareh-Nazari B., Bachelerie F., and Dargemont C. 1997. Evidence for a role of CRM1 in signal-mediated nuclear protein export. *Science* **278:** 141.

Pemberton L.F., Blobel G., and Rosenblum J.S. 1998. Transport routes through the nuclear pore complex. *Curr. Opin. Cell Biol.* **10:** 392.

Rout M.P., Blobel G., and Aitchison J.D. 1997. A distinct nuclear import pathway used by ribosomal proteins. *Cell* **89:** 715.

Rout M.P., Aitchison J.D., Suprapto A., Hjertaas K., Zhao Y.M., and Chait B.T. 2000. The yeast nuclear pore complex: Composition, architecture, and transport mechanism. *J. Cell Biol.* **148:** 635.

Sarkar S. and Hopper A.K. 1998. tRNA nuclear export in *Saccharomyces cerevisiae*: In situ hybridization analysis. *Mol. Biol. Cell* **9:** 3041.

Stade K., Ford C.S., Guthrie C., and Weis K. 1997. Exportin 1 (Crm1p) is an essential nuclear export factor. *Cell* **90:** 1041.

Stage-Zimmermann T., Schmidt U., and Silver P.A. 2000. Factors affecting nuclear export of the 60S ribosomal subunit in vivo. *Mol. Biol. Cell* **11:** 3777.

Venema J. and Tollervey D. 1999. Ribosome synthesis in *Saccharomyces cerevisiae*. *Annu. Rev. Genet.* **33:** 261.

Verschoor A., Warner J.R., Srivastava S., Grassucci R.A., and Frank J. 1998. Three-dimensional structure of the yeast ribosome. *Nucleic Acids Res.* **26:** 655.

Warner J.R. 1999. The economics of ribosome biosynthesis in yeast. *Trends Biochem. Sci.* **24:** 437.

Wen W., Meinkoth J.L., Tsien R.Y., and Taylor S.S. 1995. Identification of a signal for rapid export of proteins from the nucleus. *Cell* **82:** 463.

Yang Q., Rout M.P., and Akey C.W. 1998. Three-dimensional architecture of the isolated yeast nuclear pore complex: Functional and evolutionary implications. *Mol. Cell* **1:** 223.

The Ribosome in the 21st Century:
The Post-structural Era

P.B. MOORE

*Departments of Chemistry, and Molecular Biology and Biochemistry, Yale University,
New Haven, Connecticut 06520-8107*

The last time protein synthesis was the topic of a Cold Spring Harbor Symposium was 1969, not long after it had been established that the ribosome is an enzyme and that its RNA does not carry genetic information. The decision to make protein synthesis the focus of the Symposium again in 2001 was well-timed. Nine months before the Symposium opened, atomic resolution structures were published for both a large ribosomal subunit (Ban et al. 2000) and a small one (Schluenzen et al. 2000; Wimberly et al. 2000), and about a month before the Symposium started, a 5.5 Å resolution structure appeared for a 70S ribosome (Yusupov et al. 2001). Thus, the 66th Symposium was the first international meeting on protein synthesis to occur after the crystallographic "unveiling" of the ribosome, and the talks given provide a glimpse of how knowledge of the structure of the ribosome crystal is going to affect protein synthesis research in the decades to come.

Since the reader can rapidly discover what transpired at the Symposium by examining the papers in this volume, it is pointless to review the proceedings here. What follows instead is an essay about the impact of ribosome crystal structures on the protein synthesis field that expands on the short talk I gave at the end of the Symposium.

SEQUENCE CONSERVATION IN RIBOSOMAL RNA

At a time when the excitement surrounding crystal structures is high, it is easy to forget how important sequences have been to the ribosome field. The challenge of sequencing the 55 proteins and two large rRNAs found in the *E. coli* ribosomes was taken up in the early 1970s at a time when the sequencing of a small protein could consume years and the largest RNA that had been sequenced was less than a tenth the size of 16S rRNA. The most important first fruits of this heroic effort were the secondary structure diagrams for rRNAs that materialized in the early 1980s, shortly after the first rRNA sequences were completed (Noller and Woese 1981). They provided the first accurate indications of how rRNA is structured, and guided much of the research done in the decades that followed.

All authors cited here without dates refer to papers in this volume.

In addition to telling us about rRNA secondary structure, rRNA sequence comparisons revealed the existence of positions in rRNA chains where little, if any, species-to-species sequence variation is observed. A great deal of attention has been given to these invariant residues subsequently, because conservation is proof positive of functional and/or structural importance. In the last decade, my group has worked on two highly conserved sequences in 23S rRNA: the sarcin/ricin loop and the peptidyl transferase region. Our experiences with both raise questions about the significance of conservation in rRNAs.

Phenotypically Silent Bases in the Sarcin/Ricin Loop

The sarcin/ricin loop (SRL) includes the longest run of conserved residues in the entire ribosome. It is the apex of helix 95 in 23S rRNA, and is the site attacked by two families of protein toxins, the archetypes of which are α-sarcin and ricin, respectively. Each catalyzes the cleavage of a specific covalent bond in the SRL, and ribosomes so modified do not interact with elongation factors properly and hence fail to catalyze protein synthesis effectively (Wool et al. 2000). In the early 1990s, my group launched a study of the solution structure of an oligonucleotide containing the SRL sequence in collaboration with Ira Wool and his colleagues, who had earlier demonstrated that it is a substrate for both toxins in vitro. Ultimately its structure was solved both by nuclear magnetic resonance (NMR) (Szewczak et al. 1993; Szewczak and Moore 1995) and by X-ray crystallography (Correll et al. 1998, 1999). Both the solution structure and the crystal structure of the SRL oligonucleotide in question are fundamentally the same as the structure of the SRL sequence in the intact ribosome (Ban et al. 2000).

The most conspicuous feature of the SRL is the S-turn on the 5′ side of its terminal loop. It is stabilized by a base triple involving a bulged G (G2655 in *E. coli*), which is both critical structurally and conserved. Nevertheless, not only are *E. coli* strains containing G2655A ribosomes viable, but the G2655A mutation is virtually silent phenotypically (Macbeth and Wool 1999). Only after ~100 generations of growth in competition with wild-type strains does it become apparent that G2655A ribosomes are disadvantageous. Another mutation in the conserved region of the SRL, G2661C, which converts the R base in the

GNRA tetraloop at the top of the SRL into a pyrimidine, is also almost devoid of phenotype (Tapprich and Dahlberg 1990; Wool et al. 2000). Clearly, at least some of the bases in the SRL have been conserved because they confer functional advantages so small they can hardly be detected. Conserved bases such as G2655 could be described as "quiet."

Quiet Bases in the Peptidyl Tranferase Center

We discovered at the Symposium that there are conserved bases in the peptidyl transferase center of the large ribosomal subunit which are remarkably quiet, at least as far as peptidyl transferase activity is concerned. A generation's worth of biochemical and molecular biological experiments have identified nucleotides belonging to the central loop of the fifth domain of 23S rRNA as components of the ribosome's peptidyl transferase site (Noller 1991). Those nucleotides form a compact assembly at the small subunit end of the exit tunnel where electron microscopic studies have shown the peptidyl transferase site is located (Frank et al. 1995; Stark et al. 1997b; Ban et al. 2000). Furthermore, crystal structures of large subunits with peptidyl transferase substrate and transition state analogs bound have demonstrated that the nucleotides in question account entirely for the peptidyl transferase site (Nissen et al. 2000).

These crystal structures have focused attention on a single nucleotide in the peptidyl transferase site, A2451 (*E. coli*), which is 100% conserved in cytoplasmic ribosomes. A2451 is located in the middle of the peptidyl transferase site, and my colleagues and I have proposed that it promotes peptide-bond formation by acting as a general acid/base (Muth et al. 2000; Nissen et al. 2000). Experiments reported at this Symposium, most of them done in vitro, show that mutations of A2451 appear to have only modest effects on the capacity of ribosomes to catalyze peptide-bond formation (Polacek et al. 2001; Kim et al.) and to make proteins (Gregory et al.). Furthermore, mutants of G2447, a highly conserved base that helps position A2451 and which we had suggested might enhance the catalytic activity of A2451 by polarizing it (Nissen et al. 2000), are almost without phenotype. Whether either of these bases actively catalyzes peptide-bond formation or not, it is remarkable that mutations of residues so centrally located have so little effect on activity.

Here, it is necessary to point out that unlike the SRL mutations discussed above, A2451 mutations are not quiet in vivo. Despite the minimal effect they seem to have on the rate of peptide formation in vitro, they are dominant lethal in vivo (Muth et al. 2000). This paradox might be resolved if the role of A2451 in catalyzing the hydrolysis that releases completed polypeptides from tRNAs were far more important than its role in promoting peptide-bond formation (Hansen et al.).

In light of such observations, what conclusion can biochemists draw from the observation that some base in rRNA is conserved? Apparently tiny functional advantages, perhaps 1% advantages, suffice to ensure conservation, but biochemists are incapable of measuring *anything* with 1% accuracy. Furthermore, the making or breaking of a hydrogen bond, which is about the smallest molecular change a structural biologist is likely to be able to identify with confidence, has an associated free-energy change corresponding to at least a tenfold change in rates or binding constants. The structural correlate of a mutation that leads to a 1% change in rate or binding constant is likely to be very subtle! Thus, the observation that a particular rRNA base is conserved does not mean that its role is something a biochemist/structural biologist can easily understand.

It would be helpful if geneticists/molecular biologists could devise an experimental strategy for distinguishing RNA bases that are conserved for 1% reasons—quiet bases—from bases that are conserved for tenfold or greater reasons—important bases. The latter are clearly worth biochemical/structural investigation because their contributions to function are likely to be understandable as well as significant. Most quiet bases will be bit players functionally, but it could occasionally happen that the role played by a conserved base is important, but that it is quiet nonetheless because its role can be performed by *any* other base almost as well. (The possibility that a nonconserved base could contribute directly to function is dismissed on the grounds that the probability that the cell would be totally indifferent to base substitution at a functionally important position is infinitesimal. The vexing issue of how conserved a base must be in order that its conservation be considered functionally meaningful will not be pursued here.)

ON THE MEANING OF RIBOSOME CRYSTAL STRUCTURES

Nonimaging Approaches to Ribosome Structure

There was panic in the air in 1999, at the last international meeting of the ribosome fraternity in Helsingor, Denmark. Its cause was concern about the direction in which the field was headed. Atomic resolution structures of ribosomes had not yet been obtained, but lower resolution electron density maps of both ribosomal subunits were presented at the meeting, and it was clear that atomic resolution structures would soon follow. This was a problem for many because investigation of the ribosome by nonimaging techniques had been an important activity for 30 years. The papers and posters presented at this Symposium show how the community reacted. Almost no such work was presented. At any similar meeting held in the preceding three decades, it would have been a major topic.

It is easy now to dismiss the literature generated using the techniques just alluded to, and perhaps even to question the wisdom of those who contributed to it. However, it is important to remember that at the time the first such studies were done, and for decades thereafter, it was not obvious that atomic resolution structures of intact ribosomes or ribosomal subunits would *ever* be obtained. If they had not, the information that literature contains, which was gleaned so painfully, would be all we know about ribosome structure today. Furthermore, data of this kind are useful even for molecules whose structures can be, or have been, determined. On more than one occasion, comparison of protein crystal structures with noncrystallographic data has led to the realization that electron density maps had been misinterpreted.

The Neutron Map of the 30S Subunit

The 30S crystal structure is particularly interesting to me personally because of my multi-year involvement in the determination of the positions in proteins in the 30S subunit by neutron scattering (Capel et al. 1987, 1988). The technique we used measures distances between protein centers of mass a pair at a time. Hence it is sensible to compare the neutron map of the small subunit with its crystal structure by plotting neutron map distances as a function of center-to-center distance in the crystal structure. If the neutron map were perfect, all the points in such a plot would fall on the diagonal. Figure 1 displays the results for the 79 inter-protein distances measured by neutron scattering that can be compared directly with the crystal structures. The correlation is obvious; the neutron experiment did indeed measure distances. There are six egregious outliers, however: five where the distance estimate that emerged is much too short, and a sixth where the distance estimate is much larger than it should have been. Three of the outliers involve S20. Two relate to S16, and one to S14. Since the data for S20 are internally consistent, it cannot be ruled out that the position of S20 in the reconstituted subunits examined was not the same as its position in the crystals. It is not obvious why anomalous distances were obtained for S16 and S14.

The neutron map can also be compared with the crystal structure by superimposing on the locations found for protein centers of mass crystallographically. The centers of mass of 13 of the 19 proteins the two maps describe (S2, S3, S4, S5, S6, S7, S8, S9, S10, S11, S12, S15, S17) superimpose on each other with a root mean square (rms) error of about 18 Å, which is less than the radius of a small spherical protein. Furthermore, within that set, the least accurately positioned protein is only 32 Å from where it belongs. Of the remaining 6 proteins, 3 (S14, S16, S20) are poorly placed because the distances measured for them include some that are wildly inaccurate. For the other 3 proteins (S13, S18, S19), the distances measured are as accurate as those that positioned the 13 proteins whose locations are well-determined. The reason their positions are incorrect is that the distances that placed them in the neutron map relate them to sets of proteins that are nearly co-planar. Thus, the data are almost equally consistent with two positions for each of these proteins, one above the plane determined by the reference proteins and the other below, and the wrong choice was the one favored. All the proteins mislocated in the neutron map have low molecular weights, which means that the signal-to-noise ratio was poor in the data measured for all of them. They are also proteins for which only a small number of distances were measured, and that made them particularly vulnerable to the ambiguity just described.

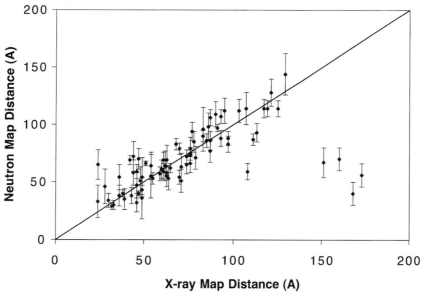

Figure 1. Comparison of inter-protein distances from the neutron map of the small subunit with those from the X-ray structure of the small subunit. The position of the center of mass of each protein in the 30S subunit of *T. thermophilus* was computed using the coordinates reported by Ramakrishnan and coworkers (Wimberly et al. 2000), and inter-protein distances were computed therefrom. Those distances are compared with the neutron map center-of-mass distances listed in Capel et al. (1988). The errors assigned neutron center-to-center distances are standard deviations, and they were computed from the standard deviations reported for center-of-mass coordinates in the neutron map. Distances are plotted for only the 79 protein pairs whose relationship was determined directly by a neutron measurement. (*Note:* The *T. thermophilus* small subunit structure used lacks S1 and S21.)

Other Retrospective Analyses

Not surprisingly, I sympathize with others who are interested in comparing noncrystallographic results with what has been learned from the atomic resolution crystal structures (Sergiev et al.). Which experiments produced reliable results? How should the data have been interpreted? Cells contain ribonucleoproteins that are a lot less tractable than the ribosome; e.g., the spliceosome. It makes sense to investigate their structures by non-imaging methods at this point for the same reasons it made sense to do similar experiments on the ribosome in the past. However, it would be silly for anyone to do so using methods that failed with the ribosome or to interpret new data unguided by the ribosome experience. Finally, the ribosome may be more dynamic than the average all-protein enzyme (see below), and careful examination of the noncrystallographic structure data could lead to insights into this aspect of its behavior.

On the Value of the Atomic-resolution Structures Now in Hand

One of the shibboleths of ribosome crystallographers is the sentiment that their field is still in its infancy. There is truth to it, even if it may seem self-serving to non-crystallographers. It is unlikely that translocation or proofreading will be understood mechanistically in the absence of atomic-resolution structures for many more ribosome/substrate/factor complexes than have been solved so far. What impresses me, however, is not how little we have at this point, but how much.

Because we have a small number of atomic-resolution structures in hand, a lot of valuable information can be gleaned from structural data the resolution of which is less than atomic. The Santa Cruz structure of the *Thermus thermophilus* 70S ribosome is a case in point (Yusupov et al. 2001; Noller et al.). The crystals from which it is derived diffract to a resolution insufficient for the determination of macromolecular structures de novo: 5.5 Å. Nevertheless, an atomic-resolution interpretation has emerged. As it turns out, the chain trace of 16S rRNA was worked out directly from the electron density maps obtained by the Santa Cruz group, an achievement that echoes the success the MRC group had in deducing the structure of the central domain of 16S rRNA from their 5.5 Å resolution electron density map of the small subunit in 1999 (Clemons et al. 1999). The Santa Cruz interpretation of the large subunit part of the *T. thermophilus* 70S structure, however, rests in large part on the atomic-resolution structure of the *Haloarculi marismortui* large ribosomal subunit because the middle of the large subunit is too complicated to interpret de novo at 5.5 Å resolution. The structural similarity between the large subunits of the two species is so high that RNA features in the large subunit part of the 70S electron density map that correspond to structures not seen in the *H. marismortui* large subunit could be modeled. However, it is difficult to model protein sequences using electron density maps having resolutions much worse than 3 Å, and for that reason, the Santa Cruz interpretation of proteins in the 70S structure derives entirely from prior information, and is fragmentary in the large subunit because of the lack of homology between the ribosomal proteins in *H. marismortui* and *T. thermophilus*.

The electron microscopists have demonstrated that they too can use atomic-resolution crystal structures to interpret low-resolution electron density maps (Beckmann et al.; Frank and Agrawal). The analysis of eukaryotic ribosome structures presented at the Symposium provided a dramatic example. In addition, the tiny samples required to obtain three-dimensional reconstructions give electron microscopists a huge advantage when it comes to the analysis of translational intermediates that are difficult to prepare (Frank and Agrawal). Most of what we know today about where factors bind to the ribosome comes from electron microscopy, for example, and its edge as a tool for understanding intermediate states is unlikely to disappear anytime soon. The information electron microscopy can provide is limited only by resolution, and relief in this area may be forthcoming in the future also (van Heel).

How Flexible is the Ribosome?

Over the years, many experiments have hinted that the ribosome may be more flexible than the typical, proteinaceous enzyme (see, e.g., Moore 1987). At this Symposium, Yonath and coworkers presented a structure for the large ribosomal subunit from *H. marismortui* obtained from crystals containing high concentrations of K^+. Significant differences are reported to exist between it and the Yale structure, in which Na^+ is the predominant cation (Bashan et al.). If substantiated, these results would imply that the ribosome is remarkably flexible.

Had the two kinds of crystals studied been crystals of the same protein prepared under different conditions, the expectation would have been that the two structures would be identical in the interiors of their domains, within error, but might differ in relative domain locations and in the placement of surface side chains. The differences between the two *H. marismortui* large subunit structures are reportedly much bigger than that (Bashan et al.). Not only are all the projections from the body of the large subunit that are disordered in the Yale structure reported to be more visible in the new structure (which may be a resolution artifact), but the tails of at least some of the proteins that penetrate the subunit are in different positions in the two structures, which is hard to understand given the intricate networks of interactions with neighboring RNA sequences that stabilize their conformations. (Evidence for a similar plasticity of protein tail conformation has been advanced for the 30S subunit [Schluenzen et al. 2000].) In addition to these local differences, the new *H. marismortui* large subunit structure is reported to be more compact than the Yale structure, a finding that suggests a sensitivity of ribosome structure to ionic conditions that has no parallel in the protein world.

In this context, it is worth pointing out that the backbone trace for *T. thermophilus* 23S rRNA reported by Noller and coworkers (Yusupov et al. 2001) is essentially identical to that reported by the Yale group for the *H. marismortui* large ribosomal subunit (Ban et al. 2000) over almost the entire length where the two are homologous. Furthermore, the *T. thermophilus* structure in question is that of the 70S ribosomes, of course, not the large subunit in isolation, and the crystals from which it was obtained were stabilized in a solvent of much lower ionic strength than that used for the stabilization of either of the two *H. marismortui* large subunit crystals. If the conformation of the large ribosomal subunit is as sensitive to conditions as the differences in the two *H. marismortui* structures suggest, why isn't the *T. thermophilus* structure different from both?

Interestingly, the structures we have obtained for *H. marismortui* 50S subunit from crystals in which K^+ is the dominant cation (J. Hansen, unpubl.) are identical to those obtained from crystals in which Na^+ predominates. The visibility of projecting RNA stem-loops is no different, and there are no indications of alterations in protein conformation or location. Thus, the distinctive features of the structure reported by the Yonath group have to be ascribed entirely to a difference in K^+ concentration of about a factor of two. Finally, there is no reason to believe that the Yale structure is that of an inactive conformation of the large ribosomal subunit. The peptidyl transferase activity of the Yale crystals has been verified experimentally (Hansen et al.), and since electron density corresponding to L1, L10, and L11 is observed in Yale electron density maps, the allegation that these components are absent from the Yale crystals is without merit.

However the crystallographic issues just discussed are resolved, it remains that electron microscopic studies have shown significant motions within the large subunit during the elongation cycle (Frank and Agrawal 2000). Evidence has also been advanced that the peptide tunnel can open and close, an intriguing finding that may turn out to be the first indication that nascent proteins are transported down the tunnel by a peristaltic mechanism (Gabashvili et al. 2001; Frank and Agrawal). The conformational plasticity of the small subunit is even more obvious (Gabashvili et al. 1999). There is overwhelming evidence that protein synthesis is accompanied by substantial motions of its three domains, and intradomain motion has been documented also (Brodersen et al. 2000; Carter et al. 2000, 2001; Schluenzen et al. 2000; Ogle et al. 2001; Pioletti et al. 2001).

Is the ribosome fundamentally different from a protein enzyme in its plasticity? The hydrophobic centers of protein domains are almost as densely packed as crystals of small organic molecules (Richards 1974; Chothia 1984), which tends to reduce their flexibility. RNAs are different. An RNA domain is, to first order, an array of double helices, and there is no way to pack double helices so that they fill space with protein-like efficiency. Thus, there are solvent-filled voids within RNA domains that could enable motions much larger than those seen in proteins. Nevertheless, it would be premature to conclude that RNA domains are easy to deform. The helices in RNA domains are rod-like objects held in place by networks of specific interactions with their neighbors. Geodesic domes are built the same way, and in a geodesic dome the ratio of the volume enclosed to the volume devoted to structural elements is much larger than it is in the ribosome. If the RNA in the ribosome is geodesic-dome-like, significant motions will be accompanied by the making and breaking of inter-helix interactions, and its domains will be protein-like in their plasticity. Clearly, more experiments need to be done.

RIBOSOME FUNCTION IN THE POST-STRUCTURAL ERA

Satisfying as they are, macromolecular structures are not an end in themselves. Structures are determined because without them it is almost impossible to understand the chemical basis of macromolecular function. It is appropriate to ask, therefore, what the structures discussed above have taught us about ribosome function. The answer should be a comfort to those who feared that the ribosome field was about to end in 1999, for in many instances, they have heightened the mystery surrounding many aspects of ribosome function, not resolved them.

tRNA Binding and Mis-translation

The aspect of ribosome function that has been most interestingly illuminated by structures so far is its decoding mechanism (Brodersen et al; Noller et al.; see also Brierly and Pennell; Stahl). For example, it is now clear that 16S rRNA bases around C1400 act to test whether codon–anticodon interactions are cognate, and we can begin to understand why mutations in S12 affect fidelity. It is equally clear there is much we do not understand. The A-site tRNA/ribosome interaction that has been visualized crystallographically is the one that exists after the tRNA in the A site has been accepted by the proofreading process. Electron microscopic reconstructions have shown that aminoacyl tRNAs are delivered to the ribosome by EF-Tu at a site that overlaps the A site only at the anticodon end, and that a dramatic change in tRNA position accompanies (Stark et al. 1997a,b). What does the tRNA–mRNA interaction look like when tRNA is bound in the delivery site, and how does the ribosome hold onto tRNAs accepted into the delivery site while the enormous motion occurs that results in their entering the A site? A specific proposal was advanced at the Symposium that should be easy to test (Simonson and Lake).

Several talks at the Symposium dealt with experiments done on systems that exploit what an engineer might call "ribosome failure modes," and they provide additional evidence of how meager our understanding really is. It has long been known that translocation does not always advance ribosomes down mRNAs by the standard three nucleotides. The failure to do so leads to frameshifting and other more dramatic forms of misreading (Atkins et al.; Brierley and Pennell). Ribosomes can be induced to hop, skip, and jump their way down mRNAs by mRNA sequences and structures that are still in the process of being identified. How can people who do not understand the mechanism that normally enables ribosomes to maintain reading frame explain phenomena like these?

IRES sequences in mRNAs, which enable eukaryotic ribosomes to initiate protein synthesis in a cap-independent manner, can be thought of as a ribosome failure mode specific to eukaryotic ribosomes that has become functionally important in some contexts. A tantalizing electron microscopic image of an IRES sequence bound to a ribosome was provided at the Symposium (Kieft et al.), which hints at how those sequences work, but the remaining challenge is formidable. By prokaryotic standards (see Gualerzi et al.), initiation is astonishingly complex in eukaryotes; the details are still being worked out (Ali and Jackson; Pestova and Hellen; Gross et al.; Asano et al.). Again, until we understand what happens normally, how are we to understand what IRESs do?

The L7/L12 Mystery

Then there is the interesting puzzle posed by L7/L12. L7/L12 was the first ribosomal protein sequenced, and its carboxy-terminal domain was the first bit of ribosomal protein to have its structure determined crystallographically (Leijonmarck et al. 1980). It has been known for years that L7/L12 plays an important role in elongation factor/ribosome interactions (Moller and Maassen 1986). Electron microscopic experiments have revealed both where factors bind to the ribosome and where L7/L12 is located (Agrawal et al. 1999; Stark et al. 2000). L7/L12 is associated with the same segment of 23S rRNA that binds L11, and there is persuasive evidence that L11 contacts EF-G during elongation (Frank and Agrawal). The interaction of L7/L12 with the stem of the stem-loop that is capped by L11 is mediated by L10, and the L10/L7/L12 assembly points out into solution, away from the factor-binding site. The L10/L7/L12 assembly looks gratuitous; it looks as though the ribosome ought to work perfectly well without them. Yet it does not. What is the explanation of these observations, let alone the recent report that L7/L12 stimulates the GTPase activity of EF-G in vitro (Savelsbergh et al. 2000)?

Translocation

Translocation has long looked likely to be the most difficult single aspect of protein synthesis to understand mechanistically. It clearly involves coordinated conformational changes within the two subunits that result in relative motions as well as in motions of ribosome-bound mRNA and tRNAs (Stark et al. 1995; Agrawal et al. 1999; Frank and Agrawal 2000). It is probably related to interactions between tRNA-binding sites (Blaha and Nierhaus). What the new structures provide is a far more precise understanding of what has to occur during translocation, but little insight into how and why it occurs. A few years ago, when it was discovered that EF-G resembles the EF-Tu/aminoacyl tRNA/GTP ternary complex, there was a brief period when it seemed that translocation would soon be understood conceptually (Nissen et al. 1995). Although data obtained since then support the hypothesis that EF-G acts as a ternary complex analog during translocation (Andersen and Nyborg), why and

how this enables it to catalyze translocation remains as puzzling as ever. The molecular mimicry idea seems very robust, however. It appears that release factors also resemble the ternary complex (Kaji et al.).

Higher-order Functions

Ribo-centrics tend to think of ribosomes as objects that act in a vacuum, save for the aminoacyl tRNAs, mRNAs, and protein factors supplied to them by the rest of the cell. They do not, of course, and the aspect of their interactions with other cellular components that is best understood is that which relates to the production of proteins that must be inserted into or pass through membranes. It has long been known that ribosomes engaged in the synthesis of such proteins associate with membranes by binding to a secretory apparatus that is part of the membrane. Symposium participants were treated to EM images of 80S ribosomes complexed with Sec61, the protein that mediates the docking of 80S ribosomes with the membrane-associated components of the secretory apparatus (Beckmann et al.), and a discussion of how the signal recognition particle selects ribosomes for binding to membranes (A.E. Johnson et al.) This rich area of ribosome-related biochemistry awaits the attention of crystallographers, and its elucidation will doubtless be accompanied by insights into an aspect of ribosome function about which we are now totally ignorant: how nascent peptides pass through the exit tunnel in the large ribosomal subunit.

Peptide-bond Formation

Finally, the peptidyl transferase site provides a vivid example of how far we are from a real understanding of the ribosome. On paper, peptide-bond formation is simple nucleophile/electrophile chemistry, and ever since the fragment reaction was discovered in the 1960s, it has appeared to be the simplest of the tasks performed by the ribosome. The large subunits the Yale group has studied crystallographically are active in poly(U)-directed polyphenylalanine synthesis both before and after crystallization, and catalyze a reaction related to the fragment reaction in the crystalline state. Furthermore, all the structures obtained for substrates and products bound to the peptidyl transferase site of those subunits suggest that the N3 of A2451 may be protonated at neutral pH and are consistent with its playing an active role in peptide-bond formation (Hansen et al.). Nevertheless, as noted earlier, the mutagenesis data discussed above apparently do not support that hypothesis, and we now know that the chemical reactivity data reported last year that was taken as indicating that A2451 has a pK_a around 7.5 (Muth et al. 2000) were misleading. The pH-dependent reactivity of that base can only be demonstrated in *E. coli* ribosomes that are incapable of peptide-bond formation because they have been inactivated (Gregory et al.). The pK_a of A2451 cannot be examined directly in activated ribosomes because it is unreactive to low-molecular-weight reagents.

The disconnect that appears to exist between the structural data and the genetic and biochemical data about the peptidyl transferase site reminds one how reluctant the ribosome has always been to give up its secrets. One is also reminded how often the discourse in the field has been a dialectic. Almost no significant proposal gets advanced that does not immediately elicit an evidence-backed counterproposal.

In this instance, one cannot help wondering whether the fragment assay for peptidyl transferase activity itself is a source of confusion. Bacterial ribosomes make peptide bonds at an average rate around 20 s^{-1} at 37°C (Kjeldgaard and Gaussing 1974), and the fraction of each elongation cycle devoted to peptide-bond formation itself is small (Pape et al. 1998). For the sake of argument, suppose that once the P and A sites of the ribosome contain appropriate substrates, peptide-bond formation occurs at a rate of 10^2 s^{-1}. The rate could be much faster than this, but it is unlikely to be much slower. At 37°C, fragment reaction proceeds in isolated 50S subunits at rates around 10^{-2} s^{-1}, which is 4 orders of magnitude slower. Maybe the low-molecular-weight substrate/ribosome structures available do not capture the whole truth about the peptidyl transferase site.

All of these difficulties notwithstanding, the A2451 hypothesis should not be abandoned prematurely. Every base has an oxygen or a nitrogen in exactly the same position as the N3 of adenine, and if A2451 is replaced by a C, for example, it can make almost exactly the same interactions with surrounding bases (J. Hansen, pers. comm.). This being the case, what does it mean that mutations of A2451 don't have a big effect on peptidyl transferase activity?

CONCLUSION

No one aware of the impact of protein crystal structures on enzymology will be in the least surprised by what atomic-resolution ribosome structures have done for the ribosome field so far. Almost all the questions about ribosome function that were on the table before structures were obtained remain on the table today. The reason is that structures seldom lay fundamental functional problems to rest. What structures do instead is to reformulate those problems in atomic terms, and by so doing make it possible to do experiments that address them in ways that would otherwise have been impossible. That process is under way today in the ribosome field, and I think we can look forward to a rich harvest of new insight.

ACKNOWLEDGMENTS

I thank my colleagues T.A. Steitz, J. Hansen, M. Schmeing, and D. Klein for many useful discussions, but hasten to emphasize my responsibility for all errors of fact or interpretation this paper may contain. This work was supported by a grant from the National Institutes of Health (P01-GM-22778).

REFERENCES

Agrawal R.K., Heagle A.B., Penczek P., Grassucci R.A., and Frank J. 1999. EF-G-dependent GTP hydrolysis induces translocation accompanied by large conformational changes in the 70S ribosome. *Nat. Struct. Biol.* **6**: 643.

Ban N., Nissen P., Hansen J., Moore P.B., and Steitz T.A. 2000. The complete atomic structure of the large ribosomal subunit at 2.4 Å resolution. *Science* **289**: 905.

Brodersen D.E., Clemons W.M., Carter A.P., Morgan-Warren R.J., Wimberly B.T., and Ramakrishnan V. 2000. The structural basis for the action of the antibiotics tetracycline, pactamycin and hygromycin B on the 30S subunit. *Cell* **103**: 1143.

Capel M.S., Kjeldgaard M., Engelman D.M., and Moore P.B. 1988. The positions of S2, S13, S16, S17, S19, and S21 in the 30S ribosomal subunit of *Escherichia coli*. *J. Mol. Biol.* **200**: 65.

Capel M.S., Engelman D.M., Freeborn B.R., Kjeldgaard M., Langer J.A., Ramakrishnan V., Schindler D.G., Schneider D.K., Schoenborn B.P., Sillers I.-Y., Yabuki S., and Moore P.B. 1987. A complete mapping of the proteins in the small ribosomal subunit of *E.coli*. *Science* **238**: 1403.

Carter A.P., Clemons W.M., Brodersen D.E., Morgan-Warren R.J., Wimberly B.T., and Ramakrishnan V. 2000. Functional insights from the structure of the 30S ribosomal subunit and its interactions with antibiotics. *Nature* **407**: 340.

Carter A.P., Clemons W.M., Broderson D.E., Morgan-Warren R.J., Hartsch T., Wimberly B.T., and Ramakrishnan V. 2001. Crystal structure of an initiation factor bound to the 30S ribosomal subunit. *Science* **291**: 498.

Chothia C. 1984. Principles that determine the structure of proteins. *Annu. Rev. Biochem.* **53**: 537.

Clemons W.M., Jr., May J.L.C., Wimberly B.T., McCutcheon J.P., Capel M.S., and Ramakrishnan V. 1999. Structure of a bacterial 30S ribosomal subunit at 5.5 Å resolution. *Nature* **400**: 833.

Correll C.C., Wool I.G., and Minishkin A. 1999. The two faces of the *Escherichia coli* 23S rRNA sarcin/ricin domain: The structure at 1.11 Å resolution. *J. Mol. Biol.* **292**: 275.

Correll C.C., Munishkin A., Chan Y., Ren Z., Wool I.G., and Steitz T.A. 1998. Crystal structure of the ribosomal RNA loop essential for binding both elongation factors. *Proc. Natl. Acad. Sci.* **95**: 13436.

Frank J. and Agrawal R.K. 2000. A rachet-like inter-subunit reorganization of the ribosome during translocation. *Nature* **406**: 318.

Frank J., Zhu J., Penczek P., Li Y., Srivastava S., Verschoor A., Rademacher M., Grassucci R., Lata R.K., and Agrawal R.K. 1995. A model for protein synthesis based on cryo-electron microscopy of the *E. coli* ribosome. *Nature* **376**: 441.

Gabashvili I.S., Agrawal R.K., Grassucci R., Squires C.L., Dahlberg A.E., and Frank J. 1999. Major rearrangements in the 70S ribosomal 3D structure caused by a conformational switch in 16S ribosomal RNA. *EMBO J.* **18**: 6501.

Gabashvili I.S., Gregory S.T., Valle M., Grassucci R., Worbs M., Wahl M., Dahlberg A.E., and Frank J. 2001. The polypeptide tunnel system in the ribosome and its gating in erythromycin resistance mutants of L4 and L22. *Mol. Cell* **8**: 181.

Kjeldgaard N.O. and Gaussing K. 1974. Regulation of biosynthesis of ribosomes. In *Ribosomes* (ed. M. Nomura et al.), p. 369. Cold Spring Harbor Laboratory, Cold Spring Harbor, New York.

Leijonmarck M., Eriksson S., and Liljas A. 1980. Crystal structure of a ribosomal component at 2.6Å resolution. *Nature* **286**: 824.

Macbeth M.R. and Wool I.G. 1999. The phenotype of mutations of G2655 in the sarcin/ricin domain of 23S ribosomal RNA. *J. Mol. Biol.* **285**: 965.

Moller W. and Maassen J.A. 1986. On the structure, function and dynamics of L7/L12 from *Escherichia coli* ribosomes. In *Structure, function, and genetics of ribosomes* (ed. B. Hardesty and G. Kramer), p. 309. Springer-Verlag, New York.

Moore P.B. 1987. On the modus operandi of the ribosome. *Cold Spring Harbor Symp. Quant Biol.* **52**: 721.

Muth G.W., Ortoleva-Donnelly L., and Strobel S.A. 2000. A single adenosine with a neutral pKa in the ribosomal peptidyl transferase center. *Science* **289**: 947.

Nissen P., Ban N., Hansen J., Moore P.B., and Stetiz T.A. 2000. The structural basis of ribosome activity in peptide bond synthesis. *Science* **289:** 920.

Nissen P., Kjeldgaard M., Thirup S., Polekhina G., Reshetnikova L., Clark B.F., and Nyborg J. 1995. Crystal structure of the ternary complex of Phe-tRNAPhe, EF-Tu, and a GTP analog. *Science* **270:** 1464.

Noller H.F. 1991. Ribosomal RNA and translation. *Annu. Rev. Biochem.* **60:** 191.

Noller H.F. and Woese C.R. 1981. Secondary structure of 16S ribosomal RNA. *Science* **212:** 403.

Ogle J.M., Brodersen D.E., Clemons W.M., Tarry M.J., Carter A.P., and Ramakrishnan V. 2001. Recognition of cognate transfer RNA by the 30S ribosomal subunit. *Science* **292:** 897.

Pape T., Wintermeyer W., and Rodnina M.V. 1998. Complete kinetic mechanism of elongation factor tu-dependent binding of aminoacyl-tRNA to the A site of the *E. coli* ribosome. *EMBO J.* **17:** 7490.

Pioletti M., Schlunzen F., Harms J., Zarivach R., Glumann M., Avila H., Bashan A., Bartels H., Auerbach T., Jacobi C., Hartsch T., Yonath A., and Franceschi F. 2001. Crystal structures of complexes of the small ribosomal subunit with tetracycline, edeine and IF3. *EMBO J.* **20:** 1829.

Polacek N., Gaynor M., Yassin A., and Mankin A.S. 2001. Ribosomal peptidyl transferase can withstand mutations at the putative catalytic nucleotide. *Nature* **411:** 498.

Richards F.M. 1974. The interpretation of protein structures: Total volume, group volume dsitributions and packing density. *J. Mol. Biol.* **82:** 1.

Savelsbergh A., Mohr D., Wilden B., Wintermeyer W., and Rodnina M.V. 2000. Stimulation of the GTPase activity of translation elongation factor G by ribosomal protein L7/L12. *J. Biol. Chem.* **275:** 890.

Schluenzen F., Tocilj A., Zarivach R., Harms J., Gluehmann M., Janell D., Bashan A., Bartles H., Agmon I., Franceschi F., and Yonath A. 2000. Structure of functionally activated small ribosomal subunit at 3.3 Å resolution. *Cell* **102:** 615.

Stark H., Rodnina M.V., Wieden H.-J., van Heel M., and Wintermeyer W. 2000. Large-scale movement of elongation factor G end extensive conformational change of the ribosome during translocation. *Cell* **100:** 301.

Stark H., Rodnin M.V., Rinke-Appel J., Brimacombe R., Wintermeyer W., and van Heel M. 1997a. Visualization of elongation factor Tu on the *Escherichia coli* ribosome. *Nature* **389:** 403.

Stark H., Mueller F., Orlova E.V., Schatz M., Dube P., Erdemir T., Zemlin F., Brimacombe R., and van Heel M. 1995. The 70S *Escherichia coli* ribosome at 23 Å resolution: Fitting the ribosomal RNA. *Structure* **3:** 815.

Stark H., Orlova E.V., Rinke-Appel J., Junke N., Mueller F., Rodnina M., Wintermeyer W., Brimacombe R., and van Heel M. 1997b. Arrangement of tRNAs in pre- and posttranslocational ribosomes revealed by electron cryomicroscopy. *Cell* **88:** 19.

Szewczak A.A. and Moore P.B. 1995. The sarcin/ricin loop, a modular RNA. *J. Mol. Biol.* **247:** 81.

Szewczak A.A., Moore P.B., Chan Y.-L., and Wool I.G. 1993. The conformation of the sarcin/ricin loop from 28S ribosomal RNA. *Proc. Natl. Acad. Sci.* **90:** 9581.

Tapprich W.E. and Dahlberg A.E. 1990. A single base mutation at position 2661 in *E. coli* 23S ribosomal RNA affects the binding of ternary complex to the ribosome. *EMBO J.* **9:** 2649.

Wimberly B.T., Brodersen D.E., Clemons W.M., Morgan-Warren R.J., Carter A.P., Vonrhein C., Hartsch T., and Ramakrishnan V. 2000. Structure of the 30S ribosomal subunit. *Nature* **407:** 327.

Wool I.G., Correll C.C., and Chan Y.-L. 2000. Structure and function of the sarcin-ricin domain. In *The Ribosome: Structure, function, antibiotics, and cellular interactions* (ed. R.A. et al.), p. 461. ASM Press, Washington, D.C.

Yusupov M.M., Yusupova G.Z., Baucom A., Lieberman K., Earnest T.N., Cate J.H. D., and Noller H.F. 2001. Crystal structure of the ribosome at 5.5 Å resolution. *Science* **292:** 883.

Author Index

Subject Index

A

αε model, 136, 140
A0 site cleavage, 582
A2BP1, 298
A1492 and A1493
  decoding and, 23–24
  IF1 and, 364
  requirements for adenines, 25
A2451
  acid-base mechanism of catalysis
    proposal, 104–105
  chloramphenicol-resistance muta-
    tion, 104–105
  extent of DMS reactivity, 111–112
  IF2 interaction and, 365
  mutant analysis, 123–125
  peptide-bond formation and, 119,
    612–613
  peptidyl transferase study and, 110
  pH-dependent conformational
    change and, 113–115
  quiet bases in peptidyl transferase
    center, 608
  role in ribosome function, 112
A2476 and A2478, 364
A2486, 39–40
A2660, 364
aaRS. See Aminoacyl-tRNA synthetase
aa-tRNA. See Aminoacyl-tRNA
Abf1p, 570
Abundant in neuroepithelium area
  protein (ANA), 298
Acid-base mechanism of catalysis,
  104–105, 115
Acid-urea polyacrylamide gel electrophoresis, 195
Actin, 440
Actin cytoskeleton and eEF1A
  actin alterations effect on protein
    synthesis, 445
  actin's link to translation/mRNA
    metabolism, 445
  eEF1A and actin binding, 445
  genetic link between actin and
    eEF1A, 446
Active site analysis by site-directed
  mutagenesis, 60
  A2451 mutant analysis, 123–125
  chloramphenicol resistance of mu-
    tants, 124
  components arrangement, 119
  G2447 mutant analysis, 124
  interpretation of in vivo proper-
    ties, 124–125

materials and methods, 120
mutant ribosome populations analysis, 120–121
peptide-bond formation sequence,
  119, 122–123
pH-dependent conformational
  changes, 119–120
ribosome role in catalyzing peptide-bond formation,
  125
single-turnover peptidyl transferase assays, 121–122
Ad. See Adenovirus
Adenines, 25, 36
Adenovirus (Ad) translational control
  gene expression, 259
  inhibition of cellular protein syn-
    thesis
    by displacement of Mnk from
      eIF4G, 261–263
    by eIF4E targeting, 260–261
  ribosome shunting, 260
    facilitated by *cis* elements,
      263–265
    mechanism of direction by tri-
      partite leader, 265–266
  translational inhibition model,
    259–260
AdT (Amidotransferase), 177, 179
Aeropyrum pernix, 164
Affinity labeling, 532
A-form RNA helices, 130
Amber codon suppression, 197–198
Amidotransferase (AdT), 177, 179
Amino acids recognition by syn-
  thetases. See Amino-
  acyl-tRNA synthetase
Amino acid sufficiency control by
  TOP mRNAs, 478
Aminoacyl-RNA synthesis in ancient
  RNA, 208–209
Aminoacyl-tRNA (aa-tRNA)
  peptide growth study and, 97
  ternary complex of eEF1A and,
    430–432
  transorientation and, 127
  tRNA conformity with EF-Tu,
    185–187
Aminoacyl-tRNA synthetase (aaRS)
  binding role of RaiA, 512–513
  classes, 167
  class I
    amino acids recognition,
      168–170
    domain architecture, 168
    lysyl-tRNA, 175–177

  tRNA recognition, 170–172
CysRS synthesis
  alternate routes considerations,
    180–181
  dual specificity ProCysRS oc-
    currence, 181
  ProCysRS and aaRS evolution,
    181–182
  recognition of ProCysRS, 181
  substrate specificity of pro-cys-
    tRNA, 181
  reaction steps, 167
  recognition of nucleotide residues,
    167
  tRNA-dependent amidations
    function of GatCAB and
      GatDE subunits,
      177–179
    glutamine and ATP hydrolysis
      coupling, 179
    pathway of amide AA-tRNA
      formation, 180
    sequestering of misaminoacy-
      lated tRNAs, 179–180
    structure and roles, 177
    substitution for two aaRSs, 177
Aminoglycoside antibiotics, 249
Aminoglycosides, 139
A-minor motif, 36
ANA (abundant in neuroepithelium
  area protein), 298
Anchor sites, 73
Animal pole blastomeres, 348
Antibiotics
  aminoglycoside, 249
  codon-anticodon base-pairing
    recognition, 24–26
  complexes, 28–29
  decoding site location, 23–24
  inhibition patterns and the E-site,
    139
  interference with 30S initiation
    process, 47–48
Antisense oligonucleotides, 270
Anti-Shine-Dalgarno region, 45, 46f
Antizyme frameshifts and recoding,
  224
Apontic, 332
Aquifex aeolicus, 164, 521, 595–596
Arabidopsis thaliana, 305
Archaea
  biosynthesis mechanism, 154, 155t
  eIF2 subunits in, 413–414,
    420–421
  ribosomal subunit export and, 604
  transcription machinery, 562

Page numbers followed by f indicate a figure and by t indicate a table.

617